PHYSICAL CONSTANTS

Speed of light	$c\ = 3.00 \times 10^8$ m/sec
Gravitational constant	$G\ = 6.67 \times 10^{-11}$ N $\cdot$ m^2/kg^2
Avogadro's number	$N_A = 6.023 \times 10^{23}$ particles/ g $\cdot$ atom
Boltzmann's constant	$k\ = 1.3806 \times 10^{-23}$ J/K
Gas constant	$R\ = 8.314$ J/g $\cdot$ mole $\cdot$ K = 1.9872 cal/g $\cdot$ mole $\cdot$ K
Planck's constant	$h\ = 6.63 \times 10^{-34}$ J $\cdot$ sec
Electron charge	$e\ = 1.60 \times 10^{-19}$ C
Electron rest mass	$m_e = 9.11 \times 10^{-31}$ kg = 5.4858 $\times 10^{-4}$ u
Proton rest mass	$m_p = 1.67 \times 10^{-27}$ kg = 1.00727 u
Neutron rest mass	$m_n = 1.67 \times 10^{-27}$ kg = 1.00866 u
Coulomb's law constant	$k\ = 9.0 \times 10^9$ N $\cdot$ m^2/C^2
Permeability constant	$\mu_0 = 4\pi \times 10^{-7}$ N/A^2
Standard gravitational acceleration	$g\ = 9.81$ m/s^2 = 32.17 ft/s^2
Mass of earth	5.98×10^{24} kg
Average radius of earth	6.38×10^6 m
Average density of earth	5570 kg/m^3
Average earth-moon distance	3.84×10^8 m
Average earth-sun distance	1.496×10^{11} m
Mass of sun	1.99×10^{30} kg
Radius of sun	7×10^8 m

PHYSICS

PHYSICS

A General Introduction

ALAN VAN HEUVELEN
NEW MEXICO STATE UNIVERSITY

Little, Brown and Company

BOSTON   TORONTO

Library of Congress Catalog Card No. 81-81010

ISBN 0-316-89710-8

9 8 7 6 5 4 3 2

RM

Published simultaneously in Canada
by Little, Brown & Company (Canada) Limited

Printed in the United States of America

Cover photo: Lissajous figure from a light source. Lissajous figures are created by two perpendicular oscillating forces acting on an object whose changing position can be recorded. In this example, oscillations of a light pendulum trace a Lissajous figure that is then distorted by mirrors and recorded on film.

To my parents,
teachers by example

Preface

I have been guided by four goals in writing this text: (1) to provide a complete, logical, and clear introduction to the basic principles of physics for use in the algebra and trigonometry (precalculus) introductory physics course taken by students in biology, the health professions, geology, engineering technology, and other fields of science; (2) to develop the student's ability to use physics principles in a systematic way to solve problems; (3) to make both the physics principles and the problem-solving techniques accessible and meaningful to the student by *integrating* into the presentation a wide variety of interesting, quantitative, real-world examples and problems; and (4) to develop the analytical skills of students in a way that will be useful to them in their scientific and personal endeavors, not just in solving physics problems. Let me mention briefly how each of these goals has shaped the text.

1. *A complete, logical, and clear introduction to the basic principles and concepts of physics.* Two strategies have been used to attain this goal. First, new principles are carefully developed through a series of logical steps ending with a statement of the principle in a general form. Whenever possible, the presentation and development of a new principle is related to the student's experiences and interests. Once a principle has been introduced, it is used repeatedly throughout the book in examples and problems.

Second, although the text is organized in a conventional way (beginning with mechanics and ending with modern physics), I have introduced a few changes in the usual order to make it easier for students to grasp the basic ideas underlying and unifying various fields of physics. Because of these changes, the presentation in each part develops logically from a small number of basic ideas, thus avoiding a proliferation of unrelated concepts and equations. For example, Part I, Newtonian Mechanics, emphasizes one idea: forces, impulses, or torques cause an object to experience an acceleration, a change in momentum, or an angular acceleration, respectively. Newton's second law ($\Sigma \mathbf{F} = m\mathbf{a}$) or equations derived from it ($\Sigma \mathbf{F} \, \Delta t = \Delta \mathbf{p}$ and $\Sigma \boldsymbol{\tau} = I\boldsymbol{\alpha}$) are combined with kinematics to describe a variety of phenomena. Students can easily grasp this concept of cause and effect and learn to use the Newtonian approach in problem solving.

The unifying idea in Part II, Energy and Its Transformations, is the conservation-of-energy principle. Energy can be transferred between a system and its environment or converted among different forms within a system; yet with all these transformations, the total energy of the universe is unchanged. This conservation principle, as expressed by the work-energy equation or the first law of thermodynamics, can be introduced and easily understood as a second important problem-solving technique.

Parts I and II, then, show two fundamental concepts and two ways that physicists perform calculations when dealing with natural phenomena. Each subsequent part of the book not only increases the student's repertoire of concepts but also combines new analytical tools with those learned earlier. By emphasizing the basic principles, the presentation of new concepts unfolds in a logical manner from a relatively small number of important ideas, key equations, and central analytical techniques.

2. *The development of the student's ability to use physics principles to solve problems.* As just mentioned, the book is organized to emphasize basic principles. Careful attention is also given to typical ways in which these principles are used to solve problems. For example, the construction of force diagrams and the resolution of forces into components are two procedures that are emphasized in the presentation of Newton's laws. I have found that if students systematically use these techniques whenever appropriate, they can routinely work challenging, sophisticated problems that they might otherwise find difficult or discouraging.

To allow students ample opportunity to develop their problem-solving skills, the text includes 303 in-chapter solved example problems that illustrate problem-solving procedures. These procedures are reinforced by applying them to different subjects throughout the book. Also included are 420 questions and 1560 end-of-chapter problems; all of them have been worked independently and checked for accuracy by at least two people. The problems, which are grouped according to the corresponding chapter sections, are ordered from easy to difficult within their groupings. Most are of moderate difficulty, since these are the problems one is most likely to assign frequently. Many chapters also have general problems at the end that may be slightly more difficult than the regular ones or that require students to integrate ideas from several sections of the chapter or from more than one chapter. Answers to the odd-numbered problems are provided at the back of the book. The SI metric system of units is used throughout the text except at the beginning where English-system units are occasionally used.

3. *Physics concepts and problem-solving procedures are brought to life by applying them to interesting, realistic situations.* Teaching and learning are much more enjoyable and effective if the principles can be illustrated with relevant, thought-provoking examples. In recent years, textbooks have attempted to motivate students in physics by providing applications to other subjects, notably biology. These applications have usually been presented in separate sections or essays that qualitatively discuss the ways in which physics relates to other subjects. But because there is seldom enough time to learn all of the important physics concepts and how to use them, optional sections, no matter how interesting, are often omitted. To avoid this problem I have *integrated* many applications into the presentation of the physics principles themselves. Although standard physics examples appear, physics is also applied to biology, technology, the earth and its climate, and other fields of science as well as to everyday life. *All of these examples involve quantitative calculations and are presented with physical analogs that are easily understood by students and physicists alike.* The integration of these examples in the presentation allows the student to learn physics principles while applying them to a variety of interesting situations.

4. *Development of analytical skills.* I have long felt that students should be encouraged to develop analytical thinking skills that carry over into their per-

sonal and professional lives. As a means to this end, I have included a number of estimation problems in which students are not given all the information they need to work the problem. Instead, they must decide what quantities are needed to make the estimate, what principles and equations must be used to make the estimate, and on approximate values for the quantities to be used. Imaginative situations have been chosen for these estimation problems so that students will be curious to learn the results of their estimates.

Also included are short sections called Interludes in each of the seven parts of the text. Interludes are designed to encourage analytical thinking by exposing the student to specific techniques (for example, the uses of scaling or of exponential functions) and showing how they can be applied in a real-world context. Interludes, which include computational examples and problems, can be the subject of one lecture period.

Ancillary Aids

Several useful items have been prepared to accompany *Physics: A General Introduction*. Lois M. Kieffaber's *Study Guide and Workbook* is available for students. For each chapter of the text, it includes a list of important terms and equations, many helpful problem-solving hints related directly to the kinds of problems discussed in the chapter, and many additional worked example problems plus three or four exercise problems (with complete solutions in the appendix) that can be used by students to test their mastery of the material covered in that chapter. Also, I have prepared for the instructor a *Test Bank* which will be made available to adopters of the text by writing to Little, Brown and Company. The *Test Bank* contains over 500 problems for use in tests or as further homework assignments. An *Instuctor's Manual with Solutions* also accompanies the text.

Acknowledgments

I have incurred more debts in writing this book than I can hope to acknowledge. I am grateful for the many comments and suggestions by the following professors who reviewed parts of the manuscript: Angelo Armenti (Villanova University), John Botke (Effects Technology), Frank O. Clark (University of Kentucky), Robert Beck Clark (Texas A and M), Peter G. Debrunner (University of Illinois-Urbana), Stanley Hertzbach (University of Massachusetts, Amherst), Konrad Mauersberger (University of Minnesota), Joseph McCauley (University of Houston), Robert B. Prigo (University of California, Santa Barbara), John W. Robson (University of Arizona), Lawrence Rowan (University of North Carolina, Chapel Hill), Radah R. Roy (Arizona State University), Melvin Schwartz (St. John's University), Ernest Urvater (Colorado State University), Richard Whitlock (University of North Carolina, Greensboro), Charles A. Whitten, Jr. (University of California, Los Angeles), George Williams (University of Utah), Gordon G. Wiseman (University of Kansas). A very special thanks to Professors Robert Bearse (University of Kansas), W. M. Hartmann (Michigan State University), D. L. Rutledge (Oklahoma State University), and Noel Yeh (State University of New York, Binghamton), who reviewed all or major portions of the text and provided many valuable comments, and to Professor Lois M. Kieffaber (University of New Mexico) who made many valuable suggestions in addition to preparing the *Study Guide and Workbook*. I was also aided by personal discussion with many of my colleagues at New Mexico State University, including George Burleson, Yugal Behl, Horace Coburn, Ralph Dressel, Robert Liefeld, August Miller, Budd Ram, Thor Stromberg, Otto Theimer, and Charles Young.

Three students, Tim Meier, Karla Ojanen, and Pam Murray, reviewed preliminary versions of the manuscript and provided many excellent suggestions for change. Several graduate students including Sam Crow, David DePoister, David Lynch, Greg Quist, and Forrest Rennick checked the answers and wording of the problems and questions. Thanks to Josie Vigil, who typed some of the problems, and to Marian Baker, who edited most of the original version of the manuscript. A special thanks to two very efficient typists: Nancy Denzler, who typed most of the original manuscript, and Pat Paciano, who typed the final version.

Finally, I am grateful to the very competent people at Little, Brown and Company for their professional, yet personable, help in all stages of the project. I am especially indebted to Ian Irvine for his excellent guidance throughout the project, to Jane Tufts for her careful review and revision of the original manuscript, and to Elizabeth Schaaf and her colleagues in production for their professional work in putting the book together.

Summary Contents

Introduction   xxv

PART I
Newtonian Mechanics   1

1 Scalars and Vectors   3
2 Statics   16
3 Describing Motion: Kinematics   35
4 Newton's Laws of Motion   59
5 Momentum   80
       INTERLUDE Making Rough Estimates   94
6 Circular Motion at Constant Speed   98
7 Rotational Motion with Angular Acceleration   115

PART II
Energy and its Transformations   131

8 Work and Energy   133
9 Thermal Energy, Temperature, and Heat   160
10 Heat Transfer   182
       INTERLUDE Using Proportions   203
11 Entropy and Thermodynamics   206

PART III
Gases, Liquids and Solids 225

12 Gases 227
13 Fluid Statics 252
14 Fluid Dynamics 271
15 Elastic Properties of Solids 289

INTERLUDE Scaling 302

PART IV
Vibrations and Waves 307

16 Vibration 309
17 Traveling Waves 326

INTERLUDE Dimensional Analysis 347

18 Standing Waves and Sound 352

PART V
Light and Optics 369

19 Reflection and Refraction 370
20 Optical Instruments 387
21 Wave Interference 416

INTERLUDE The Difficulty of Seeing Small Objects 434

PART VI
Electricity and Magnetism 441

22 Electric Force 443
23 Electrical Energy 461

24 Electric Current and Ohm's Law   478
25 Direct-Current Electric Circuits   495
26 Capacitors   510
27 Magnetic Forces and Fields   528
28 Induced Voltage   547
29 Alternating-Current Circuits   565
        INTERLUDE Impedance Matching   589

PART VII
Modern Physics   593

30 Special Relativity   595
31 Wave-Particle Duality   613
32 The Bohr Model of the Atom   633
33 Quantum Mechanics and Atoms   649
34 The Nucleus and Radioactivity   664
        INTERLUDE The Exponential Function   691
35 Fission, Fusion, and Elementary Particles   698
    Appendixes
    Mathematics Review   718
    Graphing   721
    Selected Isotopes   728

    Answers to Odd-Numbered Problems   732
    Index   739

Contents

Introduction   xxv

PART I
Newtonian Mechanics   1

1 Scalars and Vectors   3

1.1 Physical Quantities and Standards   3
1.2 Systems of Units   4
1.3 Scalar and Vector Quantities   5
1.4 Graphical Representation of Vectors   5
1.5 Graphical Addition and Subtraction of Vectors   6
1.6 Vector Addition by Components (Qualitative)   8
1.7 Components of a Vector   10
1.8 Vector Addition by Components   12
   Summary   13   Questions   14   Problems   14

2 Statics   16

2.1 First Condition of Equilibrium   16
2.2 Problem-Solving Techniques and Examples   18
2.3 Torques   22
2.4 Second Condition of Equilibrium   24
2.5 Gravitational Torques   27
   Summary   30   Questions   30   Problems   31

3 Describing Motion: Kinematics   35

3.1 Types of Motion   35
3.2 Average Speed and Velocity   36
3.3 Instantaneous Speed and Velocity   38

3.4 Acceleration 40
3.5 Graphical Analysis of Velocity and Acceleration 43
3.6 Linear Motion with Constant Acceleration 45
3.7 Gravitational Acceleration 50
3.8 Projectile Motion 52
Summary 54   Questions 55   Problems 55   General Problems 58

4 Newton's Laws of Motion 59

4.1 Newton's First Law of Motion 59
4.2 Newton's Second Law of Motion and Mass 60
4.3 Units for Newton's Second Law 62
4.4 Weight 64
4.5 Newton's Third Law 65
4.6 Applications of Newton's Second Law 66
4.7 Friction Forces 71
Summary 76   Questions 76   Problems 77   General Problems 79

5 Momentum 80

5.1 Momentum 80
5.2 The Impulse-Momentum Equation 81
5.3 Conserved Quantities 85
5.4 Conservation of Momentum 87
Summary 90   Questions 91   Problems 91   General Problems 93

INTERLUDE **Making Rough Estimates** 94

6 Circular Motion at Constant Speed 98

6.1 Angular Position Coordinate 98
6.2 Units to Describe Angular Position 99
6.3 Tangential Speed and Angular Velocity 101
6.4 Centripetal Acceleration 102
6.5 Centripetal Force 105
6.6 Newton's Law of Gravitation 107
6.7 Weightlessness 109
Summary 111   Questions 111   Problems 112   General Problems 114

7 Rotational Motion with Angular Acceleration 115

7.1 Angular Acceleration 115
7.2 Tangential Acceleration 116
7.3 Kinematic Equations for Constant Angular Acceleration 117
7.4 Torque and Angular Acceleration 118
7.5 Moment of Inertia 120
7.6 Solving Problems in Rotational Motion 122
7.7 Angular Momentum and Its Conservation 125
Summary 128   Questions 128   Problems 129

PART II
Energy and Its Transformations   131

8 Work and Energy   133

8.1   Work   133
8.2   Energy   135
8.3   Kinetic Energy (KE)   136
8.4   Gravitational Potential Energy (PE_g)   138
8.5   Elastic Potential Energy (PE_s)   140
8.6   Conservation-of-Energy Principle   142
8.7   Work-Energy Calculations   143
8.8   Thermal Energy and Friction Forces   149
8.9   Inelastic and Elastic Collisions   151
8.10 Power   152
      Summary   154   Questions   155   Problems   155   General Problems   159

9 Thermal Energy, Temperature, and Heat   160

9.1   Thermal Energy   160
9.2   Temperature   161
9.3   Heat and the Conservation-of-Energy Principle   163
9.4   Specific Heat Capacity   164
9.5   Calorimetry   166
9.6   Melting and Freezing   168
9.7   Boiling and Condensation   172
9.8   Thermal Expansion   174
      Summary   177   Questions   178   Problems   178   General Problems   180

10 Heat Transfer   182

10.1 Heat Transfer by Conduction   182
10.2 Heat Transfer by Convection   184
10.3 Heat Transfer by Radiation   187
10.4 Heat Transfer by Evaporation   189
10.5 Temperature Control and the First Law of Thermodynamics   193
      Summary   199   Questions   200   Problems   200   General Problems   202

      INTERLUDE   Using Proportions   203

11 Entropy and Thermodynamics   206

11.1 Irreversible Processes   207
11.2 Statistical Definition of Entropy   208
11.3 Favored Forms of Energy   210
11.4 Thermodynamic Definition of Entropy   211
11.5 Second Law of Thermodynamics   213

11.6 Heat Engines 214

11.7 Heat Pumps and Refrigerators 217

11.8 Recurrence Paradox 219

Summary 221 Questions 221 Problems 221

PART III
Gases, Liquids, and Solids 225

12 Gases 227

12.1 Description of a Gas 227

12.2 Pressure 228

12.3 Gauge Pressure 229

12.4 Equations of State 230

12.5 Ideal Gas Law 232

12.6 Kinetic Theory of Gases 235

12.7 Work Done by a Gas 239

12.8 *P-V* Diagrams and Thermodynamics of Gases 242

12.9 Diffusion 244

12.10 Osmosis 245

Summary 247 Questions 248 Problems 249

13 Fluid Statics 252

13.1 Pressure 252

13.2 Density 253

13.3 Pressure Variation with Depth 255

13.4 Externally Applied Pressure 257

13.5 Archimedes' Principle 259

13.6 Surface Tension 262

13.7 Capillary Action 265

Summary 266 Questions 267 Problems 267 General Problems 270

14 Fluid Dynamics 271

14.1 Types of Fluid Flow 271

14.2 Flow Rate and Equation of Continuity 272

14.3 Bernoulli's Equation 273

14.4 Applications of Bernoulli's Equation 277

14.5 Viscous Fluid Flow 280

14.6 Reynolds Number and Turbulence 283

14.7 Drag Force 284

Summary 285 Questions 286 Problems 286 General Problems 288

15 Elastic Properties of Solids 289

15.1 Elastic Deformations 289
15.2 Types of Solid Deformations 290
15.3 Stress and Strain 290
15.4 Length Deformation 292
15.5 Shear Deformation 294
15.6 Volume Deformation 296
15.7 Elastic Membranes 297
    Summary 298   Questions 299   Problems 300   General Problems 301

    INTERLUDE **Scaling** 302

PART IV
Vibrations and Waves 307

16 Vibration 309

16.1 Vibratory Motion 309
16.2 Energy of Simple Harmonic Motion 311
16.3 Frequency of Simple Harmonic Motion 312
16.4 The Pendulum 315
16.5 Kinematics of Simple Harmonic Motion 317
16.6 Damping 318
16.7 Energy Transfer Between Oscillators 319
    Summary 322   Questions 322   Problems 323   General Problems 324

17 Traveling Waves 326

17.1 Properties of Waves 326
17.2 Types of Waves 327
17.3 Amplitude, Speed, Frequency, and Wavelength 329
17.4 Reflection and Transmission 332
17.5 Principle of Superposition 334
17.6 Complex Waves 335
17.7 Beat Frequencies 338
17.8 Doppler Effect 339
    Summary 343   Questions 343   Problems 344   General Problems 346

    INTERLUDE **Dimensional Analysis** 347

18 Standing Waves and Sound 352

18.1 Standing Waves on Strings 352
18.2 Standing Waves in Air Columns 356
18.3 Pitch 359

18.4 Quality of Sound 359
18.5 Intensity and Intensity Level 360
18.6 The Human Ear 364
       Summary 365   Questions 365   Problems 366   General Problems 367

PART V
Light and Optics   369

19 Reflection and Refraction   370

19.1 The Nature of Light 371
19.2 Coherence, Wavefronts, and Rays 372
19.3 Reflections 373
19.4 Index of Refraction 374
19.5 Refraction 375
19.6 Total Internal Reflection 378
19.7 Dispersion 381
       Summary 384   Questions 384   Problems 385

20 Optical Instruments   387

20.1 Lenses 387
20.2 Formation of Images 388
20.3 Thin-Lens Equation 391
20.4 The Human Eye 394
20.5 Angular Magnification and Magnifying Glasses 397
20.6 Combinations of Lenses 400
20.7 Aberrations 406
20.8 Mirrors 408
       Summary 411   Questions 412   Problems 413

21 Wave Interference   416

21.1 Huygens' Principle 416
21.2 Young's Double-Slit Experiment 417
21.3 Gratings 421
21.4 Thin-Film Interference 423
21.5 Single-Slit Diffraction 426
21.6 Holography 428
       Summary 430   Questions 431   Problems 431

       INTERLUDE **The Difficulty of Seeing Small Objects** 434

PART VI
Electricity and Magnetism   441

22 Electric Force   443

22.1 Electric Charge   443
22.2 Coulomb's Law   445
22.3 Static Electric Forces   446
22.4 Dielectric Constant   449
22.5 Electric Dipoles   451
22.6 Electric Fields   452
22.7 Lines of Force   455
22.8 Are Electric Fields Real?   456
  Summary   457   Questions   458   Problems   458   General Problems   460

23 Electrical Energy   461

23.1 Electrical Potential Energy (PE_q) 461
23.2 Electrical Potential Energy Involving Multiple Charges   465
23.3 Work-Energy Calculations   466
23.4 Electrical Potential Difference and Potential   469
23.5 Potential Difference and Electric Field   473
  Summary   474   Questions   475   Problems   475   General Problems   477

24 Electric Current and Ohm's Law   478

24.1 Electric Power Sources   478
24.2 Electric Current (I) and Drift Velocity   479
24.3 Ohm's Law and Resistance   482
24.4 Applications of Ohm's Law   486
24.5 Resistance and Temperature   489
24.6 Electrical Power   490
  Summary   492   Questions   492   Problems   492   General Problems   494

25 Direct-Current Electric Circuits   495

25.1 Kirchhoff's Loop Rule   495
25.2 Kirchhoff's Junction Rule   497
25.3 Use of Kirchhoff's Rules   498
25.4 Equivalent Resistance   500
25.5 Measuring Current and Voltage   503
  Summary   505   Questions   505   Problems   505

26 Capacitors 510

26.1 Capacitors 510
26.2 Capacitance 511
26.3 Dielectrics 514
26.4 Dielectric Breakdown 515
26.5 Energy in a Charged Capacitor 518
26.6 Charging and Discharging a Capacitor Through a Resistor 519
26.7 Capacitors in Series and in Parallel 522
Summary 524   Questions 525   Problems 525

27 Magnetic Forces and Fields 528

27.1 Magnetic Fields 528
27.2 Magnetic Force on a Moving Charge 529
27.3 Circular Motion in a Magnetic Field 532
27.4 Torque on a Current Loop 535
27.5 A Direct-Current Electric Motor 536
27.6 The Cause of a Magnetic Field 537
27.7 Calculating the Magnetic Field Caused by Moving Charges 539
27.8 Magnetic Materials 541
Summary 543   Questions 543   Problems 544   General Problems 546

28 Induced Voltage 547

28.1 Electromagnetic Induction 547
28.2 Faraday's Law 548
28.3 The Polarity of Induced Voltage 550
28.4 The Electric Generator 552
28.5 Mutual Inductance 554
28.6 Self-Inductance 556
28.7 Transformers 558
Summary 560   Questions 560   Problems 561   General Problems 563

29 Alternating-Current Circuits 565

29.1 AC Generator 565
29.2 Alternating Current Through Resistors 566
29.3 RMS Voltage and Current 567
29.4 Apparent Resistance of a Coil 569
29.5 Apparent Resistance of a Capacitor 572
29.6 Phase Angles of Voltages and Currents in AC Circuits 574
29.7 RLC Series Circuit 577
29.8 Power in AC Circuits 580
29.9 Resonance in RLC Series Circuit 582
Summary 584   Questions 585   Problems 585   General Problems 587

INTERLUDE Impedance Matching 589

PART VII
Modern Physics   593

30 Special Relativity   595

30.1 Events and Inertial Reference Frames   596
30.2 Postulates of Special Relativity   596
30.3 Time Dilation   597
30.4 Length Contraction   601
30.5 Relativistic Mass   603
30.6 Relativistic Energy   605
30.7 Equivalence of Mass and Energy   608
Summary   609   Questions   610   Problems   610   General Problems   611

31 Wave-Particle Duality   613

31.1 Electromagnetic Waves   613
31.2 Electromagnetic Spectrum   615
31.3 Photons   616
31.4 Photon Experiments   618
31.5 The Wave-Particle Nature of EM Radiation   621
31.6 The Wave-Particle Nature of Matter   622
31.7 The Uncertainty Principle   624
31.8 Electron Microscopes   627
Summary   629   Questions   629   Problems   630   General Problems   632

32 The Bohr Model of the Atom   633

32.1 The Nuclear Model of the Atom   633
32.2 Atomic Spectra   634
32.3 The Bohr Model   637
32.4 De Broglie's Quantum Condition   644
Summary   645   Questions   646   Problems   646   General Problems   648

33 Quantum Mechanics and Atoms   649

33.1 Quantum Mechanics   649
33.2 Quantum Numbers for Atomic Wave Functions   650
33.3 Multielectron Atoms   652
33.4 The Periodic Table   654
33.5 Absorption and Emission of Radiation   656
33.6 Stimulated Emission and Lasers   657
33.7 X-Ray Emission   660
Summary   661   Questions   662   Problems   662   General Problems   663

34 The Nucleus and Radioactivity 664

34.1   The Nucleus   664
34.2   Nuclear Force   665
34.3   Binding Energy per Nucleon   666
34.4   Nuclear Reactions   669
34.5   Radioactive Decay   671
34.6   Half-Life and Decay Rate   675
34.7   Radioactive Dating   677
34.8   Radioactive Decay Series   679
34.9   Ionizing Radiation   679
34.10 Sources and Effects of Ionizing Radiation   682
34.11 Radiation Detection Devices   685
     Summary   687   Questions   687   Problems   688   General Problems   690

     INTERLUDE   **The Exponential Function**   691

35 Fission, Fusion, and Elementary Particles 698

35.1 Nuclear Fission   698
35.2 Nuclear Power Plants   699
35.3 Breeder Reactors   705
35.4 Fusion   705
35.5 Elementary-Particle Physics   707
35.6 Classification Schemes and Conservation Laws   711
35.7 Quarks   713
     Summary   715   Questions   715   Problems   715   General Problems   717

Appendixes

A.   Mathematics Review   718
B.   Graphing   721
C.   Selected Isotopes   728

Answers to Odd-Numbered Problems   732

Index   739

Introduction

Physics is a fundamental science encompassing subjects ranging from atoms and the subatomic particles of which atoms are made to galaxies and their constituents—pulsars, black holes, neutron stars, white dwarfs, and even the sun and our own earth. Physicists search for general rules or laws that bring understanding to the chaotic behavior of our surroundings. The laws, once discovered, often seem obvious, yet their discovery usually requires years of theorizing and experimentation.

For example, Apollonius of Perega in 200 B.C. adopted the concept that the earth occupied the center of a revolving universe. Three hundred years later Ptolemy provided a theory to explain the complicated motion of the planets in that earth-centered universe. Ptolemy's theory, which predicted with surprising accuracy the changing positions of the planets, was accepted for the next fourteen hundred years. Then Copernicus, who studied astronomy at the time Columbus sailed to America, developed a theory for the motion of the heavenly bodies in which the sun resided at the center of the universe and the earth moved in orbit about it just as did the other planets. More than a hundred years later the theory was confirmed by careful observation by Galileo Galilei. Finally, fifty years after Galileo's death, Isaac Newton formulated three simple laws of motion and the universal law of gravitation, which together provided the basic understanding needed to explain the orbital motion of the earth and the other planets.

Newton's simple laws also give us the understanding needed to guide rockets to the moon, to build skyscrapers, and to realize why we should not lift heavy weights when in a bent position. Newton's inspiration provided not only the basic resolution of an eighteen-hundred-year-old problem but also a general framework for analyzing the mechanical properties of nature. Now, three hundred years after Newton's death, we are able to learn these basic laws of motion and how to use them during the first few weeks of our physics course. It is hard to appreciate the great struggle of our predecessors who developed the new understanding that now seems routine.

Today, those same struggles occur in most branches of science. Only the subject matter has changed. How does the brain work? What causes the earth's magnetic field? What is the nature of the pulsating sources of x-ray radiation in our galaxy? Are protons and neutrons really made of three smaller particles called quarks held together by exchanging other particles called gluons—or is this quark theory a misconception such as Ptolemy's complicated model of an earth-centered universe?

The pursuit of basic understanding often seems greatly removed from the activities of daily living. How could knowing that a proton is made of three smaller particles called quarks possibly affect our lives? Contemporaries of J. J. Thomson could just as easily have asked him a similar question in 1897 after his discovery of the electron. He probably could not have provided a satisfactory answer. Yet, less than a century later, the electron plays an integral part in our lives. Moving electrons in electric circuits produce light for reading, heat for cooking, waves for transmitting news, stock quotations, and information about bank accounts, and much of the information provided by radio, television, and high-fidelity systems. Could quarks, if discovered, possibly bring such a technological revolution in the one hundred years following their discovery? The answer is almost certainly yes, but the revolution cannot be envisioned now, just as Thomson could not envision the technology initiated by his discovery.

In this book we will concentrate on developing an understanding of the important, basic laws of physics. These laws encompass subjects such as mechanics, energy, waves, light, optics, electricity, magnetism, and the more modern topics of atoms and nuclei. Just as important, we will try to foster the ability to use these laws to analyze relevant problems of current interest in physics and in other fields of science and technology such as biology, medicine, geology, architecture, engineering, agriculture, and anthropology. You will learn about techniques used by archeologists to determine the age of bones, about electron microscopes, airport metal detectors, ways in which heat is gained and lost in homes, the development of stresses and tensions in body muscles, and a variety of other subjects. Thus, you will see how the intellectual struggles of scientists in the past have led to relevant endeavors in the present. I hope that you enjoy your exposure to these endeavors and that they provide part of the basis for your own future explorations of our interesting world.

PHYSICS

Newtonian Mechanics

I

We often hear that knowledge in science increases rapidly. We should note, however, that many of the principles used in physics today were developed over a hundred years ago. Our present subject, Newtonian mechanics, was introduced over three hundred years ago by men such as Galileo Galilei and Isaac Newton.

In Newtonian mechanics we consider the way in which forces affect the motion of objects. Under what conditions do objects move faster or slower, and under what conditions do they remain at rest? Newtonian mechanics helps us answer many down-to-earth questions. For example: What is the compression in the discs of your backbone when you lift a 50-lb weight? What is the minimum stopping distance in a car collision that allows the passengers to avoid injury? How does a ballerina spin so fast during a pirouette? How far from the earth must a communications satellite orbit if it is to remain above the same point on the earth's rotating surface? Besides helping us analyze such questions, Newtonian mechanics provides the foundation for much of the rest of our development of physics.

Scalars and Vectors

1

Physicists seek to describe nature in terms of a small number of basic laws. These laws involve equations that relate quantities such as force, mass, acceleration, momentum, and energy. Many of these quantities are said to be *scalars*—quantities having only a magnitude and a unit of measure. Examples are time, distance, and energy. The normal rules of algebra are used to add, subtract, multiply, and divide scalar quantities.

Other quantities, such as force and velocity, have a magnitude, unit of measure, and a direction. These quantities with direction are called *vectors*. To add and subtract them we must use special rules, which are developed in this chapter. This mathematical background plays an essential role in subsequent chapters, where we develop and use some of the important laws of physics.

1.1 Physical Quantities and Standards

In the physical sciences, nature is described using a variety of precisely defined quantities. These quantities differ from the descriptive words used in everyday language—a person is *nice* or a flower is *pretty*. The meanings of such everyday words depend on the previous experiences of the persons using them. The word *nice* may be used by one person to rate physical appearance and by another to rate personality. For this reason we cannot rate a person exactly using such words as *nice*.

In physics and in some other branches of science, the quantities used to describe nature are defined so that they have the same meaning for any person using them. Quantities such as time, length, speed, and pressure can be measured and given specific values. Each quantity, when measured, has a magnitude (a number) and a unit of measure. For example, a basketball player 2.1 m tall is 2.1 times longer than a unit of length called the meter. If we all agree upon the definition of a meter, then the player's height is described unambiguously.

An essential component of measurement is having precise standards that define units such as the meter. Standards have not always been defined precisely. At one time a "foot" was the length of the king's foot in England. Early in the present century the "foot" was defined in several different ways. Property evaluations in Brooklyn were made using the United States foot, the Bushwick foot, the Williamsburg foot, and the foot of the 26th Ward. Some strips of

Brooklyn property were untaxable, because after surveys made using different "feet," these strips of land legally did not exist!

We can see that it is important to have standards that are precisely defined and that are used in common by people involved in trade, science, and industry. The present unit of length used in science is called the *meter*. It was originally defined as the distance between two lines inscribed on a bar of metal stored in a special vault in a small town near Paris, France. Replicas made from the standard were distributed around the world. A meter in France had the same length as meters in Canada, the United States, Peru, or in any other country.

A metal bar such as used for the length standard can deteriorate with time and is accessible to only a few. In 1960, therefore, the meter was redefined as the length of 1,650,763.73 wavelengths of light emitted by a particular atomic transition of krypton atoms. This new standard equals the length of the old but is more accurate than the old, is accessible to scientists in any laboratory in the world, and does not deteriorate. The current standard for time also depends on an atomic phenomenon.

1.2 Systems of Units

We might ask ourselves how many of these quantities such as distance and time need to have standards. For example, if we have standards for length and for time, then we do not need a separate standard for speed, since its unit (meters per second) is already defined by the length and time standards. We find that we can describe most scientific observations using the units of measure of only the seven basic quantities listed in Table 1.1. They are called **basic quantities** because each unit has a well-defined standard.

Scientists use many more than seven quantities to describe their observations, but the units of these other quantities, called **derived quantities,** are just combinations of basic units. Speed is an example of a derived quantity; its units are the ratio of meters and seconds. Often, a special name is given to a frequently used derived unit. These names usually honor a noteworthy physicist. For example, the derived unit of force—the kg m/s^2—is called a *newton* in honor of Isaac Newton.

A *system of units* consists of a set of standards for some or all of the basic units. In a system of units that evolved in English-speaking countries, the so-called English system, the units for length, mass, and time are the foot, slug, and second. In the metric system the units of length, mass, and time are the meter, kilogram, and second. The modernized version of the metric system, based on atomic standards, is called the **SI system** from the French words *Système International d'Unités* (International System of Units).

Today, SI metric units are used officially in over 75 percent of world trade. The United States and Canada are the only industrialized countries not using this system as a basis of trade, although they are adopting it. One big advantage of the metric system is that large and small units are all related to the basic units by powers of ten. For example, 1 kilometer = 1000 meters, and 1 centimeter = $\frac{1}{100}$ meter. Thus, it is much easier to convert meters to kilometers than to convert feet to miles. The prefixes used for these multiples are listed in a table on the inside back cover.

TABLE 1.1 Basic Units of the SI System

Quantity	Unit name	Unit symbol
Length	meter	m
Mass	kilogram	kg
Time	second	s
Electric current	ampere	A
Temperature	degree kelvin	K
Amount of substance	mole	mol
Luminous intensity	candela	cd

1.3 Scalar and Vector Quantities

We have said that quantities in science can be measured and specific values given to them. Each quantity can be represented by a symbol indicating its numerical value and unit of measure. For example, the symbol E could represent an energy of 100 calories, or t could represent a time of 3.5 seconds. These quantities are examples of what are called scalars.

A **scalar** is a quantity that has a magnitude and usually a unit of measure. Scalars are given algebraic symbols such as E, t, and v. The familiar rules of algebraic addition, subtraction, multiplication, and division are used to mathematically manipulate scalars.

Often we need to indicate the direction as well as the magnitude of a quantity. Perhaps you have been told that a hidden treasure has a "displacement" from you of 100 miles in a direction 50° north of west. *Displacement* is a vector quantity indicating (1) the distance from one point to another and (2) the direction of an arrow pointing from the first to the second point. *Distance* is the corresponding scalar quantity; it indicates only the separation of the points, not the direction from one to the other.

A **vector** is a quantity with both magnitude and direction. Vectors are represented by boldface algebraic symbols such as **F, d,** and **a.** * Special rules will be developed to add, subtract, and multiply vectors.

The **magnitude** of a vector is a number and unit of measure that indicates its numerical value. The magnitude is represented by the same symbol as the vector, except that it is not in boldface. For example, a 30-m displacement vector **A** that points west has a magnitude $A = 30$ m. The magnitude is always either a positive number or zero.

Several examples of scalar and vector quantities are listed in Table 1.2. Notice that each vector in Table 1.2 is specified by a magnitude and a direction, whereas scalars are specified only by magnitudes.

1.4 Graphical Representation of Vectors

Vector quantities are often represented by arrows. The arrow's direction indicates the direction of the vector, and the arrow's length indicates the vector's magnitude. An appropriate scale must be used when drawing the length of the arrow. For example, in Fig. 1.1, 1 cm represents 50 km. Thus, vector **A,** which is 3 cm long, represents a 150-km displacement toward the west. Vector **B** represents a 100-km displacement 30° north of east.†

*In handwritten work, vectors are usually represented by the symbol with an arrow over it ($\vec{F}$, $\vec{d}$, and $\vec{a}$).

†A displacement whose direction is 30° north of east makes a 30° angle with an axis pointing toward the east. The vector is oriented above the axis toward the north rather than below the axis toward the south.

TABLE 1.2 Examples of Scalar and Vector Quantities

Scalars		
Symbol	Name	Example
d	distance	30 m
v	speed	50 m/s
t	time	15 s
E	energy	2000 cal

Vectors		
Symbol	Name	Example
d	displacement	30 m north
v	velocity	50 m/s west
F	force	100 lb up
a	acceleration	12 m/s² down

FIG. 1.1. Two displacement vectors. **A** is 3 cm long and represents a 150-km displacement toward the west. **B** is 2 cm long and represents a 100-km displacement 30° north of east.

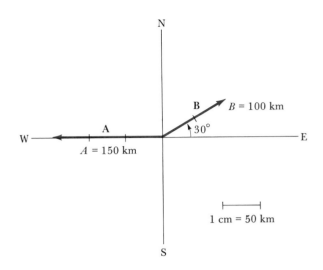

Forces are another familiar quantity represented by vectors. Forces are often caused by a push or pull of some object on another. Sometimes, but not always, the objects are clearly in contact with each other. A book pushing down on a table or a boxer's glove punching an opponent are forces, but so too is the earth's gravitational pull on a skydiver (see Fig. 1.2). The unit of force in the English system of units is the *pound*. In the metric system, the unit of force is the *newton*. One newton is approximately the weight of a medium-sized apple, a little less than one-fourth of a pound.

1.5 Graphical Addition and Subtraction of Vectors

How are vector quantities added and subtracted? The normal rules of algebraic addition and subtraction are not satisfactory. Suppose, for example, that you take a two-day trip. The first day you travel 500 km toward the east along a straight road. The second day involves another 500-km displacement, but not necessarily in the same direction. In what direction and how far from your starting position are you now located? The answer, of course, depends on the direction of the second day's trip. You could be 1000 km east of your home if you continued traveling east during the second day (Fig. 1.3a). However, if you traveled west on the second day, your net displacement would be zero—you would be back where you started (Fig. 1.3b). If you traveled north during the second day, your net displacement would be a little over 700 km to the northeast of your home (Fig. 1.3c). When adding displacement vectors or vectors of any type, we are concerned only with their net result. For displacement vectors, we are concerned with the magnitude and direction of our final displacement from the starting position. If we travel around the world and end back where we started, our net displacement is zero. We get no credit for the manner in which we reached that destination.

Vector addition is visualized most easily using a graphical technique. Suppose we want to add the two vectors **A** and **B** shown in Fig. 1.4a. These could be displacement vectors, force vectors, or whatever. To add them graphically, we

(a)

(b)

(c)

FIG. 1.2. (a) A boxer receiving a punch, and (b) the vector representation of the force caused by the punch on the boxer's head. (c) Shown here is the gravitational force of the earth pulling down on a skydiver.

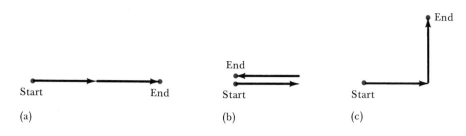

FIG. 1.3. Two separate 500-km trips are each represented by a displacement vector. The net displacement from the starting position depends on the direction of the second day's trip.

place the tail of **B** at the head of **A,** as in Fig. 1.4b. We can move vectors from one place to another for addition, but we cannot change the magnitude or direction of a vector while moving it. Having moved vector **B,** we draw another vector, **R,** from the tail of **A** to the head of **B,** as in Fig. 1.4c. This vector **R** represents the net displacement (or net force if **A** and **B** are force vectors) and is called the **resultant** of vectors **A** and **B.** Written as a mathematical equation, we say that $\mathbf{R} = \mathbf{A} + \mathbf{B}$. Because **A** and **B** are vectors, we must use a technique such as illustrated in Fig. 1.4c to find **R.** We cannot simply add the magnitudes of **A** and **B** to find **R**!

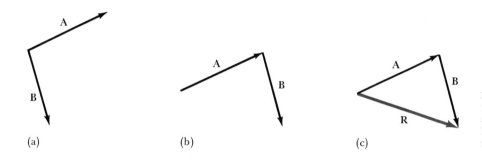

FIG. 1.4. (a) Two vectors. (b) To add these vectors, the tail of one vector is placed at the head of the other. (c) The resultant, $\mathbf{R} = \mathbf{A} + \mathbf{B}$, goes from the tail of the first to the head of the last vector.

In order to add three or more vectors graphically, we place them head to tail one at a time, as shown in Fig. 1.5b. The magnitude and direction of the vectors must not be changed, but the order in which we add the vectors makes no difference. Then, draw vector **R** from the tail of the first vector to the head of the last, as in Fig. 1.5c. This resultant vector **R** equals $\mathbf{A} + \mathbf{B} + \mathbf{C}$.

Usually, the graphical technique for adding vectors is used to provide a rough check on another vector addition technique introduced later. However, if done with care using a ruler and protractor, the graphical vector addition technique provides a fairly accurate method for determining the resultant of several vectors.

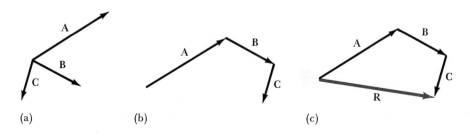

FIG. 1.5. Three vectors shown in (a) are added by joining them head to tail in (b). (c) The resultant, $\mathbf{R} = \mathbf{A} + \mathbf{B} + \mathbf{C}$, goes from the tail of the first to the head of the last vector.

FIG. 1.6. (a) Three displacement vectors; (b) the resultant displacement.

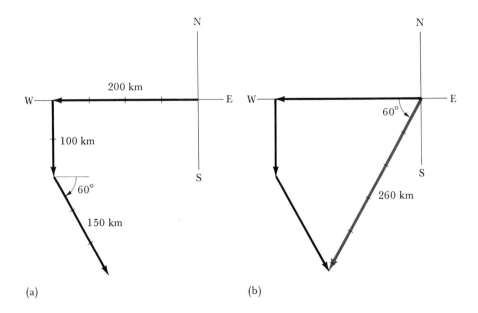

(a)   (b)

EXAMPLE 1.1 A car travels 200 km west, 100 km south, and finally 150 km at an angle 60° south of east. What is the net displacement of the car?

SOLUTION A vector diagram representing the three displacements drawn head to tail is shown in Fig. 1.6a. To find the resultant displacement **R** we draw an arrow from the tail of the first displacement to the head of the last. We measure the magnitude of the resultant with a ruler and find that its length is 5.2 cm (Fig. 1.6b). Since each centimeter represents 50 km, the magnitude of the resultant displacement is 5.2 × 50 km = 260 km. Using a protractor, we find that the direction of the resultant is 60° south of west. ∎

Occasionally we will need to subtract two vectors. Consider a general case of vector subtraction illustrated in Fig. 1.7, where vector **B** is subtracted from **A**. We first multiply vector **B** by −1, which reverses the direction of **B** and produces the vector −**B** (Fig. 1.7b). We then add **A** and −**B**, as shown in Fig. 1.7c. The resultant goes from the tail of **A** to the head of −**B** (Fig. 1.7d) and is just the difference of **A** and **B**:

$$R = A + (-B) = A - B.$$

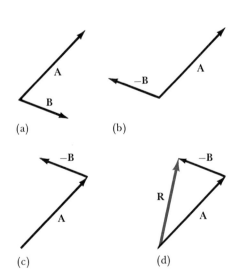

FIG. 1.7. (a) Two vectors **A** and **B**; (b) **A** and −**B**; (c) the addition of **A** and −**B**; and (d) the resultant **R** = **A** + (−**B**) = **A** − **B**.

1.6 Vector Addition by Components (Qualitative)

The graphical method of adding and subtracting vectors takes considerable time if done with accuracy. A different vector addition technique developed in this section is more accurate and is also faster and more convenient for solving a variety of interesting problems involving vectors. This *component method,* as it is called, depends upon our ability to resolve a vector into components.

Consider the displacement vector shown in Fig. 1.8a. A coordinate system has been drawn that has an *x* axis pointing east and a *y* axis pointing north. The **x component** of **A** is a vector $\mathbf{A}_x$ that points along the *x* axis and whose length equals the projection of **A** on that axis, as shown in Fig. 1.8b. The **y component** of **A** is the projection of **A** on the *y* axis and is labeled $\mathbf{A}_y$ in Fig. 1.8b. Notice that the sum of $\mathbf{A}_x$ and $\mathbf{A}_y$ equals **A**, as shown in Fig. 1.8c.

In the component method of adding vectors, each vector being added is first resolved into its components. Then, instead of adding the vectors, we add their components. This seems reasonable, since the components of a vector, when added, equal the vector itself. Thus, it should not matter whether we add the vectors or their components.

The component method of adding vectors is illustrated in Fig. 1.9. In (a), vectors **A** and **B** are added graphically to produce a resultant vector **R** (the dashed line). In (b), the double crossed lines through vectors **A** and **B** indicate that they have now been replaced by their components. Instead of adding **A** and **B**, we add their components $\mathbf{A}_x$, $\mathbf{A}_y$, $\mathbf{B}_x$, and $\mathbf{B}_y$ as illustrated in Fig. 1.9b. The resultant is the same as when adding **A** and **B** directly, as shown in Fig. 1.9a.

We can, if we wish, change the order of addition of vectors. This has been done in Fig. 1.9c, where $\mathbf{A}_x$ and $\mathbf{B}_x$ are added first to find what is called the *x* component of the resultant vector, called $\mathbf{R}_x$. Note that since $\mathbf{B}_x$ points to the left, it cancels some of the effect of $\mathbf{A}_x$, which points to the right. After finding $\mathbf{R}_x$, we then add $\mathbf{A}_y$ and $\mathbf{B}_y$ to find the *y* component of the resultant, $\mathbf{R}_y$. Then, when $\mathbf{R}_x$ and $\mathbf{R}_y$ are added, as shown in (c), we once again get the same resultant vector **R**.

Notice that the triangle made of **R**, $\mathbf{R}_x$, and $\mathbf{R}_y$, and shown in Fig. 1.9d, is a right triangle. According to the Pythagorean theorem the sum of the squares of its sides ($R_x^2 + R_y^2$) equals the square of the triangle's hypotenuse (R^2). Or,

$$R = \sqrt{R_x^2 + R_y^2}.$$

(a)

(b)

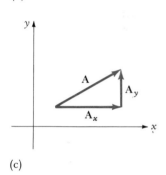

(c)

FIG. 1.8. (a) A displacement vector **A** and (b) its *x* and *y* components ($\mathbf{A}_x$ and $\mathbf{A}_y$). (c) The components when added equal **A**.

(a)

(c)

(b)

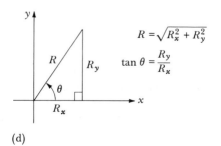

$$R = \sqrt{R_x^2 + R_y^2}$$

$$\tan \theta = \frac{R_y}{R_x}$$

(d)

FIG. 1.9. (a) Vectors **A** and **B** are added graphically to produce **R**. (b) **A** and **B** are replaced by their components, and then the components are added to produce **R**. (c) The order of component addition is rearranged to determine $\mathbf{R}_x$ and $\mathbf{R}_y$. (d) The magnitude R and direction θ of **R** can be determined from R_x and R_y.

Thus, we can determine the magnitude of **R** if R_x and R_y are known. Also, using the fact that $\tan \theta = R_y/R_x$, we can calculate the angle θ that **R** makes with the x axis.

We see then that our component technique for adding vectors involves three steps: (1) determine the components of each vector being added; (2) add the x components of these vectors to find R_x and the y components to find R_y; and (3) determine the magnitude R and direction θ of the resultant vector from the values of R_x and R_y. The details of these calculations are outlined with examples in the following sections.

1.7 Components of a Vector

The component of a vector does not need to be specified as a vector. Its subscript—for example, the x on A_x—indicates the axis along which the component lies. If the component points in the positive direction of that axis, it is given a plus sign. If it points toward the negative direction of the axis, it is given a negative sign. For instance, the x component of the force shown in Fig. 1.10 is negative, since F_x points in the negative x direction, whereas the y component is positive, since it points upward in the positive y direction.

Next we need to determine the magnitude of the components. For this, a brief review of trigonometry is helpful. Consider the right triangle shown in Fig. 1.11. Recall the definitions of the sine, cosine, and tangent functions:

$$\sin \theta = \frac{o}{h},$$

$$\cos \theta = \frac{a}{h},$$

$$\tan \theta = \frac{o}{a}.$$

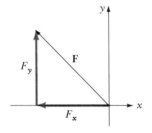

FIG. 1.10. F_x is negative, F_y positive.

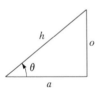

FIG. 1.11. The sides of the triangle are labeled h (hypotenuse), o (side opposite the angle θ), and a (side adjacent to the angle θ).

Using these relationships, we can find the magnitude of the force components shown in Fig. 1.12. The hypotenuse is the magnitude of the force and the sides are the magnitudes of the x and y components of **F**. Thus

$$\sin \theta = \frac{F_y}{F} \quad \text{or} \quad F_y = F \sin \theta, \tag{1.1}$$

$$\cos \theta = \frac{F_x}{F} \quad \text{or} \quad F_x = F \cos \theta, \tag{1.2}$$

$$\tan \theta = \frac{F_y}{F_x}.$$

Equations (1.1) and (1.2) are used to calculate the magnitudes of F_x and F_y if F and θ are known. A table of sine, cosine, and tangent functions appears on the inside back cover. Most calculators also give the sines and cosines of different angles.

The technique, then, for finding the components of a vector may be summarized as follows.

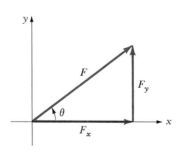

FIG. 1.12. If F and θ are known, F_x and F_y can be found using trigonometry.

The x and y **components** of a vector $\mathbf{F}$ are

$$F_x = \pm F \cos \theta, \tag{1.3}$$
$$F_y = \pm F \sin \theta, \tag{1.4}$$

where F is the magnitude of the vector and θ is the angle (90° or less) that $\mathbf{F}$ makes with the x axis. F_x is positive if it points in the positive x direction and negative if it points in the negative x direction. F_y is either positive or negative, depending on its direction relative to the y axis.

Be sure to note that the angle θ appearing in Eqs. (1.3) and (1.4) is the angle that the vector makes with the positive or negative x axis and not the angle it makes with the y axis.

EXAMPLE 1.2 A carpenter pulls a load of lumber on a cart with a 60-lb force directed 40° above the horizontal (Fig. 1.13a). Calculate the horizontal and vertical components of the force.

SOLUTION The force and a set of axes used to describe it are shown in Fig. 1.13b. Both force components shown in (c) are positive. Since $F = 60$ lb and $\theta = 40°$, we find using Eqs. (1.3) and (1.4) that

$$F_x = F \cos \theta = (60 \text{ lb})(\cos 40°) = \underline{46 \text{ lb}},$$
$$F_y = F \sin \theta = (60 \text{ lb})(\sin 40°) = \underline{39 \text{ lb}}. \qquad \blacksquare$$

EXAMPLE 1.3 You fly 4200 km in a direction 47° north of west while traveling from Minneapolis to Fairbanks. Using a coordinate system such as shown in Fig. 1.14a with the x axis pointing east and the y axis north, calculate the x and y components of your displacement.

SOLUTION The x component of the trip is negative, since it points west in the negative x direction, whereas the y component is positive. The components are then

$$A_x = -A \cos \theta = -(4200 \text{ km})(\cos 47°) = \underline{-2860 \text{ km}},$$
$$A_y = +A \sin \theta = +(4200 \text{ km})(\sin 47°) = \underline{3070 \text{ km}}. \qquad \blacksquare$$

(a)

(b)

(c)

FIG. 1.13. (a) A carpenter pulling a load of lumber. (b) The force of the rope on the cart. (c) The horizontal and vertical components of the force.

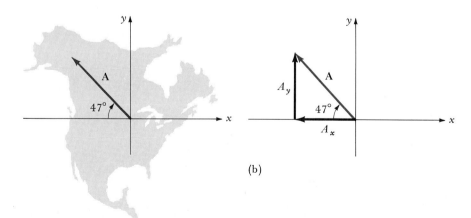

(a)

(b)

FIG. 1.14. (a) A 4200-km trip from Minneapolis to Fairbanks. (b) The x component is negative and the y component positive.

(a)

(b)

(c)

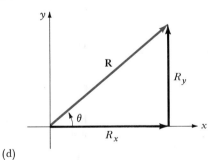

(d)

FIG. 1.15. (a) Two displacement vectors. (b and c) When added, the vectors produce a resultant displacement whose components equal the sum of the components of **A** and **B**. (d) The magnitude and direction of the resultant vector are determined from the values of its components.

1.8   Vector Addition by Components

Consider the two displacement vectors shown in Fig. 1.15a. **A** has a magnitude of 50 km and points 20° above the positive x axis. **B** has a magnitude of 40 km and points 65° above the positive x axis. The vectors are added graphically in Fig. 1.15b. Notice that the sum of A_x and B_x equals R_x, and that the sum of A_y and B_y equals R_y. The specific values of the components are determined using Eqs. (1.3) and (1.4):

$$R_x = A_x + B_x = A \cos 20° + B \cos 65°$$
$$= (50 \text{ km})(0.94) + (40 \text{ km})(0.42) = 64 \text{ km},$$

and

$$R_y = A_y + B_y = A \sin 20° + B \sin 65°$$
$$= (50 \text{ km})(0.34) + (40 \text{ km})(0.91) = 53 \text{ km}.$$

We can now use these resultant components, R_x and R_y, to find **R** as shown in Fig. 1.15d. The magnitude of **R** is determined using the Pythagorean theorem:

$$R = \sqrt{R_x^2 + R_y^2} = \sqrt{64^2 + 53^2} = 83 \text{ km}.$$

The angle θ that **R** makes with the x axis is determined using the expression for the tangent of that angle:

$$\tan \theta = \frac{R_y}{R_x} = \frac{53 \text{ km}}{64 \text{ km}} = 0.83.$$

From the table of tangents on the inside back cover, we see that a 40° angle has a tangent of approximately 0.83.* Thus, **R** points 40° above the positive x axis and is 83 km long.

This vector addition technique can be used for any type of vector; it is summarized below for force vectors.

Vector addition technique using components: To add a group of vectors to find the resultant $\mathbf{R} = \mathbf{F}_1 + \mathbf{F}_2 + \mathbf{F}_3 + \cdots$, we first add the x components of these vectors to find R_x and then add the y components to find R_y:

$$R_x = F_{1x} + F_{2x} + F_{3x} + \cdots, \tag{1.5}$$
$$R_y = F_{1y} + F_{2y} + F_{3y} + \cdots. \tag{1.6}$$

Equations (1.3) and (1.4) are used to find the components of each vector. Having found R_x and R_y, the magnitude of the resultant is

$$R = \sqrt{R_x^2 + R_y^2}. \tag{1.7}$$

The direction of **R** is determined from a picture of **R** constructed from R_x and R_y. The exact angle, θ, that **R** makes with the x axis is determined from the equation

$$\tan \theta = \left| \frac{R_y}{R_x} \right| \tag{1.8}$$

or

$$\theta = \arctan \left| \frac{R_y}{R_x} \right|. \tag{1.9}$$

*The angle can also be determined by calculating the arctangent of 0.83 using a calculator.

The lines on either side of R_y/R_x in Eqs. (1.8) and (1.9) are absolute value signs and mean that only the positive values of R_x and R_y are substituted into these equations. When solving for θ, we then determine the angle of 90° or less that the resultant vector makes with either the positive or negative x axis. The quadrant in which the vector points depends on the signs of R_x and R_y. Consider the next example.

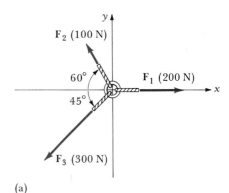

EXAMPLE 1.4 Three ropes attached to a ring pull it as shown in Fig. 1.16a. Calculate the resultant of these three forces.

SOLUTION You should check carefully the sign and angle used when calculating the components of each force. The resultant components are:

$$R_x = F_{1x} + F_{2x} + F_{3x} = F_1 \cos 0° - F_2 \cos 60° - F_3 \cos 45°$$
$$= (200 \text{ N})(1.0) - (100 \text{ N})(0.5) - (300 \text{ N})(0.71) = -62 \text{ N},$$
$$R_y = F_{1y} + F_{2y} + F_{3y} = F_1 \sin 0° + F_2 \sin 60° - F_3 \sin 45°$$
$$= (200 \text{ N})(0) + (100 \text{ N})(0.87) - (300 \text{ N})(0.71) = -126 \text{ N}.$$

(a)

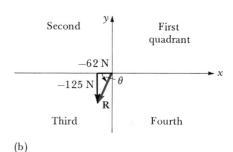

For this problem, the resultant has an x component of -62 N and a y component of -126 N. A drawing of **R** based on these component values appears in Fig. 1.16b. We see that **R** is in the third quadrant. The rough graphical addition of F_1, F_2, and F_3 shown in Fig. 1.16c confirms that **R** does point into the third quadrant (a good check on our math). The magnitude of **R** and the exact angle that it makes with the x axis are calculated using Eqs. (1.7) and (1.8):

$$R = \sqrt{62^2 + 126^2} = \underline{140 \text{ N}},$$
$$\tan \theta = \frac{126 \text{ N}}{62 \text{ N}} = 2.0.$$

(b)

According to our table of tangents, a 63.5° angle has a tangent of 2.0. Since **R** points into the third quadrant, its direction must be 63.5° below the negative x axis.

We should note that this single force **R** has exactly the same effect on the ring as the three forces shown in Fig. 1.16a. The resultant force equals the vector sum of these three forces mathematically and in terms of its effect on the ring. ∎

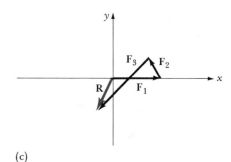

(c)

FIG. 1.16. (a) Three forces pulling on a ring. (b) The components of the resultant indicate that it points into the third quadrant. (c) A graphical addition of the three vectors confirms the component calculations.

Summary

1. **Measurable quantities:** In physics we describe nature in terms of measurable quantities. Most quantities are given in a unit of measure based on a standard or a unit derived from standardized units. The SI metric system provides most of the standards for scientific measurement. There are two types of measurable quantities, scalars and vectors.

Scalars are quantities that have only a magnitude, including a unit of measure. The normal rules of algebraic addition, subtraction, multiplication, and division are used with scalars.

Vectors are quantities that have a magnitude and a direction. Special mathematical rules are needed to manipulate vectors.

2. **Vector addition and subtraction by the graphical technique:** To add vectors graphically, we join the vectors head to tail. The resultant vector goes from the tail of the first vector to the head of the last and is the sum of these vectors ($\mathbf{R} = \mathbf{A} + \mathbf{B} + \mathbf{C} + \cdots$). Vectors may be moved from one place to another for this type of addition, but their magnitudes and directions cannot be changed. Vectors are subtracted graphically by adding the negative of the vector to be subtracted; that is, $\mathbf{R} = \mathbf{A} + (-\mathbf{B}) = \mathbf{A} - \mathbf{B}$. A negative vector, for example $-\mathbf{B}$, is the vector with its direction reversed.

3. **Vector addition by components:** To add vectors by the com-

ponent technique, we must first find the **x and y components** of each vector being added. They are given by the equations

$$F_x = \pm F \cos \theta, \qquad (1.3)$$
$$F_y = \pm F \sin \theta, \qquad (1.4)$$

where F is the magnitude of the vector and θ is the angle (90° or less) that **F** makes with either the positive or negative x axis. The signs of F_x and F_y depend upon the direction of the projection of **F** on the x and y axes, respectively. We then calculate the x and y components of the resultant vector by adding the x and y components of each vector being added:

$$R_x = F_{1x} + F_{2x} + F_{3x} + \cdots, \qquad (1.5)$$
$$R_y = F_{1y} + F_{2y} + F_{3y} + \cdots. \qquad (1.6)$$

Having determined R_x and R_y, we calculate the magnitude and direction of the resultant vector using the equations

$$R = \sqrt{R_x^2 + R_y^2} \qquad (1.7)$$
$$\tan \theta = \left| \frac{R_y}{R_x} \right| \quad \text{or} \quad \theta = \arctan \left| \frac{R_y}{R_x} \right|, \quad \textbf{(1.8) or (1.9)}$$

where θ is the angle, 90° or less, that **R** makes with either the positive or negative x axis. The quadrant in which **R** points is determined from a rough drawing of **R** based on its component values.

Questions

1. Many years ago in England, the standard of weight was based on the weight of 32 wheat kernels. In what ways does this seem like a good standard compared to our present one—a metal cylinder stored near Paris? In what ways does it seem like a bad standard?

2. Most quantities used in physics can be measured in units with well-defined standards. Why are there not similar standards for descriptive words used in everyday language such as *pretty, nice, lovely,* and *bad?*

3. Choose some word, such as mentioned in the previous problem, and invent a standard for it. For example, develop a standard for the word *friendly.* Your standard for this or some other everyday word should allow us to use the word with a more precise meaning. We should, for instance, be able to use your standard to assign numerical values of friendliness for different people. Indicate the reasons why your definition may be deficient.

4. A medium-sized apple weighs one newton. (a) Name five other objects that weigh about one newton (± 0.5 N). (b) Name several objects that you estimate weigh between 5 and 15 newtons; (c) between 50 and 150 newtons.

5. Which of the following quantities are scalars and which are vectors: (a) wind velocity, (b) population of a town, (c) force of a cable pulling an elevator, (d) weight of a book, (e) number of leaves on a tree, (f) age of the universe, and (g) flow rate of blood in the circulatory system?

6. (a) Describe five vector quantities in your immediate surroundings. Be sure to give the estimated magnitude and direction of each. They should not all be forces. (b) Describe five scalars in your immediate surroundings.

7. Can two forces of unequal magnitude add to zero? Explain.

8. Can a nonzero displacement vector pointing north and one of equal magnitude pointing east add to zero? Explain.

Problems

1.5 Graphical Addition and Subtraction of Vectors

All problems in this section must be worked using graphical techniques.

1. A person travels 30 km east and then 50 km in a direction 60° south of east. Using a scale of 10 km to 1 cm, determine the magnitude and direction of the final vector displacement of the person.

2. A bull is pulled with ropes by two cowboys. One exerts a force of 150 lb in a direction toward the east. The other exerts a 180-lb force toward the south. Calculate the magnitude and direction of the net force on the bull.

3. A hot air balloon drifts 12 km in a direction 70° south of east. An abrupt change in air movement then carries the balloon 8.0 km in a direction 40° south of west. Use graphical analysis to determine the balloon's resultant displacement.

4. A marching band while on parade travels 1.5 km south, 1.0 km east, and then 1.0 km north. How far must the band members walk, using the shortest possible path, to return to their cars parked at the starting position of the parade?

5. A traveling salesman during the first three days of the week travels 200 km west, 300 km 53° north of west, and finally 200 km 30° north of east. Calculate the salesman's resultant displacement after the three days of travel.

6. On January 16, 1954, a South African sharp-nosed frog named "Leaping Lena" (later found to be a male) set the world's record distance for three consecutive jumps: 4.0 m 30° N of E, 3.2 m 15° S of E, and 3.6 m 25° S of E. Determine his record displacement from the starting position after the three jumps (magnitude and direction).

7. Calculate the sum of forces F_1 and F_2 shown in Fig. 1.17.

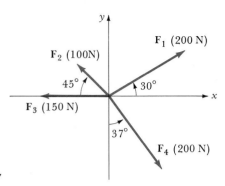

FIG. 1.17

8. Calculate the sum of forces F_3 and F_4 shown in Fig. 1.17.

9. Calculate the sum of forces F_1, F_3, and F_4 shown in Fig. 1.17.

10. Calculate the sum of displacements **A, B,** and **C** shown in Fig. 1.18.

11. Calculate $F_1 - F_2$ for the forces shown in Fig. 1.17.

12. Calculate $F_4 - F_3$ for the forces shown in Fig. 1.17.

13. Calculate $A - C$ for the displacements shown in Fig. 1.18.

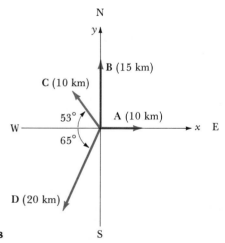

FIG. 1.18

14. Show for the forces in Fig. 1.17 that $F_1 + F_2 + F_3 = F_3 + F_2 + F_1$.

15. For vectors **C** and **D** shown in Fig. 1.18, calculate the following vectors: (a) $C + D$, (b) $D + C$, (c) $C - D$, and (d) $D - C$.

16. A force vector of magnitude 100 N is directed west, and a second force of magnitude 200 N is directed 30° east of north (or 60° north of east). Find the magnitude and direction of a third force which, when added to the first two, gives a resultant whose magnitude is zero.

1.7 Components of a Vector

17. Calculate the x and y components of displacements **A** and **C** shown in Fig. 1.18.

18. Calculate the x and y components of displacements **B** and **D** shown in Fig. 1.18.

19. Calculate the x and y components of F_1 and F_3 shown in Fig. 1.17.

20. Calculate the x and y components of F_2 and F_4 shown in Fig. 1.17.

21. Using a coordinate system with the x axis pointing east and the y axis pointing north, calculate the components of each displacement of the hot air balloon discussed in Problem 3.

22. Using a coordinate system with the x axis pointing east and the y axis pointing north, calculate the components of each displacement of the traveling salesman discussed in Problem 5.

1.8 Vector Addition by Components

All vector addition in this section must be done by the component technique.

23. Calculate the resultant displacement of the person described in Problem 1.

24. Calculate the resultant force pulling the bull described in Problem 2.

25. Calculate the resultant displacement of the balloon described in Problem 3.

26. Calculate the resultant displacement of the marching band described in Problem 4.

27. Calculate the resultant displacement of the salesman described in Problem 5.

28. Calculate the resultant displacement of the frog described in Problem 6.

29. Add forces F_1 and F_2 shown in Fig. 1.17.

30. Add forces F_3 and F_4 shown in Fig. 1.17.

31. Add forces F_1, F_3, and F_4 shown in Fig. 1.17.

32. Add displacements **A, B,** and **C** shown in Fig. 1.18.

33. In order to reach a person's house from your present location, you must travel 0.80 km east and 0.30 km south. Calculate the magnitude and direction of a displacement vector to the person's house.

34. Find the magnitudes and directions of forces whose x and y components are: (a) 10 N and −10 N; (b) −20 N and −30 N; (c) −5 N and +10 N; and (d) 20 N and −10 N.

35. Find the magnitudes and directions of displacements whose x and y components are: (a) −7 km and +4 km; (b) +20 km and −30 km; (c) 0 and 10 km; and (d) −20 km and 0.

36. Two displacement vectors when added produce a resultant of magnitude 100 km pointing 37° S of E. One displacement vector is 200 km in a direction 60° N of E. Determine the magnitude and direction of the other displacement.

37. Three forces when added produce a resultant that is 100 N pointing 63° above the positive x axis. If one of the forces is 50 N in the positive y direction and another is 100 N in a direction 45° below the positive x axis, what are the magnitude and direction of the third force?

2 Statics

Statics is the study of forces acting on objects that are either at rest or moving at constant speed in a straight path. Both groups of objects are said to be in equilibrium. Objects at rest are in static equilibrium and those moving in a straight path at constant speed are in dynamic equilibrium. In this chapter we consider only objects at rest.

For an object to remain in equilibrium, it must satisfy two conditions: (1) The vector sum of all forces acting on the object must be zero. If not zero, the object begins to move in the direction of the resultant force and hence no longer remains at rest. (2) Also, for the object to be rotationally stable, the forces must act on it in a way that prevents it from rotating, tipping, or falling over.

Each of these conditions can be written in the form of a mathematical equation. These equations are used to analyze a variety of situations. For example, an engineer can estimate the safe load that the girders of a building can carry. Or we can estimate the tension force in back muscles when a person is bending over to lift a weight. Statics also gives us an interesting and visual introduction to forces and the way in which we manipulate them in vector equations.

2.1 First Condition of Equilibrium

The first condition of equilibrium says that if an object is to remain at rest, then the vector sum of all forces acting on it must be zero. A young man is holding a flowerpot in Fig. 2.1a. Experience tells us that if the man lets go of the pot, it falls to the ground. If he pushes or lifts hard in an upward direction, the pot rises. Evidently, two forces act on the pot: one pushing up and one pulling down. If the two forces balance, the pot remains stationary. If the upward force becomes greater than the downward force, the pot rises, and vice versa.

The two forces are represented by vectors $\mathbf{F}_c$ and $\mathbf{w}$ in Fig. 2.1b. $\mathbf{F}_c$ is called a **contact force** and represents the force of the man's hand pushing up on the pot. Two objects exert contact forces on each other whenever they touch. The direction of a contact force is always perpendicular to the surface of contact. The force labeled $\mathbf{w}$ is the pot's **weight** and is caused by the gravitational pull of the earth on the pot. The magnitude of the weight force can be determined by measuring the pot's weight on a scale or balance.*

(a)

(b)

FIG. 2.1. (a) A man holding a flowerpot. (b) The forces acting on the pot.

*Weight is defined more carefully in Chapter 4.

16

For the pot to remain at rest, the magnitudes of $\mathbf{F}_c$ and $\mathbf{w}$ must be equal, and they must point in opposite directions. This condition is stated mathematically by the equation $\mathbf{F}_c = -\mathbf{w}$. The minus sign means simply that $\mathbf{F}_c$ points in the opposite direction of $\mathbf{w}$. Adding $\mathbf{w}$ to both sides of this equation produces the desired equation:

$$\mathbf{F}_c + \mathbf{w} = 0.$$

The forces acting on the block add to zero, causing it to be in equilibrium. Be sure to note that this is an equation involving vectors, and vector addition must be used when adding the forces.

In general, we state this equilibrium condition as follows.

First condition of equilibrium: If an object is in equilibrium, then the vector sum of all forces acting on the object must be zero. This can be written as a mathematical equation as follows:

$$\Sigma\, \mathbf{F} = \mathbf{F}_1 + \mathbf{F}_2 + \mathbf{F}_3 + \cdots = 0. \tag{2.1}$$

The Greek letter sigma, Σ, means summation. Thus, $\Sigma\, \mathbf{F}$ represents the summation of all the forces acting on the object, and $\mathbf{F}_1$, $\mathbf{F}_2$, $\mathbf{F}_3$, ... represent these forces.

The first condition of equilibrium can also be applied to the situation shown in Fig. 2.2a. A person pushes with force $\mathbf{F}$ on a wagon resting on the side of a hill. How large must $\mathbf{F}$ be to prevent the wagon from rolling down the hill?

Two other forces in addition to $\mathbf{F}$ act on the wagon, as shown in Fig. 2.2b. First, the contact force $\mathbf{F}_c$ of the ground pushes up on the tires. The contact force is directed perpendicular to the surface of contact between the tires and the road. You might note that the tires also exert a contact force on the ground. Since we are interested only in forces acting on the wagon, we do not show the force of the wagon on the ground.

The third force acting on the wagon is the downward force of gravity $\mathbf{w}$ equal in magnitude to the wagon's weight.

In Fig. 2.2c all of the force vectors have been moved so that their tails are at the origin of a coordinate system. Also, the weight force $\mathbf{w}$ has been resolved into its components, $\mathbf{w}_x$ and $\mathbf{w}_y$. If $\mathbf{w}_x$ is greater in magnitude than $\mathbf{F}$, the wagon will start to roll down the hill in the negative x direction. In order for the wagon to remain stationary, $\mathbf{F}$ must just balance $\mathbf{w}_x$. Also, $\mathbf{F}_c$ and $\mathbf{w}_y$ must balance each other. If they do not, the wagon "sinks" into the hill ($\mathbf{w}_y > \mathbf{F}_c$) or rises into the air ($\mathbf{F}_c > \mathbf{w}_y$). We know from experience that this does not happen, which implies that $\mathbf{F}_c$ and $\mathbf{w}_y$ balance each other.

This example illustrates an alternate way of expressing the first condition of equilibrium. For an object to be in equilibrium, the summation of the x components ($F_{1x}, F_{2x}, F_{3x}, \ldots$) of all forces acting on the object must be zero, and the sum of the y components ($F_{1y}, F_{2y}, F_{3y}, \ldots$) of the forces must also equal zero. This can be written mathematically as:

Alternate form of **first condition of equilibrium:**

$$\Sigma\, F_x = F_{1x} + F_{2x} + F_{3x} + \cdots = 0. \tag{2.2a}$$
$$\Sigma\, F_y = F_{1y} + F_{2y} + F_{3y} + \cdots = 0. \tag{2.2b}$$

In physics the first condition of equilibrium is used most often as stated in Eqs. (2.2a) and (2.2b). With the assistance of these equations, we can determine the

(a)

(b)

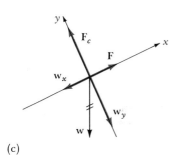

(c)

FIG. 2.2. (a) A man pushing a wagon in order to hold it at rest. (b) The forces acting on the wagon, and (c) the forces resolved into components.

(a)

(b)

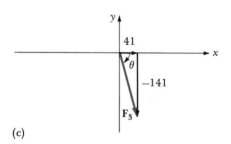

(c)

FIG. 2.3. (a) Three ropes exert forces on a metal ring. (b) The three forces acting on the ring relative to a set of coordinates. (c) The unknown force F_3 constructed from its x and y components.

magnitude and direction of an unknown force acting on an object if all the other forces acting on the object are known. This technique is used in the next example to determine the unknown force of a rope pulling on a metal ring.

EXAMPLE 2.1 Three people pull on ropes attached to a ring as shown in Fig. 2.3a. One force F_1 has a magnitude of 100 N and pulls along the positive x axis. The second rope exerts a force F_2 of 200 N in a direction 45° above the negative x axis. A third rope pulls with an unknown force F_3 that causes the ring to remain in equilibrium. Calculate the magnitude and direction of F_3.

SOLUTION For the ring to remain at rest, the sum of the x components of all forces acting on the ring must be zero, and the sum of the y components of these forces must also be zero:

$$\Sigma F_x = 0 \quad \text{or} \quad F_{1x} + F_{2x} + F_{3x} = 0.$$
$$\Sigma F_y = 0 \quad \text{or} \quad F_{1y} + F_{2y} + F_{3y} = 0.$$

Using Eqs. (1.3) and (1.4), we substitute the components of F_1, F_2, and F_3 into the above equations, obtaining

$$100 \cos 0° - 200 \cos 45° + F_{3x} = 0,$$

and

$$100 \sin 0° + 200 \sin 45° + F_{3y} = 0.$$

Substituting the sines and cosines into these equations and rearranging, we find that $F_{3x} = 41$ N and $F_{3y} = -141$ N. Next, F_3 is drawn from its components in Fig. 2.3c. The magnitude of F_3 is determined using Eq. (1.7):

$$F_3 = \sqrt{F_{3x}^2 + F_{3y}^2} = \sqrt{41^2 + (-141)^2} = 147 \text{ N}.$$

Our picture constructed from the components of F_3 shows that it points below the positive x axis into the fourth quadrant. The tangent of the angle that F_3 makes with the x axis is given by Eq. (1.9):

$$\theta = \arctan \left| \frac{F_{3y}}{F_{3x}} \right| = \arctan \left(\frac{141}{41} \right) = \arctan 3.44 = 73.8°.$$

Thus, F_3 has a magnitude of 147 N and points 73.8° below the positive x axis. ∎

2.2 Problem-Solving Techniques and Examples

Often we use statics to calculate one or two unknown forces acting on an object if all other forces acting on it are known. Suppose, for example, that you bend over to lift a box of books. We can identify many of the forces acting on your body: the force of gravity, the books pulling down on your arms, and the floor pushing up on your feet. Statics, however, enables us to estimate with reasonable accuracy two difficult-to-measure forces acting inside your body: the tension in your back muscles and the compression force on discs that separate vertebrae in the backbone. If we can calculate these forces when using different lifting techniques, then perhaps we can learn what is the best lifting technique in order to avoid back injury.

How do we use statics and the equations of equilibrium to calculate these unknown forces? A useful procedure is outlined below. We take one or two small parts of a complex situation and apply the equations of equilibrium to these parts. If the following technique is mastered, complex statics problems can be solved routinely. The procedure is illustrated for the simple example of a plant that weighs 50 N and hangs at the end of a rope attached to the ceiling of a room. We wish to find the force of the rope on the plant.

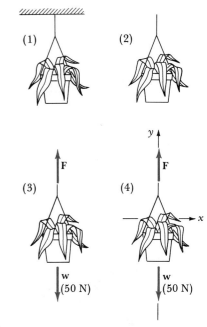

Techniques for Solving Problems in Statics

1. Make a drawing of the whole structure being considered.

2. Choose one part or object in this structure and draw it separately. The object chosen is usually a part of the structure on which many of the known forces and at least one of the unknown forces act. Both conditions of equilibrium must be satisfied by this object, since it is at rest.

3. After choosing an object on which to focus attention, draw arrows to represent all forces acting on that object. These forces should be identified symbolically (for example, $\mathbf{P}, \mathbf{w}, \mathbf{T}, \mathbf{F}_c$). This is called a force diagram or a free-body diagram. Be sure to show all forces acting on the object but do not show forces that this object exerts on other parts of the structure.

4. Next, superimpose an x axis and a y axis on your force diagram. These axes may be oriented in any direction. Once you decide on the placement of the axes and the origin of these coordinates, all future calculations must be consistent with that choice.

5. Write the two equations for the first condition of equilibrium [Eqs. (2.2a) and (2.2b)]. To do this, you must first resolve the known forces into their components as described in Section 1.7. You will also have to include in your equations symbols for the unknown components. When finished, you will have two equations that you can use to solve for unknowns. (Later in the chapter the second condition of equilibrium provides a third equation to use in solving for unknowns.)

6. Rearrange the equations from Part 5 to solve for the unknowns.

In some problems you may have to complete the six steps outlined above several times, each time using a different object for your force diagram. For easy problems you may not need to follow this procedure. However, it is difficult to learn the procedure on difficult problems, so try to use it for both easy and difficult ones.

Most of the following examples that illustrate this technique involve what are called **tension forces,** the force with which a string, rope, or cable pulls on an object to which it is attached. The tension force always pulls in the direction of the cable and is exerted uniformly along its entire length.

EXAMPLE 2.2 In Fig. 2.4a a 100-N weight hangs by a system of three cords oriented as shown. Calculate the tension in each cord.

SOLUTION The first step in solving the problem involves drawing the whole structure, already done for us in Fig. 2.4a. Next we choose one part of this

(5) $\quad \Sigma F_x = F_x + w_x$
$\quad\quad = 0 + 0 = 0$
$\quad \Sigma F_y = F_y + w_y$
$\quad\quad = F - w$
$\quad\quad = F - 50\,\text{N} = 0$

(6) $\quad F = \underline{50\,\text{N}}$

(a)

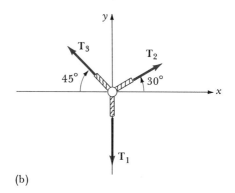

(b)

FIG. 2.4. (a) A weight suspended from cords. (b) A force diagram of the knot at the junction of the three cords.

(a)

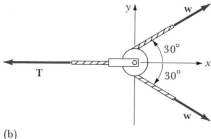

(b)

FIG. 2.5. (a) A traction device for the leg. (b) The forces acting on the pulley of the traction device.

structure and draw a force diagram for it. For this problem we choose the knot at the junction of the three cords as our object of interest for a force diagram (Fig. 2.4b). The knot is the only part of the whole structure on which the three tension forces act. Note that the object for the force diagram does not have to be separate from other objects but can be attached to adjacent objects, such as the three cords.

Three tension forces act on the knot as shown in Fig. 2.4b. The tension in the vertical cable (T_1) equals the weight hanging at its end (100 N). If it did not, the weight would either rise or fall, depending on whether T_1 was greater or less than 100 N. For the weight to remain stationary, T_1 must equal the 100-N weight of the block.

A coordinate system has been drawn in the force diagram shown in Fig. 2.4b. We can now apply the equilibrium equations ($\Sigma F_x = 0$ and $\Sigma F_y = 0$) in order to find T_2 and T_3:

$$\Sigma F_x = 0 \quad \text{or} \quad T_{1x} + T_{2x} + T_{3x}$$
$$= +T_1 \cos 90° + T_2 \cos 30° - T_3 \cos 45° = 0.$$

$$\Sigma F_y = 0 \quad \text{or} \quad T_{1y} + T_{2y} + T_{3y}$$
$$= -T_1 \sin 90° + T_2 \sin 30° + T_3 \sin 45° = 0.$$

If we substitute $T_1 = 100$ (we omit the newton units until the end to keep the equations simple) and the values of the sines and cosines into the above two equations, we find that

$$0.0T_1 + 0.87T_2 - 0.71T_3 = 0, \qquad \textbf{(2.3a)}$$

$$-100 + 0.5T_2 + 0.71T_3 = 0. \qquad \textbf{(2.3b)}$$

Each of the above equations contains two unknowns. A common technique for solving such equations is to rearrange one of them so that one unknown is equal to some function of the other unknown. From Eq. (2.3a), we find that $T_2 = (0.71/0.87)T_3 = 0.82T_3$. Substituting this expression for T_2 into Eq. (2.3b), we find that $-100 + 0.5(0.82T_3) + 0.71T_3 = 0$. This equation can be solved for T_3:

$$T_3 = \frac{100}{0.5(0.82) + 0.71} = \underline{89\ \text{N}}.$$

Since $T_2 = 0.82T_3$, we find then that $T_2 = 0.82 \times 89 = \underline{73\ \text{N}}$. ∎

As you may have noted in the previous example, the object considered in statics problems (such as the knot) is often some small part of the whole structure shown in the original picture. The object chosen for a force diagram usually has most of the important forces, including the unknown force or forces, acting on it. Note carefully the object chosen in the next example. Once again, it is the one part of the whole structure on which the known and unknown forces act.

EXAMPLE 2.3 Calculate the magnitude of the tension force **T** pulling on the foot in the traction device shown in Fig. 2.5a. The weight **w** hanging from the cable attached to the wall is 40 N, or about 9 lb.

SOLUTION When several ropes or cables are joined together at a junction or a knot, as in Example 2.2, the tension in each cable is usually different. However, the tension along a single cable is the same at every point. This is true even if the cable changes directions as it passes around a pulley. Thus, the tension at all points along the hanging cable shown in Fig. 2.5a is the same and equals the 40-N weight hanging at its end.*

We choose the pulley attached to the foot as the object of interest for a force diagram (Fig. 2.5b). To visualize more easily the forces acting on the pulley, we include in the force diagram the small sections of rope attached to or passing around the pulley. In this way it becomes apparent that the hanging cable exerts two forces on the pulley, each of magnitude **w** and oriented as shown in Fig. 2.5b. The cable attached to the foot exerts an unknown tension force **T** on the pulley. The equations of equilibrium for the pulley are:

$$\Sigma F_x = 0 \quad \text{or} \quad -T\cos 0° + w\cos 30° + w\cos 30° = 0.$$
$$\Sigma F_y = 0 \quad \text{or} \quad 0 + w\sin 30° - w\sin 30° = 0.$$

Since $w = 40\,\text{N}$ and $\cos 0° = 1$, the x equation allows us to solve for T:

$$T = 2w\cos 30° = 2(40\,\text{N})0.87 = \underline{69\,\text{N}}.\quad\blacksquare$$

EXAMPLE 2.4 A young couple parked in a car on a moonlit evening find that they are stuck in the sand. The young woman, having studied physics, ties a rope to the car's front bumper and stretches it around a tree in front of the car as shown in Fig. 2.6a. She then pushes with a 300-N (equal to 70-lb) force against the center of the rope, causing it to deflect as shown. The car starts to move. Calculate the force of the rope on the car.

SOLUTION The problem is worked easily by choosing as the object for a force diagram the short section of rope being pushed by the young woman (Fig. 2.6b). This is the one part of the whole situation shown in Fig. 2.6a on which the known force acts. Three forces act on this center section of rope (Fig. 2.6b): the woman pushes it with a 300-N force, and the rope is pulled from each side with an unknown tension force caused by the adjacent pieces of rope.

Using the coordinate axes shown in Fig. 2.6b, we find that the equations of equilibrium for the center section of rope are:

$$\Sigma F_x = T_{1x} + F_x + T_{2x} = T_1\cos 5° + F\cos 90° - T_2\cos 5° = 0,$$
$$\Sigma F_y = T_{1y} + F_y + T_{2y} = -T_1\sin 5° + F\sin 90° - T_2\sin 5° = 0.$$

The first equation above implies that $T_1 = T_2$; that is, the tension is the same on each side of the rope. Substituting $T_1 = T_2 = T$ into the second equation and solving for T produces the equation

$$T = \frac{F\sin 90°}{2\sin 5°} = \frac{(300\,\text{N})(1)}{2(0.087)} = 1720\,\text{N}.$$

The young woman, using ingenuity, has succeeded in applying a force of almost 400 lb to the car. $\blacksquare$

(a)

(b)

FIG. 2.6. (a) A technique for pulling a car out of the sand. (b) A force diagram for the section of rope being pushed by the woman.

---

*This is true only if the cable has negligible weight compared to the weight hanging at its end and if the pulleys around which it passes are frictionless. In all of our problems we will assume that these two conditions hold.

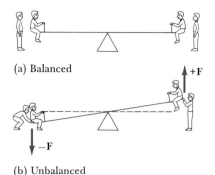

(a) Balanced

(b) Unbalanced

FIG. 2.7. A seesaw rotationally unbalanced by two forces that add to zero.

2.3 Torques

We have found that when an object is in equilibrium, the sum of all forces acting on it is zero. This first condition of equilibrium guarantees that the object, if at rest, will not start moving to a new location. However, this condition does not guarantee that the object will not tip over or rotate. Another condition is needed to insure rotational equilibrium.

Consider a seesaw in equilibrium (Fig. 2.7a). Since it is in equilibrium, the sum of the forces acting on it must be zero. We can disturb this equilibrium by applying two forces of equal magnitude, one up, the other down, at opposite ends of the seesaw (Fig. 2.7b). The seesaw begins to rotate and is no longer in rotational equilibrium. However, the sum of all forces acting on the seesaw is still zero, since we have added a net force of zero [$\mathbf{F} + (-\mathbf{F}) = 0$]. Evidently, another condition besides balanced forces is needed to maintain rotational equilibrium. This condition concerns the torques caused by the forces.

Consider the ability of the force of your hand to turn a wrench attached to a rusted bolt. In Fig. 2.8a, the hand pulls up directly over the head of the wrench. When applied in this position, the force will not turn the wrench. If we apply the force halfway along the handle, as in Fig. 2.8b, it may turn the bolt, although with difficulty. If we apply the same force at the end of the handle as in Fig. 2.8c, the bolt will turn more easily. We see that the place where a force is applied to an object is important in causing the object to rotate.

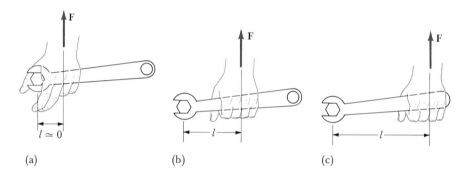

FIG. 2.8. A force's ability to turn a wrench depends on where the force is applied: the farther from the head, the better the turning ability.

A quantity called **torque** is a measure of the ability of a force to cause rotation. Torque is given the symbol τ (the Greek letter tau). In the simplest cases, when the force is directed perpendicular to an object such as a wrench, torque depends on the magnitude of the force and on the distance between the point of application of the force and the **fulcrum,** the point about which the object rotates. This distance is called the moment arm or lever arm of the force and given the symbol l. The torque τ of a force is the product of the magnitude of the force F and the moment arm l; that is, $\tau = (F)(l)$.

Often the forces that produce rotation are not directed perpendicular to the object. Consider the force $\mathbf{F}$ of the rope pulling on the beam in Fig. 2.9. The beam can rotate about a bearing placed at its left end. We will choose this position as the origin of coordinates. To calculate the torque caused by the force $\mathbf{F}$ about this origin, we first need to find the moment arm. To do so draw an extended, dashed line that passes through the point where the force acts on the object and that points in the direction of $\mathbf{F}$. This line is called the **line of action** of $\mathbf{F}$. Next, draw a dotted line from the origin of coordinates to the line of action. This second line is drawn so that it is perpendicular to the line of action, forming

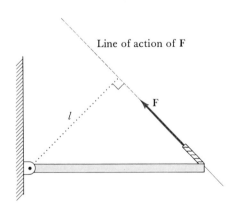

Line of action of F

FIG. 2.9. The line of action of a force pulling on a beam, shown together with the moment arm l of that force.

a right angle at their intersection (see Fig. 2.9). The distance from the origin to this intersection of the two lines is the moment arm *l* of the force. Thus, a general definition of the **moment arm** is the perpendicular distance from the origin of the coordinates to the line of action of the force. Having found the moment arm, we can then define the torque:

The **torque** τ caused by a force **F** is the product of the magnitude of the force and its moment arm *l*:

$$\tau = \pm(F)(l). \qquad (2.4)$$

The torque is positive if the force tends to rotate the object counterclockwise about the origin and is negative if it tends to rotate the object clockwise.

EXAMPLE 2.5 Calculate the torque caused by the 40-lb weight hanging from the beam shown in Fig. 2.10a. For this problem place the origin of coordinates at the pin on the left side of the beam.

SOLUTION The line of action and moment arm of this force are shown in Fig. 2.10b. We see that $l = 5$ ft, $F = 40$ lb, and the torque is negative, since it tends to rotate the beam clockwise about the origin. The torque caused by the force is $\tau = -(5\text{ ft})(40\text{ lb}) = \underline{-200\text{ ft} \cdot \text{lb}}$. ∎

EXAMPLE 2.6 A rope pulls on the end of a beam as pictured in Fig. 2.10a. The tension in the rope is 800 N. Calculate the torque about the pin at the end of the beam due to this tension force.

SOLUTION The line of action of **T** passes directly through the pin where the origin of coordinates is located (Fig. 2.10c). Consequently, the moment arm of this force is zero, and the torque is therefore zero: $\tau = (l)(T) = (0)(800\text{ N}) = \underline{0}$. ∎

EXAMPLE 2.7 Calculate the torque on the beam shown in Fig. 2.11a due to the 500-N force **F**. The length of the beam is 3 m.

SOLUTION We are given that $F = 500$ N and $L = 3$ m. The moment arm of **F** is shown in Fig. 2.11b and can be calculated by taking the sine of the 30° triangle shown in that figure:

$$\sin 30° = \frac{\text{opposite side}}{\text{hypotenuse}} = \frac{l}{L}$$

(a)

(b)

(c)

FIG. 2.10. (a) A beam being pulled from its end by a rope tension force and supporting a 40-lb weight. (b) The line of action and moment arm of the weight force. (c) The line of action of the tension force passes through the origin, so that its torque is zero. Another force (not shown) is needed to keep the beam in equilibrium.

FIG. 2.11. (a) A force acting on a beam; (b) the moment arm *l* of the force; and (c) resolution of the force into components in order to calculate the torque by the component method.

(c)

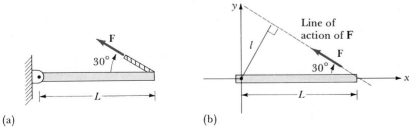

(a)   (b)

or $l = L \sin 30°$. The torque due to **F** is positive, since it tends to rotate the beam counterclockwise, and is

$$\tau = +(F)(l) = (F)(L \sin 30°) = (500 \text{ N})(3 \text{ m} \times 0.5) = \underline{750 \text{ N} \cdot \text{m}}.$$

Another way to calculate the torque caused by **F** is shown in Fig. 2.11c. The force is first resolved into its x and y components, $\mathbf{F}_x$ and $\mathbf{F}_y$. We then calculate separately the torque caused by each component. As can be seen in Fig. 2.11c, the moment arm of $\mathbf{F}_y$ is L and that of $\mathbf{F}_x$ is zero because its line of action passes through the origin. The total torque caused by **F** is the sum of the torques of its components:

$$\tau = (F_y)(l_y) + (F_x)(l_x) = (F \sin 30°)(L) + (F \cos 30°)(0)$$
$$= (500 \text{ N} \times 0.5)(3 \text{ m}) + 0 = \underline{750 \text{ N} \cdot \text{m}}.$$

This is the same answer as calculated using the technique shown in Fig. 2.11b. Often it is easier to calculate a torque by replacing a force with its components and then adding the torques caused by each component. ∎

2.4   Second Condition of Equilibrium

We learned in the previous section that the torque exerted by a force is a measure of its ability to disturb an object's rotational equilibrium. In order for an object to remain in rotational equilibrium, the positive torques that tend to rotate it counterclockwise must be balanced by equal magnitude negative torques that tend to rotate it clockwise. This is the second condition of equilibrium:

Second condition of equilibrium: For an object to remain in rotational equilibrium, the sum of all torques acting on the object must be zero:

$$\Sigma \tau = \tau_1 + \tau_2 + \tau_3 + \cdots = 0, \tag{2.5}$$

where $\tau_1, \tau_2, \tau_3, \ldots$ represent the torques caused by all forces $\mathbf{F}_1, \mathbf{F}_2, \mathbf{F}_3, \ldots$ acting on the object.

The two conditions that must be satisfied for an object to remain in equilibrium are summarized below:

$$\text{First condition:} \quad \Sigma F_x = 0, \tag{2.2a}$$
$$\Sigma F_y = 0. \tag{2.2b}$$
$$\text{Second condition:} \quad \Sigma \tau = 0. \tag{2.3}$$

In the following examples these conditions are used to calculate unknown forces acting on objects in equilibrium.

EXAMPLE 2.8   A woman shown in Fig. 2.12a supports the weight of a 200-lb crate resting on a lever. Calculate the upward force of her hand on the lever and the force of the fulcrum on the lever. Her hand is 15 ft from the fulcrum, and the crate is 3 ft from the fulcrum.

SOLUTION The lever is a logical object for a force diagram, since both unknown forces act on it as does the known weight force of the crate pushing down on it (see Figs. 2.12b and c).

The origin of the x and y coordinates may be placed anywhere. However, once chosen, the torques must then be calculated relative to that origin even though it may not be the real fulcrum of the lever. If the object is in equilibrium, the torques must add to zero no matter what origin is used. For example, in Fig. 2.12b the origin is located at the point where the woman's hand pushes on the beam. We calculate the tendency of forces to cause rotation about that point. The weight force **w** tends to rotate the beam counterclockwise about this point and causes a torque equal to $\tau_w = +(w)(l) = +(200\text{ lb})(12\text{ ft}) = 2400\text{ lb}\cdot\text{ft}$. The actual fulcrum tends to rotate the beam clockwise about the origin shown in Fig. 2.12b and causes a torque given by $\tau_F = -(F)(l_F) = -(F)(15\text{ ft})$. Since **P** has a zero moment arm about the origin shown in (b), its torque is zero. The sum of these three torques must be zero if the beam is in rotational equilibrium:

$$\Sigma\,\tau = \tau_F + \tau_w + \tau_P = -(15\text{ ft})F + 2400\text{ lb}\cdot\text{ft} + 0 = 0.$$

Rearranging this equation, we find that

$$F = \frac{2400\text{ lb}\cdot\text{ft}}{15\text{ ft}} = \underline{160\text{ lb}}.$$

Let us rework this problem, only this time with the origin of coordinates located at the real fulcrum, as shown in Fig. 2.12c. The moment arms of the forces are different about that origin as are the torques. But since the beam is in equilibrium, the torques must still add to zero:

$$\Sigma\,\tau = \tau_F + \tau_w + \tau_P = +(F)(0\text{ ft}) - (200\text{ lb})(3\text{ ft}) + (P)(15\text{ ft}) = 0.$$

Rearranging this equation, we find that

$$P = \frac{(200\text{ lb})(3\text{ ft})}{(15\text{ ft})} = \underline{40\text{ lb}}.$$

We have been able to solve for both unknown forces by simply applying the torque equation for two different coordinate systems. ∎

When using this torque equation to solve for unknown forces, it is often useful to place the origin of coordinates at the point where one of the unknown forces acts on the object, as was done in Fig. 2.12b and c. At that point the torque caused by this unknown force is zero because it has a zero moment arm. Consequently, that particular unknown force does not appear in the torque equation, and the equation can be used to solve for the other unknown force.

Finally, we should note that the results of these calculations using the second condition of equilibrium are consistent with the first condition of equilibrium. Referring to Fig. 2.12b, we see that Eq. (2.2b) should be written:

$$\Sigma\,F_y = F_y + w_y + P_y = F - w + P = 0.$$

Substituting the results of our previous calculations using the torque equations, $F = 160\text{ lb}$, $w = 200\text{ lb}$, and $P = 40\text{ lb}$, in the above equation produces a net force of zero on the left side of the equation. Our torque calculations are consistent with the first condition of equilibrium.

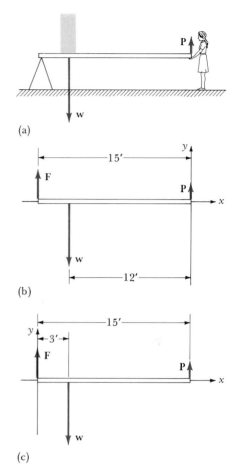

(a)

(b)

(c)

FIG. 2.12. (a) A woman applying a small force to the end of a beam in order to support a large weight force near the beam's fulcrum. (b) and (c) Force diagrams of the beam for two different coordinate systems.

(a)

(b)

FIG. 2.13. (a) The foot as a lever. (b) A model of the foot as a lever while supporting the leg by tension in the Achilles tendon.

EXAMPLE 2.9 The foot and Achilles tendon muscle provide a nice example of a body lever (Fig. 2.13a). The ball of the foot touching the ground is the fulcrum, the bones in the foot are the arm of the lever, and the Achilles tendon and its associated muscles provide the tension force needed to lift the load (the calf bone in this case). The ground pushes up on the ball of the foot with a force of half the person's weight. The heel of the foot is just off the ground. Find the tension **T** in the Achilles tendon and the compression force **C** at the joint between the leg and foot for a person weighing 180 lb.

SOLUTION The object of interest is the lever consisting of the foot bones. The coordinates used are shown in Fig. 2.13b. The conditions of equilibrium for this lever are:

$$\Sigma F_x = 0 \quad \text{or} \quad w_x + C_x + T_x = 0 + 0 + 0 = 0.$$
$$\Sigma F_y = 0 \quad \text{or} \quad w_y + C_y + T_y = 180\,\text{lb}/2 - C + T = 0.$$
$$\Sigma \tau = 0 \quad \text{or} \quad -\frac{w}{2}(l_w) + C(l_c) + T(l_T)$$

or

$$-(180\,\text{lb}/2)(5.5\,\text{in}) + C(0) + T(1.6\,\text{in}) = 0.$$

T can be calculated using the last equation:

$$T = \frac{(180\,\text{lb}/2)(5.5\,\text{in})}{1.6\,\text{in}} = \underline{310\,\text{lb}}.$$

Using the y-force equation, we find that

$$C = \frac{180\,\text{lb}}{2} + T = 90\,\text{lb} + 310\,\text{lb} = \underline{400\,\text{lb}}.$$

A person's Achilles tendon becomes ripped or torn if the tension exceeds about 1500 lb. This tension is exceeded occasionally when athletes with powerful muscles in their lower leg run or jump. ∎

EXAMPLE 2.10 Calculate the tension in the cable supporting the beam shown in Fig. 2.14a. The beam's weight can be ignored. A plant weighing 100 N hangs from the beam as shown. Also calculate the force of the wall support on the beam.

SOLUTION A force diagram for the beam is shown in Fig. 2.14b. The origin of the coordinates is placed at the location of the unknown force **F** of the wall support on the beam. The horizontal component of **F** balances the horizontal component of **T**. The vertical component of **F** is needed to prevent the left edge of the beam from falling. To calculate **F,** it is easiest to consider its components F_x and F_y as unknowns (see Fig. 2.14c). These components plus the unknown cable tension T can be found using the three equations of equilibrium:

$$\Sigma F_x = 0 \quad \text{or} \quad F_x + w_x + T_x = F_x + 0 - T\cos 37° = 0.$$
$$\Sigma F_y = 0 \quad \text{or} \quad F_y + w_y + T_y = F_y - w + T\sin 37° = 0.$$
$$\Sigma \tau = 0 \quad \text{or} \quad \tau_F + \tau_w + \tau_T$$
$$= (F)(0) - (w)(0.6\,m) + \underbrace{(T\sin 37°)(0.9\,\text{m})}_{T_y} = 0.$$

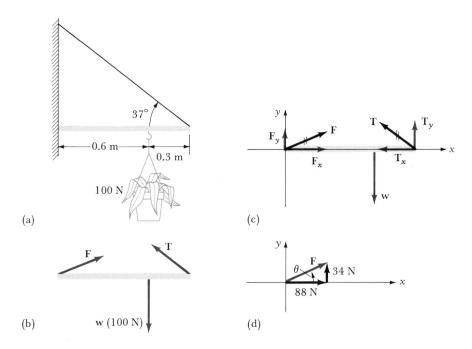

Note that the torques due to $\mathbf{F}_x$, $\mathbf{F}_y$, and $\mathbf{T}_x$ (shown in Fig. 2.14c) are zero, since these component forces have zero moment arms.

The last equation can be solved for T after substituting $w = 100\,\text{N}$:

$$T = \frac{w(0.6\,\text{m})}{\sin 37°(0.9\,\text{m})} = \frac{(100\,\text{N})(0.6\,\text{m})}{0.6(0.9\,\text{m})} = \underline{110\,\text{N}}.$$

Using the x-force equation, we find that

$$F_x = T\cos 37° = (110\,\text{N})(0.8) = 88\,\text{N}.$$

Using the y-force equation, we find that

$$F_y = w - T\sin 37° = 100\,\text{N} - (110\,\text{N})(0.6) = 34\,\text{N}.$$

The magnitude of the force of the wall support is determined using Eq. (1.7):

$$F = \sqrt{F_x^2 + F_y^2} = \sqrt{88^2 + 34^2} = 94\,\text{N}.$$

F is drawn from its component values in Fig. 2.14d. It points into the first quadrant. The tangent of the angle that it makes with the x axis is

$$\tan\theta = \left|\frac{F_y}{F_x}\right| = \frac{34\,\text{N}}{88\,\text{N}} = 0.39.$$

A 21° angle has a tangent of 0.39. Thus, **F** has a magnitude of 94 N and points 21° above the positive x axis. ∎

2.5 Gravitational Torques

The force of gravity often produces a torque on an object. For example, when a person bends over to pick something off the ground, the force of gravity on the upper body creates a torque that tends to rotate the upper body about the hips,

$\mathbf{F}_{\text{muscle}}$

w

FIG. 2.15. A man picking up an empty can. The weight of his upper body creates a torque about the hips that must be balanced by the torque produced by his back muscles.

as shown in Fig. 2.15. To prevent this the back muscles must exert an opposing torque. The line of action of this muscle force is along the muscle itself. However, it is not obvious what the line of action of the weight force is, since it is exerted approximately uniformly over the whole upper body. Gravity acts on the head, shoulders, arms, chest, and so forth. Each weight force on each body part has a different moment arm. This is true for any **extended body**—that is, for any body whose mass is distributed in space. It would be cumbersome to calculate separately the torques caused by the weights of each small part of an extended body.

Fortunately, we do not need to do this. Every extended body has what is called a **center of gravity**—a point at or near the center of the body at which we can pretend that all of the weight force acts when calculating torques. A variety of experimental and theoretical techniques can be used to measure or calculate the center of gravity of different objects, but we will not consider these techniques here. In most of the problems in this book the position of the center of gravity of an object is given. To calculate the gravitational torque acting on the object, we pretend that all of its weight force acts at the center of gravity. The usual rules for calculating torques are used.

EXAMPLE 2.11 A 150-lb rock climber hangs out from a ledge, as shown in Fig. 2.16a. Calculate the force $\mathbf{F}$ of the ledge on the climber's feet and the horizontal tension force $\mathbf{T}$ on the climber's arms. The climber's center of gravity is 1.5 ft out from the side of the cliff.

SOLUTION A force diagram for the climber is shown in Fig. 2.16b. We place the origin of coordinates at the climber's feet so that the unknown force F does not enter into the torque equation. The equations of equilibrium are

$$\Sigma F_x = 0 \quad \text{or} \quad F_x + T_x + w_x = F_x - T + 0 = 0.$$
$$\Sigma F_y = 0 \quad \text{or} \quad F_y + T_y + w_y = F_y + 0 - w = 0.$$
$$\Sigma \tau = 0 \quad \text{or} \quad \tau_F + \tau_T + \tau_w$$
$$= (F)(0) + (T)(4.5 \text{ ft}) - (w)(1.5 \text{ ft}) = 0.$$

Notice that, when calculating the weight torque, the whole weight force $\mathbf{w}$ is assumed to act at the center of gravity. Substituting $w = 150$ lb into the last equation, we find that

$$T = \frac{(150 \text{ lb})(1.5 \text{ ft})}{(4.5 \text{ ft})} = \underline{50 \text{ lb}}.$$

We find from the x-component equation that $F_x = T = 50$ lb and from the y-component equation that $F_y = w = 150$ lb. The magnitude of $\mathbf{F}$ is determined using Eq. (1.7):

$$F = \sqrt{F_x^2 + F_y^2} = \sqrt{50^2 + 150^2} = 160 \text{ lb}.$$

The components of $\mathbf{F}$ are both positive. Thus, $\mathbf{F}$ must point into the first quadrant. The tangent of the angle it makes with the x axis is

$$\tan \theta = \frac{150}{50} = 3 \quad \text{or} \quad \theta = 72°.$$

Thus, $\mathbf{F}$ has a <u>magnitude of 160 lb and points 72° above the positive x axis.</u> ∎

Center of gravity

(a)

$\mathbf{T}$

4.5 ft

$\mathbf{F}$ $\mathbf{w}$

1.5 ft

(b)

FIG. 2.16. (a) A climber hanging on a ledge. (b) The force diagram for the climber.

The final example in this chapter illustrates an important problem related to muscle tensions and to the compression forces on body joints. The actual calculations are a little difficult, but the background gained from other examples in this chapter should allow us to understand the principles. *

EXAMPLE 2.12 Lower-back problems have been with humans since our prehistoric ancestors assumed an upright position. In the following example we calculate the large compression force produced on the backbone when improper lifting techniques are used and the large tension forces in back muscles needed to do this lifting.

Figure 2.17 shows a person lifting a 50-lb weight from a leaning position. One group of back muscles is the primary "cable" holding the person in this position. The backbone serves as a "beam" to support this lifting activity. The backbone is made of many small bones called vertebrae, which are separated by discs. A disc is similar to a flexible, fluid-filled bag. As the back flexes, the discs change shape (Fig. 2.18).

Figures 2.19a and b show a mechanical model of the back of a 180-lb person lifting a 50-lb weight. We want to solve for the tension T in the cable (the back muscles) and the compressional force **F** on the rotary joint at the bottom of the beam (this compression will be present in much of the spine and can cause serious disc problems, particularly in the lower spine). The cable supporting the beam attaches to the beam (spine) one-third of the way from its end. The weight of the trunk w_1 at its center of gravity is 72 lb; w_2 (86 lb) is the combined weight of the head, arms, and 50-lb object being lifted, and w_2 acts at the center of gravity of these three objects.

SOLUTION The equations for equilibrium for this system are given below.

$$\Sigma F_x = -T \cos 18° + F_x = 0.$$
$$\Sigma F_y = -T \sin 18° - 72 - 86 + F_y = 0$$
$$\Sigma \tau = -72\left(\frac{L}{2}\cos 30°\right) - 86(L \cos 30°) + T\left(\frac{2L}{3}\sin 12°\right) = 0.$$

F_x and F_y are the unknown components of the compression force **F**. Solving the last equation, we find that $T = \underline{760\text{ lb}}$! We then find from the x-force equation that $F_x = 730$ lb and from the y-force equation that $F_y = 390$ lb. Thus, the magnitude of F is

$$F = \sqrt{730^2 + 390^2} = \underline{830\text{ lb}}.$$

*This example is adapted from a paper by L. A. Strait, V. T. Inman, and H. J. Ralston, *Am. J. Phys.* **15:**375–382 (1947).

FIG. 2.17. Lifting a weight incorrectly by bending the back rather than the legs.

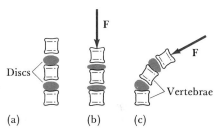
FIG. 2.18. (a) Vertebrae separated by discs under normal load; (b) larger compressional forces causing the discs to flatten; and (c) change of shape of the disc to accommodate changes in direction of the spine.

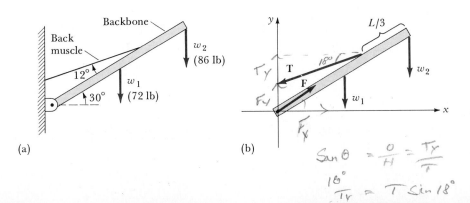
(a) (b)
FIG. 2.19. (a) A mechanical model of the back of a person lifting a weight. The beam is the spine, the cable is the spinal extensor muscles, w_1 is the weight of the trunk, and w_2 is the weight of the head, arms, and the 50-lb weight being lifted. (b) The forces on the beam.

The direction of **F** is determined from the equation

$$\tan \theta = \frac{F_y}{F_x} = \frac{390}{730} = 0.53 \quad \text{or} \quad \theta = \underline{28°}.$$

The back muscles and backbone experience forces greater than one-third ton when lifting from a bent position as shown in Fig. 2.17. These forces are reduced to less than half these values if the 50-lb weight is lifted by bending the knees while keeping the back vertical! ∎

Summary

1. **Conditions for equilibrium:** Objects in equilibrium are either at rest or moving with constant speed in a straight line. We have considered only objects at rest. For an object to be in equilibrium, all the forces and torques acting on the object, when added, must be zero:

First condition: $\Sigma F = 0$ **(2.1)** or
$$\Sigma F_x = 0. \quad \textbf{(2.2a)}$$
$$\Sigma F_y = 0. \quad \textbf{(2.2b)}$$

Second condition: $\Sigma \tau = 0.$ **(2.5)**

If the first condition of equilibrium is satisfied, an object at rest remains at rest; if the object is moving, its state of motion is unchanged. If the second condition is satisfied, an object is rotationally stable.

2. **Torques:** Torque indicates the ability of a force to cause a rotation. The torque τ of a force **F** acting on an object is

$$\tau = \pm(F)(l), \quad \textbf{(2.4)}$$

where l is the moment arm of the force. The moment arm is the perpendicular distance from the line of action of the force to the origin of coordinates. The torque is positive if the force tends to rotate the object counterclockwise about the origin and is negative if it tends to rotate the object clockwise.

3. **Problem-solving techniques:** To solve problems in statics:

    a. Make a drawing of the whole structure being considered.

    b. Choose one part or object of this structure and draw it separately. The part chosen usually has the unknown forces acting on it.

    c. Draw arrows representing all the forces acting on this object. Identify each force with letter symbols.

    d. Superimpose an x axis and a y axis on your diagram.

    e. Write the conditions for equilibrium [Eqs. (2.2) and (2.5)] for this object.

    f. Solve these equations for the unknown forces.

    g. Occasionally, the whole procedure should be repeated for a different object in order to find other unknown forces.

4. **Gravitational torques:** The gravitational force is exerted on every part of an extended body. To calculate the gravitational torque, however, we can pretend that all the weight force acts at a point near the center of the body called the center of gravity.

Questions

1. A hammock is tied with ropes between two trees. Are its ropes more or less likely to break if stretched tightly or loosely between the trees? Explain.

2. Something is wrong with the orientation of the ropes shown in Fig. 2.20. Using the first condition of equilibrium for the hanging pulley, explain this error and redraw the system as you would expect to see it.

3. Is it possible for an object to be in equilibrium when only one force acts on it? Explain.

4. Give three examples of an object that is not in rotational equilibrium even though the sum of the forces acting on the object is zero.

5. The tension in body muscles needed to lift various objects is usually five to ten times greater than the weight being lifted. Explain. A simple example would be helpful.

6. A ladder leans against a wall. Construct a force diagram showing the direction of all forces acting on the ladder. Be sure to identify the cause of each force.

7. Two books lie on top of each other on a table. Construct a force diagram showing the forces acting on the bottom book. Identify each force.

FIG. 2.20

8.   A strong wind blows on a young tree. Construct a force diagram for the tree. Explain which force prevents the tree from tipping over.

9.   A tee-shaped support with a plant hanging at the end is attached to the wall, as in Fig. 2.21. Three bolts can be used to attach it to the wall. When doing this, is it better to place two bolts at the top and one at the bottom or one at the top and two at the bottom? Explain.

10.   Is it more difficult to do a sit-up with your hands stretched in front of you or with them behind your head? Explain.

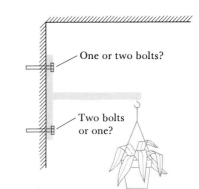

FIG. 2.21

Problems

2.1 and 2.2   First Condition of Equilibrium and Problem-Solving Techniques

1.   Three boys pull at the ends of three ropes tied together in a knot. The first boy pulls with a 20-lb force in the positive x direction, while the second pulls with a 40-lb force at an angle $45°$ above the positive x axis. With what force $\mathbf{F}_3$ (magnitude and direction) must the third boy pull so that the sum of these three forces is zero?

2.   A force table used in student laboratories in physics classes has a ring to which three strings are tied. The first two strings exert the following forces on the ring: $\mathbf{T}_1$ (2 N directed $37°$ below the positive x axis) and $\mathbf{T}_2$ (4 N directed $80°$ above the positive x axis). Calculate the force of the third string on the ring if it remains in equilibrium.

3.   Three ropes tied to a ring pull on it with the following forces: $\mathbf{T}_1$ (50 N in positive y direction), $\mathbf{T}_2$ (20 N $25°$ above negative x axis), and $\mathbf{T}_3$ (70 N $70°$ below negative x axis). Calculate the force that a fourth rope must exert on the ring to keep it in equilibrium.

4.   Calculate the tensions T_1, T_2, and T_3 in the ropes shown in Fig. 2.22 for the case where $\theta_2 = 63°$, $\theta_3 = 45°$, and $w = 100$ N.

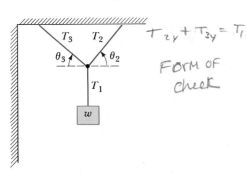

FIG. 2.22

5.   Calculate the tensions in the three ropes shown in Fig. 2.22 for the case where $\theta_2 = 37°$, $\theta_3 = 0°$, and $w = 10$ lb.

6.   Calculate the tensions in the three ropes shown in Fig. 2.22 for the case where $\theta_2 = 30°$, $\theta_3 = 50°$, and $w = 200$ N.

7.   Redraw Fig. 2.22 with $\theta_2 = 50°$ and $\theta_3 = 0$. Rope two is found to have a tension T_2 equal to 100 N. Calculate the values of T_1, T_3, and w.

8.   A sign weighing 15 lb hangs from a post as shown in Fig. 2.23. If the tension in the horizontal rope is 6 lb, calculate the magnitude and direction of the force caused by the other rope on the sign.

FIG. 2.23

9.   A monkey sits in a tire attached to a rope that passes over a pulley fastened to the ceiling and then back down to the monkey. With what downward force must a monkey weighing 100 N pull on the free end of the rope in order to support his weight while sitting in the tire? Explain. Note that the rope pulls up on the monkey with a force equal in magnitude to that with which he pulls down on the rope.

10.   A crate weighing 2000 N hangs by a single rope from the ceiling. A second rope attached to a pulley pulls the hanging rope to the side as shown in Fig. 2.24. If $\theta_1 = 60°$, what is the magnitude and direction of the tension force $\mathbf{T}_2$?

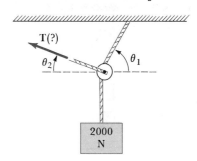

FIG. 2.24

11. The 2000-N hanging crate shown in Fig. 2.24 is pulled to the side by a 200-N tension force T_2. Write the two equations for the first condition of equilibrium using the pulley as the object of interest for your force diagram. Calculate θ_1 and θ_2. You may need to use the identity $\sin^2\theta + \cos^2\theta = 1$.

12. Find the tension T and weight w needed to keep the 100-N weight shown in Fig. 2.25 in equilibrium.

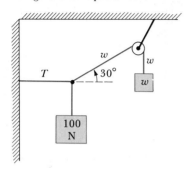

FIG. 2.25

13. The strap of a backpack causes the shoulder muscle to be deflected by $4°$ (see Fig. 2.26) when the strap pulls down on this muscle with a force of 20 lb. Calculate the tension in the muscle.

FIG. 2.26

14. A 150-lb tightrope walker stands at the center of a tightrope that is 20 ft long. The tightrope will break if its tension exceeds 1500 lb. Calculate the smallest angle at which the rope can bend up from the horizontal on either side of the walker in order to avoid breaking.

2.3 Torques

15. A 260-lb mutineer on Captain Bligh's ship is made to "walk the plank." The plank, which extends 9 ft beyond its support, will break if subjected to a torque greater than 2200 lb · ft. Will the crewman break the plank before stepping off its end? Explain.

16. A high diver weighing 700 N stands at the end of a diving board 3.5 m from its support. Calculate the magnitude of the torque on the board about the point of support.

17. Calculate the torques about point O caused by forces $\mathbf{F}_1$, $\mathbf{F}_2$, $\mathbf{F}_3$, and $\mathbf{F}_4$ in Fig. 2.27. All forces have magnitudes of 100 N and are applied a distance of $2m$ from O.

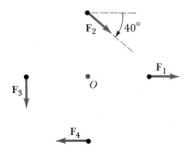

FIG. 2.27

18. Your hand holds a quart of milk while your arm is bent at the elbow in a 90° angle. *Estimate* the torque caused by the milk on your arm about the elbow joint. Indicate all numbers used in your calculations. This is an estimate and your answer may differ by 10 to 50 percent from the answers of others.

19. A person's bent arm is shown in Fig. 2.28. The person holds a ball whose weight is 40 N. Calculate the torques on the forearm about the elbow joint caused by the downward force of the ball and the upward forces of the biceps and triceps muscles. $F_T = 100$ N and $F_B = 340$ N.

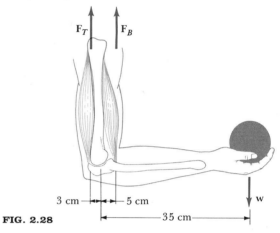

FIG. 2.28

20. Three forces are applied to the beam shown in Fig. 2.29. (a) Calculate the torque about the pivot point on the left caused by forces $\mathbf{F}_1$ and $\mathbf{F}_2$. (b) At what distance must $\mathbf{F}_3$ be applied to cause a torque that balances those produced by $\mathbf{F}_1$ and $\mathbf{F}_2$? All forces have magnitudes of 200 N.

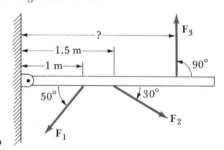

FIG. 2.29

21. The torque caused by the tension force $\mathbf{T}$ shown in Fig. 2.19 is $T(2L \sin 12°/3)$. Justify this statement.

22. A 50-lb girl sits 4 ft from the fulcrum of a seesaw. (a) Calculate the magnitude of the torque exerted by the girl on the seesaw. (b) At what distance from the fulcrum on the other side should the girl's 80-lb friend sit in order that his torque is equal in magnitude to her torque?

2.4 Second Condition of Equilibrium

23. The fulcrum of a seesaw 4 m long is located 2.5 m from one end. A person weighing 300 N sits on the long end. What must the weight of a person sitting at the short end be in order to balance the seesaw? Ignore the weight of the seesaw.

24. A rod 1 m long hangs from the ceiling with a string attached to its middle. A 200-N weight hangs from one end of the rod and a 100-N weight from the other end. Where should a third 150-N

weight hang relative to the center string in order to balance the rod?

25. Two people carry a stretcher 4 m long. A lion weighing 1000 N sits on the stretcher 1 m from one side. (a) Place an origin of coordinates at the position of the person nearest the lion and calculate the torque exerted by the lion on the stretcher. (b) What must the force of the other person on the stretcher be to produce a torque that balances that caused by the lion? (c) Repeat this procedure with the origin at the person farthest from the lion and find the force exerted on the stretcher by the person nearest the lion.

26. The biceps muscle of the person's arm shown in Fig. 2.28 provides the positive torque needed to support the 50-N ball held in the person's hand. Calculate the force of the biceps muscle on the arm. Ignore the triceps muscle and the arm's weight.

27. The person's arm shown in Fig. 2.28, instead of holding a ball, pushes down on a table with a force of 20 N at the fingers. Calculate the force of the triceps muscle needed to balance the torque caused by the upward force of the table on the fingers. Ignore the biceps muscle and the arm's weight.

28. Find the force of the biceps muscle on the arm in Fig. 2.28 when a person lifts 30 lb with the hand. Also calculate the force of the bone in the upper arm on the bone in the forearm at the elbow joint. Ignore the triceps muscle and the weight of the arm.

29. Calculate the magnitude of the tension force **T** exerted by the hamstring muscles (in the back of the thigh) and the compression force **F** at the joint between the thigh bone and calf bone when a 20-lb force pulls on the foot as shown in Fig. 2.30.

FIG. 2.30

30. A 130-lb person stands on a diving board (Fig. 2.31). Calculate the forces of the two supports on the board. Ignore the weight of the board.

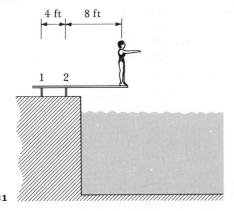

FIG. 2.31

31. The weightless beam shown in Fig. 2.32a is analogous to a human body when lifting a weight in a bent position (Fig. 2.32b). Calculate the tension in the cable and the force of the wall support on the beam when the hanging weight is 400 N (90 lb.).

FIG. 2.32. (a) A beam with hanging weight, and (b) its biological analog.

32. Calculate the tension in a system such as shown in Fig. 2.32a only with an 800-N hanging weight and a beam tilted at 30° rather than 15°.

33. Find the tension T in the cable and the force **F** (magnitude and direction) of the wall support on the weightless beam shown in Fig. 2.33.

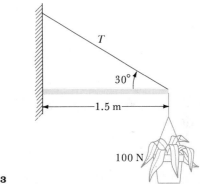

FIG. 2.33

34. A plank 10 m long is supported by cables from each end. The left cable has a tension of 300 N and the right cable 400 N. Calculate the weight and location of a person sitting on the plank. Ignore the weight of the plank.

35. The object marked A in the mobile in Fig. 2.34 weighs 10 N. Calculate the weight of object B. The numbers in Fig. 2.34 indicate the relative lengths of the rods on each side of their supporting cord. Ignore the weight of the rods.

FIG. 2.34

2.5 Gravitational Torques

36. Repeat Problem 23 concerning a seesaw, but this time include the seesaw's 200-N weight in the calculations. The seesaw's center of gravity is at its geometrical center.

37. The diving board shown in Fig. 2.31 weighs 60 lb, and its center of gravity is at the board's geometrical center. Calculate the forces of the two supports on the board when a 120-lb person stands on the end.

38. The person's hand shown in Fig. 2.28 supports a 100-N weight. The forearm weighs 10 N and its center of gravity is 18 cm from the elbow joint. Calculate the force of the biceps muscle on the person's arm. Ignore the tension in the triceps muscle.

39. Repeat Problem 27 involving the triceps muscle but include the 10-N weight of the forearm in your calculations. The arm's center of gravity is 18 cm from the elbow joint. Ignore the force of the biceps muscle.

40. Repeat Problem 33 for a beam weighing 100 N whose center of gravity is at the beam's geometrical center.

41. Repeat Problem 34, but include the plank's 100-N weight in your calculations. Its center of gravity is at the center of the beam.

42. A football player's head, shown in Fig. 2.35, can be considered as a lever. The vertebra at the bottom of the skull serves as the fulcrum. The torque caused by the weight w of the head and helmet (18 lb) is balanced by the torque caused by the trapezius muscle tension T in the neck. (a) Calculate T. (b) If a 40-lb downward force is applied to the face mask, what muscle tension T is required to keep the head in equilibrium?

FIG. 2.35

Describing Motion: Kinematics

3

Having learned in the last chapter about objects at rest, we now begin the study of moving objects. Galileo Galilei (1564–1642), the Italian astronomer, physicist, and mathematician, was the first to introduce quantitative techniques for describing motion. He developed the concept of acceleration as the rate at which an object's velocity changes. Using his own heartbeat as a clock, Galileo discovered that the time taken by a pendulum to complete one back-and-forth vibration did not depend on the size of the swing. With this information, he invented a pendulum clock, versions of which are still sold today.

Galileo became most famous for his studies of the planets and stars. These studies led to a book supporting the theory that the earth was not at the center of the universe but was just another planet moving about the sun. Removing the earth from its cherished position at the center of the universe conflicted with religious views; and Galileo, at age 69 and in poor health, was forced by a tribunal of the Inquisition to confess the error of his thinking. After his confession, Galileo was forbidden from further work in science. While under guard in a small stone house, Galileo managed to prepare his last book, *Dialogues Concerning Two New Sciences,* summarizing his earlier research in the study of motion and mechanics. Galileo must be considered the father of **kinematics,** a subject in which an object's changing motion is described quantitatively.

In this chapter we will develop techniques for predicting the future position and motion of objects such as race cars, golf balls, rockets, and skydivers. These techniques rely on careful definitions of quantities such as velocity and acceleration. Equations will be formulated that relate an object's acceleration to its changing position and velocity at different times. These kinematical equations appear frequently in our future discussion of physics.

3.1 Types of Motion

We are surrounded by a variety of objects in motion—people, cars, the moon, the leaves vibrating on a tree, blood rushing through our arteries, and electrons speeding toward our television screen. This great variety of motion, when analyzed closely, is just a combination of three basic types:

1. **Translational motion,** which consists of the movement of an object from one place to another (for example, the movement of a car traveling along a highway or of a baseball speeding toward a batter);

2. **Rotational motion,** which consists of the turning or rotating of an object about an axis (such as the rotation of the hands of a clock or the earth's rotation around its north-south axis);

3. **Vibrational motion,** which consists of the back-and-forth movement of an object about a center of vibration (for instance, the back-and-forth movement of a swing or of a vibrating violin string).

For now, we will consider only translational motion. Rotational and vibrational motions are examined in Chapters 6, 7, and 16.

3.2 Average Speed and Velocity

Average Speed

The concept of speed is well known to us all, primarily because of our use of automobiles. A car that travels 80 km in one hour has an average speed of 80 km/hr.

Average speed, $\bar{v}$, is defined as the distance traveled divided by the time required to travel that distance:

$$\bar{v} = \frac{d}{t} = \frac{\text{distance traveled}}{\text{time}}. \tag{3.1}$$

A line over a symbol (for example, the line over the v) indicates the average value of the quantity appearing under the line. Units of speed include kilometers per hour, meters per second, miles per hour, or any unit of distance divided by time.

EXAMPLE 3.1 Calculate the average speed of an automobile that travels 60 km east and then 80 km north in two hours (Fig. 3.1).

SOLUTION The total distance traveled by the car is $60 + 80 = 140$ km. The time required for this trip is 2 hours. Thus,

$$\bar{v} = \frac{d}{t} = \frac{140 \text{ km}}{2 \text{ hr}} = \underline{70 \text{ km/hr}}. \qquad \blacksquare$$

Unit Conversions

Occasionally, we need to convert a quantity from one system of units to another. For example, suppose that we wish to convert a speed of 50 miles/hr to a speed in units of m/s. The conversion table inside the front cover of the book indicates that 1 mile = 1609 m. Also, we know that 3600 s = 1 hr. We rewrite these equalities as $1 = (1609 \text{ m}/1 \text{ mile})$ and $1 = (1 \text{ hr}/3600 \text{ s})$. Now to convert 50 miles/hr to m/s, we multiply by one, twice, substituting the expressions in the above equalities for the ones:

$$50 \frac{\text{miles}}{\text{hr}} = \left(50 \frac{\text{miles}}{\text{hr}}\right)(1)(1) = \left(50 \frac{\text{miles}}{\text{hr}}\right)\left(\frac{1609 \text{ m}}{1 \text{ mile}}\right)\left(\frac{1 \text{ hr}}{3600 \text{ s}}\right) = \underline{22.3 \frac{\text{m}}{\text{s}}}.$$

FIG. 3.1

The mile in the numerator cancels the mile in the denominator, just as algebraic quantities cancel. The hours also cancel.

If a unit to be converted appears to the second or higher power, then each power must be converted. For example, suppose that the cross-sectional area of the bone in your ankle is 0.5 in^2, and you need to know its area in units of cm^2. According to the conversion table inside the front cover, $1 \text{ in} = 2.54 \text{ cm}$ or $1 = (2.54 \text{ cm}/1 \text{ in})$. Each inch of the 0.5-in^2 area must be converted, so we multiply by one, twice:

$$0.5 \text{ in}^2 = (0.5 \text{ in}^2)(1)(1) = (0.5 \text{ in}^2)\left(\frac{2.54 \text{ cm}}{1 \text{ in}}\right)\left(\frac{2.54 \text{ cm}}{1 \text{ in}}\right) = 3.2 \text{ cm}^2.$$

The next example involves our definition of average speed plus a relatively easy unit conversion.

EXAMPLE 3.2 A boy runs along a straight road for 10 minutes at a speed of 3 m/s. The boy tires and walks for 5 minutes at 1 m/s. (a) Calculate the total distance traveled by the boy. (b) What is his average speed?

SOLUTION (a) The total distance traveled will be the sum of the distances traveled during each part of the trip. We can rearrange Eq. (3.1) to calculate these distances. While running at 3 m/s, the boy travels a distance

$$d = \bar{v}t = \left(3 \frac{\text{m}}{\text{s}}\right)(10 \text{ min})$$

$$= \left(3 \frac{\text{m}}{\text{s}}\right)\left(10 \text{ min} \times \frac{60 \text{ s}}{1 \text{ min}}\right) = \left(3 \frac{\text{m}}{\text{s}}\right)(600 \text{ s}) = 1800 \text{ m}.$$

While walking at 1 m/s, the boy travels a distance

$$d = \bar{v}t = \left(1 \frac{\text{m}}{\text{s}}\right)\left(5 \text{ min} \times \frac{60 \text{ s}}{1 \text{ min}}\right) = 300 \text{ m}.$$

Thus, the total distance traveled is $1800 \text{ m} + 300 \text{ m} = \underline{2100 \text{ m}}$.

(b) The average speed of the boy is calculated by dividing the total distance traveled by the total time $(10 \text{ min} + 5 \text{ min} = 15 \text{ min})$ used to travel this distance. Hence,

$$\bar{v} = \frac{d}{t} = \frac{2100 \text{ m}}{15 \text{ min}} = \frac{2100 \text{ m}}{15 \text{ min}\left(\dfrac{60 \text{ s}}{1 \text{ min}}\right)} = \underline{2.3 \text{ m/s}}. \qquad \blacksquare$$

Average Velocity

Speed, a scalar quantity, indicates only how far an object moves in a certain time but does not indicate the direction of that motion. Velocity, a vector quantity, depends on both the speed of an object and its direction of motion.

Average velocity, $\bar{\mathbf{v}}$, is defined as an object's displacement (a vector quantity) divided by the time required for this displacement:

$$\bar{\mathbf{v}} = \frac{\mathbf{d}}{t} = \frac{\text{displacement}}{\text{time}}. \qquad (3.2)$$

The idea of average velocity is seldom used in our future work. However, a slightly different quantity called instantaneous velocity, or simply velocity, is very important in the development of kinematics.

3.3 Instantaneous Speed and Velocity

We found in Example 3.1 that the average speed of a car that traveled 140 km in a time of 2 hr was 70 km/hr. It is unlikely that the car traveled at a constant 70 km/hr during the whole trip because most driving involves many starts and stops. The car probably traveled slower than 70 km/hr for part of the trip and faster during other portions. The car may even have moved backward for part of the trip. To have complete knowledge of how the trip was made, we need to know the car's velocity at each instant of time during the trip.

Instantaneous velocity indicates how fast an object moves at each instant of time and the direction of that motion. *Instantaneous speed* indicates only how fast an object is moving at each instant of time. (The speedometer of an automobile indicates its instantaneous speed.) For most of this chapter, we will only consider one-dimensional motion—motion along a single coordinate axis. In this case we need not use vector symbols for velocity. If an object moves in the positive direction along the axis, its velocity is positive; if it moves in the negative direction, its velocity is negative. The object's speed is the magnitude of the velocity.

In order to develop a more precise definition for instantaneous velocity and speed in one dimension, consider the motion of a drag racer shown in Fig. 3.2. The car starts from rest and, when given a signal, accelerates in the positive direction along a 200-m track, arriving at the finish line 4.0 s later. Hence, the average velocity of the car during the race is $\bar{v} = +200 \text{ m}/4.0 \text{ s} = +50.0 \text{ m/s}$. (We omit the vector symbol for $\bar{v}$ since the motion is restricted to one dimension.)

In order to obtain a more complete profile of the changing velocity of the car, suppose we photograph the racer at 0.5-s time intervals from the start of the race to the finish. Distance markers beside the track will show us where the car is located at the end of each successive time interval (Fig. 3.2). Because the car moves in only one direction, we can use the coordinate x to represent the car's position at different times t. In Table 3.1, the values of x are listed for different times. Notice that at a time of 0.0 s the race car is at the starting line at position 0 m. During the next 0.5 s, the car moves to position $x = 3$ m. Its average

TABLE 3.1 The Position and Speed of a Drag Racer at Different Times

$t(s)$	$x(m)$	$\Delta x(m)$	$\Delta x/\Delta t = \bar{v}$ (m/s)
0.0	0		
		3	3/0.5 = 6
0.5	3		
		10	10/0.5 = 20
1.0	13		
		16	32
1.5	29		
		22	44
2.0	51		
		27	54
2.5	78		
		34	68
3.0	112		
		41	82
3.5	153		
		47	94
4.0	200		

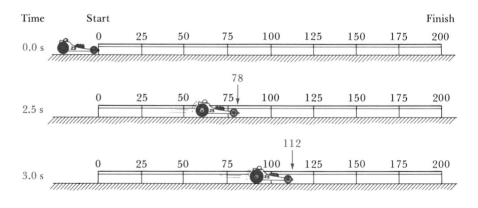

FIG. 3.2. The position of a drag racer at different times is determined by photographs showing its position relative to markers in the background.

velocity during that time is its change in position Δx, divided by the change in time Δt required for the position change:*

$$\bar{v} = \frac{\Delta x}{\Delta t} = \frac{3\,\text{m} - 0\,\text{m}}{0.5\,\text{s} - 0\,\text{s}} = +6\,\text{m/s}.$$

During the next 0.5 s, from $t = 0.5$ s to $t = 1.0$ s, the car moves from position 3 m to 13 m. Thus, its average velocity during that time is

$$\bar{v} = \frac{\Delta x}{\Delta t} = \frac{13\,\text{m} - 3\,\text{m}}{1.0\,\text{s} - 0.5\,\text{s}} = \frac{+10\,\text{m}}{0.5\,\text{s}} = +20\,\text{m/s}.$$

A list of the calculated values of average velocity for each successive 0.5-s time interval appears in the last column of Table 3.1. The velocities are positive because the car moves in the positive x direction.

Since the car's velocity is continually increasing, it almost certainly moves faster at the end of each 0.5-s time interval than at the beginning. In order to calculate its instantaneous velocity, we must divide the trip into much smaller time intervals. We might require, for example, that pictures be taken so often that the change in velocity between successive time intervals can be no more than 0.1 percent. If this condition is satisfied, then we consider the velocities we calculate from these frequent pictures to be instantaneous velocities. In practice, such accuracy is seldom necessary.

This example demonstrates that, in order to calculate an instantaneous velocity, we must know the displacement of an object during a very short time interval. For one-dimensional motion, instantaneous velocity and speed are defined as follows:

The **instantaneous velocity** of an object moving along the x axis is its change in position Δx divided by the short time Δt required for this change:

$$\text{Instantaneous one-dimensional velocity} = v = \frac{\Delta x}{\Delta t}. \tag{3.3}$$

If the object moves in the positive x direction, Δx is positive, as is the velocity. If the object moves in the negative x direction, Δx and v are both negative. The **instantaneous speed** of the object is just the magnitude of v and is always a positive number.†

EXAMPLE 3.3 An automobile moves along a straight track beside which there are markers indicating the car's position at different times. If the car is at

---

*The symbol Δ, delta, means a change in the quantity following the Δ. Thus, Δx (read "delta x") is a change in the position coordinate x; Δt (read "delta t") is a change in time; Δx and Δt are single symbols, not the product of Δ times x or Δ times t. The units of Δx are the units of x (for example, meters) and the units of Δt are the units of t.

†Using calculus, one-dimensional velocity is defined as

$$v = \lim_{\Delta t \to 0} \frac{\Delta x}{\Delta t};$$

that is, the instantaneous velocity equals the limiting value of the change in x divided by the change in t as the time interval approaches zero. As Δt approaches zero, so does Δx. However, the ratio of the two small numbers, which is the velocity, can be quite large, depending on how fast the object moves. This limiting value of $\Delta x/\Delta t$ is called the derivative of x with respect to t and is written as $v = dx/dt$.

position $x_1 = 51.2$ m at a time $t_1 = 5.3$ s and is at position $x_2 = 49.8$ m at a time 5.4 s, what is the car's velocity during that time interval?

SOLUTION The change in position of the car is $\Delta x = x_2 - x_1 = 49.8$ m $- 51.2$ m $= -1.4$ m. The negative sign means that the car has moved to the left, toward the negative x direction. The time interval during which this displacement occurred is $\Delta t = t_2 - t_1 = 5.4$ s $- 5.3$ s $= 0.1$ s. Hence, the car's velocity at that time is

$$v = \frac{\Delta x}{\Delta t} = \frac{x_2 - x_1}{t_2 - t_1} = \frac{-1.4 \text{ m}}{0.1 \text{ s}} = \underline{-14 \text{ m/s}}.$$

The velocity is negative because the car is moving in the negative x direction. The car's speed at that time is 14 m/s, the positive (absolute) value of -14 m/s.

◼

For the more general case where the object's motion is not restricted to one dimension, its instantaneous velocity is defined as its displacement $\Delta \mathbf{d}$ (a vector) divided by the very short time Δt needed for this displacement:

$$\text{instantaneous velocity} = \mathbf{v} = \frac{\Delta \mathbf{d}}{\Delta t}. \tag{3.4}$$

Suppose that a car undergoes a displacement of 2.3 m in a direction 30° N of E during a 0.1-s time interval; the car's velocity is 2.3 m/0.1 s = 23 m/s and is directed 30° N of E. The car's speed is simply the magnitude of the velocity or, in this case, 23 m/s.

Hereafter, we use the word *speed* to mean "instantaneous speed" and *velocity* to mean "instantaneous velocity."

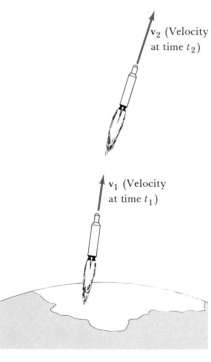

$\mathbf{v}_2$ (Velocity at time t_2)

$\mathbf{v}_1$ (Velocity at time t_1)

FIG. 3.3. The acceleration of a rocket during a time $t_2 - t_1$ is the ratio of its change in velocity ($\mathbf{v}_2 - \mathbf{v}_1$) and the time change ($t_2 - t_1$).

3.4 Acceleration

Whenever the velocity of an object changes, we say that the object is **accelerating.** A car waiting at a red light accelerates after the stoplight changes to green. Its velocity increases from zero as it moves through the intersection. A car also accelerates when it comes to a stop at a stop sign. We often call this a **deceleration**—a decrease in the speed of an object.

Acceleration can be defined more precisely with reference to Fig. 3.3. Suppose that at time t_1 an object is moving with a velocity $\mathbf{v}_1$ and at a later time t_2 its velocity has changed to $\mathbf{v}_2$.

The **average acceleration, $\overline{\mathbf{a}}$,** of the object is its change in velocity divided by the time required for that change:

$$\overline{\mathbf{a}} = \frac{\mathbf{v}_2 - \mathbf{v}_1}{t_2 - t_1} = \frac{\text{change in velocity}}{\text{time}}. \tag{3.5}$$

Hence, acceleration indicates the change in velocity and not how fast something is moving.

The **units of acceleration** are the units of velocity (for example, m/s or ft/s) divided by the unit of time (s). Thus, acceleration has units of length divided by time squared (for instance, m/s^2 or ft/s^2).

Three factors affect acceleration: (1) the change in the magnitude of the object's velocity—that is, its change in speed; (2) the change in the direction of the object's velocity; and (3) the time required for these changes. The following examples involving motion in one direction illustrate the first and third of these factors. For one-dimensional motion we do not need to use vector signs if we agree to indicate directions by $+$ and $-$ instead. A positive velocity indicates motion in the positive x direction; a negative velocity, motion in the negative x direction.

EXAMPLE 3.4 An automobile travels east in the positive x direction at $+20$ m/s and needs to pass a truck (Fig. 3.4). The automobile's velocity is increased to $+25$ m/s in 2 s; it still moves toward the east. Calculate its acceleration.

SOLUTION Since the motion is in the positive x direction, both the initial and final velocities are positive. The change in velocity is $v_2 - v_1 = (+25 \text{ m/s}) - (+20 \text{ m/s}) = +5$ m/s. This change occurs in a time $t_2 - t_1 = 2$ s. The average acceleration is then

$$\bar{a} = \frac{v_2 - v_1}{t_2 - t_1} = \frac{+5 \text{ m/s}}{2 \text{ s}} = \underline{+2.5 \text{ m/s}^2}. \quad\blacksquare$$

EXAMPLE 3.5 A car traveling east in the positive x direction at $+20$ m/s comes up behind a large truck and is unable to pass (Fig. 3.5). The car slows to $+15$ m/s in a time of 2 s. Calculate the car's acceleration.

SOLUTION The change in the car's velocity is $v_2 - v_1 = (+15 \text{ m/s}) - (+20 \text{ m/s}) = -5$ m/s. Note that the change in velocity is negative because the car's velocity in the positive x direction decreased by 5 m/s. This decrease occurred in a time $t_2 - t_1 = 2$ s. The average acceleration is then

$$\bar{a} = \frac{v_2 - v_1}{t_2 - t_1} = \frac{-5 \text{ m/s}}{2 \text{ s}} = \underline{-2.5 \text{ m/s}^2}.$$

The acceleration is negative because the car's velocity along the positive x direction decreased. It lost 2.5 m/s of speed in this direction each second. $\blacksquare$

Acceleration also depends on the time required for the velocity change. If the change in velocity had occurred in one second instead of two, the car's acceleration would have been twice as great.

As with speed and velocity, we are often interested in an object's instantaneous acceleration.

An object's **instantaneous acceleration, a,** is defined as its change in velocity, **Δv,** divided by the very short time, Δt, needed for that change:

$$\mathbf{a} = \frac{\mathbf{\Delta v}}{\Delta t}. \quad (3.6)$$

Time (t_1): 0
Velocity (v_1): 20 m/s, east

Time (t_2): 2 s
Velocity (v_2): 25 m/s, east

FIG. 3.4. The acceleration of the car is positive, since its velocity in the x direction increases.

Time (t_1): 0
Velocity (v_1): 20 m/s, east

Time (t_2): 2 s
Velocity (v_2): 15 m/s, east

FIG. 3.5. The acceleration of the car is negative, since its velocity in the positive x direction decreases.

The time interval Δt should be so short that the acceleration varies little from one time interval to the next.

We can estimate the instantaneous acceleration of an object if we know its velocity at different times. Consider an example of this calculation for the case of linear motion—motion along a straight line. The skydiver in Fig. 3.6 undergoes linear motion when falling toward the earth. She leaves the plane at time zero with an initial velocity of zero. As time progresses, her downward velocity increases; that is, she accelerates in the downward direction. Her velocities at different times are listed in Fig. 3.6. We have chosen to call the downward direction *positive* so that the velocities are all positive.

In order to calculate the skydiver's acceleration during different time intervals, we divide her change in velocity by the time required for that change. For example, at a time of 2 s her velocity is $+58$ ft/s. At 4 s her velocity has increased to $+98$ ft/s, an increase of $+40$ ft/s in the 2-s time interval. Thus, her average acceleration during that time is

$$\overline{a} = \frac{\Delta v}{\Delta t} = \frac{+40 \text{ ft/s}}{2 \text{ s}} = +20 \text{ ft/s}^2.$$

The calculated accelerations of the woman during other time intervals are listed in the last column of Fig. 3.6.

The technique we have used illustrates the way in which instantaneous acceleration can be estimated. In order to calculate more accurately the instantaneous acceleration at different times, we would need to know the woman's change in velocity over much shorter time intervals—perhaps every 0.1 s instead of every 2 s.

Time	Velocity	Δv (ft/s)	$a = \Delta v/\Delta t$ (ft/s^2)
0 s	0 ft/s		
		58	29 = 58/2
2 s	58 ft/s		
		40	20 = 40/2
4 s	98 ft/s		
		28	14
6 s	126 ft/s		
		20	10
8 s	146 ft/s		
		14	7
10 s	160 ft/s		
		8	4
12 s	168 ft/s		

FIG. 3.6. The changing velocity of a skydiver at different times.

3.5 Graphical Analysis of Velocity and Acceleration

Scientists often perform experiments in which one quantity is measured as another one changes. For example, we might photograph the changing position of a falling skydiver as a function of time. The information from these measurements can be used to determine a variety of things, such as the variation of the skydiver's velocity and acceleration with time, and the effect of air resistance on his or her fall.

The measured data are usually recorded in a table. This information often becomes easier to interpret when displayed in the form of a graph. But even graphs require some skill to understand. In this section we will develop a better understanding of the slope of a graph and how the slope relates to other quantities. This knowledge is especially useful when studying motion, since the changing velocity and acceleration of an object can be determined from the slopes of graphs in which position and velocity are plotted as a function of time.

As an example, consider the changing position of BART, the San Francisco Bay Area Rapid Transit commuter train, as it leaves a station. Its position at different times is tabulated in Fig. 3.7a. These measured data are plotted in a graph in Fig. 3.7b. BART's position appears on the vertical axis and the time on the horizontal axis. Each point on the graph corresponds to a measurement of position at one particular time. A smooth curve runs through these data points.

The slope of the curve shown in Fig. 3.7b—that is, how fast the curve rises or falls—indicates the velocity of the object. We can estimate the slope at a point by drawing a triangle, such as shown in Fig. 3.7b. The horizontal side of the triangle is called the *run* of the curve, and in this case it equals a change in time (10 s). The vertical line is called the *rise* of the curve; in this graph it equals the change in position (270 m) that occurs during the run time. The ratio of these

Time (s)	0	10	20	30	40	50	60
Position (m)	0	50	180	370	600	870	1200

(a)

(b)

FIG. 3.7. (a) A table showing the position of a BART train at different times. (b) These same data displayed in a graph.

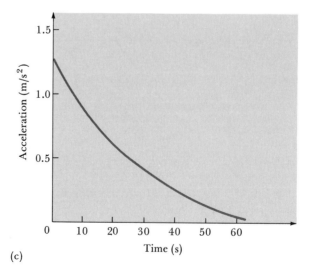

FIG. 3.8. (a) The slope of an x-versus-t graph is the velocity. (b) The slope of a v-versus-t graph is the acceleration. (c) An acceleration-versus-time graph.

two changes—the rise divided by the run—equals approximately the slope of the curve:

$$\text{slope} \cong \frac{\text{rise}}{\text{run}}.$$

For the graph plotted in Fig. 3.7b, the slope is just the velocity of the train:

$$\text{slope} = \frac{\text{rise}}{\text{run}} = \frac{\Delta x}{\Delta t} = \text{velocity}.$$

Notice that the slope is small at early times when the train begins its departure from the station. As it gains speed, the slope of the curve increases.

In the last paragraph we used the word *approximately*. The slope equals the ratio of the rise and run only if the run is infinitesimally short. To calculate exactly the slope of the curve at a particular point, we draw a line tangent to the curve at that point, as shown in Fig. 3.8a. The **slope** equals exactly the rise divided by the run of the tangent line.

We are now able to construct a velocity-versus-time graph, as shown in Fig. 3.8b. The velocity at each time is just the slope of the position-versus-time curve (Fig. 3.8a) at that time. Notice that after about 50 s the train's velocity changes little; it has nearly reached its maximum velocity.

Having constructed a velocity-versus-time graph, we can now use the same procedure to construct an acceleration-versus-time graph. Notice that the slope of the velocity-versus-time graph is given by

$$\text{slope} = \frac{\text{rise}}{\text{run}} = \frac{\Delta v}{\Delta t}.$$

which is just the acceleration. Thus, the slope of each point on the velocity-versus-time graph equals the train's acceleration at that point (plotted in Fig. 3.8c). The train's acceleration is greatest as it leaves the station. Eventually, after reaching a speed of about 30 m/s, the train's velocity no longer increases; its acceleration is zero.

3.6 Linear Motion with Constant Acceleration

The techniques described in previous sections for calculating the velocity and acceleration of an object require that we know an object's position at many different times. In this section we derive four simple equations that relate the position, velocity, and acceleration of an object at different times. With one of these equations we can calculate an object's acceleration if its position at two different times is known. By using other equations, we can calculate an object's future position and velocity if its acceleration is known.

Although these simple equations are easier to use than the techniques developed in the last sections, they have one major disadvantage: *the equations apply only to objects whose acceleration is constant.* Few situations in life involve constant acceleration, but many come close (for example, the acceleration of a falling object or of a car skidding to a stop). These equations will often be used in such situations even though the acceleration is not strictly constant.

The four equations are derived from the definitions of average velocity and acceleration. Because the derivations are done for linear motion, we will omit the vector signs and keep track of directions using + and − signs only.

EQUATION 1   The object's velocity at time zero is represented by the symbol v_0, and at a later time t its velocity is v. Since the object's velocity has changed by an amount $v - v_0$ during a time $t - 0$, its acceleration during that time, according to Eq. (3.5), is

$$a = \frac{v - v_0}{t - 0}.$$

Rearranging this equation to solve for v gives us our first equation of motion for constant acceleration:

$$v = v_0 + at. \tag{3.7}$$

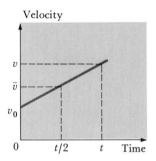

FIG. 3.9.   A velocity-versus-time graph is a straight line for constant acceleration.

EQUATION 2   Our next equation is derived by equating two different expressions for the average velocity of an object that starts at time zero with velocity v_0 and ends at time t with velocity v. To develop the first expression, we plot velocity on the ordinate of a graph versus time on the abscissa, as shown in Fig. 3.9. The straight line represents the equation $v = v_0 + at$. At $t = 0$, the velocity equals v_0. Since the acceleration is constant, the velocity increases in direct proportion to t and is represented by the straight line shown in Fig. 3.9. The average value of the velocity during that time interval is just the arithmetic average of the initial velocity and the final velocity:

$$\bar{v} = \frac{v + v_0}{2}.$$

The average velocity also equals the displacement of the object divided by the time required for the displacement. If the object starts at position x_0 at time zero and ends at position x at time t, its displacement is $x - x_0$. Thus, its average velocity is

$$\bar{v} = \frac{\text{displacement}}{\text{time}} = \frac{x - x_0}{t}.$$

We can now equate the two expressions for average velocity:

$$\frac{x - x_0}{t} = \frac{v + v_0}{2}.$$

Rearranging the above produces our second equation of motion for constant acceleration:

$$x - x_0 = \left(\frac{v + v_0}{2}\right)t. \tag{3.8}$$

EQUATION 3   The third equation is obtained by substituting $v = v_0 + at$ [Eq. (3.7)] for v in Eq. (3.8):

$$x - x_0 = \frac{(v + v_0)}{2}t = \frac{[(v_0 + at) + v_0]}{2}t.$$

When rearranged, we have our third equation:

$$x - x_0 = v_0 t + \tfrac{1}{2} a t^2. \tag{3.9}$$

EQUATION 4 The last equation is obtained by solving Eq. (3.7) for t; that is, $t = (v - v_0)/a$. This expression for t is then inserted into Eq. (3.8):

$$x - x_0 = \left(\frac{v + v_0}{2}\right) t = \left(\frac{v + v_0}{2}\right)\left(\frac{v - v_0}{a}\right) = \frac{v^2 - v_0^2}{2a}.$$

Rearranging, we obtain the fourth equation of motion:

$$2a(x - x_0) = v^2 - v_0^2. \tag{3.10}$$

In summary, then, we have four equations of motion for constant acceleration that relate the object's acceleration a, initial position x_0 and initial velocity v_0 at time zero to the object's final position x, and final velocity v at time t:

EQUATIONS FOR LINEAR MOTION WITH CONSTANT ACCELERATION

$$v = v_0 + at, \tag{3.7}$$

$$x - x_0 = \left(\frac{v + v_0}{2}\right) t, \tag{3.8}$$

$$x - x_0 = v_0 t + \tfrac{1}{2} a t^2, \tag{3.9}$$

$$2a(x - x_0) = v^2 - v_0^2. \tag{3.10}$$

To use these equations, list all of the known information about the position and motion of an object at different times; then look for one of the four equations that can be used to solve for a desired unknown using the given information. Note that each equation uses only three of the four variables x, v, a, and t.

EXAMPLE 3.6 A drag racer accelerates from an initial velocity of zero to a final velocity of 40 m/s in a distance of 80 m. Calculate its acceleration, which is assumed constant.

SOLUTION The following information is given:

$$v_0 = 0,$$
$$v = 40 \text{ m/s},$$
$$x - x_0 = 80 \text{ m},$$

and we need to determine a, the car's acceleration. Equations (3.7) and (3.9) both contain two unknowns, time and acceleration, and cannot be used to determine a. Equation (3.8) does not contain the acceleration and hence cannot be used to calculate a. However, acceleration is the only unknown in Eq. (3.10), and this equation can be used to solve for a. Rearranging, we find that

$$a = \frac{v^2 - v_0^2}{2(x - x_0)} = \frac{(40 \text{ m/s})^2 - (0)^2}{2(80 \text{ m})} = \underline{10 \text{ m/s}^2}. \qquad \blacksquare$$

EXAMPLE 3.7 A sled on an air track at Holloman Air Force Base near Alamogordo, New Mexico (called the Daisy Decelerator) is used to test the effect of high accelerations on the body. (a) If the vehicle starts from rest and accelerates at 50 m/s^2 (a little greater than five times the acceleration of a freely falling body), what is its speed after traveling 100 m? (b) How much time is required for this 100-m trip? (c) If the vehicle is stopped abruptly by an air compression ram in 10 m, what is its acceleration (deceleration) when stopping? Assume that the acceleration is constant.

SOLUTION   (a) The initial conditions are

$$x_0 = 0,$$
$$v_0 = 0,$$
$$a = 50 \text{ m/s}^2.$$

We want to find v when $x = 100$ m. Equation (3.10) can be solved for v:

$$v^2 = v_0^2 + 2a(x - x_0) = 0^2 + 2(50 \text{ m/s}^2)(100 \text{ m}) = 10,000 \text{ m}^2/\text{s}^2,$$

or

$$v = \sqrt{10,000 \text{ m}^2/\text{s}^2} = \underline{100 \text{ m/s}}.$$

(b) Equation (3.7) can be used to find the time t needed for this change in velocity:

$$t = \frac{v - v_0}{a} = \frac{100 \text{ m/s} - 0}{50 \text{ m/s}^2} = \underline{2.0 \text{ s}}.$$

(c) We must consider this part as a separate problem because the acceleration changes abruptly as the car hits the air compression ram. The final conditions in parts (a) and (b) are the initial conditions for this part. Thus, $x_0 = 100$ m and $v_0 = 100$ m/s. We want to find the acceleration that reduces the velocity to zero ($v = 0$) at position $x = 110$ m. Rearranging Eq. (3.10), we find

$$a = \frac{v^2 - v_0^2}{2(x - x_0)} = \frac{0^2 - (100 \text{ m/s})^2}{2(110 \text{ m} - 100 \text{ m})} = \underline{-500 \text{ m/s}^2}.$$

The sled's acceleration is opposite in direction to its velocity, which is positive; therefore, the sled is decelerating.

This deceleration is about 51 times the acceleration of gravity. If a human being undergoes an acceleration more than 30 times the acceleration of gravity (called 30g's), it is likely that some form of injury will occur. At 50g's, a passenger in an automobile collision has a greater than 50 percent chance of serious injury—perhaps death. ■

When working problems in kinematics, often we must consider the motion of two objects. For example, what length of time is required for a faster car to overtake a slower car? In these problems the most difficult part is arranging the known information so that we can use equations. Consider carefully the way in which the following problem is worked.

EXAMPLE 3.8   A hobo wishes to catch a freight train that stands in front of a train station. He waits 50 ft behind the train whose last car is 50 ft from the station. As the train starts to leave the station with an acceleration of 2 ft/s^2, the hobo runs toward the train at a constant speed of 15 ft/s. Will he catch the train

(a)

(b)

FIG. 3.10. (a) A hobo waits to catch a train. (b) Later, after the train has started, the hobo and the last car reach the same position at some unknown time.

before the last car reaches the station, or will he have to run past the station to catch the train?

SOLUTION To express the known information quantitatively, we must first choose a coordinate system, as shown in Fig. 3.10. The x axis points in the direction of motion. The origin can be placed anywhere; for convenience, we will put it at the hobo's original position. The initial conditions at time zero are, then:

Hobo:	$x_0 = 0$ ft,	Train:	$x'_0 = 50$ ft,
	$v_0 = 15$ ft/s,		$v'_0 = 0$,
	$a = 0$.		$a' = 2$ ft/s^2.

Our problem is to find the position where the hobo reaches the train. Let us use Eq. (3.9) to write general expressions for the position of the hobo (x) and of the train (x') at some future time:

Hobo: $\quad x = x_0 + v_0 t + \tfrac{1}{2}at^2 = 0 + 15t + (\tfrac{1}{2})(0)t^2 = 15t$ **(3.11)**

Train: $\quad x' = x'_0 + v'_0 t + \tfrac{1}{2}a't^2 = 50 + (0)t + (\tfrac{1}{2})(2)t^2 = 50 + t^2$. **(3.12)**

We have omitted units to keep the equations simple. These two equations give the position of the hobo and of the train at any time in the future. However, we are interested in the specific time when the hobo and the train are at the same position; that is, when $x = x'$. If we set Eqs. (3.11) and (3.12) equal to each other, we find that

$$15t = 50 + t^2$$

or

$$t^2 - 15t + 50 = 0. \qquad \textbf{(3.13)}$$

This equation, when solved for t, will tell us the time when the hobo and the back of the train are in the same position; that is, when $x = x'$. After finding that time, we substitute back into either Eq. (3.11) or Eq. (3.12) to find the position where this occurs.

Equation (3.13) is a quadratic equation of the form

$$at^2 + bt + c = 0. \qquad \textbf{(3.14)}$$

Two values of t are solutions of this equation:

$$t = \frac{-b \pm \sqrt{b^2 - 4ac}}{2a}. \qquad (3.15)$$

For our particular problem, $a = 1$, $b = -15$, and $c = 50$. Thus, the times when the hobo and train are in the same position are

$$t = \frac{-(-15) \pm \sqrt{(-15)^2 - (4 \times 1 \times 50)}}{2} = 5 \text{ s or } 10 \text{ s.}$$

The hobo and the last car of the train will be at the same position 5 s after the train starts and also 10 s after the train starts. The first of these solutions corresponds to the time the hobo first reaches the train. Evidently he is moving fast enough to run past the last car and farther up along the train. By 10 s, the train has accelerated to such a speed that the last car catches up to and passes him.

If the hobo joins the train at the earlier time, he will be at position $x = 15t = 15(5 \text{ s}) = 75$ ft or 25 ft before the station. ■

3.7 Gravitational Acceleration

One important example of linear motion with approximately constant acceleration is the vertical motion of objects thrown into the air or of objects falling through the air. In fact, if it were not for the resistive force of air on these objects, their acceleration would be constant. This acceleration is caused by the gravitational force of the earth pulling down on the objects. This force—their weight—causes them to accelerate toward the center of the earth with what is called **gravitational acceleration.** The magnitude of this acceleration, which we give the symbol g, at the earth's surface has a value

$$g = 9.8 \text{ m/s}^2 = 32.2 \text{ ft/s}^2. \qquad (3.16)$$

The value of g varies slightly with elevation and latitude and also depends on the geological formations at different parts of the earth's surface. In all of our calculations, we will use the rounded-off value, $g = 9.8$ m/s^2 or 32 ft/s^2.

We have said that this gravitational acceleration is a constant only if we ignore air resistance. For what situations can air resistance be ignored? The answer to this question depends on several factors:

1. If you desire great accuracy, then air resistance must be considered.

2. How fast is the object moving? The faster the object moves, the more air resistance affects its acceleration.

3. The shape and surface area of an object greatly affect air resistance. Air resistance on an open parachute is much greater than that on a parachute still enclosed in its bag.

4. Air resistance is much more important for a light object, such as a Ping-Pong ball, than for a heavier object of the same size, such as a golf ball.

Obviously, no general rule tells us when air resistance can be ignored. In our examples and problems the objects will be moving through the air at moderate speeds and will be relatively heavy and/or streamlined. We will ignore air resistance, even though its effect may be significant.

Let us use the set of four equations derived in the previous section to relate the position, velocity, and acceleration, at different times, of objects moving in the vertical direction under the influence of the gravitational acceleration. We will use a y coordinate to indicate the vertical position of the object. For this vertical motion, Eqs. (3.7)–(3.10) become

$$v = v_0 + at, \tag{3.7y}$$

$$y - y_0 = \left(\frac{v + v_0}{2}\right)t, \tag{3.8y}$$

$$y - y_0 = v_0 t + \tfrac{1}{2}at^2, \tag{3.9y}$$

$$2a(y - y_0) = v^2 - v_0^2. \tag{3.10y}$$

These equations relate the object's acceleration, a, initial vertical position, y_0, and initial vertical velocity, v_0, at time zero, to its final vertical position, y, and final vertical velocity, v, at time t. If the positive y axis points up, then the acceleration due to gravity is down; that is, $a = -g$. If the y axis points down, then $a = +g$, since the acceleration is downward in the positive y direction. Let us consider some examples.

EXAMPLE 3.9   A high-jumper leaps into the air with an initial speed of 12 ft/s. How high will he be at the top of his flight when $v = 0$?

SOLUTION   If we choose the ground level as the origin of our y-coordinate axis and the y axis points up, then the known quantities for this problem are:

$$y_0 = 0 \text{ ft,}$$
$$v_0 = 12 \text{ ft/s,}$$
$$a = -g = -32 \text{ ft/s}^2.$$

We wish to know the value of y when $v = 0$.* Equation (3.10y) can be used to calculate the jumper's height at the top of the jump:

$$y - y_0 = \frac{v^2 - v_0^2}{2a} = \frac{0^2 - (12 \text{ ft/s})^2}{2(-32 \text{ ft/s}^2)} = \underline{2.3 \text{ ft.}} \quad \blacksquare$$

EXAMPLE 3.10   A person can test your reaction time by dropping a ruler between your fingers, as shown in Fig. 3.11. The distance the ruler falls before you catch it indicates the time the ruler falls and hence your reaction time. What is your reaction time if the ruler falls 20 cm before you catch it?

SOLUTION   Let us choose the initial position of the lower edge of the ruler as the origin of the coordinate system. We will follow the motion of this lower edge. If we orient the y axis so that it points down (Fig. 3.11b), then the known quantities are:

$$y_0 = 0,$$
$$v_0 = 0,$$
$$a = g = 9.8 \text{ m/s}^2.$$

FIG. 3.11.   (a) A person drops a meter stick. (b) A second person catches it. The distance the stick falls is related to the second person's reaction time.

---

*At the top of a person's jump, his velocity is neither up nor down. The person has stopped for just an instant while changing directions.

We wish to find the time t when the bottom edge of the ruler has fallen to $y = 20\,\text{cm} = 0.2\,\text{m}$. It is important to use units all from the same system of units. Thus, if g is in m/s², then y should be in meters and not in centimeters. Equation (3.9y) can be used to find t:

$$y - y_0 = v_0 t + \tfrac{1}{2}at^2,$$

or

$$(0.2\,\text{m}) - (0) = (0)t + \tfrac{1}{2}(9.8\,\text{m/s}^2)t^2.$$

Thus,

$$t^2 = \frac{2(0.2\,\text{m})}{(9.8\,\text{m/s}^2)} = 0.041$$

or

$$t = \underline{0.20\,\text{s}}.$$

This is about the amount of time needed for information to travel from your eye to your brain and then to your hand so that it can grab the ruler. Do not be discouraged if your time is longer than this; 0.2 s is a fast reaction time.

3.8 Projectile Motion

A **projectile** is an object that moves through space under the influence of the earth's gravitational force. Two coordinates must be used to describe the projectile's motion, since it moves horizontally as well as vertically. Examples of projectile motion are the path followed by a baseball having been hit into the air by a bat or an ice skater jumping over some barrels. Often a strong, abrupt force initiates the motion of a projectile. Following this force, the projectile moves through the air and is influenced only by the gravitational force of the earth pulling down on it and by air resistance. If the effect of air resistance is ignored, it is quite easy to use equations, such as those developed in the previous sections, to analyze the motion of a projectile—how far it will travel, how high it will go, and so forth.

When the four linear motion equations are used to analyze projectile motion, the horizontal and vertical motions of the projectile are treated independently, as illustrated in Figs. 3.12 and 3.13. An ejector seat is placed in a car. When the car is at rest, the seat is adjusted so that, when released, the driver shoots straight up into the air and returns again to the seat (Fig. 3.12). Without readjusting the seat, the driver next travels at constant speed toward an underpass (Fig. 3.13). Just before the car reaches the underpass, the ejector seat shoots the driver straight up into the air. Since the driver and car are both moving horizontally at the same speed before the driver is released, both continue this horizontal motion as though nothing had happened. Thus, when the

FIG. 3.12. An ejector seat shoots a person straight up into the air. The seat is adjusted so that the person returns to the car when it is at rest.

FIG. 3.13. Without readjusting the seat, a person shot from the moving car also returns to the seat. Since the car is moving, the driver has an initial x component of velocity equal to that of the car. This velocity is unaffected by the vertical motion.

driver comes back down to the level of the car, he again finds himself back in the front seat, having flown over the underpass. The point is that both objects had the same horizontal motion, which was unaffected by the vertical flight of the driver.

Projectile motion, then, is described in terms of motion in the horizontal or x direction, and independently by motion in the vertical or y direction. If air resistance can be ignored, the projectile's horizontal motion continues at constant velocity because its horizontal acceleration is zero. The vertical motion, however, will have acceleration due to gravity. We can therefore restate the four linear motion equations as two separate sets of equations: one set describes the horizontal or x motion of the projectile, for which $a_x = 0$; the other set describes the vertical or y motion, for which $a_y = -g$ (the y axis is oriented upward).

PROJECTILE MOTION

x Equations $(a_x = 0)$		y Equations $(a_y = -g)$	
$v_x = v_{x_0},$	(3.7x)	$v_y = v_{y_0} - gt,$	(3.7y)
$x - x_0 = \left(\dfrac{v_{x_0} + v_x}{2}\right)t,$	(3.8x)	$y - y_0 = \left(\dfrac{v_{y_0} + v_y}{2}\right)t,$	(3.8y)
$x - x_0 = v_{x_0}t,$	(3.9x)	$y - y_0 = v_{y_0}t - \frac{1}{2}gt^2,$	(3.9y)
$0 = v_x^2 - v_{x_0}^2.$	(3.10x)	$-2g(y - y_0) = v_y^2 - v_{y_0}^2.$	(3.10y)

Most often a projectile leaves the ground at some angle θ with a speed v_0 as shown in Fig. 3.14. In such a case, the initial speed in the x direction is the x component of $\mathbf{v_0}$, or $v_{x_0} = v_0 \cos \theta$. The initial speed in the y direction is the y component of $\mathbf{v_0}$, or $v_{y_0} = v_0 \sin \theta$ [see Eqs. (1.3) and (1.4)].

Often, the y equations are used to determine the length of time the projectile remains in the air and the x equations to determine how far the projectile will have traveled horizontally during that time. For example, if a projectile leaves ground level and returns to ground level, Eq. (3.9y) with $y = 0$ and $y_0 = 0$ can be solved for t. Equation (3.9x) can then be used to calculate the distance the projectile travels in that time.

FIG. 3.14. When fired at an angle, a projectile has an initial motion in both the x and the y directions.

EXAMPLE 3.11 A football player kicks off from the 40-yd line. How far will the ball travel before hitting the ground if its initial speed v_0 is 80 ft/s and the ball leaves the ground at an angle of 30°?

SOLUTION First, the y equations are solved to find the time of flight of the ball. The known conditions for the y motion are

$$y_0 = 0 \quad \text{and} \quad v_{y_0} = v_0 \sin 30°, \quad \text{where } v_0 = 80 \text{ ft/s.}$$

We need to find the time t when $y = 0$—that is, when the ball returns to ground level. Equation (3.9y) can be used:

$$0 - 0 = v_0 \sin 30° \, t - \tfrac{1}{2}gt^2$$

or

$$t = \frac{2v_0 \sin 30°}{g} = \frac{2(80 \text{ ft/s})(0.5)}{(32 \text{ ft/s}^2)} = 2.5 \text{ s.}$$

Next, we find the horizontal distance $(x - x_0)$ the ball travels during this time. Equation (3.9x) provides this information:

$$x - x_0 = v_{x_0}t = (v_0 \cos 30°)t = \left(80 \frac{\text{ft}}{\text{s}}\right)(0.87)(2.5 \text{ s})$$

$$= \underline{173 \text{ ft}}. \qquad \blacksquare$$

Summary

1. **Speed, velocity, and acceleration: Kinematics** is the study of the way in which an object's motion changes with time. The description of motion is based on the definitions of the vector quantities' displacement, velocity, and acceleration. The scalar magnitudes of displacement and velocity are called distance and speed, respectively. Speed is the distance traveled divided by the time needed to travel that distance; velocity is the displacement of an object (magnitude and direction) divided by the time needed for the displacement; acceleration is the change in an object's velocity divided by the time needed for that change. If we are interested in the values of these quantities over a relatively long period of time, we find their average values, which are given by the following equations:

$$\text{average speed} = \bar{v} = \frac{d}{t}, \qquad (3.1)$$

$$\text{average velocity} = \bar{\mathbf{v}} = \frac{\mathbf{d}}{t}, \qquad (3.2)$$

$$\text{average acceleration} = \bar{\mathbf{a}} = \frac{\mathbf{v}_2 - \mathbf{v}_1}{t_2 - t_1}, \qquad (3.5)$$

where d is the distance traveled, $\mathbf{d}$ is the displacement, and t is the time.

The instantaneous values of velocity and acceleration are defined as

$$\text{instantaneous velocity} = \mathbf{v} = \frac{\Delta \mathbf{d}}{\Delta t}, \qquad (3.4)$$

$$\text{instantaneous acceleration} = \mathbf{a} = \frac{\Delta \mathbf{v}}{\Delta t}, \qquad (3.6)$$

where $\Delta \mathbf{d}$ is the displacement and $\Delta \mathbf{v}$ the change in velocity during a very short time Δt.

The above definitions apply to motion in any direction. For motion along one axis or dimension, called linear motion, the following definitions apply:

$$\text{average velocity} = \bar{v} = \frac{x_2 - x_1}{t_2 - t_1},$$

$$\text{instantaneous velocity} = v = \frac{\Delta x}{\Delta t}, \qquad (3.3)$$

$$\text{average acceleration} = \bar{a} = \frac{v_2 - v_1}{t_2 - t_1},$$

$$\text{instantaneous acceleration} = a = \frac{\Delta v}{\Delta t},$$

where x indicates the axis along which motion occurs, and the two possible directions along this axis are indicated by $+$ and $-$.

2. **Equations of linear motion with constant acceleration:** When acceleration is constant during linear motion, a set of four simple equations can be used to relate position, velocity, and acceleration at different times. These equations are

$$v = v_0 + at, \qquad (3.7)$$

$$x - x_0 = \left(\frac{v + v_0}{2}\right)t, \qquad (3.8)$$

$$x - x_0 = v_0 t + \tfrac{1}{2}at^2, \qquad (3.9)$$

$$2a(x - x_0) = v^2 - v_0^2. \qquad (3.10)$$

In these equations, x_0 and v_0 represent the object's position and velocity at time zero; x and v represent its position and velocity at a later time t; and a represents the object's constant acceleration. These equations can be used to calculate an object's acceleration if its changing position and velocity at different times are known. If the acceleration is known, the equations can be used to calculate the future position and velocity of the object.

3. **Gravitational acceleration:** The vertical motion of objects moving freely in the air comes close to being linear motion with constant acceleration. If we ignore the effects of wind and air resistance, then objects experience constant acceleration caused by the earth's gravitational force. The acceleration is directed toward the center of the earth and has a magnitude

$$\text{gravitational acceleration} = g = 9.8 \text{ m/s}^2 = 32 \text{ ft/s}^2.$$

4. **Projectile motion:** Projectiles are objects moving both horizontally and vertically through the air. If we ignore the resistive force of the air on the projectiles, then the horizontal and vertical components of their motion can be described independently using two separate sets of equations derived from Eqs. (3.7)–(3.10). For the horizontal-motion equations, the acceleration is zero ($a_x = 0$). For the vertical-motion equations, the acceleration is that caused by gravity ($a_y = -g$), where, for the vertical component of motion, the y axis points up.

Questions

In several of the questions and problems in this chapter and in most subsequent chapters, you will be asked to make an estimate. In Question 6, for example, in order to estimate the child's deceleration while sliding to a stop on grass, you will need to use an equation or perhaps more than one equation. You will also need to make a reasonable guess at certain numbers (a child's typical running speed, the distance the child slides when coming to a stop, or the time required to slide to a stop). Obviously, your answers will not match exactly those of some other person or those of the grader. There is no single correct answer to these estimation questions and problems. The numbers, though, should be reasonable. The grader will probably approve your efforts if your estimates are within a factor of about 2 or 3 of his or hers. Such estimates are called *order-of-magnitude* estimates. You should justify and explain clearly the basis on which you have made your estimates.

1. Design and describe an experiment to allow you to measure the speed of a baseball when thrown by a major league third baseman toward first base.

2. Design and describe an experiment to measure the speed of sound.

3. How much speed does your body gain in 2 seconds when it undergoes an acceleration of $2 \ m/s^2$? of $4 \ m/s^2$? What is its final speed in each case if its initial speed is $1 \ m/s$?

4. (a) Can a car traveling at $30 \ m/s$ have the same acceleration as a bicycle rider traveling at $2 \ m/s$? Explain. (b) Give an example of an object with zero speed that has a nonzero acceleration.

5. Give an example of an object that has a nonzero acceleration but moves with a constant speed.

6. A child running on a thick grass lawn slides to a stop. *Estimate* the child's acceleration. Indicate all numbers used in your estimate and your estimation technique.

7. You jump from a fence 2 m high and land stiff-legged in soft soil. *Estimate* your deceleration while being stopped by the soil. Indicate the numbers and technique used in your estimation.

8. Is it true that a rock dropped from rest falls three times farther in the second second than in the first? Explain.

9. A ball is thrown up and caught at the same level. If we ignore air resistance, its time while moving up is the same as its time moving down. Which time is longer if we do not ignore air resistance, or are they still the same? Explain.

10. *Estimate* the maximum speed with which you can throw a ball straight up in the air. Use any technique you wish.

11. For novice tennis players who have trouble getting their serve into the service court, it is advantageous to stand toward the corner of the court (at B in Fig. 3.15), rather than near the center line (at A). The ball is in the air longer when it reaches the net when served from B than from A. Why does this help get it into the service court?

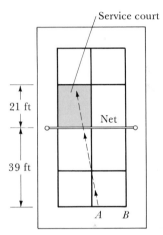

FIG. 3.15

Problems

3.2 Average Speed and Velocity

1. Four seconds after seeing the lightning in a distant cloud, you hear thunder. How far is the cloud in kilometers and in miles? Sound travels at about $330 \ m/s$.

2. You are traveling in an automobile at $56 \ km/hr$ (35 mph). You glance for one second at a person standing on the sidewalk. Calculate the distance your car travels while you are looking toward the person.

3. Light and radio waves travel at a speed of $3 \times 10^8 \ m/s$. (a) Distances to stars are measured in light-years. A light-year is the distance light travels in one year. Calculate the number of meters in one light-year. (b) If there were intelligent beings on a planet of Alpha Centauri, the nearest star, how much time would be required to receive a response to a message sent there? Alpha Centauri is $4.1 \times 10^{16} \ m$ from the earth.

4. If the continents are drifting apart at a constant speed of $3 \ cm/yr$ and Africa and South America are separated by $6.4 \times 10^8 \ cm$, how long has this drift been occurring?

5. A high-speed camera takes pictures of the position of a car at a rate of 2400 pictures per minute. Calculate the distance traveled by a car from one picture frame to the next if the car is moving at $100 \ km/hr$.

6. A complete picture on a television screen is produced in $\frac{1}{30}$ second. Each picture is formed by 525 traverses of the electron beam across the screen, each crossing at a slightly lower level. If the width of the screen is $40 \ cm$, how fast is the electron beam traveling across the screen?

7. You drive for 40 minutes at $90 \ km/hr$ and then stop for 10 minutes. You then drive 30 minutes at $100 \ km/hr$. Calculate your average speed for the entire trip.

8. Tommy Tortoise drives his car at 55 mph for 4 hours. Hank Hare drives at 70 mph for 45 minutes of each hour and then has a 15-minute coffee break. Hank also travels for 4 hours. Will Tommy or Hank travel farther? By how much?

9. A bat flying toward a wall in a cave sends sound waves toward the wall. The waves travel at 330 m/s, and the wall is 2 m from the bat. If the bat is flying at 20 m/s, approximately how far will the bat travel during the time required for the sound waves to make a round trip from the bat to the wall?

10. In 1926, Johnny Weissmuller set the world 400-m swimming record in a time of 4 min, 57.0 s. In 1966, Martha Randall set the women's record in 4 min, 38.0 s. By how many meters would Martha have beaten Johnny if they had raced each other?

11. What is the speed in meters per second of a car traveling (a) 65 mph? (b) 25 mph? (c) What is the speed in miles per hour of a car traveling at 100 km/hr?

12. Two cities are separated by 250 miles. What is their separation (a) in meters? (b) in kilometers? (c) If it takes 4.6 hours to travel between these cities, how many seconds are required for the trip?

13. The surface area of a person's body is 1.6 m². Calculate the area (a) in square centimeters, (b) in square feet.

14. The volume of a 130-lb person is about 1.8 ft³. Calculate the person's weight in newtons and his volume in cubic meters and in cubic centimeters.

15. (a) An electron involved in energy conversion in photosynthesis travels about 100 μm in 3 ms. What is the average speed of the electron in units of meters per second? (b) What is the electron's speed in miles per hour?

3.3 Instantaneous Speed and Velocity

16. A skydiver falls from an airplane. His positions below the plane at different times are given below. (a) Calculate the diver's average speed during the first 5 seconds. (b) Calculate the diver's average speed during each one-second time interval.

time (s)	0	1	2	3	4	5	6
position (m)	0	5	19	42	74	112	154

17. A sample of red blood cells is placed in a tube of water. The cells start to fall slowly to the bottom. The distance of one cell below its starting position as a function of time is listed in the table below. Calculate the cell's average speed (a) during the first second, (b) during the time interval from 0.4 to 0.6 s. (c) After one second of falling, the cell falls at the same speed it had during the time from 0.8 to 1.0 s. How long will it take the cell to fall 3 cm further to the bottom of the tube?

time (s)	0.0	0.2	0.4	0.6	0.8	1.0
position (μm)	0.000	0.096	0.232	0.384	0.544	0.710

18. A weather balloon continuously radios its elevation back to a base station. Its elevations at several different times are given below:

time (min)	0	5	10	15	20	25
elevation (km)	1.2	1.6	1.9	2.0	1.8	1.1

Calculate its average velocity for the time interval (a) from 5 to 10 min, (b) from 20 to 25 min, (c) from 0 to 25 min. (d) If the balloon's velocity after 25 min is the same as it was from 20 to 25 min, when will it hit the ground?

19. The positions of a girl at different times during a potato race are listed below. (a) Calculate the child's average velocity during each 2-s time interval. (b) Do you think she moved with a greater speed at 1 s or at 9 s? Explain.

t (s)	0	2	4	6	8	10	12
x (ft)	0	9	26	36	28	11	0

20. A young woman wishes to run 440 yards in 60 s. Her positions on the track at 3-s time intervals for the first 15 s are listed below. (a) Calculate her average speed during each time interval. (b) If she continues to run for the last 338 yards at the same speed as she ran during the time from 12 to 15 s, what will be her time for the full 440 yards?

time (s)	0	3	6	9	12	15
distance (yd)	0	16	35	57	79	102

3.4 Acceleration

21. An airport runway is 1000 m long. An airplane must attain a speed of 66 m/s in order to take off. (a) If the average speed of the airplane down the runway is 33 m/s, for what length of time is the plane on the runway? (b) What is the average acceleration of the plane?

22. A young boy running with a speed of 7 m/s slides into second base. If the boy's deceleration is -4 m/s², how much time is required for the boy to stop?

23. The Stanford high-energy linear accelerator is 2 miles long. If, during the first half-mile, electrons are accelerated from rest to a speed of 3×10^7 m/s, what is the average acceleration of the electrons? Assume that the average speed of an electron is half its speed at the end of the half-mile.

24. The velocity of a drag racer at different times after the start of a race is given below. Calculate (a) the racer's acceleration during each time interval and (b) the average acceleration for the first 4 s.

t (s)	0.0	0.5	1.0	1.5	2.0	2.5	3.0	3.5	4.0
v (m/s)	0.0	5.4	14.3	25.7	38.0	49.2	58.1	67.3	74.4

25. A rock falling near the surface of the moon has velocities at different times toward the moon as given below. Calculate the rock's average acceleration (a) for the first 10 s and (b) for each 2-s time interval listed.

t (s)	0.0	2.0	4.0	6.0	8.0	10.0
v (m/s)	8.3	11.5	14.6	18.0	21.1	24.3

3.5 Graphical Analysis of Velocity and Acceleration

26. Make a position-versus-time graph of the data presented in Problem 16. Draw a smooth curve that passes through these data points. Estimate the slope of the curve at $t = 0, 2, 4,$ and 6 s. From this information construct a velocity-versus-time graph.

27. Construct a position-versus-time graph using the data presented in Problem 18. Estimate the slope of the graph at $t = 0$, 10, 15, and 25 s. Using this information, construct a velocity-versus-time graph.

28. Using the data in Problem 24, construct a velocity-versus-time graph. Estimate the slope of the graph at $t = 0.0$, 1.0, 2.0, and 3.0 s. From this information, construct an acceleration-versus-time graph.

29. Using the data presented in Problem 19, construct a graph of position versus time. Using this graph, estimate roughly the girl's velocity at each time listed. From this information construct a velocity-versus-time graph. Using this graph, make a rough graph of acceleration versus time.

3.6 Linear Motion with Constant Acceleration

30. A Triumph TR-7 starts at rest and accelerates to a speed of 27 m/s (equivalent to 60 mph) in 11.8 s. (a) Calculate the car's acceleration. (b) Calculate the distance traveled during that acceleration.

31. A freight train leaves a railroad station with an acceleration of 0.06 m/s^2. (a) Calculate the time needed for the train to attain a speed of 24 m/s (54 mph) if starting from rest. (b) How far has the train traveled?

32. A Triumph TR-7 accelerates from rest at 2.27 m/s^2, while a Fiat X1/9 undergoes a 2.22 m/s^2 acceleration. How much farther does the Triumph travel in 10 s than the Fiat?

33. A Jaguar undergoes constant acceleration from rest and reaches a speed of 37 m/s in a distance of 400 m. Calculate its acceleration and the time needed to reach this speed.

34. A car traveling at 35 mph is stopped in a distance of 2.5 ft when it hits a tree. (a) Calculate the average deceleration of the car. (b) Calculate the time needed to stop the car (assuming constant deceleration).

35. Lt. Col. John L. Stapp was a pioneer in testing the effects of large accelerations on the human body. He himself rode the Daisy Train at Holloman Air Force Base many times. During one trip, a sled propelled by nine rockets attained a speed of 630 mph in 5 s. The sled was stopped in 1.5 s. Calculate the average acceleration of the sled (a) during the first 5 s and (b) during the last 1.5 s. (c) How far did the sled travel during the first 5 s and (d) during the last 1.5 s? (e) A freely falling body has an acceleration of 32 ft/s^2 (called 1g). How many g's is the acceleration in part (b)?

36. A bus driver forgets to stop at a bus stop. Just as she passes the stop, she decelerates. Five seconds later the bus is 80 m beyond the stop and has slowed to 10 m/s. (a) Calculate the speed of the bus as it passed the bus stop. (b) What is the deceleration of the bus (assumed constant)?

37. A spaceship, in order to make a long voyage, must travel as fast as possible. Relativistic considerations limit the ship's speed to about one-tenth the speed of light (the speed of light is 3×10^8 m/s). In order to avoid physiological problems for the ship's occupants, the ship can accelerate at only 10 m/s^2. (a) What length of time in years is required for the ship to attain its final speed? (b) How far does it travel in that time?

38. A mother and daughter are to race for 100 m. The mother can run at 4 m/s and the daughter at 5 m/s. To make the race fair, the daughter lets her mother run a shorter distance so that the race should end in a tie. What distance should the mother run if they both start at the same time? Ignore acceleration; assume that they run at constant speed.

39. A herd of bulls is stampeding at a speed of 7 m/s. A boy, at rest 20 m in front of the bulls, accelerates at 3 m/s^2. Will the herd trample the boy? Explain.

40. Most driver's manuals show the distances required to stop a car when traveling at various speeds. The table below is taken from the 1976 *California Driver's Handbook:*

STOPPING DISTANCE: From Eye to Brain to Foot to Wheel to Road (With Perfect 4-Wheel Brakes and Ideal Conditions)

MPH	Thinking Distance	Braking Distance	Total Distance Traveled
25	27 ft	34.4 ft	61.4 ft
35	38 ft	67.0 ft	105.0 ft
45	49 ft	110.0 ft	159.0 ft
55	60 ft	165.0 ft	225.0 ft
65	71 ft	231.0 ft	302.0 ft

(a) Calculate the reaction time associated with the "thinking distance" for three different speeds. The car is traveling at a constant speed during that time. (b) Calculate the deceleration associated with the "braking distance" for three different speeds.

3.7 Gravitational Acceleration

41. A stone that is initially at rest is dropped from the top of a building. It falls for 5 s before hitting the ground. How tall is the building?

42. Divers in Acapulco leap from a cliff that is 36 m above the water below. Estimate their speed when they hit the water.

43. A ball becomes stuck in the branch of a tree. With what upward velocity must a rock be thrown so that it reaches the ball with a speed of 3 m/s? The ball is 10 m above the thrower's arm.

44. A stone is dropped into a well that is 50 m deep. How long after the stone is dropped will the sound of the splash reach the top of the well? Sound travels at 330 m/s.

45. A basketball player can jump 28 inches off the floor from a standing position. Calculate the player's speed when leaving the floor.

46. A rock is dropped from a cliff. (a) What is its instantaneous speed after falling 1.0 s? (b) What is its average speed during the first second? (c) How far does it travel in that second?

47. Do you recall Galileo's famous experiment at the Leaning Tower of Pisa (the Tower is 55 m high and deviates from the vertical by 4 m)? If Galileo had accidentally dropped one of the rocks 0.5 s before the other, with what velocity would the second rock have to be thrown downward in order to reach the ground at the same time as the first rock?

48. A building is 50 m tall. A jealous lover wishes to drop a tomato from the top of the building onto the head of his girlfriend's suitor. If the suitor walks at a speed of 1 m/s, how many meters should the suitor be from the position below the tomato when it is dropped? (See Fig. 3.16.)

FIG. 3.16

$\rightarrow\mid \mid\leftarrow d?$

3.8 Projectile Motion

49. Daring Darless wishes to cross the Grand Canyon of the Snake River by being shot from a cannon. She wishes to be launched at 60° to the horizontal so that she can spend more time in the air waving to the crowd. With what speed must she be launched to cross the 520-m gap?

50. A car drives off a horizontal embankment and lands 11 m from the edge of the embankment in a field that is 3 m lower than the embankment. With what speed was the car traveling when it left the embankment?

51. An ice skater in a barrel-jumping contest leaves the ground at a 28° angle with a speed of 10 m/s. If a barrel is 0.8 m in diameter, how many barrels can the skater jump? Ignore the problem of the barrel's height.

52. The unofficial record for a single jump by a frog was set by a South African sharp-nosed frog that leaped 5 m when being retrieved for placement in its container. If this frog left the ground at an angle of 40°, what was its initial speed?

53. Robin Hood wishes to split an arrow already in the bull's-eye of a target 40 m away. If he aims directly at the arrow, by how much will he miss? The arrow leaves the bow horizontally at 140 m/s.

54. A football punter wants to kick the ball so that it hits the ground 50 m from where it is kicked and so that the ball is in the air 4 s. At what angle and with what initial speed should the ball be kicked? Assume that the ball leaves his foot from an elevation of 1 m.

55. A baseball leaves the bat of Henry Aaron with a speed of 110 ft/s at an angle of 37° above the horizontal. The ball is 4 ft off the ground when it leaves the bat. The ball must clear a fence that is 10 ft high and 350 ft from the batter. (a) At what times after being hit will it reach the height of the fence? (b) How far from the batter will the ball be at these times? (c) Will Henry have another home run? Explain.

56. A tennis ball is served from position A in Fig. 3.15 such that it leaves the racket 8 ft above the ground in a horizontal direction with a speed of 73 ft/s (50 mph). (a) Will the ball be able to cross a 3-ft-high net 39 ft in front of the server? (b) Will the ball land in the service court, which is within 21 ft of the net on the other side of the net?

General Problems

G1. The average of any quantity that changes uniformly is the sum of its initial and final values divided by two. We used this fact to derive Eq. (3.8). Try it for the following examples: (a) Your semester grade-point averages increase uniformly as follows: 2.4, 2.7, 3.0, 3.3. Calculate your average for the four semesters, first by taking the average of all four numbers, second by taking the average of the first and last numbers. (b) Your weight decreases uniformly each month: 150, 143, 136, 129, 122 lb. Calculate your average weight during the five months using both techniques. (c) Invent some other example of a uniformly changing quantity and present similar calculations for it.

G2. Use the photos of the ball shown in Fig. 3.17 to answer the following questions. The photos were taken at $\frac{1}{30}$-s intervals. (a) Is the speed of the ball greater at the left or right? Explain. (b) Is the ball accelerating as it moves right? Explain. (c) Make measurements on the photo and estimate the ball's acceleration.

G3. A visitor to Mars drops a rock and notes its position at different times, as given below. Calculate the acceleration due to gravity on Mars.

time (s)	0.0	0.5	1.0	1.5	2.0
distance fallen (m)	0.00	0.48	1.92	4.33	7.68

FIG. 3.17. (Photo from *PSSC Physics,* 2nd ed., 1965. D. C. Heath & Co. Education Development Center, Newton, MA.)

Newton's Laws of Motion

4

In the summer of 1665 the Great (bubonic) Plague was devastating England. Ten percent of London's population died. When the universities were closed to prevent its spread, Isaac Newton, a 22-year-old student at Trinity College, Cambridge, returned to his mother's farm in Woolsthorpe, Lincolnshire. The next eighteen months of his life must be considered the most productive period of achievement for any individual in the history of science. During that time Newton invented differential and integral calculus, discovered the universal law of gravitation, developed a new theory of light, and put together the ideas for his three laws of motion, the subject of this chapter.

These three laws still serve, three hundred years after their formulation, as the basis of **dynamics,** a subject in which we analyze the effect of force in causing accelerated motion. How much force is required to stop your body during a car collision? to catch a child jumping from a wall? to hit a baseball for a 360-ft home run? Every dynamical process in nature involves in one way or another these three laws.

4.1 Newton's First Law of Motion

Newton's description of motion represented a dramatic change from the ideas of Aristotle, the Greek scientist who two thousand years earlier had written extensively about the mechanical behavior of nature. At Newton's birth, Aristotle's work was still considered the ultimate authority on this subject.

Aristotle distinguished two types of motion: natural and nonnatural. The motion of heavenly bodies in a circle was considered natural, as was the motion of a stone falling by itself toward the earth. These objects seemed to move without a visible source of force pulling or pushing them. The motion was considered natural because it continued unaided by human activity.

On the other hand, the motion of an object being pulled or pushed by a visible source of force, such as a rope pulling a sled, was considered nonnatural. The object's motion continued only if acted upon by a force. For example, if you push a book lying on a table, it will slide. When the force stops, so does the book. According to Aristotle, it was nonnatural for the book to slide; hence it stopped.

Newton saw the situation differently. A moving book slides to a stop not because it is involved in a nonnatural motion but because a friction force opposes its motion. Without friction, the book would continue sliding in a straight

(a)

(b)

FIG. 4.1. Accelerated or decelerated motion caused by an unbalanced force.

line until it fell off the table. Situations such as this are addressed by the first of his three laws of motion:

Newton's first law of motion: Every body continues in its state of rest or of constant speed in a straight line unless it is compelled to change that state because of forces acting on it.

This law is equivalent to our first condition of equilibrium introduced in Chapter 2. Recall that if the vector sum of the forces acting on an object is zero, then the object, if at rest, remains at rest. If moving, the object continues its motion at constant velocity. But what if the forces acting on the object do not add to zero? Newton's second law of motion addresses this subject.

4.2   Newton's Second Law of Motion and Mass

According to Newton, an object accelerates if acted on by an unbalanced force. Consider the situations depicted in Fig. 4.1. Each object is subjected to an unbalanced force. In (a), a person steps off a high diving board. The gravitational force of the person's weight w is not balanced by another force, and the person accelerates toward the water. In (b), a moving car hits a wall, and the unbalanced force of the wall causes the car to slow down and stop—that is, to decelerate.

These examples point to a general rule: an unbalanced force causes an object's motion to change. If the force is in the direction of motion, the object moves faster; it accelerates. If the force opposes the motion, the object will slow down; it decelerates.

The acceleration, according to Newton, depends on two factors: (1) the resultant force acting on the object, and (2) the mass or amount of matter of which the object is made. Figure 4.2 illustrates the relationship between force

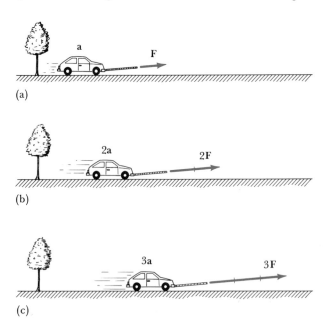

(a)

(b)

(c)

FIG. 4.2. Acceleration of a car is proportional to the net force applied.

and acceleration. In (a), a force **F** causes a car to have an acceleration of **a.** If the force is doubled, as in (b), the acceleration doubles. If the force is tripled, as in (c), the acceleration triples. The acceleration increases in direct proportion to the resultant force applied to the object:

$$\mathbf{a} \propto \mathbf{F}.$$

The symbol $\propto$ means "is proportional to."

An object's acceleration also depends on its mass.

Mass is often defined qualitatively as the amount of matter of which an object is made. Mass is also defined as a measure of the object's inertia. The greater its inertia, the less its acceleration when pushed or pulled by a force.

The idea of inertia is illustrated by considering defensive linemen on a football team. Most of these young men have relatively large mass, and so they are difficult to move. A large force causes little acceleration. A large mass is also advantageous for a running back on a football team. The larger his mass, the less he decelerates or slows down when a force is applied.

The quantitative relation between mass and acceleration is illustrated in Fig. 4.3. In (a), a force exerted on a single cart having mass m causes the cart to have an acceleration a. In Fig. 4.3b, two identical carts attached together and having a total mass $2m$ are pulled by the same force. However, their acceleration is only $a/2$ instead of a. Four carts of total mass $4m$ shown in Fig. 4.3c experience an acceleration $a/4$ when pulled by the same force. The acceleration caused by a force is inversely proportional to the mass of an object or objects on which the force acts. This inverse relationship between acceleration and mass is expressed as follows:

$$a \propto \frac{1}{m}.$$

By choosing the units of force in an appropriate fashion, we can combine the two proportionality equations ($a \propto F$ and $a \propto 1/m$) into a single equation ($a = F/m$) called Newton's second law. The units for this equation are discussed in the next section.

Newton's second law of motion: An unbalanced force acting on an object will cause the object to accelerate in the direction of the force. The acceleration is directly proportional to the resultant force acting on the object and inversely proportional to the object's mass. If a group of forces act on an object of mass m, the vector sum of these forces ($\Sigma \mathbf{F}$) causes the object to have an acceleration **a** given by

$$\mathbf{a} = \frac{\Sigma \mathbf{F}}{m} \quad \text{or} \quad \Sigma \mathbf{F} = m\mathbf{a}. \tag{4.1}$$

The direction of **a** is in the direction of the resultant force.

Newton's second law is often written as **F** = m**a.** The force **F** on the left side of the equation, however, represents the net or resultant force acting on the object—that is, the vector sum of all forces acting on the object, as Eq. (4.1) makes clear (see Fig. 4.4). Note that if the forces acting on an object add to zero, as in Fig. 4.4b, its acceleration is zero. If the object is at rest, it remains at rest. If

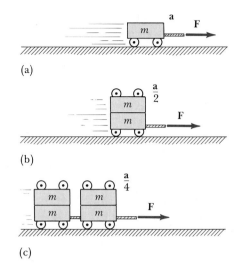

(a)

(b)

(c)

FIG. 4.3. The acceleration of objects upon which the same force acts is less for objects with greater mass.

FIG. 4.4. An object's acceleration depends on the net force acting on it.

Applied forces | Net force

(a) 5 N / 5 N → 10 N

(b) 5 N ← | 5 N → 0 N

(c) 5 N ← | 10 N → 5 N

moving, its velocity remains constant. Thus, the first condition of equilibrium discussed in Chapter 2 is just a special example of Newton's second law for the case when the forces acting on an object add to zero.

4.3   Units for Newton's Second Law

When using Newton's second law, the units of force, mass, and acceleration must all be from the same system of units. In this text SI metric units are used almost exclusively. However, English system units will be introduced and used occasionally because of their widespread usage in the past.

SI Metric System

The unit of acceleration in the SI system is the **meter/second^2** (m/s^2). Both meters and seconds are basic units with well-defined standards. The SI unit of mass is the **kilogram** (kg). One kg is a basic unit defined as the mass of a solid platinum-iridium cylinder stored in a special vault in the small town of Sèvres near Paris, France.

The SI unit of force is called the newton (N). A medium-sized apple weighs one newton; it is a little less than one-fourth of a pound. The **newton** is defined not in terms of a basic standard but by using Newton's second law. A force of one newton is an unbalanced force that will cause a 1-kg mass to experience an acceleration of 1 m/s^2. Substituting these units into Newton's second law ($\Sigma \mathbf{F} = m\mathbf{a}$) produces an expression relating newtons to these other units:

$$1 \text{ N} = (1 \text{ kg})(1 \text{ m/s}^2). \tag{4.2}$$

This expression is used often in the text for making unit conversions. To remember it, just remember $F = ma$ and the units for each quantity.

English System

The unit of acceleration in the English system is the **foot/second^2** (ft/s^2) and the unit of force is the **pound** (lb), which is defined in terms of the newton:

$$1 \text{ lb} = 4.45 \text{ N}$$

or

$$1 \text{ N} = 0.225 \text{ lb} \simeq \tfrac{1}{4} \text{ lb}.$$

The unit of mass in the English system is called the slug and is defined using Newton's second law in the form $m = F/a$. One **slug** is that mass which experiences an acceleration of 1 ft/s^2 when acted on by an unbalanced force of 1 lb:

$$1 \text{ slug} = \frac{1 \text{ lb}}{1 \text{ ft/s}^2}. \tag{4.3}$$

A one-slug mass is equal to a 14.6-kg mass:

$$1 \text{ slug} = 14.6 \text{ kg}$$

or

$$1 \text{ kg} = 0.0685 \text{ slug}.$$

One slug equals approximately the mass of four gallons of milk.

Warning: When using any of the laws or equations of physics, express all quantities in the same system of units. For example, do not substitute mass in units of kg and acceleration in units of ft/s^2 into Newton's second law in order to calculate force. The ft/s^2 must first be converted to m/s^2. Also, each quantity must be expressed using the appropriate unit for the system of units being used. For example, when using the metric system, mass must be in kilograms and not in grams or milligrams. If mass is given in grams, first convert to kilograms before substituting into an equation.

In the examples that follow in this section and the next, Newton's second law is applied only to one-dimensional motion, in which case we need not use vector notation. In Section 4.6 we will consider more carefully the vector nature of Newton's second law.

EXAMPLE 4.1 A net force of 1000 N acts on a 1500-kg car when it is pulled by a tow truck. Calculate the car's acceleration.

SOLUTION We are given that the net force acting on the car $\Sigma F = 1000$ N and that its mass $m = 1500$ kg. Since $\Sigma F = ma$, we find that

$$a = \frac{\Sigma F}{m} = \frac{1000 \text{ N}}{1500 \text{ kg}} = 0.67 \, \frac{\text{N}}{\text{kg}}.$$

If Eq. (4.2) is rearranged, we see that 1 N/kg = 1 m/s^2. Thus, the automobile's acceleration is $\underline{0.67 \text{ m/s}^2}$. ∎

EXAMPLE 4.2 A tennis ball with a mass of 57 grams is accelerated at a rate of 200 m/s^2 when hit by a tennis racket. What force did the racket exert on the ball?

SOLUTION First, the numbers must be converted to appropriate units for the SI system. In particular, the ball's mass must be converted from g to kg:

$$m = 57 \text{ g} = 57 \text{ g} \left(\frac{1 \text{ kg}}{10^3 \text{ g}}\right) = 5.7 \times 10^{-2} \text{ kg}.$$

The force on the ball will then be

$$F = ma = (5.7 \times 10^{-2} \text{ kg})(200 \text{ m/s}^2)$$

$$= 11.4 \, \frac{\text{kg} \cdot \text{m}}{\text{s}^2} = \underline{11.4 \text{ N}}. \quad ∎$$

EXAMPLE 4.3 The caboose of a train accelerates from rest to 30 miles per hour in 200 s while a 200-lb force pulls it. Calculate (a) the acceleration of the caboose (assumed constant) and (b) the mass of the caboose.

SOLUTION The following information is known:

$$\text{At } t = 0, \quad v_0 = 0.$$

$$\text{At } t = 200 \text{ s}, \quad v = 30 \frac{\text{miles}}{\text{hr}} = 30 \frac{\text{miles}}{\text{hr}} \left(\frac{5280 \text{ ft}}{1 \text{ mile}}\right)\left(\frac{1 \text{ hr}}{3600 \text{ s}}\right) = 44 \frac{\text{ft}}{\text{s}},$$

$$F = 200 \text{ lb}.$$

Notice that the speed must be converted from miles/hr to ft/s before using it in any of our equations. Newton's second law cannot be solved directly for acceleration or mass, since both are unknowns. We can, however, find the acceleration of the caboose using Eq. (3.7) from kinematics:

$$a = \frac{v - v_0}{t} = \frac{44 \text{ ft/s} - 0}{200 \text{ s}} = \underline{0.22 \text{ ft/s}^2}.$$

(b) Newton's second law can now be used to calculate the mass of the caboose:

$$m = \frac{F}{a} = \frac{200 \text{ lb}}{0.22 \text{ ft/s}^2} = 909 \frac{\text{lb}}{\text{ft/s}^2} = \underline{909 \text{ slugs}}.$$

The conversion of lb/(ft/s²) to slug was made using Eq. (4.3). ■

4.4 Weight

The force of gravity affects almost all of nature—from the energy needed to climb stairs or rise from a chair to the tides caused by the gravitational attraction of the sun and moon on the oceans. Weight is just one example of a gravitational force. When standing on the earth's surface, your weight is the gravitational force of the earth's mass pulling on your body. If standing on the moon, your weight is reduced by one-sixth. The moon's mass is smaller and exerts less force on your body. If floating in outer space, you are weightless because there are no nearby masses exerting observable gravitational forces on your body.

Weight, then, is defined as the gravitational force exerted on an object because of its gravitational attraction to some other mass such as the earth.

An object's weight when on the earth's surface is related by a simple equation to its mass. Suppose that an object with mass m is falling freely (we ignore air resistance). The cause of its acceleration is the object's weight w—the force of gravity pulling down on it. The object experiences a downward acceleration of magnitude g equal to 9.8 m/s² or 32 ft/s². Substituting weight for force and g for acceleration into Newton's second law ($\Sigma F = ma$), we see that

$$w = mg. \tag{4.4}$$

If an object's mass is known, Eq. (4.4) can be used to calculate the object's weight when near the earth's surface, and vice versa.

Weight and mass are often incorrectly substituted for each other in Newton's second law. *Weight, the force of gravity, is in units of newtons or pounds, and it appears with other forces on the left side of Newton's second law* ($\Sigma F = ma$). Mass represents an inertial property of an object—how effective forces are in changing the object's motion. *An object's mass appears in units of kilograms or slugs on the right side of Newton's second law.*

EXAMPLE 4.4 Calculate the weight of a 0.510-kg jar of Smucker's natural peanut butter.

SOLUTION Mass is in units of kg. Thus, the peanut butter's mass must be 0.510 kg. Its weight is

$$w = mg = (0.510 \, \text{kg})(9.8 \, \text{m/s}^2) = 5.0 \, \text{kg} \frac{\text{m}}{\text{s}^2} = \underline{5.0 \, \text{N}}.$$ ■

EXAMPLE 4.5 Calculate the mass of an 80-lb bag of cement.

SOLUTION Weight is a force in units of lb or N. The cement's mass is

$$m = \frac{w}{g} = \frac{80 \, \text{lb}}{32 \, \text{ft/s}^2} = 2.5 \, \frac{\text{lb}}{\text{ft/s}^2} = \underline{2.5 \, \text{slugs}}.$$ ■

4.5 Newton's Third Law

The great potential for analyzing interesting and complex situations by using Newton's second law is not apparent from the relatively simple examples presented so far. As situations become more complex and interesting, greater care must be used to identify which forces act on an object to cause its acceleration. The words *act on* are especially important.

Newton indicated in his third law of motion that forces always come in pairs—an action force and its reaction force. The action force is that exerted by one object on another. The reaction force is that exerted by the other object on the first. For example, rest your forearm against the edge of the pages of this book, as illustrated in Fig. 4.5. Your arm exerts a force into and against the book. But the book also exerts a reaction force into and against your arm, as is apparent from the indentation of the skin and muscle at the region of contact. According to Newton's third law of motion, these forces have equal magnitudes but point in opposite directions and act on different objects. (For this example, one force acts on the book; the other, on the arm.)

Newton's third law of motion: Whenever one object exerts a force on a second object, the second object exerts a reaction force of equal magnitude but opposite direction on the first object.

When using Newton's second law of motion to calculate an object's acceleration, we must consider only the forces acting on the object. The reaction forces that it exerts on other objects do not affect its acceleration. These reaction forces, however, affect the other objects and can be considered if we wish to study their acceleration.

Suppose, for example, that a person throws a ball in the air, as shown in Fig. 4.6a. If we are interested in the speed of the ball as it leaves the person's

(a)

(b)

$\mathbf{F}_{\text{arm on book}} = -\mathbf{F}_{\text{book on arm}}$

FIG. 4.5. The force (a) of the arm on the book is equal in magnitude but opposite in direction to the reaction force (b) of the book on the arm.

(a)

(b)

(c)

FIG. 4.6. (a) A man throws a ball. (b) A force diagram of the ball and (c) of the man's forearm.

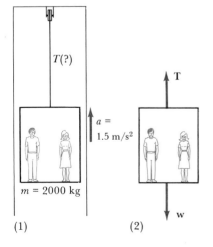

(1)   (2)

hand, we might construct a force diagram for the ball, as shown in Fig. 4.6b. The force F_H shown in this picture is the upward force of the hand on the ball. However, if our concern is with the muscle tension in the thrower's arm, we construct a force diagram of the forearm as shown in Fig. 4.6c. Notice that the ball exerts a reaction force F'_H down on the thrower's hand. The magnitude of the reaction force equals F_H but points in the opposite direction.

We see that we must define carefully the object whose motion we are considering. Then, when constructing a force diagram for that object, we consider only the forces acting on it and not the reaction forces that it exerts on other objects.

4.6 Applications of Newton's Second Law

Most problems in dynamics involve several forces that act on an object. The forces usually point in different directions. We can use vector addition to find the resultant force. The object's acceleration then equals this resultant force divided by the object's mass ($\mathbf{a} = \Sigma \, \mathbf{F}/m = \mathbf{R}/m$). Usually, it is better to use Newton's second law in component form to find the components of the acceleration along different axes:

The sum of the x components of all forces acting on an object equals the product of its mass and acceleration along the x axis. Similarly, the sum of the y components of the forces acting on the object equals the product of its mass and acceleration along the y axis. **Newton's second law,** rewritten in component form, is as follows:

$$\Sigma \, F_x = ma_x, \tag{4.5}$$
$$\Sigma \, F_y = ma_y. \tag{4.6}$$

The procedure for solving problems in dynamics is similar to the procedure used in statics. The only difference is that the sum of the forces is no longer zero; consequently, the object on which the forces act can accelerate. A summary of this problem-solving procedure follows. A diagram illustrates each step for the problem of calculating the tension in a cable needed to raise a 2000-kg elevator with an acceleration of 1.5 m/s^2.

Problem-Solving Procedure

1. Make a drawing of the whole situation being considered. Identify as precisely as possible the unknown quantity you wish to determine.

2. Construct a force diagram for a particular object in your drawing. The force diagram allows you to focus your attention on one small part of a complicated situation. All, or all but one, of the forces acting on this object are usually known. Arrows should be drawn to represent each known and unknown force acting on the object (as was done in Chapter 2 for objects in equilibrium). The object's acceleration is a consequence of the forces shown

in the diagram and is not itself a force. Hence, the $m\mathbf{a}$ (or ma_x and ma_y) of Newton's second law should not appear in the force diagram.

3. Superimpose a coordinate system on your force diagram. The coordinate system provides a set of axes along which the force components are calculated when solving Eqs. (4.5) and (4.6). You may wish to move your vectors so that their tails are at the origin. This makes it easier to calculate their components.

4. Write Newton's second law in component form:

$$\Sigma F_x = ma_x$$
$$\Sigma F_y = ma_y$$

The components of the forces are calculated using Eqs. (1.3) and (1.4).

5. Solve these equations for the unknowns. Frequently, kinematical equations such as Eqs. (3.7) through (3.10) will also be used along with Newton's second law when solving for an unknown.

The last step in this problem-solving procedure usually proceeds in one of two ways. If all the forces acting on an object are known, its acceleration can be calculated. Having found the acceleration, we can use kinematics to determine the future velocity and position of the object. This procedure is summarized below:

$$\frac{\Sigma \mathbf{F}}{m} \xrightarrow[\text{law}]{\text{second}} \mathbf{a} \xrightarrow{\text{kinematics}} \begin{array}{c}\text{changing velocity}\\\text{and position}\end{array}$$

On the other hand, if the changing position and velocity of an object are known, kinematical equations can be used to calculate its acceleration. Newton's second law is then used to determine an unknown force responsible for that acceleration:

$$\begin{array}{c}\text{changing velocity}\\\text{and position}\end{array} \xrightarrow{\text{kinematics}} \mathbf{a} \xrightarrow[\text{law}]{\text{second}} \frac{\Sigma \mathbf{F}}{m}$$

EXAMPLE 4.6 A 63-kg roller skater is pulled by a rope that exerts on the skater a 50-N force in the horizontal direction. (a) If resistive friction forces are small and ignored, calculate the skater's acceleration. (b) If the skater starts at rest, calculate her speed after being pulled 10 m.

SOLUTION (a) The whole situation described in the problem appears in Fig. 4.7a. The force diagram for the skater (Fig. 4.7b) shows all the forces acting on the skater: $\mathbf{w}$ is the skater's weight, $\mathbf{F}_c$ is the contact force of the skating surface up on the skater, and $\mathbf{T}$ is the rope tension force pulling the skater. A set of coordinate axes are superimposed on this force diagram.

We next write Newton's second law in component form; the force components are calculated using Eqs. (1.3) and (1.4).

$$\Sigma F_x = w_x + F_{c_x} + T_x$$
$$= w \cos 90° + F_c \cos 90° + T \cos 0° = ma_x,$$
$$\Sigma F_y = w_y + F_{c_y} + T_y$$
$$= -w \sin 90° + F_c \sin 90° + T \sin 0° = ma_y.$$

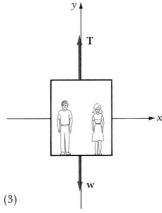

(3)

(4) $\Sigma F_x = T_x + w_x = 0 + 0 = ma_x$

$\Sigma F_y = T_y + w_y = T - w = ma_y$

(5) $T = ma_y + w = ma_y + mg$

$= 2000 \text{ kg } (1.5 \text{ m/s}^2 + 9.8 \text{ m/s}^2)$

$= \underline{2.26 \times 10^4 \text{ N}}$

(a)

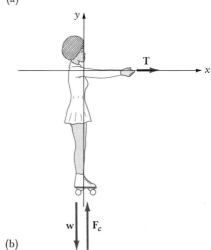

(b)

FIG. 4.7. (a) A roller skater being pulled by a rope. (b) A force diagram of the roller skater.

We learn from the x-component equation that

$$a_x = \frac{T}{m} = \frac{50\,\text{N}}{63\,\text{kg}} = \underline{0.79\,\text{m/s}^2}.$$

After substituting $\sin 90° = 1$ into the y-component equation, we find that

$$-w + F_c = ma_y.$$

If the skater does not move up or down, then $a_y = 0$, and we see that the weight and contact forces balance each other. This latter information is not needed in this problem.

(b) Having found the forward acceleration, we are now left with a problem in kinematics. Since a_x is constant, Eq. (3.10) can be used to calculate the skater's speed after traveling a distance $x - x_0 = 10$ m. The skater's initial speed v_0 is zero, and we find that

$$v^2 = v_0^2 + 2a_x(x - x_0)$$
$$= 0^2 + 2(0.79\,\text{m/s}^2)(10\,\text{m}) = 15.9\,\text{m}^2/\text{s}^2$$

or

$$v = \underline{4.0\,\text{m/s}}. \qquad \blacksquare$$

In this last example known forces were used to calculate an acceleration. The acceleration was then used to calculate the changing velocity of the skater. In the next example the procedure is reversed: an object's changing motion is used to calculate its acceleration. Then, using Newton's second law, we determine an unknown force causing the acceleration.

EXAMPLE 4.7 An 80-kg person driving a car runs into the back end of a parked cement mixer. The car and its driver, held in place by a seat belt and shoulder strap, stop in a distance of 1 m from the point of impact. If the car's original speed was 18 m/s, calculate the average force of the seat belt and shoulder strap on the person in the car.

SOLUTION The situation is pictured in Fig. 4.8a. A force diagram for the person is shown in Fig. 4.8b. In this problem we are concerned primarily with the forward motion in the positive x direction and will ignore the y-component equation. The x-component equation of Newton's second law is written below:

$$\Sigma F_x = F_{c_x} + w_x + F_{B_x}$$
$$= F_c \cos 90° + w \cos 90° - F_B \cos 0° = ma_x$$

or

$$-F_B = ma_x,$$

where F_B is the unknown magnitude of the force of the seat belt on the person. The negative sign appears in this equation because $\mathbf{F}_B$ points in the negative x direction opposite the car's motion. Without this opposing force of the seat belt, the person would continue forward through the car's window during the collision.

The above equation can be used to calculate F_B after a_x is calculated using kinematics. Remember that the car's initial speed $v_0 = 18$ m/s, and its final

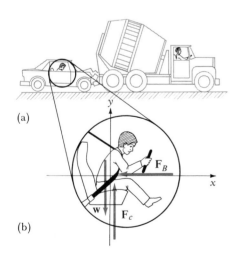

(a)

(b)

FIG. 4.8. (a) A car crashes into a cement mixer. (b) A force diagram of the car's driver.

speed is zero after decelerating to a stop in a distance of 1 m. Thus, using Eq. (3.10), we find that

$$a_x = \frac{v^2 - v_0^2}{2(x - x_0)} = \frac{0^2 - (18 \text{ m/s})^2}{2(1.0 \text{ m})} = -160 \text{ m/s}^2.$$

Substituting for m and a_x into the equation for the seat-belt force, we find that

$$-F_B = ma_x = (80 \text{ kg})(-160 \text{ m/s}^2) = \underline{-12{,}800 \text{ N}}.$$

The magnitude of the force of the seat belt and shoulder strap on the person is 12,800 N or about 1.4 tons of force. This force is great enough to break the person's ribs or cause other chest injuries. ∎

EXAMPLE 4.8 A 50-kg person on a skate board rolls down a short hill inclined at 30° with the horizontal. Calculate the person's acceleration if friction is ignored.

SOLUTION Figure 4.9a shows the situation described in the problem. In Fig. 4.9b, a force diagram of the person is shown. Problems involving inclines are solved more easily if the x axis points parallel to the plane and the y axis points perpendicular to the plane. With these coordinates, all motion is along the x axis parallel to the incline. No motion occurs perpendicular to the incline, and $a_y = 0$.

For the forces and axes shown in Fig. 4.9b, the x-component equation for the second law is

$$\Sigma F_x = F_{c_x} + w_x = F_c \cos 90° + w \cos 60°$$
$$= F_c(0) + w(0.5) = ma_x$$

or

$$a_x = \frac{0.5w}{m}.$$

But, since the magnitude of the weight force equals mg, we find that

$$a_x = \frac{0.5w}{m} = \frac{0.5(mg)}{m} = 0.5(9.8 \text{ m/s}^2) = \underline{4.9 \text{ m/s}^2}.$$

The y-component equation is not needed for this problem. ∎

FIG. 4.9. (a) A person on a skateboard rolling down a hill. (b) A force diagram of the person. The x axis is parallel to the incline and the y axis is perpendicular.

EXAMPLE 4.9 A fire breaks out in a hotel. A woman in a third-floor room ties several sheets together to make a rope. One end is tied to a radiator and the other is thrown out the window (Fig. 4.10a). The woman weighs 140 lb, but the sheets can support only 100 lb before tearing. (a) At what rate must she accelerate down the sheets in order to avoid tearing them? (b) If the distance from the window to the ground is 30 ft, how fast is she moving when she reaches the ground?

SOLUTION A force diagram of the woman appears in Fig. 4.10b. Two forces act on her: a downward weight force **w**, and the upward force $\mathbf{F}_s$ of the sheets (note that she pulls down on the sheets and the sheets pull up on her). If she were

FIG. 4.10. (a) A woman using sheets to lower herself from a burning building. (b) A force diagram of the woman.

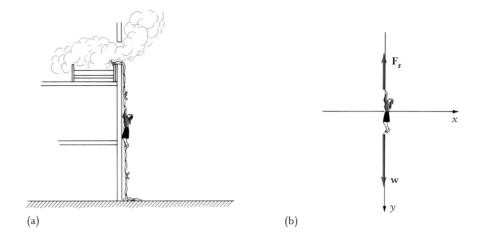

(a)   (b)

to hang at rest from the sheets, $\mathbf{F}_s$ and $\mathbf{w}$ would have to balance (both would be 140 lb). However, since the sheets will rip if F_s is greater than 100 lb, she must accelerate downward to keep from ripping the sheets. This will allow F_s to be less than 140 lb.

If we choose the y axis so that it points downward, then Newton's second law is

$$\Sigma F_y = w - F_s = ma_y.$$

Substituting $m = w/g$ into the above and rearranging, we find that

$$a_y = \frac{w - F_s}{m} = \frac{w - F_s}{(w/g)} = \left(\frac{w - F_s}{w}\right)g.$$

Given that $F_s = 100$ lb and $w = 140$ lb, then

$$a_y = \left(\frac{140\ \text{lb} - 100\ \text{lb}}{140\ \text{lb}}\right)(32\ \text{ft/s}^2) = \underline{9\ \text{ft/s}^2}.$$

(b) We can use kinematic Eq. (3.10) to estimate her speed at the bottom of the sheets. We are given: $v_{y_0} = 0$, $a_y = 9$ ft/s^2, and $y - y_0 = 30$ ft. Thus v_y, her speed at the bottom of the sheets, is

$$v_y^2 = v_{y_0}^2 + 2a_y(y - y_0)$$
$$= 0^2 + 2(9\ \text{ft/s}^2)(30\ \text{ft})$$

or

$$v_y = \underline{23\ \text{ft/s}}. \qquad \blacksquare$$

EXAMPLE 4.10 Two lovers are parked 10 m from the edge of a cliff in a sports car whose mass including that of the occupants is 1000 kg. A jealous suitor ties a rope to the car's bumper and a 50-kg rock to the other end of the rope. He then lowers the rock over the cliff and the car, in neutral, accelerates toward the cliff (we will ignore all friction forces). (a) Calculate the car's acceleration. (b) How much time do the lovers have before the car is pulled over the cliff?

SOLUTION The situation is shown in Fig. 4.11a. In this problem, it is necessary to make a force diagram for the car and a separate one for the rock (Figs.

4.11b and c). We have chosen the x axis in each figure to be the direction of motion because the car's acceleration a_x will then have the same algebraic value as that of the rock. The x-component equation of the second law for the car is

$$\Sigma F_x = T = m_{car} a_x \qquad (4.7)$$

The x-component equation for the rock is

$$\Sigma F_x = w_{rock} - T = m_{rock} a_x$$

or, substituting $w_{rock} = m_{rock} g$ into the above, we find that

$$T = m_{rock} g - m_{rock} a_x. \qquad (4.8)$$

If we substitute Eq. (4.7) into Eq. (4.8) and rearrange to solve for a_x, we find that

$$a_x = \left(\frac{m_{rock}}{m_{car} + m_{rock}} \right) g = \left(\frac{50 \text{ kg}}{1050 \text{ kg}} \right) 9.8 \text{ m/s}^2 = \underline{0.47 \text{ m/s}^2}.$$

(b) Kinematics is now used to calculate the time before the lovers plunge over the cliff. We know the following information about the car: $v_{x_0} = 0$, $a_x = 0.47 \text{ m/s}^2$, and $x - x_0 = 10 \text{ m}$. Equation (3.9) $(x - x_0 = v_{x_0} t + \frac{1}{2} a_x t^2)$ can be used to calculate the time taken by the car to travel 10 m.

Since $v_{x_0} = 0$, we find that

$$t^2 = \frac{2(x - x_0)}{a_x} = \frac{2(10 \text{ m})}{0.47 \text{ m/s}^2} = 42.6 \text{ s}^2$$

or

$$t = \underline{6.5 \text{ s}}.$$

This should be sufficient time for the lovers to apply the brakes and stop the car! ∎

4.7 Friction Forces

So far, our examples in this chapter have involved three kinds of force: (1) a tension force transmitted from one point to another along the length of a rope or cord; (2) a contact force exerted perpendicular to the surface of an object where it touches another object; and (3) a weight force caused by the gravitational pull of the earth on an object. In this section we examine yet another kind of force—friction force. A **friction force** opposes the motion of an object across a surface on which the object rests or slides and is directed parallel to the surface of contact. We consider here two of the most common types of friction: static friction and kinetic friction.

Static Friction

Static friction exists when an object does not slide along a surface on which it rests even though a force is exerted to make it slide. For example, if we push a large crate but it does not slide, we say that a static friction force resists our efforts. This friction force is static because the crate remains at rest, or "static," even though we push it.

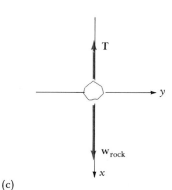

FIG. 4.11. (a) A car being pulled toward a cliff by a rope attached to a hanging rock. A force diagram (b) for the car and (c) for the rock.

FIG. 4.12. A static friction force F_s resists the tendency of an object to start sliding with whatever force is needed up to a maximum value given by F_s (max) $= \mu_s F_c$.

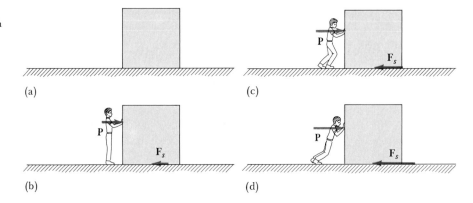

(a)   (c)

(b)   (d)

This static friction force is illustrated in Fig. 4.12. In (a), a large crate rests on a floor. No force pushes the crate; hence, there is no static friction force opposing the push. In (b), a person exerts a gentle push **P** against the crate. However, the crate does not move because a static friction force $\mathbf{F}_s$ equal in magnitude to the push opposes the motion. In (c) and (d), even larger pushing forces are unable to make the crate slide. As **P** increases, so does the static friction force resisting the motion.

If several people pushed the crate, the force **P** might become so large that the static friction force could no longer prevent the crate from sliding. The maximum force that a static friction force can exert depends on two factors:

1. The relative roughness of the two surfaces. A measure of the roughness of two surfaces is indicated by the coefficient of static friction μ_s of the surfaces. The larger the value of μ_s, the rougher the surfaces, and the harder it is to make the object start sliding.

2. The magnitude of the contact force between the object and the surface on which it rests. The larger the contact force, the harder it is to make the object start sliding.*

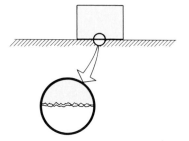

FIG. 4.13. An enlargement of a section of contact between a block and a floor shows many indentations from the two surfaces, which hook and bond to each other.

Figure 4.13, which shows an enlarged view of two surfaces in contact with each other, may make these two factors clearer. The figure shows that there are many indentations on any surface; if viewed through a microscope, no surface is perfectly smooth. The microscopic indentations and bulges in one surface tend to interlock with bulges and indentations in the other surface. If we try to pull one of these objects across the other, these bulges and indentations "hook on" to each other and resist the motion. The coefficient of static friction describes this relative roughness of surfaces. The coefficient of static friction between a rubber car tire and a concrete highway ranges from about 0.6 to 1.0, depending on the condition of the road; between two pieces of Teflon it is 0.04, and it is even less between the cartilage and bone in a healthy joint. The smaller the value of μ_s, the easier it is for the surfaces to slide across each other. Table 4.1 gives the values of μ_s for several surfaces.

TABLE 4.1 Coefficients of Static μ_s and Kinetic μ_k Friction

Surfaces	μ_s	μ_k
Rubber on concrete	0.6–1.0	0.8
Steel on steel	0.74	0.57
Aluminum on steel	0.61	0.47
Glass on glass	0.94	0.4
Wood on wood	0.25–0.5	
Ski wax on wet snow		0.1
Ski wax on dry snow		0.04
Teflon on Teflon	0.04	0.04
Greased metals	0.1	0.06
Surfaces in healthy joint		0.015

*Often, the magnitude of a contact force between an object and the surface on which it rests equals the object's weight. Thus, the greater its weight, the harder it is to make the object slide. However, the magnitude of the contact force does not always equal an object's weight, as we will see in some of the examples and problems.

Figure 4.13 also indicates why the maximum value of the static friction force depends on the contact force between the surfaces. As the contact force increases, the tiny bulges and indentations of one surface press more closely into those of the other surface, making it harder for the object to slide. A heavy crate is therefore harder to slide than a light crate.

As a result of many experiments, it has been determined that the magnitude of the **static friction force F$_s$** is always less than or equal to the product of the coefficient of static friction μ_s and the magnitude of the contact force F_c between the two surfaces:

$$F_s \leq \mu_s F_c \tag{4.9}$$

where the symbol $\leq$ means "less than or equal to." The direction of the static friction force **F$_s$** is parallel to the surface of contact and opposes the tendency to move. Note that an object sitting at rest begins to slide if pulled or pushed by a force directed parallel to the surface whose magnitude exceeds the maximum possible static friction force.

EXAMPLE 4.11 A 150-lb box sits on a floor. The coefficient of static friction between the block and floor is 0.4. What magnitude of horizontal pushing force **P** is needed to make the box slide?

SOLUTION The situation is shown in Fig. 4.14a. A force diagram of the box is shown in Fig. 4.14b. We have moved all the vector arrows so that their tails are at the origin of coordinates.

In order for the box to start sliding, P must be greater than the maximum possible value of F_s. Or,

$$P > F_s \text{ (maximum)} = \mu_s F_c.$$

We can see from the force diagram that the contact force F_c must balance the box's weight w. Thus, $F_c = w = 150$ lb. Therefore,

$$P > \mu_s F_c = (0.4)(150 \text{ lb}) = \underline{60 \text{ lb}}. \qquad \blacksquare$$

(a)

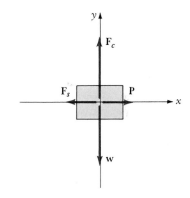

(b)

FIG. 4.14. (a) A pushing force **P** on a block is resisted by a static friction force **F$_s$**. (b) The force diagram of the block.

Kinetic Friction

If an object slides across a surface, a **kinetic friction force** opposes its sliding. The word *kinetic* implies that the object is moving. A road, for example, exerts a kinetic friction force on a car that skids to a stop.

After an object that is pushed or pulled starts to slide, less force is usually required to keep it sliding than was needed to start it sliding; that is, the kinetic friction force is less than the maximum static friction force. This is probably because the two surfaces cannot settle into each other as easily when the object moves as when it is at rest.

This effect of the motion of an object on the friction force is accounted for by defining a coefficient of kinetic friction, μ_k, which is usually less than the coefficient of static friction. Table 4.1 gives the values of μ_s and μ_k that have been determined experimentally for different surfaces in contact with each other. Notice that μ_s and μ_k are dimensionless numbers; they have no unit of measure.

Actually, the coefficient of kinetic friction μ_k is not constant for two types of surfaces. It can depend on the condition of the surfaces and also on the speed with which one object moves across the other. Usually, the value of μ_k decreases slightly as the speed of the object increases. This effect of velocity on μ_k is small, though, and will be ignored in our examples and problems.

The sliding or kinetic friction force $\mathbf{F}_k$ depends not only on μ_k but also on the magnitude of the contact force between the two surfaces, just as the static friction force did. From many experiments we know that the magnitude of the **kinetic friction force $\mathbf{F}_k$** acting on an object sliding across a surface is given by

$$F_k = \mu_k F_c. \tag{4.10}$$

The friction force $\mathbf{F}_k$ is directed parallel to the surface of contact and opposes the motion.

Surprisingly, kinetic and static friction forces usually do not depend on the area of contact between the surfaces. When the surfaces are hard and cannot fold into the little hills and valleys of each other, then friction forces are independent of area. If the two surfaces can conform to each other's shape, however, the friction does depend on area. Drag racers underinflate the rear tires of their cars so that the tires can fold into the little hills and valleys in a road. The back tires are also very wide to increase the area of contact between the tires and the road. This folding of wide tires into the indentation of the road causes the friction between the road and the tire to increase, enabling the racer to start faster. In our examples we will ignore the effect of contact area.

EXAMPLE 4.12 Police, examining the scene of an accident, observe skid marks 30 m long left by a 1200-kg car. The car skidded to a stop on a concrete highway having a coefficient of kinetic friction with the tires of 0.8. Estimate the car's speed at the beginning of the skid.

SOLUTION The situation described in the problem is shown in Fig. 4.15a, and a force diagram of the car appears in Fig. 4.15b. The kinetic friction force $\mathbf{F}_k$ is parallel to the surface of contact between the road and tires, and it points opposite the direction of motion. Before calculating $\mathbf{F}_k$, we need to calculate the contact force between the road and tires. To find F_c, we use the y-component equation of Newton's second law:

$$\Sigma F_y = F_c - w = ma_y.$$

But since no vertical motion occurs, $a_y = 0$ and $F_c - w = 0$. Thus, for this problem

$$F_c = w = mg = (1200 \text{ kg})(9.8 \text{ m/s}^2) = 1.18 \times 10^4 \text{ N}.$$

We can now calculate the kinetic friction force using Eq. 4.10:

$$F_k = \mu_k F_c = (0.8)(1.18 \times 10^4 \text{ N}) = 0.94 \times 10^3 \text{ N}$$

The x-component equation of Newton's second law can then be used to determine the car's acceleration:

$$\Sigma F_x = -F_k = ma_x$$

(a)

(b)

FIG. 4.15. (a) A car skidding to a stop. (b) A force diagram of the car.

or

$$a_x = -\frac{F_k}{m} = \frac{0.94 \times 10^3 \, \text{N}}{1200 \, \text{kg}} = -7.8 \, \text{m/s}^2.$$

Finally, kinematics Eq. (3.10) can be used to calculate the car's initial speed v_0 if its final speed is zero after skidding 30 m:

$$v^2 - v_0^2 = 2a_x(x - x_0)$$

or

$$0^2 - v_0^2 = 2(-7.8 \, \text{m/s}^2)(30 \, \text{m}).$$

Thus,

$$v_0 = \underline{21.6 \, \text{m/s}} \quad \text{or} \quad 48 \, \text{miles/hr.} \qquad \blacksquare$$

EXAMPLE 4.13 A 30-kg block is pulled along a floor by a 200-N tension force directed 37° above the floor. If the kinetic coefficient of friction μ_k between the block and floor is 0.3, what is the acceleration of the block?

SOLUTION The situation described in the problem is shown in Fig. 4.16a, while Fig. 4.16b shows a force diagram of the block. In order to know the magnitude of the friction force, we must first find the contact force F_c. The y-component equation of Newton's second law can be used to calculate F_c:

$$\Sigma F_y = T_y + F_{c_y} + F_{k_y} + w_y = ma_y$$

or

$$T \sin 37° + F_c \sin 90° + F_k \sin 0° - w \sin 90° = ma_y.$$

Since there is no motion along the y axis, $a_y = 0$ and the above equation becomes

$$0.6T + F_c - w = 0$$

or

$$F_c = w - 0.6T = mg - 0.6T = (30 \, \text{kg})(9.8 \, \text{m/s}^2) - 0.6(200 \, \text{N}) = 174 \, \text{N}.$$

We can now calculate the magnitude of the kinetic friction force:

$$F_k = \mu_k F_c = 0.3(174 \, \text{N}) = 52 \, \text{N}.$$

The x-component equation of Newton's second law will allow us to calculate the block's acceleration:

$$\begin{aligned}\Sigma F_x &= T_x + F_{c_x} + F_{k_x} + w_x \\ &= T \cos 37° + F_c \cos 90° - F_k \cos 0° + w \cos 90° \\ &= ma_x \end{aligned}$$

or

$$0.8T - F_k = ma_x.$$

Thus,

$$a_x = \frac{0.8T - F_k}{m} = \frac{0.8(200 \, \text{N}) - 52 \, \text{N}}{30 \, \text{kg}} = \underline{3.6 \, \text{m/s}^2.}$$

(a)

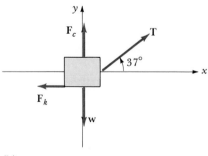

(b)

$\blacksquare$ **FIG. 4.16.** (a) A block pulled by a rope. (b) A force diagram of the block.

Summary

1. **Newton's laws of motion:** In the study called dynamics, we consider the way in which forces affect the motion of objects. An understanding of Newton's three laws of motion is fundamental to dynamics.

   Newton's first law: Every body continues in its state of rest or of constant speed in a straight line unless it is compelled to change that state because of forces acting on it.

   Newton's second law: An unbalanced force acting on an object will cause the object to accelerate in the direction of the force. The acceleration is proportional to the vector sum of the forces acting on the object and inversely proportional to the object's mass:

$$\Sigma \mathbf{F} = m\mathbf{a} \quad \text{or} \quad \begin{array}{l} \Sigma F_x = ma_y \\ \Sigma F_y = ma_y \end{array} \quad (4.1) \quad \text{or} \quad \begin{array}{l} (4.5) \\ (4.6) \end{array}$$

   Newton's third law: Whenever one object exerts a force on a second object, the second object exerts a reaction force of equal magnitude but opposite direction on the first object.

   Newton's second law relates the forces acting on an object to its acceleration. Kinematics is often used to relate an object's acceleration to its changing velocity and position. Procedures for using the second law and kinematics to solve problems in dynamics are outlined in Section 4.6.

2. **Mass and weight: Mass** is a measure of an object's inertia. The greater its inertia, the less its acceleration when pushed or pulled by a force. Mass is also an indication of the amount of matter of which the object is made.

An object's **weight** is the gravitational force exerted on the object due to its gravitational attraction to other masses such as the earth. Near the earth's surface, an object's weight and mass are related by the equation

$$w = mg, \qquad (4.4)$$

where $g = 9.8 \text{ m/s}^2$.

3. **Friction forces:** Friction forces oppose the motion of one object across another surface. If an object, when pushed or pulled, does not move because of the opposing friction force, the friction is then said to be static friction. If the object slides across a surface, it is opposed by a kinetic friction force.

   The **static friction force** $\mathbf{F}_s$ has whatever value is needed to keep an object from sliding, but does not exeed a limiting value given by the equation:

$$F_s \le \mu_s F_c, \qquad (4.9)$$

where μ_s is the coefficient of static friction between the object and the surface on which it rests and F_c is the contact force between the two surfaces.

   The **kinetic friction force** $\mathbf{F}_k$ has a magnitude

$$F_k = \mu_k F_c, \qquad (4.10)$$

where μ_k is the coefficient of kinetic friction. Both static and kinetic friction forces are directed parallel to the surface of contact and oppose the motion.

Questions

1. (a) A force is seen acting on an object, but the object does not move. Explain. (b) A 200-N force acts on a 10-kg block but causes it to have only an acceleration of 5 m/s^2. Explain.

2. A box full of lead and one of the same size full of feathers are floating weightless near a spaceship. Develop an experiment to estimate the mass of one compared to that of the other.

3. (a) *Estimate* the weight of this book in newtons, using the fact that one medium-sized apple weighs 1 N. (b) Using this estimate, calculate its weight in pounds and its mass in kilograms and slugs.

4. A horse pulls on a wagon and the wagon pulls back on the horse with an equal reaction force. How can the horse accelerate?

5. Why is it easier to pull a lawn mower than to push it?

6. A person jumps from a wall and lands stiff-legged. Use kinematics and Newton's second law to explain why the person is less likely to be injured when landing in soft sand than on concrete.

7. A person is able to push down harder on a bicycle pedal when pulling up hard on the handlebars. Explain, using a force diagram.

8. A person lifts a box from a bent position. Why is there more chance of back injury if the box is lifted quickly rather than slowly?

9. Two balls of the same material, one of mass m and the other of mass $2m$, are dropped simultaneously from the Leaning Tower of Pisa. Will the balls hit the ground at the same time? Explain carefully, using Newton's second law. Ignore air resistance.

10. If we ignore friction, which will roll down a hill faster—a sports car or a truck? Explain.

11. (a) A person standing on a rug remains stationary while pushing on a wall. Show the forces acting on the person and indicate which forces are equal in magnitude. (b) Why doesn't a person on roller skates remain stationary when pushing on a wall while standing on a concrete floor?

12. A person starts to run. (a) Show the forces acting on the person. (b) Why is the person able to accelerate in the horizontal direction?

13. A car accelerates along a road. (a) Show the forces of the car on the road. (b) Show the reaction forces of the road on the car. (c) Which force is responsible for the acceleration?

Problems

4.2 and 4.3 Newton's Second Law and Units

1. (a) Calculate the net force needed to provide a 1.5 m/s^2 acceleration for a 60-kg person on ice skates. (b) The same net force is applied to a 40-kg skater. What is her acceleration?

2. A rope will break if its tension exceeds 3000 N. Calculate the maximum possible acceleration of a 1400-kg car pulled by the rope. Ignore friction.

3. An earth satellite is being constructed in orbit above the earth. A 200-N force is used to push a steel beam into place. The beam accelerates at 0.2 m/s^2 when acted on by this force. Calculate the mass of the beam.

4. A farmer pushes a 500-kg wagon with a horizontal force of 125 N. (a) Calculate the acceleration of the wagon (ignore friction forces). (b) If the wagon starts at rest, how fast is it moving after being pushed for a time of 5 s?

5. The lower left chamber of the heart (the left ventricle) pumps blood into the aorta. The left ventricular contraction lasts 0.2 s, during which time a mass of 88 g of blood is accelerated from rest to a speed of about 0.45 m/s. (a) What is the average acceleration of the blood during this time? (b) What is the average net force of the left ventricle on the blood during this contraction?

6. A 1000-kg sports car accelerates from rest to 60 miles/hr in 11 s. Calculate the net force acting on the car.

7. A 40-g arrow traveling at 50 m/s hits a target backed by a bale of hay. The arrow stops after entering 0.5 m into the bale. (a) Calculate the deceleration of the arrow. (b) What is the average retarding force of the target on the arrow during this deceleration?

8. A football halfback weighing 210 lb, running at a speed of 16 ft/s, is stopped by a head-on collision with a defensive linebacker. The halfback is stopped 2 ft from where the tackle started. (a) Calculate the deceleration of the halfback. (b) What is his mass in slugs? (c) What is the average force acting on the halfback?

9. A car traveling at 20 m/s hits an embankment and is stopped in a distance of 1.6 m. (a) What is the deceleration of the car? (b) If the car's mass is 1800 kg, what is the average force of the embankment on the car?

10. During a head-on collision, the person in the front seat of a car decelerates from 13.3 m/s (30 miles/hr) to rest in 0.1 s. (a) What is the deceleration of the person? (b) If the person's mass is 75 kg (165 lb), calculate the average force of the seat belt and shoulder strap on the person.

11. A 0.145-kg baseball enters the glove of a catcher on a baseball team at a speed of 102 miles/hr. The baseball is stopped in a distance of 8.0 inches. Calculate the force in units of pounds and newtons of the glove on the ball if the deceleration is assumed constant.

4.4 Weight

12. (a) The mass of a gallon of milk is 3.8 kg. Calculate its weight in newtons. (b) A cement truck when filled weighs 4.2 × 10^5 N. Calculate its mass in kilograms.

13. Calculate the mass of a 3-lb bag of oranges in units of slugs; in kilograms. Calculate its weight in newtons.

14. The liquid in a can of Coca-Cola has a mass of 0.35 kg. Calculate your change in weight in units of newtons and pounds immediately after drinking the contents of one can.

15. (a) Indicate your weight in pounds and calculate your mass in slugs. (b) If a 50-lb unbalanced force acted on you, what would be your acceleration?

16. (a) Indicate your weight in pounds and convert to newtons. (b) Calculate your mass in kilograms. (c) Calculate your mass in kilograms on the moon. (d) Calculate your weight in newtons on the moon. The acceleration of gravity on the moon is one-sixth its value on the earth.

4.5 Newton's Third Law

17. A book lies on top of a second book, which rests on a table. Construct a force diagram for each book and identify in words the cause of each force. Indicate which, if any, of the forces shown in your diagrams are action-reaction pairs.

18. Describe in words and in pictures three different forces and the reaction to each force.

19. An car exerts a forward force on a trailer, and the trailer exerts an equal magnitude backward force on the car. Construct a force diagram for each vehicle and explain what force causes the car to accelerate when pulling the trailer.

20. A person is stranded in the middle of a frictionless sheet of ice. To get off the ice, the person removes his shoes and throws them simultaneously in the same horizontal direction. The shoes, whose total mass equals 0.6 kg, fly across the ice at a speed of 10 m/s. The 60-kg person slides in the opposite direction at a speed of 0.1 m/s. Explain carefully the force that caused the person to start moving.

4.6 Applications of Newton's Second Law

21. A rope connected to a 50-kg sled pulls it along a frictionless sheet of ice. The tension in the rope is 100 N, and the rope is oriented at an angle of 30° above a line drawn parallel to the ice. Calculate the horizontal acceleration of the sled.

22. A 70-kg calf running at a speed of 2.0 m/s is roped from behind. The rope exerts on the calf a steady force of 200 N in a direction of 24° above the horizontal and from the rear. Calculate the distance the calf runs after being roped and before it is stopped. Ignore all other forces with horizontal components.

23. A 50-kg woman in a crouched position pushes against the ground with a constant force while raising the mass of her body 0.2 m. The woman's speed as she leaves the ground is 2.7 m/s. Calculate her constant pushing force. (Do not forget to include her weight in your calculations.)

24. An 80-kg man jumps from a wall and lands stiff-legged in soft sand. The man's downward speed just as he touches the sand is 6.0 m/s. He is stopped after sinking 0.05 m into the sand. Calculate the force, assumed constant, of the sand on the man while he is being stopped.

25. A 90-kg businessman leaps from a burning hotel window to

be caught by a fire net. His speed as he reaches the net is 22 m/s, and he is stopped by the net after sinking 1.2 m into it. Calculate the force of the net on the man (the force is assumed constant).

26. A 130-lb rock climber must be rescued from a mountain ledge by a helicopter. After a rope from the helicopter is attached to the climber, the helicopter accelerates upward at 24 ft/s². What is the force of the rope on the climber?

27. A 3.1-g penny thrown from the top of the Sears building in Chicago lands on the head of a passerby 460 m below. Just before the penny hits the person's head, it is falling at a speed of 25 m/s. It is stopped by the head in a distance of 1 mm (10^{-3} m). (a) Calculate the deceleration of the penny. (b) Calculate the force of the penny on the head. (c) Parts of the skull bones will be fractured by forces of about 900 N (200 lb) or more. Is there a chance of fracturing the skull with the penny?

28. (a) An elevator of mass 10,000 kg accelerates upward at a rate of 1.5 m/s². What is the tension in the cable supporting the elevator? (b) Same as (a) only the elevator accelerates downward at 1.5 m/s².

29. (a) A 100-lb person stands on a bathroom scale in an elevator that is accelerating upward at 8 ft/s². What does the scale read? (b) What does the scale read if the elevator is accelerating down at 8 ft/s²? (c) If the elevator cable breaks and falls freely toward the bottom of the elevator shaft, what does the scale read? Explain.

30. An elevator whose mass is 8,000 kg is supported by a cable that breaks if subjected to a tension greater than 1.2×10^5 N. Calculate the maximum acceleration the elevator can have.

31. A car sitting at rest is hit from the rear by a car moving at 13 m/s. The car lurches forward with an acceleration of about 300 m/s². Figure 4.17 shows the neck muscle that causes the head to accelerate forward with the body. If the head has a mass of 4.5 kg, what is the horizontal component of **F** required to accelerate the head? If **F** is directed 37° below the horizontal, what is the magnitude of **F**? (*Note:* Whiplash victims often experience sore neck muscles because of these unusually large tension forces.)

FIG. 4.17

32. A 10-kg box is pulled up a frictionless plane inclined at an angle of 37° with the horizontal by an 80-N force directed parallel to the plane. Calculate the acceleration of the block.

33. A string with one 10-g die on the end is attached to the rear view mirror of a car. When the car leaves an intersection, the string makes an angle of 5° with the vertical. What is the acceleration of the car? (*Hint:* Choose the die as the object for your force diagram. Two forces act on the die, which is accelerating in the horizontal direction but not vertically. First use the vertical-com-

ponent equation of Newton's second law to find the tension in the string. The horizontal component of **T** causes the die to accelerate.)

34. A 30-kg boy hangs from a rope, which passes over a pulley. A 40-kg mass hangs at the other end of the rope. Construct separate force diagrams for the boy and the 40-kg mass. Write Newton's second law for each object. Determine the acceleration of the boy and the tension in the cable (it is not 40 kg times g).

35. A 20-kg block rests on a frictionless table. A cord attached to the block extends horizontally to a pulley at the edge of the table. A 10-kg mass hangs at the end of the cord after it passes over the pulley. Calculate the acceleration of the block and mass, and the tension in the cable.

36. Calculate the acceleration of the system shown in Fig. 4.18, and the tension in the cable. There is no friction force on the 10-kg block.

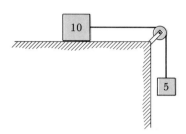

FIG. 4.18

37. The system shown in Fig. 4.19 accelerates to the left. Find the magnitude of this acceleration and the tension in the cable. There is no friction between the block and the plane.

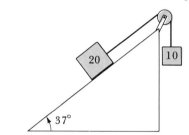

FIG. 4.19

4.7 Friction Forces

38. A 200-lb crate lies on the floor. The coefficient of static friction between the crate and floor is 0.6. What is the minimum force required to start the crate sliding?

39. A 40-kg child wearing socks does not slide across a hardwood floor until pulled by a 100-N horizontal force. Calculate the coefficient of static friction between the socks and floor. Assume that the child remains oriented vertically.

40. A block rests on an inclined plane. Its angle with the horizontal is increased. When the angle becomes 40°, the block starts to slide. Calculate the coefficient of static friction between the block and plane.

41. A 170-lb runner near the beginning of a race is leaning forward at an angle of 50° from the horizontal. His foot pushes against the track with a force **F** of magnitude 300 lb that is directed along his body. (a) What is the horizontal component of the reaction force of the ground on his foot? This is also the static friction force of the ground on his foot. (b) What is the horizontal

acceleration of the runner? (c) What is the vertical acceleration of the runner?

42.   A bus leaves a corner with an acceleration of 1.2 m/s^2. A 75-kg person standing in the bus must lean forward in order to avoid falling. At what angle with the horizontal must the person lean if unaided by any hand supports? For this problem you can assume that the net force **F** of the ground on the person (the sum of the contact force and the friction force) is directed along the person's slanted legs. The horizontal component of **F** caused by friction is responsible for the person's acceleration. The vertical component of **F** balances the person's weight.

43.   A boy pushes horizontally with a force of 60 N against a 30-kg wagon. The wagon accelerates at 0.8 m/s^2. (a) Calculate the net force causing the acceleration. (b) How much of the 60-N push was needed to overcome friction? (c) When the boy stops pushing, what is the wagon's acceleration?

44.   A train engine with one car accelerates from a station at 0.50 m/s^2. (a) What is the net force on the car whose mass is 10,000 kg? (b) A 1000-N friction force resists the motion of the car. What is the force of the coupling that pulls on the car?

45.   A 50-kg block rests on the floor. The coefficients of static and kinetic friction are 0.7 and 0.5, respectively. (a) What is the minimum force needed to start the box sliding? (b) If the same force continues to push the block after it starts sliding, what is the acceleration of the block?

46.   A 50-kg block is pulled at a constant speed by a horizontal force of 250 N. Calculate the kinetic coefficient of friction between the block and the surface on which it slides.

47.   A 1500-kg car travels at 80 km/hr. The brakes are applied abruptly, and the car skids to a stop. The kinetic coefficient of friction between the car and road is 0.7. (a) Calculate the friction force on the car. (b) Calculate the car's acceleration. (c) How far will it skid before stopping?

48.   A 110-g hockey puck leaves a hockey stick with a speed of 8.0 m/s. It slides on the ice a distance of 40 m before stopping. What is the kinetic coefficient of friction between the puck and the ice?

49.   A 3000-lb auto traveling at 59 ft/s (40 mph) coasts to a stop in 2200 ft when in gear and in 3000 ft when in neutral. Calculate the net resistive force slowing the auto in each case. The difference in the two situations is the extra friction force caused by the transmission and drive shaft when in gear.

General Problems

G1.   A football player is asked to jerk a 135-lb barbell from the ground to his chest. The initial acceleration of the barbell is 25 ft/s^2. (a) What is the initial force of the player on the barbell? (b) In Example 2.12 in Chapter 2, we found that the back-muscle tension and disc-compression forces are more than seven times the weight being lifted when in a bent position. Roughly what is the tension in the football player's back muscle?

G2.   A 20-kg crate is lowered down a plane inclined at an angle of 37° with the horizontal by a rope that exerts on the crate a 50-N force directed parallel to the plane. A 40-N friction force, also parallel to the plane, opposes the downward motion. What is the acceleration of the crate?

G3.   A 100-lb box accelerates at a rate of 2 ft/s^2. Its motion is opposed by a 40-lb kinetic friction force. (a) How large is the horizontal component of the force pulling the block? (b) If this pulling force is directed 30° above the horizontal, what is its magnitude?

G4.   Calculate the acceleration of a 100-kg crate when pulled by a 300-N force directed 45° above the horizontal. The crate slides on a surface whose coefficient of kinetic friction with the crate is 0.20.

5 Momentum

The motion of objects and the way in which this motion changes are again our subjects. Now, instead of considering the acceleration caused by forces, we will consider an object's change in momentum caused by the impulse of a force. The analysis is very similar to that used in the last chapter, and an equation relating impulse and momentum can be derived easily from Newton's second law of motion. The impulse-momentum equation is useful for studying collisions of automobiles, impacts that occur during athletic events, demolition of buildings, and many less dramatic examples involving the changing motion of an object.

Momentum is a measure of the difficulty encountered in bringing an object to rest. The greater the object's mass and velocity, the greater is its momentum. *Impulse* is the product of a force acting on an object and the time that the force acts. An impulse, such as the force of a bat on a ball, causes the object's momentum to change.

Momentum has a special significance: it is one of a small number of quantities that does not change during the many interactions of life and is called a *conserved quantity*. The total momentum of two cars involved in a crash is the same just after the crash as it was just before the crash. The fact that momentum is conserved provides a useful tool for analyzing a variety of physical processes ranging from the study of rocket propulsion to the discovery of new particles produced by high-energy accelerators.

5.1 Momentum

Momentum is a property related to an object's motion and mass. The faster the object moves and the larger its mass, the greater is its momentum. A truck moving at 30 mph has more momentum than a Ping-Pong ball flying at the same speed or than another truck moving at a slower speed. Qualitatively, *momentum reflects the tendency of a moving object to continue moving and the difficulty encountered in reducing that motion.* Momentum is defined quantitatively as follows:

The **momentum p** of an object is defined as the product of its mass m and velocity **v**:

$$\text{momentum} = \mathbf{p} = m\mathbf{v}. \tag{5.1}$$

The units of momentum are kg $\cdot$ m/s.

Since velocity is a vector quantity having magnitude and direction, momentum is also a vector quantity with magnitude and direction. The direction of an object's momentum equals the direction of its velocity.

EXAMPLE 5.1   A cement truck full of cement has a mass of 42,000 kg and travels north at a speed of 18 m/s, or about 40 mph. (a) Calculate its momentum. (b) How fast must a 900-kg Honda Civic travel to have the same momentum?

SOLUTION   (a) The truck's momentum points north in the same direction as its velocity. The magnitude of its momentum is $p = mv = (4.2 \times 10^4\,\text{kg})$ $(18\,\text{m/s}) = 7.6 \times 10^5\,\text{kg} \cdot \text{m/s}$. (b) For the momentum of the Honda to equal that of the truck, it must drive north in the same direction as the truck. The magnitude of the Honda's momentum must also equal that of the truck. Thus,

$$m_{\text{Honda}} v_{\text{Honda}} = 7.6 \times 10^5\,\text{kg} \cdot \text{m/s}$$

or

$$v_{\text{Honda}} = \frac{7.6 \times 10^5\,\text{kg} \cdot \text{m/s}}{m_{\text{Honda}}} = \frac{7.6 \times 10^5\,\text{kg} \cdot \text{m/s}}{900\,\text{kg}} = \underline{840\,\text{m/s}}.$$

This is approximately 1900 mph! It is unlikely that the momentum of the Honda or of any small car will equal that of a cement truck moving at 40 mph. ■

5.2   The Impulse-Momentum Equation

What causes an object's momentum to change? Force must be involved. Force is needed to start or stop a car. The length of time that the force acts also affects the change in momentum. For instance, a small force acting for a long time can stop a car (Fig. 5.1a) just as a large force acting for a short time can stop the car (Fig. 5.1b). These two quantities, the force and the time the force acts, determine what is called the impulse of the force.

(a)

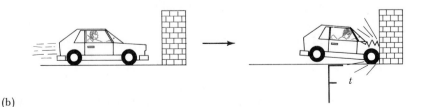

(b)

FIG. 5.1.   (a) A small force acting for a long time can cause a large change in momentum. (b) A large force acting for a short time can also cause a large change in momentum.

The **impulse** of a force **F** acting on an object for a time t is defined as the product of the force and time:

$$\text{impulse} = \mathbf{F}t. \tag{5.2}$$

Impulse has units of newton-seconds (N·s) and is the cause of an object's change in momentum.

An equation relating the impulse of a force and the subsequent change in momentum of the object on which the force acts is derived easily using Newton's second law of motion ($\Sigma\,\mathbf{F} = m\mathbf{a}$) and kinematics. We replace the acceleration in Newton's second law by the acceleration given by Eq. (3.7) from kinematics,

$$\mathbf{a} = \frac{\mathbf{v} - \mathbf{v}_0}{t},$$

to obtain

$$\Sigma\,\mathbf{F} = m\mathbf{a} = m\,\frac{\mathbf{v} - \mathbf{v}_0}{t}.$$

Rearranging the above equation produces the **impulse-momentum equation:**

$$(\Sigma\,\mathbf{F})t = m\mathbf{v} - m\mathbf{v}_0. \tag{5.3}$$

Notice that the right side of the equation equals the change in an object's momentum. The left side is the cause of momentum change and equals the sum of the impulses of all forces acting on the object during the time of the momentum change.

The impulse-momentum equation provides a nice interpretation for a variety of phenomena ranging from impulses occurring in athletics and car collisions to the impulse of a molecule of air bouncing off the inside wall of a balloon or off your skin. Most of our examples and problems are restricted to one-dimensional motion in which case we will not use vector symbols in Eq. (5.3). Instead, the directions of forces and velocities are indicated by plus and minus signs (depending on the orientation of the force or velocity relative to the axis along which motion occurs).

EXAMPLE 5.2 A golf club strikes a golf ball (Fig. 5.2). The club and ball remain in contact for 0.6×10^{-3} s. The 45-g ball leaves the club with a speed of 70 m/s. Calculate the average force of the club on the ball.

SOLUTION The force of the club on the ball is directed along the line of motion of the club and is the only force acting on the ball in that direction. We omit vector signs and consider only the magnitude of this force and the ball's change in momentum in that direction. The ball's initial speed v_0 is zero and its final speed v equals 70 m/s. Rearranging Eq. (5.3), we find that

$$F_{\text{club}} = \frac{mv - mv_0}{t} = \frac{(4.5 \times 10^{-2}\,\text{kg})(70\,\text{m/s}) - 0}{0.6 \times 10^{-3}\,\text{s}} = \underline{5.3 \times 10^3\,\text{N}}. \quad \blacksquare$$

FIG. 5.2. A golf club transferring momentum to a golf ball.

The force calculated in the last example is the average force exerted by the club on the ball during its contact time with the ball. The actual force varies—

being less than the average force during portions of the contact time and greater at other times.

In many applications of the impulse-momentum equation, the impulse time lasts less than 1 second. For example, during a head-on car collision, the car may stop in less than 0.1 s. In order to calculate the force exerted on an object during an impulse, we must know the impulse time.

Kinematics provides a moderately reliable method for estimating the impulse time. Suppose that a car crashes into a large tree. Examination of the car after the collision indicates that its front end was smashed in a distance of about 0.5 m during the collision. This 0.5 m represents the distance the car traveled from the beginning to the end of the impact and is called the car's stopping distance.

Equation (3.8) from kinematics can be used to relate this stopping distance $x - x_0$ to the stopping or impulse time t:

$$t = \frac{2(x - x_0)}{v + v_0}, \tag{3.8}$$

where v is the car's final speed (zero for our example) and v_0 is its initial speed. This equation is exact only if the car's deceleration is constant—a very improbable collision. However, the equation does provide us with a technique for estimating the stopping time if the stopping distance is known.

EXAMPLE 5.3 A car and its driver traveling at a speed of 18 m/s (40 mph) are stopped abruptly by a collision with the rear end of a parked moving van (Fig. 5.3a). The car's front end is pushed in 0.55 m by the collision; the stopping distance of the car, then, is also 0.55 m. A 60-kg driver, who is restrained by a seat belt and shoulder harness, is also stopped in 0.55 m. Calculate (a) the time required to stop the car, (b) the average force acting on the driver (Fig. 5.3b), and (c) the deceleration of the driver.

SOLUTION (a) Choosing the positive x axis to point in the direction of initial motion, the driver's initial velocity in the x direction is 18 m/s and the final velocity is zero. The time required to stop the car is calculated using Eq. (3.8):

$$t = \frac{2(x - x_0)}{v + v_0} = \frac{2(0.55 \text{ m})}{0 + 18 \text{ m/s}} = 0.061 \text{ s} = \underline{61 \text{ ms}},$$

where 1 ms = 1 millisecond = 10^{-3} s.

(b) The average force of the seat belt and shoulder strap on the driver is determined from the impulse-momentum equation:

$$F = \frac{m(v - v_0)}{t} = \frac{(60 \text{ kg})(0 - 18 \text{ m/s})}{0.061 \text{ s}} = \underline{-1.77 \times 10^4 \text{ N}}.$$

The negative sign implies that the force points in the negative x direction—opposite the direction of motion. The magnitude of the force is equivalent to 3600 lb (almost the weight of a rhinoceros).

(c) The driver's deceleration can be calculated using Eq. (3.7):

$$a = \frac{v - v_0}{t} = \frac{0 - (18 \text{ m/s})}{0.061 \text{ s}} = \underline{-295 \text{ m/s}^2}.$$

It is common to write acceleration or deceleration during impact in terms of

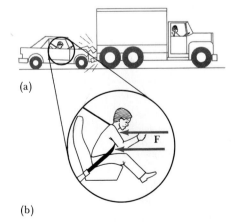

(a)

(b)

FIG. 5.3. (a) A car crashes into the rear of a moving van. (b) The force of the seat belt and shoulder strap restrains the driver of the car as it stops abruptly.

gravitational acceleration—that is, as so many g's. Since $g = 9.8 \text{ m/s}^2$, we find that

$$a = (-295 \text{ m/s}^2)\left(\frac{1g}{9.8 \text{ m/s}^2}\right) = -30g.$$

The deceleration is 30 times greater than that of a freely falling body. ∎

Will the driver in Example 5.3 be injured? In October 1976 a preliminary report of the European Experimental Vehicle Committee (EEVC)* estimated that chest injuries were improbable if shoulder-strap tension was less than about 8000 N. If one-half of the 17,700-N force on the driver in Example 5.3, or 8850 N, is caused by the shoulder strap, then there is a reasonable chance of injury. Common chest injuries in collisions include broken ribs, lacerations of the lungs, bruises of the heart, and even rupture of the aorta. This last injury can occur if the heart is pushed or pulled away from the large vessels that normally anchor it to the body. To avoid such injuries, shoulder straps can be constructed that extend or loosen slightly if the tension reaches some maximum value (for example, 8000 N). This increases the stopping time and thus decreases the magnitude of the force during the impulse.

EXAMPLE 5.4 A 30-kg boy falls from the roof of a house. After falling freely for 2.5 m, he lands on the branches of a bush that break his fall (Fig. 5.4). He is stopped after sinking 0.8 m into the bush. Estimate the average force of the bush on the boy.

SOLUTION The solution requires three steps: (1) calculate the boy's velocity as he enters the bush, (2) calculate the time needed to reduce his velocity to zero after entering the bush, and (3) use the impulse-momentum equation to calculate the average force of the bush on the boy.

The velocity and forces are described conveniently using a coordinate system with the y axis pointing down (Fig. 5.4). Since all motion and the forces point along the y axis, we can denote direction simply by using plus and minus signs (plus for down and minus for up).

Equation (3.10) from kinematics is used to calculate the boy's velocity as he reaches the bush after falling freely for 2.5 m:

$$v^2 = v_0^2 + 2a(y - y_0) \tag{3.10}$$
$$= 0^2 + 2(9.8 \text{ m/s}^2)(2.5 \text{ m}) = 49 \text{ m}^2/\text{s}^2$$

or

$$v = 7.0 \text{ m/s}.$$

The rest of the problem involves his interaction with the bush. His initial velocity as he enters the bush is 7.0 m/s and his final velocity after being stopped is zero. We calculate the stopping time using Eq. (3.8):

$$t = \frac{2(y - y_0)}{v + v_0} = \frac{2(0.8 \text{ m})}{0 + 7 \text{ m/s}} = 0.23 \text{ s}.$$

FIG. 5.4. A force diagram for a boy falling into a bush.

*Report on the Sixth International Technical Conference on Experimental Safety Vehicles, (Washington, D.C.: U.S. Department of Transportation, 1976), pp. 20–40.

While the boy is falling into the bush, two forces act on him: his downward weight force **w** and the upward force of the bush $\mathbf{F}_B$ (Fig. 5.4). $\mathbf{F}_B$ points up in the negative y direction, so it is given a negative sign; **w** points down in the positive y direction and has a positive sign. Both of these forces act on the boy along his line of motion as he falls into the bush and must be included in the left side of the impulse-momentum equation:

$$(w - F_B)t = mv - mv_0,$$

where F_B is the magnitude of the force of the bush on the boy. Rearranging to solve for F_B, we find that

$$F_B = -\left(\frac{mv - mv_0}{t}\right) + w = \frac{m(v_0 - v)}{t} + mg$$

$$= \frac{(30 \text{ kg})(7 \text{ m/s} - 0)}{0.23 \text{ s}} + (30 \text{ kg})(9.8 \text{ m/s}^2)$$

$$= \underline{1200 \text{ N}}. \qquad\blacksquare$$

EXAMPLE 5.5 A raindrop hits the head of a man with no hair (Fig. 5.5). The mass of the raindrop is 4×10^{-6} kg, its diameter is 2 mm, and its speed is 8 m/s. (a) Estimate the impulse time—that is, the time required to stop the raindrop. (b) Estimate the average force of the raindrop on his head.

SOLUTION (a) The impulse time can be calculated using Eq. (3.8) if we know the stopping distance. As a rough estimate, let us assume that as the drop lands on the head, it flattens somewhat, as shown in Fig. 5.5b. Thus the stopping distance is approximately equal to the diameter of the raindrop (2 mm). If the downward direction is chosen as the positive y direction, then according to Eq. (3.8),

$$t \simeq \frac{2(y - y_0)}{v + v_0} = \frac{2(2 \times 10^{-3} \text{ m})}{0 + 8 \text{ m/s}} = 0.5 \times 10^{-3} \text{ s}.$$

(b) The average impulse force of the head on the raindrop is

$$F = \frac{mv - mv_0}{t} = \frac{(4 \times 10^{-6} \text{ kg})0 - (4 \times 10^{-6} \text{ kg})(8 \text{ m/s})}{0.5 \times 10^{-3} \text{ s}}$$

$$= \underline{-0.064 \text{ N}}.$$

The reaction force of the raindrop on the head has the same magnitude but points down.

Can you justify the fact that we ignored the weight of the raindrop in this example, but did not ignore the boy's weight in the Example 4.4? (*Hint:* You might compare the raindrop's weight to the 0.064-N force needed to reduce the drop's momentum to zero.) $\qquad\blacksquare$

FIG. 5.5. A raindrop hitting a person's head is stopped in a time t of about 0.3×10^{-3} s.

5.3 Conserved Quantities

Scientists believe that the amount of certain quantities in the universe never changes. These quantities are called **conserved quantities.** Their amount in the universe at this particular time should be the same as at any time in the future.

Momentum is one example of a conserved quantity, as is electric charge and energy.

It may seem presumptuous to talk about quantities that are conserved in a universe, most of which we have never seen. The idea of conserved quantities has evolved from experiments done with very small, isolated parts of the universe. For example, a chemist might study a group of chemicals that are reacting in a closed vessel. The chemicals can be isolated from their environment by assuring that no heat or matter can cross the walls of the vessels. After studying many of these isolated reactions, chemists found that the number of atoms of a certain type (for example, carbon atoms or hydrogen atoms) is always the same before the reaction as after the reaction. Out of these observations has grown a conservation principle: the number of atoms of a certain type is the same in the molecules present before a reaction as after the reaction. This conservation law is the basis for the technique we learn in elementary chemistry courses—the technique of balancing chemical reactions. For example, when sugar and oxygen are "burned" to form carbon dioxide and water, we write the chemical reaction as

$$C_6H_{12}O_6 + 6\,O_2 \longrightarrow 6\,CO_2 + 6\,H_2O$$

The number of C, O, and H atoms is the same on each side of the chemical equation. The number of atoms of each type is conserved in the reaction.*

Many other physical quantities seem to be conserved in processes that take place in the isolated confines of a laboratory apparatus. Since no exception to the conservation of these quantities in isolated systems has ever been found, we assume that they are conserved in any isolated system, the universe being the largest of these systems.

But what is an isolated system? Fig. 5.6 is a schematic representation of the universe. We have drawn dashed lines around one part of the universe, which we call the system. The **system** can be any part of the universe we choose—a single atom, a container in which chemicals are reacting, a human body, the whole earth, or whatever. All the rest of the universe outside the system is called the **environment.** The dashed lines can be an actual boundary between the system and its environment, or it can be an imaginary boundary that we use to help identify the system. The system is said to be **isolated** if nothing (energy, matter, light) crosses the boundary of the system to or from the environment, and if the net sum of all external forces acting on the system is zero. Within an isolated system, the amount of a conserved quantity such as momentum will not change.

Often, an isolated system is constructed in the laboratory. The walls containing the system prevent energy and matter from entering or leaving it. At other times, we can consider a system approximately isolated by simply choosing its boundaries in an appropriate manner. Then we can apply conservation laws to learn more about elements within the system. When two cars collide, for instance, we can choose the system to be the region of space that includes both cars. The interaction of each car with the other is much greater than any other force affecting the cars. As a result, we consider the system as being approximately isolated, and we can then use conservation principles to learn more about the motions of the cars during their collision. In the next section we introduce the law of conservation of momentum and see how it can be used.

(a)

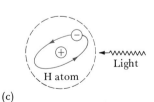

(b)

(c)

FIG. 5.6. (a) A "system" is some part of the universe in which we have a special interest. The rest of the universe not in the system is called the environment. (b) The earth is the system, and the sun and moon are its immediate environment. (c) A hydrogen atom is the system, and it is being approached by light from the environment.

*The conservation law is true for most chemical processes that occur on the earth. Exceptions to this rule involve radioactive nuclei and nuclear reactions that occur in power plants, on the sun, and in particle accelerators.

5.4 Conservation of Momentum

We can derive the conservation-of-momentum principle by looking at the simple case of a collision between two objects. Figure 5.7 shows two objects before, during, and after they collide. The objects could be balls, spherically shaped molecules, or whatever. Let us suppose that they are balls.

First, consider ball one. What is its change in momentum? Before the collision, its velocity is $\mathbf{v}_{1_0}$; following the collision, its velocity is $\mathbf{v}_1$. Its change in momentum is then $m\mathbf{v}_1 - m\mathbf{v}_{1_0}$. This change in momentum is caused by the impulsive force of ball two on ball one, $\mathbf{F}_{2\,\text{on}\,1}$, which we will assume is constant and lasts for a time t. Using Eq. (5.3), we can then write for ball one that

$$\mathbf{F}_{2\,\text{on}\,1}t = m_1\mathbf{v}_1 - m_1\mathbf{v}_{1_0}.$$

Similarly, the momentum of ball two is changed because of the impulsive force of ball one on ball two, $\mathbf{F}_{1\,\text{on}\,2}$. Applying Eq. (5.3) for ball two, we obtain

$$\mathbf{F}_{1\,\text{on}\,2}t = m_2\mathbf{v}_2 - m_2\mathbf{v}_{2_0}.$$

Adding these two equations, we find that

$$(\mathbf{F}_{2\,\text{on}\,1} + \mathbf{F}_{1\,\text{on}\,2})t = (m_1\mathbf{v}_1 - m_1\mathbf{v}_{1_0}) + (m_2\mathbf{v}_2 - m_2\mathbf{v}_{2_0}). \qquad (5.4)$$

Newton's third law, however, says that the action-reaction pair of forces, $\mathbf{F}_{2\,\text{on}\,1}$ and $\mathbf{F}_{1\,\text{on}\,2}$, must be equal in magnitude but oppositely directed. Thus their sum must be zero:

$$\mathbf{F}_{1\,\text{on}\,2} = -\mathbf{F}_{2\,\text{on}\,1}$$

or

$$\mathbf{F}_{2\,\text{on}\,1} + \mathbf{F}_{1\,\text{on}\,2} = 0.$$

As a result, the left side of Eq. (5.4) must be zero. We find, then, that the change in momentum of ball one plus the change in momentum of ball two must add to zero:

$$0 = (m_1\mathbf{v}_1 - m_1\mathbf{v}_{1_0}) + (m_2\mathbf{v}_2 - m_2\mathbf{v}_{2_0}). \qquad (5.5)$$

Equation (5.5) can be rewritten as

$$(m_1\mathbf{v}_1 + m_2\mathbf{v}_2) = (m_1\mathbf{v}_{1_0} + m_2\mathbf{v}_{2_0}). \qquad (5.6)$$

The first term in parentheses is the total momentum of the two balls after the collision. The second parentheses term is their momentum before the collision. There has been no change in the total momentum during the collision, although each ball has changed momentum.

This is an example of the general principle called the **conservation-of-momentum principle:**

When the net sum of the external forces acting on a system of particles is zero, the vector sum of the momenta of these particles is a constant, both in magnitude and direction.

External forces are forces originating from outside the system that act on objects in the system. If no such external forces act on the system, or if the external forces add to zero, then we can say that the system is isolated. In such cases, the total momentum of objects in the system will not change. There can, of course,

FIG. 5.7. Two balls before, during, and after a collision.

be changes in the momentum of individual particles due to the forces they exert on each other, but when all these momentum changes are added, the net change is zero.

Equations (5.5) and (5.6) are mathematical statements of this conservation principle for the case of two particles in a system. In applying this principle to problems, we must first carefully identify the system being considered. For example, in Fig. 5.7, we could draw dashed lines around our two objects. This imaginary boundary would identify the two objects as being in the system.

EXAMPLE 5.6 A 1000-kg car traveling east at a speed of 15 m/s has a head-on collision with a 4000-kg truck traveling west at 10 m/s. If the vehicles are locked together after the collision, what is their final velocity v? (This is a one-dimensional problem, and directions will be indicated using plus and minus signs.)

FIG. 5.8. The external forces acting on the vehicles in the system.

SOLUTION Figure 5.8 shows the system we have chosen. A downward gravitational force $\mathbf{w}$ and an upward contact force $\mathbf{F}_c$ act on each vehicle. These forces originate from outside the system and are thus external forces. But since $\mathbf{w} = -\mathbf{F}_c$ and $\mathbf{w}' = -\mathbf{F}_c'$, the net sum of these external forces is zero. We can therefore consider the system isolated.

Next we choose a coordinate system. If we choose the positive x axis as pointing toward the east, then v_0 (car) $= 15$ m/s and v_0 (truck) $= -10$ m/s. We are also given that $m_{car} = 1000$ kg and $m_{truck} = 4000$ kg. The final velocity v of the car and truck along the x direction is an unknown. We do not yet know whether v is positive or negative. Equation (5.5) becomes

$$0 = (1000 \text{ kg})(v - 15 \text{ m/s}) + (4000 \text{ kg})[v - (-10 \text{ m/s})]$$

or

$$(5000 \text{ kg})v = -25,000 \text{ kg} \cdot \text{m/s}$$

Thus,

$$v = -5 \text{ m/s}.$$

The negative sign means that the car and truck move in the negative x direction after the collision—that is, toward the west.

We can look at these results from another point of view. Initially, the total momentum of the car and truck was the sum of their individual momenta, or

$$\text{initial momentum} = (1000 \text{ kg})(+15 \text{ m/s}) + (4000 \text{ kg})(-10 \text{ m/s})$$
$$= -25,000 \text{ kg} \cdot \text{m/s}.$$

Following the collision, the car and truck are locked together and have a mass of 5000 kg. They are moving to the left at 5 m/s (in the negative x direction). Thus,

$$\text{final momentum} = (5000 \text{ kg})(-5 \text{ m/s}) = -25,000 \text{ kg} \cdot \text{m/s}.$$

There has been no change in momentum. ∎

Rocket propulsion is another interesting phenomenon involving momentum conservation. It is often mistakenly thought that fuel ejected from the rocket pushes against the air outside the rocket and thereby propels the rocket. But a force exerted against the air would propel the air, not the rocket. Also, when flying to or from the moon, there is no air for the ejected fuel to push against. Rocket propulsion can be most easily understood as an example of momentum conservation. Since no outside forces act on the rocket-fuel system, the momentum of the fuel ejected in the backward direction must be balanced by forward momentum gained by the rocketship.

The situation is similar to that of a person stranded at the center of a frictionless sheet of ice. To get off the ice, the person can remove his or her shoes and throw them parallel to the surface of the ice. The momentum gained by the shoes flying through the air is equal in magnitude but opposite in direction to the momentum the person gains while sliding off the ice. The ability to gain forward momentum by ejecting mass in the backward direction is illustrated in the next example.

EXAMPLE 5.7 The 5000-kg final stage of a rocket ejects the 10,000-kg second stage. Before ejection, the two stages together travel with a velocity of 200 m/s. If after ejection the second stage travels at 50 m/s opposite the direction of its initial motion, what is the velocity of the final stage? (See Fig. 5.9.)

SOLUTION The momentum of the two stages together before ejection must equal the sum of the momenta of the two stages separately after ejection:

$$(m_1 + m_2)v_0 = m_1 v_1 + m_2 v_2$$

$$\underset{\text{momentum}}{\text{initial}} = \underset{\text{momentum}}{\text{final}}$$

where m_1 and m_2 are the masses of the first and second stage, respectively, v_0 is their initial velocity (200 m/s), v_1 is the unknown final velocity of the first stage, and v_2 is the final velocity of the second stage (-50 m/s). * Solving for v_1, we find

$$v_1 = \frac{(m_1 + m_2)v_0 - m_2 v_2}{m_1}$$

$$= \frac{(15{,}000 \text{ kg})(200 \text{ m/s}) - (10{,}000 \text{ kg})(-50 \text{ m/s})}{5000 \text{ kg}}$$

$$= 700 \text{ m/s}. \qquad \blacksquare$$

Before ejection

$v_0 = 200$ m/s

After ejection

$v_2 = 50$ m/s $v_1 = ?$

FIG. 5.9. The ejection of the second stage of a rocket helps the first stage move faster.

The previous examples involved changes in momenta occurring along only one axis. Often, the collision of one object with another results in changes in direction and must be described using two or more coordinates. We must then rewrite Eqs. (5.5) and (5.6) in their component forms and consider momentum conservation along each axis separately. Equation (5.6) becomes:

$$m_1 v_{1_{0x}} + m_2 v_{2_{0x}} = m_1 v_{1_x} + m_2 v_{2_x}, \qquad (5.6x)$$

$$m_1 v_{1_{0y}} + m_2 v_{2_{0y}} = m_1 v_{1_y} + m_2 v_{2_y}. \qquad (5.6y)$$

The components of momentum along the x axis must be conserved as must the components along the y axis.

EXAMPLE 5.8 A cue ball traveling with a speed of 2 m/s along the x axis (see Fig. 5.10) hits the eight ball. After the collision, the cue ball travels with a

---

*We are calling velocities to the right in Fig. 5.9 positive and those to the left negative.

Before collision

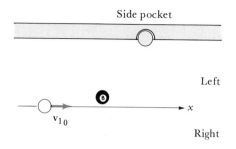

Side pocket

Left

x

v_{1_0}

Right

After collision

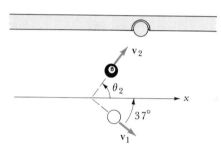

v_2

θ_2

x

$37°$

v_1

FIG. 5.10. A coordinate system for describing the motion of pool balls.

speed of 1.6 m/s in a direction $37°$ to the right of the x axis. Will the eight ball fall in the side pocket, which is oriented along a line $53°$ to the left of the x axis? The two balls have equal mass.

SOLUTION We are given that $m_1 = m_2 = m$, $v_{1_0} = 2$ m/s, $v_1 = 1.6$ m/s, and $\theta_1 = 37°$, where θ_1 is the direction of the cue ball after the collision. We wish to calculate the direction of the eight ball after the collision. This is done most easily by using the two equations for the conservation of the x and y components of momentum to solve for v_{2_x} and v_{2_y}, the x and y components of the velocity of the eight ball after the collision. We have

$$\text{momentum before} = \text{momentum after}$$

x component: $\qquad m_1 v_{1_0} = m_1 v_1 \cos 37° + m_2 v_{2_x}$

y component: $\qquad 0 = -m_1 v_1 \sin 37° + m_2 v_{2_y}$

Substituting $m_1 = m_2 = m$ and the values of v_{1_0}, v_1, $\cos 37°$, and $\sin 37°$ into the above and rearranging, we find that

$$v_{2_x} = v_{1_0} - v_1 \cos 37° = (2 \text{ m/s}) - (1.6 \text{ m/s})0.8 = 0.72 \text{ m/s}$$

and

$$v_{2_y} = v_1 \sin 37° = (1.6 \text{ m/s})0.6 = 0.96 \text{ m/s}.$$

We can then determine the direction of v_2 using the fact that

$$\tan \theta_2 = \frac{v_{2_y}}{v_{2_x}} = \frac{0.96 \text{ m/s}}{0.72 \text{ m/s}} = 1.33$$

or

$$\theta_2 = \underline{53°}.$$

The ball falls in the side pocket. ∎

Summary

1. **Momentum and impulse:** The momentum of an object is defined as the product of its mass and its velocity **v**:

$$\text{momentum} = \mathbf{p} \equiv m\mathbf{v}. \qquad (5.1)$$

A force **F** acting on an object for a time t is said to exert an impulse on the object. The impulse of the force is the product of **F** and t, or

$$\text{impulse} \equiv \mathbf{F}t. \qquad (5.2)$$

The sum of the impulses of all forces acting on an object for a certain time is equal to the change in momentum of the object during that time. This is called the **impulse-momentum equation** and is written as

$$(\Sigma \mathbf{F})t = m\mathbf{v} - m\mathbf{v}_0. \qquad (5.3)$$

When using the impulse-momentum equation to analyze one-dimensional collisions, the impulse time is often estimated using

Eq. (3.8) from kinematics:

$$t = \frac{2(x - x_0)}{v + v_0}, \qquad (3.8)$$

where $(x - x_0)$ is the distance the object moves during the impulse, v_0 is its initial velocity at the start of the impulse, and v is its velocity at the end of the impulse.

2. **Conservation of momentum:** Momentum is one of a group of conserved quantities. The amount of a conserved quantity in the universe never changes. When a small part of the universe, which we call the **system,** is isolated from the rest of the universe, the total momentum of objects in the system is constant. For a system to be **isolated,** the sum of external forces (those originating from outside the system) that act on objects in the system must add to zero. Also, no matter, energy, or light should enter or leave the system.

For an isolated system with two objects, the conservation-of-momentum equation can be written as

$$0 = (m_1\mathbf{v}_1 - m_1\mathbf{v}_{1_0}) + (m_2\mathbf{v}_2 - m_2\mathbf{v}_{2_0}). \tag{5.5}$$

The change in momentum of object one plus the change in momentum of object two must add to zero. Equation (5.5) can also

be written as

or
$$(m_1\mathbf{v}_1 + m_2\mathbf{v}_2) = (m_1\mathbf{v}_{1_0} + m_2\mathbf{v}_{2_0}) \tag{5.6}$$

$$\begin{pmatrix} \text{final momentum} \\ \text{of two objects} \end{pmatrix} = \begin{pmatrix} \text{initial momentum} \\ \text{of two objects} \end{pmatrix}$$

In an isolated system, the initial and final momenta are equal.

Questions

Several of these questions require rough estimates. Before working them, you might read the introduction to the questions in Chapter 3 or the estimation interlude following this chapter.

1. A 0.4-kg hammerhead hits a nail and pounds it partway into a board. (a) *Estimate* the speed of the hammerhead before hitting the nail. (b) Estimate the stopping distance. (c) Calculate the stopping time. (d) Use these numbers to estimate the average force of the hammerhead on the nail (which will be equal in magnitude but opposite in direction to the force of the nail on the hammerhead).

2. A heavyweight boxer hits an opponent on the head. (a) *Estimate* the speed of the glove just before it hits the head. (b) Estimate the total mass of the glove, arm, and other body parts that are moving at this effective speed. (c) Estimate the stopping distance for the glove. (d) From your estimates in (a) and (c), calculate the stopping time. (e) Use the previous results to estimate the average force of the punch.

3. A hand lies palm up on a table top. A 1-kg ball strikes the palm with a speed of 5 m/s after falling from a height of 1.25 m. (a) *Estimate* the stopping distance—that is, the extra distance the ball falls after first touching the hand. (b) Use this distance to estimate the stopping time. (c) Use this and Eq. (5.3) to calculate the estimated force of the hand on the ball (which is equal in magnitude to the force of the ball on the hand). (d) Why would the force be somewhat less if the hand caught the ball in midair?

4. A 0.21-kg pool ball, moving at a speed of 1.2 m/s, hits the cushion on the side of the pool table. If it rebounds in the opposite direction with the same speed, *estimate* the average force of the ball on the cushion during the collision.

5. A 30-kg girl jumps from a garage roof 3.5 m high onto a concrete driveway. *Estimate* the force on the girl's leg (a) if she lands on her heels with her legs stiff, and (b) if she lands on her toes and lets her knees bend while stopping.

6. A parachutist with his parachute open falls at about the same speed as a person who has jumped from a wall 3.0 m high. Because of the high speed with which he reaches the ground, the parachutist is taught to land on his toes. His ankles bend during initial contact with the ground. As his heels touch the ground, his knees begin to bend, and his body falls to the side until his thighs and then his upper body finally reach the ground. Why does this elaborate falling technique reduce injury when landing?

7. Two ice skaters stand facing each other on the ice. The first, holding a 1-lb ball, throws it to the second skater, who is initially at rest and one foot away from the first. The second throws the ball back to the first. Describe their expected motion if they continue throwing the ball back and forth to each other. Assume that the ice exerts little friction force on the skates.

8. Two cars of unequal mass but moving at the same speed collide head-on. Explain why a passenger in the smaller car is more likely to be injured than one in the larger car. Justify your reasoning with physical principles.

Problems

5.1 Momentum

1. A 1000-kg car travels at a speed of 20 m/s. (a) At what speed must a 3000-kg truck travel to have the same momentum? (b) At what speed must a 7.3-kg bowling ball travel to have the same momentum?

2. (a) Calculate the momentum of a 30-kg dog running at a speed of 8 m/s toward the west. (b) How fast must a 70-kg person run to have the same momentum as the dog?

3. *Estimate* the magnitude of your momentum in units of kg · m/s when walking to physics class. Justify verbally the numbers used in your estimate.

4. What is the change in momentum of a 0.0567-kg tennis ball traveling at 25 m/s if it hits a practice wall and rebounds with the same speed? Is the vector change in momentum in the direction of the initial velocity, the direction of the final velocity, or neither? Explain.

5. (a) A 145-g baseball traveling at 35 m/s is hit by a bat and rebounds at 40 m/s. What is the change in momentum of the ball? (b) A 0.055-kg golf ball starts at rest and leaves the tee with a speed of 50 m/s. What is the change in momentum of the golf ball?

6. A 2100-lb car traveling at 30 mph hits a truck and rebounds in the opposite direction at a speed of 10 mph. What is the car's change in momentum (magnitude and direction)?

5.2 The Impulse-Momentum Equation

7. An impulse of 7.5×10^4 N · s stops a car in 0.1 s. (a) Calculate the average force on the car. (b) If the impulse time is tripled, what is the average force on the car?

8. A (2×10^5)-N friction force caused by the brakes of a subway train stops it in 50 s. (a) Calculate the force needed to stop the train in 30 s. (b) Calculate the stopping time if a (1.5×10^5)-N friction force acts on the train.

9. A 0.057-kg tennis ball experiences a change in velocity of 30 m/s when hit by a tennis racket. The average force of the racket on the ball is 50 N. For how long are the ball and racket in contact?

10. About 80 g of blood is pumped from the heart each heartbeat. The blood starts at rest and has a speed of about 0.6 m/s in the aorta. If the pumping takes 0.17 s, what is the average force on this blood?

11. A car sitting at a stop sign is hit from the rear by a truck. The car lurches forward so that its speed just after the collision is 10 m/s. (a) If the collision time is 70 ms, what constant force is needed to accelerate the driver's 4.5-kg head? (b) the driver's 1.3-kg brain? Assume that the head and brain experience the same motion change as the car.

12. A bat is in contact with a 10^{-2}-slug baseball for 1.3×10^{-3} s. The average force of the bat on the ball is 2000 lb. The ball has an initial velocity of 120 ft/s toward the bat. What is the ball's final velocity (magnitude and direction)?

13. A 100-kg astronaut pushes against the inside back wall of a 2000-kg spaceship, causing his speed toward the front to increase from zero to 0.8 m/s. (a) If the push lasts 0.3 s, what is the average force of the astronaut on the spaceship? (b) With what speed does the spaceship recoil if it is initially at rest? (*Note:* A reaction force acts on the spaceship, the magnitude of which equals the force on the astronaut (Newton's third law).)

14. The bone in the upper part of the cheek (the zygomatic bone) can be fractured by a 900-N force lasting 6 ms or longer. A hockey puck can easily provide such a force when hitting the unprotected face. (a) What velocity change of a 0.11-kg hockey puck is needed to provide that impulsive force? (b) A padded face mask doubles the impact time. By how much does it change the force on the head behind the mask? Explain.

15. An impulse of 150 N · s is required to stop a person's head in a car collision. (a) If the head stops in 0.1 s when the cheek bone hits the steering wheel, what is the average force on the cheek bone? (b) Will the cheek bone be fractured (see Problem 14)? (c) What is the shortest impact time in order to avoid breaking the bone?

16. When a 58-g tennis ball hits a wall, it becomes flattened. Assume that the ball undergoes a constant deceleration from an initial speed of 20 m/s to a speed of zero when flying into the wall. During this deceleration the ball moves into the wall a distance of 2.5 cm. (a) Calculate the time required for this deceleration (assumed constant). (b) What is the average force of the wall on the ball while being stopped?

17. A 55-kg woman falls from a ladder and lands stiff-legged on concrete. Her speed just before hitting the cement is 7 m/s. She is stopped in a distance of 2.0 cm. (a) Calculate the stopping time. (b) Calculate the average stopping force.

18. A 180-lb man stands stiff-legged in an elevator whose cable has just broken. The elevator is falling at 60 ft/s when it starts to hit a shock-absorbing device at the bottom of the elevator. The stopping force on the man should be no more than about 18,000 lb. If more, the man's ankles or thigh bones will fracture. (a) Use the impulse-momentum equation to calculate the man's minimum stopping time in order to avoid breaking a leg. (b) Calculate the minimum stopping distance—the distance the elevator must travel while being stopped by the shock-absorbing device.

19. The majority of the European Experimental Vehicle Committee has recommended that the knee, thigh bone (femur), and hip should be subjected to a force no greater than 4000 N during car collisions. Greater forces could cause dislocations or fractures. Assume that a knee is stopped when it hits the instrument panel of the car. Assume also that the effective mass carried along with the knee is 20 percent of the body mass. (a) What is the minimum stopping time to avoid injury to the knee of an 80-kg person traveling at 15 m/s (34 mph)? (b) What is the minimum stopping distance?

20. An impact-absorbing bumper is built to stop a car of mass m moving with an initial speed v_0. The bumper exerts a force F on the car as it retracts a distance Δx during the collision. (a) Using Eqs. (3.8) and (5.3), show that the magnitude of the force and retraction distance are related by the equation

$$F = \frac{m v_0^2}{2 \, \Delta x}.$$

This general equation applies to all collisions that stop an object with initial speed v_0. Notice that the net force required to stop the car in a fixed distance increases in proportion to the square of the car's initial velocity. (b) What force is required to stop a 1500-kg car moving at 5 mph in a distance of 0.2 m? (c) Repeat (b) for a car moving at 10 mph.

21. A shot-putter can put a 16-lb shot 60 ft by applying a force of 100 lb for a time of 0.20 s. If the distance the shot is put is proportional to the impulse of the force, how far can he put the shot if the force is applied for 0.25 s rather than 0.20 s? Explain.

22. Water sprays from a firehose against a burning wall at a rate of 60 kg/s. The water's speed before hitting the wall is 40 m/s and its speed after hitting the wall is zero. Calculate the force of the water against the wall.

23. Air blowing with a speed of 10 m/s hits a sign, causing the air to stop and move to the side past the sign. The mass of air hitting the sign each second is 13 kg. (a) Calculate the change in momentum of the air each second. (b) Calculate the steady force of the air on the sign.

24. A 6 m/s wind blows against one side of one of the Twin Towers at the World Trade Center in New York City. The area of the side is 63 m × 411 m, and 2.0×10^5 kg of air hits the wall each second. If the air's speed is reduced by one-half when hitting the wall (part of the air is deflected and its speed is changed little), calculate the average force of the air on the wall.

5.4 Conservation of Momentum

25. A 3-kg gun mounted on frictionless wheels shoots a 5-g bullet in the horizontal direction at a speed of 300 m/s. What is the recoil speed of the gun?

26. A cannon mounted on the back of a ship fires a 50-kg cannon ball in the horizontal direction at a speed of 150 m/s. If the cannon and ship to which it is firmly attached have a mass of 4000 kg and are initially at rest, what is the speed of the ship just after shooting the cannon? Ignore water resistance.

27. A 90-kg space mechanic is drifting away from a spaceship at a speed of 0.2 m/s. The mechanic has no rocket to help in getting back to the ship—only a 0.5-kg wrench. With what speed and in what direction relative to the spaceship must he throw the wrench in order to acquire a speed of 0.1 m/s toward the spaceship? Explain.

28. An 80-kg spacewoman is floating at rest a distance of 10 m from a spaceship when she runs out of oxygen and fuel to power her back to the spaceship. She removes her oxygen tank (3 kg) and flings it into space away from the ship with a speed of 15 m/s. (a) At what speed does she recoil toward the spaceship? (b) How long must she hold her breath before reaching the ship?

29. A radioactive nucleus at rest decays to form an alpha particle and a product nucleus. With what speed does the product nucleus move following decay if the alpha particle moves at 2.0×10^7 m/s? The mass of the product nucleus is 54 times greater than the mass of the alpha particle.

30. There are approximately 230 million people in the United States, each with an average mass of about 60 kg. If all simultaneously jumped upward from the earth with an initial speed of 6 m/s, what would be the recoil velocity of the earth? The mass of the earth is 5.98×10^{24} kg.

31. A group of people in Quebec are playing ice baseball. A 160-lb man initially at rest catches the 0.32-lb ball traveling at 40 mph. If his skates are frictionless, what is his speed after catching the ball?

32. A boy coasts on a level surface on his tricycle at a speed of 2 m/s. The boy and tricycle together have a mass of 30 kg. What is the speed of the tricycle when a 20-kg friend, initially at rest, steps on the back of the tricycle?

33. A 10,000-kg coal car coasts under a coal storage bin with a speed of 2 m/s. As it goes under the bin, 1000 kg of coal is dropped into the car. What is the final speed of the loaded car?

34. An 80-g arrow shot from a bow passes through an apple balanced on William Tell's head. The arrow has a speed of 50 m/s before passing through the apple and 40 m/s after. The apple has a mass of 100 g. Calculate the final speed of the apple.

35. A 500-g gun sliding on ice at a speed of 4 m/s accidentally fires a 5-g bullet opposite its direction of motion. If the bullet travels at 300 m/s, what is the velocity of the gun following the shot?

36. A 900-kg car moving west at 20 m/s hits a 1200-kg car moving north at 10 m/s. The cars become locked together and move toward the northwest just after the collision. Calculate their exact direction and speed after the collision.

37. A 3000-kg spaceship travels toward the moon at a speed of 1.5×10^4 m/s. The captain of the ship wishes to change direction by 5°. Rockets eject a short burst of fuel at a speed of 3×10^6 m/s perpendicular to the ship's initial direction. Calculate the mass of fuel that must be ejected to change course.

General Problems

G1. A 40-kg child is dropped 3.0 m from the window of a burning building into the arms of a person below. The person's arms give with the child as it is being caught. What is the average force on the child if the stopping distance is (a) 0.3 m and (b) 1.0 m?

G2. A 2100-lb sports car traveling east at 20 mph has a head-on collision with a 4000-lb station wagon traveling west at 30 mph. (a) If the vehicles remain locked together, what is their final speed? (b) If the collision time is 0.1 s, what is the force needed to restrain a 150-lb person in the sports car? (c) in the station wagon?

G3. A 3-g bullet traveling at 300 m/s hits and embeds itself in a 1-kg wooden block resting on a frictionless surface. (a) What is the final speed of the block and bullet? (b) If the block after being hit slides on a surface that exerts a 3-N friction force on the block, how far will it slide?

The ability to make estimates and rough calculations is useful for a variety of reasons. (1) You need to decide whether a goal is worth pursuing; for example, can you make a living as a piano tuner in a town that already has a certain number of tuners? (2) You need to know roughly the amount of material needed for some activity—for instance, the food needed for a party or the number of bags of fertilizer needed for your lawn. (3) You want to know roughly a number before it is measured; for example, what scale should be used when measuring a voltage or how rapid a time-measuring device is needed to detect laser light reflected from a distant object? (4) You wish to check whether a measurement you have made is reasonable—for instance, the measurement of the time for light to travel to a mountain and back or the oxygen consumed by a hummingbird. (5) You can determine an unknown quantity only by an estimation procedure; for example, you estimate the number of cats in the United States or the compression force on the discs in your back when lifting a weight in different ways. *

The procedure for making rough estimates usually means selecting some basic numbers whose values are known or can be estimated, and then combining these numbers in some mathematical framework that leads to the desired answer. In the last chapter the mathematical framework involved the use of the impulse-momentum equation and equations of kinematics. Sometimes, no physics equation is needed; basic numbers are simply multiplied and divided to determine the desired result. This procedure is used in the next example and in problems at the end of the estimation interlude.

EXAMPLE 1   Estimate the fraction of the United States population that eats at McDonald's each day. (*Hint:* In 1980 a sign outside their buildings said that they had served 35 billion meals. You might try making a wild guess before examining our estimation technique.)

SOLUTION   For this type of problem, we need a scheme that allows us to calculate the unknown by using other numbers or quantities that can be estimated more reliably. For instance, suppose we knew the number of meals served by McDonald's each day. This number equals roughly the number of people eating at McDonald's each day. Then, dividing by the United States population (about 230 million) would give us the fraction of the United States population eating at McDonald's each day:

$$\frac{\text{number of meals served each day}}{\text{U.S. population}} \simeq \frac{\text{fraction of population}}{\text{eating at McDonald's each day}} .$$

---

*The subject of making rough estimates has been considered in two interesting articles in the *Physics Teacher:* (1) Mark St. John, "Thinking Like a Physicist in the Laboratory," *Physics Teacher* **18**:436–443 (1980), and (2) J. D. Memory and A. W. Jenkins, Jr., "Estimating Orders of Magnitude," *Physics Teacher* **15**:43–44 (1977).

To estimate the number of meals served by McDonald's each day we divide the total number of meals served since they started in business (35 billion according to their sign) by the total number of days McDonald's has served meals. We can probably make a rough estimate of this latter number. If you are in your twenties, then you probably remember McDonald's for most of your life. My first recollection of McDonald's goes back about twenty years to the early 1960's. Let us assume that the company has been in business for twenty years as of 1980. Since there are 365 days/year, they have served meals for about

$$20 \text{ years} \times 365 \frac{\text{days}}{\text{year}} = 7300 \text{ days.}$$

We could easily be in error by 1000 or 2000 days. Thus, we will round our estimate of 7300 days to 7000 days. Retaining the 3 in 7300 days implies greater accuracy than we can justify.

Next we divide the total number of meals served by the number of days in business to get an estimate of the average number of meals served per day:

$$\frac{35,000,000,000 \text{ meals}}{7,000 \text{ days}} = 5,000,000 \text{ meals/day.}$$

It seems likely that more meals are served per day now than during the earlier days in their history. Thus we should probably fudge this number up to 6 or 7 million per day. We will settle for 6 million.

Now then, we can return to our first equation to determine the fraction of people eating at McDonald's:

$$\frac{6 \text{ million meals/day}}{230 \text{ million people}} \simeq 0.026 = \frac{1}{38}$$

or about one of every 40 people eats at McDonald's each day.

This whole procedure can be summarized in a chart as shown below:

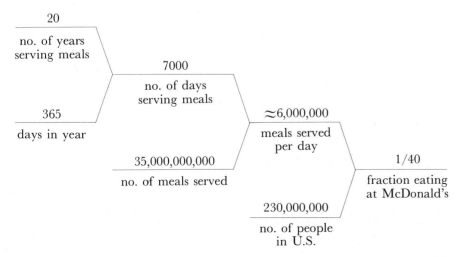

TABLE 1 The Percent Uncertainties of Numbers Used in McDonald's Estimate

Quantity	Number and uncertainty	Percent uncertainty
Years in service	20 ± 10	$\dfrac{10}{20} \times 100 = 50\%$
Days in year	365 ± 0	$\dfrac{0}{365} \times 100 = 0\%$
No. meals served	35 bil $\pm$ 5 bil	$\dfrac{5}{35} \times 100 = 14\%$
People in U.S.	230 mil $\pm$ 20 mil	$\dfrac{20}{230} \times 100 = 10\%$

Two quantities on the left are combined by multiplying or dividing to get the next quantity on the right. Each number on the right is rounded off.

All of the four numbers on the left side of the chart are numbers we have supplied to make the estimate. Some of them are known with considerable accuracy—for example, the number of days in the year. Others are known with less accuracy—the number of years that McDonald's has served meals. The uncertainty in our final answer can be estimated using the **weakest-link rule**—the calculated value is no more precise than the least precise value used to make the calculation. We first estimate the percentage uncertainty in each of the numbers used to make the estimate. The final answer is assigned the same percentage uncertainty as the least uncertain of the estimated numbers. Consider Table 1. Since the number with greatest percentage uncertainty (the number of years of service) is about 50 percent uncertain, we assign our final answer a 50 percent uncertainty:

fraction eating at McDonald's = $0.026 \pm 0.5(0.026) = 0.026 \pm 0.013$.

Our fraction could vary from about 0.013 to 0.039, or from 1/77 to 1/26. We will round these numbers, and estimate that somewhere between 1/80 to 1/30 of population eats at McDonald's each day.

A check with the national organization in August 1980 indicated that they served approximately 3 billion people in 1979 or an average of 8 million people per day, of which 7 million were served in the United States. Thus, the fraction eating at McDonald's each day in 1979 was 0.030—or 1 out of every 33 United States citizens. (This example was written and the estimate completed before calling McDonald's). ■

Problems assigned in most science and mathematics courses usually include "input information" that allows you to calculate a precise answer. Yet in much of real-life problem solving you must provide the input information yourself, and this information is often estimated or measured imprecisely. It should not upset you to find answers that are 50 or 100 percent uncertain. You still know more about the desired solution than if you made a wild guess.

The following problems should help you develop this ability to make rough estimates. To determine a solution for each problem, use a technique similar to that used in the McDonald's example. Also, using the weakest-link rule, estimate the uncertainty in your answer. Solutions provided to your instructor are either values determined by contacting an appropriate reference or the author's estimate. The latter may be significantly in error, so do not be upset if your answer differs somewhat from that estimated by another person. You should, however, justify your estimation procedure with a reasonable plan, and your input numbers should be justified with moderate accuracy.

Problems

1. How long would it take to sail around the earth nonstop?

2. How many times does the average person's heart beat during his or her lifetime? By comparison, a four-stroke car engine traveling at 60 mph completes about 10^7 strokes in one lifetime.

3. Estimate the average number of people in the United States who ride on a bicycle each day.

4. Estimate the volume of your body in cubic centimeters.

5. Estimate the number of piano tuners that work in Atlanta, Georgia.

6. Estimate the cost of gasoline used for all purposes by people who work at and attend your college or university each day.

7. Estimate the average number of miles jogged by students at your college or university each day.

8. Estimate the volume of water consumed during the day by people in your physics class, and estimate the three greatest uses of the water and the amount used for each purpose.

6

Circular Motion at Constant Speed

Kinematics and Newton's second law provided the basis in the first five chapters for analyzing the translational motion of objects and the forces that cause the motion to change. Similar techniques are used in this and the next chapter to analyze circular and rotational motion. In this chapter we restrict our attention to objects moving in a circular path at constant speed: a ball swinging in a horizontal circle at the end of a string, the moon circling the earth, or a race car moving at constant speed around a circular track.

We will find that a force is needed to prevent an object from leaving its circular path, even if the object moves at constant speed. The force pulls the object toward the center of the circle and does not change the object's speed, only its direction of motion as it moves in a circle. Examples of these forces that exert an inward force on an object as it moves in a circle are the earth's gravitational pull on the moon and the tension force of a string pulling on a ball swung at its end.

If an object's speed, while in orbit, increases or decreases, then another force is responsible for the change in speed. These forces that cause changing speed are directed tangent to the circle of motion and are considered in Chapter 7.

6.1 Angular Position Coordinate

When describing linear motion in Chapter 3, we used an x coordinate or a y coordinate to indicate an object's position at different times. To describe circular motion, such as the movement of a satellite about the earth, we need to use some other type of coordinate to indicate the object's position.

One way of doing this might be to place markers along the circumference of the circular path. The position of the object would then be given by its distance along the circumference from some reference point—like the furlong markers indicating the position of a horse from the starting line of a horse race. * Usually, in physics, the reference point for one of these circular coordinate systems is the position where the positive x axis crosses the circle (see Fig. 6.1). This position coordinate along the arc of the circle is called the **arc-length coordinate** and is given the symbol s. By convention, s is positive in the counterclockwise direction from the reference point and negative in the clockwise direction.

FIG. 6.1. The arc length s is used to indicate the location of an object on a circular path along which it moves. The race car is at the 0.7-km position—that is, $s = 0.7$ km.

*One furlong is a unit of length equal to 220 yards.

While the arc-length coordinate s is convenient for describing the motion of some objects, such as the position of horses during a horse race, it is not useful for describing the position of an earth satellite. We cannot place markers along the satellite's orbit to show its position. We can, however, indicate its angular position above the earth using telemetry. This angular coordinate technique is illustrated in Fig. 6.2. A race car travels around a circular track. To indicate the car's position by an angle, we draw a radial line from the center of the circle to the car. The angle from the x axis to that radial line, as in Fig. 6.2, is called the **angular-position coordinate** and is given the symbol θ (theta). The angular position for the race car shown in Fig. 6.2 is about 75 degrees—that is, $\theta = 75°$. θ is positive when measured in a counterclockwise direction from the positive x axis and is negative in a clockwise direction.

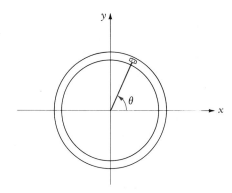

FIG. 6.2. The angular position coordinate θ is the angle from the positive x axis to a radial line drawn from the center of the circle to the object moving in a circle.

6.2 Units to Describe Angular Position

Three different units are used to measure the angular position of an object: (1) degrees, (2) revolutions, and (3) radians. The unit degree is the most familiar. There are 360° in a circle. If an object is at the top of the circle, its position is 90°. When at the bottom, its position is 270°.

Revolution is also a familiar unit. One revolution (or rev) corresponds to a complete rotation about the circle and equals 360°. An object that had made one-fourth of a revolution would be at the top of the circle. Revolutions are usually used to indicate the change in angular position per unit time. For example, an object that rotates in a complete circle 120 times each minute is said to have an angular speed of 120 revolutions per minute (120 rpm). Motorcycle engines rotate at about 500 rpm. The turntable of a high-fidelity system rotates with an angular speed of 33.3 rpm.

The unit of angular position most useful in physics is the radian. It is defined in terms of the two lengths shown in Fig. 6.3. One of these, the arc length s that was defined earlier, is the distance along the circumference of the circle from the positive x axis to the position of the object. The other length used in defining the radian is the radius r of the circle. The angle θ in units of **radians** (rad) is the ratio of s and r:

$$\theta \text{ (in radians)} = \frac{s}{r}. \tag{6.1}$$

Since the circumference of a circle is $2\pi r$, one complete rotation around a circle corresponds to a change in arc length s of $2\pi r$ and a change in angular coordinate of

$$\theta = \frac{s}{r} = \frac{2\pi r}{r} = 2\pi. $$

Thus, in a complete circle, there are 2π radians.

We can now relate the three units of angular position as follows:

$$360° = 1 \text{ rev} = 2\pi \text{ rad.} \tag{6.2}$$

This equation is useful when converting between different angle units, as in the following example.

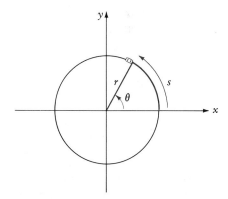

FIG. 6.3. The angle θ in units of radians is the ratio of s, the arc length, and r, the radius of the circle.

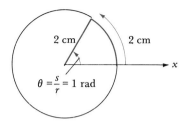

FIG. 6.4. An angle of one radian has an arc length that equals the radius of the circle.

EXAMPLE 6.1 Convert an angular position of 50° to radians and to revolutions.

SOLUTION Using Eq. (6.2), we find that $2\pi \, \text{rad}/360° = 1$ and that $1 \, \text{rev}/360° = 1$. Thus,

$$50° = (50°)\left(\frac{2\pi \, \text{rad}}{360°}\right) = \underline{0.87 \, \text{rad}}$$

and

$$50° = (50°)\left(\frac{1 \, \text{rev}}{360°}\right) = \underline{0.14 \, \text{rev}}.$$

Note that the radian unit has no dimensions; it is the ratio of two lengths. Consider, for instance, an angle of 1 rad, shown in Fig. 6.4. Its arc length equals 2 cm, as does the radius of the circle. Thus,

$$\theta = \frac{s}{r} = \frac{2 \, \text{cm}}{2 \, \text{cm}} = 1 \, \text{rad}.$$

But $2 \, \text{cm}/2 \, \text{cm} = 1$ and is unitless. We see that one radian is just the number 1. We can remove the word *radian* from an expression or multiply an expression by 1 radian with no adverse consequences. The unit radian is usually kept in an expression that involves angles in order to remind us that we are using units of radians and not degrees or revolutions.

The radian is useful in science because it can be used to relate angle positions or changes in angular position to arc lengths. For example, if a car travels 2.0 rad around a circular race track of radius 300 m, the distance along the arc that the car has traveled is determined by rearranging Eq. (6.1):

$$s = r\theta = (300 \, \text{m})(2.0 \, \text{rad}) = 600 \, \text{m}.$$

We dropped the rad unit in the answer because we are calculating a length, and the radian is therefore not needed as a reminder. Note that we could not have calculated this arc length using degrees or revolutions unless we first converted these units to radians.

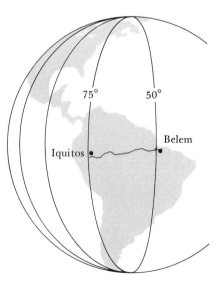

FIG. 6.5. The angular separation of cities on the equator is the difference in their longitudes.

EXAMPLE 6.2 Ocean vessels can sail up the Amazon River from Belem, Brazil, at its mouth to Iquitos, Peru—a twisting, curved trip along the river of 3700 km. Calculate the distance as the crow flies between these cities, both of which are near the equator—Belem at 48° longitude and Iquitos at 73° longitude (Fig. 6.5).

SOLUTION The arc length s of a portion of a circle is given by Eq. (6.1)—that is, $s = r\theta$. For this problem, r is the radius of the earth ($6.38 \times 10^6 \, \text{m}$). The angular separation of the two cities on the equator in units of degrees is the difference in their longitude: $73° - 48° = 25°$. To use Eq. (6.1), we must first convert 25° to units of radians:

$$\theta = 25° \left(\frac{2\pi \, \text{rad}}{360°}\right) = 0.44 \, \text{rad}.$$

Then, according to Eq. (6.1),

$$s = r\theta = (6.38 \times 10^6 \text{ m})(0.44 \text{ rad})$$
$$= 2.8 \times 10^6 \text{ m} = \underline{2800 \text{ km}}.$$

■

6.3 Tangential Speed and Angular Velocity

In Section 6.1 we used two ways to indicate the location of an object moving in a circular path: (1) the distance s of the object along the circumference of the circle from some reference point, and (2) the angular-position coordinate of the object. Figure 6.6a shows both ways of describing an object's position.

We can also use two ways to describe how fast an object is moving. Using s, we can define its speed v along the circumference of the circle as the ratio of its change in position Δs and the time Δt required for the change (see Fig. 6.6b). Thus,

$$\text{tangential speed} = v = \frac{\Delta s}{\Delta t}. \qquad \textbf{(6.3)}$$

This ratio is the object's **tangential speed**—how fast the object actually moves along the arc of the circle. In the case of a horse race, the tangential speed is the actual speed of the horse along the track. The word *tangential* is used because the motion is directed tangent to the arc of the circle.

If an object is moving, then θ is changing. We say that the object has an angular velocity. **Angular velocity** is defined as the ratio of the change in angular position $\Delta\theta$ to the time Δt required for this change, and is given the symbol ω (omega):

$$\text{angular velocity} = \omega = \frac{\Delta\theta}{\Delta t}. \qquad \textbf{(6.4)}$$

If, for example, the race car travels 90° around a circular track in 30 s, then its angular velocity is

$$\omega = \frac{90°}{30 \text{ s}} = \frac{3°}{\text{s}} = 3° \text{ s}^{-1}.$$

Notice that angular velocity indicates a change in angular position per unit time and not a change in distance per unit time. You might think of the radial line from the center of the circle to the moving object as the spoke of a bicycle tire that rotates in a circle. The angular velocity is the spoke's change in angular position or orientation per unit time. The word *velocity* is used rather than angular *speed* because in more advanced treatments ω is often assigned a direction. If the object moves counterclockwise, the direction of **ω** is out of the paper perpendicular to the plane of motion. If the object moves clockwise, **ω** is directed into the paper. This seemingly strange choice of direction for **ω** is really the only reasonable choice, since **ω** is then directed along the axis of rotation and defines clearly the plane in which rotation occurs. We will not concern ourselves in the future with the direction of **ω**.

In many examples of circular motion we will need to determine the tangen-

(a)

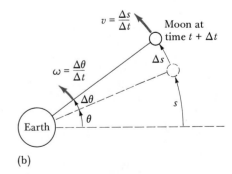

(b)

FIG. 6.6. (a) At time t the moon is at angular position θ and arc-length position s. (b) A short time, Δt, later its angular position has changed by $\Delta\theta$ and the moon has moved a distance Δs along the arc.

tial speed of an object from its angular velocity, or vice versa. For example, how fast does the groove on a high-fidelity record travel under the needle when the record turns with a certain angular velocity? To derive an expression relating v and ω, we note that when θ is in units of radians, the tangential distance s and the angular position θ are related by the equation $s = r\theta$ [Eq. (6.1)]. We can use this equation to relate v and ω. Consider the motion of the moon depicted in Fig. 6.6b. It rotates a distance Δs along its circular path in a time Δt. During the same time the moon's angular position coordinate has changed by $\Delta\theta$. From Eq. (6.1) we see that Δs and $\Delta\theta$ are also related by

$$\Delta s = r\,\Delta\theta.$$

If we divide both sides of this equation by Δt, we find that

$$\frac{\Delta s}{\Delta t} = r\frac{\Delta\theta}{\Delta t}.$$

But $\Delta s/\Delta t$ is just the tangential speed v of the moon, and $\Delta\theta/\Delta t$ is the moon's angular velocity ω. Thus, we find that tangential speed and angular velocity are related by the equation

$$v = r\omega. \tag{6.5}$$

When using this equation, the angular velocity must be in units of radians, since $\Delta s = r\,\Delta\theta$ applies only for $\Delta\theta$ in units of radians.

We should note that $v = \Delta s/\Delta t$ and $\omega = \Delta\theta/\Delta t$ are instantaneous values if Δt is very short and are average values if Δt is long. Equation (6.5) relates v and ω in either case.

EXAMPLE 6.3 The earth makes one rotation about its axis in 24 hours. (a) Calculate its angular velocity in units of rad/s. (b) Calculate the tangential speed of an object at the equator due to this rotational motion. The radius of the earth is 6.38×10^6 m.

SOLUTION (a) According to Eq. (6.4), we find that the earth's angular velocity is

$$\omega = \frac{\Delta\theta}{\Delta t} = \frac{1\text{ rev}}{24\text{ hr}} = \left(\frac{1\text{ rev}}{24\text{ hr}}\right)\left(\frac{2\pi\text{ rad}}{1\text{ rev}}\right)\left(\frac{1\text{ hr}}{3600\text{ s}}\right) = \underline{7.3 \times 10^{-5}\text{ rad/s}}.$$

(b) Using Eq. (6.5), we find that the tangential speed of an object on the earth's equator is

$$v = r\omega = (6.38 \times 10^6\text{ m})(7.3 \times 10^{-5}\text{ rad/s}) = \underline{460\text{ m/s}}.$$

Thus, a person standing on the equator travels in a circular path at a speed of 460 m/s or a little over 1000 mph. Fortunately, the earth's atmosphere rotates at about the same speed; otherwise the earth would be very windy! ∎

6.4 Centripetal Acceleration

We might think that an object moving with constant speed in a circular path is not accelerating. However, an object that is not accelerating has constant velocity, not constant speed. Notice in Fig. 6.7 that the velocity changes as a ball

FIG. 6.7. The velocity of an object moving at constant speed in a circular path is continually changing because its direction of motion is changing.

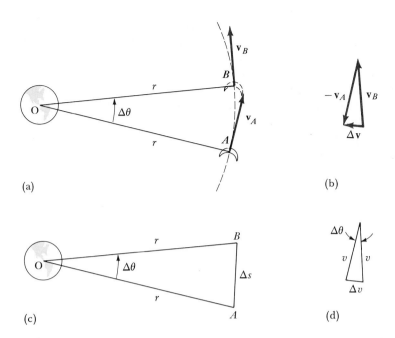

(a)

(b)

(c)

(d)

FIG. 6.8. (a) As the moon moves from position A to position B, its velocity changes from $\mathbf{v}_A$ to $\mathbf{v}_B$. (b) The change in velocity $\Delta \mathbf{v} = \mathbf{v}_B - \mathbf{v}_A$. (c) A triangle showing the distance Δs traveled by the moon; and (d) a similar velocity triangle.

swings around a circle at the end of a string. The speed may not change, but the direction of the velocity changes. Since the velocity changes direction, the ball is accelerating.

Let us derive an expression for this acceleration by considering the motion of the moon in its orbit about the earth (Fig. 6.8a). During a time Δt the moon moves from position A to position B, and the moon's velocity changes from $\mathbf{v}_A$ to $\mathbf{v}_B$. The magnitude of the velocity remains the same; only the direction changes. The acceleration of the moon equals the change in velocity divided by the change in time:

$$\mathbf{a} = \frac{\Delta \mathbf{v}}{\Delta t} = \frac{\mathbf{v}_B - \mathbf{v}_A}{\Delta t}.$$

The change in velocity $\Delta \mathbf{v} = \mathbf{v}_B - \mathbf{v}_A$ is shown in Fig. 6.8b and points toward the center of the circle.* The acceleration must also be toward the center of the circle. This acceleration of objects moving in circular paths at constant speed is called **centripetal acceleration** after the Latin word *centripetus* meaning "toward the center."

Centripetal acceleration is always a little difficult to understand. If the moon is accelerating toward the earth, why does it remain at a constant distance away rather than "falling" into the earth? In fact, if the moon did not accelerate, it would move straight ahead, as in Fig. 6.9, and leave its orbit around the earth. Because of its centripetal acceleration the moon does, in a sense, fall toward the earth, but while it falls, it continues to move forward. The combination of the two motions keeps it on a circular path of constant radius.

To find the magnitude of the centripetal acceleration, we apply some basic geometry. In Fig. 6.8c a triangle is constructed whose sides are radial lines from the earth to the moon when at positions A and B. The third side of the triangle is the distance Δs that the moon travels from position A to position B.† In Fig. 6.8d

Path if no acceleration

Path if moon started with zero velocity

Actual path

FIG. 6.9. The earth pulls the moon down toward the earth. While this happens, the moon travels to the left. The net result is a circular path of constant speed.

*The graphical subtraction of vectors was discussed at the end of Section 1.5.

†This straight line is slightly shorter than the actual curved arc along which the moon travels. The error introduced is negligible when $\Delta \theta$ is small.

another triangle is constructed from $\mathbf{v}_B$, $-\mathbf{v}_A$, and $\Delta\mathbf{v} = \mathbf{v}_B - \mathbf{v}_A$. Both $\mathbf{v}_B$ and $-\mathbf{v}_A$ have the same magnitude, equal to the moon's speed v.

Comparing these triangles and the vectors shown in Fig. 6.8a, we see that $-\mathbf{v}_A$ is perpendicular to line OA and $\mathbf{v}_B$ is perpendicular to line OB. Thus, the angle $\Delta\theta$ between OA and OB equals the angle between $-\mathbf{v}_A$ and $\mathbf{v}_B$ (Fig. 6.8b). The corresponding triangles shown in Fig. 6.8c and d must have equal angles and are similar triangles. Because they are similar triangles, the ratio of the length of the side opposite $\Delta\theta$ and the side adjacent to $\Delta\theta$ for one triangle must equal that same ratio for the other triangle:

$$\Delta\theta = \frac{\Delta s}{r} = \frac{\Delta v}{v}. \tag{6.6}$$

Since the moon travels with speed v, the distance traveled in a time Δt is

$$\Delta s = v\,\Delta t.$$

When we substitute this into Eq. (6.6), we find that

$$\frac{v\,\Delta t}{r} = \frac{\Delta v}{v}.$$

After rearranging, we have

$$\frac{\Delta v}{\Delta t} = \frac{v^2}{r}.$$

But $\Delta v/\Delta t$ is just the magnitude of the acceleration and equals v^2/r. Since v and r are the same at each point on the circle, then the magnitude of the acceleration is also the same at each point. We can summarize what we have learned about centripetal acceleration by the following statements:

An object moving in a circular path of radius r with a constant speed v has an acceleration, called **centripetal acceleration** $\mathbf{a}_c$, of magnitude

$$a_c = \frac{v^2}{r}. \tag{6.7}$$

The acceleration is directed toward the center of the circle.

The speed of an object moving in a circular path is related to its angular velocity by the equation

$$v = r\omega. \tag{6.5}$$

Thus, the magnitude of the centripetal acceleration can also be written as

$$a_c = r\omega^2. \tag{6.8}$$

EXAMPLE 6.4 Calculate the centripetal acceleration of a car traveling on a circular race track of radius 1000 m at a speed of 180 km/hr.

SOLUTION We first convert the speed to units of meters and seconds:

$$v = 180\,\frac{\text{km}}{\text{hr}}\left(\frac{1000\text{ m}}{1\text{ km}}\right)\left(\frac{1\text{ hr}}{3600\text{ s}}\right) = 50\,\frac{\text{m}}{\text{s}},$$

Then, using Eq. (6.7), we find that

$$a_c = \frac{v^2}{r} = \frac{(50\text{ m/s})^2}{1000\text{ m}} = 2.5\,\frac{\text{m}}{\text{s}^2}.$$ ∎

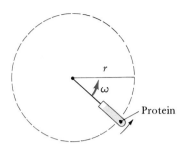

FIG. 6.10. The path of a red blood cell in an ultracentrifuge.

A centrifuge provides an important example involving centripetal acceleration. Suppose that we have a tube containing a liquid in which some red blood cells, large molecules, or grains of sand are suspended. If we set the tube in a vertical position on a table and leave it for a week, the more dense particles such as the red blood cells or grains of sand will slowly sink to the bottom of the tube. Eventually, the suspended particles will be separated at the bottom—the most dense particles on the bottom. The separation may require only minutes if done in a centrifuge.

In the centrifuge, the solution in a tube whirls in a circular path at high speed (see Fig. 6.10). The red blood cell or particle of sand feels much like a person riding in a car when turning a sharp corner. In the car you feel as though you are being thrown out against the side door. Actually, your body wants to move straight ahead but the car turns so that the door intercepts your path and pulls you around the corner with it. The same phenomenon occurs to the red blood cell or particle of sand in the centrifuge. It wants to move straight ahead but the whirling tube intercepts its path. The cell falls quickly to the bottom of the tube where it is forced by the tube to continue its motion in a circular path. The cell falls to the bottom of the tube at a rate proportional to its centripetal acceleration ($r\omega^2$). This acceleration may be hundreds of thousands of times faster than if the cell sits in a solution on a lab bench.

EXAMPLE 6.5 A red blood cell suspended in a liquid in a tube rotates in an ultracentrifuge in a circular path of radius 0.1 m (see Fig. 6.10). The tube makes 800 revolutions per second. Calculate the centripetal acceleration of the cell.

SOLUTION The angular velocity of the red blood cell in units of rad/s will be

$$\omega = 800 \frac{\text{rev}}{\text{s}} \left(\frac{2\pi \text{ rad}}{1 \text{ rev}} \right) = 1600\,\pi \frac{\text{rad}}{\text{s}}.$$

Thus, its centripetal acceleration will be

$$a_c = r\omega^2 = (0.1 \text{ m})(1600\,\pi \text{ rad/s})^2 = \underline{2.5 \times 10^6 \text{ m/s}^2}.$$

This is 260,000 times the acceleration of gravity, which is 9.8 m/s^2. The red blood cell should reach the bottom of the tube 260,000 times faster than if sitting in the same tube on the lab bench.

6.5 Centripetal Force

We have seen that an object moving in a circular path at constant speed experiences a centripetal acceleration. In this section we consider examples of forces that cause circular motion. Any force that pulls on an object causing it to move in a circle is said to exert a centripetal force on the object. Look at the ball swinging at the end of a string shown in Fig. 6.11a. The tension force **T** of the string on the ball (Fig. 6.11b) is the only force acting on the ball in the radial direction. The horizontal component of this force T_h points toward the center of the circle and causes the ball to remain in its circular orbit. Newton's second law, $\Sigma \mathbf{F} = m\mathbf{a}$, relates the horizontal component of the tension force and the acceleration it causes in the radial direction:

$$T_h = ma_c,$$

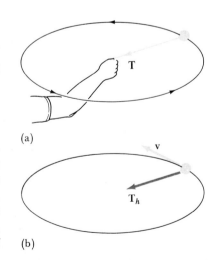

(a)

(b)

FIG. 6.11. A ball swung at the end of a string is acted on by a tension force, which causes it to move in a circular path. The force is responsible for the ball's centripetal acceleration.

FIG. 6.12. The tension force in a child's arms keeps him moving in a circular orbit.

where m is the ball's mass. The horizontal component of tension in the string depends on the ball's speed and the radius of the circle, as is apparent by substituting for a_c using either Eq. (6.7) or (6.8):

$$T_h = m\frac{v^2}{r} = mr\omega^2.$$

Consider an example in which the tension force is provided by the arms of a person.

EXAMPLE 6.6 A 20-kg child is swung by his sister, as shown in Fig. 6.12. The radius of the circular path of the child is 1.2 m. The child makes one rotation in one second. Calculate the horizontal component of tension in each of the child's arms. Ignore the child's weight, which must be balanced by an upward component of **T**.

SOLUTION We are given the following information:

$$m = 20 \text{ kg}, \qquad r = 1.2 \text{ m}, \qquad \omega = 1 \text{ rev/s} = 2\pi \text{ rad/s}.$$

The angular velocity ω is related to the horizontal component of tension pulling the child by the equation

$$T_h = ma_c = mr\omega^2 = (20 \text{ kg})(1.2 \text{ m})(2\pi \text{ rad/s})^2$$
$$= 950 \frac{\text{kg} \cdot \text{m}}{\text{s}^2} = 950 \text{ N}.$$

The horizontal component of tension in each arm will be one-half this amount, or <u>480 N</u> (about 110 lb). ∎

In some situations several forces act along the radial direction to cause centripetal acceleration. If the force points inward, it is given a positive sign; if outward, a negative sign. The centripetal acceleration points inward and has a positive sign. The next example is an interesting biological manifestation of centripetal forces.

(a)

(b)

FIG. 6.13. (a) An airplane doing a loop and (b) the forces acting on the pilot at the bottom of the loop.

EXAMPLE 6.7 (Blackout and redout) Figure 6.13 shows a pilot in an airplane moving in a vertical loop. At the bottom of the loop, the pilot may experience a phenomenon called "blackout." Two forces act on the pilot (Fig. 6.13b): his downward weight force **w** and the upward force of the airplane's seat $\mathbf{F}_{\text{seat}}$. Because the pilot moves in a circular path, these forces along the radial direction must, according to Newton's second law ($\Sigma \mathbf{F} = m\mathbf{a}$), equal the pilot's mass times his centripetal acceleration, where $a_c = v^2/r$. We find that

$$F_{\text{seat}} - w = m\frac{v^2}{r}.$$

Substituting $w = mg$ and rearranging, we find that the force of the airplane seat on the pilot is

$$F_{\text{seat}} = m\left(\frac{v^2}{r} + g\right)$$

If the v^2/r part of the above expression is large compared to g, the pilot will feel very heavy. The seat pushes up on the pilot with a large force. Blood in the pilot's body may be "left behind" in his legs. The vessels in the lower body can easily absorb a large amount of excess blood. If an upward force of three or four times that caused by gravity lasts for 20 seconds or more, blood will be unable to reach the pilot's brain. The pilot will lose consciousness, and blackout occurs. Special flightsuits are made that exert considerable pressure on the legs during these accelerations. This pressure prevents blood from draining into the veins of the legs.

If a pilot initiates a power dive from a horizontal position, the opposite effect can occur. The pilot accelerates downward and blood rushes to the head. Vision may be blurred. Blood rushing to the head fills the vessels in the retina of the eye and the pilot literally "sees red." This phenomenon is called a "redout." ∎

6.6 Newton's Law of Gravitation

Another important force that causes centripetal acceleration is the force of gravity. The quantitative description of this force was discovered by Isaac Newton during the eighteen months in 1665–1666 that Cambridge University was closed because of the Great Plague. Newton, then 22, spent that time at the family farm where he lived as a child. In the serenity of this rural environment Newton spent hours each day in intense concentration. He was putting together the ideas for his three laws of motion. He also wondered about the circular motion of the moon about the earth and the circular motion of the planets about the sun. What prevented these bodies from leaving their circular paths?

According to one of his biographers, Newton was sitting under an apple tree one day when an apple fell. Newton was struck by the idea—but not by the apple—that the force causing the apple to fall might also cause the moon to fall toward the earth. However, since the moon moved forward as it "fell," the moon followed a circular path rather than falling into the earth as the apple had done. According to Newton, he "began to think of gravity extending to the orb of the moon. . . ." Using observations about planetary motion made by Kepler, Newton deduced that the force that kept planets in orbit could also account for the earth's gravitational force on an apple. Newton had developed a logical and basic understanding of an important aspect of the structure of our universe.

This gravitational force that one mass exerts on another and that is responsible for the circular motion of the moon, earth satellites, and the planets is described below.

Newton's universal law of gravitation: The gravitational force exerted by a mass m_1 on another mass m_2 is proportional to the products of their masses and inversely proportional to the square of their separation r:

$$F = \frac{Gm_1m_2}{r^2},$$

(6.9)

where G is a constant called the universal gravitational constant and has a value

$$G = 6.67 \times 10^{-11} \, \text{N} \cdot \text{m}^2/\text{kg}^2.$$

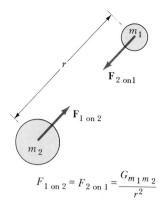

$$F_{1 \text{ on } 2} = F_{2 \text{ on } 1} = \frac{Gm_1 m_2}{r^2}$$

FIG. 6.14. Two masses exert attractive gravitational forces on each other. The forces are equal in magnitude but oppositely directed.

Figure 6.14 shows the direction of the forces that the two masses exert on each other. $\mathbf{F}_{1 \text{ on } 2}$ is the force that m_1 exerts on m_2. $\mathbf{F}_{2 \text{ on } 1}$ is the force that m_2 exerts on m_1. As can be seen, each mass exerts an attractive force on the other. The magnitude of each force is the same and can be calculated using Eq. (6.9). The separation of the masses used in these calculations must be from their centers of gravity and not from their edges.

EXAMPLE 6.8   The sun exerts a gravitational force of 3.6×10^{22} N on the earth. If the earth is 1.5×10^{11} m from the sun and has a mass of $m_e = 5.98 \times 10^{24}$ kg, what is the mass m_s of the sun?

SOLUTION   Since

$$F = \frac{Gm_e m_s}{r^2},$$

we find that

$$m_s = \frac{Fr^2}{Gm_e} = \frac{(3.6 \times 10^{22} \text{ N})(1.5 \times 10^{11} \text{ m})^2}{(6.67 \times 10^{-11} \text{ N} \cdot \text{m}^2/\text{kg}^2)(5.98 \times 10^{24} \text{ kg})}$$
$$= \underline{2.0 \times 10^{30} \text{ kg}}. \qquad \blacksquare$$

Newton's gravitational force also accounts for the weight of an apple or of any object at the earth's surface. The gravitational force caused by the earth on an object of mass m at the earth's surface is

$$F = \frac{Gm_e m}{r_e^2}$$

where r_e is the radius of the earth and m_e is its mass. Substituting for G, r_e, and m_e, we find that

$$F = \frac{(6.67 \times 10^{-11} \text{ N} \cdot \text{m}^2/\text{kg}^2)(5.98 \times 10^{24} \text{ kg})}{(6.38 \times 10^6 \text{ m})^2} m$$
$$= (9.8 \text{ m/s}^2)m = mg.$$

Note that 9.8 m/s^2 is just the value of g, the earth's gravitational acceleration at its surface, and is given by

$$g = \frac{Gm_e}{r_e^2}. \qquad (6.10)$$

An object's weight is the gravitational force of the earth on the object's mass m and equals

$$w = \frac{Gm_e m}{r_e^2} = \left(\frac{Gm_e}{r_e^2}\right) m = mg.$$

A similar procedure could be used to calculate an object's weight when on the moon or on any of the planets. We simply substitute into Eq. (6.10) the mass and radius of the planet and calculate the gravitational acceleration g on that planet. The object's weight on the planet is $w = mg$.

Consider another important application of Newton's law of gravitation and of our ideas in the last section about circular motion.

EXAMPLE 6.9 (Synchronous earth satellite) Most communication satellites remain in orbit above the same point on the earth. This requires that they circle the earth once in 24 hours—that is, their motion is synchronous with the earth's rotation. (a) At what distance from the earth should we place a satellite so that it remains synchronous with the earth's rotation? (b) What is the speed of the satellite? The earth's mass is 5.98×10^{24} kg.

SOLUTION (a) The earth's gravitational force keeps the satellite moving in a circular path. Substituting this gravitational force and the expression for centripetal acceleration, given by Eq. (6.8), into Newton's second law $\Sigma \mathbf{F} = m\mathbf{a}$, we find that

$$\frac{Gm_e m}{r^2} = m(r\omega^2), \qquad (6.11)$$

where m is the mass of the satellite, m_e the mass of the earth, r the separation of the satellite from the center of the earth, and ω the satellite's angular velocity.

Since the satellite is to make one revolution about the earth each 24 hours, its angular velocity is

$$\omega = \frac{1 \text{ rev}}{24 \text{ hr}} = \left(\frac{1 \text{ rev}}{24 \text{ hr}}\right)\left(\frac{2\pi \text{ rad}}{1 \text{ rev}}\right)\left(\frac{1 \text{ hr}}{3600 \text{ s}}\right) = 7.27 \times 10^{-5} \text{ rad/s}.$$

Solving Eq. (6.11) for r, we find that

$$r^3 = \frac{Gm_e}{\omega^2} = \frac{(6.67 \times 10^{-11} \text{ N} \cdot \text{m}^2/\text{kg}^2)(5.98 \times 10^{24} \text{ kg})}{(7.27 \times 10^{-5} \text{ rad/s})^2} = 7.55 \times 10^{22} \text{ m}^3 *$$

or

$$r = \sqrt[3]{7.55 \times 10^{22} \text{ m}^3} = \underline{4.23 \times 10^7 \text{ m}}.$$

This is almost seven times the radius of the earth.

It is interesting that our result did not depend on the mass of the satellite. Thus, any satellite placed at this distance from the earth's center will remain synchronous with the earth; it rotates with the earth and always stays above the same part of the earth's surface.

(b) The satellite's speed, according to Eq. (6.5), is

$$v = r\omega = (4.23 \times 10^7 \text{ m})(7.27 \times 10^{-5} \text{ rad/s}) = \underline{3080 \text{ m/s}},$$

or about 7000 miles/hr. ∎

6.7 Weightlessness

An interesting application of centripetal force and acceleration is the experience called weightlessness. This phenomenon is associated most often with the motion of astronauts floating in spaceships. Let us consider a more down-to-earth situation.

In Fig. 6.15a a person on the surface of the earth supports a metal sphere on his hand. Two forces act on the sphere, as shown in Fig. 6.15b. The downward weight force **w** is the actual weight of the sphere and is caused by its gravita-

*We have used the fact that $1 \text{ N} = 1 \text{ kg} \cdot \text{m/s}^2$ to convert units.

(a)

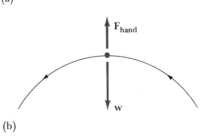

(b)

FIG. 6.15. (a) A person holds a metal sphere; since the earth is rotating, the sphere moves in a circle. (b) The forces acting on the sphere.

tional attraction to the earth. The second force acting on the sphere, $\mathbf{F}_{\text{hand}}$, is the upward force of the person's hand.

In Fig. 6.15a we see that the sphere moves in a circular path as the earth rotates. In order for any object to move in a circular path, there must be a net force pulling the object toward the center of the circle. This centripetal force causes the object's path to bend in a circle rather than move straight ahead. For the sphere, the net centripetal force is merely the difference of w and $\mathbf{F}_{\text{hand}}$. Newton's second law becomes

$$w - F_{\text{hand}} = ma_c$$
$$= m(r\omega^2) \qquad (6.12)$$

We see that F_{hand} must be less than w if the mass experiences a centripetal acceleration.

The earth's angular velocity is so small that, for practical purposes, $r\omega^2$ can be ignored for objects on the earth's surface. If $r\omega^2$ is very small, then according to Eq. (6.12) the object's weight w is equal in magnitude to the force F_{hand} needed to support that weight. However, if the earth's angular velocity is greater, or if the object circles the earth in an orbit of greater radius, then its centripetal acceleration is not negligible. All of its weight force may be needed to keep it in orbit about the earth, and no force F_{hand} is needed to support its weight. The object seems to have no weight.

Suppose that the earth's angular velocity could be increased. Imagine it rotating so fast that a mass would fly up out of your hand, much like a drop of water being spun out of wet clothes in a clothes dryer. At this point, a mass would appear to be weightless. It would require no force to support its weight.

EXAMPLE 6.10 At what angular velocity must the earth and a metal ball rotate so that the ball seems weightless?

SOLUTION The condition of weightlessness occurs when the force F_{hand} needed to support the sphere's weight is zero. Substituting $F_{\text{hand}} = 0$ into Eq. (6.12), we find that

$$w - 0 = mr\omega^2$$

or

$$\omega = \sqrt{\frac{w}{mr}} = \sqrt{\frac{mg}{mr}} = \sqrt{\frac{g}{r}}$$
$$= \sqrt{\frac{9.8 \text{ m/s}^2}{6.38 \times 10^6 \text{ m}}} = \underline{1.2 \times 10^{-3} \text{ rad/s.}} \qquad \blacksquare$$

We see from the last example that if the earth rotated with an angular velocity of 1.2×10^{-3} rad/s, or about 17 times the earth's actual angular velocity, we would need no force to support the ball's weight. If we stood on a bathroom scale, the scale would read zero. If the earth's angular velocity were even greater, we would have to grab hold of a tree or something else attached to the earth just to stay on its surface. The metal ball we were holding would rise up out of our hand and we would rise up with it if we did not anchor ourselves.

An astronaut in orbit about the earth is in a similar situation. The whole gravitational weight force caused by the attraction between the earth's mass and that of the astronaut is needed to keep the astronaut in the circular orbit. If the astronaut stood on a bathroom scale, it would exert no outward contact force on the astronaut. The astronaut would appear to have no weight.

We see that the expression *weightlessness* is inappropriate: a weight force does act on the astronaut. It would be better to say that the astronaut has no apparent weight because no outward force is needed to support the astronaut's actual weight when moving in the circular orbit about the earth.

Summary

1. **Angular position:** The angular position of an object moving in a circular path is specified by an **angular position coordinate, θ**. θ is the angle from the positive x axis to a radial line drawn from the center of the circle to the object. The units of θ can be degrees, revolutions, or radians. These units are related by the equation,

$$360° = 1 \text{ rev} = 2\pi \text{ rad}. \tag{6.2}$$

The angle θ in units of **radians** is defined as the arc length s along the circumference of the circle from the x axis to the object divided by the radius r of the circle:

$$\theta \text{ (in radians)} = \frac{s}{r}. \tag{6.1}$$

2. **Angular velocity and tangential speed:** The **angular velocity** ω of an object moving in a circular path is the ratio of the change in the object's angular position $\Delta\theta$ and the time Δt required for this change:

$$\text{angular velocity} = \omega = \frac{\Delta\theta}{\Delta t}. \tag{6.4}$$

The object's **tangential speed** is the ratio of its change in position Δs along the arc of the circle and the time Δt required for this change:

$$\text{tangential speed} = v = \frac{\Delta s}{\Delta t}. \tag{6.3}$$

The two quantities, angular velocity and tangential speed, are related by the equation

$$v = r\omega, \tag{6.5}$$

where r is the radius of the circle.

3. **Centripetal acceleration:** When an object moves in a circular path at constant speed, the direction of its velocity changes continually even though the magnitude of its velocity is constant. The object experiences a **centripetal acceleration $\mathbf{a}_c$** because of this changing direction of motion. The magnitude of the acceleration is given by

$$a_c = v^2/r \tag{6.7}$$
$$= r\omega^2. \tag{6.8}$$

The direction of $\mathbf{a}_c$ is toward the center of the circle.

Forces cause acceleration. When an object moves in a circular path, a **centripetal force** is needed to pull in on the object and keep it from leaving the circle. Any force responsible for keeping the object in its circular path is said to be a centripetal force. Newton's second law relates these forces that act along the radial direction to the centripetal acceleration they cause:

$$\Sigma F \text{ (in radial direction)} = ma_c.$$

4. **Gravitational force.** The gravitational force of attraction exerted by mass m_1 on mass m_2 is proportional to the product of their masses and inversely proportional to the square of their separation r:

$$F = \frac{Gm_1m_2}{r^2},$$

where $G = 6.67 \times 10^{-11} \text{ N} \cdot \text{m}^2/\text{kg}^2$ is the universal gravitational constant.

Questions

1. In the following examples, identify the centripetal force that causes the object to move in a circular path: (a) a ball rotating in a horizontal circle at the end of a string, (b) a ball rotating in a vertical circle at the end of a string when at the top of the circle, (c) a car rounding a curve on a highway, and (d) the moon circling the earth.

2. A pilot is performing a vertical loop-the-loop at constant speed. Why is it more likely that the pilot will black out at the bottom of the loop than at the top?

3. *Estimate* the force needed to keep a 10-inch-long section of the outside of a car tire tread on the tire when it rotates at 15 rpm. Justify any numbers used in your estimate.

4. *Estimate* the centripetal force needed to hold your foot on your leg when swinging the leg to kick a football. Justify any numbers used in your estimate.

5. A rope is tied to the handle of a pail of water. Why does the water stay in the pail when the pail is swung in a vertical circle?

6. Explain how the spin-dry cycle on a washing machine causes water to be removed from clothes. Be specific.

7. Why do we feel we are being thrown upward, out of our seats, when going over a hump on a rollercoaster?

8. Where, on the earth's surface, would you expect to experience the greatest centripetal acceleration as a result of the earth's rotation? Explain.

9. (a) If the mass of the earth and moon were both doubled, by how much would the radius of the moon's orbit about the earth change if it had the same velocity? (b) By how much would its velocity change if it had the same radius? (c) Calculate the time between new moons for each situation.

Problems

6.1 and 6.2 Angular Position Coordinate and Units

1. Perform the following conversions: 50° to radians, 160° to radians, 1.5π radians to degrees, 0.4 radians to degrees, 2.1 revolutions to radians, 3 revolutions to degrees.

2. Make the following conversions: 500 rpm to rad/s, 1 rev/27 days to rad/s, 50 rad/s² to rev/min².

3. (a) Calculate the arc length of a 20° portion of a circle whose radius is 5 m. (b) Repeat (a) for an angle of 0.7 rev.

4. The earth has a radius of 6.38×10^3 km. Two cities on the equator are separated by 1000 km. Calculate their angular separation in radians, degrees, and revolutions.

5. When punting a football, a person's leg swings through an angular displacement of 120°. The person's leg is 0.9 m long. Calculate the distance moved by the foot.

6. A pair of car headlights have an angular separation of 0.4° when viewed by a distant observer. *Estimate* the distance of the car from the observer. Justify your estimation technique.

7. A circular piece of land of diameter 350 m is divided into seven equal pie-shaped lots. Calculate the length of the front line on each lot and the angular separation of the lines at the side of each lot.

8. An insect on the edge of a 12-inch record completes one rotation in 0.56 s. (a) Calculate the change in angular position of the insect in 1.0 s. (b) Calculate the distance traveled by the insect in 1.0 s.

9. A bicycle with a 0.68-m diameter tire travels 2000 km before the tire becomes worn and must be replaced. Calculate the number of turns made by the tire in traveling that distance.

6.3 Tangential Speed and Angular Velocity

10. The earth circles the sun in a nearly circular orbit of radius 1.5×10^{11} m. (a) Calculate the angular velocity of the earth about the sun. (b) Calculate the earth's speed while moving in this orbit.

11. The tooth of a blade in a circular saw travels with a speed of 1200 m/s. The blade has a radius of 0.1 m. Calculate the angular velocity of the blade in rad/s and in rpm.

12. Calculate the angular velocity of a passenger car traveling at 50 mph on the circular turn of a highway. The radius of the turn is 400 m.

13. The tire on a bicycle rotates four times in one second. (a) Calculate the angular velocity of the tire. (b) Calculate the average tangential speed of the valve stem 0.33 m from the axis of rotation.

14. The moon is an average distance of 3.8×10^8 m from the earth. It circles the earth once each 27.3 days. (a) Calculate its angular velocity in rad/s. (b) Calculate the distance it travels in its orbit in one day.

15. The electron in a hydrogen atom rotates about its proton nucleus with an angular velocity of 4.15×10^{16} rad/s. (a) Calculate the number of revolutions the electron makes per second. (b) Calculate the electron's speed. The average distance of the electron from the nucleus is 0.53×10^{-10} m.

6.4 Centripetal Acceleration

16. Calculate the centripetal acceleration of the electron in the hydrogen atom discussed in Problem 15.

17. An airplane makes a loop of radius 500 m. The centripetal acceleration of the airplane is 40 m/s². Calculate the angular velocity and tangential speed of the airplane.

18. A red blood cell rotating in an ultracentrifuge experiences a centripetal acceleration of 150,000 *g*'s. If the cell is 0.15 m from the axis of rotation, what is the angular velocity of the ultracentrifuge in rpm?

19. Jupiter rotates once about its axis each 9 hr 56 min. Its radius is 7.15×10^4 km. Calculate the centripetal acceleration in m/s² and in *g*'s of a person standing at the equator of Jupiter.

20. Extremely dense neutron stars rotate with an angular velocity of about 1 rev/s. (a) If the radius of the star is 2.5×10^4 m, what is the speed of a mass on the star's equator? (b) Calculate the centripetal acceleration in m/s² and in *g*'s of a point on the equator.

21. *Estimate* the centripetal acceleration of a person riding on a Ferris wheel at an amusement park. Justify your choice of numbers used in the estimate.

22. A tube containing colloidal clay particles is placed in a centrifuge. The time needed for the particles to settle to the bottom of the tube is inversely proportional to the centripetal acceleration of the particles. Seven days are required for the particles to settle when the acceleration equals *g*. Calculate the time required for the clay to settle in the tube of the centrifuge whose radius is 0.3 m and angular velocity is 10,000 rpm.

6.5 Centripetal Force

23. Trout weighing less than 16 oz must be returned to the stream. A disappointed angler, after catching a 12-oz trout, decides to weigh it by swinging it at the end of a string attached to a scale in a circle of 1.5 m radius. With what angular velocity must the fish be swung to get the scale to read 16 oz? Assume that it swings in a horizontal circle.

24. A 0.5-kg mass is swung at the end of a string that is 1.2 m

long. The string breaks if the tension exceeds 400 N. What is the maximum speed with which the mass can travel in its circular path?

25. Calculate the centripetal force acting on an 1100-kg car that travels around a highway curve of radius 150 m at 27 m/s.

26. The blades on a rotary lawn mower turn 50 times per second. Calculate the force needed to hold the outer 2 cm of the blade to the inner portion of the blade if the outer part is 21 cm from the center of the blade and the mass of the outer portion is 7.0 g.

27. A car speeds around the 80-m radius curved exit ramp of a freeway. A 70-kg passenger holds the arm rest of the car door with a 220-N force in order to prevent sliding across the front seat of the car (ignore friction). Calculate the car's speed in km/hr.

28. The fire department attaches a 10-m long rope to a tree and the other end goes to a dormitory window to serve as a fire escape. A 68-kg student while swinging from the window can hold the rope with a force no greater than 1200 N. What is the maximum speed of the swinger at the bottom of the swing? Do not forget to include the swinger's weight in your force diagram and equations.

29. A 50-kg young man rides in a carnival car that rotates at the end of a beam in a vertical circle of 20 m radius with an angular velocity of 0.4 rev/s. (a) Draw a force diagram for the man at the top of the circle (he is inverted). (b) Use Newton's second law to determine the magnitude of the force of the car seat on the man at the top of the circle.

30. A person revolves at the end of a swing shown in Fig. 6.16. (a) Draw a force diagram of the swing and person whose combined mass is 80 kg. (b) Use the vertical component equation of Newton's second law to find the tension in the cable attached to the swing. (c) The horizontal component of this tension force provides the centripetal force. Use this fact to calculate the angular velocity of the swing.

FIG. 6.16

6.6   Newton's Law of Gravitation

31. Calculate the gravitational force (a) of the sun on the moon, and (b) of the earth on the moon. $m_{sun} = 2.0 \times 10^{30}$ kg, $m_{earth} = 6.0 \times 10^{24}$ kg, $m_{moon} = 7.3 \times 10^{22}$ kg, $r_{sun-moon} = 1.5 \times 10^{11}$ m, and $r_{earth-moon} = 3.8 \times 10^8$ m.

32. A black hole exerts a 50-N gravitational force on a spaceship. The black hole is 10^{14} m from the ship. Calculate the force on the ship when it moves to one-half that distance from the black hole. (*Note:* One-half of 10^{14} m is not 10^7 m.)

33. Calculate the radius of the orbit of an earth satellite that makes two rotations per day.

34. The average radius of the earth's orbit around the sun is 1.5×10^8 km. The mass of the earth is 5.98×10^{24} kg, and it makes one revolution in 365 days. (a) Calculate the speed of the earth. (b) Estimate the sun's mass, using Newton's law of gravitation and his second law of motion.

35. The moon travels in an orbit about the earth that is nearly circular with radius of 3.8×10^5 km. The earth's mass is 5.98×10^{24} kg. Calculate the period T for one moon-orbit about the earth using Newton's law of gravitation and his second law of motion.

36. Calculate the period of rotation of an earth satellite that is just above the earth's surface. Use a force diagram and Newton's second law as the starting point for your calculations.

37. Calculate the ratio of the earth's gravitational force on an 80-kg person when at the earth's surface and when 1000 km above the earth's surface. The radius of the earth is 6370 km.

38. Calculate your weight on the surface of Mars. The mass of Mars is 6.6×10^{23} kg and its radius is 3380 km.

39. A person weighing 760 N on the earth moves to Saturn. (a) What is the person's mass on Saturn? (b) Calculate the person's weight on Saturn. The mass of Saturn is 5.7×10^{26} kg, and its radius is 6.0×10^4 km.

40. The acceleration of gravity on the surface of Jupiter, the most massive planet, is 6.47 m/s², and Jupiter's radius is 7.1×10^4 km. Calculate Jupiter's mass.

6.7   Weightlessness

41. A 50-kg person stands on a bathroom scale on the earth's surface. (a) Draw a force diagram for the person. (b) With the assistance of this diagram write an expression using Newton's second law for the forces and acceleration along the radial direction. (Remember that the person rotates daily at a constant speed in a circular path, so that $a_c = r\omega^2$.) (c) If the earth's rotational speed about its axis increased by a factor of 10, what would the bathroom scale read?

42. Mars has a mass of 6.58×10^{23} kg and a radius of 3.38×10^6 m. With what angular velocity should Mars rotate about its axis so that a person standing on Mars would feel weightless?

43. People living on a space platform in outer space would normally feel weightless because there are no large masses nearby exerting gravitational forces on them. To stimulate the effect of gravitation, people could live on the edge of the platform with their heads toward the center and their feet on an outside rim (see Fig. 6.17). If the platform rotates, the outside rim exerts an inward centripetal contact force on the space persons. At what angular velocity must a platform of radius 100 m rotate in order to provide for the space persons an apparent weight of mg? Explain.

FIG. 6.17.   A space platform.

General Problems

G1. A 50-lb child sits on the edge of a merry-go-round at a distance of 10 ft from its center. If the coefficient of static friction between the child and merry-go-round is 0.7, what is the maximum angular velocity at which the merry-go-round can rotate before the child slides off?

G2. A 2000-kg truck travels around a highway curve whose radius is 250 m. The truck travels at 30 m/s. (a) Calculate the centripetal force acting on the truck. (b) This centripetal force is provided by static friction, which prevents the truck from sliding off the highway. What is the minimum coefficient of static friction needed to keep the truck from sliding? The road is flat.

G3. A 60-kg young woman is a passenger in a "rotor ride" at an amusement park. A drum of radius 5 m is spun with an angular velocity of 30 rpm. The young woman is pressed against the wall of the rotating drum as shown in Fig. 6.18. (a) Calculate the contact force of the drum on the woman (this is the centripetal force that keeps her rotating). (b) While the drum rotates, the floor is lowered. A vertical static friction force supports the woman's weight. What must the coefficient of friction be in order to support her weight?

G4. A person stands on the edge of a merry-go-round which rotates at an angular velocity of 0.50 rad/s. The 60-kg person

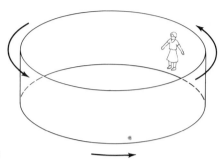

FIG. 6.18

stands 5 m from the center of the merry-go-round. (a) What must the friction force be on the person in order to prevent his or her slipping from the merry-go-round? (b) The net force of the merry-go-round on the person is the vector sum of F_c and F_s, the contact and static friction forces. At what angle θ is the vector sum of these forces directed?

G5. Plants grown on a rotating platform tilt toward the axis of rotation. (a) Explain why you think this happens. (b) *Estimate* the angle of tilt for plants rotating with an angular velocity of 1.5 rev/s at a distance of 0.1 m from the axis of rotation.

Rotational Motion with Angular Acceleration

7

We learned in Chapter 6 that a centripetal force is needed to keep an object moving in a circular path. The force points toward the center of the circle, perpendicular to the object's motion, and does not change the object's speed, only its direction of motion. In order to make the object move faster or slower, a force must be exerted parallel to its velocity—that is, tangent to the circle along which the object moves. Tangential forces can give the object a push, causing it to move faster, or can slow it by exerting an opposing force. This chapter examines these tangential forces and their effect on objects moving in circular paths.

Most of our examples in the last chapter had to do with motion in which a single object moved in a circle whose radius was much larger than the object: a racetrack is much larger than a car moving around it, and the moon's orbit about the earth is much larger than the moon. This motion in circles becomes slightly more complicated when we consider the rotational motion of extended bodies about an axis within the body—for example, the earth's rotation about its north–south axis or the rotation of your leg about the hip joint when kicking a ball. To understand this we must consider the torques caused by forces and the subsequent change in angular motion caused by the torques. The treatment is very similar to that used in Chapter 4, only Newton's second law and the kinematical equations of that chapter are replaced by similar equations appropriate for describing rotational motion. We will also introduce a rotational analog of linear momentum. These rotational equations should help us understand phenomena such as the torque needed by a motor to turn the wheels of a car and the technique used by ballerinas to perform a pirouette.

7.1 Angular Acceleration

Much of our analysis in this chapter relies on a quantity called angular acceleration. If an object's speed around the circle increases, then its angular velocity also increases. We say that the object experiences an angular acceleration:

The **angular acceleration** α of an object moving in a circle or rotating about an axis is defined as the ratio of the object's change in angular velocity $\Delta\omega$ and the time Δt required for that change:

$$\alpha = \frac{\Delta\omega}{\Delta t}. \tag{7.1}$$

The units of angular acceleration are rad/s^2.

If the time interval Δt is very small, then α, as defined above, is an instantaneous angular acceleration. Often we are interested only in the average angular acceleration over a longer time interval:

$$\bar{\alpha} = \frac{\omega - \omega_0}{t}, \tag{7.2}$$

where ω_0 is the initial angular velocity at time zero and ω is the final angular velocity at time t.

EXAMPLE 7.1 A bicycle tire starts from rest and 10 s later has an angular velocity of 2.5 rev/s. Calculate its average angular acceleration during that time.

SOLUTION We know that the initial angular velocity $\omega_0 = 0$, and the final angular velocity $\omega = 2.5$ rev/s at $t = 10$ s. Using Eq. (7.2), we find that

$$\bar{\alpha} = \frac{\omega - \omega_0}{t} = \frac{2.5 \text{ rev/s} - 0}{10 \text{ s}} = 0.25 \frac{\text{rev}}{\text{s}^2}$$

$$= 0.25 \frac{\text{rev}}{\text{s}^2} \left(\frac{2\pi \text{ rad}}{\text{rev}} \right) = 1.6 \frac{\text{rad}}{\text{s}^2}. \qquad \blacksquare$$

In more advanced treatments of angular acceleration, α is a vector quantity pointing along an axis perpendicular to the plane of the circular motion—that is, along the axis about which rotation occurs. We will not adhere strictly to conventions about the direction and signs of angular acceleration and angular velocity. For our problems and examples, α is positive if the magnitude of the angular velocity increases and negative if it decreases.

7.2 Tangential Acceleration

If an object rotates faster—if its angular velocity increases—then the speed of each point on the object also moves faster. The object experiences a tangential acceleration; its speed increases:

Tangential acceleration a_t is defined as the change in speed Δv of an object moving in a circle, or of a point on a rotating object, divided by the time Δt required for that change in speed:

$$a_t = \frac{\Delta v}{\Delta t}. \tag{7.3}$$

The units of tangential acceleration are m/s².

If Δt in the above definition is small, then an instantaneous tangential acceleration has been defined. The average tangential acceleration can be defined as

$$\bar{a}_t = \frac{v - v_0}{t}, \tag{7.4}$$

where v_0 is the object's tangential speed at time zero, and v is its speed at time t.

Angular acceleration and tangential acceleration can be related easily. Recall from Section 6.3 that tangential speed and angular velocity are related by the equation $v = r\omega$. Substituting for v and v_0 into Eq. (7.4), we find that

$$\bar{a}_t = \frac{r\omega - r\omega_0}{t} = r\left(\frac{\omega - \omega_0}{t}\right). \tag{7.5}$$

But $(\omega - \omega_0)/t$ is the average angular acceleration [Eq. (7.2)]. Thus,

$$a_t = r\alpha. \tag{7.6}$$

The tangential acceleration (change in speed) is the product of an object's angular acceleration and the distance r of the object from its center of rotation. The average signs have been dropped because Eq. (7.6) applies either to average values of a_t and α or to instantaneous values.

EXAMPLE 7.2 A bicycle tire undergoes an angular acceleration of 1.50 rad/s². (a) Calculate the tangential acceleration of its valve stem, which is 0.68 m from the axis of rotation. (b) Calculate the stem's tangential speed after 10 s, if it starts at rest.

SOLUTION (a) We are given that $r = 0.68$ m, and $\alpha = 1.50$ rad/s². Thus,

$$a_t = r\alpha = (0.68)(1.50 \text{ rad/s}^2) = \underline{1.02 \text{ m/s}^2}.$$

(b) Rearranging Eq. (7.4), we find that

$$v = v_0 + a_t t$$
$$= 0 + (1.02 \text{ m/s}^2)(10 \text{ s}) = \underline{10 \text{ m/s}}. \qquad \blacksquare$$

We now have equations relating angular position, velocity, and acceleration (θ, ω, and α, respectively) to an object's position, speed, and acceleration (s, v, and a_t) along its circular path:

$$s = r\theta,$$
$$v = r\omega,$$
$$a_t = r\alpha.$$

Several kinematical equations can be derived that relate these angular and tangential quantities to each other at different times for the special case when the angular acceleration is constant.

7.3 Kinematic Equations for Constant Angular Acceleration

In Chapter 3 we derived four equations (3.7–3.10) that relate an object's position, velocity, and acceleration at two different times. The equations apply to linear motion with constant acceleration. They apply also to the tangential motion of an object along a circular path with constant tangential acceleration. The position coordinate x is replaced by an arc-length coordinate s, the velocity v by tangential speed v, and the acceleration a by tangential acceleration a_t.

A similar set of equations can be derived to relate the angular position coordinate θ, the angular velocity ω, and the angular acceleration α at time t to its angular position θ_0 and angular velocity ω_0 at time zero. *These equations apply only if angular acceleration is constant.* The two sets of equations are written below:

Tangential	Angular	
$v = v_0 + a_t t,$	$\omega = \omega_0 + \alpha t,$	(7.7)
$s - s_0 = \left(\dfrac{v + v_0}{2}\right)t,$	$\theta - \theta_0 = \left(\dfrac{\omega + \omega_0}{2}\right)t,$	(7.8)
$s - s_0 = v_0 t + \frac{1}{2}a_t t^2,$	$\theta - \theta_0 = \omega_0 t + \frac{1}{2}\alpha t^2,$	(7.9)
$2(s - s_0)a_t = v^2 - v_0^2,$	$2\alpha(\theta - \theta_0) = \omega^2 - \omega_0^2.$	(7.10)

EXAMPLE 7.3 The switch for the turntable on a 33-rpm high-fidelity system is turned off. The table makes 20 complete rotations before slowing to a stop. Calculate its angular acceleration, assumed constant.

SOLUTION The turntable makes 20 complete turns or 20 revolutions while stopping. Thus,

$$\theta - \theta_0 = 20 \text{ rev} = 20 \text{ rev}\left(\frac{2\pi \text{ rad}}{1 \text{ rev}}\right) = 126 \text{ rad}.$$

Its initial angular velocity is

$$\omega_0 = 33 \frac{\text{rev}}{\text{min}} = 33 \frac{\text{rev}}{\text{min}}\left(\frac{2\pi \text{ rad}}{1 \text{ rev}}\right)\left(\frac{1 \text{ min}}{60 \text{ s}}\right) = 3.5 \frac{\text{rad}}{\text{s}},$$

and its final angular velocity $\omega = 0$. Equation (7.10) can be used to determine the turntable's angular acceleration:

$$\alpha = \frac{\omega^2 - \omega_0^2}{2(\theta - \theta_0)} = \frac{0^2 - (3.5 \text{ rad/s})^2}{2(126 \text{ rad})} = \underline{-4.9 \times 10^{-2} \text{ rad/s}^2}.$$

The negative sign implies that the turntable's angular velocity decreased. ∎

7.4 Torque and Angular Acceleration

What causes an object to experience an angular acceleration? Recall from Chapter 2 that for an object to remain in rotational equilibrium, the vector sum of the torques acting on it must be zero. When the torques do not add to zero, the rotational equilibrium is disturbed; the object undergoes an angular acceleration. The angular acceleration depends on the magnitude and sign of the torques exerted by forces acting on the object.

Figure 7.1 provides an example. A metal ball is attached to the end of a rod, and the other end of the rod pivots freely about a pin. The ball and the rod can rotate without friction in a horizontal plane. For the ball's speed to increase and to have greater angular velocity, a force must be applied. As we saw in Chapter 2, the magnitude of the torque τ caused by the force depends on both the magnitude of the force and the position at which the force acts on the object.

$\tau = lF$

(a)

$\tau = lF$

(b)

FIG. 7.1. The torque caused by a force depends on where the force acts on an object. The greater the moment arm l of the force, the greater the torque.

The **torque** is the product of the force F and its moment arm l, the perpendicular distance from the line of action of the force to the pivot point or origin of coordinates. Thus,

$$\tau = \pm lF.$$

FIG. 7.2. The force causes a torque and angular acceleration.

The torque is positive if it tends to rotate the object counterclockwise and negative if it rotates the object clockwise. The nearer the force is applied to the pivot plane (small l), the less effective it will be in accelerating the ball.

Newton's second law can be used to find a precise relation between the torque of a force and the resulting angular acceleration. Consider the simple example shown in Fig. 7.2. The ball's mass is m; the rod's mass is assumed small compared to that of the ball and can be ignored. A single force $\mathbf{F}$ acts on the ball in the tangential direction. Thus, Newton's second law applied to the tangential direction becomes

$$F = ma_t.$$

The moment arm of the force is just r, the distance of the ball and force from the pivot point, as shown in Fig. 7.2. Let us multiply both sides of the above equation by r:

$$rF = mra_t.$$

The left side of this equation equals the torque τ caused by $\mathbf{F}$. Substituting for a_t, using Eq. (7.6), $a_t = r\alpha$, we find that

$$rF = mr(r\alpha)$$

or

$$\tau = (mr^2)\alpha.$$

Often several forces act on a rotating object. In these cases we must add the torques caused by all forces acting on the object in order to find the angular acceleration. We find that

$$\Sigma \tau = \tau_1 + \tau_2 + \cdots = (mr^2)\alpha, \tag{7.11}$$

where τ_1, τ_2, ... are the torques caused by forces $\mathbf{F}_1$, $\mathbf{F}_2$, ... Notice that if the torques add to zero, the angular acceleration will also be zero. The object will be in rotational equilibrium.

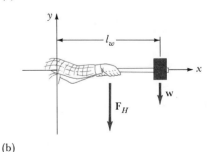

EXAMPLE 7.4 A sledgehammer with a 7-kg head is used to pound a stake. The hammerhead is 0.8 m from the body of the person swinging it. The person exerts a 50-N · m torque on the hammer (see Fig. 7.3a). (a) Calculate the angular acceleration of the hammerhead when in the position shown in Fig. 7.3a. (b) If the hammerhead's tangential speed at A is 2.0 m/s, calculate its speed 0.1 s later, assuming the acceleration remains constant. Ignore the mass of the hammer's handle.

FIG. 7.3. (a) A person exerts a torque while swinging a sledgehammer. (b) A force diagram for the sledgehammer.

SOLUTION (a) Two forces cause torques (see Fig. 7.3b). The person's hands exert a torque $\tau_H = 50\,\text{N} \cdot \text{m}$, and the hammerhead's weight causes a torque given by the expression

$$\tau_w = l_w w = l_w(mg)$$
$$= (0.8 \text{ m})(7 \text{ kg} \times 9.8 \text{ m/s}^2) = 55 \text{ N} \cdot \text{m}.$$

Rearranging Eq. (7.11), we can determine the hammerhead's angular acceleration caused by these torques:

$$\alpha = \frac{\tau_H + \tau_w}{mr^2} = \frac{50 \text{ N} \cdot \text{m} + 55 \text{ N} \cdot \text{m}}{(7 \text{ kg})(0.8 \text{ m})^2}$$
$$= 23.4 \text{ N/kg} \cdot \text{m}.$$

Using the fact that $1 \text{ N} = 1 \text{ kg} \cdot \text{m/s}^2$, we find that the angular acceleration is

$$\alpha = \underline{23.4 \text{ rad/s}^2}.$$

The radian unit is included because α is an angular quantity.

(b) The rest of the problem involves kinematics. At time 0, the hammerhead's speed is 2.0 m/s. Later, at $t = 0.1$ s, the speed has increased to $v = v_0 + a_t t$, where

$$a_t = r\alpha = (0.8)(23.4 \text{ rad/s}^2).$$

Thus,

$$v = (2.0 \text{ m/s}) + (0.8 \text{ m})(23.4 \text{ rad/s}^2)(0.1 \text{ s})$$
$$= \underline{3.9 \text{ m/s}}. \qquad \blacksquare$$

7.5 Moment of Inertia

Equation (7.11), $\Sigma \tau = (mr^2)\alpha$, is the rotational equivalent of Newton's second law. There is a strong analogy between each of the three quantities in this equation and the three quantities in Newton's second law:

	Rotational motion	Linear motion
Cause of acceleration	$\Sigma \tau$	$\Sigma \mathbf{F}$
Acceleration	α	$\mathbf{a}$
Inertia of accelerated object	mr^2	m

For linear motion, mass is a measure of an object's inertia—that is, how much force is needed to cause the object to experience a certain acceleration. For rotational motion the mass times the square of the distance of the mass from the axis of rotation (mr^2) is a measure of the object's rotational inertia. The quantity mr^2 is called the object's **moment of inertia** and is given the symbol I. Moment of inertia indicates the torque needed to cause the object to experience a certain angular acceleration. The moment of inertia measures the difficulty in starting or stopping rotation. Equation (7.11) can now be written as

$$\Sigma \tau = \tau_1 + \tau_2 + \cdots = I\alpha. \qquad (7.12)$$

Notice that the moment of inertia depends not only on the mass m of the object but also on the distance r of the mass from the point about which it rotates. The farther the mass is from the point about which it rotates, the harder it is to change its rotational motion. A heavy mallet is harder to swing when held at the end of the handle than when held in the middle of the handle, nearer the mallet. This dependence of rotational inertia on the distance r is part of our everyday experience. For example, if a baseball player has trouble getting the bat around in time to hit the ball, he or she is told to "choke up" on the bat. The result is a reduction of the moment of inertia of the bat, since its mass is brought closer to the hands. With the same torque, then, the batter can increase the angular acceleration of the bat because its moment of inertia has been reduced.

It is simple to calculate the moment of inertia of a single mass. Most objects, however, have a continuous distribution of mass in a certain shape. To get a feel for how we might calculate the moment of inertia of these objects, let us look first at several simple objects with more than one mass. Consider first the barbell shown in Fig. 7.4. Two masses are at the ends of a rod of length L. The rod is assumed to have a negligible weight compared to the weights on the end. Each mass is a distance $L/2$ from the axis of rotation and thus contributes an amount $mr^2 = m(L/2)^2$ to the total moment of inertia. The moment of inertia of the barbell about an axis through its center is therefore

FIG. 7.4. A barbell.

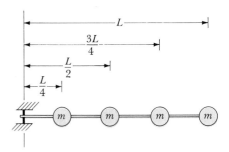

FIG. 7.5. Each mass contributes differently to the moment of inertia of this rod about an axis through its end.

$$I = m\left(\frac{L}{2}\right)^2 + m\left(\frac{L}{2}\right)^2 = 0.5 \text{ mL}^2.$$

In Fig. 7.5 we see a different rod that pivots about one end. Four equal masses are attached to the rod at quarter-length intervals along the rod. The rod itself has negligible mass. The moment of inertia of this rod is

$$I = m(\tfrac{1}{4}L)^2 + m(\tfrac{2}{4}L)^2 + m(\tfrac{3}{4}L)^2 + m(L)^2$$
$$= 1.88mL^2.$$

Once again each mass contributes to the moment of inertia of the rod. However, masses farther from the axis of rotation contribute much more than those near the axis. The mass at the right end of the rod in Fig. 7.5 contributes more to the rod's moment of inertia than the other three masses combined. This one mass makes the rod harder to rotate than the other three masses do.

The rotational inertia of what is called an extended body is calculated in a similar fashion. An **extended body** is one with mass distributed continuously in a certain shape. The disc-shaped flywheel on a car is an extended body. Your leg is an extended body, as is a pencil or a bat.

To calculate the moment of inertia of an extended body, we pretend that it is made of many small masses (Fig. 7.6). Each small mass contributes to the moment of inertia in proportion to its mass and the square of its distance from the axis about which we calculate the moment. For example, mass 7 in Fig. 7.6 contributes an amount $m_7(r_7)^2$ to the moment of inertia of a person's leg. The moment of inertia of the whole leg is then the sum of all of these small terms and is written

$$I = m_1 r_1^2 + m_2 r_2^2 + m_3 r_3^2 + \cdots. \tag{7.13}$$

Often this summation process is done using integral calculus; sometimes I is determined experimentally. In most of our examples and problems we will either be given the moment of inertia of an extended body or will use the values given in Table 7.1 for objects with common shapes.

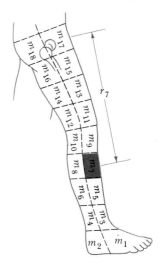

FIG. 7.6. To calculate the moment of inertia of a leg, add the moment caused by each small mass that is part of the leg.

TABLE 7.1 Moments of Inertia of Various Bodies

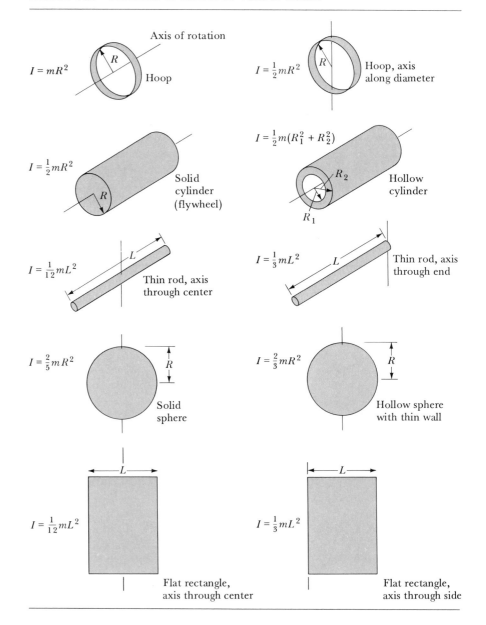

7.6 Solving Problems in Rotational Motion

The technique for solving problems involving rotational motion is very similar to that used in dynamics and statics.

1. Draw a picture of the whole situation described in the problem.

2. Choose one particular object and construct a force diagram showing all the forces acting on the object. It is important to place force arrows in the force

diagram at the position where they exert their force on the object. This position will affect the magnitude of the torque that the force exerts.

3. Superimpose a coordinate system on the force diagram. This coordinate system may be placed anywhere; however, all subsequent calculations must be consistent with this placement. Usually the origin is placed at the center of rotation.

4. Write the torque-angular acceleration equation ($\Sigma \tau = I\alpha$) for this object and substitute all the known information. The moment of inertia must be calculated using the same axis as that used to calculate the torques.

5. Solve for the unknown. Kinematical equations may also be needed to calculate the angular acceleration or some other angular property of the object.

The following examples should make this procedure clear.

EXAMPLE 7.5 Suppose that a force **F** is applied at the earth's equator and tangent to its surface for a time of one year. How large must the force be in order to stop the earth's rotation? The earth's mass is 5.98×10^{24} kg and its radius is 6.38×10^6 m.

SOLUTION The situation described in the problem is shown in Fig. 7.7. The earth itself is the object for our force diagram. The origin of coordinates is placed at the earth's center. The force **F** slows the earth's rotation by exerting a torque $\tau = rF$ on the earth, where r is the earth's radius. Substituting for the torque in Eq. (7.12), $\Sigma \tau = I\alpha$, and rearranging, we find that

$$F = \frac{I\alpha}{r}.$$

An expression for the earth's moment of inertia, assumed to be a solid sphere of uniform density, is given in Table 7.1:

$$I = \tfrac{2}{5}mr^2 = \tfrac{2}{5}(5.98 \times 10^{24} \text{ kg})(6.38 \times 10^6 \text{ m})^2$$
$$= 9.74 \times 10^{37} \text{ kg} \cdot \text{m}^2.$$

The earth's angular acceleration while slowing to a stop in a time of one year is solved using Eq. (7.7):

$$\alpha = \frac{\omega - \omega_0}{t} = \frac{0 - (2\pi \text{ rad}/24 \text{ hr})}{1 \text{ yr}}$$

$$= -\frac{(2\pi \text{ rad})}{\left(24 \text{ hr} \times \dfrac{3600 \text{ s}}{1 \text{ hr}}\right)\left(1 \text{ yr} \times \dfrac{365 \text{ days}}{1 \text{ yr}} \times \dfrac{24 \text{ hr}}{1 \text{ day}} \times \dfrac{3600 \text{ s}}{1 \text{ hr}}\right)}$$

$$= -2.3 \times 10^{-12} \text{ rad/s}^2.$$

The negative sign implies that the force and torque oppose the initial rotation.

The magnitude of the stopping force can now be found by substituting for I and α in the equation for the force:

$$F = \frac{I\alpha}{r} = \frac{(9.74 \times 10^{37} \text{ kg} \cdot \text{m}^2)(-2.3 \times 10^{-12} \text{ rad/s}^2)}{(6.38 \times 10^6 \text{ m})}$$
$$= -3.5 \times 10^{19} \text{ N}.$$

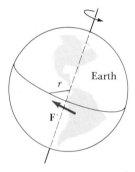

FIG. 7.7. A force slowing the earth's rotation.

(a)

(b)

FIG. 7.8. (a) A boy being chased by a bull exerts a torque on a gate to close it before the bull escapes. (b) A force diagram for the gate.

By comparison, the earth exerts a gravitational force on the moon about five times greater than the above force. ∎

EXAMPLE 7.6 A boy chased by a bull runs through a gate. The bull will escape if the gate is not closed in three seconds. The gate has a 400-kg · m² moment of inertia, and the boy pushes the gate at its open end 2.8 m from the hinges with a 100-N force that is directed perpendicular to the gate. (a) Calculate the angular acceleration of the gate. (b) Calculate the time required to close the gate if it starts at rest and must rotate through 50° before being closed.

SOLUTION (a) The situation is shown in Fig. 7.8a. A force diagram for the gate is shown in Fig. 7.8b. The origin of coordinates is located along a line at the gate's hinges. The torque caused by the boy's force is

$$\tau = lF = (2.8 \text{ m})(100 \text{ N}) = 280 \text{ N} \cdot \text{m}.$$

Substituting this and the gate's moment of inertia ($I = 400 \text{ kg} \cdot \text{m}^2$) into Eq. (7.12), and rearranging to solve for α, we find that

$$\alpha = \frac{\Sigma \tau}{I} = \frac{280 \text{ N} \cdot \text{m}}{400 \text{ kg} \cdot \text{m}^2} = \underline{0.70 \text{ rad/s}^2}.$$

By now you should be able to make the above unit conversion. Try it. Remember that radians are unitless and can be placed in an expression or removed at will. The radian is included here because α, the angular acceleration, is an angular quantity.

(b) To determine the time needed to close the gate, we use Eq. (7.9) from kinematics:

$$\theta - \theta_0 = \omega_0 t + \tfrac{1}{2}\alpha t^2,$$

where, for this problem, $\theta - \theta_0 = 50°$, $\omega_0 = 0$, and $\alpha = 0.70 \text{ rad/s}^2$. Substituting these values, and rearranging, we find that

$$t^2 = \frac{2(\theta - \theta_0)}{\alpha} = \frac{2\left(50° \times \dfrac{2\pi \text{ rad}}{360°}\right)}{0.70 \text{ rad/s}^2} = 2.49 \text{ s}^2$$

or

$$t = \underline{1.6 \text{ s}}.$$

As expected, our young hero has time to close the gate and escape. ∎

FIG. 7.9. A person's head rotates toward the floor after a faint. The rotating block serves as a model of this rotating head as it is stopped by the floor.

EXAMPLE 7.7 A person faints and falls to the floor. Just before the person's head hits the floor, it is rotating about a pivot point at the shoulders with an angular velocity $\omega_0 = 8 \text{ rad/s}$ (see Fig. 7.9). The head is stopped by the floor in 0.005 s as it hits 0.2 m from the pivot point. The moment of inertia of the rotating part of the body is 0.25 kg · m². Calculate (a) the average torque of the floor on the head, and (b) the magnitude of the average force of the floor on the head. Ignore the torque caused by the head's weight.

SOLUTION The torque caused by the force of the floor on the head is given by Eq. (7.12), $\tau = I\alpha$, where $I = 0.25 \text{ kg} \cdot \text{m}^2$. We must determine α, the head's angular acceleration using Eq. (7.7) from kinematics:

$$\alpha = \frac{\omega - \omega_0}{t} = \frac{0 - 8 \, \text{rad/s}}{0.005 \, \text{s}} = -1.6 \times 10^3 \, \text{rad/s}^2.$$

The negative sign implies that the angular velocity is decreasing. Substituting for α and I in Eq. (7.12), we find that

$$\tau = I\alpha = (0.25 \, \text{kg} \cdot \text{m}^2)(-1.6 \times 10^3 \, \text{rad/s}^2) = \underline{-400 \, \text{N} \cdot \text{m}}.$$

The negative sign implies that the torque opposes the initial motion.

(b) The moment arm l of the floor's force on the head is 0.2 m. Since the torque caused by the force is given by the expression

$$\tau = lF$$

and equals $-400 \, \text{N} \cdot \text{m}$, we find that

$$F = \frac{\tau}{l} = \frac{-400 \, \text{N} \cdot \text{m}}{0.2 \, \text{m}} = \underline{-2000 \, \text{N}}.$$

This force, equal to 450 lb, could cause a concussion. ■

7.7 Angular Momentum and Its Conservation

There are many analogies between linear motion and rotational motion. One of these analogies exists between linear momentum and what is called **angular momentum**—the tendency of a rotating object to continue its rotational motion without change. Angular momentum and its conservation principle play an essential role in describing phenomena ranging from atomic and molecular physics to the techniques used by ice skaters and ballet dancers to initiate a rapid spin or pirouette.

We can write an expression for angular momentum based on the analogous expression for linear momentum. Linear momentum p is the product of an object's mass and its linear velocity.

$$\text{linear momentum} = \mathbf{p} = m\mathbf{v}.$$

The rotational quantities analogous to mass and linear velocity are moment of inertia I and angular velocity ω, respectively. Angular momentum L can be defined in analogy to linear momentum as follows:

The **angular momentum** L of an object is the product of its moment of inertia I and its angular velocity ω:

$$L = I\omega. \tag{7.14}$$

The units of angular momentum are $\text{kg} \cdot \text{m}^2/\text{s}$.

Angular momentum is one of the important group of quantities that are conserved.* The total angular momentum in the universe never changes. To develop the principle of conservation of angular momentum, let us start with

*Conserved quantities were introduced in Section 5.3.

the rotational equivalent of Newton's second law:

$$\Sigma \tau = I\alpha.$$

Using Eq. (7.7) from kinematics to substitute for α in the above equation, we find that

$$\Sigma \tau = I\frac{\omega - \omega_0}{t}$$

or

$$(\Sigma \tau)t = I\omega - I\omega_0. \qquad (7.15)$$

The right side of this equation is the change in angular momentum of an object with moment of inertia I in a time t. The left side is the cause of this change—it is the rotational impulse caused by the torques acting on the object during that time. Equation (7.15) is exactly analogous to the impulse-momentum equation $[(\Sigma F)t = mv - mv_0]$ used in Chapter 5 to describe linear motion and impulsive collisions.

We found in Chapter 5 that if two (or more) objects are involved in a collision and if no external forces act on these objects during the collision, or if these external forces add to zero, then the linear momentum of the objects is conserved. The linear momentum remains the same before the collision as after. If one object gains linear momentum, the other object loses the same amount. A similar principle, the **conservation-of-angular-momentum principle,** applies to rotational motion:

If no torques of external origin act on a system, or if these torques add to zero, then the angular momentum of the objects in the system is constant.

We can see from Eq. (7.15) that this principle is true for a single object. If no torques act on the object, then

$$\Sigma \tau = 0,$$

and the right side of the equation must also be zero; or

$$0 = I\omega - I_0\omega_0.$$

Thus,

$$I\omega = I_0\omega_0, \qquad (7.16)$$

where I_0 and I are the object's initial and final moments of inertia, respectively, and ω_0 and ω are its initial and final angular velocities, respectively.

Let us apply this principle to two examples. In each, no torques of external origin act on a single object. When the object's moment of inertia changes, its angular velocity must also change so that $I\omega$ remains constant.

EXAMPLE 7.8 A figure skater with her arms and a leg extended, as shown in Fig. 7.10a, spins with an angular velocity of 1.0 rev/s. Her moment of inertia is $4 \text{ kg} \cdot \text{m}^2$. As she draws her arms and leg in close to the axis of rotation, as shown in Fig. 7.10b, her moment of inertia decreases to $0.8 \text{ kg} \cdot \text{m}^2$.* At what

*Remember that moment of inertia depends on mr^2. As her arms are brought close to her body, the value of r for the hands and arms decreases, as does the moment of inertia. Because moment of inertia depends on r^2, a change in r can have a dramatic effect on the value of I.

(a)

(b)

FIG. 7.10. (a) A figure skater spinning into a pirouette shown in (b).

angular velocity does she now rotate? Ignore any friction forces acting on her skates.

SOLUTION We are given that $I_0 = 4$ kg $\cdot$ m², $\omega_0 = 1.0$ rev/s, and her final moment of inertia $I = 0.8$ kg $\cdot$ m². Since no external torques act on her body, the skater's angular momentum must remain constant. Her final angular velocity can be calculated using Eq. (7.16):

$$\omega = \frac{I_0\omega_0}{I} = \frac{(4.0 \text{ kg} \cdot \text{m}^2)(1 \text{ rev/s})}{0.8 \text{ kg} \cdot \text{m}^2} = \underline{5.0 \text{ rev/s}}. \qquad \blacksquare$$

EXAMPLE 7.9 A 40-kg child stands at the edge of a circular platform that can rotate freely in a horizontal plane (Fig. 7.11a). The child and platform are initially at rest. The platform has a radius of 2 m and a moment of inertia of 500 kg $\cdot$ m². The child starts running around the edge of the platform in a counterclockwise direction and attains a speed of 3 m/s relative to the ground (Fig. 7.11b). What is the angular velocity of the platform?

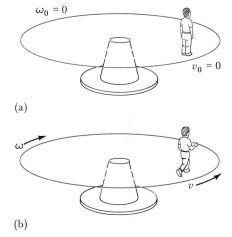

(a)

(b)

FIG. 7.11. (a) A child stands on a platform. (b) When the child runs counterclockwise, the platform must rotate clockwise to conserve angular momentum.

SOLUTION The child's counterclockwise angular momentum will be equal and opposite to the clockwise momentum of the platform, or

$$\underbrace{L_{\text{child}} - L_{\text{platform}}}_{\substack{\text{final angular} \\ \text{momentum}}} = \underbrace{0.}_{\substack{\text{initial} \\ \text{angular} \\ \text{momentum}}} \qquad \textbf{(7.17)}$$

The final angular momentum of the child will be

$$L_{\text{child}} = I\omega = (mr^2)\left(\frac{v}{r}\right) = mrv,$$

where r is the distance of the child from the center about which he or she is running, m is the child's mass, and v is the child's speed in the circular path. The final angular momentum of the platform will be

$$L_{\text{platform}} = I_p\omega_p,$$

where I_p is the moment of inertia of the platform and ω_p is its unknown angular velocity. Substituting these expressions for L into Eq. (7.17) and rearranging, we find that

$$\omega_p = \frac{mrv}{I_p} = \frac{(40 \text{ kg})(2 \text{ m})(3 \text{ m/s})}{(500 \text{ kg} \cdot \text{m}^2)}$$
$$= \underline{0.5 \text{ rad/s}}. \qquad \blacksquare$$

Before leaving this subject, we should point out that the expression for the child's angular momentum derived in the last example is a general expression for the angular momentum of a mass m when moving at speed v in a circle of radius r. This expression for angular momentum,

$$L = mrv, \qquad \textbf{(7.18)}$$

is very important in atomic physics when considering the motion of electrons in orbit about a nucleus.

Summary

By now you may feel that a bewildering variety of equations must be used to work problems involving rotational acceleration. Basically, everything we have done in this chapter is analogous to what was done when considering linear motion in Chapters 4 and 5. You should find that this correlation between linear motion and rotational motion is helpful in remembering and applying rotational equations. The important equations and concepts used in this chapter and the equivalent equations used to describe linear motion are summarized at the right.

In addition to these equations, we have also used several kinematical equations. Four of these are presented on p. 118; they relate the angular quantities θ, ω, and α at different times. Four other equations on the same page relate the quantities s, v, and a_t—quantities that describe motion along the arc of the circular path followed by the rotating object. These tangential quantities are related to their angular analog by multiplying by r; $a_t = r\alpha$, $v = r\omega$, and $s = r\theta$.

Quantity or relationship	Linear motion	Angular motion
Velocity	$v = \Delta x/\Delta t$	$\omega = \Delta\theta/\Delta t$ (7.1)
Acceleration	$a = \Delta v/\Delta t$	$\alpha = \Delta\omega/\Delta t$
Newton's second law	$\Sigma F = ma$	$\Sigma \tau = I\alpha$ (7.12)
Cause of acceleration	F	τ The torque of a force is the product of the force and its moment arm ($\tau = lF$).
Inertia of an object	m	I The rotational moment of inertia of a single object is the product of its mass and the square of its distance from the axis of rotation ($I = mr^2$). For an extended body, we divide it into many small masses and add the moments due to each of these masses ($I = m_1 r_1^2 + m_2 r_2^2 + \cdots$). The larger the value of I, the harder it is to change its state of rotation.
Momentum of an object	$p = mv$	$L = I\omega$ (7.14)
For an isolated system, linear and angular momenta are conserved.	$\Delta p_1 + \Delta p_2 + \cdots = 0$	$\Delta L_1 + \Delta L_2 + \cdots = 0$

Questions

1. Compare the force needed to open a heavy door when the door is pushed in the middle to that needed when the door is pushed at the side away from the hinges. Explain.

2. A child is given a push on a swing. Why doesn't the child continue in a vertical loop over the top of the swing? Explain in terms of torques.

3. *Estimate* the moment of inertia of a quarter about (a) an axis perpendicular to and through the center of its face and (b) an axis passing sideways through the center of the coin.

4. *Estimate* the moment of inertia of a baseball bat about an axis passing perpendicular to and through its handle.

5. *Estimate* the average force of a car door when closing on a person's fingers. Indicate all numbers you use (the moment of inertia of the door, the angular velocity of the door just as it reaches the fingers, the stopping time, and so on).

6. When riding a ten-speed bicycle up a hill, a cyclist shifts the chain to a larger-diameter gear attached to the back wheel. Why is this gear preferred rather than a smaller gear?

7. If all the people on the earth took elevators to the tops of high buildings in their communities, how would the length of the day be affected? Explain.

8. The Mississippi River carries sediment from higher latitudes toward the equator. How does this affect the length of the day? Explain.

9. A spinning raw egg, if stopped momentarily and then released by the fingers, will resume spinning. Explain. Will this happen with a hard-boiled egg? Explain.

10. When we start an automobile engine, the fan and many other parts of the car start to rotate. If angular momentum is conserved, where does this angular momentum originate?

11. If you could accelerate from zero to 8 mph anywhere on the earth's surface, where and in which direction would you run in order to increase the length of the day most?

12. Compare the magnitude of the earth's angular momentum about its axis to that of the moon about the earth. The tides exert a torque on the earth and moon so that eventually they will both rotate with the same period. The object with the greatest angular momentum will experience the smallest percentage change in the period of rotation. Will the earth's solar day increase more than the moon's period of rotation decreases? Explain.

Problems

7.1, 7.2, and 7.3 Angular Acceleration, Tangential Acceleration, and Kinematics

1. (a) When a record player is stopped, the turntable, initially rotating at 33.3 rpm, slows down and stops in 60 s. Calculate the angular acceleration of the turntable. (b) A grinding wheel, when turned on, accelerates with an angular acceleration of 190 rad/s². How much time is required to attain a final angular velocity of 1800 rpm?

2. (a) A centrifuge accelerates from rest with an angular acceleration of 30 rad/s². What is the angular velocity of the centrifuge after 30 s? (b) The angular velocity of an automobile engine accelerates from 800 rpm to 3000 rpm in 15 s. Calculate the engine's angular acceleration.

3. During a tennis serve, the top of the racquet, which is 1.5 m from its pivot point, accelerates from rest to a speed of 20 m/s in a time of 0.1 s. Calculate the average tangential acceleration of the top of the racquet and also its angular acceleration.

4. A record player is switched from 33 rpm to 45 rpm. The turntable changes speed in 2 s. Calculate its average angular acceleration and the average tangential acceleration of a point on the turntable that is 15 cm from the axis of rotation.

5. An ant clings to the outside edge of the tire of an exercise bicycle. The ant's speed increases from zero to 10 m/s in 2.5 s. (a) Calculate the average tangential acceleration of the ant. (b) If the wheel's angular acceleration is 13 rad/s², how far is the ant from the axis of rotation?

6. When a person punts a football, the foot accelerates from a speed of zero to a final speed of 11 m/s just before it hits the football. This acceleration occurs in a time of 0.3 s. (a) Calculate the average tangential acceleration of the foot. (b) If the foot rotates at the end of a leg that is 1.0 m long, what is the angular velocity of the leg just before hitting the ball? (c) What is the average angular acceleration of the leg during the 0.3 s?

7. A car tire spinning in the mud starts from rest and accelerates for 1.5 s with an angular acceleration of 40 rad/s². (a) Calculate the angular velocity of the tire at the end of the 1.5 s. (b) Calculate the tangential speed of a point on the outside of the tire 0.3 m from the axis of rotation at the end of the 1.5 s.

8. *Estimate* the angular acceleration of a car tire as you leave an intersection after a red light has turned green. Discuss the choice of numbers used in your estimate.

9. A grinding wheel takes 10 s to accelerate from rest to a final angular velocity of 360 rpm. Calculate the wheel's angular acceleration in units of rad/s².

10. The turntable on a high-fidelity system starts at rest and reaches its final angular velocity of 33 rpm in a time of 1.2 s. Calculate (a) the turntable's angular acceleration and (b) the angle through which it rotates during this time.

11. According to the speedometer on an exercise bicycle, a point on the rim of its 0.30-m radius wheel travels in a circle a distance of 200 m while its speed increases from 0 to 10 m/s. Calculate (a) the average tangential acceleration of the rim and (b) the wheel's angular acceleration.

12. Calculate what the earth's angular acceleration would be in rad/s² if the length of a day increased from 24 hr to 48 hr during the next 100 years.

13. A person on a Ferris wheel starts at rest and reaches a full speed of 5 m/s in one turn. The wheel's radius is 12 m. Calculate (a) the person's tangential acceleration and (b) the wheel's angular acceleration.

14. A father pushes the edge of a disc-shaped platform of 2.0-m radius on which his children sit. The platform starts at rest and experiences an angular acceleration of 0.3 rad/s². Calculate the distance the father must run while pushing the platform in order to increase its speed at the edge to 7 m/s.

7.4 Torque and Angular Acceleration

15. A 1.2-kg ball is attached at one end of a 0.8-m stick of negligible mass. The other end of the stick has a pin through it that allows the stick to rotate freely in a vertical circle. (a) Calculate the gravitational torque on the ball when the stick is held in a horizontal position. (b) Calculate the stick's angular acceleration just as the ball is released.

16. A sledgehammer with a 4-kg head and a 0.7-m handle of negligible mass is accelerated in an arc from rest to a final angular velocity of 6 rad/s in a time of 0.3 s. Calculate (a) the angular acceleration of the hammerhead and (b) the net torque causing that acceleration.

17. A machine used to test golf balls accelerates a golf club with a 0.24-kg head at the end of a 0.5-m handle of negligible mass. The head swings in an arc, starting from rest and attaining a final speed of 40 m/s in a time of 0.2 s. Calculate (a) the average angular acceleration of the golf club and (b) the net torque required to produce this acceleration.

18. A 0.30-kg ball moves in a horizontal circle of radius 0.90 m at the end of a stick of negligible mass. A torque of 0.036 N · m is needed to keep the stick and ball rotating at constant angular velocity. Calculate the magnitude of the air-resistance force opposing the ball's motion.

7.5 Moment of Inertia

19. Calculate the moment of inertia of the four balls shown in Fig. 7.12 about an axis perpendicular to the paper and passing through point A. The mass of each ball is 2 kg. Ignore the mass of the rods to which the balls are attached.

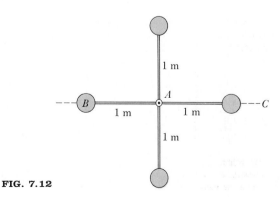

FIG. 7.12

20. Repeat Problem 19 for an axis perpendicular to the paper through point B.

21. Repeat Problem 19 for axis BC, which passes through two of the balls.

22. *Estimate* the moment of inertia of a 200-g meter stick about an axis at its end and perpendicular to the meter stick. To do this divide the meter stick into five equal sections and add the contributions to the moments of inertia caused by each section.

23. A 1-m stick can rotate in a horizontal circle about a pin in one end of the stick. A 1-kg mass sits on top of the pin and another at the end of the stick, 1 m from the pivot point. If you now put both masses together, where on the stick should they be placed so that the moment of inertia is the same as in the original situation?

7.6 Solving Problems

24. A grinding stone, whose moment of inertia is $0.1 \, kg \cdot m^2$, is rotating at 1800 rpm. What constant frictional force must be applied to its edge to bring it to rest in 60 s? The radius of the stone is 0.15 m.

25. A high-fidelity turntable is rotating with an angular velocity of 33.3 rpm when it is turned off. The turntable, whose moment of inertia is $10^{-2} \, kg \cdot m^2$, stops in 50 s. Calculate the resistive torque that slows the turntable.

26. A person applies a 120-N force while pulling a rope wrapped around a flywheel. The flywheel is a disc whose mass is 20 kg and radius is 0.2 m. Calculate the angular acceleration of the flywheel.

27. As the tides slide across the earth, they exert a friction force that opposes the earth's rotation. Because of this, the time required for one earth rotation increases by 0.0016 s every 100 years. (a) Calculate the angular deceleration of the earth. (b) With the assumption that the earth is a uniform solid sphere, what torque do the tides exert on the earth? (c) If the friction force caused by the tides were all concentrated at the equator, how large would this force be?

28. Calculate the average force needed to cause a $1 \, m \times 1 \, m$ gate of mass 30 kg and initially at rest to swing open 90° in a time of 1.8 s. The force is applied 1 m from the hinges and perpendicular to the door.

29. A merry-to-round, considered as a uniform disc of radius 5 m and mass 25,000 kg, must have a new motor that will accelerate it from rest to 1.5 rad/s in 8 s. Calculate the torque that the motor must provide to the merry-go-round.

30. A rope wrapped around a flywheel is attached to a 3-kg pail hanging at the end. The flywheel has a radius of 0.1 m and a mass of 8 kg. (a) Make a force diagram for the pail and another for the flywheel. (b) Use Newton's second law and the pail force diagram to obtain an expression relating the rope tension and the acceleration of the pail. (c) Use Eq. (7.12) and the flywheel force diagram to obtain an expression relating the rope tension to the angular acceleration of the flywheel. (d) The pail acceleration a and flywheel angular acceleration α are related by the equation $a = r\alpha$, where r is the flywheel radius. Combine this with your equations in parts (b) and (c) to obtain the value of the rope tension and pail acceleration.

7.7 Angular Momentum and Its Conservation

31. (a) Calculate the angular momentum of a 1500-kg race car traveling with a speed of 40 m/s around a circular racetrack whose radius is 800 m. (b) An electron rotating about a proton in a hydrogen atom has an angular momentum of $1.05 \times 10^{-34} \, kg \cdot m^2/s$. The electron's mass is $9.1 \times 10^{-31} \, kg$, and it moves in a circular path of $0.53 \times 10^{-10} \, m$ radius. Calculate the average speed of the electron.

32. (a) Calculate the angular momentum of a 10-kg disc-shaped flywheel of radius 9 cm when rotating with an angular velocity of 320 rad/s. (b) With what angular velocity must a 10-kg solid sphere of 9-cm radius rotate to have the same angular momentum as the flywheel?

33. A net torque of $5.0 \times 10^{-3} \, N \cdot m$ retards the motion of a grinding wheel whose moment of inertia is $6 \times 10^{-3} \, kg \cdot m^2$. How much time is needed to slow the wheel's angular velocity from 360 rpm to 120 rpm?

34. A 0.1-kg ball swings at the end of a string whose length is 0.8 m. (a) Calculate the angular momentum of the ball if it has a tangential speed of 5 m/s. (b) If the string winds around a stick at the axis of rotation, what must the ball's speed be when the string is reduced to a length of 0.4 m?

35. A physics professor sits on a chair on a circular platform of negligible mass that rotates freely on an air table. The professor initially extends his arms while holding weights in each hand. A student gives the professor a push so that he has an angular velocity of 3 rad/s. The professor pulls the weights in towards his body and holds them above his head so that his moment of inertia is $5.0 \, kg \cdot m^2$. He now rotates with an angular velocity of 8.0 rad/s. What was his initial moment of inertia with arms extended? Explain.

36. A turntable, whose moment of inertia is $1.0 \times 10^{-3} \, kg \cdot m^2$, rotates on a frictionless aircushion with an angular velocity of 2.0 rev/s. A 1.0-g beetle falls to the center of the turntable and then walks 0.15 m to its edge. Calculate the angular velocity of the turntable with the beetle on the edge.

37. A large disc-shaped platform can rotate about a center axle. A 60-kg woman stands at rest at the edge of the platform 4.0 m from its center. The platform is also at rest. She starts running around the edge of the platform and attains a speed relative to the ground of 2.0 m/s. Calculate the angular velocity of the platform. Its mass is 120 kg.

38. At present, the motion of people on the earth is fairly random; the number moving east equals the number moving west. Assume that we could get all the earth's inhabitants lined up on the land at the equator. If they all ran to the west at 3.0 m/s, *estimate* the change in angular velocity of the earth and the change in the length of the earth's day. (*Note:* You have not been given all the information you need. Justify any other numbers used in your calculations.)

Energy and Its Transformations

II

In the first part of the book we calculated the changing motion of an object using Newton's second law and kinematics. This approach becomes quite difficult when we try to analyze the motion of more than two objects that exert forces on each other.

In this part of the book we develop a new technique for solving problems— a technique based on energy. This energy approach allows us to calculate a future property of a group of objects if we know their present condition and how other properties of the objects are changing.

In Chapter 8 we will be primarily interested in developing the energy concept as a problem-solving tool. As we progress further into the subject, we will consider many interesting, everyday examples that depend upon energy transformations. These include the energy changes in our bodies, in the earth and its climate, and in all aspects of nature.

Work and Energy

Concepts related to energy are perhaps the most important in all of science. Energy appears in many different forms: the energy an object has because of its motion or its elevation above the earth, the energy of a compressed or stretched spring, the energy of positive and negative electric charges when separated, the thermal energy of hot substances, the mass energy of nuclei on the sun, the chemical energy of an octane molecule, and so forth. Every living and dynamic process in nature involves the conversion of energy from one form to another. Energy transformations are important for powering the machines built and used by humans. Energy transformations cause ocean currents, rain, wind, temperature variations, and all other phenomena related to the earth's climate. The ability to capture, transform, and store various forms of energy is the essential feature of living organisms. Energy has been called the *vis viva*—the force of life. It is an important subject that merits careful attention.

In this chapter we will learn how to calculate changes of several different forms of energy. We will also consider one way in which energy is transferred between different objects—that is, by doing work. Perhaps the most important topic of this chapter is the development of the idea of the conservation-of-energy principle, one of the basic laws of science. Let us start by analyzing one way in which energy is transferred between different objects.

8.1 Work

Our development of energy principles in this chapter evolves from the definition of a quantity called work. We will find that this definition of work differs somewhat from everyday expressions involving the word *work*—"I'm going to work," or "I have to work a problem."

In physics, work is done when a force acts on an object as the object moves from one place to another. Because of the work, energy is transferred from the object that causes the force to the object on which the force acts, or vice versa. For example, consider a bowler who exerts a force on a bowling ball causing it to accelerate forward. The ball gains energy of motion as its speed increases. The bowler, in turn, loses a little chemical energy in his or her muscles while pushing the ball. The energy transfer from bowler to ball has occurred because of the work done by the bowler on the ball.

FIG. 8.1. The work done by the rope pulling the block is $W = F \Delta r \cos \theta$, where θ is the angle between **F** and Δ**r.**

We see that work is defined in terms of the force exerted on an object as it undergoes a displacement:

If a constant force **F** acts on an object while the object undergoes a dislacement Δ**r** (Fig. 8.1), the **work** done by the force on the object during the displacement is

$$W = F \Delta r \cos \theta, \tag{8.1}$$

where θ is the angle ($180°$ or less) between the direction of **F** and the direction of Δ**r.** *

The F and Δr that appear in Eq. (8.1) are both scalar quantities—the magnitudes of the force and displacement vectors, respectively. Since work is defined as the product of three scalars (F, Δr, and $\cos \theta$), work is also a scalar quantity.

We also see from our definition of work that an object must experience a displacement before any work is done on it. According to the physics definition of work, holding an 80-lb bag of cement all day is no work, since Δr, the cement's displacement while being held, is zero.

Note also that the work done on an object by a force depends very much on the direction of the force relative to the object's displacement. The angle θ appearing in the definition of work (see Fig. 8.1) is the *angle between the vectors* **F** *and* Δ**r** and is *not* the angle that either **F** or Δ**r** makes with some axis of a set of coordinates. Consider carefully the angle θ used in the next two examples.

(a)

(b)

FIG. 8.2. (a) A carpenter pulls a load of wood. (b) The rope tension force makes an angle of $37°$ with the displacement. $W = T \Delta r \cos 37°$.

EXAMPLE 8.1 A carpenter pulls a cart load of lumber 10 m across a floor (Fig. 8.2a). The tension in the rope is 200 N and is directed $37°$ above the horizontal. Calculate the work done by the carpenter in pulling the lumber.

SOLUTION The tension force **T** and displacement Δ**r** make an angle of $37°$ with respect to each other (Fig. 8.2b). Hence, the work done by the rope tension force in moving the load of lumber is

$$W = T \Delta r \cos 37° = (200 \text{ N})(10 \text{ m})(0.8)$$
$$= \underline{1600 \text{ N} \cdot \text{m}}. \qquad \blacksquare$$

EXAMPLE 8.2 A woman jumps from a high wall and lands in soft sand (Fig. 8.3a). The sand exerts an average upward force of 50,000 N while stopping her. The woman stops after sinking 5 cm into the sand. How much work is done by the force of the sand on her?

SOLUTION The force of the sand on the woman points upward, whereas the woman's displacement, while stopping, is 5 cm downward (Fig. 8.3b); hence there is a $180°$ angle between the direction of the force and the direction of the displacement. The work done by the force of the sand on the woman is

$$W = F \Delta r \cos 180° = (50,000 \text{ N})(0.05 \text{ m})(-1) = \underline{-2500 \text{ N} \cdot \text{m}}. \qquad \blacksquare$$

*This definition of W is sometimes written in an abbreviated form as $W = \mathbf{F} \cdot \Delta \mathbf{r}$, and is called the "dot or scalar product" of vectors **F** and Δ**r.** In general, the scalar product of any two vectors, **A** and **B,** is

$$\mathbf{A} \cdot \mathbf{B} \equiv AB \cos \theta,$$

where θ is the angle ($180°$ or less) between the directions of **A** and **B.**

If negative work is done on an object, as in the last example, the object's energy decreases. For example, the woman discussed in Example 8.2 lost a form of energy called *kinetic energy* (energy of motion) because the sand's force opposed her motion. The sand did a negative amount of work on the woman and caused her to stop moving.

Units of Work

We see from the previous examples that the units of work are the product of the units of force (for example, newtons or pounds) and the units of displacement (for example, meters or feet). In fact these are the units of all forms of energy. Thus, work and energy units are newton-meters or pound-feet.

Because work and energy are such important quantities, their unit has been given a special name in the SI metric system; it is called the **joule** (J) in honor of the English physicist, James Joule (1818–1889):

$$1 \text{ joule} \equiv 1 \text{ newton-meter} \qquad \textbf{(8.2)}$$

or

$$1 \text{ J} = 1 \text{ N} \cdot \text{m}.$$

Joule, a physicist whose scientific interests were financially supported by a family brewery, spent his honeymoon near a waterfall. During his honeymoon, Joule measured with great accuracy the temperature of the water at the top of the falls and at the bottom after the water had dashed onto the rocks below. This and other experiments eventually led to the very important conservation-of-energy theorem discussed later in the chapter. Such devotion to physics surely deserves a unit named in Joule's honor.

8.2 Energy

Whenever work is done on an object, it gains energy. If 100 J of work is done, the object gains 100 J of energy. The energy may be stored by the object in a variety of forms:

Kinetic energy—the energy an object has because it is moving. For example, a moving hammerhead has more energy than one at rest.

Gravitational potential energy—the energy an object has because of its vertical separation from the earth. For instance, a pile driver high above a post has more energy to drive a post into the ground than one at a lower elevation.

Elastic potential energy—the energy stored in a stretched or compressed elastic material such as a spring. For example, the spring on the handle of a pinball machine has more energy when compressed than when in a relaxed position.

Thermal energy—the kinetic energy of atoms and molecules due to their random motion. For instance, a glass of hot water has more thermal energy than a glass of cold water. The atoms and molecules in the hot water move about faster. Their motion is random because the atoms and molecules experience frequent collisions and continually change direction.

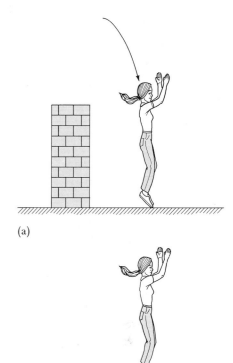

(a)

(b)

FIG. 8.3. (a) A woman jumps from a wall and is stopped while landing in sand. (b) The force of the sand on the woman is opposite her displacement as she sinks in the sand. Thus, negative work is done on her by the sand.

These forms of energy are introduced in this chapter. Other forms of energy, including electrical energy and nuclear energy, will be discussed in later chapters. We will derive equations that can be used to calculate the change in each form of energy under a variety of conditions. These equations are very general and will be used extensively throughout the book. The derivations rely on the fact that work done on an object causes its energy to increase (positive work) or decrease (negative work).

8.3 Kinetic Energy (*KE*)

Translational Kinetic Energy

Kinetic energy is the energy an object has because it is moving. A moving car has more energy than one sitting at rest. To derive an equation for calculating the change in kinetic energy of an object whose speed changes, we combine the equation for work with Newton's second law and kinematics.

Consider a small meteoroid floating in outer space (Fig. 8.4). If we apply a force to the meteoroid, it accelerates, increasing its kinetic energy. The work done by the force **F** that acts on the meteoroid for a distance $x - x_0$ is

$$W = F \Delta r \cos \theta = F(x - x_0) \cos 0° = F(x - x_0). \tag{8.3}$$

The meteoroid's acceleration is related by Newton's second law to the force pushing it:

$$F = ma, \tag{4.1}$$

where m is the meteoroid's mass. The displacement is related to the acceleration and change of speed by Eq. (3.10) from kinematics:

$$x - x_0 = \frac{v^2 - v_0^2}{2a}, \tag{3.10}$$

where v_0 is the initial speed of the meteoroid and v is its final speed. Substituting Eqs. (4.1) and (3.10) into Eq. (8.3), we find that

$$W = F(x - x_0) = (ma)\left(\frac{v^2 - v_0^2}{2a}\right) = \frac{1}{2}mv^2 - \frac{1}{2}mv_0^2.$$

In this situation, all the work done by the force has been converted to kinetic energy of the meteoroid. Thus, the change in its kinetic energy must equal $\frac{1}{2}mv^2 - \frac{1}{2}mv_0^2$. This is a general expression for calculating the change in an object's kinetic energy when its speed changes.

The **kinetic energy** *KE* of a mass m moving with a speed v is $KE = \frac{1}{2}mv^2$. The change in kinetic energy when the speed of the mass changes from an initial value v_0 to a final value v is

$$\Delta KE = KE - KE_0 = \tfrac{1}{2}mv^2 - \tfrac{1}{2}mv_0^2. \tag{8.4}$$

EXAMPLE 8.3 A 0.1-kg stone is thrown from the edge of an ocean cliff with an initial speed of 20 m/s (Fig. 8.5). When it strikes the water below, it is traveling at 45 m/s. What is the change in kinetic energy of the stone?

FIG. 8.4. A force **F** acting on an object floating in outer space causes it to accelerate.

SOLUTION   We are given that $m = 0.1$ kg, $v_0 = 20$ m/s, and $v = 45$ m/s. Thus,

$$\Delta KE = \frac{1}{2}mv^2 - \frac{1}{2}mv_0^2 = \frac{1}{2}(0.1 \text{ kg})(45 \text{ m/s})^2 - \frac{1}{2}(0.1 \text{ kg})(20 \text{ m/s})^2$$

$$= 100 \text{ kg}\frac{\text{m}^2}{\text{s}^2} - 20 \text{ kg}\frac{\text{m}^2}{\text{s}^2} = 80 \text{ kg}\frac{\text{m}^2}{\text{s}^2} = 80\left(\text{kg}\frac{\text{m}}{\text{s}^2}\right)\text{m}$$

$$= 80 \text{ N} \cdot \text{m} = \underline{80 \text{ J}}.$$

Notice that the change in kinetic energy depends only on the object's speed and not on its direction of motion. ◼

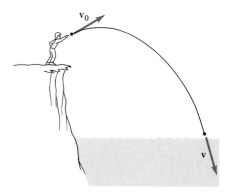

FIG. 8.5. The change in kinetic energy of the rock depends on the initial and final speed of the rock but not on its direction.

Rotational Kinetic Energy

The expression just derived for kinetic energy applies primarily to objects moving from one place to another—that is, to translational motion. An analogous equation can be used to calculate the kinetic energy of a rotating object such as a yo-yo or a potter's wheel.

Recall from Chapter 7 that the rotational equivalent of mass is moment of inertia I, and that the rotational equivalent of speed is angular velocity ω. If we replace the m in the equation for translational kinetic energy ($KE = \frac{1}{2}mv^2$) by I, and the v by ω, we have an expression for an object's rotational kinetic energy—$KE_r = \frac{1}{2}I\omega^2$:

If the angular velocity of a rotating object with moment of inertia I changes from an initial value ω_0 to a final value ω, then its change in **rotational kinetic energy** is

$$\Delta KE_r = \tfrac{1}{2}I\omega^2 - \tfrac{1}{2}I\omega_0^2. \tag{8.5}$$

A flywheel is a device built to store energy by its rotational motion. Usually it consists of a heavy, disc-shaped mass that spins about an axis through its center. By spinning faster, a flywheel stores more energy; when spinning slower, it stores less energy. In the past flywheels have been used primarily to reduce variations in energy provided by intermittent sources. For example, the engine in a conventional car runs on energy provided by a series of small explosions in its cylinders. A heavy flywheel rotating on the main shaft smooths out the bursts of energy released by the explosions. A potter's wheel serves the same purpose; the flywheel stores and smooths the irregular pedaling energy supplied by the potter.

In recent years the flywheel has drawn considerable interest as a mechanism to store energy that is normally wasted. Consider the device shown schematically in Fig. 8.6. A slightly more complicated device, tested on subway cars in New York City, reduced energy consumption by 20 to 30 percent. It works as follows. As the moving subway car approaches a station, the traction wheel rolls against the track, causing a set of gears to turn a motor and thereby increase the angular velocity of a flywheel. The traction wheel rolling against the track slows the car, reducing its translational kinetic energy. However, instead of being wasted as heat generated in the brake drums, the energy is used to increase the flywheel's angular velocity and rotational kinetic energy.

FIG. 8.6. A flywheel system stores translational kinetic energy lost by a subway car as it comes to a stop.

When the train is ready to leave the station, the motor system is reversed. The flywheel now turns the motor, causing it to drive the traction wheels, which cause the train to move forward. As the flywheel loses rotational energy, the train's speed and translational energy increase.

The U.S. Department of Energy supports several other interesting flywheel projects. They include a battery-powered vehicle for the postal service that has a flywheel braking system; a flywheel to store energy captured by wind turbines; and large flywheels to store energy from power plants during periods of low demand for use later during periods of high demand—a process called load-leveling.

EXAMPLE 8.4 The Department of Energy and the U.S. Post Office have contracted with industry for the development of a flywheel regenerative braking system for a 900-kg battery-powered jeep. The flywheel has a 0.1-kg · m² moment of inertia and can store a maximum of 3.6×10^5 J of energy. (a) Calculate the maximum angular velocity of the flywheel. (b) If all of the flywheel's rotational kinetic energy can be converted to translational kinetic energy, calculate the maximum speed the jeep can attain from a stop.

SOLUTION (a) In the example we are given that $I = 0.1$ kg · m² and $KE_r = 3.6 \times 10^5$ J. Rearranging the equation for an object's rotational kinetic energy ($KE_r = \frac{1}{2}I\omega^2$), we find the flywheel's angular velocity when storing this energy:

$$\omega^2 = \frac{2KE_r}{I} = \frac{2(3.6 \times 10^5 \text{ J})}{0.1 \text{ kg} \cdot \text{m}^2} = 7.2 \times 10^5 \text{ s}^{-2}$$

or

$$\omega = \underline{2.7 \times 10^3 \text{ rad/s}}.$$

This equals 430 revolutions per second! (b) If all of the rotational kinetic energy of the flywheel is converted to the jeep's translational kinetic energy, then $KE = \frac{1}{2}mv^2 = 3.6 \times 10^5$ J, where $m = 900$ kg is the jeep's mass. Thus,

$$v^2 = \frac{2KE}{m} = \frac{2(3.6 \times 10^5 \text{ J})}{900 \text{ kg}} = 800 \text{ m}^2/\text{s}^2$$

or

$$v = \underline{28 \text{ m/s}}.$$

This equals 63 mph. We see that a flywheel can store considerable energy. ∎

8.4 Gravitational Potential Energy (PE_g)

Gravitational potential energy is a form of energy an object has because of its vertical separation from the earth. Consider the pile driver shown in Fig. 8.7. When high above the earth, as in Fig. 8.7b, the pile driver has the potential to drive a post into the ground. At a lower elevation, as in Fig. 8.7a, the pile driver has less potential for driving a post into the ground. The only difference in the two situations is the vertical separation of the pile driver from the earth. When separated farther, it has more gravitational potential energy.

(a)   (b)

FIG. 8.7. A pile driver has more gravitational potential energy when higher above the earth as shown in (b) rather than as in (a).

We can calculate the extra gravitational potential energy of the pile driver in Fig. 8.7b compared to that in (a) by calculating the work required to lift the pile driver from position y_0 to position y. We will lift it in such a way that there are no changes in any other form of energy. To do this, we lift the pile driver with a force **F** that just balances the downward force of gravity (Fig. 8.8). This will assure that the pile driver's kinetic energy does not increase.* The work done by the cable while lifting the pile driver is

$$W = F(y - y_0)\cos 0°.$$

Since the average force of the cable on the pile driver is equal in magnitude to its weight—that is, $F = w = mg$, the work done by F is

$$W = F(y - y_0)\cos 0° = mg(y - y_0).$$

Because all the work has been used to increase the pile driver's gravitational potential energy, we find that $W = \Delta PE_g = mg(y - y_0)$. This equation can be used to calculate the change in gravitational potential energy of any mass whose vertical position relative to the earth changes from an initial position y_0 to a final position y.

If the vertical separation of a mass m from the earth is changed from an initial position y_0 to a final position y, then there is a change in **gravitational potential energy** ΔPE_g given by

$$\Delta PE_g = mg(y - y_0) = w(y - y_0). \tag{8.6}$$

Two points about Eq. (8.6) need emphasizing. (1) Gravitational potential energy depends only on the vertical separation (the height) of a mass from the earth, not on any horizontal change in position. If a mass m is pulled a distance d up a plane inclined at an angle θ, as in Fig. 8.9, the change in the block's gravitational potential energy is $mg(d \sin \theta)$, where $d \sin \theta$ is the additional vertical separation of the mass and the earth. The change in gravitational potential energy is not mgd. (2) Changes in gravitational potential energy depend only on the initial and final heights of the mass, not on the path taken to change heights. In Fig. 8.10, the net change in gravitational potential energy is $mg(y - y_0)$. The change does not depend on the fact that the car went up and down several times between its initial and final positions.

EXAMPLE 8.5 A 400-kg roller coaster starts 48 m above the ground and is 3 m above the ground at the end of the ride (Fig. 8.10). Calculate the change in gravitational potential energy of the roller coaster.

SOLUTION For this problem, $y_0 = 48$ m and $y = 3$ m. Hence, the change in gravitational potential energy is

$$\Delta PE_g = mg(y - y_0)$$
$$= (400 \text{ kg})(9.8 \text{ m/s}^2)[(3 \text{ m}) - (48 \text{ m})] = (400 \text{ kg})(9.8 \text{ m/s}^2)(-45 \text{ m})$$
$$= -180{,}000 \text{ (kg} \cdot \text{m/s}^2)\text{ m} = -180{,}000 \text{ N} \cdot \text{m} = \underline{-180{,}000 \text{ J}}.$$

At the end of the ride the roller coaster is closer to the earth, and so it has less

*To get the pile driver started moving up, we must make F just a little larger than the pile driver's weight. When we get close to position y, we can make F slightly less than its weight so that the pile driver stops. On the average, the magnitude of F just equals the pile driver's weight.

FIG. 8.8. The work done in lifting the pile driver from y_0 to y by the force **F** is $W = F(y - y_0)$.

FIG. 8.9. The change in gravitational potential energy when a mass is moved a distance d up the incline is $\Delta PE_g = mgd \sin \theta$ and not mgd.

FIG. 8.10. The change in gravitational potential energy depends on the initial and final vertical positions of an object and not on the path followed in moving between those positions.

FIG. 8.11. A compressed spring has elastic potential energy.

gravitational potential energy. This energy change is not affected by the fact that the roller coaster went up and down several times on its trip. It depends only on its initial and final positions and on its mass. ■

8.5 Elastic Potential Energy (PE_s)

Gravitational potential energy is caused by the relative position of two masses, the mass of the earth and the mass of some object near its surface. Potential energy can also be stored in an object because of its shape. In Fig. 8.11 the energy in a compressed spring can be used to shoot a ball from a cannon. A relaxed spring does not have this potential. The elastic potential energy of a spring depends on its shape—how much it is compressed or expanded.

To find the energy stored in a compressed spring, we calculate the work required to compress it. In Fig. 8.12 a force **F** slowly pushes on a spring to compress it from its equilibrium position at $x = 0$ to some other position x. (The equilibrium position is the place where the end of the spring rests when not compressed or stretched.) The force needed to compress a spring increases as the spring becomes more compressed. In fact, this force is directly proportional to the amount the spring is already compressed, an experimental observation made by the English scientist, Robert Hooke, three hundred years ago. This variation of force with the distance the spring is already compressed is illustrated in Fig. 8.12. When the end of the spring is at $x = 0$, as in (a), no force is needed to hold the spring. When the spring is compressed to position x', as in (b), the force needed to hold it there is $F' = kx'$. If compressed farther to position x, as in (c), the force needed to hold the spring in place is $F = kx$. The proportionality constant k (called the *force constant*) depends on how stiff the spring is. A stiff spring has a large value of k. A spring that is easily stretched or compressed has a small value of k.

Since the compression force is zero when $x = 0$ and is $+kx$ with a displacement from equilibrium of x, the average force F_{av} needed to compress the spring from position zero to position x is

$$F_{av} = \frac{k0 + kx}{2} = \frac{1}{2}kx.$$

The displacement of the spring is in the same direction as F_{av} and has a magnitude.

$$\Delta x = (x - 0).$$

The work done in compressing the spring is

$$W = F_{av}\,\Delta x = (\tfrac{1}{2}kx)(x - 0) = \tfrac{1}{2}kx^2.$$

This work causes the elastic potential energy of the spring to increase. Equation (8.7) below summarizes the calculation of elastic potential energy change for springs or for other elastic objects.

The **elastic potential energy** of a spring with force constant k that is stretched or compressed a distance x from its equilibrium position is $PE_s = \tfrac{1}{2}kx^2$. The

Equilibrium
$x = 0$

(a)

x'

$F' = kx'$

(b)

x

$F = kx$

(c)

FIG. 8.12. A force **F** is used to compress a spring. In (a) no force is required to hold the spring at its equilibrium position. In (b) and (c) progressively larger forces are needed to hold the spring or to slowly compress it farther. The magnitude of the force F varies with the compression distance x according to the equation $F = +kx$; k is the force constant of the spring.

change in elastic energy when the spring is stretched or compressed from an initial position x_0 to a final position x is

$$\Delta PE_s = PE_s - PE_{s_0} = \tfrac{1}{2}kx^2 - \tfrac{1}{2}kx_0^2. \tag{8.7}$$

The equilibrium position of the spring is assumed to be at $x = 0$.

Equation (8.7) applies equally well to springs that are compressed and to springs that are stretched from their equilibrium position. The equation has general validity because most materials such as bones, metal rods, wire, string, and the membranes of blood vessels show an elastic tendency described nicely by Eq. (8.7). The subscript "*s*" is used in the symbol for elastic potential energy because of the similarity in behavior of elastic materials and springs. Also, if we used the subscript "el" for elastic rather than "*s*" for spring, confusion might occur later when we discuss electrical potential energy, which also begins with "el."

EXAMPLE 8.6 The spring in a BB gun has a force constant of 1.8×10^4 N/m. When loaded, the spring is compressed a distance of 1.2 cm. (a) Calculate the change in elastic potential energy of the spring when it is released, thus returning to its equilibrium position. (b) If all the elastic potential energy is converted to the kinetic energy of a 0.36-g BB fired from the gun, what is the final speed of the BB?

SOLUTION (a) The initial and final displacements of the spring are $x_0 = 0.012$ m and $x = 0.000$ m (the spring ends at its equilibrium position). The change in elastic potential energy is calculated using Eq. (8.7):

$$\Delta PE_s = \tfrac{1}{2}kx^2 - \tfrac{1}{2}kx_0^2$$
$$= \tfrac{1}{2}(1.8 \times 10^4\,\text{N/m})(0.000\,\text{m})^2 - \tfrac{1}{2}(1.8 \times 10^4\,\text{N/m})(0.012\,\text{m})^2$$
$$= 0 - 1.3\,\text{N} \cdot \text{m} = \underline{-1.3\,\text{J}}.$$

(b) If the BB gains in kinetic energy all of the elastic energy lost by the spring, then the BB's kinetic energy as it leaves the gun is $+1.3$ J. Thus, according to Eq. (8.4), we have

$$\Delta KE = 1.3\,\text{J} = \tfrac{1}{2}mv^2 - \tfrac{1}{2}mv_0^2.$$

But since the ball's original speed v_0 is zero, we find that

$$v^2 = \frac{2(\Delta KE)}{m}$$

$$= \frac{2(1.3\,\text{J})}{0.36 \times 10^{-3}\,\text{kg}} = 7.2 \times 10^3\,\frac{\text{m}^2}{\text{s}^2}$$

or

$$v = \underline{85\,\text{m/s}}. \qquad\blacksquare$$

This last example is representative of the type of calculations we will use in the rest of the chapter to determine some unknown property of a system involved in energy change. In this example we were able to determine the increase in kinetic energy of the BB and its final speed by calculating the decrease in elastic potential energy of the spring. This procedure is based on a very important principle called the conservation-of-energy principle.

8.6 Conservation-of-Energy Principle

One of the most important and useful aspects of energy is the fact that energy is a conserved quantity. Supposedly, the amount of energy in the universe is the same now as it was millions of years ago. How can we make such a universal statement when we cannot even measure the amount of coal and oil in the earth's crust? For over one hundred years scientists have performed a variety of controlled experiments that allow them to monitor the amount of energy involved in various processes. In these experiments the form of the energy changes, but the total amount always remains constant. Since energy is conserved under a great variety of conditions on the earth, we assume that energy is conserved in the whole universe.

Occasionally experiments have challenged this faith in the energy-conservation principle. In the early 1930s, for example, some experiments were performed involving nuclear reactions. During these nuclear processes, energy seemed to disappear. A physicist, Wolfgang Pauli, proposed that a new, unseen particle carried away the missing energy. This particle, called a neutrino, was discovered twenty-three years later in 1956, and it exactly accounted for the missing energy. Pauli's faith in the energy-conservation principle had been justified.

The basic philosophy underlying conservation principles was outlined in Section 5.3 and is reviewed here with special emphasis on energy conservation. Consider Fig. 8.13, which represents the whole universe. We choose one small part, called the **system,** in which we have a special interest. The system can be a single atom, a solution of chemicals, a human body, a desert, a lake, a supernova, or whatever. Everything outside the system is called the **environment.** The **conservation-of-energy** principle says that the total energy in the system, E_{system}, plus that in the environment, $E_{\text{environment}}$, must be a constant:

$$E_{\text{system}} + E_{\text{environment}} = \text{CONSTANT}.$$

Energy can, however, be transferred between the system and the environment and can also be converted between different forms while in the system.

Energy can be transferred between the system and the environment in several different ways: as heat, Q; by the process defined earlier in the chapter called work, W; as energy carried by mass that crosses the boundary of the system; and as electromagnetic energy such as sunlight that crosses the boundary of the system. These energy-transfer processes are represented in Fig. 8.13 by arrows. If energy is transferred from the environment to the system as heat Q, by doing work W, or in any of these other ways, the change in the energy of the system (ΔE_{system}) is

$$Q + W + \cdots = \Delta E_{\text{system}}. \tag{8.8}$$

The environment loses just as much energy as the system gains. Thus, the change in energy of the environment is

$$-Q - W - \cdots = \Delta E_{\text{environment}}.$$

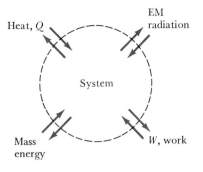

FIG. 8.13. A system, any part of the universe, and its environment have a total energy that is constant. However, energy can be transferred between the system and environment by doing work, by heat transfer, as electromagnetic (EM) radiation, and by mass transfer.

There can also be changes in the type of energy already in a system. These changes appear on the right side of Eq. (8.8). The right side, ΔE_{system}, is the sum of the changes of all the different types of energy already in a system—such as its kinetic energy, gravitational potential energy, elastic potential energy, and thermal energy. Equation (8.8) then becomes

$$\underbrace{Q + W + \cdots}_{\substack{\text{Energy cross-}\\\text{ing boundary}\\\text{into or out}\\\text{of the system}}} = \underbrace{\Delta KE + \Delta PE_g + \Delta PE_s + \Delta E_{th} + \cdots.}_{\substack{\Delta E_{system}, \text{ the sum of the}\\\text{changes of different}\\\text{forms of energy in the}\\\text{system}}} \qquad (8.9)$$

Equations (8.8) and (8.9) are two different forms of the **conservation-of-energy principle.** This equation serves as the basis of most energy calculations in science. It is used to monitor energy changes in a system in much the same way as we might monitor our own wealth. We are paid money for our work and we pay money for various services and products. This incoming and outgoing money appears on the left side of an equation such as Eq. (8.9). Our wealth increases if more comes in than goes out, and vice versa.

The different forms of wealth we already have appear on the right side of the equation—art works, stocks and bonds, real estate holdings, cars, and so forth. However, since Eq. (8.9) considers only changes in energy (or in wealth), a quantity on the right is nonzero only if it changes. If we sell a stock and buy a piece of art, our stock wealth decreases and our art wealth increases. A negative change in one form of wealth is balanced by a positive change in another form.

In the rest of this chapter we will use a restricted form of the conservation-of-energy principle to develop this energy-balancing idea. We consider situations in which energy is transferred to or from the system only by doing work. In this case Eq. (8.9) becomes

$$\underbrace{W}_{\substack{\text{Energy cross-}\\\text{ing boundary}\\\text{into or out}\\\text{of the system}}} = \underbrace{\Delta KE + \Delta PE_g + \Delta PE_s + \Delta E_{th} + \cdots.}_{\substack{\Delta E_{system}, \text{ the sum of the}\\\text{changes in different}\\\text{forms of energy in the}\\\text{system}}} \qquad (8.10)$$

and is called the **work-energy equation.**

8.7 Work-Energy Calculations

Earlier we used the definition of work to derive mathematical equations for calculating changes in different types of energy. Now we use these equations and the work-energy equation to solve for some unknown quantity involved in an energy change. A procedure for solving these problems is outlined below. You should be warned that physicists use a variety of techniques to solve energy problems, and your instructor may use a different technique than is outlined here. Each technique is correct and has advantages and disadvantages. The particular technique we use can be applied throughout the book and need not be altered later when we discuss thermodynamics or other subjects dealing with energy.

Technique for Using the Work-Energy Equation

1. Choose a system. Draw a picture of the system. The system is anything you choose; everything outside it is called the environment. It is often useful to show a picture of the system's initial state and another picture of its final state. This makes it easier to identify the types of energy changes the system undergoes.

2. Identify the external forces acting on objects in the system. External forces are forces that originate from outside the system. Then, calculate the work done on the system by these external forces as the system changes from its initial to its final state. Remember that work is a way to transfer energy between objects that exert forces on each other. When calculating work, we are interested only in work-type energy transfer across the boundary of the system. We do not need to calculate the work done by the force of one object in the system on another object in the system. We calculate only the work done on the system by forces originating in the environment. The system gains energy if the work is positive and loses energy if the work is negative.

3. Calculate the changes in the various forms of energy (ΔKE, ΔPE_g, ΔPE_s, ...) of the system as it moves from its initial state to its final state.

4. Put all of these work and energy calculations into the *work-energy equation*

$$W = \Delta KE + \Delta PE_g + \Delta PE_s + \cdots$$

and solve for the unknown.

The equations used to calculate work and the different types of energy change are summarized here:

Work:	$W = F\,\Delta r \cos \theta.$	(8.1)
Kinetic energy:	$\Delta KE = \frac{1}{2}mv^2 - \frac{1}{2}mv_0^2.$	(8.4)
	$\Delta KE_r = \frac{1}{2}I\omega^2 - \frac{1}{2}I\omega_0^2.$	(8.5)
Gravitational potential energy:	$\Delta PE_g = mg(y - y_0).$	(8.6)
Elastic potential energy:	$\Delta PE_s = \frac{1}{2}kx^2 - \frac{1}{2}kx_0^2.$	(8.7)

Perhaps the greatest difficulty in using the work-energy equation is deciding what types of energy change have occurred in a system and whether or not work is done on the system. Consider carefully two different ways of calculating the work and energy changes in the next example.

EXAMPLE 8.7 A ball of mass m is dropped from the top of a building, a height h above the ground. If the ball starts at rest, derive an expression for its speed just before hitting the ground. Ignore the resistive force of the air. Determine the speed if $h = 15$ m.

SOLUTION The problem will be worked two ways, each using a different choice of system. First we choose a system consisting of the ball and the earth (Fig. 8.14). No forces originating outside the system affect the ball's descent; thus,

$$W = 0.$$

FIG. 8.14

The ball's kinetic energy increases as its speed increases. Since it started at rest ($v_0 = 0$), its change in kinetic energy is

$$KE = \tfrac{1}{2}mv^2 - \tfrac{1}{2}mv_0^2 = \tfrac{1}{2}mv^2 - \tfrac{1}{2}m0^2 = \tfrac{1}{2}mv^2,$$

where v is the ball's unknown speed just before hitting the ground. As the ball's elevation above the earth decreases, it loses gravitational potential energy. Since its change in elevation is $y - y_0 = 0 - h = -h$, its change in gravitational potential energy is

$$\Delta PE_g = mg(y - y_0) = mg(-h) = -mgh.$$

These are the only forms of energy change that occur. There are no elastic objects present that are being stretched or compressed. No mention is made of rotational motion, so we will assume that the ball is not rotating.

Having decided on the values of W and the energy changes, we substitute these values into the work-energy equation,

$$W = \Delta KE + \Delta PE_g + \Delta PE_s,$$

and find that

$$0 = \tfrac{1}{2}mv^2 - mgh + 0. \tag{8.11}$$

Notice that the decrease in gravitational potential energy must be balanced by an equal magnitude increase in kinetic energy. This fact allows us to calculate the ball's unknown speed after falling a distance h. After cancelling the m that occurs in each term of the above equation and rearranging, we find that the ball's speed after falling 15 m is

$$v = \sqrt{2gh} = \sqrt{2(9.8 \text{ m/s}^2)(15 \text{ m})} = \underline{17 \text{ m/s}}.$$

This problem can also be solved using a completely different system that includes only the ball (Fig. 8.15). Now, the gravitational force **w** of the earth pulling on the ball is a force originating outside the system. The work done by this downward-pointing force as the ball falls in the same direction is

$$W_g = F \, \Delta r \cos \theta = wh \cos 0° = mgh,$$

where w, the ball's weight, equals mg. The changes in energy of the ball are

$$\Delta KE = \tfrac{1}{2}mv^2 - \tfrac{1}{2}m0^2 = \tfrac{1}{2}mv^2$$

and

$$\Delta PE_g = 0.$$

The gravitational potential energy change for this system equals zero, since the ball alone is the system and its position has not changed relative to any other object in the system. Changes in gravitational potential energy occur only if an object's position changes relative to the position of the earth, which is also in the system. If the earth is not in the system, its effect is included as work done by the external force of gravity.

Substituting the expression for the work done by gravity and the energy changes into the work-energy equation,

$$W = \Delta KE + \Delta PE_g + \Delta PE_s,$$

we find that

$$mgh = \tfrac{1}{2}mv^2 + 0 + 0.$$

FIG. 8.15

Before

(a)

After

(b)

FIG. 8.16. (a) Before the gun has fired, all the energy is elastic potential energy of the compressed spring. (b) The kinetic and gravitational energies increase after firing.

This is the same as Eq. (8.11), except that the gravitational term appears on the other side. Since the signs are reversed, this shift has no effect on the outcome of the equation. If we define the system carefully, the work-energy equation provides consistent results, no matter what system we choose. However, what may be called a change in energy for one system may be called work for another. ∎

EXAMPLE 8.8 A spring gun, shown in Fig. 8.16, is loaded by compressing a spring that has a force constant $k = 1.2 \times 10^4 \, \text{N/m}$. A 1-kg ball rests against the spring. When the spring is released, the ball is shot out of the end of a barrel 2 m long. If the ball leaves the end of the barrel with a speed of 10 m/s, how far was the spring compressed before the gun was fired? Ignore friction.

SOLUTION If we choose the ball, gun, and earth as the system, then there are no objects outside the system that exert forces on objects in the system; all energy changes relating to this problem occur in the system. Since no external forces affect the problem, the energy transferred to or from the system by doing work is zero:

$$W = 0.$$

The changes in kinetic energy, gravitational potential energy, and elastic potential energy are*

$$\Delta KE = \tfrac{1}{2}mv^2 - \tfrac{1}{2}mv_0^2 = \tfrac{1}{2}(1)10^2 - \tfrac{1}{2}(1)0^2 = 50,$$
$$\Delta PE_g = mg(y - y_0) = (1)(9.8)(2 \sin 30°) = 9.8,$$
$$\Delta PE_s = \tfrac{1}{2}kx^2 - \tfrac{1}{2}kx_0^2 = \tfrac{1}{2}(1.2 \times 10^5)0^2 - \tfrac{1}{2}(1.2 \times 10^4)x_0^2 = -6000x_0^2.$$

Note that the spring is initially compressed a distance x_0, the unknown we wish to determine. As the ball leaves the barrel, the spring returns to its equilibrium position ($x = 0$).

We now substitute these work and energy calculations into the work-energy equation [Eq. (8.10)],

$$W = \Delta KE + \Delta PE_g + \Delta PE_s,$$

and find that

$$0 = 50 + 9.8 - 6000x_0^2.$$

Thus,

$$x_0^2 = \frac{50 + 9.8}{6000} = 0.01$$

and

$$x_0 = \underline{0.1 \, \text{m}}. \qquad ∎$$

EXAMPLE 8.9 A cable attached to a motor pulls a 1.2×10^4-kg cable car up a hill in San Francisco (Fig. 8.17). If the car starts at rest and reaches a speed

---

*As problems become more complex, the unit conversion also becomes more complex and may occasionally cause more confusion than the physics involved. We have omitted units in this example to avoid confusion, but we will use them in the other examples in this chapter.

of 8.0 m/s in 25 m, calculate the average tension in the cable. The street is inclined at an angle of 10° with the horizontal.

SOLUTION   In this example, we choose as our system the cable car and the earth, excluding the motor pulling the car. Thus, the cable between the motor and car is the source of an external force acting on the car. The work done by the cable while pulling the car 25 m is

$$W = T \, \Delta r \cos 0° = T(25 \text{ m})(1),$$

where T is the unknown tension in the cable. The changes in the other forms of energy are

$$\Delta KE = \tfrac{1}{2}mv^2 - \tfrac{1}{2}mv_0^2 = \tfrac{1}{2}(1.2 \times 10^4 \text{ kg})(8.0 \text{ m/s})^2 = 3.8 \times 10^5 \text{ J},$$

$$\Delta PE_g = mg(y - y_0) = (1.2 \times 10^4 \text{ kg})(9.8 \text{ m/s}^2)(25 \text{ m} \sin 10°) = 5.1 \times 10^5 \text{ J} *$$

$$\Delta PE_s = 0 \quad \text{(since there are no springs in the system)}.$$

Substituting these calculations into the work-energy equation, we find that

$$T(25 \text{ m}) = 3.8 \times 10^5 \text{ J} + 5.1 \times 10^5 \text{ J} + 0$$

or

$$T = \frac{8.9 \times 10^5 \text{ J}}{25 \text{ m}} = \underline{3.6 \times 10^4 \text{ N}}. \qquad \blacksquare$$

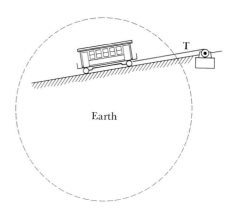

FIG. 8.17

You might try working this last example using Newton's second law and kinematics. Your answer should be the same. When using Newton's second law, though, you must use vector quantities, so the effort required is somewhat greater. The energy approach to problem solving is usually much easier. The next example would be very difficult to solve using Newton's second law because it involves the variable force of a spring. The energy approach is fairly simple.

EXAMPLE 8.10   A soapbox derby race car starts at rest at the top of a track that has a vertical drop of 15 m. The car is to be stopped at the end of the track by colliding with and compressing a spring. In order to avoid injury to the driver, the spring must be compressed 3.0 m before the car stops. What should be the force constant of the spring? The car and driver together have a mass of 113 kg. Ignore friction forces.

SOLUTION   Let us choose the car, spring, and earth as our system. There are, then, no external forces acting on the system, and hence $W = 0$. The changes in the other forms of energy are

$$\Delta KE = \tfrac{1}{2}mv^2 - \tfrac{1}{2}mv_0^2 = \tfrac{1}{2}m0^2 - \tfrac{1}{2}m0^2 = 0,†$$

$$\Delta PE_g = mg(y - y_0) = (113 \text{ kg})(9.8 \text{ m/s}^2)(0 \text{ m} - 15 \text{ m}) = -1.66 \times 10^4 \text{ J},$$

$$\Delta PE_s = \tfrac{1}{2}kx^2 - \tfrac{1}{2}kx_0^2 = \tfrac{1}{2}k(3.0 \text{ m})^2 - \tfrac{1}{2}k0^2 = \tfrac{1}{2}k(3.0 \text{ m})^2.$$

*Notice that the change in vertical elevation is 25 m sin 10° and not 25 m.

†Notice that the car starts at rest and ends at rest. Thus, its net change in kinetic energy is zero. We need not concern ourselves with the fact that the car was moving during the race; we are interested only in its initial and final values of kinetic energy.

Substituting these results into the work-energy equation, we find that

$$0 = 0 - 1.66 \times 10^4 \, \text{J} + \tfrac{1}{2}k(3.0 \, \text{m})^2.$$

The force constant of the spring should be

$$k = \frac{2(1.66 \times 10^4 \, \text{J})}{9.0 \, \text{m}^2} = 3.7 \times 10^3 \, \text{J/m}^2 = \underline{3.7 \times 10^3 \, \text{N/m}}. \qquad \blacksquare$$

EXAMPLE 8.11 Several reports have been made of parachutists whose parachutes did not open and who survived their impact with the earth by landing in a snowbank. One such event occurred in 1955 during a large airborne parachute jump exercise in Alaska. A paratrooper dropped 370 m from a C-119 airplane and landed in a snowbank. "The impact looked like a mortar round exploding in the snow. When the airmen reached the spot, they found a young paratrooper flat on his back at the bottom of a $3\frac{1}{2}$ foot crater in the snow . . . . He could talk and did not appear injured."* He did suffer several minor injuries, including a fractured clavicle and bruises, but soon rejoined his unit. Calculate the average force of the snow in stopping his fall. We will assume that the paratrooper's mass was 90 kg and that he was falling at a speed of 54 m/s, or 120 mph, as he entered the snow.

SOLUTION As our system, we include only the falling paratrooper (Fig. 8.18). Two external forces act on him: the downward force of gravity **w,** and the upward force **F** of the snow, the unknown that we wish to determine. The paratrooper dropped a distance d of 3.5 ft or 1.07 m into the snow. Thus, the work done by each force while stopping the paratrooper is

$$W_g = (w)(d)(\cos 0°) = (mg)(d)(\cos 0°)$$
$$= (90 \, \text{kg} \times 9.8 \, \text{m/s}^2)(1.07 \, \text{m})(1) = 944 \, \text{J},$$
$$W_F = (F)(d)(\cos 180°) = F(1.07 \, \text{m})(-1) = -(1.07 \, \text{m})F.$$

The changes in the energy of the paratrooper while being stopped are

$$\Delta KE = \tfrac{1}{2}mv^2 - \tfrac{1}{2}mv_0^2 = -\tfrac{1}{2}(90 \, \text{kg})(54 \, \text{m/s})^2 = -131{,}220 \, \text{J},$$
$$\Delta PE_g = 0,$$
$$\Delta PE_s = 0.$$

Notice that the gravitational potential energy change is zero since the earth is not part of the system. The earth's effect has been included by considering the work done by gravity.

Putting these work and energy calculations into the work-energy equation, we find that

$$944 \, \text{J} - (1.07 \, \text{m})F = -131{,}220 \, \text{J} + 0 + 0$$

or

$$F = \frac{131{,}220 \, \text{J} + 944 \, \text{J}}{1.07 \, \text{m}} = \underline{1.2 \times 10^5 \, \text{N}}.$$

This is an average force of about 12 tons! $\qquad \blacksquare$

FIG. 8.18. The forces acting on a paratrooper as he falls into a snowbank after his parachute fails to open.

*R. G. Snyder, "Terminal Velocity Impacts Into Snow," *Military Medicine,* **131:**1290–1298 (1966).

8.8 Thermal Energy and Friction Forces

The work-energy technique can be used to solve problems involving many other forms of energy. In this section we introduce a new type of energy and consider one way in which it is produced.

Thermal energy is the random kinetic energy of the atoms and molecules of objects in a system. Imagine two glasses of water, one whose temperature is 60°C and the other, 30°C. The glasses look identical but the warmer water has more thermal energy. Its water molecules are translating, rotating, and vibrating more than those in the cooler water. This random atomic and molecular kinetic energy, called thermal energy, is given the symbol E_{th}.

Let us consider one way in which the thermal energy of a system increases. Friction forces between objects that rub against each other cause the thermal energy of these objects to increase. This is why your hands get warmer when rubbed together.

To derive an expression for the thermal energy produced by friction, consider the block shown in Fig. 8.19a. The block is pulled along a level surface by a force **F**. We choose **F** so that it just balances the kinetic friction force F_k that opposes the block's motion. Once moving, the block continues moving at constant speed.

We choose the block and the surface on which it slides as our system. The force pulling the block originates from outside the system and does work on the block. If the block is pulled a distance Δx, the work done by **F** is

$$W_F = F\,\Delta x \cos 0° = F\,\Delta x.$$

But since we have arranged our experiment so that the magnitude of **F** equals the magnitude of the friction force F_k, we can substitute F_k for F in the above equation:

$$W_F = F_k\,\Delta x.$$

Since the block moves at constant speed, its kinetic energy does not change. The only change in energy caused by the work is a change in thermal energy. Using the work-energy equation, with a term included for thermal energy,

$$W = \Delta KE + \Delta PE_g + \Delta PE_s + \Delta E_{th},$$

we find that

$$F_k\,\Delta x = 0 + 0 + 0 + \Delta E_{th}.$$

The atoms and molecules in the block and table are now moving faster and have more random kinetic energy—that is, thermal energy—as a result of the block's rubbing against the surface on which it slides.

The **thermal energy** of an object is the random kinetic energy of the atoms and molecules of which the object is made. The change in thermal energy caused by friction of a system when a block moves a distance Δx across another surface in the system is given by

$$\Delta E_{th} = F_k\,\Delta x. \tag{8.12}$$

F_k is the magnitude of the kinetic friction force between the block and the surface on which it slides. The thermal energy is shared by the moving object and the surface.

(a)

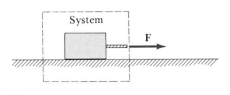

(b)

FIG. 8.19. (a) A force **F** pulls a block to the right. The force is balanced by a kinetic friction force F_k that resists the motion of the block. (b) If the system is the block and the surface on which it slides, then the only external force that does work on the block is **F**.

Note that Δx is the actual distance the block travels while in contact with the surface.

EXAMPLE 8.12 A 1100-kg car traveling at 18 m/s skids to a stop on a level road in a distance of 24 m. Calculate the average kinetic friction force acting on the car as it stops.

SOLUTION If we choose the car, road, and earth as the system, no external forces act on the system. Thus

$$W = 0.$$

The only energy changes that occur are a decrease in the car's kinetic energy and a corresponding increase in the thermal energy of the tires and road. The changes are

$$\Delta KE = \tfrac{1}{2}mv^2 - \tfrac{1}{2}mv_0^2 = -\tfrac{1}{2}(1100\text{ kg})(18\text{ m/s})^2 = -1.8 \times 10^5\text{ J}^*$$

and

$$\Delta E_{th} = F_k\,\Delta x = F_k(24\text{ m}).$$

Substituting into the work-energy equation,

$$W = \Delta KE + \Delta PE_g + \Delta PE_s + \Delta E_{th},$$

we find that

$$0 = -1.8 \times 10^5\text{ J} + 0 + 0 + F_k(24\text{ m}).$$

Thus

$$F_k = \frac{1.8 \times 10^5\text{ J}}{24\text{ m}} = 7.5 \times 10^3\text{ N}. \qquad \blacksquare$$

EXAMPLE 8.13 You wish to build a ski lift for the small hill next to your living quarters. The cable is to pull a 100-kg skier up a hill inclined at an angle of 20° to the horizontal. The skier's speed is zero at the bottom and is 6 m/s at the top. The hill is 50 m long, and a 150-N friction force opposes the skier's motion. Calculate the average tension in the cable pulling the skier.

SOLUTION We choose the skier, hill, and earth as the system. However, the motor and cable are outside the system and do work on the skier. The work done by the cable is

$$W = T\,\Delta r \cos 0° = T(50\text{ m})(1),$$

where T is the unknown cable tension. The changes in energy of the system are

$$\Delta KE = \tfrac{1}{2}mv^2 - \tfrac{1}{2}mv_0^2 = \tfrac{1}{2}(100\text{ kg})(6\text{ m/s})^2 - 0 = 1800\text{ J},$$
$$\Delta PE_g = mg(y - y_0) = (100\text{ kg})(9.8\text{ m/s}^2)(50\text{ m sin }20°) = 16{,}800\text{ J},$$
$$\Delta PE_s = 0,$$
$$\Delta E_{th} = F_k\,\Delta x = (150\text{ N})(50\text{ m}) = 7500\text{ J}.$$

---

*The energy released in this stop is about equal to the chemical energy released to your body when you eat a small slice of bread.

Putting all of these calculations into the work-energy equation,

$$W = \Delta KE + \Delta PE_g + \Delta PE_s + \Delta E_{th}.$$

we find that

$$T(50 \text{ m}) = 1800 \text{ J} + 16,800 \text{ J} + 0 + 7500 \text{ J}$$

or

$$T = \frac{1800 \text{ J} + 16,800 \text{ J} + 7500 \text{ J}}{50 \text{ m}} = \underline{520 \text{ N}}.$$ ∎

8.9 Inelastic and Elastic Collisions

Energy and momentum are both conserved when two objects collide. However, the form of the energy almost always changes, some of the original kinetic energy being converted during the collision to other forms such as thermal energy. This happens when a basketball bounces on the floor. Each bounce produces a little thermal energy as the ball's leather or rubber skin flexes and bends. The ball leaves the floor with less speed and less kinetic energy than it had when hitting the floor. This event is called an **inelastic collision**—one in which kinetic energy is converted to other forms of energy. A car collision is inelastic since a considerable portion of the car's initial kinetic energy is lost as its shape is changed during the collision. Even though kinetic energy is lost during inelastic collisions, momentum is conserved.

EXAMPLE 8.14 A ballistic pendulum is used to measure the speed of a fast-moving projectile. The projectile hits and imbeds in a wooden block hanging from strings, as shown in Fig. 8.20. Derive an expression for the height h that the block and projectile swing up into the air after the collision.

SOLUTION The collision is inelastic, since kinetic energy is lost as the projectile splits the wood fibers while becoming imbedded in the block. But momentum is conserved:

initial momentum = final momentum

or

$$mv_0 = (m + M)v, \qquad (8.13)$$

where v_0 is the initial speed of the projectile of mass m, and v is the speed of the block of mass M and projectile just after the collision has occurred.

Following the collision, the block and projectile swing up into the air. Kinetic energy is converted to gravitational potential energy:

$$0 = \Delta KE + \Delta PE_g$$

or

$$0 = -\tfrac{1}{2}(M + m)v^2 + (M + m)gh.$$

Rearranging the above and substituting from Eq. (8.13) for v, we find that

$$v_0 = \sqrt{2gh}\left(\frac{m + M}{m}\right).$$ ∎

FIG. 8.20. A ballistic pendulum.

FIG. 8.21. A ball swings down and hits a second ball, initially hanging at rest.

Often, when atoms and molecules collide with each other, the total kinetic energy of the particles before the collision equals the sum of their kinetic energy afterward. If one atom or molecule gains kinetic energy, the other loses an equal amount. These events are called **elastic collisions**—ones in which both kinetic energy and momentum are conserved. Most collisions involving macroscopic objects are inelastic, but occasionally they can be approximated as elastic, as in the next example.

EXAMPLE 8.15 You may be familiar with a toy that was developed years ago for demonstrations in physics classes and is now sold in stores. The device consists of several identical steel balls, each hanging from two strings as shown in Fig. 8.21. Consider an experiment in which one ball swings down and hits a second ball hanging at rest. Confirm that the first ball stops when it hits the second, and that the second moves off at about the same speed that the first ball had before the collision. We assume that the thermal energy created by the collision is small and that the collision is elastic.

SOLUTION Since the collision is assumed elastic, both kinetic energy and momentum are conserved:

$$\text{Kinetic energy:} \quad \tfrac{1}{2}mv_{1_0}^2 + \tfrac{1}{2}mv_{2_0}^2 = \tfrac{1}{2}mv_1^2 + \tfrac{1}{2}mv_2^2, \tag{8.14}$$

$$\text{Momentum:} \quad mv_{1_0} + mv_{2_0} = mv_1 + mv_2, \tag{8.15}$$

where m is the mass of each ball, v_{1_0} and v_{2_0} are the initial velocities of balls one and two, and v_1 and v_2 are their final velocities. In this problem, $v_{2_0} = 0$. Substituting for v_{2_0} in Eq. (8.15), cancelling the m's, and rearranging, we find that

$$v_1 = v_{1_0} - v_2. \tag{8.16}$$

Substituting Eq. (8.16) into Eq. (8.14) leads to the equation

$$\tfrac{1}{2}mv_{1_0}^2 + \tfrac{1}{2}m0^2 = \tfrac{1}{2}m(v_{1_0} - v_2)^2 + \tfrac{1}{2}mv_2^2.$$

Cancelling the m's and rearranging to solve for v_2, we find that

$$v_2 = v_{1_0}.$$

After the collision, ball two has the same velocity as ball one had before the collision. Substituting this result into Eq. (8.16), we see now that $\underline{v_1 = 0}$. The first ball is stopped by the collision. ∎

8.10 Power

We have been considering a variety of processes involving energy changes but have not discussed the time required for these changes. Power is an important quantity used to describe the rate of doing work or the rate of the transformation of one form of energy into another.

Power is defined as the ratio of the work ΔW done and the time Δt required to do that work:

$$\text{power} = P = \frac{\Delta W}{\Delta t}. \tag{8.17}$$

Power can also be defined as the ratio of the change in a certain form of energy ΔE, such as electrical or chemical energy, and the time Δt required for that energy change:

$$\text{power} = P = \frac{\Delta E}{\Delta t}. \tag{8.18}$$

The SI metric unit of power is the **watt,** where 1 watt $= 1$ J/s.

A 60-watt lightbulb converts 60 J of electrical energy to other forms of energy (heat and light) each second. A cyclist while pedaling a bicycle at moderate speed typically converts chemical energy in the body to other uses at a rate of about 400–500 watts, or 400–500 joules per second.

Another power unit commonly used to describe the power of engines is the **horsepower:**

$$1 \text{ horsepower (hp)} = 746 \text{ watts.}$$

A half-horsepower engine converts electrical energy to other forms of energy at a rate of 0.5×746 watts $= 373$ watts, or 373 J/s.

EXAMPLE 8.16 A crane lifts a 300-kg load at a constant rate of 1.0 m per second. Calculate the rate at which the crane is performing the work.

SOLUTION The work done by the crane per unit time is

$$P = \frac{\Delta W}{\Delta t} = \frac{F(\Delta h)\cos 0°}{\Delta t} = \frac{(mg)(\Delta h)(1)}{\Delta t}$$

$$= \frac{(300 \text{ kg} \times 9.8 \text{ m/s}^2)(1 \text{ m})(1)}{(1 \text{ s})} = \frac{2900 \frac{\text{kg} \cdot \text{m}}{\text{s}^2} \text{m}}{\text{s}}$$

$$= 2900 \frac{\text{N} \cdot \text{m}}{\text{s}} = 2900 \frac{\text{J}}{\text{s}} = \underline{2900 \text{ watts.}} \qquad \blacksquare$$

Metabolic Rate

The energy for bodily activity is provided by the chemical energy of the foods we eat. The body breaks the larger molecules of these foods into smaller molecules having less chemical energy. The absolute value of the rate of conversion of this chemical energy $|\Delta E/\Delta t|$ into other forms of energy by our bodies is called our **metabolic rate.** The metabolic rate depends on many factors, including one's weight, physical activity, and the efficiency of bodily processes. Table 8.1 provides a list of the metabolic rates of people under several different conditions and in several different units of measure. The "kcal" is a unit of energy related to the joule as follows:

$$1 \text{ kcal} = 1000 \text{ calories} = 4186 \text{ joules.}$$

Dieticians call a kcal simply a Cal. A piece of bread provides about 70 kcal of metabolic energy.

As we can see from Table 8.1, a 150-lb person when at rest uses energy at about the same rate as a 100-watt bulb. During a day of rest, the 150-lb person

TABLE 8.1 Energy Usage in Various Activities

System	$|\Delta E/\Delta t|$, The metabolic rate		
	Watts	kcal/hr	kcal/day
100-lb person at rest	80	70	1,600
150-lb person at rest	100	90	2,100
200-lb person at rest	120	110	2,600
Person walking at 3 mph	280	240	5,800
Moderate exercise	470	400	10,000
Heavy exercise	700	600	14,000

would use about 2100 kcal of food. To keep his or her weight constant, the person must eat roughly 2100 kcal of food. A 200-lb person would require about 2600 kcal per day. Exercise or other activity increases the caloric usage. A person loses about 1 lb of body weight for each 3500 kcal of food omitted from the diet. A person can also lose a pound of weight by increasing the metabolic rate by exercising so that 3500 extra kcal of energy is used.

Summary

1. **Work:** Work is one way of transferring energy to or from an object. If a constant force $\mathbf{F}$ acts on the object as it undergoes a displacement $\Delta \mathbf{r}$, the work done by the force on the object during the displacement is

$$W = F \Delta r \cos \theta, \tag{8.1}$$

where θ is the angle (180° or less) between the direction of $\mathbf{F}$ and the direction of $\Delta \mathbf{r}$. The object gains energy if the work done on it is positive and loses energy if the work is negative.

2. **Energy changes:** The objects in a system can have a variety of different forms of energy. Four different types of energy are described below, along with equations used to calculate the change in each form.

Kinetic energy (KE): Kinetic energy is energy an object has because of its motion. The change in kinetic energy of a mass m whose speed changes from v_0 to v is

$$\Delta KE = \tfrac{1}{2}mv^2 - \tfrac{1}{2}mv_0^2. \tag{8.4}$$

The change in an object's **rotational kinetic energy** is

$$\Delta KE_r = \tfrac{1}{2}I\omega^2 - \tfrac{1}{2}I\omega_0^2, \tag{8.5}$$

where I is its moment of inertia, and ω_0 and ω are its initial and final angular velocities.

Gravitational potential energy (PE_g): Gravitational potential energy is energy an object has because of its vertical separation from the earth's surface. The change in gravitational potential energy when a mass m moves from a vertical position y_0 to vertical position y is

$$\Delta PE_g = mg(y - y_0). \tag{8.6}$$

Elastic potential energy (PE_s): Elastic potential energy is energy an elastic object such as a spring has because it is com-

pressed or stretched from its equilibrium position. The change in elastic potential energy when the spring is stretched or compressed from position x_0 to position x is

$$\Delta PE_s = \tfrac{1}{2}kx^2 - \tfrac{1}{2}kx_0^2, \tag{8.7}$$

where k is the force constant of the spring.

Thermal energy (E_{th}): Thermal energy is the kinetic energy of the atoms and molecules in a system because of their random motion. The increase in thermal energy when a block slides a distance Δx on a surface while being opposed by a kinetic friction force F_k is

$$\Delta E_{th} = F_k \Delta x. \tag{8.12}$$

The thermal energy is shared by the block and the surface on which it slides.

3. **Work-energy equation:** Energy is a conserved quantity. If a small part of the universe called the system gains or loses energy from its environment because of work done on the system by forces originating in the environment, then the change in the system's energy is

$$W = \Delta E_{\text{system}}$$

or

$$W = \Delta KE + \Delta PE_g + \Delta PE_s + \Delta E_{th} + \cdots. \tag{8.10}$$

This equation, called the work-energy equation, applies to situations in which the only type of energy transfer from the environment to the system is that done by work.

4. **Power:** Power is defined as the rate of doing work or the rate of energy conversion from one form to another:

$$P = \frac{\Delta W}{\Delta t} \quad \text{or} \quad \frac{\Delta E}{\Delta t}. \tag{8.17 or 8.18}$$

Questions

1. Several activities are described below. For each, show the force exerted by the person and the displacement of the object on which the force acts. Using the definition of work, explain whether the force does positive, negative, or zero work. (a) A person slowly lifts a box from the floor to a table. (b) A person slowly lowers a box from a table to the floor. (c) A person carries a bag horizontally from one location to another. No lifting or lowering takes place. (d) A Dutch boy holds his thumb in a hole in the dike, preventing water from rushing through the hole and flooding Holland.

2. A person rides a bicycle along a horizontal road. Consider the following forces and decide, by showing the direction of the force and the direction of the displacement, whether the work done is positive, negative, or zero. (a) The force of the pedal *on* the person as the pedal and foot move down. (b) The force of the bike seat *on* the person as the person moves forward 100 m. (c) The gravitational force of the earth *on* the person as he or she moves forward 100 m.

3. Give three examples of activities that are often called work in everyday life but that, according to the physics definition, are not work. Explain why, in each case, no physics work is done.

4. *Estimate* the change in gravitational potential energy when you rise from bed to a standing position each morning (or afternoon).

5. A pendulum is displaced from the vertical and released. (a) As it swings down toward the vertical axis, its kinetic energy increases. From where does this energy come? (b) Eventually the pendulum will stop swinging and hang at rest, straight down. Where has the energy gone?

6. The heart does about 1 J of work each heartbeat. (a) *Estimate* the total work done by the heart in an average lifetime. (b) Mt. Everest is 9400 m high. Calculate the mass that would have to be raised from sea level to the top of Mt. Everest in order to cause an increase in gravitational potential energy equal to the work done by the heart in a lifetime.

7. A car with standard transmission can roll farther in neutral than when in gear. In terms of energy, why is this possible?

8. Describe a process in which gravitational potential energy is converted mostly into (a) kinetic energy, (b) elastic potential energy, and (c) thermal energy caused by friction.

9. Estimate the maximum horsepower you can exert when raising your body mass as fast as possible up a flight of ten stairs. Justify any numbers you use in your estimate. The only energy change you should consider is the change in gravitational potential energy—that is, $P = \Delta PE_g/\Delta t = mg(y - y_0)/\Delta t$.

10. Calculate the lifting power exerted by a person during each of the following activities: (a) the person lifts a 10-kg mass 1 m higher in 4 s, (b) the person lifts the 10-kg mass 1 m higher in 2 s, (c) the person lifts the 10-kg mass 2 m higher in 4 s, and (d) the person lifts a 20-kg mass 1 m higher in 4 s.

11. *Estimate* the horsepower exerted by a weight lifter who lifts 150 kg from the ground to his chest. Justify any numbers you use in your estimate.

Problems

8.1 Work

1. (a) A boy pulls a wagon 30 m along a sidewalk with a rope tension force of 80 N. The force is directed 25° above the horizontal. How much work is done by the boy? (b) How much work is done if the same tension force is directed parallel to the ground?

2. Calculate the work done by a woman (a) when slowly lifting a 15-kg suitcase 0.8 meter upward, (b) when lowering it 0.8 meter, and (c) when holding it at rest.

3. (a) An 80-N rope tension force is used to slowly pull a wagon 50 m up a hill inclined at an angle of 20°. Calculate the work done by the rope if it pulls parallel to the hill. (b) Repeat (a), except that the rope slowly lowers the wagon 50 m down the hill.

4. A car is stopped in 0.8 m as it runs into a pile of sand. The work done on the car by the sand is -6×10^5 J. (a) Calculate the average force of the sand on the car. (b) Why was the work done by the sand negative? (c) How is the average force altered if the stopping distance is doubled? Justify your answer.

5. A rope attached to a truck pulls a water skier along an irrigation canal. The tension in the rope is 400 N, and the rope makes an angle of 15° with the canal. Calculate the work done by the rope in pulling the skier 300 m.

6. The cable in an elevator exerts an average 20,000-N tension force on the elevator while lifting it 35 m. (a) Calculate the work done by the cable. (b) Calculate the work done if the cable exerts the same force while lowering the elevator.

7. A man does 5800 J of work while pulling a rope attached to a cart loaded with bricks. The tension force in the rope is 320 N and the cart moves 20 m along the level ground. Draw a picture showing the orientation of the rope relative to the ground. Justify with appropriate calculations the orientation shown in the picture.

8. A rope tow at a ski area exerts an average force of 60 lb while pulling a skier 500 ft up a hill inclined at an angle of 18° with the horizontal. The force is exerted parallel to the hill. Calculate the work done by the rope on the skier.

9. A lever 1.1 m long is used to raise the back bumper of a car. The fulcrum of the lever is 0.1 m from the bumper. (a) How much work is done by a man pushing with a 500-N force on the long end of the lever while it is displaced downward a distance of 0.5 m? (b) How far up is the bumper raised? (c) Calculate the work done by the lever on the car.

10. A bellhop lifts a 20-kg suitcase 0.4 m, then slowly carries it 30 m along a hall, and finally sets it back down on the floor.

Calculate the work done by the bellhop on the suitcase during each part of this trip.

11. A person slowly pulls the free end of the rope shown in Fig. 8.22 with a force of 90 N. The rope moves a distance of 1.0 m. (a) Calculate the work done by the person on the rope. (b) Calculate the work done by the hanging pulley on the 180-N weight. Justify carefully this latter calculation.

$T = 90$ N

$w = 180$ N

FIG. 8.22

8.3 Kinetic Energy

12. The kinetic energy of a 1500-kg car increases by 2.6×10^5 J. Its final speed is 20 m/s. Calculate its initial speed.

13. A 3-kg rabbit and a 10-kg Irish setter have the same kinetic energy. If the setter is running at a speed of 8 m/s, how fast is the rabbit running?

14. Calculate the change in kinetic energy when a 1600-kg car accelerates (a) from 0 to 35 mph, (b) from 35 mph to 70 mph.

15. *Estimate* your kinetic energy when walking to physics class.

16. A 1500-kg car travels 15 km along a city road. Instead of driving at a steady speed of 15 m/s and synchronizing his motion with the stoplights, the driver accelerates the car from rest to 30 m/s and then stops for the next stoplight (the lights are 1 km apart). (a) Calculate the kinetic energy the car uses each time it accelerates from zero to 30 m/s. (b) Calculate the total energy it uses for these accelerations during the 15 km trip. (c) If the car is only 5 percent efficient, it uses 20 times more gasoline energy than the kinetic energy generated by its accelerations. How much extra energy does the car use because of the accelerations? (d) If one piece of bread releases 2.5×10^6 J of energy, how many pieces of bread, in terms of energy, does this start-and-stop driver waste?

17. A 10,000-kg railroad car, moving at 1.0 m/s, bumps into and couples with an identical car at rest. The coupled cars move at 0.5 m/s. (a) Calculate the net change in kinetic energy. (b) If $\Delta KE \neq 0$, speculate about where the energy went (ΔKE negative) or came from (ΔKE positive).

18. Huge amounts of air circulate across the earth in regular patterns. A tropical air current called the Hadley cell carries about 2×10^{11} kg of air per second past a cross section while moving toward the equator. The average air speed is about 1.5 m/s. (a) Calculate the kinetic energy of this moving air. (b) The United States in 1980 used about 7×10^{16} J of energy. Calculate the ratio of the kinetic energy of the 2×10^{11} kg of air in the Hadley cell that passes toward the equator each second and the energy consumed in the United States each second.

19. Calculate the change in rotational kinetic energy when the angular velocity of the turntable of a high-fidelity system increases from 0 to 33 rpm. Its moment of inertia is 6×10^{-3} kg $\cdot$ m^2.

20. A grinding wheel with a (4.0×10^{-3})-kg $\cdot$ m^2 moment of inertia gains 1.2×10^3 J of rotational kinetic energy after starting from rest. Calculate the flywheel's final angular velocity.

21. The U.S. Department of Energy at one time had plans for a 1500-kg automobile to be powered completely by the rotational kinetic energy of a flywheel. (a) If the 300-kg flywheel had a 6.0-kg $\cdot$ m^2 moment of inertia and could turn at a maximum angular velocity of 3600 rad/s, calculate the energy stored in the flywheel. (b) How many accelerations from a speed of zero to 15 m/s could the car make before the flywheel's energy was dissipated, assuming 100 percent energy transfer and no flywheel regeneration during braking?

22. The angular velocity of a flywheel increases by 40 percent. By what percent does its rotational kinetic energy increase?

8.4 Gravitational Potential Energy

23. A 5200-kg trolley car in San Francisco is pulled a distance of 360 m up a hill inclined 12° from the horizontal. Calculate the change in gravitational potential energy of the car.

24. *Estimate* the change in gravitational potential energy of a person who successfully climbs Mt. Everest from sea level.

25. The 150-kg frame for a waterbed is lowered 1.2 m straight down from the back of a U-haul truck to the ground. Calculate the change in gravitational potential energy.

26. (a) Calculate the change in gravitational potential energy of a 50-kg mountain climber after moving 30 m up a vertical slope. (b) If the muscles in the body can convert chemical energy into gravitational potential energy with an efficiency of no more than 5 percent, calculate the least muscle chemical energy used to climb this slope.

27. A ski slope drops at an angle of 37° to the horizontal and is 500 m long. (a) Calculate the change in gravitational potential energy of a 60-kg skier who goes down this slope. (b) If 30 percent of the gravitational potential energy lost is converted into kinetic energy, how fast is the skier traveling at the bottom of the slope?

28. The racetrack for a soapbox derby is 300 m long and has a vertical drop of 28 m. (a) Calculate the change in gravitational potential energy of a racecar and its driver (a total mass of 113 kg) from start to finish of a race. (b) If all this energy is converted to kinetic energy, how fast is the car moving at the end of the race?

8.5 Elastic Potential Energy

29. A spring with a (1.2×10^4)-N/m force constant is stretched from its equilibrium position to a position $x = 6$ cm. (a) Calculate the change in elastic potential energy of the spring. (b) Calculate its change in elastic potential energy as it returns from the 6-cm position to a position 4 cm from equilibrium. (c) Calculate its elastic potential energy change as it moves from the 4-cm position back to its equilibrium position.

30. A spring-loaded car bumper is built to be compressed 0.3 m when it hits a wall while moving at a speed of 9 m/s. The energy the spring must absorb is 80,000 J. What should be the force constant of the bumper?

31. An elevator cable breaks. The elevator's fall is stopped by compression of a spring at the bottom of the shaft. The spring, whose force constant is 2×10^5 N/m, must absorb 3×10^4 J of energy when being compressed by the elevator. How far is the spring compressed?

32. A partially compressed spring is compressed farther, causing

its elastic potential energy to increase by a factor of 1.5. Calculate the factor by which the compression distance is increased.

33. A 700-kg rocket sled traveling at 50 m/s is stopped by compression of a spring. The force constant of the spring is 1.2×10^6 N/m. (a) Calculate the change in kinetic energy of the sled as it stops. (b) How far will the spring be compressed in order that its change in elastic energy balances the kinetic-energy change of the sled?

34. The force required to stretch a slingshot by different amounts is graphed in Fig. 8.23. (a) Calculate the force constant k of the slingshot. (b) How much energy is required to stretch the sling from $x = 0$ to 15 cm? (*Hint:* For part (a), use the equation $F = kx$ for one or two known points on the F-versus-x curve in order to calculate k.)

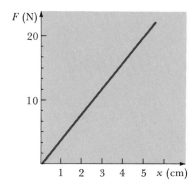

FIG. 8.23. A graph of the force F needed to compress a spring versus the compression x of the spring.

8.7 Work-Energy Calculations

(In solving these problems be sure to use the work-energy equation.)

35. With what speed must a 90-kg basketball player leave the ground in order to raise his or her center of mass 0.8 m in order to make a slam dunk?

36. A 1000-N force pointing in the horizontal direction pushes a 200-kg cart on a level, frictionless track. The cart has an initial speed of 20 m/s. What is its speed after being pushed for 50 m?

37. A 3000-lb car crashes into a pile of sand and stops in 4 ft. If the car was initially moving at 30 mph, what is the average force of the sand on the car? Ignore friction forces.

38. Calculate the speed of a rock after falling 55 m from the top of the Leaning Tower of Pisa. Ignore the resistive forces of the air.

39. A 3-kg mass rests on a 30° frictionless, inclined plane (Fig. 8.24). The mass rests against a spring having a 4200-N/m force constant which is compressed 15 cm. When the spring is released,

FIG. 8.24

the mass shoots up toward the end of the plane, a distance of 1.8 m from where it started. Will the mass reach the end of the plane? Explain. If so, what is its speed as it leaves the plane?

40. A daredevil driver is shown in Figs. 3.12 and 3.13 being ejected from a spring-loaded car ejector seat. If an 80-kg daredevil wishes to be able to fly to a height of 7 m after leaving the seat, what force constant is required for the spring in the ejector seat? When loaded, the spring will be compressed 0.3 m. (*Note:* The driver will rise a total distance of 7.3 m after the spring is released.)

41. A vehicle built to test the effects of high accelerations on people is moving at a speed of 20 m/s. The vehicle runs into a piston that compresses the air in a cylinder and stops the car in 0.3 m. A 70-kg person is also stopped in the same distance by the force of shoulder straps and seatbelts on her body. Calculate the average force of these belts on the person during the stop.

42. A 55-kg woman jumps from the window of a burning building. (a) Calculate her speed just as she reaches a fire fighter's net 8.0 m below. Ignore air resistance. (b) She is stopped by the net after falling 1.2 m into it. Calculate the average force of the net on her body.

43. A 0.20-kg egg is dropped from a ladder a vertical distance of 4.0 m. (a) Calculate the speed of the egg at the bottom of its fall. (b) The egg will break if subjected to an impulsive force greater than 80 N. Over what minimum distance must a constant stopping force be exerted in order to avoid breaking the egg?

44. A person on a bicycle traveling at 10 m/s on a horizontal surface stops pedaling as he starts up a hill inclined at an angle of 3°. How far along the incline will the person travel before stopping? Ignore friction forces.

45. An inventive young houseperson has developed a spring-loaded clothes shoot. This device, when released, shoots a pile of folded clothes from the basement up to the second floor of a home. The device is designed to hurl 2 kg of clothes upward a distance of 5.0 m (counting the 0.3 m the spring is initially compressed). What must the force constant of the spring be if 70 percent of the spring's energy is converted to the gravitational potential energy of the clothes when their elevation increases (the other 30 percent being converted to thermal energy by friction forces)?

46. A rope exerts a 40-N force while lowering a 20-kg mass down a frictionless plane inclined at an angle of 15° (the force acts parallel to the plane). If the speed of the mass is initially 2.0 m/s, calculate its final speed after moving 10 m down the plane.

47. You are hired by a company to design a net to be used by fire fighters for catching people who have jumped from burning buildings. The net is designed to catch a 70-kg person falling at a speed of 15 m/s. (a) Calculate the stopping distance—that is, the distance the net stretches downward while stopping the person's fall. This distance should be such that the average force of the net on the person is 6000 N or less. A greater force may cause bone dislocations. (b) Calculate the effective force constant k of the net. (*Hint:* You will probably want to use a different system in part (b) than in part (a).)

48. One technique used to gain height in high jumping is to "throw" the arms and one leg up into the air. This extra kinetic energy is converted into additional gravitational potential energy, thus increasing the height of the jump. Assume that the arms and one leg represent 30 percent of the body weight and that these three body parts have an upward velocity 50 percent greater than

that of the other body parts. (a) Calculate the ratio of the jumper's kinetic energy with this extra arm and leg motion and his kinetic energy if the arms and legs have the same speed as the rest of the body. (b) Based on the calculations in part (a), by what percent should a jumper be able to increase his height using this arm and leg motion?

8.8 Thermal Energy and Friction Forces

(Use the work-energy technique to solve these problems.)

49. A 50-kg crate is pulled 10 m across a floor by a rope directed horizontally. If the coefficient of friction between the crate and floor is 0.6, how much thermal energy is created?

50. A 1500-kg car traveling at 20 m/s is brought to a stop by the opposing forces of friction. If these opposing forces have an average magnitude of 800 N, how far will the car travel before stopping?

51. A 1500-kg car is initially coasting at 20 m/s down a hill inclined at 2° with the horizontal. If an 800-N friction force opposes the car, how far will it travel down the hill before stopping?

52. Consider the soapbox derby race car discussed in Problem 28. If the speed of the car at the end of the race is 18 m/s, calculate the average resistive force due to friction acting on the car during the race.

53. A spring with force constant k of 10,000 N/m is compressed 0.20 m, and a 2.0-kg block rests against the spring. The block is shot up a plane 3.0 m long, inclined at 37°. A 10-N friction force opposes the block as it slides up the plane. Calculate the speed with which the block leaves the plane.

54. An 80-N friction force opposes the motion of a bicycle rider. (a) How much thermal energy is created because of friction by the cyclist in traveling 1 mile? (b) How far must the cyclist travel to produce 3×10^5 J of thermal energy? By comparison, this is the energy released to the body after one eats a slice of bread.

55. A 6.0-kg block slides from position A down a frictionless curve to position B (Fig. 8.25). At B, a friction force opposes the motion of the block so that it comes to a stop 2.5 m from B. Calculate the coefficient of kinetic friction between the block and the surface after position B.

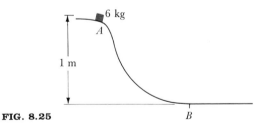

FIG. 8.25

56. A spring with a (1.2×10^4)-N/m force constant is compressed 6.0 cm, and a 10-kg block is placed against it. When the spring is released, the block shoots forward along a horizontal surface whose coefficient of friction is 0.20. How far will the block travel before coming to a stop?

8.9 Inelastic and Elastic Collisions

57. An 80-g arrow moving at 80 m/s hits and imbeds in a 10-kg block resting on ice. (a) Use the conservation-of-momentum principle to determine the speed of the block and arrow just after the collision. (b) Use the work-energy principle to determine how far

the block slides on the ice following the collision. The coefficient of kinetic friction between it and the ice is 0.20.

58. A person sitting on a sled (a total mass of 60 kg) catches a 4.0-kg jug of wine thrown by a friend. The sled, initially at rest, coasts 4.0 cm following the catch before coming to a stop. The coefficient of friction between the ice and sled is 0.05. Calculate the initial speed in the horizontal direction of the jug of wine.

59. An 1800-kg station wagon moving at 10 m/s hits a 900-kg stationary compact car from the rear. The two vehicles become locked together. (a) Calculate their speed just after the collision. (b) What fraction of their total initial kinetic energy is lost during the collision?

60. A billiard ball at rest is hit head on by a second billiard ball moving with a speed of 1.5 m/s. If the collision is elastic, and we ignore rotational motion, calculate the final speeds of each ball.

61. A ball of mass 4.0 kg moving at speed 2.0 m/s has an elastic, head-on collision with a second, stationary ball of mass 1.0 kg. Calculate the final velocity of each ball.

62. A ball of mass 1.0 kg moving at speed 2.0 m/s has an elastic, head-on collision with a second, stationary ball of mass 4.0 kg. Calculate the final velocity of each ball.

8.10 Power

63. A horse, according to James Watt, can do work at a rate of one horsepower. At what speed can a horse pull a wagon on which the horse exerts a 400-N force?

64. *Estimate* the power in watts and in horsepower generated by the Russian weight lifter, Vasily Alexeev, during his record jerk at the 1976 Olympics. As part of that lift, he raised 255 kg from his chest to above his head (a distance of about 0.7 m). This was done in approximately one second.

65. A large tree can evaporate 500 kg of water a day. (a) If the water is raised 8 m, how much energy must be provided by the tree to lift the water this distance? (b) If the evaporation occurs over a 12-hour period, what is the average power in watts needed to provide this increase in gravitational energy?

66. How much time is needed for a motor whose output is 3.0 hp to lift a 1000-kg elevator a distance of 30 m?

67. Calculate the output power needed for the pump of a fire engine that must lift 30 kg of water a vertical distance of 20 m each second.

68. (a) Calculate the power output of the motor of a 1500-kg car traveling on a level road at a speed of 90 km/hr if the car is opposed by a total friction force of 750 N. (b) If the car travels at the same speed and is opposed by the same friction force while climbing a 4° hill, what power must the engine now generate? (*Hint:* Add the rates at which the motor does work against friction and against gravity.)

69. A 150-lb person wishes to lose 10 lb of body weight in 30 days. Estimate the number of kilocalories the person should have in his or her diet to achieve this goal (see Table 8.1). Remember that omission of 3500 kcal from the diet results in a loss of 1 lb of weight.

70. A 150-lb person walks at 3 mph one hour a day for one year. (a) Use the numbers in Table 8.1 to calculate the extra number of kilocalories of energy used because of the walking. (b) The person should expect to weigh how much less at the end of the year because of the walking?

71. In the past, salmon would swim over seven hundred miles to spawn at the headwaters of the Salmon River in central Idaho. The trip took about 22 days, and the fish consumed energy at a rate of 2.0 watt for each kilogram of body weight. (a) What is the total energy used by a 3-kg salmon while making this 22-day trip? (b) About 80 percent of this energy is released by burning fat and the other 20 percent by burning protein. How many grams of fat are burned? One gram of fat releases 3.8×10^4 J of energy. (c) If the salmon is about 15 percent fat by weight at the beginning of the trip, how many grams of fat does it have left at the end of the trip?

General Problems

G1. James Joule spent a portion of his honeymoon measuring the difference in temperature of water at the top and the bottom of the Sallanches waterfall in the valley of Chamonix in France. His experiments confirmed the conservation-of-energy principle. We will consider the results expected for such experiments if performed at Niagara Falls which are about 48 m high. (a) If we ignore air resistance, what is the speed of a mass m of water just before hitting stones at the bottom of the falls? (b) 4180 J of energy is required to warm 1 kg of water by 1 C°. If all of the kinetic energy of the water just before hitting bottom is converted to thermal energy when the water hits the rocks, by how much is the temperature of the water increased?

G2. A spring with force constant $k = 1.2 \times 10^4$ N/m is compressed 0.10 m, and a 3-kg mass is placed against it. The spring is released and the block is shot up a 37° inclined plane. When the block returns to the spring, it compresses the spring only 0.05 m before coming to a stop. Approximately how far up the plane did the block move before returning to the spring? The coefficient of kinetic friction between the plane and block is 0.46.

G3. A car driving at a steady speed of 55 mph on a level road uses 1 gallon of gas to go 20 miles. About 80 percent of the energy is used to overcome internal friction in moving parts in the car; the other 20 percent is used to overcome external forces (drag forces due to air resistance, friction forces of the road, and flexion of the tire while it rotates). Use this information to estimate the net sum of these external forces on the car. One gallon of gasoline provides 1.4×10^8 J of energy.

9 Thermal Energy, Temperature, and Heat

In Chapter 8 we used the work-energy equation ($W = \Delta E_{\text{system}}$) to analyze energy transformations involving macroscopic objects, objects easily seen with our eyes. In this and the next three chapters we look at energy transformations involving the atoms and molecules of which matter is made. This atomic and molecular energy of a system, called *internal energy,* consists of (1) thermal energy—the random kinetic energy of the atoms and molecules, and (2) the potential energy of these tiny particles resulting from their bonds and interactions with each other.

Temperature is an important indicator of thermal energy, and we will define it along with different temperature scales. We will also define a quantity called heat—a mechanism, like work, for transferring energy from one object to another. Several consequences of heat transfer are discussed, including temperature changes; changes between the solid, liquid, and gaseous states of matter; and changes in the dimensions of objects as they become warmer or cooler.

Processes involving thermal energy and heat are very important, affecting our metabolic rates, the climate of the earth, and every sensation of comfort and discomfort we feel that is related to temperature. For example, a 3 percent increase in body temperature causes our metabolic rate to double. A 1 percent increase in the earth's temperature would cause the polar ice caps to melt, and several of our large cities would be flooded.* Certainly, thermal energy and heat are important subjects that deserve careful attention.

9.1 Thermal Energy

The **thermal energy** in a substance is the kinetic energy associated with the random motion of all its atoms and molecules. Consider a cup of water. The water molecules in the cup experience three different types of random motion, as depicted in Fig. 9.1. They move or translate from one position to another (Fig. 9.1a). Collisions with other water molecules frequently interrupt this motion, causing the molecule to change direction and speed—hence the use of the word *random* to describe the motion. The molecules also vibrate (Fig. 9.1b). The hydrogen atoms attached to the oxygen atom (H$\diagup$O$\diagdown$H) vibrate in and out

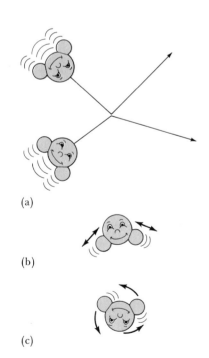

FIG. 9.1. Types of motion of a water molecule: (a) translational, interrupted by a collision with another water molecule; (b) vibrational; and (c) rotational.

(a)

(b)

(c)

*These percentage changes in temperature are on the absolute or Kelvin scale, to be introduced soon.

160

and bend back and forth. The molecules also spin or rotate as they move through space (Fig. 9.1c). Because all three types of motion—translational, vibrational, and rotational—occur simultaneously, there is much variation in the motion of different molecules at any instant in time. Some may be rotating more and translating less, while some move up and others down.

Each molecule's motion changes almost one hundred billion billion (10^{14}) times per second because of collisions with other molecules. A molecule moving slowly before a collision may move fast afterward. However, when averaged for one second or more, we find that the average motion of every molecule of water in the glass is about the same—even though there is great variation in the motion of the molecules at each instant of time.

If the water becomes warmer, the molecules translate from one place to another faster, become more distorted in shape as they vibrate, and rotate more rapidly. The average energy associated with each type of motion increases. If the water becomes cooler, the translational, vibrational, and rotational motions decrease, as does the average energy associated with each type of motion. We see that thermal energy is related closely to the idea of temperature.

9.2 Temperature

Temperature is an indicator of the average thermal energy of the atoms or molecules in a substance whose temperature is being measured. As the temperature increases, so does the average thermal energy of its atoms or molecules. Thermal energy and temperature, then, are closely related quantities—but there is an important distinction. Temperature is a measure of the average thermal energy *per* molecule or atom, whereas thermal energy is the total random kinetic energy of *all* the atoms or molecules in the substance. If, for example, the ocean and a glass of water are at the same temperature, then the average thermal energy of one water molecule in the ocean is the same as that of one in the glass. However, the ocean contains much more thermal energy because there are so many more water molecules in the ocean than in the glass.

Temperature is measured using a variety of substances that change in a predictable way as their temperatures change. For example, the length of a metal rod increases as it warms. This variation of length with temperature is called a **thermometric property**—a property of matter that varies in a predictable manner with temperature. Other thermometric properties are the increase in the pressure of an enclosed gas as it warms, the increase in the volume of space occupied by a liquid such as mercury as it warms, the change in color of light emitted by a flame as the flame becomes hotter, and the increase in electrical resistance of a metal wire as its temperature increases. Every thermometer uses one of these or some other thermometric property as the basis for its temperature measurement.

One of the most familiar thermometers is the liquid mercury thermometer (Fig. 9.2). A narrow tube is attached to a bulb containing liquid mercury. As the bulb becomes warm, the mercury expands farther into the tube. Marks are placed on the outside of the tube at points to which the mercury rises when the bulb is placed in a temperature bath whose temperature is fixed. For example, one mark can be placed at the point to which the mercury rises when the bulb is placed in a solution of freezing water. Another mark can be inscribed at the point to which the mercury rises when the bulb is in boiling water at atmospheric pressure.

FIG. 9.2. A liquid mercury thermometer. As mercury in the bulb is warmed, it expands farther up into the capillary tube.

Temperature scales are constructed by assigning numbers to these fixed temperatures. On the Celsius temperature scale, the number zero is assigned to the freezing point of water and the number 100 to its boiling point. The space between these two marks is divided into 100 equal divisions or intervals. A temperature is expressed as a **point** on the scale in units of degrees Celsius (°C). For example, the average temperature of a person's body is 37.0°C. The difference of two temperatures is called an **interval** whose unit is Celsius degrees (C°). For instance, if the outdoor temperature increases from 20°C to 26°C, the change in temperature, an interval, equals 6 C°.*

Another temperature scale used commonly in the United States is the Fahrenheit scale. This scale, developed by Gabriel Fahrenheit (1686–1736), a German physicist, has the number 32 assigned to the freezing point of water and 212 to its boiling point. Fahrenheit wanted a scale in which all temperatures commonly measured would be positive. The lowest temperature he could obtain with a freezing mixture of water and salt, and the lowest outdoor temperature he expected to measure in Holland where he lived, were both greater than 0°F if he assigned a value of 32°F to the freezing point of water. Thus, he felt that all temperatures that he would need to measure would be positive numbers.

Fahrenheit was mistaken about the range of temperatures that scientists would someday want to measure. We often measure temperatures below 0°F (the coldest outdoor temperature measured on the earth was −127°F on August 24, 1960, at Vostok Station, Antarctica). There is, however, a lower limit to temperature. The coldest any object can possibly become is −273.15°C or −459.67°F. At this temperature the random motion of the atoms and molecules in a substance is at a minimum, and the temperature cannot be lowered further by decreasing this motion. This lowest possible temperature is called *absolute zero*.

Fahrenheit chose 212°F as the boiling point of water so that the normal body temperature would be near 100°F. There is supposedly no upper limit to temperature. Atoms near the center of stars can be several million degrees Fahrenheit.

On the Fahrenheit scale the boiling and freezing points of water are separated by 180 F° (that is, 212°F − 32°F), while on the Celsius scale these same two points are separated by only 100 C°. The ratio of these two numbers, 180 F°/100 C° = 9 F°/5 C°, can be used to relate temperature changes on one scale to those on the other. To calculate a Fahrenheit temperature T_F from a Celsius temperature T_C we must first multiply the Celsius temperature by $\frac{9}{5}$. We then add 32 to account for the fact that the freezing temperature of water on the Fahrenheit scale is 32°F and not zero. Thus,

$$T_F = \tfrac{9}{5}T_C + 32. \tag{9.1}$$

To calculate an equivalent Celsius temperature from a Fahrenheit temperature, Eq. (9.1) can be rearranged to give

$$T_C = \tfrac{5}{9}(T_F - 32). \tag{9.2}$$

Let us consider two examples of these temperature conversions.

EXAMPLE 9.1 If the temperature of a room is 20°C, what is its temperature in degrees Fahrenheit?

---

*This scale is named in honor of its inventor, the Swedish astronomer Anders Celsius (1701–1744). It was formerly called the centigrade scale, the Latin prefix *centi* meaning 1/100th.

FIG. 9.3. Several common temperatures on the Kelvin, Celsius, and Fahrenheit temperature scales.

SOLUTION   Using Eq. (9.1), we find that

$$T_F = \tfrac{9}{5}(20) + 32 = 36 + 32 = \underline{68\,°\text{F}}.$$ ■

EXAMPLE 9.2   The normal core temperature of the body is 98.6°F. What is the core temperature in °C?

SOLUTION   Using Eq. (9.2), we find that

$$T_C = \tfrac{5}{9}(98.6 - 32) = \tfrac{5}{9}(66.6) = \underline{37.0\,°\text{C}}.$$

Note that we first subtracted 32°F from 98.6°F to get 66.6°F. This is the number of Fahrenheit degrees that the body is above the freezing point of water. Since each Fahrenheit degree is $\tfrac{5}{9}$ of a Celsius degree, we find that the normal core temperature of the body is $(\tfrac{5}{9})66.6\,°\text{F} = 37.0\,°\text{C}$.   ■

Another very important temperature scale in physics is called the Kelvin or absolute temperature scale. In this scale, the temperature of absolute zero is assigned the value 0 K.* Thus, 0 K is equivalent to $-273.15\,°\text{C}$ or to $-459.67\,°\text{F}$. Temperature intervals on the Kelvin scale are the same as on the Celsius scale. The freezing point of water on the Kelvin scale is 273.15 K and the boiling point is 100 degrees higher or 373.15 K. Figure 9.3 shows the relationship among the three scales.

9.3 Heat and the Conservation-of-Energy Principle

The thermal energy of a system can change in many ways. The engine of an idling car becomes warm because much of the energy released when gasoline burns in the engine is converted into thermal energy. If you rub your hands together vigorously, they become warm because work done by the muscles in your arms is converted into thermal energy caused by friction as your hands rub together.

---

*The Kelvin degrees are abbreviated K, and not °K, according to a convention established by the international standards organization, the General Conference of Weights and Measures.

FIG. 9.4. When the "hot" molecules in the flame hit the glass, kinetic energy is transferred to the atoms in the glass. The glass and its contents become warmer. This is a form of heat transfer.

FIG. 9.5. Energy can enter or leave a system by heat transfer Q or by work W done on or by the system. The change in the energy of the system ΔE_{system} is given by the equation $Q + W = \Delta E_{\text{system}}$.

Another common way in which the thermal energy in a system is changed is by a process called heat transfer. **Heat transfer** occurs when two objects at different temperatures are brought into contact with each other. Thermal energy in the hotter object is transferred to the cooler one. Consider the effect of the flame of a bunsen burner on a glass of water (Fig. 9.4). The gas molecules in the flame have a temperature of about 900°C. Since they are so hot, the random kinetic energy of a molecule in the flame is much greater than the random kinetic energy of an atom in the glass. When a fast-moving gas molecule in the flame collides with the glass, an atom in the glass gains kinetic energy, and the gas molecule of the flame loses kinetic energy. The atom in the glass is warmed. There has been an energy transfer from the flame to the glass: we call it a heat transfer.

Heat is energy transferred to or from an object because of a difference in its temperature and that of some other object it contacts in its environment. The cooler object gains energy and the warmer body loses energy. Heat is given the symbol Q and has units of energy (such as joules, calories, and Btu).

The terms *heat* and *thermal energy* are often used synonymously. They are, however, different quantities, although the difference is subtle. The thermal energy in an object is the amount of random kinetic energy of its atoms and molecules.

We will often be casual in our use of the term *heat,* but in a strict sense, heat must be defined relative to a system (see Fig. 9.5). Heat is energy transferred between a system and its environment because of a difference in their temperatures. If the system gains energy, Q is positive; if the system loses energy, Q is negative.

When we use the work-energy principle ($W = \Delta E_{\text{system}}$) to keep track of the energy in a system, the changes in thermal energy ΔE_{th} or of other forms of internal energy belong on the right-hand side of the equation because thermal energy is one of many different forms of the energy in a system. Heat Q, however, belongs on the left-hand side of the equation because heat Q, like work, is energy added to or removed from a system. Thus, the work-energy equation becomes

$$
\begin{aligned}
W + Q &= \Delta E_{\text{system}} \\
&= \Delta KE + \Delta PE_g + \Delta E_{th} + \Delta E_{\text{chem}} + \cdots.
\end{aligned}
\tag{9.3}
$$

Equation (9.3) is based on the conservation-of-energy principle (see Section 8.6) and will be used extensively in future calculations in this and later chapters.

9.4 Specific Heat Capacity

To calculate the heat energy added to or removed from an object, we need to introduce a new quantity. The **specific heat capacity** c of a substance is the amount of heat that must be added to a unit mass of that substance to raise its temperature by 1 C°. The units of specific heat capacity are energy per mass per degree temperature—for example, J/kg · C°, cal/g · C°, Btu/slug · F°. Table 9.1 lists the values of the specific heat capacity of some common substances. The specific heat capacity of water is 4180 J/kg · C°. This means that 4180 J of energy must be added to 1 kg of water to raise its temperature 1 C°. If we know

TABLE 9.1 Specific Heat Capacities of Various Substances

Substance	c (J/kg · C°)	Liquid substances	c (J/kg · C°)	Gaseous substances[a]	c (J/kg · C°)
Iron or steel	450	Water	4180	Steam	1970
Copper	390	Methanol	2550	Oxygen	910
Aluminum	900	Ethanol	2480	Nitrogen	1040
Lead	130	Ethylene glycol	2380	Dry air	~1000
Glass	840	Benzene	1720		
Sodium chloride	880				
Ice	2090				
Wood	1680				
Sand	820				
Brick	840				
Concrete	880				
Human body	3470				

[a] At constant pressure

the specific heat capacity of a substance, we can find the amount of heat required to raise the temperature of a mass m of that substance by an amount ΔT.

EXAMPLE 9.3 You wish to take a bath and will need to warm 160 kg of water by 14 C°. How much heat is required?

SOLUTION Note first that to warm 1 kg of water by 1 C° requires 4180 J. To warm 160 kg by 1 C° would require 160×4180 J $= 670{,}000$ J. However, we wish to warm the water 14 C° instead of 1 C°. We will need 14 times as much energy or $14 \times 160 \times 4180$ J $= \underline{9.4 \times 10^6 \text{ J}}$.* ∎

We see from this last example that the heat Q needed to raise the temperature of a mass m of substance with specific heat capacity c by a temperature ΔT is the product of m, c, and ΔT:

$$Q = mc\,\Delta T = mc(T - T_0). \qquad (9.4)$$

The ΔT in Eq. (9.4) is the final temperature T of the object minus its initial temperature T_0 before the heat was added or removed. When the object's temperature increases, ΔT and Q are positive numbers (energy is added to the object). When the object's temperature decreases, ΔT and Q are negative numbers (the object loses energy).

EXAMPLE 9.4 A 30-kg child has a temperature of 102.2°F. How much heat must be removed from the child's body to lower his temperature to 98.6°F? (See Fig. 9.6.)

SOLUTION From Table 9.1 we see that the specific heat capacity of biological material $c = 3470$ J/kg · C°. This means that 3470 J of energy must be re-

*Electricity in southern California in 1981 cost about 1 cent for 6.0×10^5 J (or 6 cents per kilowatt hour). Our bath cost about 16 cents.

FIG. 9.6. Heat transfer occurs from a hot child to a cool tub of water.

FIG. 9.7. A massive wall is heated by sunlight. Air that circulates along the wall is warmed.

moved from 1 kg of biological material to lower its temperature by 1 C°. The child's mass is $m = 30$ kg. The desired temperature change is

$$\Delta T = T - T_0 = (98.6°\text{F} - 102.2°\text{F}) = -3.6\,\text{F}° = (-3.6\,\text{F}°)\left(\frac{5\,\text{C}°}{9\,\text{F}°}\right)$$

$$= -2.0\,\text{C}°.$$

Using Eq. (9.4), we find that the amount of heat energy that must be removed to lower the child's temperature is

$$Q = \left(3470\,\frac{\text{J}}{\text{kg}\cdot\text{C}°}\right)(30\,\text{kg})(-2.0\,\text{C}°) = \underline{-2.1 \times 10^5\,\text{J}}.$$

The negative sign reminds us that heat was transferred out of the child's body.

■

As we see in Table 9.1, the specific heat capacity for different materials varies considerably. This variation affects our lives in many ways. Notice, for example, that the specific heat capacity of water is about five times greater than that of sand. This is one reason why the temperatures of cities near large bodies of water remain more stable on a daily and seasonal basis than the temperatures of cities inland. During the daylight hours, absorption of sunlight causes a temperature increase roughly five times greater in sandy terrain than in a large body of water. At night, as heat is lost, the temperature of the sandy terrain drops more rapidly than that of land near a body of water.

Specific heat capacity is also important to people interested in solar heating and cooling. They must carefully consider the specific heat capacity of meterials used to store solar energy. During the day, energy gained from sunlight is used to warm this storage material. At night the energy is released to warm the home. It is important to have a storage material with a high specific heat capacity, one that will store large amounts of energy per unit mass.

In passive solar heating, sunlight coming through a window is absorbed by some surface in the house (Fig. 9.7). If the surface has a low specific heat capacity and a small mass, it will warm quickly, making the daytime temperature uncomfortably high. At night it will cool quickly, resulting in uncomfortably low temperatures. It is better to have the sunlight absorbed by a massive wall with a high heat capacity. The resulting temperature change $\Delta T = Q/mc$ will be small because both m and c are large.

9.5 Calorimetry

In order to calculate the heat Q required to raise an object's temperature by a certain amount, we must know the object's specific heat capacity, c. *Calorimetry* is an experimental technique used to measure specific heat capacity. Calorimetry is also used to measure the energy required to melt or boil a substance, to measure the metabolic rates of animals, and to calculate the chemical energy released during chemical reactions and even during explosions.

Calorimetry measurements are made using a device called a *calorimeter*. Calorimeters vary considerably in construction, depending on the type of measurement to be made. A calorimeter used to measure the specific heat capacity of a solid material is shown in Fig. 9.8. It consists of an insulated can containing a

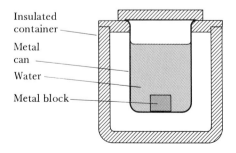

FIG. 9.8. A schematic representation of a calorimeter.

known amount of water. The insulation around the can (for example, styrofoam or a vacuum jacket) prohibits the transfer of heat into or out of the can.

The can and water are set at some known initial temperature, and the object whose specific heat capacity we wish to measure is warmed or cooled to a temperature different from that of the water and can. The object is then placed in the water. The can, water, and object eventually reach the same final temperature. If the object starts at a higher temperature, then it cools by transferring heat to the water and can. The water and can, in turn, are warmed. The net heat change is zero since heat gained by the water and can equals that lost by the object, or vice versa.

We are able to write this heat-transfer process as follows:

$$Q_{\text{can}} + Q_{\text{water}} + Q_{\text{object}} = 0, \qquad (9.5)$$

where Q_{can}, Q_{water}, and Q_{object} are the heats gained or lost by the can, water, and object, respectively. If heat is gained, Q is positive; if heat is lost, Q is negative.

We can use Eq. (9.4) ($Q = cm\,\Delta T$) to calculate the heat change for each term in Eq. (9.5). We find, then, that

$$c_{\text{can}} m_{\text{can}} \Delta T_{\text{can}} + c_{\text{water}} m_{\text{water}} \Delta T_{\text{water}} + c_{\text{object}} m_{\text{object}} \Delta T_{\text{object}} = 0.$$

The only unknown in the above equation is c_{object}. Thus, we are able to solve for the unknown specific heat capacity.

EXAMPLE 9.5 An insulated aluminum can whose mass is 0.20 kg contains 0.30 kg of water at 25°C. A 0.10-kg metal block whose temperature is 80°C is lowered into the water. The final temperature of the water, can, and block is 30°C. What is the specific heat capacity of the metal block?

SOLUTION In this example the metal block gives up heat as its temperature is lowered from 80°C to 30°C:

$$\begin{aligned} Q_{\text{block}} &= (c_{\text{block}})(m_{\text{block}})(T - T_0)_{\text{block}} \\ &= (c_{\text{block}})(0.10\ \text{kg})(30°\text{C} - 80°\text{C}) \\ &= (c_{\text{block}})(0.10\ \text{kg})(-50\ \text{C}°) \\ &= -(c_{\text{block}})(5\ \text{kg} \cdot \text{C}°). \end{aligned}$$

The water and can both gain energy as their temperatures rise from 25°C to 30°C:

$$Q_{\text{water}} = (c_{\text{water}})(m_{\text{water}})(T - T_0)_{\text{water}}$$

$$= \left(4180\ \frac{\text{J}}{\text{kg} \cdot \text{C}°}\right)(0.30\ \text{kg})(30°\text{C} - 25°\text{C}) = +6270\ \text{J},$$

$$Q_{\text{can}} = (c_{\text{can}})(m_{\text{can}})(T - T_0)_{\text{can}}$$

$$= \left(900\ \frac{\text{J}}{\text{kg} \cdot \text{C}°}\right)(0.20\ \text{kg})(30°\text{C} - 25°\text{C}) = +900\ \text{J}.$$

Note that we used the specific heat of aluminum when calculating Q_{can}.

Since no heat has left the calorimeter, the sum of the three heat terms must be zero [Eq. (9.5)], or

$$900\ \text{J} + 6270\ \text{J} - (c_{\text{block}})(5\ \text{kg} \cdot \text{C}°) = 0.$$

After rearranging the equation, we find that

$$c_{block} = \frac{(900 + 6270)\,J}{5\ kg \cdot C^\circ} = 1430\ \frac{J}{kg \cdot C^\circ}.$$ ∎

Calorimeters have been built to measure the resting or basal metabolic rate (BMR) of humans and animals. The person or animal rests in a calorimeter, which is insulated so that heat cannot be transferred through its walls. Air at a temperature lower than the body temperature is circulated into the container and then out of it. The air is warmer when it leaves the container than when it enters; the energy gained by the air is energy lost by the person's or animal's body. The energy gained by the air per unit time equals the basal metabolic rate of the person or animal. An actual human calorimeter is a more complicated device that measures moisture lost by evaporation and radiant energy lost by the body. Conceptually, however, it works as we have described.

EXAMPLE 9.6 A rabbit, when placed in an animal calorimeter, each second causes 2.0×10^{-3} kg of circulating air to increase in temperature by 2.7 C°. Calculate the basal metabolic rate of the rabbit.

SOLUTION The basal metabolic rate is the rate at which heat is added to the air, or is $Q_{air}/\Delta t$. Using this and Eq. (9.4), we find that

$$BMR = \frac{Q_{air}}{\Delta t} = \frac{mc_{air}\Delta T}{\Delta t} = \frac{m}{\Delta t}c_{air}\Delta T.$$

We are given that the mass of air transported through the calorimeter per unit time is

$$\frac{m}{\Delta t} = 2.0 \times 10^{-3}\ \frac{kg}{s}.$$

From Table 9.1, the specific heat capacity of the air $c_{air} = 1000$ J/kg · C°. The temperature change of the air $\Delta T = +2.7$ C°. Thus,

$$BMR = \left(2.0 \times 10^{-3}\ \frac{kg}{s}\right)\left(1000\ \frac{J}{kg \cdot C^\circ}\right)(2.7\ C^\circ)$$

$$= 5.4\ \frac{J}{s} = \underline{5.4\ watts}.$$ ∎

9.6 Melting and Freezing

When heat is added to or removed from matter, other changes besides a temperature change can occur. For example, the heat can cause a change in state of the matter. Most forms of matter can exist in three different states—solid, liquid, and gas. Water, for example, can exist as ice, as liquid water, or as gaseous water vapor. In this section we consider the energy involved in melting a solid or in freezing a liquid. The subject of the next section is boiling and condensation—transitions between the liquid and gaseous states of matter.

Melting Solids

The atoms and molecules in a solid are usually bonded to each other in a well-defined "crystalline" structure. They can vibrate about an equilibrium position but are unable to rotate or move to new positions in the solid. The organization of water molecules in ice is depicted in Fig. 9.9a. The oxygen atom of a water molecule shares hydrogen atoms with each of four neighboring water molecules.

If heat is added to a solid such as ice, its temperature increases. The atoms and molecules vibrate with greater amplitude. As more and more heat is added, eventually the vibrations become so violent that the atoms or molecules break away from their neighbors and move about more freely. The matter has melted: it has changed from the solid state to the liquid state. When water melts, neighboring water molecules no longer share hydrogen atoms (Fig. 9.9b). Each water molecule has its own two hydrogens. This change of state usually occurs at a well-defined temperature called the **melting temperature,** which varies from substance to substance—the melting temperature of water is 0°C, whereas that of copper is 1083°C.

The amount of energy required to break the crystalline structure of a solid to form a liquid also differs for different substances. A quantity called the heat of fusion (or more formally, the latent heat of fusion) is a measure of the energy required to melt a solid. The **heat of fusion** L_f of a solid substance is defined as the energy needed to melt 1 kg of the solid into the liquid state at the melting temperature of the substance. The units of L_f are J/kg. For example, 3.35×10^5 J of energy is required to melt 1 kg of ice at 0°C; thus, the heat of fusion of ice is 3.35×10^5 J/kg. A list of melting temperatures and heats of fusion of different substances appears in Table 9.2.

The amount of energy needed to melt larger or smaller amounts of matter is directly proportional to the mass m of the matter involved. For example, twice as much energy is required to melt 2 kg of ice as to melt 1 kg of ice.

In general, we can say that the heat Q needed to melt a mass m of solid is

$$Q = mL_f. \tag{9.6}$$

This is always a positive number since energy must be added to the material to melt it.

H$_2$O solid

H$_2$O liquid

(b)

FIG. 9.9. (a) The crystalline structure of the water molecules in ice. (b) The more random arrangement of the molecules in liquid water.

TABLE 9.2 Heats of Fusion and Vaporization

Substance	Melting		Boiling	
	Melting temperature (°C)	L_f, Heat of fusion (J/kg)	Boiling temperature (°C)	L_v, Heat of vaporization (J/kg)
Water	0.00	3.35×10^5	100.00	2.256×10^6
Ethanol	−114	10.42×10^5	78	0.854×10^6
Hydrogen (H$_2$)	−259.31	0.586×10^5	−252.89	0.452×10^6
Oxygen (O$_2$)	−218.83	0.138×10^5	−182.97	0.213×10^6
Nitrogen (N$_2$)	−210.10	0.257×10^5	−195.81	0.201×10^6
Aluminum	658.7	3.984×10^5	2300	10.5×10^6
Copper	1083	1.34×10^5	2600	3.1×10^6
Iron	1530	2.721×10^5	2500	6.364×10^6

Freezing Liquids

Freezing is the opposite of melting. If energy is removed from a liquid, its temperature will decrease; the vibrational, rotational, and translational motions of the atoms and molecules in the liquid are reduced. When the liquid reaches what is called its freezing temperature, the random motion of the atoms and molecules is slow enough that they start to fuse or bond with each other; the solid state begins to form. This transition from the liquid to the solid state is called freezing and occurs at the same temperature at which the solid melts. Thus, the freezing and melting temperatures of a substance are the same.

As the atoms or molecules fuse together into the crystalline structure of the solid state, energy is released. The amount of energy released equals that needed to break the bonds when the solid melts.

The energy Q released when a mass m of liquid freezes to form the solid state is

$$Q = -mL_f. \tag{9.7}$$

The minus sign reminds us that the liquid loses or releases energy when it freezes.

It is important to note that a substance does not change temperature while it is melting or freezing. For example, when heat is added to ice at $0°C$, no temperature change occurs in the ice-water mixture until all of the ice melts. Similarly, if heat is removed from water at $0°C$, it does not change temperature until all of the water has been converted to ice.

EXAMPLE 9.7 (a) Calculate the heat that must be added to 500 g of ice at $0°C$ to melt it. (b) Calculate the extra energy needed to raise the temperature of the melted ice from $0°C$ to $22°C$.

SOLUTION (a) We use Eq. (9.6) to find the heat that must be added to the ice to melt it:

$$Q = mL_f = (0.5 \text{ kg})(3.35 \times 10^5 \text{ J/kg}) = \underline{1.7 \times 10^5 \text{ J}}.$$

It has now been converted from ice to water at $0°C$. (b) To calculate the heat required to raise the temperature of 0.5 kg of water from $0°C$ to $22°C$, we use Eq. (9.4). The specific heat capacity of water is $4180 \text{ J/kg} \cdot \text{C}°$ (Table 9.1). Thus,

$$Q = mc \, \Delta T = (0.5 \text{ kg})(4180 \text{ J/kg} \cdot \text{C}°)(22° - 0°C)$$
$$= \underline{0.46 \times 10^5 \text{ J}}.$$

The total energy required to change 0.5 kg of ice at $0°C$ to 0.5 kg of water at $22°C$ is $1.7 \times 10^5 \text{ J}$ plus $0.46 \times 10^5 \text{ J}$ or $2.2 \times 10^5 \text{ J}$. ∎

EXAMPLE 9.8 Ten grams of ice at $0°C$ is placed in a styrofoam cup containing 300 g of coffee initially at $90°C$. Calculate the final temperature T of the coffee after all the ice has melted. Assume that no heat leaves the styrofoam cup. The specific heat capacity of coffee is $4180 \text{ J/kg} \cdot \text{C}°$, the same as that of water.

SOLUTION Since no heat leaves the cup, the coffee must lose an amount of heat equal to that gained by the ice, and the net heat change is zero:

$$Q_{\text{ice}} + Q_{\text{coffee}} = 0. \tag{9.8}$$

The heat gained by the ice has two parts: (1) that needed to melt it; and (2) that needed to raise the melted ice's temperature from 0°C to its final temperature T. Thus,

$$Q_{\text{ice}} = m_{\text{ice}}L_f + m_{\text{ice}}c_{\text{water}}(T - 0°\text{C}).$$

The heat lost by the coffee is given by

$$Q_{\text{coffee}} = m_{\text{coffee}}c_{\text{coffee}}(T - 90°\text{C}).$$

Notice that the final temperature T is less than 90°C, hence Q_{coffee} is negative. Substituting these two expressions into Eq. (9.8), we find that

$$\overbrace{\underbrace{m_{\text{ice}}L_f}_{\substack{\text{to melt} \\ \text{ice}}} + \underbrace{m_{\text{ice}}c_{\text{water}}(T - 0°\text{C})}_{\substack{\text{to raise } T \text{ of} \\ \text{melted ice}}}}^{\text{heat gained by ice}} + \overbrace{m_{\text{coffee}}c_{\text{coffee}}(T - 90°\text{C})}^{\text{heat lost by coffee}} = 0.$$

When solved for T, we find that

$$
\begin{aligned}
T &= \frac{m_{\text{coffee}}c_{\text{coffee}}(90°\text{C}) - m_{\text{ice}}L_f}{m_{\text{ice}}c_{\text{water}} + m_{\text{coffee}}c_{\text{coffee}}} \\
&= \frac{(0.3\,\text{kg})(4180\,\text{J/kg} \cdot \text{C}°)(90°\text{C}) - (0.01\,\text{kg})(3.35 \times 10^5\,\text{J/kg})}{(0.3\,\text{kg})(4180\,\text{J/kg} \cdot \text{C}°) + (0.01\,\text{kg})(4180\,\text{J/kg} \cdot \text{C}°)} = \underline{85°\text{C}}. \blacksquare
\end{aligned}
$$

EXAMPLE 9.9 (Making ice cream) An ice cream maker is shown in Fig. 9.10. The ingredients of the ice cream at 0°C are placed in the inner, metal container. Unfortunately, the ingredients do not solidify until their temperature is lowered several degrees below zero. How can we lower the temperature of these ingredients below zero using ice whose temperature is approximately 0°C?

SOLUTION The freezing temperature of ice is lowered by the presence of salt. For example, a 10 percent (by weight) solution of salt water freezes at −6.5°C. If we surround the metal container holding the ice cream ingredients with a mixture of salt and ice, the salt that contacts the ice causes it to melt. Remember that the salt ice is above its freezing temperature. But energy is needed to melt a solid. The energy is drawn out of the ice cream ingredients and ice that does not contact the salt, thus lowering their temperatures. In summary, energy is added to the salt ice, causing it to melt. The energy is removed from the ice cream ingredients and other parts of the ice, causing their temperature to decrease. We can write a heat-balance equation in the following way:

Insulated container

Ice cream

Ice and salt

Stirring motor

$$\overbrace{\underbrace{m_i L_f}_{\substack{\text{to} \\ \text{melt} \\ \text{ice}}} + \underbrace{m_i c_i(T - 0°\text{C})}_{\substack{\text{to lower } T \\ \text{of ice}}}}^{\substack{\text{net heat gained} \\ \text{by ice}}} + \overbrace{m_{ic}c_{ic}(T - 0°\text{C})}^{\substack{\text{heat lost by} \\ \text{ice cream}}} = 0,$$

where m_i is the mass of the salt ice, L_f is its heat of fusion, and c_i is its specific heat capacity; T is the final temperature of the ice cream and salt ice (less than 0°C), m_{ic} is the mass of the ice cream, and c_{ic} is its specific heat capacity. Solving for T,

FIG. 9.10. An ice cream maker.

we find that

$$T = -\frac{m_i L_f}{m_i c_i + m_{ic} c_{ic}}.$$

We see that the final temperature will be below zero, since the final temperature equals a group of positive numbers having a negative sign in front of them.

■

9.7   Boiling and Condensation

Boiling Liquids

Even in the liquid state, there are forces of attraction between neighboring atoms and molecules. In order for a molecule to leave the surface of a liquid, it must have enough energy to break away from the attracting forces of its neighbors (Fig. 9.11). When heat is added to the liquid, its temperature increases. The atoms and molecules move about faster and bump into each other harder. Eventually, some molecules in the liquid have enough energy to break away from their neighbors and leave the surface to become free molecules in the gaseous atmosphere above the liquid. This is called **evaporation.** As the liquid's temperature increases, the rate of evaporation increases because more molecules have enough energy to break away from their neighbors.

If enough heat is added to a liquid, eventually so many molecules will be leaving the liquid that all the extra energy added to the liquid is used by these escaping molecules. No energy is left to increase the temperature of the molecules remaining behind, and the liquid is said to be **boiling.**

The **boiling temperature** of a liquid is the temperature at which any heat added to the liquid causes the liquid to evaporate but does not cause an increase in temperature. Like freezing temperature, boiling temperature varies from substance to substance; at atmospheric pressure water boils at 100°C, and the boiling temperature of copper is 2300°C.

The amount of energy required to convert a liquid to a gas depends on the particular liquid. A quantity called the heat of vaporization (or the latent heat of vaporization) is a measure of the energy needed to boil a liquid. The **heat of vaporization** L_v of a substance is the energy required to convert 1 kg of the substance from the liquid to the gaseous state at the liquid's boiling temperature. The units of L_v are J/kg. The heat of vaporization of water is 2.256×10^6 J/kg. Table 9.2 lists the boiling temperatures and heats of vaporization for a variety of substances.

More energy is required to boil away a large mass of liquid than a small mass.

The amount of heat Q required to boil a mass m of substance whose latent heat of vaporization is L_v is

$$Q = mL_v. \tag{9.9}$$

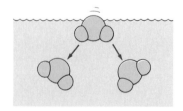

FIG. 9.11. A water molecule at the surface of the water is prevented from leaving the water by its attraction to neighboring molecules. Only the most energetic molecules can escape.

EXAMPLE 9.10 How much heat is required to evaporate 20 g of ethanol?

SOLUTION The latent heat of vaporization of ethanol is $L_v = 8.5 \times 10^5$ J/kg. Thus, the heat required to evaporate 20 g is

$$Q = (0.020 \text{ kg})(8.5 \times 10^5 \text{ J/kg}) = \underline{1.7 \times 10^4 \text{ J}}.$$

Ethanol (rubbing alcohol) is often applied to a person's skin in order to cool the body. Heat is transferred from the body and skin to the ethanol, causing it to evaporate. ∎

Condensation

Condensation, the reverse of evaporation, occurs when gas molecules are cooled to their boiling temperature. The further removal of energy causes the gas molecules to come together (that is, condense) to form a liquid. The **condensation temperature** is the same as the boiling temperature.*

The heat given up when 1 kg of gas condenses to form 1 kg of liquid is called the **heat of condensation** and equals the heat of vaporization.

The heat energy Q released when a mass m of gas condenses to the liquid state is

$$Q = -mL_v. \tag{9.10}$$

The negative sign reminds us that the gas releases energy when it liquefies.

The temperature at which a liquid boils or a gas condenses depends on the pressure of the gas above the liquid. Table 9.3 shows the variation of the boiling temperature of water as a function of pressure. As the air pressure above the water decreases, so does its boiling temperature. This variation of boiling temperature with pressure has a significant effect on cooking times. In Denver, Colorado, the air pressure is about 610 millimeters of mercury (mm Hg) compared to a pressure of 760 mm Hg at sea level. We see from Table 9.3 that water boils at about 94°C at Denver. It takes longer to boil an egg in Denver because the cooking temperature is lower.

The moon has no atmosphere, and its air pressure is nearly zero. On the moon, water boils at a very low temperature—below 0°C. An astronaut standing on the moon must wear a pressurized suit to prevent his or her blood from boiling.

EXAMPLE 9.11 (Electric power plants) An electric power plant is illustrated in Fig. 9.12. Water is heated by burning coal, oil, or gas (a fossil fuel plant) or by energy released by nuclear fusion (a nuclear boiling water reactor). The water is heated under very high pressure from about 80°C to 260°C; it does not boil until reaching about 250°C. High-pressure steam produced above the water pushes through a steam turbine, causing its blades to rotate. An electric generator attached to the turning blades produces electricity. After leaving the turbine, the steam is cooled in a condenser and converts back to liquid water, which returns to the boiler for another cycle.

Let us make three calculations: (a) Estimate the energy transfer rate $Q/\Delta t$ to the water in the boiler needed to convert 8 kg/s of water to steam. (b) If 35 percent of the energy transferred to the water results in production of electrical

TABLE 9.3 Boiling Temperature of Water as a Function of Pressure

Pressure (mm of Hg)	Boiling T (°C)
1500	120.3
1000	108.0
760 (atmospheric pressure at sea level)	100.0
700	97.8
600	93.6
400	83.0
100	51.6
50	38.1
20	22.2
10	11.3
4.6	0.0
2.0	−10.8

*Some materials pass directly from the gaseous to the solid state or vice versa without becoming a liquid. When this occurs, the solid is said to **sublime**. The heat required to make this transition is called the **heat of sublimation.** Dry ice, the solid form of carbon dioxide, is such a substance. It sublimes from the solid, dry-ice state to the gaseous state at a temperature of about −79°C.

FIG. 9.12. A schematic representation of a fossil fuel electric power plant.

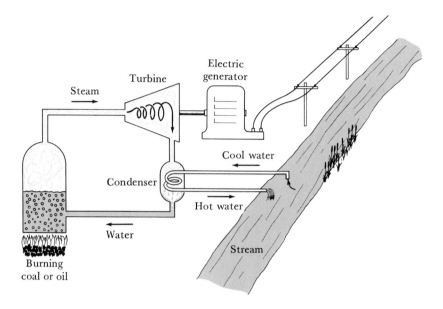

energy, what is the rate of electric power production? (c) If the other 65 percent of the energy is carried away by cooling water at the condenser, what is the rate of heat transfer to the environment?

SOLUTION (a) We wish to convert 8 kg/s of water at 80°C into steam at 260°C. Because of the high pressure in the boiler, water does not boil until about 250°C. The rate of heat transfer will be

$$
\frac{Q}{\Delta t} = \underbrace{\frac{\Delta m}{\Delta t} c_{\text{water}}(250°\text{C} - 80°)}_{\substack{\text{to raise water} \\ \text{to } 250°\text{C}}} + \underbrace{\frac{\Delta m}{\Delta t} L_v}_{\substack{\text{to boil} \\ \text{water}}} + \underbrace{\frac{\Delta m}{\Delta t} c_{\text{steam}}(260° - 250°\text{C})}_{\substack{\text{to raise } T \text{ of steam} \\ \text{from } 250°\text{C to } 260°\text{C}}}
$$

$$
= \left(8\,\frac{\text{kg}}{\text{s}}\right)\left(4180\,\frac{\text{J}}{\text{kg} \cdot \text{C}°}\right)(170\,\text{C}°) + \left(8\,\frac{\text{kg}}{\text{s}}\right)\left(2.3 \times 10^6\,\frac{\text{J}}{\text{kg}}\right)
$$

$$
+ \left(8\,\frac{\text{kg}}{\text{s}}\right)\left(2000\,\frac{\text{J}}{\text{kg} \cdot \text{C}°}\right)(10\text{C}°)
$$

$$
= 24 \times 10^6\,\frac{\text{J}}{\text{s}} = 24 \times 10^6\,\text{watts} = \underline{24\,\text{megawatts}}.
$$

(b) The electric power produced is 0.35×24 megawatts = 8.4 megawatts. (c) The remaining 65 percent of the power, or 0.65×24 megawatts = 15.6 megawatts is transferred to the environment as thermal pollution. This huge amount of heat is one of the environmental problems of most types of electric power plants. ∎

9.8 Thermal Expansion

In addition to causing changes in the temperature and the state of a substance, the transfer of heat can also cause a change in the volume of a substance. As its temperature increases, its atoms and molecules vibrate with greater amplitude.

FIG. 9.13. A road that buckled because of expansion during the summer heat. (Courtesy of the New York State Dept. of Transportation)

If free to translate, they move faster and have more energetic collisions. Because of their increased motion, the atoms and molecules occupy slightly more space than when moving less vigorously. They are, on the average, spread farther apart when the substance is hot than when cool. Therefore, almost all forms of matter expand when heated and contract when cooled.

Architects and engineers must consider thermal expansion and contraction of metal beams when designing tall buildings and bridges. Expansion joints are usually placed in sidewalks, bridges, and highways to prevent their buckling during summer heat. The road shown in Fig. 9.13 did not have an expansion joint, and as its length increased on a hot summer day, the road buckled. Dentists must use filling materials that expand and contract with temperature change at the same rate as a tooth; otherwise, the filling will become loose.

The change in length—that is, **linear expansion**—ΔL of an object that is warmed or cooled depends on its change in temperature ΔT, on its original length L, and on the type of material of which it is made. These quantities are related by the equation

$$\Delta L = \alpha L \, \Delta T. \tag{9.11}$$

The quantity α is called the **coefficient of thermal expansion.** From Eq. (9.11) we see that $\alpha = \Delta L / L \, \Delta T$ and has units of inverse degrees ($C^{\circ -1}$), since the length units cancel. Table 9.4 lists the values of α for various materials. The coefficient α is not a constant for a particular material but changes slightly with temperature. However, because the change is usually small, we will ignore it in our problems and examples.

EXAMPLE 9.12 A steel beam in a bridge extends 25 m across a small stream. What is its change in length from the winter, when its temperature is $-20°C$, to the summer, when it is $38°C$?

TABLE 9.4 Coefficients of Thermal Expansion at 20°C

Substance	Linear α ($C^{\circ -1}$)	Volume γ ($C^{\circ -1}$)
Aluminum	25×10^{-6}	72×10^{-6}
Steel and iron	12×10^{-6}	36×10^{-6}
Glass (pyrex)	3×10^{-6}	9×10^{-6}
Glass (jar)	$\sim 10 \times 10^{-6}$	$\sim 30 \times 10^{-6}$
Brick and concrete	$\sim 10 \times 10^{-6}$	$\sim 30 \times 10^{-6}$
Rubber	$\sim 80 \times 10^{-6}$	$\sim 240 \times 10^{-6}$
Ethanol	250×10^{-6}	750×10^{-6}
Methanol	400×10^{-6}	1200×10^{-6}
Gasoline	$\sim 300 \times 10^{-6}$	$\sim 900 \times 10^{-6}$
Air		3670×10^{-6}

SOLUTION The change in temperature of the bridge is

$$\Delta T = 38°C - (-20°C)$$
$$= 58\ C°.$$

Using Eq. (9.11), we find that the bridge's change in length is

$$\Delta L = \alpha L\ \Delta T = (12 \times 10^{-6}\ C°^{-1})(25\ m)(58\ C°)$$
$$= 0.017\ m = \underline{1.7\ cm}.\qquad\blacksquare$$

EXAMPLE 9.13 The aluminum lid of a jar of dill pickles is stuck to the jar. To loosen the lid so it can be opened, hot water is poured over the lid, causing it to expand. If the temperature increase of the lid and glass is 40 C°, calculate the change in circumference of the lid and of the glass on which it is screwed. The diameter of the lid before heating is 22 cm.

SOLUTION The lid and glass expand uniformly in all directions. Thus, as each warms, its diameter d increases by an amount Δd equal to

$$\Delta d = \alpha d\ \Delta T,$$

where α is the coefficient of linear expansion of the aluminum lid $(25 \times 10^{-6}\ C°^{-1})$ or glass rim $(10 \times 10^{-6}\ C°^{-1})$, and ΔT is the temperature change. The circumference of the circle changes from πd to $\pi(d + \Delta d)$ or by an amount $\pi\ \Delta d = \pi\alpha d\ \Delta T$. For the aluminum lid, the circumference change is

$$\pi\alpha_{al}d\ \Delta T = \pi(25 \times 10^{-6}\ C°^{-1})(22\ cm)(40\ C°)$$
$$= 0.069\ cm = \underline{0.69\ mm}.$$

The circumference of the glass changes by

$$\pi\alpha_{glass}d\ \Delta T = \pi(10 \times 10^{-6}\ C°^{-1})(22\ cm)(40\ C°)$$
$$= 0.028\ cm = \underline{0.28\ mm}.$$

Since the aluminum expands more than the glass, we should now be able to open our jar of pickles. $\blacksquare$

Because a material usually expands uniformly in all directions as it becomes warmer, its volume increases. The change in volume ΔV of a substance depends on its change in temperature, its original volume V, and on the material of which it is made:

$$\Delta V = \gamma V\ \Delta T. \qquad (9.12)$$

The quantity γ (the Greek letter gamma) is called the **coefficient of volume expansion;** its value depends on the substance that is expanding or contracting. The units of γ are $C°^{-1}$, the same as those of the coefficient of linear expansion. A list of the values of γ for a variety of substances appears in Table 9.4.

The coefficient of volume expansion γ of most substances is about three times the coefficient of linear expansion α of that same substance. This seems reasonable since the volume of an object is the product of three linear dimensions, such as length, height, and width, and each expands as it is heated. Although the mathematics needed to prove this statement is a little more difficult than this simple argument implies, we can see that it is true by comparing the values of α and γ in Table 9.4.

EXAMPLE 9.14 A bunsen burner heats the air in a cylinder whose original volume is 250 cm³ when at 20°C. A movable piston at the top of the cylinder allows the volume to change. Calculate the volume of the air after it has warmed to 40°C.

SOLUTION The coefficient of volume expansion of air listed in Table 9.4 is $\gamma = 3670 \times 10^{-6}$ C°⁻¹. The volume change then will be

$$\Delta V = \gamma V \Delta T = (3670 \times 10^{-6} \text{ C}°^{-1})(250 \text{ cm}^3)(40°\text{C} - 20°\text{C}) = 18 \text{ cm}^3.$$

The volume of the gas will now be $250 + 18 = \underline{268 \text{ cm}^3}$. ■

FIG. 9.14. Since water at 4°C is denser than water at 0°C, the warmer water sinks. The cold water freezes at the top of a lake rather than at the bottom.

We have said that most substances expand as they warm and contract as they cool. There is one notable and important exception. When water is warmed from 0°C to 4°C, it contracts and becomes more dense; that is, it weighs more per unit volume. Notice in Fig. 9.14 that the dense 4°C water has sunk to the bottom of the pond. The colder and less dense 0°C water floats at the top. If the water begins to freeze, the cold water at the top freezes first. Consequently, a lake freezes from the top down rather than from the bottom up. During the winter, fish remain near the bottom of the lake in the slightly warmer water.

The change in volume of matter as it warms and cools also affects our atmosphere. We see from Table 9.4 that gases such as air, when heated, expand much more than solids and liquids. As a gas expands, it becomes less dense; its molecules on the average are spaced farther apart than in a cool gas. For this reason a hot (less dense) gas will rise and float on a cool (more dense) gas. This property of gases is very important in cleansing the atmosphere surrounding our large cities. Air warmed at the earth's surface rises and carries with it air contaminants. Cool, clean air sinks down from above to replace the warm air.

Occasionally, an inversion occurs. For some reason, the air at the earth's surface remains cool and is covered by warmer air above. The temperature is said to be inverted. The cool, dense air at the surface does not rise, and pollution accumulates until the inversion breaks.

Summary

1. **Temperature scales:** An object's temperature is a measure of the average random kinetic energy of its atoms and molecules. Three temperature scales are used to measure temperature: the Celsius scale, the Fahrenheit scale, and the Kelvin scale. Temperatures in the Celsius and Fahrenheit scales are related by the equations $T_F = \frac{9}{5}T_C + 32$ or $T_C = \frac{5}{9}(T_F - 32)$. The Kelvin temperature is calculated by adding 273 degrees to the Celsius temperature.

2. **Thermal energy** E_{th} is the random kinetic energy of all of the atoms and molecules in a substance. **Heat** Q is the energy transferred to or from a substance because of a difference in its temperature and that of its environment. When calculating the energy changes of a system using the conservation-of-energy principle,

$$Q + W = \Delta E_{\text{system}}$$
$$= \Delta KE + \Delta PE_g + \Delta E_{th} + \cdots, \quad (9.3)$$

heat appears on the left side because it is energy added to or removed from the system. Thermal energy is on the right side because it is a form of energy in the system.

3. **Specific heat capacity:** The amount of heat Q that must be added to a mass m of substance in order to change its temperature by ΔT is calculated using the equation

$$Q = mc \, \Delta T, \quad (9.4)$$

where c is called the specific heat capacity of the substance.

4. The amount of energy consumed or released by a mass m of substance when it **changes state** can be calculated using the equations below:

To melt a solid	$Q = mL_f$.	(9.6)
To freeze a liquid	$Q = -mL_f$.	(9.7)
To vaporize a liquid	$Q = mL_v$.	(9.9)
To condense a gas	$Q = -mL_v$.	(9.10)

L_f and L_v are the latent **heats of fusion and vaporization**, respectively.

5. The **linear expansion** ΔL of an object with length L caused by

a temperature change ΔT is calculated using the equation

$$\Delta L = \alpha L \, \Delta T. \tag{9.11}$$

where α is called the **coefficient of thermal expansion.**

    The **change in volume** ΔV of a substance caused by a temper-ature change ΔT is given by

$$\Delta V = \gamma V \, \Delta T, \tag{9.12}$$

where V is the original volume of the substance, and γ is called the **coefficient of volume expansion.**

Questions

1.   We have said that a water molecule in a glass of water has 10^{14} collisions with other molecules in the glass each second. Estimate the number of years a college football player would have to play (24 hours a day) in order to have the same number of collisions. Explain all the assumptions used in your estimate.

2.   If you double the absolute temperature of a glass of ice water, what will its new temperature be in degrees Celsius? Explain.

3.   Deserts often have much greater daily temperature variations than vegetated land at the same latitude. Explain.

4.   A person has a solar collector that requires 2 m^3 of water to store the solar energy. Approximately what volume of rocks would store the same amount of energy (assuming the same temperature change)?

5.   Dublin in Ireland and Edmonton in Alberta, Canada, are at the same latitude ($53.5°$). The January and July average temperatures in Dublin are $40°F$ and $60°F$, respectively, while in Edmonton they are $10°F$ and $70°F$. Why is the temperature variation so much greater in Edmonton than in Dublin?

6.   A tray of hot water and a tray of cold water are placed in the freezer of a refrigerator. Which tray will freeze first? Explain.

7.   Why does spraying fruit trees with water before a frost help prevent the fruit from freezing? (*Note:* Because of the sugar in fruit, it freezes at about $-1°C$.)

8.   A farmer's fruit storage cellar is unheated. To prevent the fruit from freezing, the farmer places a barrel of water in the cellar. Why does this help prevent freezing?

9.   The melting temperature of ice decreases by about $1 \text{ C}°$ for each 1200 N/cm^2 of pressure on the ice. *Estimate* the melting temperature of ice under a 60-kg ice skater. Explain any assumptions you make.

10.   Your property must be surveyed, and measurements will be made with a metal tape. Your tax rate will be based on the amount of land you have. For tax purposes, is it better to have the survey done in the summer or in the winter? Explain.

Problems

9.2   Temperature

1.   (a) The body temperature of a chicken is about $107°F$. Calculate its temperature in degrees Celsius. (b) The temperature of a pig is $39.4°C$. Calculate its temperature in degrees Fahrenheit.

2.   The moon's temperature on its bright side is about $100°C$ and on its dark side, about $-173°C$. Calculate both these temperatures in degrees Fahrenheit and in degrees Kelvin.

3.   (a) The highest temperature recorded for the earth was $136°F$ on September 13, 1922, in Azizia, Libya. The United States record is $56.7°C$ on July 10, 1913, in Death Valley, California. By how much in Fahrenheit degrees did the United States record miss the earth's record? (b) The earth's record cold temperature was $-127°F$ on August 24, 1960, at Vostok Station, Antarctica. Calculate this temperature in degrees Celsius and Kelvin.

4.   The temperature of the center of the earth is about $10,000°F$. Calculate this temperature in degrees Celsius and Kelvin.

5.   A frog can survive for long periods with a body temperature anywhere between $-1.2°C$ and $30°C$. Calculate these extremes in degrees Fahrenheit.

6.   The ocean temperature off the shore of the state of Washington varies from about $15°C$ in the summer to $10°C$ in the winter. This is a $5\text{-C}°$ variation. What is the temperature variation in Fahrenheit and Kelvin degrees?

7.   Nitrogen changes from a liquid to a gas at $-196°C$. Calculate this temperature in degrees Fahrenheit and Kelvin.

9.3 and 9.4   Heat, Specific Heat Capacity, and the Conservation-of-Energy Principle

8.   A person will die if his or her body core cools from $37°C$ to about $31°C$. Calculate the heat that must be removed from the body of a 70-kg person to cause this temperature change.

9.   A 60-kg person consumes about 2000 kcal of food in one day. If 10 percent of this food energy is converted to thermal energy and cannot leave the body, calculate the temperature change of the person. (*Note:* 1 kcal = 4180 J.)

10.   Calculate the amount of heat required to raise the temperature (a) of 0.5 kg of water by $10 \text{ C}°$, (b) of 0.5 kg of ethanol by $10 \text{ C}°$, (c) of 0.5 kg of iron by $10 \text{ C}°$.

11.   The addition of $10,000 \text{ J}$ of heat to a container of water causes its temperature to increase by $7 \text{ C}°$. Calculate the mass of the water.

12.   When a 3-kg piece of metal gains $1.75 \times 10^4 \text{ J}$ of energy, its temperature increases by $15 \text{ C}°$. Calculate its specific heat capacity.

13.   A tree leaf has a mass of 0.8 g and a specific heat capacity of $3700 \text{ J/kg} \cdot \text{C}°$. The energy of the sunlight striking the leaf per

second is 2.8 J/s. If all the solar energy striking the leaf is absorbed, calculate the energy absorbed by the leaf (a) in 1 second, (b) in 1 minute. (c) If this energy were not removed from the leaf, calculate the temperature change of the leaf in 1 minute. (*Note:* Do not be surprised if your answer is large. A leaf clearly needs ways to remove heat absorbed from the sun. These mechanisms are discussed in the next chapter.)

14. Temperature is one of the most dynamic properties of soil. The daily temperature fluctuation of the top 10 cm of bare, clay soil on a summer day is about 14 C°. Calculate the net heat gained by 1 m² of soil that is 0.1 m deep when it is warmed 14 C°. The density of the clay soil is approximately 2.4×10^3 kg/m³, and its specific heat capacity is about 900 J/kg · C°.

15. (a) Calculate the energy required to increase the temperature of a 200-kg cast-iron car engine from 30°C to 1000°C (1000°C is about the melting temperature of iron). (b) If the engine of this car when idling generates thermal energy at a rate of about 8000 J/s, and if none of the energy is removed from the engine, how long will the car run before the block and engine melt?

16. The Superdome in New Orleans holds 76,791 people for a sporting event. The volume of air in the dome is about 3×10^6 m³. If all the seats in the dome are filled and each person transfers his metabolic thermal energy to the air in the dome at a rate of 100 watts (100 J/s), calculate the temperature change of the air in 2 hours. Assume that no heat leaves the air through the walls, floor, or ceiling of the dome. The density of air is 1.3 kg/m³ and its specific heat capacity is about 1000 J/kg · C°.

17. It is recommended for purposes of ventilation that the inside air in a home be replaced with outside air once every 2 hours. This air infiltration occurs naturally by leakage through tiny cracks around doors and windows, even in well-caulked and weather-stripped homes. The home's volume is 450 m³. The density of air is 1.3 kg/m³, and its specific heat capacity at constant pressure is about 1000 J/kg · C°. (a) Calculate the mass of air lost per 2 hours. (b) Calculate the mass of air lost per second. (c) Calculate the heat lost per second by this air if its temperature is changed by 38 C°, the difference in the temperature of the inside and outside air. (This figure is the energy per second needed to warm outside air leaking into the house.)

18. A 0.5-kg hammerhead traveling at 5 m/s hits a 7-g nail (a tenpenny nail) whose specific heat capacity is 470 J/kg · C°. (a) If the hammerhead stops when hitting the nail, what is the change in kinetic energy of the hammerhead? (b) Assume that 50 percent of the kinetic energy lost by the hammerhead is converted to thermal energy in the nail. Calculate the temperature increase of the nail. (c) If another 20 percent of the kinetic energy lost by the hammerhead is converted into thermal energy in the hammerhead, how much energy is available to drive the nail into the wood? Explain.

19. A 5-g lead bullet traveling at 300 m/s penetrates a wooden block and stops. If 50 percent of the initial kinetic energy of the block is converted into thermal energy in the bullet, by how much does the bullet's temperature increase?

20. (a) The great Arizona crater was created by the impact of a meteorite whose mass was estimated to be 5×10^8 kg. The meteorite's speed before impact was about 10,000 m/s. Calculate the kinetic energy of the meteorite before the collision. (b) If 20 percent of this energy is converted to thermal energy in the meteorite

whose specific heat capacity is 900 J/kg · C°, by how much will its temperature increase? (*Note:* Large amounts of rock found near the crater had melted on impact and then resolidified, indicating that the temperature of the rock during impact reached at least 1700°C.)

9.5 Calorimetry

21. A 300-g insulated aluminum calorimeter holds 150 g of water. The water and aluminum are initially at 30°C; 200 g of ethanol at 55°C is added to the water. Calculate the final temperature of the mixture.

22. A 250-g copper calorimeter holds 200 g of water. The water and copper are initially at 30.0°C. A 150-g metal block whose temperature is 90.0°C is placed in the calorimeter. The final temperature of the mixture is 34.1°C. Calculate the specific heat of the metal block.

23. To 200 g of coffee (essentially water) at a temperature of 70°C in an insulated cup, 25 g of milk at 10°C is added. If the specific heat capacity of milk is 3800 J/kg · C°, by how much is the coffee temperature lowered when the milk is added? Ignore the specific heat capacity of the cup.

24. A 500-g aluminum calorimeter contains 300 g of water. The water and aluminum are initially at 40°C. A cold 200-g iron block with a temperature of 0°C is added to the water. Calculate the final temperature of the mixture.

25. Bath water of 150 kg mass is at a temperature of 44°C. How much cool water at 20°C must be added to lower the temperature to 38°C? Ignore heat lost to the tub and air.

26. A small lake initially at 16°C and containing 5.0×10^9 kg of water absorbs 1.0×10^4 kg/s of 18°C water from a stream used to cool an electric power plant. Calculate the time needed to warm the lake 0.5 C°, assuming no heat is lost to the ground or air.

27. A man whose basal metabolic rate is 120 watts lies in a human calorimeter. Each minute 10.4 kg of air flows through the calorimeter. Calculate the change in temperature of the air if all the heat leaving the man's body is absorbed by the air.

28. A woman lies in a human calorimeter through which 1.0 m³ of air flows each minute. The air leaving the calorimeter is 5 C° warmer than air entering. Calculate the basal metabolic rate of the woman. The density of air is 1.3 kg/m³.

9.6 Melting and Freezing

29. Calculate the amount of energy needed to change a 0.5-kg block of ice at 0°C into water at 20°C.

30. When 1.4×10^5 J of energy is removed from 0.6 kg of water initially at 20°C, will all the water freeze? If not, how much remains unfrozen?

31. An electric heater warms a large 0°C block of ice at a rate of 200 J/s. Calculate the mass of ice that melts in 10 minutes.

32. A 300-g aluminum calorimeter containing 200 g of water has 100 g of ice at 0°C placed in it. The water and calorimeter are initially at 40°C. Calculate the final temperature of the calorimeter and its contents.

33. A 400-g copper calorimeter containing 250 g of water has 100 g of ice at 0°C placed in it. The final temperature of all the contents is 15°C. Calculate the initial temperature of the calorimeter and the 250 g of water.

34. A tub containing 50 kg of water is placed in a farmer's canning cellar, initially at 10°C. On a cold evening the cellar loses heat through the walls at a rate of 1200 J/s. Without the tub of water, the fruit would freeze in 4 hours (the fruit freezes at −1°C, because the sugar in the fruit lowers the freezing temperature). By how much time will the presence of the water delay the freeze? (*Hint:* Calculate the energy released in cooling the water to 0°C and then freezing it. This energy is made available to the room. The time delay is the extra time required to lose this energy through the walls.)

9.7 Boiling and Condensation

35. How much energy is required to convert (a) 0.1 kg of water at 100°C to steam at 100°C? (b) 0.1 kg of liquid ethanol at 78°C to ethanol vapor at 78°C?

36. Calculate the amount of energy required to convert 2 kg of water at 80°C to steam at 100°C.

37. A kettle containing 1.5 kg of boiling water absorbs heat from a gas stove at a rate of 600 J/s. How long must the water boil before it is converted entirely to water vapor?

38. During a back rub, 80 g of ethanol (rubbing alcohol) is converted from a liquid to a gas. Calculate the heat removed from your body for this conversion. Ignore the ethanol's change in temperature before it changes state.

39. An enclosed, insulated container holds 50 g of ethanol. To this, 50,000 J of heat is transferred. Calculate the final state and temperature of the ethanol if it is initially at 28°C. The heat capacity of ethanol in the gaseous state is 1800 J/kg · C°.

40. A nuclear power plant generates 1000 MW (megawatts) of waste heat. If this heat is disposed of in an evaporative cooling tower, how much water must be evaporated (a) per second? (b) per day?

9.8 Thermal Expansion

41. If one steel beam extended the entire length of the Sears Building in Chicago (1424 ft), by how much would the beam contract from its summer high temperature of 37°C to its winter low of −20°C?

42. The main steel span of the Golden Gate Bridge is about 1350 m long. By how much will it expand if its temperature is increased from 10°F to 100°F?

43. An iron steam-heating pipe extends 70 ft from the basement of a building to the fifth floor. What is the change in length of the pipe from the time when no steam flows ($T = 20$°C) to a time when it carries steam at 100°C?

44. An interstate highway is made of concrete slabs 25 m long placed end to end. (a) What expansion gap must be left between the slabs to prevent buckling when the temperature changes from 20°C to 50°C? (b) Starting with the gap calculated in (a), find the gap when the temperature decreases to −20°C.

45. Two concrete roadway slabs 25 m long are accidently laid without expansion gaps (Fig. 9.15). How high will the slabs buckle up if their temperatures increase from 10°C to 50°C?

FIG. 9.15

46. Two steel rails, each 50 ft long, are butted firmly together when at a temperature of 0°C. Their other ends are anchored to ties. How high will the butted section rise if the temperature increases to 100°C? The situation is similar to that pictured in Fig. 9.15 for the concrete highway slabs.

47. The length of a steel measuring tape 30 m long is calibrated under a specified tension at a temperature of 20°C. (a) How long is the tape when it is at 35°C? (b) What is the 20°C length of a piece of land that measures 252.50 m according to this tape when at a temperature of 35°C?

48. The volume of an iron ball of radius r is given by $\frac{4}{3}\pi r^3$. A particular ball has a diameter of 10.00 cm and is 0.01 cm too large to fit through the hole in a metal plate. What temperature change of the ball will allow it to fit through the hole?

49. A 50-gallon (1.9×10^5 cm³) steel drum at 5°C is filled to the brim with gasoline. The drum and its contents are warmed to 40°C. How much gasoline is lost through the open cap of the drum?

50. Suppose that the coefficient of volume expansion of a slender person were 10^{-3} C⁻¹. By how much would his or her temperature need to increase in order for the body volume to increase by 10 percent? Describe two problems you might have if this technique were used to increase your dimensions.

51. The radiator of a car is filled to the overflow level with 5.4 liters of water. The water is warmed from 25°C to 95°C. How much water spills out of the overflow tube? The coefficient of volume expansion of water in this temperature range is 550×10^{-6} C⁻¹.

General Problems

G1. A lightning flash releases about 10^{10} J of electrical energy. If all this energy is added to 50 kg of water (the amount of water in a 165-lb person) at 37°C, what is the final state and temperature of the water?

G2. A substance called Glauber salt ($Na_2SO_4 \cdot 10\,H_2O$) can be used as an energy-storage material for a solar collector. Energy collected from the sun raises the temperature of the salt, causing it to melt at 32°C; additional absorbed energy raises the tempera-

ture of the liquid salt. At night the energy is released as the salt cools and returns to the solid state. (a) Calculate the energy required to change the temperature of 1 kg of Glauber salt from 25°C to 50°C. (b) Does 1 kg of Glauber salt or 1 kg of water store more energy over the same temperature change? Explain. For Glauber salt, c (solid) = 1920 J/kg · C°, c (liquid) = 2840 J/kg · C°, and $L_f = 2.4 \times 10^5$ J/kg.

G3. (a) Calculate the energy that must be removed from 50 kg of water (about the amount in a person) to cool it by 2 C°. (b) Suppose that the heat removed in (a) were used to change the state of a small fraction of the liquid water to the gaseous state. How many kilograms of water would change state? (c) Describe in words the analogous mechanism for removing heat from a human body.

G4. Moisture in the air in the gaseous state condenses to form a cloud. A rainstorm follows, resulting in 2 cm of rain over an area of 2 km × 2 km. Estimate the energy released as the water con-

denses and drops to the earth. Note that the mass of 1 m³ of water is 1000 kg.

G5. A vacuum jar contains 100 g of water initially at 30°C. A vacuum pump lowers the pressure in the jar. (a) Using Table 9.3, estimate the pressure at which the water will start to boil. (b) If the latent heat of vaporization of water at this temperature is about 2.7×10^6 J/kg, how much heat is removed from the water when 1 g of the water boils away? (c) Calculate the temperature change in the remaining 99 g of water when this heat is removed.

P. 21, 39, 45, 49
33, 25, 18

10 Heat Transfer

In Chapter 9 we learned that heat addition or removal causes matter to change temperature or state and to expand or contract. In this chapter we consider the actual mechanisms by which heat is transferred to or from an object. These heat-transfer mechanisms include (1) conduction, (2) convection, and (3) radiation. We will also discuss a fourth mechanism, evaporation, which is really not a form of heat transfer but has the effect of cooling a body. These four mechanisms, either singly or in combination, are involved in the temperature control of a wide variety of systems such as our bodies, the buildings and homes in which we live, and the earth. Later in the chapter we will discuss the means by which the temperature of these systems is controlled.

10.1 Heat Transfer by Conduction

If we place a metal pan over the burner of a gas stove, the handle soon becomes too hot to hold (Fig. 10.1). Heat entering the pan from the flame is transferred from the pan to the handle by a process called *conduction*. The atoms in the hot part of the pan have more thermal energy than those in the cooler handle. When these hot atoms jiggle about and collide with their slower-moving neighbors, kinetic energy is transferred. The atoms that have just gained energy collide with their neighbors, and more thermal energy passes along the pan toward the handle. Eventually, atoms at the outer end of the handle are warmed by this bumping-type motion that passes from atom to atom along the pan. Heat transfer by **conduction** is a transfer of thermal energy from atom to atom through a material and is caused by the collision of hotter atoms with cooler, slower-moving atoms. The atoms themselves do not move from one end of the material to the other; only their random kinetic energy is transferred.

In most of the examples in this chapter we will consider the rate at which heat is transferred. The **heat-transfer rate** H is the amount of heat ΔQ that passes from one region of an object to another region divided by the time Δt needed for that heat transfer.

$$\text{heat-transfer rate} = H = \frac{\Delta Q}{\Delta t}. \tag{10.1}$$

The units of H are watts, since 1 watt is the ratio of a unit of heat, the joule, and a unit of time, the second (1 watt = 1 J/s).

(a)

(b)

"Ouch!"

(c)

FIG. 10.1. (a) A pan is placed on a stove. (b) The stove, on high, warms the pan. (c) Eventually the handle, heated by conduction, becomes too hot to hold.

OFF

HIGH

HIGH

The heat-conduction rate across an object, such as shown in Fig. 10.2, can be determined using the following equation:

$$\left(\frac{\Delta Q}{\Delta t}\right)_{conduction} = H_{cd} = \frac{KA(T_2 - T_1)}{L} \qquad (10.2)$$

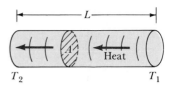

FIG. 10.2. Heat transfer by conduction occurs along a metal rod of cross-sectional area A and length L. The rate of heat transfer from region 1 at temperature T_1 to region 2 at temperature T_2 is given by Eq. (10.2).

H_{cd} is the **conductive heat-transfer rate** from region 1 at temperature T_1 to another part of the object, region 2, at temperature T_2; L is the distance the heat must travel from region 1 to region 2; and A is the cross-sectional area across which the heat travels. The quantity K, called the **thermal conductivity,** indicates the ease or difficulty of heat transfer by conduction through different types of materials. A list of the thermal conductivities of several materials appears in Table 10.1. Metals, which are good heat conductors, have large values of K; poor heat conductors such as air, wood, and styrofoam have small values of K. These latter materials are often called heat **insulators** and are used to prevent the escape of heat from ice chests, thermos bottles, and homes. In the SI metric system of units, K is in units of watts/m $\cdot$ C°, A is in m², L is in m, T_1 and T_2 are in °C, and H_{cd} is in watts.

In our heat-transfer calculations in this chapter we will be interested in the heat transfer to or from one particular region of some object. For example, we may want to know the rate of heat lost from our skin to the surrounding air or the rate at which heat passes from inside a room to the outside air. Equation (10.2) and *all of our heat-transfer equations are written so that we can calculate the rate at which heat is entering or leaving region 1 at temperature* T_1. The signs in Eq. (10.2) and in equations we develop later are such that, if $\Delta Q/\Delta t$ is positive, heat is entering region 1. If $\Delta Q/\Delta t$ is negative, heat is leaving region 1.

EXAMPLE 10.1 One end of a copper rod 1.2 m long and with a cross-sectional area of 3.0×10^{-4} m² is in boiling water at 100°C. The other end is in ice at 0°C. (a) Calculate the conductive heat-transfer rate to the ice. (b) How much time is required to melt 10 g of ice?

SOLUTION (a) We are given the following information: $L = 1.2$ m, $A = 3.0 \times 10^{-4}$ m², $T_2 - T_1 = T_{water} - T_{ice} = 100°C - 0°C = 100$ C°. We have chosen to call region 1 the ice side of the rod and region 2 the boiling-water side. Our calculations will then tell us the heat-transfer rate to the ice, since this is what we call region 1.

The conduction coefficient for copper is $K_{copper} = 385$ watts/m $\cdot$ C°. Substituting in Eq. (10.2), we find that

$$H_{cd} = \left(385 \frac{\text{watts}}{\text{m} \cdot \text{C}°}\right)(3.0 \times 10^{-4} \text{ m}^2)\frac{(100 \text{ C}°)}{(1.2 \text{ m})} = \underline{9.6 \text{ watts.}}$$

(b) We must now determine the amount of heat ΔQ needed to melt 0.01 kg of ice. Using Eq. (9.7) and the value of the heat of fusion L_f for ice that is given in Table 9.2, we find that

$$\Delta Q = mL_f = (0.01 \text{ kg})(335{,}000 \text{ J/kg}) = 3350 \text{ J.}$$

If we call the unknown time required to melt this ice Δt, then the heat-transfer rate must be

$$\frac{\Delta Q}{\Delta t} = \frac{3350 \text{ J}}{\Delta t}$$

TABLE 10.1 Thermal Conductivity K of Different Materials

Material	K (watts/m $\cdot$ C°)
Metals	
Aluminum	205
Brass	109
Copper	385
Steel	50.2
Other solids	
Dry soil	0.2
Red brick	0.6
Concrete	0.84
Glass	0.8
Ice	1.6
Styrofoam insulation	0.016
Wood	0.04
Fat	0.021
Muscle	0.042
Bone	0.042
Other materials	
Air	0.024
Water	0.59

Front view

Side view

FIG. 10.3. An earth house. The roof and three walls are covered with earth. A fourth wall of windows provides a view and collects solar energy in the winter.

and is equal to the actual conductive heat-transfer rate calculated from part (a), or

$$\frac{3350 \text{ J}}{\Delta t} = 9.6 \text{ watts.}$$

Therefore,

$$\Delta t = \frac{3350 \text{ J}}{9.6 \text{ watts}} = \frac{3350 \text{ J}}{9.6 \text{ J/s}} = \underline{350 \text{ s}}.$$ ∎

EXAMPLE 10.2 (a) A mobile home has walls that are 0.03 m thick and have a thermal conductivity of 0.1 watt/m · C°. The total area of the walls is 45 m². Calculate the heat-transfer rate from the inside of the home to the outside if the inside temperature is 20°C and the outside temperature is −5°C. (b) Compare the conductive heat-transfer rate of the mobile home walls to that of the roof of a large 45-m² room in an earth house covered with 0.5 m of dry soil (Fig. 10.3).

SOLUTION (a) In this problem the heat travels a short distance $L = 0.03$ m across a large area $A = 45$ m². The small value of L and large value of A both contribute to a high heat transfer by conduction. The temperature difference across which the heat travels is

$$T_2 - T_1 = T_{\text{outside}} - T_{\text{inside}} = (-5°C) - (20°C) = -25 \text{ C}°.$$

Substituting these values into Eq. (10.2), we find that

$$H_{cd} = \left(0.1 \frac{\text{watt}}{\text{m} \cdot \text{C}°}\right)(45 \text{ m}^2)\frac{(-25 \text{ C}°)}{0.03 \text{ m}} = \underline{-3750 \text{ watts}}.$$

The negative sign indicates that the heat flows away from region 1—that is, the region inside the house. The house is losing energy to the outside by heat conduction at a rate of 3750 watts. (b) Dry soil according to Table 10.1 has a thermal conductivity $K_{\text{soil}} \simeq 0.2$ watt/m · C°. Substituting $L = 0.5$ m, $A = 45$ m², $T_2 - T_1 = -25$ C°, and the value of K_{soil} into Eq. (10.2), we find that the conduction heat-transfer rate from the roof of the room up through the earth is

$$H_{cd} = \left(0.2 \frac{\text{watt}}{\text{m} \cdot \text{C}°}\right)(45 \text{ m}^2)\frac{(-25 \text{ C}°)}{0.5 \text{ m}} = \underline{-450 \text{ watts}}.$$

The extra thickness of the soil provides a barrier that greatly reduces the heat-loss rate from the house compared to the −3750-watt heat-transfer rate through the thin walls of the mobile home. ∎

10.2 Heat Transfer by Convection

Remember that conductive heat transfer means the transfer of kinetic energy from atom to atom: the atoms do not move from one place to another. **Convective heat transfer** is the transfer of thermal energy from one place to another by mass that moves between these places and carries the energy with it. An example of *forced convective heat transfer* is the forced-air heating system used to warm homes in the winter. A furnace warms air, the air is blown or forced by a fan through ducts to different parts of the house, and then cold-air ducts return cool air to the furnace for reheating. The thermal energy is carried by the air from the furnace to other parts of the house; hence this is a form of convective heat transfer.

FIG. 10.4. A car engine is cooled by convection. Moving water carries thermal energy away from the engine to the radiator.

The engine in an automobile is cooled by forced convection (Fig. 10.4). Water cooled in the radiator is pumped through the engine, where it absorbs heat generated by the engine. This warmed water then returns to the radiator for cooling. Heat is carried away from the engine by the flowing water.

A bicycle rider can lose energy as cool air moves across the skin (Fig. 10.5). The air carries heat away from the skin. Since the air mass is moving relative to the skin, we call this convective heat transfer. The examples just discussed are called *forced convection* because the fluid's motion past the object is caused by a mechanical force (a fan in the furnace pushes the air, a radiator pump pushes the water, and a bicycle rider pedals the bicycle through the air).

Natural convection occurs when a fluid such as air or water moves because of its change in temperature and density caused by absorption of heat from another object. During the day, for example, air near the earth's surface is warmed by its contact with the earth. As it warms, its volume increases. The air becomes less dense and rises, allowing cooler, more dense air from above to come down and replace the warm air. The air movement is a natural convection current. In a home air that is warmed near a radiator rises, and cool air near the ceiling falls. This natural convection current causes the room to become uniformly heated.

FIG. 10.5. The body of a cyclist is cooled convectively by air moving across the skin.

The calculation of the rate of convective heat transfer is a complex subject and depends on many factors. We can approximate the convective heat-transfer rate using the following equation:

$$\left(\frac{\Delta Q}{\Delta t}\right)_{\text{convection}} = H_{cv} = hA(T_2 - T_1) = hA(T_{\text{fluid}} - T_{\text{surface}}), \quad \textbf{(10.3)}$$

where ΔQ is the heat transferred convectively in a time Δt from or to a surface of area A across which a fluid such as air or water moves. T_{surface} (that is, T_1) is the temperature of the surface to or from which heat is being transferred, and T_{fluid} (that is, T_2) is the temperature of the fluid such as air or water that moves across that surface. The quantity h in Eq. (10.3) is called the **convection coefficient** and accounts for many of the other variables on which the convective heat-transfer rate depends, such as the type of fluid that crosses the surface, the speed of the fluid, and the shape and texture of the surface across which the fluid moves. The convection coefficient must be determined for each unique situation. Tables and empirical equations are often used to estimate the value of h for different situations. Figure 10.6 gives the value of h for dry air crossing a person's bare skin. We see that the value of h varies considerably with air speed. For example, if the air speed is 1 m/s, we see from Fig. 10.6 that $h \simeq 15$ watts/m^2 C°. If the air speed is 3 m/s, then $h \simeq 26$ watts/m · C°.

It is important when using Eq. (10.3) to consider the sign of the convective heat-transfer rate. If the fluid is warmer than the surface across which it moves, then $T_2 - T_1$ is greater than zero, and H_{cv} is positive. Heat is transferred from the warm, moving fluid to the cooler surface. For example, hot air from a hair

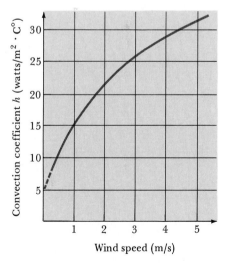

FIG. 10.6. The convection coefficient h as a function of air speed for dry air moving across bare skin. (Adapted from G. Brengelmann, *Physiology and Biophysics III,* 20th ed., T. C. Ruch and H. D. Patton, editors (Philadelphia: W. B. Saunders Company, 1973), p. 108.)

dryer warms the skin as the air blows across it. If, on the other hand, the moving fluid is cooler than the surface across which it moves, then $T_2 - T_1$ is less than zero, and H_{cv} is negative. The cool fluid removes heat from the warmer surface.

EXAMPLE 10.3 A young woman rides a bicycle at a speed of 4 m/s in still air with a temperature of 30°C. Her skin temperature is 35°C and her surface area is 1.4 m². If one-half her skin area is exposed to the air, what is the rate of convective heat loss from her skin?

SOLUTION From Fig. 10.6 we see that the convection coefficient h when $v = 4$ m/s is about 29 watts/m² · C°. Although the rider's surface area is 1.4 m², only half that area is exposed. For the half that is covered, the air speed across the skin is zero, and no convection occurs. Thus in our calculations we use only one-half the rider's area or 0.7 m². Substituting into Eq. (10.3), we find that

$$H_{cv} = \left(29 \, \frac{\text{watts}}{\text{m} \cdot \text{C}°}\right)(0.7 \text{ m}^2)(30°\text{C} - 35°\text{C}) = \underline{-102 \text{ watts}}.$$

The negative sign means that the rider is losing energy from her skin at a rate of 102 watts. Under these conditions, convection would account for about 30 percent of her heat loss. ∎

Wind-Chill Temperature

Our skin temperature is normally about 33°C but can vary from about 30°C to 35°C, depending on the conditions of the environment. In the winter when the air temperature is low, there is convective heat loss from our bodies to the cool air that moves across the skin. As we see from Eq. (10.3) and Fig. 10.6, the rate of this heat loss increases as the speed of the cool air increases. We feel colder on a cold, windy day than on a cold, still day.

A quantity called the wind-chill temperature accounts for this effect of air speed on how cold we feel. The **wind-chill temperature** is the temperature of still, cold air that would cause the same heat-loss rate from our bodies as occurs when cool air is moving at a particular speed. Table 10.2 gives the wind-chill temperatures for different conditions of temperature and wind speed. For example, if the air temperature is 0°C and the wind speed is 10 m/s, the body loses

TABLE 10.2 Wind-Chill Temperatures[a]

Air temperature, °C	Wind velocity, m/s (mph)				
	2 (4.5)	5 (11.2)	10 (22.4)	15 (33.6)	20 (44.7)
2	1.0	−6.6	−12	−16	−18
0	−1.3	−8.4	−15	−18	−20
−5	−7.0	−15	−22	−26	−29
−10	−12	−21	−29	−34	−36
−20	−23	−34	−44	−50	−52

[a] The numbers in color indicate the temperature of still air at which the heat-loss rate is the same as when warmer air whose temperature is indicated on the left moves past the body at different speeds.

heat at the same rate as if the temperature were $-15°C$ with no wind speed. One of the functions of clothing is to produce a layer of still air around the skin, decreasing air movement and heat losses by convection.

10.3 Heat Transfer by Radiation

A third method of heat transfer involves a form of energy known as **electromagnetic radiation,** which is produced by vibrating and accelerating electric charges. For example, when electrons vibrate back and forth in the antenna of a radio station, radio waves, a form of electromagnetic radiation, leave the antenna much as water waves move away from a beach ball pushed up and down in a pond. These waves can travel in a vacuum, in air, and in many other media. Their speed in a vacuum is 3.0×10^8 m/s (186,000 miles per second). A small fraction of the waves leaving the radio station can be absorbed by the antenna of a car radio as electrons in the antenna are forced to move up and down on the passing wave. An energy transfer has thus occurred from the antenna of the radio station to that of the car. We usually do not call this particular process a heat transfer, but since energy moves from one place to another with the waves, it is an important energy-transfer mechanism.

Another process, commonly termed **heat transfer by radiation,** is analogous to energy transfer by radio waves. Consider a log burning in a fire. Because of their high temperature, many of the molecules present in and near the fire vibrate somewhat more violently than usual. These rapid molecular vibrations produce a form of radiation called *infrared waves*. The waves travel from the fire in all directions at the same speed as radio waves. Infrared waves falling on your skin are absorbed and cause the molecules in your skin to vibrate with greater amplitude. Your skin feels warmer. This process is called heat transfer by radiation.

There are many different forms of electromagnetic radiation, the most familiar being light—the visible radiation detected by our eyes. Radio waves, microwaves, infrared and ultraviolet radiation, x-rays, and gamma rays are all forms of electromagnetic radiation. All these types of radiation travel at the same speed (3.0×10^8 m/s in air or a vacuum), are produced by electric charges that vibrate or experience some other type of accelerated motion, and are absorbed by matter that lies in their path. Because energy is carried by the waves from their source to the place where the waves are absorbed, we call this process heat transfer by radiation.

The sun emits all these forms of radiation, although most of the radiative energy leaving the sun consists of light and infrared radiation. The earth is bathed by approximately 1.7×10^{17} J of this radiative energy *each second*—more than the total energy consumed in the United States *each year*. If the earth had absorbed this energy for five million years (about the age of man) without losing an equal amount of energy in some other manner, its temperature would now be well over $1000°C$. However, our planet has managed to remain at a constant temperature (within a few degrees) by emitting into space each second an amount of electromagnetic radiation equal to that absorbed from the sun (Fig. 10.7). The earth is not special in this ability to emit radiation. Every object with a temperature above absolute zero emits electromagnetic radiation. A fire, a wall heater, a table, our bodies, and even ice cubes emit radiative energy.

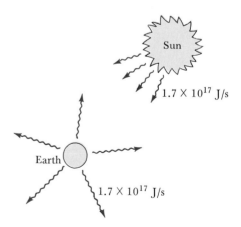

FIG. 10.7. The earth absorbs 1.7×10^{17} J/s of radiative energy from the sun and emits energy at the same rate into outer space.

The rate at which radiation is emitted from an object depends on its temperature T_1, its surface area A, and the type of surface it has. The surface is characterized by a quantity called emissivity e. **Emissivity** is a unitless number ranging from 0 to 1. A good emitter of radiation, such as a dark or black surface, has an emissivity close to 1. For a poor emitter, such as a white or shiny surface (a mirror), e is close to zero. For the earth the average value of e is about 0.95. For the human body it is 0.98.

The rate of emission of electromagnetic radiation from an object depends on these three quantities—temperature T_1, area A, and emissivity e—according to the following equation:

$$\left(\frac{\Delta Q}{\Delta t}\right)_{\text{emission}} = R_e = e\sigma A T_1^4, \tag{10.4}$$

where $\sigma = 5.67 \times 10^{-8}$ watts/m² · K⁴ is a constant called the Stefan-Boltzmann constant. Equation (10.4) is called Stefan's law. Note that the rate of emission of electromagnetic radiation from an object depends on the *fourth power of its temperature*. When this law is used, the *temperature must be in degrees Kelvin* (K)!

Every object emits electromagnetic radiation at a rate in proportion to the fourth power of its temperature. Our bodies, for example, emit radiation at a rate of about

$$R_e \simeq (0.98)\left(5.67 \times 10^{-8}\,\frac{\text{watts}}{\text{m}^2 \cdot \text{K}^4}\right)(1.7\ \text{m}^2)[\overbrace{(273 + 35)}^{\text{skin }T\text{ in K}}\ \text{K}]^4 = 850\ \text{watts}.$$

$$\underset{\substack{\uparrow \\ \text{skin }T \\ \text{in }°\text{C}}}{}$$

This is about eight times our resting metabolic rate. We would certainly cool to very low temperatures if we did not gain energy from some other external energy source. Our bodies gain considerable energy by absorbing electromagnetic energy emitted by objects in our environment. When an object, such as your body, is in an environment that has a uniform temperature T_2, the environment emits radiation in proportion to T_2^4. Some of this radiation is absorbed by the object. The rate of absorption by the object is given by

$$\left(\frac{\Delta Q}{\Delta t}\right)_{\text{absorption}} = R_a = e\sigma A T_2^4, \tag{10.5}$$

where A is the object's area, and e its emissivity. If the object is a good emitter with an emissivity near 1, it will also be a good absorber.

For the object, then, the **net radiative heat-transfer rate** is the radiation absorbed minus the radiation emitted:

$$H_r = R_a - R_e$$

or

$$H_r = e\sigma A T_2^4 - e\sigma A T_1^4 = e\sigma A\,(T_2^4 - T_1^4). \tag{10.6}$$

FIG. 10.8. A stove emits more radiation per unit area to the room than it absorbs because its surface temperature is higher than that of the room.

EXAMPLE 10.4 A black potbellied stove has a surface area of 1.5 m² and a surface temperature of 90°C (Fig. 10.8). Calculate the net radiative heat-transfer rate of the stove in a room whose wall temperatures are 15°C. The emissivity of the black surface of the stove is about 1.0.

SOLUTION First, we must convert the temperatures to degrees Kelvin. We do so by adding 273 to the temperature in Celsius degrees:

$$T_1 = T_{\text{stove wall}} = 273 + 90 = 363 \text{ K} \quad \text{and} \quad T_2 = T_{\text{room wall}} = 273 + 15$$
$$= 288 \text{ K}.$$

The net radiative heat-transfer rate from the stove, then, will be

$$H_r = R_a - R_e = e\sigma A \left(T_2^4 - T_1^4\right)$$
$$= (1.0)\left(5.67 \times 10^{-8}\frac{\text{watts}}{\text{m}^2 \cdot \text{K}^4}\right)(1.5 \text{ m}^2) \times [(288 \text{ K})^4 - (363 \text{ K})^4]$$
$$= \underline{-890 \text{ watts}}.$$

The negative sign indicates that the stove is losing energy to its surroundings.

■

Thermography

The fact that objects emit radiation in proportion to the fourth power of temperature provides a useful technique for locating warmer objects surrounded by slightly cooler objects. Suppose, for example, that an object is 10 percent warmer than the temperature of its surroundings. If neighboring objects or material has a temperature T, then the warmer object has a temperature $1.1T$. The rate of radiative emission from the cool objects is proportional to T^4, whereas the rate of emission from the hotter object is proportional to $(1.1T)^4$. But $(1.1T)^4 = 1.1^4 T^4 = 1.46T^4$. We see that the warmer object emits 1.46 times more radiative energy per unit surface area (that is, 46 percent more) than neighboring objects, even though its temperature is only 10 percent hotter. Infrared cameras can easily detect the difference in radiative emission and in temperature from slightly warmer objects in cool surroundings.

Figure 10.9 shows an infrared picture called a **thermogram.** The change in temperature of a person's hand is apparent as he starts smoking a cigarette. Smoking causes the circulation to decrease and, hence, lowers the skin temperature. Less infrared radiation is emitted. Thermography is also used to detect cancerous growths near the skin's surface. These growths are usually warmer than their surroundings because of increased chemical activity in them.

This type of thermal photography is also used for other purposes. Figure 10.10 shows an infrared film of a home. Hot spots, such as the chimney and windows, appear white, indicating where the most energy is being lost from the home.

10.4 Heat Transfer by Evaporation

The molecules in a liquid continually move about and collide with each other. At any instant there is much variation in their speed—some move fast and others slow. Slow-moving, less energetic molecules may reach the water surface, but they are usually prevented from escaping because of their attraction to neighboring molecules at the surface. However, energetic, fast-moving molecules that reach the surface can escape (Fig. 10.11). The continual escape of these energetic molecules from a liquid into the gas above, called **evaporation,** is a process that occurs at all temperatures.

(a)   0 min

(b)   3 min

FIG. 10.9. The general reduction in heat emission from the hand as a result of cigarette smoking. Commercial thermographic systems, consisting of an infrared detector, camera and display unit, can measure temperatures in the range 15°–45°C with a minimum detectable temperature difference of 0.1 C° at 30°C object temperature. (Courtesy AGA Corporation, Pine Brook, N.J.)

FIG. 10.10. Thermograph of a home showing the "hot spots"—regions where heat is leaving the home. (Courtesy of the National Bureau of Standards)

Evaporation tends to cool a liquid because only the energetic, "hot" molecules escape. Those that remain in the liquid are less energetic and cooler. Therefore, the average thermal energy and temperature of molecules that remain in the liquid is less because of evaporation.

Often, energy from some other source is added to the liquid to compensate for the decrease in thermal energy caused by evaporation. When you perspire, a layer of moisture covers your skin, and as the moisture evaporates, it becomes cooler. However, thermal energy created by metabolic processes inside your body is carried to the skin by the blood. The layer of moisture, then, is cooled by evaporation and warmed by heat removed from inside your body. Sweating is one of the important mechanisms for maintaining a constant body temperature when air temperatures are high or when we exercise.

The rate of heat loss by evaporation depends on the rate of evaporation—that is, the ratio of the mass Δm of liquid evaporated and the time Δt required for this evaporation:

$$\text{evaporation rate} = \frac{\Delta m}{\Delta t}. \tag{10.7}$$

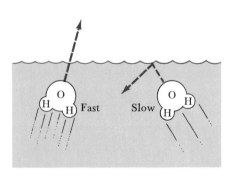

FIG. 10.11. A fast-moving water molecule can escape from a liquid when it hits the surface. A slow-moving one cannot escape.

The evaporation rate is affected by many factors. (1) The evaporation rate of water depends on the relative humidity of the air. If the humidity is high, the air contains many water molecules, which may hit the liquid surface and join the liquid. This process, *condensation*, cancels the effect of evaporation. This is why it is more difficult to remain cool on a hot, humid day than on a hot, dry day. On a humid day, water condenses on our skin almost as fast as it evaporates, and there is no net cooling effect. (2) The rate of evaporation depends also on whether the air near the body is moving. For instance, when water is evaporated from a lake on a still day, the water vapor remains above the water's surface (Fig. 10.12a). Many molecules return to the lake from this vapor. This condensation cancels the effect of evaporation. Should a breeze blow, this gaseous water vapor moves away and fewer molecules return to the lake's surface (Fig. 10.12b). Thus, the net evaporation rate is greater when a breeze blows. (3) The evaporation rate depends also on the temperature of the liquid. As it is warmed, more molecules gain the energy needed to leave the liquid surface. (4) The rate of evaporation depends also on the surface area of the liquid. If the liquid is spread over a thin layer, as it is when the body sweats, evaporation is faster than when the same amount of liquid is held in a compact container.

With all of these variables to consider, it is difficult to write a general equation for calculating the rate of evaporation. What we can say is that the heat ΔQ lost by a liquid when a mass Δm of the liquid evaporates is given by

$$\Delta Q \simeq -L_e \, \Delta m,$$

where L_e is the heat of vaporization of the liquid at the temperature and pressure of the liquid that is evaporating. The heat of vaporization changes a little as the temperature and pressure change. For example, the heat of vaporization of water from a person's skin when at $35°C$ is approximately $2.4 \times 10^6 \, \text{J/kg}$ whereas the boiling temperature heat of vaporization is $2.256 \times 10^6 \, \text{J/kg}$. We will use this approximate skin temperature value in our examples and problems. Thus, the heat lost by water when a mass Δm of the water evaporates is approximately

$$\Delta Q = -(2.4 \times 10^6 \, \text{J/kg}) \, \Delta m \qquad \text{(for water)}.$$

The minus sign indicates that the liquid remaining after evaporation has lost energy because of evaporation.

The evaporative heat transfer rate—the heat lost per unit time—is calculated using the following equation:

$$\left(\frac{\Delta Q}{\Delta t}\right)_{\text{evaporation}} = H_e = -L_e \frac{\Delta m}{\Delta t} \tag{10.8}$$

or

$$\left(\frac{\Delta Q}{\Delta t}\right)_{\text{evaporation}} = H_e = -(2.4 \times 10^6 \, \text{J/kg}) \frac{\Delta m}{\Delta t} \qquad \text{(for water),} \tag{10.9}$$

where $\Delta m / \Delta t$ is the evaporation rate. Several representative values of the evaporation rate for a person in various desert conditions are listed in Fig. 10.13. For example, with the outside temperature at $37.8°C$, an average person sitting nude in the sun evaporates about 0.2 g of water each second.

EXAMPLE 10.5 Calculate the heat-loss rate by evaporation for an individual walking clothed in the desert sun at $100°F$.

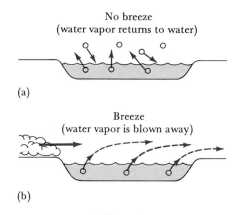

FIG. 10.12. (a) When there is no breeze, evaporated water remains above the lake and some water vapor returns to the lake. This reduces the net evaporation rate. (b) When a breeze blows, evaporated water is removed from the lake and cannot return. The net evaporation rate is greater with a breeze.

FIG. 10.13. Evaporation rate under different desert conditions. The temperature is 37.8°C. (Adapted from Gosselin in E. F. Adolph et al., eds., *Physiology of Man in the Desert*. New York: Hafner Press, 1969.)

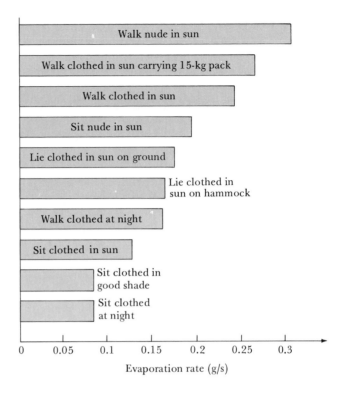

SOLUTION A temperature of 100°F equals 37.8°C. Thus, we can use the evaporation-rate numbers given in Fig. 10.13. We see that the person's evaporation rate ($\Delta m/\Delta t$) when walking clothed on a desert is about 0.24 g/s. From Eq. (10.9) we find that

$$H_e = -(2.4 \times 10^6 \, \text{J/kg})(0.24 \, \text{g/s} \times 10^{-3} \, \text{kg/g}) = -580 \, \text{J/s}$$
$$= \underline{-580 \, \text{watts}}. \quad \blacksquare$$

Evaporation is an extremely important heat-loss mechanism on the desert (and elsewhere). There is a net gain of radiative energy when walking in the sun of almost 500 watts. This energy, along with thermal energy produced by metabolism (about 200 J/s when walking), must be removed from the body. Evaporation is responsible for most of the heat removal. We would be unable to live outdoors on warm days if we could not sweat.

EXAMPLE 10.6 The sun irradiates the earth's outer atmosphere with about 1.7×10^{17} J of energy each second. Of this energy 47 percent is actually absorbed by the earth's surface, and the other 53 percent is absorbed or reflected by the atmosphere. Of the 47 percent absorbed, almost half is used to evaporate water. The water that evaporates recondenses in the atmosphere to form clouds. How much water is removed from the earth each second by this evaporation?

SOLUTION We have been told that the amount of energy used for evaporation per second is half of 47 percent of 1.7×10^{17} J. The evaporative heat-loss rate will be

$$H_e = \frac{(0.5)(0.47)(1.7 \times 10^{17}\,\text{J})}{1\,\text{s}}.$$

In general, however, the evaporative heat-loss rate is

$$H_e = -(2.4 \times 10^6\,\text{J/kg})\frac{\Delta m}{\Delta t}. \tag{10.8}$$

Setting these two expressions equal to each other, we find that

$$\frac{\Delta m}{\Delta t} = -\frac{(0.5)(0.47)(1.7 \times 10^{17}\,\text{J/s})}{(2.4 \times 10^6\,\text{J/kg})} = \underline{-1.7 \times 10^{10}\,\text{kg/s}}.$$

This is equivalent to the evaporation each second of all the water in a lake one-half mile in diameter and over 100 feet deep!   ■

10.5   Temperature Control and the First Law of Thermodynamics

The conservation-of-energy principle $(Q + W = \Delta E_{\text{system}})$ provides a useful framework for analyzing systems whose temperature is controlled by various heat-transfer and energy-conversion mechanisms. In this section we use the principle to study temperature control in our bodies, in our homes, and on the earth. We will be interested in the heat ΔQ added to or removed from a system, the work ΔW done on the system, and the subsequent energy change of the system ΔE_{system} in a time Δt. Dividing each term in the conservation-of-energy equation by Δt results in the equation

$$\frac{\Delta Q}{\Delta t} + \frac{\Delta W}{\Delta t} = \frac{\Delta E_{\text{system}}}{\Delta t}. \tag{10.10}$$

Let us consider briefly each of these three terms to be used in our analysis.

Rate of Change of a System's Energy: $\Delta E_{\text{system}}/\Delta t$

A system's energy can change with time in many ways. In this and the next chapter we will consider primarily changes in the system's internal energy, and in particular (1) its change in thermal energy with time $(\Delta E_{\text{th}}/\Delta t)$, and (2) its change in chemical energy with time $(\Delta E_{\text{chem}}/\Delta t)$. This latter form of energy, E_{chem}, is a form of electrical potential energy that holds atoms and molecules together. Examples of chemical energy change are the melting of ice, the conversion of natural gas and oxygen to combustion products, and the conversion by our bodies of sugar and oxygen to carbon dioxide and water. Whenever bonds between atoms in a molecule or bonds between different molecules are broken or formed, chemical energy changes.

For now, we consider only the changes in these two forms of internal energy and not the changes in mechanical energy of a system (for example, in its gravi-

(a)

F′ (force of foot on pedal)

(b)

F (force of pedal on foot)

(c)

FIG. 10.14. (a) A rider pedaling a bike. (b) The rider pushes down on the pedal with a force **F′**. (c) The pedal pushes up on the rider with a force **F**.

tational potential energy or its kinetic energy). Thus,

$$\frac{\Delta E_{\text{system}}}{\Delta t} = \frac{\Delta E_{\text{th}}}{\Delta t} + \frac{\Delta E_{\text{chem}}}{\Delta t} \qquad \text{(when there are no changes} \quad \textbf{(10.11)} \\ \text{in mechanical energy).}$$

The energy conservation equation $Q + W = \Delta E_{\text{system}}$ when applied only to changes in a system's internal atomic and molecular energy is called the **first law of thermodynamics.**

Power: $\Delta W / \Delta t$

The work done on a system per unit time is called *power*. When calculating work, we consider only forces originating from outside the system that act on the system. The work done by one of these external forces **F** as the system or some part of the system undergoes a displacement $\Delta \mathbf{r}$ is $\Delta W = F \Delta r \cos \theta$, where θ is the angle between the direction of **F** and $\Delta \mathbf{r}$.

Consider, for instance, the work done on a bicycle rider pictured in Fig. 10.14a. We choose the rider as the system. The rider pushes down on the pedal with a force **F′** (Fig. 10.14b), the force of the rider on his environment. To calculate the work done on the system, we must recognize the work done by the force **F** of the pedal on the rider's foot (Fig. 10.14c). This force points upward, whereas the displacement of the foot is down. Since **F** and $\Delta \mathbf{r}$ point in opposite directions, we find that the work done by the pedal on the system is a negative number:

$$\Delta W = F \Delta r \cos \theta = F \Delta r \cos 180° = -F \Delta r.$$

A negative value for work means simply that the system lost energy by doing work. Thus, part of the energy change of the bicycle rider was used to do work, and the rider has less energy because of the work that was done. Often, when we choose an engine or a person as our system, W and $\Delta W / \Delta t$ will be negative numbers. The engine or person does work and hence has less energy. *

Heat-Transfer Rate: $\Delta Q / \Delta t$

In this chapter we have discussed four ways to transfer heat into or out of a system: conduction, convection, radiation, and evaporation. The net heat-transfer rate $\Delta Q / \Delta t$ is the sum of the four heat-transfer rates:

$$\frac{\Delta Q}{\Delta t} = \left(\frac{\Delta Q}{\Delta t}\right)_{\text{cond}} + \left(\frac{\Delta Q}{\Delta t}\right)_{\text{conv}} + \left(\frac{\Delta Q}{\Delta t}\right)_{\text{rad}} + \left(\frac{\Delta Q}{\Delta t}\right)_{\text{ev}}$$
$$= H_{cd} + H_{cv} + H_r + H_e. \qquad \textbf{(10.12)}$$

Heat transfer is positive if energy is added to the system, negative if energy is lost by the system.

---

*The first law of thermodynamics is sometimes written as $Q - W' = \Delta E_{\text{system}}$. When written this way, W' is defined as the work done by the system on the environment. For the bicycle-rider example, $W' = F' \Delta r \cos \theta = F' \Delta r \cos 0° = +F' \Delta r$. The rider did positive work on the environment. The negative sign in front of W' in the first law of thermodynamics indicates that the system's energy decreased by $-W' = -F' \Delta r$ because of this work. Since the magnitudes of **F** and **F′** are equal, both sign conventions produce the same result.

First Law of Thermodynamics

Substituting Eqs. (10.11) and (10.12) into Eq. (10.10) provides us with a general equation for describing the rate of a system's internal energy change:

$$\underbrace{H_{cd} + H_{cv} + H_r + H_e}_{\dfrac{\Delta Q}{\Delta t}} + \dfrac{\Delta W}{\Delta t} = \underbrace{\dfrac{\Delta E_{th}}{\Delta t} + \dfrac{\Delta E_{chem}}{\Delta t} + \cdots}_{} \qquad (10.13)$$

$$\dfrac{\Delta Q}{\Delta t} + \dfrac{\Delta W}{\Delta t} = \dfrac{\Delta E_{system}}{\Delta t}$$

We will apply Eq. (10.13), the time rate of change of quantities in the first law of thermodynamics, to three special situations involving temperature control: (1) body temperature control; (2) home temperature control; and (3) temperature control of the earth.

Body Temperature Control

The temperature of the central core of the body is maintained at about 37.0°C (98.6°F), even though the conditions outside and inside the body vary dramatically. Our metabolic rates, for instance, can increase by a factor of twenty during vigorous exercise. The extra thermal energy generated by the exercise must be removed from our body. If not removed, the body's core temperature would increase by 6°C in about 15 minutes, and brain damage and convulsions would occur.

Not only must we be able to remove great excesses of energy, we must be able to do this in a variety of external conditions. We live and work in outdoor environments that range in temperature from well below freezing to over 40°C (104°F). The ability of the body to adapt to different conditions is illustrated in the next example.

EXAMPLE 10.7 A runner with a metabolic rate of 800 watts (eight times the resting rate) runs on an indoor track at a speed of 3 m/s. The runner's surface area is 1.7 m² and skin temperature is 35°C. Air resistance and friction forces oppose the runner and do negative work on the runner at a rate of −100 watts (the runner loses energy by doing work against friction and air resistance). The runner's core temperature remains constant. The building in which the track is located has air and wall temperatures of 28°C. (a) Calculate the rate of heat transfer by convection and radiation (we ignore conductive heat transfer, since it is small compared to the other forms of heat transfer). (b) Using Eq. (10.13), estimate the heat-loss rate by evaporation. (c) Use the result of part (b) to calculate the runner's evaporation rate, $\Delta m/\Delta t$.

SOLUTION (a) First, we use Eq. (10.3) to calculate the convective heat-transfer rate. According to Fig. 10.6, the convection coefficient for a person moving at 3 m/s is $h = 25$ watts/m² · C°. Thus,

$$H_{cv} = hA(T_{air} - T_{skin}) = \left(25 \, \frac{\text{watts}}{\text{m}^2 \cdot \text{C}°}\right)(1.7 \, \text{m}^2)(28°\text{C} - 35°\text{C})$$

$$= \underline{-300 \, \text{watts}}.$$

The radiative heat-transfer rate is determined using Eq. (10.6):

$$H_r = e\sigma A (T^4_{walls} - T^4_{skin})$$

$$\simeq (0.98)\left(5.67 \times 10^{-8} \frac{watts}{m^2 \cdot K^4}\right)(1.7 \text{ m}^2)[(301 \text{ K})^4 - (308 \text{ K})^4]$$

$$= \underline{-75 \text{ watts}}.$$

Notice that we used an emissivity for the runner's body of 0.98 and that we converted temperatures to degrees Kelvin before substituting in Eq. (10.6). (b) We are now ready to substitute numbers into Eq. (10.13). We must remember that the change in chemical energy of the runner $(\Delta E_{chem}/\Delta t)$ is just the negative of his metabolic rate. The runner loses chemical energy at a rate of -800 watts due to the "burning" of sugar-type molecules with oxygen to form carbon dioxide and water. Also, since the runner's temperature is constant, $\Delta E_{th}/\Delta t$ is zero. We find, then, that Eq. (10.13)

$$H_{cd} + H_{cv} + H_r + H_e + \frac{\Delta W}{\Delta t} = \frac{\Delta E_{th}}{\Delta t} + \frac{\Delta E_{chem}}{\Delta t}$$

becomes

$$0 - 300 \text{ watts} - 75 \text{ watts} + H_e - 100 \text{ watts} \simeq 0 - 800 \text{ watts}$$

or

$$H_e = \underline{-325 \text{ watts}}.$$

(c) The evaporative heat-loss rate, as expressed by Eq. (10.9), equals -325 watts:

$$H_e \simeq -(2.4 \times 10^6 \text{ J/kg})\frac{\Delta m}{\Delta t} = -325 \text{ watts}$$

or

$$\frac{\Delta m}{\Delta t} = \frac{325 \text{ watts}}{2.4 \times 10^6 \text{ J/kg}} = 0.14 \times 10^{-3} \text{ kg/s} = 0.14 \text{ g/s} = \underline{490 \text{ g/hr}}.$$

The runner will lose about one-half liter of water in one hour. A normal individual can easily sweat over one liter per hour, so our runner will be able to maintain a constant temperature by moderate sweating. Notice that 13 percent (100 watts) of the runner's energy was used to do work, 38 percent (300 watts) was lost by convection, 10 percent (75 watts) by radiation, and 40 percent (325 watts) by evaporation—that is, sweating. ■

Home Temperature Control

Equation (10.13) provides a satisfactory basis for examining home temperature control. We can, for example, estimate the rate at which energy must be supplied to a home by a furnace, wood stove, or other heating system in order to maintain the home's temperature. The changes in energy for two homes are summarized in Table 10.3, along with anticipated heating costs for three different heating systems.

Heat losses from the homes are primarily by conduction through the ceiling, walls, windows, and floors and by convection leakage of warm air out of and cold air into the house. The energy required to evaporate water for humidification purposes is also a source of heat loss. This humidified air leaks out of the house through door openings, cracks around windows, and other small openings.

In Home One, with a minimum of insulation, 22,280 watts of power must be provided to compensate for conductive, convective, and evaporative heat losses. Of the total energy needed, 800 watts are supplied radiatively by sunlight entering windows; 2000 watts are provided by energy generated by the home's occupants (about 100 watts per person) and by thermal energy made available by the use of electric and gas appliances; the remaining 19,480 watts must be supplied by the heating system. *

Three types of heating systems are commonly used: (1) forced-air gas furnaces; (2) electric radiative heating systems; and (3) heat pumps (discussed in Chapter 11). Table 10.3 gives estimates of the heating costs per year for each of these systems. Of course, the cost of energy varies substantially in different parts

*The $\Delta E_{chem}/\Delta t$ term is chemical in origin. For example, gas that is burned in a furnace has less chemical energy after burning than before, hence the negative sign for this quantity.

TABLE 10.3 Heat-Loss Rates and Heating Costs for Homes of about 1500-ft^2 Area^a

		Home One (Minimum insulation)	Home Two (Adequate insulation and storm windows)
Conduction	Walls	−4,580	−1,330
	Ceiling	−2,240	−1,440
	Floor	−3,670	−1,220
	Windows	−7,750	−3,720
+			
Convection	Ventilation (one air change per two hours)	−2,940	−2,940
+			
Radiation	Solar through windows	+800	+800
+			
Evaporation	Humidification	−1,100	−1,100
‖		‖	‖
$\frac{\Delta E_{chem}}{\Delta t}$	Energy-loss rate by people and appliances	−2,000	−2,000
	Energy-loss rate from heating system	−19,480	−8,950

	Yearly heating costs	
1. Gas furnace	$610	$310
2. Electric radiative	$2,260	$1,160
3. Heat pump	$1,140	$590

^{a} An inside-to-outside temperature difference of 39°C is assumed. Home One has 4-inch insulation in the ceiling and none in the walls. Home Two has 6-inch insulation in the ceiling, 4-inch in the walls, a foil with air gap under the floor, and storm windows. All units are in watts except the heating costs, which are in dollars. The yearly heating costs are calculated for locations with an annual degree day heating requirement of about 5700, including these cities: Boise, Boston, Columbus, Indianapolis, Lincoln, Pittsburgh, and Providence. The costs are based on a charge of $0.06 per kilowatt hour for electricity and $0.28 per therm for natural gas.

FIG. 10.15. The temperature of the earth during the last 500,000 years. (Adapted from J. D. Hays et al., "Variations in Earth's Orbit: Pace Maker of the Ice Ages," *Science,* **194:**1121–1132 (10 December 1976). Copyright 1976 by the American Association for the Advancement of Science.)

of the country, and these costs increase each year. Costs also depend on the efficiency with which the heating system operates. The numbers in Table 10.3 are representative of energy costs in Ohio for the year 1980.

Home Two is insulated better than Home One and has storm windows. Notice that heat losses through the walls have been reduced by 70 percent for Home Two compared to Home One. The storm windows have reduced heat loss through windows by 50 percent. Overall, one-half as much energy is needed to heat Home Two as Home One.

We should look a little more closely at the convective heat losses caused by warm air leaking out of the house and by cool air leaking through cracks around windows and doors and through other small openings. These leaks into and out of the house, called **infiltration,** can be reduced by placing weatherstripping around doors and windows and by caulking small cracks in the walls. Some infiltration is desirable for purposes of ventilation. It is recommended that there be one complete air change every two to five hours. Unfortunately, most homes change air every hour or less, and office buildings change air in even less time. The homes in our example are assumed to make one air change every two hours.

Temperature Control of the Earth

The earth's temperature has fluctuated between 19°C and 27°C for the last five hundred thousand years (Fig. 10.15), during which there have been about five major changes in climate. When the earth's temperature was low, continental ice sheets 260 m thick extended into the middle latitudes. The most recent period of glaciation ended about six thousand years ago, when the Wisconsin glacier melted.

In recent years concern has grown that increasing amounts of carbon dioxide in the atmosphere will absorb radiation normally leaving the earth and cause a small increase in the earth's temperature. A rise of several Celsius degrees could supposedly melt the polar ice caps and cause major changes in the earth's climate.

Let us apply the first law of thermodynamics to the system pictured in Fig. 10.16 to see why this change might occur. The system inside the dashed lines consists of the earth and its atmosphere. The system gains radiant energy from the sun, R_{sun}, and loses radiant energy emitted by the earth and its atmosphere to outer space, R_{earth}. Thus, for this system, Eq. (10.13) becomes

$$R_{sun} - R_{earth} = \frac{\Delta E_{th}}{\Delta t} + \frac{\Delta E_{chem}}{\Delta t}.$$

If $R_{sun} - R_{earth} = 0$, then the internal energy and temperature of the system remain constant. If $R_{sun} - R_{earth} > 0$, the system's internal energy increases, and if $R_{sun} - R_{earth} < 0$, its internal energy decreases.

The radiant energy entering the system from the sun consists primarily of visible and short-wavelength infrared radiation. Energy emitted by the earth and its atmosphere is at much longer wavelengths—long-wavelength infrared radiation. Because carbon dioxide is a strong absorber in this long-wavelength region, increased carbon dioxide in the atmosphere would absorb radiation normally leaving the earth, R_{earth} would be reduced, and $R_{sun} - R_{earth}$ would be greater than zero. The internal energy of the earth would increase, as would its temperature. As the earth's temperature increased, it would radiate energy at a

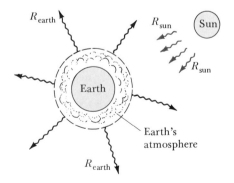

FIG. 10.16. The earth and its atmosphere are a system that gains energy by radiation from the sun (R_{sun}) and loses it by radiation to outer space (R_{earth}).

faster rate. Eventually its radiation rate would again balance energy coming from the sun, and the earth's temperature would stabilize at a new, higher temperature.

It has been estimated that doubling the CO_2 concentration in the atmosphere would cause about a 3-C° increase in the earth's temperature. If present trends continue, the CO_2 concentration will double by the year 2020. The CO_2 concentration is increasing for two reasons: (1) when fossil fuels such as oil, coal, and gas are burned, CO_2 is emitted into the atmosphere; and (2) forests normally absorb CO_2 as part of their photosynthetic growth process, but many forests on the earth are being cut for other uses (paper, building materials, and so forth).

It is difficult to calculate the changes expected because of the decreased radiation leaving the earth. We can see, however, that if $R_{sun} - R_{earth} > 0$, then $\Delta E_{th}/\Delta t$ and $\Delta E_{chem}/\Delta t$ for the earth system will be greater than zero. A positive value of $\Delta E_{th}/\Delta t$ would cause the earth's temperature to increase. A positive value for $\Delta E_{chem}/\Delta t$ would cause ice to melt. Remember that chemical energy is needed to break the bonds of water molecules in ice when converting ice to liquid water. Thus, our simple model predicts a temperature increase and some melting of ice.

This analysis is greatly simplified. Supposedly higher temperatures would increase the rate of evaporation from the oceans, causing an increased cloud cover over the earth, which in turn, would reduce the radiation absorbed by the earth from the sun (thus, R_{sun} would decrease). At present, there is no theory that scientists feel is adequate to predict the consequences of changing concentrations of atmospheric gases.

Summary

Heat is transferred between objects or between different parts of an object by four mechanisms: conduction, convection, radiation, and evaporation.

1. **Conduction:** Heat is transferred from the hot to the cold part of an object by the transfer of thermal energy from atom to atom through the material. The atoms do not leave their equilibrium positions. Only thermal energy moves through the material. The rate of heat transfer is calculated using the equation

$$H_{cd} = \frac{KA(T_2 - T_1)}{L}. \tag{10.2}$$

2. **Convection:** Heat is transferred from one place to another by mass that moves between these places. The mass carries thermal energy with it. The rate of heat transfer by convection, when a fluid (such as air) crosses the body, is calculated using the equation

$$H_{cv} = hA(T_2 - T_1). \tag{10.3}$$

3. **Radiation:** Heat is transferred by electromagnetic radiation, such as light and infrared waves. This radiation is emitted by any object with a temperature above absolute zero. The net rate of radiative heat transfer from an object is the difference of its rate of absorption and its rate of emission of radiation:

$$H_r = R_a - R_e = e\sigma A T_2^4 - e\sigma A T_1^4, \tag{10.6}$$

where T_2 is the temperature of the surroundings in degrees Kelvin, and T_1 is the object's temperature in degrees Kelvin.

4. **Evaporation:** When a liquid evaporates, the liquid that remains behind is cooled. The rate of heat transfer by evaporation is

$$H_e = -L_e \frac{\Delta m}{\Delta t}, \tag{10.8}$$

where L_e is the latent heat of evaporation of the liquid and $\Delta m/\Delta t$ is the rate of evaporation (the mass of liquid evaporated per unit time). For water near 35°C,

$$H_e = -(2.4 \times 10^6 \text{ J/kg}) \frac{\Delta m}{\Delta t}. \tag{10.9}$$

5. **Temperature control and the first law of thermodynamics:** When these heat-transfer mechanisms are incorporated into the first law of thermodynamics, we have an equation that is useful in the analysis of a wide variety of heat-transfer systems:

$$\underbrace{\frac{\Delta Q}{\Delta t}} + \underbrace{\frac{\Delta W}{\Delta t}} = \underbrace{\frac{\Delta E_{system}}{\Delta t}}$$

or

$$H_{cd} + H_{cv} + H_r + H_e + \frac{\Delta W}{\Delta t} = \frac{\Delta E_{th}}{\Delta t} + \frac{\Delta E_{chem}}{\Delta t} + \cdots \tag{10.13}$$

This equation can be used to calculate an unknown heat-transfer rate, work rate, or internal energy change if all other terms in the equation are known.

Questions

1. A potato into which several nails have been pushed bakes faster than a similar potato with no nails. Explain.

2. Explain carefully why storm windows are effective at reducing heat losses in a home.

3. The water in a paper cup can be boiled by placing the cup directly over a flame (such as from a candle or bunsen burner). Explain why this is possible without burning a hole in the cup. (*Note:* The flame may even "touch" the cup!)

4. Explain two reasons why blowing across hot soup or coffee helps to lower its temperature.

5. Placing a moistened finger in the wind can help identify the wind's direction. Explain why.

6. Joggers often accumulate large amounts of water on their skin when running with the wind. When running against the wind, their skin may seem almost dry. Give two explanations.

7. Why does covering a keg of beer with wet towels on a warm day help keep the beer cool?

8. Explain why dogs can cool themselves by panting.

9. Why is it more difficult to work vigorously on a hot, humid day than on a hot, dry day?

10. The water in an outdoor swimming pool is heated faster and remains warmer if covered by a transparent, plastic cover than if not covered. Give two reasons.

11. Drapes and other forms of window treatment (curtains, blinds, and so on) help conserve energy if closed on a cold winter night. Explain why.

12. A burning candle is placed between two cans containing equal amounts of water. One can is black; the other is silver. In which can will the water be heated most rapidly? Explain.

13. A woman has a cup of hot coffee and a small container of milk. The milk is at room temperature and will be added to the coffee before she drinks it. If the woman must wait ten minutes before drinking the coffee, and she wants it to be as hot as possible at that time, should she add the milk and wait ten minutes or wait ten minutes and add the milk? Explain.

14. Look carefully at your surroundings for one or more days. Make four recommendations of ways in which energy lost by heat transfer might be reduced. For one of these recommendations, make a rough estimate of the percentage of energy now lost that you think would be saved if your recommendation were accepted. Show your calculations.

15. When the body is exposed to very cold temperatures, thermal energy must be conserved so that the body core temperature is not lowered. To do this, blood circulation to the skin and extremities is reduced, resulting in lower skin and extremity temperatures. Explain why this helps conserve body energy. (*Note:* One possible consequence of this strategy is frostbite of the extremities. The body sacrifices the extremities rather than the core!)

Problems

10.1 Heat Transfer by Conduction

1. One end of a steel bar with cross-sectional area of 0.060 m² and length of 0.50 m is in the flame of a bunsen burner, and the other end is on a block of ice. The hot end is at 250°C. (a) Calculate the heat transferred from the hot end to the ice in 30 minutes. (b) Calculate the number of kilograms of ice that will melt.

2. A solid brass rod of cross-sectional area 0.10 m² is stuck 1.0 m into the ground. The temperature at the bottom of the rod is 12°C. If the ground surface temperature is −10°C, at what rate will the rod carry heat to the surface? Ignore heat transfer into or out of the sides of the rod.

3. The wall of an adobe home in Mexico is 20 in thick and has an area of 30 m² and a thermal conductivity of 0.4 watt/m · C°. (a) If the inside temperature is 15°C and the outside temperature is −5°C, what is the rate of energy loss through this wall? (b) How much energy is lost in 24 hours?

4. (a) Calculate the rate of heat conduction through a concrete wall that is 0.20 m thick (8.0 in) and has an area of 10 m². A 20-C° temperature difference exists across the wall. (b) Repeat the calculation for a styrofoam block of the same dimensions.

5. A windowpane is 1.0 m × 1.3 m and is 3.0 mm thick. Its inner surface is at 10.0°C and its outer surface at 9.5°C. Calculate the conductive heat-transfer rate through the glass.

6. A goosedown sleeping bag has an exposed area of 2.5 m² and is insulated with 6.0 cm of goosedown whose thermal conductivity is 2.5×10^{-2} watts/m · C°. If a person in the bag is at 37°C and the outside temperature is −10°C, will the person lose heat through the bag faster than it is produced by metabolic processes (about 100 watts)? Explain.

7. A sheep is covered with wool that is about 2 cm thick and has a thermal conductivity of about 0.050 watt/m · C°. If its inside temperature is 37°C and it lies on snow that is at 0°C, calculate the heat lost by conductivity to the snow. The area of contact is about 0.15 m². How does this heat-loss rate compare to the sheep's metabolic rate of 50 watts?

10.2 Heat Transfer by Convection

8. Air at a temperature of −20°C blows across a 1.0-m × 1.5-m window whose outside temperature is constant at 5°C. If the coefficient of convection, h, is 10 watts/m² · C°, what is the convective heat loss rate from the window?

9. A nude person stands in front of a fan that blows 25°C air at a speed of 2.5 m/s. If the person's surface area is 1.6 m² and skin temperature is 35°C, estimate the convective heat-loss rate from the person's body.

10. A woman runs with a speed of 3 m/s at the beach. Her skin

temperature is 36°C, and her surface area is 1.4 m². The air is at a temperature of 30°C and is blowing at a speed of 2 m/s (the air moves across her body at a relative speed of 3 + 2 = 5 m/s). (a) Estimate her heat loss by convection. (b) What is the heat-loss rate by convection if she runs with the wind? Explain.

11.   The temperature of a leaf is 5 C° above the air temperature. The surface area of the leaf is 2 cm² (1 cm² per side). When the air is moving at 5 mph, the convection coefficient is 24 watts/m² · C°. What is the convective heat-loss rate from the leaf? (*Note:* The leaf gains heat from the sunlight at a rate of 0.14 watts. Thus, the leaf must have other heat-loss mechanisms besides convection.)

12.   A person blows 35°C air across the top of a bowl of soup with a temperature of 90°C. The air moves at a speed of 3 m/s, and the top surface area of the soup is 80 cm². (a) *Estimate* the convective heat-loss rate of the soup. (b) How much heat is lost in 10 minutes? (c) If the soup has a mass of 160 g and a specific heat capacity of 4000 J/kg · C°, what is its temperature change due to convection in 10 minutes?

10.3   Heat Transfer by Radiation

13.   The emissivity of tungsten is about 0.36. If the tungsten filament in a light bulb is at a temperature of 3000 K and emits radiant heat (light and infrared radiation) at a rate of 60 watts, what is the area of the filament?

14.   A carbon rod, when heated to 1000 K, emits radiant energy at a rate of 20 watts. Calculate the radiant energy emission rate if the rod is (a) at 1200 K, and (b) at 1500 K.

15.   The radiant energy emitted by the earth's surface per unit of area, R_e/A, is about 420 watts/m² and its emissivity is 0.95. (a) Estimate the temperature of the earth. (b) The moon's average emittance is about 320 watts/m². Estimate its temperature. (*Note:* Without our atmosphere, the earth's emittance rate and temperature would be the same as the moon's. Our atmosphere absorbs and reflects back to the earth much of the radiant energy that would otherwise leave. Our atmosphere is a "blanket" that helps keep the earth warm.)

16.   (a) Calculate the radiative heat-transfer rate of a woman with a surface area of 1.5 m² and emissivity 1.0 while standing in a bikini in a room at 25°C. Her skin temperature is 35°C. (b) Repeat the calculations for a woman in a sauna where the wall and air temperature is 45°C (113°F).

17.   A young man in shorts has a surface area of 1.7 m², emissivity 0.95, and rides a bicycle on a cloudy day at a speed of 5 m/s. The air temperature is 30°C, and his skin temperature is 36°C. Calculate the man's net radiative heat-transfer rate.

18.   A leaf with an area of 1 cm² on each side and emissivity 1.0 absorbs radiation on each side at a rate of about 700 watts/m². What should the leaf's temperature be if it must lose radiative energy at the same rate?

19.   The average sunlight incident on a photoelectric solar collector on a factory roof for the 8-hour working day is 700 watts/m². What must the area of the collector be in order to provide 3000 watts of electricity? Assume the radiant energy is converted to electricity with an efficiency of 15 percent and that the collector's emissivity is 1.0.

20.   The walls, floor, and ceiling of a room are at 20°C. By how many degrees Celsius must the ceiling temperature be increased in order to provide 1.2 kilowatts of extra radiant energy to the room

compared to what it would provide if at 20°C? The ceiling has an area of 50 m² and an emissivity of 0.95.

10.4   Heat Transfer by Evaporation

21.   A bowl of hot soup loses 0.4 g of water by evaporation in 1 minute. Calculate the average evaporative heat-transfer rate of the soup during that minute, assuming the soup is primarily water.

22.   (a) *Estimate* the evaporative heat-transfer rate of a person walking clothed in the desert sun while carrying a 15-kg pack. (b) How much heat does the person lose in 30 minutes of walking?

23.   Assume that the average radiative heat-transfer rate per unit area to a lake is 60 watts/m². If the solar energy absorbed by 1 m² of the lake's surface is used to evaporate water, how many grams of water are evaporated from 1 m² of surface per second? per day?

24.   A marathon runner loses heat by evaporation at a rate of 380 watts. Calculate the moisture lost while running for 3 hours.

25.   A keg of beer gains heat from its surroundings at a rate of 20 watts. (a) At what rate in grams per second must water evaporate from a towel placed over the keg in order to cool the keg at the same rate that heat is being absorbed? (b) How much water in grams is lost in 2 hours?

10.5   Temperature Control and the First Law of Thermodynamics

26.   Air from a hair dryer blows across a person's head (0.1 m² area) at a speed of about 5 m/s. The air temperature is 60°C, and the temperature of the head is 38°C. (a) Estimate the convective heat-transfer rate. (b) If 65 percent of the heat added to the head by convection is used to evaporate water, calculate the evaporation rate of the water in grams per second and in grams per minute.

27.   A man with surface area of 2.1 m² lies in the sun on a well-insulated blanket such that there is no conductive heat transfer from the surface on which he lies. Only 1.1 m² of his skin area is exposed to the sun and wind. His skin temperature is 38°C. (a) Solar radiation is absorbed by the exposed skin at a rate of 1320 watts/m². What is the net rate of radiative heat transfer to this 1.1 m² area of skin (ignore the radiative heat transfer to the rest of the skin)? (b) If a 3-m/s breeze blows across the body (air temperature 33°C), what is the convective heat-transfer rate? (c) If the man's metabolic rate is 120 watts and no heat is lost by conduction, what must his evaporative heat-transfer rate be in order to maintain a constant temperature? (d) What is his perspiration rate, $\Delta m/\Delta t$?

28.   A cyclist with an exposed surface area of 1.5 m² and emissivity 0.95 rides on an indoor track in a building with walls and air at a temperature of 28°C. The cyclist's metabolic rate is 600 watts, and her skin temperature is 35°C. The bicycle and rider travel at a speed of 4 m/s. The cyclist expends work energy at a rate of 100 watts (that is, $\Delta W/\Delta t = -100$ watts). (a) Estimate the rate of heat transfer by convection and radiation. (We will ignore the small rate of conductive heat transfer.) (b) Use the time rate of change of the first law of thermodynamics to estimate the heat-loss rate by evaporation. (c) Calculate the expected rate of evaporation.

29.   A marathon runner does 150 J of work each second and uses 1000 J of chemical energy each second. (a) How much heat must be removed from the runner's body per second to keep his temper-

ature constant? (b) If 50 percent of this heat loss is caused by evaporation, how much water does the runner lose per second? (c) How much water does the runner lose in 3 hours (a respectable time for a marathon run)?

30. A perfectly insulated bowl contains 250 g of soup. The exposed upper surface of the soup has an area of 0.03 m² and an emissivity of 1.0. The soup is initially at 50°C, and the air temperature is 20°C. The evaporation rate $\Delta m/\Delta t = 4.0 \times 10^{-3}$ g/s. Air blows across the soup at a speed of 1.0 m/s. (a) Calculate the sum of the evaporative, convective, and radiative heat-loss rates from the soup (ignore the conductive loss). (b) How much heat will the soup lose in 5 minutes? (c) What will the soup's temperature be after 5 minutes? The specific heat capacity of the soup is 4000 J/kg · C°.

31. A man wearing jogging shorts stands in a room whose temperature is 20°C. The man's skin temperature is 33°C, his skin area 1.8 m², and his emissivity 1.0. Air circulates past his skin so that the convection coefficient is 6.0 watt/m² · C°. (a) Calculate the rates of heat transfer by convection and radiation. We ignore conductive heat transfer and we assume that no evaporative heat loss occurs, since the man needs to conserve thermal energy. (b) Using the results of (a) and Eq. (10.13), estimate the man's metabolic rate $(-\Delta E_{chem}/\Delta t)$ if his temperature is to remain constant. (*Note:* The man can double or triple his resting metabolic rate if he shivers. Does this man need to shiver to maintain constant temperature?)

32.* On an average winter day (38°F or 3°C average temperature) a "standard" house requires heat because the energy already there is lost at the following rates: (i) 2.1 kilowatts (kW) are lost through partially insulated walls and roof by conductance,

*Problems 32 through 35 are reprinted from Joan Ross and Marc Ross, *Physics Teacher*, **16**:272 (May 1978). Copyright 1978 by The American Association of Physics Teachers.

(ii) 0.3 kW are lost through the floor by conductance, and (iii) 1.9 kW are lost through the windows (no storm windows) by conductance. Additional heat is also needed at the following rates: (iv) 2.3 kW to heat the air entering the house through cracks, flues, and other openings (infiltration losses), and (v) an additional 1.1 kW to humidify the incoming air (because warm air must contain more water vapor than cold air for people to be comfortable). What is the total rate at which energy is lost from this house?

33. On the same average winter day some heat energy is supplied to the standard house (see Problem 32) in the following amounts (on a day with average cloudiness): (i) sunlight through windows—0.5 kW, (ii) people's body heat—0.2 kW, and (iii) heat from appliances—1.2 kW. How many kilowatts must be supplied to the standard house in Problem 32 by the heating system of the house in order to keep it at constant temperature?

34. Suppose that the following design changes are made in the house in Problem 32: (i) added insulation of walls, roof, and floors, cutting the losses incurred there by 60 percent; (ii) tightly fitting double-glazed windows with selective coatings to reduce the passage of infrared light, or with special shutters, cutting conductance losses by 70 percent; (iii) elimination of cracks, closing of flues, and so on, cutting infiltration losses by 70 percent. What is the total rate at which energy is lost from this house?

35. The "conservation" house of Problem 34 should include other design changes, such as shifting windows away from the north to the south sides, and improvements in appliances (which might involve a microwave oven, venting of dryer heat internally in winter, substitution of new types of fluorescent bulbs for incandescent bulbs, and a well-insulated hot water tank). Heat energy is supplied to this house at the following rates: (i) sunlight through windows—1.0 kW, (ii) people's warmth—0.2 kW, and (iii) appliance warmth—0.8 kW. How many kilowatts must be supplied by the heating system of the conservation house to keep it at constant temperature?

General Problems

G1. A canteen is covered with wet canvas. If 15 g of water evaporates from the canvas and if 50 percent of the heat used to evaporate the water is supplied by the 400 g of water in the canteen, calculate the temperature change of the water in the canteen.

G2. Insulation rated as R-12 will conduct $\frac{1}{12}$ Btu/hr of heat through each square foot of surface if there is a 1-F° temperature difference across the material. We say that R-12 insulation has a thermal resistance R of 12 hr · ft² · F°/Btu. The heat-conduction rate through a material of area A, across which there is a temperature difference $T_2 - T_1$, is

$$\left(\frac{\Delta Q}{\Delta t}\right)_{cond} = \frac{1}{R} A(T_2 - T_1).$$

(a) What is the rate of conductive heat flow through 1000 ft² of R-12 insulation across which there is a temperature difference of 60 F°? (b) Convert this to units of watts. (c) Repeat parts (a) and (b) for R-19 insulation.

G3. A resting man loses heat at a rate of 100 watts. (a) If all this

heat is transferred to a cup of tea, initially at 20°C, approximately how long is required to warm the tea to 30°C? The tea contains 250 g of water. (b) Devise an apparatus to transfer heat from the man to the tea. There is no correct answer. Let your imagination and hard thinking guide you. It is improbable that your device will be able to transfer 100 percent of the heat to the tea!

G4. Assume that because of increasing CO_2 concentration in the atmosphere, the net radiation heat-transfer rate for the earth and its atmosphere is $+0.02 \times 1350$ watts/m², corresponding to a 2 percent decrease in radiation leaving the earth. (a) Calculate the extra heat added to the earth and its atmosphere in 10 years. (*Note:* The radiation falls on an area equal to πr_{earth}^2, where $r_{earth} = 6.4 \times 10^6$ m.) (b) If 30 percent of this energy is used to melt the polar ice caps, how many kilograms of ice will be melted in 10 years? (c) How many cubic meters of ice will be melted? (*Note:* The density of water is 10^3 kg/m³.) (d) By how much will the level of the ocean rise in 10 years? (*Note:* The area of the ocean surface is 4.0×10^{14} m².)

A proportion is defined as the equality of two ratios; for instance, $a/b = c/d$ is a proportion. Proportions can be used to determine the expected change in one quantity when another quantity changes. For example, suppose that the cable for an elevator is replaced by a new cable whose diameter is 50 percent larger. By what percentage does the weight that can be lifted increase? (It is not 50 percent.) Proportions can be used in everyday life to answer questions such as whether or not a large container of a product is a better buy on a cost-per-unit-weight basis than a smaller container.

To illustrate the use of proportions, consider the pricing of pizza. Suppose that a small pizza costs a certain amount. How much should a larger pizza cost? The cost of pizza should increase in proportion to its area, not its diameter. Thus,

$$\text{COST} = k(\text{AREA}) = k(\pi r^2), \tag{1}$$

where r is the radius of the pizza and k is a constant that depends on the price of ingredients per unit area. If the area of the pizza doubles, the cost should double, but k remains unchanged.

Let us rearrange Eq. (1) so that the two variable quantities (COST and radius) are on the right side of the equation, and the constants are on the left:

$$k\pi = \frac{\text{COST}}{r^2}.$$

This equation should apply to any size pizza. If r increases, the cost should increase in such a way that the ratio COST/r^2 remains constant. Thus, we can write a proportion for pizzas of two different sizes:

$$k\pi = \frac{\text{COST}}{r^2} = \frac{\text{COST}'}{r'^2}. \tag{2}$$

EXAMPLE 1 A 7-inch pizza costs \$2.50. How much should a 10-inch pizza cost?

SOLUTION Substituting COST = \$2.50 for a pizza of radius $r = 7''/2$, we can then use Eq. (2) to calculate the COST of a pizza having radius $r' = 10''/2$:

$$\text{COST}' = \frac{r'^2}{r^2}\,\text{COST} = \frac{(10''/2)^2}{(7''/2)^2}\,\$2.50 = \underline{\$5.10}. \qquad \blacksquare$$

This procedure can be used for any equation that relates two quantities that change while all other quantities remain constant. We simply place all the constant quantities on one side of the equation and the variables on the other side. If one variable changes, the other variable must change in a manner that keeps the variable side of the equation equal to the side with constants.

EXAMPLE 2 The resistive drag force of air on a moving car F_D depends on the car's speed v according to the following equation:

$$F_D = \frac{1}{2} C_D \rho A v^2, \qquad (3)$$

where A is the car's cross-sectional area, ρ is the density of air, and C_D is a unitless coefficient called the drag coefficient whose value depends on the car's shape and is approximately equal to 1. For a particular car, C_D, ρ, and A remain nearly constant as the car's speed changes. If the drag force is 130 N when the car travels at 15 m/s, what is the drag force at 25 m/s?

SOLUTION We first rearrange Eq. (3) so that the constants are on one side and the variables (F_D and v) on the other:

$$\frac{1}{2} C_D \rho A = \frac{F_D}{v^2}.$$

If v changes, F_D must change so that the ratio F_D/v^2 remains equal to the constants on the left. Thus, for two different speeds, v and v', the corresponding drag forces, F_D and F'_D, must be related by the proportion

$$\frac{1}{2} C_D \rho A = \frac{F_D}{v^2} = \frac{F'_D}{v'^2}$$

or

$$F'_D = \frac{v'^2}{v^2} F_D = \frac{(25 \text{ m/s})^2}{(15 \text{ m/s})^2} \, 130 \text{ N} = \underline{360 \text{ N}}. \qquad \blacksquare$$

Sometimes we are told that a quantity changes by a certain percent. For example, if the temperature decreases by 5 percent, then the new temperature T is related to the initial temperature T_0 by the equation $T = 0.95 T_0$. If the temperature increases by 5 percent, then $T = 1.05 T_0$. A small percentage change in one quantity, such as temperature, can sometimes cause a much larger percentage change in another quantity, such as the radiative emission from a surface.

EXAMPLE 3 The temperature of the surface of a Franklin stove becomes 5 percent hotter when a new log is added. Calculate the percentage increase in the radiative power emission from the surface of the stove.

SOLUTION The radiative power emission from a surface is given by Eq. (10.4):

$$R_e = \sigma e A T^4,$$

where σ is a constant, e is the emissivity of the surface, A is the area of the surface, and T is its temperature in degrees Kelvin. Placing the constants on the left side of an equation and the variables, R_e and T, on the right, we have

$$\sigma e A = \frac{R_e}{T^4}.$$

As the temperature changes, the radiative emission must change so that R_e/T^4 remains equal to the constants on the left. Hence,

$$\sigma e A = \frac{R_e}{T^4} = \frac{R'_e}{T'^4}.$$

If the temperature increases by 5 percent, then $T' = 1.05\,T$ and

$$\frac{R'_e}{R_e} = \frac{T'^4}{T^4} = \frac{(1.05\,T)^4}{T^4} = \frac{1.05^4 T^4}{T^4} = 1.05^4 = 1.22.$$

The radiative emission rate is 1.22 times its original rate, an increase of 22 percent. ∎

Problems

1. The distance y that an object falls in a time t if starting at rest is given by the equation $y = \frac{1}{2}at^2$. On the moon a rock falls 10 m in 3.50 s. How much time does it take to fall 15 m, assuming that the acceleration is constant?

2. A circular, wool quilt of 1.2-m diameter costs $82. What should the price of a 1.6-m diameter quilt be if it is to cost the same per unit area?

3. A 0.8-cm diameter rope will support a 5000-N weight. If the strength of a rope increases in proportion to its cross-sectional area, how much weight will a 1.0-cm diameter rope support?

4. The outside wall of a chimney emits radiative energy at a rate 25 percent greater than a neighboring wall of the same area. If the temperature of the adjacent wall is 0°C, what is the temperature of the chimney wall in degrees Celsius?

5. A box of wood 1.0 m on each side costs $56. What should be the price of a box of wood that is 1.8 m on each side if its cost per unit volume equals that of the smaller box?

6. By what fraction should the driver of a car reduce its speed in order to decrease the drag force by 30 percent?

7. You decide to open a pizza parlor. The ingredients and cost factors require that you charge $3.10 for a pizza 7 inches in diameter. How large should you make a pizza whose price is 50 percent more than that of the 7-inch pizza?

8. The radiant heat emission rate from a person's skin decreases from 850 to 820 watts. If the skin's initial temperature was 37.0°C, what is its final temperature? Be sure to convert to degrees Kelvin before working the problem.

11

Entropy and Thermodynamics

Thermodynamics, the study of processes involving energy transformations and heat, originated in the early part of the nineteenth century, about sixty years after the start of the industrial revolution. During the industrial revolution, there was considerable interest in developing machines to do mechanical work. The steam engine, for instance, was built to pump water from mine shafts in England. Early models of the steam engine were extremely inefficient, but as they were gradually improved and their efficiency increased, scientists began to wonder if these machines had an ultimate level of efficiency. This wondering led to the formulation of thermodynamic principles that allowed scientists to calculate the maximum efficiency of a machine such as a steam engine.

Thermodynamics has remained an important tool for analyzing the efficiency or lack of efficiency of all types of mechanical devices such as steam engines, electric power plants, and automobiles. In recent years the laws of thermodynamics have also been used to analyze energy transformations in biological and geological systems. These thermodynamic laws can be used to study energy transformations occurring in living cells; energy transformations of global concern; and even philosophical ideas about the expected distribution of energy billions of years from now.

We have already seen in Chapter 10 how the first law of thermodynamics, based on the idea that energy is neither created nor destroyed, helps us understand a variety of processes such as the temperature control of our bodies, our homes, and the earth. In this chapter we will focus on the second law of thermodynamics, which deals with the degradation or conversion of energy from useful to less useful forms.

In thermodynamics the usefulness of energy is rated according to the amount of work that can be done with it. The chemical energy stored in the bonds of octane molecules can do more work and is more useful than an equal amount of thermal energy in hot pavement or in sea water. The usefulness of the energy is rated using a quantity called **entropy.** Entropy, like golf, has an inverted scale: the lower the score, the better is the energy in terms of usefulness. We will find that entropy is related closely to the idea of order and disorder. Organized, useful energy has low entropy. The organized bonds between atoms in an octane or sugar molecule is an ordered way of storing energy. Less useful energy has high entropy and is more disordered. The thermal energy of a water molecule due to its random motion is a disordered and less useful form of energy.

206

According to the second law of thermodynamics, the entropy of the universe increases during any process in nature. For example, the ordered form of the energy in octane molecules is converted to thermal energy as octane molecules combine with oxygen while burning. The burning of these molecules causes our entropy score to go up. Your body increases the entropy of the universe as you read this book: the chemical energy of sugar molecules in your body is being converted to thermal energy. The universe is not losing energy—the energy is just being converted to a less useful form.

The consequences of the second law of thermodynamics extend from the practical to the profound. The law affects the efficiency with which machines can operate, and it also influences the costs of maintaining a comfortable environment in which to live. If it were not for nature's observance of this law, it seems likely that we could find a fountain of youth to prevent our aging. Even philosophical ideas about predestination and the destiny of the universe have been viewed in terms of entropy and the second law of thermodynamics.

To begin our discussion, we look at why the first law does not account for all aspects of energy transformation, which leads to an introduction to entropy and the second law.

11.1 Irreversible Processes

Our discussions of energy until now have been based on the conservation-of-energy principle and the first law of thermodynamics. If a part of the universe called the system gains energy because of work done on it or by heat transfer into it, then the environment surrounding the system loses an equal amount of energy, or vice versa. If a system's kinetic energy increases, there must be an equal decrease of some other form or forms of energy.

While this principle was being developed, scientists observed that most processes in nature move in only one direction—people grow only older, not younger; a log can burn with oxygen to produce thermal energy, carbon dioxide, and ashes, but ashes in a fireplace do not spontaneously combine with thermal energy and carbon dioxide to produce a log and oxygen; a rock will roll down a hill, but it will not spontaneously roll up a hill. None of these reverse processes is prohibited by the conservation-of-energy principle. Consider the rock.

As a rock rolls down a hill, some of its kinetic energy is converted by friction into thermal energy in the rock and in the ground. When the rock stops at the bottom of the hill, all its kinetic energy has been converted into thermal energy—the rock and ground are slightly warmer. Why is the rock unable to extract thermal energy from the ground when at the bottom of the hill and convert it to an equal amount of kinetic energy to roll up the hill? Conservation of energy does not prohibit this. It just says that the amount of thermal energy lost must equal the kinetic energy gained. Yet even though this process of rolling up a hill is allowed by energy conservation, we never see it happen (except in cartoons, such as Fig. 11.1).

Our experiences in life indicate that some processes are irreversible—they move only in one direction. There must be some explanation to indicate why these processes are irreversible. The explanation is embodied in the second law of thermodynamics and is related to the idea of entropy. One way to understand entropy is based on a statistical approach.

FIG. 11.1. This could happen to you, according to the first law of thermodynamics.

11.2 Statistical Definition of Entropy

To arrive at a statistical definition of entropy, imagine a tray on which four identical coins rest. We will calculate the number of ways that a particular distribution of heads and tails can occur following a shake of the tray. We can obtain five different distributions, by shaking four coins. These different distributions are distinguished by the symbol i, which can assume the integer values from zero through four. We see in Table 11.1 that the $i = 0$ state has zero tails and four heads, the $i = 1$ state has one tail and three heads, and so forth. Each of these five states is called a **macrostate** of the system of four coins. Each macrostate has a different number of tails n_t and of heads n_h. In thermodynamics we call n_t and n_h **state variables,** since they are used to distinguish one particular macrostate of a system from other macrostates.

Next, we want to determine the number of ways in which the ith macrostate can occur. This is called the **thermodynamic probability** P_i of the ith state. Macrostate zero with zero tails and four heads can occur in only one way—if all coins lie with heads up. For macrostate zero, $P_0 = 1$. Macrostate one, with one tail and three heads, can be attained in four different ways, since there are four coins, and we have four choices for the coin that will be the tail. For macrostate one, $P_1 = 4$. Column 3 in Table 11.1 lists the number of ways of getting the different macrostates. Each of these ways is called a *microstate* of that particular macrostate. For example, macrostate one has four microstates. Macrostate two has six; that is, there are six unique ways in which the four coins can be arranged to get two tails and two heads.

In general, for any number of coins, we can calculate the number P_i of

TABLE 11.1 Different Ways in Which Four Coins Can Be Arranged

Macrostate label, i	Different arrangements of four coins ● = tail O = head	Thermodynamic probability, P_i	Entropy $S_i = k \ln P_i$
0	O O O O	1	0
1	● O O O O ● O O O O ● O O O O ●	4	$1.4\,k$
2	● ● O O ● O ● O ● O O ● O ● ● O O ● O ● O O ● ●	6	$1.8\,k$
3	● ● ● O ● ● O ● ● O ● ● O ● ● ●	4	$1.4\,k$
4	● ● ● ●	1	0

microstates of the ith macrostate by using the equation

$$P_i = \frac{n!}{n_t! n_h!}, \tag{11.1}$$

where n is the total number of coins, and n_t and n_h are the number of tails and heads for the ith state.* Using Eq. (11.1), we find that the number of microstates of macrostate two is

$$P_2 = \frac{4!}{2!2!} = \frac{4 \cdot 3 \cdot 2 \cdot 1}{(2 \cdot 1)(2 \cdot 1)} = 6.$$

If we add all the numbers in the third column of Table 11.1, we find that there are 16 microstates. This means that there are 16 unique ways in which our four coins can be arranged, and each is equally likely to occur if the tray is given a good shake. We see that our chance of getting all heads after a shake is 1 in 16, our chance of getting three heads and one tail is 4 in 16, and our chance of getting two heads and two tails is 6 in 16. Obviously, the distribution with two heads and two tails is most likely to occur, and the distributions with all heads or all tails are least likely to occur. These latter states are said to be more **or-dered**—each coin is the same in the all-heads or all-tails macrostates. These ordered states have the least probability of occurring, and we will find that they also have the least entropy. The two-heads, two-tails macrostate is more **disor-dered**—there is more variation from coin to coin. Also, this macrostate can be formed in six different ways. When the tray is being shaken, a particular coin might jump from heads to tails and another from tails to heads. The distribution of coins is still in the $i = 2$ macrostate if it started in that state. This probable state allows more variation and randomness. In the statistical ap-proach to thermodynamics, we measure the randomness or disorder of a state in terms of its probability of occurring—*the greater the probability of the state, the greater is its disorder or randomness.*

Entropy also increases with disorder. The entropy S_i of the ith macrostate can be defined in terms of the thermodynamic probability P_i of the state:

$$\text{entropy of } i\text{th macrostate} = S_i = k \ln P_i, \tag{11.2}$$

where $\ln P_i$ is the abbreviation for the natural logarithm of the number P_i (the use of logarithms is reviewed briefly in Appendix A); k is a proportionality constant called **Boltzmann's constant** and has a value

$$k = 1.3805 \times 10^{-23} \, \text{J/K.†}$$

The fourth column of Table 11.1 lists the values of entropy S_i for the different macrostates. The values are listed in units of k.

The *second law of thermodynamics* says that the universe, as it changes, moves toward states with the greatest thermodynamic probability, and therefore to-ward states of maximum entropy. Consequently, if we start with a tray of four

---

*The exclamation marks (!) in Eq. (11.1) mean that we multiply the number before the exclamation mark by all the positive integers smaller than the number. Thus, for example, $8! = 8 \cdot 7 \cdot 6 \cdot 5 \cdot 4 \cdot 3 \cdot 2 \cdot 1$ and $2! = 2 \cdot 1$. This mathematical shorthand is called "factorial." 8! is said to be eight factorial, and $n!$ is said to be n factorial. Zero factorial is defined as one: $0! = 1$.

†Ludwig Boltzmann (1844–1906) made many significant contributions to the statistical develop-ment of thermodynamics. His grave is marked with a stone that carries this important defining equation for entropy: $S = k \ln P$.

coins showing all heads—entropy zero—and shake the tray, supposedly each shake will produce a distribution with greater entropy. Eventually, the entropy should reach a maximum possible value, $1.8\,k$, the entropy of the two-heads, two-tails distribution.

This law does not work well for distributions with small numbers of objects. For the tray with four coins, there is an excellent chance that after one shake the system will not be in a state with greater entropy. There is even a 1-in-16 chance of getting all heads or all tails, states with zero entropy. With a large number of coins, though, the situation is quite different. With 100 coins, for instance, the state with zero tails and 100 heads is about 10^{29} less likely to occur than the state with 50 heads and 50 tails. We would have to shake the tray once per second for about 3×10^{22} years (about seven trillion times longer than the age of the earth) in order to get one shake resulting in zero tails and 100 heads.

If we increase the number of coins to one million, states with equal numbers of heads and tails are overwhelmingly more probable than states with an unequal number of heads and tails. There is, for example, almost a 10^{87} better chance of getting an equal number of heads and tails than of getting 51 percent heads and 49 percent tails after one shake of one million coins. If we started with one million coins having all heads up, we would eventually, after many shakes, expect to have a 50–50 distribution with perhaps imperceptible fluctuations about this equal distribution. This 50–50 distribution is the state with highest probability and highest entropy.

In general, for systems with large numbers of objects, the move toward high-probability, high-entropy states is inevitable. Once the system is in the state of maximum entropy, it cannot leave that state. The system can perhaps bump into neighboring states of high probability. But usually these states are so close to the maximum-probability state that we do not know the system has changed.

This final state of maximum entropy is often called the **equilibrium state.** If the system is in the most probable state, it does not change. The system has reached equilibrium. A chemical reaction moves toward the equilibrium state as time progresses. Once the state is reached, the reaction stops and we see no more changes in the chemical properties of the system.

11.3   Favored Forms of Energy

The statistical definition of the second law of thermodynamics says that during any process of nature, the universe moves into more probable states—states that are favored by statistics. These favored states have greater entropy and disorder than those from which they evolve. The move toward statistically favored, high-entropy states causes energy to be converted from ordered to disordered forms. The sequence shown in Fig. 11.2 depicts the progression of nature toward more probable states. Unfortunately, this progression results in the conversion of energy to less useful forms. Organized energy is more useful for running engines or for use as energy for our bodies. Random thermal energy with hot and cold reservoirs is next most useful—it is used in power plants to produce electricity. Random thermal energy at uniform temperature is least useful—heat does not flow and work cannot be done with this energy.

It used to be thought that the evolution of living animals, such as humans, was perhaps an exception to the second law of thermodynamics (the molecules

Organized forms of energy ⟶ Hot and cold reservoirs of thermal energy ⟶ Uniformly distributed thermal energy

Oil

Hot water   Cool water

| Least probable but most useful | Medium probable and medium useful | Most probable but least useful |

FIG. 11.2. The transformation of energy into forms favored by statistics. Unfortunately, when in statistically favored forms, the energy is least useful. (Charles Moore/Black Star)

in living animals are very organized—see Fig. 11.3). It is now understood that the growth of order in these living beings must be considered relative to the great disorder produced in their environment to allow for this growth. A person consumes and converts complicated molecules with low entropy, such as carbohydrates, fats, and proteins, into much smaller molecules (CO_2, H_2O). A healthy adult consumes about 500 kg or a half-ton of these foods each year. The degradation of this food results in the transfer to the environment of two billion joules of thermal energy. The human may remain ordered, but our environment is greatly disordered. A UNESCO document on environmental pollution gives these figures and concludes:

> Maintenance of life is an expensive process in terms of generation of disorder, and no one can understand the full implications of human ecology and environmental pollution without understanding that first.

The figures just mentioned represent only the human entropy increase due to food consumption. In industrialized countries the entropy increase is almost 100 times greater per person, owing to the degradation of high-quality fuels that are used to maintain our mechanized society.

The scientific basis of the energy crisis rests with the second law of thermodynamics and ideas of entropy.* The problem is the rate of degradation of energy to a less useful form rather than the actual loss of energy.

11.4 Thermodynamic Definition of Entropy

The statistical approach to entropy makes sense—it seems reasonable that nature moves toward more probable distributions. For most real-life problems, though, it is not practical to determine statistically the probability that a system will be in different possible states. Even small systems are made of huge numbers of atoms and molecules. A simple bacterial cell consists of roughly 10^{11} atoms.

Fortunately, there is another, more practical, approach to entropy and the second law of thermodynamics, which was developed in the 1800s, before the atomic nature of matter was understood and years before the statistical definition of entropy was determined. Unfortunately, entropy as defined thermody-

*There are also many social and political ramifications of the energy crisis.

FIG. 11.3 Is this organized being a mistake—a contradiction of the second law of thermodynamics?

namically is difficult to perceive. It does not affect our senses the way temperature or force does. In this book we will not fully develop this thermodynamic approach to entropy; we will, however, define entropy and use this definition to analyze some representative energy transformations that occur in machines and in other parts of nature.

In thermodynamics, the **change in entropy** ΔS of a system as it moves from an initial state to some arbitrary final state is defined as

$$\Delta S = \frac{\Delta Q}{T}, \qquad (11.3)$$

where ΔQ is the heat added to or removed from the system as it moves between these two states, and T is the temperature of the system *in degrees Kelvin* during this heat transfer. This definition is valid only during reversible changes in the system—changes involving no degradation of energy. No friction forces can be acting on a system or on its environment during the reversible process. All entropy changes are due only to heat transfer. We will not pursue here this subject of reversibility; it receives considerable attention in more comprehensive treatments of thermodynamics.

The thermodynamic definition of entropy change [Eq. (11.3)] is useful for analyzing transformations of energy involving macroscopic objects—that is, objects large enough to be seen easily with our unaided eyes.

EXAMPLE 11.1 Calculate the change in entropy of a block of ice to which 10^4 J of heat is added causing part of the ice to melt. The temperature of the ice is 0°C.

SOLUTION When using Eq. (11.3) to calculate changes in entropy, temperatures must be in degrees Kelvin. The temperature of the block of ice is 0°C or 273 K. Thus, the change in entropy is

$$\Delta S = \frac{\Delta Q}{T} = \frac{10^4 \text{ J}}{273 \text{ K}} = \underline{37 \text{ J/K.}}^* \qquad \blacksquare$$

EXAMPLE 11.2 Calculate the change in entropy of a pot of boiling water to which 10^4 J of energy is added. The water's temperature is 100°C.

SOLUTION The temperature of the boiling water in degrees Kelvin is $100 + 273 = 373$ K. Thus, the change in entropy of the water is

$$\Delta S = \frac{\Delta Q}{T} = \frac{10^4 \text{ J}}{373 \text{ K}} = \underline{27 \text{ J/K.}} \qquad \blacksquare$$

The water discussed in Example 11.2 received the same heat as the ice discussed in Example 11.1, yet their changes in entropy were different. A hot object is already quite disordered (water is less ordered than ice). Hence, when heat is added to a hot object, its change in disorder is more difficult to notice and is less than the entropy change of a cold object to which heat is added. This

---

*A J/K is sometimes called an entropy unit (eu). Thus, 1 J/K = 1 eu.

dependence of entropy change on temperature is very important when we consider the overall entropy change of the universe during processes involving heat transfer, as we learn in the next section.

11.5 Second Law of Thermodynamics

The **second law of thermodynamics** can now be stated in the following fashion:

During any process of nature, the entropy change of the universe—the entropy change of a system plus that of its environment—must be greater than or equal to zero. This law can be written in mathematical symbols as

$$\Delta S_{\text{system}} + \Delta S_{\text{environment}} \geq 0. * \qquad (11.4)$$

If a system is isolated from its environment, there is no interaction between the two. The change in entropy of the isolated system during some process, then, is independent of changes occurring in the environment. In such a case, a system's entropy must increase or, at best, remain constant. Thus,

$$\Delta S_{\text{system}} \geq 0 \text{ (isolated system).} \qquad (11.5)$$

The situation in which entropy remains constant—when $\Delta S = 0$—is called equilibrium: the system is not changing. For any dynamic process of nature, changes are occurring, entropy is increasing, and the inequalities in Eqs. (11.4) and (11.5) must be used. If Eq. (11.4) [or Eq. (11.5) for an isolated system] is not satisfied, then the process being described cannot occur.

EXAMPLE 11.3 A 10-g cube of ice at 0°C lies on ground whose temperature is 20°C. Calculate the change in entropy that results from heat transfer from the ground to the ice, causing the ice to melt and become 10 g of water at 0°C. The ground is considered a very large reservoir whose temperature remains constant at 20°C.

SOLUTION We will calculate the entropy changes of the ice and ground separately, then add them to find the net entropy change. First, the ice.

The heat needed to melt 10 g of ice is given by Eq. (9.6):

$$\Delta Q = mL_f = (0.010 \text{ kg})(3.35 \times 10^5 \text{ J/kg}) = +3350 \text{ J}.$$

The change in entropy of the ice when it melts is

$$\Delta S_{\text{ice}} = \frac{\Delta Q}{T_{\text{ice}}} = \frac{+3350 \text{ J}}{273 \text{ K}} = +12.3 \text{ J/K}.$$

The ground loses just as much heat as the ice gained. Thus, the entropy change of the ground is

$$\Delta S_{\text{ground}} = \frac{\Delta Q}{T_{\text{ground}}} = \frac{-3350 \text{ J}}{(273 + 20) \text{ K}} = -11.4 \text{ J/K}.$$

---

*The symbol $\geq$ means "greater than or equal to"; that is, the left side of the equation must be greater than or equal to the right side.

The net change in entropy is

$$\Delta S = \Delta S_{\text{ice}} + \Delta S_{\text{ground}} = +12.3 \, \text{J/K} - 11.4 \, \text{J/K} = \underline{+0.9 \, \text{J/K}}.$$

Since the entropy change is positive, the process is allowed by the second law of thermodynamics. ∎

EXAMPLE 11.4 Calculate the entropy change when a container holding 0.5 kg of water at 30°C contacts a second container holding 0.50 kg of water at a temperature of 70°C. Heat is transferred from the hot to the cool water until both containers are at 50°C.

SOLUTION To calculate the change in entropy, we consider separately the entropy change of each container of water, then add these changes to find the net change. The heat added to the cooler water as it warms from 30°C to 50°C is given by Eq. (9.4):

$$\Delta Q = mc \, \Delta T = (0.50 \, \text{kg})(4180 \, \text{J/kg C}°)(50°\text{C} - 30°\text{C}) = 41,800 \, \text{J}.$$

The warmer water loses an equal amount of energy when cooling from 70°C to 50°C.

We now have a problem. To calculate the entropy change using the equation $\Delta S = \Delta Q/T$, we must associate some temperature with the material gaining or losing the heat. But the temperature of each container changes as heat flows into or out of it. To do the calculation correctly, we must transfer the heat in very small increments, each of which causes a small change in temperature. The total change in entropy is just the sum of these small entropy changes.

As a rough approximation, we will pretend that all the heat was transferred from the warm water at its average temperature (half of 70°C + 50°C, or 60°C) to the cold water at its average temperature (40°C). The total change in entropy is, then,

$$\Delta S = \Delta S_{\text{cool}} + \Delta S_{\text{hot}} \simeq \frac{\Delta Q}{\overline{T}_{\text{cool}}} - \frac{\Delta Q}{\overline{T}_{\text{hot}}}$$

$$= \frac{41,800 \, \text{J}}{(273 + 40) \, \text{K}} - \frac{41,800 \, \text{J}}{(273 + 60) \, \text{K}}$$

$$= +134 \, \text{J/K} - 126 \, \text{J/K} = \underline{+8 \, \text{J/K}}.$$

The negative sign appears in front of the entropy change for the hot container because it lost heat. An exact calculation using integral calculus indicates that the entropy change was actually 8.04 J/K. Our approximate method using average temperatures provides satisfactory results when temperature changes are not too great. Since the net change in entropy is positive, the process is allowed by the second law of thermodynamics. ∎

11.6 Heat Engines

Most of the world's electricity is produced by heat engines. These engines generally operate at about 35 percent efficiency, which means that only 35 percent of the energy input to the engine results in the production of electrical energy. The other 65 percent is lost as waste heat to the environment. The first and second laws of thermodynamics are very useful for determining the best possible efficiency of different energy-transduction devices. We should be able to decide if we can generate electricity at 100 percent efficiency using a heat engine, or if the 35 percent now achieved is the best we can expect.

A heat engine used to generate electricity is represented schematically in Fig. 11.4a and b. Heat enters the engine on the left side, where water is continually warmed. The heat is produced by burning gas, oil, or coal or by nuclear fission. The temperature of this water on the hot side is typically 560 °C. In Fig. 11.4b this heat source (the burning coal or whatever) is represented by a box on the left labeled "Input reservoir," whose temperature is T_{hot}. An amount of heat Q_{hot} flows from this reservoir into the heat engine and represents the input energy used to operate the engine.

In the engine the water under high pressure boils and is converted to gaseous water vapor (Fig. 11.4a), which causes the blades of a turbine to rotate. As the turbine rotates, an electric generator turns, and electricity is produced. The electricity can then be used to do work. In our block diagram (Fig. 11.4b) the electrical energy leaving the heat engine is symbolized by W for work and represents the useful output of the generator.

Finally, after leaving the turbine (Fig. 11.4a again), the steam is condensed back to the liquid state and returns to the boiler for another trip around the loop. When the steam is condensed, heat is released. This heat is carried away by evaporating water in evaporative cooling towers or by passage of the steam through coils immersed in the cool water of a lake or stream. In Fig. 11.4b the temperature at the condenser coils is T_{cold}, and an amount of heat Q_{cold} leaves the engine at the condenser. The condenser temperature is typically about 100 °C.

Let us now use the first and second laws of thermodynamics to determine the efficiency of this engine. The **efficiency** e of a device is defined as the ratio of

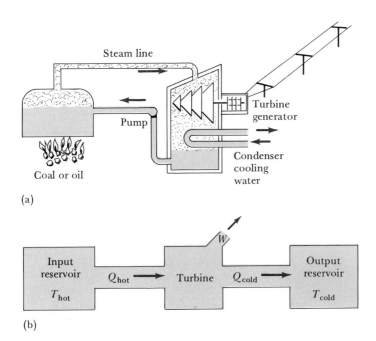

(a)

(b)

FIG. 11.4. (a) A heat engine to produce electricity. (b) A thermodynamic representation of the engine.

the useful output of the device divided by its input:

$$\text{efficiency} = e = \frac{\text{useful output}}{\text{input}}. \tag{11.6}$$

For a heat engine, its useful output is the work it does, and its input is the heat entering the engine from the hot reservoir:

$$\text{efficiency of heat engine} = e = \frac{W}{Q_{\text{hot}}}. \tag{11.7}$$

Once operating, a heat engine remains in a stable or steady state, and its internal energy does not change. Thus, according to the first law of thermodynamics, the energy entering the engine must equal that leaving the engine, or

$$Q_{\text{hot}} - Q_{\text{cold}} - W = 0.$$

This equation has been written so that Q_{hot}, Q_{cold}, and W are all positive numbers. W is the work done *by* the engine.

Solving the above equation for W and then substituting in Eq. (11.7), we find that

$$e = \frac{Q_{\text{hot}} - Q_{\text{cold}}}{Q_{\text{hot}}} = 1 - \frac{Q_{\text{cold}}}{Q_{\text{hot}}}. \tag{11.8}$$

If no heat is emitted from the engine—if $Q_{\text{cold}} = 0$—then the efficiency is 1. This means that all the input heat is converted to work, and the engine is 100 percent efficient. We will find, however, that this is not possible.

Using the second law of thermodynamics, we learn more about the efficiency of the engine by calculating its change of entropy. The input reservoir to the engine, where the fuel is burning, is at a high temperature, T_{hot}. The entropy change due to the heat removal from this reservoir is

$$\Delta S_{\text{hot}} = \frac{-Q_{\text{hot}}}{T_{\text{hot}}}.$$

The output reservoir of the engine at the condenser is at a lower temperature, T_{cold}. The entropy change due to heat input into this reservoir is

$$\Delta S_{\text{cold}} = \frac{Q_{\text{cold}}}{T_{\text{cold}}}.$$

The total entropy change is the sum of the entropy change at the input and the entropy change at the output. This total change must be greater than, or equal to, zero according to the second law of thermodynamics:

$$\Delta S_{\text{total}} = \Delta S_{\text{hot}} + \Delta S_{\text{cold}} = -\frac{Q_{\text{hot}}}{T_{\text{hot}}} + \frac{Q_{\text{cold}}}{T_{\text{cold}}} \geq 0.$$

This last inequality can be rewritten as

$$\frac{Q_{\text{cold}}}{Q_{\text{hot}}} \geq \frac{T_{\text{cold}}}{T_{\text{hot}}}. \tag{11.9}$$

When this is substituted into Eq. (11.8), we find that

$$e \leq 1 - \frac{T_{\text{cold}}}{T_{\text{hot}}}. \tag{11.10}$$

Thus, the **maximum efficiency of a heat engine,** as limited by the second law of thermodynamics, depends on the temperatures of the input and output reservoirs of the engine. For example, if $T_{hot} \simeq 560°C \simeq 833\,K$ and if $T_{cold} = 100°C = 373\,K$, then the efficiency can be no more than

$$ e \leq 1 - \frac{373\,K}{833\,K} = 0.55. $$

The engine can be at best 55 percent efficient. The efficiency in practice of a modern coal-fired power plant is less than 40 percent.

11.7 Heat Pumps and Refrigerators

Heat flows naturally from hot objects to cool objects. A heat pump is a device that reverses the direction of flow; heat is transferred from a cold reservoir to a warmer one. A heat pump can be used to warm your house in the winter by extracting thermal energy from cool air outside and transferring it into the house to keep the house warm. A refrigerator is another example of a heat pump; heat is transferred from cool air inside the refrigerator to warmer air outside the refrigerator.

A heat pump used to warm a house in the winter is depicted in Fig. 11.5a. Heat, removed from cold air at temperature T_{cold} outside the house, is absorbed by a fluid circulating through the heat pump. This input heat is labeled Q_{cold} in Fig. 11.5b. The heat pump cannot operate without some outside source of energy that does work W circulating and compressing the fluid in the heat pump. The input work is usually done by an electric motor and is represented by the symbol W in Fig. 11.5b. Finally, heat Q_{hot} is transferred from the heat pump into the house, whose temperature is T_{hot}.

The useful output of a heat pump is the heat Q_{hot} transferred into the house. The energy needed to do this is the work W done by its motor. To be efficient, a heat pump should transfer a lot of heat while doing little work. The heat pump is rated using an efficiency quantity called the **coefficient of performance,** which is the ratio of the heat transferred into the hot reservoir and the work needed to transfer that heat:

$$ \text{coefficient of performance of heat pump} = \eta = \frac{Q_{hot}}{W}. \qquad \textbf{(11.11)} $$

The larger the value of η, the better is the heat pump. Typical heat pumps used for home heating have coefficients of performance of 2 or 3; in other words, two or three times more heat is transferred into the house than the work that is done transferring the heat. We can use the first and second laws of thermodynamics to see if this is the best we can do.

The heat pump, while operating, remains in a stable situation. Its internal energy does not change. According to the first law of thermodynamics, the energy entering the heat pump must equal that leaving it:

$$ Q_{cold} + W - Q_{hot} = 0. $$

Once again, all quantities are positive. Solving the above equation for W and substituting into Eq. (11.11), we find that

$$ \eta = \frac{Q_{hot}}{Q_{hot} - Q_{cold}}. \qquad \textbf{(11.12)} $$

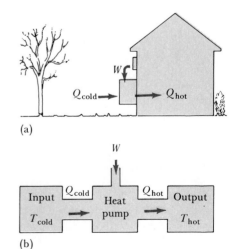

FIG. 11.5. (a) A heat pump to warm a house in the winter. (b) A schematic representation of a heat pump.

This equation can be rewritten in terms of temperatures, using the second law of thermodynamics. The entropy change of the input reservoir is

$$\Delta S_{\text{cold}} = \frac{-Q_{\text{cold}}}{T_{\text{cold}}},$$

and that of the output reservoir is

$$\Delta S_{\text{hot}} = \frac{+Q_{\text{hot}}}{T_{\text{hot}}}.$$

Thus, according to the second law of thermodynamics, the net entropy change of this heat-pump system, including the two reservoirs, is

$$-\frac{Q_{\text{cold}}}{T_{\text{cold}}} + \frac{Q_{\text{hot}}}{T_{\text{hot}}} \geq 0. \tag{11.13}$$

With a little manipulation of Eqs. (11.12) and (11.13), we find that the maximum possible value for the **coefficient of performance of a heat pump** is limited by the following equation:

$$\eta = \frac{Q_{\text{hot}}}{W} \leq \frac{T_{\text{hot}}}{T_{\text{hot}} - T_{\text{cold}}}. \tag{11.14}$$

EXAMPLE 11.5 A heat pump deposits 14,000 J of heat into a home. The temperature of the outside air is $-10°C$ (263 K), and the air in the house is at $20°C$ (293 K). (a) Calculate the maximum possible coefficient of performance of the heat pump. (b) Calculate the amount of electrical work done in operating the heat pump at this performance level.

SOLUTION (a) We are given that $T_{\text{cold}} = 263$ K and $T_{\text{hot}} = 293$ K. According to Eq. (11.14), the coefficient of performance of a heat pump operating between these two temperatures is limited to the following value:

$$\eta \leq \frac{T_{\text{hot}}}{T_{\text{hot}} - T_{\text{cold}}} = \frac{293}{293 - 263} = \underline{9.8}.$$

A heat pump could in principle deposit almost ten times more heat into the house than the work used to extract the heat. (b) Rearranging our definition of coefficient of performance, we find that

$$W = \frac{Q_{\text{hot}}}{\eta} = \frac{14,000 \text{ J}}{9.8} = \underline{1430 \text{ J}}.$$

Only 1430 J of work is needed to transfer 14,000 J of heat. Unfortunately, most real heat pumps operate with coefficients of performance of only 2 or 3. ∎

Refrigerator

A refrigerator works much like a heat pump, except that heat is withdrawn from inside the refrigerator and deposited in the warmer air outside, as depicted in Fig. 11.6. The useful function achieved by a refrigerator is the removal of heat Q_{cold} from inside the refrigerator, and an amount of work W is expended to

FIG. 11.6. (a) A refrigerator is a heat pump. (b) A schematic representation of the energy transformations.

achieve this removal. The efficiency quantity used to rate refrigerators, also called a coefficient of performance, is

$$\text{coefficient of performance of refrigerator} = \eta = \frac{Q_{\text{cold}}}{W}. \quad \textbf{(11.15)}$$

A calculation similar to that performed for a heat pump shows that the **coefficient of performance (COP) of a refrigerator** is limited by the equation

$$\text{COP (refrigerator)} = \eta = \frac{Q_{\text{cold}}}{W} \leq \frac{T_{\text{cold}}}{T_{\text{hot}} - T_{\text{cold}}}. \quad \textbf{(11.16)}$$

EXAMPLE 11.6 How much electrical energy is required to remove 700,000 J of heat from the freezer of a refrigerator operating at its maximum coefficient of performance? The freezer temperature is $-5°C$, and the room temperature is $20°C$.

SOLUTION We are given the following information: $T_{\text{cold}} = -5°C = 268$ K, $T_{\text{hot}} = 20°C = 293$ K, $Q_{\text{cold}} = 700,000$ J. We wish to find the work done by the electrical compressor of the refrigerator. Using Eq. (11.16), we find that

$$Q_{\text{cold}} \frac{T_{\text{hot}} - T_{\text{cold}}}{T_{\text{cold}}} \leq W$$

or

$$(700,000 \text{ J}) \frac{293 \text{ K} - 268 \text{ K}}{268 \text{ K}} = 65,000 \text{ J} \leq W.$$

Thus, at least 65,000 J of electrical work is needed to remove the heat. Most refrigerators use much more than the optimum electrical work needed. ■

11.8 Recurrence Paradox

In this chapter we have considered entropy and the second law of thermodynamics from two points of view: (1) the statistical view, and (2) a view based on heat and the tendency for energy to be dispersed as thermal energy rather than as other, more useful forms. In the statistical view the universe moves toward those states that are most probable. In terms of energy, states in which the energy is dispersed randomly in the form of thermal energy are most probable. Thus, it seems that the two views are consistent.

Yet several interesting controversies have evolved from a supposed conflict between these two views of the second law of thermodynamics. One of these, the **recurrence paradox,** is based on the fact that, statistically, if the universe consists of a finite number of particles, then there is a finite although extremely small chance for every arrangement of these particles to occur. If we wait long enough, the particles should at some time in the future be arranged again just as they are now. It is like shaking a tray full of coins. If we do it long enough, at some time in the future we will certainly get a distribution of all heads or of any other arrangement of heads and tails. The German philosopher Friedrich Nietzsche stated this idea as follows:

If the universe may be conceived as a definite quantity of energy, as a definite number of centres of energy—and every other concept remains indefinite and therefore useless—it follows therefrom that the universe must go through a calculable number of combinations in the great game of chance which constitutes its existence. In infinity [of time], at some moment or other, every possible combination must once have been realized; not only this, but it must have been realized an infinite number of times.

On the basis only of the ideas of statistical probability and random distributions, we would expect that at some future time all the atoms and molecules of the universe will be back in their present arrangement. It seems reasonable, then, that history should move forward in about the same way as it does now—the drama of life will be repeated, not once, but many times.

This idea seems to conflict with the approach to the second law based on heat. In this approach energy continues to be dispersed, and there seems to be no way in which it can be shaken back into order.

Recent theories of cosmology have been addressing this conflict, not so much in terms of entropy but by using ideas of mechanics and the origin of the universe. The universe supposedly started its evolution about twenty billion years ago as a giant ball of mass. The huge gravitational force of this mass caused it to crush down on itself. The intense pressure and heat produced a chain of nuclear reactions and an explosion called the **big bang.** The mass of the universe was thrown into motion away from this explosion. It is now generally accepted that the universe is expanding—that all parts are moving apart—as if in consequence of a big bang. Cosmologists are considering whether this expansion will continue forever or will eventually stop. By analogy, consider a rocket leaving the earth. If its velocity is great enough, it will be able to escape the earth's gravitational pull; if its velocity is too small, the earth's gravitation will eventually pull it back.

Cosmologists are trying to calculate the density of mass in the universe to decide whether it is great enough eventually to halt this expansion. If it is, then the halt will occur, and thereafter the mass of the universe will be drawn back toward its center, eventually resulting in the **big crunch** and a likely recurrence of the big bang. Some theories predict that such a crunch will occur in about fifty billion years. It is speculated that during this process of contraction, the entropy of the universe may be decreasing. Natural processes will be reversed in time—like a clock running backward. It would seem reasonable that if the universe does oscillate in and out, as predicted by these theories, perhaps Nietzsche's idea about the recurrence of events in history might be valid. There would, however, be an extremely long waiting time between similar events.

Another group of cosmologists have said that the measurements of the universe indicate that its mass density is too small to halt its expansion. In this case the increase in entropy is expected to continue until all mass and energy are distributed randomly in space. Toward the end, neither living beings nor mechanical machines will be able to operate—there will be no food or sources of high-quality energy available for their operation. The universe will have reached its so-called **heat death.** All matter will be distributed uniformly. Energy will be in the form of thermal energy (or heat). Since the matter and energy are distributed over so much space, the temperature will be cold—several degrees Kelvin. Unfortunately, this fate seems a little boring. Hopefully, we can look forward to a big crunch.

FIG. 11.7 Flammarion's conception of the "heat death" of the universe. From the novel, *Le Fin du Monde* (1894).

Summary

1. **Irreversible processes:** Many processes of nature are observed to move in only one direction and are said to be irreversible. Examples are aging, the burning of a log, and the rolling of a rock down a hill. The reverse of these processes is allowed by the first law of thermodynamics, yet we do not see the reverse processes occur. The concept of entropy has been developed to help us understand why these reverse processes do not occur. *Entropy* is a measure of disorder and the randomness of the distribution of a system's energy. The greater the disorder, the greater is the entropy.

2. **Statistical development of entropy and the second law of thermodynamics:** Entropy can be defined quantitatively by using either statistics or a thermodynamic definition involving heat and temperature. In statistics, the **entropy** S_i of the ith macrostate of a system is defined as

$$S_i = k \ln P_i, \tag{11.2}$$

where P_i is the **thermodynamic probability** of the ith state. The thermodynamic probability is the number of different ways the objects in the system can be arranged to form that macrostate. Each of these different ways is called a microstate. The second law of thermodynamics says that the universe continually moves toward states of greater entropy. Since entropy increases as the probability of a state increases, we see that the universe moves toward more probable states. This movement toward probable states causes thermal energy to be favored over more organized forms of energy such as the chemical energy of molecular bonds. Most processes in nature move toward increasing thermal energy, while the amount of organized, useful energy decreases.

3. **Thermodynamic development of entropy and the second law of thermodynamics:** In thermodynamics, the **change in entropy** ΔS of a system or object at temperature T (in degrees Kelvin) to which heat ΔQ is added or removed is

$$\Delta S = \frac{\Delta Q}{T} \tag{11.3}$$

The **second law of thermodynamics** says that during any process, the entropy change of a system plus that of its environment must be greater than zero:

$$\Delta S_{\text{system}} + \Delta S_{\text{environment}} \geq 0. \tag{11.4}$$

The first and second laws of thermodynamics are especially useful for analyzing the efficiency of heat engines, heat pumps, and other systems involved in the transfer of heat for useful purposes.

Questions

1. Describe five everyday examples of processes that involve entropy increase. Be sure to state all the parts of nature involved in these entropy increases.

2. A cup of hot coffee sits on your desk. Assume that some of the thermal energy in the coffee can be converted to gravitational potential energy—enough so that the coffee rises up above the rim of the cup. *Estimate* the drop in temperature needed to accomplish this energy conversion. Why does this conversion not occur?

3. In terms of the statistical definition of entropy, why is there a good chance that the entropy of five coins being shaken on a tray will decrease, whereas this chance for 10^6 coins is negligible?

4. The entropy of the molecules that form leaves on a tree decreases in the spring of each year. Is this a contradiction of the second law of thermodynamics? Explain.

5. The efficiency of nuclear power plants used to produce electricity is about 40 percent. Under what condition could this conversion efficiency be increased to 100 percent? What difficulty might be encountered in doing this?

6. There is a huge amount of thermal energy in the oceans of the earth. Indicate two problems that limit the usefulness of this energy as a source of power for a heat engine.

7. An air conditioner transfers heat from a cold reservoir to a hot reservoir. How can this heat transfer be reconciled with the second law of thermodynamics?

8. Would it be possible to use a refrigerator with its door open as a device to cool a house in the summer? Explain.

9. Would it be possible, in terms of the first and the second laws of thermodynamics, for a refrigerator to remove as much heat from inside the refrigerator as it gave off into the air outside? Explain.

10. An air conditioner and a refrigerator both use work (electrical energy) to remove heat from a cold reservoir and deposit it in a hot reservoir. Why does an air conditioner cool a house and a refrigerator warm the house (even with its door open)?

11. Assume that there is a finite probability that at some future time all the atoms and molecules of the universe will be in a situation such as exists now. History should then repeat itself. What, however, will happen if the atoms and molecules are in the same position but moving in the opposite directions to those in which they now move? Explain.

12. List five ways in which the universe would be different if it had reached its final equilibrium state.

Problems

11.2 Statistical Definition of Entropy

1. (a) Identify, in terms of the numbers of heads and tails, all the different macrostates of a system of five coins. (b) Calculate the number of microstates of each macrostate. (c) Calculate the entropy of each state.

2. Repeat Problem 1 for a system with ten coins.

3. (a) Calculate the ratio of the thermodynamic probabilities of a system of twenty coins when in a macrostate with 10 heads and 10 tails and when in a macrostate with 18 heads and 2 tails. (b) Do the same for a system with ten coins and for the states with

5 heads and 5 tails, compared to 9 heads and 1 tail. (c) When you compare your answers to parts (a) and (b), what do you infer about a similar ratio for a system with 10,000 coins?

4. Nine numbered balls are dropped randomly into three boxes. The number of balls falling into each box are labeled n_1, n_2 and n_3 (the numbers must add to nine). (a) Identify five possible arrangements or macrostates of the balls (there are many more than five). (b) Compare the probability of the equal distribution ($n_1 = 3$, $n_2 = 3$, and $n_3 = 3$) to that of a 0, 0, 9 distribution. (c) Compare the probability of an equal distribution to that of a 2, 3, 4 distribution. (*Note:* The thermodynamic probability is given by $P = n!/(n_1!n_2!n_3!)$, where n is the total number of balls.)

5. Twenty balls are dropped in four boxes. Calculate the ratio of the probabilities of the following distributions: (a) 5, 5, 5, 5 to 3, 4, 6, 7; (b) 5, 5, 5, 5 to 0, 0, 0, 20; and (c) 5, 5, 5, 5 to 1, 3, 7, 9. The equation provided in Problem 4, if extended to four boxes, can be used to calculate the probabilities.

6. Two dice are rolled. Macrostates of these dice are distinguished by the total number for each roll (that is, 2, 3, 4, . . . , 12). (a) Calculate the number of microstates for each macrostate. For example, there are three microstates for the macrostate 4: 2, 2; 3, 1; and 1, 3. (b) What is the macrostate with greatest entropy? (c) What is the macrostate with least entropy? Explain.

7. Three dice are rolled. (a) Identify the different possible macrostates. (b) Indicate the possible microstates for the macrostate that shows 17 on the dice; (c) for the macrostate that shows 16. (d) How many more times probable is the 16 state than the 17 state?

8. The probability that a lost person wandering about an island will be on the north part is one-half and on the south part is also one-half. (a) Calculate the probability that three lost people wandering about independently will all be on the south half. (b) Repeat part (a) for six people all on the south half of the island.

11.4 Thermodynamic Definition of Entropy

9. (a) Calculate the change in entropy when 0.50 kg of water freezes at 0°C. (b) Calculate the change in entropy when 0.50 kg of water boils at 100°C.

10. (a) Calculate the change in entropy when 2.0 kg of ice at 0°C melts. (b) Calculate the change in entropy when 2.0 kg of steam at 100°C condenses to water at 100°C.

11. Heat is added to 1.0 kg of ice at −10°C. (a) *Estimate* the entropy change of the ice as it warms to 0°C, (b) as it melts at 0°C, (c) as it warms from 0°C to 100°C, (d) as it boils at 100°C, and (e) as the steam warms from 100°C to 120°C. (f) What is the total entropy change?

12. *Estimate* the entropy changes of (a) 200 g of water that is warmed from 20°C to 50°C, (b) 200 g of ethanol that is warmed from 20°C to 50°C, and (c) 200 g of copper that is warmed from 20°C to 50°C.

13. A towel at 30°C, while drying on a line, absorbs 1000 J of energy from the sun thus causing water to evaporate. Calculate the entropy change of the evaporated water. The towel does not change temperature and the latent heat of vaporization at this temperature is 2.4×10^6 J/kg.

11.5 Second Law of Thermodynamics

14. *Estimate* the total change in entropy of two containers of water. One container holds 0.10 kg of water at 70°C and is warmed to 90°C by heat emitted from the other container. This other container, also holding 0.10 kg of water, cools from 30°C to 10°C. Is this heat-transfer process allowed by the first law of thermodynamics? the second? Explain.

15. (a) 0.10 kg of ice at 0°C is added to 0.30 kg of tea at 30°C. Calculate the final temperature of the mixture after reaching equilibrium. The heat capacity of ice tea is the same as that of water. (b) Estimate the entropy change of the universe during this process. Is is allowed by the second law of thermodynamics?

16. A home at 25°C loses heat at a rate of 14,000 watts to the outside air, whose temperature is −10°C. Calculate the rate of entropy change due to this heat transfer.

17. A vat containing 200 kg of water is kept in a fruit cellar during the winter. On a cold night the water freezes, releasing heat to the room. The heat passes from the cellar to the outside air, whose temperature is −20°C. Calculate the entropy change for this process if the cellar remains at 0°C.

18. A 250-g aluminum calorimeter cup containing 200 g of water is at a temperature of 10°C. A 300-g block of copper at 60°C is placed in the cup. (a) Calculate the final equilibrium temperature. (b) Estimate the change in entropy of the cup, water, and copper block during this process.

19. A 5.0-kg block sliding at 8.0 m/s on a level surface at 20°C is stopped by friction after sliding 10 m. Calculate the change in entropy of the universe for this process. (*Hint:* Use the heat definition of entropy. The heat that would be added to the system is the same as the thermal energy gained by friction.)

20. A 5.0-kg block slides from an initial speed of 8.0 m/s to a final speed of zero. It travels 12 m down a plane inclined at 15° with the horizontal before being stopped by friction. Calculate the change in entropy of the universe for this process if the block and plane are at a temperature of 27°C. (See the hint with Problem 19.)

11.6 Heat Engines

21. A heat engine operates between a heat reservoir at temperature 700 K and another at 300 K. Calculate the maximum amount of work the engine can do if 100,000 J of heat is transferred from the hot reservoir to the engine. Calculate the maximum possible efficiency of the engine.

22. An inventor claims to have developed a heat engine that can be attached to the exhaust system of a car. It operates on heat emitted from the exhaust ($T = 150°C$). The output side of his engine is at air temperature (10°C). (a) Calculate the maximum efficiency of the engine. (b) If the car exhaust system deposits 2.0×10^5 J/s of heat into this heat engine, what is the maximum power of the engine? (Power is work/time.)

23. If the maximum efficiency of an automobile engine is 0.56 and the exhaust temperature is 120°C, what is the temperature of the burning gas in the engine? How much work could this engine produce when burning 1 gallon of gasoline? (One gallon releases 1.4×10^8 J of heat.) (*Note:* There are many inefficiencies other than those of the engine that reduce the overall car efficiency.)

24. A thermal hot spring, 400 m deep in the ground, is at a temperature of 220°C. The average temperature of the earth at its

surface is 15°C. (a) Calculate the maximum possible efficiency of a heat engine operating between the hot springs and the earth's surface. (b) How much heat must be transferred from the hot springs to an engine each second in order to provide a work output of 100,000 watts?

25. Near Bermuda, ocean water at the surface has an average temperature of about 24°C. At a depth of 800 m, the water temperature is about 10°C. (a) Calculate the maximum efficiency of a heat engine operating between these two heat reservoirs. (b) How much heat per second would have to be transferred down from the surface water to the engine and then exhausted to the water below in order to produce 10,000 kilowatts of electrical power?

26. A nuclear power plant operates between a high-temperature heat reservoir at 560°C and a low-temperature stream at 20°C. (a) Calculate the maximum possible efficiency of this heat engine. (b) How much heat per second must be put into the engine to produce 1000 megawatts of electrical power (work/time)? (c) How much heat per second is dumped into the environment? (d) If this heat is absorbed by the Niagara River with a flow rate of 2×10^6 kg/s, by how much will the river's temperature change?

27. A 60-kg woman walking on level ground at 1 m/s metabolizes energy at a rate of 230 watts. When she walks up a 5-degree incline at the same speed, her metabolic rate increases to 370 watts. Calculate her efficiency at converting chemical energy into gravitational potential energy.

11.7 Heat Pumps and Refrigerators

28. (a) Calculate the maximum possible coefficient of performance of a heat pump that operates between a room at 20°C and metal plates 10 m in the ground at a temperature of 15°C, and one that operates between a room at 20°C and outside air at −10°C. (b) If heat is pumped into the house at a rate of 14,000 watts, what is the least work that must be done by each pump during 1 second? Assume that the heat pump operates at the maximum coefficient of performance allowed by nature.

29. A heat pump collects heat from outside air at 5°C and delivers it into a house at a duct at a temperature of 40°C (at this temperature the air is slightly warmer than body temperature and feels warm if moving past your body). (a) Calculate the maximum possible coefficient of performance of the heat pump. (b) If the heat pump's motor does 1000 J of electrical work during a certain time, how much heat is delivered into the house, assuming the heat pump works at maximum coefficient of performance? (c) Repeat part (c) for a heat pump whose coefficient of performance is 2.5 (the efficiency of most real-life heat pumps).

30. The freezer of a refrigerator operates between a temperature of −15°C and room temperature (22°C). (a) Calculate the maximum coefficient of performance of the freezer. (b) How much energy is required to freeze 0.20 kg of water (an ice-cube tray) at 0°C and then lower its temperature to −15°C? (c) How much work must the freezer do to extract this much heat? (d) How much heat is deposited into the room?

31. An ice-making machine is to convert 2.0 kg of water at 0°C into 2.0 kg of ice at 0°C. The room temperature surrounding the ice machine is 20°C. (a) How much heat must be removed from the water? (b) Calculate the minimum work in the form of electrical energy needed to extract this heat by the ice-making machine. (c) How much heat is deposited in the room?

Gases, Liquids, and Solids

III

In Parts I and II we have concentrated on two of the basic theories of science—Newton's laws of motion and the conservation-of-energy principle. In Part III we will apply these theories to the study of the physical properties of gases, liquids, and solids. A knowledge of the properties of these three forms of matter will allow us to better understand the physiology of our bodies, the factors in our environment that affect our lives (for example, the climate), the evolving structure of the earth, and many other interesting properties of nature.

Gases 12

A gas is a form of matter in which the atoms and molecules are relatively far apart. The atoms and molecules are in constant motion, and they collide frequently and violently with each other. Collisions also occur when gas molecules hit solid and liquid surfaces in their environment. The collisions cause a force and pressure to be exerted on these surfaces.

Gases are described in terms of quantities such as pressure, temperature, the volume of space occupied, and the amount of gas in that space. These quantities are called **state variables** because they describe the state or status of a gas—how hot it is, the pressure it exerts on a surface, and so forth. State variables are related to each other by equations called **equations of state.** Scientists use these equations of state in conjunction with the laws of thermodynamics to study phenomena such as climate changes in the atmosphere, the operation of gas-driven heat engines, and the change in pressure in a person's skull when the space inside the skull changes.

12.1 Description of a Gas

Suppose for a moment that you have been reduced to the size of a molecule—about 10^{-10} times your normal size. What would life be like if you, in your reduced state, became a molecule in the air? Other molecules would be constantly moving past you in all directions at speeds of about 1000 mph (see Fig. 12.1); some would be moving faster, others slower.

On the average, you would be separated from the nearest molecule by a distance of about ten times your own size, a situation comparable to that of twenty people spread over the area of a football field. You are usually separated by some distance from the other air molecules, during which time you feel little effect from their presence. Nevertheless, your isolation is interrupted by frequent collisions. The collisions are violent, since the molecules move at speeds of about 1000 mph. Each time a molecule crashes into you, you are knocked randomly in a new direction. After a collision you move only about 10^{-7} meter before being hit by another molecule. During the course of a minute, you experience about 300 billion collisions, each causing your direction of motion to change. Because of the random nature of these collisions, you find after one minute that you have moved only about 4 cm from your starting position, even though your speed during that time has averaged about 1000 mph. During this time, you have had

FIG. 12.1. Life as you would see it if you were an air molecule.

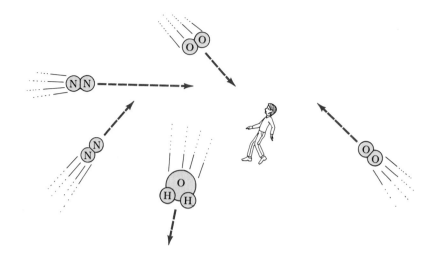

no meaningful relations with any particular air molecule, just a succession of abrupt collisions.

This picture of air is characteristic of all gases. The space between atoms and molecules in any gas is large compared to their size (see Fig. 12.2). The molecules move fast, but because frequent collisions interrupt and change their direction of motion, they undergo little net displacement during any short period.

Besides colliding with each other, molecules and atoms in a gas also bounce off solid objects and liquid surfaces in their path. If a molecule of air hits your skin, it rebounds. Although we cannot sense these individual collisions, the molecule does exert a small force on the skin during the collision. This force is the cause of what we call *pressure*.

12.2 Pressure

When molecules of air or of any gas collide with a solid or liquid surface, they exert force against that surface. The force is of an impulsive nature, much like the force of a tennis ball hitting a practice board. Our skin cannot sense the force of individual collisions, but we can sense the average effect of a large number of the collisions. Each second about 10^{23} molecules of air strike each square centimeter of our skin. The net effect of these collisions is a force that seems constant in nature. Later in this chapter we discuss the magnitude and atomic nature of the force caused by these abrupt collisions. For now, we turn to only the pressure caused by the collisions of the atoms and molecules in a gas with a solid or liquid surface.

FIG. 12.2. This dumbbell-shaped object is about 20 million times the size of an oxygen or nitrogen molecule. On the same scale the nearest air molecule would be near the right edge of this page.

Pressure P is defined as the force F exerted perpendicularly on a surface divided by the area A over which the force is exerted:

$$P = \frac{F}{A}. \tag{12.1}$$

At sea level, for example, collisions of air molecules with a solid surface cause an average force of $1.01 \times 10^5\,\text{N}$ to be exerted on each square meter of surface.

Thus, atmospheric pressure at sea level is $1.01 \times 10^5 \, \text{N/m}^2$. In terms of the English system of units, the pressure of air at sea level is $14.7 \, \text{lb/in}^2$ (or 14.7 psi—pounds per <u>s</u>quare <u>in</u>ch).

Units of Pressure

For measuring pressure there are many kinds of instruments and nearly as many units. One device used to measure gas pressure, a **manometer,** consists of a U-shaped tube containing mercury or water (Fig. 12.3). If the tube is open on both ends, air pushes down on each column with equal pressure, and the columns on each side of the tube remain at the same elevation (Fig. 12.3a). However, if one end of the tube is closed and the air removed from it, creating a vacuum, the air pressure forces the mercury down on the open side and up into the evacuated side, as shown in Fig. 12.3b. The height h of the mercury column in the evacuated side above that in the side open to the air is a measure of the pressure of the air. At sea level the mercury column will be about 760 millimeters or 29.9 inches higher on the left side than on the right side of the tube. Thus, we say that the pressure of the earth's atmosphere at sea level is 760 **millimeters of mercury** (760 mm Hg) or 29.9 **inches of mercury** (in Hg).

These pressure units are based entirely on a measuring technique. Since pressure is defined as the ratio of a force to an area, one would expect units such as N/m^2 or lb/ft^2. However, various other hybrid units, some based on particular measuring techniques, are in common use today: one **atmosphere** equals the pressure of air at sea level, the **bar** is a unit of pressure slightly less than one atmosphere, and the **pascal** (Pa) is the SI unit for pressure and equals $1 \, \text{N/m}^2$. Factors needed to convert between these and other units of pressure are listed in Table 12.1.

12.3 Gauge Pressure

The pressure we have been discussing is called *absolute pressure,* or simply *pressure.* It is the pressure we would measure with a manometer having one arm evacuated (Fig. 12.3b). This device compares the pressure of the gas pushing down on the mercury in the open side of the tube to the zero pressure in the evacuated side of the tube.

(a)

(b)

FIG. 12.3. (a) When both tubes of the manometer are open to the atmosphere, the mercury on each side is at the same elevation. (b) If one side of the manometer is closed and evacuated, the atmospheric pressure on the open side pushes the mercury down and up into the evacuated side. The height of the mercury column in the evacuated side above the column in the open side is a measure of the atmospheric pressure.

TABLE 12.1 Conversion Factors for Common Units of Pressure. Each number gives the value of 1 unit of the pressure named at the left in terms of the unit named at the top; for example, 1 lb/in² equals 51.7 mm Hg

	N/m^2 (pascal)	lb/in^2	atm	bar	mm Hg (torr)	cm H_2O
N/m^2 or Pa[a]	1	1.45×10^{-4}	9.87×10^{-6}	10^{-5}	7.50×10^{-3}	1.02×10^{-2}
lb/in^2	6.89×10^3	1	6.80×10^{-2}	6.89×10^{-2}	51.7	70.3
atm	1.01×10^5	14.7	1	1.01	760	1.03×10^3
bar	10^5	14.5	0.987	1	750	1.02×10^3
mm Hg (torr)	133	1.93×10^{-2}	1.32×10^{-3}	1.33×10^{-3}	1	1.36
cm H_2O	98.1	1.42×10^{-2}	9.68×10^{-4}	9.81×10^{-4}	0.736	1

[a] A N/m^2 is also called a pascal (Pa).

FIG. 12.4. When a person blows into the right side of an open-tube manometer, the extra pressure on this side forces water down and up into the left arm.

FIG. 12.5. When a person sucks on the right side of the manometer, the reduced pressure causes the water to rise on this side.

Many actual pressure-measuring devices do not have an evacuated side but are open on both sides. An open-tube manometer compares an unknown pressure of gas above one side of the tube to the atmospheric pressure above the other side. If both sides are exposed to air as in Fig. 12.3a, the pressure is the same on each side, and the mercury or water in the tube will be at the same level on each side.

If, however, one side of the manometer is attached to a container of gas at a pressure different from atmospheric pressure, the column of mercury or water on that side will be displaced either upward or downward, depending on whether the pressure is less or greater than atmospheric pressure. For example, if you blow into a rubber hose attached to the right arm of a manometer containing water, as in Fig. 12.4, the water column will be displaced down on the right, owing to the extra pressure above it. * The difference in the height h of the water column on the left compared to the right indicates the pressure difference. This pressure difference is called **gauge pressure.** The gauge—that is, the measuring device—compares the unknown air pressure on the right to atmospheric pressure on the left. The absolute pressure P on the right is just the gauge pressure plus atmospheric pressure P_{atm}:

$$P = P_{atm} + P_{gauge}. \qquad (12.2)$$

For the situation shown in Fig. 12.4, the gauge pressure is positive.

If a person sucks on a manometer as shown in Fig. 12.5, the gauge pressure will be negative—that is, less than atmospheric pressure. The column of water is forced up into the partially evacuated right side by the atmospheric pressure on the left side.

In this and later chapters we will usually need to use absolute pressures in our equations. For simplicity we will use the word *pressure* to mean "absolute pressure" from here on. When necessary, Eq. (12.2) can be used to convert gauge pressure to (absolute) pressure and vice versa.

EXAMPLE 12.1 The gauge pressure inside a straw when you begin to suck on it while drinking a soda is −15 mm Hg. What is the (absolute) pressure in the straw?

SOLUTION

$$P = P_{atm} + P_{gauge} = 760 \text{ mm Hg} - 15 \text{ mm Hg}$$
$$= \underline{745 \text{ mm Hg}}. \qquad \blacksquare$$

12.4 Equations of State

Pressure is one of several measurable characteristics of gases. Others are temperature, the volume of space occupied by the gas, and the amount of gas in that space. These quantities, called **state variables,** are used to describe the condition or state of the gas: is the gas hot or cold? compressed under high pressure into a small space or spread over a large volume under low pressure?

*You would not want to blow or suck on the air above a mercury manometer; mercury is a highly toxic substance!

In the case of gases, scientists have found that state variables are related to each other in the form of equations called **equations of state.** They help us determine how changes in one state variable will affect other state variables. For instance, we know that if we close the nozzle of a bicycle pump so that air cannot leave it and then try to compress the gas in the pump by pushing down on the handle, the force with which we must push increases as the gas becomes more compressed. The pressure of the gas increases as its volume decreases. An equation of state is a mathematical equation that can be used to calculate this expected change in pressure when the volume of gas changes.

Robert Boyle (1627–1691), a contemporary of Newton, developed the equation of state that relates the pressure of a gas to its volume. Boyle used a manometer closed on one side (Fig. 12.6). A fixed amount of air was trapped in the closed side by a column of mercury, and the pressure on the air could be increased by adding mercury to the open column. Boyle found that if the temperature and amount of air in the tube remained constant, the volume of space occupied by the air decreased in inverse proportion to the pressure acting on the air; that is, $V \propto 1/P$. If the pressure doubled, the volume decreased to one-half its original volume. Boyle described this relationship between pressure and volume with the following equation, now called **Boyle's law:**

PV = constant (for a fixed amount of gas at constant temperature)   **(12.3)**

For the product of P and V to remain constant, one of the quantities must decrease if the other increases, and vice versa. Absolute pressures must be used in this equation.

We often use Boyle's law by comparing the product of the pressure and volume in a gas before some process occurs and after it is over. If the amount of gas and the temperature of the gas remain constant, then the product of the final values of pressure and volume must equal the product of the initial values, since both products equal the same constant:

$$PV = P_0 V_0 = \text{constant.} \qquad \textbf{(12.4)}$$

Once again, absolute pressures must be used in this equation.

EXAMPLE 12.2 Suppose that a person's skull is filled with air at a gauge pressure of $+8$ mm Hg. A skull fracture reduces the skull volume and, in turn, increases the gauge pressure in the skull to $+16$ mm Hg. Assuming that the air temperature remains constant, calculate the ratio of the final skull volume to the initial skull volume.

SOLUTION By converting from gauge to absolute pressures and then solving Eq. (12.4) for V/V_0, we get the desired ratio:

$$\frac{V}{V_0} = \frac{P_0}{P} = \frac{(8 \text{ mm Hg} + 760 \text{ mm Hg})}{(16 \text{ mm Hg} + 760 \text{ mm Hg})} = \frac{768}{776} = \underline{0.99}. \qquad \blacksquare$$

The observed pressure change in the last example is caused by only a 1 percent decrease in volume. In terms of pressure, a person's skull behaves much like a skull filled with air. The gauge pressure is normally about 8 mm Hg. Pressures exceeding 16 to 20 mm Hg are cause for concern.

The calculation in the last example shows that pressure measurements eas-

(a)

(b)

FIG. 12.6. (a) A small column of mercury creates little pressure, and the gas occupies a large volume. (b) A larger pressure is created by a high column of mercury; the gas occupies less volume.

ily reveal changes in skull volume and can therefore help in the detection of a number of head problems including brain disorders, tumors, hemorrhage, infections, swelling of the brain, and injuries to the skull. A sinus headache, for instance, occurs when sinus fluids occupy space in the skull, reducing the volume of space available for its usual contents. The reduction in volume increases the pressure on the brain, causing a headache.

Boyle's law is just one of many empirical equations developed to calculate expected changes in one state variable caused by changes in others. For example, with Charles' law ($P/T = $ constant) we can calculate expected changes in the pressure of a constant volume of gas when the temperature of the gas changes. Perhaps the most famous and most used of these equations of state is the ideal gas law.

12.5   Ideal Gas Law

The **ideal gas law** is an equation of state relating the variables pressure, temperature, volume, and amount of gas. For most gases at temperatures somewhat above their condensation temperature, these variables are related by the equation

$$PV = NkT, \tag{12.5}$$

where P is the gas pressure, N is the number of molecules of gas in a volume of space V, T is the gas temperature *in degrees Kelvin,* and k is a constant, called **Boltzmann's constant,** that equals 1.38×10^{-23} J/K.

EXAMPLE 12.3   A home whose volume equals 80 m^3 contains air at a pressure of one atmosphere and at a temperature of 20°C (Fig. 12.7). How many molecules of air are in the home? If the average mass of one molecule is 4.8×10^{-26} kg, what is the mass of all of the air in the home?

SOLUTION   When using the ideal gas law, we must convert temperatures to degrees Kelvin [$T = 20°C = (273 + 20)$ K $= 293$ K]. Also, pressure must be in units consistent with other quantities in the equation. In this example we convert from atm to N/m^2: $P = 1$ atm $= 1.01 \times 10^5$ N/m^2. Substituting in Eq. (12.5), we find that the number of molecules in the house is

$$N = \frac{PV}{kT} = \frac{(1.01 \times 10^5 \text{ N/m}^2)(80 \text{ m}^3)}{(1.38 \times 10^{-23} \text{ J/K})(293 \text{ K})} = \underline{2.0 \times 10^{27}}. *$$

Since the average mass of each molecule equals 4.8×10^{-26} kg, the total mass of the air in the house is

$$m = (2.0 \times 10^{27} \text{ molecules})(4.8 \times 10^{-26} \text{ kg/molecule}) = \underline{96 \text{ kg}}.$$

The air weighs about 200 lb. ∎

EXAMPLE 12.4   The gauge pressure of air in a car tire is 30.0 lb/in^2 and its temperature is 27°C. As the car is driven along a highway, the temperature of the air in the tire increases to 57°C and the volume increases by 5 percent. What is the final air gauge pressure in the tire?

FIG. 12.7.   A typical house contains about 100 kg of air.

*Remember that $1 \text{ N} \cdot \text{m} = 1 \text{ J}$.

SOLUTION This problem can be worked using Eq. (12.5): $PV/T = Nk$. Since the number of air molecules in the tire does not change, the right side of this equation is unchanged as the gas warms. Thus, the final value of PV/T after the air has warmed must equal the initial value:

$$\frac{PV}{T} = \frac{P_0 V_0}{T_0} = Nk.$$

The above equation can be rearranged to determine the final pressure:

$$P = \left(\frac{V_0}{V}\right)\left(\frac{T}{T_0}\right) P_0. \qquad (12.6)$$

Since the final volume is 5 percent greater than the initial volume, the ratio V_0/V is $V_0/1.05V_0 = 1/1.05$. When we use the ideal gas law, the temperatures must be in degrees Kelvin. Thus,

$$\frac{T}{T_0} = \frac{(57 + 273)\,\text{K}}{(27 + 273)\,\text{K}} = \frac{330}{300} = 1.10.$$

The pressures substituted into Eq. (12.6) must be absolute pressures and not gauge pressures. Equation (12.2) is used to find the initial air pressure in the tire: $P_0 = P_{\text{atm}} + P_{\text{gauge}_0} = 14.7\,\text{lb/in}^2 + 30.0\,\text{lb/in}^2 = 44.7\,\text{lb/in}^2$. We now substitute these quantities into Eq. (12.6) to find the final air pressure in the tire:

$$P = \left(\frac{1}{1.05}\right)(1.10)(44.7\,\text{lb/in}^2) = 46.8\,\text{lb/in}^2.$$

The final gauge pressure is the final pressure minus the atmospheric pressure, or

$$P_{\text{gauge}} = P - P_{\text{atm}} = 46.8 - 14.7 = \underline{32.1\,\text{lb/in}^2}.$$

We were able to use mixed units in this example (temperature in the metric system unit of degrees Kelvin, and pressure in the English system unit of pounds per square inch) because we used ratios, and the units canceled each other. ∎

Avogadro's Number, Moles, and the Ideal Gas Law

Most containers of gas, when near atmospheric pressure, hold huge numbers of atoms or molecules of gas. A cubic centimeter of air contains about 3×10^{19} molecules. This is such a large number that it has little meaning.

Sir James Jeans (1877–1946), the British astrophysicist, offered an interesting example of the enormous number of molecules involved in normal activities. If George Washington's last breath, which contained more than 10^{22} molecules, is now distributed uniformly in the air of the earth's atmosphere, then there are about five molecules from his last breath in each 2000 cm^3 of the atmosphere. Since your lungs hold about 2000 cm^3 of air, it is likely that five of the molecules now in your lungs were also part of George Washington's last breath.

We can avoid using large numbers in the ideal gas law by defining a quantity, the mole, and a related number, called Avogadro's number. One **mole** of a substance (solid, liquid, or gas) consists of 6.023×10^{23} atoms or molecules of that substance. The number, 6.023×10^{23}, is called **Avogadro's number** and is given the symbol N_A. One mole of any form of matter contains a mass in grams equal to the atomic or molecular mass of that form of matter. For example, one mole of atomic hydrogen (H) has a mass of 1 gram, one mole of molecular hydrogen (H_2) has a mass of 2 grams, one mole of helium (He) has a mass of 4 grams, one mole of molecular oxygen (O_2) has a mass of 32 grams, and so on.

Using the definition above, we can now define the number of molecules of a substance in terms of Avogadro's number and the number of moles of that substance. If we have 0.4 mole of oxygen in a container, the number of molecules of oxygen will be 0.4 times Avogadro's number, or $0.4(6.023 \times 10^{23}) = 2.4 \times 10^{23}$ molecules of oxygen. In general, the number N of atoms or molecules of a particular substance is related to the number n of moles of that substance and to Avogadro's number N_A by the equation

$$N = nN_A. \tag{12.7}$$

If we substitute $N = nN_A$ into Eq. (12.5), we find that

$$\frac{PV}{T} = Nk = nN_A k = nR,$$

where $R \equiv N_A k = 8.314$ J/mole $\cdot$ K and is called the **universal gas constant.** Using this constant, we can rewrite the **ideal gas law** in terms of the number of moles of gas, a much more manageable number than the number of gas molecules:

$$PV = nRT. \tag{12.8}$$

Equation (12.8) is a good description of the relationship between the state variables of most real gases of low density at temperatures well above the temperature at which the gas condenses to a liquid. As before, the temperature in this equation must be in degrees Kelvin, and absolute pressures must be used.

EXAMPLE 12.5 A balloon containing 18 g of helium has a volume of 0.10 m³. The pressure inside the balloon is 1.2×10^5 N/m². Calculate the temperature of the gas in the balloon.

SOLUTION We can solve the ideal gas law [Eq. (12.8)] to find the temperature. We must first, however, determine the number of moles of helium in the balloon. One mole of helium has a molecular weight of 4.0 g. Thus,

$$18 \text{ g} = (18 \text{ g})\left(\frac{1 \text{ mole}}{4.0 \text{ g}}\right) = 4.5 \text{ mole}.$$

After rearranging the ideal gas law, we find that

$$T = \frac{PV}{nR} = \frac{(1.2 \times 10^5 \text{ N/m}^2)(0.10 \text{ m}^3)}{(4.5 \text{ mole})(8.314 \text{ J/mole K})} = \underline{320 \text{ K}}. \qquad \blacksquare$$

EXAMPLE 12.6 A scuba diver breathes 600 cm³ of air per breath and takes 30 breaths/min. (a) What volume of air does the diver breathe in 20 minutes? (b) The diver is under the water at a depth such that this air is inhaled at a pressure of 2.5 atm. The tank from which the air is released is at 240 atm. What must be the volume of the tank so that the air will last 20 minutes? (c) Calculate the mass of the air in the tank at a temperature of 290 K. One mole of air has a mass of 29 g.

SOLUTION (a) The volume of air inhaled in 20 minutes is

$$V = (600 \text{ cm}^3/\text{breath})(30 \text{ breaths/min})(20 \text{ min})$$

$$= 3.6 \times 10^5 \text{ cm}^3 = (3.6 \times 10^5 \text{ cm}^3)\left(\frac{10^{-6} \text{ m}^3}{1 \text{ cm}^3}\right) = \underline{0.36 \text{ m}^3}.$$

(b) This volume V of air at a pressure of 2.5 atm is to be compressed to a tank volume V_t at a presssure of 240 atm. We can find V_t by using Boyle's law:

$$PV = P_t V_t$$

or

$$V_t = \frac{P}{P_t} V = \left(\frac{2.5 \text{ atm}}{240 \text{ atm}}\right)(0.36 \text{ m}^3) = \underline{3.8 \times 10^{-3} \text{ m}^3}.$$

(c) By Eq. (12.8), the number of moles of air in the tank is

$$n = \frac{P_t V_t}{RT}$$

where, for the tank, $V_t = 3.8 \times 10^{-3} \text{ m}^3$. The pressure must be converted from atm to N/m²:

$$P_t = (240 \text{ atm})\left(\frac{1.01 \times 10^5 \text{ N/m}^2}{1 \text{ atm}}\right) = 2.4 \times 10^7 \text{ N/m}^2.$$

We find, then, that

$$n = \frac{(2.4 \times 10^7 \text{ N/m}^2)(3.8 \times 10^{-3} \text{ m}^3)}{(8.314 \text{ J/mole} \cdot \text{K})(290 \text{ K})} = 38 \text{ moles}.$$

Since each mole has a mass of 29 g, the mass of the air is $m = 38$ moles (29 g/mole) = $\underline{1100 \text{ g}}$. ∎

A Note about Units

Often we can work a problem using ratios. For example, in part (a) of the previous example the volume of air in the tank depended on the ratio of the pressure at the diver's lungs to the pressure in the tank. In that situation any absolute pressure unit can be used, since the units cancel. In part (b), though, when the number of moles of gas is calculated, the units do not cancel. It is essential when making this type of calculation to *use a common system of units for all quantities*. The SI metric system is highly recommended and will be used almost exclusively in this text.

12.6 Kinetic Theory of Gases

We have seen that quantities such as pressure, temperature, volume, and number of molecules are related by equations such as Boyle's law and the ideal gas law. We can substitute numbers into these equations and find the value of some unknown quantity. Doing this gives us little extra insight into the reason gases behave as they do. Why, for example, does the pressure of a gas double when its volume is reduced to one-half its original volume? Boyle's law describes this behavior but does not help us understand why this happens.

This variation of gas pressure with volume is really very surprising. Remember that a gas is made of molecules separated by distances much greater than their own size (Fig. 12.2). Almost 99.9 percent of the volume of a container of air is empty. It should be easy to "crowd" these air molecules into a smaller space. If we reduce the volume by half, the volume occupied by the gas is still about 99.8 percent empty. The molecules are still not rubbing elbows with each other. Yet, when we do reduce the volume by a factor of one-half, the molecules exert twice as much pressure on the walls of their container. How can this be?

To answer such questions, scientists have proposed models concerning the nature and behavior of gases. Boyle proposed that a gas might be made of tiny particles that were springy—like little balls of curly wool piled together. The "wool," if coarse and springy enough, could resist compression much as air does. Isaac Newton, on the other hand, proposed that air was made of tiny particles somewhat farther apart. His particles resisted compression because of a repulsive force between the particles which increased as their separation decreased.

Although both these models seem to explain the relationship between volume and pressure, neither explains the dramatic dependence of pressure on temperature. If the temperature of a constant volume of gas is doubled, its pressure doubles. In Newton's model the pressure depends only on the separation of the molecules, which does not explain the dependence of pressure on temperature. Similarly, Boyle's model does not account for variations in gas pressure with temperature.

A successful model for gases was not developed until the nineteenth century, about one hundred fifty years after Boyle discovered his law. This theory, the kinetic theory of gases, will be used to derive an equation of state for gases, which, when compared to the ideal gas law, will result in new insights into the meaning of temperature and thermal energy. The **kinetic theory of gases** states that gases consist of atoms and molecules in constant and rapid motion. The pressure caused by these atoms and molecules is a result of collisions of the atoms and molecules with the walls of their container (Fig. 12.8).

This model of gases and Newton's second law are used now to derive an expression for the pressure caused by a gas against the walls of its enclosure. To complete this derivation, we will make several assumptions that simplify our calculations without affecting the general nature of the results. These assumptions are needed because of the huge number of atoms and molecules in a gas and because of the complicated motion of the individual atoms and molecules. We assume that individual atoms and molecules are point masses having no size, that they all move at the same speed v, and that one-third of the molecules move parallel to the x axis, one-third move parallel to the y axis, and one-third move parallel to the z axis. Also, we assume that the container holding the gas has length L on each side, and that the x, y, and z axes are oriented parallel to the walls of the container, as shown in Fig. 12.9.

Our assumption of zero size allows us to ignore the internal structure of a molecule—the fact that it is made of nuclei surrounded by electrons. We can also ignore the vibrational and rotational motion of the molecules and the collisions of one molecule with another; they are too small to collide. The only collisions that occur are with the walls of the container. These collisions are assumed to be elastic; that is, there is no loss of kinetic energy during a collision.

Let us now consider the pressure of the gas on a wall lying in the y-z plane. When a molecule hits this wall, it rebounds in the opposite direction with the same speed (Fig. 12.9) because the collision is assumed to be elastic. The wall

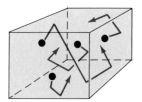

FIG. 12.8. Molecules of gas in constant motion cause a pressure because of their collisions with the walls of their container.

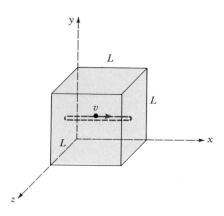

FIG. 12.9. A molecule that hits the wall, moving with speed v, rebounds and moves with a speed v in the opposite direction. The impulse of the force is $f \Delta t = 2mv$.

must therefore cause a force f' on the molecule given by Eq. (5.3),

$$f' \Delta t = mv_x - mv_{x_0}, \qquad (5.3)$$

where Δt is the time of duration of the collision, f' is the average force of the wall on the molecule during that time, m is the molecule's mass, v_{x_0} is its velocity in the x direction before the collision, and v_x is its velocity in the x direction after collision. For a molecule moving with a speed v in the x direction, its x components of velocity before and after a collision with the wall are $v_{x_0} = v$ and $v_x = -v$. Thus,

$$f' = \frac{m(-v) - mv}{\Delta t} = \frac{-2mv}{\Delta t}.$$

The force f of the molecule on the wall is the reaction to the force f' of the wall on the molecule. Since action-reaction forces have equal magnitudes but opposite directions, we find that

$$f = \text{force of molecule on wall} = +\frac{2mv}{\Delta t}. \qquad (12.9)$$

After colliding with the wall, the molecule rebounds and flies across the box to hit the other wall. Rebounding from this wall, it returns again for another collision with the original wall. Each collision causes an impulse $f \Delta t$, the result of an abrupt force acting for a very short time. Let us multiply and divide this impulse by τ, the time needed for one round trip across the box and back. Since the box has length L, the distance traveled at speed v is $2L$. This transit time for one round trip is

$$\tau = \frac{2L}{v}. \qquad (12.10)$$

The impulse can now be written as

$$f \Delta t = f \Delta t \frac{\tau}{\tau} = \left(f \frac{\Delta t}{\tau} \right) \tau = (\bar{f}) \tau.$$

We can think of $\bar{f}$ as an average force, smaller than f by a factor $\Delta t/\tau$, that acts during the whole time of transit τ. The impulse caused by this small average force $\bar{f}$ spread over the longer time τ has the same effect—that is, impulse—as the larger, abrupt force f acting for a short time Δt. According to Eq. (12.9), $f \Delta t = 2mv$. But since $f \Delta t = \bar{f} \tau$, then $\bar{f} \tau$ must also equal $2mv$. Rearranging and substituting for τ from Eq. (12.10), we find that

$$\bar{f} = \frac{2mv}{\tau} = \frac{2mv}{(2L/v)} = \frac{mv^2}{L}.$$

If there are N molecules in the box and one-third of them are hitting the wall (the other two-thirds move perpendicular to the wall and do not hit it), then the average force due to all of these molecules is

$$F = \frac{N}{3} \bar{f} = \frac{N}{3} \left(\frac{mv^2}{L} \right).$$

The pressure exerted on the wall by this force equals the force divided by the area of the wall. For a cubical box, the area is L^2. Thus,

$$P = \frac{F}{A} = \frac{1}{L^2} \left(\frac{N}{3} \right) \left(\frac{mv^2}{L} \right) = \frac{1}{3} \left(\frac{Nmv^2}{L^3} \right).$$

The volume of a cube of length L on each side is L^3. After substituting V for L^3, and rearranging, we find that

$$PV = \frac{1}{3} Nmv^2.$$

If we multiply and divide the right side by 2, we can see the explicit dependence of the product PV on the kinetic energy of a molecule:

$$PV = \frac{2}{3} N \left(\frac{1}{2} mv^2 \right) \tag{12.11}$$

Now let us compare the ideal gas law ($PV = NkT$) to the above. Both equations have PV on the left. Thus, for both equations to be true simultaneously, we find that

$$\frac{1}{2} mv^2 = \frac{3}{2} kT.$$

The translational kinetic energy of a molecule in the gas equals a constant times the temperature of the gas.

Although our derivation was for molecules moving at constant speed v, the result holds in general for "real" molecules that at any one time move with a variety of speeds. For this real gas we replace v^2 by the average of the square of a molecule's speed $(\overline{v^2})$. Thus,

$$\frac{1}{2} m\overline{v^2} = \frac{3}{2} kT.$$

If a container holds N molecules at temperature T, the average kinetic energy due to the random motion of all molecules is $N(\frac{3}{2})kT$. This energy due to the random motion of molecules is called thermal energy.

The **average thermal energy** per molecule in a gas at temperature T is

$$E_{\text{th}}/\text{molecule} = \tfrac{1}{2} m\overline{v^2} = \tfrac{3}{2} kT. \tag{12.12}$$

The **thermal energy of N molecules** is

$$E_{\text{th}} = N(\tfrac{3}{2})kT. \tag{12.13}$$

Temperature must be in degrees Kelvin when we use these equations.

We see that temperature is a measure of thermal energy!

The above expressions for thermal energy include the random kinetic energy due to translational motion only. We have ignored a molecule's thermal energy because of its rotational and vibrational motions. In our examples and problems in this chapter we assume that Eq. (12.13) includes all the thermal energy of a gas of N atoms. In a more rigorous treatment of gases consisting of diatomic molecules at room temperature, the $\frac{3}{2}$ in front of kT should be replaced by $\frac{5}{2}$ in order to include the molecule's rotational thermal energy.

EXAMPLE 12.7 A cup contains about 5.0×10^{21} gaseous helium atoms at a temperature of $27°C$. Calculate the thermal energy of these atoms.

SOLUTION The thermal energy of these atoms is

$$E_{\text{th}} = N \left(\frac{3}{2} kT \right) = (5.0 \times 10^{21}) \frac{3}{2} \left(1.38 \times 10^{-23} \frac{\text{J}}{\text{K}} \right) (300 \text{ K}) = \underline{31 \text{ J}}. \quad \blacksquare$$

Root-Mean-Square Speed

Equation (12.12) allows us to estimate the speeds of different types of atoms or molecules in a gas. If we solve the equation for $\overline{v^2}$, we find that

$$\overline{v^2} = \frac{3kT}{m}.$$

This is the average of the square of the speed of a molecule in the gas. The square root of $\overline{v^2}$ is called the **root-mean-square speed:**

$$v_{rms} = \text{root-mean-square speed} = \sqrt{\overline{v^2}} = \sqrt{\frac{3kT}{m}}, \qquad \textbf{(12.14)}$$

where T is the temperature of the gas and m is the mass of the atom or molecule whose root-mean-square speed is being calculated.

We should remember that at any instant there is great variation in the speeds of the different atoms and molecules. The equation allows us to estimate, roughly, the average speed of an atom or molecule in the gas.

Equation (12.14) applies not only to single atoms and molecules in a gas, but also to small particles consisting of many atoms or molecules, such as specks of dust, grains of pollen, and particles of smog (see Fig. 12.10).

FIG. 12.10. Particles of gas have the same average kinetic energy. If their mass is large, their speed is small; if their mass is small, their speed is large.

EXAMPLE 12.8 Calculate the root-mean-square speed of an air molecule at a temperature of 27°C. One mole of air has a mass of 29 g. How does v_{rms} of air molecules compare to the speed of sound in air (340 m/s)?

SOLUTION Since one mole (6.023×10^{23} molecules) of air has a mass of 29 g, the average mass of a molecule is

$$m = \frac{29 \text{ g}}{6.023 \times 10^{23}} = 4.8 \times 10^{-23} \text{ g} = 4.8 \times 10^{-26} \text{ kg}.$$

We find, then, that

$$v_{rms} = \sqrt{\frac{3kT}{m}} = \sqrt{\frac{3(1.38 \times 10^{-23} \text{ J/K})(27 + 273) \text{ K}}{4.8 \times 10^{-26} \text{ kg}}} = \underline{510 \text{ m/s.}}$$

Sound waves are caused by air molecules oscillating in and out of regions of varying air density. Since the molecules cannot move into or out of these regions faster than their random speed, the speed with which sound waves are formed and travel is limited by this random speed. The speed of sound in a gas is about two-thirds the root-meant-square speed of its molecules. ∎

12.7 **Work Done by a Gas**

We have learned that the thermal energy of a gas depends only on its temperature and the number of molecules in the gas. This relationship between thermal energy and temperature is important when we use the laws of thermodynamics to analyze energy-transformation processes involving gases, such as those occurring in the cylinders of automobile engines or in the earth's atmosphere. These processes involve work, and our equation for calculating work must be modified slightly when we think of the work done on or by a gas as its volume changes.

FIG. 12.11. The work done on the gas by the force of the piston when displaced by a distance Δr is $W = F\,\Delta r = P\,\Delta V$, where ΔV is the volume change of the gas and P is its pressure.

As you may remember from Chapter 8, work is defined as

$$W = F\,\Delta r \cos\theta, \tag{8.1}$$

where F is the magnitude of a force acting on an object as it undergoes a displacement of magnitude Δr, and θ is the angle between the direction of **F** and $\Delta \mathbf{r}$. Let us apply this definition to determine the work needed to change the volume of a cylinder containing gas. In Fig. 12.11 a force is shown pushing against a movable piston that forms one wall of the cylinder. The work done by the force as the piston undergoes a small displacement Δx is $W = F\,\Delta x \cos 0° = F\,\Delta x$. If we multiply and divide this expression for work by A, the cross-sectional area of the piston and cylinder, we find that

$$W = F\,\Delta x = F\,\Delta x \frac{A}{A} = \frac{F}{A}(\Delta x A) = P\,\Delta V,$$

where F/A is the pressure exerted by the gas against the piston, and ΔV is the change in volume of the gas (the cross-sectional area of the cylinder times its change in length). This is a general expression for the work done on a gas during any change in volume. But if we use the convention that ΔV is negative when the volume decreases, then a negative sign should be placed in front of the $P\,\Delta V$ so that the work done on the gas is positive when its volume decreases.

The **work** W done **on a gas** when at pressure P as its volume changes by an amount ΔV is

$$W = -P\,\Delta V. \tag{12.15}$$

If the volume decreases, ΔV is negative and W is positive. If the volume increases, ΔV is positive and W is negative.

In our earlier use of the first law of thermodynamics and conservation of energy principle $(Q + W = \Delta E)$, W was defined as the work done *on* an object in the system. If the gas in a cylinder is the system, then W is the work done *on the gas* and is defined as $W = -P\,\Delta V$. In some treatments of thermodynamics, the first law is written as $Q - W' = \Delta E$, where W' is defined as the work done *by* the system on its environment. For a gas, W', the work done *by the gas,* is defined as $W' = +P\,\Delta V$. Both conventions produce the same result, since a minus sign appears in front of W' in the first law when we use the latter convention.

EXAMPLE 12.9 A cylinder of gas has a movable piston of cross-sectional area $0.12\,\text{m}^2$ (Fig. 12.12). The piston can move freely up or down and exerts a constant pressure of $2.40 \times 10^5\,\text{N/m}^2$ on the gas. If the gas is heated, it expands and the piston moves 4.0 cm upward. Calculate the work done by the piston on the gas.

SOLUTION The volume of the gas increases by $\Delta V = A\,\Delta x = +(0.12\,\text{m}^2)(0.04\,\text{m}) = +4.8 \times 10^{-3}\,\text{m}^3$. The work done by the piston on the gas is

$$W = -P\,\Delta V = -(2.40 \times 10^5\,\text{N/m}^2)(4.8 \times 10^{-3}\,\text{m}^3) = \underline{-1.15 \times 10^3\,\text{J}}.$$

The gas in turn does work $W' = +1.15 \times 10^3\,\text{J}$ on the piston while lifting it.

FIG. 12.12. The weight of a piston causes a constant pressure to be exerted on the gas.

Temperature Change
When the Volume of a Gas Changes

A change in volume of a gas can affect its temperature. Suppose that the volume of a gas decreases by an amount $-|\Delta V|$, and that no heat is added to or removed from the gas. The work done on the gas is $W = -P(-|\Delta V|) = P|\Delta V|$ and is a positive number. Substituting into the first law of thermodynamics ($Q + W = \Delta E$) for the case when $Q = 0$, we find that

$$0 + P|\Delta V| = \Delta E.$$

The internal energy ΔE of the gas must increase. Most of this increase results in increased thermal energy. Since thermal energy is proportional to temperature [Eq. (12.13)], the temperature of the gas increases. By similar reasoning we can show that the temperature of a gas decreases if its volume increases—that is, if no heat is added to or removed from the gas.

This change in temperature with volume change has many interesting consequences. For instance, when air moves up into the atmosphere from the earth below, it expands because of the decrease in pressure. If the change in elevation and expansion are rapid, the gas exchanges little heat with surrounding air, and its temperature is lowered because of the expansion. If the temperature is lowered below the dew point, water condenses and rain falls.

Another example in which the change in volume of air affects our climate occurs in the fall in California. Warm air known as the Santa Ana wind blows west from the desert toward the ocean. As the wind moves down off the mountains at the west coast and piles up against dense ocean air, its volume contracts, and its temperature increases. The air, already warmed by its passage across the desert, becomes even warmer. The change in temperature can be significant, as is apparent from the next example.

EXAMPLE 12.10 Estimate the change in temperature of air if its volume V contracts by 1 percent with no heat addition or removal. The number of molecules per unit volume (N/V) in the air is approximately 2.4×10^{25} molecules/m^3 and its average pressure is one atmosphere (1.01×10^5 N/m^2).

SOLUTION The volume decreases by 1 percent, or $\Delta V = -0.01V$, where V is the original volume of the gas. The work done in compressing the gas is

$$W = -P\,\Delta V = -P(-0.01\ V) = 0.01\ PV.$$

Heat is neither added to nor removed from the gas, so that

$$Q = 0.$$

The change in internal thermal energy of the gas is

$$\Delta E_{\text{th}} = N(\tfrac{3}{2})k(T - T_0) = N(\tfrac{3}{2})k\,\Delta T,$$

where ΔT is the temperature change of the gas, and N is the number of gas molecules.

Substituting for W, Q, and ΔE into the first law of thermodynamics, we find that

$$0 + 0.01PV = N(\tfrac{3}{2})k\,\Delta T$$

or

$$\Delta T = \frac{2(0.01)PV}{3Nk} = \frac{0.02}{3}\left(\frac{V}{N}\right)\frac{P}{k}$$

$$= \frac{0.02}{3}\left(\frac{1}{2.4 \times 10^{25}\ \text{molecules/m}^3}\right)\frac{(1.01 \times 10^5\ \text{N/m}^2)}{(1.38 \times 10^{-23}\ \text{J/K})} = \underline{2.0\ \text{K}}.$$

The temperature will increase by 2 K (by 2 C° or 3.8 F°) with only a 1 percent volume decrease. A 5 percent volume decrease would cause a 10-C° temperature increase.* ∎

12.8 P-V Diagrams and Thermodynamics of Gases

P-V Diagrams

When the pressure in a gas remains constant as it expands or contracts, the work calculation is quite easy: we multiply the pressure times the change in volume. However, if the pressure changes as the volume changes, the calculation is more difficult. In this case, a technique using P-V diagrams is helpful for making the work calculation.

A P-V diagram consists of a graph in which the pressure of a gas is plotted on the vertical axis for different values of volume plotted on the horizontal axis. When a gas expands under constant pressure, the P-V diagram consists of a straight, horizontal line as shown in Fig. 12.13a. The magnitude of work done on the gas during this expansion is $P(V - V_0)$ and equals the shaded area under the curve, since the area equals the product of the height P of the rectangle and its width $(V - V_0)$. The sign of the work is negative if $V > V_0$ and is positive if $V < V_0$.

If pressure changes while a gas expands or contracts, as represented by the P-V diagram shown in Fig. 12.13b, *the magnitude of the work done on the gas still equals the shaded area under the curve.* The expansion $(V - V_0)$ can be divided into many small expansions, each of magnitude ΔV. The work done during each small expansion is $P\,\Delta V$, where P is the average pressure during the small expansion and ΔV is the small change in volume. The area of each narrow rectangle in Fig. 12.13b equals the product of P and ΔV and therefore equals the work done during each small expansion. As the pressure decreases, the area of the rectangles becomes smaller, as does the work done during that small expansion. By adding the area of all of these rectangles, we have the total area under the curve and the total work done during the volume change.

Gases and the First Law of Thermodynamics

P-V diagrams are used to represent many thermodynamic processes involving gases, such as the change of state of a gas in the cylinder of an engine as it moves through one of its cycles of expansion and contraction, or the change in state of

(a)

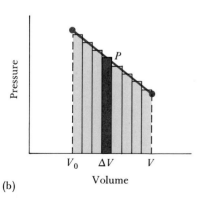

(b)

FIG. 12.13. (a) A change in volume at constant pressure. (b) A change in volume as the pressure changes. The area under the curve equals the magnitude of the work done on the gas during the volume change.

*Since air consists mostly of diatomic molecules whose thermal energy at room temperature equals $\frac{5}{2}kT$ rather than $\frac{3}{2}kT$ (the extra kT is present because of their rotational motion), our estimates are a little high. A 5 percent volume decrease should lead to a 6-C° temperature increase.

air as it rises up into the atmosphere on a hot, summer day. The succession of points on the diagram represents the changing pressure and volume of the gas. The magnitude of the work done on the gas during the process equals the area under the *P-V* curve. If the amount of gas undergoing change is known, the ideal gas law or some other gas law can be used to calculate the temperature change of the gas:

$$\Delta T = T - T_0 = \frac{PV}{nR} - \frac{P_0 V_0}{nR}$$

where *P* and *V* are the final pressure and volume of the gas and P_0 and V_0 are the initial values. Having determined the temperature change, we can then use Eq. (12.13) to calculate the change in internal energy of the gas. The first law of thermodynamics can then be used to calculate the heat added to or removed from the gas.

Alternatively, we can calculate the work done on or by the gas using the area under the *P-V* curve. If we know the heat added to the gas or removed from it, we can use the first law of thermodynamics to calculate the change in internal energy of the gas, and using Eq. (12.13) we can then calculate its change in temperature $(\Delta E_{\text{th}} = \frac{3}{2} Nk \, \Delta T)$.

EXAMPLE 12.11 As $0.90 \, \text{m}^3$ of air at temperature $37°\text{C}$ rises on a hot summer day, the pressure of surrounding air decreases from $1.00 \, \text{atm}$ to $0.70 \, \text{atm}$ and the volume of the air increases to $1.15 \, \text{m}^3$ (represented by the *P-V* diagram in Fig. 12.14). (a) Calculate the work done on the air. (b) If the air gains $0.70 \times 10^4 \, \text{J}$ of heat, by how much does its internal energy change? (c) Calculate the final temperature of the gas.

SOLUTION (a) The work done on the gas during the expansion can be calculated by counting the squares under the line representing this expansion in the *P-V* diagram shown in Fig. 12.14. Each square of height $0.1 \times 10^5 \, \text{N/m}^2$ and width $0.05 \, \text{m}^3$ corresponds to work of $(0.1 \times 10^5 \, \text{N/m}^2)(0.05 \, \text{m}^3) = 500 \, \text{J}$. Since there are 43 squares under the line, the total work done on the gas during the expansion is

$$W = -(43 \text{ squares})(500 \, \text{J/square}) = \underline{-2.15 \times 10^4 \, \text{J}}.$$

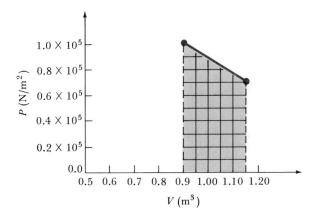

FIG. 12.14. The line represents a process in which a gas expands from $0.9 \, \text{m}^3$ at pressure $1.0 \times 10^5 \, \text{N/m}^2$ to $1.3 \, \text{m}^3$ at pressure $0.7 \times 10^5 \, \text{N/m}^2$.

The work is negative, since $W = -P\Delta V$ is negative if ΔV is positive. (b) To determine the air's change in internal energy, we substitute $W = -2.15 \times 10^4$ J and $Q = 0.70 \times 10^4$ J into the first law of thermodynamics ($Q + W = \Delta E$). Thus,

$$\Delta E = Q + W = 0.70 \times 10^4 \text{ J} - 2.15 \times 10^4 \text{ J} = \underline{-1.45 \times 10^4 \text{ J}}.$$

(c) To determine the temperature change of the gas, we can first determine the number of moles of gas that have expanded by using the ideal gas law and the initial state of the gas:

$$n = \frac{P_0 V_0}{R T_0} = \frac{(1.01 \times 10^5 \text{ N/m}^2)(0.90 \text{ m}^3)}{(8.314 \text{ J/mole} \cdot \text{K})(310 \text{ K})} = 35.3 \text{ mole}.$$

We can then calculate the final temperature of the gas using the ideal gas law and the final state of the gas:

$$T = \frac{PV}{nR} = \frac{(0.7 \times 1.01 \times 10^5 \text{ N/m}^2)(1.15 \text{ m}^3)}{(35.3 \text{ mole})(8.314 \text{ J/mole} \cdot \text{K})} = 277 \text{ K} = \underline{4 \text{ °C}}.$$

We could also have calculated the temperature using the calculated change in internal energy (-1.45×10^4 J) and Eq. (12.13) with the substitution $N = nN_A = (35.3 \text{ mole})(6.023 \times 10^{23} \text{ molecules/mole}) = 2.13 \times 10^{25}$ molecules:

$$\Delta T = T - T_0 = \frac{2}{3}\frac{\Delta E_{th}}{Nk} = \frac{2(-1.45 \times 10^4 \text{ J})}{3(2.13 \times 10^{25})(1.38 \times 10^{-23} \text{ J/K})} = -33 \text{ K}$$

or a change in temperature of -33 C°. Since $T_0 = 37$ °C, $T = \underline{4 \text{ °C}}$. It is interesting to note that the gas gained energy by heat transfer, but as it expanded, its temperature still decreased because of energy lost from the work done. ■

12.9 Diffusion

Earlier, we pictured the molecules in a gas as moving at high speed and having frequent collisions. A molecule follows a zig-zag path, its course changing abruptly at each collision. If we followed the motion of one molecule of air, it would be displaced only several centimeters in one minute even though traveling with an average speed of about 500 m/s. During the minute, it would experience over 10^{11} collisions, each collision causing its speed and direction of motion to change.

If a group of special molecules, such as those from a bottle of perfume, enter the air at one location, they also experience this random motion. As time progresses, the molecules slowly move or diffuse away from their point of origin into the surrounding space (Fig. 12.15). **Diffusion** is the slow spreading of atoms and molecules away from areas in which they are concentrated into areas in which they are less concentrated.

Diffusion plays an important role in biological processes. Oxygen released in capillaries diffuses from the blood to cells near the capillaries, and muscle fibers receive oxygen that diffuses from surrounding material into the fibers. Since a fiber needs oxygen to twitch, the action of muscles may be limited by the rate of diffusion of oxygen into the fiber.

$t = 0$

Later

FIG. 12.15. As time passes, molecules of perfume diffuse farther from their point of origin.

12.10 Osmosis

Osmosis is a form of diffusion that occurs across a membrane. The membrane allows small molecules to pass but prohibits the passing of larger molecules. The skin of a wrinkled prune will allow water to pass, but not the sugar molecules dissolved inside the skin. When the prune is placed in water, it absorbs water through its skin until it becomes swollen. It does not, however, lose sugar or any of its nutritious ingredients through the skin. A red blood cell, if placed in water, swells owing to absorption of extra water through its membrane, but the cell does not lose hemoglobin or other large molecules from inside.

During osmosis, *there is a net fluid flow across the membrane into the region with the greatest concentration of dissolved molecules or ions*. This is why water flows into a prune toward the region of high sugar concentration and why water flows into a red blood cell toward the region of high hemoglobin concentration. If a person soaks an infected sore in epsom salts, fluid flows out of the sore into the water with high salt concentration.

What causes this flow of fluid through a membrane toward a region of higher concentration of dissolved molecules? Consider the two solutions pictured in Fig. 12.16a. The solution on the left of the membrane is a mixture of water and sugar molecules; that on the right is pure water. The solutions are separated by a semipermeable membrane that allows only water to pass, not the sugar molecules. The two types of molecules (water and sugar) act much like molecules of gas. When the water molecules hit the membrane, they can pass through, but the sugar molecules cannot: they bounce off the membrane much as gas molecules bounce off the walls of a container. These sugar molecules create a pressure on the membrane that happens to be the same as the pressure P we would expect from a group of atoms of gas hitting a wall:

$$P = \frac{NkT}{V} = \frac{nRT}{V},$$

where N is the number of sugar molecules in the left chamber of volume V at temperature T, and n is the number of moles of sugar in this chamber.

Let us next consider the pressure caused by the water molecules. More water molecules per unit volume are on the right side of the membrane than are on the left because some of the space that could have been occupied by water in the left chamber is occupied by sugar molecules. Thus, more water molecules will be hitting the membrane per unit time from the right side than from the left. Since water can pass through the membrane, we find that a net movement of water occurs from the right side to the left. As this continues, the water level

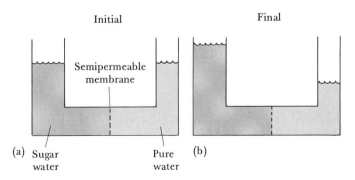

Initial   Final

Semipermeable membrane

(a) Sugar water   Pure water   (b)

FIG. 12.16. (a) A semipermeable membrane allows water but not sugar molecules to pass. More water molecules pass through the membrane from the right side, where they are most concentrated, to the left, where sugar molecules occupy the space normally occupied by water. (b) Eventually the water level rises until this osmosis stops. The extra pressure of the solution on the left compared to that on the right is called an osmotic pressure.

on the left will rise and the level on the right will fall (Fig. 12.16b). This flow continues until the added pressure on the left causes the frequency of collisions of water molecules from the left to balance that from the right.

Because the diffusion stops when the number of water molecules hitting the membrane is the same on each side, we can see that the net water pressure on the membrane is zero. There is, however, the extra pressure on the membrane from the left caused by the dissolved sugar molecules. This pressure, called the **osmotic pressure,** equals the difference in pressure across the membrane, the higher pressure existing on the side of the membrane containing the dissolved molecules. The value of the osmotic pressure is

$$\text{osmotic pressure} = P_{\text{os}} = \frac{nRT}{V}, \tag{12.18}$$

where n is the number of moles of dissolved molecules of any type in the chamber that is at the higher pressure.*

When working with solutions, it is common to define the **molar concentration** c of the solution—the number of moles of solute n (such as sugar) divided by the volume V of the solution in which it is dissolved:

$$\text{molar concentration} = c = \frac{n}{V}. \tag{12.19}$$

The osmotic pressure across the membrane is, then,

$$P_{\text{os}} = cRT. \tag{12.20}$$

EXAMPLE 12.12 A balloon-shaped membrane containing sugar water of concentration 0.20 mole/1000 cm³ and at a temperature of 27°C is placed in a container of pure water. If the membrane is permeable to water but not to sugar, what is the pressure difference between the inside and outside of the balloon when osmosis stops?

SOLUTION Water will move toward the solution with the highest concentration of solute—in this case, into the membrane. When osmosis has stopped, the water pressure on each side of the membrane is equalized. However, the sugar exerts a net outward pressure equal to

$$P_{\text{os}} = cRT = \left(\frac{0.2 \text{ mole}}{1000 \text{ cm}^3}\right)\left(8.314 \frac{\text{J}}{\text{mole K}}\right)(300 \text{ K})$$

$$= 0.50 \frac{\text{J}}{\text{cm}^3} = 5.0 \times 10^5 \text{ J/m}^3 = \underline{5.0 \times 10^5 \text{ N/m}^2}.$$

This is a pressure difference of about five atmospheres and could easily cause the balloon to burst. A similar fate is suffered by red blood cells stored in water.

■

Osmosis is important in the transport of water up the stalks of plants. If the water in the plant stalk contains dissolved molecules, such as sugar in sap, then

---

*The situation in which molecules are dissolved on both sides of the membrane is considered in Example 12.14.

ground water flows across the membranes into plant roots, forcing water higher in the plant stalk. However, if the ground water contains too much salt concentration, osmosis causes water to flow out of the plant into the ground water. This is one reason for watering plants with distilled water: salt does not accumulate in the soil. Salt can be leached from the soil of house plants by running water through the soil for several minutes.

EXAMPLE 12.13 A 1.5-m tall Norfolk pine is planted in soil that contains 0.100 mole/liter of salt ions (Na^+, K^+, Cl^-, and so forth). Calculate the concentration of dissolved molecules in the tree fluid that causes water to move by osmosis to the top of the tree. The plant grows at a temperature of 27°C or 300 K.

SOLUTION From Table 12.1, we see that a pressure difference of one atmosphere causes water to rise 1030 cm or 10.3 m higher on the evacuated side of a manometer than on the high-pressure side. In this example, the water is to rise 1.5 m up the tree. Thus, the osmotic pressure must be

$$P_{os} = \frac{1.5 \text{ m}}{10.3 \text{ m}} (1 \text{ atm}) = 0.15 \text{ atm} = 0.15 \times 10^5 \text{ N/m}^2.$$

The salt in the soil draws water out of the plant, whereas molecules dissolved in the tree sap draw water into the plant. These dissolved molecules must be greater in concentration than the salt. The c in Eq. (12.20) is just the difference in concentrations of dissolved molecules or ions across the membrane of the plant roots. The net osmotic pressure in the desired direction is calculated as follows:

$$P_{net} = P_{in} - P_{out} = c_{sap}RT - c_{salt}RT = (c_{sap} - c_{salt})RT$$

or

$$c_{sap} - c_{salt} = \frac{P_{net}}{RT} = \frac{0.15 \times 10^5 \text{ N/m}^2}{(8.314 \text{ J/mole K})(300 \text{ K})} = 6 \frac{\text{moles}}{\text{m}^3}$$

$$= 6 \frac{\text{moles}}{\text{m}^3} \left(\frac{1 \text{ m}^3}{10^6 \text{ cm}^3}\right)\left(\frac{10^3 \text{ cm}^3}{1 \text{ liter}}\right) = 6 \times 10^{-3} \frac{\text{moles}}{\text{liter}}.$$

Therefore

$$c_{sap} = 0.006 \text{ mole/liter} + c_{salt}$$
$$= (0.006 + .1) \text{ mole/liter}$$
$$= \underline{0.106 \text{ mole/liter}}.$$
∎

Summary

1. **Gases** consist of atoms and molecules in constant motion that are separated far from each other except during frequent collisions. When the atoms and molecules collide with a solid or liquid surface, a small impulsive force is exerted on the surface. As a result of the many collisions per unit time, fairly uniform pressure is exerted by the gas on the surface.

2. **Pressure** is defined as the force that is exerted by a fluid such as air perpendicularly to a surface divided by the area of the surface:

$$P = \frac{F}{A}. \tag{12.1}$$

3. **Equations of state:** In addition to pressure, gases are described using other quantities such as temperature T, the volume of space occupied by the gas V, the number of atoms or molecules N in that space, and the number of moles n in the space. These quantities are related to each other by equations of state such as **Boyle's law:**

$$PV = \text{constant (if } T \text{ and } N \text{ are constant),} \qquad (12.3)$$

and the **ideal gas law:**

$$PV = NkT \qquad (12.5)$$

or

$$PV = nRT, \qquad (12.8)$$

where

$$n = \text{number of moles of gas} = \frac{N}{N_A}, \qquad (12.7)$$

and $N_A = 6.023 \times 10^{23}$ is called **Avogadro's number;** $k = 1.38 \times 10^{-23}$ J/K is called **Boltzmann's constant;** and $R \equiv N_A k = 8.314$ J/mole $\cdot$ K is the **universal gas constant.**

4. **Thermal energy:** For a gas at temperature T (in degrees Kelvin) the average thermal energy due to translational motion of one of its molecules is

$$E_{th}/\text{molecule} = \tfrac{1}{2}m\overline{v^2} = \tfrac{3}{2}kT. \qquad (12.12)$$

The **root-mean-square velocity** of a particle of mass m in the gas is

$$v_{rms} = \sqrt{\overline{v^2}} = \sqrt{\frac{3kT}{m}}. \qquad (12.14)$$

The **thermal energy** due to translational motion of N molecules is

$$E_{th} = N(\tfrac{3}{2})kT. \qquad (12.13)$$

5. **Work done on a gas:** The work W done *on a gas* as its volume changes is

$$W = -P\,\Delta V, \qquad (12.15)$$

where ΔV is positive for a volume increase and negative for a decrease. This expression for work can be used with the first law of thermodynamics $(Q + W = \Delta E)$ to analyze energy-transformation processes involving gases. Sometimes it is more convenient to consider the work done *by a gas,*

$$W' = +P\,\Delta V,$$

in which case the first law of thermodynamics is written as $Q - W' = \Delta E$.

6. **Osmosis** is the flow of a liquid such as water across a semipermeable membrane toward a region that contains the highest concentration of dissolved ions and molecules. The **osmotic pressure** caused by these dissolved ions and molecules is

$$P_{os} = \frac{n}{V}RT = cRT, \qquad \textbf{(12.18) and (12.20)}$$

where $c = n/V$ is the **molar concentration** of the dissolved ions and molecules.

Questions

1. A skin diver when 20 m under the water is exposed to a pressure from the water of about 3×10^5 N/m^2. *Estimate* the net force of the water pushing in on his chest (equal to the force pushing in on his back). Why doesn't the diver's chest collapse under this large force?

2. *Estimate* the force of air pushing down on your head and shoulders in units of newtons and pounds. Does this downward force of the air on your body affect the reading of a bathroom scale when you weigh yourself? Explain.

3. One gram of air contains about 2×10^{22} molecules. If these molecules were distributed uniformly on the earth's surface, estimate the number that would be under your feet right now. The radius of the earth is 6.38×10^6 m. (*Hint:* First calculate the number that would lie on each square meter of the earth's surface.)

4. The weight of the earth's atmosphere on each 1 m^2 of surface is about 10^5 N. *Estimate* the thickness of the atmosphere if it condensed to a liquid with each 1000 kg occupying 1 m^3 of space.

5. Why do air bubbles increase in size as they rise?

6. A paper towel is crumpled and stuffed in the bottom of a tall glass. The glass is pushed, open end down, into some water until it is completely immersed. Will the towel remain dry? Explain.

7. Place a flat piece of cardboard over the top of a glass *filled* with water. Hold the glass over a sink and turn it upside down while holding the cardboard in place. If you remove your hand from the cardboard, it should remain in place and prevent water from falling out of the glass. Explain why this happens.

8. One mole of chicken feathers is spread uniformly over the surface of the earth. *Estimate* the thickness of this layer of feathers. Justify any assumptions made in your calculations. (*Note:* The radius of the earth is about 6×10^6 m.)

9. The temperature of the air in a room is doubled from 15°C to 30°C. By what fraction does the thermal energy of the air molecules increase?

10. A room is at a temperature of 20°C, and the speed of sound is 340 m/s. To about what temperature must the room air be increased in order to double the speed of sound? Explain.

11. Ten grams of nitrogen molecules have 1400 J of thermal energy, as do 10 g of oxygen molecules (held in a separate container). Are both containers at the same temperature? If not, which is hotter? Explain.

12. Why is air as it leaves a cylinder in which it has been compressed cooler than when in the cylinder?

13. Ammonia molecules move with a speed of about 600 m/s. Why do we not smell ammonia throughout a room the instant it is spilled on a table?

14. Why do not air molecules diffuse from near the earth's surface (where they are concentrated) into outer space (where they are not concentrated)?

Problems

12.1 Description of a Gas

1. There are about 3×10^{25} molecules of air in each cubic meter of the atmosphere. If each molecule had its own cube of space, what is the length of one side of the cube? How many times greater is the length of a side of the cube than the length of a molecule that is about 3×10^{-10} m long?

2. What is the length of the side of a cube that holds 4×10^9 molecules of air (equal in number to the people on the earth)? There are 3×10^{25} molecules in each cubic meter.

3. An air molecule traveling at 500 m/s has about 10^{10} collisions per second with other air molecules. (a) What distance does the molecule travel between collisions? (b) If the average distance between the nearest molecules is 3×10^{-9} m, how many other molecules does a molecule pass on the average between collisions?

12.2 Pressure

4. Calculate the force of the earth's atmosphere on the cover of your physics book.

5. Calculate the force of the earth's atmosphere (a) on one side of an 8×11-inch piece of paper, (b) on one side of a 3×5-inch index card.

6. Air is withdrawn from inside a cube. The air outside at a pressure of 1 atm exerts a force of 2.0×10^4 N on each wall of the cube. What is the volume of the cube?

7. A cylindrical iron plunger is held against the ceiling, and the air is pumped from inside it. A 140-lb person hangs by a rope from the plunger (Fig. 12.17). What minimum radius must the plunger have so that air pressure keeps it forced against the ceiling? Atmospheric air pressure is 14.7 lb/in².

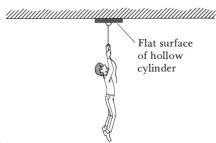

Flat surface of hollow cylinder

FIG. 12.17

8. The air pressure drops by 0.3 bar when you climb from sea level to an elevation of 3000 m. Convert this pressure change to units of (a) mm Hg, (b) lb/in², and (c) N/m².

9. The pressure of a boxer's glove on his opponent's head during a knockout punch is about 5×10^5 N/m². Convert this pressure to units of (a) atm, (b) lb/in², (c) mm Hg, and (d) cm H_2O.

12.3 Gauge Pressure

10. (a) The pressure in a cylinder of compressed air is 52.0 lb/in². Calculate its gauge pressure. (b) An open-tube manometer with one side attached to a spinal tap has mercury that is 11 mm higher on the side open to the atmosphere. What is the gauge pressure in the spine? The absolute pressure?

11. Air is sucked from a straw, reducing the absolute pressure in the straw by 10 percent. What is the gauge pressure in units of N/m² in the straw after sucking? The absolute pressure?

12. The gauge pressure of blood in the aorta increases from 80 to 120 mm Hg when blood is pumped into the aorta from the heart. Calculate the percentage increase in the absolute blood pressure in the aorta.

13. During a hurricane, the gauge pressure in a house is zero and that above the roof is -8 mm Hg. If the area of the roof is 200 m², what is the net force of the air on the roof—that is, the difference of the force caused by the air from below and that exerted from above?

12.4 Equations of State

14. A pressure of 50 lb/in² is required to blow a cork out of the end of a cork gun. Air inside the barrel is at a pressure of 1 atm when a piston is 1.5 ft from a cork placed in the end of the barrel. As the piston is pushed toward the cork, the pressure increases. How far is the piston from the cork when it is dislodged from the gun? Assume that the air does not change temperature.

15. The middle ear has a volume of 0.6 cm³ when at sea level. Assuming that no air leaves the middle ear and that its temperature remains constant, what will be its volume at 1500 m above sea level where the air pressure is 0.83 atm?

16. A cylindrical diving bell, open at the bottom and closed at the top, is 4 m tall. When at the ocean's surface, the bell is filled with air at atmospheric pressure. The pressure increases by one atmosphere for each 10 m the bell is lowered below the ocean surface. If the bell is lowered 36 m below the ocean surface, how many meters of air space are left at the top of the inside of the bell?

17. When surrounded by air at a pressure of 1 atm, a basketball has a radius of 0.12 m. (a) Calculate its volume. (b) Calculate its volume when held 15 m under the surface of water where the absolute pressure is 2.5 atm. (c) Calculate its radius when at this pressure.

18. The inside of a student's skull has a volume of 1130 cm³, and the gauge pressure in the skull is $+6$ mm Hg. While cramming for an exam, the volume of the student's brain increases, causing an increase in gauge pressure to $+25$ mm Hg. Estimate the increased volume of the brain. (*Note:* This increase equals the decrease in space available inside the skull.)

12.5 Ideal Gas Law

19. During the compression stroke of the cylinder of a diesel engine, the air pressure in the cylinder increases from 1 atm to 50 atm and the temperature increases from 26°C to 517°C. Calculate the ratio of the initial volume to the final volume of the cylinder.

20. (a) Calculate the number of molecules per unit volume (N/V) in the atmosphere at the top of Mt. Everest. The pressure is 0.31 atm and the temperature is -30°C. (b) Calculate the number of molecules per unit volume at sea level, where the pressure is 1.0 atm and the temperature is 20°C.

21. Homemade beer is capped into a bottle at a temperature of 27°C and pressure of 1.2 atm. The cap will pop off if the pressure

inside the bottle exceeds 1.5 atm. What temperature must the gas inside the bottle reach in order to pop the cap?

22. A cork gun is shot by heating the gas in a cylinder with both ends closed (one end is closed by the cork). The gas in the cylinder starts at a temperature of 20°C with a pressure of 1.2 atm. The cork explodes from the gun when the temperature reaches 200°C. What is the pressure in the gun when the cork leaves the gun?

23. A bubble of water rises from the bottom of a lake where the pressure is 1.8 atm to the lake surface where the pressure is 1.0 atm. If the bubble's volume at the bottom of the lake is 0.1 m^3, what is its volume at the surface? The water temperature changes from 10°C at the bottom to 22°C at the surface.

24. (a) What are the approximate molecular masses (in grams) of lithium (Li), argon (Ar), and water (H_2O)? (b) Approximately how many moles is 33.6 g of water (H_2O)? (c) 33.6 g of oxygen (O_2)? (d) 33.6 g of helium (He)?

25. (a) How many atoms of carbon form a mass of 1.70 g? (b) How many molecules of oxygen (O_2) have a mass of 11.3 g? (c) What is the mass of 4.0×10^{25} molecules of nitrogen (N_2)?

26. A 600-m^3 home contains air at 17°C and a pressure of 1 atm. If the air pressure remains at 1 atm, how many moles of air are removed from the home as the air warms to 27°C?

27. (a) The pressure in a helium balloon is 1.3×10^5 N/m^2 and its temperature is 27°C. How many helium atoms are in the balloon whose volume is 0.8 m^3? (b) Calculate the number of moles of helium.

28. Oxygen is withdrawn from a tank of volume 0.04 m^3. The tank's pressure drops from 10 atm to 2 atm, and the temperature is lowered from 300 K to 290 K. (a) How many moles of oxygen are removed from the tank? (b) How many grams of oxygen (O_2) are removed?

29. A 0.05-m^3 tank holds 90 g of carbon dioxide (CO_2) of molecular mass 44 g at a temperature of 30°C. (a) Calculate the number of moles of CO_2 in the tank. (b) Calculate the pressure in the tank.

30. The air in the middle ear has a volume of 0.6 cm^3, pressure of 1 atm, and temperature of 37°C when at sea level. After the person climbs to an altitude of 3000 m, the volume is still 0.6 cm^3, the pressure is now 0.66 atm, and the temperature is 35°C. How many moles of air have been removed from the middle ear?

12.6 Kinetic Theory of Gases

✗ 31. A (5.0×10^{-3})-m^3 container holds 1.2×10^{23} oxygen molecules at a pressure of 0.9×10^5 N/m^2. If the mass of each molecule is 5.3×10^{-27} kg, what is the root-mean-square speed of the molecules?

32. The hydrogen atoms on the sun's surface are at a temperature of 6000 K. Each atom has a mass of 1.66×10^{-27} kg. Calculate the root-mean-square speed of these atoms. Would the speed of sound be faster or slower on the sun than on the earth? Explain.

33. A nitrogen molecule $(4.7 \times 10^{-26}$ kg), a virus $(1.6 \times 10^{-20}$ kg), and a golf ball (0.045 kg) are all in a room at a temperature of 300 K. Calculate the root-mean-square speed of each object.

34. The root-mean-square speed of a gas of hydrogen molecules (H_2) is 1700 m/s. Calculate the molecular mass of a smoke particle suspended in the gas if its root-mean-square speed is 0.14 m/s $(m_{H_2} = 3.35 \times 10^{-27}$ kg).

35. The speed that a projectile must have to escape from the earth is 1.12×10^4 m/s. (a) At what temperature would an average molecule of oxygen in the earth's atmosphere be able to escape from the earth $(m_{O_2} = 5.31 \times 10^{-26}$ kg)? (b) Repeat part (a) for hydrogen $(m_{H_2} = 3.35 \times 10^{-27}$ kg). Comment on the earth's ability to retain hydrogen and oxygen.

36. (a) If the total mass of a car is 1500 kg and an average atom in the car has a mass of 1.0×10^{-25} kg, how many atoms does the car contain? (b) If each atom has an average thermal energy of $\frac{3}{2}kT$ and the temperature of the car is 30°C, what is the total thermal energy of the car? (c) At what speed must the car drive so that its nonrandom kinetic energy along a highway ($\frac{1}{2}mv^2$) equals its thermal energy?

37. (a) Show that the root-mean-square speed of gas molecules at temperature T_1 is related to that at temperature T_2 by the equation

$$\frac{v_1}{v_2} = \sqrt{\frac{T_1}{T_2}}.$$

(b) The speed of sound is roughly proportional to the rms speed of the gas molecules. If sound's speed is 331 m/s at 0°C, what is its speed at 20°C?

38. The root-mean-square speed of atom A is 1.5 times that of atom B when the atoms are at the same temperature. Calculate the ratio of the mass of atom A to that of atom B.

39. If the speed of sound is proportional to the root-mean-square speed of a molecule, show that the speed of sound in helium (He) is approximately 2.6 times the speed in nitrogen (N_2).

40. A box is divided in the middle by a wall. Hydrogen molecules (H_2) moving with a root-mean-square speed of 300 m/s are placed on the left side, and nitrogen molecules (N_2) moving at 100 m/s are on the right. Calculate the ratio of the temperature of the hydrogen gas compared to the temperature of the nitrogen gas.

41. A hailstorm causes an average pressure of 1.4 N/m^2 on the 200-m^2 flat roof of a house. The hailstones, each of mass 7.0×10^{-3} kg, are traveling at an average speed of 10 m/s and rebound after hitting the roof with the same speed. How many hailstones hit the roof each second?

42. A 0.058-kg tennis ball, traveling at 25 m/s, hits a wall, rebounds with the same speed in the opposite direction, and is hit again by a tennis player, causing the ball to return to the wall at the same speed. The ball returns to the wall once each 0.6 s. (a) Calculate the force of the ball on the wall averaged over the time between collisions. (b) If ten people are practicing against the same wall, whose area is 30 m^2, calculate the average pressure of these ten tennis balls against the wall.

12.7 and 12.8 Work Done by a Gas, P-V Diagrams, and Thermodynamics

43. A gas expands from a volume of 2.0×10^3 cm^3 to a volume of 3.4×10^3 cm^3 while at a pressure of 1.2×10^5 N/m^2. Calculate (a) the work done on the gas during the expansion, and (b) the work done by the gas.

44. One cubic meter of air containing 2.4×10^{25} molecules expands to 1.2 times its original volume as it moves up in the atmosphere. Suppose that the average pressure acting on the gas is 0.9 atm (the pressure actually decreases as the gas expands).

(a) Calculate the work done on the gas during its expansion. (b) If no heat enters or leaves the gas, calculate its change in temperature using the first law of thermodynamics and the expression for the thermal energy of a gas.

45. In a certain process involving a gas, $-1000\,J$ of work is done on the gas as it absorbs $3000\,J$ of heat. Calculate its change in internal energy. Does the gas expand or contract? Does its temperature increase or decrease? Explain.

46. An ideal gas at a constant pressure of $0.90 \times 10^5\,N/m^2$ expands from an initial volume of $0.50\,m^3$ to $0.60\,m^3$. As it expands, the gas absorbs $3000\,J$ of heat. (a) Calculate the work done on the gas. (b) Calculate its change in thermal energy. Does the temperature of the gas increase, decrease, or remain unchanged? Explain.

47. An ideal gas is compressed slowly at constant temperature to 0.7 times its original volume. While the gas is being compressed, $5000\,J$ of heat is removed from it. (a) Calculate the change in thermal energy of the gas. (b) Calculate the work done on the gas during compression.

48. Ten moles of an ideal gas has an initial volume of $0.30\,m^3$ at 1 atm pressure. Heat is added to the gas, causing its temperature to increase by $200\,C°$ while its pressure remains constant at 1 atm. Calculate (a) the increase in internal energy of the gas, (b) the initial and final temperatures of the gas, (c) its final volume, (d) the work done on the gas, and (e) the heat added to the gas.

49. A $0.035\text{-}m^3$ cylinder containing 1.5 moles of an ideal gas at $40°C$ is closed at one end by a piston. The piston is compressed to a final volume of $0.0031\,m^3$. The compression occurs rapidly so that no heat is lost during compression (called an adiabatic compression). The final temperature of the gas is $540°C$. How much work is done during compression?

50. Calculate the work done on a gas as it moves (a) from A to B for the P-V diagram shown in Fig. 12.18, (b) from B to C, (c) from C to D, and (d) from D to A.

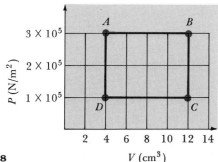

FIG. 12.18

51. Estimate the work done on a gas during the expansion (a) from A to B shown in Fig. 12.19, (b) from B to C, and (c) from C to A.

52. The changes of internal thermal energy during each portion of the cycle shown in Fig. 12.18 are: $A \to B$ $(5.6\,J)$, $B \to C$ $(-8.2\,J)$, $C \to D$ $(-4.8\,J)$, and $D \to A$ $(+7.4\,J)$. Calculate the work done on the gas and the heat added to or removed from the gas during each step of the cycle.

53. The changes in internal thermal energy of an ideal gas moving along each step of the cycle shown in Fig. 12.19 are: $A \to B$ $(10\,J)$, $B \to C$ $(-9\,J)$, $C \to A$ $(-1\,J)$. Calculate the work done on

the gas during each step of the cycle and the heat added to or removed from the gas during each step.

FIG. 12.19

54. The burning of 1 gallon of gasoline in a car engine releases $1.2 \times 10^8\,J$ of heat. The car, while traveling at $20\,m/s$, does work at a rate of $7 \times 10^3\,J/s$. If the car travels 20 miles on 1 gallon of gasoline, calculate its overall efficiency (the ratio of the work done and the energy expended).

12.10 Osmosis

55. Seawater with $400\,moles/m^3$ of salt ions dissolved in it and distilled water at $27°C$ are placed on opposite sides of a semipermeable membrane. Calculate the osmotic pressure across the membrane. In which direction does the net flow of water across the membrane occur?

56. A rigid balloon with semipermeable walls contains $27°C$ water of volume $5\,cm^3$. Dissolved in the water is $0.20\,g$ of glucose with molecular weight 180. (a) What is the molar concentration of glucose in the balloon? (b) The balloon is placed in a glass of water containing salt with dissolved ions at a molar concentration of $1.6 \times 10^2\,mole/m^3$. Calculate the osmotic pressure due to the glucose and due to the salt. (c) Calculate the net osmotic pressure, the difference of the pressures calculated in (b). Will the balloon expand or shrink? Explain.

57. The osmotic pressure at the base of a house plant kept at $27°C$ is $120\,mm\,Hg$. (a) Calculate the molar concentration of dissolved molecules in the plant's vessels if there are no dissolved molecules in the soil water. (b) Suppose that the soil water contains $0.010\,mole/liter$ of dissolved molecules and ions and that the concentration of dissolved molecules in the plant is that calculated in (a). What is the osmotic pressure across the plant roots?

58. A typical red blood cell has a volume of $95\,\mu m^3$ $(95 \times 10^{-18}\,m^3)$ and holds about 3.4×10^8 molecules of hemoglobin that are impermeable to the cell walls. Most other molecules in the blood can cross the cell walls. If a red blood cell is placed in pure water, what osmotic pressure is developed in the cell relative to the water? Actually, the cell stretches and its walls become permeable to hemoglobin before this pressure is attained. Assume that the temperature is $17°C$.

59. The osmotic pressure in a prune when placed in water at $300\,K$ is about $15\,atm$. (a) Calculate the molar concentration of sugar in the prune when at room temperature. (b) *Estimate* the volume of a prune, and then calculate the number of moles of sugar in the prune. (c) If the molecular weight of sugar is 180, about how many grams of sugar are in the prune?

13 Fluid Statics

Any form of matter that flows, such as a liquid or a gas, is called a **fluid.** In this chapter we use the ideas of statics to describe static fluids—or fluids at rest. In the next chapter we consider fluid dynamics, or fluids in motion.

Fluids are of paramount importance to living organisms. The human body is two-thirds water, and liquids in the body not only transport nutrients to its 10^{14} living cells but also carry waste products from these cells. The oxygen we need for combustion of foods flows in and out of the respiratory system along with other molecules in the air.

Fluids are also responsible for most of the climatic conditions on the earth. Air and ocean currents moderate the climates of different parts of the globe, while convection currents in the atmosphere carry pollutants away from large cities. Even the drift of the continents on the earth's mantle can be considered a form of flow.

Because fluids are not rigid, they change shape when exposed to external forces. If you walk into a wall, its rigid structure opposes your motion. If you walk into air, it simply flows past your body. Because of this lack of rigidity, it is convenient to use the quantities pressure and density rather than force and mass when studying fluid statics and dynamics. We will use these quantities with Newton's laws to develop some of the important principles concerning static fluids.

13.1 Pressure

When we study fluids, the quantity pressure assumes a role of primary importance. What force is to solids in the study of statics and dynamics, pressure is to fluids.

Pressure P was defined in Chapter 12 as the ratio of the magnitude of a force applied perpendicular to a surface and the area over which the force is exerted:

$$P = \frac{F}{A}. \tag{12.1}$$

Pressure is a scalar quantity and has no direction. However, the force caused by the pressure of a fluid has a magnitude and a direction. The magnitude of the

force on a surface of area A is

$$F = PA.$$

The direction of the force caused by the pressure is perpendicular to and toward the surface. For example, the molecules of air colliding with a person's head cause a pressure of about $10\,\mathrm{N/cm^2}$, so the magnitude of the force on a 1-cm² area is 10 N. The direction of the force depends on the orientation of the skin, as depicted in Fig. 13.1. The air hitting a 1-cm² area at the top of the man's head causes a 10-N force directed downward, whereas a 1-cm² area under his chin experiences a 10-N force pointing up. A horizontal force acts on the sides of the face.

In summary, we see that a static fluid at pressure P exerts a force **F** against a surface. The magnitude of the force against a surface of area A is

$$F = PA. \tag{12.1}$$

The direction of the force is perpendicular to and toward the surface.

FIG. 13.1. The pressure of the earth's atmosphere causes a force on the skin that is always perpendicular to the skin.

EXAMPLE 13.1 Estimate the force of the earth's atmosphere on the skin under your chin when your head is held upright.

SOLUTION The area of the skin under the chin is about $5\,\mathrm{cm} \times 8\,\mathrm{cm} = 40\,\mathrm{cm^2} = 40 \times 10^{-4}\,\mathrm{m^2}$. The pressure of the earth's atmosphere at sea level is $1.0 \times 10^5\,\mathrm{N/m^2}$. Thus, the magnitude of the air's force on the skin is

$$F = PA \simeq (1.0 \times 10^5\,\mathrm{N/m^2})(40 \times 10^{-4}\,\mathrm{m^2}) = \underline{400\,\mathrm{N}}.$$

The direction of the force is perpendicular to and toward the surface. For this example the force points up (approximately). ∎

13.2 Density

Just as pressure is a more convenient quantity than force when we discuss fluids, density is more useful than mass to describe their inertial property. The mass density—that is, **density**—of a substance is given the Greek letter ρ (rho) and is defined as the ratio of the mass m of the substance and its volume V:

$$\text{density} = \rho = \frac{m}{V}. \tag{13.1}$$

The units of density are $\mathrm{kg/m^3}$, $\mathrm{g/cm^3}$, $\mathrm{slug/ft^3}$, and so on. The densities of various substances are listed in Table 13.1 in the SI unit $\mathrm{kg/m^3}$. Notice that water has a density of $1000\,\mathrm{kg/m^3}$. Hence, 1000 kg of water occupies 1 cubic meter. Platinum has a density of $21{,}450\,\mathrm{kg/m^3}$, whereas the density of air is about $1.3\,\mathrm{kg/m^3}$.

Densities are often quoted in units of $\mathrm{g/cm^3}$ rather than $\mathrm{kg/m^3}$. A density of $1\,\mathrm{g/cm^3}$ equals a density of $1000\,\mathrm{kg/m^3}$, as we can see by the following conversion:

$$1\,\frac{\mathrm{g}}{\mathrm{cm^3}} = 1\,\frac{\mathrm{g}}{\mathrm{cm^3}}\left(\frac{1\,\mathrm{kg}}{10^3\,\mathrm{g}}\right)\left(\frac{10^6\,\mathrm{cm^3}}{1\,\mathrm{m^3}}\right) = 10^3\,\frac{\mathrm{kg}}{\mathrm{m^3}}.$$

TABLE 13.1 Densities of Various Substances

Solids		Liquids		Gases	
Substance	Density (kg/m³)	Substance	Density (kg/m³)	Substance	Density (kg/m³)^a
Aluminum	2,700	Acetone	792	Dry air 0°C	1.29
Copper	8,920	Ethyl alcohol	791	10°C	1.25
Gold	19,300	Methyl alcohol	810	20°C	1.21
Iron	7,860	Gasoline	660–690	30°C	1.16
Lead	11,300	Mercury	13,600	Helium	0.178
Platinum	21,450	Milk	1,028–1,035	Hydrogen	0.090
Silver	10,500	Seawater	1,025	Oxygen	1.43
Bone	1,700–2,000	Water, 0°C	999.87		
Brick	1,400–2,200	3.98°C	1,000.00		
Cement	2,700–3,000	20°C	1,001.80		
Clay	1,800–2,600	Blood plasma	1,030		
Glass	2,400–2,800	Blood, whole	1,050		
Ice	917				
Balsa wood	120				
Oak	600–900				
Pine	500				
Planet Earth	5,252				
Moon	3,340				
Sun	1,410				
Universe (average)	10^{-27}				
Pulsar	10^8–10^{11}				

^{a} At a 1-atm pressure and 0°C temperature

Equation (13.1) can be used to calculate the unknown mass or unknown volume of a fluid whose density is known.

EXAMPLE 13.2 A swimming pool has dimensions 20 m × 10 m, and the water is 1.8 m deep. Calculate the mass of the water in the pool.

SOLUTION The water's volume is

$$V = 20\,\text{m} \times 10\,\text{m} \times 1.8\,\text{m} = 360\,\text{m}^3,$$

and its density is 1000 kg/m³. Rearranging Eq. (13.1), we find that the water's mass is

$$m = \rho V = (1000\,\text{kg/m}^3)(360\,\text{m}^3) = \underline{3.6 \times 10^5\,\text{kg}}. \qquad \blacksquare$$

EXAMPLE 13.3 Calculate what the volume of a 100-kg woman would be if her density equaled that of a pulsar (10^{11} kg/m³), a small but extremely dense rotating star.

SOLUTION Rearranging Eq. (13.1), we find that the woman's volume would be

$$V = \frac{m}{\rho} = \frac{100\,\text{kg}}{(10^{11}\,\text{kg/m}^3)} = \underline{10^{-9}\,\text{m}^3}.$$

Note that $1 \, \text{mm} = \frac{1}{1000} \, \text{m} = 10^{-3} \, \text{m}$. Thus, $(1 \, \text{mm})^3 = (10^{-3} \, \text{m})^3 = 10^{-9} \, \text{m}^3$. We see that the woman would be the size of a cube that is 1 mm long on each edge. ■

Mass densities are rarely used in the English system of units. It is more common to use a quantity called **weight density** D, the ratio of an object's weight w and its volume V:

$$\text{weight density} = D = \frac{w}{V}.$$

Since $w = mg$, we find that mass density and weight density are related by the equation $D = w/V = mg/V = (m/V)g = \rho g$, or

$$\rho = \frac{D}{g}.$$

To convert a weight density to a mass density, we simply divide by g, the acceleration due to gravity.

Many tables in handbooks list what is called the **relative density** or **specific gravity** of a substance instead of its density. The relative density of a substance is defined as the ratio of its density to the density of water at $3.98 \, °\text{C}$ ($1000 \, \text{kg/m}^3$). For example, since the relative density of lead is 11.34, its density is 11.34 times greater than that of water, or $11{,}340 \, \text{kg/m}^3$. Relative density is a unitless quantity since it is the ratio of two quantities having the same units.

13.3 Pressure Variation with Depth

The pressure exerted by a fluid varies with depth; the deeper an object is in a fluid, the greater the pressure acting on the object. You may have felt the increase in water pressure on your ears while swimming under water, or the decrease in air pressure when driving or hiking from lower to higher elevations in the mountains, when taking off in an airplane, or while riding up in a skyscraper's elevator. When our ears "pop," air is released from the middle ear. This release of air causes a reduction in air pressure inside the middle ear that compensates at least partially for the reduced pressure outside the ear when at a higher elevation.

The pressure at a particular depth in a fluid is due to the weight of the fluid above. For instance, the blood pressure in the veins of a person's feet exceeds the pressure in the head because the blood in the feet supports the weight of the blood above it. This increased pressure may cause one's feet to swell, especially if one must stand while working.

The variation of pressure with depth is somewhat like stacking books on a table (Fig. 13.2). Imagine that each book is a layer of air. The only pressure on the top book is that due to the air pushing down from above. However, the second book must support the weight of the top book plus the atmospheric air above it. The sixth book in the stack must support the weight of the five books above it plus the atmospheric air, and so forth.

We can derive an equation that allows us to calculate the difference in pressure at different depths in a fluid. Consider the shaded portion of fluid shown in Fig. 13.3. The bottom surface is at vertical elevation y_1 and the top

FIG. 13.2. The pressure is greatest on books deepest in the stack because they must support the weight of all the books above them.

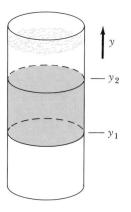

FIG. 13.3. A cylinder of fluid.

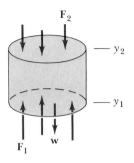

FIG. 13.4. The forces acting on the shaded section of fluid shown in Fig. 13.3.

surface is at elevation y_2. The difference in pressure between y_1 and y_2 should depend on the extra weight of fluid that must be supported at elevation y_1 compared to that at elevation y_2.

The vertical forces acting on the shaded volume are shown in Fig. 13.4. The fluid above the shaded volume pushes down with a force $\mathbf{F}_2$. If the pressure at elevation y_2 is P_2 and the cross-sectional area of the cylinder is A, then the magnitude of the force from above is

$$F_2 = P_2 A.$$

Similarly, fluid from below the shaded section of fluid pushes up with a force $\mathbf{F}_1$ equal in magnitude to

$$F_1 = P_1 A.$$

The third force acting on the shaded volume is its weight $\mathbf{w}$, a downward-directed force.

Since the fluid is not moving and therefore is in equilibrium, these three vertically directed forces must add to zero. With the y axis pointing upward, this equilibrium condition is written as

$$F_1 - F_2 - w = 0,$$

or

$$P_1 A - P_2 A - w = 0. \tag{13.2}$$

The weight of the shaded section of fluid depends on its density and volume. The volume is

$$V = A(y_2 - y_1),$$

where $(y_2 - y_1)$ is the height of the shaded section of fluid. The weight of this part of the fluid is

$$w = mg = (\rho V)g = \rho A(y_2 - y_1)g.$$

Substituting for w in Eq. (13.2), we find that

$$P_1 A - P_2 A - \rho g A(y_2 - y_1) = 0.$$

After dividing the equation by A, we find that

$$P_1 - P_2 = \rho g(y_2 - y_1)$$

or

$$P_1 = P_2 + \rho g(y_2 - y_1). \tag{13.3}$$

This equation allows us to calculate the pressure at one elevation in a fluid in terms of the pressure at a second elevation. The equation has been derived with the assumption that the y axis points upward and that the density of the fluid does not change with elevation.

EXAMPLE 13.4 Calculate the pressure on a skin diver who is 10 m below the surface of the water. The density of seawater is $1025 \, \text{kg/m}^3$ and the air pressure at the water's surface is $1.013 \times 10^5 \, \text{N/m}^2$.

SOLUTION Let $y_2 = 0$ m at the water's surface. Then, $P_2 = 1.013 \times 10^5$ N/m². The elevation of the skin diver is then $y_1 = -10$ m, and the pressure of the water on the skin diver at that elevation is

$$P_1 = P_2 + \rho g(y_2 - y_1)$$
$$= 1.013 \times 10^5 \text{ N/m}^2 + (1025 \text{ kg/m}^3)(9.8 \text{ m/s}^2)[0 - (-10 \text{ m})]$$
$$= 1.013 \times 10^5 \text{ N/m}^2 + 1.005 \times 10^5 \text{ N/m}^2 = \underline{2.018 \times 10^5 \text{ N/m}^2}.$$

This is equivalent to a pressure of 2 atm. The pressure increases by 1 atm for each 10 m under the water's surface. For example, if the pressure at the surface is 1 atm, the pressure 30 m below the surface is 1 atm + 3 atm = 4 atm. ∎

EXAMPLE 13.5 Calculate the change in air pressure when you climb a mountain that is 500 m high. Assume that the density of air is uniform and has a value of 1.3 kg/m³. Actually, the density of air decreases slowly with increasing elevation but not enough to significantly affect our calculations.

SOLUTION Let $y_1 = 0$ represent the elevation of the starting position, and $y_2 = 500$ m is the position of the top of the mountain. Then the change in pressure as we climb the mountain is

$$P_2 - P_1 = \rho g(y_1 - y_2)$$
$$= (1.3 \text{ kg/m}^3)(9.8 \text{ m/s}^2)(0 - 500 \text{ m}) = \underline{-6370 \text{ N/m}^2}.$$

The pressure has decreased by 6370 N/m², or about 0.06 atm, as we ascended the mountain. ∎

13.4 Externally Applied Pressure

We have seen that fluid pressure varies with the elevation. Another important property of fluids is the uniform increase in pressure throughout an enclosed volume of fluid that is caused by increased pressure at one location in the fluid. This consequence is predicted by Eq. (13.3):

$$P_1 = P_2 + \rho g(y_2 - y_1).$$

If the pressure P_2 at elevation y_2 increases, there must be an equal increase in pressure P_1 at elevation y_1. This property was first recognized in the seventeenth century by the French mathematician and philosopher, Blaise Pascal (1623–1662). The principle, named in his honor, is stated below:

Pascal's principle: An external pressure exerted on a static, enclosed fluid is transmitted uniformly throughout the volume of the fluid.

For example, if the pressure at the pumping station of a city water system is increased by 10 pounds per square inch (psi), then the pressure at all the homes connected by water lines to the pumping station also increases by 10 psi. The pressure varies in different homes, depending on their elevation relative to the pumping station, but an increase in pressure at the station causes an equal increase throughout the water system.

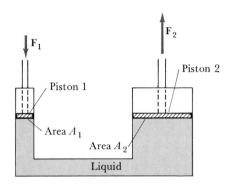

FIG. 13.5. A hydraulic press. A force on piston 1 causes an increase in pressure that is transmitted through the fluid. This increased pressure causes a large force on piston 2 because of its larger area.

FIG. 13.6. A hydraulic can crusher.

Pascal's principle has many applications. The hydraulic press is a form of simple machine that converts small forces into larger forces, or vice versa. Hydraulic presses are used in automobile service stations to lift cars and by dentists and barbers to raise and lower the chairs on which their clients sit. The hydraulic brakes of an automobile are also a form of hydraulic press. Most of these devices work on the simple principle illustrated in Fig. 13.5, although the actual devices are usually more complicated in construction.

A force F_1 is applied to a piston that has a cross-sectional area A_1. The pressure in the fluid just under the piston is

$$P = \frac{F_1}{A_1}.$$

Because the pressure is transmitted uniformly throughout the fluid, the pressure of the fluid under piston 2 is also $P = F_1/A_1$.* But since the area of piston 2 is greater than that of piston 1, we find that the upward force of the fluid on piston 2 is greater than the force causing the pressure at piston 1. The upward force of the fluid on piston 2 is

$$F_2 = PA_2 = \left(\frac{F_1}{A_1}\right)A_2 = \left(\frac{A_2}{A_1}\right)F_1. \tag{13.4}$$

Since A_2 is greater than A_1, F_2 is also greater than F_1.

EXAMPLE 13.6 How large a force is needed on a small piston of area $2\,\text{cm}^2$ in order to support a 1000-N weight resting on a piston of area $20\,\text{cm}^2$?

SOLUTION We are given that $A_1 = 2\,\text{cm}^2$, $A_2 = 20\,\text{cm}^2$, $F_2 = 1000\,\text{N}$, and are asked to find F_1. After rearranging Eq. (13.4), we find that

$$F_1 = \left(\frac{A_1}{A_2}\right)F_2 = \left(\frac{2\,\text{cm}^2}{20\,\text{cm}^2}\right)1000\,\text{N} = \underline{100\,\text{N}}. \qquad \blacksquare$$

EXAMPLE 13.7 A hydraulic can crusher is shown in Fig. 13.6. The large piston has an area of $8\,\text{m}^2$ and exerts a force $\mathbf{F}_2$ of magnitude $2 \times 10^6\,\text{N}$ on the cans. Calculate the magnitude of the force $\mathbf{F}_1$ exerted by the small piston (area $10\,\text{cm}^2$) on the fluid. Do not ignore the fact that the large piston is 1 m higher than the small piston.

SOLUTION The pressure P_2 in the fluid at the top of the crusher is

$$P_2 = \frac{F_2}{A_2} = \frac{2.0 \times 10^6\,\text{N}}{8.0\,\text{m}^2} = 2.5 \times 10^5\,\text{N/m}^2.$$

Then, according to Eq. (13.3), the pressure on the fluid at the small piston is

$$P_1 = P_2 + \rho g(y_2 - y_1)$$
$$= 2.5 \times 10^5\,\text{N/m}^2 + (1000\,\text{kg/m}^2)(9.8\,\text{m/s}^2)(1.0\,\text{m} - 0)$$
$$= 2.5 \times 10^5\,\text{N/m}^2 + 0.098 \times 10^5\,\text{N/m}^2 \simeq 2.6 \times 10^5\,\text{N/m}^2.$$

*We are assuming that piston 2 is at the same elevation as piston 1. If not, Eq. (13.3) will tell us the pressure differences under the two pistons because of their differences in elevation.

Since the small piston has an area of 10 cm² (10^{-3} m²), the magnitude of the force $\mathbf{F}_1$ on this piston is

$$F_1 = P_1 A_1 = (2.6 \times 10^5 \, \text{N/m}^2)(10^{-3} \, \text{m}^2) = \underline{260 \, \text{N}}.$$ ∎

Pascal's principle also has important consequences in biology and medicine. For example, glaucoma, an eye disease, involves Pascal's principle. A clear fluid called aqueous humor fills two chambers in the front of the eye (Fig. 13.7). In the normal eye new fluid is continually secreted into these chambers while old fluid drains from the eye through small sinus canals. When a person has glaucoma, these drainage canals close, fluid accumulates in the front of the eye, and the pressure inside the eye increases. This pressure increase is transmitted to the retina at the back of the eye and causes degenerative changes in the retina that can eventually lead to blindness.

FIG. 13.7. When extra fluid cannot drain out of the eye's cornea, pressure increases throughout the eye, possibly causing blindness.

13.5 Archimedes' Principle

Fluids exert upward buoyant forces on objects in the fluid. These buoyant forces either completely or partially support the object's weight. For example, the buoyant force of water supports a person's weight while floating in a swimming pool. The buoyant force of seawater supports the million-pound weight of icebergs. The atmosphere exerts a buoyant force that helps support the weight of a balloon.

To derive an equation for calculating the buoyant force of a fluid on an object, consider a metal block immersed in a liquid as shown in Fig. 13.8. The fluid, a liquid in this case, exerts forces on all six surfaces of the block. The forces exerted on opposite sides of the block cancel each other. However, the upward force $\mathbf{F}_1$ due to the fluid pressure from below the block and the downward force $\mathbf{F}_2$ due to that from above the block do not cancel, since the pressure of the liquid on the bottom is greater than that on the top. The net force of the fluid on the block $\mathbf{F}_1 - \mathbf{F}_2$, points upward, since $\mathbf{F}_1$ is greater in magnitude than $\mathbf{F}_2$. This net fluid force is called the **buoyant force B** of the fluid, and its magnitude is

$$B = F_1 - F_2 = P_1 A - P_2 A = (P_1 - P_2)A, \tag{13.5}$$

where A is the area of the bottom and top surfaces of the block, and P_1 and P_2 are the fluid pressures at the bottom and top of the block, respectively.

The difference in pressure at two different elevations in the fluid is given by Eq. (13.3):

$$P_1 - P_2 = \rho g(y_2 - y_1).$$

Substituting in Eq. (13.5), we find that the buoyant force is

$$B = \rho g(y_2 - y_1)A.$$

The product $(y_2 - y_1)A$ is the volume V of the block—its height times its area—and also equals the space taken up by the block in the fluid. This space is called the **volume V of fluid displaced** by the object in the fluid. Substituting V in place of $(y_2 - y_1)A$ results in an expression for the buoyant force of the fluid:

$$B = \rho g V.$$

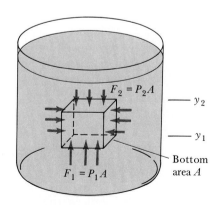

FIG. 13.8. The forces acting on a metal block immersed in a liquid.

It is important to note that the density ρ in this equation is the density of the fluid and not of the object in the fluid. According to Eq. (13.1), the product ρV is the mass of a volume V of fluid having density ρ. When we multiply mass times g, we convert mass to weight. Thus, $\rho g V$ is the weight of fluid whose volume equals V. From now on we will place subscript f on ρ and on V to remind us that these quantities refer to "fluid" density and "fluid" volume.

The **buoyant force B** exerted by a fluid on an object in the fluid is equal in magnitude to the weight of fluid displaced by the object. If the displaced fluid has density ρ_f and volume V_f, then the buoyant force is

$$B = \rho_f g V_f. \tag{13.6}$$

If an object is completely immersed in a fluid, the volume of the displaced fluid equals the volume of the object. If the object floats, then the volume of the displaced fluid equals the volume of space taken up by the object below the fluid's surface.

The derivation of Eq. (13.6) was for a solid cube, but the results apply to objects of any shape. The pressure on the lower sides of these irregular-shaped objects exceeds that on the upper sides, and the resultant buoyant force is given by Eq. (13.6).

The principle just derived, called **Archimedes' principle** was supposedly discovered by the Greek mathematician Archimedes while sitting in his bath. He had been asked by the king to determine whether a crown was made of pure gold or of less valuable metals made to look like gold. In the bath Archimedes thought of a technique he could use to measure the density of the crown: he would weigh the crown first in air and then when immersed in water. Using these two measurements and a small amount of calculation, he would be able to tell whether the crown's density was that of gold or of a less dense and less valuable metal. Archimedes was so excited by his flash of brilliance that he ran from his house shouting, "Eureka!"—the Greek for "I have found [it]!" Whether he was naked or wore a towel is still unknown.

EXAMPLE 13.8 A crown is found to weigh 25.0 N when suspended in air and 22.6 N when suspended in water. Calculate the density of the crown to see if it is made of gold, as claimed by its donor. The density of water is $1000 \, \text{kg/m}^3$.

SOLUTION The weight of the crown is related to its density ρ and its volume V by Eqs. (4.4) and (13.1):

$$w = mg = \rho V g. \tag{13.7}$$

We know the crown's weight (25.0 N) but not its volume; hence we cannot determine its density. We learn more by considering the forces shown in Fig. 13.9 that act on the crown when in water: **B** is the upward buoyant force of the water on the crown, **w** is the crown's weight (25.0 N), and **T** is the upward tension force of the string that is attached from the crown to the spring balance. According to the scale, $T = 22.6 \, \text{N}$. Since the crown hangs at rest, these three factors must add to zero:

$$T + B - w = 0. \tag{13.8}$$

FIG. 13.9. A force diagram for a crown hanging in water.

We substitute into the above equation for the buoyant force:

$$B = \rho_f g V,$$

where V is the volume of displaced water, which equals the volume of the crown, since it is completely immersed, and ρ_f is the density of the water. After substituting for B in Eq. (13.8) and rearranging, we find that

$$V = \frac{w - T}{\rho_f g}.$$

Rearranging Eq. (13.7) to solve for the crown's density ρ and substituting for the crown's volume from above, we find that

$$\rho = \frac{w}{gV} = \frac{w}{g} \frac{\rho_f g}{(w - T)} = \frac{w}{(w - T)} \rho_f$$

$$= \frac{25.0 \text{ N}}{(25.0 \text{ N} - 22.6 \text{ N})} \ 1000 \ \frac{\text{kg}}{\text{m}^3} = \underline{10{,}400 \text{ kg/m}^3}.$$

We see from Table 13.1 that $10{,}400 \text{ kg/m}^3$ is less than the density of gold. Thus, the crown is not made of pure gold. ■

Archimedes' principle is the basis of a therapy procedure for persons with motor difficulty. If a person exercises while partially submerged in water, the buoyant force has the effect of reducing the person's weight, and less muscle force is needed to move body parts. The person may be able to walk while partially submerged in water but not when out of the water. The exercise and movement of the limbs when in water prevents disuse atrophy.

EXAMPLE 13.9 Calculate the triceps muscle tension needed to slowly rotate the forearm downward when it is submerged in water (Fig. 13.10). The density of the forearm is 970 kg/m^3 and that of the water is 1000 kg/m^3. The volume of the forearm is $1.4 \times 10^{-3} \text{ m}^3$.

SOLUTION A force diagram for the forearm is shown in Fig. 13.10: $\mathbf{T}$ is the tension in the triceps muscle, $\mathbf{F}$ the force of bone in the upper arm on the lower arm, $\mathbf{w}$ the weight of the arm acting at its center of gravity, and $\mathbf{B}$ the buoyant force of the water on the arm.

If the arm slowly rotates downward, then the sum of the torques acting on the arm must be zero (assuming no angular acceleration):

$$\Sigma \tau = 0$$

or

$$-l_T(T) + 0(F) - l_w(w) + l_w(B) = 0.$$

Thus,

$$T = \frac{l_w(B - w)}{l_T}. \tag{13.9}$$

The arm's weight is given by Eqs. (4.4) and (13.1):

$$w = mg = \rho_{\text{arm}} V g.$$

FIG. 13.10. A force diagram for a forearm lowered under water.

The buoyant force of the water on the arm is

$$B = \rho_{\text{water}} V g,$$

where V is the volume of fluid displaced, equal in this case to the arm's volume. Substituting in Eq. (13.9), we find that

$$T = \frac{l_w(\rho_{\text{water}} - \rho_{\text{arm}})Vg}{l_T}$$

$$= \frac{(0.15 \text{ m})(1000 \text{ kg/m}^3 - 970 \text{ kg/m}^3)(1.4 \times 10^{-3} \text{ m}^3)(9.8 \text{ m/s}^2)}{0.03 \text{ m}} = \underline{2.1 \text{ N}}.$$

The tension would be over thirty times greater for the same movement if the arm were out of the water. ∎

EXAMPLE 13.10 It is believed that the earth's crust is composed of about thirty continent-sized pieces of rock that "float" on a semifluid "plastic" mantle below the earth's surface (Fig. 13.11). If the average density of the continental mass is 2800 kg/m³ and that of the mantle is 3300 kg/m³, what fraction of the continental crust is floating above the mantle (like an iceberg floating above water)?

SOLUTION Since the crust is floating, its weight w must be balanced by the upward buoyant force of the mantle:

$$B - w = 0 \quad \text{or} \quad B = w. \tag{13.10}$$

The continental weight is related to its density ρ_c and volume V_c by Eqs. (4.4) and (13.1):

$$w = mg = \rho_c g V_c.$$

The buoyant force of the mantle (considered as a fluid) is

$$B = \rho_m g V_m,$$

where ρ_m is the density of the mantle and V_m is the volume of space in the mantle that is occupied by the floating continent. For example, if 60 percent of the continent is in the mantle, then $V_m = 0.6V_c$.

Substituting for B and w in Eq. (13.10), we find that

$$\rho_m g V_m = \rho_c g V_c$$

or

$$\frac{V_m}{V_c} = \frac{\rho_c}{\rho_m} = \frac{2800 \text{ kg/m}^3}{3300 \text{ kg/m}^3} = 0.85.$$

We see that 85 percent of the continent is submerged in the mantle. Only <u>15 percent</u> is above the mantle, and much of that is covered with water. ∎

FIG. 13.11. A continent "floats," partially submerged, on the earth's mantle.

Ocean

Crust 2800 kg/m³

Upper mantle 3300 kg/m³

Water?

FIG. 13.12. Could water remain "piled" on a road without any support from the sides?

13.6 Surface Tension

Most of us would be surprised to see a "pile" of water shaped like a giant drop sitting on a road (Fig. 13.12) or an unsupported wall of water around a home. Whenever we see large amounts of water, a container is always holding it in place, such as the concrete walls of a swimming pool, the plastic mattress of a water bed, or a bathtub.

Yet it is possible to form small, unsupported walls and piles of water. We have all seen small drops of water on an oil-covered driveway, and children routinely form small, thin walls of water when blowing soap bubbles. What supports these tiny walls and piles of water?

A bubble of soap water is not supported by a membrane at its surfaces. The water itself forms a very thin network of molecules at its surface that holds the bubble together. This surface network is held in place by attractive forces between adjacent molecules. These forces are caused by the nonuniform distribution of positive and negative electric charges on the molecules. The molecules arrange themselves so that the part of one molecule with a small positive charge is, on the average, nearer the part of a neighboring molecule with a small negative charge, resulting in a small attractive force between the molecules.

The net force on a molecule inside a liquid is zero, since the forces of other molecules pull it from all directions (see Fig. 13.13). However, molecules at the surface do not have attractive forces pulling on them from outside the water. They are attracted only toward molecules at the side and to those deeper in the solution (Fig. 13.13). Consequently, a small layer of molecules on the liquid surface is held together and pulled back into the liquid. This surface layer acts much like a plastic membrane that helps to contain the water.

The strength of this liquid surface layer is measured in terms of a quantity called surface tension. In order to define surface tension, consider a canvas bag with a slash of length l across its surface (Fig. 13.14a). When we sew the cut together and fill the bag with water, we find that a force pulls on the stitches from each side of the cut (Fig. 13.14b). If we represent the total force of the stitches in one direction by the symbol F, then the surface tension γ is defined as the ratio of F and l:

$$\text{surface tension} = \gamma = \frac{F}{l}. \quad (13.11)$$

Surface tension is the force per unit length needed to hold the surface together. The force is tangent to the liquid surface and is caused by the mutual attraction of the surface molecules to each other and to other molecules below the surface. The force has the same effect as a thin membrane holding the liquid in place (see Fig. 13.14b). Table 13.2 lists the surface tension of several liquids in units of newtons per meter.

EXAMPLE 13.11 An inverted U-shaped wire bounded on the bottom by a slide wire that can be moved up or down surrounds a thin film of soap water (Fig. 13.15a). If the surface tension of the soap water is 0.025 N/m, and the slide wire weighs 1.0×10^{-3} N and is 0.03 m long, how much weight w must be hung from the slide wire to keep it in equilibrium?

SOLUTION Let us choose the slide wire, hanging weight, and a small section of soap film as the object of interest for a force diagram (Fig. 13.15b). Since this group of objects is at rest, the upward forces must be equal in magnitude to the downward forces. The downward forces are the weight of the slide wire w' and the weight w of the hanging mass.

The upward force F is caused by the surface tension on each side of the soap

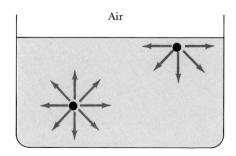

FIG. 13.13. A molecule under the surface of a liquid is attracted from all sides by other molecules. A molecule at the surface is attracted toward molecules at its sides and to others deeper in the liquid. These forces create a tension in the liquid's surface, which has the same effect as a membrane containing the liquid.

(a)

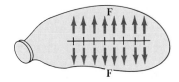

(b)

FIG. 13.14. (a) A water bag with a cut of length l. (b) When sewed together, a net surface force **F** pulls in each direction on the stitches.

TABLE 13.2 Surface Tensions of Several Liquids

Surface	In contact with	Temperature (°C)	Surface tension γ (N/m)
Acetone	Air	20	0.0237
Ammonia	Air	34	0.0181
Benzene	Air	20	0.0289
Ethyl alcohol	Air	20	0.0228
Glycerol	Air	20	0.0634
Methyl alcohol	Air	20	0.0226
Water	Air	0	0.0756
		20	0.0728
		60	0.0662
		100	0.0589
Soap water	Air	20	0.025
Mercury	Air	20	0.465

film (Fig. 13.15c). Since both sides of the film help support the weights, the total surface tension force is twice the tension force on one surface of the film:

$$F = 2\gamma l.$$

Since the sum of the vertical forces is zero, we find that

$$2\gamma l - w' - w = 0$$

or

$$w = 2\gamma l - w' = 2(0.025 \text{ N/m})(0.03 \text{ m}) - 1.0 \times 10^{-3} \text{ N} = \underline{0.5 \times 10^{-3} \text{ N}}.$$

∎

Liquid surface tension is important in a variety of situations. For example, bugs such as water striders are supported by the surface tension of water. The water sags under their weight, but the surface does not "tear." The striders float even though they are more dense than the water.

Surface tension is of importance in the fabric and cleansing industries. The spaces between threads in tent and umbrella material are much larger than the size of water molecules, yet these materials are waterproof because the surface tension in water is so great that individual water molecules are unable to leave the water to pass through the material. The water molecules hang together and bridge the gaps in the fabric. Cleansing agents such as soaps are forms of matter that, when added to water, reduce its surface tension, allowing the water to

FIG. 13.15. (a) A soap film supported in a wire frame. (b) The forces acting on the slide wire. **F** is the force caused by the tension in the soap film. (c) Side view of the film supporting the weights.

break apart and penetrate more easily into the fabric of a material, where it dissolves dirt.

Disinfectants often are solutions of low surface tension. The solution, instead of forming drops, spreads easily over the surface of a wound and over the surface of bacteria cells in the wound. Insect sprays also have low surface tension. The spray will spread easily into the insect trachea, causing it to become clogged.

13.7 Capillary Action

The force of attraction between similar type molecules is called **cohesion.** Cohesive forces are responsible for the surface tension in liquids. There can also be an attractive force between different types of molecules; this force is called **adhesion.** Adhesive tape sticks to the surface of the skin because of the adhesion between the molecules of the tape and those of the skin. The meniscus that you observe when water partially fills a glass cylinder is caused by the adhesive force between water molecules and the atoms and molecules in the glass (Fig. 13.16). The water molecules crawl up the side of the glass because of this adhesion.

Capillary action is a result of both adhesion and cohesion. A capillary is a hole or bore of small radius in a solid material. If the entrance to the capillary is in contact with a liquid that adheres to the solid capillary walls, the liquid will be drawn up into the capillary. The rise of liquid into a capillary is called **capillary action.** Examples are the rise of water in the bore of a glass capillary tube, the rise of water up the small capillaries in plants, the absorption of water into the small pores of a paper towel, and the movement of kerosene up the wick of a kerosene lamp.

The height to which a liquid will rise into a capillary is limited not by the adhesion of the water to the sides of the capillary but by the ability of the thin film of liquid at the sides of the meniscus to support the weight of liquid that has been lifted. Consider the rise of water in the glass capillary pictured in Fig. 13.17a. The thin film of water at the sides of the meniscus pulls like suspenders on the water below (Fig. 13.17b). The surface tension in this meniscus supports the weight of fluid that has been lifted up the capillary tube. The upward surface tension force exerted by the meniscus on the liquid below is given by Eq. (13.11):

$$\text{UP:} \quad F = \gamma 2\pi r,$$

where $2\pi r$ is the circumference of the capillary tube—that is, the length around the edge of the meniscus that is supporting the liquid below.*

The downward weight force on the liquid that has been lifted is

$$\text{DOWN:} \quad w = mg = (\rho V)g = \rho(\pi r^2 h)g,$$

where h is the height of the liquid that has risen in the capillary, and V is the volume of liquid that has risen in the capillary.

---

*We have assumed that the liquid at the top of the meniscus is parallel to the surface of the capillary. This is usually the case when water contacts glass. However, often the liquid at the top of the capillary makes an angle θ with the side of the capillary wall (see Fig. 13.18). The upward component of the surface tension force is then

$$F = \gamma 2\pi r \cos \theta.$$

FIG. 13.16. Water molecules attracted to atoms in glass cause a meniscus at the top edge of a glass cylinder.

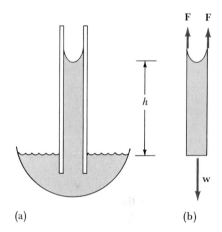

FIG. 13.17. (a) Capillary action causes water to rise in the bore of a capillary tube. (b) A force diagram of this raised water.

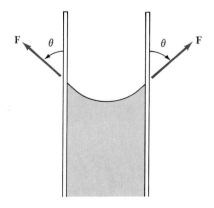

FIG. 13.18

Setting the upward and downward forces equal,

$$\gamma 2\pi r = \rho(\pi r^2 h)g,$$

we find that

$$h = \frac{2\gamma}{\rho g r} \qquad (13.12)$$

If r is small, the liquid can be raised a considerable height by capillary action.

EXAMPLE 13.12 The small capillary tubes in a plant have a radius of about 0.005 mm. To what height will 20°C water be raised in these tubes by capillary action?

SOLUTION According to Tables 13.1 and 13.2, the density of water is 1000 kg/m^3, and its surface tension is 0.073 N/m. The height to which the water will rise by capillary action is

$$h = \frac{2\gamma}{\rho g r} = \frac{2(0.073 \text{ N/m})}{(1000 \text{ kg/m}^3)(9.8 \text{ m/s}^2)(5 \times 10^{-6} \text{ m})} = \underline{3.0 \text{ m}}.\qquad \blacksquare$$

Summary

1. A **fluid** is a form of matter, such as a liquid or gas, that can flow. A static fluid—a fluid that is not moving—exerts a force perpendicular to and toward a surface it contacts. The magnitude of the force is

$$F = PA, \qquad (12.1)$$

where P is the pressure in the fluid contacting the surface, and A is the area of the surface.

2. The **density** ρ of a substance is defined as the ratio of its mass m and volume V:

$$\rho = \frac{m}{V}. \qquad (13.1)$$

3. The pressure in a fluid increases with depth. The pressure P_1 at a vertical elevation y_1 is related to the pressure P_2 at vertical elevation y_2 by the equation

$$P_1 = P_2 + \rho g(y_2 - y_1), \qquad (13.3)$$

where ρ is the density of the fluid. The equation applies when the positive y axis is directed upward. This variation of pressure with depth occurs because the fluid at a certain depth must support the weight of fluid above it. At greater depths more weight must be supported.

4. **Pascal's principle** says that if external pressure is exerted on an enclosed static fluid, the pressure increase will be transmitted uniformly through the whole volume of the fluid. Equation (13.3) predicts this pressure increase, since if P_2 is increased by an external pressure, then there must be an equal increase in the pressure, P_1.

5. A fluid exerts an upward **buoyant force** on an object that is partially or totally submerged in the fluid. According to **Archimedes' principle,** the magnitude of this buoyant force equals the weight of fluid displaced by the object. If the density of the fluid is ρ_f and the volume of fluid displaced by the object is V_f, then the buoyant force B is

$$B = \rho_f g V_f. \qquad (13.6)$$

6. **Surface tension:** The mutual attraction of molecules at the surface of a liquid to each other and to molecules deeper in a fluid gives rise to a surface tension that has the effect of a membrane containing the fluid. The surface tension γ is the force per unit length needed to hold these surface molecules together:

$$\gamma = \frac{F}{l}. \qquad (13.11)$$

The force is directed tangent to the surface.

7. **Capillary action** occurs when liquids are drawn up a small capillary tube because of the adhesion of the liquid to the surface of the tube and because of cohesion of molecules of liquid to each other. The height to which the liquid is drawn is given by the equation

$$h = \frac{2\gamma}{\rho g r}, \qquad (13.12)$$

where r is the radius of the capillary, γ is the surface tension of the liquid, and ρ is its density.

Questions

1. Give two examples in which a fluid exerts an upward force on an object, and two examples in which it exerts a downward force.

2. A milk carton is filled mostly with milk but also with some air at atmospheric pressure. What would happen to the carton (a) if it were immersed about 50 m under the surface of the water? (b) if it were taken many thousands of meters above the earth's surface?

3. If the atmosphere had a uniform density of 1.3 kg/m^3 from the earth's surface up to the "top" of the atmosphere, how high would the atmosphere be? The pressure at the earth's surface is $1.0 \times 10^5 \text{ N/m}^2$. Explain why there is air even at an elevation of $2.5 \times 10^4 \text{ m}$ above the earth's surface.

4. *Estimate* the increase in water pressure in a water bed when a 150-lb person lies on the bed.

5. When placed in a lake, an object either floats on the surface or sinks. It does not float at some intermediate location between the surface and the bottom of the lake. However, a weather balloon floats at some intermediate elevation between the earth's surface and the top of its atmosphere. Explain.

6. *Estimate* the force needed to hold a basketball under water. Justify any numbers used in your estimate.

7. A rubber raft floating in a swimming pool holds a heavy rock. Does the level of water in the pool rise or fall when the rock is thrown out of the raft and into the water? Explain.

8. Will a boat float higher or lower in salt water or in fresh water? Explain.

9. A hollow sphere hanging from one arm of an equal-arm balance is balanced exactly by a solid weight of the same material on the other arm. The air is then removed from the room holding the balance. Will the arms remain balanced? If not, which arm moves lower? Explain.

10. A glass with one ice cube in it is filled to the *rim* with water. Will the water level remain unchanged, decrease, or overflow when the ice melts? Explain. (*Hint:* You must consider the density of the ice and the volume of space displaced by the ice before melting, compared to the volume of the melted ice.)

11. A helium balloon floats in a car. In which direction will the balloon move when the car accelerates from a stop sign? Explain.

12. Your skin's surface area in square meters can be estimated using the equation

$$A = 0.20m^{0.425}h^{0.725},$$

where m is your mass in kilograms, and h is your height in meters. (a) Use this equation to estimate your own surface area. (b) If an atom on the surface of your skin is about $1.2 \times 10^{-10} \text{ m}$ in radius, roughly how many atoms are in the top layer of skin on your surface?

13. Hot water leaks more easily from small holes in faucets than cold water at the same pressure. Explain.

14. Sprinkle pepper over the surface of a dish of water. What do you think will happen to the pepper when you gently touch a bar of soap to the water surface? Try it and describe the results. Why do you think this happened? (*Hint:* Think of two membranes fastened together, one stretched somewhat more tightly than the other.)

15. How can capillary action be used to estimate the radii of glass capillary tubes?

Problems

13.1 Pressure

1. You hold your physics book in the air so that its back cover faces the ground. Approximately how large a force in newtons and in pounds does the atmosphere exert on the cover? Why does this force not support the book's weight?

2. Calculate the downward force of the earth's atmosphere on the state where you attend college.

3. Suppose that you could remove all the air from inside your bedroom. Calculate the force of the earth's atmosphere on the outside of the door leading into the bedroom in units of newtons, pounds, and tons.

4. The atmosphere at a pressure of 1 atm exerts a downward force of $0.8 \times 10^5 \text{ N}$ on a circular table. Calculate the radius of the table.

13.2 Density

5. A single-level home has a floor area of 200 m^2 with ceilings that are 2.6 m high. Determine the mass of the air in the house when at $20°C$.

6. About two-thirds of your body mass consists of water. Calculate the volume of water in a 70-kg person.

7. A 2.0-lb fish has a volume of 0.029 ft^3. Calculate the fish's (a) weight density, and (b) mass density.

8. Calculate the average mass density of the earth. The earth's mass is $5.98 \times 10^{24} \text{ kg}$, and its radius is $6.38 \times 10^6 \text{ m}$.

9. A person's mass while dieting decreases by 5 percent. Exercise creates muscles and reduces fat, causing the person's density to increase by 2 percent. Calculate the percentage change in the person's volume.

13.3 Pressure Variation with Depth

10. A car sinks 6 m below the surface of a lake. Calculate the net force pushing on its 0.7-m^2 door. Atmospheric air pressure pushes out on the door from inside the car and down on the water at its surface.

11. The gravity-fed water reservoir for a city is located 40 m above the city. Calculate the water gauge pressure at homes in the

city in units of newtons per square meter, pounds per square inch, and atmospheres.

12. Estimate the average blood gauge pressure in an artery of the foot, 1.2 m below the heart. The average blood gauge pressure in the heart is 100 mm Hg.

13. A man blows into a hose connected to a U-shaped tube containing water. Before being blown, the water on each side of the tube is at the same level. While being blown, the water is forced 0.2 m higher on the side opposite the man. Calculate the gauge pressure in the man's lungs when blowing.

14. A woman sucks water 0.3 m up a straw. Calculate the gauge pressure in her lungs while sucking.

15. You can develop a gauge pressure of -30 mm Hg in the lungs while sucking on a straw. Up what length straw can you suck a fruit drink with a density of 1200 kg/m³?

16. A man climbs 500 m up a mountain. (a) Assuming a constant air density of 1.2 kg/m², calculate the change in air pressure acting on him. (b) The pressure on the inside and outside of the man's eardrum is balanced when he starts the climb. What is the net force acting on the eardrum at the top of the mountain if there is no pressure change inside the ear? The area of the eardrum is 0.2 cm².

17. A device known as Hare's apparatus (Fig. 13.19) is used to measure the relative density of a fluid. Suction in a common, center tube causes water to rise in the tube on the right and a different type of fluid of unknown density to rise in the tube on the left. When the valve in the center tube is closed, the water has risen to a level of 10.5 cm, and the fluid on the left has risen to a level of 9.2 cm. Calculate the density of the fluid on the left. The density of water is 1000 kg/m³.

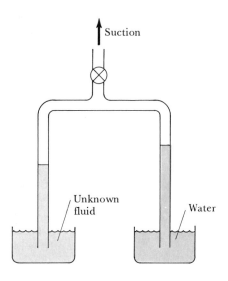

FIG. 13.19. Hare's apparatus for determining relative density in liquids.

18. Olive oil is poured into a U-tube containing mercury with the result shown in Fig. 13.20. Both sides of the tube are open to the atmosphere. Calculate the density of the olive oil. The density of mercury is 13,600 kg/m³.

FIG. 13.20

13.4 Externally Applied Pressure

19. A hydraulic lift is used to raise a 900-kg car. Compressed air equal in pressure to 4.0×10^5 N/m² is applied to a small cylinder. Calculate the area of a large piston supporting the car's weight.

20. A 10-kg mass is placed on top of a piston of a hydraulic lift; the piston's area is 3.0 cm². A 100-kg woman wrestler sits on a piston of 60-cm² area. Will she cause the 10-kg mass to rise? Explain.

21. As shown in Fig. 13.21, a 100-N force pulling up on a lever causes an increase in pressure under a small piston of 8.0-cm² area. (a) Calculate the pressure created under the small piston. (b) Calculate the upward force **F** on the large, 160-cm² piston of the hydraulic press.

FIG. 13.21

22. A person is poked in the eye with a finger that exerts a 60-N force over an area of 0.3 cm² at the front of the eye. The increased pressure in the eye is transmitted to the back of the eye. (a) Calculate the force on the 0.05-cm² area from which the optic nerve

leaves the eye. (b) Calculate the force at the back of the eye on a single rod whose diameter is about 10^{-6} m.

13.5 Archimedes' Principle

23. A person with a density of 980 kg/m³ floats in seawater of density 1025 kg/m³. What fraction of the person is submerged?

24. When a 70-kg person steps on an ice floe, the top surface of the ice is just barely submerged. Calculate the mass of the ice floe. The density of seawater is 1025 kg/m³ and that of ice is 920 kg/m³.

25. Calculate the buoyant force on an ice cube of volume 12 cm³ when placed in ethyl alcohol of density 810 kg/m³. Will the ice cube float? Explain.

26. Logs of density 600 kg/m³ are used to build a raft. What is the weight of the maximum load that can be supported by a raft built from 300 kg of logs?

27. A life preserver is manufactured to support a 70-kg person with 20 percent of his volume out of the water. The person's density and that of the water are both about 1000 kg/m³. If the density of the life preserver is 100 kg/m³ and it is completely submerged, what must its volume be?

28. A gold crown weighs 40 N when in air. What will it weigh when suspended (a) in water, and (b) in ethyl alcohol?

29. Calculate the upward buoyant force on a person's brain that is immersed in a fluid in the skull whose density is 1007 kg/m³. The brain's mass is 1.0 kg, and its density is 1040 kg/m³.

30. A protein molecule of mass 1.1×10^{-22} kg and density 1.3×10^3 kg/m³ is placed in a vertical tube of water of density 1000 kg/m³. Calculate the initial acceleration of the protein.

31. A 30-kg rock of density 2400 kg/m³ is immersed in a lake and must be moved from near a boat dock. Calculate the force needed to lift the rock while it is under the water.

32. While analyzing a sample of ore, a geologist finds that it weighs 2.00 N in air and 1.13 N when immersed in water. Calculate the density of the ore.

33. A 10-kg weather balloon with a volume of 18 m³ is filled with 4.0 kg of helium. (a) Calculate the buoyant force on the balloon if the air density is 1.3 kg/m³. (b) Calculate the initial upward acceleration of the balloon.

13.6 Surface Tension

34. If the apparatus with a slide wire 0.05-m long shown in Fig. 13.15 supports a total weight of 4.2×10^{-3} N, what is the surface tension of the liquid?

35. A horizontal, platinum wire having a 0.015-g mass and 5-cm length is dipped in ethyl alcohol. Calculate the force needed to pull the wire slowly from the alcohol.

36. The surface tension of a liquid can be determined by measuring the force F needed to lift a horizontal, circular platinum ring of radius r and weight w from the liquid's surface. (a) Derive an equation for γ in terms of F, r, and w. (b) If $F = 12.2 \times 10^{-3}$ N, $r = 2.0$ cm, and $w = 5.4 \times 10^{-3}$ N, calculate the value of γ for a particular fluid.

37. A thin circular wire of 1.12-cm radius is supported horizontally in a liquid and withdrawn slowly by an upward force of 5.53×10^{-3} N. If the surface tension of the liquid is 0.032 N/m, what is the mass of the wire? Note that the ring is pulled down by two surfaces, one inside the ring and one outside.

38. One leg of a water strider standing on 20°C water makes a depression of radius 2.0 mm. The water surface tension force is exerted at an angle of 53° with the horizontal (Fig. 13.22). (a) Calculate the weight supported by surface tension at the sides of this depression. (b) Calculate the weight and mass of the water strider if each of its six legs causes a similar depression.

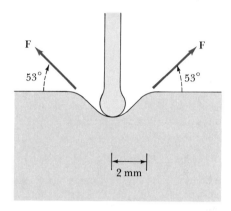

FIG. 13.22. A water strider's leg being supported by surface tension.

39. The tension in the skin of the left ventricle of the heart is 290 N/m. A 4-cm cut is made in the left ventricle during open-heart surgery. How many stitches must be used to repair the cut? Each stitch will safely support a tension of 0.4 N.

13.7 Capillary Action

40. Calculate the heights to which water at 20°C will rise in glass capillary tubes of radii 0.2 mm, 0.5 mm, 1.0 mm, and 2.0 mm.

41. Water rises to a height of 15.6 cm in a glass capillary tube. Estimate the radius of the tube. To what height would ethyl alcohol rise in the same tube?

42. If capillary action were responsible for lifting water to the tops of redwood trees 100 m tall, what would be the maximum radii of capillaries in these trees?

43. It has been calculated that capillary action will lift water to the following heights in various types of soils: (a) 0.1 m in fine gravel, (b) 0.5 m in coarse sand, (c) 2 m in fine sand, (d) 10 m in silt, and (e) 50 m in clay. Estimate the radii of pores in each type of soil.

44. Water is raised 14 cm in the bore of a capillary tube. How high will soap water be raised in the same tube? Explain.

45. Two large parallel glass plates separated by a distance of 0.5 mm are dipped into water. To what height will water rise between the plates?

General Problems

G1.   A 60-kg woman having a density of 980 kg/m^3 stands on a bathroom scale. Calculate her weight reduction because of the buoyant force of air on her body.

G2.   A hollow, cylindrical rod with an outer radius of 0.12 cm is closed at the bottom (Fig. 13.23). Weights dropped into the rod cause it to float vertically in 20°C water with the weighted end down. The rod and weights together have a mass of 0.18 g. (a) If the water contacts the rod parallel to its surface, calculate the water surface tension force **F** that supports the rod. (b) If the bottom of the rod is a distance h below the surface, calculate the buoyant force on the rod in terms of h. (c) Combine the forces in (a) and (b) and the known weight of the rod and its contents to calculate the distance h.

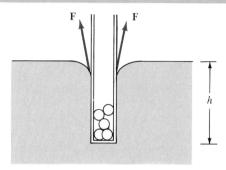

FIG. 13.23.   A cylinder supported by surface tension and a buoyant force.

Fluid Dynamics

14

In the last chapter we studied fluid statics—fluids at rest. In this chapter we will learn about fluid dynamics—moving fluids. The study of fluid dynamics will permit us to answer a variety of questions concerning moving fluids: What causes the roof to be lifted off a house during a strong wind? Why does air moving through the bronchial passages cause us to snore? How does the speed of the flow of blood affect the chance of one's having a heart attack? How does air support the weight of massive jumbo jets? These and countless other questions can be answered using one basic equation developed in 1738 by the Swiss physicist Daniel Bernoulli (1700–1782). Bernoulli's equation allows us to calculate variations in fluid pressure caused by a fluid's motion.

Friction in fluids, a subject that is ignored in Bernoulli's equation, becomes important when fluids move through narrow vessels. This friction in moving fluids and the equally important friction or drag force of a fluid on a moving object are considered toward the end of the chapter.

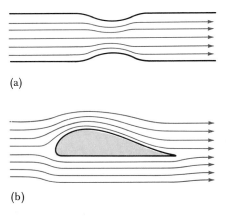

FIG. 14.1. Laminar flow (a) of water through a constriction in a pipe and (b) of air over an airplane wing.

14.1 Types of Fluid Flow

When the pressure is lower on one side of a fluid than on the other, the fluid will flow toward this low-pressure region. For instance, large masses of air in the earth's atmosphere move from regions of high into regions of low pressure. Blood flows from the arterial side of the circulatory system at a pressure of about 100 mm Hg to the venous side at a pressure of about 5 mm Hg.

Fluid flow is characterized by two main types of flow, streamline flow and turbulent flow. In **streamline flow,** sometimes called laminar flow, every particle of fluid that passes a particular point follows the same path as particles that passed that point before it. The particles move in an unchanging line called a streamline. The streamlines of water molecules passing through a constriction in a pipe are depicted in Fig. 14.1a. The streamlines of air moving past the wing of an airplane are shown in Fig. 14.1b.

Turbulent flow is characterized by agitated, disorderly motion. Instead of following a given path, molecules of a fluid swirl and form whirlpool patterns called eddies. The turbulent flow of water moving through a narrow constriction in a pipe is illustrated in Fig. 14.2a. The turbulent motion of air behind an inclined airplane wing is shown in Fig. 14.2b. Usually, turbulent flow occurs in

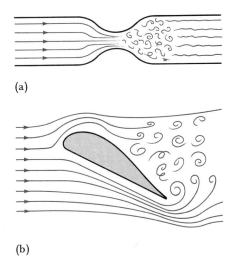

FIG. 14.2. Turbulent flow (a) of water in a pipe and (b) of air behind a tilted airplane wing.

fluids moving at high speed. Friction forces are normally much greater during turbulent flow than during streamline flow. In this chapter we concentrate mostly on situations involving streamline flow.

14.2 Flow Rate and Equation of Continuity

We are often interested in the volume V of fluid that flows past some cross section of a vessel in a time t. The ratio of this volume and the time required for it to flow is called the **flow rate** of the fluid and is given the symbol Q:

$$Q \equiv \frac{V}{t}. \tag{14.1}$$

The units of flow rate are expressed in m^3/s, ft^3/s, ft^3/min, and so forth.

Often we cannot directly measure the actual volume of fluid that passes along a vessel per unit time. For example, we cannot directly measure the volume of blood pumped through an artery in a certain time without somehow collecting and measuring that volume. Doubtless, a person would object to having his blood collected for a routine flow-rate measurement, but there are indirect techniques for measuring the flow rate. One of these methods, discussed in Chapter 17, utilizes a noninvasive measurement of the blood's speed and the cross-sectional area of the vessel carrying the blood.

Consider the fluid moving along the pipe pictured in Fig. 14.3a. We are interested in the volume of fluid that crosses a cross section of area A at some position along the pipe. For the pipe shown in Fig. 14.3a, the darkened portion of the fluid moves forward and passes the cross section A in a time t. Thus, after a time t, the back part of this fluid has moved forward to the position shown in Fig. 14.3b.

The volume V of fluid in this darkened portion of the cylinder is the product of its length l and the cross-sectional area of the pipe:

$$V = lA.$$

We find, then, that

$$Q = \frac{V}{t} = \frac{lA}{t} = \frac{l}{t}A.$$

However, l is also the distance the fluid moves in a time t. Thus, l/t is the average speed $\bar{v}$ of the fluid. Substituting $\bar{v} = l/t$ into the above equation, we find that

$$Q = \bar{v}A. \tag{14.2}$$

This is a general equation that allows us to calculate the **flow rate** of fluid past any cross-sectional area along a pipe or vessel.

EXAMPLE 14.1 The heart pumps blood at a flow rate of $80\ cm^3/s$ into the aorta whose diameter is $1.5\ cm$. Calculate the average speed of blood in the aorta.

SOLUTION The cross-sectional area of the aorta is

$$A = \pi r^2 = \pi(1.5\ cm/2)^2 = 1.8\ cm^2.$$

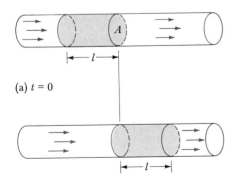

(a) $t = 0$

(b) Time, t

FIG. 14.3. The darkened section of fluid requires a time t to pass the cross section A. The volume of fluid that flows in this time is $V = lA$.

The average speed of blood in the aorta is determined by rearranging Eq. (14.2):

$$\bar{v} = \frac{Q}{A} = \frac{80 \text{ cm}^3/\text{s}}{1.8 \text{ cm}^2} = 44 \text{ cm/s}.$$ ∎

FIG. 14.4. A fluid flows through a vessel with changing cross-sectional area. For incompressible fluids, the flow rate across A_1 must equal that across A_2.

Often, the cross-sectional area of a vessel through which fluid flows varies from one part of the vessel to another. If the fluid is incompressible—if its volume undergoes negligible change when the pressure acting on the fluid increases—then the volume of fluid that passes one cross section of the vessel equals that passing another cross section. During any period of time, the fluid entering the region between these cross sections (for example, A_1 and A_2 in Fig. 14.4) must equal that leaving the region. The flow rate in must equal the flow rate out. In terms of the cross sections A_1 and A_2 pictured in Fig. 14.4 we can say that

$$Q = \bar{v}_1 A_1 = \bar{v}_2 A_2, \qquad (14.3)$$

where $\bar{v}_1$ is the average speed of fluid passing cross section A_1, and $\bar{v}_2$ is the average speed of that passing A_2. In a narrow section of pipe where A is small, the speed of the fluid will have to be great in order for the flow rate to remain constant.

Equation (14.3) is called the **continuity equation** and is used to relate the cross-sectional areas and speeds of fluid flow in different parts of a vessel carrying the fluid. It will be an especially useful equation when discussing applications of Bernoulli's equation later in this chapter.

EXAMPLE 14.2 Blood normally flows with an average speed of about 10 cm/s in the large arteries whose radii are about 0.3 cm. Suppose that a small section of artery is reduced in radius by one-half because of thickening of its walls (called atherosclerosis). Calculate the blood speed past the constriction.

SOLUTION The blood speed past the constriction $\bar{v}_2$ will be related to the speed $\bar{v}_1$ in the clear artery by the equation

$$\bar{v}_2 = \frac{A_1}{A_2}\bar{v}_1 = \left(\frac{\pi r_1^2}{\pi r_2^2}\right)\bar{v}_1$$

$$= \left(\frac{r_1}{r_2}\right)^2 \bar{v}_1 = \left(\frac{r_1}{0.5 r_1}\right)^2 \bar{v}_1 = 4\bar{v}_1.$$

Thus, $\bar{v}_2$ will be 40 cm/s if $\bar{v}_1$ is 10 cm/s. ∎

14.3 Bernoulli's Equation ✗

In 1738, Daniel Bernoulli formulated an equation that could be used to determine the variation of pressure in fluids as a function of the speed of the fluid and the elevation of the vessel through which the fluid passed. This equation can be used to explain why airplane wings have lift, how to avoid having the roof of your house blown off during a tornado or hurricane, why plaque in arteries becomes dislodged (possibly causing a stroke), and many other phenomena involving moving fluids.

If Bernoulli had developed this equation in the late 1800s or in this century, it is unlikely that the equation would have been named for him because it is just the work-energy equation (Eq. 8.10) applied to a special type of fluid flow. Bernoulli, however, devised the equation long before thermodynamics and ideas about energy had been introduced to the world of science. Since these topics are now well understood, we can use the work-energy equation to derive Bernoulli's equation.

A pipe through which fluid flows is shown in Fig. 14.5a. We will assume that the fluid (1) undergoes streamline flow and (2) is incompressible and that (3) no friction forces act on the fluid. We will use the work-energy equation to describe the behavior of the fluid as it moves a short distance along the vessel. We choose as our system the darkened section of fluid pictured in Figs. 14.5a and b.

The fluid flows to the right. Fluid behind the system at position 1 pushes the system to the right with a pressure P_1, while the fluid ahead of the system at position 2 pushes back on the system with a pressure P_2. The speed of the fluid varies from position to position according to the equation of continuity $(A_1 v_1 = A_2 v_2)$. As a result, the fluid moves faster at position 2 than at position 1, since the vessel is narrower at 2 than at 1.

The net effect of the movement of the fluid a short distance to the right is summarized in Fig. 14.5c. A volume of fluid initially at position 1 moving at a

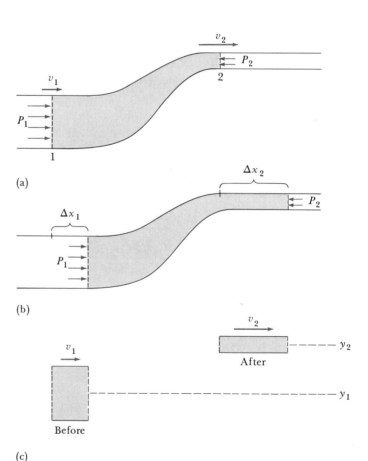

FIG. 14.5. (a) A vessel carrying fluid. The darkened part of the fluid is the system—the part to which we apply the work-energy equation. (b) The system a short time later after it has moved along the vessel. (c) The net effect of this movement is that part of the system has been transferred from the *before* region to the *after* region.

speed v_1 has now been transferred, in effect, to position 2, where it moves at speed v_2. Both the kinetic and gravitational potential energies of the fluid in the system have changed. These changes have been caused by the work done on the fluid by the forward and backward pressures at positions 1 and 2, respectively. In terms of the work-energy equation, the energy changes are related by the equation

$$W = \Delta KE + \Delta PE_g.$$

Let us now calculate each of these terms in the above equation.

Work: Work is done on the system by the force of the fluid pushing from the left with pressure P_1 and of the fluid on the right pushing back on the system with pressure P_2. If the fluid moves a distance Δx_1 on the left side (Fig. 14.5b), it will have to move a farther distance Δx_2 on the right side, since the vessel is narrower on the right. Using Eq. (8.1), we find the net work done on the fluid to be

$$W = F_1 \Delta x_1 \cos 0° + F_2 \Delta x_2 \cos 180°$$
$$= F_1 \Delta x_1 - F_2 \Delta x_2 = P_1 A_1 \Delta x_1 - P_2 A_2 \Delta x_2.$$

The volume of fluid ΔV that has moved from the left to the right is $\Delta V = A_1(\Delta x_1) = A_2(\Delta x_2)$. The above expression for work becomes

$$W = (P_1 - P_2) \Delta V.$$

Change in kinetic energy: The mass of fluid that has moved is related to its density and volume by the equation

$$m = \rho \Delta V.$$

As the system moves, a small amount of fluid originally moving with speed v_1 at position 1 has, in effect, been transferred to position 2, where it moves with speed v_2. Thus, the kinetic energy change of the mass m of fluid shown in Fig. 14.5c is

$$\Delta KE = \tfrac{1}{2}mv_2^2 - \tfrac{1}{2}mv_1^2$$
$$= \tfrac{1}{2}(\rho \Delta V)v_2^2 - \tfrac{1}{2}(\rho \Delta V)v_1^2.$$

Change in gravitational potential energy: The gravitational potential energy of the system has also changed, since a mass m of fluid has effectively moved from elevation y_1 to elevation y_2. The change in gravitational potential energy is

$$\Delta PE_g = mg(y_2 - y_1) = (\rho \Delta V)g(y_2 - y_1).$$

If we put these three energy changes into the work-energy equation and cancel the common ΔV in each term, we find that

$$P_1 - P_2 = \tfrac{1}{2}\rho(v_2^2 - v_1^2) + \rho g(y_2 - y_1). \qquad \textbf{(14.4)}$$

This is one form of **Bernoulli's equation.** It relates the pressures, speeds, and elevations of a fluid at any two points in the fluid (our positions 1 and 2 in Fig. 14.5 were completely arbitrary). Equation 14.4 applies to the nonturbulent flow of a frictionless, incompressible fluid. The term on the left side of Eq. (14.4) is a consequence of the work done because of the difference in fluid pressure pushing forward and that pushing back. The two terms on the right result from the changes in kinetic and gravitational potential energies of the fluid.

Bernoulli's equation is often written in a form that is easier to use and perhaps easier to remember. A rearrangement of Eq. (14.4) yields

$$P_1 + \tfrac{1}{2}\rho v_1^2 + \rho g y_1 = P_2 + \tfrac{1}{2}\rho v_2^2 + \rho g y_2. \tag{14.5}$$

We see that the sum of the three terms on the left for position 1 in the fluid must equal the sum of the same three terms for another position, position 2. Since the choice of positions is arbitrary, the sum of these three terms must be constant at all points in the fluid:

$$P + \tfrac{1}{2}\rho v^2 + \rho g y = \text{constant} \tag{14.6}$$

These three terms must add to the same value at any point along a streamline.

EXAMPLE 14.3 Water flows through a pipe of radius 0.10 m at a speed of 5.0 m/s. (a) Calculate the flow rate in this pipe. (b) The water goes down a hill with a vertical drop of 4.0 m and then flows along a pipe of radius 0.080 m. Calculate the pressure of the fluid in the small pipe minus the pressure in the large pipe. The density of water is 1000 kg/m^3.

SOLUTION (a) The situation described in the problem is shown in Fig. 14.6. The flow rate is

$$Q = v_1 A_1 = v_1 (\pi r_1^2) = (5.0 \text{ m/s})\pi(0.10 \text{ m})^2 = \underline{0.16 \text{ m}^3/\text{s}}.$$

(b) We wish to find the pressure difference, $P_2 - P_1$, using Bernoulli's equation. The speed of the fluid at position 2 in Fig. 14.6 can be determined using the continuity equation:

$$v_1 A_1 = v_2 A_2$$

or

$$v_2 = \frac{A_1}{A_2} V_1 = \frac{\pi(0.10 \text{ m})^2}{\pi(0.080 \text{ m})^2}(5.0 \text{ m/s}) = 7.8 \text{ m/s}.$$

From Bernoulli's equation, we find that

$$\begin{aligned}
P_2 - P_1 &= \tfrac{1}{2}\rho v_1^2 - \tfrac{1}{2}\rho v_2^2 + \rho g(y_1 - y_2) \\
&= \tfrac{1}{2}(1000 \text{ kg/m}^3)(5.0 \text{ m/s})^2 - \tfrac{1}{2}(1000 \text{ kg/m}^3)(7.8 \text{ m/s})^2 \\
&\quad + (1000 \text{ kg/m}^3)(9.8 \text{ m/s})(4.0 \text{ m}) = \underline{+21{,}000 \text{ N/m}^2}. \quad\blacksquare
\end{aligned}$$

FIG. 14.6. Water flows from position 1 toward position 2.

When using Bernoulli's equation, we learn some surprising consequences of the effect of motion on fluid pressure. Consider the air above the roof of a house during a windstorm. If the air's kinetic energy increases, some other term in Eq. (14.6), such as the pressure, must decrease so that the sum of the three terms remains constant. We find that the pressure above the roof of a house during a windstorm is less than normal. If the windows are closed, the normal pressure inside the house may blow the roof off. The unbalanced pressure can be corrected by opening the windows. It is cheaper to repair soiled furniture and carpet than to replace a roof.

A simple experiment can help convince you that the pressure of air on a surface *decreases* because the air moves. Hang two sheets of paper straight down as shown in Fig. 14.7. When you blow air between the sheets, the pressure is

FIG. 14.7. Two pages held apart are pulled together because of the reduced pressure of air blown between them.

reduced and the sheets move together. After you stop blowing, the sheets separate.

The reed of an oboe closes in a similar fashion. When air is blown into the reed, the pressure is reduced compared to pressure outside the reed where the air is not moving. The reed closes just as the papers in Fig. 14.7 come together. When closed, air flow stops, the pressures inside and outside the reed equalize, and the reed reopens causing air flow to start. The pressure is again lowered inside the reed due to the moving air, and it again closes. The rhythmic opening and closing of the reed initiates the sound we hear coming from the oboe.

Snoring is caused in a similar way when the bronchial passage alternately opens and closes. As air flows through the passage, air pressure inside is reduced and air pressure outside causes the passage to close. When air flow stops, the pressures equalize and the passage reopens. The rhythmic opening and closing of the passage resembles the vibration of an oboe reed and creates the snoring sound.

14.4 Applications of Bernoulli's Equation

When using Bernoulli's equation, we choose two different positions in a fluid. One position is located at a place where we want to find an unknown property of the fluid—for example, the pressure. The second position is chosen at a place where the pressure, density, speed, and elevation of the fluid are known. When all of this known information is substituted into Bernoulli's equation, we can solve for the unknown.

EXAMPLE 14.4 (a) Calculate the speed with which water flows from a hole in the dam of a large irrigation canal. The hole is 0.80 m below the surface of the water (Fig. 14.8). (b) If the hole has a radius of 2.0 cm, what is the flow rate of water from the hole?

SOLUTION We choose position 1 as the place where the water leaves the hole (see Fig. 14.8). Position 2 can be chosen at a place where the pressure, elevation, and speed of the water are known. A convenient place is at the top of the water behind the dam (see Fig. 14.8). There is so much water behind the dam that its level moves downward very slowly. Thus, the water speed at position 2 is nearly zero ($v_2 \simeq 0$). Since position 2 is 0.80 m higher than position 1, $y_2 - y_1 = 0.80$ m. The water pressure at both positions 1 and 2 is atmospheric pressure: $P_1 = P_2 = P_{atm}$.

We next substitute all of this information into Eq. (14.4):

$$P_{atm} - P_{atm} = \tfrac{1}{2}\rho(0^2 - v_1^2) + \rho g(y_2 - y_1).$$

Solving for v_1, we find that

$$v_1 = \sqrt{2g(y_2 - y_1)}.$$

This is called **Torricelli's theorem.** Substituting for g and $(y_2 - y_1)$, we find that

$$v_1 = \sqrt{2(9.8 \text{ m/s}^2)(0.80)} = \underline{4.0 \text{ m/s}}.$$

(b) The flow rate of the water from the hole is determined using Eq. (14.2),

FIG. 14.8

FIG. 14.9. Air blowing across a roof causes the pressure to be reduced. If the wind blows fast enough, the internal air pressure can blow the roof off.

$Q = A_1 v_1$, where A_1 is the area of the hole: $A_1 = \pi r_1^2 = \pi (0.020 \text{ m})^2 = 1.3 \times 10^{-3} \text{ m}^2$. The flow rate is, then,

$$Q = A_1 v_1 = (1.3 \times 10^{-3} \text{ m}^2)(4.0 \text{ m/s}) = \underline{5.0 \times 10^{-3} \text{ m}^3/\text{s}.} \quad \blacksquare$$

EXAMPLE 14.5 A 45 m/s (100 mph) wind blows across the top of a flat-roofed house (Fig. 14.9). Calculate the net force on the roof due to the pressure differential from inside the house to the outside. Assume that the pressure in the house is atmospheric pressure and that the roof has an area of 200 m². The density of air equals 1.3 kg/m^3.

SOLUTION Air above the house pushes down with a force, $F_1 = P_1 A$, where P_1 is the air pressure above the house and A is the area of the roof. The air inside the house pushes up with a force, $F = P_{atm} A$, where P_{atm} is the atmospheric pressure of the air inside the house. The net upward force is

$$F_{net} = F - F_1 = P_{atm} A - P_1 A = (P_{atm} - P_1)A. \quad (14.7)$$

We can use Bernoulli's equation to find the pressure P_1. To do this, we choose a second point in the air that is at the same elevation as air above the roof, but where the air is not blowing. This point may be miles away. We must be able to move freely from position 1 to position 2 without passing across or through any artificial barrier. If the house is completely closed, there is an artificial barrier between position 1 and the inside of the house, so we cannot choose position 2 inside the house.

If we go far from the house, we will find another place where the wind is not blowing and which is at the same elevation as position 1. At this place, if we are at sea level, the pressure is also atmospheric pressure. Thus,

$$P_2 = \text{atmospheric pressure} = P_{atm}.$$

The air is at rest at this distant position ($v_2 = 0$) and its elevation is the same as at position 1 ($y_2 = y_1$). From Eq. (14.5),

$$P_1 + \tfrac{1}{2}\rho v_1^2 + \rho g y_1 = P_{atm} + \tfrac{1}{2}\rho(0)^2 + \rho g y_1.$$

Solving for P_1, we find that

$$P_1 = P_{atm} - \tfrac{1}{2}\rho v_1^2.$$

Substituting for P_1 into Eq. (14.7), we find that

$$\begin{aligned} F_{net} = (P_{atm} - P_1)A &= [P_{atm} - (P_{atm} - \tfrac{1}{2}\rho v_1^2)]A = \tfrac{1}{2}\rho v_1^2 A \\ &= \tfrac{1}{2}(1.3 \text{ kg/m}^3)(45 \text{ m/s})^2(200 \text{ m}^2) \\ &= \underline{2.6 \times 10^5 \text{ N},} \end{aligned}$$

or an upward force of about 30 tons.

This huge upward force is caused by the reduced pressure over the roof compared to the atmospheric pressure in the house. To avoid having the roof lifted off the house by this pressure imbalance, open some windows so that the pressure has a chance to equalize. $\quad \blacksquare$

FIG. 14.10. Blood must accelerate as it moves past a plaque. The reduced blood pressure of this fast-moving blood may cause the plaque to become dislodged.

Bernoulli's principle can play a villainous role in causing heart attacks. A person's arteries may become constricted by the growth of an atherosclerotic plaque such as shown in Fig. 14.10. If dislodged, the plaque moves downstream and may block a smaller vessel so that no blood flows through it. Should this happen to one of the arteries supplying blood to the heart muscles, a heart attack may occur. Bernoulli's principle helps us understand how a plaque becomes dislodged, as described in the simplified model of a plaque in the next example.

FIG. 14.11. A simplified version of a plaque.

EXAMPLE 14.6 Blood of density 1030 kg/m^3 flows through the unobstructed part of the blood vessel shown in Fig. 14.11 at a speed of 0.5 m/s. The cross-sectional area through which blood flows past the plaque is one-sixth the normal area of the vessel. The bottom surface of the plaque facing down toward position 2 has an area of 0.60 cm^2, as do the top surfaces in the channels facing position 1. Calculate the net downward force that tends to dislodge the plaque.

SOLUTION The blood pressure at position 2 is less than at position 1 because the blood must flow at high speed through the constricted artery, whereas it sits at rest in the channels at position 1 ($v_1 = 0$). We determine the blood speed at position 2, using the continuity equation, by comparing the speed of the blood and area of the blood vessel in the unconstricted artery (v and A) to that in the constricted artery (v_2 and A_2):

$$v_2 = \frac{Av}{A_2} = \left(\frac{A}{A_2}\right)v = (6)(0.5 \text{ m/s}) = 3 \text{ m/s}.$$

Let us next calculate the difference in pressure from position 1 to position 2 in Fig. 14.11. We will also assume that positions 1 and 2 are at approximately the same elevation; that is, $y_1 \cong y_2$. According to Eq. (14.4),

$$P_1 - P_2 = \tfrac{1}{2}\rho v_2^2 - \tfrac{1}{2}\rho v_1^2 + \rho g(y_2 - y_1)$$
$$= \tfrac{1}{2}(1030 \text{ kg/m}^3)(3 \text{ m/s})^2 - 0 + 0 = 4640 \text{ N/m}^2.$$

Since the pressure is greater from above the plaque at position 1 than from below at position 2, a net downward force acts on the plaque. If the top and bottom areas against which the fluids press is $0.60 \text{ cm}^2 = 6.0 \times 10^{-5} \text{ m}^2$, then the net force trying to dislodge the plaque is

$$F_{\text{net}} = F_1 - F_2 = (P_1 - P_2)(\text{area})$$
$$= (4640 \text{ N/m}^2)(6.0 \times 10^{-5} \text{ m}^2) = \underline{0.28 \text{ N}}.$$

This is about the weight of one-fourth of a medium-sized apple pulling on the plaque. ∎

Bernoulli's principle also plays a major role in flight by helping to support the weight of an airplane. An airplane wing (Fig. 14.12) is curved so that air must travel farther across the top of the wing than across the bottom. Because of this, the speed of air across the top is greater than that across the bottom, causing a reduction in the pressure above the wing. This reduced pressure is partially responsible for the lift of the wing.

EXAMPLE 14.7 Suppose that the speed of air over the top of the wing is 180 m/s, that the air speed under the wing is 165 m/s, and that the density of air at the altitude where the plane flies is 0.80 kg/m^3. (a) Calculate the pressure difference between the bottom and top of the wing. (b) If the plane has a mass of

FIG. 14.12. Air travels faster over the top of a wing than under the bottom. This causes a greater pressure reduction over the wing and a net lift force.

2.0×10^4 kg, what must be the area of the wings so that the lift can support the airplane's weight?

SOLUTION (a) Let us take position 1 under the wing and position 2 over the wing (Fig. 14.12). We will assume that the difference in gravitational potential energies is negligible because the vertical separation of the points is very small, as is the air density. We find, then, using Bernoulli's equation, that

$$P_1 - P_2 = \tfrac{1}{2}\rho v_2^2 - \tfrac{1}{2}\rho v_1^2$$
$$= \tfrac{1}{2}(0.80 \text{ kg/m}^3)(180 \text{ m/s})^2 - \tfrac{1}{2}(0.80 \text{ kg/m}^3)(165 \text{ m/s})^2 = \underline{2070 \text{ N/m}^2}.$$

(b) The net lift force on the wing is

$$F_1 - F_2 = (P_1 - P_2)A.$$

This must balance the airplane's weight, $w = mg$. Thus,

$$mg = (P_1 - P_2)A.$$

Rearranging the above equation to find the area of the wing, we find that

$$A = \frac{mg}{P_1 - P_2} = \frac{(2.0 \times 10^4 \text{ kg})(9.8 \text{ m/s}^2)}{2070 \text{ N/m}^2} = \underline{95 \text{ m}^2}. \qquad \blacksquare$$

14.5 Viscous Fluid Flow

In our previous discussion and examples we have ignored the role of friction forces. This approximation is not appropriate in many practical cases. In fluids, just as in solids, friction is often responsible for the generation of thermal energy, but the way in which this thermal energy is created differs for liquids compared to solids. When one solid slides across another, the microscopic ridges and trenches on one surface hook and bond to the trenches and ridges on the other surface, and the thermal energy is created at the surface between the solids.

When a fluid moves across a solid, thermal energy is not generated at their interface. A thin layer of fluid, perhaps just a few molecules thick, does not move at all. The molecules in this layer continually encounter the microscopic ridges and trenches of the solid surface. However, layers of fluid above this interface do slide across each other, much as pictured in Fig. 14.13. This sliding of one layer of fluid across another causes the friction force in fluids. Consequently, the friction in fluids is generated within the fluid rather than at the interface of the fluid and a solid.

The net effect of this friction is known as the **viscosity** of the fluid. The more tightly one layer of fluid is held to an adjacent layer, the greater is the viscosity of the fluid, and the greater is the amount of thermal energy generated as the fluid moves. Molasses and honey are very viscous fluids, whereas water is less viscous.

If a fluid is pushed through a cylindrical pipe, a viscous retarding force F_v opposes its motion. The magnitude of this **viscous force** is related to the speed v of the fluid at the center of the pipe by the equation

$$F_V = 4\pi\eta l v,$$

where l is the length of the pipe, and the Greek letter eta, η, is the **coefficient of viscosity** of the fluid at a particular temperature. Its magnitude indicates the

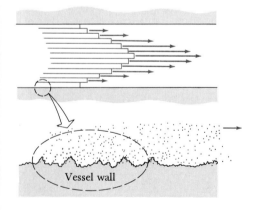

FIG. 14.13. The fluid moving down the center of a vessel moves fastest. Fluid at the edge does not move at all. The molecules of fluid in this surface layer stick to the molecules in the ridges and trenches of the solid surface of the vessel.

difficulty with which a fluid flows—or how viscous the fluid is. The coefficient of viscosity for honey is larger than that for water. A list of the coefficients of viscosity (or simply viscosities) of several fluids appears in Table 14.1. The SI unit of viscosity is N · s/m². Another commonly used viscosity unit is the **poise** (P), where 1 N · s/m² = 10 P.

For a fluid to move through a pipe at constant speed, a driving force must balance the opposing viscous retarding force. If the pressure at the entrance of a pipe is P_1, then the force pushing the fluid forward is $F_1 = P_1 A$, where A is the cross-sectional area of the pipe. The pressure force opposing the motion at the exit of the pipe at pressure P_2 is $F_2 = P_2 A$. The net driving force is, then,

$$F_1 - F_2 = (P_1 - P_2)A.$$

If the fluid moves at constant speed, this driving force must be equal in magnitude to the viscous retarding force. Thus,

$$(P_1 - P_2)A = 4\pi \eta l v.$$

If the pipe has a radius r, then its cross-sectional area is $A = \pi r^2$. Substituting this for A and rearranging, we find that

$$P_1 - P_2 = \left(\frac{4\eta l}{r^2}\right) v. \tag{14.8}$$

This equation relates the entrance and exit pressure difference to properties of the fluid and to properties of the vessel carrying the fluid. Remember that v is the speed of the fluid along the center of the pipe.

TABLE 14.1 Viscosities of Liquids and Gases

Substance	Viscosity (N · s/m²)
Gases	
Air (30°C)	1.9×10^{-5}
Water vapor (100°C)	1.25×10^{-5}
Liquids	
Water (0°C)	1.8×10^{-3}
(20°C)	1.0×10^{-3}
(40°C)	0.66×10^{-3}
(60°C)	0.47×10^{-3}
(80°C)	0.36×10^{-3}
(100°C)	0.28×10^{-3}
Alcohol (20°C)	1.2×10^{-3}
Blood plasma (37°C)	1.5×10^{-3}
Blood, whole (37°C)	$\sim 4 \times 10^{-3}$
Oil, light machine (37°C)	0.113
Oil, SAE No. 10	0.20
Glycerin	0.629

EXAMPLE 14.8 Blood flows with a speed of 0.5 m/s down the center of the aorta. If the length of the aorta is 0.4 m and its radius is 0.8 cm, what is the drop in pressure needed to force the blood through the aorta?

SOLUTION We have been given the following information: $l = 0.4$ m, $r = 0.8$ cm $= 8 \times 10^{-3}$ m, and $v = 0.5$ m/s. The coefficient of viscosity of blood, according to Table 14.1, is $\eta = 4 \times 10^{-3}$ N · s/m². Substituting these numbers into Eq. (14.8), we find that

$$P_1 - P_2 = \frac{4(4 \times 10^{-3} \text{ N} \cdot \text{s/m}^2)(0.4 \text{ m})}{(8 \times 10^{-3} \text{ m})^2} (0.5 \text{ m/s}) = \underline{50 \text{ N/m}^2}.$$

Since 133 N/m² = 1 mm Hg, the pressure drop along the aorta due to friction in units of mm Hg is

$$P_1 - P_2 = (50 \text{ N/m}^2)\left(\frac{1 \text{ mm Hg}}{133 \text{ N/m}^2}\right) = 0.4 \text{ mm Hg.}$$

This is a very small part of the 100-mm Hg drop in blood pressure from the exit of the heart, through the circulatory system, and back again to the entrance of the heart. Most of the blood-pressure drop in the circulatory system occurs across the capillaries and the small vessels, called arterioles, which feed the capillaries. These vessels have small radii, and according to Eq. (14.8), the smaller the value of r, the greater is the pressure drop. ∎

We are often interested in the pressure needed to maintain a certain fluid flow rate. The flow rate, defined as the volume of fluid that passes through a vessel per unit time, is related to the average speed $\bar{v}$ and cross-sectional area A of the vessel by the equation

$$Q = \frac{V}{t} = \bar{v}A.$$

For a cylindrical pipe, the fluid flows fastest at the center of the pipe and not at all at the edge of the pipe (the fluid is stuck on the microscopic ridges and trenches of this surface). Overall, the average speed $\bar{v}$ of the fluid is just half of its speed v at the center of the pipe ($\bar{v} = v/2$). For a pipe of cross-sectional area $A = \pi r^2$,

$$Q = \bar{v}A = \frac{v}{2}(\pi r^2). \tag{14.9}$$

Solving Eq. (14.8) for v, we find that

$$v = \left(\frac{r^2}{4\eta L}\right)(P_1 - P_2).$$

Substituting the above expression for v into Eq. (14.9) produces the equation,

$$Q = \left(\frac{\pi r^4}{8\eta l}\right)(P_1 - P_2). \tag{14.10}$$

This is called **Poiseuille's law,** named in honor of the French physician Jean Louis Marie Poiseuille (1799–1869). Poiseuille established this law empirically with remarkable accuracy.

Notice that the flow rate is proportional to the fourth power of the radius of the vessel carrying the fluid. If the radius of a vessel carrying fluid is reduced by a factor of 2, the flow rate will be reduced by a factor of 16—that is, 2^4.

EXAMPLE 14.9 A well has a 1.0-cm radius pipe that carries water from the ground to a reservoir. Calculate the factor by which the flow rate increases if a 1.5-cm radius pipe replaces the 1.0-cm pipe. The pressure across and the length of the pipe are unchanged.

SOLUTION The ratio of the flow rate through the 1.5-cm pipe and that through the 1.0-cm pipe is

$$\frac{Q(1.5\text{-cm pipe})}{Q(1.0\text{-cm pipe})} = \frac{\left[\dfrac{\pi(1.5 \text{ cm})^4}{8\eta l}\right](P_1 - P_2)}{\left[\dfrac{\pi(1.0 \text{ cm})^4}{8\eta l}\right](P_1 - P_2)} = \frac{1.5^4}{1.0^4} = 1.5^4 = \underline{5.1}.$$

The flow rate is increased by a factor of 5.1 with only a 1.5 increase in the radius of the pipe! ■

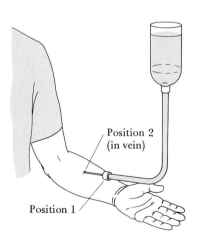

FIG. 14.14. The flow rate of blood into a vein depends on the radius of the needle.

EXAMPLE 14.10 Blood is to be transferred from the reservoir shown in Fig. 14.14 into a person's vein. The flow rate is regulated by the length and radius of the needle, since all other vessels carrying the blood have large radii and cause little resistance to the flow. The flow rate is to be $1.0 \times 10^{-6}\,\text{m}^3/\text{s}$ (about one liter each 15 minutes). The pressure at the entrance of the needle (position 1) is $9000\,\text{N/m}^2$ greater than that in the vein at position 2. The viscosity of blood is $4 \times 10^{-3}\,\text{N} \cdot \text{s/m}^2$. If the length of the needle is 0.03 m, calculate the radius of the needle needed to get the desired flow rate.

SOLUTION When Poiseuille's law is solved for the radius of the needle, we find that

$$r^4 = \frac{8\eta l Q}{\pi(P_1 - P_2)} = \frac{8(4 \times 10^{-3}\,\text{N} \cdot \text{s/m}^2)(0.03\,\text{m})(1.0 \times 10^{-6}\,\text{m}^3/\text{s})}{\pi(9000\,\text{N/m}^2)}$$

$$= 3.73 \times 10^{-14}\,\text{m}^4$$

or

$$r = 4.3 \times 10^{-4}\,\text{m} = \underline{0.43\,\text{mm}}. \qquad \blacksquare$$

14.6 Reynolds Number and Turbulence

Poiseuille's law applies only to laminar, or streamline, flow. As the speed of laminar flow increases, an abrupt transition to turbulent flow may occur. Once started, greater pressure is needed to maintain turbulent flow because the viscous friction force increases during turbulent flow.

The abrupt onset of turbulence is characterized by a number called the **Reynolds number** (Re):

$$\text{Re} = \frac{2\bar{v}r\rho}{\eta}, \qquad (14.11)$$

where $\bar{v}$ is the average speed of the fluid, ρ is its density, η is the fluid's coefficient of viscosity, and r is the radius of the vessel that carries the fluid. The Reynolds number is dimensionless, since all of its units cancel. It has been determined experimentally that if Re is less than 2000, laminar flow occurs; if Re is greater than 3000, turbulent flow occurs; and if Re is between 2000 and 3000, the flow is unstable and may be either laminar or turbulent.

EXAMPLE 14.11 Estimate the maximum rate of laminar flow of water in a garden hose whose radius is 1.0 cm. The water temperature is 20°C.

SOLUTION When the Reynolds number exceeds 2000, flow may become turbulent. Thus, the maximum average speed for laminar flow of water having density $\rho = 1000\,\text{kg/m}^3$ and viscosity $\eta = 1.0 \times 10^{-3}\,\text{N} \cdot \text{s/m}^2$ is, according to Eq. (14.11),

$$\bar{v} = \frac{\eta \cdot \text{Re}}{2r\rho} \simeq \frac{(1.0 \times 10^{-3}\,\text{N} \cdot \text{s/m}^2)(2000)}{2(1.0 \times 10^{-2}\,\text{m})(1000\,\text{kg/m}^3)} = 0.10\,\text{m/s}.$$

The flow rate is

$$Q = \bar{v}A = \bar{v}(\pi r^2) = (0.10\,\text{m/s})(\pi)(1.0 \times 10^{-2}\,\text{m})^2 = \underline{3.1 \times 10^{-5}\,\text{m}^3/\text{s}}. \qquad \blacksquare$$

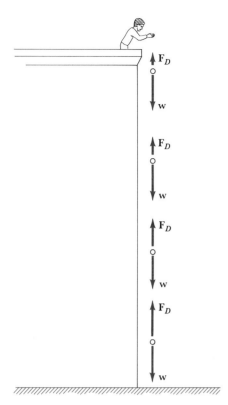

FIG. 14.15. As a ball falls faster, the magnitude of the drag force increases. Eventually, $\mathbf{F}_D$ and $\mathbf{w}$ are equal in magnitude and the ball falls at a constant terminal speed, v_T.

14.7 Drag Force

In the last two sections we analyzed the effect of viscosity on a fluid flowing through a tube. In this section we discuss the viscous friction force that opposes the motion of an object relative to a fluid. Examples include the resistive force of water on a swimmer and the resistive force of air on a car, bicycle, or skydiver. These viscous friction forces are often called **drag forces.**

A Ping-Pong ball dropped from a building is illustrated in Fig. 14.15. If the ball falls slowly, laminar flow occurs. Streamlines of air pass the ball as it falls. As the ball's speed increases, the flow of air past the ball becomes turbulent. The onset of turbulence is predicted by a different form of the Reynolds number:

$$\mathrm{Re} = \frac{vL\rho}{\eta}, \qquad (14.12)$$

where v is the object's speed, L is the object's length, ρ is the density of the fluid, and η is the fluid's viscosity. This number is used for an entirely different purpose than the Reynolds number introduced in the last section, and the two should not be confused. When Re, given by Eq. (14.12), is less than about one, laminar flow occurs past an object moving through a fluid, and when Re is greater than one, the flow past the object is turbulent. The other form of the Reynolds number is used to predict the type of flow of fluid through a vessel.

The drag force acting on an object moving relative to a fluid depends on whether the flow past the object is laminar or turbulent. If laminar, the drag force increases approximately in proportion to the object's speed relative to the fluid:

$$F_D = Dv,$$

where D is a constant. For the situation where a circular object of radius r falls at speed v through a liquid with viscosity η, the drag force for laminar flow is

$$F_D = 6\pi\eta r v. \qquad (14.13)$$

This equation is called **Stokes' law.**

If fluid flow past the moving object is turbulent, the drag force increases approximately in proportion to the square of the object's speed relative to the fluid:

$$F_D = kv^2.$$

For instance, for objects moving through air with turbulent flow,

$$F_D \cong \tfrac{1}{2}C_D\rho Av^2, \qquad (14.14)$$

where ρ is the density of air, C_D is a unitless number called the drag coefficient, and A is the cross-sectional area of the object as seen along its line of motion.

EXAMPLE 14.12 Estimate the air drag force on a 1130-kg compact car when moving at a speed of 27 m/s (60 mph). The cross-sectional area of the car is roughly 2.0 m^2 and the drag coefficient C_D is approximately 0.5 for a well-designed car. The density of air is 1.3 kg/m^3.

SOLUTION From Eq. (4.14) we find that

$$F_D = \tfrac{1}{2}(0.5)(1.3 \text{ kg/m}^3)(2.0 \text{ m}^2)(27 \text{ m/s})^2 = 470 \text{ kg m/s}^2 = \underline{470 \text{ N}}.$$

This is a force of about 100 lb. ∎

Terminal Speed

An object falling through a fluid eventually reaches a speed where the force pulling it in one direction (for example, the weight force pulling it down) is balanced by the opposing drag force. The net force on the object is zero; it no longer accelerates and continues to move at a constant speed called its **terminal speed** v_T. The terminal speed of a skydiver is approximately 54 m/s (120 mph), whereas that of a falling raindrop is about 12 m/s. A red blood cell falls in a beaker of water with a terminal speed of about 10^{-7} m/s. The next example shows how an object's terminal speed can be calculated.

EXAMPLE 14.13 Estimate the terminal speed of a Ping-Pong ball weighing 2.2×10^{-2} N whose radius is 1.9×10^{-2} m. The ball experiences a turbulent drag force with a drag coefficient of 0.6.

SOLUTION The ball reaches terminal speed when its weight is balanced by the upward drag force:

$$w - F_D = 0.$$

Substituting Eq. (14.14) for F_D and rearranging, we find that

$$v_T = \sqrt{\frac{2w}{C_D \rho A}} = \sqrt{\frac{2w}{C_D \rho \pi r^2}} = \sqrt{\frac{2(2.2 \times 10^{-2} \text{ N})}{(0.6)(1.3 \text{ kg/m}^3)(\pi)(1.9 \times 10^{-2} \text{ m})^2}}$$

$$= \underline{7.1 \text{ m/s}}.$$ ∎

Summary

1. **Fluid flow:** When the pressure is greater in one region of a fluid than in another, the fluid flows from the high-pressure to the low-pressure region. If the fluid flows smoothly in streamlines (each molecule of fluid following the one in front of it), then the flow is said to be **laminar.** If the fluid swirls and forms eddies, the flow is said to be **turbulent.**

2. **Flow rate:** The flow rate Q is defined as the volume of fluid V that flows past a cross section of the vessel carrying the fluid divided by the time t for the fluid to pass:

$$Q \equiv \frac{V}{t}. \tag{14.1}$$

When fluid flows through a pipe of cross-sectional area A at an average speed $\bar{v}$, then the flow rate is

$$Q = \bar{v}A. \tag{14.2}$$

If the fluid is incompressible, then the flow rate past one cross section in a pipe equals that past another cross section. Thus,

$$Q = v_1 A_1 = v_2 A_2, \tag{14.3}$$

where v_1 is the speed of the fluid past a cross section of area A_1, and v_2 is its speed past a cross section of area A_2. Equation (14.3) is called the **continuity equation.**

3. **Bernoulli's equation** results from applying the work-energy equation to a section of incompressible, frictionless fluid flowing along a pipe. We find that the pressure, speed, and elevation of the fluid at one position are related to that at a second position in the fluid by the equation

$$P_1 + \tfrac{1}{2}\rho v_1^2 + \rho g y_1 = P_2 + \tfrac{1}{2}\rho v_2^2 + \rho g y_2, \tag{14.5}$$

where ρ is the density of the fluid.

4. **Friction in fluids:** When fluids flow through vessels, the layer of fluid nearest the vessel does not move. But layers of

fluid away from the surface do move and slide across layers nearer the vessel. This sliding of one layer across another causes a friction force in fluids that opposes their motion. This friction is called **viscosity. A coefficient of viscosity** η characterizes the degree of viscosity in fluids.

Viscous forces oppose the motion of all fluids. We can often ignore these viscous forces when fluids move through vessels with large cross-sectional areas and when the flow is laminar. However, the flow rate can be greatly affected by viscous forces when the fluid flows through vessels of small cross-sectional areas. For such vessels the flow rate Q is given by **Poiseuille's equation**

$$Q = \left(\frac{\pi r^4}{8\eta l}\right)(P_1 - P_2), \tag{14.10}$$

where $P_1 - P_2$ is the pressure change across a length l of vessel whose radius is r.

5. **Drag force:** A viscous drag force opposes the motion of an object through a fluid. If turbulence is caused in the fluid as the object passes, the drag force is proportional to the object's speed squared. If the object does not create turbulence, the drag force is usually proportional to the first power of the object's speed. We can calculate the **Reynolds number** [Eq. (14.12)] to decide whether turbulence does or does not occur.

Questions

1. Explain why a wind blowing past the window of a building causes the window to be blown out rather than in.

2. The top of a convertible bulges when it is driving along a highway. Explain why this happens.

3. How does Bernoulli's principle help explain the fact that air is drawn up the chimney of a house?

4. Air blowing through a tube connected to an inverted funnel will support a Ping-Pong ball as shown in Fig. 14.16. Use Bernoulli's principle to explain this phenomenon.

FIG. 14.16

5. A person can open an envelope by cutting the end and blowing air past the cut end, thus causing the envelope to bulge open. Explain using Bernoulli's principle.

6. If the friction in moving fluids occurs in the fluid rather than at the boundary of the fluid, why doesn't blood get very hot? Explain in detail.

7. An air gun blowing air over a Ping-Pong ball will support it, as shown in Fig. 14.17. Construct a force diagram for the ball. Explain in terms of these forces how the ball can remain in equilibrium.

FIG. 14.17

8. A ball of mass m and another of the same material, only with mass $2m$, are dropped simultaneously from the Leaning Tower of Pisa. Will the balls hit the ground at the same time? If not, which will hit first? Explain. Do not ignore air resistance.

Problems

Unless stated otherwise, assume in these problems that atmospheric pressure is $1.01 \times 10^5 \, \text{N/m}^2$ and that the densities of water and air are $1000 \, \text{kg/m}^3$ and $1.3 \, \text{kg/m}^3$, respectively.

14.2 Flow Rate and Equation of Continuity

1. (a) Water flows through a hose at a rate of $100 \, \text{cm}^3/\text{s}$. If the radius of the hose is $1.0 \, \text{cm}$, at what speed does the water travel? (b) If a constriction reduces the radius to $0.30 \, \text{cm}$, what is the water speed through the constriction, assuming the flow rate is unchanged?

2. (a) A small river channel has a rectangular cross section of 5.0-m width and 1.2-m depth. If water flows at a speed of $0.80 \, \text{m/s}$, what is its flow rate? (b) If the width of the stream is reduced to $3.0 \, \text{m}$ and the depth increases to $1.5 \, \text{m}$ as the water passes an embankment, what is the speed of the water past the embankment?

3. Each second $0.070 \, \text{m}^3$ of water flows through a fire hose. If the water is to leave the nozzle at a speed of $25 \, \text{m/s}$, what should its diameter be?

4. A main water line for a housing project must deliver water at a maximum flow rate of $0.010 \, \text{m}^3/\text{s}$. If the speed of the water at

this flow rate is 0.30 m/s, calculate the diameter of the pipe carrying the water.

5. Water flows at a speed of 2.0 m/s through a pipe of radius 0.10 m. At a junction the water divides and travels through two pipes, each of radius 0.060 m. How fast does the water travel in these smaller pipes?

6. The flow rate of blood in the aorta is 80 cm^3/s. Beyond the aorta this blood travels through about 6×10^9 capillaries. If the radii of these capillaries is 8×10^{-4} cm, what is the speed of flow of blood through the capillaries?

14.3 and 14.4 Bernoulli's Equation and Applications

7. Water flows through a pipe of radius 0.10 m at a speed of 2.0 m/s. The gauge pressure in the pipe is 5.0×10^4 N/m^2. The pipe goes up a hill 3.0 m high before it levels again. If the pipe on the hill has a reduced radius of 0.080 m, (a) what is the speed of water in this pipe, and (b) what is the gauge pressure in the pipe?

8. Water flows through a fire hose up a staircase in a burning hotel. Water leaves the hose from a nozzle of radius 2.0 cm at a speed of 4.0 m/s. The nozzle is 15 m above the fire truck. If the hose at the fire truck has a radius of 3.0 cm, what is the pressure of the water at the truck? (*Hint:* The pressure on the water when it leaves the nozzle is atmospheric pressure.)

9. A garden hose of 0.80 cm radius is connected to one of 1.0 cm radius. The smaller hose is held on the roof of a house 4.0 m above the larger hose. Water leaves the smaller hose at a speed of 6.0 m/s and at atmospheric pressure. Calculate the pressure in the larger hose on the ground.

10. A U-shaped tube, open at both ends, contains water. If air blows across the top of one end at a speed of 10 m/s and not across the other, how much higher will the water be in the side below the moving air compared to that below the air at rest?

X11. A straw extends 3.0 cm out of a glass of water. How fast must air blow across the top of the straw in order to draw water to the top of the straw?

12. How fast must air blow across the bottom of the tube shown in Fig. 14.18 in order to draw water up the tube so that flow begins?

0.1 m

FIG. 14.18 v?

13. A (1.1×10^4)-kg airplane has wings of area 50 m^2. (a) Calculate the lift force needed to support the airplane. (b) Calculate the pressure difference between the bottom and top of the wing needed to support the airplane. (c) If air of density 0.9 kg/m^3 travels under the wing at a speed of 200 m/s, how fast must air travel over the wing to provide the desired lift?

14. A large, open barrel of water has a hole of 0.010 m diameter in the bottom a distance of 0.40 m down from the top of the water.

Calculate the speed with which water leaves the hole and the water flow rate.

15. Air of density 1.3 kg/m^3 blows past a 1.2 m $\times$ 2.2 m window of a skyscraper at a speed of 25 m/s. The air inside the building is at atmospheric pressure. (a) Calculate the difference in pressure between the inside of the window and the outside. (b) Calculate the net force of the air on the window (magnitude and direction).

16. A large, closed barrel of water has a 0.01-m diameter opening at the bottom through which water leaves at a speed of 2.0 m/s. Calculate the pressure at the top of the water level in the barrel, 0.50 m above the hole.

17. Water sits at rest behind an irrigation dam. The water is 1.2 m above the bottom of a gate that, when lifted, allows water to flow under the gate. To what height h from the bottom of the dam should the gate be lifted in order to allow a flow rate of 1.0×10^{-2} m^3/s? The gate is 0.5 m wide.

18. The gauge pressure outside a vessel carrying blood is 3000 N/m^2. Inside the vessel the gauge pressure is 5000 N/m^2 and blood flows at a speed of 0.50 m/s. The normal vessel has a radius of 1.0 cm. A plaque reduces the radius of the inside of a small section of the vessel. To what radius must the vessel be reduced at this constriction so that the outside pressure is greater than that inside, thus causing the vessel to close?

14.5 Viscous Fluid Flow

19. A horizontal water line of 5.0 cm radius is 500 m long. Water at 20°C flows at a rate of 1.0×10^{-2} m^3/s. (a) Calculate the pressure difference from the beginning to the end of the line. (b) If the pressure difference is kept constant and the flow rate is to be doubled, what must be the radius of the pipe?

20. There is a 700-lb/ft^2 pressure drop along a horizontal, 100-ft water line of radius 0.5 inch. Calculate the flow rate of water along the line. The viscosity of water in the English system of units is 2.3×10^{-5} lb $\cdot$ s/ft^2.

21. A 5.0-cm diameter firehose 60 m long carries 20°C water at a speed of 12 m/s. Calculate the drop in pressure along the hose due to viscous friction.

22. Your heart pumps blood at a flow rate of about 80 cm^3/s. This blood flows through about 9×10^9 capillaries, each of radius 4×10^{-4} cm, and 0.1 cm long. Calculate the pressure drop across a capillary, assuming a blood viscosity of 4×10^{-3} N/s $\cdot$ m^2.

23. Repeat Problem 22 for the arterioles. There are about 5×10^8 arterioles, each of radius 1×10^{-3} cm and of average length 1.0 cm.

24. Water flows in a solar collector through a copper tube of 0.60 cm radius that is 20 m long. The average temperature of the water is 60°C and the flow rate is 200 cm^3/s. Calculate the pressure drop along the tube, assuming the water does not change elevation.

25. The average radius of an arteriole, a small vessel carrying blood to the capillaries, is 1×10^{-2} mm. (a) If its radius is reduced by 10 percent and there is no change in pressure across the vessel, by what factor is the flow rate changed? (b) If the radius is reduced by 10 percent and the flow rate remains constant, by what factor must the pressure change?

26. A piston pushes 20°C water through a horizontal tube of 0.20 cm radius and 3.0 m long. One end of the tube is open and at atmospheric pressure. (a) Calculate the force needed to push the piston so that the flow rate is 100 cm³/s. (b) Repeat the problem if the piston pushes SAE 10 oil instead of water.

14.6 Reynolds Number and Turbulence

27. Blood flows at an average speed of 0.40 m/s in an artery whose radius is 0.50 cm. Based on a calculation of the Reynolds number, determine whether the flow is laminar or turbulent.

28. The radius of a vessel carrying blood at a flow rate of 20 cm³/s is slowly reduced at a constriction until the flow, initially laminar, becomes turbulent. At the constriction, the Reynolds number now equals 2500. Calculate the radius of the vessel when turbulence starts. (*Note:* The blood speed changes as the radius changes.)

29. Air flows from a furnace through a circular duct having a diameter of 0.30 m. Do you think the flow is laminar or turbulent? Explain.

30. Water at a temperature of 40°C flows in a copper tube whose radius is 0.30 cm. Calculate the maximum flow rate before turbulence starts.

14.7 Drag Force

31. A 2300-kg "luxury" car has a 0.6 drag coefficient and an effective frontal area of 2.8 m². Calculate the air drag force on the car at 55 miles/hr (24 m/s) and at 70 miles/hr (31 m/s). The density of air is 1.3 kg/m³.

32. *Estimate* the drag force opposing your motion when you ride a bicycle at 5 m/s.

33. *Estimate* the drag force on a baseball thrown by a major league pitcher.

34. Calculate the drag force on a red blood cell having a radius of 1.0×10^{-5} m and moving through 20°C water at a speed of 1.0×10^{-5} m/s. (Assume that laminar flow occurs.)

35. A protein of radius 3×10^{-9} m falls through a tube of water whose viscosity η is 1.0×10^{-3} N · s/m². A constant downward force of 3×10^{-22} N acts on the protein. (a) Use Stoke's law and the other information provided to estimate the terminal speed of the protein. (b) How many hours would be required for the protein to fall 0.1 m?

36. Colloidal clay particles each have a constant downward force of 3×10^{-17} N acting on them. The particles settle 10 cm in water in 10^2 minutes. Calculate the radii of these particles. The viscosity of water is 1.0×10^{-3} N · s/m².

37. A sphere falls through a fluid. A constant downward force of 0.50 N acts on the sphere. The drag force opposing the sphere's motion is given by

$$F_D = 2v,$$

where F_D is in newtons if v is in meters per second. What is the terminal speed of the sphere?

38. A balloon floats down through the air, pulled by a constant downward force of 0.050 N. A drag force opposing the balloon's motion is given by the equation

$$F_D = 0.03v^2,$$

where F_D is in newtons if v is in meters per second. Calculate the terminal speed of the balloon.

39. A parachutist, whose mass including that of the parachute is 80 kg, falls toward the earth with parachute open at a constant terminal speed of 8.5 m/s. If the drag force on the parachutist is given by Eq. (14.14) and the drag coefficient $C_D = 0.5$, what is the area of the parachute?

40. A balloon, whose radius when inflated is 0.2 m, falls slowly through air with a terminal speed of 0.4 m/s. Its drag coefficient C_D is approximately 0.5. (a) What is the drag force on the balloon? (b) What is the balloon's weight (remember that it is falling at terminal speed)? (c) What is the balloon's mass?

General Problems

G1. We can include the effect of friction in Bernoulli's equation by adding a term for the thermal energy generated because of the viscous retarding force acting on the fluid. Show that the term to be added to Eq. (14.4) for flow in a vessel of uniform cross-sectional area A is

$$\frac{\Delta E_{th}}{\Delta V} = \frac{4\pi \eta l v}{A},$$

where v is the speed of the fluid of viscosity η along the center of a pipe whose length is l.

G2. A glucose solution of viscosity 2.2×10^{-3} N · s/m² and of density 1030 kg/m³ flows from an elevated bag into a vein. The needle into the vein has a radius of 0.30 mm and is 2.0 cm long. All other tubes leading to the needle have much larger radii, and viscous forces in them can be ignored. The pressure in the vein is 1000 N/m² above atmospheric pressure. (a) Calculate the gauge pressure needed at the entrance of the needle in order to maintain a flow rate of 0.10 cm³/s. (b) To what elevation should the bag

containing the glucose be raised in order to maintain this pressure at the needle?

G3. (a) Show that the work done per unit time by viscous friction in a fluid with a flow rate Q across which there is a pressure drop P is

$$\frac{W}{t} = PQ = Q^2 R = \frac{P^2}{R},$$

where $R = 8\eta l / \pi r^4$ is called the **flow resistance** of the fluid moving through a vessel of radius r. (b) By what percentage must the work per unit time increase if the radius of a vessel decreases by 10 percent and all other quantities including the flow rate remain constant (the pressure does not remain constant)?

G4. A round grain of sand with a radius of 0.15 mm and density of 2300 kg/m³ is placed in a 20°C lake. Calculate the terminal velocity of the sand as it sinks into the lake. Do not forget to include the buoyant force acting on the sand.

Elastic Properties of Solids

15

In the last three chapters we have been studying fluids—gases and liquids. Now we come to solids, a form of matter in which bonds between neighboring atoms and molecules are strong. Whereas both liquids and gases flow and take on the shape of their container, solids tend to have definite shapes that they retain unless acted on by external forces. When opposing external forces are applied, the atomic and molecular bonds may be stretched, compressed, bent, or changed in other ways. When the forces are removed, however, the bonds between the atoms often return to their original orientation and length, and the solid returns to its original shape. When this happens, we say that the solid is **elastic**—it returns to its original shape after the external force deforming it has been removed.

Not all solids have this elastic property. Clay, when deformed by a force, remains deformed. This permanent alteration of the shape of clay is actually another form of flow, such as we studied in the last chapter. Here we will be concerned primarily with elastic deformations of solids, such as compressing a bone, stretching a blood vessel, and shearing a continental plate as it moves on the earth's surface. We will develop and use equations that relate the forces causing these deformations to the actual magnitude of the deformations. Let us start by considering a simple and familiar example—stretching a spring.

15.1 Elastic Deformations

Stretching and compressing a spring illustrates many of the more complex types of elastic deformation we will encounter later on. When no force acts on the spring and mass shown in Fig. 15.1, the end of the spring rests at its **equilibrium position.** As the force pulling the mass increases, the spring stretches and moves farther from its equilibrium position. We find by experiment that the spring's displacement x from equilibrium is proportional to the applied force F:

$$F \propto x.$$

For example, if a force of magnitude F_0 causes a displacement x_0, then a force $2F_0$ will cause a displacement $2x_0$. If the force is removed from the spring, it returns to its equilibrium position ($x = 0$).

The proportionality between the magnitude of the applied force and the

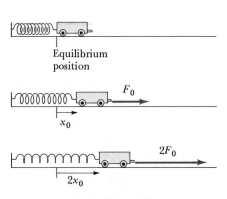

FIG. 15.1. As the force pulling on a spring increases, its displacement from equilibrium increases.

Tension   Compression

(a)

Shear

(b)

Change in volume

(c)

FIG. 15.2. The three basic types of deformation of solids. (a) A tensile or compressive force causes a change in an object's length. (b) A shear force causes a change in an object's shape. (c) A pressure change causes a change in an object's volume.

displacement from equilibrium can be written as an equation, called **Hooke's law:**

$$F = kx. \qquad (15.1)$$

The proportionality constant k, called the **force constant,** depends on how stiff the spring is: the value of k is large for a stiff spring and small for an easily stretched spring.

All types of elastic deformations can be described in an analogous manner. The description has three elements: (1) a cause of the deformation such as the force pulling the spring in Fig. 15.1; (2) a measure of the deformation such as the x coordinate used to indicate a spring's displacement from equilibrium; and (3) a proportionality constant that relates the deformation and its cause in an equation such as Eq. (15.1). For deformations of elastic solids, the cause of the deformation is called a **stress,** the resulting deformation is called a **strain,** and the proportionality constant relating stress and strain is called a **modulus of elasticity.** The definition of each of these three quantities depends on the particular type of deformation being described.

15.2   Types of Solid Deformations

Although a solid can be deformed in many ways, such as compression, stretching, twisting, bending, folding, and volume changes, there are only three basic types of deformations. All others are combinations of these three types. As illustrated in Fig. 15.2, the three basic deformations and the forces causing them are (a) a **change in length** caused by a tension or compression, (b) an **angular change** in the orientation of the surfaces of a solid caused by a shearing force, and (c) a **volume change** caused by a uniform change in pressure on all of the surfaces of a solid.

Other types of deformation are either a special example of one of these deformations or a combination of two or three of them. For example, bending is a combination of a stretch and a compression. The fibers in the top half of the board shown in Fig. 15.3 are being stretched; those in the bottom half are being compressed. The fibers along the center of the board (the dashed line in Fig. 15.3) are unchanged in length.

Stress and strain are two quantities used to describe the nature of the applied force and the resulting deformation, respectively. Our next task is to define these terms.

15.3   Stress and Strain

When a force acts on a solid to produce a change in its dimensions or shape, the cause of the deformation is called a stress.

Stress is defined as the force acting on a solid divided by the area over which this force acts:

$$\text{stress} = \frac{\text{force}}{\text{area}} = \frac{F}{A} \qquad (15.2)$$

FIG. 15.3. Bending is a combination of tension in the upper portion of the board and compression in the bottom portion.

The units of stress are N/m^2, $lb/in.^2$, and so forth.

When defined as force per unit area, stress is proportional to the deformation it causes. If defined simply as force, stress would not be proportional to the deformation. Consider the groups of identical rubber bands pictured in Fig. 15.4. Group A consists of one rubber band elongated a distance ΔL by a force F. Group B consists of two rubber bands that are also elongated the distance ΔL, but a force $2F$ is needed to produce this elongation, since there are two rubber bands instead of one. A force $4F$ is needed to elongate the four rubber bands in group C by a distance ΔL.

Although the elongation is the same for each group of rubber bands, the force causing the elongation is different. Thus, the force is not directly proportional to the deformation. If we define the stress as the force per cross-sectional area of the rubber bands, however, then the stress is the same for each group and is proportional to the deformation. For example, if the cross-sectional area of one rubber band is A, the ratio of force and area for each group is the same:

$$\frac{F}{A} = \frac{2F}{2A} = \frac{4F}{4A}.$$

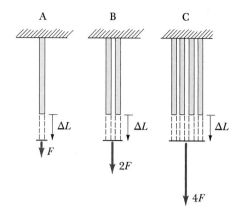

FIG. 15.4. Three groups of rubber bands are each displaced the same distance. If stress is defined as force/area, then in each case the stress is the same as is the elongation caused by the stress.

Each group of rubber bands was stretched the same amount because the force per area (the stress) acting on each group was the same.

The quantity **strain** is a measure of the deformation caused by a stress. The specific definition of strain depends on which of the three types of deformations shown in Fig. 15.2 are being described. In most cases we can think of the strain as being the actual change in an object divided by the original value of the quantity that has changed:

$$\text{strain} = \frac{\text{change in dimension}}{\text{original dimension}} \tag{15.3}$$

For instance, if a bone is 100 mm long when unstressed and is 99 mm long when under stress, the strain is the change in length divided by the original length, or

$$\text{strain} = \frac{1 \text{ mm}}{100 \text{ mm}} = 0.01.$$

If a uniform external pressure causes the volume of a rock to change from 10.00 cm^3 to 9.98 cm^3, then the volume strain is the change in volume divided by the original volume, or

$$\text{strain} = \frac{0.02 \text{ cm}^3}{10.00 \text{ cm}^3} = 0.002.$$

Shearing strain is a little more complicated and will be defined later in the chapter.

For elastic materials the strain increases in proportion to the stress placed on the material. The proportionality constant relating the stress and strain is called a **modulus of elasticity:**

$$\text{stress} = (\text{modulus of elasticity})(\text{strain}). \tag{15.4}$$

The modulus of elasticity plays the same role as the force constant k in the equation relating the force F pulling a spring and its displacement from

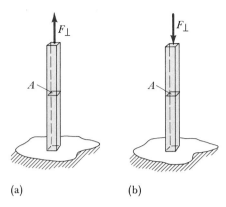

FIG. 15.5. (a) A tensile stress and (b) a compressive stress.

equilibrium x:

$$F = kx.$$

In the next three sections the stress, strain, and modulus of elasticity are defined carefully for each of the three basic types of solid deformations.

15.4 Length Deformation

Forces that stretch a solid produce what is called **tensile stress,** since the force causes tension in the stretched object. Its atoms and molecules are pulled farther apart. Pulling a rope and stretching a tendon are examples of tensile stresses (Fig. 15.5a).

A **compressive stress** is caused by a force pushing against one side of an object while it is supported by an equal force on the other side (Fig. 15.5b). Compressing a bone and pounding a nail are examples of compressive stresses.

To define stress caused by tension and compression forces, we choose a cross-sectional area through some plane of the object (see Fig. 15.5). The stress across that plane is the component of the applied force that is perpendicular or normal to that plane $F_\perp$ divided by the area A of the plane:

$$\text{tensile or compressive stress} = \frac{F_\perp}{A}.$$

The strain caused by a tensile or compressive stress is defined as the change in length ΔL of the object divided by its original length L:

$$\text{tensile or compressive strain} = \frac{\Delta L}{L}.$$

We have included the object's original length in our definition because strain, when defined this way, depends only on molecular properties of the material being stretched or compressed and not on its dimensions. Consider the rubber bands shown in Fig. 15.6. If an equal force is applied to each, the rubber band of length $2L$ will be displaced twice as far as the one of length L because twice as many molecules are in the longer rubber band, each being stretched in the same direction with the same tension. We see, however, that the strain produced is the same for each rubber band:

$$\text{strain} = \frac{\Delta L}{L} = \frac{2\,\Delta L}{2L}$$

Strain is independent of the dimensions of the object and depends only on its atomic structure.

Elastic materials are ones for which the stress and strain are proportional to each other. If the stress is doubled, the strain is doubled. The ratio of the stress and strain of an object is called the *elastic modulus* of the object. It is a measure of how difficult it is to deform the material of which the object is made. For the case of tensions or compressions, the elastic modulus is called **Young's modulus** and is defined as

$$\text{Young's modulus} = Y = \frac{\text{tensile or compressive stress}}{\text{tensile or compressive strain}} = \frac{F_\perp/A}{\Delta L/L}. \quad \textbf{(15.5)}$$

FIG. 15.6. The distance an elastic object is stretched by a force is proportional to the object's length.

TABLE 15.1 Young's Modulus and Ultimate Strengths of Various Materials

Material	Young's modulus (N/m^2)		Ultimate strength (N/m^2)	
	Tension	Compression	Tension	Compression
Aluminum	7.1×10^{10}	7×10^{10}	2.0×10^8	3.5×10^8
Steel	20×10^{10}	20×10^{10}	5.2×10^8	
Brick	2×10^{10}	2×10^{10}	0.4×10^8	
Glass	7×10^{10}	7×10^{10}	0.7×10^8	3.5×10^8
Bone	1.6×10^{10}	0.9×10^{10}	1.2×10^8	1.7×10^8
Oak	1.4×10^{10}	10^{10}		10^8
Pine	0.9×10^{10}			
Silk	6×10^9			
Spider's thread	3×10^9			
Catgut	3×10^9			
Hair	10^{10}		10^8	
Tendon	10^9			
Rubber	10^6			
Tooth (human crown)				1.5×10^8

Some representative values of Young's modulus are listed in Table 15.1. Notice that for some materials, Y is different for tension than for compression. This is because the attractive force between atoms when stretched does not equal the repulsive force when compressed. We see that Young's modulus for steel is approximately 10^5 times greater than that for rubber; that is, steel is 10^5 times harder to stretch than rubber. Bone is about 10^4 times more difficult to stretch than rubber, and a tendon about 10^3 times more difficult.

EXAMPLE 15.1 The compressive force on the bone in the lower leg of a person standing on the balls of his feet with heels slightly elevated is about 1800 N (see Fig. 15.7). Calculate the change in the length of the bone caused by this compressive force. The bone is a hollow cylinder with an inner radius of 0.60 cm, an outer radius of 1.4 cm, and a length of 0.42 m.

SOLUTION The change in length can be determined by rearranging Eq. (15.5):

$$\Delta L = \frac{F_\perp}{A} \frac{L}{Y}.$$

The compression force and length of the bone are given in the problem statement. Young's modulus for bone that is being compressed is given in Table 15.1: $Y = 0.9 \times 10^{10} \, N/m^2$. The cross-sectional area of the bone is the area of a circle of radius 1.4 cm minus the area of the hollow core—a circle of radius 0.6 cm. Thus,

$$A = \pi r_{outer}^2 - \pi r_{inner}^2 = \pi(0.014 \, \text{m})^2 - \pi(0.006 \, \text{m})^2 = 5.0 \times 10^{-4} \, \text{m}^2.$$

We find, then, that

$$\Delta L = \frac{(1800 \, \text{N})(0.42 \, \text{m})}{(5.0 \times 10^{-4} \, \text{m}^2)(0.9 \times 10^{10} \, \text{N/m}^2)} = 1.7 \times 10^{-4} \, \text{m} = \underline{0.17 \, \text{mm}}. \quad \blacksquare$$

FIG. 15.7. The force **F** on the tibia when a person stands as shown is about 1800 N.

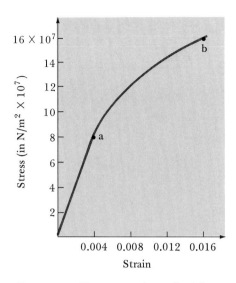

FIG. 15.8. The stress-strain graph of the humerus. The values of stress plotted on the vertical axis should be multiplied by 10^7.

It is common when testing materials to make a stress-versus-strain graph of the material. A compressive stress-versus-strain graph for the bone in the upper arm (the humerus) is shown in Fig. 15.8. The slope of the straight-line portion of this graph from the origin to point a (the elastic region) is Young's modulus for bone. Beyond point a, the material may still try to return to its original shape, but nonelastic structural changes have occurred. At point b, the material breaks or fractures. The maximum stress that the material can experience before fracture occurs is called the **ultimate tensile** or **compressive strength.** Some representative values of ultimate strength of different materials appear in Table 15.1.

15.5 Shear Deformation

A second type of solid deformation is caused by a **shear force.** If you place your hand on the top cover of a book and gently push it to the side, your hand generates a shear stress (Fig. 15.9). The resulting distortion of the book as the pages slide across each other is a shear strain. The **shear stress** is defined as

$$\text{shear stress} = \frac{F_{\|}}{A},$$

where $F_{\|}$ is the force exerted parallel to a given cross-sectional area. In this example, the area is that of the book's top surface.

This choice of area may seem strange, since the force of your hand is not spread over the whole top cover of the book. The cover, however, is held together as a unit by the forces between its atoms, and the amount of displacement caused by the force depends on this area. We may think of adjacent pages in the book as being bonded loosely to each other by frictional and molecular forces. The strength of the bonding between adjacent pages increases as the area increases. It is harder to make adjacent layers of a solid "slide" across each other when the layers have large cross-sectional areas. Our definition of stress, then, accounts not only for the force but also for this effect of area on the shearing of the object exposed to the force.

The result of the stress is a **shear strain,** defined as the distance Δs that the top surface is displaced in the direction of the force divided by the thickness L of the book or whatever is undergoing a **shear strain** (see Fig. 15.9c):

$$\text{shear strain} = \frac{\Delta s}{L}.$$

Note carefully the symbols and corresponding geometrical dimensions used in this definition compared to those used to define tensile and compressive strain.

Shear strain is independent of the dimensions of the object undergoing the strain. Imagine two books equal in every way, except that one is twice as thick as the other. If an equal shear stress is applied to each book (Fig. 15.10), the thicker book will be displaced twice as far as the thinner book. But since shear strain is defined as $\Delta s/L$, we find that the strain is the same for both books ($\Delta s/L = 2\,\Delta s/2L$). Shear strain, then, is an indicator of the displacement per unit thickness.

If the object being sheared is elastic, there is a proportionality between the shear strain and the shear stress. The ratio of the shear stress and the shear strain in the elastic region is called the **shear modulus** and is given the symbol S:

(a)

(b)

(c)

FIG. 15.9. (a) A book having an area A. (b) A force parallel to the cover causes a stress $F_{\|}/A$. (c) The strain caused by this stress is the ratio of Δs and L. Note carefully the meaning of the symbols!

$$\text{shear modulus} = S = \frac{\text{shear stress}}{\text{shear strain}} = \frac{F_{\|}/A}{\Delta s/L}. \qquad \textbf{(15.6)}$$

The shear modulus is roughly one-third the value of Young's modulus for similar materials. Some representative values of shear modulus are listed in Table 15.2.

EXAMPLE 15.2 According to the geological theory of plate tectonics, the surface of the earth consists of about twenty rigid plates that move slowly past each other on the mantle of the earth. One way in which earthquakes can occur is that the edges of the plates, originally stuck together, slip, and a cleavage in the earth's surface occurs. Let us estimate the size of the shear force that results in an earthquake.

SOLUTION Two plates, as seen from above, are pictured in Fig. 15.11a. They move slowly past each other in opposite directions. Friction forces prevent the sides of each of the plates that are touching from moving. As each plate moves forward (Fig. 15.11b), a shearing strain is caused by the force of each plate holding the other's edge in place.

Let us estimate roughly the magnitude of the shear force of one plate on the other. We use the following numbers, which represent, at best, the order of magnitude of quantities appearing in Eq. (15.6).

Area: A fault may be several hundred kilometers long (we will use a distance of 100 km) and extends perhaps 10 km into the ground. The area of contact between the two plates is roughly

$$(1 \times 10^5 \, \text{m})(1 \times 10^4 \, \text{m}) = 10^9 \, \text{m}^2.$$

Length: The plate is sheared for perhaps 10 km on each side of the edge between the two plates: $L \simeq 10^4 \, \text{m}$.

Displacement Δs: The plates move at a speed of about 4 cm/year. Since major earthquakes occur roughly every 50 to 100 years, we estimate that the unimpeded part of each plate away from the fault has moved a distance of about (75 years)(0.04 m/year) $\simeq 3$ m.

Shear modulus: A representative value of the shear modulus for rock is $3 \times 10^{10} \, \text{N/m}^2$.

(a)

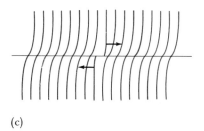

(b)

FIG. 15.10. The shear strain is the same for both books, since $\Delta s/L = 2 \Delta s/2L$.

TABLE 15.2 The Shear Modulus of Various Materials[a]

Material	Shear modulus (N/m²)
Aluminum	3×10^{10}
Steel	8×10^{10}
Glass	$(0.8\text{–}2.3) \times 10^{10}$
Bone	$\sim 10^{10}$
Rubber (soft, vulcanized)	$(0.8\text{–}1.6) \times 10^6$
Quartz	$(0.8\text{–}3) \times 10^{10}$
Wood	$(0.8\text{–}1) \times 10^{10}$

[a] The value of the shear modulus can vary somewhat between different samples of the same material.

(a)

(b)

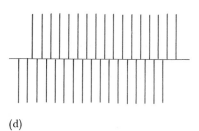

(c)

(d)

FIG. 15.11. (a) Two plates on the earth's crust moving in opposite directions are held together at a surface of contact (the center line). (b) Strain develops as the plates away from this contact move in opposite directions. When the strain becomes too great, the plates slip [(c) and (d)], causing a cleavage in the earth at the interface between the two plates.

Solving Eq. (15.6) for shear force $F_\parallel$, we find that

$$F_\parallel = \frac{AS\,\Delta s}{L} = \frac{(10^9\,\text{m}^2)(3 \times 10^{10}\,\text{N/m}^2)(3\,\text{m})}{10^4\,\text{m}} = \underline{0.9 \times 10^{16}\,\text{N}.}^*$$

When this force becomes too great, the rocks will slip at their boundary, causing an earthquake. After the slip, the plates start shearing again, as depicted in Fig. 15.11b. Another earthquake can be expected at some future time when the force again becomes too great to restrain the rocks.

The strain that occurs in a plate stores a great amount of elastic potential energy—like the energy of a stretched or compressed spring. This energy is released when the plates slip. A large earthquake can release about 10,000 times more energy than that released by the first atomic bomb! Such great releases of energy are avoided if the faults slip more frequently. Much less energy is released per earthquake in a larger number of small earthquakes. ∎

15.6 Volume Deformation

A volume deformation is caused by a uniform change in the pressure exerted perpendicular to all surfaces of a solid. The **volume stress** is defined as

$$\text{volume stress} = \frac{\Delta F}{A} = \Delta P,$$

where ΔF is the magnitude of the increased or decreased force acting perpendicular to a portion of the surface of area A (Fig. 15.12), and ΔP is the corresponding pressure change.

The **volume strain** is defined as the change in volume ΔV of the solid divided by the original volume:

$$\text{volume strain} = \frac{\Delta V}{V}.$$

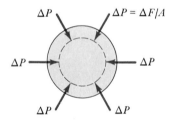

If the solid is elastic and the stress occurs in the elastic region, the solid will return to its original volume when the stress is removed. The negative ratio of the stress and strain in this elastic region is a constant called the **bulk modulus** B:

$$\text{bulk modulus} = B = -\frac{\text{volume stress}}{\text{volume strain}} = -\frac{\Delta P}{\Delta V/V}. \qquad (15.7)$$

The negative sign is included in the definition because an increase in the pressure acting on the solid causes a decrease in volume; that is, ΔV is negative. With the use of this negative sign convention, the bulk modulus is then a positive number.

The inverse of the bulk modulus is called the **compressibility**:

$$\text{compressibility} = \frac{1}{B} = -\frac{1}{V}\frac{\Delta V}{\Delta P} = -\frac{1}{\Delta P}\frac{\Delta V}{V}. \qquad (15.8)$$

The compressibility is the fractional change in volume $\Delta V/V$ caused by a pressure change, ΔP. The easier it is to compress a material, the greater is its compressibility. Representative values of the bulk modulus and compressibility of various solids and liquids are listed in Table 15.3.

*This is about equal to the weight of the top 3 m of land in the whole state of New York.

FIG. 15.12. The change in volume of a solid is caused by a uniform change in pressure ΔP (equal to the change in force ΔF per area A).

TABLE 15.3 The Bulk Modulus and Compressibility of Various Materials

Material	Bulk modulus (N/m²)	Compressibility (m²/N)
Aluminum	7.7×10^{10}	0.13×10^{-10}
Copper	14×10^{10}	0.07×10^{-10}
Steel	16×10^{10}	0.06×10^{-10}
Marble	7×10^{10}	0.14×10^{-10}
Granite	4.7×10^{10}	0.21×10^{-10}
Mercury	27×10^9	0.37×10^{-10}
Water	2.0×10^9	4.9×10^{-10}
Ethyl alcohol	0.9×10^9	11×10^{-10}

EXAMPLE 15.3 Calculate the density of water near the bottom of the ocean where the pressure is about 500 atm.

SOLUTION The density of water at the ocean's surface, where the pressure is 1 atm, is about 1025 kg/m³. Thus, 1 m³ of water at the surface has a mass of 1025 kg. Let us calculate the change in volume ΔV of this water when at a pressure of 500 atm. According to Eq. (15.7),

$$\Delta V = -\frac{(\Delta P)V}{B}.$$

The pressure change is (500 atm − 1 atm) = 499 atm, which is equivalent to 499×10^5 N/m². Thus the volume change is

$$\Delta V = -\frac{(499 \times 10^5\,\text{N/m}^2)(1\,\text{m}^3)}{2.0 \times 10^9\,\text{N/m}^2} = -0.025\,\text{m}^3.$$

The density of the water is then

$$\rho = \frac{m}{V + \Delta V} = \frac{1025\,\text{kg}}{1000\,\text{m}^3 - 0.025\,\text{m}^3} = \frac{1025\,\text{kg}}{0.975\,\text{m}^3} = \underline{1051\,\text{kg/m}^3}.$$

The water is compressed only 2 to 3 percent at the ocean bottom where the pressure is several hundred atmospheres. Liquids and solids are not easily compressed. ∎

15.7 Elastic Membranes

One important example of elastic tensile stress and strain is the stretching that occurs in the vessels of the human circulatory system. The volume of the aorta and large arteries increases as blood is pumped into these vessels once each heartbeat. The tension in the walls of the vessels also increases as they are stretched. Let us consider the stress-versus-strain curve for one of these blood vessels.

A vessel of length l is pictured in Fig. 15.13a. In (b), it has been cut lengthwise along the side and hung from a support. If we hang weights from the bottom, the vessel stretches much as a sheet of rubber would stretch. As more weight is added, the vessel stretches farther.

The stress on the vessel is the force F hanging from its end, divided by its cross-sectional area A. The strain caused by this stress is the change in length ΔL of the vessel, divided by its original length L. The solid curve in Fig. 15.14 is a typical stress-versus-strain curve for a human artery. The varying slope of this curve equals the varying Young's modulus for the vessel.

Two different types of elastic fibers in human arteries are responsible for the elasticity of these vessels. One fiber, elastin, is abundant in arteries of the circulatory system. The elastin is easily stretched—about six times more easily than rubber. The stress-versus-strain curve of an elastin fiber is shown by the dashed line in Fig. 15.14.

Collagen is a second type of fiber found in arteries. Because it is much stiffer than elastin, a small number of collagen fibers can give an artery a high degree of resistance to distension. Usually, the collagen in the walls of an artery is somewhat slack. It does not start stretching until the vessel has already been stretched somewhat beyond its normal size. A stress-versus-strain curve for collagen is shown by the dotted line in Fig. 15.14.

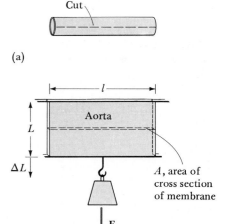

FIG. 15.13. (a) A cylindrical vessel is cut and (b) hung from a support. As weights are added, the membrane stretches.

FIG. 15.14. The stress-versus-strain curve of a human aorta and its individual elastic constituents.

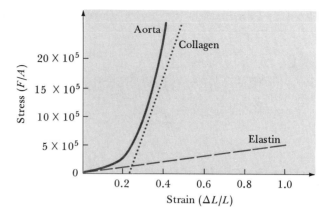

An actual artery is a combination of elastin and collagen. Its stress-versus-strain curve is a composite of the two. The initial portion of the curve is quite flat, indicating an easily extended membrane. As the membrane is stretched further, more and more collagen fibers lose their slack and contribute to the total membrane tension. The final steep slope is representative of a membrane whose resistance to extension is limited primarily by collagen fibers.

As a person ages, the blood vessels acquire a greater amount of collagen. Also, these collagen fibers lose their slack. The stress-versus-strain curves of blood vessels of different age groups are shown in Fig. 15.15, indicating that the vessels become somewhat stiffer with age.

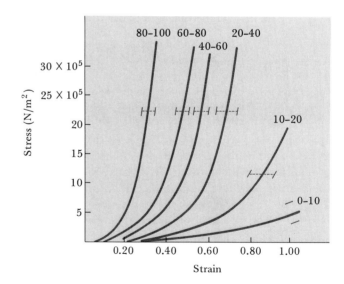

FIG. 15.15. The stress graphs of the human aorta for different age groups are shown. As a person ages, the aorta contains more collagen. The collagen also loses its slack. Both effects cause a greater slope in the curve (a greater Young's modulus). (Reproduced by permission of the National Research Council of Canada from the *Canadian Journal of Biochemistry and Physiology* **37:**557–569 [1959])

Summary

1. An **elastic deformation** is one in which an object returns to its original shape when the force causing the deformation is removed. The force causing the deformation is often proportional to the deformation itself. For example, for a spring, the stretching force F is proportional to the spring's displacement x from equilibrium ($F = kx$). The proportionality constant k, called the **spring's force constant,** depends on the stiffness of the spring.

The forces causing deformations of solids are characterized in terms of a quantity called stress. **Stress** is defined as the ratio of the force acting on the solid and the area over which the force is exerted:

$$\text{stress} = \frac{\text{force}}{\text{area}}. \tag{15.2}$$

The deformation of a solid caused by a stress is characterized in terms of a quantity called strain. **Strain** is defined as the change in some dimension of a solid (its length, orientation, or volume) divided by an original dimension of the object:

$$\text{strain} = \frac{\text{change in dimension}}{\text{original dimension}}. \tag{15.3}$$

For an elastic solid, the stress is proportional to the strain. The proportionality constant is called a **modulus of elasticity:**

$$\text{stress} = (\text{modulus of elasticity})(\text{strain}) \tag{15.4}$$

All possible elastic deformations of solids can be analyzed in terms of a combination of three basic types (length, shear, and volume deformations).

2. **Length deformation**—caused by compression or tensile forces:

$$\text{stress} = \frac{F_\perp}{A},$$

$$\text{strain} = \frac{\Delta L}{L},$$

$$\text{Young's modulus } Y = \frac{\text{stress}}{\text{strain}} = \frac{F_\perp/A}{\Delta L/L}. \tag{15.5}$$

FIG. 15.16

3. **Shear deformation:**

$$\text{stress} = \frac{F_\parallel}{A},$$

$$\text{strain} = \frac{\Delta s}{L},$$

$$\text{shear modulus } S = \frac{\text{stress}}{\text{strain}} = \frac{F_\parallel/A}{\Delta s/L}. \tag{15.6}$$

FIG. 15.17

4. **Volume deformation:**

$$\text{stress} = \frac{\Delta F}{A} = \Delta P,$$

$$\text{strain} = \frac{\Delta V}{V},$$

$$\text{bulk modulus } B = -\frac{\text{stress}}{\text{strain}} = -\frac{\Delta F/A}{\Delta V/V} = -\frac{\Delta P}{\Delta V/V}. \tag{15.7}$$

FIG. 15.18

Questions

1. *Estimate* the effective force constant of the springs of a car. (Assume that the car has only one spring that is compressed when a force is applied.) Make any rough measurements needed and indicate clearly your estimation technique.

2. A spring with force constant k is cut in half. What is the force constant of one of the halves? Explain.

3. Why do mountain climbers use ropes that have Young's moduli less than those of most other ropes? (*Hint:* Consider the distance needed to stop a falling climber as the rope becomes taut and the force of the rope on the climber.)

4. A parachutist whose parachute has not opened lands in snow. Using your knowledge of the ultimate compressive strength of bones, *estimate* the minimum stopping distance for the parachutist in order to avoid breaking a bone. The terminal speed of a freely falling parachutist is about 50 m/s.

5. Based on your own experimental measurements, *estimate* the Young's modulus of a single strand of a rubber band. Justify any numbers used and explain your procedure carefully.

6. *Estimate* the shear modulus of this book. Justify any numbers used and explain your procedure carefully.

7. How does the shear stress change (a) when the force acting on a cube of Jell-O parallel to its top surface is doubled, (b) when the force remains constant but each dimension of the Jell-O is doubled? (c) How does the strain change for these two situations?

8. The pressure acting on the outside surface of an inflated balloon increases by exactly 20 percent. Calculate the percentage change in the balloon's volume, assuming that no temperature change occurs and that the initial pressure on the balloon is 1.0 atm.

Problems

15.1 Elastic Deformations

1. A spring having a force constant of 1.5×10^3 N/m is stretched 0.12 m from its equilibrium position. (a) Calculate the force pulling on the end of the spring. (b) If the force calculated in (a) is increased by 60 N, how far is the spring now displaced from equilibrium?

2. A force of 20 N causes a spring to be displaced from equilibrium by 2.5 cm. (a) Calculate the force constant of the spring. (b) What is its displacement when a force of 45 N pulls on the spring?

3. A 1.5-kg mass hangs at the end of a spring. When 0.5 kg more is added, the spring is displaced by an extra 7.0 cm. How far was the spring displaced when only 1.5 kg hung at its end?

15.3 Stress and Strain

4. (a) A brick wall, 2 m tall, 0.1 m thick, and 20 m long, has a total weight of 8.0×10^4 N. Calculate the stress on the foundation under the bottom row of bricks. (b) A rope of radius 0.50 cm supports a weight of 400 N. Calculate the stress in the rope.

5. (a) A table with four identical legs, each of cross-sectional area 1.0×10^{-3} m^2, supports a weight of 600 N. Calculate the stress in each leg. (b) A wire supporting a 50-N metal ball has a stress of 6.4×10^7 N/m^2. Calculate the cross-sectional area and radius of the wire.

6. A hollow, cylindrical bone, with an external radius of 1.2 cm and an internal radius of 0.5 cm, supports a load that causes a stress of 2.7×10^6 N/m^2. Calculate the weight of the load.

7. (a) A rope 10 m long is stretched 3.0 cm when a person hangs at its end. Calculate the strain in the rope. (b) A rope 2.4 m long, used to pull a car from a mud hole, has a strain of 0.05. How much is it stretched when pulling the car?

8. (a) A violin string is stressed to produce a strain of 2.0×10^{-4}. If its length is increased by 1.4×10^{-2} cm, what is the unstretched length of the string? (b) Oil in a barrel of volume 0.20 m^3 expands by 10 cm^3 when heated. What is the volume strain of the oil?

15.4 Length Deformation

9. A 3.0-m long steel wire of radius 0.60 mm supports a weight of 50 N. By how much is the wire elongated because of the weight?

10. A 2.5-m long upright beam supports a load of 3000 N. The beam is compressed 0.20 cm by the load. Young's modulus of the beam is 1.0×10^{10} N/m^2. Calculate its cross-sectional area.

11. An elevator weighing 1.0×10^5 N is supported by two steel cables, each with a cross-sectional area of 5.0 cm^2. Calculate the elongation of the cables if the elevator hangs 30 m from its support.

12. A nylon rope used in mountain climbing stretches 1.2 m when supporting a 75-kg climber. The rope is 40 m long and has a diameter of 0.80 cm. (a) Calculate Young's modulus of the rope. (b) How far will it stretch when supporting a 90-kg climber?

13. A tendon, originally 7.5 cm long, is stretched 0.30 cm. (a) Calculate the strain in the tendon. (b) Calculate the stress causing this strain. (c) If the cross-sectional area of the tendon is 0.30 cm^2, how great is the force causing the stress?

14. (a) If the ultimate tensile strength of hair is 10^8 N/m^2 and each hair has a diameter of 0.10 mm, how many hairs are needed to support the weight of a 60-kg person? (b) Estimate what fraction of the hair on your head this number is (assuming you have a full head of hair). Justify any assumptions you make.

15. The ultimate tensile strength of the bone in your lower leg (the tibia) is 1.4×10^8 N/m^2 and its Young's modulus is 1.8×10^{10} N/m^2. Assuming that the stress and strain are proportional up to the point of fracture, calculate the strain in the bone at the point of fracture.

16. When in tune, a steel banjo string has a tension of 1300 N. If the ultimate tensile strength of the string is 6.0×10^8 N/m^2, what must be the area of the string so that it does not break? Allow a factor of five [that is, it should be able to support a tension of $5 \times (1300$ N)].

17. The ultimate compressive strength of a tooth is about 1.5×10^8 N/m^2. Estimate the force a molar can support before it fractures. Justify any numbers used in your calculation.

18. A glass plate 0.1 m $\times$ 0.2 m in area and 0.3 cm thick lies on a flat surface. (a) Calculate the maximum load the glass will support before it breaks. (b) Calculate the distance the glass is compressed when one-half of this load is added to the plate.

15.5 Shear Deformation

19. A wooden block, 0.20 m wide, 0.40 m long, and 0.50 m high and with a shear modulus 9.0×10^9 N/m^2, is glued to a floor. A shearing force of 2.5×10^5 N is applied to the top surface. Calculate the distance that the top edge of the block is displaced by the shear force.

20. A horizontal cylindrical beam made of wood and having a radius of 0.07 m and a shear modulus of 1.0×10^{10} N/m^2 projects out a distance of 0.1 m from the wall of an adobe house in which it is embedded. A 100-kg mass hangs from its end. Estimate the downward shear displacement of the end of the beam.

21. A soft rubber block of shear modulus 0.8×10^6 N/m^2 is used to stop cars at the end of a soapbox derby track (Fig. 15.19). A car bumper hits the side of the block at its top, 0.5 m above the ground. The shearing force of the car against the block is 6.4×10^4 N when the top of the block is displaced 0.1 m. What is the top area of the block?

FIG. 15.19

22. A 3.0-cm cube of gelatin has a force of 0.40 N exerted parallel to its top surface. This surface is displaced 0.50 cm. Calculate the shear modulus of the gelatin.

23. A plate of the earth's crust is sheared because one edge cannot move while the rest of the plate moves. Energy released during an earthquake indicates that the force retarding the motion at the edge is 10^{16} N. Following the earthquake, the plate edge slips 1.5 m. The length of the plate perpendicular to the surface that

bends as the plate moves is 15 km. Estimate the area of the side of the plate that is prevented from moving.

24. The elastic potential energy change of a solid object that is sheared a distance Δs is given by

$$\Delta PE_{el} = \frac{1}{2}\left(\frac{AS}{L}\right)\Delta s^2$$

(see Sec. 15.5 for the meaning of the symbols). Estimate the elastic energy that will be released when the plate discussed in Example 15.2 "slips" during an earthquake.

15.6 Volume Deformation

25. Calculate the volume changes of 0.5 m^3 of each of the following types of materials when exposed to a pressure increase of 10 atm: (a) steel, (b) aluminum, (c) water, and (d) ethyl alcohol.

26. The pressure on a granite rock 100 km below the earth's surface is approximately 3.7×10^4 atm. Estimate the percentage increase in the density of granite at this depth compared to its density at the earth's surface.

27. At what depth below the surface of water will the density of the water increase by 0.1 percent because of the increased pressure? (*Note:* The pressure increases by one atmosphere for each 10 m of depth below the surface.)

28. A head-shrinker decides to mechanize his operation by applying a high-pressure gas to the heads. Estimate the pressure needed if the head is to be decreased in volume by 30 percent. Assume that the bulk modulus of the head is one-half that of water.

29. The oil in a hydraulic press occupies a volume of 0.500 m^3 when the pressure is 2 atm. Calculate the change in volume when the pressure is increased to 12 atm. The bulk modulus of oil is about 1.5×10^{10} N/m^2.

15.7 Elastic Membranes

30. (a) Estimate the Young's modulus of elastin and of collagen in their elastic regions using the stress-versus-strain curves shown in Fig. 15.14. (b) Which type of material is more difficult to stretch and by what factor?

31. Use the equation derived in Problem G3 below to estimate the elastic potential energy stored in a section of aorta that is 50 cm long when its radius increases from 1.00 cm to 1.18 cm. The thickness of the aorta wall is about 0.2 cm, and its Young's modulus in the elastic region is about 8×10^5 N/m^2. (*Hint:* The initial length of the membrane being stretched is the circumference of the aorta.)

General Problems

G1. An 80-kg person hangs from a 10 m long weightless crossbar that is supported by steel cables (Fig. 15.20). The cable on the left has a radius of 1.0 mm and that on the right a radius of 2.0 mm. Calculate the distance x from the left cable that a person should hang so that both cables stretch the same amount.

FIG. 15.20

G2. A 70-kg person jumps from a wall and lands stiff-legged on one heel. The person is stopped after sinking 0.010 m into the ground. The person is falling at a speed of 9.0 m/s just before hitting the ground. (a) Calculate the person's deceleration (assumed constant) after hitting the ground. (b) Calculate the constant force of the ground on the person. (c) If the person's thigh

bone has an ultimate compressive strength of 1.7×10^8 N/m^2 and a cross-sectional area of 3×10^{-4} m^2, will the bone shatter? Explain.

G3. In Chapter 8 we learned that the elastic potential energy of a stretched spring is $PE_s = \frac{1}{2}kx^2$, where x is the displacement of the spring from equilibrium, and k is the spring constant. Use a derivation similar to that in Section 8.4 to show that the elastic potential energy of a stretched wire is

$$PE_s = \frac{1}{2}\left(\frac{AY}{L}\right)\Delta L^2,$$

where A is the cross-sectional area of the wire, L is its length, Y is the Young's modulus of the wire, and ΔL is the distance it is stretched.

G4. A 75-kg mountain climber falls at a speed of 10 m/s when the rope attached to her begins to become taut. The climber falls an extra 1.2 m before the rope stops her. (a) Calculate the change in kinetic energy of the climber. (b) Calculate the change in her gravitational potential energy while being stopped by the rope. (c) Based on your answers to parts (a) and (b), what is the change in elastic potential energy of the rope? (d) If the Young's modulus of the rope is 4.5×10^7 N/m^2 and its length is 10 m, what is the cross-sectional area of the rope (see the equation derived in Problem G3)?

When we watch ants pursuing their daily activities, the environment in which they move seems similar to ours, except smaller. For an insect like an ant, a patch of grass must look like a forest, a puddle of water like a lake, and a pebble like a boulder. If our bodies could be reduced to the size of an ant's, would life go on as usual, or would we be unable to adapt to this small world because of some basic physiological problem caused by our new size?

Scaling is used in science to determine the effect of size on function. Can we live normal, active lives if reduced to the size of an ant or if increased to the size of King Kong? Scaling is also used to construct models of larger or smaller objects. For example, a geologist can build a model whose changes in a short time represent structural changes in the earth that take millions of years to occur. Engineers and architects use scaling to make models for testing designs of buildings, airplane wings, power plants, and so forth.

To start our discussion of scaling, look at the two groups of blocks shown in Fig. 1. The first group consists of two cubes, one sitting on the other. Each cube has sides of length 1 (we omit the units). The two blocks together have dimensions $1 \times 1 \times 2$.

The second group of blocks has the same shape as the first but is three times longer in each direction. Its dimensions are $3 \times 3 \times 6$. The scaling factor of the second group of blocks compared to the first is three, since each dimension of the second group is three times longer than the similar dimension of the first group.

The **scaling factor** λ is the ratio of a dimension d' of one object and the similar dimension d of another object whose shape is the same as the first:

$$\text{scaling factor} = \lambda = \frac{d'}{d}.$$

If the objects have similar shapes, any dimension can be used to calculate the scaling factor, just as long as the same dimension is used for each object.

How does the area of similar shaped objects depend on the scaling factor? Picture in your mind the bottom surface area of the blocks shown in Fig. 1. For the small set of blocks, the bottom area A is $1 \times 1 = 1$, whereas for the larger set of blocks the area A' is $3 \times 3 = 9$. The ratio of the two areas, $A'/A = 9/1 = 9$ is the square of the scaling factor. The ratio of the areas of the side surfaces of the blocks also equals the scaling factor squared, since, for the side surfaces, $A'/A = (3 \times 6)/(1 \times 2) = 18/2 = 9$.

Areas scale as the square of the scaling factor:

$$\frac{A'}{A} = \lambda^2.$$

This seems reasonable. Each dimension has increased by λ, and area is the product of two dimensions.

How do the volumes of these two similarly shaped groups of blocks com-

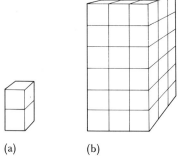

(a)   (b)

FIG. 1. Each dimension in the large group of blocks is three times longer than the corresponding dimension in the small group.

pare? The volume of the small group of blocks is $1 \times 1 \times 2 = 2$, whereas that of the larger groups is $3 \times 3 \times 6 = 54$. The ratio of their volumes, $V'/V = 54/2 = 27$, is the cube of the scaling factor (note that $3^3 = 27$). This also seems reasonable. Each dimension has increased by λ, and the volume is the product of three dimensions. Thus, the volume of one object should be λ^3 times that of the other object.

Volumes scale as the cube of the scaling factor:

$$\frac{V'}{V} = \lambda^3.$$

If two objects have equal and uniform density, then their masses and weights also scale in proportion to the cube of the scaling factor because mass and weight are proportional to volume [recall Eq. (13.1)—$m = \rho V$]. If the volume of one object is λ^3 bigger than that of another, its mass is also λ^3 bigger, assuming their densities are equal.

EXAMPLE 1 During a one-year period, a young girl grows so that each dimension in her body increases by 5 percent. By what percentage does her weight increase, assuming that her density remains unchanged?

SOLUTION One of the dimensions of her body, such as her height, increases in one year from a value d to a value d', where, for a 5 percent increase, $d' = 1.05d$; the scaling factor is 1.05. Her weight increases in proportion to the scaling factor cubed, or

$$\frac{w'}{w} = \lambda^3 = 1.05^3 = 1.16.$$

Thus, her weight increases by 16 percent. ∎

EXAMPLE 2 (King Kong's Achilles tendon) A giant such as the popular movie figure, King Kong, would have many structural problems if shaped like a man or woman. Suppose the giant is 60 ft tall and stands on the balls of his feet with his heels slightly elevated. Will his Achilles tendon tear? The tension in the Achilles tendon of a 6-ft tall, 180-lb man of similar shape and stance is 310 lb. The man's tendon tears if the tension exceeds 1500 lb.

SOLUTION The giant is 10 times taller than the man; thus, $\lambda = 10$. The giant's mass and weight are $\lambda^3 = 10^3 = 1000$ times greater than that of the man. However, the strength of the giant's muscles and tendons is only λ^2, or 100 times greater than that of the man, since strength depends on the cross-sectional area of muscles, tendons, and bones. Thus, the giant can exert 100 times more force with his muscles and tendons than can the man and can support 100 times more force on his bones. But, unfortunately, he is 1000 times heavier.

In Example 2.9, we found that the tension in the Achilles tendon of a 180-lb man when his heels are slightly elevated is

$$T = \frac{(w/2)(5.5 \text{ in})}{(1.6 \text{ in})} = \frac{(180 \text{ lb}/2)(5.5 \text{ in})}{(1.6 \text{ in})} = 310 \text{ lb}.$$

For the giant, the tension in the Achilles tendon will be

$$T' = \frac{(\lambda^3 w/2)(\lambda\, 5.5 \text{ in})}{(\lambda\, 1.8 \text{ in})} = \lambda^3 (310 \text{ lb}) = 310{,}000 \text{ lb}.$$

A man's Achilles tendon will rip when the tension exceeds about 1500 lb. Since the giant's tendon is λ^2, or 100 times larger in cross section, it will rip if the tension exceeds λ^2 1500 lb, or 150,000 lb. But to stand with his heel slightly off the ground requires a tension of 310,000 lb. *Merely standing on the ball of his foot will rip the tendon.* If the giant runs or jumps, the tension is even greater. The compression force on his bones and joints will cause fractures when standing. A 60-ft giant shaped like a man would be unable to stand, walk, or run. ∎

EXAMPLE 3   Parachutists have survived impacts in snow banks even though their parachutes did not open. Their speed at impact was about 120 mph, the terminal speed of a falling skydiver. Will a giant skydiver ten times the length of an average-sized woman ($\lambda = 10$) or an elf one-tenth the length of an average-sized woman ($\lambda = 0.1$) have a greater terminal speed as they reach the snow bank?

SOLUTION   At terminal speed the weight of the falling person is balanced by the opposing drag force of the air $\mathbf{F}_D$ (see Fig. 2):

$$w - F_D = 0.$$

Using Eq. (14.14) to substitute for F_D, we find that

$$w - \frac{1}{2} C_D \rho A v_T^2 = 0$$

or

$$v_T = \sqrt{\frac{2w}{C_D \rho A}},$$

where C_D is the drag coefficient (the same roughly for all three divers), ρ is the density of the air, A is the cross-sectional area of the diver, and v_T is the diver's terminal speed.

The ratio of the terminal speed of a giant or elf and that of a woman with a similar shape is

$$\frac{v_T'}{v_T} = \frac{\sqrt{2w'/C_D \rho A'}}{\sqrt{2w/C_D \rho A}} = \sqrt{\frac{2w'}{C_D \rho A'} \frac{C_D \rho A}{2w}} = \sqrt{\frac{w'}{w} \frac{A}{A'}} = \sqrt{(\lambda^3) \frac{1}{(\lambda^2)}} = \sqrt{\lambda}.$$

FIG. 2. Forces on a falling skydiver.

Note that weight scales as the third power of the scaling factor and area as the second power. We see that the ratio of the terminal speeds of similarly shaped falling objects scales as the square root of the scaling factor. Using $v_T(\text{woman}) = 120\,\text{mph}$, we find that

$$v_T'(\text{giant}) = \sqrt{10} \times 120\,\text{mph} = \underline{380\,\text{mph}}$$

whereas

$$v_T'(\text{elf}) = \sqrt{0.1} \times 120\,\text{mph} = \underline{38\,\text{mph}}.$$

The elf moves at a much lower speed when arriving at the snow bank. An extension of calculations such as this indicates that the force per unit area needed to stop the giant is over 100 times greater than that needed to stop the elf, if they stop in the same distance. ∎

Gravity is a less important factor for small animals than for large ones. An insect can hang easily from a ceiling, but a giant such as King Kong will break bones and rip tendons while walking because of the huge mass that must be supported by his legs. It appears that gravity limits the size of animals living on the earth. Had life evolved on a less massive planet, animals might have been larger.

Geologists have used scaling to construct models of geological formations. The materials for these models are chosen so that their change with time represents similar structural changes expected in the earth. The time scale can be reduced so that changes in the model occurring in several hours represent similar changes in the earth that may occur in millions of years. To build these models, the scaling of distance, time, and mass must be considered carefully, as must the effect of these choices of scale on other quantities such as density, force, pressure, and flow rate. The following example illustrates the effect of scale on a derived quantity such as flow rate.

EXAMPLE 4 A reservoir fills with water at a rate of $1000\,\text{m}^3/\text{min}$. A model of the reservoir is built so that each dimension in the model is 10^{-3} times the similar dimension of the reservoir, and the filling of water in the model takes 10^{-2} less time than filling the actual reservoir. At what rate should water flow into the model?

SOLUTION The scaling factors for distance and for time are:

$$\text{distance scaling factor} = \lambda = \frac{d_{\text{model}}}{d_{\text{reservoir}}} = 10^{-3},$$

$$\text{time scaling factor} = \tau = \frac{t_{\text{model}}}{t_{\text{reservoir}}} = 10^{-2},$$

where t_{model} represents the time between events happening in the model (such as the start and completion of filling the model with water) and $t_{reservoir}$ represents the time for the same events in the actual reservoir.

The flow rate of water into the model Q_{model} is the volume V_{model} of water that flows into the model divided by the time t_{model} required for this flow:

$$Q_{model} = \frac{V_{model}}{t_{model}}.$$

Since volume scales as the cube of the distance scaling factor, we find that

$$Q_{model} = \frac{V_{model}}{t_{model}} = \frac{\lambda^3 V_{reservoir}}{\tau\, t_{reservoir}} = \frac{\lambda^3}{\tau} Q_{reservoir}$$

$$= \frac{(10^{-3})^3}{(10^{-2})}(1000 \text{ m}^3/\text{min}) = 10^{-4}\text{ m}^3/\text{min} = \underline{100 \text{ cm}^3/\text{min}}. \quad \blacksquare$$

Problems

1. A 55-kg boy is 1.5 m tall. What will be his mass when his height increases 0.2 m, assuming that he retains the same body shape and density?

2. A watermelon 30 cm long costs $3 and a similarly shaped watermelon 24 cm long costs $1.80. Which watermelon is the better buy based on a price-per-unit-volume basis? Explain.

3. A 36-kg girl who is 1.40 m tall and her 1.80-m tall mother have exactly the same shape and density. Each lies on a board of nails for a neighborhood carnival. (a) If the girl is supported by 70 nails, how many nails support the mother? They lie on boards having an equal number of nails per unit area. (b) What is the mother's weight? (c) If the pain experienced depends on the average force per nail, which person experiences the greater pain? Explain.

4. A 70-kg man and an 80-kg man have similar builds. (a) Calculate the scale factor of the heavier man compared to the lighter man. (*Note:* It is not 1.14.) (b) If the smaller man is 1.60 m tall, how tall is the larger man? (c) A person's strength depends primarily on the cross-sectional area of his muscles. If the smaller man can lift a 300-N weight with his forearm, how much weight can the larger man lift with his forearm?

5. A person's strength is proportional to the cross-sectional area of his or her muscles. Why is it reasonable that an ant can lift many times its own weight, while a man or woman cannot?

6. A girl and her mother (heavier than her daughter) each use identical parachutes for a 1-mile jump. Who will reach the ground in the shortest time? Explain.

7. In Example 2.9 we found that a 310-lb tension in the Achilles tendon was needed to lift the heels of a 6-ft, 180-lb person slightly off the ground. This tension is directly proportional to a person's weight. The cross-sectional area of a 180-lb person's Achilles tendon is approximately 0.25 in^2. (a) What is the tension per unit area in the person's tendon? (b) An Achilles tendon will tear if the tension per area exceeds 6000 lb/in.^{2}. What is the height of the tallest giant who can lift his heels off the ground (stand on his toes) without tearing a tendon?

8. A model of a dam is built. The model is 0.5×10^{-3} times smaller than the actual dam in every dimension. A time-scale factor of 10^{-2} is chosen for the model. (a) If the water flow rate into the actual dam is 20 m^3/s, what should it be for the model? (b) If the model fills in six hours, what time is needed to fill the actual dam?

9. You are to make a geological model of a mountain. The modeling material must be chosen such that its compression strength and other properties are related by scaling as closely as possible to the properties of the actual mountain material (limestone). The differential compression strength of limestone (the unbalanced weight that it will support per unit area without breaking) is approximately 40,000 N/cm^2. If our model is to be 10^{-3} times the size of the mountain, what should the compression strength of the model material be? Explain. (*Note:* Weak, wet clay is often used.)

Vibrations and Waves

IV

Much of our study of physics has involved moving objects: a golf ball flying through the air, a car skidding to a stop, blood rushing through our arteries. In this section we analyze a different type of motion called wave motion. A wave consists of a disturbance that moves from one place to another. Examples are the disturbance in the level of water caused by a water wave, the disturbance of the pressure and density of air as a sound wave passes, and the disturbance of the compression and shear strain in the earth as an earthquake passes.

Waves are of utmost importance for the transmission of information. Our senses of sight and hearing depend upon light and sound waves. Our ability to think depends upon information transferred by nerve impulses, a form of wave motion. X-rays and ultrasound are used in medicine for diagnostic purposes. Geologists use radio waves and sound waves to explore the earth's crust. Microscopes bend light waves so that we see tiny objects such as body cells. Telescopes gather light and other waves from distant galaxies.

Waves are also responsible for energy transfer. Most of the earth's energy comes from the sun via light and infrared waves. A surfer derives his energy of motion from a water wave.

Our analysis of wave motion begins in Chapter 16 by looking at the physics of vibration. This is a natural starting place, since waves are produced and detected by vibrating objects. Following this introduction, we take up in Chapter 17 the general properties of all waves, with particular reference to mechanical type waves such as sound waves and water waves. A special kind of wave important in music (called a standing wave) is considered in Chapter 18.

Vibration 16

Vibration, also called oscillation, is a word used to describe the motion of an object that moves back and forth in a repetitive fashion. The world is full of objects that vibrate: banjo strings, electrons in a television antenna, our eardrums, the atoms in molecules, and ducks on a rough sea. Vibrations are not always desirable. Tall buildings must be constructed so that they do not sway excessively, and shock absorbers in your car dampen its up-and-down vibrations while traveling on a bumpy road. At other times much effort is made to produce vibrations. The hills and valleys on a record groove initiate mechanical and electrical vibrations in a high-fidelity system, and when amplified, the electrical vibrations cause the diaphragm of a speaker to vibrate. The sound waves produced by the speaker cause our eardrums to vibrate, and we hear sound. Sound, like most forms of communication, is created and detected by objects that vibrate.

The study of vibrations involves no new laws or principles of physics. Instead, we use Newton's second law and energy principles to analyze vibratory motion. The analysis starts with the study of a single mass vibrating at the end of a spring. The mathematical equations developed for this simple system can be easily extended to describe the motion of other vibrators such as a pendulum, a swinging leg, or a string on a musical instrument. We conclude our study by considering the conditions under which one vibrator can transfer energy to another.

(a)

(b)

16.1 Vibratory Motion

All objects that vibrate have an **equilibrium position** and are subject to a **restoring force** that always tries to return the object to its equilibrium position. Consider a cart that is attached to the end of a spring (Fig. 16.1). The spring shown in (a) is neither compressed nor extended, and the cart sits at rest at its equilibrium position. If the cart is pulled to the right, as in (b), the spring exerts a restoring force on the cart toward the left. If the cart is pushed to the left as in (c), the spring exerts a restoring force toward the right. The spring always exerts a force on the cart toward the equilibrium position.

The restoring force F_r of the spring on the cart is an example of a **Hooke's law-type force** that can be written as

$$F_r = -kx, \tag{16.1}$$

(c)

FIG. 16.1. (a) An object is at its equilibrium position. (b) If displaced to the right, the spring exerts a force to the left. (c) If displaced to the left, the spring exerts a restoring force to the right.

309

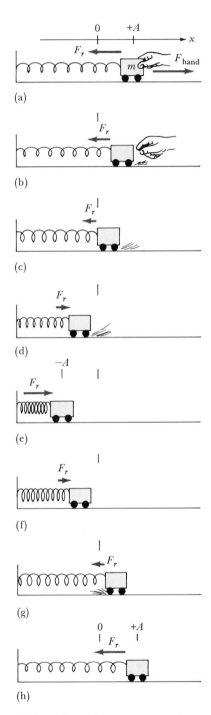

(a)

(b)

(c)

(d)

(e)

(f)

(g)

(h)

FIG. 16.2. A time sequence showing the effect of the restoring force on an oscillator during one cycle of motion. The speed of the cart is at maximum as it passes through the equilibrium position.

where x is the displacement of the cart from equilibrium and k is the force constant for the spring. The negative sign means that the force points opposite the direction of the displacement. The **force constant** k is a measure of a spring's stiffness. *

Why does a restoring force cause an object to vibrate? Another cart is shown in Fig. 16.2. If we pull the cart to the right as in (a) and release it as in (b), the spring's force on the cart causes it to accelerate toward the left. When it reaches the equilibrium position, the cart has momentum and overshoots it. As the cart passes the equilibrium position, the spring's restoring force starts pushing in the opposite direction. The cart decelerates and is finally stopped for an instant on the left (e), and then begins to move back to the right. It once again overshoots the equilibrium position and finally stops for an instant at its starting position. We say that the cart has completed one **cycle of motion;** it has made one **vibration.** If the friction forces in the system are small, the cart will continue to vibrate back and forth many times. The vibratory motion is called **simple harmonic motion** if the restoring force is proportional to the object's displacement from equilibrium.

This vibrational motion is described using several quantities that are defined below.

Amplitude A: The amplitude of a vibration is the magnitude of the maximum displacement of the object from equilibrium. For the cart shown in Fig. 16.2, the amplitude of vibration is the distance that the cart is displaced to the right at the start of its vibration.

Period τ (Greek letter tau): The period of a vibration is the time required for one vibration. For the cart shown in Fig. 16.2, the period is the time needed for the cart to move from position A at the right to position $-A$ at the left and back again to position A.

Frequency f: The frequency of vibration is the number of vibrations (also called cycles) that occur in a given time period. The most commonly used frequency unit is called the **hertz** (Hz), where one hertz equals one vibration per second:

$$1 \text{ Hz} = 1 \frac{\text{vib}}{\text{s}} = 1 \frac{\text{cycle}}{\text{s}}.$$

The words *vibration* and *cycle* indicate what is being counted but are not really units; these words can be removed from the units of equations involving frequency. Thus, an appropriate unit for frequency is $1/\text{s}$ or s^{-1}.

If an object's frequency of vibration is known, its period can be calculated, and vice versa. Suppose, for example, that the mass shown in Fig. 16.2 completes 5 vibrations in one second. Its frequency is 5 vib/s. The time needed to complete one vibration—that is, the period of vibration—is $\frac{1}{5}$ second. If the vibration frequency is 10 vib/s, then the period is $\frac{1}{10}$ second. The frequency and period are the inverse of each other:

$$f = \frac{1}{\tau}. \tag{16.2}$$

*In Chapters 8 and 15, when discussing springs, we used an equation $F = kx$. The force in that case was the external force needed to displace the cart and spring from equilibrium. In this section we are talking about the restoring force of the spring on the cart. This force opposes an external force and hence has the opposite sign.

EXAMPLE 16.1 The Empire State Building sways back and forth at a vibration frequency of 0.125 Hz. Calculate its period of vibration.

SOLUTION Rearranging Eq. (16.2) to solve for τ, we find that

$$\tau = \frac{1}{f} = \frac{1}{0.125 \text{ Hz}} = \frac{1}{0.125 \text{ vib/s}} = \underline{8 \text{ s/vib.}}$$

Each vibration requires eight seconds. ■

16.2 Energy of Simple Harmonic Motion

When a system vibrates, its energy continually interchanges between kinetic energy and some form of potential energy. Consider the system shown in Fig. 16.3. The only forms of energy involved in the vibration of the system, which includes the cart and the spring, are the kinetic energy of the cart given by $\frac{1}{2}mv^2$ (see Section 8.3) and the elastic potential energy of the spring, $\frac{1}{2}kx^2$ (see Section 8.5). Because no forces from outside the system do work on the system as it vibrates (we are ignoring friction), the **total energy** of the cart and spring must remain constant. If the cart is displaced from equilibrium at position x and moves with speed v, as shown in Fig. 16.3a, then its energy is

$$E = \frac{1}{2}mv^2 + \frac{1}{2}kx^2. \tag{16.3}$$

When the cart moves to its maximum displacement from equilibrium, as shown in Fig. 16.3b, the elastic potential energy is at its maximum and the kinetic energy is zero (the cart has stopped for an instant):

$$E = \frac{1}{2}m0^2 + \frac{1}{2}k(\pm A)^2 = \frac{1}{2}kA^2.$$

When the cart passes its equilibrium position, its elastic potential energy is zero because $x = 0$. However, its kinetic energy is then a maximum:

$$E = \frac{1}{2}mv_{\text{max}}^2 + \frac{1}{2}k0^2 = \frac{1}{2}mv_{\text{max}}^2.$$

Note that all three of these energy expressions are equal—the energy is constant. Only the relative amounts of the kinetic energy of the mass and the elastic potential energy of the spring change.

EXAMPLE 16.2 A spring with a 1.6×10^4 N/m force constant and a 0.10-kg mass at its end has a total vibrational energy of 3.2 J. (a) Calculate its amplitude of vibration. (b) Calculate its maximum speed. (c) Calculate its speed when displaced 0.010 m from equilibrium. (d) What is its amplitude of vibration if its energy is doubled? Ignore friction forces.

SOLUTION (a) When at its maximum displacement from equilibrium, the speed of the mass is zero (it has stopped for an instant to change directions), and

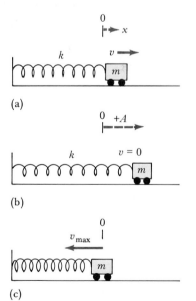

FIG. 16.3. (a) The total energy of a harmonic oscillator is the sum of its kinetic energy and its elastic energy. (b) When the cart is displaced a maximum distance from equilibrium, the energy is all elastic energy of the spring. (c) When the cart is passing through the equilibrium position, the energy is all kinetic energy of the cart.

its displacement from equilibrium is either $+A$ or $-A$. The energy of the system is

$$E = \frac{1}{2}m0^2 + \frac{1}{2}kA^2.$$

Solving for A, we find that

$$A = \sqrt{\frac{2E}{k}} = \sqrt{\frac{2(3.2\text{ J})}{1.6 \times 10^4\text{ N/m}}} = \underline{0.020\text{ m}}.$$

(b) The maximum speed occurs when the mass is passing its equilibrium position (when $x = 0$):

$$E = \frac{1}{2}mv_{\text{max}}^2 + \frac{1}{2}k0^2$$

or

$$v_{\text{max}} = \sqrt{\frac{2E}{m}} = \sqrt{\frac{2(3.2\text{ J})}{0.10\text{ kg}}} = \underline{8.0\text{ m/s}}.$$

(c) The system's energy when displaced 0.010 m from equilibrium is a combination of kinetic energy and elastic potential energy:

$$E = \frac{1}{2}mv^2 + \frac{1}{2}kx^2,$$

or

$$v^2 = \frac{2E}{m} - \frac{k}{m}x^2 = \frac{2(3.2\text{ J})}{0.10\text{ kg}} - \frac{1.6 \times 10^4\text{ N/m}}{0.10\text{ kg}}(0.010\text{ m})^2 = 48\text{ m}^2/\text{s}^2.$$

Thus,

$$v = \underline{6.9\text{ m/s}}.$$

(d) When the energy is doubled so that $E' = 6.4$ J, the vibrational amplitude of the mass is

$$A' = \sqrt{\frac{2E'}{k}} = \sqrt{\frac{2(6.4\text{ J})}{1.6 \times 10^4\text{ N/m}}} = \underline{0.028\text{ m}}.$$

(Notice that its amplitude of vibration did not double when its energy doubled.)

∎

16.3 Frequency
of Simple Harmonic Motion

The period of vibration of a simple harmonic oscillator can be estimated using a little reasoning along with the ideas of energy developed in the last section. If the period is known, the frequency can be calculated, since $f = 1/\tau$.

Recall the period of vibration for the cart pictured in Fig. 16.2. It moves a total distance $4A$ during one vibration: $2A$ to the left and $2A$ back to the right.

If the average speed $\bar{v}$ of the mass is known, the time τ needed to move the distance $4A$ is $\tau = 4a/\bar{v}$.* Estimating the average speed of the oscillator to be

*Remember that speed = distance/time. Therefore, time = distance/speed, or $\tau = 4A/\bar{v}$.

about half its maximum speed, then

$$\tau = \frac{4A}{\bar{v}} \sim \frac{4A}{(v_{max}/2)} = 8\,\frac{A}{v_{max}}. \tag{16.4}$$

The expression is approximate, since $\bar{v}$ is not exactly equal to $v_{max}/2$.

We can use our knowledge of the oscillator's energy to rewrite this equation for τ in terms of such quantities as k and m—quantities that are intrinsic properties of the oscillator. Recall that the energy of the oscillator when displaced a maximum distance A from its equilibrium position is

$$E = \frac{1}{2}kA^2.$$

This also equals the kinetic energy of the mass as it passes through its equilibrium position at $x = 0$:

$$E = \frac{1}{2}mv_{max}^2.$$

Since the energy is the same in each position, these two expressions for energy are equal:

$$\frac{1}{2}kA^2 = \frac{1}{2}mv_{max}^2.$$

Rearranging, we find that

$$\frac{A}{v_{max}} = \sqrt{\frac{m}{k}}.$$

The above can be substituted into Eq. (16.4):

$$\tau \sim 8\sqrt{\frac{m}{k}}.$$

This is a surprisingly good estimate. An exact calculation using calculus shows that the period of vibration of a mass at the end of a spring is

$$\tau = 2\pi\sqrt{\frac{m}{k}}. \tag{16.5}$$

Since the frequency of vibration is inversely proportional to the period ($f = 1/\tau$), the **frequency of vibration of a simple harmonic oscillator** is

$$f = \frac{1}{2\pi}\sqrt{\frac{k}{m}}. \tag{16.6}$$

Notice that the frequency is greater for objects fastened to stiff springs (large values of k) and less for objects with large mass.

EXAMPLE 16.3 A 0.20-kg mass hangs at the end of a rubber band that vibrates at a frequency of 1.2 Hz. Calculate the force constant of the rubber band.

SOLUTION Equation (16.6) can be rearranged to solve for k:

$$k = 4\pi^2 f^2 m = 4\pi^2 (1.2 \text{ s}^{-1})^2 (0.20 \text{ kg}) = 11.4 \text{ kg/s}^2 = \underline{11.4 \text{ N/m}}.\,^*$$ ■

A Mass Hanging at the End of a Spring

The vibrational motion of a mass hanging at the end of a spring is slightly more complicated than its motion when vibrating horizontally on a frictionless surface. We must now include the force of gravity in our analysis.

When no mass hangs at the end of a spring, as in Fig. 16.4a, it has a length L; when the mass is added to the spring, its length increases by ΔL. The equilibrium position of the mass is now a distance $L + \Delta L$ from the spring's support (b). The forces acting on the mass, shown in (c), are the upward force of the spring, equal in magnitude to $k\,\Delta L$, and the downward force of gravity, mg. Since the mass is in equilibrium, the sum of the forces acting on the mass add to zero, or

$$k\,\Delta L - mg = 0.$$

What happens if we displace the mass a small distance above this new equilibrium position, and then release it? Let us choose the x axis pointing up, with $x = 0$ at the new equilibrium position of the mass. If we displace the mass a distance x above this equilibrium position (d) and then let go (e), the mass will accelerate downward because the upward force of the spring on the mass has been reduced from $k\,\Delta L$ to $k(\Delta L - x)$, whereas the downward gravitational force remains unchanged. Newton's second law for motion along the x direction ($\Sigma F_x = ma_x$) becomes

$$k(\Delta L - x) - mg = ma_x.$$

But since $k\,\Delta L - mg = 0$, we find that

$$-kx = ma_x,$$

which is exactly the same force and acceleration equation we get for a horizontally vibrating mass when displaced a distance x from its equilibrium position. The frequency of vibration, then, must be the same for a hanging mass as for

*Note that $1\,\dfrac{\text{N}}{\text{m}} = \left(1 \text{ kg}\,\dfrac{\text{m}}{\text{s}^2}\right)\dfrac{1}{m} = 1\,\dfrac{\text{kg}}{\text{s}^2}.$

FIG. 16.4 (a) (b) (c) (d) (e)

one vibrating on a horizontal surface at the end of the spring:

$$f = \frac{1}{2\pi} \sqrt{\frac{k}{m}}.$$

EXAMPLE 16.4 A mass hanging at the end of a spring vibrates at frequency f. When additional mass is added, the system vibrates at a frequency f' that is 15 percent less than f. By what percentage did the mass at the end of the spring increase?

SOLUTION Since the frequency decreases by 15 percent, $f' = 0.85f$. Taking the ratio of f' and f and substituting Eq. (16.6) for the frequencies, we find that

$$\frac{f'}{f} = 0.85 = \frac{(1/2\pi)\sqrt{k/m'}}{(1/2\pi)\sqrt{k/m}} = \sqrt{\frac{m}{m'}},$$

where m is the mass when the system vibrates at frequency f, and m' is the mass when vibrating at f'. Squaring and rearranging, we find that

$$\frac{m'}{m} = \frac{1}{0.85^2} = \frac{1}{0.72} = 1.38.$$

Since $m' = 1.38\,m$, the mass must have been increased by <u>38 percent</u>. ∎

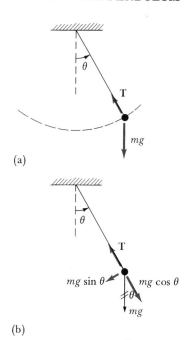

(a)

(b)

FIG. 16.5. (a) Two forces act on the pendulum ball: the rope-tension force **T** and the ball's weight mg. (b) The component of the weight force tangent to the arc along which the ball travels serves as the restoring force for the ball.

16.4 The Pendulum

A **simple pendulum** consists of a mass hanging at the end of a cord of negligible weight. Pulled to the side and released, the mass swings back and forth in approximate simple harmonic motion.

A pendulum, like all simple harmonic oscillators, has a surprising property. The same time is taken for one swing of small amplitude as for one of larger amplitude. This property was first observed by Galileo, who used his heartbeat to time the period of a pendulum. He realized that a pendulum could be used as a standard of time for a clock. Galileo pendulum clocks are still sold today, over three hundred years after their invention.

An expression for the frequency at which a pendulum swings can be determined by considering the forces acting on the mass at its end (Fig. 16.5a). The cord attached to the mass exerts a tension force **T** in the direction of the cord and perpendicular to the arc along which the mass moves. The downward weight force mg has two components (Fig. 16.5b). One component of magnitude $mg \cos \theta$ is perpendicular to the arc of motion, and is balanced by **T**, the tension force. The other component is tangent to the arc and restores the pendulum toward its equilibrium position. This component serves as a restoring force and is given by the equation

$$F_r = -mg \sin \theta.$$

(The negative sign indicates that the restoring force points opposite the direction of the displacement.)

For small angles, the value of θ, when using radian units, is approximately equal to $\sin \theta$ (see Table 16.1). Notice, for example, that when $\theta = 0.3491$ radi-

TABLE 16.1 Values of θ and $\sin \theta$ for Small Angles

θ radians	(degrees)	$\sin \theta$
0	(0)	0
0.0872	(5)	0.0872
0.1745	(10)	0.1736
0.3491	(20)	0.3420
0.5236	(30)	0.5000

FIG. 16.6. The linear displacement x of a pendulum from its equilibrium position is related to its angular displacement θ by the equation $x = L\theta$, where θ must be in units of radians.

ans (equal to $20°$), $\sin \theta = 0.3420$. They differ by only two percent, even for this moderately large angle.

Because $\theta \simeq \sin \theta$ for small angles, the restoring force can be written as

$$F_r \simeq -mg\theta.$$

The angle θ is related to x, the displacement of the mass from its equilibrium position (the heavy line on the arc in Fig. 16.6), by a variation of Eq. (6.1):

$$x = L\theta,$$

where L is the length of the pendulum. Thus, we find that the restoring force is

$$F_r = -mg\theta = -mg\,\frac{x}{L} = -\frac{mg}{L}x.$$

The above equation for restoring force is very similar to that for a mass at the end of a spring ($F_r = -kx$), the only difference being that the force constant k for a spring is replaced by mg/L for a pendulum. The quantity mg/L is called the effective force constant of the pendulum and can be substituted in place of k for all the equations derived to describe a mass vibrating at the end of a spring. Consequently, the frequency of a pendulum is given by Eq. (16.6) with mg/L replacing k:

$$f = \frac{1}{2\pi}\sqrt{\frac{k}{m}} = \frac{1}{2\pi}\sqrt{\frac{mg/L}{m}}$$

or

$$f = \frac{1}{2\pi}\sqrt{\frac{g}{L}}. \tag{16.7}$$

We see that the frequency of a pendulum depends only on its length and on the acceleration of gravity, not on the mass of the object hanging at its end nor on the amplitude of vibration.

Equation (16.7) was, before 1950, the basis of a technique used by geologists to locate salt domes, which are likely spots for oil. Since the density of these geological structures is less than that of surrounding parts of the earth's crust, the value of g is less above a salt dome than above surrounding parts of the earth, so a pendulum should swing slower when hanging above a salt dome.

EXAMPLE 16.5 Use the pendulum-frequency equation to estimate the number of steps that is "natural" for a leg to take per unit time while walking—in other words, the natural frequency of vibration of the leg.*

SOLUTION The center of mass of the leg is approximately at the knee, and the length L that is used in Eq. (16.7) is about half the actual length of the leg. If

---

*The leg is not a simple pendulum but is an example of what is called a **physical pendulum**—an extended body that swings back and forth about a pivot point. The frequency of swinging of a physical pendulum is $f = (1/2\pi)\sqrt{mgL/I}$, where L is the distance from the pivot point to its center of mass, m is its mass, and I is the moment of inertia of the pendulum about its pivot point. Since our calculations in this example are approximate, we will treat the leg as a simple pendulum.

a leg is 1 m long, then $L = 0.5$ m and its natural swinging frequency is approximately

$$f = \frac{1}{2\pi}\sqrt{\frac{g}{L}} \simeq \frac{1}{2\pi}\sqrt{\frac{9.8 \text{ m/s}^2}{0.5 \text{ m}}} = \underline{0.7 \text{ s}^{-1}}.$$

The period or time for one step is the inverse of f, or

$$\tau = \frac{1}{f} \simeq 1.4 \text{ s}. \qquad\qquad \blacksquare$$

According to our equation, then, it is natural to take one step with your right leg (or your left leg) each 1.4 s or so. To test this estimate, you can stand on the toes of one foot and swing the free leg back and forth at what seems like a natural frequency (Fig. 16.7). Count the number of swings in 60 s. When you divide this number into 60 s, you will have the period for one swing. It should be about 1.4 s.

If your legs are short, the value of L is less than 0.5 m and the period is shorter than 1.4 s. Similarly, longer legs have longer periods of vibration. Giraffes take relatively few steps per unit time because of their long legs. A Chihuahua dog, on the other hand, has very short legs and a high "walking frequency."

FIG. 16.7. The natural frequency of vibration of your leg is about the same as a comfortable walking speed.

16.5 Kinematics of Simple Harmonic Motion

We now turn to the kinematics of simple harmonic motion. How does the position of a simple harmonic oscillator vary with time? The variation of the position of a vibrating object with time can be determined with the aid of the experiment shown in Fig. 16.8. A wheel with a peg fastened to its edge rotates next to a mass vibrating at the end of a spring. Light from a projector shines on both objects, and the shadows from the vibrating mass and the rotating peg move up and down on a screen. The mass is caused to vibrate so that the amplitude of its vertical motion equals the radius of the rotating circle. If the rotational motion of the circle is adjusted to make one complete rotation each time the mass completes one vibration, we find that their shadows move up and down on the screen in exactly the same pattern of motion. The vertical position of the peg on the rotating wheel changes with time in the same fashion as a vibrating object's position changes with time.

If we use the coordinate x to represent the vertical position of the peg, we see

FIG. 16.8. The shadow of the peg on a screen moves up and down in the same motion as a mass vibrating at the end of a spring.

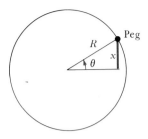

FIG. 16.9. The distance of the peg above the horizontal is $x = R \sin \theta$.

from Fig. 16.9 that

$$x = R \sin \theta, \tag{16.8}$$

where R is the radius of the circle and θ is the angle between a horizontal line and a radial line drawn from the center of the circle to the peg. If we start our clock running when the peg is at the side of the circle—that is, $\theta = 0$ when $t = 0$—then at a later time t the angular position θ of the peg is $\theta = \omega t$, where ω is the angular velocity of the rotating wheel. [Remember from Eq. (6.4) that $\omega = \Delta\theta/\Delta t = (\theta - \theta_0)/(t - t_0) = (\theta - 0)/(t - 0) = \theta/t$]. Since the wheel completes one rotation through 2π radians in a time of one period τ, we see that $\omega = 2\pi/\tau$. Thus,

$$\theta = \omega t = \frac{2\pi}{\tau} t = 2\pi f t,$$

where we have substituted f for $1/\tau$ in the last part of this equation. The frequency f represents the number of times the peg rotates in a circle each second and also represents the vibrational frequency of the mass attached to the spring. Substituting for θ in Eq. (16.8), we find that the position of the vibrating mass changes with time according to the equation

$$x = A \sin 2\pi \frac{t}{\tau} = A \sin 2\pi f t. \tag{16.9}$$

The radius of the circle has been replaced by the amplitude of vibration. The position x of the vibrating mass is plotted as a function of time t in the graph shown in Fig. 16.10. The general pattern is called a sinusoidal type curve, and an object whose motion can be described using Eq. (16.9) is said to undergo **sinusoidal motion.**

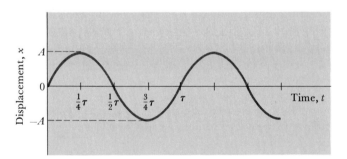

FIG. 16.10. A position-versus-time graph for a simple harmonic oscillator.

16.6 Damping

In the preceding sections, we have ignored the effect of friction on vibrating objects. Yet friction is very important in real-world vibrations. If friction were not present in the elastic suspension system of a car, for instance, the car would continue vibrating up and down for miles after crossing a bump on a road. Tall buildings rely on friction to reduce their vibrational energy when they sway in the wind.

Friction is present to varying degrees in every object that vibrates or oscillates. The vibrations are said to be **damped** by the friction. A **weakly damped oscillator** has little friction and continues to vibrate for many periods, but the amplitude of each successive vibration is slightly reduced because of friction (Fig. 16.11a).

If the friction force opposing the vibration is increased, the vibrational energy is transferred more quickly to thermal energy. The amplitude of vibration decreases more rapidly, as shown in Fig. 16.11b. In an **overdamped oscillator,** the damping is so great that all of the elastic energy of the stretched oscillator is converted to thermal energy as the oscillator makes its first move toward its equilibrium position. When overdamped, the oscillator takes a very long time to return to the equilibrium position.

A **critically damped oscillator** is one in which the oscillator returns to equilibrium in the shortest possible time but does not overshoot the equilibrium position (Fig. 16.11d). The springs and suspension system of a car should be critically damped to avoid continued vibrations when traveling on a bumpy road.

16.7 Energy Transfer Between Oscillators

Much of our contact with the world around us is a result of the transfer of energy between different oscillators. Vibrational energy created in one person's throat is transmitted as sound waves to another person's ears. The emission of light and its absorption in our eyes is a result of energy transfer between atomic and molecular oscillators.

A dramatic example of the effects of energy transfer between different types of oscillators is shown in Fig. 16.12. The Tacoma Narrows Bridge near Tacoma, Washington, was opened for traffic on July 1, 1940. It was constructed in a way that allowed several vibrations to occur with little damping. If gusts of wind hit the bridge at intervals of time about equal to the period of these vibrations, the wind would induce large amplitude vibrations in the bridge. On November 7, 1940, a strong wind caused a twisting vibration whose amplitude at the edge of the road was over 3 m (Fig. 16.12a). The bridge withstood these twisting, up-and-down vibrations for about one hour before it collapsed (Fig. 16.12b). The Tacoma Narrows Bridge, a vibrator, was excited by absorbing energy from another vibrating source of energy—the fluctuating pressure of the air blowing against the bridge.

(a)

(b)

(c)

(d)

FIG. 16.11. The oscillators shown have varying degrees of damping. With damping, the oscillator energy is converted into thermal energy, thus reducing or even stopping the vibrations.

(a)

(b)

FIG. 16.12. The Tacoma Narrows Bridge is an unusual example of energy transfer between oscillators. Gusts of wind at about the same frequency as a twisting vibration in the bridge caused large amplitude vibrations. The bridge oscillated for one hour before collapsing. (Wide World)

FIG. 16.13. The swinger gains energy if positive work is done by the force—that is, if F and Δr are in the same directions.

FIG. 16.14. The swinger loses energy if negative work is done by the force—that is, if F and Δr are in opposite directions.

The ability of one vibrator to transfer energy to another depends on two factors: (1) *the frequency of one vibrator relative to the frequency of the other vibrator, and* (2) *the contact or coupling between the vibrators.* Consider first how the frequencies of two vibrators affect their ability to exchange energy. For efficient energy transfer from one to the other, they should vibrate at nearly the same frequency. Imagine a person pushing a child on a swing. To push the child higher, the person must push so that positive work is done on the child—that is, the force must be exerted in the direction of the child's displacement, as shown in Fig. 16.13. To transfer additional energy to the child, the next push must occur one period later when the child is again swinging forward.

If the pushes are at a frequency different from that at which the child swings, the next push may occur as the child swings back, in which case the push opposes the child's motion (Fig. 16.14). The push will reduce the amplitude of vibration and the child will have less energy. Maximum energy transfer occurs when the pushing is at the same frequency as the swinging.

The importance of having the vibrator frequencies equal in order for efficient energy transfer to occur is illustrated by another experiment involving the speaker of a sound system and a champagne glass (Fig. 16.15). We will make the speaker vibrate at different frequencies by connecting it to an electrical oscillator. The sound energy leaving the speaker is kept constant as the sound frequency is varied. The relative amplitude of vibration of the glass at each frequency is detected by using a small pick-up coil attached to the glass.

As frequency varies, the amplitude of vibration of the glass also varies, as shown in Fig. 16.16. The graph of vibration amplitude versus frequency is called the **response function** of the glass and indicates the ability of the glass to vibrate at different frequencies. The peak of the curve at frequency f_0 is called the **resonant frequency** of the glass—that is, the frequency at which the glass vibrates with greatest amplitude. For the vibration of the speaker to transmit energy to the glass, the speaker's frequency must overlap with the resonant curve of the glass; the greater the overlap, the more energy is transferred from one to the other.

For example, if the speaker is vibrating at a frequency of 250 Hz and the resonant frequency of the glass is 700 Hz, there is no overlap and no transfer of energy from the speaker to the glass (Fig. 16.17a). As the speaker's frequency is

FIG. 16.15. A system to record the transfer of vibrational energy from a speaker to a champagne glass at different frequencies.

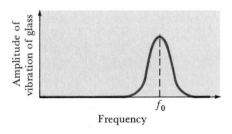

FIG. 16.16. The resonant vibration of a glass occurs at frequency f_0.

FIG. 16.17. The amount of energy transfer between vibrators depends on the overlap of their frequencies.

moved closer to that of the glass, the speaker frequency begins to overlap with the glass resonant curve, and energy is transferred to the glass. Maximum energy is transferred when both frequencies are the same, as shown in Fig. 16.17d. Singers have supposedly fractured glassware by inducing large amplitude vibrations at the resonant frequency of the glass.

The second condition needed to transfer energy from one vibrating object to another is that there must be a coupling mechanism through which energy can travel from one oscillator to the other. The stronger the coupling is, the greater the amount of energy that is transferred in a certain time. For the speaker and glass, the medium coupling the speaker vibrations to the glass is the air between the speaker and the glass. Pressure variations caused by the speaker travel as sound waves through the air to the glass. The coupling can be reduced by moving the glass farther from the speaker so that the sound from the speaker is diffused more and is less intense at the glass. If the experiment is done in a vacuum, no energy transfer occurs at any frequency because the coupling medium (the air) has been removed.

These conditions for energy transfer affect our ability to communicate information by sound and by light. They also affect a variety of daily occurrences that may or may not have been part of your experience. For example, your car may vibrate when traveling at a certain speed. The car probably has a wheel that is slightly imbalanced. When rotating at a frequency equal to the frequency at which some other part of the car can vibrate, energy is transferred from the imbalanced wheel to that other vibrator. As the car moves slower or faster, the wheel's rotational frequency differs from that of the car part and no energy transfer occurs. Electric motors on devices such as refrigerators can cause some other part of the refrigerator to vibrate if the armature of the motor rotates at the same frequency that the other part of the refrigerator vibrates. Your arm swings when you walk if the "walking frequency" of your legs equals the "swinging frequency" of your arms. Raising or lowering the arms changes their resonant frequency and may cause them to stop swinging.

Summary

1. **Restoring force:** Objects that vibrate are acted on by a restoring force when displaced from their equilibrium position. For simple harmonic motion, the restoring force $\mathbf{F}_r$ is proportional to the object's displacement x from equilibrium. For a mass at the end of a spring,

$$F_r = -kx, \qquad (16.1)$$

where k is the **force constant** of the spring.

2. **Vibration:** The vibrational motion of an object is characterized in terms of three quantities:

The **amplitude** A of the vibration is the magnitude of the maximum displacement from equilibrium of the vibrating object.

The **period** τ is the time required for one complete vibration.

The **frequency** f is the number of vibrations per unit time. Frequency and period are related by the equation

$$f = \frac{1}{\tau}. \qquad (16.2)$$

Frequency is measured most often in units of hertz (Hz):

$$1 \text{ Hz} = 1 \text{ vib/s} = 1 \text{ s}^{-1}.$$

3. The **energy** of a freely vibrating, simple harmonic oscillator that is acted on by negligible friction force is

$$E = \frac{1}{2}mv^2 + \frac{1}{2}kx^2. \qquad (16.3)$$

4. The **frequency of vibration** of different devices depends on intrinsic properties of the devices. For a mass m at the end of a spring with force constant k,

$$f = \frac{1}{2\pi}\sqrt{\frac{k}{m}}. \qquad (16.6)$$

For a pendulum with a mass at the end of a cord of length L, the vibration frequency is

$$f = \frac{1}{2\pi}\sqrt{\frac{g}{L}}, \qquad (16.7)$$

where g is the acceleration of gravity (9.8 m/s^2).

5. **Damping** occurs when friction converts the vibrational energy of an oscillator to thermal energy, causing the amplitude of vibration to decrease.

6. **Energy transfer** occurs between two vibrators if (1) they can vibrate at approximately the same frequency, and (2) there is a coupling medium or mechanism through which energy can transfer from one oscillator to the other.

Questions

1. A banjo string, a pendulum, and water sloshing in a glass all undergo vibratory motion. Draw a force diagram indicating the cause of the restoring force for each type of motion.

2. At what position(s) during simple harmonic motion is an oscillator's speed zero? Where is its acceleration zero? Explain.

3. By how much must we increase the amplitude of vibration of a mass at the end of a spring in order to double its maximum speed during a vibration?

4. If you double the amplitude of vibration of a mass at the end of a spring, how does this affect the values of τ, k, E, and v_{max}?

5. A known mass placed at the end of a spring causes it to stretch a distance ΔL. Explain how you can calculate the frequency of vibration of this mass if you do not know the value of the spring's force constant k.

6. Describe two different ways to measure the force constant of a rubber band using only a 100-g mass, a stopwatch, and a meter stick.

7. (a) *Estimate* the effective force constant of the suspension system of a car. Describe carefully your technique. (b) Estimate the frequency at which the car is likely to vibrate.

8. A pendulum clock is running too fast. How should the length be adjusted? Explain.

9. The length of the rod on a pendulum clock that sits in an outdoor area is carefully adjusted to give the correct time in the summer. Will it run fast or slow in the winter when the temperature is colder? Explain.

10. A pendulum clock is moved from the Mississippi Delta region to the Rocky Mountains. Will the clock run faster or slower in the Rockies? Why? How can it be adjusted to compensate for this change?

11. Oil is often found in a geological structure called a salt dome. The rock in the salt dome is less dense than surrounding rock and is pushed up so that it "floats" on surrounding rock. Explain how a sensitive pendulum can detect the location of a salt dome.

12. By experiment and by calculation *estimate* the natural period and frequency of vibration of your arm. Explain carefully the numbers used in each estimate.

13. A pendulum and a mass at the end of a spring are both carefully adjusted to make one vibration each second when on the earth's surface. Will the period of vibration of the pendulum or of the mass be affected if they are placed on the moon? Explain.

14. Will the frequency of vibration of a swing be greater or less when you stand on the swing than when sitting? Explain.

15. *Estimate* the time required for one step with one leg of a Chihuahua and of an elephant. Assume that they walk at their natural frequency.

16. The amplitude of vibration of a swing slowly decreases to zero if you do not pump your legs while swinging. Explain carefully what happens to the vibrational energy.

17. If you walk with your arms hanging down, they often oscillate forward and back at the same frequency as your stepping frequency. However, if your arms are bent at the elbow so that the forearms are horizontal, they often do not oscillate. Why?

Problems

16.1 Vibratory Motion

1. (a) Musicians in an orchestra tune their instruments to what is called "concert A," a frequency of 440 Hz. Calculate the period for one vibration. (b) The atoms in an oxygen molecule complete one vibration in a time of 2.11×10^{-14} s. Calculate the frequency of vibration of O_2.

2. A mass at the end of a spring makes one vibration in 0.2 s. (a) Calculate the frequency of vibration. (b) If the period is doubled by increasing the mass, what is the frequency?

3. (a) The period of the lowest-frequency sound the normal ear can hear is 0.05 s. Calculate its vibrational frequency. (b) The highest-frequency sound heard by a normal ear is about 20,000 Hz. Calculate the time for one vibration.

4. An FM radio station broadcasts at a frequency of 90 MHz. What distance does a radiowave leaving the antenna travel in one vibration if the wave travels at a speed of 3.0×10^8 m/s?

5. A spring stretches 5.0 cm when pulled by a 100-N force. Calculate the force constant of the spring.

6. A spring is stretched by a 100-N force. The displacement of the end of the spring increases 5.0 cm beyond its original stretched position when the force increases 20 percent to 120 N. Calculate the force constant of the spring and its original displacement.

16.2 Energy of Simple Harmonic Motion

7. A simple harmonic oscillator (SHO) vibrates with an amplitude A. (a) What fraction of the total energy of the oscillator is elastic potential energy and what fraction kinetic energy when the oscillator is at position $x = A/2$? (b) At what position is the oscillator when its kinetic energy equals its elastic potential energy?

8. A spring with a force constant 1200 N/m has a 55-g mass at its end. (a) If the vibrator energy is 6.0 J, what is the amplitude of vibration? (b) What is the maximum speed of the mass? (c) What is the speed when the mass is at a position $x = A/2$?

9. A 15-N force applied to a mass attached to a spring causes the mass to move 0.060 m from its equilibrium position. When the force is removed, the mass undergoes simple harmonic motion. (a) Calculate the force constant of the spring. (b) Calculate the energy of the oscillator.

10. A spring with force constant 2.5×10^4 N/m has a 1.4-kg mass at its end. (a) If its amplitude of vibration is 0.030 m, what is the total energy of the oscillator? (b) What is the maximum speed of the mass? (c) If the energy is tripled, what is the new amplitude? (d) What is the maximum speed of the oscillator?

11. When exposed to the weakest sound it can hear, the ear's energy of vibration is about 10^{-19} J. If the force constant of the ear is 20 N/m, what is the amplitude of vibration of the eardrum? How does this compare to the size of an atom ($\sim 5 \times 10^{-10}$ m)?

12. A 0.10-kg mass at the end of a spring of force constant 120 N/m undergoes horizontal vibrations of amplitude 0.080 m. As the mass reaches its maximum amplitude of vibration, a 0.04-kg piece of clay is added to the mass. Calculate the ratio of the maximum speed of the oscillator before the mass is added and that of the oscillator after the mass is added.

16.3 Frequency of Simple Harmonic Motion

13. A 1.6-kg mass attached to a spring undergoes simple harmonic motion of amplitude 0.12 m and has an energy of 72 J. Calculate the force constant of the spring and the frequency of vibration.

14. A 0.10-kg mass vibrating at the end of a spring has an extra mass added to it when its displacement is $x = A$. How large a mass should be added to reduce the frequency to half its initial value?

15. The vibrations of a 30-g mass attached to a spring are to serve as the time standard of a clock. Calculate the desired force constant of the spring if the mass makes one vibration each second.

16. A spring with unknown force constant hangs from a support. When a 0.20-kg mass is placed at the end of the spring, the spring stretches 0.10 m below its starting position. (a) Calculate the force constant of the spring. (b) Calculate the frequency of vibration of the mass.

17. A 70-kg gymnast sitting on a trampoline causes it to sag 0.10 m at the center. (a) Calculate an effective force constant for the trampoline. (b) The trampoline is pulled downward an extra 0.05 m by a strap sewed under the center of the trampoline. When the strap is released, what is the energy and frequency of vibration of the trampoline? Ignore the mass of the material that is stretched.

18. A 0.10-kg mass sliding at a speed of 5.0 m/s on a frictionless, horizontal surface runs into and sticks to a spring of force constant 250 N/m. Calculate the amplitude of vibration and frequency of vibration of the mass.

19. A spring-type oscillator of force constant 1200 N/m vibrates with an amplitude of 0.15 m. At the end of one vibration when $x = A$, half of the mass attached to the spring falls off. Calculate

the ratio of the final to initial values of energy (E_f/E_0), amplitude (A_f/A_0), and frequency (f_f/f_0).

20. A sound wave traveling through water must vibrate approximately 1000 times more mass than one traveling through air. Calculate the ratio of the vibration amplitude of sound in air compared to that in water, assuming equal frequency and energy of the vibration.

21. Low-frequency vibrations (less than 5 Hz) are annoying to humans if the product of the amplitude and the frequency squared (Af^2) equals 0.5×10^{-2} m·s^{-2} (or more). Determine the frequency and amplitude of a carnival vibrator that produces these annoying vibrations if the device vibrates a total mass of 120 kg and has a vibrational energy of 12 J.

16.4 The Pendulum

22. In 1851, the French physicist Jean Foucault hung a large iron ball on a wire about 61 meters (200 ft) long to show that the earth rotates. The pendulum appears to swing in different directions as the earth turns under it. Calculate the swinging frequency of this pendulum.

23. You wish to build a clock whose pendulum makes one swing back and forth per second. (a) Calculate the desired length of the rod (assumed to have negligible mass) holding the weight at its end. (b) Will the rod need to be shorter or longer if you include the mass of the rod in your calculations? Explain.

24. You are designing a pendulum clock whose period can be adjusted by 10 percent by changing the length of the pendulum. By what percent must you be able to change the length to provide this flexibility of period? Explain.

25. Calculate the maximum speed of a child on a swing 3.0 m long when the amplitude of vibration is 1.2 m.

26. (a) Calculate the change $\Delta\tau$ in the period of a pendulum caused by a small change Δg in the acceleration of gravity. *Hint:* The period will be related to the acceleration of gravity by the equation

$$\tau + \Delta\tau = \frac{1}{2\pi}\sqrt{\frac{g+\Delta g}{L}}.$$

To find $\Delta\tau$, expand $(g+\Delta g)^{1/2}$ using the binomial theorem:

$$(g+\Delta g)^{1/2} \simeq g^{1/2} + \frac{1}{2}g^{-1/2}\Delta g + \cdots.$$

(If Δg is small, we can ignore higher power terms of Δg.) (b) Use your results to find the fractional change in period $\Delta\tau/\tau$ in terms of $\Delta g/g$. (c) What is $\Delta\tau/\tau$ for a 0.2 percent variation in g that occurs when moving across the surface of a salt dome?

16.5 Kinematics of Simple Harmonic Motion

27. A graph of position versus time for an object undergoing simple harmonic motion is shown in Fig. 16.18. Estimate from the graph the amplitude and period of the motion, and calculate the object's frequency.

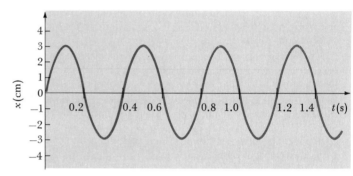

FIG. 16.18

28. A mass at the end of a spring undergoes simple harmonic motion of amplitude 6 cm and frequency 4 Hz. Calculate the position of the mass at (a) 0.10 s and (b) 0.15 s.

29. Draw a graph showing the position-versus-time curve of a simple harmonic oscillator (a) with twice the frequency of that shown in Fig. 16.18 and (b) with the same frequency but twice the amplitude.

General Problems

G1. A 0.12-kg calculator is to be tested for its ability to withstand vibration and acceleration. It is strapped to a cart of negligible mass at the end of a spring. The vibration amplitude is to be 5.0 cm. (a) Calculate the force constant of the spring if the cart is to experience an initial acceleration of 10 g. (b) Calculate the energy of vibration.

G2. Calculate the frequency of vibration of the mass shown in Fig. 16.19.

G3. Derive an expression for the frequency of oscillation of the mass shown in Fig. 16.20.

G4. A 500-kg ball at the end of a 30-m cable suspended from a crane is used to demolish a building. (a) If the ball has an angular displacement of 15° from the vertical, calculate its linear displacement from the vertical along the arc of its motion. (b) Calculate the kinetic energy of the ball at the bottom of the arc.

FIG. 16.19

FIG. 16.20

G5. The Sears Building in Chicago has a mass of about 5×10^8 kg and sways back and forth at a frequency of about 0.1 Hz. (a) Estimate the force constant for this swaying motion. (b) A gust of wind hitting the building exerts a force of about 4×10^6 N. By approximately how much is the top of the building displaced by a steady force equal in magnitude to that of the wind?

G6. A hole is drilled through the center of the earth. The gravitational force on a mass in that hole is mgr/R, where r is the distance of the mass from the earth's center and R is the radius of the earth (6.4×10^6 m). A mass dropped into the hole executes simple harmonic motion. Find the period of the motion. How does one half this time compare to the time needed to fly in an airplane halfway around the earth?

17 Traveling Waves

Vibrations, such as discussed in the last chapter, are essential for the creation, transmission, and detection of waves; and without waves the earth would be a quiet, dull place—no sound, no light, and no nerve impulses transferring information between your brain and other body parts. Nor would there be communication by radio, television, or telephone. Even the ocean would be a less enchanting place.

All these different types of waves have common features, the most important of which is their ability to transfer energy and information. In this chapter we will analyze general principles that apply to all waves and will use these principles to study mechanical waves such as sound waves, waves on stretched strings, shock waves that pass through the earth following an earthquake, and water waves.

17.1 Properties of Waves

The word *wave* brings to mind the picture of water waves moving across the ocean. The wave consists of repeated disturbances in the level of the water, and these disturbances move one after the other toward the shore. A sound wave is a moving disturbance of the pressure and density of matter. If you clap your hands, for instance, air molecules move rapidly from between your hands, causing a compression of nearby air. Following the compression, the molecules return slowly to their original positions. However, the compression spreads rapidly at a speed of about 340 m/s to more distant parts of the air. In general, a **wave** can be defined as a disturbance that moves from one place to another.

The sound wave produced by clapping your hands is an example of an **impulsive wave,** as is the shock wave caused by an earthquake. An impulsive water wave occurs when you throw a rock into a pond, as shown in Fig. 17.1. A single disturbance with a few trailing ripples moves at constant speed away from the source of the disturbance.

A **periodic wave** is caused by a repeated disturbance. For instance, the diaphragm of a high-fidelity speaker vibrates back and forth, and the air pressure in front of the diaphragm alternately increases and decreases. This repeated or periodic disturbance of air pressure moves away from the speaker in the form of a sound wave. A periodic water wave can be created in a pond by pushing a large beach ball up and down in the water (Fig. 17.2). The disturbances created

FIG. 17-1. A rock thrown into a pond causes a single impulsive wave to spread in all directions from the splash.

FIG. 17.2. A periodic wave is produced by a repetitive disturbance—a vibration.

A water wave

Wave direction

Water molecules move in circular orbits when wave passes by

by each up-and-down motion of the ball move one after the other, at constant speed, away from the source. The peaks of the waves are called **crests,** and the depressed regions between crests are called **troughs.**

All waves have several features in common. One important property is that *a wave does not transfer matter from one place to another.* A wave is a moving disturbance, but no net displacement of matter occurs. Consider again water waves on a pond. The molecules of water at the surface undergo a circular, vibratory motion about an equilibrium position at the center of the circle (Fig. 17.3). However, the water as a whole does not move from one side of a pond to the other. There is no transfer of matter across the pond.

A second property of waves is that they *transfer energy.* A wave in water passing under a boat is momentarily able to lift the boat—increase its gravitational potential energy. The boat also gains a little upward kinetic energy as the wave passes. The wave could not momentarily increase the boat's energy if it did not carry energy. Waves traveling on the earth's surface after an earthquake can carry enough energy to level a building miles from the quake. Such a wave does indeed transfer energy.

A third important property of waves is their ability to *transmit information.* A child on one side of a pond might arrange with a friend on the other side that two impulsive waves in short succession constitute a signal for the friend to return home. Much of the information in the human nervous system is transmitted in a similar way—that is, by coded electrical impulses. Periodic waves can also carry information by the frequency and amplitude of the waves. Our ability to recognize different sounds depends on the frequency and amplitude of sound waves.

We conclude, therefore, that waves are traveling disturbances that can transfer energy and information but not matter.

17.2 Types of Waves

Waves are classified in a variety of ways. In the last section we distinguished impulsive waves from periodic waves. Waves can also be categorized according to the type of disturbance that moves through space. A sound wave consists of a fluctuation in the pressure and density of matter. A water wave consists of a variation in the level of water at its surface. A radio wave consists of a fluctuating electric and magnetic disturbance that moves through space, causing electric charges in the path of the wave to vibrate.

Waves are also classified according to the direction in which objects in the path of the wave vibrate relative to the direction the wave travels. A **transverse wave** causes an object to vibrate perpendicular to the direction the wave travels, whereas a **longitudinal wave** causes the object to vibrate parallel to the direction

FIG. 17.4. This is an example of a transverse wave; the rope fibers vibrate perpendicular to the direction the wave travels.

the wave travels. Consider the transverse wave shown in Fig. 17.4. The fibers of the rope vibrate up and down perpendicular to the direction of travel of the wave. Water waves, light waves, and radio waves are other examples of transverse waves.

For longitudinal waves, the particles in the medium vibrate back and forth in the same direction that the wave travels. A longitudinal wave created on a slinky is shown in Fig. 17.5. As the person's hand pushes forward and pulls back on one end of the slinky, the coils in front of the hand are alternately compressed or crowded together, and then expanded or made less crowded. This latter situation is called a **rarefaction**—a decrease in the density or crowding of a part of a medium. Each region of compression and rarefaction moves to the right as new compressions and rarefactions are produced by the hand's vibration. Individual coils do not move from one end of the slinky to the other; only the disturbance of coil spacing moves. The individual coils do, however, vibrate back and forth about their equilibrium position. Since their vibrational motion is parallel to the direction in which the disturbance travels, the wave is called a longitudinal wave. Sound waves are also longitudinal waves.

FIG. 17.5. A longitudinal wave is created on a slinky by vibrating the hand back and forth. The coils vibrate in the same direction that the wave travels.

Longitudinal waves are often pictured by plotting their density or pressure at one instant of time at different positions in space. Two examples are shown in Fig. 17.6. In (a) the number of slinky coils per unit length (linear density) at one instant of time is plotted on the vertical axis versus position on the horizontal axis. In (b) the density of air as a sound wave passes is plotted on the vertical axis as a function of position on the horizontal axis. Notice that the crests of the waves in (a) and (b) are regions of greater than normal density while the troughs are regions of reduced density. The graphs shown in Fig. 17.6 are called **wave-**

FIG. 17.6. (a) The spacing between coils on a slinky is plotted versus coil position. (b) The density of air as a sound wave passes is plotted versus position in space.

(a)

(b)

forms and represent the disturbance caused by the wave at one instant of time at different positions. Since the wave moves, the waveform also indicates the changing pressure or density at one location at different times.

17.3 Amplitude, Speed, Frequency, and Wavelength

Periodic waves of all types are described using quantities such as amplitude, speed, frequency, and wavelength.

Amplitude

The **amplitude** of a wave is the magnitude of the maximum disturbance caused by the wave. Consider the longitudinal wave moving along the slinky shown in Fig. 17.6a. The number of coils in each centimeter length of slinky is plotted for different positions along the slinky. When no wave is present, there are 5 coils per cm. This is the equilibrium number of coils per cm. As the wave passes, the number fluctuates between 4 coils/cm and 6 coils/cm. The amplitude of the wave equals the maximum disturbance from equilibrium, or 1 coil/cm.

Speed

The **speed** v of a wave is the distance that a disturbance travels per unit time. The speed of a particular type of wave depends on properties of the medium in which the wave travels. Consider an impulse sent along the row of springs and masses shown in Fig. 17.7. The speed of the impulse depends on how stiff the springs are. If they are very stiff, an impulse on one side of a spring is transmitted quickly to the other side. We expect that an impulse should travel faster along a row of masses connected together by stiff springs than along a row of masses connected by easily flexed springs. In general, the speed of a wave in a medium is proportional to the square root of some elastic property of the medium, a property that measures the "stiffness" of the bonds or interactions between neighboring atoms and molecules in the medium:

$$v \propto \sqrt{\text{elastic property}}.$$

The mass of the spheres shown in Fig. 17.7 also affects the speed of the wave. When the first spring on the left pushes the second mass toward the right, its acceleration depends on how large its mass is. The greater its mass, the less its acceleration is and the longer the time needed to compress the next spring. The impulse travels slowly if the mass of the spheres is large and vice versa. In general, the speed of a wave is inversely proportional to the square root of a quantity that depends on the density of mass in a medium—that is, on an inertial property:

$$v \propto \frac{1}{\sqrt{\text{inertial property}}}.$$

When these two factors that determine a wave's speed are combined, we see that speed depends on an elastic property and on an inertial property of a

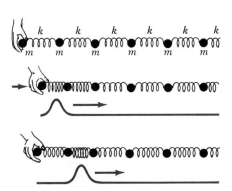

FIG. 17.7. The speed at which a pulse travels along this chain of spheres and springs depends on the force constant of the springs and the mass of the spheres.

medium in the following way:

$$v \propto \sqrt{\frac{\text{elastic property}}{\text{inertial property}}}. \tag{17.1}$$

We will not derive the exact equations for the speed of different types of waves. Several equations are listed in Table 17.1 along with representative values of the speed of these waves in different media. Notice, for example, that the speed of sound in a gas or liquid is given by

$$v = \sqrt{\frac{B}{\rho}},$$

where B is the bulk modulus of the gas or liquid (an elastic property) and ρ is its density (an inertial property).

EXAMPLE 17.1 An explosion is detonated at the surface of the ocean over the Japanese trench, the deepest part of the ocean. The impulse travels to the bottom of the trench, is reflected, and returns to a ship at the surface in 14.6 s. Calculate the depth of the ocean trench.

SOLUTION We first calculate the speed of sound in seawater using Eq. (17.2) in Table 17.1. The density of seawater is about $1030 \, \text{kg/m}^3$. The bulk modulus of water from Table 15.3 is $2.0 \times 10^9 \, \text{N/m}^2$. Thus, the speed of sound in seawater is

$$v = \sqrt{\frac{B}{\rho}} = \sqrt{\frac{2.0 \times 10^9 \, \text{N/m}^2}{1030 \, \text{kg/m}^3}} = 1.4 \times 10^3 \, \text{m/s}.$$

TABLE 17.1 Equations for Calculating the Speed of Different Types of Waves

Type of wave	Speed of wave[a]	Sample values		
Sound in a gas or liquid	$\sqrt{\dfrac{B}{\rho}}$	Air Water	340 m/s 1500 m/s	(17.2)
Compressional or longitudinal wave in solids (P-wave)	$\sqrt{\dfrac{B + \frac{4}{3}S}{\rho}}$	Clay Sandstone Limestone Granite Salt	1000 m/s 2000 m/s 4000 m/s 5000 m/s 6000 m/s	(17.3)
Shear or transverse wave in solids (S-waves)	$\sqrt{\dfrac{S}{\rho}}$	About 0.6 times the speed of compressional waves given above		(17.4)
Compressional wave in thin rod or bone	$\sqrt{\dfrac{Y}{\rho}}$	Bone	3000 m/s	(17.5)
Transverse wave on a string	$\sqrt{\dfrac{T}{\mu}}$	Violin A-string 288 m/s Violin G-string 128 m/s		(17.6)

[a] ρ = density, μ = mass per unit length, B = bulk modulus, S = shear modulus (also called modulus of rigidity), Y = Young's modulus, and T = tension in the string. These different elastic moduli are defined in Chapter 15.

The sound travels a distance two times the ocean depth D (down and back). This depth is related to the speed and time of travel by the equation

$$2D = vt = (1.4 \times 10^3 \,\text{m/s})(14.6 \,\text{s}) = 2.04 \times 10^4 \,\text{m}.$$

Thus the ocean depth at this trench is 1.02×10^4 m, or <u>10.2 km</u>. ∎

Frequency

The **frequency** f of a wave is the number of back-and-forth or up-and-down vibrations of a part of the medium through which the wave travels. For example, for the wave traveling along the slinky shown in Fig. 17.6a, the frequency of the wave equals the number of back-and-forth vibrations per unit time of a coil in the slinky. Usually, the frequency of a periodic wave equals the frequency of vibration of its source. *

Wavelength

The **wavelength** λ (lambda) of a periodic wave is the distance between adjacent troughs, between adjacent crests, or between any two identical parts of adjacent disturbances (see Fig. 17.6). The wavelength depends on the wave's speed and frequency and is not an independent quantity. If the hand pushes the slinky shown in Fig. 17.5 back and forth at a low frequency, for instance, then a crest created by one compression will travel a relatively long distance before the next compression is created. Thus, the crests of the wave will be separated by large distances (long wavelength). If, on the other hand, the hand vibrates in and out at a high frequency, there will be little distance between adjacent crests. A previous crest can travel only a short distance before another crest is created. The separation of crests (wavelength) is short.

We can, in fact, derive an equation that allows us to calculate the wavelength of a wave if we know its speed and frequency. Recall that the time required for one vibration is called the period of vibration and is given the symbol τ. Note that if we produce one crest by pushing in on a coiled spring, the next crest will be created one vibration or one period τ later. The original crest has traveled a distance of one wavelength. Its speed is, then,

$$v = \frac{\text{distance}}{\text{time}} = \frac{\lambda}{\tau}. \tag{17.7}$$

But the period τ and frequency are related by the equation

$$f = \frac{1}{\tau}.$$

Substituting for $1/\tau$ in Eq. (17.7), we find that

$$v = f\lambda \tag{17.8}$$

---

*We will learn later that the frequency is altered if the source of the waves is moving or if the observer is moving. We ignore these effects of motion until Section 17.8.

or

$$\lambda = \frac{v}{f}.$$

We see that if f is large (a high frequency wave), then λ is short; if f is small (a low frequency wave), then λ is long. Equation (17.8) can be used to calculate any one of the three quantities (λ, v, or f) when the other two are known.

EXAMPLE 17.2 A sound wave created by a radio speaker vibrating at 300 Hz has a wavelength in air of 1.15 m. What is the speed of sound in air?

SOLUTION Using Eq. (17.8), we find that

$$v = f\lambda = (300 \text{ s}^{-1})(1.15 \text{ m}) = \underline{345 \text{ m/s}}.$$

Notice that we substituted the frequency unit s^{-1} in place of the unit Hz.

■

17.4 Reflection and Transmission

Until now, we have been considering waves moving through a homogeneous medium, a medium with the same properties everywhere. What happens to a wave when it reaches the end of a medium, or when there is an abrupt change from one medium to another? At these boundaries between different media, a wave separates—part is **transmitted** to the new medium and part is **reflected** back into the old. For example, a sound wave traveling in air is partially reflected at a wall and partially transmitted into the wall. Light waves traveling in air when striking the surface of glass are partly reflected and partly transmitted into the glass.*

The amount of reflection at these interfaces and the nature of the reflected wave can provide considerable information about differences of the two media at their interface. Geologists, for instance, have determined that the antarctic ice sheet rests on earth rather than water because radiowaves reflected from the bottom of the ice sheet are characteristic of an ice-earth interface rather than an ice-water interface.

To understand why waves are partially reflected at these interfaces between media, consider two ropes woven together, as shown in Fig. 17.8. The rope on the left is thin and has a small mass per unit length, while the rope on the right has a large mass per unit length. By flicking the hand quickly up and down, a wave pulse is started in the left rope and travels toward the interface.

When the pulse reaches the interface, only part of it is transmitted to the large rope. Because of its large mass per unit length, the large rope is harder to accelerate upward than the light rope. Only a small-amplitude upward pulse is initiated in the heavy rope (Fig. 17.8g). This is called the **transmitted** pulse.

FIG. 17.8. A pulse traveling along a thin rope is partially reflected and partially transmitted at the interface with a large rope. The reflected pulse is inverted—there is a 180° change in phase.

*Some absorption of a wave's energy also takes place when it reaches the boundary between two media. In fact, **absorption** can occur at any part of a medium through which the wave travels. The coordinated vibrations of the atoms in the medium are turned into random kinetic energy—that is, to thermal energy. The rate of this conversion varies greatly. The slower the conversion, the longer the wave lives.

Because the large rope does not "give," it exerts a downward force on the small rope (Fig. 17.8f). The small rope is "jerked" below its equilibrium position by the downward force, starting an inverted, reflected pulse back toward the left. We say that there has been a 180° **change in phase** of the reflected wave. An upward pulse is converted to a downward pulse. In the case of sound waves, a compression of air molecules would be converted into a rarefaction—that is, decreased density—and vice versa.

A change of phase does not always occur at an interface. A pulse traveling along a heavy rope toward an interface with a light rope is depicted in Fig. 17.9. When the pulse reaches the interface, the small rope restrains the large rope less than usual. The large rope overshoots and initiates a large transmitted pulse in the small rope. A small, upright, reflected pulse is also created by this overshoot. In this situation the incident and reflected pulses are both upright. There is no phase change.

The degree to which waves are reflected and transmitted at the boundary between different media can be analyzed using a quantity called impedance. The **impedance** of a medium is a measure of the difficulty a wave has in distorting the medium and depends on the elasticity, density, and viscosity of the medium and on the frequency of the wave passing through it.

The subject of impedance is complex, and we will not consider the mathematical details here. Nevertheless, we can get a qualitative feel for the idea of impedance by comparing the difficulty you would have in vibrating the oar of a boat back and forth in air and then in water. The water offers much more resistance to the oar than does the air, and the water has greater impedance. A sound wave causes matter to vibrate back and forth; the water offers more resistance or impedance to a sound wave than does air.

Whenever a wave strikes the boundary between two media having different impedances, the wave is partially reflected and partially transmitted. If the impedance increases, the phase of the reflected wave is inverted; an incident compression becomes a reflected rarefaction. If the impedance decreases, the reflected pulse has the same phase as the incident pulse.

Sound waves are used routinely in medicine to probe different organs of the body. In most instances the reflected waves are used only to determine the location of some interface between two media and not to determine the impedance of the reflecting medium. For example, a high frequency sound pulse reflected from a piece of metal lodged in a person's eye is shown in Fig. 17.10. The horizontal axis indicates the time required for the pulse to travel from the eye's

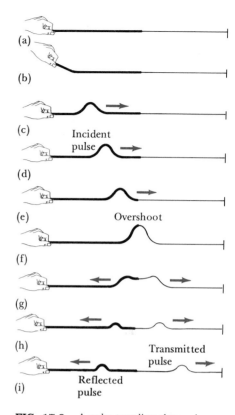

FIG. 17.9. A pulse traveling along a heavy rope is partially reflected and partially transmitted at the interface with a thin rope. The reflected pulse in this case is upright, as was the incident pulse. There is no change in phase.

FIG. 17.10. An ultrasound pulse originating at the eye's surface is reflected back to the detector by a piece of metal lodged in the eye. An echo also comes from the back of the eye at the retina.

FIG. 17.11. An ultrasound emulsifier. The tip vibrates at the frequency 2.3×10^4 Hz and can emulsify cataracts and other types of tissues. (Courtesy of CooperVision Systems/ Cavitron)

FIG. 17.12. Ultrasound echoes are detected by a transmitter and receiver on (a) the right side and (b) the left side of the skull. Notice in trace (a) that the echo from the midline when the pulse starts from the right side is longer than the midline echo when starting from the left side (b). The midline has been shifted left.

surface to the metal and back again. If we know the speed of sound in the eye, we can then calculate the distance to the metal.

Ultrasound scanners can produce a two-dimensional image of internal interfaces in the body by using an array of receiving elements to detect high-frequency sound reflected from these surfaces. The scanners are ideal for detecting and diagnosing tumors, cysts, foreign bodies, and the altered location of interfaces. The ultrasound emulsifier shown in Fig. 17.11 can remove cataracts, cysts, and tumors by fragmenting tissue when the surgical tip vibrates at high frequency against it.

Ultrasound, this high-frequency sound, has several other applications. It is used routinely to determine if injury has caused swelling of the brain. Two sound generators are placed on each side of a patient's skull. Sounds leaving the generators are reflected at three places: one on each side of the skull, and a third from the tough midline membrane of the brain (Fig. 17.12). If the midline membrane is not centered, it is likely that swelling on one side of the brain has caused it to be displaced. An operation is probably necessary to stop a brain hemorrhage.

17.5 Principle of Superposition

In the past sections, we have been studying the behavior of a single wave or pulse. A wave characterized by a single frequency and wavelength is often called a **sinusoidal wave.** These waves are initiated by simple harmonic oscillators. Most periodic or repetitive disturbances of a medium are not sinusoidal waves but are combinations of two or more different frequency waves traveling through the same medium at the same time. The sound coming from a violin, for instance, is a combination of almost twenty waves, each of different frequency and wavelength.

In much of the rest of this chapter and in later chapters on waves, we will be concerned with what happens in a medium when two or more waves simultaneously pass through a medium. The waves or pulses are added according to the principle of superposition:

Principle of superposition: When two or more waves pass through the same medium at the same time, the net displacement of any point in the medium is the sum of the displacements that would be caused by each wave if alone in the medium at that time.

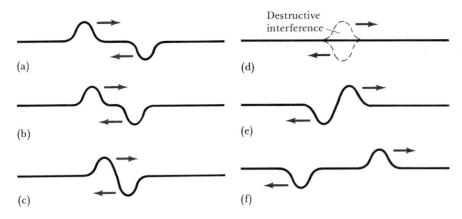

FIG. 17.13. Two wave pulses, one upright, the other inverted, pass at the center of a rope. The two pulses interfere destructively to form a resultant wave smaller than either of the component pulses.

An example will enable us to understand this principle more clearly. In Fig. 17.13a two pulse waves are shown moving toward each other along a string. The pulses pass through each other and continue moving in the same direction with the same amplitude, as shown in (f). When the pulses are at the same part of the string, however, the individual pulses cancel the effect of each other and the string does not move. The pulse moving to the left causes a negative displacement of the string that cancels the positive displacement caused by the pulse moving to the right. The resultant pulse shown in (d) is an example of **destructive interference**—the resultant displacement is smaller than the displacement caused by either wave if moving alone through the medium.

If two pulses that are displaced in the same direction (as shown in Fig. 17.14) pass each other on the string, the resultant displacement is greater than either pulse (part d). This is called **constructive interference**—the resultant displacement has greater amplitude than both of its components.

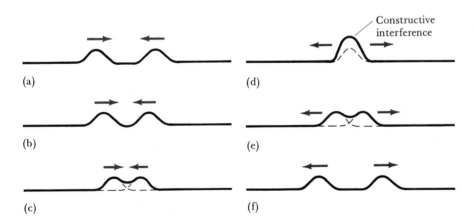

FIG. 17.14. Two upright pulses pass at the center of a rope. The resultant pulse is larger than either component pulse. They are said to interfere constructively.

17.6 Complex Waves

One important application of the principle of superposition is the analysis of what are called complex waves. A **complex wave** consists of two or more sinusoidal waves of different frequency and wavelength. Most sound waves that we hear are complex waves. Engineers transmit information along telephone lines and from television and radio antennas in the form of complex waves. Psychologists use complex wave analysis to analyze information collected by electro-encephalographic recordings of brain waves.

Let us first learn how a complex wave is formed from two or more simple waves. In Fig. 17.15a, two sound waves of different frequencies are sent simultaneously down a hallway. The variation in air pressure caused by the first wave at one instant of time at different positions along the hall is shown in (b). The graph is like a snapshot of the change in air pressure at each position along the hall if only the first wave were present. The variation of air pressure caused by the second wave at the same instant is shown in (c).

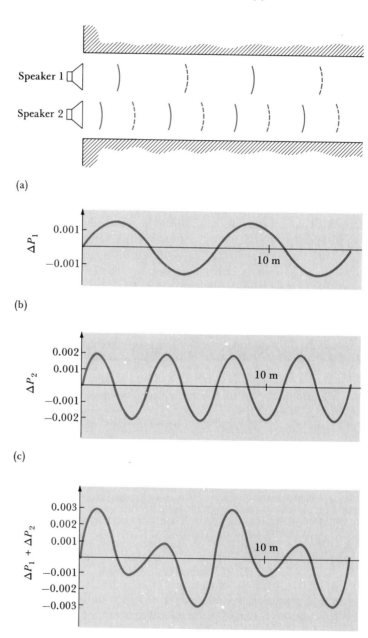

FIG. 17.15. (a) Two sound waves emitted by different speakers are sent down a hall. The variation in pressure of the air caused by each sound wave is depicted in (b) and (c). (d) The complex pressure wave is produced by adding these two waves.

If both waves travel down the hall at the same time, we do not see either pattern. Instead, we see a complex wave pattern that results from the combination of the pressure variations of both waves. This complex wave is calculated using the principle of superposition and can be constructed on a graph by adding the pressure variations caused by each wave at each point in space. At a position 10 m down the hall, wave 1 in Fig. 17.15 causes a pressure increase of 0.001 N/m²—that is, $\Delta P_1 = +0.001$ N/m². Wave 2 causes a pressure decrease of 0.002 N/m²—$\Delta P_2 = -0.002$ N/m². Thus, the actual pressure change at this point due to both waves together is $\Delta P_1 + \Delta P_2 = (0.001 - 0.002)$ N/m² or -0.001 N/m². By adding the pressures at each point along the hall, we are able to construct the complex wave shown in (d). Since both sinusoidal waves move to the right at the speed of sound, the complex wave also moves to the right at the speed of sound. If a microphone were placed in the hall, it would detect a fluctuating pressure such as shown in (d), a pattern called the **waveform** of the complex wave.

Practical reasons arise for wanting to construct a complex wave. Suppose that you wish to build an electric organ that can produce sounds like those of a clarinet or violin. The organ must be able to produce complex waves that look similar to those from the actual instruments. This can be done by combining sinusoidal waves to obtain the appropriate complex wave. In order to do this, we must first determine what frequency waves are part of a complex wave coming from a clarinet or violin when playing a particular pitch sound.

A variety of techniques are used to separate a complex wave into its sinusoidal components. For example, a sound wave detected by a microphone can be fed into a device called a spectrum analyzer. This device electronically analyzes the wave pattern and plots on a screen the amplitude of waves at different frequencies that are present in the complex wave. This amplitude-versus-frequency graph is called a **frequency spectrum** of the complex wave. A frequency spectrum of the complex wave we constructed in Fig. 17.15 is shown in Fig. 17.16. The height of the line at each frequency is proportional to the amplitude of the wave at that frequency. Notice that wave 2 has a greater amplitude than wave 1.

Examples of actual sound wave pressure variations and the corresponding frequency spectra are shown in Fig. 17.17. The complex wave shown in (a) is that of a piano when playing concert A at 440 Hz. The wave pattern is repetitive and represents a complex wave made mostly of a sinusoidal wave of frequency 440 Hz. The complex wave also consists of components of smaller amplitude at 880 Hz, 1320 Hz, and higher frequency, as is indicated by the frequency spectrum of the wave shown in (b). The complex wave from a violin (Fig. 17.17c) when playing the same pitch sound has many more peaks and valleys than that from the piano. The frequency spectrum of the violin complex wave is shown in (d). Notice that many more sinusoidal waves have been combined to form the violin complex wave. A component at 2200 Hz is especially prominent in the violin and is responsible for much of the bumpy appearance of the complex wave shown in (c).

The presence or absence of various sinusoidal waves in complex waves can be detected by our ears. This is one reason a violin sounds different from a piano, even though they play the same pitch sound. Our ears in conjunction with our brains have the remarkable ability to serve as a spectrum analyzer for sound waves. How this is done is still a matter of active research.

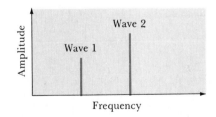

FIG. 17.16. A frequency spectrum for the complex wave shown in Fig. 17.15d.

FIG. 17.17. (a) The waveform of sound coming from a piano playing concert A (440 Hz) and (b) the frequency spectrum of the waveform. (c) and (d) are the waveforms and frequency spectrum of a violin when playing concert A.

(a)

(b)

(c)

(d)

17.7 Beat Frequencies

Another interesting application of the principle of superposition is the formation of beats, the complex wave that results when two sinusoidal waves of slightly different frequency are combined. The amplitude of the complex wave alternately increases and decreases, as shown in Fig. 17.18c. Each segment of rising and falling amplitude is called a **beat.**

To see why a beat pattern is formed, consider the varying pressure of sound produced at the ears of a person who is equidistant from two different sound sources, one of which vibrates at frequency 20 Hz and the other at 22 Hz. The fluctuating pressure at each ear as a function of time for each wave and for their resultant is shown in Fig. 17.18. At time $t = 0$, wave 1 is initiating an increase in air pressure whereas wave 2 is initiating decreased pressure. The two pressure variations cancel and the complex wave has zero amplitude at this time. At $t = 0.25$ s, wave 1 has completed 5.0 vibrations and is again initiating increased

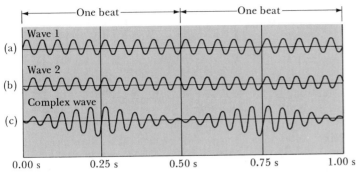

One beat — One beat

(a) Wave 1

(b) Wave 2

(c) Complex wave

0.00 s   0.25 s   0.50 s   0.75 s   1.00 s

FIG. 17.18. The fluctuating pressure as a function of time for (a) a 20-Hz sound wave and (b) a 22-Hz sound wave. (c) If both waves are present at the same time, the resultant complex wave has a fluctuating amplitude. Each region of rising and falling amplitude is called a beat.

pressure at the ear. Wave 2, vibrating at higher frequency, has completed 5.5 vibrations and instead of causing decreased pressure, now causes increased pressure. At this time, the two waves are said to be *in phase,* and the resultant complex wave has a large amplitude. At $t = 0.50$ s, the two waves are again out of phase—that is, wave 1 causes increased pressure that is canceled by the decreased pressure of wave 2. We have now completed our beat pattern, the rise and fall of the amplitude of the complex wave. During the time from $t = 0.50$ s to $t = 1.00$ s, another beat occurs.

One beat pattern occurs every 0.50 s, and the beat frequency is 1 beat/0.50 s or 2 beats/s. This equals the frequency difference of the waves used to construct the complex wave (22 Hz − 20 Hz). In general, the number of beats produced per unit time, called the **beat frequency** f_{beat}, equals the magnitude of the difference of the frequencies of the two waves used to construct the beat pattern:

$$f_{\text{beat}} = |f_1 - f_2|. \tag{17.9}$$

Beats are useful in precise frequency measurements. For example, a piano tuner can easily set the frequency of middle C on a piano to 262 Hz by observing the beat frequency produced when the piano and a 262-Hz tuning fork are sounded simultaneously. If a beat frequency of 2 Hz is observed, then the piano must be vibrating either at 260 Hz or at 264 Hz. The string is then tightened or loosened until the beat frequency is reduced to zero.

17.8 Doppler Effect

Until now, we have been studying situations in which the source of a wave and the observer are both stationary. In this case, the frequency of a wave as it leaves its source is the same as the frequency of the wave when detected by an observer. If either the source or observer is moving, though, a change in frequency occurs. Suppose you are standing near a racetrack as fast-moving race cars pass. While a car approaches, a high-pitched, whirring sound is heard, but as the car passes, the frequency or pitch of the sound suddenly and noticeably drops. A similar phenomenon occurs when one listens to the sound from the horn of a passing car or the whistle of a passing train. The change from a higher pitch as the sound source approaches to a lower pitch as it moves away is an example of the **Doppler effect**—any change in the observed frequency of a wave that is caused by the motion of the wave source or the observer.

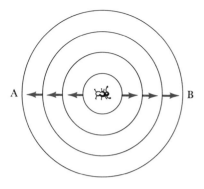

FIG. 17.19. The circular lines represent wave crests on the water caused by a beetle bobbing up and down at constant frequency in the same place.

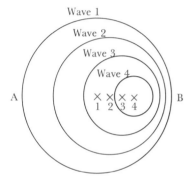

FIG. 17.20. If a water beetle moves to the right as it creates new waves, the crests are crowded closer together in front of the beetle and are spread farther apart behind the beetle.

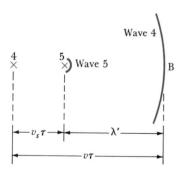

FIG. 17.21. A source emits a wave from position 4. During the period of time τ before the next wave is emitted, the source moves a distance $v_s\tau$ to the right. Wave 4 has moved a distance $v\tau$ from position 4. The separation λ' of wave 4 and the new wave 5 is $\lambda' = v\tau - v_s\tau$.

We can see why the observed frequency changes by examining the waves created by a water beetle bobbing up and down in the water at a constant frequency. If the beetle bobs up and down in the same place, the pattern of wave crests appears as shown in Fig. 17.19. The crests move away from the source at a constant speed. The large circle represents a crest produced by the beetle some time ago and the small circle is a crest produced more recently. The distance between adjacent crests is one wavelength and is the same toward observer A as toward observer B. The frequency of waves reaching both observers is the same.

Now suppose the bobbing beetle moves to the right at a slower speed than the speed at which the waves move. Each new wave produced by the beetle originates from a point farther to the right (Fig. 17.20). Wave crest 1, the first to be created, is a large circle centered at position 1 where the beetle started bobbing. Wave crest 2 is a smaller circle centered a step to the right at position 2, and so forth.

The waves in front of the beetle are more closely spaced than those behind the beetle. A Doppler frequency shift occurs because observer B in front of the beetle encounters the waves at a higher than usual frequency, and observer A behind the beetle encounters waves at lower than usual frequency.

To derive an equation for calculating this shift, we first calculate the separation of crests in front of the beetle. Take a look at the separation of wave 4 from the wave initiated by the beetle's fifth step (wave crest 5 shown in Fig. 17.21). If the beetle bobs up and down at frequency f, the time between the start of wave 4 and the start of wave 5 is $\tau = 1/f$. During this time the beetle moves a distance $v_s\tau$ to the right where v_s is the speed of "source" of the waves, the beetle in this case. Wave 4 has moved a distance $v\tau$ during that same time where v is the speed of the wave through the water. We see in Fig. 17.21 that the crests of waves 4 and 5 are separated by a distance

$$\lambda' = (v - v_s)\tau.$$

These waves are incident on observer B at a frequency

$$f' = \frac{\text{wave speed}}{\text{wavelength}} = \frac{v}{\lambda'} = \frac{v}{(v - v_s)\tau}.$$

But since $f = 1/\tau$, we find that

$$f' = f\frac{v}{(v - v_s)}. \tag{17.10}$$

This equation can be used to calculate the observed frequency f' of waves coming from a moving source that emits waves at frequency f. Here v_s is the speed of the source and is positive if the source is moving toward the observer and negative if the source is moving away from the observer.

A change in the observed frequency also occurs if the observer is moving and the source is stationary. Assume, for instance, that a friend remains stationary in the water of a swimming pool and creates waves by pushing a beach ball up and down. If you swim toward the source, you will encounter the waves more frequently than if you remain stationary. This is the Doppler effect due to a moving observer. If the observer moves toward the source at speed v_0, then the waves appear to move past the observer at a speed of $v + v_0$. Since the wavelength λ of the waves from a stationary source is uniform in all directions, the observer will encounter waves at a frequency

$$f' = \frac{\text{speed relative to waves}}{\text{wavelength}} = \frac{v + v_O}{\lambda}.$$

But, for the source, $\lambda = v/f$. Thus,

$$f' = f\left(\frac{v + v_O}{v}\right). \tag{17.11}$$

This equation can be used to calculate the frequency f' detected by an observer moving at speed v_O toward (v_O positive) or away from (v_O negative) a stationary source emitting waves at frequency f.

The general equation for the **Doppler effect** for sound waves combines Eqs. (17.10) and (17.11). The frequency f' of waves encountered by the observer is related to the frequency f of waves generated by the source by the following equation:

$$f' = f\left(\frac{v + v_O}{v - v_S}\right). \tag{17.12}$$

In this equation, v is the speed of the wave through the medium in which it travels; v_O is the speed of the observer relative to the medium, and is positive ($+$) if the observer moves toward the source and negative ($-$) if the observer moves away from the source; v_S is the speed of the source relative to the medium and is positive ($+$) if the source moves toward the observer and negative ($-$) if the source moves away from the observer.*

EXAMPLE 17.3 A train traveling at 40 m/s has a horn that vibrates at a frequency of 200 Hz. Calculate the frequency of the horn's sound heard by a person standing by the track (a) as the train approaches the observer, and (b) as the train moves away from the observer. The speed of sound in air is about 340 m/s.

SOLUTION (a) Because the observer is stationary, $v_O = 0$. As the train approaches the observer, $v_S = +40$ m/s (v_S is positive because the train approaches the observer). Thus, the frequency heard by the observer is

$$f' = f\left(\frac{v + v_O}{v - v_S}\right) = (200 \text{ Hz})\left(\frac{340 \text{ m/s} + 0}{340 \text{ m/s} - 40 \text{ m/s}}\right) = \underline{227 \text{ Hz}}.$$

(b) As the train departs, v_S equals -40 m/s (v_S is negative because the train is moving away from the observer). Thus, the frequency heard by the observer as the train moves away is

$$f' = (200 \text{ Hz})\left(\frac{340 \text{ m/s} + 0}{340 \text{ m/s} - (-40 \text{ m/s})}\right) = 200 \text{ Hz}\left(\frac{340 \text{ m/s}}{380 \text{ m/s}}\right) = \underline{179 \text{ Hz}}.$$

The frequency of the horn's sound decreases from 227 Hz to 179 Hz as the train passes. ∎

*This equation applies only to motion of the source and observer along the direction of a line joining them.

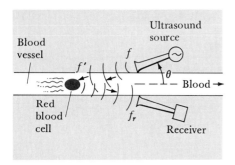

FIG. 17.22. Sound waves leave a source, are reflected from a moving red blood cell, and are detected by a receiver. The frequency shift of the detected waves indicates the speed of the blood.

The Doppler effect is the basis of a technique used to measure the speed of flow of blood. High-frequency sound waves called ultrasound are directed into an artery, as shown in Fig. 17.22. These waves are reflected by red blood cells moving toward the sound source. If the angle θ is small, the frequency of sound f' striking a red blood cell is related to the frequency f emitted from the source by the equation

$$f' = f\left(\frac{v + v_c}{v - 0}\right) = f\left(\frac{v + v_c}{v}\right), \tag{17.13}$$

where v is the speed of sound in blood and v_c is the speed of the red blood cell.

The cells reflect this sound to a receiver, which detects the sound. The cells act as a moving source at frequency f'. The frequency detected at the receiver f_r is related to f' by the equation

$$f_r = f'\left(\frac{v + 0}{v - v_c}\right) = f'\left(\frac{v}{v - v_c}\right).$$

Substituting for f' from Eq. (17.13), we find that the frequency at the receiver is

$$f_r = f\left(\frac{v + v_c}{v}\right)\left(\frac{v}{v - v_c}\right)$$

or

$$f_r = f\left(\frac{v + v_c}{v - v_c}\right). \tag{17.14}$$

EXAMPLE 17.4 A source of ultrasound emits waves at a frequency of 100,000 Hz. The waves are reflected by red blood cells moving toward the source at a speed of 0.3 m/s. Calculate the frequency of sound detected at a receiver next to the source. The speed of sound in the blood is 1500 m/s.

SOLUTION According to Eq. (17.14), the frequency of sound detected at the receiver is

$$f_r = (100,000\text{ Hz})\left(\frac{1500\text{ m/s} + 0.3\text{ m/s}}{1500\text{ m/s} - 0.3\text{ m/s}}\right) = \underline{100,040\text{ Hz}}.$$

There is a 40-Hz shift in frequency. The shift is usually detected as a beat frequency between the detected sound and a part of the 100,000-Hz sound emitted by the source. In this case, the beat frequency is 40 Hz. The beat frequency is easily measured and provides an accurate indicator of the blood speed. ∎

Doppler radar systems used by police to measure the speed of automobiles are very similar to the system just described for measuring the speed of red blood cells. Microwaves sent from a source in the police car reflect from a moving car and back to a receiver in the police car. If the car moves toward the police car, the microwaves are shifted to higher frequency. The greater the car's speed, the larger the frequency shift.

The equation used to calculate the Doppler frequency shift for electromagnetic waves such as microwaves, radio waves, and light waves is different from that used for sound waves. Conceptually, however, the frequency shifts observed when the source of a wave moves relative to the observer is the same for all types of waves.

Some of the most interesting applications of the Doppler effect are in astronomy. Astronomers have found that, like the frequency of sound from a train moving away from us, the frequency of light coming from distant galaxies is shifted to lower frequency than light from similar sources in our own galaxy. Based on calculations using the Doppler effect, it appears that nearby galaxies are moving away from us at speeds of about 250,000 m/s. Distant galaxies are moving away at speeds up to 90 percent the speed of light. The universe is moving apart and expanding in all directions.

Summary

1. **Waves:** A **wave** is a disturbance that travels from one place to another. A single disturbance moving away from a source is called an **impulse.** A repetitive disturbance is called a **periodic wave.** A **transverse wave** is one in which the disturbance causes objects in its path to vibrate perpendicular to the direction in which the wave travels. A **longitudinal wave** is one in which the disturbance causes vibration along the same direction in which the wave travels.

The motion of waves is characterized in terms of the following quantities:

Amplitude A is the maximum displacement from equilibrium caused by the wave motion.

Speed v is the distance that some part of the wave travels per unit time and is an intrinsic property of the medium in which the wave travels. The speed is the square root of the ratio of an elastic property of the medium and of an inertial property. For example, for sound waves in air or liquid, $v = \sqrt{B/\rho}$, where B is the bulk modulus of the medium (an elastic constant), and ρ is its density (an inertial constant).

Frequency f is the number of vibrations per unit of time of some point in the medium along which the wave travels and is usually equal to the frequency of the source causing the wave.

Wavelength λ is the distance between two identical points of adjacent waves (for example, the distance between two crests). The wavelength depends on the frequency and speed according to the following equation:

$$v = \lambda f. \tag{17.8}$$

2. **Reflection and transmission:** Whenever a wave passes the boundary between two different media, part of the incident wave is reflected and part is transmitted. The amplitude of the reflected wave depends on the difference in impedance of the two media,

where impedance is a measure of the difficulty a wave has in causing the medium to vibrate. If a wave moving in a medium of low impedance (such as air) strikes a boundary with a medium of higher impedance (such as water), the phase of the wave is inverted; a compression in the incident wave becomes a rarefaction in the reflected wave. If the wave passes from a medium of higher to lower impedance, the phase of the reflected wave is unchanged.

3. **Principle of superposition:** When two or more waves pass through the same medium at the same time, the net disturbance of any point in the medium is the sum of the disturbances that would be caused by each wave if alone in the medium at that time. The principle of superposition has many applications. **Complex waves** are produced by adding sinusoidal waves, each of different frequency and wavelength.

4. **Beats:** When the frequencies of two waves are slightly different, their addition results in a beat pattern. The beat frequency is related to the frequencies of the two waves by the equation

$$f_{\text{beat}} = |f_1 - f_2|. \tag{17.9}$$

5. **Doppler effect:** The frequency at which we observe a wave usually equals the frequency of vibration of the wave's source. However, if either the source or the observer is moving, the observed frequency f' will differ from the source frequency f. The Doppler frequency shift for sound waves is given by the equation

$$f' = f\left(\frac{v + v_0}{v - v_s}\right). \tag{17.12}$$

Here v is the speed of the wave; v_0, the observer speed, is positive $(+)$ if the observer moves toward the source and negative $(-)$ if the observer moves away from the source; and v_s, the source speed, is positive $(+)$ if the source moves toward the observer and negative $(-)$ if the source moves away from the observer.

Questions

1. Describe three phenomena that show that waves transfer energy from one place to another.

2. When a bell is rung, what happens to the vibrational energy initially present in the bell?

3. A television antenna has horizontal rods that are approximately equal in length to half the wavelength of the waves carrying the television signal. Use this fact to estimate the frequency of the waves. Their speed is 3×10^8 m/s.

4. Invent and describe an experiment to *estimate* the speed of sound in air. You can use only everyday items found in a drugstore or general store.

5. When an earthquake occurs, you feel the tremor before hearing the sound. Explain. If you are 20 km from the earthquake, *estimate* by calculation the time delay between the tremor and the sound.

6. Explain why the speed of sound is greater in humid air than in dry air at the same pressure and temperature.

7. How can you use the time delay between a lightning flash and the sound of thunder to estimate the distance to the place where the flash occurred? Justify any assumptions you make by using an example calculation.

8. Describe two useful types of information a geologist can obtain by detecting sound reflected from different materials under the earth's surface.

9. Two speakers hang from racks placed in an open field. When sound of the same frequency comes from both speakers, there are places on the field where no sound is heard and other places at about the same distance from the speakers where the sound is loud. Explain.

10. Two identical sound waves are sent down a long hall. Does the resultant wave necessarily have an amplitude twice that of the constituent waves? Explain. Under what conditions would we hear no sound?

11. Sound waves of all frequencies in the audio-frequency range (20–20,000 Hz) travel at the same speed in air. Explain how some common experience supports this statement.

12. How can we use the phenomena of beats to tune a piano string to exactly the same frequency as an electronic oscillator?

13. A child traveling on a merry-go-round blows a whistle. Describe the variation in sound frequency heard by a stationary observer some distance from the merry-go-round. Be sure to indicate what direction the child is moving relative to you at different times in this frequency variation. How is this similar to the variation of the frequency of light we detect coming from a spinning galaxy?

Problems

The speed of sound in air is assumed to be 340 m/s for all of these problems unless stated otherwise.

17.3 Amplitude, Speed, Frequency, and Wavelength

1. People can hear sounds ranging in frequency from about 20 Hz to 20,000 Hz. Calculate the wavelengths for each of these sounds.

2. (a) An orchestra tunes to concert A at 440 Hz. What is the wavelength of this sound? (b) In order for a bat's sonar to be able to detect small objects, such as insects, the wavelength of the sound the bat emits must be about the size of the object. What is the frequency if the wavelength is 3.0 mm?

3. A dolphin has a sonar system that emits sounds having a frequency of 2.0×10^5 Hz. Calculate the wavelength of the sound. Remember that the dolphin is not in the air.

4. A car drives at 22 m/s along a road that has posts at the side. A "wooshing" sound is heard every 0.7 s as the car passes a post. What distance separates the posts?

5. (a) The speed of sound in water, which has a density of 1000 kg/m³, is 1450 m/s. Calculate the bulk modulus of water. (b) Calculate the speed of sound in a long, thin bone whose Young's modulus Y is 1.6×10^{10} N/m² and whose density is 1800 kg/m³.

6. A telephone lineman is told to stretch the wire between poles to a tension of about 800 N. Not having a tension scale, the lineman decides to measure the speed of a pulse created in the wire when he hits it with a wrench. The pulse travels from one pole 60 m to another pole and back again in 2.6 s. The 60-m wire has a mass of 15 kg. Should the wire be tightened or loosened? Explain.

7. The salt in a salt dome in the earth is about as dense as clay, yet compressional sound waves travel about five times faster in a salt dome than in clay. Calculate the ratio of the elastic modulus expression $(B + \frac{4}{3}S)$ for salt compared to that for clay.

8. Radio waves travel at a speed of 1.7×10^8 m/s through ice. A radio wave pulse sent into the antarctic ice reflects off the earth at the bottom and returns to the surface in 32.9×10^{-6} s. How deep is the ice?

9. A flash of lightning is seen and 2.4 s later, thunder is heard coming from the same location. (a) Why is there this long delay? (b) How far away did the lightning flash occur? Justify any assumptions you make in your calculations.

10. A robot is built with tiny sound detectors at the ends of its arms. The sound is transmitted through a 1.2-m long tungsten rod to a recorder. (a) Calculate the delay time for the sound to pass through the rod. Tungsten has a density of 18,800 kg/m³ and a Young's modulus of 35.5×10^{10} N/m². (b) How does this delay time compare roughly (one-tenth, one hundred times, or whatever) to the delay time for a nerve impulse to travel that distance? Explain.

11. An earthquake produces a longitudinal pressure wave called a P-wave* that travels through the body of the earth at a speed $v = \sqrt{(B + \frac{4}{3}S)/\rho} = 7.8$ km/s. The quake also produces a transverse wave called an S-wave that travels at speed $v = \sqrt{S/\rho} = 4.5$ km/s. If the earth's density ρ equals 2400 kg/m³, use the above information to calculate the bulk modulus B and shear modulus S of the earth.

17.4 Reflection and Transmission

12. The echo of an ultrasound impulse from a piece of metal lodged in the eye is shown in Fig. 17.10. Calculate the distance between the object and the eye's surface. Ultrasound travels at about 1500 m/s in the eye.

*The P stands for primary. These waves travel faster than transverse waves and are the first ones we feel following an earthquake. Slower-traveling S-waves (secondary) are felt second.

13. A bat receives a reflected sound wave from a fly. If the reflected wave is returned to the bat 0.042 s after it is sent, how far is the fly from the bat? Would you expect the reflected wave pulse to be in or out of phase with the incident wave pulse? Explain.

14. A sound wave created by an explosion at the earth's surface is reflected by a discontinuity of some type under the earth's surface (Fig. 17.23). Calculate the distance from the surface to the discontinuity. Does it have a greater or lesser impedance to sound than the earth above it? Explain. Sound travels at about 3000 m/s through the top layer of the earth.

FIG. 17.23   Time (s)

15. Ultrasound pulses transmitted from left to right and from right to left across the skull are shown in Fig. 17.12. Estimate the distance that the midline of the brain is displaced. The speed of sound in the skull is approximately 1500 m/s.

17.5   Principle of Superposition

16. The pulses shown in Fig. 17.24 (time zero) are moving toward each other at speeds of 10 m/s. Draw graphs showing the resultant pulse at times of 0.1, 0.2, and 0.3 s.

FIG. 17.24   Position (m)

17. Repeat Problem 16 for the case where the pulse on the right in Fig. 17.24 is inverted rather than upright.

18. Two waves shown in Fig. 17.25 at time zero move toward each other at a speed of 10 m/s. Draw graphs of the resultant wave at times of 0.1, 0.2, and 0.3 s.

FIG. 17.25   Position (m)

19. Carefully add the waves shown in Fig. 17.26 to get the resultant complex waveform.

FIG. 17.26

20. Carefully add the waves shown in Fig. 17.27 to get the resultant complex waveform.

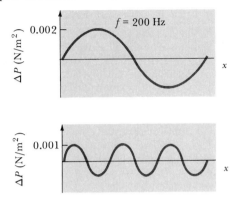

FIG. 17.27

21. (a) Make a frequency spectrum for the complex wave constructed in Problem 19 and (b) for the complex wave constructed in Problem 20.

22. A complex wave is made of two sinusoidal waves of equal amplitude. One wave has a frequency of 175 Hz and the other a frequency of 525 Hz. (a) Make a frequency spectrum of the complex wave. (b) Construct the waveform of the complex wave. It should include two full wavelengths of the lowest frequency wave. Start both waves with zero amplitude on the left and rising.

23. Repeat Problem 22 for a complex wave made of three sinusoidal waves: wave 1 at 175 Hz and amplitude 0.002 N/m², wave 2 at 350 Hz and amplitude 0.001 N/m², and wave 3 at 525 Hz and amplitude 0.001 N/m².

17.7   Beat Frequencies

24. (a) Two violins play sounds at frequencies of 440 Hz and 443 Hz. What beat frequency is heard? (b) The higher-frequency violin is tuned down in pitch so that a beat frequency of 1 Hz is heard. At what frequencies might this violin now be playing (there are two)? Explain.

25. A 30-Hz sound is annoying or dissonant to the ear. If one clarinet plays a sound at a frequency of 360 Hz, at what frequencies (more than one) might another clarinet simultaneously play so that dissonant 30-Hz beats are created? Explain.

26. A sound of wavelength 1.2 m and another of wavelength 1.0 m are played simultaneously. What beat frequency is heard?

27. A tuning fork vibrates at a frequency of 440 Hz. When it and another tuning fork vibrate together, a beat frequency of 2 Hz is produced. (a) What are the possible frequencies at which the second fork is vibrating? (b) Will a small piece of wax added to one of the tongs of the second fork cause its frequency to increase or decrease? Explain. (c) When the wax is added and both forks are sounded, the beat frequency decreases. Which of the frequencies in (a) was the actual frequency of vibration of the second tuning fork? Explain.

28. A hyperactive being from the planet Krypton is found to have cycles of activity triggered by hormones. One cycle has a period of eight days, and the other cycle a period of ten days. (a) Calculate the beat frequency—that is, the difference in frequency of these two cycles. (b) Calculate the beat period, the inverse of the beat frequency. This is the time from one large

maximum of hyperactivity (both hormones working at their maximum) to the next.

17.8 Doppler Effect

29. A car horn vibrates at a frequency of 250 Hz. (a) Calculate the frequency a stationary observer hears as the car approaches at a speed of 20 m/s and (b) departs at 20 m/s. (c) If the car is stationary, what frequency is heard by an observer approaching the car at 20 m/s and (d) by an observer departing from the car at 20 m/s?

30. A car drives at a speed of 25 m/s along a road parallel to a railroad track. A train traveling at 15 m/s sounds a horn that vibrates at 300 Hz. (a) If the train and car are moving toward each other, what frequency sound is heard by a person in the car? (b) If the train and car are moving away from each other, what frequency of sound is heard in the car?

31. A whistle is placed at the end of a stick that rotates in a horizontal plane. A stationary observer hears a sound coming from the whistle that varies from a minimum value of 395 Hz (when the whistle is rotating away from the observer) to a maximum of 405 Hz one-half turn later when the whistle is rotating toward the observer. When at rest, the whistle vibrates at 400 Hz. Calculate the speed of the whistle (a) when moving away from the observer and (b) when moving toward the observer. (*Note:* Astronomers use a similar technique to determine the angular velocity of galaxies. The speed v due to rotational motion, the angular velocity ω, and the radius of the galaxy are related by the equation $v = r\omega$. They use the Doppler shift of light to calculate v.)

32. A red blood cell travels at a speed of 0.40 m/s in a large artery. A sound with a frequency of 100,000.0 Hz enters the blood opposite the direction of flow. (a) Calculate the frequency of sound reflected from the cell and detected by a receiver. (b) If the emitted and received sounds are combined in the receiver, what beat frequency is measured?

33. A bat emits short pulses of sound at a frequency of 3.90×10^5 Hz. As the bat swoops toward a flat wall at a speed of 30 m/s, this sound is reflected from the wall back to the bat. (a) What is the frequency of sound incident on the wall? (b) Consider the wall as a sound source at the frequency calculated in (a). What frequency sound does the bat hear coming from the wall?

34. Suppose that police used an ultrasound speed detector to determine the speed of moving cars. (a) If a stationary sonar system emits sound at 1.00×10^5 Hz, what frequency of reflected sound is detected by the sonar system from a car moving toward the sonar at 40 m/s. (b) If the emitted and detected sounds are mixed in the detector, what beat frequency is heard?

35. A hungry student working in a cafeteria decides to eat plates of food that pass on a conveyor belt. The plates are separated by 3 m, and the belt moves at a speed of 9 m/min. (a) How many plates of food does the student eat per minute? (b) As the student's hunger is appeased, he moves with the belt at a speed of 6 m/min. Calculate the number of plates of food that now reach the student each minute. (This shift is similar to the Doppler shift for a moving listener.)

General Problems

G1. A record rotates at an angular velocity of 3.5 rad/s (33.3 rev/min). (a) At what speed does a groove move under the needle if it is 10 cm from the center of the record? (b) If the little hills and valleys in the groove cause the needle to vibrate at 500 Hz, what is the separation of two adjacent hills?

G2. A rock is dropped down a well. A splash is heard 3.0 s after the rock is dropped. (a) If you ignore the time required for the sound to travel up the well and the effect of the drag force of the air on the rock, how deep is the well? (b) Use the answer from part (a) to estimate the time required for the splash sound to travel up the well. (c) Recalculate the depth of the well after subtracting the time for the sound to travel.

G3. A basketball player's teammate shouts at him to catch a ball. *Estimate* the time required for the basketball player to respond—for a sound to travel a distance of 10 m, be detected by the ear, travel as a nerve impulse to the brain, be processed, then travel as a nerve signal for muscle action back to an arm, and finally cause a muscle contraction. Nerve impulses travel at a speed of about 120 m/s in humans. You will have to make reasonable guesses for any other numbers not known.

G4. Two sound speakers separated by 100 m face each other and vibrate in unison at a frequency of 85 Hz. Determine three places between the speakers where the sound is almost zero.

G5. A Doppler speed meter operating at exactly 1.02×10^5 Hz emits sound waves and detects the same waves after they are reflected from a baseball thrown by the pitcher, Nolan Ryan. The receiver "mixes" the reflected wave with a small amount of the emitted wave and measures a beat frequency of 0.30×10^5 Hz. How fast is the ball moving?

INTERLUDE   Dimensional Analysis

You may recall from Chapter 1 that a small number of physical quantities, seven to be exact, are called basic quantities. Since beginning our study of physics, we have used five of these quantities—length, time, mass, temperature, and quantity of matter. In the SI metric system the units of these quantities are, respectively, the meter, second, kilogram, kelvin, and mole. Other derived quantities such as speed, acceleration, and force can be defined in terms of these basic quantities. The **dimensions** of a derived quantity are the basic quantities used in its definition. For example, the dimensions of speed are length divided by time (l/t or lt^{-1}). The dimensions of acceleration are lt^{-2} and those of force are mlt^{-2}.

   Dimensional analysis involves the use of the dimensions of quantities to derive mathematical equations relating the quantities. A person unfamiliar with dimensional analysis will be surprised at the simplicity of deriving these equations. To illustrate, we will introduce the technique using an example.

EXAMPLE 1   Derive an equation for the speed of a wave on a stretched string.

SOLUTION   Dimensional analysis relies to a certain extent on our physical intuition. We must, therefore, make an educated guess at what quantities affect the wave's speed. Fortunately, if we guess wrong, the error is often made apparent by our subsequent calculations. For this example, let us assume that the speed of a wave on a stretched string depends on its tension T, on its mass per unit length μ, and on the amplitude of the wave A. Actually, the speed does not depend on A, but we may not realize that before starting our derivation. In this example our error is corrected for us by the techniques of dimensional analysis.

   We wish now to determine the type of equation relating these quantities to speed. Is the equation of the form $v = T/\mu A$, $T^2 A/\mu$, T/A, $\sqrt{T/\mu}$, or of some other form? We can write a general equation that encompasses all of the above expressions as follows:

$$v = CT^x \mu^y A^z, \tag{1}$$

where $x, y,$ and z are unknown exponents of the quantities T, μ, and A, respectively, and C is an unknown constant that has no dimensions (such as π, 8, or some other number). If the speed is given by the equation $v = T^{1/2}\mu^{-1/2}$, then $x = \frac{1}{2}, y = -\frac{1}{2}, z = 0$, and $C = 1$. Dimensional analysis can be used to determine the values of $x, y,$ and z but not the value of C, which must be determined by experiment or by a more detailed derivation using basic principles.

   Next, we consider the dimensions of all quantities appearing in Eq. (1):

Physical quantity	Symbol	Dimensions
Speed of wave	v	lt^{-1}
Tension in string (a force)	T	mlt^{-2}
Mass per unit length	μ	ml^{-1}
Amplitude of wave	A	l

347

Substituting the dimensions for each quantity in Eq. (1), we find that

$$lt^{-1} = (mlt^{-2})^x (ml^{-1})^y (l)^z,$$
$$lt^{-1} = (m^x l^x t^{-2x})(m^y l^{-y})(l^z),$$
$$lt^{-1} = m^{x+y} l^{x-y+z} t^{-2x}. \tag{2}$$

Remember that C has no dimensions. In order for the last equation to be dimensionally correct, each dimension must appear to the same power on the left side and on the right side of the equation. Length, for instance, appears to power one on the left side of the equation and to power $x - y + z$ on the right side. For the equation to be dimensionally correct, we require that

$$1 = x - y + z.$$

For time to have the same dimensions on each side of the equation, we require that

$$-1 = -2x.$$

Mass does not appear on the left side of Eq. (2). We say that mass appears to the power zero, since $m^0 = 1$. For mass to have the same dimensions on each side of Eq. (2), we require that

$$0 = x + y.$$

We now have three equations relating x, y, and z:

$$1 = x - y + z,$$
$$-1 = -2x,$$
$$0 = x + y.$$

Solving these equations, we find that $x = \frac{1}{2}, y = -\frac{1}{2}$, and $z = 0$. We have determined that the speed of a wave on a string is calculated using an equation of the form

$$v = CT^{1/2} \mu^{-1/2} A^0 = \underline{CT^{1/2} \mu^{-1/2}}.$$

The constant C cannot be determined by dimensional analysis. A derivation of this same equation using Newton's second law indicates that $C = 1$ and that

$$v = T^{1/2} \mu^{-1/2} = \sqrt{T/\mu},$$

which is just Eq. (17.6) given in Table 17.1. ∎

We see that deriving equations through dimensional analysis involves the following steps: (1) Choose a set of quantities on which a quantity such as wave speed depends. If you choose wrong, dimensional analysis may indicate your error by either showing that the quantity should be to the power zero (it does not belong in the expression), or you may find that there is no satisfactory set of exponents for the set of quantities you have chosen. In the latter case, you will have to try again with a different set of quantities. (2) Having selected the quantities, construct an equation of the form $v = CT^x\mu^y A^z$ and substitute the dimensions for each quantity in this equation. (3) For the equation to be dimensionally correct, the exponent of each dimension must be the same on each side of the equation. This requirement allows us to write several equations relating x, y, and z. (4) Solve these equations for x, y, and z.

EXAMPLE 2   Use dimensional analysis to derive an equation for the magnitude of the centripetal acceleration of a mass m moving in a circle of radius r at constant speed v.

SOLUTION   We will assume that the acceleration depends on all three of these quantities: m, r, and v. The equation relating centripetal acceleration to these quantities is written in the general form

$$a_c = Cm^x r^y v^z.$$

Substituting the dimensions into the above equation (remember that C is dimensionless), we find that

$$lt^{-2} = (m)^x (l)^y (lt^{-1})^z,$$
$$lt^{-2} = m^x l^{y+z} t^{-z}.$$

In order for the equation to be dimensionally consistent, we find that

$$0 = x,$$
$$1 = y + z,$$
$$-2 = -z.$$

These equations are solved to determine the values of the exponents: $x = 0$, $y = -1$, and $z = 2$. Thus,

$$a_c = Cm^0 r^{-1} v^2 = \underline{Cv^2/r}.$$

A derivation such as done in Chapter 6 indicates that $C = 1$ and that $a_c = v^2/r$. ∎

EXAMPLE 3   Suppose that a hole is drilled through the earth (Fig. 1). Use dimensional analysis to determine an equation for the time τ needed for a person to fall down the hole from one side of the earth to the other and then back again. Ignore the drag force of air on the falling person.

SOLUTION   His acceleration as he starts to fall is the acceleration due to gravity g, but as he approaches the center of the earth, the acceleration decreases

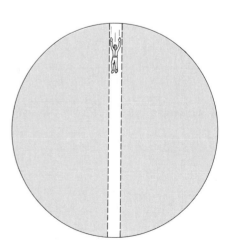

FIG. 1

to zero. Even though the acceleration changes, the time for the trip must depend on his initial acceleration g and also on the distance of the trip (four times the earth's radius R). Our previous experience with falling masses leads us to believe that all masses should fall with the same acceleration if we ignore drag forces. So we will not include the person's mass as a quantity affecting the time for the trip. The effect of the earth's mass M is included in the value of g. To be sure that the earth's mass does not affect the time for the trip in some other way, we will include it in our dimensional analysis.

The equation for the time τ for the round trip is written in terms of the acceleration of gravity g, the radius of the earth R, and the earth's mass M as follows:

$$\tau = CR^x g^y M^z.$$

Substituting the dimensions for each quantity, we find that

$$t = (l)^x (lt^{-2})^y (m)^z,$$
$$t = l^{x+y} t^{-2y} m^z.$$

To make this equation dimensionally correct, we require that

$$0 = x + y,$$
$$1 = -2y,$$
$$0 = z.$$

Solving for x, y, and z, we find that $z = 0, y = -\frac{1}{2}$, and $x = \frac{1}{2}$. The equation for the time of one trip through the earth and back again is of the form

$$\tau = CR^{1/2} g^{-1/2} M^0 = C\sqrt{R/g}.$$

A calculation using Newton's second law and techniques developed in Chapter 16 leads to the equation $\tau = 2\pi\sqrt{R/g}$. Substituting $R = 6.4 \times 10^6$ m and $g = 9.8$ m/s^2, we find that $\tau = 2\pi\sqrt{(6.4 \times 10^6 \text{ m})/(9.8 \text{ m/s}^2)} = 5.1 \times 10^3$ s $= 1.4$ hr. Actually, the drag force of air would greatly increase the time needed for the trip. ∎

The examples of dimensional analysis presented here and in the problems that follow are fairly simple. In the hands of an expert, dimensional analysis provides guidance in solving much more complex problems. Lord Rayleigh, a brilliant and versatile physicist of the early twentieth century, combined a keen insight into the essentials of a physics problem with the techniques of dimensional analysis to provide neat and simple solutions to a wide variety of problems. He and others have found dimensional analysis to be an important technique for guiding and checking their work in physics. *

*A particularly good dimensional analysis reference for the beginning student is H. E. Huntley, *Dimensional Analysis* (New York: Holt, Rinehart and Winston, 1951).

Problems

1. Use dimensional analysis to determine the swinging frequency of a simple pendulum consisting of a mass m hanging in the earth's gravitational field g at the end of a string of length L.

2. Use dimensional analysis to determine the dependence of the speed of sound v in air on the density ρ and bulk modulus B of the air.

3. A string of length L and of mass m when stretched with tension T vibrates at frequency f. Use dimensional analysis to determine the functional dependence of f on these other quantities.

4. An object falls from rest a distance h in a time t. Use dimensional analysis to determine the functional form of an equation for calculating h at different times.

Chapter opening illustration with the number 18.

Standing Waves and Sound

Onis Morrison pulls his bow across a catgut string while Gerti Zinn sings a refrain of "Old Joe Clark":

> Round and around, Old Joe Clark,
> Round and around I say,
> He'd follow me ten thousand miles,
> To hear my fiddle play.

The sounds of an old-time fiddler or a concert violinist start as the vibration of a string fixed at both ends. The sound is different from the whine of truck tires on a highway or the groan of a furnace. It is called music.

Much of the uniqueness of musical sounds depends upon what physicists call *standing-wave vibrations*. For a fiddler playing "Old Joe Clark," a standing-wave vibration is the unique vibration of the string that occurs when it is excited by a bow. The string's vibration is the result of waves moving back and forth in opposite directions along the string. These waves interfere with each other and produce a vibration of the string as a whole that looks like a wave that is not moving; hence the name *standing wave*. The standing-wave vibration of a string affects the pitch, quality, and loudness of the violin's sound. The study of standing waves is a natural place to start this chapter, much of which is devoted to musical types of sound.

FIG. 18.1. Onis Morrison and Gerti Zinn in concert. (Courtesy of the Ozark Folk Center, an Arkansas State Park)

18.1 Standing Waves on Strings

To begin our study of standing waves on a string, let us consider the rope shown in Fig. 18.2. One end of the rope is attached to a fixed support while the other end is held in a woman's hand. The woman wishes to shake the rope up and down at a frequency that will produce large amplitude vibrations on the rope.

To start, she sends a pulse down the rope (b–d). When it reaches the fixed end, the upward pulse is reflected and inverted to a downward pulse traveling back toward her hand (e–g). When it reaches her hand, it is once again reflected and inverted to an upward pulse (h).

To make this pulse bigger, she can give her hand another upward shake, making sure to time the shake so that it adds to the previous upward pulse that has just been reflected (h). If her upward shake is too soon, the new pulse she creates will interfere destructively with the inverted pulse that is returning to her hand. To make each new pulse interfere constructively with a previous pulse, then, she must shake her hand upward each time a previous pulse returns to her hand. If she continues to synchronize her shaking frequency with the start of reflected pulses from her hand, the amplitude of the disturbance traveling along the rope grows and looks like the vibration shown in part (l).

The frequency at which she should vibrate her hand to cause the rope vibration shown in (l) depends on the length L of the rope and the speed v of the pulse along the rope. The time for one complete round trip of a pulse down the rope and back is

$$\tau = \frac{2L}{v}.$$

Thus, the frequency $f_1{}^*$ of the upward shakes should be one shake per time τ, or

$$f_1 = \frac{1}{\tau} = \frac{1}{2L/v} = \frac{v}{2L}. \qquad (18.1)$$

If the frequency is less than this, each new pulse will interfere destructively with previous pulses and the rope will vibrate little. Should the woman raise and lower her hand at a frequency $f_1 = v/2L$, each new pulse will add to the last and create large-amplitude vibrations.

The vibration shown in (l) is called the **fundamental frequency vibration** or the **first harmonic vibration** of the string or rope with fixed ends. It is the lowest frequency of the allowed large-amplitude vibrations on the rope. If you pluck a guitar string of length L, the sound you hear coming from the guitar is at the fundamental frequency given by Eq. (18.1). The speed that appears in this equation is the speed of a pulse on the string and not the speed of sound in air.

EXAMPLE 18.1 Calculate the fundamental frequency of vibration of the A-string on a violin. Its mass is 0.30 g, it is 33 cm long, and it is stretched with a tension of 77 N.

SOLUTION The speed of a wave on a string is given by $v = \sqrt{T/(m/L)}$ (Eq. 17.6), where m is the mass of the vibrating portion of the string and L is its length. For the violin A-string

$$v = \sqrt{\frac{77 \text{ N}}{0.30 \times 10^{-3} \text{ kg}/0.33 \text{ m}}} = 290 \text{ m/s}.$$

The fundamental frequency of vibration is

$$f_1 = \frac{v}{2L} = \frac{290 \text{ m/s}}{2(0.33 \text{ m})} = 440 \text{ s}^{-1} = \underline{440 \text{ Hz}}. \qquad \blacksquare$$

A string will vibrate with large amplitude at other frequencies besides the fundamental. Consider again the woman shaking a rope. If she shakes the rope at a frequency twice the fundamental frequency, two pulses travel along the

FIG. 18.2. Pulses generated by shaking a rope up and down cause the vibration pattern shown in (l).

*The subscript "1" on f_1 indicates the first or lowest frequency at which she can shake the rope to get large-amplitude vibrations. There are higher frequencies at which large-amplitude vibrations also occur, as we will soon learn.

(a)

(b)

(c)

(d)

(e)

(f)

FIG. 18.3. If two pulses are generated on a stretched rope in the time required for one round trip of a pulse, the second harmonic vibration pattern shown in (f) is produced.

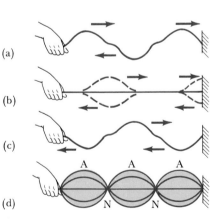

(a)

(b)

(c)

(d)

FIG. 18.4. If we initiate three pulses in the time needed for a pulse to make one round trip along the rope, we get the third harmonic vibration pattern as shown in (d).

rope as shown in Fig. 18.3a. Each pulse that returns to her hand after reflection from the opposite end is reinforced by another upward shake of her hand.

However, as a downward, reflected pulse passes the middle of the rope, it interferes destructively with an upward pulse that has just left the hand (c), and the middle of the rope never becomes displaced. As the inverted, reflected pulse approaches the hand and the upright pulse approaches the fixed end, as shown in (d), the rope's shape appears inverted compared to its shape in (a). After the pulses are reflected from the hand and from the fixed end of the rope, as in (e), the rope returns to its original shape as shown in (a).

The vibration pattern of the rope when shaken at a frequency f_2 equal to twice the fundamental frequency,

$$f_2 = 2f_1 = 2\left(\frac{v}{2L}\right),$$

is shown in (f) and is called the **second harmonic vibration** or the **first overtone vibration.** The place where the rope undergoes no vibration is called a **node** and is labeled by an N in (f). The places where the vibration amplitude is greatest are called **antinodes** (A in Fig. 18.3f).

If the rope is shaken at three times the fundamental frequency,

$$f_3 = 3f_1 = 3\left(\frac{v}{2L}\right),$$

three pulses travel along it. Each pulse is reinforced by an upward motion of the hand at just the right time and large amplitude vibrations are again produced. The vibration pattern that results is shown in Fig. 18.4d and is called the **third harmonic vibration** or the **second overtone.**

In general, a rope or string will vibrate with large amplitude at any frequency such that pulses originating at the source (the hand) reinforce pulses that have been reflected from the source after traversing the string. This condition is met when the string's *standing-wave vibration frequency* is

$$f_n = n\left(\frac{v}{2L}\right) \qquad \text{where } n = 1, 2, 3, 4, \ldots. \qquad (18.2)$$

Thus, there are many frequencies at which we can shake the rope or string and get large amplitude vibrations. Each frequency corresponds to a different value of the positive integer n. The wave patterns that are produced on the rope for these different values of n do not appear to travel or move. They appear to be stationary vibrations, called **standing waves,** and are caused by the interference of waves moving in opposite directions. The vibration patterns for several standing waves are shown again in Fig. 18.5. Notice that they are all integral multiples of the fundamental frequency.

The wavelengths of these standing-wave vibrations are calculated using Eq. (17.8), $\lambda = v/f$. For example, the wavelength of the fundamental vibration is

$$\lambda_1 = \frac{v}{f_1} = \frac{v}{(v/2L)} = 2L.$$

For the second harmonic, the wavelength is

$$\lambda_2 = \frac{v}{f_2} = \frac{v}{2(v/2L)} = L.$$

FIG. 18.5. The first four standing-wave vibration patterns on a string fixed at both ends.

Frequency	Wavelength	A	Name
$f_1 = \dfrac{v}{2L}$	$\lambda_1 = 2L$		First harmonic (fundamental)
$f_2 = 2\left(\dfrac{v}{2L}\right)$	$\lambda_2 = \dfrac{2L}{2}$		Second harmonic (first overtone)
$f_3 = 3\left(\dfrac{v}{2L}\right)$	$\lambda_3 = \dfrac{2L}{3}$		Third harmonic (second overtone)
$f_4 = 4\left(\dfrac{v}{2L}\right)$	$\lambda_4 = \dfrac{2L}{4}$		Fourth harmonic (third overtone)

In general, the wavelengths of the string's harmonic vibrations are expressed by the equation

$$\lambda_n = \frac{v}{f_n} = \frac{v}{n(v/2L)} = \frac{2L}{n} \qquad \text{where } n = 1, 2, 3, 4, \ldots . \qquad \textbf{(18.3)}$$

Remember that these wavelengths are for waves traveling on the string and not for the sound waves in the air caused by the string's vibrations.

EXAMPLE 18.2 If tuned correctly, the fundamental frequency of the A-string of a violin is 440 Hz. List the frequencies of several other standing-wave vibrations of the A-string.

SOLUTION The string, when bowed or plucked, vibrates simultaneously at many different frequencies. These frequencies are integral multiples of the fundamental and include 880 Hz, 1320 Hz, 1760 Hz, and so forth. ∎

EXAMPLE 18.3 A banjo D-string is 0.69 m long and has a fundamental frequency of 587 Hz. The string vibrates at a higher fundamental frequency if its length is reduced by holding the string down over a fret. Where should a fret be placed to play the note F-sharp, which has a fundamental frequency of 734 Hz?

SOLUTION The fundamental frequency of vibration f_1 is related to the string length L by the equation $f_1 = v/2L$, or $L = v/2f_1$. We cannot solve immediately for the unknown length when the string is vibrating at 734 Hz since we do not know v, but we can use the fact that the string is 0.69 m long and vibrates at 587 Hz to solve for v:

$$v = f_1(2L) = (587 \ s^{-1})(2 \times 0.69 \ \text{m}) = 810 \ \text{m/s}.$$

Since the speed is constant as the length is changed, we can solve for the unknown length L' when the string is vibrating at the higher frequency f_1' equal to 734 Hz:

$$L' = \frac{v}{2f_1'} = \frac{810 \text{ m/s}}{2 \times 734 \text{ s}^{-1}} = 0.55 \text{ m}.$$

The fret must be located a distance 0.69 m − 0.55 m = $\underline{0.14 \text{ m}}$ from the end of the string. ∎

18.2 Standing Waves in Air Columns

Standing-wave vibrations occur in many objects besides strings. They are manifested in the swaying of tall buildings, the twisting vibrations of bridges, the motion of a drumhead, and the vibration of the gas in an air column. This last type of vibration is important for the generation of sound by brass and woodwind musical instruments.

The formation of standing-wave vibrations in air columns (pipes) is very similar in concept to the formation of such vibrations in strings. A pressure pulse is initiated at one end of the pipe. This might be done by blowing air across an open-hole tube, such as in a flute. Or it might be done by blowing air through a vibrating reed, such as in a clarinet. Little swirls or puffs of air enter the air column in a repetitive fashion.

Consider the fate of one of these pressure pulses while moving in an air column with both ends open, called an **open pipe.** The pressure pulse moves along the pipe as shown in Fig. 18.6a–c. At the open end, part of the pulse is transmitted out of the pipe and part is reflected. It may surprise you that sound

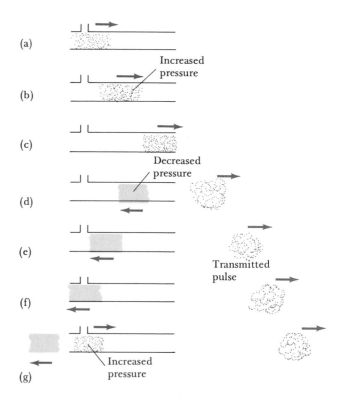

FIG. 18.6. A pulse of increased pressure traveling along a pipe (a–c) is partially transmitted and partially reflected at the end (d). After being reflected from the end where it started, a pulse of increased pressure again moves to the right (g).

is reflected at the open end of a pipe. The pipe confines the movement of air molecules and causes the air inside the pipe to have a different impedance to sound waves than the air outside the pipe. Whenever there is an abrupt change of impedance, such as occurs at the open end, waves are partially reflected.

The reflected pressure pulse has an inverted phase. If the incident pulse caused increased pressure, the reflected pulse causes decreased pressure.* This pulse of decreased pressure moves back toward the end of the pipe where it started (d–f) and is reflected at the open end on the left (g). After reflections from both open ends, the pulse again has increased pressure and is moving toward the right. If a new puff of air is initiated at this time, it interferes constructively with the reflected pulse, and its amplitude is increased. If the new puff of air is too early or too late, it interferes destructively with the reflected pulse, and the amplitude is reduced.

The frequency of standing-wave vibrations in pipes depends on the time needed for a pulse to travel down the pipe and back again. If the sound pulse travels at a speed v along a pipe of length L, then the time τ needed to travel a distance $2L$ is $\tau = 2L/v$. The puffs should occur once each time period τ or at a frequency $f_1 = 1/\tau = v/2L$. This equation is the same as that for the fundamental frequency of vibration of a string. However, in this case, v is the speed of a pulse through the air inside the pipe (about 340 m/s).

Just as a string can be excited to vibrate at multiples of the fundamental frequency, the pipe also vibrates with large amplitude if excited by puffs of air at integral multiples of the fundamental frequency. In general, an *open pipe* can vibrate with large amplitude at the following *standing-wave frequencies:*

$$f_n = n\left(\frac{v}{2L}\right) \quad \text{where } n = 1, 2, 3, \ldots, \quad \textbf{(18.2)}$$

where v is the speed of air in the pipe, and L is its length. Musical instruments that function as open pipes include the flute, recorder, oboe, and bassoon.

Next, let us turn to a pipe called a **closed pipe,** which is closed at one end and open at the other end. Musical instruments that behave somewhat like closed pipes include the clarinet, trumpet, trombone, and human voice. The closed end of these instruments is at the reed, mouthpiece, or vocal cords, respectively. The determination of the standing-wave vibration frequencies of closed pipes is slightly more complicated than for open pipes.

Suppose that a pulse of increased pressure is initiated near the open end of the pipe (Fig. 18.7a). When the pulse is reflected at the closed end, its phase is unchanged (d). When the reflected pulse travels back to the open end, it is once again reflected. However, this time there is a change in phase; it becomes a pulse of decreased pressure (g). If a new pulse of increased pressure is initiated, it cancels the effect of the reflected pressure pulse. To add to the reflected pulse in a constructive way, we must wait for it to make an additional trip down the pipe and back again. During each round trip, its phase is inverted once. After two round trips it is again a pulse of increased pressure (m), and a new pulse adds constructively to increase its amplitude. In order, then, for a large-amplitude

*The pressure pulse changes phase at the end of an open tube. A displacement pulse, which indicates the motion of air molecules, does not change phase at the boundary of an open tube. We are discussing pressure pulses, not displacement pulses.

FIG. 18.7. A pulse of increased pressure must make two trips up the pipe and back before it again becomes a pulse of increased pressure.

standing wave to be produced, the time τ between pulses must be $4L/v$. The frequency of the fundamental is $f_1 = 1/\tau = v/4L$.

The closed pipe is unique in that overtone vibrations do not occur at all frequencies that are integral multiples of the fundamental frequency. For example, at twice the frequency of the fundamental, a pulse makes only one trip up and down the pipe before another pulse is started. The reflected pulse, which is inverted, interferes destructively with the new pulse. The pipe cannot vibrate at twice the fundamental frequency. A closed pipe can, however, vibrate at *odd* integral multiples of the fundamental frequency. The *standing-wave vibration frequencies of a closed pipe* are

$$f_n = n\left(\frac{v}{4L}\right) \qquad \text{where } n = 1, 3, 5, 7, \ldots \qquad \textbf{(18.4)}$$

EXAMPLE 18.4 The fundamental frequency of vibration of a hole in a bed of rocks is 35 Hz. (a) How deep is the hole? (b) At what other frequencies would the hole resonate if air were blown across the open end?

SOLUTION (a) The hole is closed at the bottom so that its fundamental frequency is given by $f_1 = v/4L$. Thus,

$$L = \frac{v}{4f_1} = \frac{340 \text{ m/s}}{4(35 \text{ s}^{-1})} = \underline{2.4 \text{ m}}.$$

(b) A closed pipe vibrates only at odd multiples of the fundamental frequency. If the speed of the air blowing across the top of the hole is increased, higher harmonics are excited. The next higher is at three times the fundamental frequency, or $f_3 = 3f_1 = \underline{105 \text{ Hz}}$. The next allowed frequency is $f_5 = 5f_1 = \underline{175 \text{ Hz}}$. ∎

EXAMPLE 18.5 A flute made from a straw behaves like an open pipe and has six holes cut in it (Fig. 18.8). The length of the flute's air column (called the bore length) is the distance from the mouth to the nearest open hole. Opening a hole is like cutting the end off the straw and decreasing its length.

With all holes closed, the bore length is 16 cm and the flute vibrates with a fundamental frequency f_1. With the third hole open, the fundamental frequency f_1' equals $\frac{4}{3}f_1$. What is the flute's bore length with the third hole open?

SOLUTION The ratio of f_1' and f_1 equals 4/3. Substituting Eq. (18.2) for the fundamental frequencies and canceling the speed of sound, which is the same for both vibrations, we find that the fundamental frequencies and bore lengths are related by the equation

$$\frac{4}{3} = \frac{f_1'}{f_1} = \frac{v/2L'}{v/2L} = \frac{L}{L'}$$

or

$$L' = \frac{3}{4}L = \frac{3}{4}(16 \text{ cm}) = \underline{12 \text{ cm}}.$$ ∎

FIG. 18.8. A flute made by cutting holes in a straw. The length of the air column is the distance from the mouth to the nearest open hole.

18.3 Pitch

The word **pitch** is used to indicate a person's subjective impression about the frequency of a sound. A high-pitch sound like that from a piccolo has a high vibration frequency; while a low-pitch sound like that from a foghorn has a low vibration frequency. A normal ear is sensitive to sounds in the audio frequency range from about 20 Hz to 20,000 Hz. As a person grows older, the limits of this hearing range shrink.

Most sounds we hear are complex waves consisting of sinusoidal waves of many different frequencies. The sound we call noise includes waves of a broad range of frequencies, and we seldom associate a pitch with the noise. However, if the complex sound consists of a fundamental frequency vibration and higher harmonics whose frequencies are multiples of the fundamental frequency, then we usually identify the pitch of the sound as being that of the fundamental frequency. This is true even when the amplitudes of the higher harmonics are greater than that of the fundamental.

Several physiological aspects of our hearing system can cause variations in the pitch we hear even though the frequency of the sound is unchanged. For example, a loud, high-frequency sound may seem to have a higher pitch than the same-frequency sound played softly. Under some conditions these variations may be considerable. Fortunately, these conditions do not often arise in music, and the pitch change can usually be ignored.

18.4 Quality of Sound

A violin playing concert A at 440 Hz does not sound like an oboe or French horn playing the same note. Why do musical instruments sound differently even when they play equally loudly at the same pitch? We say that the sounds of the instruments have different **quality** or timbre, the characteristic that distinguishes one sound from others of equal pitch and loudness.

Differences in the quality of various sounds are related to their frequency spectra. When a musical instrument is played, more than one of its standing-wave vibrations is excited. For example, when bowing a violin string, ten or twenty standing-wave vibrations are excited simultaneously. The pitch of the sound is that of the lowest standing-wave frequency present in the complex wave. *The quality of the sound depends on the number and relative amplitudes of the other standing-wave frequencies that make up the complex sound wave and on the way that this spectrum of tones changes with time.*

The frequency spectrum of a sound source is usually influenced by several parts of the source. For instance, when a person is talking or singing, air rushes through the vocal cords. The cords vibrate open and closed at a frequency that depends on their tension. The puffs of air entering the throat (Fig. 18.9a) produce a sound wave such as shown in Fig. 18.9b. A frequency spectrum of this complex wave is shown in Fig. 18.9c.

The sound coming from a person's mouth does not have the sound spectrum shown in (c). A person's throat, mouth, and nasal passage (called the vocal tract) are cavities that have their own resonant frequencies (Fig. 18.10). These cavities, taken together, form an unusually shaped air column that is approximately 17 cm long and is like a closed pipe (closed at the vocal cords and open at

(a)

(b)

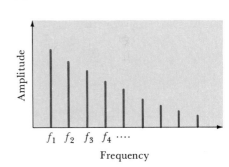

(c)

FIG. 18.9. (a) As the vocal cords open and shut, bursts of air move along the throat. (b) The pressure wave that moves is a complex wave comprising many individual frequency components. (c) The frequency spectrum of the complex wave in (b).

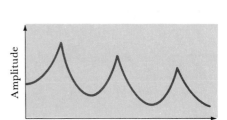

FIG. 18.10. The frequency response of the vocal tract. The first broad resonance frequency is characteristic of a closed pipe that is 17 cm long. The other resonant frequencies are higher odd harmonics.

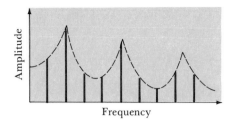

FIG. 18.11. The sound radiated from the mouth is a combination of the vocal-cord frequency spectrum (Fig. 18.9c) as altered by the vocal-tract resonator (Fig. 18.10).

FIG. 18.12. Differences in the frequency spectra for the words *who'd* and *heed* allow us to distinguish these words.

the mouth). The first four resonant frequencies of this "pipe" are at approximately 500, 1500, 2500, and 3500 Hz.* If the cavity is longer, its resonant frequencies are at lower frequency.

This vocal tract is excited at the many frequencies of the complex wave coming from the vocal cords (Fig. 18.9c). Those frequencies from the vocal cords that are closest to the resonant peaks of the vocal tract are accentuated; the others are attenuated. As a result, the sound that leaves the mouth includes waves at frequencies initiated by the vocal cords, but the amplitudes of these waves are increased or decreased in proportion to the vocal tract resonance curve at each frequency. For the frequency spectra of the vocal cords and vocal tract shown in Figs. 18.9c and 18.10, respectively, the frequency spectrum of the sound leaving the mouth appears as shown in Fig. 18.11.

To change the quality of sound, the shape and size of the vocal tract are altered, causing variations in the frequency spectrum of the vocal tract and thus of the radiated sound. Assume for instance, that a person says the words *who'd* and *heed* such that the first resonant frequency of the vocal tract for each word occurs at about 300 Hz (Fig. 18.12). The second resonant frequency in *who'd* is at about 900 Hz, where it is expected. When saying *heed*, the tongue is raised in the middle of the mouth. This attenuates the 900-Hz resonance. In fact, the next strong resonant frequency we hear when saying *heed* is near 2000 Hz. The word *heed* has a quality associated more with higher-frequency sounds because of the attenuation of these lower-frequency vocal tract resonances (Fig. 18.12). The regions where the vocal tract reinforces the vibrations from the vocal cords are called **formants.**

The subject of quality is even more complex than we have described. No single explanation exists for the sound quality produced by all instruments or by our voices, but the characteristics we've mentioned can help us understand some aspects of this subject.

18.5 Intensity and Intensity Level

We have distinguished musical and spoken sounds by two properties: pitch and quality. Sounds are also distinguished by a third property, loudness. The loudness of a sound depends partly on the amplitude of the vibrations the sound induces in our eardrums, which in turn depends on the magnitude of the pressure variation caused by the sound wave. The pressure fluctuation of a barely audible sound is called the **threshold of audibility.** For the normal ear of a young person, the threshold pressure is about $2 \times 10^{-5} \, \text{N/m}^2$, which is less than one billionth of atmospheric pressure. A sound that is painful or harmful to the ear is about 10^6 times greater in pressure, or about $20 \, \text{N/m}^2$. The pressures just mentioned are fluctuations about the normal pressure of the atmosphere $(1.0 \times 10^5 \, \text{N/m}^2$ at sea level).

Instruments built to measure the loudness of a sound do not measure the sound's pressure amplitude but instead measure two other quantities called intensity and intensity level. The **intensity** I of a wave of any type is defined as the wave energy that crosses an area A perpendicular to the wave's direction in a

*Notice that $f_1 = v/4L = (340 \text{ m/s})/4(0.17 \text{ cm}) = 500$ Hz. Also, because it is a closed pipe, only the odd harmonics are present.

time t divided by the area and time:

$$I = \frac{E}{tA} = \frac{P}{A}.$$ (18.5)

Intensity can also be defined in terms of power P since power is the energy that flows per unit time. The units of intensity are the units of energy (joule) divided by the product of the units of area (m^2) and time (s), or $J/m^2 \cdot s$. The unit J/s is called a watt; thus, the units of intensity are often given as watt/m^2.

EXAMPLE 18.6 The intensity of sound at a construction site is 0.1 watt/m^2. Assuming that the area of the eardrum is 0.2 cm^2, how much sound energy is absorbed by one ear in an eight-hour work day?

SOLUTION Rearranging Eq. (18.5) to solve for energy, we find that

$$E = IAt = \left(0.1\,\frac{J}{s \cdot m^2}\right)(0.2 \times 10^{-4}\,m^2)\left(8\,hr \times \frac{3600\,s}{1\,hr}\right) = \underline{0.06\,J}. \qquad \blacksquare$$

The intensities of some familiar sounds are listed in Table 18.1. Our environment includes sounds that differ in intensity from about 10^{-12} watts/m^2 (barely audible) to nearly 1 watt/m^2 (a painful sound). The noise in a classroom is about 10^{-7} watt/m^2.

Because of the wide variation in the range of sound intensities, a different quantity for measuring sound intensities has been developed called intensity level. Intensity level is not a measure of a sound's intensity but is a comparison

TABLE 18.1 Intensities and Intensity Levels of Common Sounds

Source of sound	Intensity (watt/m²)	Intensity level (dB)	Description
Large rocket engine (nearby)	10^6	180	
Jet takeoff (nearby)	10^3	150	
Pneumatic riveter; machine gun	10	130	
Rock concert with amplifiers (2 m); jet takeoff (60 m)	1	120	Pain threshold
Construction noise (3 m)	10^{-1}	110	
Subway train	10^{-2}	100	
Heavy truck (15 m); Niagara Falls	10^{-3}	90	Constant exposure endangers hearing
Noisy office with machines; average factory	10^{-4}	80	
Busy traffic	10^{-5}	70	
Normal conversation (1 m)	10^{-6}	60	
Quiet office	10^{-7}	50	Quiet
Library	10^{-8}	40	
Soft whisper (5 m)	10^{-9}	30	Very quiet
Rustling leaves	10^{-10}	20	
Normal breathing	10^{-11}	10	Barely audible
	10^{-12}	0	Hearing threshold

of the intensity of one sound and the intensity of a reference sound. **Intensity level** β is defined on a logarithmic scale as follows:

$$\beta = 10 \log \frac{I}{I_0} \tag{18.6}$$

where I_0 is a reference intensity to which other intensities I are compared (for sound, $I_0 = 10^{-12}$ watt/m^2). The log in Eq. (18.6) represents the logarithm to the base ten of the ratio I/I_0.* The unit of β in Eq. (18.6) is the decibel (dB). The intensity levels β of several familiar sounds are listed in Table 18.1.

EXAMPLE 18.7 The sound in an average classroom has an intensity of 10^{-7} watt/m^2. (a) Calculate the intensity level of that sound. (b) If the sound intensity is doubled, what is the new sound intensity level?

SOLUTION (a) The ratio of the sound intensity I and the reference intensity I_0 is

$$\frac{I}{I_0} = \frac{10^{-7} \text{ watt/m}^2}{10^{-12} \text{ watt/m}^2} = 10^5.$$

Since $\log 10^5$ is 5, the sound intensity level is

$$\beta = 10 \log 10^5 = 10(5) = \underline{50 \text{ dB}}.$$

(b) If I is doubled, the ratio of I and I_0 becomes

$$\frac{I}{I_0} = \frac{2 \times 10^{-7} \text{ watt/m}^2}{10^{-12} \text{ watt/m}^2} = 2 \times 10^5.$$

The sound intensity level is, then,

$$L = 10 \log (2 \times 10^5) = 10(5.3) = \underline{53 \text{ dB}}. \qquad \blacksquare$$

The intensity levels listed in Table 18.1 can be calculated easily by considering the power of ten by which the sound's intensity exceeds the reference intensity. For example, the sound intensity in a noisy office is about 10^{-4} watt/m^2, which is 10^8 times more intense than the reference intensity of 10^{-12} watt/m^2. Since the noisy office is 8 powers of ten greater than I_0, the intensity level is $10(8) = 80$ dB.

The intensity levels β_1 and β_2 of two different sounds of intensities I_1 and I_2, respectively, are compared by taking the difference, $\beta_2 - \beta_1$, of their intensity levels. Using the properties of logarithms listed in the footnote below,† we find that

$$\beta_2 - \beta_1 = 10 \log (I_2/I_0) - 10 \log (I_1/I_0)$$
$$= (10 \log I_2 - 10 \log I_0) - (10 \log I_1 - 10 \log I_0)$$
$$= 10 \log I_2 - 10 \log I_1$$

or

$$\beta_2 - \beta_1 = 10 \log (I_2/I_1). \tag{18.7}$$

*The use of logarithms is reviewed in Appendix A.

†Note that $\log (ab) = \log a + \log b$ and that $\log (a/b) = \log a - \log b$.

EXAMPLE 18.8 A chicken farmer decides to increase the size of her flock from 500 chickens to 2000 chickens. By how much does the intensity level of the sound of the chickens increase?

SOLUTION We will represent the sound intensity of 500 chickens by the symbol I. When the number of chickens is quadrupled, the intensity increases to $4I$. The difference in the intensity levels is

$$\beta_{2000} - \beta_{500} = 10 \log (I_{2000}/I_{500}) = 10 \log (4I/I) = 10 \log 4 = \underline{6 \text{ dB}}. \quad \blacksquare$$

Our ears do not sense that the loudness of a sound increases in direct proportion to either the intensity or the intensity level of the sound. If the intensity of a sound doubles, we do not think that the sound is twice as loud. Only when the sound intensity is ten times greater do we sense that the sound is twice as loud. If a sound is 10^2 times more intense than another sound, it seems 2^2 times louder to our ears; if 10^3 times more intense, it seems 2^3 times louder.

Since the intensity level of a sound increases by 10 dB each time the intensity increases by a factor of ten, the loudness of a sound seems to double each time the intensity level increases by 10 dB. A sound that is 30 dB greater than another sound seems 2^3 times louder.

In the last example, the number of chickens belonging to the farmer increased by a factor of four, as did the intensity of the sound coming from the chickens. However, the sound would seem less than two times louder. The intensity would have to increase by a factor of ten and the intensity level by 10 dB before the loudness would seem to double.

When two sounds of different intensity levels are heard simultaneously, the intensity level of the two sounds is not much more than that of the louder sound. The loud sound can sometimes mask or hide the less loud sound. In order to find the intensity level of two sounds when heard simultaneously, we must first add their intensities. To do this, we must be able to calculate their intensities from their intensity levels. This is done by dividing the intensity level by ten. This is the factor of ten by which the sound intensity exceeds the reference intensity. For example, suppose that a shout produces an intensity level of 75 dB. The intensity of the sound is

$$I = 10^{75/10} I_0 = 10^{7.5} I_0 = (10^{0.5})(10^7) I_0$$
$$= 3.2 \times 10^7 I_0 = 3.2 \times 10^7 (10^{-12} \text{ watt/m}^2) = 3.2 \times 10^{-5} \text{ watt/m}^2. ^*$$

EXAMPLE 18.9 The noise from a vacuum cleaner is 85 dB while that of a television set is 78 dB. What is the intensity level of the two sounds together?

SOLUTION The intensities of the two sounds are first calculated and then added to find the total sound intensity. This total intensity is then converted to intensity level in the usual fashion. The intensity of the sound from the vacuum cleaner is

$$I = 10^{85/10} I_0 = (3.2 \times 10^8)(10^{-12} \text{ watt/m}^2) = 3.2 \times 10^{-4} \text{ watt/m}^2.$$

The intensity of the sound from the television is

$$I' = 10^{78/10} I_0 = (6.3 \times 10^7)(10^{-12} \text{ watt/m}^2) = 0.6 \times 10^{-4} \text{ watt/m}^2.$$

*Remember that $10^{a+b} = (10^a)(10^b)$. Thus, $10^{7.5} = 10^{0.5+7.0} = (10^{0.5})(10^{7.0}) = 3.2 \times 10^7$.

The total intensity of both sounds is

$$I_{total} = 3.2 \times 10^{-4}\,watt/m^2 + 0.6 \times 10^{-4}\,watt/m^2 = 3.8 \times 10^{-4}\,watt/m^2.$$

The intensity level of both sounds together is

$$\beta_{total} = 10\,\log\frac{I_{total}}{I_0} = 10\,\log\frac{3.8 \times 10^{-4}\,watt/m^2}{10^{-12}\,watt/m^2}$$

$$= 10\,\log\,(3.8 \times 10^8) = 10(8.6) = \underline{86\,dB}. \qquad \blacksquare$$

18.6   The Human Ear

A simplified sketch of the human ear is shown in Fig. 18.13. The part of the ear at the side of our head, called the pinna, serves to gather and guide sound energy into the auditory canal. Because of the pinna, the pressure variation at the eardrum due to a sound wave is about two times the pressure variation that would occur without the pinna and auditory canal.

The pressure variation at the eardrum causes it to vibrate. A large fraction of the vibrational energy is transmitted through three small bones in the middle ear known collectively as the ossicles and individually as the hammer, anvil, and stirrup. These three bones constitute a lever system that increases the pressure that can be exerted on a small membrane of the inner ear known as the oval window. The pressure increase is possible for two reasons: (1) The hammer feels a small pressure over a large area of its contact with the eardrum whereas the stirrup exerts a large pressure over the small area of its contact with the oval window. This difference in areas causes the pressure to be increased by a factor of about fifteen to thirty. (2) The three bones act as a lever that causes increased force of the stirrup against the oval window, although the displacement caused by the force is less than if the bones were not present. The pressure can be increased by up to three times by this lever action. The overall pressure fluctuation of the stirrup against the oval window can be 180 times greater than the pressure fluctuation of a sound wave in air if incident directly on the window (a factor of two due to the pinna and auditory canal, a factor of up to thirty because of the area change from the eardrum to the oval window, and a factor of three from the lever action of the ossicles).

The increased pressure fluctuation against the oval window causes a fluctuating pressure in the fluid inside the cochlea of the inner ear. The fluctuating pressure is sensed by nerve cells along the basilar membrane. Nerves nearest the

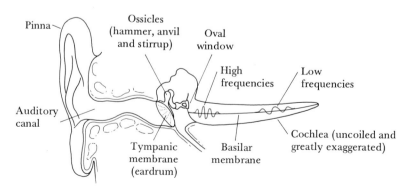

FIG. 18.13. A simplified version of the ear.

oval window respond to high-frequency sounds whereas nerves farther from the window respond to low-frequency sounds. Thus, our ability to distinguish the frequency of a sound depends on the variation in sensitivity of different cells along the basilar membrane to different frequencies. This ability is remarkable in that the whole membrane is only about one centimeter long, and yet the human hearing system responds to sounds ranging in frequency from about 20 Hz to 20,000 Hz. A normal ear can distinguish sounds that differ in frequency by 0.3 percent (a frequency difference of 3 Hz can be distinguished for two sounds near 1000 Hz).

Summary

1. **Standing waves** are produced by the interference of waves reflecting back and forth on or in an object of length L. For a **string** and for an **open pipe,** the frequencies of the standing-wave vibrations are

$$f_n = n(v/2L) \qquad \text{where } n = 1, 2, 3, 4, \ldots . \qquad \textbf{(18.2)}$$

For strings, v is the speed of the wave on the string and equals $\sqrt{T/(m/L)}$. For air columns or pipes, v is the speed of sound in the pipe. For a **closed pipe,** the *resonant frequencies* are

$$f_n = n(v/4L) \qquad \text{where } n = 1, 3, 5, 7, \ldots . \qquad \textbf{(18.4)}$$

2. The **pitch** of a sound is our subjective impression of how high or low its frequency is. The pitch of a musical sound is very nearly the same as the lowest standing-wave frequency of the sound's source.

3. The **quality** of a sound is the characteristic of a sound that distinguishes it from other sounds of the same pitch and loudness. Quality depends on the frequency spectra of the complex waves coming from the sound source.

4. The **loudness** of a sound depends on the energy in a sound wave and the fluctuations in air pressure it causes against our ear. The relationship between loudness and sound energy is quantified using the quantities intensity and intensity level.

Sound **intensity** I is defined as the wave energy E that passes a cross section of area A in a time t, divided by that area and time:

$$I = E/At = P/A. \qquad \textbf{(18.5)}$$

Since E/t is defined as power, intensity is also the sound power P passing a cross section of area A.

The **intensity level** β of a sound of intensity I is defined as

$$\beta = 10 \log (I/I_0) \qquad \textbf{(18.6)}$$

in units of dB, where I_0 is the least intense sound heard by the normal ear (10^{-12} watt/m^2) and log is the logarithm to the base ten of the ratio I/I_0. A 10-dB change in intensity level corresponds roughly to a doubling of the loudness of the sound.

Questions

1. The low-frequency strings on a piano are about the same length as those used for the middle frequencies. The low-frequency strings, however, are more massive. Why?

2. How are the fundamental frequencies of the strings of a stringed instrument affected by increasing their tension? Explain.

3. The wavelength of the wave on a string of fixed length is unchanged when the string is tightened. However, the sound wave coming from the string does have a different frequency and wavelength. Explain.

4. Standing-wave vibrations analogous to those on a vibrating string can occur on a bridge. What would the second harmonic bridge vibration look like? How might it be excited?

5. A mechanical device produces vibrations at one end of a string of fixed length. (a) Show qualitatively on a graph the expected vibration amplitude-versus-frequency curve that you expect to observe if the frequency of the exciting device is increased from zero to some arbitrary high frequency. (b) If you lightly rest your finger one-fourth of the way from one end of the string, how is the curve in (a) modified? Explain and show the expected curve.

6. If you look at the third harmonic of a vibrating string, you will see two nodes between the ends. The string vibrates in three sections. How does kinetic energy get across nodes from one antinode to another?

7. After inhaling helium, the frequency at which you speak is increased so that you sound like Donald Duck. Why does this occur?

8. A hot-air duct blows warm air on the pipes of an organ. Explain how this affects their vibration frequency.

9. Assume that the members of an orchestra did not warm their instruments before a concert. Compare the change in pitch of the stringed instruments to that of the woodwinds as the instruments and air around them warm. Be sure to justify your conclusions.

10. Estimate the sound frequencies that resonate between the ceiling and tile floor of a typical classroom.

11. You fill a pop bottle with water. Describe and explain the changing sound you hear as the water nears the top.

12. If you touch the middle of a violin string very lightly while it is being bowed, you hear a sound twice the normal frequency. What has your finger done to cause the change in pitch of the sound?

13. The pitch of notes played by instruments in an orchestra ranges from about 50 Hz to 4000 Hz. Why do high-fidelity systems amplify sounds up to 20,000 Hz? How would the sound of an orchestra be altered if the system reproduced sounds to only 4000 Hz?

14. Why is the quality of the human voice affected by a head cold?

15. Sound can travel from inside your mouth to your ear via bones in your head. How might this affect your perception of the sound of your voice compared to the sound you hear from a tape recorder? (*Note:* Bones transmit low-frequency sounds better than high.)

16. The intensity of a sound is 10^{-2} watt/m^2. (a) What power of sound crosses a square area whose edges are each 1 cm long? (b) 2 cm long? (c) 10 cm long? (d) 100 cm long?

17. When a drummer crashes the cymbals, what happens to the sound energy created by the crash?

18. A sound intensity is increased by 10^4. (a) About how much louder does the sound seem? (b) By how much has the sound intensity level increased?

19. The intensity level of a sound is decreased by 20 dB. (a) About how much more quiet does it seem? (b) By what factor has the sound intensity changed?

Problems

For all problems, unless stated otherwise, assume that the speed of sound in air is 340 m/s.

18.1 Standing Waves on Strings

1. A banjo G-string is 0.69 m long and has a fundamental frequency of 392 Hz. (a) Calculate the speed of a wave or pulse on the string. (b) Identify three other frequencies at which the string can vibrate.

2. The D-string on a banjo is 0.69 m long, has a mass of 0.30 g, and vibrates at a fundamental frequency of 587 Hz. (a) Calculate the speed of a wave on the string. (b) Calculate the tension in the string.

3. How far from the end of the banjo string discussed in Problem 1 must a fret be placed so that its fundamental frequency is increased to 490 Hz when the string is held down at the fret?

4. When shaken at one end at a frequency of 15 Hz, a thin rubber hose vibrates in three segments with nodes separating each segment. Identify its fundamental frequency of vibration and the frequencies of its second and fifth harmonics (assuming it is held with constant tension).

5. A cello string is 0.68 m long, has a mass of 1.3 g, and is pulled by a tension of 170 N. At what frequency does the string vibrate?

6. A violin A-string 0.33 m long vibrates at a fundamental frequency of 440 Hz. Where should your finger press the string against the finger board so that its decreased length causes it to vibrate at (a) five-fourths times the original frequency and (b) four-thirds times the original frequency?

7. Two wires on a piano are the same length and are pulled by the same tension. Wire 1 vibrates at a frequency 1.5 times that of wire 2. Calculate the ratio of their masses (m_2/m_1).

8. By what percentage must the tension in a violin string be increased in order to increase its frequency by a factor of 1.10 (that is, by 10 percent)?

9. A canary sits 10 m from the edge of a clothesline 30 m long, and a grackle sits 5 m from the other end. At what frequency can the line be vibrated so that the grackle is dislodged by the vibra-

tion while the canary sits undisturbed? The tension in the line is 200 N and the mass per unit length is 0.10 kg/m.

10. Perform an *estimation* calculation to determine approximately the fundamental frequency of vibration of a telephone line near where you live. You will need to estimate the distance between poles, the mass per unit length of the wire, and the tension in the wire.

18.2 Standing Waves in Air Columns

11. (a) Calculate the first three standing-wave frequencies of an open pipe 40 cm long. (b) Do the same for a closed pipe 40 cm long.

12. A clarinet (considered a closed pipe) has an effective bore length of 0.60 m. Calculate the first three standing-wave frequencies of the clarinet when all the finger holes are closed.

13. The 2779-m Brooklyn-Battery Tunnel, connecting Brooklyn and Manhattan, is one of the world's longest underwater car tunnels. (a) Calculate its fundamental frequency of vibration. (b) What harmonic must be excited so that it resonates in the audio region at 20 Hz or greater?

14. The first three standing-wave frequencies of a straw whistle are 330 Hz, 660 Hz, and 990 Hz. (a) Is the whistle similar to an open or closed pipe? Explain. (b) How long is the whistle?

15. Think of the outer ear of a human being as a tube that is 2.6 cm deep and closed at the eardrum. (a) Calculate the fundamental resonance frequency of this tube. (b) At what portion of the audio frequency range is this resonance located? (c) How might this resonance affect our hearing?

16. A person hums in a well and finds strong resonances at frequencies of 60 Hz, 100 Hz, and 140 Hz. (a) What is the fundamental frequency of the well? Explain. (b) How deep is the well?

17. (a) Use the dimensions of a 12-oz pop bottle to *estimate* its fundamental resonant frequency when empty. (b) Calculate the depth of water that must be added to increase its frequency by a factor of 4/3.

18. Assume that a friend has a hole in his head extending from one ear out the other. *Estimate* the frequency at which air in this hole would vibrate when excited by moving air as the friend jogs.

19. The speed of sound can be measured using an apparatus such as the one shown in Fig. 18.14. A 440-Hz tuning fork vibrating above a tube partially filled with water initiates sound waves in the tube. The air inside the tube vibrates in sympathy with the tuning fork when the water in the tube is lowered 0.20 m and 0.60 m from the top of the tube. Use this information to calculate the speed of sound in air.

FIG. 18.14

18.4 Quality of Sound

20. A bassoon has a strong formant in the region 450 to 500 Hz. (a) Which harmonic is strongest when a person is playing a note at a frequency of 120 Hz? (b) At a frequency of 160 Hz? Explain.

21. If you close your lips around an open tube and sing with a slowly rising "glissando" from low to high frequency, you find that at certain pitches it is very hard to produce a sound. For a tube 1.0 m long, explain why the pitches of no sound are near 85 and 255 Hz. In what way is this experiment similar to the attenuation of some sounds that occurs in the human vocal tract?

22. The formant region of a musical instrument extends from 800 to 1000 Hz. (a) Which harmonics are loudest when playing a note whose fundamental frequency is 150 Hz? (b) 200 Hz? (c) 300 Hz?

18.5 Intensity and Intensity Level

23. The sound intensity at a gasoline station next to a turnpike averages 10^{-3} watt/m^2. The owner decides to collect this energy and convert it to thermal energy for heating his building. If his conversion is 100 percent efficient, what is the length of the side of a square sound collector that is needed to provide thermal energy at a rate of 500 watts? Is this practical? Explain.

24. The sound intensity three miles from the takeoff point of a Concord supersonic jet is 0.6 watt/m^2. What area sound collector would absorb enough energy to run a 40-watt lightbulb?

25. A wall transmits only 0.10 percent of the sound energy incident on it. (a) What sound power is transmitted into an adjacent room through a 3 m $\times$ 5 m wall? The incident sound intensity is 2.5×10^{-4} watt/m^2. (b) What radius hole would allow the same power of sound to pass?

26. A rock band plays music with an average intensity of 0.10 watt/m^2. *Estimate* the sound energy striking your face during a two-hour concert.

27. In music a very soft sound called "triple piano" (ppp) has an intensity of about 10^{-8} watt/m^2. A very loud sound called "triple forte" (fff) has an intensity of about 10^{-2} watt/m^2. Convert these intensities to intensity levels (units of dB).

28. Convert the following intensities to intensity levels: (a) 3.0×10^{-8} watt/m^2 in a quiet living room, (b) 2.0×10^{-5} watt/m^2 inside a car, (c) 6.0×10^{-3} watt/m^2 near a passing truck, and (d) 5.0×10^{-1} watt/m^2 at a construction site where an air hammer is operating.

29. The noise from a Concord supersonic jet is 0.20 watt/m^2 at a distance of 5.6 km. (a) Calculate the intensity level of this sound. (b) At a distance of 11.2 km, the sound intensity is reduced by one-fourth. What is the intensity level at this distance?

30. Calculate the change in intensity level when a sound intensity is increased by (a) a factor of eight ($I_2 = 8I_1$), (b) a factor of eighty, and (c) a factor of 800.

31. One violin creates a sound whose intensity level is 60 dB. Calculate the intensity level of sixteen violins, each playing at this intensity.

32. A swarm of 10,000 locusts causes a sound of intensity level 75 dB. Calculate the intensity level caused by one locust.

33. Music from a rock band is at an average intensity level of 106 dB. How long must the band play to cause one joule of acoustical energy to strike your body, whose cross-sectional area is 1.2 m^2.

34. A person has a 34-dB hearing loss. By what factor must a hearing aid amplify the sound intensity in order to restore normal hearing?

35. Soundproofing reduces the noise in an office at a factory from 90 dB to 75 dB. Calculate the ratio of the original sound intensity and the final sound intensity.

36. A high-fidelity speaker is said to have a 3-dB flat response to sounds from 100 Hz to 10,000 Hz. This means that if electrical signals of equal power are sent to the speaker at any two frequencies in this range, the sound intensity levels coming from the speaker at these frequencies will differ by 3 dB or less. What is the maximum variation of the sound intensities produced by the speakers?

37. Two separate sounds are incident on the ear. The intensity level of one sound is 55.0 dB and of the other is 70.0 dB. What is the intensity level of the two sounds together?

General Problems

G1. The beat frequencies produced by several like instruments that are slightly out of tune with each other cause a pleasing sensation called the "chorus effect." By what factor must the tension of the A-string of one violin initially vibrating at 440 Hz be increased in order to produce 2-Hz beats with another violin with which it was initially in tune?

G2. Inhabitants of a planet in the galaxy Ursa Minor have hearing systems that consist of cylindrical holes in their chests. Each hole is a different length and has a fixed membrane of nerves at the bottom for detecting sound. (a) If the holes vary in length from 5 cm to 50 cm and the speed of sound on the planet is 900 m/s, what range of frequencies can these people detect? (b) If there is a total of 51 tubes that vary in length by ±1 cm from the neighboring tubes, what is the ratio of the frequency of the longest tube to that of the next shorter tube? (c) How does this hearing system compare in range and discrimination (ability to distinguish sounds of different frequency) to our own? The human ear can distinguish two sounds that differ in frequency by about 0.3 percent.

G3. While camping, you record a thunderclap whose intensity is 10^{-2} watt/m^2. The clap reaches you three seconds after a flash of lightning that caused the thunder. Calculate the total acoustical power generated by the lightning. (*Hint:* Assume that the sound intensity is the same over all parts of the surface of a sphere centered at the place where the lightning flash occurred.)

G4. A window whose area is 1.5 m^2 opens onto a street. Sound at an intensity level of 75 dB enters the open window. How long would it take to collect enough sound energy to warm a cup of tea (0.20 kg of water) from 20°C to 40°C?

Light and Optics

In Part IV we examined several general properties of waves and their applications in the study of mechanical waves such as sound waves. Now we introduce several new ideas about waves. These ideas and those developed earlier will be used in the study of light waves. In Chapter 19 we learn about the reflection of light at the boundary between two media and about the refraction (bending) of the light transmitted into the new medium. This information is important in Chapter 20, where we study the bending of light that passes through lenses and the reflection of light from mirrors. We will discuss how lenses and mirrors change the visual appearance of objects viewed with cameras, projectors, eye glasses, microscopes, and telescopes. In Chapter 21 we encounter some of the interesting phenomena that depend on the interference of light waves coming from different sources. We will learn how the wavelength of light is measured in spectrographs, about the diffraction or spreading of light as it passes through narrow openings, and about a new type of three-dimensional photography called holography.

Reflection and Refraction

19

In our previous discussions of waves, we learned that a wave traveling in one medium is partly reflected and partly transmitted at its boundary with a second medium. We limited our discussion to waves that strike a boundary perpendicular to its surface. In this chapter we examine the more general situation in which the directions of the incident, reflected, and transmitted waves are not perpendicular to the boundary between media. We will find that if the incident wave does not strike perpendicular to the surface, then the transmitted wave moves in a different direction from the incident wave; the transmitted wave is said to be *refracted*.

Refraction is a general property of all waves (sound, water, light, and so forth). The refraction of light waves is especially important because it affects the ability of glass lenses in microscopes, telescopes, and eyeglasses to focus and change the direction of light. Most of our examples in this chapter involve the reflection and the refraction of light, although the principles to be used apply to all types of waves. Before turning to the reflection and refraction of waves, we will discuss briefly the nature of light waves.

19.1 The Nature of Light

The study of the nature of light has extended over thousands of years. Four hundred years before Christ, during the Golden Age of Enlightenment in Greece, men like Socrates and Plato speculated that light rays left our eyes, were reflected from objects, and then returned, thereby allowing us to see. Euclid's book, *Optics,* published around 300 B.C., contained fifty-eight theorems about the nature of light, especially its geometrical behavior. According to legend, Archimedes, a scientist who studied Euclid's work, built an optical device to defend his hometown in Sicily against the Romans. The device, a huge concave mirror, would reflect sunlight on enemy ships and cause them to catch fire and burn. According to a popular myth, the device during one battle supposedly burned several of the sixty ships beseiging the town. The other ships fled in panic.

In more recent history, our understanding of the nature of light has centered on two conflicting theories. During the 1700s, a **corpuscular theory** was championed by Newton and widely accepted by other scientists. According to this theory, light streaming from objects consisted of huge numbers of tiny particles or corpuscles of light. The corpuscular theory was supported by the fact that light seemed to move in straight lines instead of spreading out as waves do. Because Newton also observed that sunlight passing through a prism was broken into a rainbow of different colors, he felt that white light consisted of many kinds of tiny particles, each a different color. When the different light corpuscles were brought back together, white light was reformed.

Not everyone believed the corpuscular theory. A conflicting theory of light was proposed by the Dutch scientist Christian Huygens (1629–1695), a contemporary of Newton. According to Huygens, light was a **wave** that traveled through an invisible substance called ether (not to be confused with the liquid used as an anesthetic). As light passed, the ether vibrated. In 1801, an English scientist, Thomas Young (1773–1829), showed that light beams from two sources interfered with each other in the same way that two sound waves interfere. This constructive and destructive interference of light was inconsistent with the corpuscular theory. It was, however, consistent with the wave theory, and so, during the 1800s, light was regarded as a wave phenomenon. The nature of light waves was eventually described in detail in the 1860s by the British physicist James Clerk Maxwell (1831–1879). Maxwell's four equations embraced all of the previously known behavior of electrical and magnetic phenomena. His equations predicted the existence of light waves (and other types of electromagnetic waves to be discussed in Part VII) and the speed at which they travel. Maxwell's equations left little doubt about the wave nature of light.

The reign of the wave theory was short. New experiments performed in the latter part of the nineteenth century and early in this century produced results inconsistent with the wave theory. In 1905 Albert Einstein published a theory concerning what was called the photoelectric effect. According to this theory, light absorbed by atoms consisted of individual, massless bundles of energy (like Newton's corpuscles) that are now called **photons.** The light absorption was inconsistent with the wave theory of light.

Since the early 1900s, scientists have viewed light as a phenomenon with a dual nature, part particle and part wave. For now, all phenomena we discuss can be explained in terms of the wave nature of light. In Part VII we will examine in greater detail the particle nature of light.

19.2 Coherence, Wavefronts, and Rays

Like all waves, light can be viewed as a series of crests and troughs moving away from a source. If several sources produce waves simultaneously, the surrounding space is disturbed by the interference of the waves from each source. If the different sources synchronize their vibrations, the resultant wave disturbance can have an orderly appearance—like a marching band on parade. This is called a **coherent wave.** A wave created by random disturbances (such as by rocks thrown randomly into a small pond) has a disorganized appearance that changes unpredictably. This is called an **incoherent wave.** In this chapter, and the next, we will consider the direction of travel of all types of waves (sound waves, water waves, light waves, and so on). Our attention will be restricted to coherent waves because they are easiest to picture.

Coherent waves can be drawn as one crest following another at a distance of one wavelength. These crests are called **wavefronts.** The disturbance caused by the wave is the same at every point on the wavefront. Wavefronts moving across a large body of water are pictured in Fig. 19.1.

FIG. 19.1. A series of water waves moves toward a shore. The crests are examples of wavefronts. A ray is an arrow drawn perpendicular to the wavefronts and pointing in the direction the wave travels.

A **ray** is an arrow drawn perpendicular to a wavefront that points in the direction of the wave's motion (Fig. 19.1). Wavefronts and rays are convenient geometrical devices to help visualize the changing direction of a wave as it is reflected and transmitted at the boundary between two media.

19.3 Reflection

As we learned in Chapter 17, when a wave is incident on a boundary between two media that have different properties, part of the wave is reflected. For example, when light traveling in air strikes glass, a fraction of the light is reflected and the remaining portion is transmitted into the glass. The fraction that is reflected depends on the type of glass and on the angle at which the light meets the surface of the glass. If the light strikes perpendicular to the glass, about 4 percent is reflected. If it strikes at other angles, a larger percent is reflected.

The direction of travel of a reflected light beam depends on the roughness of the reflecting surface relative to the wavelength of the reflected wave. If the wave is incident on a rough surface, the reflected wave travels in many directions (Fig. 19.2). This is called **diffuse reflection** and occurs if the irregularities in the surface are about the same size or larger than the wavelength of the wave striking the surface. The walls of concert halls are often made a "fluted" shape (Fig. 19.3a) so that sound waves striking the walls are reflected in different directions. A person in the audience receives a small amount of reflected sound from many parts of the wall (Fig. 19.3a) rather than a larger amount of reflected sound from one part of the wall (Fig. 19.3b). Light is being reflected diffusely from the pages of this book because the irregularities in its surface are larger than the wavelength of the light waves hitting the page.

If, however, a surface is flat, the reflected wave travels in only one direction after reflection. For example, if the wall of a concert hall is flat, as in Fig. 19.3b, sound from the stage will be reflected from only one part of each wall onto your ears. It may sound as if the performing group is actually located at these parts of the walls. The reflection of waves from surfaces whose irregularities are smaller than the wavelength of the wave is called **specular reflection.** Examples are the reflection of sound from flat walls and the reflection of light from a mirror.

The directions of the incident and reflected wave for specular reflection are shown by drawing rays. In Fig. 19.4a, a narrow beam of light is reflected from a smooth surface such as a mirror. The incident ray shown in Fig. 19.4b makes an

FIG. 19.2. Diffuse reflection from an uneven surface.

(a)

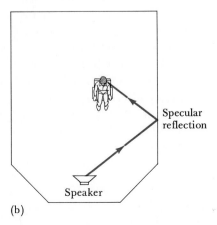

(b)

FIG. 19.3. (a) With fluted walls, sound is deflected from each small section of a wall to a listener. (b) If the walls are flat, an intense reflected sound comes from only one part of the wall.

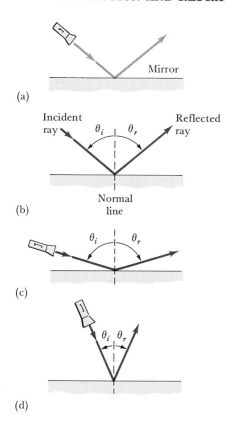

(a)

(b)

(c)

(d)

FIG. 19.4. (a) A narrow beam of light from a flashlight is reflected from a mirror. (b) The directions of the incident and reflected rays are indicated by the angles θ_i and θ_r that they make with a normal line. (c) and (d) As the angle of incidence increases, so does the angle of reflection.

angle θ_i between itself and a dashed line drawn perpendicular to the surface (called the normal line). The reflected ray makes an angle θ_r with respect to the normal line. According to the **law of reflection,** the angle of incidence θ_i equals the angle of reflection θ_r:

$$\theta_i = \theta_r. \tag{19.1}$$

The incident and reflected rays are on opposite sides of the normal line and in the same plane with it.

The law of reflection is easily confirmed by shining a flashlight or laser beam on a mirror. As the incident angle is changed, the reflected angle also changes (Fig. 19.4b, c, and d).

19.4 Index of Refraction

When light or any wave strikes the interface between two media, the transmitted wave usually travels in a different direction from that of the incident wave. The transmitted wave is said to be **refracted**—that is, its direction changes. Refraction is caused by a change in speed of the wave as it passes from one medium to another. Geophysicists usually analyze the refraction of sound waves at the boundaries between different media in the earth in terms of the speeds of sound in these media. In the case of light waves, another quantity called the index of refraction is used.

The **index of refraction** n of a medium is the ratio of the speed of light in a vacuum (given the symbol c) and the speed of light v in the medium:

$$n = \frac{c}{v}. \tag{19.2}$$

The index of refraction has no units, since it is the ratio of two quantities with identical units. A list of the speeds of light in several different media and of their indices of refraction appears in Table 19.1. Notice that in a vacuum, light travels at a speed

$$c = 2.997925 \times 10^8 \text{ m/s} \simeq 3.00 \times 10^8 \text{ m/s}.$$

Unless great accuracy is desired, the rounded-off value of 3.00×10^8 m/s will be used in our examples and problems. The speed of light c in a vacuum is always greater than its speed v in other media. Since c is greater than v for all media, the index of refraction $n = c/v$ will always be greater than one. However, the speed of light in air is so close to its speed in a vacuum that the index of refraction of air for practical purposes is 1.000.

EXAMPLE 19.1 Light travels at a speed of 2.25×10^8 m/s in water. Calculate the index of refraction of the water.

SOLUTION

$$n = \frac{c}{v} = \frac{3.00 \times 10^8 \text{ m/s}}{2.25 \times 10^8 \text{ m/s}} = \underline{1.33}. \qquad \blacksquare$$

TABLE 19.1 Speed and Index of Refraction of Yellow Light in Various Substances

Material	Speed ($\times 10^8$ m/s)	Index of refraction (c/v)
Vacuum	$c = 2.997925 \times 10^8$	1.00000
Air	$v = 2.99706$	1.00029
Carbon dioxide	2.99658	1.00045
Helium	2.99782	1.000034
Water (20°C)	2.2490	1.3330
Ethyl alcohol	2.2016	1.3617
Methyl alcohol	2.2555	1.3292
Benzene	1.9968	1.5014
Carbon disulfide	1.8415	1.6279
Sugar solution, 25%	2.1846	1.3723
Sugar solution, 50%	2.1112	1.4200
Sugar solution, 75%	2.0292	1.4774
Glass, light crown	1.976	1.517
Glass, dense crown	1.888	1.588
Glass, light flint	1.899	1.579
Glass, heavy flint	1.820	1.647
Canada balsam	1.959	1.530
Fluorite	2.091	1.434
Diamond	1.240	2.417

19.5 Refraction

As we said in the last section, the direction of a wave usually changes as it passes from one medium to another; the transmitted wave is said to be refracted. Refraction is caused by the change in speed of a wave as it passes from one medium to another. This phenomenon is illustrated by considering a marching band that moves from pavement onto a field of soft mud as shown in Fig. 19.5. Each row of the band represents a wave crest. As each band member reaches the mud, his speed decreases, and his direction of travel moves closer to the normal line. If the band moves out of the mud onto the pavement, its speed increases

Pavement

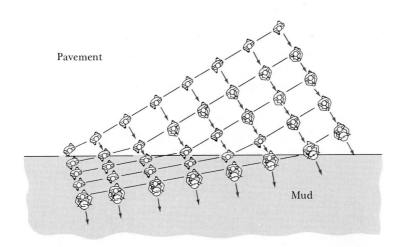

Mud

FIG. 19.5. The direction of travel of a marching band changes as its speed decreases when moving from pavement onto a mud field.

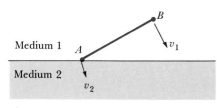

Medium 1

Medium 2

(a)

(b)

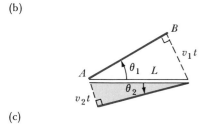

(c)

FIG. 19.6. (a) A wavefront moves from Medium 1 toward Medium 2. (b) During a time t, Point B travels a distance $v_1 t$, while Point A travels a distance $v_2 t$. (c) By comparing these triangles, we can derive Snell's law.

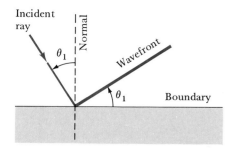

FIG. 19.7. A ray makes the same angle with the normal that the wavefront makes with the interface.

and its direction of travel bends away from the normal line. The change in direction is similar for waves. If a wave enters a medium in which its speed decreases, its direction bends toward the normal; if it enters a medium in which its speed increases, its direction bends away from the normal.

To derive an equation for calculating the expected change in direction, consider the wavefront of the wave shown in Fig. 19.6a as it approaches the boundary between two media. The wavefront could represent the crest of a sound wave moving from air into a wall or the wavefront of light moving from air into glass or water. Point B on the wavefront requires a time t to move forward to the boundary between the two media. Since the wave's speed in medium 1 is v_1, point B moves a distance $v_1 t$ during that time (Fig. 19.6b). Point A, which has just entered medium 2, moves a distance $v_2 t$ in the second medium during that same time. If v_2 is less than v_1, then point A moves a shorter distance than point B, and the wave refracts toward the normal line. However, if v_2 is greater than v_1, point A moves a longer distance than point B, and the wave refracts away from the normal line.

The change in direction can be calculated by comparing the triangles shown in Fig. 19.6c. They each have the same hypotenuse—the length L along the boundary between the media. But the angles θ_1 and θ_2 differ because the sides of the triangles opposite θ_1 and θ_2 are different lengths. Consider the sines of these angles:

$$\sin \theta_1 = \frac{v_1 t}{L} \quad \text{and} \quad \sin \theta_2 = \frac{v_2 t}{L}.$$

If we take the ratio of these two sines, we find that

$$\frac{\sin \theta_1}{\sin \theta_2} = \frac{v_1 t / L}{v_2 t / L} = \frac{v_1}{v_2}$$

or

$$\frac{\sin \theta_1}{v_1} = \frac{\sin \theta_2}{v_2}. \tag{19.3}$$

This equation, known as **Snell's law,** was determined experimentally in a different form in 1621 by the Dutch scientist Willebrord Snell (1591–1626).

We have derived the equation using the angles that the wavefronts make with the boundary between the media. However, it is customary when considering refraction to use the angle that a ray moving in the direction of travel of the wave makes with the normal line. We see from Fig. 19.7 that the angle θ_1 between the ray and the normal line equals the angle θ_1 between the wavefront and the boundary. The same applies to angle θ_2. Thus, Eq. (19.3) can be used to calculate the change in direction of wave rays relative to normal lines. In the future we will always refer to angle θ_1 as the direction of an incident ray relative to the normal line and θ_2 as the direction of the transmitted or refracted ray relative to the normal line. Snell's law is then written as

$$\frac{\sin \theta_1}{v_1} = \frac{\sin \theta_2}{v_2}, \tag{19.3}$$

where v_1 is the wave's speed in the medium of the incident wave and v_2 is its speed in the medium of the refracted wave. Equation (19.3) is used by geophysicists when analyzing the refraction of sound waves passing between different layers of the earth.

A different form of Snell's law is used to analyze the refraction of light. In this form, Eq. (19.2) is used to replace the wave's speed by the indices of refraction. Substituting $v_1 = c/n_1$ and $v_2 = c/n_2$ into Eq. (19.3), Snell's law becomes

$$\frac{\sin \theta_1}{c/n_1} = \frac{\sin \theta_2}{c/n_2}$$

or

$$n_1 \sin \theta_1 = n_2 \sin \theta_2. \tag{19.4}$$

This form of Snell's law will be used in all of our calculations involving light. (See Fig. 19.8 for the meaning of the symbols.)

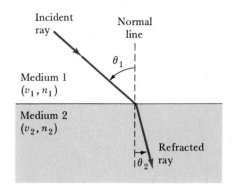

FIG. 19.8. Light is refracted at an interface between two media.

EXAMPLE 19.2 Light traveling in air strikes an interface with benzene at an incident angle of 35°. Calculate the direction of the refracted wave.

SOLUTION The following information is known: $\theta_1 = 35°$, $n_1 = 1.00$ (air), and $n_2 = 1.50$ (benzene). The sine of the angle of refraction is given by Eq. (19.4):

$$\sin \theta_2 = \frac{n_1 \sin \theta_1}{n_2} = \frac{1.00 \sin 35°}{1.50} = 0.382$$

or

$$\theta_2 = \underline{22.5°}.$$

Notice that the wave slowed as it entered the benzene. This decrease in speed caused it to bend toward the normal. ∎

EXAMPLE 19.3 A person fishing with a spear gun while standing on a shore sees a fish in the water. The fish appears along a line of sight making an angle of 50° with the normal (Fig. 19.9a). (a) At what angle does light coming from the fish to the person strike the water-air interface from below? (b) Should the person aim above or below the line of sight in order to hit the fish?

SOLUTION (a) Because light moves faster in air than in water, the ray bends away from the normal as it leaves the water (Fig. 19.9b). We wish to calculate the unknown incident angle θ_1. The refracted wave in the air makes an angle, $\theta_2 = 50°$, with the normal. Thus, the following information is known: $n_1 = 1.33$ (water), $n_2 = 1.00$ (air), and $\theta_2 = 50°$ (in air). The sine of the angle θ_1 is determined from Eq. (19.4):

$$\sin \theta_1 = \frac{n_2 \sin \theta_2}{n_1} = \frac{1.00 \sin 50°}{1.33} = 0.576$$

or

$$\theta_1 = \underline{35.2°}.$$

(a)

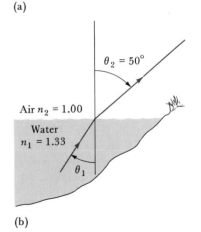

(b)

FIG. 19.9. (a) A person sees light coming from a fish in the water. (b) The actual light path is shown.

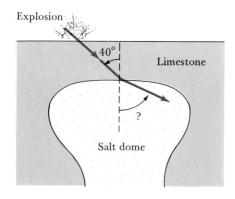

FIG. 19.10. A ray of sound is refracted at the interface between limestone and the top of a salt dome.

(b) Because the ray bends away from the normal as it leaves the water (Fig. 19.9b), the person should shoot <u>below</u> his or her line of sight in order to hit the fish. ∎

EXAMPLE 19.4 Sound caused by an explosion travels at 4000 m/s in limestone and strikes a boundary at the top of a salt dome (Fig. 19.10). If part of the sound strikes the boundary at an angle of incidence of 40°, calculate the angle of refraction. The speed of sound in salt is about 6000 m/s.

SOLUTION In this problem, speeds rather than indices of refraction are known. Thus, Eq. (19.3) will be used rather than Eq. (19.4). The known information is $\theta_1 = 40°$, $v_1 = 4000$ m/s, and $v_2 = 6000$ m/s. The sine of the refracted wave is

$$\sin \theta_2 = v_2 \frac{\sin \theta_1}{v_1} = (6000 \text{ m/s}) \frac{\sin 40°}{4000 \text{ m/s}} = 0.964$$

or

$$\theta_2 = \underline{74.6°}.$$

Because the wave traveled from a slower to a faster medium, the wave bent away from the normal as it entered the salt (Fig. 19.10). ∎

Mirages

One of the interesting consequences of the refraction of light waves is the formation of mirages. On a hot day there may be a layer of very hot air just above the ground. This hot air is less dense than cooler air farther above the surface. Light travels slightly faster through the less dense, hot air than through the cooler air. Wavefronts of light leaving a distant object are bent as shown in Fig. 19.11a because the lower part of a wavefront travels faster than the upper part. Some of the light originally slanting downward is refracted up through the hot air and into our eyes; the light never hits the ground. The refracted light from a distant tree (Fig. 19.11b) appears to come from the surface in front of the tree, like the reflection of light off a pond of water. A mirage is caused not by a loss of mental facility on a hot, dry desert, but by refracted light that appears to be reflected light from the smooth surface of a pond in front of an object.

Occasionally, a motorist observes what appears to be a wet spot on a distant part of a hot highway. This is just light that originates in the sky and travels toward the highway. Before reaching the pavement, it is bent or refracted up into your eye by the hot air above the road. You are seeing light from the sky that curved down toward the highway but missed.

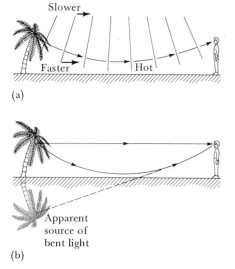

FIG. 19.11. (a) Wavefronts are bent because the light travels faster in the hot air than in cooler air. (b) Rays enter a person's eye as though coming from an inverted object in the ground. The mirage looks like the reflection of the tree off a pond.

19.6 Total Internal Reflection

When the speed of light increases as it passes from one medium to another, the light bends away from the normal as shown in Fig. 19.12a. As the angle of the incident ray increases, the refracted ray bends farther from the normal. For a

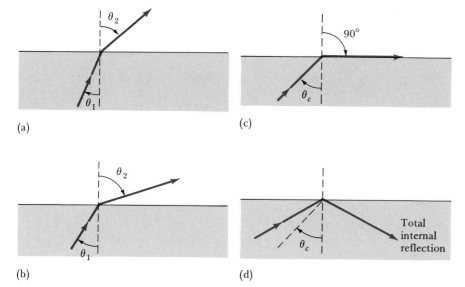

(a)

(b)

(c)

(d)

FIG. 19.12. (a) and (b) If the speed of a wave increases as it moves from one medium to another, its direction bends away from the normal. (c) At an incident angle called the critical angle θ_c, the refracted wave moves parallel to the boundary between the media ($\theta_2 = 90°$). (d) If θ_1 is greater than θ_c, total internal reflection occurs and there is no refracted wave.

particular incident angle called the **critical angle** θ_c (Fig. 19.12c), the refracted ray makes an angle of 90° with the normal and skims along the boundary between the two media. The critical angle is determined by setting $\theta_1 = \theta_c$ and $\theta_2 = 90°$ in Snell's law:

$$\sin \theta_c = \frac{n_2}{n_1} \sin 90° = \frac{n_2}{n_1}. \tag{19.5}$$

EXAMPLE 19.5 Calculate the critical angle for light moving from water to air (Fig. 19.12c).

SOLUTION The refracted ray makes an angle $\theta_2 = 90°$ when $\theta_1 = \theta_c$. The index of refraction of water, the medium of the incident ray, is $n_1 = 1.33$. The refracted wave moves in air. Thus, $n_2 = 1.00$. Using Snell's law, we find that

$$\sin \theta_c = \frac{n_2 \sin \theta_2}{n_1} = \frac{1.00 \sin 90°}{1.33} = 0.752$$

or

$$\theta_c = \underline{48.8°}. \qquad \blacksquare$$

If a ray strikes a surface at an angle less than θ_c, part of the wave is refracted into the second medium. However, for incident angles greater than θ_c, we find using Snell's law that the sine of the refracted angle ($\sin \theta_2$) is greater than one. The sine of an angle can never be greater than one. This impossible result is avoided because there is no refracted wave for incident angles greater than θ_c; the wave is *totally reflected* (100 percent) at the boundary between the media (Fig. 19.12d). This phenomenon, called **total internal reflection,** occurs when a wave travels at an incident angle greater than the critical angle while trying to move into a second medium in which its speed increases—that is, into a medium with a smaller index of refraction.

FIG. 19.13. A ray of light in glass strikes an interface with air. Is the light totally reflected?

EXAMPLE 19.6 Light travels inside a block of light crown glass (Fig. 19.13). If the beam strikes a surface between the glass and air at an incident angle of 51°, will part of the beam be refracted into the air?

SOLUTION In Fig. 19.13, we call medium 1 the glass and medium 2 the air. We see from Table 19.1 that the index of refraction of light crown glass is 1.517 and that of air is 1.000. Thus, according to Snell's law, the sine of the refracted angle is

$$\sin \theta_2 = \frac{n_1 \sin \theta_1}{n_2} = \frac{1.517 \sin 51°}{1.000} = 1.18.$$

If, when using Snell's law, we find that the sine of the refracted angle is greater than 1.00, we interpret it as meaning that no refraction occurs. The ray experiences total internal reflection.

Our result can be checked by calculating the critical angle for light striking an interface between crown glass and air:

$$\sin \theta_c = \frac{n_2 \sin 90°}{n_1} = \frac{1.000(1.00)}{1.517} = 0.659$$

or

$$\theta_c = 41.2°.$$

Since 51° is greater than θ_c, the light is totally reflected. ∎

Fiber Optics

Suppose that the glass discussed in the last example was a long thin fiber of glass. If light moving through the glass strikes the top edge of the glass at an angle of 51° (or at any angle greater than θ_c), the light is totally reflected. The reflected light moves toward the bottom surface of the glass and strikes it at an incident angle of 51° (see Fig. 19.14). The light is once again totally reflected. The light continues to move along the fiber bouncing from one side to the other, each bounce resulting in total reflection. The fiber, which is called a **light pipe,** carries light along the fiber much the way water flows along a pipe. No light is lost from the sides.

These fibers that carry light have become an important new tool in technology. The communications industry is developing laser communications systems in which telephone conversations and other forms of information are transmitted by light waves traveling along flexible glass fibers. These glass fibers are also used to make ornamental lamps consisting of a large number of flexible fibers that carry light upward along the fibers from a light source at the base.

Fiber optic devices have been built to view places that are normally inaccessible to our eyes, such as the inside of car radiators or the inside of a person's stomach. In these devices, a bundle of tiny glass fibers transmits a view of an inaccessible region at one end of the bundle to a person's eye looking at the other end of the bundle. If a lens is placed at the entrance end of the bundle, a larger field of view can be transmitted along the rods. A picture taken inside the knee joint is shown in Fig. 19.15a. An outer group of fibers carries light into the joint to illuminate it; an inner group carries an image of the illuminated surfaces of the joint to the viewer.

FIG. 19.14. After being totally reflected at the top of the glass, light is again totally reflected at the bottom. This glass is a "light pipe."

(a)

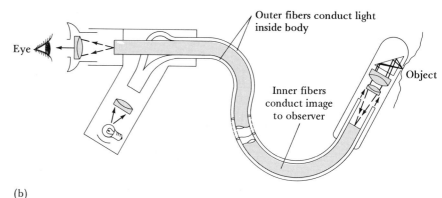

(b)

FIG. 19.15. (a) A photograph of cartilage in a knee joint. A thin metal suction device is cleaning damaged areas. (b) A schematic representation of the fiber optic device used to view the joint. (Photograph courtesy of Dr. William H. Dunbar)

Prisms

Since the critical angle for total internal reflection from a glass-air interface is less than 45°, glass prisms with 45° and 90° angles are used to reflect light through 90° or 180° angles in many optical instruments such as telescopes and binoculars. Examples of the reflecting ability of prisms are shown in Fig. 19.16.

The reflective ability of prisms is preferred over mirrors for several reasons. First, prisms reflect almost 100 percent of the light incident on the prism, whereas mirrors reflect somewhat less than 100 percent. Second, mirrors tarnish and lose their reflective ability with age, whereas prisms retain their reflective ability. Finally, prisms can invert an image—that is, make it appear upside down (Fig. 19.16c). This inversion may seem like a disadvantage. However, the

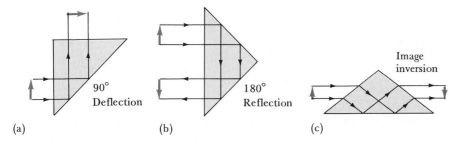

(a)

(b)

(c)

FIG. 19.16. A prism is used in three different ways to change either the direction of light from an object [(a) and (b)] or the orientation of light from an object (c).

lenses in optical devices such as telescopes and binoculars cause an object viewed through the telescope to appear inverted. In binoculars, a form of telescope, prisms are placed in the light path so that the inverted image of the telescope lenses is reinverted and appears right side up.

19.7 Dispersion

Light waves vary in wavelength from about 700 nm (red light) to 400 nm (violet light). In a vacuum the different wavelengths of light all travel at the same speed, but in media such as glass and water the speed of the different colors and

TABLE 19.2 Speed and Index of Refraction for Different Wavelengths of Light in Silicate Flint Glass

Color	Wavelength (nm)	v(m/s)	n
Red	650	1.860×10^8	1.613
Yellow	580	1.851×10^8	1.621
Green	530	1.843×10^8	1.628
Blue	470	1.834×10^8	1.636
Violet	400	1.806×10^8	1.661

FIG. 19.17. White light is separated into its many wavelength components by refraction through a prism.

wavelengths of light vary. For example, red light having a wavelength of 650 nm travels at 1.860×10^8 m/s in silicate flint glass, whereas violet light of 400-nm wavelength travels at 1.806×10^8 m/s (Table 19.2). Any medium that causes waves of different wavelength to move at different speed is said to be a **dispersive medium.** The medium can cause white light that consists of all different wavelengths of light to be separated or dispersed into its different wavelength components. A prism such as that shown in Fig. 19.17 causes the dispersion of white light.

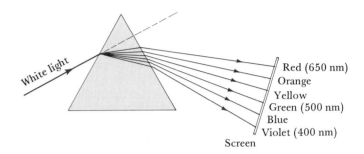

To understand why the dispersion shown in Fig. 19.17 occurs, consider a wavefront and ray of light entering the prism shown in Fig. 19.18. A wavefront speeds through the air until its right edge strikes the glass (Fig. 19.18a). This edge slows down and bends as it enters the glass, while the left edge that is still in the air continues speeding forward (Fig. 19.18b and c). The slowing of the wave causes it to bend and change direction. As the wavefront reaches the second surface (Fig. 19.18e), the left edge of the wavefront moves into the air first. It speeds forward while the right edge is still shuffling along through the glass at slower speed. As the wavefront emerges from the prism, it is deflected even farther to the right.

All colors of light undergo a similar change in direction, but the change is greatest for those wavelengths (colors) that are slowed most in the glass. Hence, violet is deflected farthest to the right and red the least.

This nonuniform deflection of light composed of many wavelengths is useful for analyzing the composition of light from different sources. Light from the

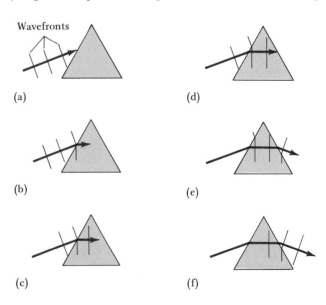

FIG. 19.18. Wavefronts of light are bent at the surfaces of a prism.

sun, for example, is made of all different wavelengths. Thus, deflection of sunlight through a prism produces a rainbow of all the different colors (Fig. 19.17). On the other hand a hot gas of hydrogen atoms emits light at only certain discrete wavelengths: red light at about 656 nm and blue light at 486 nm, 434 nm, and 410 nm. If the light given off by hydrogen atoms is passed through the prism and then projected on a screen, separate lines for each discrete wavelength are seen. The 410 nm blue light is deflected most; the red light is deflected least. If light from a star has lines at these four hydrogen wavelengths, then an astronomer knows that hydrogen is one type of atom on the star. Other atoms can be identified in a similar manner.

Rainbows

Rainbows are a dramatic example of the effect of dispersion. Water, like glass, causes the dispersion of light. When white light from the sun enters a spherical raindrop as shown in Fig. 19.19, the light is refracted or bent. The different colors of light are bent by different amounts. After reflection off the back surface of the drop, the different colors of light are refracted again as they leave the front surface.

Each drop disperses a full spectrum of colors. Yet, an observer on the ground facing away from the sun sees at most only one color of light coming from a particular drop. If the person sees red light from a drop, the violet light is deflected above her head. Or if the person looks at drops lower in the sky, she may see violet light coming from the drops. The red light from these drops is deflected below her eyes onto the ground. She sees red light when her line of view makes an angle of 42° with the beam of sunlight and violet light when the angle is 40°. Other colors of light are seen at intermediate angles.

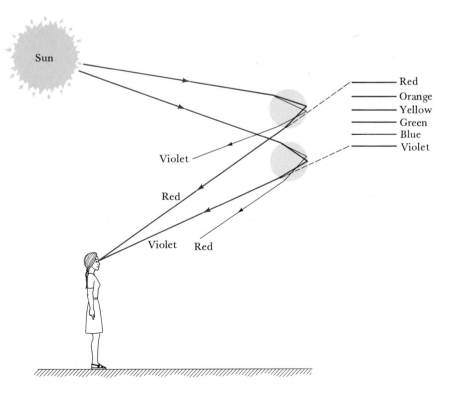

FIG. 19.19. Red light is refracted and reflected from a raindrop at a different angle from violet light. As a result, we see different colors of light coming from different parts of the sky. A rainbow is formed.

Summary

1. **Coherence:** A **coherent wave** is caused by the interference of waves coming from several sources whose vibrations are synchronized with each other. A coherent wave usually has an orderly appearance, like a marching band on parade. An **incoherent wave** is caused by the interference of waves coming from different sources whose disturbances are random and unsynchronized. The incoherent wave has a disorganized appearance that changes unpredictably.

2. **Wavefronts and rays:** A coherent wave can be drawn as one crest following another. These moving crests are called **wavefronts,** and the disturbance caused by the wave is the same at every point on the wavefront. A **ray** is an arrow drawn perpendicular to the wavefront that points in the direction of travel of the wave.

3. **Reflection: Diffuse reflection** results in a wave being reflected in many directions and occurs when a wave strikes a surface whose irregularities are larger than the wavelength of the wave. **Specular reflection** results in a reflected wave that moves in only one direction and occurs when the wave strikes a smooth surface. The angle that the reflected wave makes with a normal line drawn perpendicular to the reflecting surface equals the angle that the incident ray makes with that line (see Fig. 19.20).

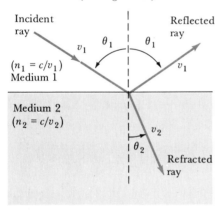

FIG. 19.20. The angles used to describe the directions of the incident, reflected, and refracted waves.

4. **Index of refraction:** The index of refraction n of a medium is defined as the speed of light in a vacuum c (3.00×10^8 m/s) divided by the speed of light in the medium v:

$$n = \frac{c}{v}. \tag{19.2}$$

5. **Refraction and Snell's law:** When a wave strikes the interface between two media, the transmitted wave usually travels in a different direction from the incident wave. The transmitted wave is said to be **refracted**—that is, to change direction—and is called the refracted wave. The direction of the refracted ray θ_2 relative to the normal line (see Fig. 19.20) is determined using Snell's law in one of two forms:

$$\frac{\sin \theta_1}{v_1} = \frac{\sin \theta_2}{v_2} \tag{19.3}$$

or

$$n_1 \sin \theta_1 = n_2 \sin \theta_2, \tag{19.4}$$

where θ_1 is the angle of the incident ray relative to the normal line, v_1 and v_2 are the speeds of the wave in the first and second media, respectively, and n_1 and n_2 are the indices of refraction of these media.

6. **Total internal reflection:** When the speed of a wave increases as it moves from one medium to another, the refracted ray bends farther away from the normal line. If the incident ray makes an angle θ_c, called the **critical angle,** with the normal line, the refracted ray is at $90°$ and skims along the surface between the media. If the incident angle is greater than θ_c, the wave is **totally reflected**—there is no refracted wave. The critical angle is determined by substituting $\theta_1 = \theta_c$ and $\theta_2 = 90°$ into Snell's law.

7. **Dispersion:** Any medium that causes waves of different wavelengths to move at different speeds is said to be a **dispersive medium.** A glass prism causes different colors and wavelengths of light to travel at different speeds and is able to separate white light into a rainbow of colors by the varying angular refraction of the different colors of light.

Questions

1. (a) Estimate the thickness of a page of this book. With a little ingenuity you should be able to do this within about 10 percent. (b) Calculate the number of 500-nm wavelengths of light that equal this thickness.

2. The Cathedral of Girgenti in Sicily is the shape of half an ellipse that has been rotated about its long axis. The confessional was unknowingly placed at one focus of the ellipse. A person standing at the other focus can easily overhear the penance of a sinner speaking in private to a priest. Explain, using a ray diagram.

3. A wave passes from one medium to another. Why does the wave's frequency not change, whereas its wavelength does?

4. A person sits in the balcony of a concert hall. Does more sound reach the person when the air is warm near the ground floor and cool near the ceiling, or vice versa? Explain.

5.   Lumps of sugar are added to a container of water (Fig. 19.21). The water is not stirred while the sugar dissolves slowly at the bottom. A narrow laser beam enters the container, bends downward, is reflected from the bottom, and then bends again as it leaves the container. Explain the bending of the laser beam. [Adapted from W. M. Strouse, *Am. J. Phys.*, **40**:913 (1972).]

FIG. 19.21

6.   Why does the air shimmer above a hot highway?

7.   A rod partly immersed in water is shown in Fig. 19.22. Use ray diagrams to explain the apparent break in the rod at the place where it enters the water.

8.   Place a penny in the bottom of an empty cup whose walls are not transparent to light (a styrofoam cup works well). Now, look down into the cup at an angle that puts the penny just out of sight. When you slowly add water to the cup, the penny comes

FIG. 19.22

into view, even though you have not moved your eyes or the penny. Explain why this occurs.

9.   A fish looks upward along a line making a 45° angle with the water-air interface. What does it see? Explain.

Problems

19.3   Reflection

1.   Two mirrors, each 10 m long and separated by 0.4 m, are parallel and face each other. A light ray enters the space between the mirrors at a 45° angle and reflects back and forth while traversing the 10-m length of the mirrors. (a) Estimate the distance traveled by the ray while between the mirrors. (b) Calculate the time needed to travel this distance.

2.   Two mirrors are oriented at right angles. A ray strikes the horizontal mirror at an incident angle of 65°, reflects from it, and then hits the vertical mirror. (a) Calculate the angle of incidence at the vertical mirror. (b) Show that the ray leaves the vertical mirror parallel to its original direction.

3.   A spherical mirror is made by silver-coating half of the inside of a sphere of radius R = 20 cm. A light ray moves horizontally 5.0 cm above the axis of the mirror as in Fig. 19.23. When reflected from the mirror, how far from the mirror does the ray cross the axis? You will need to use some trigonometry and geometry to solve this problem.

FIG. 19.23

19.4   Index of Refraction

4.   (a) The speed of light in acetone is 2.210×10^8 m/s. Calculate the index of refraction of acetone. (b) The index of refraction of ice is 1.305. Calculate the speed of light in ice.

5.   (a) Light travels 0.700 times as fast in Jell-O as it does in a vacuum. Calculate the index of refraction of Jell-O. (b) If it travels 1.10 times faster in bouillon soup than in Jell-O, what is the soup's index of refraction? Be sure to justify your calculation technique.

6.   The index of refraction for radio waves passing through ice is 1.305. If the ice is 2400 m thick, how much time is needed for radio waves to travel from the surface of the antarctic ice sheet to the bottom and back to the surface?

7.   A water molecule vibrating at a frequency of 9.66×10^{13} Hz in a glass of water emits infrared radiation. (a) Calculate the wavelength of the wave when in the water. (b) Calculate the wavelength after it leaves the water and is in the air.

8.   Light travels from the sun to the earth in 500 seconds. By approximately how much time would the trip be changed if the earth had no atmosphere? Indicate any approximations used in your calculations.

9.   When a wave passes from one medium to another, its frequency remains the same but its wavelength changes. Derive an expression for the ratio of the wavelength in medium 1 to that in medium 2 as a function of the indices of refraction of the two media ($\lambda_1/\lambda_2 = ?$).

19.5   Refraction

10.   (a) Light passes from air into a 25 percent sugar solution at an incident angle of 40°. Calculate the angle of the refracted wave. (b) Light passes from ethyl alcohol to air at an incident angle of 18°. Calculate the angle of the refracted wave in the air.

11. (a) Light traveling through air strikes a piece of light crown glass at an angle of 72°. Calculate the angle of refraction of the ray in the glass. (b) A ray passing from air to cyclohexane is incident at 48° and has an angle of refraction of 31°. Calculate the index of refraction of the cyclohexane.

12. A light ray traveling in air is incident on an interface with another medium at an angle of 57°. The reflected ray and the refracted ray make an angle of 90° with respect to each other. Calculate the index of refraction of the second medium.

13. A sound wave traveling in air ($v = 340$ m/s) is incident on a surface of water at an incident angle of 5.0°. Calculate the direction of the refracted sound wave in the water ($v_{\text{water}} = 1400$ m/s).

14. A ray of light traveling in air is incident at an angle of 42° on an interface with a water-sugar solution. If the refracted wave travels at an angle of 27°, what is the speed of light in the water-sugar solution?

15. A sound wave traveling in limestone at a speed of 4000 m/s is incident at an angle of 38° on an interface with an unknown material. The refracted sound leaves the interface at an angle of 50°. Calculate the speed of sound in the unknown material, and use Table 17.1 to guess what this material is.

16. A bagpipe marching band moving at a speed of 0.80 m/s on a hard surface enters a mud field and slows to 0.50 m/s. If the front row enters the mud at an incident angle of 30°, calculate its direction in the mud.

17. Light traveling in air enters a sheet of glass ($n = 1.58$) at the side of an aquarium and then passes into the water ($n = 1.33$). If the angle of incidence at the air-glass interface is 22°, calculate the angle at which the light wave travels in the glass and in the water.

18. Sound traveling in limestone enters the top surface of a cubical salt dome at an angle of incidence of 35°. After passing through the salt dome, the sound leaves the right side and reenters the limestone. Calculate its final direction in the limestone and show the direction in a figure. ($v_{\text{limestone}} = 4000$ m/s and $v_{\text{salt}} = 6000$ m/s).

19. Light moving up and toward the right in air enters the side of a cube of gelatin ($n = 1.30$) at an incident angle of 80°. Calculate the angle at which the light leaves the top surface of the cube.

19.6 Total Internal Reflection

20. Calculate the critical angle for light inside a diamond incident on an interface with air.

21. (a) Calculate the critical angle of reflection for sound moving at 340 m/s in air that strikes a surface of water where its speed is 1500 m/s. (b) Will music played by a rock band on the shore annoy fish in the water? Explain.

22. Sound travels at a speed of 330 m/s in cool air near the earth's surface. It strikes an interface with overlying warmer air, in which its speed is 350 m/s. Indicate the angular region of incidence for which total internal reflection occurs.

23. A scuba diver swimming under water at night shines a beam of light so that it hits the water-air interface at an incident angle of 52°. Will the light be seen above the water? Explain.

24. Light is incident on the boundary between two media at an angle of 30°. If the refracted light makes an angle of 42°, what is the critical angle for light incident on the same boundary?

25. A refractometer is a device used to measure the index of refraction of an unknown liquid. Light generated in the liquid is refracted into a standard material whose index of refraction is known and is greater than that of the unknown liquid. The incident beam is adjusted for total internal reflection and Eq. (19.5) is used to calculate the unknown index. Invent a design for such a device and perform a sample calculation to show how it might work.

26. (a) Rays of light are incident on a glass-air interface (Fig. 19.24a). Calculate the critical angle for total internal reflection ($n_{\text{glass}} = 1.58$). (b) If there is a thin, horizontal layer of water ($n = 1.33$) on the glass, will a ray incident on the glass-water interface at the critical angle calculated in (a) be able to leave the water (Fig. 19.24b)? Justify your answer.

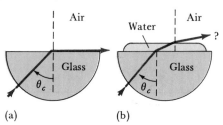

(a)   (b)

FIG. 19.24

27. Rays of light enter the end of a light pipe from air at an angle θ_1 (Fig. 19.25). The index of refraction of the pipe is 1.72. Calculate the greatest angle θ_1 for which the ray is totally reflected at the top surface of the glass-air interface inside the pipe.

FIG. 19.25

28. An optic fiber ($n = 1.72$) is often coated with a protective covering of glass ($n = 1.50$). (a) Calculate the critical angle for the fiber-glass interface. (b) Calculate the critical angle for the glass-air interface. (c) Calculate the critical angle for a fiber-air interface (no glass covering). (d) Will a ray incident on the fiber-glass interface at the angle calculated in (c) eventually enter the air from the glass-air interface of a coated fiber? Explain.

29. What must be the minimum value of the index of refraction of the prism shown in Fig. 19.16a in order that light be totally reflected where indicated? Is the index of refraction of most types of glass greater than this value?

19.7 Dispersion

30. Red light of wavelength 650 nm strikes the side of the prism ($n_{\text{red}} = 1.613$) shown in Fig. 19.17 at an incident angle of 53.8°. (a) Calculate the refracted angle relative to a line perpendicular to the first air-glass interface. (b) If the prism has three 60° angles, calculate the incident angle at the second glass-air interface. (c) Calculate the refracted angle relative to a line perpendicular to this surface. (d) Calculate the net angular change in direction of the light.

31. Repeat Problem 30 for blue light for which $n_{\text{blue}} = 1.636$.

Optical Instruments

20

In 1590 Zacharias Janssen of Holland made the first microscope, and eighteen years later another Dutchman, Hans Lippershey, constructed the first telescope. Science was embarking on a new era that would eventually allow us to see the tiny details of living organisms and the grand formations in the sky.

Microscopes and telescopes are only two of many optical instruments used to alter the way we see things. Other instruments that will be considered in this chapter, besides the human eye, include the camera, slide projector, and eyeglasses. These devices depend on glass lenses that gather and focus light from an object to form an image seen by our eyes. We will learn how these images are formed to make things look bigger or smaller, closer or farther away, upright or inverted. We will also find out how curved mirrors can be used to produce images that are different (for example, bigger) than the object being viewed.

20.1 Lenses

Our eyes perceive an object as being located at a point from which rays of light leave the object. A **lens** is an optical device that changes the apparent location of an object by altering the path of light rays coming from the object.

The effect of a lens on light is easily visualized by constructing a lens from two prisms, as shown in Fig. 20.1a. When parallel rays of light pass through the prism lens, they bend toward an axis through the center of the lens. Because the rays converge or move towards each other, this lens is called a **converging lens** (also called a convex lens). The rays of light, after passing through the prism lens, intersect at different places. If, instead of using prisms, we make a converging lens whose surfaces are properly curved, all the rays will pass through one point (Fig. 20.1b). This point where *parallel* rays meet after passing through a

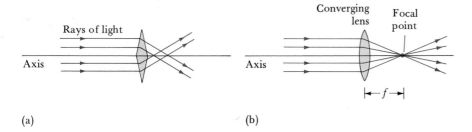

(a)   (b)

FIG. 20.1. (a) A converging lens built from two prisms causes rays of light to cross at different points after passing through the lens. (b) A lens with properly curved surfaces causes parallel rays to converge at the focal point.

(a)

(b)

FIG. 20.2. (a) Two prisms join to form a diverging lens. Rays of light move apart or diverge after passing through the prisms. (b) A diverging lens with properly curved surfaces causes parallel rays to appear to diverge from a single point called the focal point.

converging lens is called the **focal point** of the lens. (The word *parallel* is emphasized because parallel rays are the only ones that converge at the focal point. Rays not parallel to the axis cross it at other points.) The **focal length** f is the distance of this focal point from the lens. For a converging lens, the focal length is always a positive number.

Another type of lens commonly used in optical instruments is a concave or **diverging lens.** Instead of bringing rays of light together, a diverging lens causes them to diverge or move apart. This effect can be illustrated using two prisms, as shown in Fig. 20.2a. If parallel rays of light strike one side of the prism, the refracted rays diverge when leaving the other side. If the surfaces of a lens are properly curved, all parallel rays will appear to diverge from the same point on the left side of the lens (Fig. 20.2b). The point from which parallel rays appear to diverge after leaving a diverging lens is called the **focal point** of the lens. The **focal length** is the distance from this focal point to the lens. For diverging lenses, the focal length is a negative number.

20.2 Formation of Images

The use of lenses in optical instruments, such as cameras, movie projectors, microscopes, and telescopes, can be illustrated using **ray diagrams** that show the path of light along several carefully chosen rays. The intersection of these rays after passing through a lens indicates the effect of the lens on our perception of an object—does it appear bigger, smaller, closer to our eye, farther away, inverted, or upright?

To construct a ray diagram, draw a lens with a faint line passing through the center of the lens and perpendicular to its surface (Fig. 20.3a). The line is called the principal axis or the **axis** of the lens. Place dots on the axis on each side of the lens to indicate the focal points of the lens. The distance from the lens to the focal points represents the focal length of the lens, a distance that is usually given for a particular lens.

Next, draw an arrow to the left side of the lens perpendicular to and resting on the axis, as shown in Fig. 20.3a. The arrow represents the object being viewed through the lens. The distance between the arrow and the lens in the drawing represents the separation of the actual object from the lens. The convention of always placing the object on the left side of the lens simplifies our later mathematical calculations. Finally, draw two or three carefully chosen rays of light that pass from the object through the lens. The rays used for this purpose are described below and pictured in Fig. 20.3.

Ray 1 leaves the object and moves parallel to the axis. As the ray passes through a converging lens, it is refracted toward the axis and passes through the focal point on the right (Fig. 20.3a). If it passes through a diverging lens, ray 1 is refracted away from the axis and appears to come from the focal point on the left (the dashed line in Fig. 20.3b indicates the ray's apparent origin).

FIG. 20.3. Three special rays passing through converging and diverging lenses. Two or three of these rays can be used to locate the image of any object.

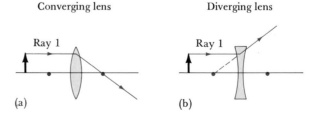

Converging lens   Diverging lens

Ray 1   Ray 1

(a)   (b)

Ray 2 (Fig. 20.3c and d) leaves the object and passes directly through the middle of the lens. Although the ray may be refracted slightly at the front and back surfaces of the lenses, the two refractions cancel each other. If the lens is thin, the ray will move straight through its center.

(c)   (d)

For converging lenses, *Ray 3* leaves the object and passes through the focal point on the left (Fig. 20.3e). The lens refracts the ray so that it moves parallel to the axis on the right. For a diverging lens, ray 3 leaves the object and moves toward the focal point on the right (Fig. 20.3f). Before reaching the focal point, the ray is refracted parallel to the axis on the right. The dashed line indicates the ray's path if the lens were not present.

(e)   (f)

If two or three of these rays are drawn on the same diagram, they may converge on the right side of the lens at a point called the *real image* of the object. If the rays diverge on the right side, the point from which they seem to diverge is called the *virtual image* of the object.

Formation of a Real Image

To illustrate the formation of real images, we will locate an object to the left side of a converging lens at a distance of about two times the focal length of the lens (Fig. 20.4a). Light rays leave the object in all directions, but only some of these rays pass through the lens. The paths of our special three rays from the tip of the object are shown in Fig. 20.4b. We find that the rays converge on the right side of the lens at point *I*.

Rays of light are also coming from other points on the object. For example, three rays leaving the middle of the arrow are brought together at point *I′*, a little above point *I* (Fig. 20.4c). When a ray diagram is constructed for rays leaving each point on the object, we find that an inverted image is formed at a distance *s′* on the right of the lens (Fig. 20.4d). If a screen is placed at a distance *s′* to the right of the lens, we will see a real, inverted picture of the object on the screen. This is called a **real image,** since the rays actually come together to form an image of the object. The distance *s′* is called the **image distance** and the distance *s* from the object to the lens is called the **object distance.**

A sharp, clear image of an object appears in only one place. If a screen is placed closer to the lens than distance *s′* (Fig. 20.5a), where rays leaving the tip of the arrow have not yet converged to a point, the image on the screen will be blurry and out of focus. If the screen is placed to the right of *s′* (Fig. 20.5b),

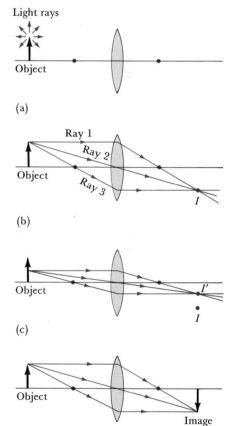

FIG. 20.4. (a) Rays of light leave the tip of the arrow in all directions. (b) Three special rays from the tip converge at *I* after passing through the lens. (c) Three rays from the middle of the arrow converge at *I′* above *I*. (d) A real inverted image of the object is formed at a distance *s′* to the right of the lens.

FIG. 20.5. If a screen is placed at any distance from the lens other than s', a blurry image of the object is formed.

(a)

(b)

where rays from the tip are diverging after having converged at I, a blurry image is again the result. Only at distance s' from the lens is a sharp, clear, focused image seen.

The distance s' of the focused image from the lens can be determined by experiment, by drawing a ray diagram, or by calculation using a thin-lens equation (to be introduced later). The ray diagram technique is the least accurate of these methods and is used mainly as a check on calculations using the thin-lens equation. The image position depends on where the object is located and on the focal length of the lens. Notice in Fig. 20.6a and b that the image distance

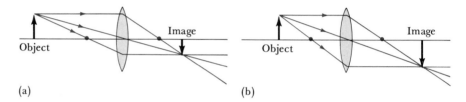

FIG. 20.6. As the object moves toward the focal point from the left, the image moves farther toward the right.

(a)

(b)

increases as the object distance decreases. However, if an object is placed between the focal point and the lens, a real image is no longer formed, as we will see next.

Virtual Image

When the rays from an object diverge after passing through a lens (Fig. 20.7a), a real image is not formed. We cannot place a screen on the right side of the lens and see an image of the object. But, if we look with our eyes at light passing through the lens, the rays appear to diverge from a position behind the lens (Fig. 20.7b). At this position we see the **virtual image** of the object.

FIG. 20.7. (a) The three special rays leaving an object placed inside the focal point of a converging lens. (b) The rays, after passing through the lens, appear to come from point I. This is a virtual image, since light does not actually come from this point.

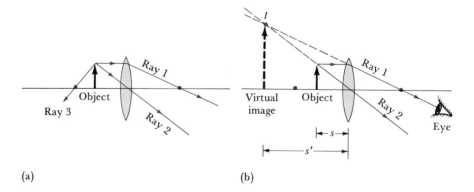

(a)

(b)

If an object is placed between a converging lens and its focal point, a virtual image is formed. To locate the image, consider the rays of light shown in Fig. 20.7a. Ray 3 does not pass through the lens and cannot be used to locate the image. However, rays 1 and 2 do pass through the lens and diverge on the right side. These rays from the tip of the arrow appear to our eyes to diverge from point I on the left side of the lens (where the dashed lines in Fig. 20.7b cross). Since the directions of these rays imply that light is actually coming from I, we think the tip of the arrow is located at I. This apparent location of the arrow is said to be a virtual image of the arrow.

A virtual image is usually drawn using a dashed arrow rather than a solid arrow. The image is always located on the same side of the lens as the object. Examples of virtual images of objects seen through diverging lenses are shown in Fig. 20.8.

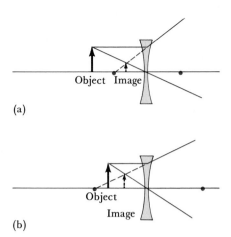

(a)

(b)

FIG. 20.8. Virtual images formed by diverging lenses.

20.3 Thin-Lens Equation

As we saw in the last section, a ray diagram can be used to locate the image of light rays passing through a lens from an object. The ray-diagram technique provides a means of determining the type of image (real or virtual, inverted or upright) and its approximate location. The image location can be determined more accurately using an equation called the **thin-lens equation:**

$$\frac{1}{f} = \frac{1}{s} + \frac{1}{s'}, \tag{20.1}$$

where f is the focal length of the lens, s is the object distance, and s' is the image distance. This equation can be solved for any one of the three quantities (f, s, or s') if the other two quantities are known. The equation is used in the design of optical instruments such as telescopes, microscopes, slide projectors, and cameras. The equation is also the basis for choosing lenses for eyeglasses.

Several **sign conventions** are important when using the thin-lens equation. These conventions apply for light rays moving from left to right.

1. The **focal length** f is positive for converging lenses and negative for diverging lenses.

2. The **object distance** s is positive if the object is to the left of the lens and negative if the object is to the right of the lens.*

3. The **image distance** s' is positive for real images formed to the right of the lens and negative for virtual images formed to the left of the lens.

The thin-lens equation is derived geometrically by considering the triangles formed by the special rays shown in Fig. 20.9a. Look at the two triangles formed by ray 2 and the axis of the lens (Fig. 20.9b). The angle θ is the same for each triangle, as is the tangent of each angle. Thus,

$$\tan \theta = \frac{h}{s} = \frac{h'}{s'}.$$

(a)

(b)

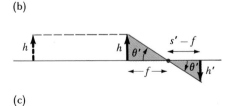

(c)

FIG. 20.9. (a) A ray diagram for a converging lens that forms a real image of an object. (b) and (c) Two pairs of similar triangles are used to derive the thin-lens equation.

---

*This rule is useful for systems containing more than one lens. The object of the second lens is the image of the first. Occasionally, the object of the second lens may actually be to the right of the lens. We will not need to use this rule until Section 20.7. For now, the object distance is always positive.

The right side of ray 1 also forms two triangles with the axis of the lens (Fig. 20.9c). The angle θ' is the same for each of these triangles, as is the tangent of each angle. Thus,

$$\tan \theta' = \frac{h}{f} = \frac{h'}{s' - f}.$$

The last two equations can be rearranged so that the quantity h'/h appears on one side of each equation. We find then that

$$\frac{s'}{s} = \frac{h'}{h}$$

and

$$\frac{s' - f}{f} = \frac{h'}{h}.$$

Since the right sides of these two equations are identical, their left sides must be equal:

$$\frac{s'}{s} = \frac{s' - f}{f}.$$

Dividing each term in this equation by s' yields

$$\frac{1}{s} = \frac{1}{f} - \frac{1}{s'}$$

or

$$\frac{1}{s} + \frac{1}{s'} = \frac{1}{f}.$$

The thin-lens equation can also be derived for diverging lenses and for converging lenses in which virtual images are formed.

EXAMPLE 20.1 An object is placed 20 cm to the left of a diverging lens whose focal length is -10 cm (the negative sign is used for a diverging lens). Where is the image of this object located and is the image real or virtual?

SOLUTION A ray diagram for this system is shown in Fig. 20.10. On the right side of the lens, light appears to come from a virtual image between the lens and the left focal point—about 7 cm left of the lens. Thus, we expect that $s' \simeq -7$ cm. The negative sign indicates that the image is virtual and to the left of the lens. Let us see if the same result is obtained using the thin-lens equation.

Rearranging Eq. (20.1) to solve for s', we find that

$$\frac{1}{s'} = \frac{1}{f} - \frac{1}{s} = \frac{1}{-10 \text{ cm}} - \frac{1}{20 \text{ cm}}$$

$$= -0.100 \text{ cm}^{-1} - 0.050 \text{ cm}^{-1} = -0.150 \text{ cm}^{-1}$$

or

$$s' = \frac{1}{-0.150 \text{ cm}^{-1}} = \underline{-6.67 \text{ cm}}.$$

FIG. 20.10. A ray diagram for a diverging lens.

Remember that to find s', you must invert the value of $1/s'$ obtained from Eq. (20.1). Forgetting to invert is a common error in optics problems! Our thin-lens calculation ($s' = -6.67\,\text{cm}$) checks well with the estimated solution ($s' \simeq -7\,\text{cm}$) determined using a ray diagram. ■

FIG. 20.11. A ray passing through the center of the lens forms two similar triangles that are used to calculate the linear magnification of the lens.

Linear Magnification

For some optical devices, the height of the image compared to the height of the object must be known. You might, for instance, want to know whether the image of a person 2 m tall will fit on a film that is 35 mm high. The ratio of the height of the image h' and the height of the object h is called the **linear magnification** m:

$$m = \frac{h'}{h}.$$

The linear magnification is a unitless number that simply tells how much taller the image is than the object; for example, if $m = 1$, the image and object are the same height. By convention, the height is a positive number for an upright object or an upright image and a negative number for an inverted object or image. For example, in Fig. 20.11, h' has a negative value because the image is inverted. If m is a negative number, the image is inverted relative to the object.

The linear magnification of an optical system can be calculated using the object and image distances, s and s', rather than their heights. In Fig. 20.11, the two shaded triangles are similar. Thus,

$$\frac{-h'}{s'} = \frac{h}{s}.$$

The negative sign in the above equation is needed so that both ratios have the same sign. Recall that h' shown in Fig. 20.11 is a negative number. Thus, $-h'$ is a positive number as is the ratio $-h'/s'$. This ratio equals the positive ratio h/s. We can rearrange the above equation as follows: $h'/h = -s'/s$. Therefore, the linear magnification m of an optical system can be calculated using either ratio:

$$m = \frac{h'}{h} = -\frac{s'}{s}. \tag{20.2}$$

Equation (20.2) applies to any type of lens and to real and virtual images. If m is negative, the image is inverted relative to the object.

Cameras

A camera, shown in Fig. 20.12, has a lens of fixed focal length. To focus sharply on objects at different distances, the film moves forward and back relative to the lens. Actually, the lens is moved relative to the film, but the effect is the same. As the object approaches the lens (decreasing object distance s), the film must be moved farther from the lens (increasing image distance s'). For a camera, then, the focal length f is a constant, and the image distance s' is varied to accommodate different object distances s.

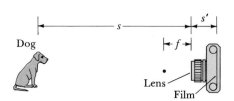

FIG. 20.12. A real image of a dog is formed by a lens on the film of a camera.

EXAMPLE 20.2 A camera has a converging lens with a focal length of 5.00 cm. A dog 1.0 m tall, standing 3.0 m from the lens, is photographed. (a) Where must the film be located relative to the lens so that a sharp image is produced on the film? (b) How large is the image?

SOLUTION The optical system appears as shown in Fig. 20.12. The known information is: $f = 5.00$ cm, $s = 300$ cm, and $h = 1.0$ m. (a) To find where the film should be placed relative to the camera lens, we must solve Eq. (20.1) for the image distance s':

$$\frac{1}{s'} = \frac{1}{f} - \frac{1}{s} = \frac{1}{5.00 \text{ cm}} - \frac{1}{300 \text{ cm}}$$

$$= 0.2000 \text{ cm}^{-1} - 0.0033 \text{ cm}^{-1} = 0.1967 \text{ cm}^{-1}$$

or

$$s' = \frac{1}{0.1967 \text{ cm}^{-1}} = \underline{5.08 \text{ cm}}.$$

(b) To calculate the height h' of the image, we must first solve for the linear magnification of the lens using Eq. (20.2):

$$m = -\frac{s'}{s} = -\frac{5.08 \text{ cm}}{300 \text{ cm}} = -0.0169.$$

The negative sign means the image is inverted. Again, using Eq. (20.2), we can find the image height:

$$h' = mh = (-0.0169)(1.0 \text{ m}) = -0.0169 \text{ m} = \underline{-16.9 \text{ mm}}.$$

The inverted image of the dog should occupy about one half the height on a 35-mm film. ∎

20.4 The Human Eye

The human eye resembles in many ways an expensive camera. The eye is equipped with a built-in cleaning and lubricating system, an exposure meter, an automatic field finder, and a continuous supply of film. Light from an object enters the cornea (Fig. 20.13), a transparent covering over the surface of the eye, and passes through a transparent lens held in place by ciliary muscles. An iris in front of the lens opens or closes like the shutter on a camera in order to regulate the amount of light entering the eye. The cornea and lens together act as a lens of variable focal length that focuses light from an object to form a real image on the back surface of the eye, called the retina.

The retina acts like the film of a camera. It contains about 130 million light-sensitive cells called cones and rods. Light absorbed by these cells initiates photochemical reactions that cause electrical impulses in nerves attached to the cones and rods. The signals from individual cones and rods are combined in a complicated network of nerve cells and transferred from the eye to the brain via the optic nerve. What we see depends on which cones and rods are excited by absorbing light and on the way in which the electrical signals from different cones and rods are combined and interpreted by the brain.

The cones are concentrated in one part of the retina called the fovea. The

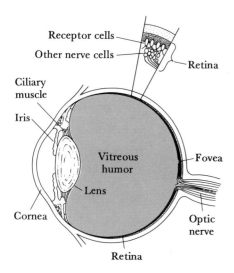

FIG. 20.13. The human eye.

fovea is about 0.3 mm in diameter and contains 10,000 cones and no rods. Each cone in this region has a separate nerve fiber that leads to the brain along the optic nerve. Because of the large number of nerves coming from this small area, the fovea is the best part of the retina for resolving the fine details of a bright object. Besides providing a region of high visual acuity, the cones in the fovea and in other parts of the retina are specialized for detecting different colors of light.

The concentration of cones decreases outside the fovea. In these peripheral regions, the rods predominate. Their density in the retina (about 150,000 per square millimeter) is about the same as that of the cones in the fovea region. However, the light signals from perhaps 100 adjacent rods are brought together into a single nerve cell that leads to the brain. This combining of the rod signals reduces our ability to see the fine details of an object but helps us see dimly lit objects, since many small signals are combined to produce a larger signal.

Optics of the Eye

A normal eye can focus on objects located anywhere from about 25 cm to hundreds of miles away. This ability to focus on objects at different distances is called **accommodation.** Unlike the camera, which uses a fixed focal length lens and a variable image distance to accommodate different object distances, the eye has a fixed image distance (about 2.1 cm) and a variable focal length lens system.

The changing focal length of the eye's lens is illustrated in Fig. 20.14. Looking at distant objects, the ciliary muscle attached to the lens of the eye relaxes and the lens becomes less curved (Fig. 20.14a). When less curved, the focal length increases and an image is formed at the retina. If the lens remains flattened, and the object moves closer to the lens, the image will then move back behind the retina causing a blurred pattern of light on the retina. To avoid this, the ciliary muscles contract and cause an increase in the curvature of the lens, reducing its focal length (Fig. 20.14b). With reduced focal length, the image moves forward and again forms a sharp, focused image on the retina. If your eyes become tired after reading for many hours, it is because the ciliary muscles have been tensed to keep the lenses of your eyes curved.

The **far point** of the eye is the greatest distance to an object upon which the relaxed eye can focus. The **near point** of the eye is the closest distance of an object upon which the tensed eye can focus. For the normal eye, the far point is effectively infinity (we can focus on the moon and on distant stars) and the near point is about 25 to 50 cm. We turn next to the changes in the eye that cause nearsightedness and farsightedness.

Nearsightedness

A nearsighted person's eyes can focus clearly on objects that are close to the eye but not on those that are distant. The far point of a nearsighted person may be only a few meters rather than infinity.

Nearsightedness occurs if a person's eyeball is larger than the usual diameter (variations in diameter of 1 mm can easily cause a person to be nearsighted).

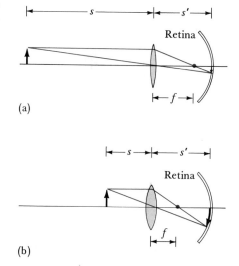

(a)

(b)

FIG. 20.14. The eye has a fixed image distance s' and a variable focal length f to accommodate different object distances s.

Relaxed eye

Distant
object

(a)

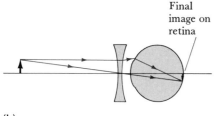

Final
image on
retina

(b)

Image formed by
eyeglass lens
at eye's far point

(c)

FIG. 20.15. (a) When a nearsighted person views a distant object, the image is formed in front of the retina. (b) A diverging eyeglass lens causes rays to diverge so that a final image is formed on the retina. (c) The corrective lens forms a virtual image at the eye's far point.

In such cases, the image of a distant object is formed in front of the retina (Fig. 20.15a), even when the eye lens is relaxed. Tensing the ciliary muscles increases the curvature of the lens and causes the image to move even farther in front of the retina. Nearsightedness also occurs if the cornea and relaxed lens are too curved—the image of a distant object is again formed in front of the retina.

In order to correct nearsighted vision, a diverging eyeglass lens is placed in front of the cornea. The rays from an object diverge slightly while passing through the eyeglass lens so that when they pass through the lens of the eye, an image is formed farther back in the eye (Fig. 20.15b). You can think of the eyeglass lens as performing a trick on the eye. If an object is very distant (for example, $s \simeq \infty$), then the focal length of the eyeglass lens is chosen so that its virtual image is formed at the far point of the eye (Fig. 20.15c). Light passing through the eyeglass lens appears to come from the image at the far point, and not from the more distant object. To calculate the desired focal length of the eyeglass lens, we set $s = \infty$ and $s' = -$(far point). The negative sign accounts for the fact that the image of the object as seen through the lens of the eyeglass is virtual and to the left of the lens. The value of f is then calculated using the thin-lens equation.

EXAMPLE 20.3 The far point for a nearsighted person is 2.0 m from the eye. What focal length lens should the person use in order to focus on a very distant object ($s = \infty$)?

SOLUTION We are given that $s = \infty$ and that the image of this distant object should be at the far point of the eye—that is, at $s' = -2.0$ m. Inserting these values into Eq. (20.1), we can calculate the focal length of the lens for the eyeglasses:

$$\frac{1}{f} = \frac{1}{s} + \frac{1}{s'} = \frac{1}{\infty} + \frac{1}{-2.0 \text{ m}} = 0 + \frac{1}{-2.0 \text{ m}} = -\frac{1}{2.0 \text{ m}}$$

or

$$f = \underline{-2.0 \text{ m}}.$$

The focal length of the lens for a nearsighted person equals the negative value of the person's far point. ∎

Farsightedness

Unlike nearsighted people, farsighted persons are able to see distant objects but cannot focus on those nearby. Whereas the normal eye has a near point of 25 to 50 cm, a farsighted person may have a near point several meters from the eye. If his vision is uncorrected by glasses, a farsighted person will have to hold a book or newspaper at his nearpoint, several meters from the eye, in order to focus on the print. If the book is held closer to the eye, the image of the book is formed behind the retina, as shown in Fig. 20.16a. Light striking the retina is blurred and out of focus. Farsightedness may occur if the diameter of a person's eyeball is smaller than usual or if the lens is unable to curve enough when the ciliary muscles contract.

Converging eyeglass lenses are used to correct farsighted vision. The eyeglass lens bends light rays from an object toward the axis of the lens. The rays converge even farther as they enter the eye's cornea and lens, and the image is formed on the retina (Fig. 20.16b). Without the bending caused by the eyeglass lens, the image is formed behind the retina.

The converging lens used to correct farsightedness must have a focal length that is greater than the near point of a normal eye, which by convention is said to be 25 cm. When an object such as a book is held at the normal near point and viewed through the eyeglass lens, a virtual image of the object is formed behind the object by the glasses (Fig. 20.16c). The eye looking through the eyeglass lens perceives light to be coming from the virtual image. An eyeglass lens of the correct focal length will cause the virtual image to be formed at the actual near point of the farsighted person's eye. To determine the focal length of the eyeglass lens, we need to solve the thin-lens equation for f, given that $s = 25$ cm (the actual distance of the book or nearby object from the eyeglass lens) and $s' = -$near point (the position of the virtual image produced by the eyeglass lens). Light entering the eye appears to come from s'. The negative sign means that the virtual image is on the same side of the lens as the object.

EXAMPLE 20.4 A farsighted person has a near point of 1.5 m. Calculate the focal length of lenses needed for her eyeglasses.

SOLUTION We know that $s = 25$ cm (the position at which an object is held when viewed from close range) and $s' = -150$ cm (the position from which we want her eye to perceive the light as coming). Substituting into Eq. (20.1) and solving for f, we find that

$$\frac{1}{f} = \frac{1}{s} + \frac{1}{s'} = \frac{1}{25 \text{ cm}} + \frac{1}{-150 \text{ cm}} = 0.0333 \text{ cm}^{-1}$$

or

$$f = +30 \text{ cm}. \qquad \blacksquare$$

(a)

(b)

(c)

FIG. 20.16. (a) The image of a near object is formed behind the retina for a farsighted person. (b) A converging eyeglass lens will cause the rays to converge on the retina. (c) The corrective lens forms a virtual image at the eye's near point.

Diopter

An optometrist prescribes lenses for eyeglasses in terms of the power P of the lens in diopters rather than in terms of the focal length. The **power** of a lens is defined as the inverse of its focal length in meters:

$$P \text{ (diopters)} = \frac{1}{f \text{ (meters)}}. \qquad (20.3)$$

For example, the power of a diverging lens of focal length $f = -50$ cm is $P = 1/(-0.50 \text{ m}) = -2$ diopters. A lens of high power has a short focal length and causes rays to converge rapidly after passing through the lens.

20.5 Angular Magnification and Magnifying Glasses

The linear magnification of an optical system compares only image and object heights, but the apparent size of an object as judged by the eye depends not only on its height but also on its location relative to the eye. For example, a pencil

FIG. 20.17. The size of the image of a pencil on the retina depends upon its distance from the retina.

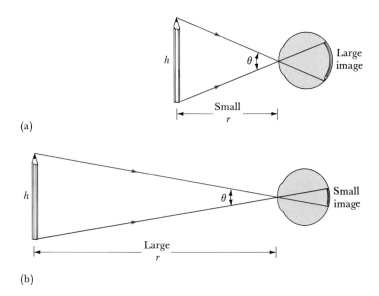

(a)

(b)

held 25 cm from your eye appears longer than one held 100 cm away (Fig. 20.17). In fact, the pencil may appear longer than a 100-story building that is several miles away, even though the pencil is much smaller in size.

A person's impression of an object's size is proportional to the angle θ shown in Fig. 20.17. This angle, called the **angular size,** is related to an object's height h and its distance r from the eye by the equation

$$\theta = \frac{h}{r}. \tag{20.4}$$

Remember that an angle θ, in units of radians, is the ratio of an arc length (h in Fig. 20.17) and the radius of a circle (r in Fig. 20.17).*

If a person looks at an object through a system of lenses, he or she sees light that appears to come from the final image of the system of lenses. If the person's eye is close to the last lens, then the angular size θ' as seen through the lenses is

$$\theta' = \frac{h'}{s'}, \tag{20.5}$$

where h' is the height of the final image and s' is the distance of the final image from the last lens. The **angular magnification** M of an optical system is defined as the ratio of θ' and θ:

$$M = \frac{\theta'}{\theta} = \frac{\text{angular size of final image of optical system}}{\text{angular size of object as seen by unaided eye}}. \tag{20.6}$$

A magnifying glass is the simplest optical device that provides angular magnification. The device consists of a single converging lens, as shown in Fig. 20.18a. If the object is held between the focal point and the lens, a magnified virtual image is formed behind the object. When looking through the magnifying glass, the viewer thinks the light is coming from this enlarged image. By

FIG. 20.18. (a) A magnifying glass produces an enlarged virtual image of an object held near the glass. (b) The object appears much smaller if viewed directly.

(a)

(b)

*This expression for θ is exact only if h is the curved length on the arc of a circle ($s \circlearrowright$) and not the straight line between the corners of the triangle ($h \triangleright$). For small angles, $h \simeq s$ and Eq. (20.4) is a good approximation.

comparison, if one looks directly at the object without the aid of the magnifying glass, as shown in Fig. 20.18b, the object appears much smaller.

To calculate the angular magnification of the magnifying glass, we compare the angular size θ' of the image seen through the magnifying glass and the angular size θ of the object as seen with the unaided eye. The angular size of the image viewed through the magnifying glass is $\theta' = h'/s'$ (see Fig. 20.18a). By considering ray 2 in Fig. 20.18a, we see that $h'/s' = h/s$. Thus,

$$\theta' = \frac{h}{s},$$

where h is the actual height of the object, and s is its distance from the magnifying glass.

Next, we consider the angular size of the object as seen with the unaided eye. The angular size θ of an object of height h held a distance r from the eye is $\theta = h/r$ (Fig. 20.19a). As the object is brought closer to the eye (Fig. 20.19b and c), its angular size increases. Should the object be brought closer than the near point, a blurred image is formed on the retina (Fig. 20.19d). The maximum angular size θ_{max} of an object of height h, when focused on the retina, is

$$\theta_{max} = \frac{h}{\text{near point}}. \tag{20.7}$$

(a)

(b)

(c)

(d)

FIG. 20.19. The angular size θ of an object increases as it is brought closer to the eye. If brought too close, as in (d), an image is formed behind the eye, causing a blurred image on the retina.

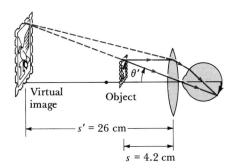

FIG. 20.20. A magnifying glass forms a virtual image of a stamp held near the glass.

The **angular magnification of the magnifying glass** (called **magnifying power**) is the ratio of the angular size θ' as seen through the magnifying glass and the maximum angular size θ_{max} as seen with the unaided eye. Thus,

$$M = \frac{\theta'}{\theta_{max}} = \frac{h/s}{h/(\text{near point})} = \frac{\text{near point}}{s}. \qquad (20.8)$$

EXAMPLE 20.5 A stamp is held 4.2 cm from a magnifying glass whose focal length is +5.0 cm. Locate the virtual image of the stamp and calculate the angular magnification of the magnifying glass. The near point of the eye is 25 cm.

SOLUTION The image distance s' is calculated using the thin-lens equation:

$$\frac{1}{s'} = \frac{1}{f} - \frac{1}{s} = \frac{1}{5.0\text{ cm}} - \frac{1}{4.2\text{ cm}} = -0.0381\text{ cm}^{-1}$$

or

$$s' = -26\text{ cm}.$$

The image is virtual and is located 26 cm behind the lens, as shown in Fig. 20.20. Since 26 cm is greater than the near point of the eye, a clear image is formed on the retina. The eye perceives rays of light as coming from this image.

The angular magnification of the magnifying glass is

$$M = \frac{\text{near point}}{s} = \frac{25\text{ cm}}{4.2\text{ cm}} = \underline{6.0}. \qquad \blacksquare$$

20.6 Combinations of Lenses

Many optical instruments that have more than one lens (such as microscopes and telescopes) form an image of an object that is magnified more than is possible using only a single lens. The technique for locating the final image and calculating its magnification is not difficult but requires that several factors be carefully considered.

Consider the two-lens system shown in Fig. 20.21a. The lenses are separated by a distance d and have focal lengths f_1 and f_2. The object is a distance s_1 from lens 1. The location of the final image of the two-lens system can be determined as follows:

1. Using Eq. (20.1), determine the location s_1' of the image formed by the first lens (Fig. 20.21b). If s_1' is positive, the image is to the right of lens 1. If s_1' is negative, the image is to the left of lens 1. In our example, s_1' is positive.

2. The image of lens 1 is now the object for lens 2. The object distance s_2 (Fig. 20.21c) is

$$s_2 = d - s_1', \qquad (20.9)$$

where d, the separation of lenses, is a positive number and s_1' is either positive or negative, depending on the value calculated in step 1. If s_1' is a larger positive number than d, the image of lens 1 is formed to the right of lens 2; s_2 will then be a negative number (be sure to retain this negative sign in subsequent calculations).

(a)

(b)

(c)

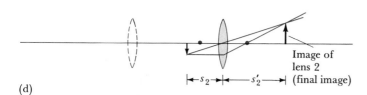

(d)

FIG. 20.21. A step-by-step scheme for locating the final image of the two-lens system shown in (a) is illustrated. (b) First locate the image of the first lens. (c) Calculate the object distance for the second lens. (d) Locate the final image formed by the second lens.

3. Use the thin-lens equation to calculate the image distance s_2' of lens 2 (Fig. 20.21d):

$$\frac{1}{s_2'} = \frac{1}{f_2} - \frac{1}{s_2}.$$

This is the location of the final image relative to the second lens. If s_2' is positive, the final image is real and located to the right of lens 2. If s_2' is negative, the final image is virtual and to the left of lens 2. In our example, the final image is real and to the right of lens 2.

4. The total magnification of the two-lens system is the product of the magnifications of each lens.

The above procedure is simple to use, but care must be taken to follow the appropriate sign conventions.

EXAMPLE 20.6 An object is located 50 m from a +15.0-cm focal length converging lens (Fig. 20.22). A diverging lens with a −4.20-cm focal length is placed 10.0 cm to the right of the first lens. Locate the final image of the two-lens system.

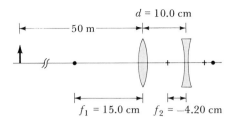

$d = 10.0$ cm

50 m

$f_1 = 15.0$ cm $f_2 = -4.20$ cm

FIG. 20.22. A two-lens system that is similar to an opera glass.

SOLUTION The following information is given: $f_1 = +15.0$ cm, $f_2 = -4.20$ cm, $d = 10.0$ cm, and $s_1 = 5000$ cm. First, locate the image of the first lens using Eq. (20.1):

$$\frac{1}{s_1'} = \frac{1}{f_1} - \frac{1}{s_1} = \frac{1}{+15.0 \text{ cm}} - \frac{1}{5000 \text{ cm}} = 0.06647 \text{ cm}^{-1}$$

or

$$s_1' = \frac{1}{0.06647 \text{ cm}^{-1}} = +15.05 \text{ cm}.$$

Next, locate the object of the second lens using Eq. (20.9):

$$s_2 = d - s_1' = 10.0 \text{ cm} - 15.05 \text{ cm} = -5.05 \text{ cm}.$$

The negative sign indicates that the object of the second lens is to its right. Finally, using Eq. (20.1), we can locate the image of the second lens:

$$\frac{1}{s_2'} = \frac{1}{f_2} - \frac{1}{s_2} = \frac{1}{-4.20 \text{ cm}} - \frac{1}{-5.05 \text{ cm}} = -0.0401 \text{ cm}^{-1}$$

or

$$s_2' = \frac{1}{-0.0401 \text{ cm}^{-1}} = -25.0 \text{ cm}.$$

The negative sign indicates that the final image is 25.0 cm to the left of the second lens. Since it is to the left, the final image is virtual. ■

Telescopes

A telescope (Fig. 20.23a) is a two-lens optical system that helps us see distant objects. The most common telescope consists of two converging lenses separated by a distance approximately equal to the sum of their focal lengths. When observing a distant object, the first lens produces a real image just beyond its own focal length, as shown in (b). The second lens is located so that this image of the first lens is just inside the focal point of the second lens. The image of the second lens is a magnified virtual image, as shown in (c). Our eye perceives that light as coming from this enlarged virtual image. The magnification of the telescope is the product of the magnification of each lens.

EXAMPLE 20.7 A lion 1.2 m tall stands 50 m from the first lens of the telescope shown in Fig. 20.23a. (a) Locate the final image of the lion. (b) Calculate the linear magnification of the telescope and the height of the final image. (c) Calculate the angular magnification of the telescope.

SOLUTION (a) We will leave the details of calculating the final image position to you. The results of these calculations are listed below and shown in Fig. 20.23b and c.

$$s_1 = 50 \text{ m} = 5000 \text{ cm},$$
$$s_1' = 20.08 \text{ cm},$$
$$s_2 = 4.42 \text{ cm},$$
$$s_2' = -38.1 \text{ cm}.$$

(a)

(b)

(c)

The final image is 38.1 cm to the left of the second lens. (b) The total linear magnification is the product of the linear magnifications of each lens [Eq. (20.2)]:

$$m = m_1 m_2 = \left(-\frac{s_1'}{s_1}\right)\left(-\frac{s_2'}{s_2}\right) = \frac{(s_1')(s_2')}{(s_1)(s_2)} = \frac{(20.08 \text{ cm})(-38.1 \text{ cm})}{(5000 \text{ cm})(4.42 \text{ cm})} = \underline{-0.035}.$$

The negative sign indicates that the image is inverted. The height of the final image is

$$h' = mh = (-0.035)(1.2 \text{ m}) = -0.042 \text{ m} = \underline{-4.2 \text{ cm}}.$$

(c) To calculate the angular magnification, we compare the angular size of the lion as seen through the optical system to its angular size as seen with the unaided eye. Through the optical system, the lion seems to be a distance of 38.1 cm from the second lens, and its inverted height is 4.2 cm. Thus, according to Eq. (20.5),

$$\theta' = \frac{h'}{s'} = \frac{-4.2 \text{ cm}}{38.1 \text{ cm}} = -0.110 \text{ radians.}$$

To the unaided eye, the lion is 50 m away and 1.2 m tall. Its angular size to the unaided eye is

$$\theta = \frac{h}{r} = \frac{1.2 \text{ m}}{50 \text{ m}} = 0.024 \text{ radians.}$$

The angular magnification is given by Eq. (20.6):

$$M = \frac{\theta'}{\theta} = \frac{-0.110 \text{ radians}}{0.024 \text{ radians}} = \underline{-4.6}.$$

The lion appears 4.6 times taller when viewed through the telescope. The negative sign indicates that its image is inverted. Prisms, if placed between the two lenses, will reinvert the image (see Fig. 19.16) so that the final image is upright; binoculars use such prisms. ∎

The Compound Microscope

The compound microscope is another common example of a two-lens system. Microscopes have been used for over three hundred years to examine the details of small objects. In 1665 Robert Hooke published a book, *Micrographia*, which contained detailed drawings of insects, snowflakes, and other small objects that had been examined with microscopes whose angular magnifications were 100 or less. Today, the very best light microscopes provide angular magnifications of about 1000. With these instruments, scientists can distinguish objects separated by 2×10^{-7} m (about the length of 700 atoms lined in a row).

Although modern microscopes may contain as many as ten or more lenses, they work in principle much like the two-lens microscope used by Robert Hooke (depicted schematically in Fig. 20.24a). In this microscope both lenses are converging. The lens nearest the object is called the objective lens, and that nearest the eye of the observer is called the eyepiece or ocular lens. Both lenses have relatively short focal lengths (on the order of 2 cm or less) and are typically separated by 10 to 20 cm. When the object is placed a little to the left of the focal point of the objective lens (Fig. 20.24b), a real, magnified image of this object is formed in the microscope tube. If a screen were placed in the tube at the image distance s'_1, a real image of the object would be formed on the screen. This image of the objective lens has a linear magnification given by Eq. (20.2):

$$m_1 = -\frac{s'_1}{s_1}.$$

The image of the first lens is the object for the second. Since this object is located just inside the focal point of the second lens, a final, virtual image is formed outside the microscope (Fig. 20.24c). The second lens of the microscope is actually a magnifying glass used to view the real image of the first lens. The angular magnification of this magnifying glass is, according to Eq. (20.8),

$$M_2 = \frac{\text{near point}}{s_2}.$$

Since the original object has already been magnified by an amount $m_1 = -(s'_1/s_1)$, the total angular magnification M of the microscope is the product of the two magnifications:

$$M = m_1 M_2 = \left(-\frac{s'_1}{s_1}\right)\left(\frac{\text{near point}}{s_2}\right). \qquad (20.10)$$

This expression for angular magnification can be rewritten in terms of the focal lengths of the lenses and their separation d. Consider Fig. 20.25. Notice that $s_1 \simeq f_1, s'_1 \simeq d - f_2$, and $s_2 \simeq f_2$. By substituting these values in Eq. (20.10), the angular magnification becomes

$$M \simeq \left(-\frac{d - f_2}{f_1}\right)\left(\frac{\text{near point}}{f_2}\right). \qquad (20.11)$$

(a)

(b)

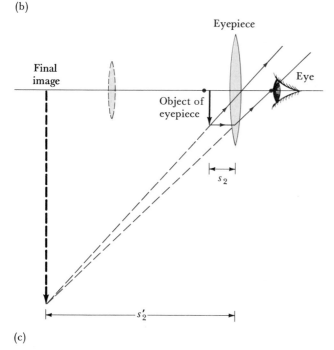

(c)

FIG. 20.24. (a) A two-lens microscope. (b) A magnified, real image of an object is formed by the objective lens. (c) This image is the object for the eyepiece which serves as a magnifying glass for examining the object.

The negative sign indicates that the final image is inverted. This expression for magnification is easier to use than Eq. (20.10) since we need not locate the image and object positions. It depends only on the focal lengths of the lenses, on their separation, and on the near point of the eye.

EXAMPLE 20.8 A compound microscope has an objective lens of focal length 0.80 cm and an eyepiece of focal length 1.25 cm. The lenses are separated by 18.0 cm. If a red blood cell is located 0.84 cm in front of the objective lens, (a) where is the final image of the cell located, and (b) what is its angular magnification? The viewer's near point is 25 cm.

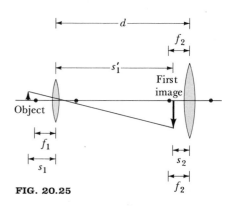

FIG. 20.25

SOLUTION   (a) We are given the following information: $f_1 = +0.80$ cm, $f_2 = +1.25$ cm, $d = 18.00$ cm, and $s_1 = +0.84$ cm. The image of the first lens is found using the thin-lens equation:

$$\frac{1}{s_1'} = \frac{1}{f_1} - \frac{1}{s_1} = \frac{1}{0.80 \text{ cm}} - \frac{1}{0.84 \text{ cm}} = 0.0595 \text{ cm}^{-1}$$

or

$$s_1' = \frac{1}{0.0595 \text{ cm}^{-1}} = +16.8 \text{ cm}$$

The image of the first lens is the object of the second. The object distance is, according to Eq. (20.9), $s_2 = d - s_1' = 18.0$ cm $- 16.8$ cm $= 1.2$ cm. The image of the second lens, which is the final image, is then found from the thin-lens equation:

$$\frac{1}{s_2'} = \frac{1}{f_2} - \frac{1}{s_2} = \frac{1}{1.25 \text{ cm}} - \frac{1}{1.20 \text{ cm}} = -0.0333 \text{ cm}^{-1}$$

or

$$s_2' = \frac{1}{-0.0333 \text{ cm}^{-1}} = \underline{-30 \text{ cm.}}$$

(b) The linear magnification of the first lens is, according to Eq. (20.2), $m_1 = -s_1'/s_1 = -(16.8 \text{ cm})/(0.84 \text{ cm}) = -20$. The angular magnification of the second lens is $M_2 = $ (near point)$/s_2 = 25$ cm$/1.2$ cm $= 21$. Thus, the total angular magnification is

$$M = m_1 M_2 = (-20)(21) = \underline{-420.}$$

The negative sign indicates that the image is inverted.

The angular magnification can be calculated approximately using Eq. (20.11):

$$M \simeq -\left(\frac{d - f_2}{f_1}\right)\left(\frac{\text{near point}}{f_2}\right) = -\left(\frac{18.0 \text{ cm} - 1.25 \text{ cm}}{0.80 \text{ cm}}\right)\left(\frac{25 \text{ cm}}{1.25 \text{ cm}}\right) = -420.$$

As we see, the above equation for approximating magnification produces the same result in this example as the more exact calculation using expressions for the magnification of each lens.   ■

Scientists using microscopes have found that the eye feels the least strain when the final image is about 100 cm from the eyepiece. This is called the position of minimum eyestrain. The final image can be shifted to this position through a small shift in the location of the eyepiece; the change in angular magnification is usually minor.

20.7   Aberrations

We have been using the thin-lens equation to locate the image produced by lenses that gather and focus light from an object. Most lenses form images that are slightly out of focus, even at the location where a sharp image is supposed to be formed. The lens defects causing this lack of focus are called **aberrations.**

Some aberrations result from improper lens construction and flaws in the material used to construct the lens. Others, such as spherical aberrations, chromatic aberrations, and astigmatisms, result from inherent refractive properties of light passing through the lens.

Spherical Aberration

Lenses are easiest to make if their surface has a spherical shape. Unfortunately, light rays moving parallel to the principal axis of a spherical lens do not all cross the axis at the same point (Fig. 20.26). Rays passing through the lens near its edge are refracted more than desired and cross the principal axis before rays that pass through the center region of the lens. Thus, there is no single focal point for all parallel rays passing through the lens. There is, however, one position along the axis where the transmitted light forms the smallest circle, called the **circle of least confusion** (Fig. 20.26). Parallel rays of light seem to focus best at this position on the axis.

Spherical aberration can be minimized by collecting light from a smaller region or aperture near the center of the lens, or by combining several lenses that provide the same net focal length as a single lens but whose spherical aberrations cancel each other.

Chromatic Aberration

Chromatic aberration is caused by the difference in the degree to which different colors of light are refracted. Blue light, for example, is refracted more than red light at the surfaces of a converging lens (Fig. 20.27a), so the blue light focuses

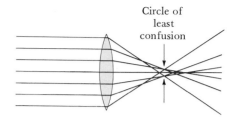

FIG. 20.26. Light rays moving parallel to the axis of a spherical lens do not all cross exactly at the focal point. The image of the rays is smallest at the circle of least confusion.

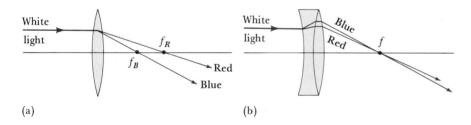

(a)

(b)

FIG. 20.27. (a) Different colors of light are focused at different distances from a lens. (b) An achromatic compound lens brings different color rays to focus at the same point.

closer to the lens than the red light. This variation in focal length for different colors of light can be corrected by making a compound lens of two different materials such as shown in Fig. 20.27b. If properly constructed, this compound lens, called an achromatic lens (from the Greek *achrōmatos*, meaning "without color"), can focus all colors of light at the same point.

Astigmatism

Astigmatism is a common defect of the eye caused by a lack of symmetry in the eye's cornea. For instance, if the cornea is flatter along a vertical axis parallel to the corneal surface than along a horizontal axis parallel to the surface, light passing through each axis is brought to focus at different distances from the cornea. Astigmatism is detected by viewing a set of lines such as shown in Fig.

FIG. 20.28. If you have astigmatism, some lines will appear fainter and less focused than others.

20.28. The lines should appear equally intense if you do not have astigmatism. If you do have astigmatism, lines in one direction may appear fainter and less focused than lines perpendicular to that direction. Astigmatism of the eye is corrected by eyeglass lenses that are less curved in one direction than in another.

Another form of astigmatism, called off-axis astigmatism, has a noticeable effect on glass lenses used in optical devices such as cameras. The image of a point O (Fig. 20.29) that is above the principal axis of a converging lens does not form a point image but instead forms a horizontal line image and a vertical line image at slightly different distances from the lens. Once again, a circle of least confusion occurs between these lines where the point is focused best. Off-axis astigmatism is reduced in photography by placing the important part of a photographed scene in the center of the field of view. Less important background that is farther from the principal axis will suffer more from off-axis astigmatism. Expensive cameras reduce off-axis astigmatism by combining several lenses whose off-axis astigmatism effects cancel each other.

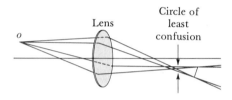

FIG. 20.29. Off-axis astigmatism results in a slightly confused image.

20.8 Mirrors

We have been discussing optical devices that focus and change the direction of light by refracting it as the light passes through glass lenses. We now consider devices such as mirrors that focus light by reflecting it. Mirrors can be used in much the same way as lenses. For example, most modern research telescopes use a mirror to focus light from distant stars and galaxies onto a film or into our eyes. The mirror of the large telescope at Mt. Palomar is 5.1 m (17 ft) in diameter. A converging lens of this size would sag under its own weight if supported at its edges.

A more commonly used mirror, such as that found on your bathroom wall, is called a **plane mirror** (the word *plane* indicates that it is flat). If an object is held in front of the mirror, as in Fig. 20.30, an upright image is formed an equal distance behind the mirror. The image is the place from which reflected rays of light seem to come—the place from which the dashed lines in Fig. 20.30 diverge. The height h' of the image is the same as the height h of the object. Thus, the linear magnification [Eq. (20.2)] of a plane mirror is $m = h'/h = 1$. The image shown in Fig. 20.30 is virtual since the light does not actually pass through the image. It just appears to come from the image.

Although plane mirrors only change the direction of light, curved mirrors can focus light to produce expanded or contracted images. Most of these curved mirrors are portions of either the inside or outside surface of a sphere. If the mirrored surface is on the inside of a spherical surface, it is said to be a **concave** or **converging mirror** (Fig. 20.31a). If the mirrored surface is on the outside of

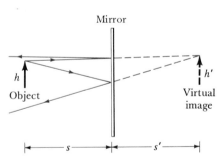

FIG. 20.30. An object and its virtual image, as formed by a plane mirror. For a plane mirror, $s = s'$.

FIG. 20.31. (a) Parallel rays cross at focal point after reflection from a converging (concave) mirror. (b) Parallel rays appear to leave the focal point after reflection from a diverging (convex) mirror.

(a)   (b)

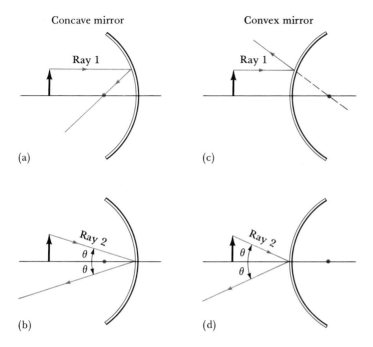

Concave mirror   Convex mirror

(a)   (c)

(b)   (d)

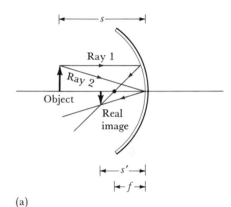

FIG. 20.32. The paths followed by two special rays after reflection from mirrors. The rays are used to construct mirror images.

(a)

the sphere, it is said to be a **convex** or **diverging mirror** (Fig. 20.31b). Two techniques are used to find the size and location of the image produced by these mirrors: (1) the ray technique, and (2) the mirror equation. Both techniques are similar to image location techniques used with lenses.

Ray Techniques

An image of an object formed by a mirror can be located by drawing two special rays that leave the object (Fig. 20.32). *Ray 1* moves parallel to the axis of the mirror. For a converging mirror, the reflected ray passes through the focal point on the axis in front of the mirror (Fig. 20.32a). For a diverging mirror, the reflected ray appears to diverge from the focal point behind the mirror (the dashed line path shown in Fig. 20.32c). *Ray 2* reflects from the center of each type of mirror at the place where the principal axis crosses the mirror (Fig. 20.32b and d). The angle of the reflected ray equals the angle of the incident ray.

The image of an object is located at the place where these rays intersect after reflecting from the mirror (a real image). If the rays diverge after leaving the mirror, the image is located at the position from which the rays appear to diverge from behind the mirror (a virtual image). Several examples of the ray technique of locating mirror images are shown in Fig. 20.33.

Mirror Equation

The location of the image can be calculated using a **mirror equation** [the same as the thin-lens equation—Eq. (20.1)]:

$$\frac{1}{f} = \frac{1}{s} + \frac{1}{s'}.$$

(b)

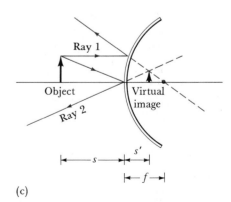

(c)

FIG. 20.33. Examples of image location for mirrors.

FIG. 20.34

The linear magnification of the image is calculated using the same equation that was used for lenses [Eq. (20.2)]:

$$m = \frac{h'}{h} = -\frac{s'}{s}.$$

The symbols in these equations are illustrated in Fig. 20.33 and described below:

1. The **object distance** s is the distance of the object from the mirror and is a positive number.

2. The **image distance** s' is the distance of the image from the mirror and is positive for real images in front of the mirror (to the left in Fig. 20.33a) and negative for virtual images behind the mirror (Fig. 20.33b and c).

3. The **focal length** f is the distance of the focal point from the mirror and is positive for converging mirrors (focal point in front of the mirror) and negative for diverging mirrors (focal point behind the mirror).

4. The **object height** h and **image height** h' are positive if upright and negative if inverted.

EXAMPLE 20.9 A dentist examines the cavity in a person's tooth using a concave mirror of focal length $+2.0$ cm (Fig. 20.34). If the cavity is 1.5 cm from the mirror's surface, (a) where is the image located, (b) is it real or virtual, and (c) what is its linear magnification?

SOLUTION The following information is given: $f_1 = +2.0$ cm (a concave mirror has a positive focal length), and $s = 1.5$ cm. (a) The mirror equation [Eq. (20.1)] is used to find the image position:

$$\frac{1}{s'} = \frac{1}{f} - \frac{1}{s} = \frac{1}{2.0 \text{ cm}} - \frac{1}{1.5 \text{ cm}} = -0.167 \text{ cm}^{-1}$$

or

$$s' = \frac{1}{-0.167 \text{ cm}^{-1}} = \underline{-6.0 \text{ cm}}.$$

(b) Since s' is negative, the image is <u>virtual</u> and appears to be 6.0 cm behind the mirror. (c) Using Eq. (20.2), we find that the image magnification is

$$m = -\frac{s'}{s} = -\frac{-6.0 \text{ cm}}{1.5 \text{ cm}} = \underline{+4.0}.$$

The cavity will appear to be four times its actual size. Since the magnification is positive, the image has the same orientation as the cavity. ∎

EXAMPLE 20.10 A young man with a loaf of French bread stands 4.0 m from a store observation mirror that has a focal length of -1.8 m (Fig. 20.35). The height of his image in the mirror is 0.45 m. Calculate (a) the image location, (b) its linear magnification, and (c) the young man's actual height.

SOLUTION The following information is given: $s = 4.0$ m, $f = -1.8$ m (a diverging mirror has a negative focal length), and $h' = 0.45$ m. (a) The image

FIG. 20.35. A diverging mirror. (Photograph by Douglas Rachele)

location is found from the mirror equation [Eq. (20.1)]:

$$\frac{1}{s'} = \frac{1}{f} - \frac{1}{s} = \frac{1}{-1.8 \text{ m}} - \frac{1}{4.0 \text{ m}} = -0.56 \text{ m}^{-1} - 0.25 \text{ m}^{-1} = -0.81 \text{ m}^{-1}$$

or

$$s' = \frac{1}{(-0.81 \text{ m}^{-1})} = \underline{-1.2 \text{ m}}.$$

The young man appears as a virtual image 1.2 m behind the mirror. (b) and (c) He is magnified by an amount

$$m = -\frac{s'}{s} = -\frac{-1.2 \text{ m}}{4.0 \text{ m}} = \underline{0.30}.$$

Since the magnification is positive, the image is <u>upright</u>. The boy's actual height h is determined by rearranging Eq. (20.2):

$$h = \frac{h'}{m} = \frac{0.45 \text{ m}}{0.30} = \underline{1.50 \text{ m}}.$$ ■

Summary

1. **A lens** is an optical device that changes the apparent location of an object by altering the path of light coming from the object. The image—that is, the apparent location of the object as seen with the lens—can be located by using ray diagrams or by using the thin-lens equation.

2. **Ray diagrams:** A **real image** can be located by determining the place where two or three rays from the object (Fig. 20.3) cross after passing through the lens. If the rays diverge after passing through the lens, we say that a **virtual image** is formed at the point from which the rays seem to originate on the same side of the lens as the object.

3. **Thin-lens equation:** The image position can be calculated using the thin-lens equation:

$$\frac{1}{f} = \frac{1}{s} + \frac{1}{s'}, \tag{20.1}$$

where f is the focal length of the lens, s is the distance of the object from the lens, and s' is the distance of the image from the lens.

If we arrange the lens so that light moves through it from left to right, then the sign conventions for the thin-lens equation are: (a) The **focal length** f is positive for converging (convex) lenses and negative for diverging (concave) lenses. (b) The **object distance** s is positive if the object is to the left of the lens and negative if to the right. (This rule is needed only for systems with two or more lenses.) (c) The **image distance** s' is positive for real images formed to the right of the lens and negative for virtual images formed to the left of the lens.

For a two-lens optical system, the image of the first lens is the object of the second. The object distance s_2 for the second lens is

$$s_2 = d - s_1', \tag{20.9}$$

where d is the separation of the lenses and s_1' is the image distance of the first lens.

4. **Linear magnification:** The linear magnification m of a single lens is the ratio of the image height h' and object height h. The height is positive if the image or object is upright and negative if inverted. The linear magnification can also be calculated using the image and object distances:

$$m = \frac{h'}{h} = -\frac{s'}{s}. \tag{20.2}$$

For a two-lens system, the total linear magnification is just the product of the individual linear magnifications of each lens.

5. The **eye's optical system** consists of a cornea and lens that together act as a converging lens to produce a real image of objects on the eye's retina. By varying the focal length of the lens, the normal eye can focus on objects from a **far point** of infinity to a **near point** of 25 to 50 cm.

6. **Angular magnification:** The **angular size** θ of an object as seen by the unaided eye is $\theta = h/r$, where h is its height, and r is its distance from the eye. The maximum angular size θ_{max} of an object is $\theta_{max} = h/(\text{near point})$, where the near point is the closest an object can be held to the eye while remaining in focus. When viewed through an optical device, the angular size is $\theta' = h'/s'$, where h' is the height of the final image of the optical system, and s' is the distance from the lens nearest the eye to the final image. The angular magnification M of distant objects is

$$M = \frac{\theta'}{\theta}. \tag{20.6}$$

The angular magnification of nearby objects when seen through a **magnifying glass** is

$$M = \frac{\theta'}{\theta_{\max}} = \frac{\text{near point}}{s}. \qquad (20.8)$$

7. **Mirrors:** The images of objects viewed in a mirror are located by the ray technique using the rays shown in Fig. 20.32 or by using the mirror equation [the same as Eq. (20.1)]. The sign conventions for the mirror equations are: (a) The **focal length** f is positive for a converging (concave) mirror and negative for a diverging (convex) mirror. (b) The **object distance** s is positive. (c) The **image distance** s' is positive for real images in front of the lens and negative for virtual images behind the mirror.

Questions

1. A bubble of air is suspended under water. Draw a ray diagram showing the approximate path followed by light rays passing from the water through the bubble and back into the water. Does the bubble act like a converging or a diverging lens?

2. A person underwater cannot focus clearly on any object. Yet if the person wears goggles, objects can be seen clearly. Explain.

3. The lens of a slide projector is adjusted so that a clear image of the slide is formed on a screen. If the screen is moved farther away, must the lens be moved toward or away from the slide? Explain.

4. The retina has a blind spot at the place where the optic nerve leaves the retina. Design and describe a simple experiment that allows you to perceive the presence of this blind spot. Why doesn't the blind spot affect your normal vision?

5. Borrow the glasses of a person who is known to be farsighted. Invent and describe a simple experiment to measure the approximate focal lengths of the lenses of his or her glasses.

6. You run toward a building with walls of a metallic, reflective material. Your speed is 1.5 m/s. How fast does your image appear to move toward you? Explain.

7. A large concave mirror of focal length 3 m stands 20 m in front of you. Describe the changing appearance of your image as you move from 20 m to 1 m from the mirror. Indicate distances from the mirror where the change in appearance is dramatic.

8. Two mirrors (one plane and one diverging) at the side of a truck are shown in Fig. 20.36. Estimate the focal length of the curved mirror. Carefully explain the reasoning behind your estimate.

9. Explain why your image in a plane mirror has its left and right sides reversed, while the image of your head and legs are not reversed.

10. You stand in the Hall of Mirrors at Versailles. You are 4 m from one wall of mirrors. What is the desired distance setting on your camera when photographing your image in the mirrors? Explain.

11. The image of three children standing in front of a "trick mirror" is shown in Fig. 20.37. Why is the face of the child on the right elongated whereas his hands seem shortened? Justify your answer in terms of the magnification produced by different types of mirrors.

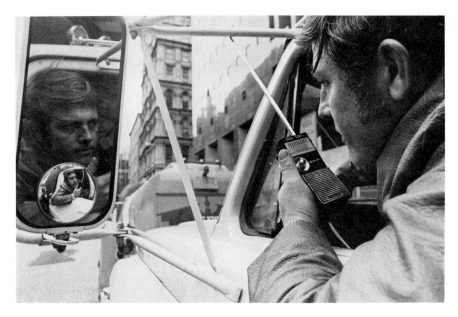

FIG. 20.36. (© 1979 by Ted Cowell/Black Star)

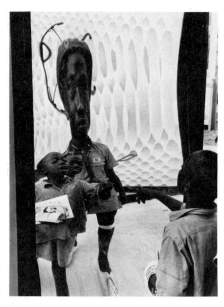

FIG. 20.37. (Marilyn Silverstone © 1964 Magnum Photos)

Problems

20.2   Formation of Images

1.   Using a ruler, carefully draw ray diagrams to locate the images of the following objects: (a) An object that is 30 cm from a converging lens of 10-cm focal length. (b) An object that is 14 cm from the same lens. (c) An object that is 5 cm from the same lens. (Choose a scale so that your drawing fills a significant portion of the width of a paper.) Measure the image location on your drawings and indicate if they are real or virtual, upright or inverted.

2.   Repeat the procedure described in Problem 1, only for the following lenses and objects: (a) An object that is 7 cm from a converging lens of 10-cm focal length. (b) An object that is 7 cm from a diverging lens of −10-cm focal length.

3.   Repeat the procedure described in Problem 1, only for the following lenses and objects: (a) An object that is 30 cm from a diverging lens of −10-cm focal length. (b) An object that is 14 cm from the same lens. (c) An object that is 5 cm from the same lens.

20.3   Thin-Lens Equation

4.   Use the thin-lens equation to locate the images of the objects described in Problem 1.

5.   Use the thin-lens equation to locate the images of the objects described in Problem 3.

6.   Light passes through a narrow slit, then through a lens, and onto a screen. The slit is 20 cm from the lens. The screen, when adjusted for a sharp image of the slit, is 15 cm from the lens. What is the focal length of the lens?

7.   A slide is located 20.0 cm from the lens of a slide projector. A focused image is produced on a screen 2.5 m from the lens. (a) Calculate the focal length of the lens. (b) Calculate the distance from the slide to the lens if the screen is moved so that it is 2.0 m from the lens.

8.   A carpenter ant being photographed is 18 cm from the film. At what places can the 4-cm converging camera lens be located so that a sharp image of the ant is produced on the film?

9.   A camera is adjusted so that a clear image of the moon is produced on the film by a lens of +8-cm focal length. How must the lens-film distance be altered to produce a focused image of a painting held 72 cm from the lens?

10.   A camera with a 5.0-cm focal length lens is used to photograph a secret document whose height is 10 cm. At what distance from the lens must the document be held so that an image 2.5 cm high is produced on the film? (Note: The real image is inverted.)

11.   An object 5 cm tall is held 50 cm from a lens with a focal length of 10 cm. (a) Calculate the height and location of the image and indicate whether it is real or virtual, upright or inverted. (b) Repeat for an object 5 cm tall held 50 cm from a lens of −10-cm focal length.

12.   (a) Make a rough graph of linear magnification versus object distance for a converging lens of 20-cm focal length as the object distance is varied from ∞ to 0. (b) Do the same for a diverging lens of −20-cm focal length. For each graph, indicate in which regions the image is real and in which regions it is virtual.

13.   A motion picture director views a scene through a diverging lens. The scene is reduced in size by the lens to the size it has on the theater screen. Calculate the desired focal length for a one-fifth reduction when the lens is held 10 m from the scene.

14.   An enlarging camera has a lens of +12.5-cm focal length. The object is a negative that is 35 mm tall. If the image to be produced on a film is 150 mm tall, how far from the lens should the object be placed?

15.   An aerial camera with a lens of 0.4-m focal length is used to photograph a landscape 2 km wide from a height of 5 km. Calculate the width of the image on the film.

16.   A slide is located 5.00 m from a screen. The view on the slide is magnified twenty times when projected on the screen. Calculate the focal length of the projector lens and the distance of the slide from the lens.

20.4   The Human Eye

17.   The image distance for the lens of a person's eye is 2.10 cm. Calculate the focal length of the eye's lens system for an object (a) at infinity, (b) 500 cm from the eye, and (c) 25 cm from the eye.

18.   (a) A woman can focus only on objects that are from 150 cm to 25 cm from her eyes. What focal length lenses does she need for her glasses? (b) A man can focus only on objects that are at least 3 m from his eyes. What focal length lenses does he need for his glasses?

19.   (a) While wearing glasses of +150-cm focal length, a man can see clearly all objects between infinity and 25 cm. What are the near and far points of his unaided eyes? (b) Repeat the calculations in (a) for a man who wears glasses of −300-cm focal length.

20.   A woman who can focus only on objects that lie from 100 to 300 cm from her eyes needs bifocals. (a) Calculate the desired focal length of the upper half of the glasses used to see distant objects. (b) Calculate the focal length of the lower half used to read a book held 30 cm from her eyes.

21.   A 35-year-old patent clerk needs glasses of 50-cm focal length in order to read patent applications held 25 cm from his eyes. Five years later, while wearing the same glasses, he must hold the patent applications 40 cm from his eyes in order to see them clearly. What should be the focal length of new glasses so that he can read again at 25 cm?

22.   Assume that the eye accommodates to objects at different distances by altering the distance from the lens system to the retina. If the lens system has a focal length of 2.10 cm, what is the lens-retina distance needed to view objects at (a) infinity, (b) 300 cm, and (c) 25 cm?

20.5   Angular Magnification and Magnifying Glasses

23.   An aphid on a plant leaf is examined with a magnifying glass of +6.0-cm focal length. The glass is held so that the final virtual image is 40 cm from the lens. Assuming a near point of 30 cm, calculate the angular magnification.

24.   A stamp is viewed through a magnifying glass of 5.0-cm focal length. Calculate the object distance for virtual images formed at

(a) negative infinity, (b) -200 cm, and (c) -25 cm. (d) Calculate the angular magnification in each case.

25. A person has a near point of 100 cm. (a) What is the nearest distance that a magnifying glass of 6-cm focal length can be held from print on a page and still form an image beyond the person's near point? (b) Calculate the angular magnification for an image at the near point?

26. A magnifying glass of 12-cm focal length is held 10 cm from a page of fine print. (a) Where is the image located? (b) How large is the image of a letter that is 1.2 mm high? (c) Assuming a near point of 30 cm, calculate the angular magnification using Eq. (20.5).

20.6 Combinations of Lenses

27. A converging lens with a 25-cm focal length is placed 50 cm in front of a diverging lens having a focal length of -40 cm. An object 2 cm tall is placed 30 cm in front of the converging lens. Determine (a) the location of the final image, (b) its orientation, and (c) whether it is real or virtual.

28. Two converging lenses, each with 20-cm focal length, are separated by 60 cm. Locate the final image of an object placed 90 cm in front of the first lens. Is the image real or virtual, upright or inverted?

29. A telephoto-lens system consists of a converging lens of 30-cm focal length and a diverging lens of -10-cm focal length. The lenses are separated by 24 cm. An object standing 100 m from the first lens is photographed. (a) Where should the film be located? (b) If the object is 2 m tall, what is the image height on the film?

30. A telephoto lens consists of a converging lens of 6-cm focal length followed by a diverging lens of -3-cm focal length. The lenses are separated by 3.5 cm. (a) Locate the final image of a whooping crane 1.5 m tall that is 120 m in front of the first lens. (b) Calculate the height of its image on the film.

31. During a laboratory experiment, the focal length of a diverging lens is measured by first forming a real image of a light source using a converging lens. The image is formed on a screen 20 cm from the lens. The diverging lens is then placed halfway between the converging lens and screen. To obtain a focused image, the screen must now be moved 15 cm farther to the right (away from the lenses). Calculate the focal length of the diverging lens.

32. Confirm that the effective focal length f of two thin lenses of focal lengths f_1 and f_2, when in contact, is

$$\frac{1}{f} = \frac{1}{f_1} + \frac{1}{f_2}.$$

33. Three lenses, each of $+10$-cm focal length, are placed in a line and separated from each other by 15 cm. If an object is placed 20 cm to the left of the first lens, where is the final image?

34. A telescope consisting of a $+3.0$-cm objective and a $+0.6$ cm eyepiece is used to view an object that is 20 m from the objective. (a) What must the distance between the objective and eyepiece be in order to produce a final, virtual image 100 cm to the left of the eyepiece? (b) What is the total angular magnification?

35. A Galilean telescope, sometimes called an opera glass, such as shown in Fig. 20.22, consists of a $+10.0$-cm objective and a -2.0-cm eyepiece. The lenses are separated by 8.0 cm. A Victoria butterfly 20 m from the objective is viewed. (a) Locate the final

image. (b) Calculate the total linear magnification. (c) Calculate the total angular magnification.

36. A Galilean telescope such as shown in Fig. 20.22 produces an upright image whereas an astronomical telescope produces an inverted image. A Galilean telescope is used to view a deer 2 m tall that is 50 m from the objective lens ($f = +100$ cm). The eyepiece of the telescope has a focal length of -5 cm. (a) Locate the image of the objective. (b) What must be the object distance s_2 for the eyepiece in order that the final image be at the eye's near point (that is, $s_2' = -25$ cm)? (c) What is the separation of the lenses? (d) What is the height of the final image? (e) What is the total angular magnification?

37. The world's largest telescope made only from lenses (that is, it has no mirrors) is located at the Yerkes Observatory near Chicago. Its objective is 1 m in diameter and has a focal length f_1 of $+18.9$ m. The eyepiece has a focal length f_2 of $+7.5$ cm. The objective and eyepiece are separated by 18.970 m. (a) What is the location of the final image of a moon crater 3.8×10^5 km from the earth? (b) If the crater has a diameter of 2 km, what is the size of its final image? (c) Calculate the angular magnification of the telescope by comparing the angular size of the image as seen through the telescope and the object as seen by the unaided eye. (*Note:* The angular magnification should be approximately f_1/f_2, as it is for most telescopes of this type.)

38. You are marooned on a tropical island. Design a telescope from a cardboard map tube and the lenses of your eyeglasses. One lens has a $+1.0$-m focal length and the other has a $+0.3$-m focal length. The telescope should allow you to view an animal 100 m from the objective with the final image being formed 1.0 m from the eyepiece. Indicate the location of the lenses and the expected angular magnification.

39. A microscope has a $+0.4$-cm objective and a $+2.5$-cm eyepiece. (a) Where must the object for the eyepiece be located in order to form a final virtual image 100 cm in front of the eyepiece? (b) If the lenses are separated by 20 cm, what is the distance between the object being viewed and the objective lens? (c) Assuming a near point of 25 cm, calculate the total angular magnification.

40. A microscope has an objective with a $+8$-mm focal length that forms an image 160 mm from the lens. The eyepiece produces an angular magnification of 9. Calculate the total magnification of the microscope.

41. The objective lens of a microscope has a focal length of 0.8 cm and an eyepiece of focal length 2.0 cm. An object is placed 0.9 cm in front of the objective. The final virtual image is 100 cm from the eyepiece at the position of minimum eyestrain. (a) Calculate the separation of the lenses. (b) Calculate the total angular magnification.

42. A microscope has a 0.5-cm objective lens and a 1.8-cm eyepiece. The lenses are separated by 20 cm. The final virtual image is 100 cm from the eyepiece. (a) Where should the object be located relative to the objective lens? (b) Calculate the total angular magnification of the microscope.

43. A dissecting microscope is designed with a larger-than-normal distance between the object and the objective lens. One of these microscopes has an objective lens of 5-cm focal length and an eyepiece of 2-cm focal length. The lenses are separated by 15 cm. The final virtual image is located 100 cm from the eyepiece.

(a) Calculate the distance of the object from the objective lens.
(b) Calculate the total angular magnification.

20.8 Mirrors

44. Locate the position, orientation, and type of image formed by an object held in front of a concave mirror of focal length 20 cm. The object distance is (a) 200 cm, (b) 40 cm, and (c) 10 cm. (d)–(f) Repeat the problem for a convex mirror of focal length −20 cm.

45. Use both a ray diagram and the mirror equation to locate the image of an object held (a) 200 cm, (b) 60 cm, and (c) 30 cm from a convex mirror of focal length −50 cm. Indicate if the image is real or virtual, upright or inverted.

46. (a) Repeat Problem 45 for objects held (a) 200 cm, (b) 60 cm, and (c) 30 cm in front of a concave mirror of focal length 50 cm.

47. The moon's diameter is 3.5×10^3 km, and its distance from the earth is 3.8×10^5 km. Calculate the position and size of the image formed by the Hale Observatory reflecting mirror, whose focal length is 16.5 m.

48. You wish to order a mirror from a scientific supply company. You want to use the mirror for shaving or makeup. The mirror should produce an image that is upright and magnified by a factor of two when held 30 cm from your face. What type and focal length mirror should you order?

49. A lamp hangs from the ceiling of a theater. You are to buy a mirror that will form an expanded real image of the lamp's flickering filament on a screen that is 5 m from the lamp. The image should be ten times larger than the filament and inverted. What focal length mirror should be purchased, and where should it be located relative to the lamp?

50. Calculate the focal length of the mirror Archimedes might have used to burn ships that were 150 m away. Justify your answer.

51. You view yourself in a large convex mirror of −1.2-m focal length from a distance of 3 m. (a) Locate your image. (b) If you are 1.7 m tall, what is your image height?

52. A fortune teller looks into her silvered-surface, crystal ball whose radius is 10 cm and whose focal length is −5 cm. (a) If her eye is 30 cm from the ball, where is the image of her eye? (b) Estimate the size of that image.

53. A clothing store orders a cylindrical mirror that magnifies a customer's width by 0.9 and his height by 1.0 when the customer stands 2.0 m from the mirror. (a) Calculate the focal length due to the horizontal curvature for two types of mirrors that will produce this 0.9 magnification. (b) Calculate the horizontal magnification for each type mirror if the person is 5.0 m away. (c) Which mirror produces a more constant magnification?

21 Wave Interference

In the last chapter, when we analyzed optical instruments using ray diagrams and the thin-lens equation, we had little need to consider the wave nature of light. However, a variety of phenomena involving light and all forms of waves can be explained only in terms of the wave nature of light. These phenomena depend on the constructive and destructive interference of waves originating from different points in space.

We learned about wave interference in Chapter 17. Our interest then was primarily with sound waves moving in one direction. If the crests (or the troughs) of two sound waves overlapped at the same place at the same time, constructive interference occurred; the two waves combined to form a new wave of increased amplitude. However, if a crest of one wave overlapped a trough of another wave, destructive interference occurred; the resultant wave had a smaller amplitude than either of the two waves.

In this chapter we will analyze the interference of waves originating from different points in space. Our analysis will allow us to understand the patterns formed when light passes through a single small opening such as a narrow slit or through several closely spaced slits. We will also learn about the interference of light reflected from thin films such as soap bubbles and the transparent films that often cover the surface of a lens. Finally, we will find out how wave interference makes possible a new technology of three-dimensional photography called holography. As with all phenomena discussed in this part of the book, the interference effects to be described apply to all types of waves, but we will restrict most of our attention to light waves for which the effects are easily demonstrated.

21.1 Huygens' Principle

From the time of Alexander the Great (350 B.C.) to that of the kings of France, the rulers of countries often hired scientists to bring prestige and fame to their courts. One of the top recruits in the seventeenth century was a Dutch mathematician, Christian Huygens, who was selected by King Louis XIV to direct all scientific research in France. During his tenure with the king, Huygens developed the best telescopes of the time, discovered the moons of Jupiter, and invented a pendulum clock so accurate that it easily detected small differences in the acceleration of gravity at different parts of the earth.

Huygens' most famous contribution to science was his wave theory of light, proposed in *Treatise on Light* (1673). This treatise contained many ideas about the nature of light that have since been confirmed. His technique for determining the future position of a wave based on its present location is the subject of this section. This technique, called *Huygens' principle,* is illustrated in Fig. 21.1a. A wavefront represented by line *AB* is shown moving at speed *v* away from a point source *S*. The location of the wavefront a short time *t* after it is at position *AB* is determined by drawing a large number of circular arcs, each centered on a different point of wavefront *AB*. Each arc represents a small wave, called a **wavelet,** that moves away from a point on the original wavefront. The radius of a wavelet *r* depends on the speed *v* of the wavefront at the point where the wavelet originates and on the time *t* that the wavelet has traveled away from that point. The three quantities are related by the equation $r = vt$. The interference of these wavelets produces a new wavefront whose position *CD* is a line drawn tangent to the edges of these wavelets, as shown in Fig. 21.1a and b. **Huygens' principle** can be summarized as follows:

Every point on a wavefront can be considered as the source of an arc-shaped wavelet that moves in the forward direction at the same speed at which the wave moves. The new wavefront is a line drawn tangent to the edges of these wavelets.

Huygens' principle is useful for determining the shape of wavefronts that pass through narrow openings or that move past obstacles. For example, Huygens' principle could be used to decide whether ocean waves during a gale will bend around a small island to strike a harbor on the lee side of the island.

The bending of waves that pass through a narrow opening or move past an obstacle is called **diffraction** and is illustrated in Fig. 21.2. In this example, the opening is about the same size as the wavelength of the wave passing through the opening. The Huygens' wavelets form a new wavefront that moves forward and spreads to the sides as shown in Fig. 21.2a. As the opening is made smaller, the wave spreads more after passing through the opening (b)—like a single arc-shaped wavelet spreading from a point. In general, the spreading of the wave increases as the size of the opening decreases.

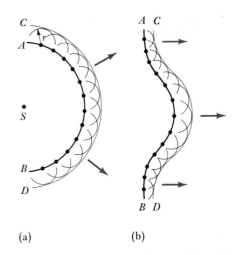

(a) (b)

FIG. 21.1. Circular wavelets move forward from wavefront *AB* to form a new wavefront *CD*. In (a) the wavelets leaving each part of *AB* move at the same speed. In (b) the wavelets move faster at the center of the wavefront than at the sides.

21.2 Young's Double-Slit Experiment

Although Christian Huygens proposed his wave theory of light in 1673, it received little support during his lifetime. None of the experiments with light at that time required an explanation based on waves, and the corpuscular theory of light was most popular. This changed, however, over a hundred years later because of the work of the brilliant English physician and physicist, Thomas Young (1773–1829). At age two, Young read fluently; by four, he had read the Bible twice; by age fourteen he knew eight languages. During his adult life, Young made many contributions to the understanding of fluids, concepts of work and energy, and the elastic properties of materials. He was also the first person to make progress in deciphering Egyptian hieroglyphics.

Young's most important contributions to science were made in the field of light and optics. For instance, he discovered that the eye's lens could change shape, allowing it to focus on objects at different distances. He also developed a

(a)

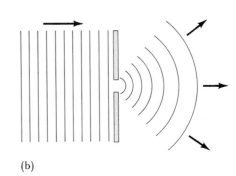

(b)

FIG. 21.2. Waves spread or diffract more after passing through a small opening (b) than through a larger opening (a).

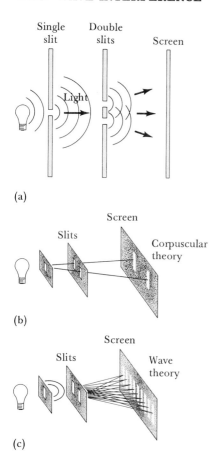

(a)

(b)

(c)

FIG. 21.3. (a) A schematic representation of Young's double-slit experiment. (b) The expected pattern of light on a screen if light is made of corpuscles. (c) The actual pattern observed by Thomas Young.

FIG. 21.4. Circular Huygens' wavelets originate at each slit. Interference of crests from each wave (the dots) produce large amplitude waves (bright light). In between, relative calm exists because of the destructive interference of a crest from one wave with a trough from the other (darkness).

successful theory explaining color vision. Young received his greatest acclaim among physicists for his double-slit interference experiment (reported in 1803), which became the cornerstone of the wave theory of light. With this experiment Young laid to rest the corpuscular theory of light that had been popular during the 1700s.

The **double-slit experiment** is illustrated in Fig. 21.3a. Light first passes through a small opening, like a narrow slit; in Young's experiment sunlight passed through a small hole in his window shutter. Wavelets produced in the single slit form wavefronts that spread and fall on two closely spaced slits. If light is made of individual little bullet-like bundles (corpuscles), then these bundles should move straight through the slits and produce a sharp image of each slit on a screen placed behind the slits (Fig. 21.3b). However, the pattern Young actually observed on the screen consisted of alternating bright and dark lines (called fringes), as shown in Fig. 21.3c.

Young explained that the pattern of fringes was caused by the interference of waves coming from each slit. Bright bands of light occurred where the light waves interfered constructively; dark bands occurred where the waves interfered destructively. This phenomenon is illustrated in Fig. 21.4. Circular wavefronts spread from each narrow slit. In some directions (indicated by the arrows in Fig. 21.4) a crest of one wave overlaps a crest of another (the dots in Fig. 21.4)—that is, they constructively interfere—and the wave amplitude is doubled. In other directions the waves interfere destructively—a crest of one wave overlaps a trough of another and the screen is dark. If you observe the waves striking a screen, you will see alternating large amplitude waves (bright light) at the positions of the arrows and no disturbance (darkness) in between.

We can predict the directions in which large amplitude waves are formed as follows (see Fig. 21.5). First, note that the same wave enters each slit from the left; the disturbances at the slits are synchronized or in phase with each other. A crest (or a trough) leaves slit 1 in Fig. 21.5b at the same time that a crest (or a trough) leaves slit 2. The waves spread in all directions, but we consider only the waves moving toward the numbered positions of bright light (large amplitude waves) on the screen shown in Fig. 21.5a. Zero, at the center of the interference pattern, is a band of bright light called the **zeroth-order interference maximum**. At this point, light has traveled an equal distance from slit 1 (S_1) and from slit 2 (S_2), as shown in Fig. 21.5b. Since position 0 is equidistant from each slit, crests (and troughs) that simultaneously leave the slits arrive simultaneously at position 0, producing constructive interference and large amplitude waves.

Constructive interference also occurs at position 1, which is one wavelength λ farther from slit 1 than from slit 2 (Fig. 21.5c). At position 2, the difference in

(a)

(c)

(b)

Slit 2

Slit 1

(d)

FIG. 21.5. (a) Light passes through a single slit. Coherent waves then pass through a pair of slits and produce a pattern of alternating bright and dark bands on a screen. (b) Constructive interference occurs at the center band because the waves from each slit travel an equal distance to the band. (c) The distance from the slits to band 1 differs by one wavelength and (d) to band 2 by two wavelengths.

distance is two wavelengths and again constructive interference occurs (Fig. 21.5d). Between the bright bands, a crest from one slit arrives at the same time as a trough from the other slit. Destructive interference occurs and creates a dark band.

At the nth bright band to the right or left of the center band, the difference in distance from the two slits is n wavelengths of light ($n\lambda$ in Fig. 21.6). This bright band is called the **nth-order interference maximum.** The angle θ of rays moving from the slits to the nth order interference maximum is calculated with reference to Fig. 21.6. The small, shaded triangle shown next to the slits has a hypotenuse d equal to the slit separation. The side of the triangle opposite the angle θ has a length $n\lambda$. Thus, we see that $\sin \theta = n\lambda/d$. The angle θ in the small triangle equals the angular deflection θ of light moving toward the **nth bright band** and is calculated using the same equation:

$$\sin \theta = \frac{n\lambda}{d}, \qquad (21.1)$$

where λ is the wavelength of the light, d is the slit separation, and n is an integer (0 for the center maximum, 1 for the first-order bright bands on each side, and so on).

FIG. 21.6. Rays are drawn from the slits to the nth bright band above the center maximum.

EXAMPLE 21.1 When red light of wavelength 650 nm passes through a pair of slits, a series of bands appears on a screen. (a) The fifth band to the left of the center maximum has an angular deflection of 8.5°. Calculate the slit separation d. (b) Calculate the angular deflections θ of the zeroth-, first-, and second-order interference maximum.

SOLUTION   (a) The following information is given: $\lambda = 650 \times 10^{-9}$ m, $n = 5$, and $\theta = 8.5°$. Equation (21.1) is solved for the slit separation d:

$$d = \frac{n\lambda}{\sin \theta} = \frac{5(650 \times 10^{-9}\,\text{m})}{\sin 8.5°} = \underline{2.2 \times 10^{-5}\,\text{m}}.$$

(b) The sines of the angular deflections of the other bands are

$$\sin \theta = \frac{n\lambda}{d} = \frac{n(650 \times 10^{-9}\,\text{m})}{2.2 \times 10^{-5}\,\text{m}} = n(0.030).$$

Thus, for $n = 0$, $\sin \theta = 0$, and $\theta = \underline{0}°$; for $n = 1$, $\sin \theta = 0.030$, and $\theta = \underline{1.7°}$; and for $n = 2$, $\sin \theta = 0.060$, and $\theta = \underline{3.4°}$.   ∎

EXAMPLE 21.2   Blue light, rather than red light, is incident on the pair of slits discussed in the previous example. The angular deflection of the fifth band to the left of the center is 6.3°. Calculate the wavelength of this light.

SOLUTION   We are given that $n = 5$, $\theta = 6.3°$, and $d = 2.2 \times 10^{-5}$ m (calculated in the previous example). Equation (21.1) is solved for the unknown wavelength:

$$\lambda = \frac{d \sin \theta}{n} = \frac{(2.2 \times 10^{-5}\,\text{m})(\sin 6.3°)}{5} = 480 \times 10^{-9}\,\text{m} = \underline{480\,\text{nm}}.   ∎$$

In the last example, known values of d, θ, and n were used to calculate the unknown wavelength of light passing through two slits. The angular deflection θ is usually small and difficult to measure directly. It can, however, be determined indirectly as illustrated in Fig. 21.7. The distance y_n of the nth bright band from the center (zeroth-order) maximum is measured, along with the distance D of the slits from the screen. The value of θ is determined from the equation

$$\tan \theta = \frac{y_n}{D}. \tag{21.2}$$

Because $\tan \theta \simeq \sin \theta$ for small angles (10° or less) such as occur in most of these experiments, we can set the right sides of Eqs. (21.1) and (21.2) equal to each other. Thus,

$$\sin \theta = \frac{n\lambda}{d} \simeq \frac{y_n}{D}. \tag{21.3}$$

If the slit separation d is known, the wavelength of light can be calculated using the above equation after measuring y_n and D.

Young's double-slit experiment is historically very important because of its confirmation of the wave nature of light, but the analysis of double-slit interfer-

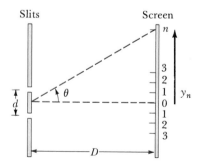

FIG. 21.7

ence patterns has little practical use. A modern descendant of a double slit, called a grating, has great practical use in the analysis of the wavelengths of light emitted and absorbed by different types of atoms and molecules.

21.3 Gratings

A **grating** consists of many closely spaced slits (for example, 10,000 in one centimeter) and can be used to separate light of different wavelengths and to measure these wavelengths.

The operation of a grating with only five slits is illustrated in Fig. 21.8. Light that strikes one of these slits spreads after passing through the slit. The waves from each of the slits interfere to produce beams of very intense light in several directions and little light elsewhere.

The first intense band to the left of the center maximum strikes the screen at position 1 (Fig. 21.8a). The extra distance of this position from adjacent slits is one wavelength (Fig. 21.8b). Rays from all slits interfere constructively to form a bright spot. The angle of deflection θ and the positions y_n of the **intensity maxima of light from a grating** are calculated using the same equations as were used for a double slit:

$$\sin \theta = \frac{n\lambda}{d} \simeq \frac{y_n}{D}, \qquad \text{where } n = 0, 1, 2, \ldots, \qquad (21.3)$$

where d is the separation of adjacent slits, λ is the wavelength of the light, D is the distance from the grating to the screen, and y_n is the distance on the screen from the center bright spot to the nth bright spot. As with double slits, the positions of bright spots are called **orders** (zeroth order for $n = 0$, first order for $n = 1$, and so forth).

The bright lines or spots are somewhat more intense and better defined than those from double slits because many more slits contribute to the interference pattern. Instead of finding alternating light and dark regions, one finds a large region of darkness followed by a narrow, bright band or spot of light.

EXAMPLE 21.3 A grating with 4000 lines per centimeter is illuminated with red light from a laser. A screen placed 2.0 m from the grating has spots of light that are separated on either side of the center maximum by 0.50 m. What is the wavelength of the light?

SOLUTION We are given that $D = 2.0$ m, $y_1 = 0.50$ m, and

$$d = \frac{1}{4000 \text{ lines/cm}} = 2.5 \times 10^{-4} \text{ cm} = 2.5 \times 10^{-6} \text{ m}.$$

Note that the separation of adjacent lines is $\frac{1}{4000}$ cm and not 4000 cm. Rearranging and substituting in Eq. (21.3) for the case where $n = 1$, we find that

$$\lambda \simeq \frac{dy_n}{nD} = \frac{(2.5 \times 10^{-6} \text{ m})(0.50 \text{ m})}{(1)(2.0 \text{ m})} = 6.3 \times 10^{-7} \text{ m} = \underline{630 \text{ nm}}.$$

The laser light illuminating the grating is probably from a helium-neon laser that emits light of 632.8-nm wavelength. ∎

(a)

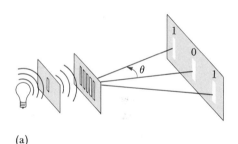

(b)

FIG. 21.8. (a) Circular wavelets of light spread from each slit in a grating. A bright band occurs on a screen at point 1 where the distance from adjacent slits differs by one wavelength as shown in (b).

Spectrograph

Wavelength measurements using gratings are made routinely in many fields of science. Astronomers, for instance, determine what types of atoms are present in distant stars by identifying the wavelengths of light coming from atoms in the stars. This is done using a **spectrograph** (Fig. 21.9). Light, such as that from a star, passes through a narrow slit S_1 and then through a lens L_1 (Fig. 21.9a) that bends and collimates the rays of light into a parallel beam. The beam of light then passes through a grating that causes the light to separate into different order beams. The zeroth-order beam is undeflected. A first-order beam is deflected both left and right. The magnitude of the deflection depends on the wavelength of the light.

Another lens L_2 is placed beyond the grating (Fig. 21.9b) to focus the deflected beams of light through another narrow slit onto a film. The lens L_2 and slit S_2 are wavelength selectors since the wavelength of light that exposes the film depends on the orientation of the lens and slit. As this collimating system is rotated, the film is exposed at different wavelengths.

Different types of atoms emit different wavelengths of light. For example, mercury emits a purple doublet line at 408 nm, a blue line at 436 nm, a greenish-yellow line at 546 nm, a yellowish doublet at 579 nm, and two weak red lines near 620 nm. The 546-nm line is being viewed by the spectrograph pictured in Fig. 21.9b. As the collimating system is rotated, an exposed film, such as shown in Fig. 21.9c, is produced. The wavelengths at which the film is exposed indicate

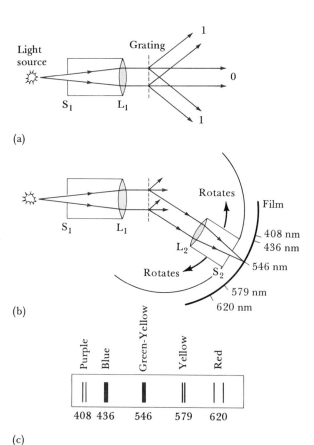

FIG. 21.9. (a) The input side of a spectrograph. (b) A spectrograph with the collimating system. (c) An exposed film from mercury gas.

the types of atoms producing the light. If a star contains mercury atoms, then the film should be exposed at those wavelengths shown in (c). It will also be exposed at the wavelengths of light emitted by other atomic constituents of the stars. The amount by which the film is exposed at different wavelengths indicates the abundance of the different types of atoms on the star. (The spectra of light emitted by several different types of atoms are shown on the color insert facing page 432.)

EXAMPLE 21.4 Recall that light or sound coming from a source moving away from us seems to have a lower frequency (and longer wavelength) than if the source is not moving relative to us. The wave is said to experience a Doppler shift. Astronomers, by observing the Doppler shift of light from distant galaxies, can calculate the speed at which the galaxies are moving away from or toward us. They do this by comparing the wavelength of a particular band of light coming from hydrogen atoms (called the H_α line). The H_α line produced by a hydrogen discharge lamp in a laboratory on the earth has a wavelength of 656 nm. (a) Calculate its angular deflection in first order using a grating spectrograph that has 5000 lines per cm. (b) This same line, when coming from hydrogen gas in the stars of the Corona Borealis galaxy, produces an angular deflection of 20.6°. Calculate the wavelength of the H_α line from this galaxy.

SOLUTION (a) The following information is given: $\lambda = 656$ nm, $n = 1$ (first order), and $d = 1$ cm/5000 $= 2.00 \times 10^{-4}$ cm $= 2.00 \times 10^{-6}$ m. Using Eq. (21.3), we find that the sine of the angular deflection is:

$$\sin \theta = \frac{n\lambda}{d} = \frac{(1)(656 \times 10^{-9} \text{ m})}{(2.00 \times 10^{-6} \text{ m})} = 0.328,$$

$$\theta = \underline{19.1°}.$$

(b) The same line from hydrogen gas in the star is deflected by 20.6°. Its observed wavelength is calculated using Eq. (21.3):

$$\lambda = \frac{d \sin \theta}{n} = \frac{(2.00 \times 10^{-6} \text{ m})(\sin 20.6°)}{1} = \underline{704 \text{ nm}}.$$

The longer wavelength (704 nm instead of 656 nm) indicates that the galaxy is moving away from the earth. The magnitude of the shift toward a longer wavelength (the red shift) when substituted in the Doppler equation for light indicates that the Corona Borealis is moving away at about 0.2×10^8 m/s, or a little less than 10 percent of the speed of light. ∎

21.4 Thin-Film Interference

Another example of wave interference of light is the rainbow of colors reflected from soap bubbles and from thin films of oil floating on water. This effect is caused by the interference of light reflected from different surfaces of the film. In Fig. 21.10a, part of an incident ray of light reflects from the top surface of a film at A and the rest of the light is transmitted into the film. Part of the transmitted ray reflects from the bottom surface at B and moves back toward the top. As a portion of this ray leaves the top surface at C and moves toward our eye, it interferes with the ray reflected at A. The interference may be either constructive

(a)

Wavefronts

(b)

(c)

FIG. 21.10. (a) Light rays are shown reflecting from the top and bottom surface of a thin film. In (b) the crests of rays leaving the surface (indicated by the horizontal lines) are in phase and constructive interference occurs. In (c) the crests are out of phase, resulting in destructive interference.

(bright light) or destructive (reduced light intensity), depending on the wavelength of the light and on the extra distance ABC traveled by the ray passing through the film. If light of many wavelengths strikes the surface, the light leaving the surface consists primarily of those wavelengths that undergo constructive interference.

In order for constructive interference to occur, the light leaving the surface at C must be in phase with light reflected at A (Fig. 21.10b); the crests and troughs of the two waves must coincide. Two factors affect the relative phase of the waves: (1) the phase of a wave can change when it is reflected from the surface of a medium with greater impedance, and (2) a phase difference is caused by the extra distance ABC traveled by the wave passing through the film.

Consider first the phase change caused by reflection. Remember from Section 17.4 that a wave changes phase when reflected from the boundary of a medium with greater impedance (an upright incident pulse on a string is converted to a downward reflected pulse). If the pulse or wave reflects from a boundary with less impedance, however, no phase change occurs.

The same type of phenomenon occurs for light. The impedance of a medium through which light travels is proportional to its index of refraction n. If light moving in air ($n = 1.00$) reflects from a soap bubble ($n \simeq 1.3$), the light changes phase. However, if light moving in a soap bubble reflects from a boundary of a medium with lesser index of refraction, as occurs at point B in Fig. 21.10a, no phase change takes place. The two rays shown leaving the surface in Fig. 21.10a differ in phase by $\frac{1}{2}\lambda$ because of the change in phase caused by reflection at A but not at B.

The phase also differs because of the extra distance ABC traveled by the wave passing through the film. If the rays move perpendicular to the surface and the thickness of the film is t, then the extra distance traveled by the ray passing through the film is $2t$. If $2t = \frac{1}{2}\lambda$, then this half-wavelength phase change is balanced by the $\lambda/2$ phase change caused by reflection at A. The rays shown in Fig. 21.10b leave the surface in phase and interfere constructively. Constructive interference also occurs if $2t = \frac{3}{2}\lambda, \frac{5}{2}\lambda, \ldots$, because when the $\frac{1}{2}\lambda$ phase change caused by reflection is added to the $\frac{3}{2}\lambda, \frac{5}{2}\lambda, \ldots$ change caused by the extra distance traveled, the waves differ by an *integral multiple* of wavelengths. Since a crest of one wave still coincides with a crest of the other, the waves are in phase and interfere constructively. Destructive interference occurs when the crest of one reflected ray overlaps the trough of the other ray, as illustrated in Fig. 21.10c.

We have ignored one important factor that affects the interference. When a light wave travels in a substance having an index of refraction n greater than 1.0, its speed decreases according to Eq. (19.2): $v = c/n$. When moving at slower speed, its wavelength decreases according to Eq. (17.8): $\lambda = f/v$. We can combine these two equations to determine the wavelength of light in air (or a vacuum) λ_{air} that will undergo constructive interference in a soap bubble or oil film with index of refraction n:

$$\lambda = \frac{v}{f} = \frac{1}{f}\left(\frac{c}{n}\right) = \left(\frac{c}{f}\right)\frac{1}{n} = \frac{\lambda_{air}}{n}.$$

Thus, the *wavelengths of light moving in air that interfere constructively with a soap bubble or oil film* must satisfy the condition that

$$2t = \frac{1}{2}\left(\frac{\lambda_{air}}{n}\right), \frac{3}{2}\left(\frac{\lambda_{air}}{n}\right), \frac{5}{2}\left(\frac{\lambda_{air}}{n}\right), \ldots \qquad (21.4)$$

where t is the thickness of the film, λ_{air} is the wavelength of the light in the air, and n is the index of refraction of the thin film.

EXAMPLE 21.5 A soap bubble appears blue in color when viewed perpendicular to its surface. Determine several possible values for the thickness of the soap film whose index of refraction is 1.38.

SOLUTION Blue light has a wavelength of about 450 nm. Rearranging Eq. (21.4), we find that the thickness could be one of several possible values:

$$t = \frac{\lambda_{air}}{4n} = \frac{450 \text{ nm}}{4(1.38)} = \underline{82 \text{ nm}},$$

$$t = \frac{3\lambda_{air}}{4n} = \underline{245 \text{ nm}},$$

$$t = \frac{5\lambda_{air}}{4n} = \underline{408 \text{ nm}}. \qquad \blacksquare$$

If we look at the surface of a soap bubble that is nearest our eyes, the light that we see travels perpendicular to the surface of the film. The extra distance traveled by light passing through the film and back is $2t$. However, light reaching our eyes from regions to the side of this nearest surface travels through the film along an angular path that is longer than $2t$ (Fig. 21.11). Thus, constructive interference occurs for longer-wavelength light coming from these regions. This is why we see a rainbow of colors coming from different parts of the thin film.

The phenomenon of thin-film interference is used to produce nonreflective coatings for glass lenses and prisms. The reduction of reflection is especially important for expensive optical devices such as microscopes that use many lenses to produce an image without aberrations. For example, if four percent of the light striking a glass-air interface is reflected from each surface, then for the twenty surfaces of an optical system with ten lenses, only $(0.96)^{20} = 0.44$ or 44 percent of the light will be transmitted through all the lenses.

To reduce the amount of light reflected, a thin layer of hard transparent material with an index of refraction less than that of glass is coated on the glass. In this case we want light reflected from the top surface of the coating at A (Fig. 21.10a) to interfere destructively with light passing through the film and back out at C. Both rays are reflected from media with greater impedance causing a $\lambda/2$ phase change at both A and B. Thus, the relative phase of the two rays is unchanged by reflection.

In order for **destructive interference** to occur, the path difference must cause a one-half wavelength phase change. The *thickness of the nonreflective film on a glass lens* must satisfy the following equation:

$$2t = \frac{1}{2}\left(\frac{\lambda_{air}}{n}\right), \frac{3}{2}\left(\frac{\lambda_{air}}{n}\right), \frac{5}{2}\left(\frac{\lambda_{air}}{n}\right), \ldots \qquad (21.5)$$

The coating on glass lenses is usually made the correct thickness to reduce reflection at a wavelength of 550 nm in the center of the visible spectrum of light. The reflection of this yellow-green light is reduced by the coating from about 4 percent to about 1 percent. However, the coating is less effective at reducing the reflection at extreme wavelengths of light (red at 700 nm and violet at 400 nm). An expensive lens with a thin-film coating has a purple hue, since the red and violet light that are reflected more produce a purple color.

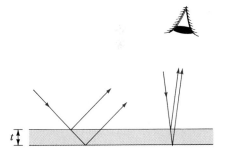

FIG. 21.11. The path length through a thin film is shorter when one is looking directly at light coming from the nearest surface than when one is looking at light from regions at the side.

EXAMPLE 21.6 Magnesium fluoride, MgF, with an index of refraction of 1.38 is used to coat a glass lens of index of refraction 1.50. What should the thickness of the coating be to reduce the reflection of 550-nm light?

SOLUTION According to Eq. (21.5),

$$t = \frac{\lambda_{air}}{4n} = \frac{550 \text{ nm}}{4(1.38)} = \underline{100 \text{ nm}}.$$

The film can also have a thickness of 300 nm, 500 nm, or other odd multiples of 100 nm. ∎

21.5 Single-Slit Diffraction

The spreading and bending of waves behind obstacles that lie in their path is called **diffraction.** The diffraction of light waves passing through a single slit is illustrated in Fig. 21.12. In (a), the slit width is large compared to the wavelength of the light, and the pattern of light cast on the screen after passing through the slit has a shape almost exactly like that of the slit. A small amount of light does bend behind the slit but is not apparent when compared to the intense light at the center of the pattern in (a). In (b) the slit width has been

(a)

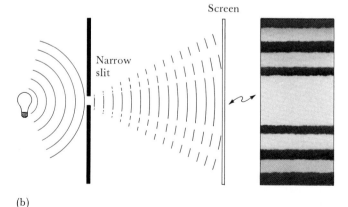

(b)

FIG. 21.12. (a) Light casts a sharp image on a screen of a slit whose width is much wider than the wavelength of light. (b) If the slit width is about the same as the wavelength of light, the diffraction or spreading of the light is easily observed.

reduced to almost the same size as the wavelength of the light. The light spreads after passing the slit and forms a broad central band of light on the center of the screen and several weaker bands or fringes of light at the sides. *The diffraction or spreading of light (or any type of wave) increases as the width of the opening through which the light passes decreases.*

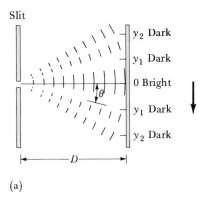

Slit

(a)

The angular spread of the light beam after passing through a slit can be calculated with reference to Fig. 21.13. The slit opening is divided into a number of smaller regions [twelve for the slit shown in (b)]. Every region vibrates in synchronization with other regions as waves strike the slit from the left. If the slit is excited by light of only one wavelength (monochromatic light), then in certain directions the light from some regions interferes destructively with light from other regions, producing a dark band of light on a screen. If light from different regions interferes constructively, a bright band is produced.

Look at the formation on the screen of the first dark band at position y_1 shown in Fig. 21.13a. Light from region 7 shown in Fig. 21.13b travels a half-wavelength farther to y_1 than does light from region 1. Therefore, the waves from these two regions interfere destructively and produce no net light at position y_1. The same is true for regions 2 and 8, 3 and 9, 4 and 10, and so on. Since the light from each region is cancelled by that from another, there is no net light intensity at position y_1.

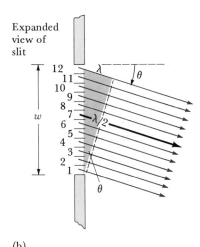

(b)

The angle θ between a line from the slit to the center of the diffraction pattern on the screen and a line from the slit to position y_1 can be determined by considering the shaded triangle shown in Fig. 21.13b. The hypotenuse of this triangle is the slit width w and the length of the side of the triangle opposite the angle θ equals the wavelength λ. We see, then, that the light from the slit produces the **first dark band** on the screen when the angular deflection θ from the slit to the dark band satisfies the equation $\sin \theta = \lambda/w$.

If the angle θ is slightly larger so that waves from region 12 travel $3\lambda/2$ farther to the screen than waves from region 1, we no longer have total destructive interference. Light from regions 1 and 5 are $\lambda/2$ out of phase and cancel each other, as does light from regions 2 and 6, 3 and 7, and 4 and 8. However, this leaves the light from regions 9, 10, 11, and 12 uncancelled, and we find a moderately intense beam of light when $\sin \theta = (3\lambda/2)/w$. The intensity at these first bright fringes on the sides is less than one-third the intensity at the center maximum.

FIG. 21.13. (a) Light incident on a narrow slit produces alternating bands of bright light and darkness on a screen. (b) The first dark band is caused by the destructive interference of light coming from different parts of the slit: Region 1 cancels 7, 2 cancels 8, and so on.

At even larger angular deflection, the light from region 12 travels 2λ farther to the screen than the light from region 1. Light from region 4 travels $\lambda/2$ farther to the screen than light from region 1; the light from the two regions interferes destructively to cancel each other as does light from regions 2 and 5, 3 and 6, 7 and 10, 8 and 11, and 9 and 12. All light is cancelled by light from another region and another dark band or fringe is formed on the screen. This band occurs when the angular deflection is calculated from the equation $\sin \theta = 2\lambda/w$.

In general, **dark bands in single-slit diffraction patterns** occur when the angular deflection θ to the *dark* band is calculated using the following equation:

$$\sin \theta = \frac{n\lambda}{w} \qquad \text{for } n = 1, 2, 3, 4, \ldots,$$

where w is the slit width, and λ is the wavelength of the light. Bright bands are formed between the dark bands. As with double-slit interference fringes, the angular deflection can also be determined using the distance y_n on a screen of the nth dark band from the central maximum for a screen that is separated from the slit by a distance D. We find that

$$\sin \theta = \frac{n\lambda}{w} \simeq \frac{y_n}{D} \qquad \text{for } n = 1, 2, 3, 4, \ldots . \tag{21.6}$$

EXAMPLE 21.7 Light of wavelength 630 nm passes through a single slit. The angular deflection to the tenth dark band on the side of the center maximum is 3.6°. (a) Calculate the slit width. (b) Calculate the angular deflection θ to the tenth dark band using light of 450-nm wavelength.

SOLUTION (a) The following information is given: $\lambda = 630$ nm, $n = 10$ (tenth dark band), and $\theta = 3.6°$. Equation (21.6) is rearranged to find the slit width:

$$w = \frac{n\lambda}{\sin \theta} = \frac{(10)(630 \times 10^{-9} \text{ m})}{(\sin 3.6°)} = 1.0 \times 10^{-4} \text{ m} = \underline{0.10 \text{ mm}}.$$

(b) When 450-nm light passes through the same 0.10-mm wide slit, the sine of the angular deflection is

$$\sin \theta = \frac{n\lambda}{d} = \frac{(10)(450 \times 10^{-9} \text{ m})}{(1.0 \times 10^{-4} \text{ m})} = 0.045,$$
$$\theta = \underline{2.6°}. \qquad \blacksquare$$

21.6 Holography

Holography is a form of photography that relies on the interference of light reflected from an object and light from another beam called the reference beam. The reflected beam and the reference beam combine to form a set of interference fringes on a holographic film. The exposed film (called a hologram), when properly viewed, produces a three-dimensional image of the object used to make the hologram even though the object is no longer present.

One procedure for making a hologram is illustrated in Fig. 21.14a. A laser provides the monochromatic (single-wavelength), coherent (an orderly pattern of moving crests and troughs) light source needed to produce the hologram. The laser light spreads after passing through a pinhole and then illuminates both the object and the holographic film. The light falling directly on the film is called the reference beam. The film is also illuminated by light reflected from the object. The interference of the reflected light with the reference beam produces concentric fringes that criss-cross over the film. The film, after processing, becomes transparent with the fringes forming a bewildering pattern that looks nothing like the object.

The hologram can now be viewed by illuminating it with only the reference beam (Fig. 21.14b). When we look through the hologram, light entering our eyes appears to come from the object used to make the hologram, even though the object is no longer present. The image seen by our eyes retains all of the three-dimensional characteristics of the original object.

To understand how this holographic process works, let us turn to the production of the hologram of a very small object such as a tiny, silvered bead (Fig.

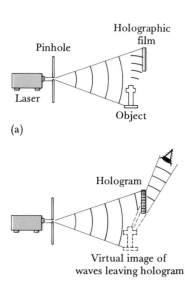

FIG. 21.14. (a) One technique for producing a hologram. (b) A hologram is viewed by illuminating it with the reference beam and viewing transmitted waves that appear to come from the object (which is not present).

21.15a). Light from the reference beam illuminates both the object and the film. Spherical scattered light waves move away from the object and also illuminate the film. A series of concentric interference fringes will form on the film (Fig. 21.15b). The fringes result from the constructive and destructive interference of the reference beam and the scattered beam of light. When a crest of one wave overlaps a trough of the other, the waves cancel. At this point, no light strikes the film; it remains unexposed (see Fig. 21.15a). Where two crests overlap, a wave of twice the amplitude forms; the film is exposed. Notice that the fringes are more closely spaced at the top of the hologram in Fig. 21.15b than at the bottom, a fact that is important when one views the hologram.

To view the hologram, the object is removed and the hologram is illuminated by the reference beam, as shown in Fig. 21.16a. Light passing through the hologram splits into three transmitted beams (analogous to the transmitted beams passing through a grating). The zeroth-order beam moves straight through the film; the first-order beams are deflected above and below the zeroth-order beam. The angular deflection θ of the rays in the upward direction (see Fig. 21.16b) depends on the spacing between the interference fringes on the film. A light ray is deflected more at the top of the hologram than at the bottom ($\theta > \theta'$) because the fringes are closer together at the top (see Fig. 21.15b). Rays 1 and 1' in Fig. 21.16b appear to diverge from the place where the object was located. While viewing the transmitted light along this first-order beam, our eye "thinks" the light is coming from the object at point A. We see the object there, even though it is not there. The other first-order beam converges below the zeroth-order beam. It forms a real image of the object. This beam could form a three-dimensional real image of the object in chalk dusk floating in the air at that region. The light moving toward this real image is usually ignored when viewing a hologram.

(a)

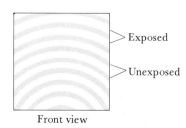

(b)

FIG. 21.15. (a) Light scattered from a silver bead interferes at a holographic film with light from a reference beam to form exposed and unexposed regions. (b) The developed film consists of a set of concentric rings.

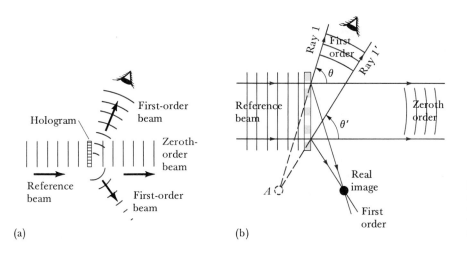

(a) (b)

FIG. 21.16. (a) A developed hologram of a silver bead is viewed using a reference beam. (b) The first-order transmitted beam moving upward (including rays 1 and 1') appears to come from the silver bead.

The hologram of a more complicated object is produced in the same way. A separate set of concentric interference fringes is produced by light reflected from each point on the object. Thus, the overlap of these fringes produces the bewildering pattern on the film. Nevertheless, when we look through the hologram, light entering our eyes appears to come from each point of the original object that produced a set of fringes. Thus, we see the object, even though it is no longer present.

FIG. 21.17. A cylindrical hologram is made by exposure from above with laser light.

The actual and potential applications of holography are many and interesting. Holograms can be used for displays in store windows. For this purpose, cylindrical holograms seem most appropriate. The film is arranged in a cylinder with the object in the middle (Fig. 21.17). The object is irradiated from above by a laser. Part of the beam is reflected from the object onto the film and another part, the reference beam, illuminates the film directly.

After development, the film is replaced below the laser and irradiated by the reference beam, but this time with no object present. As we look through the hologram, the object appears to be inside. A different view of the object is seen from each direction (Fig. 21.18).

Holography can.also be used to locate stresses in mechanical devices such as newly designed tools. In acoustical holography, the object is irradiated with ultrasound rather than light. Soft tissue in a person's body can be imaged. Thus, acoustical holography provides a noninvasive method for examining internal body parts.

FIG. 21.18. Views seen through a cylindrical hologram from different directions. (Courtesy of T. H. Jeong, Lake Forest College)

Summary

1. **Huygens' principle:** Every point on the wavefront of a wave can be considered as the source of a small, circular **wavelet** that moves in the forward direction at the same speed as the wave. The new wavefront a short time later is a line drawn tangent to the edges of these wavelets.

2. **Young's double-slit interference:** If light waves strike two narrow, closely spaced slits, the circular wavelets leaving each slit on the other side interfere constructively at some places to form bright bands of light and destructively at other places to form no light. The angular deflection θ of rays moving in the direction of **bright bands** is determined using the equation

$$d \sin \theta = n\lambda \qquad \text{for } n = 0, 1, 2, \ldots, \tag{21.1}$$

where d is the slit separation, λ is the wavelength of the light, and n is the order of the bright band (zero for the central maximum, one for the first bands on each side, and so on). If the waves are projected on a screen, the nth bright band is separated on the screen from the central maximum by a distance y_n given by

$$\sin \theta = \frac{n\lambda}{d} \simeq \frac{y_n}{D}, \tag{21.3}$$

where D is the distance from the slits to the screen.

3. **Gratings:** A **grating** consists of many closely spaced slits. Equation (21.3) can be used to locate the angular deflection to bright spots on a screen of the waves passing through a grating.

For a grating, d is the separation of adjacent slits. Gratings are used to separate a beam of light into its different colors and wavelengths.

4. **Thin-film interference:** The bright colors coming from a soap bubble or thin film of oil floating on water are caused by the constructive and destructive interference of rays reflected from the top surface of the film with rays reflected from the bottom surface of the film. For constructive interference, the two rays must differ in phase by an integral number of wavelengths; for destructive interference, their phase must differ by $\frac{1}{2}\lambda$, $\frac{3}{2}\lambda$, and so on. Phase differences occur because of phase change caused by reflection and because of differences in the path lengths of the rays.

5. **Diffraction:** When a wave passes the edge of an obstacle or through a small opening, it bends or spreads on the other side, a property called **diffraction.** Light passing through a narrow slit of width w forms a band of light in the center and alternating bands of darkness and light on each side. The angular deflection θ to

dark bands is determined using the equation

$$\sin \theta = \frac{n\lambda}{\omega} \simeq \frac{y_n}{D} \quad \text{for } n = 1, 2, 3, \ldots, \quad \textbf{(21.6)}$$

where λ is the wavelength, and n is a positive integer that labels the dark bands (1 for the first-order dark band on the sides of the central maximum, 2 for the second-order dark band on the sides, and so on). If the diffraction pattern is projected on a screen separated by a distance D from the slit, the distance y_n of the nth dark band from the central maximum is related to the angular deflection θ by Eq. (21.6).

6. **Holography** is a form of photography that relies on the interference of light reflected from an object and light from another beam called the reference beam. These two beams combine to form a set of interference fringes on a holographic film called a hologram. When properly illuminated using only the reference beam, a three-dimensional image of the object is formed.

Questions

1. Describe the appearance of the double-slit interference pattern produced by white light.

2. Devise and carefully describe a double-slit type interference experiment for sound waves. You can use one speaker, excited by an electronic oscillator, as your source of sound waves.

3. An onion skin is held in front of a light. As you move your head sideways relative to the onion skin, a rainbow spectrum of colors appears. Explain.

4. Suppose that a soap bubble has a uniform thickness. If you see green light of 520-nm wavelength when looking at the nearest part of the bubble, why will you see longer-wavelength light to the sides of this green region?

5. One edge of a flat piece of glass rests on a mirror while the other edge rests on a thin wire that separates the glass and mirror.

When looking down from above at the glass, which is illuminated with light of only one wavelength, we see alternating bright and dark fringes. Explain.

6. The width of a single slit is slowly decreased. Describe the changing pattern of monochromatic light that illuminates a screen after passing through the slit.

7. Sound passes through a doorway into an open area outdoors. Describe the changing pattern of the central intensity maximum as the sound frequency is increased.

8. A foghorn has an opening shaped like an elongated rectangle. In order to spread sound over a broad horizontal angle, should the long side of the horn be oriented up and down, or sideways? Explain.

Problems

21.1 Huygens' Principle

1. Draw a wave crest that extends about 15 cm across a paper. Wavelets that emanate from the central third of the crest travel at a speed of 6 cm/s. On the sides, the wavelets travel at 3 cm/s. (a) Using Huygens' wavelets, construct the shape of the crest $\frac{1}{4}$ second in the future. (b) Using the new crest formed in (a), construct the crest's shape as seen $\frac{1}{4}$ second later (that is, $\frac{1}{2}$ second after the initial crest).

2. A wedge-shaped wave crest, such as shown in Fig. 21.19, spreads outward. The top edge travels at a speed of 6 cm/s and

the bottom edge at 3 cm/s. Using Huygens' wavelets, construct the wave's shape $\frac{1}{4}$ and $\frac{1}{2}$ seconds in the future. Pay special attention to the "point" of the wedge.

3. The wave crest shown in Fig. 21.20 has been traveling in medium 1 at a speed of 6 cm/s. Use Huygens' wavelets to construct the shape of the crest $\frac{1}{4}$ and $\frac{1}{2}$ second after it reaches medium 2 where its speed is 3 cm/s.

FIG. 21.19

FIG. 21.20

21.2 Young's Double-Slit Experiment

4. Green light of wavelength 510 nm is incident on two slits that are separated by 0.60 mm. (a) Calculate the angular deflection of light to the zeroth-, first-, and second-order bright bands. (b) Repeat the calculations for infrared waves of wavelength 1020 nm.

5. The red light from a helium-neon gas laser has a wavelength of 630 nm. (a) Calculate the angular deflection of the light to the first three bright bands when incident on a pair of slits separated by 0.40 mm. (b) Calculate the spatial separation of the zeroth- and second-order bright bands when projected on a screen located 5.0 m from the slits.

6. Red light of wavelength 630 nm is incident on a pair of slits. The interference pattern is projected on a wall 6.0 m from the slits. The fourth bright band is separated from the central maximum by 2.8 cm. (a) Calculate the slit separation. (b) What is the angular deflection of the second-order bright band?

7. Monochromatic light passes through a pair of slits separated by 0.025 mm. On a screen 2.0 m from the slits, the third-order bright fringe is separated from the central maximum by 15.4 cm. What is the wavelength of the light?

8. A mercury lamp emits light of wavelength 436 nm. After passing through a narrow single slit, the light strikes two slits separated by 0.20 mm and located 0.50 m from the single slit. After passing through the pair of slits, the light strikes a screen 9.0 m away. (a) Make a sketch on paper of the interference pattern that you expect to see on the screen (use the correct dimensions). (b) Below the interference pattern, sketch the pattern expected if corpuscles of light cause a geometric image of the slits to be formed on the screen.

9. Two radio antennas separated by a distance $d = 600$ m emit waves of wavelength 200 m. The waves are in phase at the antennas. A row of houses is built 1000 m from the antennas and parallel to a line drawn between the antennas (the houses would be along the solid line in Fig. 21.21, where the speakers serve as antennas). A house that is an equal distance from each antenna receives a strong radio signal. Approximately what is the least distance to a house on either side (a) where the signal is strong again and (b) where the signal is weak? Explain.

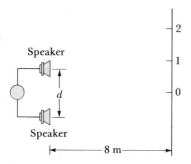

FIG. 21.21

10. Two stereo speakers separated by a distance $d = 2.0$ m play the same musical note of frequency 1000 Hz. A listener starts from position 0 (Fig. 21.21) and walks along a line parallel to the speakers. (a) Is the sound easily heard at position zero? (b) Calculate the distance from position 0 to positions 1 and 2 where intense sound is also heard. The speed of sound is 340 m/s.

21.3 Gratings

11. Light of wavelength 656 nm and 410 nm emitted from a hot gas of hydrogen atoms strikes a grating with 5300 lines per centimeter. Calculate the angular deflection of both wavelengths in the first and second orders.

12. Most light emitted by hot sodium atoms has a wavelength of 589.3 nm. When passed through a grating, the first-order angular deflection of this light is 16.1°. (a) Calculate the separation of adjacent lines in the grating and the number of lines per centimeter. (b) Calculate the angular deflection of the third-order line. (c) Calculate the distance on a screen located 2.0 m from the grating of the first-order line from the central maximum.

13. The 630-nm light from a helium-neon laser irradiates a grating. The light then falls on a screen where the first bright spot is separated from the central maximum by 0.51 m. Another wavelength of light produces its first bright spot 0.43 m from its central maximum. Calculate the wavelength of the second wavelength.

14. Green light of wavelength 518 nm is incident on a grating with 11,000 lines per centimeter. (a) What is the highest order band that can be seen? Explain your answer. (b) Calculate the distance on a screen of the first-order band from the central maximum if the screen is 1.0 m from the grating.

15. You are to order a grating to be used to detect the wavelength of infrared radiation whose wavelength is approximately 2000 nm. How many lines per centimeter should your grating have for the angular deflection in the second order of about 42°?

16. White light consisting of wavelengths from 400 to 700 nm irradiates a grating with 6000 lines/cm. (a) Calculate the range of angular deflection of the white light in the first order, and (b) in the second order. (c) Do the two orders overlap?

17. Two different wavelengths of light shine on the same grating. The third-order line of wavelength A (λ_A) has the same angular deflection as the second-order line of wavelength B (λ_B). Calculate the ratio λ_B/λ_A. Be sure to show the reasoning leading to your solution.

18. A fence consists of alternating slats and openings, the openings being separated from each other by 0.40 m. Parallel wavefronts of a single-frequency sound wave irradiate the fence from one side. A person 20 m from the fence walks parallel to it. Intense sound is heard directly in front of the fence and in another region 15 m farther along a line parallel to the fence. Calculate the wavelength and frequency of the sound ($v_{\text{sound}} = 340$ m/s).

19. A reflection grating reflects light from adjacent lines in the grating instead of allowing it to pass through slits as in a transmission grating. Interference between the reflected light waves produces different-order reflection maxima. The angular deflection of bright bands, assuming perpendicular incidence, is calculated using Eq. (22.1). White light is incident on the wing of a Morpho butterfly (whose wings act as a reflection grating). Red light of wavelength 660 nm is deflected in first order at an angle of 1.2°. (a) Calculate the angular deflection in first order of blue light (460 nm). (b) Calculate the angular deflection in third order of yellow light (560 nm).

21.4 Thin-Film Interference

20. A soap bubble of thickness 200 nm has index of refraction 1.40. Determine the longest wavelength of bright light reflected from the bubble in a direction perpendicular to its surface.

VISIBLE-REGION SPECTRA FOR SELECTED ELEMENTS

Tungsten Lamp

Iron Arc

Molecular Hydrogen

Atomic Hydrogen

Neon

Barium

Fraunhofer Lines

Courtesy of Bausch & Lomb

21. Light of 690-nm wavelength interferes constructively when reflected from a soap bubble having index of refraction 1.33. Determine two possible thicknesses of the soap bubble.

22. A film of oil with index of refraction 1.50 is spread on water whose index of refraction is 1.33. Estimate two possible thicknesses of the film in a region where the film appears red when viewed perpendicular to its surface.

23. A lens coated with a thin layer of material having an index of refraction 1.25 reflects the least amount of light at a wavelength of 590 nm. Calculate the minimum thickness of the coating.

24. The index of refraction of a nonreflective coating on a prism is 1.36. If the thickness of the coating is 100 nm, what wavelength of light is reflected least?

25. A film of transparent material 120 nm thick and having index of refraction 1.25 is placed on a glass sheet having index of refraction 1.50. Calculate (a) the longest wavelength of light that interferes destructively when reflected from the film and (b) the longest wavelength that interferes constructively.

26. Two flat glass surfaces are separated by a 160-nm gap of air. (a) Explain why 640-nm wavelength light illuminating the air gap is reflected brightly. (b) What wavelength of radiation is not reflected from the air gap?

21.5 Single-Slit Diffraction

27. Light of wavelength 630 nm is incident on a long, narrow slit. Calculate the angular deflection of the first diffraction minimum if the slit width is (a) 0.020 mm, (b) 0.20 mm, (c) 2.0 mm.

28. Water waves are incident on the 40-m opening to a harbor. If the angular deflection of the first diffraction minimum inside the harbor is 28°, what is the wavelength of the water waves?

29. Microwaves incident on a slit 11 cm wide have an angular deflection to the second diffraction minimum of 37°. Calculate the wavelength of the microwaves.

30. Babinet's principle states that the diffraction pattern of complementary objects is the same. For example, a slit in a screen (▯) produces the same diffraction pattern as a screen the same size as the slit (ı); a hair should produce the same diffraction pattern as a slit of the same width. Calculate the width of a hair that, when irradiated with laser light of wavelength 630 nm, produces a diffraction pattern on a screen with the first minimum 2.5 cm on the side of the central maximum. The screen is 2.0 m from the hair.

31. The opening of a high-fidelity speaker is shaped like a slit. Calculate the maximum width in order that the first diffraction minimum of sound is at least 45° on each side of the direction in which the speaker points. Perform the calculations for sound waves of frequency (a) 200 Hz, (b) 1000 Hz, and (c) 10,000 Hz. The speed of sound is 340 m/s.

32. Assume that the earth, its structures, and its inhabitants are all decreased in size by the same factor. *Estimate* the decrease required in order that the first-order diffraction dark band of 500-nm light entering a typical room window is at 90° (the central bright band would light most of the room). Explain all aspects of your calculations.

The Difficulty of Seeing Small Objects

Most of classical physics provides what is called a **deterministic** description of nature; if we know the present status of an object, then its future status is determined with absolute certainty. Newtonian mechanics is deterministic; if the forces acting on an object and its initial position and velocity are known, Newton's laws can be used to determine with great accuracy the future position and velocity of the object. Using Newtonian mechanics, for example, the future motions of the planets can in theory be predicted with an accuracy limited only by our ingenuity, skill, and patience in measuring their present positions and velocities and in calculating their future motions.

In theory, the future course of events in the life of a human is equally predictable. An extremely elaborate machine would have to measure and record the position and velocity of every particle in your body and of other particles in your environment that exert forces on your body. This information would then be transferred to a giant computer that would calculate the whole course of your future life.

The task of calculating your future is impossible for a variety of reasons. For one thing the body consists of over 10^{28} particles (electrons, protons, and neutrons), whereas the biggest computers can store only about 10^{20} bits of information. But the idea that a person's future is determined does not depend on whether or not instruments can actually measure and calculate that future. It depends on the supposition that a particle's position and velocity and the forces acting on it at each instant of time are precisely defined. If this is true, then the particle's future motion is also precisely defined, even though we do not know what that future will be. If the future of each particle in a person is determined, then the person's future must also be determined. Based on this idea, a philosopher might argue that free will—our ability to make choices that affect the future—is an illusion. *

The idea of determinism is hard to accept, given the large number of decisions we seem to make each day and the number of seemingly random events that affect our lives. The development early in this century of a new theory of the atom, called quantum mechanics, appeared to free electrons and other small particles from a future that was completely determined. According to the theory of quantum mechanics, a particle's position and velocity cannot simultaneously be measured with unlimited accuracy. If we measure the particle's position accurately, its velocity is uncertain. If we measure its velocity accurately, its position is uncertain. In classical physics we need to know both the particle's initial position and velocity in order to determine its future motion. Thus, according to quantum mechanics, we lack half the information needed to predict the future motion of the electrons and other small particles in our body. If the motion of these small particles is uncertain, then the future of a person must also be uncertain. Quantum mechanics seemed to restore support for the idea of free will.

*Philosophy is a complex and sophisticated subject, and the comments of a physicist as an amateur philosopher should be taken with a grain of salt.

We will have to postpone until Chapter 31 a discussion of the reason for the uncertainty in the position and velocity of small particles. We can, however, think about the problems encountered in measuring accurately the positions of small particles. These position-measurement ideas will be useful later when we develop the ideas of quantum mechanics. The ideas are also of great practical importance for persons using optical instruments such as cameras, microscopes, and telescopes. Can the fine structure of a cell be seen with a microscope? Can a reconnaissance aircraft take photographs of a missile site with sufficient detail? We will find that diffraction (studied in the last chapter) limits our ability to see the fine structure of small objects or the details of distant objects.

Consider the light passing through the small hole shown in Fig. 1a. Light leaving the hole produces a diffraction pattern on a screen behind the hole. The pattern resembles that formed by light passing through a narrow slit, except that alternating rings of bright light and darkness are formed rather than bands of light and darkness. The angle θ between a line drawn from the hole to the center of the pattern and a line drawn toward the first dark ring is calculated using the equation

$$\sin \theta = 1.22 \frac{\lambda}{d}, \qquad (1)$$

where λ is the wavelength of the wave passing through the hole and d is the diameter of the tiny hole. Equation (1) is similar to Eq. (21.6), which was used to calculate the angular deflection to dark bands in a single-slit diffraction pattern. The factor 1.22 that appears in Eq. (1) and not in Eq. (21.6) is a result of the circular geometry of the opening.

The diffraction pattern produced by a circular hole is exactly like the pattern produced by a circular disc of the same size (Fig. 1b). The only difference is that the light regions in the diffraction pattern of the hole are dark in the diffraction pattern of the disc, and vice versa.

The angular deflection θ to the first ring in either diffraction pattern shown in Fig. 1a or b depends on the diameter of the hole or disc and on the wavelength of the radiation used to form the pattern. According to Eq. (1), the sine of the angle θ (and θ itself) increases as the wavelength of the radiation increases ($\sin \theta \propto \lambda$). However, the angular deflection decreases as the diameter of the opening increases ($\sin \theta \propto 1/d$). This latter variation of angular spread with diameter provides a convenient way to measure the size of small objects. The actual diffraction patterns of laser light that diffracted around two spheres of different size is shown in Fig. 2a and b. The spheres are approximately the size of human body cells. The larger sphere produces the narrower diffraction pattern shown in (b). The diffraction pattern shown in (c) is that of a red cell of a chicken. The irregularities in this pattern provide information about distortions in the spherical geometry of the cell and also indicate the presence of the cell nucleus, which has different light-transmission properties than the rest of the cell.

(a)

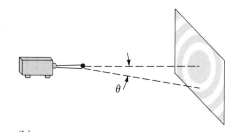

(b)

FIG. 1. (a) The diffraction pattern of light passing through a small hole has the same spread and shape as (b) the diffraction pattern of light passing a disc with the same diameter as the hole.

(a)

(b)   (c)

FIG. 2. The diffraction patterns produced by 633-nm light from an He-Ne laser passing (a) a spherical bead 6 μm in diameter, (b) a spherical bead 16 μm in diameter, and (c) a chicken's elliptical red blood cell 13 μm long by 7 μm wide. (Courtesy of Dr. Douglas Burger, Yale University School of Medicine and the Los Alamos Scientific Laboratory)

Scientists are now developing instruments that can measure the size of a large number of cells in a short time. The cells pass one by one in a narrow stream of water through a laser beam. The diffraction pattern caused by a cell as it passes through the beam is recorded by an array of light detectors behind the cell. The information from the detectors is transmitted to a computer that calculates the size of the cell. The cell size indicates the age of the cell. In some human diseases, such as cervical cancer, the cells in a sample of tissue have an abnormal size and age distribution. At the present time technicians must measure the size of a large number of cells by examining them one by one with a microscope. An automated system based on diffraction measurements should be more accurate, faster, and less expensive.

We see that diffraction provides a convenient method for measuring the size of a cell. But we find also that diffraction limits our ability to see objects such as electrons that are much smaller than cells, and this has an important impact on our earlier discussion of determinism. Consider again Eq. (1):

$$\sin \theta = 1.22 \frac{\lambda}{d}.$$

Remember that θ is the angle between a line from the particle toward the center of the diffraction pattern and another line from the particle toward the first ring to the side of the central region of the pattern. If $1.22\lambda/d$ is 1, then $\sin \theta = 1$ and $\theta = 90°$. In this case, the central region of the diffraction spreads over the whole space beyond the particle. We cannot tell where the object causing the pattern is located because all we see on a screen beyond the object is a uniform pattern of light (no rings). If $1.22\lambda/d$ is greater than 1, then according to Eq. (1), $\sin \theta > 1$. This impossible result means that the diffraction pattern is spread even farther, and we see only the center of the central portion of the diffraction pattern. It has no rings and we cannot determine the location of the object causing the pattern.

436

If a diffraction pattern with rings is to be formed, one that can be used to locate an object and determine its size, the quantity on the right of Eq. (1) must be less than 1:

$$\sin \theta = 1.22 \frac{\lambda}{d} < 1$$

or

$$\lambda < \frac{d}{1.22}. \qquad (2)$$

It is common to ignore the 1.22 in Eq. (2). In that case, we can say that *a particle's position and size can be determined using waves only if the wavelength of the radiation is less than the diameter of the particle.* The diameters of protons and neutrons are about 10^{-15} m, and the diameter of an electron is not known but is much less than that of a proton. The wavelength of light is over 10^8 times longer than the diameter of a proton and cannot be used to locate these particles. Even x-rays and gamma rays have wavelengths that are longer than the size of a proton or neutron and much longer than the size of an electron. If we do manage to produce waves whose wavelengths are shorter than these particles, the waves will disrupt the future motion of the particle in an unpredictable way, as we will see in Chapter 31. Either we cannot locate the particle because the wavelength is too long, or we disrupt its future motion in an uncertain way because the wavelength is too short. Thus, we cannot simultaneously determine an object's position and velocity with complete accuracy. The future of a small particle is uncertain.

EXAMPLE 1 Suppose that you are reduced to the size of an ant. One of your ant friends holds a tiny flashlight in front of your eyes. Over what angular region does light leaving the flashlight spread after passing through the pupil of your eye on its way toward the retina?

SOLUTION If all of your body parts are reduced proportionately, the pupil of your eye becomes very small when your body is reduced to the size of an ant. Light passing through the small opening into your eye is diffracted over a broad angular region. To calculate the angular spread, we first estimate the diameter of the eye's pupil if your body is decreased to the size of an ant. An ant is approximately 2 mm long and a person is approximately 2 m long. Thus, when shrunk, each dimension in the person's body decreases by a factor of about $2 \, \text{mm}/2 \, \text{m} = 2 \times 10^{-3} \, \text{m}/2 \, \text{m} = 10^{-3}$. The pupil in a person's eye has a diameter of about 1.5 mm. After shrinking, the diameter is reduced by 10^{-3}:

$$d = (10^{-3})1.5 \, \text{mm} = 1.5 \times 10^{-3} \, \text{mm} = 1.5 \times 10^{-6} \, \text{m}.$$

Next, consider green light of 500-nm wavelength entering this small, circular opening. The light forms a diffraction pattern on the retina with a bright region in the center and alternating dark and bright rings circling the central bright region. The angular deflection from a line drawn toward the center of the pattern to a line drawn toward the first dark ring is determined using Eq. (1):

$$\sin \theta = 1.22 \frac{\lambda}{d} = 1.22 \left(\frac{500 \times 10^{-9}\,\text{m}}{1.5 \times 10^{-6}\,\text{m}} \right) = 0.41,$$

$$\theta = 24°.$$

Two lines moving toward opposite sides of the first dark ring are separated by twice this angle, or by 48°, as shown in Fig. 1a. Thus, the bright region in the center of the pattern spreads over an angular region that is a little less than 48° wide. Most of our retina would be covered by light from this one small light source. It would be difficult for an ant-sized person to see the details of his or her environment, since light reflected from two closely spaced objects would excite the same group of cones and rods over a large region of the retina. We would see a flood of light but could not determine from where it was coming. ■

EXAMPLE 2 A biologist builds a device to detect and measure the size of insects. The device emits sound waves. If an insect passes through the beam of sound waves, it produces a diffraction pattern on an array of sound detectors behind the insect. What is the lowest-frequency sound that can be used to detect a fly that is about 3 mm in diameter? The speed of sound is 340 m/s.

SOLUTION The wavelength of a sound wave is related to its frequency f by Eq. (17.8), $f = v/\lambda$, where v is the speed of sound. In order to produce a diffraction pattern that is spread over a reasonable angular region, the biologist builds the device so that the angular deflection to the first ring of the diffraction pattern is approximately 30° or less. By combining Eqs. (17.8) and (1), we find that the frequency of the sound wave should be

$$f = \frac{v}{\lambda} = v \left(\frac{1.22}{d \sin \theta} \right) = (340\,\text{m/s}) \left[\frac{1.22}{(3 \times 10^{-3}\,\text{m})(\sin 30°)} \right]$$

$$= 2.8 \times 10^5\,\text{s}^{-1} = 2.8 \times 10^5\,\text{Hz}. \qquad ■$$

Problems

1. Laser light of wavelength 630 nm passes through a tiny hole. The angular deflection of the transmitted light is 26°. Calculate the diameter of the hole.

2. Sound leaves your mouth. (a) Estimate the diameter of your mouth when open wide. (b) Calculate the angular deflection of sound of frequency 200 Hz and of frequency 15,000 Hz. If, during your calculations, you find that sin $\theta > 1$, explain the meaning of this result. The speed of sound is 340 m/s.

3. Light of 626-nm wavelength from a helium-neon laser passes two different-size body cells. The angular deflection of the first ring in the diffraction patterns that are formed as the light passes the cells is (a) 0.060 radians and (b) 0.085 radians. Calculate the size of each cell.

4. A sound of frequency 1000 Hz passes a basketball. Estimate the angular deflection from the basketball to the first ring in the diffraction pattern. The speed of sound is 340 m/s.

5. What is the diameter of the smallest object that forms a diffraction pattern when irradiated by the (8×10^4)-Hz sound from a bat?

Electricity and Magnetism

VI

Much of our past study of physics has used Newton's second law and energy principles to analyze the motion of objects. In this part of the book these same techniques will be applied to the study of electrical and magnetic phenomena.

Electric and magnetic forces, first discovered by the ancient Greeks, were enshrouded in mystery for centuries. The practical use of electricity began only about a hundred years ago. Today, a relatively short time later, electrical and magnetic devices provide instant communication of information by television, telephone, and microwave networks that carry such things as international news and stock quotations. Electrical energy is transmitted from centralized power plants for operating electric stoves, toasters, lights, and dishwashers. Modern computer systems store information and perform calculations that help to control space flights, the operations of businesses, and charge accounts.

We cannot examine the details of all the interesting and important applications of modern electricity and magnetism. We can, however, develop a basic understanding of the underlying principles that made these applications possible. Our analysis begins with a discussion of the basic force between two charged particles. As we continue through Part VI, our interest will move toward the electronic applications of electricity and magnetism.

Electric Force

Your body contains over 10^{28} protons and nearly an equal number of electrons. The protons have what is called a positive electric charge; the electrons have a negative charge. The electrical force between these charges is one of the most important forces in nature.

Suppose for a moment that you could borrow all the electrons from a friend's body and put them into your pocket. The mass of the electrons would be about 20 grams. With no electrons, your friend would have a huge positive charge. You, on the other hand, would have a huge negative charge in your pocket. If you stood 10 m from your friend, the attractive force between the two of you would be about 10^{23} tons—over one hundred thousand times greater than the gravitational force between the earth and the sun. Obviously, your pocket would be ripped off long before you had gathered even a tiny fraction of your friend's electrons.

We never find 10^{28} protons or 10^{28} electrons gathered in one place on the earth. Equal numbers of positive and negative charges almost always combine to form atoms—tiny particles that have no net charge. So the huge electrical force just described does not occur.

There are, however, many instances in which smaller charge imbalances occur. In this chapter we will be learning how to calculate the magnitude and direction of the forces between these charges. Our concern is mostly with forces between stationary charges—that is, with **static electricity.**

22.1 Electric Charge

Many words and concepts must be defined in terms of the effects they produce. *Love,* for example, can be defined as a strong feeling that causes a person to delight in and desire the presence of another. Love, then, is defined in terms of the effects it produces—delight and desire.

Electric charge is also defined by the effect it produces. No one has ever seen electric charge; it has no known weight, color, length, or width. Yet the effects of electric charge are among the most important in nature. Almost every dynamic process in your body and in its environment involves a rearrangement of particles with electric charge, and these rearrangements cause dramatic changes in the forces acting on the particles.

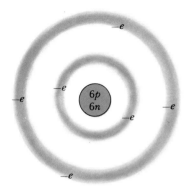

FIG. 22.1. A carbon atom consists of a nucleus with six protons (charge +6) and six neutrons that are surrounded by six electrons in different orbits.

Atoms, the basic units of which all matter is built, consist of three types of particles: electrons, protons, and neutrons. Two of these particles—electrons and protons—have electric charge while the neutron has no charge. The magnitude of the electric charge of an electron equals that of a proton (even though the electron has only $\frac{1}{2000}$ the mass of a proton). The proton is said to have a positive electric charge; the magnitude of the charge is given the symbol e. An electron has a negative charge whose value is represented by the symbol $-e$. An atom has no net charge, since it contains equal numbers of protons and electrons.

Over one hundred different types of atoms have been discovered. The atoms differ in the number of protons, neutrons, and electrons of which they are made. The central core of an atom, called the **nucleus,** consists of protons and neutrons. Electrons circle the nucleus in a variety of different orbits. Hydrogen, for example, has a nucleus of one proton surrounded by one orbiting electron. Carbon, on the other hand, has six protons and six neutrons in its nucleus, and six electrons move in orbits around the nucleus (Fig. 22.1).

As you can see in Fig. 22.1, the electrons in an atom occupy different orbits. Electrons in orbits nearest the nucleus are tightly bound to it by their electrical attraction. The electrons farthest from the nucleus (called the outer electrons) are loosely bound. It is common for an outer electron to actually leave one atom and move into an outer orbit of another type of atom. When such a charge transfer occurs, the transferred electron is usually bound more tightly to the new atom than it was to the old atom. An atom that has lost an electron has a net charge of $+e$ (it has one more proton than electron) and is called a **positive ion.** An atom that gains an extra electron has a net charge of $-e$ and is called a **negative ion.**

This type of charge transfer often occurs when materials made of different types of atoms are brought into contact with each other. The small electrical shock one receives when touching another person after shuffling across a synthetic rug is the result of removal of electrons from the rug through your shoes and onto your body. Some of these electrons leave your body quickly (the shock) when you touch another person.

The existence of electric charge and of charge transfer can be demonstrated using a balloon, a wool scarf, and a thin plastic clothes bag (Fig. 22.2). The balloon is suspended by a string. Rubbing the balloon with a plastic bag (a) causes the balloon to accumulate excess electrons from the bag and to be-

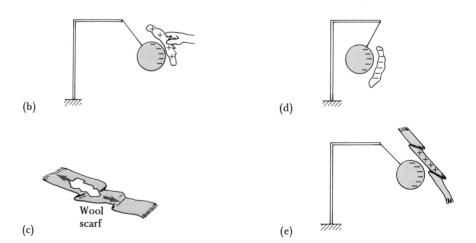

FIG. 22.2. (a) A plastic bag is rubbed against a rubber balloon. (b) This transfers negative charge from the bag to the balloon resulting in their attraction. (c) The plastic bag is now rubbed against a wool scarf. (d) The bag is now charged negatively and repels the balloon. (e) The positively charged scarf attracts the balloon.

come negatively charged. The plastic is left with an equal number of positive charges (positive ions). We find that the plastic bag and balloon attract each other (b). *Opposite charges attract.*

If the plastic bag is then rubbed against a wool scarf (c), electrons will be transferred from the scarf to the bag because electrons are more attracted to the atoms in plastic than to those in wool. The plastic bag, which started with a deficiency of electrons, gains so many electrons from the scarf that it becomes negatively charged. The plastic bag and balloon will now repel each other (d). *Like charges repel.* The scarf, which loses electrons when rubbed with plastic, becomes positively charged and will attract the balloon, since opposite charges attract (e).

Unit of Charge

The SI metric system unit of charge is called the coulomb (C) after the French engineer Charles Augustin de Coulomb (1736–1806). One **coulomb** equals the charge of an aggregate of 6.25×10^{18} protons. The charge e of a single proton, in units of coulombs, is

$$e = 1.60 \times 10^{-19} \, \text{C}.$$

An electron has a charge $-e$ equal to $-1.60 \times 10^{-19} \, \text{C}$. Balloons rubbed with plastic, as illustrated in Fig. 22.2, typically accumulate about 10^{12} excess electrons, each with a charge $-1.60 \times 10^{-19} \, \text{C}$. The total charge q on the balloon is then

$$q = 10^{12}(-e) = 10^{12}(-1.60 \times 10^{-19} \, \text{C}) = -1.60 \times 10^{-7} \, \text{C}.$$

22.2 Coulomb's Law

The electrical force exerted by one charged object on another depends on the amount of charge on each object, the sign of the charges (positive or negative), and the distance between them. The magnitude of the electrical force can be calculated using an equation determined experimentally in 1788 by Charles Coulomb. The equation is called **Coulomb's law:**

$$F = \frac{kq_1 q_2}{r^2}. \tag{22.1}$$

The amount of electric charge on the two objects is represented by the symbols q_1 and q_2, and r represents the distance separating the charges. The quantity k is a proportionality constant whose value depends on the units of the other quantities in Eq. (22.1).* When the SI metric system units of coulomb, meter, and newton are used for the quantities charge, distance, and force, respectively,

---

*An identical symbol k is used to represent the magnitude of the force constant of a spring, but there should be little confusion. The proportionality constant k in Eq. (22.1) usually appears in a mathematical expression involving electric charges, whereas the force constant k appears in expressions such as those for the restoring force and energy of a spring ($F_r = -kx$, $E = \frac{1}{2}kx^2$).

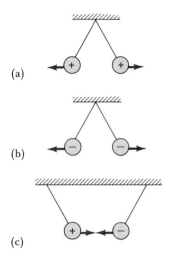

(a)

(b)

(c)

FIG. 22.3. (a) and (b) Like charges repel. (c) Unlike charges attract.

then the units of k are newton-meter^2/coulomb^2. The value of k determined by experiment is

$$k = 8.98742 \times 10^9 \, \text{N} \cdot \text{m}^2/\text{C}^2 \simeq 9.0 \times 10^9 \, \text{N} \cdot \text{m}^2/\text{C}^2.$$

The value $9.0 \times 10^9 \, \text{N} \cdot \text{m}^2/\text{C}^2$ will be used for k in all of our calculations.

The force between charges also depends on the medium in which the charges are embedded. This medium dependence is considered in Section 22.4.

The direction of the force that one charge exerts on another depends on the relative signs of the two charges. If the charges have like signs, then they exert repulsive forces on each other as shown in Fig. 22.3a and b. If the charges have opposite signs, they exert attractive forces on each other as shown in Fig. 22.3c. Coulomb's law is used to determine only the magnitude of the force; we will *ignore the signs of the charges when substituting in Coulomb's law.*

EXAMPLE 22.1 A young woman accumulates a charge q_1 of $+6.0 \times 10^{-6}$ C while sliding out of the front seat of a car. Her boyfriend, who has been waiting in the wind, has gained electrons and now has a charge q_2 of -2.0×10^{-6} C. Calculate the magnitude of the electrical force that each person exerts on the other when separated by a distance 0.50 m. Is the force attractive or repulsive?

SOLUTION The magnitude of the force exerted by one charged person on another is determined using Eq. (22.1):

$$F = \frac{kq_1q_2}{r^2} = \frac{(9.0 \times 10^9 \, \text{N} \cdot \text{m}^2/\text{C}^2)(6.0 \times 10^{-6} \, \text{C})(2.0 \times 10^{-6} \, \text{C})}{(0.50 \, \text{m})^2} = \underline{0.43 \, \text{N}}.$$

When using Eq. (22.1), remember that only the magnitude of the charges q_1 and q_2 are used to determine the magnitude of the force. Thus, the negative sign is not included when substituting the value of q_1. Because the charges have opposite signs, the woman and man are <u>attracted to each other</u> by a force of equal magnitude. ∎

22.3 Static Electric Forces

Most electrical systems consist of many charged particles, each of which exerts an electrical force on all the other charges, and vice versa. In order to calculate the net force acting on one charge due to the other charges, one must add each electrical force acting on the charge using the techniques of vector addition developed in Chapters 1 and 2. You may wish to review the summaries of these chapters in order to freshen your memory about vector problem-solving methods. The technique is illustrated qualitatively in Fig. 22.4a.

Three charges are shown: the upper positive and negative charges reside on electrodes placed by a geologist at the earth's surface and the lower charge represents a sodium ion (Na$^+$) dissolved in the ground water under the earth. In this situation we are interested in determining the direction in which the sodium ion will be pushed by the charges on the electrodes.

Whenever considering the effect of forces on an object, we always start our problems by choosing an object of interest and then drawing a force diagram showing all forces acting on that object. Since the sodium ion is the object of

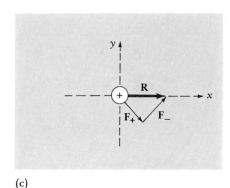

(a) (b) (c)

FIG. 22.4. (a) Positive and negative charges on electrodes near the earth's surface exert forces, shown in (b), on a sodium ion under the ground. The resultant electrical force acting on the sodium is shown in (c).

interest in the situation shown in Fig. 22.4a, we look at only the forces acting on the sodium ion. (We are not concerned with the effects of the sodium ion on other charges, or with the effects of the other charges on each other.)

The force diagram for the sodium ion (Fig. 22.4b) shows two electrical forces acting on it: (1) a repulsive force $\mathbf{F}_+$ caused by the charge of the positive electrode, and (2) an attractive force $\mathbf{F}_-$ caused by the charge of the negative electrode. We ignore for now drag forces that oppose the motion of the ion through the ground water. The net electrical force on the ion is the vector sum of the forces. They can be added graphically as in Fig. 22.4c or by using the component technique developed in Chapter 1. Examples using both techniques will be presented shortly.

In Fig. 22.4c we see that the net force on the sodium ion is toward the right. Therefore, if only the net electric force acts on the ion, it accelerates to the right. However, when it starts moving, a drag force will oppose its motion. The drag force represents the average effect of retarding collisions between the sodium ion and other atoms, ions, and molecules in its path. The effect of this drag force will be addressed in detail in a later chapter. For now, our primary concern is learning how to calculate the net electrical force on a particular charge near other charges.

EXAMPLE 22.2 From the information shown in Fig. 22.5a, calculate the net electrical force of charges q_2 $(-1.0 \times 10^{-6}\,\text{C})$ and q_3 $(+3.0 \times 10^{-6}\,\text{C})$ on q_1 $(+2.0 \times 10^{-6}\,\text{C})$.

FIG. 22.5. (a) A system of three charges. (b) A force diagram showing the forces on charge q_1 due to q_2 and q_3. (c) The graphical addition of these two forces to find the resultant force.

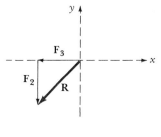

(a) (b) (c)

SOLUTION The object of interest is charge q_1. A force diagram for that charge is shown in Fig. 22.5b. $\mathbf{F}_2$ is the attractive force of q_2 on q_1 and $\mathbf{F}_3$ is the repulsive force of q_3 on q_1. The graphical addition of the forces, illustrated in Fig. 22.5c, indicates that the resultant force points below the negative x-axis into the third quadrant.

In order to calculate the magnitude and direction of the resultant force, we will add the vectors by the component technique. Before doing this, let us first calculate the magnitude of each force by using Coulomb's law [Eq. (22.1)]:

$$F_2 = \frac{kq_1q_2}{r^2} = \frac{(9.0 \times 10^9 \, \text{N} \cdot \text{m/C}^2)(2.0 \times 10^{-6} \, \text{C})(1.0 \times 10^{-6} \, \text{C})}{(1.0 \, \text{m})^2}$$

$$= 1.8 \times 10^{-2} \, \text{N}$$

and

$$F_3 = \frac{kq_1q_3}{r^2} = \frac{(9.0 \times 10^9 \, \text{N} \cdot \text{m}^2/\text{C}^2)(2.0 \times 10^{-6} \, \text{C})(3.0 \times 10^{-6} \, \text{C})}{(2.0 \, \text{m})^2}$$

$$= 1.4 \times 10^{-2} \, \text{N}.$$

The x and y components of the resultant are then calculated using Eqs. (1.3)–(1.6):

$$R_x = F_{2x} + F_{3x} = F_2 \cos 90° - F_3 \cos 0° = 0 - F_3 = -1.4 \times 10^{-2} \, \text{N}$$

and

$$R_y = F_{2y} + F_{3y} = -F_2 \sin 90° + F_3 \sin 0° = -F_2 + 0 = -1.8 \times 10^{-2} \, \text{N}.$$

The magnitude of the resultant force is found using Eq. (1.7):

$$R = \sqrt{R_x^2 + R_y^2} = \sqrt{(-1.4 \times 10^{-2} \, \text{N})^2 + (-1.8 \times 10^{-2} \, \text{N})^2}$$

$$= \underline{2.3 \times 10^{-2} \, \text{N}.}$$

In order to completely specify a vector quantity, we must also know its direction. The angle θ that the vector makes with the negative x axis is determined using Eq. (1.8):

$$\tan \theta = \left| \frac{R_y}{R_x} \right| = \frac{1.8 \times 10^{-2} \, \text{N}}{1.4 \times 10^{-2} \, \text{N}} = 1.29$$

or

$$\theta = 52°.$$

Thus, the resultant points $\underline{52° \text{ below the negative } x \text{ axis.}}$ The fact that it points below the negative x axis is apparent from the signs of the components and from the vector addition shown in Fig. 22.8c. ∎

EXAMPLE 22.3 Two balloons hang at the ends of strings that are 0.80 m long and have negligible weight. Each balloon has a mass of 1.5 g. The balloons have equal but unknown electrical charges that cause the balloons to repel each other. If the balloons are separated from each other by 0.80 m, as shown in Fig. 22.6a, what is the magnitude of the charge on each balloon?

SOLUTION To solve the problem we must first draw a force diagram for one of the balloons and use the first condition of equilibrium to determine the electrical force acting on that balloon. Although either balloon can be chosen as the object of interest for the force diagram, we will choose the balloon on the right. A force diagram for this balloon is shown in Fig. 22.6b. Three forces act on the balloon: (1) a string-tension force **T**, (2) a downward weight force **w** of magnitude mg, and (3) an electrical force **F** caused by the repulsion between the like charges on the balloons. The forces are analyzed using the coordinate system shown in (b).

Since the balloon is at rest, the sum of the x and y components of the forces must each be zero (the first condition of equilibrium). Thus, using Eqs. (1.3) and (1.5) for the x components, we find that

$$\Sigma F_x = T_x + w_x + F_x = -T\cos 60° + mg\cos 90°^{\;0} + F\cos 0° = 0.$$

Solving this last equation for F, we find that $F = 0.5T$. Next, using Eqs. (1.4) and (1.6) for the y components, we find that

$$\Sigma F_y = T_y + w_y + F_y = T\sin 60° - mg\sin 90° + F\sin 0°^{\;0} = 0.$$

Substituting for the sines and rearranging, we find that $T = mg/0.87$. Substituting this result into our earlier result using the x component equation ($F = 0.5T$), we find that

$$F = 0.5T = 0.5\left(\frac{mg}{0.87}\right) = 0.58\,mg.$$

The magnitude of the electrical force **F** is given by Coulomb's law [Eq. (22.1)]:

$$F = 0.58\,mg = \frac{kqq}{r^2},$$

where q is the unknown charge on each balloon and g is the acceleration of gravity (9.8 m/s^2). Solving for q, we find that

$$q = \sqrt{\frac{0.58\,mgr^2}{k}} = \sqrt{\frac{0.58(1.5 \times 10^{-3}\,\text{kg})(9.8\,\text{m}\cdot\text{s}^{-2})(0.80\,\text{m})^2}{(9.0 \times 10^{-9}\,\text{N}\cdot\text{m}^2/\text{C}^2)}}$$

$$= 0.79 \times 10^{-6}\,\text{C} = \underline{0.79\ \mu\text{C}}. \qquad\blacksquare$$

(a)

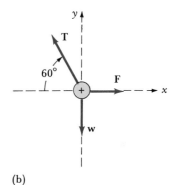

(b)

FIG. 22.6. (a) Two positively charged balloons repel each other. (b) A force diagram for the balloon on the right.

22.4   Dielectric Constant

In the previous sections of this chapter, we have ignored the effect of a medium in which two charges reside on the force exerted by the charges on each other. This effect can be significant. For example, the force between a sodium ion and a chlorine ion when in water is only $\frac{1}{80}$ the force between the same ions when in a vacuum (or, to a very good approximation, in air). The effects of the medium are particularly important in biochemical processes, many of which involve the rearrangement of electrical charges in molecules. In living organisms these rearrangements are facilitated by water, a solvent that reduces the electrical forces holding the molecules together.

To account for the effect of a medium on the magnitude of the force between two charges, a quantity called the dielectric constant is included in Coulomb's law. The **dielectric constant** K of a medium is the ratio of the electrical force between two charges when in a vacuum (called F_{vacuum}) and the force between the same two charges when kept at the same separation in the medium (called F_{medium}):

$$K = \text{dielectric constant of a medium} = \frac{F_{vacuum}}{F_{medium}}.$$

Coulomb's law as written in Eq. (22.1) applies only to charges in a vacuum (or in air): $F_{vacuum} = kq_1q_1/r^2$. Substituting this expression for F_{vacuum} into the above equation and rearranging, we find that the force F (we drop the subscript medium from now on) between the charges when separated by a distance r in a medium of dielectric constant K is

$$F = \frac{F_{vacuum}}{K} = \frac{kq_1q_2}{Kr^2}. \tag{22.2}$$

The dielectric constants of a number of media are listed in Table 22.1. Notice that, according to Eq. (22.2), the larger the value of K, the smaller is the force between the charges.*

EXAMPLE 22.4 Calculate the ratio of the electrical force between a potassium ion (K^+) and a chlorine ion (Cl^-) when separated by a distance r in ethanol and when separated by the same distance in water.

SOLUTION The ratio of the forces is

$$\frac{F_{ethanol}}{F_{water}} = \frac{\dfrac{kq_1q_2}{(K_{ethanol})r^2}}{\dfrac{kq_1q_2}{(K_{water})r^2}} = \frac{K_{water}}{K_{ethanol}} = \frac{80}{26} = \underline{3.1}. \qquad \blacksquare$$

We learned in the last example that the force between two ions in ethanol is about 3.1 times more than the force between the same ions when in water. For this reason, potassium chloride and any salt should separate into ions (dissolve) more easily in water than in ethanol. In the human body, most charged particles (ionized salts, charged macromolecules, cell membranes, and so forth) reside in water. The effect of water on the electrical interactions of these charges is great. The rate of human metabolism is very dependent on the dielectric properties of water. Nerve conduction would be impossible if salt did not dissolve in water to form ions.

The reason the electrical force between charges is reduced greatly in water and in certain other media is related to the dipole nature of these media, as we learn next.

TABLE 22.1 Dielectric Constants of Different Substances

Substance	Dielectric constant $(K)^a$
Vacuum	1.0000
Dry air	1.0006
Wax	2.25
Mica	2–7
Glass	4–7
Benzene	2.28
Paper	3.5
Axon membrane	8
Body tissue	8
Ethanol	26
Methanol	31
Water	80

^{a} At 20°C.

---

*The constant k is often written as $k = 1/4\pi\varepsilon_0$, where ε_0 is called the permittivity of free space ($\varepsilon_0 = 8.85 \times 10^{-12}\ C^2/m^2 \cdot N$). Coulomb's law is then written as

$$F = \frac{k}{K}\frac{q_1q_2}{r^2} = \frac{1}{K4\pi\varepsilon_0}\frac{q_1q_2}{r^2} = \frac{1}{4\pi\varepsilon}\frac{q_1q_2}{r^2}.$$

The product $K\varepsilon_0$ is called the permittivity of a medium with dielectric constant K. This unusual way of writing Coulomb's law allows the proportionality constant ε_0 to be related more naturally to other electromagnetic phenomena to be discussed in later chapters.

22.5 Electric Dipoles

We have seen that the electrical force between charges is less when the charges are in a medium such as water than when in air or a vacuum. This reduction in the electrical force is caused by the **polar** nature of water molecules. Although a water molecule has a net charge of zero, the charge is not distributed uniformly throughout the molecule. Excess positive charge resides on one side of the molecule, and an equal excess of negative charge is on the other side.

Many different types of molecules have this polar electric charge distribution. These molecules are characterized in terms of their electric dipole moments. Consider the equal-magnitude but opposite-sign charges shown in Fig. 22.7a. The **electric dipole moment p** of this charge distribution is a vector shown in Fig. 22.7b that points from the negative charge toward the positive charge. The magnitude of the dipole moment is the product of the positive charge q and the distance a separating the charges:

$$\text{electric dipole moment } p = qa. \qquad (22.3)$$

Let us see how the electric dipole moment of a polar molecule such as water (H_2O in Fig. 22.8a) is formed. The molecule (Fig. 22.8b) has three nuclei: an oxygen nucleus with a charge $+8e$ and two hydrogen nuclei (protons) each with charge $+e$. The ten electrons in the molecule are arranged somewhat as shown in (b). Two electrons circle close to the oxygen nucleus while the other eight electrons are located in four orbits, often called lobes. Two of the lobes have a proton and two do not. This distribution of electrons and protons results in a charge separation with two electron pairs pointing in one direction and two electron pairs with protons pointing in the other direction. This charge separation produces an electric dipole moment **p** whose magnitude $6.1 \times 10^{-30} \text{ C} \cdot \text{m}$ is the same as would be produced by a charge e separated from a charge $-e$ by a distance a equal to 0.38×10^{-10} m, or about one-fifth the size of a water molecule.

We can now understand why the electrical force between ions when in water is less than when in air. If a positive ion or any positively charged particle is placed in water, the water molecules nearest the ion are no longer randomly oriented. The negative part of the water molecules tend to rotate so that they point toward the positive ion (the dashed circle on the left in Fig. 22.9). With a negative ion in the water, the positive sides of the water molecules tend to point toward the negative ion (the right circle in Fig. 22.9). As far as the negative ion is concerned, the positive ion appears to have less charge than usual because its positive charge is reduced by the negative charges hovering near it. The charge of the negative ion is reduced in a similar way by the presence of the positive ends of the water molecules. Consequently, the force between the ions is reduced.

(a)

(b)

FIG. 22.7. (a) An electric dipole consists of a positive charge q and an equal-magnitude negative charge $-q$ separated by a distance a. (b) The electric dipole moment of these charges is a vector from $-q$ toward q whose magnitude $p = qa$.

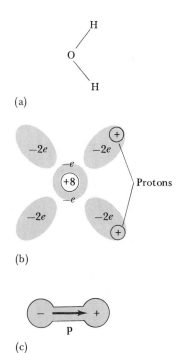

(a)

(b)

(c)

FIG. 22.8. (a) A water molecule. (b) The distribution of nuclei and electrons in a water molecule results in an effective dipole charge distribution and dipole moment, such as shown in (c).

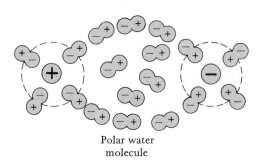

Polar water molecule

FIG. 22.9. Dipolar water molecules reduce the effective charge of ions dissolved in the water.

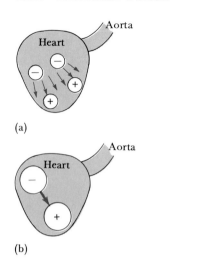

(a)

(b)

FIG. 22.10. (a) Tiny dipoles are created as heart muscle cells contract. (b) The net heart dipole at one instant is the vector sum of the many tiny dipoles.

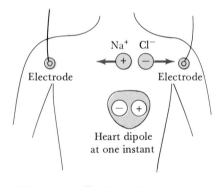

FIG. 22.11. The heart's dipole causes plus and minus charges on the body surface to move in opposite directions. This charge separation is detected by electrodes of an electrocardiograph.

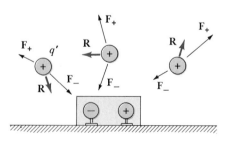

FIG. 22.12. The force on a charge q' varies from place to place, depending upon the positions and types of charges under the box.

It should be emphasized that the water molecules are continually bumping each other. The orientation of one molecule may change 10^{14} times each second. The reduction in the effective charge of an ion in water is a statistical phenomenon that depends upon the average orientation of the water molecules.

An interesting consequence of the polar nature of water molecules is related to the ability of microwave ovens to cook foods. The microwave radiation produced in these ovens interacts with the dipole-charge distribution of water molecules and exerts a torque on the dipole, causing the molecule to rotate. The heating in a microwave oven results from the extra rotational energy imparted to the water molecules by the microwaves. This rotational energy is converted to thermal energy as the molecules collide with each other while rotating. A dry paper plate placed by itself in a microwave oven will remain cool because its molecules do not have electric dipole moments and are thus unaffected by microwaves.

Probably the most important clinical application related to electric dipoles is a result of the dipole charge on the heart. As each muscle cell in the heart contracts, electric charges move in a way that produces a charge separation—an electric dipole in each muscle cell (Fig. 22.10a). The vector sum of the many tiny dipoles from all the contracting muscle cells results in a large electric dipole of the heart (Fig. 22.10b). The magnitude and direction of the heart's dipole depends on the number and orientation of muscle cells that are contracting at any one time. Since the heart repeats a cycle of contraction about once each second, it produces a dipole moment that changes magnitude and direction in a repetitive fashion—once each second.

At any instant of time the dipole charges on the heart exert forces on ions in the body's surface tissue that cause the ions to move. For example, the heart dipole shown in Fig. 22.11 forces sodium ions in the tissue to move toward the left and chlorine ions to move toward the right. Electrodes placed on the arms absorb charges of different sign. The heart acts much like a battery pushing opposite-sign charges in opposite directions in the body tissue. An electrocardiogram (ECG or EKG) is a recording of the charge separation on the body's surface caused by the electric dipole moment of the heart. Abnormalities in the size or timing sequence of different phases of the heartbeat cycle are easily detected by these electrodes.

22.6 Electric Fields

In the previous sections we used Coulomb's law and vector addition to calculate the resultant electric force acting on a charge due to the presence of other charges. What if we do not know the distribution of electric charges in a certain region of space? How might we probe the surrounding space to learn about that distribution? The answer to this question leads to the idea of electric fields and to an alternative method for describing electric forces.

To understand the concept of electric fields, suppose that your instructor asks you to determine the charges on two small metal spheres that have been placed under a box (Fig. 22.12). Your assignment is to devise a technique to determine the sign, magnitude, and location of the charges without removing the box.

One possible technique might involve building a device to measure the electric force on a small test charge held at different places outside the box. You could attach a tiny metal ball with a small positive electric charge (q') to a spring balance that measures the magnitude and direction of the electric force acting on q'. Knowing the force on q' will help you determine the charge distribution that causes the force. For example, if the spheres under the box are charged as shown in Fig. 22.12, the force on test charge q' is downward when q' is above and to the left side of the box, toward the left when q' is over the center of the box, and upward when q' is above and to the right side of the box. This information indicates that a dipole charge is under the box and that the positive charge is on the right. If the force on test charge q' is measured at every point in the space around the box, we can determine the exact charge distribution under the box.

Now suppose that another person who does not know about the charges under the box is given the device that measures the force on the test charge q'. The person discovers that the force acting on q' changes as q' is moved from one place to another. Knowing nothing of the charges under the box, the person moving the device might say that the space outside the box varies in some way. Electrical "hills" and "valleys" invisible to the naked eye seem to be detected by the force acting on the test charge q'. These hills and valleys constitute an electric field. A charge placed in the electric field will feel a force that depends on the magnitude and direction of the field.

An electric field is analogous to a gravitational field. The mass of the earth is said to create a gravitational field just as the charges under the box in Fig. 22.12 create an electric field. If another mass, such as the mass of a skydiver, is placed in the earth's gravitational field, it experiences a gravitational force. If another charge, such as q' in Fig. 22.12, is placed in an electric field, it experiences an electric force.

An electric field is defined in terms of the force acting on a charge placed in the field rather than in terms of the charges causing the field:

An **electric field E** at a point in space is the ratio of the net electric force **F** acting on a small, positive test charge q' placed at that point, divided by the value of the test charge:

$$\mathbf{E} = \frac{\mathbf{F}}{q'}. \tag{22.4}$$

The units of electric field are newtons/coulomb.

The reason for dividing the force by q' is to define the electric field in a way that is independent of the size of the test charge. The electric field is a property of the space and not of the charge testing the field. For example, if the test charge is doubled, the electric force acting on it would also double. However, the ratio $2F/2q'$ remains the same. The electric field depends only on the charges causing the field and not on the test charge used to measure it. Although the field is defined to be independent of the charge q' that measures the field, q' can affect the field if its value is too large. The value of q' must be so small that it does not cause the charges that produce the field to move.

Equation (22.4) is often used to calculate the force on a charge q that is placed in a known electric field **E.** In this case, Eq. (22.4) becomes

$$\mathbf{F} = q\mathbf{E}.$$

This equation can be used to determine the force on both positive and negative charges. If q is positive, then **F** and **E** are in the same direction; if q is negative, **F** is in the opposite direction of **E.**

Two techniques are commonly used to calculate the electric field at a point in space. In the first, we calculate using vector addition the net electrical force $\mathbf{F}$ on a small test charge q' placed at that point and then use Eq. (22.4) to find $\mathbf{E}$. In more advanced treatments, calculus is used to add the forces caused by continuous charge distributions. The second technique for finding electric fields is based on a principle developed by the German mathematician and physicist Karl Friedrich Gauss (1777–1855). This principle (called Gauss's law) is beyond the scope of this book, although we will occasionally use the results of a Gauss's law determination of $\mathbf{E}$ in an example or problem. In Example 22.5 we use the first technique described above to determine the electric field at a point.

EXAMPLE 22.5 Calculate the magnitude and direction of the electric field at point A in Fig. 22.13a. The field is caused by the dipole charges q and $-q$ ($q = 0.50 \times 10^{-6}$ C) that are separated by 0.40 m.

SOLUTION To determine the electric field at point A, a small, positive test charge q' is placed there. The electric field is given by $\mathbf{E} = \mathbf{F}/q'$, where $\mathbf{F}$ is the resultant electric force on q'. We need not specify the exact charge of q', since its value cancels out of our final result. To determine $\mathbf{F}$, we use vector addition. First, draw a force diagram for q' at position A (Fig. 22.13b). Two forces act on q': $\mathbf{F}_{\text{pos}}$ is the repulsive force caused by $+q$, and $\mathbf{F}_{\text{neg}}$ is the attractive force caused by $-q$. The resultant force $\mathbf{F}$ is the sum of $\mathbf{F}_{\text{pos}}$ and $\mathbf{F}_{\text{neg}}$. The x component of the electric field is

$$E_x = \frac{F_x}{q'} = \frac{F_{\text{pos}} \cos 0° - F_{\text{neg}} \cos 0°}{q'} = \frac{F_{\text{pos}} - F_{\text{neg}}}{q'}.$$

and the y component is

$$E_y = \frac{F_y}{q'} = \frac{F_{\text{pos}} \sin 0°\,^{0} + F_{\text{neg}} \sin 0°\,^{0}}{q'} = 0.$$

The magnitudes of the two forces are found using Eq. (22.1):

$$F_{\text{pos}} = \frac{kqq'}{r^2} = \frac{(9.0 \times 10^9 \text{ N} \cdot \text{m}^2/\text{C}^2)(0.50 \times 10^{-6}\text{ C})q'}{(1.20 \text{ m})^2} = (3.13 \times 10^3\, q')\text{N/C},$$

$$F_{\text{neg}} = \frac{kqq'}{r^2} = \frac{(9.0 \times 10^9 \text{ N} \cdot \text{m}^2/\text{C}^2)(0.50 \times 10^{-6}\text{ C})q'}{(1.60 \text{ m})^2} = (1.76 \times 10^3\, q')\text{N/C}.$$

Thus,

$$E_x = \frac{F_x}{q'} = \frac{F_{\text{pos}} - F_{\text{neg}}}{q'} = \frac{(3.13 \times 10^3\, q' - 1.76 \times 10^3\, q')}{q'} = 1.37 \times 10^3 \text{ N/C}$$

and

$$E_y = \frac{F_y}{q'} = 0.$$

The magnitude of the electric field (a vector) is determined using Eq. (1.7):

$$E = \sqrt{E_x^2 + E_y^2} = \sqrt{(1.37 \times 10^3 \text{ N/C})^2 + (0)^2} = \underline{1.37 \times 10^3 \text{ N/C}}.$$

Since $E_y = 0$ and E_x is positive, the field points in the positive x direction.

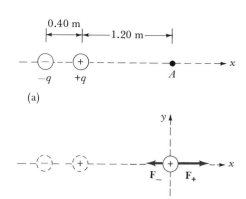

(a)

(b)

FIG. 22.13. (a) Dipole charges create an electric field at position A. (b) A force diagram for a small test charge placed at A.

EXAMPLE 22.6 An electron (charge -1.6×10^{-19} C) moves toward the screen of an oscilloscope (Fig. 22.14). On its way, it passes between two parallel plates. The upper plate has a positive charge and the lower plate has an equal negative charge. The electric field between the plates points downward and has a uniform magnitude of 1.0×10^3 N/C. Calculate the magnitude and direction of the electric force on the electron due to this charge distribution.

SOLUTION In this example, the electric field is given. The force is determined using Eq. (22.4). Because the electron has a negative charge, the force on the electron is opposite the direction of the electric field (the force points up). The magnitude of the force is

$$F = qE = (1.6 \times 10^{-19} \text{ C})(1.0 \times 10^3 \text{ N/C}) = 1.6 \times 10^{-16} \text{ N}. \qquad \blacksquare$$

FIG. 22.14. A downward electric field causes an upward force on an electron.

22.7 Lines of Force

Electric fields can be represented by drawing what are called lines of force. Lines of force were first used by an English scientist, Michael Faraday (1771–1867). Faraday, who had to leave school at thirteen to help support his destitute family, was self-educated and had no formal training in mathematics. Because of his lack of mathematical expertise, Faraday developed the notion of lines of force to help himself visualize the effects of electric and magnetic fields.

A **line of force** is a line drawn such that its direction at any point is in the direction of the electric field at that point. The separation of neighboring lines indicates the magnitude of the field in that region. If the lines are close to each other, the electric field in that region is relatively strong; if the lines are far apart, the field is weak. Lines of force always originate on a positive charge and end on a negative charge.

Several examples of lines of force are shown in Fig. 22.15. In (a) the lines

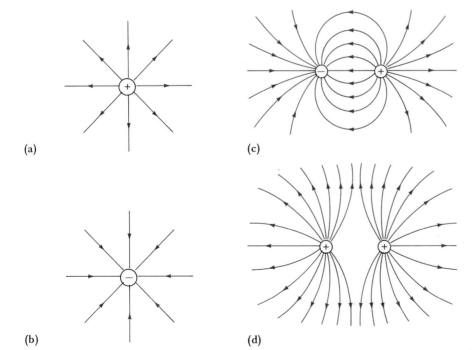

(a)

(b)

(c)

(d)

FIG. 22.15. Lines of force.

FIG. 22.16. Lines of force along a young woman's hair. (Photograph by John Womack)

leaving a single positive charge are shown. A small positive charge placed on one of the lines would feel an electric force in the direction of the line. A negative charge would feel a force in the opposite direction of the line.

Lines moving in toward a single negative charge are shown in (b). In (c), the lines of force caused by a dipole charge move from the positive to the negative charge. In (d), the lines of force leave two positive charges. The negative charges on which the lines end are far away and are not shown in the drawing.

The young woman shown in Fig. 22.16 has a large negative charge on her body and hair. The electric field (and lines of force caused by this charge) points in toward her head as in Fig. 22.15b. The force on the excess negative charge on her hair is opposite the direction of the electric field and causes her hair to point out away from her head.

22.8 Are Electric Fields Real?

If a positive electric charge sits on the desk one meter in front of you, is there actually a difference in the space between you and the charge because of its presence? Does some sort of "elastic band" stretch between you and the charge, ready to grab a negative charge entering the space or repel a positive charge?

One great virtue of the electric-field concept is that it removes the need to accept what Nobel physicist, P. W. Bridgeman, has called "that intellectual monstrosity, action at a distance." In the action-at-a-distance viewpoint (that is, using Coulomb's law), a charge at one place initiates a force that magically leaps across space, with no effect on that space. The force somehow comes to life again some distance away where it exerts a force on another charge. In the electric-field concept, the space between charges is viewed as the medium of a physical action that is passed from point to point until it finds its target, where it produces a force. The electric field can be imagined as a line of dominoes with a target at the end. If you push the first, it pushes the second, and so forth, until the last domino tips over and hits the target. In the same sense an electric field passes an electric force from point to point, eventually exerting a force on a distant charge.

The concept of an electric field is often used when calculating the effect of vibrating electric charges on other charges. Imagine, for instance, the motion of electrons in the antenna of a radio station. These electrons oscillate up and down at the broadcast frequency of the station. If, at one instant of time, an excess of electrons accumulates at the top (Fig. 22.17a), then the bottom of the antenna is left with a deficiency of electrons, and the antenna has a dipole charge distribution. If a free negative charge exists at some distance from the antenna, the dipole exerts a downward force on it, as shown in Fig. (a).

A short time later, the electrons in the antenna will have oscillated back down, thus reversing the orientation of the antenna's dipole. The force on the free electron also reverses direction (Fig. 22.17b). Thus, the free electron oscillates up and down at the same frequency as the electrons in the antenna. In the same manner the free electrons in a car radio antenna are forced to oscillate in the same pattern as those in the station's antenna.

This transfer of force does not occur instantaneously (as the action-at-a-distance view would lead us to think). There is a delay in the transfer of the

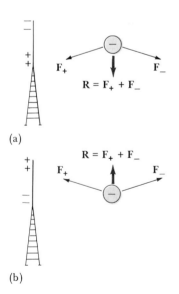

FIG. 22.17. The force on a negative charge some distance from a radio antenna oscillates up and down as the charge on the antenna oscillates.

force. It is as though the force propagates through space at a speed c equal to 3.0×10^8 m/s. We say that an electric field is created that moves at the speed of light away from the antenna. Consider Fig. 22.18. The dipole charge in the transmitting antenna shown in (a) creates an upward electric field in the space in front of it. This electrical distortion of space propagates away from the dipole at a speed c. Its effect on a distant charge is therefore delayed by the time needed for the field to travel to the distant charge. As this field moves away, the charge distribution on the antenna changes and produces a new field near the antenna. The moving electric field is called an **electromagnetic wave.** A charge some distance from the antenna oscillates up and down as the electromagnetic wave passes it. The charge bobs up and down like a boat on a rough sea.

We will be examining electromagnetic waves in more detail later on. For now, we should realize that electromagnetic waves are one important example of electric fields providing an accurate description of an electrical phenomenon.

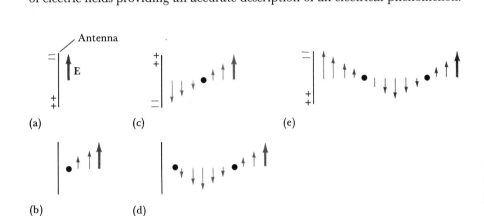

FIG. 22.18. A time sequence of the motion of an oscillating electric field leaving an antenna. The field moves at a speed of 3.0×10^8 m/s in air and is called an electromagnetic wave.

Summary

1. **Coulomb's law:** A proton has a positive electric charge e equal to 1.6×10^{-19} C; an electron has a negative charge $-e$. The magnitude of the force between a charge q_1 and a charge q_2 when separated by a distance r is given by Coulomb's law:

$$F = \frac{kq_1q_2}{Kr^2}, \qquad (22.2)$$

where $k = 9.0 \times 10^9$ N $\cdot$ m^2/C^2 and K is the dielectric constant of the material in which the charges are immersed. If the charges are in a vacuum or in dry air, $K = 1$; if in water, $K = 80$. The direction of the electrical force between two charges depends on the signs of the charges: like charges repel and unlike charges attract.

2. **Multiple charges:** When several electrical charges exert forces on one particular charge, the resultant electrical force is the vector sum of all the forces:

$$\mathbf{R} = \mathbf{F}_1 + \mathbf{F}_2 + \mathbf{F}_3 + \cdots .$$

The forces can be added graphically or by using the component technique (both techniques are explained in detail in Chapter 1):

$$R_x = F_{1x} + F_{2x} + F_{3x} + \cdots , \qquad (1.5)$$

$$R_y = F_{1y} + F_{2y} + F_{3y} + \cdots , \qquad (1.6)$$

$$R = \sqrt{R_x^2 + R_y^2}, \qquad (1.7)$$

$$\tan \theta = |R_y/R_x|. \qquad (1.8)$$

3. **Electric dipole:** An electric dipole is a positive charge q and a negative charge $-q$ of equal magnitude that are separated by a distance a. The **electric dipole moment p** points from $-q$ to q and has a magnitude

$$p = qa. \qquad (22.3)$$

4. **Electric field:** The electric field $\mathbf{E}$ at a point equals the electric force $\mathbf{F}$ acting on a small, positive test charge q' placed at that point, divided by the value of the charge:

$$\mathbf{E} = \frac{\mathbf{F}}{q'}. \qquad (22.4)$$

If a charge q is placed in a known electric field $\mathbf{E}$, the force on the charge is determined by rearranging Eq. (22.4):

$$\mathbf{F} = q\mathbf{E}.$$

Questions

1. "In all processes that occur in nature, electric charge is conserved." Explain what is meant by this statement and describe one particular example.

2. After a snowstorm or windstorm, a wire fence in the Plains states may occasionally give an intense shock to a person moving near it. Why might this happen?

3. An object acquires a positive electrical charge. Does its mass increase, decrease, or remain the same? Explain. How is its mass affected if it acquires a negative electrical charge?

4. At one time it was thought that electric charge was a weightless fluid. An excess of this fluid resulted in a positive charge; a deficiency resulted in a negative charge. Describe an experiment for which this hypothesis provides a satisfactory explanation. Describe an experiment that is not explained by this hypothesis.

5. Two charges exert a 4.0-N force on each other when separated by 1.0 m. Indicate the magnitude of the force when (a) the separation is doubled, (b) the separation is reduced by one half, (c) the magnitude of one charge is reduced by one half, and (d) both charges are doubled.

6. The electrical force of one electric charge on another is proportional to the products of their charges—that is, $F_q \propto q_1 q_2$. Invent an experiment to show that the force is not proportional to the sum of their charges ($q_1 + q_2$).

7. An electric dipole such as a water molecule is in a constant electric field. Will the force of the field cause the dipole to have a linear acceleration along a line in the direction of **E**? Explain. Will the field exert a torque on the dipole? Explain.

8. A dipole charge, such as pictured in Fig. 22.19, exists on your heart. A circular wire is placed on your chest. Where on the wire will positive charge accumulate and where will negative charge accumulate because of the force exerted on the charge by the heart's dipole? Explain.

FIG. 22.19

9. A metal sphere has no charge on it. A positive charge is brought near, but does not touch, the sphere. Show that the positive charge can exert a force on the sphere even though the sphere has no net charge. (*Hint:* Electrons move about freely on the surface of the metal.)

Problems

In some of these problems you may need to know that the mass of an electron is 9.11×10^{-31} kg.

22.1 Electric Charge

1. (a) The earth has an excess of 6.0×10^5 extra electrons on each square centimeter of surface. Calculate the electric charge in coulombs on each square centimeter of surface. (b) As you walk across a rug, 9.0×10^{-23} kg of electrons transfer to your body. Calculate the number of electrons and the total charge in coulombs on your body.

2. (a) A drop of rain accumulates a charge of -3.2×10^{-18} C. By how much does its mass increase? (b) A person standing in the wind loses 1.0×10^{-21} kg of electrons. Calculate the charge on the person after the electrons are lost.

22.2 Coulomb's Law

3. The membrane of a body cell has a positive ion of charge $+e$ on the outside wall and a negative ion of charge $-e$ on the inside wall. Calculate the magnitude of the electrical force between these ions if the membrane thickness is 0.80×10^{-9} m.

4. Calculate the electric force between two protons in the nucleus of a helium atom when separated by 2.0×10^{-15} m.

5. Calculate the number of electrons that must be transferred from the earth to the moon so that the electrical attraction between them is equal in magnitude to their gravitational attraction. What is the mass of this number of electrons?

6. Two metal spheres each with no charge are separated by one mile (1609 m). How much electric charge must be transferred from one sphere to the other so that their electrical attraction is 1.0×10^3 N?

7. Two protons are separated by 1.0×10^{-10} m. (a) What is the magnitude of the electrical repulsion force between them? (b) Use Newton's law of gravitation to find the magnitude of the gravitational attraction between the protons when separated by the same distance ($m_p = 1.67 \times 10^{-27}$ kg). (c) Calculate the ratio of these two forces.

8. Two metal spheres are held 2.0 m apart by a board that fractures when compressed by a force greater than 2.0×10^4 N. Each sphere starts with 10^{27} electrons and 10^{27} protons. Calculate the fraction of electrons that must be moved from one sphere to the other in order to fracture the board.

9. A man and woman each contain 4.0×10^{28} electrons and an equal number of protons. Calculate the electrical force between them if 1 percent of the woman's electrons are transferred to the man, who is 100 m away.

10. Two identical metal spheres are separated by 1.00×10^5 m. The spheres exert an 18.9-N repulsive force on each other. A wire is then connected between the spheres so that they become equally charged. The spheres now repel each other with a 22.5-N force. Calculate the original charges on the spheres.

22.3 Static Electric Forces

11. Two (1.0×10^{-3})-C charges are separated by 2.0 m. Calculate the force on a $-(2.0 \times 10^{-4})$-C charge that is along a line with the other charges and is 4.0 m from the nearest charge.

12. Three 1.0-C charges are equally spaced on a straight line. The separation of each charge from its neighbor is 100 m. Find the force on the center charge if (a) all charges are positive, (b) all charges are negative, and (c) the rightmost charge is negative and the other two positive.

13. Four ions (Na⁺, Cl⁻, Na⁺, Cl⁻) each separated from its neighbors by 3.0×10^{-10} m are in a row. The charge of a sodium ion is $+e$ and that of a chlorine ion is $-e$. Calculate the force on the chlorine ion at the end of the row due to the other three ions.

14. A 1.0-C charge and a 2.0-C charge are separated by 100 m. Where should a $-(1.0 \times 10^{-3})$-C charge be located on a line between the positive charges so that the net electrical force on the negative charge is zero?

15. Two charges, q and $4q$, are separated by 1.0 m. Determine the sign, magnitude, and position of a third charge that causes all three charges to remain in equilibrium.

16. Determine the magnitude and direction of the resultant force caused by charges q_1 and q_3 on charge q_2 (see Fig. 22.20).

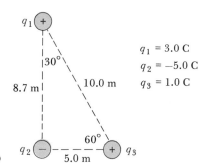

$q_1 = 3.0$ C
$q_2 = -5.0$ C
$q_3 = 1.0$ C

FIG. 22.20

17. Calculate the magnitude and direction of the resultant force caused by charges q_2 and q_3 on charge q_1 (see Fig. 22.20).

18. Calculate the resultant electrical force caused by charges q_1 and q_2 on charge q_3 (see Fig. 22.20).

19. Four (1.0×10^{-4})-C charges are located at the corners of a square whose sides are 2.0 m long. Calculate the net electrical force on the charge at the lower left corner due to the other three charges.

22.4 Dielectric Constant

20. Calculate the force between a sodium ion (charge e) and chlorine ion (charge $-e$) when separated by 2.8×10^{-10} m in air and when separated by the same distance in water.

21. Repeat Problem 3 using the fact that the dielectric constant of a cell membrane is 8.

22. A (2.0×10^{-6})-C positive charge and a $-(2.9 \times 10^{-6})$-C negative charge are separated by 0.10 m in water. Calculate the distance that the charges must be separated when in benzene so that the force between them is the same as when they were in water.

23. The chlorine ion shown in Fig. 22.21 has a charge $-e$ and the sodium ion a charge $+e$. Calculate the force acting on a hy-

droxide ion of charge $-e$ placed at position A. The charges are all immersed in water.

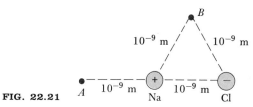

FIG. 22.21

24. An African fish called the *Gymnarchus niloticus* has a (1.0×10^{-7})-C positive charge at its head and a $-(1.0 \times 10^{-7})$-C negative charge at its tail. The charges are separated by 0.20 m. Calculate the magnitude and direction of the force on a hydroxide ion (charge $-e$) that is 0.10 m above the center of a line drawn between the two charges. Remember that the fish and charges are in water.

22.5 Electric Dipoles

25. Calculate the force on (a) a potassium ion (charge e) and (b) a hydroxide ion (charge $-e$) when placed at position A in Fig. 22.21. The charges are immersed in body tissue whose dielectric constant is 8.

26. Repeat Problem 25, only for position B in Fig. 22.21.

27. A shuttle train moves from one station to another. It is powered by equal-magnitude, opposite-sign charges on top of each station (see Fig. 22.22). If a 7000-N friction force opposes the train's initial motion, how great must the dipole charge be to induce an initial acceleration of 1.0 m/s²? The train's mass is 2.0×10^4 kg, and it has a (3.0×10^{-2})-C positive charge on its roof.

FIG. 22.22

28. The heart has a dipole charge distribution with a $+(1.0 \times 10^{-7})$-C charge separated by 4.0 cm from a $-(1.0 \times 10^{-7})$-C charge. The charges are on a horizontal line with the positive charge at the right side. Calculate the force on a sodium ion (charge e) in the tissue ($K = 8.0$) 6.0 cm above the center of a line drawn between the charges on the heart.

29. A geologist places two electrodes in the ground 800 m apart. An electric generator produces a dipole charge at the electrodes (the left electrode carries a charge of -2.0×10^{-4} C; the right electrode carries a charge of $+2.0 \times 10^{-4}$ C). Calculate the electric force acting on a hydroxide ion (charge $-e$) in the ground 300 m below the center of a horizontal line drawn from one electrode to the other. The dielectric constant of the earth is 8.0.

22.6 Electric Fields

30. The charges on the two electrodes described in Problem 29 produce an electric field in the earth. Calculate the magnitude and direction of the field halfway between the electrodes at a point just under the earth's surface.

31. Calculate the magnitude and direction of the electric field (a) 0.40 m from a $+(1.0 \times 10^{-3})$-C charge; (b) 0.80 m from a $+(1.0 \times 10^{-3})$-C charge; (c) 0.40 m from a $-(1.0 \times 10^{-3})$-C charge; and (d) 0.80 m from a $-(1.0 \times 10^{-3})$-C charge.

32. A uranium nucleus has 92 protons. (a) Calculate the magnitude of the electric field at a distance of 0.58×10^{-12} m from the nucleus (about the radius of the innermost electron orbiting the nucleus). (b) What is the magnitude of the force on an electron due to this electric field?

33. Calculate the electric field (magnitude and direction) at points A and B in Fig. 22.21. Assume the medium has a dielectric constant of 1.0.

34. Four 1.0-C charges are placed at the corners of a square whose sides are 50 m long. (a) Calculate the electric field at the center of the square (if you wish, you may use a verbal symmetry argument to justify your answer). (b) Repeat the calculations in (a), only for a square with two negative 1.0-C charges at the top and two positive 1.0-C charges at the bottom.

35. An electron in the tube of a television set starts with zero speed at the edge of a negatively charged filament. What must be the average electric field between this filament and the screen of the television set if the electron acquires a speed of 2.0×10^6 m/s in a distance of 20 cm? (*Hint:* This problem involves the use of kinematics and dynamics.)

36. A sphere weighing 10 N floats in the uniform electric field above a metal plate. The field points up and has a magnitude of 250 N/C. Calculate the magnitude and sign of the charge on the sphere.

37. An invader from a planet in the galaxy Maffei 1 shoots electrons from her ray gun. The electrons travel at a speed of 8.0×10^5 m/s. An electric-field shield is built to stop the electrons. Determine the average magnitude of an electric field that will stop the electrons in a distance of 12 cm. Does the field point toward or away from the person holding it? Explain.

38. An electron passes between two parallel plates as shown in Fig. 22.14. A 0.20-N/C electric field points down. The electron moves a horizontal distance of 4.0 cm in the field and has an initial velocity in the horizontal direction of 5.0×10^4 m/s. Calculate the change in vertical position of the electron as it leaves the 4.0-cm horizontal region.

22.7 Lines of Force

39. (a) Indicate, approximately, the direction of the net electric force on a small, positive charge placed at positions A, B, and C in Fig. 22.23. (b) Continue along the directions of these vector arrows and draw lines of force in the directions they are expected to move beyond points A, B, and C.

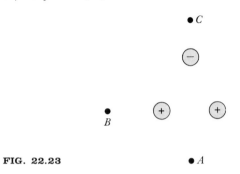

FIG. 22.23

40. Repeat the procedure outlined in Problem 39 for the charges shown in Fig. 22.24.

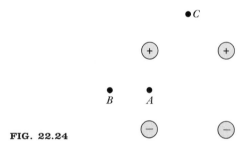

FIG. 22.24

General Problems

G1. Suppose that the electron in a hydrogen atom moves in a circular orbit of radius 0.53×10^{-10} m about the proton nucleus. (a) Calculate the magnitude of the electrical force between the charges. (b) Calculate the electron's centripetal acceleration and speed.

G2. Suppose that the earth and moon initially have zero charge. Then 1000 kg of electrons are transferred from the earth to the moon. Calculate the radius of a stable moon orbit when both the electrical and gravitational forces of attraction are acting on the moon and it completes one rotation about the earth in 29.5 days.

G3. Calculate the electrical torque on a water molecule about a point at its center when the molecule is oriented perpendicular to a 500-N/C electric field. Consider the molecule as a charge $+e$ and another charge $-e$ separated by a distance a equal to 0.38×10^{-10} m.

G4. A tiny drop of water has an excess of 50 electrons. The drop remains stationary if placed in a 50-N/C downward electric field. Calculate the mass and radius of the drop.

G5. An electron is projected with a horizontal velocity of magnitude v_0 into a constant downward electric field $\mathbf{E}$ between two parallel plates. Show that the equation of its trajectory while in the field is

$$y = \left(\frac{eE}{2mv_0^2}\right)x^2,$$

where m is the electron's mass and y is its vertical displacement when its horizontal displacement in the field is x.

G6. An atom normally has no electric dipole moment. However, when placed in an electric field, the nucleus is displaced in the direction of the field and the electrons are displaced opposite the field direction. The induced electric dipole moment is $\mathbf{p} = \alpha\mathbf{E}$, where α is called the **polarizability** of the atom. An atom with polarizability of $1.7 \times 10^{-41}\,\mathrm{C^2 \cdot m/N}$ is placed 6.0×10^{-8} m from a charge $+e$. Calculate the induced electric dipole moment.

Electrical Energy 23

In the last chapter we looked at the force that one or more electric charges exert on another charge. These electric forces, along with other types of forces, can be used in Newton's second law to calculate the acceleration of an object [$\mathbf{a} = (\Sigma\,\mathbf{F})/m$]. The force-acceleration approach to solving problems was introduced and used in Part I of the book but is seldom used in problems dealing with electricity. Instead, the energy approach introduced in Chapter 8 is preferred.

The energy problem-solving technique is based on the work-energy equation ($W = \Delta E_{\text{system}}$) or on the more general energy-conservation equation ($Q + W = \Delta E_{\text{system}}$) for problems that involve heat transfer. The expression ΔE_{system} represents the sum of the changes in the many different forms of energy (kinetic, gravitational potential, elastic potential, and thermal) that a system of particles can have. In this chapter we will introduce another form of energy to include with ΔE_{system}—the change in electrical potential energy that occurs when a group of charged particles moves closer or farther from each other. After learning how to calculate the change in electrical potential energy of a group of charges, we will use the work-energy approach to solve problems involving electric charges.

23.1 Electrical Potential Energy (PE_q)

In Chapter 8, we found that the amount of potential energy in a system always depends upon the relative position of the parts of that system. For example, a mass m and the earth have more gravitational potential energy (PE_g) when separated by several meters than when resting against each other, and a spring has more elastic potential energy (PE_s) when its coils are stretched or compressed than when relaxed. Similarly, two or more electric charges possess electrical potential energy (PE_q) and the amount depends on their positions relative to each other.

Consider the cannon in Fig. 23.1a. Two positive charges are held near each other in the barrel of the cannon. When the trigger is released, the positively charged cannonball is repelled from the other positive charge causing it to fly out the end of the barrel (Fig. 23.1b). The initial electrical potential energy [PE_q in (a)] is converted to kinetic energy (KE) and gravitational potential energy (PE_g) as the charges move apart. The situation shown in (a) is similar to that of

(a)

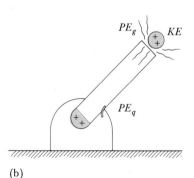

(b)

FIG. 23.1. (a) Two positive charges have considerable electrical potential energy when held near each other—like a compressed spring. (b) The electrical energy is converted to other energy forms as the charges move apart.

(a)

(b)

FIG. 23.2. (a) Opposite charges have considerable electrical potential energy when held far apart—like a stretched spring. (b) The electrical energy is converted to other forms as the charges move closer.

a compressed spring; instead of the coils of the spring being squeezed together, two like charges are squeezed or forced together. The *electrical potential energy of like charges increases as they are forced nearer each other and decreases as they move apart.*

Next, imagine the electrical potential energy of the two opposite-sign charges used as a nutcracker in Fig. 23.2. When separated by considerable distance (a), the opposite charges have more electrical potential energy. When released, the negatively charged block rushes toward the positive charge (b). The kinetic energy of the block increases and the electrical potential energy of the charges decreases. We see that the electrical potential energy of unlike charges increases as they are pulled apart and decreases as they move together.

To derive an expression for the change in electrical potential energy of two electric charges whose separation changes, let us return to the conservation-of-energy principle:

$$Q + W = \Delta E_{\text{system}}, \tag{8.8}$$

where Q is the heat flow into or out of the system, and W is the work done on objects in the system by forces that act from outside the system. The system's energy can change in many ways:

$$Q + W = \overbrace{\Delta KE + \Delta PE_g + \Delta PE_s + \Delta PE_q + \Delta E_{th} + \cdots}^{\Delta E_{\text{system}}}, \tag{8.9}$$

where ΔKE is the system's change in kinetic energy; ΔPE_g, ΔPE_s, and ΔPE_q are its changes in gravitational, elastic, and electrical potential energies, respectively; and ΔE_{th} is its change in internal thermal energy.

To derive an expression for calculating ΔPE_q, we will choose a hypothetical system so that when work is done on this system, only its electrical potential energy changes. The system shown in Fig. 23.3 consists only of charges q_1 and q_2, initially separated by a distance r_0. The like charges repel each other with an electrical force whose magnitude is given by Eq. (22.1): $F = kq_1q_2/r_0^2$. A person can prevent charge q_1 from moving away from q_2 by pushing q_1 with an equal-magnitude force $\mathbf{F}_0$ toward q_2 (Fig. 23.3a).

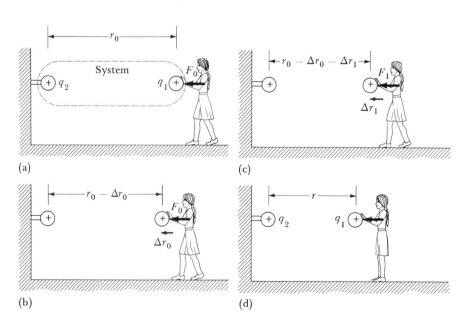

(a)

(b)

(c)

(d)

FIG. 23.3. A person does work in pushing two charges closer together.

If the person pushes q_1 just a tiny bit harder, the charge can be displaced a small distance Δr_0 toward q_2 (Fig. 23.3b). The work ΔW_0 done by the pushing force during this small displacement can be calculated using Eq. (8.1):

$$\Delta W_0 \simeq F_0\, \Delta r_0 = \frac{kq_1q_2}{r_0^2}\, \Delta r_0.$$

The calculation is approximate since the force needed to push q_1 changes as its separation from q_2 changes. If Δr_0 is small, the equation is a good approximation of the work done during that small displacement. If we wish to push the charge farther, however, more force must be exerted to move q_1 since it is now closer to q_2. The work done during the next small displacement Δr_1 shown in Fig. 23.3c is approximately

$$\Delta W_1 \simeq F_1\, \Delta r_1 = \frac{kq_1q_2}{(r_0 - \Delta r_0)^2}\, \Delta r_1.$$

For each step closer, the amount of work needed to move q_1 closer to q_2 increases.

The total work done in moving the charge from an initial separation r_0 to a final separation r (Fig. 23.3d) equals the sum of the work for each small step. This type of infinitesimal addition is done easily using calculus. The results of such a calculation are

$$W = \Delta W_1 + \Delta W_2 + \Delta W_3 + \cdots = \frac{kq_1q_2}{r} - \frac{kq_1q_2}{r_0} = kq_1q_2\left(\frac{1}{r} - \frac{1}{r_0}\right).$$

At each step there is no acceleration and, therefore, no change in kinetic energy; nor are there changes in gravitational, elastic, or thermal energies. Also no heat flows into or out of the system. Only the electrical potential energy changes because the charges are separated by less distance. By substituting for W into Eq. (8.9),

$$Q + W = \Delta KE + \Delta PE_g + \Delta PE_s + \Delta PE_q + \Delta E_{th} + \cdots,$$

we find that

$$0 + kq_1q_2\left(\frac{1}{r} - \frac{1}{r_0}\right) = 0 + 0 + 0 + \Delta PE_q + 0 + \cdots.$$

The above equation applies to any situation in which two charges in a system move closer to each other or farther apart.

The change in **electrical potential energy** ΔPE_q when the separation of two charges, q_1 and q_2, is changed from an initial separation r_0 to a final separation r is given by the equation

$$\Delta PE_q = kq_1q_2\left(\frac{1}{r} - \frac{1}{r_0}\right). \tag{23.1}$$

The above equation applies for charges in a vacuum. If the charges are in a medium with dielectric constant K, the force between the charges is reduced by a factor, $1/K$. The change in electrical potential energy is then

$$\Delta PE_q = \frac{kq_1q_2}{K}\left(\frac{1}{r} - \frac{1}{r_0}\right). \tag{23.2}$$

Electrical potential energy is measured in units of joules.

Equations (23.1) and (23.2) can be used for both positive and negative changes; *the sign of the charges must be included in calculations involving these two equations.* Notice that electrical potential energy is proportional to $1/r$ and not to $1/r^2$ as in Coulomb's force law.

EXAMPLE 23.1 An electron in a hydrogen atom is originally in an orbit separated from the proton nucleus by a distance r_0 of 2.12×10^{-10} m. Calculate the change in electrical potential energy when the electron moves into a closer orbit separated from the proton nucleus by 0.53×10^{-10} m.

SOLUTION Remember from Chapter 22 that the charge of an electron is -1.6×10^{-19} C (q_1) and that of a proton is $+1.6 \times 10^{-19}$ C (q_2). The change in electrical potential energy is given by Eq. (23.1):

$$\Delta PE_q = kq_1q_2\left(\frac{1}{r} - \frac{1}{r_0}\right)$$
$$= (9.0 \times 10^9 \text{ N} \cdot \text{m}^2/\text{C}^2)(-1.6 \times 10^{-19} \text{ C})(+1.6 \times 10^{-19} \text{ C})$$
$$\times \left(\frac{1}{0.53 \times 10^{-10} \text{ m}} - \frac{1}{2.12 \times 10^{-10} \text{ m}}\right)$$
$$= \underline{-3.3 \times 10^{-18} \text{ J}}.$$

Notice that both the negative sign for the electron charge and the positive sign for the proton charge are included in the calculation. We see that the electrical potential energy of unlike charges decreases as they move closer to each other.

■

EXAMPLE 23.2 A sodium ion (charge e) and chlorine ion (charge $-e$) are initially separated by a distance r_0 equal to 3.5×10^{-10} m. When they are dissolved in water, their separation r increases to 100×10^{-10} m. Compare the changes in electrical potential energy required for this change in separation when (a) in air ($K = 1.0$) and (b) in water ($K = 80$).

SOLUTION (a) Using Eq. (23.2), we find that the change in electrical potential energy when in air is

$$\Delta PE_q = \frac{kq_1q_2}{K}\left(\frac{1}{r} - \frac{1}{r_0}\right)$$
$$= \frac{(9.0 \times 10^9 \text{ N} \cdot \text{m}^2/\text{C}^2)(1.6 \times 10^{-19} \text{ C})(-1.6 \times 10^{-19} \text{ C})}{1.0}$$
$$\times \left(\frac{1}{100 \times 10^{-10} \text{ m}} - \frac{1}{3.5 \times 10^{-10} \text{ m}}\right)$$
$$= \underline{640 \times 10^{-21} \text{ J}}.$$

(b) In water, the energy change is $\frac{1}{80}$ the change when in air:

$$\Delta PE_q = \frac{1}{80}(640 \times 10^{-21} \text{ J}) = \underline{8.0 \times 10^{-21} \text{ J}}.$$

Approximately eighty times more energy is required to pull a sodium and chlorine ion apart in air than in water. This is the main reason why table salt (NaCl) remains crystalline in air but dissolves (separates into ions) in water.

■

23.2 Electrical Potential Energy Involving Multiple Charges

If one particular electric charge q' moves relative to several other stationary charges, the total change in electrical potential energy is the sum of the changes caused by each charge. For example, charge q', pictured in Fig. 23.4a, is initially separated from three other charges, q_1, q_2, and q_3, by distances r_{1_0}, r_{2_0}, and r_{3_0}, respectively. When q' is moved to a new location (Fig. 23.4b), its separations from these charges are now r_1, r_2, and r_3. The total change in electrical potential energy is the sum of the changes in energy caused by each charge:

$$PE_q = \frac{kq_1q'}{K}\left(\frac{1}{r_1} - \frac{1}{r_{1_0}}\right) + \frac{kq_2q'}{K}\left(\frac{1}{r_2} - \frac{1}{r_{2_0}}\right) + \frac{kq_3q'}{K}\left(\frac{1}{r_3} - \frac{1}{r_{3_0}}\right).$$

The above equation is usually written in an abbreviated form as

$$PE_q = \sum_{i=1}^{3} \frac{kq_iq'}{K}\left(\frac{1}{r_i} - \frac{1}{r_{i_0}}\right), \tag{23.3}$$

where the symbol Σ (sigma) stands for the sum of the expression following it for each possible value of i. For the example illustrated in Fig. 23.4, i can assume three values ($i = 1, 2,$ and 3); we add the changes in electrical potential energy that q' has because of its change in position relative to q_1, q_2, and q_3. If q' moves relative to five other charges, then the summation would include five terms.

(a)

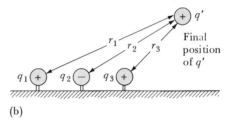

(b)

FIG. 23.4. A charge q' moves from (a) an initial position to (b) a final position.

EXAMPLE 23.3 Calculate the change in electrical potential energy when a 10-μC charge (q') is brought from infinity to position B, shown in Fig. 23.5.

SOLUTION The 10-μC charge moves relative to three fixed charges. The known information is shown in Fig. 23.5. Because q' starts infinitely far from all three charges, $r_{1_0} = r_{2_0} = r_{3_0} = \infty$. The change in electrical potential energy is calculated using Eq. (23.3):

$$\Delta PE_q = kq_1q'\left(\frac{1}{r_1} - \frac{1}{\infty}\right) + kq_2q'\left(\frac{1}{r_2} - \frac{1}{\infty}\right) + kq_3q'\left(\frac{1}{r_3} - \frac{1}{\infty}\right)$$

$$= \frac{9.0 \times 10^9 \, (6.0 \times 10^{-6})(10 \times 10^{-6})}{1.4}$$

$$+ \frac{9.0 \times 10^9 \, (-5.0 \times 10^{-6})(10 \times 10^{-6})}{1.0}$$

$$+ \frac{9.0 \times 10^9 \, (4.0 \times 10^{-6})(10 \times 10^{-6})}{1.4}$$

$$= 0.39 - 0.45 + 0.26 = \underline{0.20 \text{ J}}.$$

Notice that when the positive charge q' moved closer to the negative charge q_2, a -0.45-J energy change occurred; whenever opposite charges move closer, the electrical potential energy decreases. This decrease was more than balanced by the positive increases in PE_q as q' moved closer to q_1 and q_3 (PE_q increases as like charges move closer). Since the net electrical potential energy increased, we can assume that some external agent, not pictured, did work to push q' to position B.

FIG. 23.5

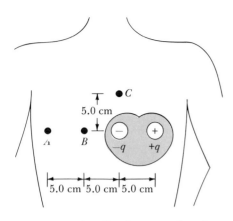

FIG. 23.6. A sodium ion moves through the body tissue.

EXAMPLE 23.4 A sodium ion (charge q' equal to $+e$) moves through body tissue from position A to position B shown in Fig. 23.6. Calculate the change in electrical potential energy of the sodium ion due to its change in position relative to the dipole charge on the heart ($q = 4.0 \times 10^{-6}$ C). The dielectric constant of the body is about 8.

SOLUTION In this example, Eq. (23.3) has two terms, one for the change in energy of the sodium ion relative to charge $+q$ and another for its change relative to $-q$. The sodium ion starts 0.15 m from charge $+q$ and 0.10 m from charge $-q$ and ends 0.10 m from $+q$ and 0.05 m from $-q$. The change in electrical potential energy is

$$\Delta PE_q = \frac{k(+q)q'}{K}\left(\frac{1}{r_+} - \frac{1}{r_{+_0}}\right) + \frac{k(-q)q'}{K}\left(\frac{1}{r_-} - \frac{1}{r_{-_0}}\right)$$

$$= \frac{(9.0 \times 10^9)(+4.0 \times 10^{-6})(1.6 \times 10^{-19})}{8}\left(\frac{1}{0.10} - \frac{1}{0.15}\right)$$

$$+ \frac{(9.0 \times 10^9)(-4.0 \times 10^{-6})(1.6 \times 10^{-19})}{8}\left(\frac{1}{0.05} - \frac{1}{0.10}\right)$$

$$= 2.4 \times 10^{-15}\,\text{J} - 7.2 \times 10^{-15}\,\text{J} = \underline{-4.8 \times 10^{-15}\,\text{J}}. \qquad \blacksquare$$

In the last example we found that a sodium ion would have less electrical potential energy at position B than at A. In moving from A to B, the sodium ion resembles a rock rolling down a hill; the rock's gravitational potential energy decreases, as does the electrical potential energy of the sodium ion.

It is natural for systems to move toward states in which their potential energy is minimized. A rock rolls down a hill, not up. A sodium ion moves from A to B—that is, in the direction of decreasing electric potential energy rather than from B to A. This tendency can be explained in two ways. (1) Forces responsible for potential energy push objects in a direction that minimizes their potential energy. For example, the net electrical force on a sodium ion if at A in Fig. 23.6 is toward B and in the direction of decreasing potential energy. (2) Our analysis of entropy in Chapter 11 indicated that nature moves toward the most probable states. In Example 23.4, as the sodium ion moves, it collides with atoms and molecules along its path causing thermal energy to be produced. The situation in which a large number of molecules have a small amount of thermal energy is more probable than one in which the sodium ion has all the energy in the form of electrical potential energy.

23.3 Work-Energy Calculations

Now that we have learned how to calculate changes in electrical potential energy, we will use the work-energy technique developed in Chapter 8 to solve some problems. We will ignore problems involving heat flow; thus, $Q = 0$. The work-energy equation with electrical potential energy included becomes

$$W = \Delta KE + \Delta PE_g + \Delta PE_s + \Delta PE_q + \Delta E_{th} + \cdots, \qquad (8.10)$$

where

$$W = F\,\Delta r \cos\theta, \qquad (8.1)$$

$$\Delta KE = \frac{1}{2} mv^2 - \frac{1}{2} mv_0^2, \tag{8.4}$$

$$\Delta PE_g = mg(h - h_0), \tag{8.6}$$

$$\Delta PE_s = \frac{1}{2} kx^2 - \frac{1}{2} kx_0^2, \tag{8.7}$$

$$\Delta PE_q = \sum_{i=1}^{N} kq_i q' \left(\frac{1}{r_i} - \frac{1}{r_{i_0}} \right), \tag{23.3}$$

$$\Delta E_{th} = F_k \, \Delta x. \tag{8.12}$$

You should review Chapter 8 for the notation or meaning of these different expressions. To use Eq. (8.10) we first calculate the work done on the system and the known energy changes as the system moves from an initial to a final state; then, we substitute these known quantities into the work-energy equation to calculate one unknown quantity.

EXAMPLE 23.5 An electron (charge $-e$) starts at rest at an infinite distance r_0 from a stationary proton (charge e). Calculate the speed of the electron after it moves unimpeded to a distance r of 10^{-9} m from the proton. The mass of an electron is 9.11×10^{-31} kg.

SOLUTION In the problem there are no springs ($\Delta PE_s = 0$), no friction forces ($\Delta E_{th} = 0$), and no work is done ($W = 0$). The change in gravitational potential energy of the electron with respect to the proton is very small and can be ignored ($\Delta PE_g = 0$). We must account for only the following energy changes:

$$\Delta KE = \frac{1}{2} mv^2 - \frac{1}{2} mv_0^2 = \frac{1}{2} (9.11 \times 10^{-31} \, \text{kg})v^2 - 0 \text{ (note that } v_0 = 0),$$

$$\Delta PE_q = kq_1 q_2 \left(\frac{1}{r} - \frac{1}{r_0} \right) = (9 \times 10^9 \, \text{N} \cdot \text{m}^2/\text{C}^2)(-1.6 \times 10^{-19} \, \text{C})$$

$$\times (1.6 \times 10^{-19} \, \text{C}) \left(\frac{1}{10^{-9} \, \text{m}} - \frac{1}{\infty} \right) = -2.3 \times 10^{-19} \, \text{N} \cdot \text{m}.$$

Substituting into the work-energy equation, we find that

$$0 = \frac{1}{2} (9.11 \times 10^{-31} \, \text{kg})v^2 + 0 + 0 - 2.3 \times 10^{-19} \, \text{N} \cdot \text{m} + 0.$$

Rearranging to solve for v, we find that

$$v^2 = \frac{2(2.3 \times 10^{-19} \, \text{N} \cdot \text{m})}{(9.11 \times 10^{-31} \, \text{kg})} = 5.05 \times 10^{11} \left(\frac{m^2}{s^2} \right)$$

or

$$v = \underline{7.1 \times 10^5 \, \text{m/s}}. \qquad \blacksquare$$

EXAMPLE 23.6 A soapbox derby race car, shown in Fig. 23.7, has a $-(2.3 \times 10^{-4})$-C negative charge (q_1) under its hood. This charge is attracted by a 0.30-C positive charge (q_2) placed beyond the end of the racetrack. Calculate the speed v of the racecar and racer (total mass of 113 kg) at the end of the race. A 40-N friction force opposes the motion of the car. The dielectric constant of air is 1.0.

FIG. 23.7

(a)

(b)

SOLUTION Our system consists of everything in Fig. 23.7. No work is done on the system by forces from outside the system, and there are no springs, so $\Delta PE_s = 0$. The changes in the other forms of energy are:

$$\Delta KE = \frac{1}{2} mv^2 - \frac{1}{2} mv_0^2 = \frac{1}{2} (113 \text{ kg})v^2,$$

$$\Delta PE_g = mg(y - y_0) = (113 \text{ kg})(9.8 \text{ m/s}^2)(0 - 28 \text{ m}) = -3.1 \times 10^4 \text{ J},$$

$$\Delta PE_q = \frac{kq_1 q_2}{K} \left(\frac{1}{r} - \frac{1}{r_0} \right)$$

$$= \frac{(9.0 \times 10^9 \text{ N} \cdot \text{m}^2/\text{C}^2)(-2.3 \times 10^{-4} \text{ C})(+0.30 \text{ C})}{1.0} \left(\frac{1}{30 \text{ m}} - \frac{1}{330 \text{ m}} \right)$$

$$= -1.9 \times 10^4 \text{ J},$$

$$\Delta E_{th} = F_k \Delta x = (40 \text{ N})(300 \text{ m}) = 1.2 \times 10^4 \text{ J}.$$

When these values are substituted in the work-energy equation to solve for v,

$$0 = \frac{1}{2}(113 \text{ kg})v^2 - (3.1 \times 10^4 \text{ J}) + 0 - (1.9 \times 10^4 \text{ J}) + (1.2 \times 10^4 \text{ J}),$$

we find that

$$v^2 = \frac{2[3.1 \times 10^4 \text{ J}) + (1.9 \times 10^4 \text{ J}) - (1.2 \times 10^4 \text{ J})]}{113 \text{ kg}} = 673 \text{ m}^2/\text{s}^2,$$

$$v = \underline{26 \text{ m/s}}.$$

23.4 Electrical Potential Difference and Potential

We learned in Sections 23.1 and 23.2 that the electrical potential energy of a system changes if one electric charge q' moves relative to other stationary charges. Equation (23.3) is used to calculate the change in electrical potential energy caused by the altered position of q' relative to the stationary charges.

This situation can be treated in another way using the concept of electrical potential difference. Instead of considering the motion of q' relative to other charges, we think of q' as moving in a space that has electrical hills and valleys that are created by the other charges. *Electrical potential difference* is a measure of the change in elevation of these electrical hills or valleys. If q' is allowed to move freely in this space, it moves toward a region where its electrical potential energy decreases (like a boulder rolling down a hill toward a region of lower gravitational potential energy).

The quantity *electrical potential difference* is defined to reflect the change in electrical energy a particle undergoes when moving between different points in space. The definition is made in terms of the space rather than in terms of the stationary charges causing the space to be modified.

The **electrical potential difference** V_{AB} from point A to point B is defined as the change in electrical potential energy $\Delta PE_{q_{A \to B}}$ that occurs when a small, positive charge q' moves from A to B, divided by the value of q':

$$V_{AB} = \frac{PE_{q_B} - PE_{q_A}}{q'} = \frac{\Delta PE_{q_{AB}}}{q'}. \tag{23.4}$$

The unit of electrical potential difference is the **volt,** where one volt is the change in electrical potential difference that occurs when a one-coulomb charge experiences a change in electrical potential energy of one joule. We see that

$$1 \text{ volt} = 1 \frac{\text{joule}}{\text{coulomb}} \quad \text{or} \quad 1 \text{ V} = 1 \frac{\text{J}}{\text{C}}.$$

We often use the less formal expression *voltage change* or simply *voltage* when referring to the potential difference between two points. For example, the potential difference or voltage across the terminals of a car battery is 12 V. For a positive charge at the negative terminal of the car battery, the positive terminal looks like an electrical hill that is 12 V high.

Electronic instruments, called voltmeters, are used to measure the voltage between two different points. For example, voltmeters of a special type can measure the potential differences between different parts of your body caused by the dipole change of your heart; geologists use voltage measurements to probe the material in the earth's crust; and technicians make voltage measurements between different points in electronic circuits of television sets, radios, and other electronic instruments.

Since voltage is measured so routinely, Eq. (23.4) is often used to calculate the change in electrical potential energy ΔPE_q that occurs when a charge q moves through a known change in voltage V_{AB}:

$$\Delta PE_{q_{AB}} = qV_{AB}.$$

This equation is usually written in a more abbreviated form as

$$\Delta PE_q = qV, \tag{23.5}$$

where V is the potential difference between the final and initial positions for which the energy change is calculated. The equation can be used for positive or negative charges. The sign of the charge must be included when using the equation.

In summary, we should think of voltage, or potential difference, as an indicator of the difference in the electrical elevation of two points as seen by a positive charge; the higher the voltage, the higher the electrical hill. A positive charge tends to roll down an electrical hill toward regions of lower voltage. However, a negative charge tends to move in the opposite direction toward regions of higher voltage where its electrical potential energy, according to Eq. (23.5), becomes less.

EXAMPLE 23.7 The potential difference V_{AB} from an electron source (called a cathode) of an x-ray tube to the positively charged anode is $+40,000$ volts (Fig. 23.8). If an electron begins at rest ($v_0 = 0$) and moves toward the anode, calculate its speed v just before reaching the anode. The mass m of an electron is 9.11×10^{-31} kg and its charge q is -1.6×10^{-19} C.

SOLUTION In the system shown in Fig. 23.8, no external forces of significance act on the system. Therefore, $W = 0$. The only energy changes are the electrical potential energy change due to the electron moving across the tube and the corresponding change in the electron's kinetic energy. The energy changes are calculated using Eqs. (23.5) and (8.5):

$$\Delta PE_q = qV = (-1.6 \times 10^{-19}\,\text{C})(40,000\,\text{V})$$
$$= -6.4 \times 10^{-15}\,\text{C} \cdot \text{V} = -6.4 \times 10^{-15}\,\text{J},$$

$$\Delta KE = \frac{1}{2}mv^2 - \frac{1}{2}mv_0^2 = \frac{1}{2}(9.11 \times 10^{-31}\,\text{kg})v^2.$$

These quantities are substituted into the work-energy equation [Eq. (8.10)], which is then solved for the final speed v:

$$0 = \frac{1}{2}(9.11 \times 10^{-31}\,\text{kg})v^2 + 0 + 0 - (6.4 \times 10^{-15}\,\text{J}) + 0,$$

$$v^2 = \frac{2(6.4 \times 10^{-15}\,\text{J})}{(9.11 \times 10^{-31}\,\text{kg})} = 1.4 \times 10^{16}\,\text{m}^2/\text{s}^2,$$

$$v = \underline{1.2 \times 10^8\,\text{m/s}}.$$

The electron is traveling at almost 0.4 times the speed of light! Whenever masses move at speeds of the order of 0.1 or more times the speed of light, theories of relativity should be used to calculate their speed. Our calculation is only an approximation; nevertheless, we can see from our calculations that the electron (a negative charge) loses electrical potential energy as it moves toward higher voltage. The loss of electrical energy is balanced by an increase in kinetic energy. ∎

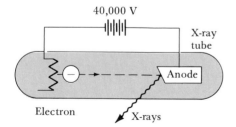

FIG. 23.8. An x-ray tube. When electrons are accelerated through a 40,000-V potential difference and crash into the anode, x-rays are produced.

EXAMPLE 23.8 An electric motor draws -260 C of electric charge (q) through a potential difference of $+120$ V. (a) Calculate the change in electric potential energy. (b) If 20 percent of this energy decrease results in the production of thermal energy, how much work can the motor do?

SOLUTION (a) The change in electrical potential energy is calculated using Eq. (23.5):

$$\Delta PE_q = qV = (-260 \text{ C})(120 \text{ V}) = \underline{-3.1 \times 10^4 \text{ J}}.$$

(b) If 20 percent of this energy is converted to thermal energy, then the change in thermal energy is

$$\Delta E_{th} = 0.20(3.1 \times 10^4 \text{ J}) = 0.6 \times 10^4 \text{ J}.$$

Substituting the values of ΔPE_q and ΔE_{th} into the work-energy equation, we can then calculate the work W done by the motor:

$$W = \Delta KE + \Delta PE_g + \Delta PE_s + \Delta PE_q + \Delta E_{th},$$
$$= 0 + 0 + 0 - (3.1 \times 10^4 \text{ J}) + (0.6 \times 10^4 \text{ J}) = \underline{-2.5 \times 10^4 \text{ J}}.$$

The negative sign means that the motor loses energy by doing work; that is, it does 2.5×10^4 J of work on the environment. The environment gains the 2.5×10^4 J of energy that the motor loses. ∎

Absolute Potential

Potential difference is an indicator of the difference in electrical elevation of two points. We are usually interested in the electrical potential difference between two points in space, but in some situations it is convenient to ascribe an absolute value to the electrical potential at each point in space. For example, one point in an electrical circuit is often called the **ground point.** The voltage at that point is defined as zero. **Absolute potential** is the value of the voltage at other points in the electrical circuit relative to this reference point of zero potential.

The word *ground* is used because in many electrical instruments and appliances the ground point actually connects via a wire to a pipe that passes under the earth. The voltage at the ground point is exactly the same as that of the earth outside a house. Since pipes in the ground maintain a constant voltage all over a city, it is a good reference level for electrical potential difference measurements.

The reference point used for potential difference measurements and calculations is completely arbitrary. For instance, when describing the electrical potential at different points surrounding the nucleus of an atom, it is common to designate a point infinitely far from the nucleus as a point of zero absolute potential or zero voltage. The absolute potential or voltage at other points relative to this reference point is determined using Eqs. (23.1) and (23.4) in the following manner. If another point is separated by a distance r from a nucleus with charge q, its potential relative to the zero-voltage reference point at infinity is

$$V = \frac{PE_{q_r} - PE_{q_\infty}}{q'} = \frac{\dfrac{kqq'}{r} - \dfrac{kqq'}{\infty}}{q'} = \frac{\dfrac{kqq'}{r} - 0}{q'} = \frac{kq}{r}. \qquad \textbf{(23.6)}$$

An electron that is infinitely far from the nucleus of charge q is at a position where the absolute voltage is kq/∞, or zero. When separated from charge q by a distance r, the voltage is kq/r.

EXAMPLE 23.9 Calculate the absolute potential at a point 0.53×10^{-10} m from a proton. This is the average distance of an electron from the proton nucleus of a hydrogen atom. Assume that the potential is zero at a point infinitely far from the proton.

SOLUTION The following information is known: $q = 1.6 \times 10^{-19}$ C (charge of proton) and $r = 0.53 \times 10^{-10}$ m. The absolute potential is calculated using Eq. (23.6):

$$V = \frac{kq}{r} = \frac{(9.0 \times 10^9 \,\text{N} \cdot \text{m}^2/\text{C}^2)(1.6 \times 10^{-19}\,\text{C})}{(0.53 \times 10^{-10}\,\text{m})}$$

$$= 27 \frac{\text{N} \cdot \text{m}}{\text{C}} = 27 \frac{\text{J}}{\text{C}} = \underline{27\,\text{V}}. \qquad \blacksquare$$

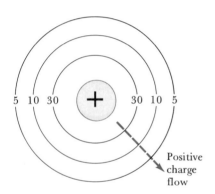

FIG. 23.9. Equipotential lines surround a positive charge.

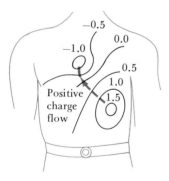

FIG. 23.10. Equipotential lines on the surface of a person's body at one time during a heartbeat cycle. The lines are caused by the dipole charge on the heart.

Equipotential Surfaces

Contour maps can be drawn to represent electrical potential. In Fig. 23.9, a large positive charge q resides at one point in space, and the circular lines surrounding the charge represent lines of constant electrical potential. A small test charge placed at one point on a line has the same electrical potential energy as at any other point on that line. The line is part of a three-dimensional surface called an **equipotential surface;** every point on the surface is at the same electrical potential. The lines drawn on Fig. 23.9 are analogous to lines drawn on a contour map that represent regions of constant vertical elevation. Just as water tends to flow from regions of higher to regions of lower elevation, positive electrical charge tends to flow perpendicular to equipotential lines toward regions of lower potential (see the dashed line in Fig. 23.9). As the voltage decreases, so does the electrical potential energy of the positive charge ($\Delta PE_q = qV$). Negative charge also tends to move toward regions where its electrical potential energy is minimized. ΔPE_q will be negative for a negative charge if V is positive. Thus, negative charges tend to move in the oppsite direction of positive charges toward regions of higher electrical potential.

The dipole charge on a person's heart creates equipotential contours on a person's body surface. An example of these contours at one instant of time is shown in Fig. 23.10 (the dipole charges on the heart causing the contours are not shown). At any point on one of these lines, a charge in the skin will have the same electrical potential energy as when located at another point on the same line.

A positive charge tends to move toward regions of lower electrical potential. The -1.0-mV (millivolt) circle in Fig. 23.10 is like a low elevation bog in which water accumulates after a rain. On the body, positive charge accumulates there. Negative charge accumulates at high "electrical elevations"—near the $+1.5$-mV peak.

23.5 Potential Difference and Electric Field

As we learned earlier, potential difference (voltage) is a measure of the change in the electrical elevation that an electric charge experiences as it moves between different positions in space. If, for example, the change in voltage from point A to point B in Fig. 23.11 is $+10$ V, then a positive electric charge will have to be pushed up this electrical hill.

Such a hill could be caused by the group of charges shown in Fig. 23.12— work must be done to move q toward B, the location of a row of positive charges.

These charges shown in Fig. 23.12 produce an electric field that points from B to A. The electric field exerts a force $\mathbf{F}_q$ on a charge q placed in the field. The force is given by Eq. (22.4):

$$\mathbf{F}_q = q\mathbf{E}.$$

In order for a positive charge q to be moved in a direction opposite the field, it must be pushed with an opposing force $\mathbf{F}$ (represented by the push of the person in Fig. 23.12) whose magnitude is at least as large as $\mathbf{F}_q$:

$$\mathbf{F} = -q\mathbf{E}.$$

The negative sign means that the person's force $\mathbf{F}$ is opposite $\mathbf{E}$, or toward the right.

The electric field pointing from B to A in Fig. 23.12 and the voltage hill from A to B in Fig. 23.11 are just two different ways of representing the electrical effect of the row of charges at A and B. These two quantities, $\mathbf{E}$ and V, are not independent but are intimately related, as we learn next.

To move the charge q a small distance Δx to the right requires work. If the charge moves at a constant speed, then all of the work is converted into electrical potential energy:

$$W = \Delta PE_q.$$

The work done by the pushing force F shown in Fig. 23.12 is

$$W = F\,\Delta x.$$

Since the charge does not accelerate, $\mathbf{F}$ exactly balances the electrical force caused by the electric field: $\mathbf{F} = -\mathbf{F}_q = -q\mathbf{E}$. Thus,

$$W = -qE\,\Delta x.$$

But according to Eq. (23.5), the change in electrical potential energy when the charge q passes through a change in voltage ΔV that occurs when it moves the distance Δx is

$$\Delta PE_q = q\,\Delta V.$$

When these values of W and ΔPE_q are substituted into the work-energy equation ($W = \Delta PE_q$), we find that

$$-qE\,\Delta x = q\,\Delta V$$

or

$$E = -\frac{\Delta V}{\Delta x}. \tag{23.7}$$

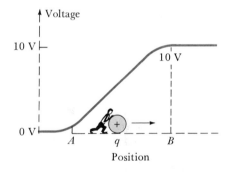

FIG. 23.11. The electrical potential energy of charge q increases as it is pushed toward the region of higher voltage at B.

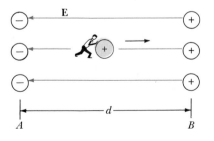

FIG. 23.12. The electrical hill shown in Fig. 23.11 could be caused by the charge distribution here. These charges at A and B produce an electric field that points to the left and opposes the motion of q.

The electric field equals the negative change in voltage ΔV divided by the distance Δx over which that voltage change occurs. If the voltage changes by amount V in distance d, then the average electric field in that region is

$$E_{av} = -\frac{V}{d}. \qquad (23.8)$$

Notice that if V is positive while moving from point A to point B, then $\mathbf{E}$ points in the opposite direction—from B to A. A positive change in voltage produces an electric field pointing in the opposite direction—hence the negative sign in Eqs. (23.7) and (23.8).

If the voltage change across a region is known, Eqs. (23.7) and (23.8) can be used to calculate the electric field in that region. The equations can also be used to calculate a change in voltage when moving in a known electric field.

EXAMPLE 23.10 A 12-V potential difference exists across a wire that is 2.0 m long, the higher voltage existing on the right side of the wire. Calculate the average electric field along the wire. Does it point toward the right or left?

SOLUTION According to Eq. (23.8), the average electric field along the wire is

$$E_{av} = -\frac{V}{d} = -\frac{12\,\text{V}}{2.0\,\text{m}} = \underline{-6.0\,\text{V}/\text{m}}.$$

The electric field points opposite the direction of increasing voltage, or toward the <u>left</u>. ∎

Summary

1. **Electrical potential energy:** The change in electrical potential energy when the separation of charges q_1 and q_2 is changed from r_0 to r is

$$\Delta PE_q = \frac{kq_1 q_2}{K}\left(\frac{1}{r} - \frac{1}{r_0}\right), \qquad (23.2)$$

where K is the dielectric constant of the material in which the charges move. If one charge q moves relative to several other fixed charges $q_1, q_2, q_3, \ldots$, the change in electrical potential energy is the sum of the energy changes for q relative to each fixed charge:

$$\Delta PE_q = \sum_{i=1} \frac{kq_i q}{K}\left(\frac{1}{r_i} - \frac{1}{r_{i_0}}\right). \qquad (23.3)$$

2. **Work-energy calculations:** Electrical potential energy is one of the many forms of energy that can change when a system moves from an initial to a final state. The work-energy equation can be used to determine an unknown change when other energy changes are known (assuming there is no heat flow):

$$W = \overbrace{\Delta KE + \Delta PE_g + \Delta PE_s + \Delta PE_q + \Delta E_{th} + \cdots}^{\text{energy changes in system}}. \qquad (8.10)$$

The different expressions in this equation are discussed in Chapter 8 (see the summary in Chapter 8).

3. **Electrical potential difference (voltage):** Electrical potential difference is an indicator of the difference in electrical "elevation" of two points in space. When a charge q moves up or down one of these hills or valleys from point A to point B, its electrical potential energy changes by an amount $PE_{q_B} - PE_{q_A} = qV_{AB}$, which is usually written in the more abbreviated form

$$\Delta PE_q = qV, \qquad (23.5)$$

where V is the potential difference (change in voltage) from A to B. The unit of potential difference called the **volt** is defined as one joule per coulomb ($1\,\text{V} = 1\,\text{J}/\text{C}$).

4. **Potential difference and electric field.** The average electric field E_{av} between two points separated by a distance d is

$$E_{av} = -\frac{V}{d}, \qquad (23.8)$$

where V is the difference in the voltage from the first to the second point. The minus sign reminds us that the electric field points from the higher toward the lower voltage.

Questions

1. A person's heart has a dipolar charge on it with the positive charge on the bottom and the negative charge on the top of the heart. Will the electrical potential energy of a sodium ion (charge $+e$) on the person's left side increase or decrease if the sodium ion moves up? The same for a chlorine ion (charge $-e$)? Justify each answer.

2. The metal sphere on the top of a Van de Graaf generator has a large negative charge. In which direction will a negatively charged ion in the air move relative to the sphere in order to experience a decrease in electrical potential energy? Justify your answer.

3. In order to get more paint from spray guns on metal car parts, the paint droplets are charged negatively as they leave the gun and a voltage source is attached to the metal car part. Should the car part be at a higher or lower voltage than the spray gun in order to attract the negatively charged paint drops? Justify your answer.

4. A bird perches on a power line that is at 100,000 V. Explain why ions in the bird's body do not flow, causing the bird to be electrocuted.

5. Your left hand touches a metal rack at zero volts, and your right hand touches the 100-V power source in a piece of electronic equipment. Indicate the direction in which sodium ions in the tissue of your right arm flow. In which direction do chlorine ions flow? Justify your answer.

6. If the voltage in a region is constant, can the electric field be nonzero? Explain your answer.

7. Can electric charges be arranged so that the electric field and absolute potential are zero at the same point? Show such an arrangement of charges and a point or points that satisfy this condition.

8. Show a charge arrangement and point in space where the absolute potential is zero, but the electric field is not zero.

9. Show a charge arrangement and point in space where the electric field is zero, but the absolute potential is not zero.

10. A person shuffles across a rug and accumulates an excess of electrons. A spark will jump from one person to another if the electric field between them exceeds about 3×10^6 N/C. Estimate the difference in voltage between the two people based on some reasonable numbers that are part of your own experience with sparks. Be sure to state your assumptions. Do not be surprised by a large calculated voltage.

11. Using the definitions of electric field, work, and potential difference and the appropriate units for each quantity, show that the units of electric field can be written as either N/C or V/m.

Problems

23.1 Electrical Potential Energy

1. Calculate the change in electrical potential energy when a sodium ion (charge e) and chlorine ion (charge $-e$) are moved from an initial separation of 2.8×10^{-9} m to 1.0×10^{-7} m when (a) in air and (b) in water.

2. Calculate the change in electrical potential energy when a hydrogen atom is ionized. Its electron moves from an orbit that is 0.53×10^{-10} m from the proton nucleus to a relatively large separation (effectively infinity).

3. The shell of a male tortoise is given a $-(3.0 \times 10^{-4})$-C charge and a female tortoise a $+(2.0 \times 10^{-4})$-C charge. Calculate the change in electrical potential energy when the tortoises move from a separation of (a) 2.00 m to 0.50 m and (b) 0.50 m to 4.00 m.

4. Two Ping-Pong balls suspended in air from strings have electric charges of $+1.0 \times 10^{-6}$ C and $+3.0 \times 10^{-6}$ C. The balls are initially separated by 20 cm. Calculate their final separation if their electrical potential energy changes by $+0.045$ J.

5. An electron is 0.10 cm from a $+3.0 \times 10^{-3}$ C electrical charge. (a) Calculate the electrical force F_0 on the electron. (b) The electron is pulled slowly by a force so that it moves to a distance of 0.11 cm from the charge. Calculate the electrical force F on the electron when at this distance. (c) Estimate the work done by the average force pulling the electron $[(F + F_0)/2] \Delta x$. (d) Compare this number to the change in electrical potential energy of the electron relative to the charge.

23.2 Electrical Potential Energy Involving Multiple Charges

6. Calculate the change in electrical potential when a (2.0×10^{-3})-C charge moves from position A to position B (Fig. 23.13).

FIG. 23.13

7. (a) Calculate the change in electrical potential energy when a positive ion with charge $+e$ moves from position B to position C (Fig. 23.13). The dielectric constant of the medium surrounding the charges is 8.0. (b) Repeat the above calculations for a negative ion with charge $-e$. (c) On the basis of these energy changes, which ion would prefer position B and which position C?

8. Calculate the change in electrical potential energy when a $+5.0$-C charge moves to position C (Fig. 23.14) from infinity: $q_1 = q_2 = q_3 = +10.0$ C.

9. Repeat the calculation described in Problem 8, only for $q_1 = q_3 = -10.0$ C and $q_2 = +10.0$ C.

10. Calculate the change in electrical potential energy when a $+2.0$-C charge moves from position C to position D (Fig. 23.14): $q_1 = q_3 = +10.0$ C and $q_2 = -20.0$ C.

FIG. 23.14

11. A chlorine ion (charge $-e$) moves from position B to position C shown in Fig. 23.6. Calculate its change in electrical potential energy. The magnitude of the dipole charges on the heart ($+q$ and $-q$) is 4.0×10^{-6} C and the charges are in tissue having an average dielectric constant of 7.0.

23.3 Work-Energy Calculations

12. Two electrons are initially separated from each other by 0.50×10^{-9} m. When released, the electrons fly apart. Calculate the speed of each electron when separated by a large distance ($r \simeq \infty$). (*Hint:* Each electron gains kinetic energy equal to one half the total electrical potential energy lost.)

13. An alpha particle (two protons and two neutrons together in one nucleus) with a mass of 6.64×10^{-27} kg and charge $+2e$ flies head-on from a large distance toward a stationary gold nucleus (charge $+79e$) at a speed of 3.0×10^{7} m/s. Calculate the distance of the alpha particle from the gold nucleus when it stops.

14. Four charges ($-e, +e, -e, +e$) in a row are each separated from their neighboring charges by 3.0×10^{-10} m. Calculate the work required to remove the positive charge on the right side to infinity.

15. Consider 10^{12} chlorine ions (each with mass 5.9×10^{-26} kg and charge $-e$) moving through the body tissue from position B (Fig. 23.6) to position A. If the ions start at rest and their speed at A is 0.10 m/s, calculate the thermal energy added to the body. The average dielectric constant of the body tissue is 7.0 and q = 4.0×10^{-6} C.

16. A 5.0-kg cart shown in Fig. 23.15 has on its roof a small metal sphere of charge -2.0×10^{-4} C. The sphere moves toward another sphere of charge $+1.5 \times 10^{-3}$ C located on the wall. As it moves, the cart pulls a cable that passes over a pulley and lifts a 10-kg mass. If the cart starts at rest 5.0 m from the wall charge, calculate its speed when 2.0 m from the wall charge. Ignore friction.

FIG. 23.15

17. Repeat the calculation for Problem 16, only this time assume that a 40-N friction force opposes the motion of the cart.

18. An ion cannon, such as shown in Fig. 23.1, consists of a positively charged ($+2.0 \times 10^{-4}$)-C metal ball compressed into a plastic barrel so that it is 0.10 m from an equal charge at the closed end of the barrel. When the ball is released, it is shot out the open end of the barrel by the repulsive force between the charges. Calculate the speed of the 10-kg ball as it leaves the end of the barrel 10.0 m from where it started. Assume that no friction force exists in the barrel. The barrel is oriented at $37°$ with respect to the horizontal.

19. Repeat Problem 18, assuming that a 90-N friction and drag force opposes the motion of the ball up the barrel.

23.4 Electrical Potential Difference and Potential

20. (a) What is the change in the electrical potential energy of an electron that passes across the $+(9.0 \times 10^{9})$-V potential difference of a high-energy linear accelerator? (b) How many electrons must pass this potential difference in order to release 1.0 J of electrical potential energy?

21. During a lightning flash, -15 C of charge move through a potential difference of 8.0×10^{7} V. How much electrical potential energy is released?

22. A spark of electricity jumps from a person's finger to your nose and through your body. While passing through the air, the spark travels across 2.0×10^{4} V and releases 3.0×10^{-7} J of electrical potential energy. How much charge in coulombs and how many electrons flow?

23. The voltage from the cathode to the screen of a television set is $+22,000$ V. If a (9.11×10^{-31})-kg electron leaves the cathode with an initial speed of zero, what is its speed just before hitting the screen?

24. An electron starting at rest accelerates across a potential difference V_0. If the voltage is doubled, by what factor does the electron's final speed increase?

25. Suppose 10^{18} electrons start at rest and move along a wire through a $+12$-V potential difference. (a) Calculate the change in electrical potential energy of all the electrons. (b) The final speed of the electrons is 0.10 m/s. Use the work-energy equation to calculate the increase in thermal energy in the wire due to collisions of the electrons with atoms in the wire. The mass of an electron is 9.11×10^{-31} kg.

26. A nerve cell is shaped like a cylinder. The membrane wall of the cylinder has a $+0.090$-V potential difference from the inside to the outside of the wall. To help maintain this potential difference, sodium ions are pumped from inside the cell to the outside. For a typical cell, 10^{9} ions are pumped each second. (a) Calculate the decrease in chemical energy each second required to produce this increase in electrical potential energy. (b) If there are roughly 7×10^{11} of these cells in the body, how much chemical energy is used in pumping sodium ions each second. (c) What fraction of an average person's metabolic rate is this energy that is used to pump ions (see Table 8.1)?

27. A battery charger forces 5.0 C of positive charge to move across a potential difference of 12 V in a battery each second. (a) Calculate the electrical energy increase of the battery in 1 s. (b) Calculate the increase in electrical potential energy of the battery in one hour of charging.

28.   A 10,000-kg shuttle bus carries a $+15$-C charge on a sphere on its top. Through what potential difference must the bus travel in order to acquire a speed of 10 m/s, assuming no friction force?

29.   Two (-3.0×10^{-6})-C charges are separated by 0.20 m. Calculate the absolute potential (assuming zero volts at infinity) at a point (a) halfway between the charges and (b) 0.20 m to the side of one of the charges along a line joining them.

30.   (a) Calculate the absolute potential at a distance of 0.53×10^{-10} m from a proton. (b) What is the change in electrical potential energy when an electron moves from infinity to 0.53×10^{-10} m from the proton and (c) when it moves from 0.53×10^{-10} m to infinity?

31.   (a) Calculate the absolute potential at a point halfway between electric dipole charges that are separated by 0.10 m and for which $q = 5.0 \times 10^{-6}$ C. (b) Calculate the absolute potential at a distance 0.20 m to the side of the positive charge along a line joining the two charges.

23.5   Potential Difference and Electric Field

32.   An average electric field of magnitude 200 V/m exists between two plates separated by 5.0 cm. How large a potential difference exists from one plate to the other?

33.   A parallel-plate capacitor is an electronic device that consists of two large parallel plates separated by a small distance. (a) Calculate the average electric field between the plates if a 120-V potential difference exists from one plate to the other and if their separation is 0.50 cm. (b) A spark will jump if the electric field exceeds 3.0×10^6 V/m. What is the closest the plates can be placed to each other to prevent sparking?

34.   (a) Calculate the electric field across a body cell membrane. A 0.085-V potential difference exists from one side to the other, and the membrane is 7.0×10^{-9} m thick. (b) Calculate the magnitude of the electrical force on a sodium ion (charge $+e$) in the membrane.

35.   If the electric field between your finger and some other object exceeds 3.0×10^6 V/m, a spark will jump. If you shuffle across a rug, a spark will jump when your finger is 0.20 cm from another person's nose. Calculate the difference in voltage between your body and the other person's body just before the spark jumps.

General Problems

G1.   Protons are accelerated from the end of an ion gun used to propel a rocketship. (a) Calculate their speed if they start at rest and pass through a potential difference of 50,000 V. The mass of a proton is 1.67×10^{-27} kg. (b) Calculate the momentum given to 3.0 kg of these protons. (c) Calculate the final speed of a 4000-kg rocketship after emitting the protons. Assume that the ship starts at rest while floating in space and that all the protons are emitted in one burst from the resting ship.

G2.   A lightning flash occurs when -40 C of charge jump from a cloud to the earth through a potential difference of 4.0×10^8 V. Calculate the change in temperature and state of 50 kg of water (the mass of water in a 150-lb person) if all the electrical potential energy is converted to thermal energy in the water. The water starts at 37°C.

G3.   Each second, -4 C of electric charge pass through $+120$-V potential difference in the coils of a 180-kg water heater. Calculate the time needed to increase the water's temperature by 10 C°.

24 Electric Current and Ohm's Law

In the last two chapters we saw how to incorporate electric force and electric energy into the general principles of physics that were introduced early in the book. With this background we now begin our study of electric charge that moves or flows from one place to another. All the electrical appliances we use, such as radios, electric stoves, and flashlights, depend on the flow of electric charge.

In this chapter we analyze some very simple electric circuits that allow electric charge to flow in a loop, such as shown in Fig. 24.1. An electric circuit of any type must include (1) a voltage source, such as a battery, that forces electric charge to move through the rest of the circuit from one terminal of the voltage source back to the other, and (2) an unbroken path through the rest of the circuit along which the charge can move. The path may consist of wires connected to light bulbs, the heating elements of toasters, or whatever. The charge that moves through this circuit is called an electric **current.**

We will learn how to determine the amount of current that flows through a circuit and the rate at which electrical energy is used by an element in the circuit, such as a light bulb. The circuits considered in this chapter consist of a voltage source and one other element through which current flows. In the next chapter, more complicated circuits are presented.

FIG. 24.1. A simple electric circuit.

FIG. 24.2. A battery.

24.1   Electric Power Sources

The first step in producing an electric current is to separate positive and negative charges in an electrical power source such as a battery. Energy is needed for this charge separation. For example, hydroelectric power generation depends on the conversion of the kinetic and gravitational potential energy of water into the electrical potential energy needed to separate charges. The chemical energy released by the burning of coal, oil, and natural gas provides the electrical potential energy used by many electric power plants.

One of the most common electrical power sources, the battery, also derives its electrical energy from the energy released by chemical reactions. These chemical reactions can be complex, yet many of the basic properties of all batteries can be understood with reference to the simple battery shown in Fig. 24.2. Two dissimilar metal rods or plates, called **electrodes,** are immersed in a dilute acidic solution called the **electrolyte.** In a typical flashlight battery, one electrode is made of carbon and the other (the outer casing of the battery) is made of zinc; the electrolyte is absorbed in a powdery, sawdustlike material that is placed between the electrodes.

The metal electrodes dissolve in the electrolyte. When zinc dissolves, positively charged zinc ions (Zn^{2+}) enter the electrolyte and leave two electrons behind on the zinc electrode. Eventually the electrode accumulates such an excess of electrons that dissolved zinc ions are attracted back to the electrode at the same rate as zinc ions dissolve and leave the electrode; the net charge on the electrode remains constant. The other metal electrode may dissolve at a slower rate and accumulates a smaller number of electrons than the first electrode. Thus, although both electrodes become negatively charged, one electrode has less negative charge and is at a higher electrical potential than the other. In a flashlight battery, the potential difference between the electrodes is 1.5 V, the carbon electrode being at a higher potential than the zinc electrode.

The parts of the electrodes that stick out of the electrolyte are called the battery **terminals.** If the battery terminals are connected by wires to a flashlight bulb, electrons leave the negative terminal and move through the wire, through the filament of the flashlight bulb, and then back along the other wire to the positive terminal. As this electric charge flows, more of the electrode dissolves to provide a steady supply of electric charge. The battery becomes "dead" when one or the other of the electrodes becomes dissolved to the extent that it can no longer supply electric charge.

Electric power sources such as batteries are represented in electric circuits by the symbol shown in Fig. 24.3. By convention, the positive terminal is on the side with the longer vertical line; the negative terminal is on the side with the shorter line.

Any device such as a battery that separates positive and negative electric charge and therefore produces a voltage is called a **source of electromotive force,** or a **source of emf.** The emf of such a device (represented by the symbol $\mathcal{E}$) is the steady voltage across its terminals when no net charge is moving onto or off of the terminals. The emf $\mathcal{E}$ of a flashlight battery is 1.5 V and of a car battery is 12 V. It would have been better to call these devices voltage sources, since they provide a potential difference or voltage across their terminals rather than force. Years of common usage have left us with this unfortunate terminology.

FIG. 24.3. The symbol $\mathcal{E}$ represents the voltage across the terminals of a source of emf when no current is flowing.

24.2 Electric Current (*I*) and Drift Velocity

In the previous section we learned how a source of emf separates positive and negative electric charges, causing a voltage to exist between the terminals of the source. If a wire is attached from one terminal to the other, electric charge flows through the wire. As the source of emf continues to separate charges, a steady flow of charge continues through the wire. The flow of charge is called an **electric current** and is given the symbol *I*.

In the case of a wire the electric current is caused by a stream of electrons moving from the negative voltage terminal toward the positive terminal. If we could look into the wire, we might see approximately 10^{20} electrons pass our line of sight each second.

Current is not always caused by electron flow. For example, if you hold the positive terminal of a car battery in your right hand and the negative terminal in your left hand, negative ions in your body such as Cl⁻ flow toward the positive terminal and positive ions such as Na⁺ and K⁺ flow toward the negative terminal. This is called an **ion current.**

In general, electric current I is defined as the number of coulombs of electric charge Δq that pass a cross section of a material divided by the time Δt required for the charge to pass. Thus,

$$I = \frac{\Delta q}{\Delta t}. \tag{24.1}$$

The unit of current called the **ampere,** or **amp** (A), is defined as the flow of one coulomb of charge each second:

$$1 \text{ A} = 1 \frac{\text{C}}{\text{s}}. \tag{24.2}$$

The direction of the current is the direction that positive electric charges flow and opposite the direction that negative charges flow.

The convention for describing the direction of an electric current is occasionally confusing because most electric currents are caused by moving electrons (negative charges), whereas the current is by definition in the opposite direction to that in which the electrons move. The confusion can be traced back to Benjamin Franklin. When Franklin rubbed a glass rod with a piece of silk, electrons moved from the rod to the silk, and the rod was left with a deficiency of electrons. Franklin chose arbitrarily to call the charge on the rod positive and the electron charge on the silk negative. The opposite choice would have been more convenient. Electric current would then have been defined as flowing in the same direction as that in which "positive" electrons move through a wire. We cannot, however, blame Benjamin Franklin for his choice. The electron causing the negative charge was not discovered until a hundred years later. Franklin could not have known that these were the charges that moved when a current flowed. In any case, since the electron was assigned a negative charge, current is now defined as flowing in the opposite direction to that in which electrons move.

EXAMPLE 24.1 Each second, 10^{17} electrons flow from right to left across a cross section of wire attached to the two terminals of a battery. Calculate the magnitude and direction of the electric current in the wire.

SOLUTION The magnitude of the charge e of an electron is 1.6×10^{-19} C. Since 10^{17} electrons pass a cross section in the wire each second, the total charge flowing each second is

$$\Delta q = 10^{17} e = 10^{17}(1.6 \times 10^{-19} \text{ C}) = 1.6 \times 10^{-2} \text{ C}.$$

Substituting for Δq and Δt in Eq. (24.1), we find that the magnitude of the electric current is

$$I = \frac{\Delta q}{\Delta t} = \frac{1.6 \times 10^{-2} \text{ C}}{1 \text{ s}} = 1.6 \times 10^{-2} \frac{\text{C}}{\text{s}} = \underline{1.6 \times 10^{-2} \text{ A}}.$$

The direction of the current is <u>from left to right</u>, opposite the direction of electron flow. ∎

An electric current produced by a battery is an example of a **direct current.** When a direct current flows, charge always moves in the same direction—electrons move away from the negative terminal of the emf source toward the positive terminal.

In many emf sources, one terminal will have first an excess of positive charge and then, a short time later, an excess of negative charge. Charges in a wire attached to the terminals of this type of electric power supply are first pushed in one direction, then back in the opposite direction, creating an **alternating current** (ac). The voltage across the wall socket in a home changes polarity at a frequency of 60 Hz (60 times each second). Charges in light bulbs and in other electronic devices that are connected to the terminals of a wall socket oscillate back and forth at 60 Hz. In this chapter we will only consider direct currents.

Drift Velocity

The flow of electric charge through a material is an interesting and in some ways a surprising process. One of the surprises is the low speed at which the charges move. Consider electrons in a copper wire.

Each atom in the copper has one outer electron that can leave the atom and moves randomly through the metal at speeds of about 10^6 m/s. The motion of these free electrons is interrupted by frequent collisions—almost 10^{14} each second. The collisions are random, and an electron changes direction at each collision so that it has little net displacement after many collisions. *

If the copper wire is placed between the terminals of a battery, part of the motion is no longer random. As a consequence of the electric field caused by the voltage between the terminals, there is a slight "drift" of electrons toward the positive terminal of the battery. Initially, as the electrons are pulled through the wire by the electric field, they accelerate. However, their forward motion is interrupted by collisions with other atoms and ions in the wire. The collisions produce the effect of a **resistive drag force** that balances the electric field force, and the electrons move at a constant terminal speed called the drift speed or drift velocity. The **drift velocity** v_d is defined as the average displacement d of a charge divided by the time t required to achieve that displacement:

$$v_d = \frac{d}{t}.$$

We note only the net displacement after many collisions and not the distance traveled randomly between collisions. The drift velocity is much slower than the speed of the random motion of the electrons.

We can derive an equation that relates the electron drift velocity to the electric current in the wire. Consider the wire of cross-sectional area A shown in

---

*The process being described is most representative of the motion of ions in solutions. The description of electron motion is more complex than we have indicated. The results of our model, however, can be applied to electron motion, and we will ignore some of the more complicated aspects of electron motion in metals.

Time zero

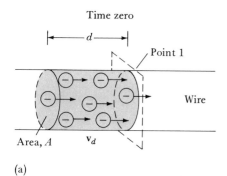

Wire

Area, A v_d

(a)

Time t later

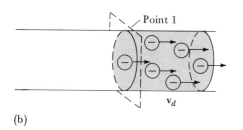

Point 1

v_d

(b)

FIG. 24.4. During a time t, the electrons in the shaded section of wire drift a distance d to the right. The drift is caused by a voltage across the wire.

Fig. 24.4a. The shaded cylinder has a length d equal to the distance that electrons drift in time t. The volume of the cylinder is $\mathrm{Vol} = (A)(d)$. All free electrons in the shaded cylinder pass point 1 at the front of the cylinder during a time t (Fig. 24.4b).

The number N of free electrons in the shaded cylinder is

$$N = n_e(\mathrm{Vol}) = n_e(Ad),$$

where n_e, the **free-electron density,** is the number of free electrons per unit volume. For copper the free-electron density equals the number of copper atoms per unit volume since each copper atom contributes one free electron to the metal. The electric current in the wire is the charge q that passes point 1 divided by the time t it takes the charge to pass. The charge q equals the number N of free electrons in the shaded cylinder times the charge of each electron $(-e)$. Thus, we find that the current is

$$I = \frac{q}{t} = \frac{N(-e)}{t} = \frac{[n_e(Ad)](-e)}{t}.$$

Substituting v_d in place of d/t, we find that

$$I = n_e A v_d(-e)$$

or

$$I = -(n_e e A)v_d. \qquad \textbf{(24.3)}$$

The negative sign means that the velocity of the electrons is opposite to the direction of the current in the wire.

The surprising result of this analysis of drift velocity occurs when we substitute numbers into Eq. (24.3). For example, when starting your car, a current of 100 A typically flows along the copper wires that connect the battery to the starter motor. The cross-sectional area A of the wire is about 30 mm² or 30×10^{-6} m², and the number of free electrons per unit volume n_e in copper is about 8×10^{28} electrons/m³. Rearranging Eq. (24.3) to solve for v_d, we find that

$$v_d = -\frac{I}{n_e A e} = -\frac{100\ \mathrm{C/s}}{(8 \times 10^{28}\ \mathrm{m^{-3}})(30 \times 10^{-6}\ \mathrm{m^2})(1.6 \times 10^{-19}\ \mathrm{C})}$$
$$= -3 \times 10^{-4}\ \mathrm{m/s}.$$

An electron moves only 3×10^{-4} m, or 0.3 mm, in one second. In an hour, an electron has moved only about 1 m. Even though electrons drift slowly along the wire, the current can be large because of the great number of electrons that are moving.

24.3   Ohm's Law and Resistance

Having developed an understanding of what electric current is, let us turn now to the two important quantities that affect the current passing through an object: (1) the potential difference—that is, the change in voltage from one end of the object to the other—and (2) the resistance offered by the object to charges flowing through it. The three quantities, current I, voltage V, and resistance R, were first related in an equation by the German physicist, Georg Simon Ohm,

in 1827. The equation,

$$V = IR, \qquad (24.4)$$

called **Ohm's law,** has been considered important enough to be included in the physics curriculum of students for over one hundred years. When Ohm presented his idea, it was received with scorn.

Ohm, at the age of 38, had served for ten years as a poorly paid mathematics and science teacher at the Jesuit College of Cologne. To qualify for a university position, he was required to produce some kind of scientific masterpiece, the value of which would bring recognition and university job offers. After many years of experimenting with electricity, during which time he published numerous short papers, Ohm produced a 250-page manuscript entitled *Mathematical Measurements of Electrical Currents.*

The paper was ignored by most of Ohm's German colleagues. One critic who did not ignore it said, "A physicist who professed such heresies was unworthy to teach science." Ohm, unfortunately, had presented his work at a time and place where experiment was disdained as a means of acquiring knowledge. He did not receive a university position and even had to resign from the Jesuit College. After six dismal years, King Ludwig I of Bavaria helped Ohm obtain a professorship at the Polytechnic School of Nuremberg. Ohm eventually received recognition for his work. In 1841 he was awarded the Copley Medal by the Royal Society of London, and in 1842 he was honored as the Society's most distinguished foreign member.*

To understand the meaning of Ohm's law and the idea of resistance, imagine a large number of skydivers falling to the earth (Fig. 24.5a). They are caught in a net from which they slide onto a conveyor belt that lifts them back to the sky, where they again fall to the net. The situation is analogous in many ways to the electric circuit shown in Fig. 24.5b. Let us compare the two by moving around each "circuit" with a skydiver or with an electron.

As the skydiver moves up the conveyor belt, his or her gravitational potential energy increases at the expense of energy used to run the conveyor's motor. Similarly, as an electron moves through a battery, its electrical potential energy increases at the expense of chemical energy in the battery.

At the top of the conveyor belt, a skydiver slides off and starts a fall to earth. Shortly after the fall begins, the diver reaches a terminal speed of about 50 m/s. At this speed the drag force of the air, which resists the diver's fall, balances the downward force of gravity; no acceleration occurs, and the diver falls at a constant speed. As the diver falls, gravitational energy is converted to thermal energy and the air surrounding the diver becomes slightly warmer.

A similar phenomenon occurs to electrons flowing in the wire attached to the battery. The battery produces an electric field **E** in the wire that pulls the electrons toward the positive terminal. This field is similar to the gravitational field pulling the skydiver. As the electron moves, it bumps into atoms and ions in the wire. These collisions have the effect of a drag force. Almost the instant the electron starts to move along the wire, the electric force pulling the electron is balanced by the drag force opposing its motion. The electron then moves at constant terminal speed (called drift velocity). As the electron is pulled through the wire by the electric field, it loses electrical potential energy. The energy is

*Jay E. Greene, ed., *100 Great Scientists* (New York: Washington Square Press, 1964), pp. 214–217.

(a)

(b)

FIG. 24.5. The motion of the skydivers around a loop (a) is analogous to the motion of an electron around the current loop (b).

(a)

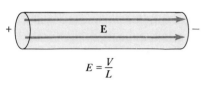

$$E = \frac{V}{L}$$

(b)

FIG. 24.6. (a) A voltage V across a material of length L causes (b) an electric field of magnitude $E = V/L$.

converted to thermal energy by the electron's collisions with atoms and ions in its path. The wire becomes hot.

With this background we can now derive Ohm's law. Consider a material of length L and cross-sectional area A across which a voltage (potential difference) V exists (Fig. 24.6a). The voltage produces in the material an electric field whose magnitude is given by Eq. (23.8):*

$$E = \frac{V}{L}.$$

The electric force on a charge q of magnitude e in the field is given by Eq. (22.4):

$$F_q = qE = eE = e\frac{V}{L}.$$

The electric force causes the charge to accelerate, but its motion is interrupted by frequent collisions (about one every 10^{-14} s) that have the same effect as a retarding drag force (see Section 14.7).†

The drag force $\mathbf{F}_D$ increases as the drift velocity of the charge increases. For most materials the **drag force** is directly proportional to the drift velocity:

$$F_D = Dv_d,$$

where D is a proportionality constant called the **drag coefficient.** D depends on the type of charge that is moving and on properties of the material through which the charge moves.

As the charge drifts faster through the metal, F_D increases until eventually the electric force pulling the charge is balanced by the drag force opposing its motion. The charge no longer accelerates, but moves at constant drift speed. This happens when

$$F_q = F_D.$$

Substituting the expressions written earlier for F_q and F_D, we find that

$$e\frac{V}{L} = Dv_d$$

or

$$v_d = \frac{eV}{LD}.$$

In the previous section, we determined that the electron current I was related to the drift velocity by Eq. (24.3):

$$I = en_e v_d A.$$

Substituting from above for v_d, we find that

$$I = en_e \left(\frac{eV}{LD}\right) A.$$

---

*We will ignore the signs of charges, fields, and voltages in our derivation.

†As we mentioned earlier, this model of the drag force is appropriate for ion currents but not for electron currents. However, the result we obtain applies quite well to electron currents in wires, although a much more complex set of reasoning must be used to reach this result.

This equation can be rearranged into a form that resembles Ohm's law ($V = IR$):

$$V = I\left[\left(\frac{D}{n_e e^2}\right)\frac{L}{A}\right]. \tag{24.5}$$

The quantity $D/n_e e^2$ is called the **resistivity** ρ (rho) of the material through which the charge flows. The resistivity depends on intrinsic properties of the material, such as the drag coefficient and the number of free electrons per unit volume in the material. Copper has a low resistivity because of its large concentration of free electrons and the relative lack of hindrance experienced by the electrons while moving through the copper (small D). On the other hand the resistivity of glass is about 10^{20} times greater than that of copper because it contains so few free electrons; since n_e is small, ρ (which is proportional to $1/n_e$) is very large. Copper is an example of a material called an electrical **conductor** (low resistivity) whereas glass is an example of an electrical **insulator** (high resistivity that prohibits the flow of electricity). The resistivities of a variety of substances are listed in Table 24.1. Substances are sometimes characterized in terms of another quantity called conductivity. The **conductivity** σ of a substance is the inverse of its resistivity ($\sigma = 1/\rho$). Thus, a material with low resistivity has high conductivity and is a good conductor of electricity, and vice versa.

Substituting $\rho = D/n_e e^2$ into Eq. (24.5), we find that

$$V = I\left(\rho\frac{L}{A}\right).$$

The quantity in parentheses is the **electrical resistance** R of an object of length L, cross-sectional area A, and resistivity ρ:

$$R = \rho\frac{L}{A}. \tag{24.6}$$

With this substitution in Eq. (24.5) we have arrived at Ohm's law:

$$V = IR.$$

It is important to note that the electrical resistance of an object depends not only on the resistivity of the material of which it is made, but also on its dimensions L and A.

The unit of electrical resistance is called the **ohm** and is given the symbol Ω (omega). One ohm is defined as the electrical resistance of an object that will allow a one-ampere (1-A) electrical current to flow when a one-volt potential difference is placed across the object. Rearranging Ohm's law to solve for resistance ($R = V/I$), we see that

$$1\ \text{ohm} = 1\ \Omega = \frac{1\ \text{V}}{1\ \text{A}}. \tag{24.7}$$

Ohm's law does not apply to all materials. For example, the current passing through a transistor does not increase in direct proportion to an increase in voltage. However, for most materials, Ohm's law adequately describes the relationship between current, voltage, and resistance.

When drawing an electric circuit to represent the path of charge flow from a source of emf through wires and other elements of the circuit such as light

TABLE 24.1 Resistivity (ρ) of a Variety of Materials

Material	Resistivity (ohm · m)[a]
Metals	
Silver	1.6×10^{-8}
Copper	1.7×10^{-8}
Aluminum	2.8×10^{-8}
Electrical insulators	
Ordinary glass	9×10^{11}
Hard rubber	1×10^{16}
Shellac	1×10^{14}
Wood	$10^8 - 10^{12}$
Body parts	
Blood	1.5
Lung	20
Fat	25
Human trunk	5
Membrane	5×10^7
Geological materials	
Igneous rocks	$10^2 - 10^7$
Sedimentary rocks	$1 - 10^5$
Ground water	~ 10

[a] At 20°C.

bulbs, we represent the electrical resistance of these elements with the symbol $\overset{R}{\sim\!\!\!\sim}$. The symbol could represent the resistance of the filament of a light bulb, the resistance of your body between electrodes held in your hands, or the resistance of a device actually called a resistor that is used routinely in electrical circuits. Wires, which have very low resistance, are usually represented in electric circuits by straight lines.

24.4 Applications of Ohm's Law

Ohm's law is used routinely in the analysis of all types of electric circuits. In this section we will encounter several fairly simple situations in which a voltage across one object causes an electric current to pass through it. Our analysis involves the equation that defines electric current ($I = \Delta q/\Delta t$), the equation for the resistance of an object ($R = \rho L/A$), and Ohm's law ($V = IR$).

EXAMPLE 24.2 (a) If an automobile starter has a resistance R of 0.10 ohm, how much current flows through the starter when it is connected to a 12-volt battery? (b) How much charge passes through the starter if it runs for 10 seconds?

SOLUTION (a) Using Eq. (24.4), we find that

$$I = \frac{V}{R} = \frac{12 \text{ V}}{0.10 \ \Omega} = 120 \ \frac{\text{V}}{\Omega} = \underline{120 \text{ A}}.$$

The conversion of units from V/Ω to A (ampere) is made using Eq. (24.7). (b) The total electric charge that flows is determined by rearranging Eq. (24.1):

$$\Delta q = I(\Delta t) = (120 \text{ A})(10 \text{ s}) = 1200 \text{ A} \cdot \text{s} = \underline{1200 \text{ C}}.$$

The unit conversion A $\cdot$ s to C is possible because 1 A (ampere) equals 1 C/s. Thus, $1 \text{ A} \cdot \text{s} = 1 \text{ C/s} \cdot \text{s} = 1 \text{ C}$. ∎

EXAMPLE 24.3 (a) Calculate the electrical resistance of a copper cylinder that is 0.50 m long and has a radius of 4.5 cm. (b) Calculate the resistance of the cylinder if it is stretched to a length of one mile (1609 m) with no change in volume. (c) Calculate the electric current through each piece of metal if a 1.5-V potential difference is placed across its ends.

SOLUTION (a) The following information is used in our calculations:

$$\text{area of metal } A = \pi r^2 = (3.14)(0.045 \text{ m})^2 = 6.4 \times 10^{-3} \text{ m}^2,$$
$$\text{length of metal } L = 0.50 \text{ m},$$
$$\text{resistivity of copper } \rho = 1.7 \times 10^{-8} \ \Omega \cdot \text{m (from Table 24.1)}.$$

The electrical resistance R of the cylinder is found using Eq. (24.6):

$$R = \rho \frac{L}{A} = (1.7 \times 10^{-8} \ \Omega \cdot \text{m})\left(\frac{0.50 \text{ m}}{6.4 \times 10^{-3} \text{ m}^2}\right) = \underline{1.3 \times 10^{-6} \ \Omega}.$$

(b) The area A' of the stretched cylinder can be determined from the fact that the volume of the original cylinder ($V = LA$) equals the volume when stretched ($V' = L'A'$). Thus, $L'A' = LA$ and

$$A' = \frac{L}{L'}A = \frac{0.50 \text{ m}}{1609 \text{ m}}(6.4 \times 10^{-3} \text{ m}^2) = 2.0 \times 10^{-6} \text{ m}^2.$$

The resistance of the wire is, then,

$$R' = \rho \frac{L'}{A'} = (1.7 \times 10^{-8} \, \Omega \cdot \text{m})\frac{1609 \text{ m}}{2.0 \times 10^{-6} \text{ m}^2} = \underline{14 \, \Omega}.$$

Even though the volume of the copper is the same, the resistance has increased by 10^7 because the copper was stretched into a long, thin wire rather than a short, thick cylinder. The resistance of the long, thin wire is analogous to the resistance that a fluid experiences when flowing through a long, thin tube. (c) The electric current in each case is given by Ohm's law:

original cylinder: $\quad I = \dfrac{V}{R} = \dfrac{1.5 \text{ V}}{1.3 \times 10^{-6} \, \Omega} = 1.2 \times 10^6 \dfrac{\text{V}}{\Omega}$

$$= \underline{1.2 \times 10^6 \text{ A}}.$$

stretched cylinder: $\quad I' = \dfrac{V}{R'} = \dfrac{1.5 \text{ V}}{14 \, \Omega} = \underline{0.11 \text{ A}}.$ ■

EXAMPLE 24.4 In the human body, most electric current is caused by the movement of sodium (Na^+), potassium (K^+), chlorine (Cl^-), and magnesium (Mg^{2+}) ions. The resistivity ρ of the body tissue depends upon the concentration of these ions and is approximately $5 \, \Omega \cdot \text{m}$. (a) Estimate the electrical resistance of the lower leg of a human. (b) Calculate the current through the leg if connected to a 120-V power source, the average voltage across a wall socket.

SOLUTION (a) Resistance depends upon the resistivity and the dimensions of the object. Let us assume that the lower leg has a length $L = 0.30$ m and a radius of about 5.0 cm $= 0.050$ m. Its cross-sectional area is, then,

$$A = \pi r^2 = \pi(0.050 \text{ m})^2 = 7.8 \times 10^{-3} \text{ m}^2.$$

Thus,

$$R = \rho \frac{L}{A} = (5 \, \Omega \cdot \text{m})\frac{0.30 \text{ m}}{7.8 \times 10^{-3} \text{ m}^2} = \underline{200 \, \Omega}.$$

(b) The current flowing through the leg when connected to 120 V is determined using Ohm's law:

$$I = \frac{V}{R} = \frac{120 \text{ V}}{200 \, \Omega} = \underline{0.6 \text{ A}}.$$

This is about six times greater than the current that will cause death to a human being (see Fig. 24.7). ■

The resistance calculated in the last example is the resistance of the internal body tissue, which is an excellent conductor of electricity. Many more deaths caused by electric shock would occur if it were not for our skin, which is a poor electrical conductor. When skin is dry, the electrical resistance across it ranges from several thousand ohms to hundreds of thousands of ohms; if the skin is moist, the electrical resistance across it is lowered. Consequently, electrical appliances, such as light switches and radios, should not be touched while you are in the bathtub. The resistance across the skin at your hands and feet may be low enough for a fatal current to pass through your body if you accidentally touch the 110-V line entering the appliance from the wall socket. The currents at which various physiological effects are produced on the human body are listed in Fig. 24.7.

FIG. 24.7. Physiological effects of electric current. [From Peter Strong, *Biophysical Measurements* (Beaverton, Ore.: Tektronix, Inc., 1970), p. 257. Reprinted by permission.]

Bilateral Impedance Rheography

It is estimated that 50,000 deaths occur each year because of clots in blood vessels. About 30 percent of the people who have orthopedic or abdominal surgery develop clots in their veins following the surgery. These clots can be detected by injecting radioactive iodine into the blood system, a procedure that is painful, costly, and not especially good for the patient's health.

An electrical-resistance measuring technique, called **bilateral impedance rheography,** has been developed to detect these same blood clots noninvasively. A voltage is applied across two electrodes attached to the region to be tested (the foreleg is shown in Fig. 24.8a). The electrical resistance of the leg between the electrodes is determined by measuring the current produced by this voltage (that is, $R = V/I$).

A cuff is then placed around the limb above the electrodes. When the cuff is inflated with air, the pressure on the blood vessels under the cuff is increased. The pressure is made less than that in the arteries entering the region but greater than that in the veins leaving the region. Blood fills the lower limb from the arteries but cannot leave by the veins. Usually, there is a 19 percent decrease in the electrical resistance of a normal leg when this blood pooling occurs, but if a

FIG. 24.8. (a) The resistance of the tissue between electrodes depends on the amount of blood in the tissue. (b) The electrical resistance decreases when the cuff in (a) is inflated so that blood can enter but not leave the lower leg. Since blood had already "pooled" in the right leg because of a clot, its resistance changed less when the cuff was inflated.

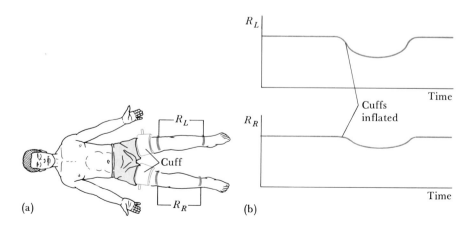

clot is present in the leg, pooling has occurred before the cuff is inflated. In this case, the electrical resistance decreases by less than 19 percent when the cuff is inflated.

Recordings of the electrical resistance of the two legs of a 56-year-old male are shown in Fig. 24.8b. The man had a clot in his right foreleg.

24.5 Resistance and Temperature

We have seen that the electrical resistance of an object depends on the resistivity of the material of which it is made and on its dimensions. The resistivity of a material is not constant; it changes as the material's temperature changes. For example, the resistance of the tungsten filament in a light bulb increases by more than a factor of ten as the filament warms after the light is turned on. At the higher temperature, the atoms and ions in the tungsten vibrate with greater amplitude. Electrons have a more difficult time finding a path through the hot material.

The resistivity of metals and their alloys varies with temperature roughly according to the equation

$$\rho_T = \rho_0(1 + \alpha \, \Delta T), \tag{24.8}$$

where ρ_0 is the resistivity at some reference temperature (usually $0°C$), ρ_T is the resistivity at a temperature interval ΔT above or below the reference temperature, and α is called the **temperature coefficient of resistance.** Representative values for α are given in Table 24.2. The units for α are $C^{°-1}$. The resistance R of a metal wire of length L and cross-sectional area A also varies in a similar fashion:

$$R_T = R_0(1 + \alpha \, \Delta T). \tag{24.9}$$

The variation of an object's electrical resistance with temperature provides a convenient method for measuring temperature. For example, the resistance of a coil of very thin wire is first measured at a reference temperature. The coil is then moved to a region of unknown temperature and its resistance is again measured. Equation (24.9) is used to calculate the difference in the unknown temperature and the reference temperature. Platinum is a common metal used for these measurements. It has a high melting temperature and is free from corrosion and oxidation.

EXAMPLE 24.5 A platinum resistance thermometer has a resistance of $100.0 \, \Omega$ when placed in an ice bath $(0°C)$. A geologist lowers the thermometer into a hot springs and the resistance increases to $123.5 \, \Omega$. What is the temperature of the hot springs?

SOLUTION The temperature coefficient of resistance of platinum is $0.00392 \, C^{°-1}$ (see Table 24.2). Equation (24.9) is rearranged to solve for the difference in temperature ΔT of the hot springs and the reference temperature:

$$\Delta T = \frac{R_T - R_0}{\alpha R_0} = \frac{123.5 \, \Omega - 100.0 \, \Omega}{(0.00392 \, C^{°-1})(100.0 \, \Omega)} = 60.0 \, C°.$$

TABLE 24.2 Temperature Coefficients of Resistance

Material	$\alpha \; (C^{°-1})$
Aluminum	0.0039
Constantan (60% Cu, 40% Ni)	0.000002
Copper (annealed)	0.00393
Iron	0.0050
Nichrome	0.0004
Platinum	0.00392
Silver	0.0038
Tungsten	0.0045
Carbon	−0.0005

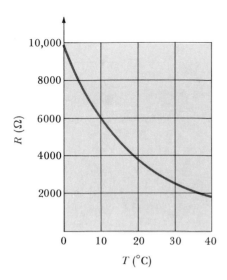

FIG. 24.9. The variation of the resistance of a thermistor with temperature.

Since the reference temperature is $0°C$, the hot springs is at a temperature of $60°C$. ∎

The resistance of some materials, known as semiconductors, does not vary as just described. **Thermistors,** for instance, are semiconductor devices whose resistance decreases dramatically as their temperature increases. The resistance of a thermistor may decrease by several percent for each $1C°$ rise in temperature (see Fig. 24.9). Thermistors provide an accurate and rapid indication of the temperature of objects in the temperature range from $-100°C$ to $300°C$.

24.6 Electrical Power

Most electrical appliances are rated in terms of **power,** the rate at which they convert electrical potential energy into other forms of energy (such as thermal energy):

$$\text{power} = P = \frac{\Delta PE_q}{\Delta t}.\qquad(24.10)$$

The unit of power is the watt, where

$$1 \text{ watt} = 1\,\frac{J}{s}.$$

For example, a 60-watt light bulb converts $60\,J$ of electrical energy into heat and light each second.

In order to calculate the electric power used by some electrical devices such as refrigerators, Eq. (24.10) is rewritten in a form that depends on the current I passing through the device and on the potential difference V across the device. Recall from Eq. (23.5) that the change in electrical potential energy when a charge Δq flows through a potential difference V is $\Delta PE_q = (\Delta q)V$. Substituting this expression for ΔPE_q in Eq. (24.10), we find that

$$P = \frac{\Delta PE_q}{\Delta t} = \frac{(\Delta q)V}{\Delta t} = \left(\frac{\Delta q}{\Delta t}\right)V = IV,$$

where $I = \Delta q/\Delta t$ is the current flowing through the device across which a potential difference V exists. If the object's resistance R is known, the above expression can be rewritten in different forms by using Ohm's law ($V = IR$) to substitute for either V or I:

$$\begin{aligned} P &= IV \\ &= I(IR) = I^2R \\ &= \left(\frac{V}{R}\right)V = V^2/R. \end{aligned}\qquad(24.11)$$

In summary, we see that the electrical power expended by a device of resistance R through which a current I flows when a potential difference V is placed across the device is given by any one of the three forms of Eq. (24.11).

EXAMPLE 24.6 A man holds the leads of a 100-V source of emf in his hands, causing a 10-mA current to flow through his body. (a) What is the resistance of his body? (b) At what rate is electrical energy converted to other forms of energy in his body? (c) What is the total amount of energy converted in 30 seconds?

SOLUTION (a) Ohm's law (Eq. 24.4) can be rearranged to solve for the resistance of his body:

$$R = \frac{V}{I} = \frac{100 \text{ V}}{10 \text{ mA}} = \frac{100 \text{ V}}{10 \times 10^{-3} \text{ A}} = \underline{10^4 \, \Omega}.$$

(b) The power can be calculated using any one of the three forms of Eq. (24.11):

$$P = VI = (100 \text{ V})(10 \times 10^{-3} \text{ A}) = \underline{1 \text{ watt}},$$
$$P = I^2R = (10 \times 10^{-3} \text{ A})^2(10^4 \, \Omega) = \underline{1 \text{ watt}},$$
$$P = V^2/R = (100 \text{ V})^2/(10^4 \, \Omega) = \underline{1 \text{ watt}}.$$

The following unit conversions have been used in the above equations:

$$VA = (J/C)(C/s) = J/s = \text{watt},$$
$$A^2\Omega = AA\Omega = A(V/\Omega)\Omega = AV = \text{watt},$$
$$V^2/\Omega = V(V/\Omega) = VA = \text{watt}.$$

(c) By rearranging Eq. (24.10), we find that the total electrical potential energy converted to other forms of energy in 30 s is:

$$\Delta PE_q = (P)(\Delta t) = (1 \text{ watt})(30 \text{ s}) = \underline{30 \text{ J}}. \qquad \blacksquare$$

Electric power companies charge their customers according to the amount of electrical potential energy the customer uses each month, but instead of charging for each joule of energy, the companies charge by an energy unit called the kilowatt-hour. A **kilowatt-hour** (kW · h) is the energy expended when a device converts electrical potential energy to other forms at a rate of 1000 watts (1 kilowatt or 1 kW) for a time of one hour. The conversion of kilowatt-hours to joules is shown below:

$$1 \text{ kW} \cdot \text{h} = 1 \text{ kW} \times 1 \text{ h} \left(\frac{1000 \text{ watt}}{1 \text{ kW}}\right)\left(\frac{3600 \text{ s}}{1 \text{ h}}\right)$$
$$= 3.6 \times 10^6 \text{ watt} \cdot \text{s} = 3.6 \times 10^6 \text{ J}.$$

In 1982 electric utility companies charged about 7 cents for 1 kW · h of electrical energy. By comparison, the chemical energy released by 14 slices of bread is about 1 kW · h and in 1982 cost approximately 50 to 60 cents.

EXAMPLE 24.7 Calculate the cost to operate a 120-watt light bulb for 24 hours if electricity costs 8 cents per kW · h.

SOLUTION The bulb uses energy at a rate of 120 watts or at a rate of 0.12 kW. Thus, the energy used in 24 hours is (24 h)(0.12 kW) = 2.9 kW · h. The cost of this amount of electrical energy at 8 cents/kW · h is:

$$\text{cost} = 2.9 \text{ kW} \cdot \text{h} \left(\frac{8 \text{ cents}}{\text{kW} \cdot \text{h}}\right) = \underline{23 \text{ cents}}. \qquad \blacksquare$$

Summary

1. A **source of electromotive force** (emf), such as a battery, uses various forms of energy to separate positive and negative electric charges. The **emf** $\mathcal{E}$ of the source is the steady voltage across its terminals when no current flows out of or into the terminals.

2. When a conducting object is attached to the terminals of a source of emf, electric charge flows through the object. The **current** I is defined as the charge Δq that flows past a cross section in the object divided by the time Δt needed for the charge to pass:

$$I = \frac{\Delta q}{\Delta t}. \tag{24.1}$$

The current is in units of **amperes** (A), where $1\text{ A} = 1\text{ C/s}$. The direction of the current is opposite the direction in which negatively charged electrons flow.

3. The magnitude of the current depends upon the voltage V across the object and upon its electrical resistance R. The **resistance** of the object is

$$R = \rho \frac{L}{A}, \tag{24.6}$$

where ρ is its **resistivity** (see Table 24.1), L is the length of the object, and A is its cross-sectional area. The unit of resistance is the **ohm** (Ω), where 1 ohm = 1 volt/amp.

4. For many objects, the current I passing through an object of resistance R when a voltage V is placed across the object is determined using **Ohm's law**:

$$V = IR. \tag{24.4}$$

5. **Power** is the rate at which electrical energy is converted to other forms of energy. When an electrical current I passes through an object of resistance R across which a voltage V exists, the power usage can be determined using any of the following expressions:

$$\text{power} = P = \frac{\Delta PE_q}{\Delta t} \tag{24.10}$$

$$= IV = I^2/R = V^2/R. \tag{24.11}$$

Questions

1. If the drift velocity of electrons in wires that connect the terminals of a battery to a light bulb is only 0.3 mm per second, why does the light turn on the instant the wires are connected, even though the wires leading to the bulb may be several meters long?

2. The wires from a high-fidelity amplifier to the speaker carry electric signals that cause the diaphragm of the speaker to move as the current reaching the speaker changes. If the wires are 5 m long and the drift velocity of electrons in the wires is less than 1 mm/s, why do the speakers respond almost instantly to changes in current occurring at the amplifier?

3. When you open the valve to a water faucet, water rushes out. Explain why electric charge does not get pushed out the end of a wire attached to the terminal of a battery.

4. Calculate the ratio of the resistances of two wires that are identical except (a) wire A is twice as long as wire B, (b) wire A has twice the radius of wire B, and (c) wire A is made of copper and wire B is made of aluminum.

5. At one time aluminum rather than copper wires were used to transmit electricity through homes. Which wire must have the larger radius if they are the same length and have the same electrical resistance? Explain.

6. Why can birds perch on a 100,000-V power line with no adverse effects?

7. When repairing electronic equipment in which high voltages are present, you should keep one hand in your back pocket and work only with the other hand. Explain carefully why this procedure is recommended.

8. Does a 60-watt light bulb have more or less resistance than a 100-watt bulb? Explain.

9. Two long wires are connected across the same voltage. Calculate the ratio of the power used in wire A to that used in B if the wires are identical except that (a) wire A is twice as long as B, (b) the radius of wire A is twice the radius of wire B, (c) wire A is made of copper and wire B of aluminum.

10. Explain the difference between a kilowatt and a kilowatt-hour in terms of the quantities for which they are units.

Problems

24.2 Electric Current and Drift Velocity

1. A 100-A current flows through a starter motor while an automobile is being started. (a) Calculate the number of electrons flowing each second past a cross section of the starter wire. (b) How many electrons move past a cross section during the 10 seconds needed to start the automobile?

2. A long wire is connected to the terminals of a battery. In 8 seconds 9.6×10^{20} electrons pass a cross section along the wire. Calculate the current in the wire. If the electrons flow from left to right, in which direction is the current?

3. A typical flashlight battery will produce a 0.5-A current for about 3 hours before losing its charge. Calculate the total number

of electrons that flow during that time past a cross section of wire connecting the battery and light bulb.

4. Suppose that all the people on the earth moved at the same speed around a circular track. Approximately how many times per second would each person have to pass the starting line in order that the "people current" be the same as the number of electrons passing a cross section in a wire when a 1-A current flows?

5. A 5.0-A current caused by moving electrons flows through a wire. (a) Calculate the number of electrons that flow past a cross section each second. (b) Suppose that the same number of water molecules flow along a tiny stream. Calculate the volume of water that flows past a cross section of the stream each second. (*Hint:* The density of water is 1000 kg/m^3, and one water molecule has a mass of 3.0×10^{-26} kg.)

6. Calculate the average drift velocity of an electron in a copper wire of 0.20-mm radius carrying a current of 1.0 A.

7. A copper wire carrying a direct current of 60 A stretches 50 km from an electric power plant to your home. Calculate the time needed for an electron to travel from the plant to your home. The wire has a radius of 1.0 mm and an electron density of 8.0×10^{28} electrons per cubic meter. (*Note:* Actually, electric power companies transmit power by alternating current rather than by direct current, as we will learn in Chapter 29.)

24.3 and 24.4 Ohm's Law and Resistance

8. (a) Calculate the current through a 0.35-Ω flashlight filament when connected across two 1.5-V batteries (a potential difference of 3.0 V). (b) Calculate the resistance of the heating elements in a toaster if a 10-A current flows when connected across a 120-V potential difference.

9. A person accidently touches a 120-V power line with one hand while touching a 0-V ground wire with the other hand. Calculate the current through the body when the hands are dry (100,000-Ω resistance) and when wet (5000-Ω resistance). Are either or both currents dangerous? Explain.

10. (a) The light in an automobile draws a current of 1.0 A when connected to a 12-V battery. What is the light's resistance? (b) What potential difference is needed to produce a current of 5.0 mA through a 2.0-MΩ resistor?

11. As the potential difference in volts across a thin platinum wire increases, the current in amperes changes as follows: $(V, I) = (0, 0), (1.0, 0.112), (3.0, 0.337),$ and $(6.1, 0.675)$. Plot a graph of potential difference as a function of current and indicate whether the platinum wire satisfies Ohm's law. If so, what is the resistance of the wire?

12. A platinum resistance thermometer consists of a coil of 0.10-mm diameter platinum wire wrapped in a coil. Calculate the length of wire needed so that the coil's resistance at 20°C is 25 Ω. The resistivity of platinum at this temperature is 1.0×10^{-7} Ω · m.

13. A copper wire carries electric current 50 km from the power plant to your home. The resistance of the wire results in a 1200-V decrease in voltage when a 120-A current flows. Calculate the diameter of the wire.

14. The BMT subway line in New York City stretches roughly 30 km from the Bronx to Brooklyn. Calculate the electrical resist-

ance of its rail that carries electric current. The rail has a cross section of about 40 cm^2 and is made of steel whose resistivity is 10×10^{-8} ohm · m.

15. A wire whose resistance is 3.6 Ω is stretched so that its length is tripled while its volume remains unchanged. Calculate the resistance of the stretched wire.

16. A human nerve cell can be considered as a long, thin cylinder of radius 5×10^{-6} m and 0.3 m length. The resistivity of the fluid inside the cell is 0.5 Ω · m. Calculate the resistance of this fluid.

17. The electrical resistance of the fluid inside a rubber tube serves as an indicator of a linear displacement of some object attached to the tube. If the tube is stretched, its length increases and its cross-sectional area decreases. Both changes lead to increased resistance. The device is called a fluid-filled rubber-tube displacement transducer. Calculate the percentage change in resistance of the fluid inside a tube if its length is increased 10 percent. (*Hint:* The fluid's volume remains constant.)

18. The wires in a home carry 15 A of current. The voltage drop across the wires because of their resistance should be 1.0 V or less. If the wires are made of aluminum and have a radius of 1.2 mm, what is their maximum length?

19. You fall from a cliff and grab a high-voltage power line to break your fall. Your hands are separated by 0.60 m. The wire carries 300 A of current through a copper wire with 3.0×10^{-4} m^2 cross-sectional area. Calculate the voltage between your hands. Would you be electrocuted by this voltage? Explain. You will be in serious trouble if you grab another line with one hand while holding the high-voltage line with the other hand. Why?

20. Estimate the earth's resistance between two electrodes separated by 1 km. Sedimentary rock of resistivity 10^3 Ω · m is between the electrodes. Explain how you arrived at your choice of area to use in the calculation.

24.5 Resistance and Temperature

21. Calculate the resistance of a nichrome wire at 60°C if its resistance at 0°C is 25.00 Ω.

22. If the copper wire of an electromagnet has a resistance of 0.300 Ω when at room temperature (20°C) with no current flowing, what is the temperature of the wire when the current is turned on and the resistance increases to 0.347 Ω?

23. Calculate the resistance of a carbon rod when at 60°C if its resistance at 0°C is 0.0180 Ω.

24. (a) A thermistor, whose resistance varies with temperature as shown in Fig. 24.9, is placed in a stream of air, causing it to have a resistance of 2420 Ω. What is the temperature of the air? (b) What is the approximate resistance of the thermistor if the temperature of the air decreases by 10 C° from the answer in (a)?

25. The tungsten filament of a light bulb has a resistance of 8.00 Ω when no current flows, and its temperature is 20°C. Estimate the filament's temperature when a 1.00-A current flows after a 120-V potential difference is placed across the filament.

24.6 Electrical Power

26. A light bulb when connected to a 110-V potential difference uses electrical energy at a rate of 60 watts. Calculate the resistance of the bulb and the current flowing through the bulb.

27. A pocket calculator draws 0.40 A of current when connected to a 6.0-V battery. At what rate does the calculator use electrical energy?

28. An electric heater is to provide 1.4 kW of power when connected to a 120-V potential difference. What should the resistance of the heater be when turned on?

29. A mercury cell has a 1.5 A · h capacity (the product of the current it supplies and the time the current is supplied before the battery loses its charge). If the potential difference across its terminals is 1.35 V, how long can the cell provide the 0.1 mW power needed for a heart pacemaker before the cell stops operating?

30. The 12-V battery of a car provides a 0.20-A current to a car radio. How much electrical energy is provided in 30 minutes?

31. A refrigerator connected to a 120-V line runs for 180 hr during one month. While on, it draws 3.0 A of current. (a) Calculate the electrical energy used by the refrigerator during the month in kilowatt-hours and in joules. (b) Calculate the cost of the electricity at a rate of 7 cents per kilowatt-hour.

32. During a lightning flash, -6 C of charge are transferred across a potential difference of about 10^7 V. The flash lasts approximately 1 ms. Calculate the average power during the flash—that is, the average rate of electrical energy decrease.

33. An x-ray tube consists of a beam of electrons accelerated across a potential difference of about 40,000 V. Approximately 3×10^{16} electrons, accelerated through this potential difference, crash into a tungsten anode each second. Calculate the average electric power dissipated by the electron beam.

34. A half-horsepower motor draws a current of 4.0 A from a 120-V line. (a) Calculate the electric power used by the motor. (b) Calculate the motor's efficiency—that is, the work it does per unit time (convert $\frac{1}{2}$ hp to watts) divided by the electrical power it uses.

35. The United States will be switching more to coal-fired electric power plants in the 1980s rather than gas- or oil-fired plants. (a) *Estimate* the mass of coal that will be burned each year if 100 percent of our electricity is provided by burning coal. (b) *Estimate* the thermal energy transferred to the environment by this coal burning. The following numbers may be helpful. One kilogram of coal provides 2.9×10^7 J of thermal energy. About 35 percent of the coal's thermal energy is converted to electrical energy and the other 65 percent is transferred to the environment. The average person in the United States uses 31 kW · h of electricity per day.

36. A 1.5-V battery provides energy to a flashlight bulb at a rate of 15 watts. (a) Calculate the current through the bulb. (b) The battery provides energy for one hour, at which time its electrical energy has been discharged. Calculate the energy provided in units of joules and in kilowatt-hours. (c) Based on the cost of a flashlight battery at a local store, *estimate* the cost per kilowatt-hour of the electrical energy. How does this number compare to the cost per kilowatt-hour of electricity to your home from the local electric utility company (assume a cost of roughly $0.05–0.10/kW · h)?

37. *Estimate* the cost of electricity used to complete the homework assignment that includes this problem. The only restriction is that the assignment is worked at night (for estimation purposes only). Justify any numbers used in your estimate.

General Problems

G1. The resistivity of nichrome wire is $1.0 \times 10^{-6} \Omega \cdot$ m. Calculate the length of wire needed for a 1200-watt electric heater that is connected across a 120-V potential difference. The wire's radius is 0.40 mm.

G2. An electric hot water heater has two resistive heating elements that together produce thermal energy at a rate of 7000 watts. You are the last in a group of people to take a shower and find that there is no more hot water available. How long will you have to wait for the 150 kg of water in the water heater to warm from 15°C to 60°C? If the electric power company charges 7 cents per kilowatt-hour, what is the cost of heating the water?

G3. An electric grill made of iron has a specific heat of 460 J/kg · C° and a mass of 2.8 kg. To cook French toast, the grill is warmed from 20°C to 350°C by resistive heating wires that produce thermal energy at a rate of 1500 watts when connected to

a 115-V potential difference. Fifty percent of the thermal energy is radiated into the room as the grill warms. How many minutes are required to warm the grill?

G4. An airplane deicer melts 0.10 kg of ice from the wings of an airplane each minute. The deicer consists of resistive heating wires connected to a 24-V battery. Calculate the current through the heating wires and their resistance. Assume that the deicer transfers 100 percent of its energy to the ice.

G5. The wires in a large electromagnet carry 150 A of current across a voltage of 40 V. The wires are cooled by water flowing near the coiled wire of the electromagnet. If water flows at 2.0×10^4 cm^3 per minute, what is the difference in temperature of water leaving the magnet compared to water entering the magnet?

Direct-Current Electric Circuits

25

Until now, we have discussed only the electric current flowing from a voltage source through a single electronic device that has resistance. Most electric circuits consist of a combination of many resistive elements, one or more voltage sources, switches, and a variety of other electronic devices. Your home, for example, can be represented crudely by the electronic circuit shown in Fig. 25.1.

In this chapter we will develop techniques to determine the electric current flowing through each element in one of these more complicated circuits and the electric power used by each element. Our attention is restricted to circuits powered by direct-current sources of emf that have a variety of different resistive elements.

25.1 Kirchhoff's Loop Rule

Two important rules used to analyze complicated circuits are Kirchhoff's loop rule and Kirchhoff's junction rule. First the loop rule. The **loop rule** is based on the conservation-of-energy principle and states that the algebraic sum of the voltage changes around any closed path or loop in a circuit is zero. The loop rule can be understood by examining the flow of electric charge from a battery through wires connected to the filament of a light bulb (Fig. 25.2a). An elec-

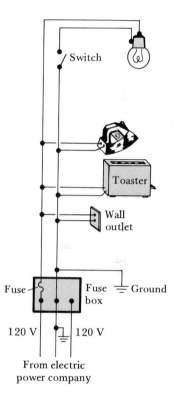

FIG. 25.1. An electric circuit for a home wiring system.

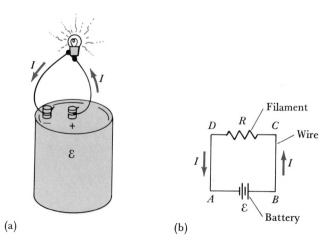

(a)   (b)

FIG. 25.2. (a) A light bulb is attached to a battery by wires. (b) An electronic circuit representing the equipment shown in (a).

tronic circuit representing this arrangement of battery, wires, and filament is shown in Fig. 25.2b.

Let us follow the motion of a positive charge as it makes one complete counterclockwise trip around the circuit (the charge moves in the same direction as the current). When using the loop rule, we will always pretend that moving positive charges cause the current. It is much easier to keep track of the signs of voltage changes caused by a positive charge moving in the direction of the current than to keep track of the signs of voltage changes caused by negative charges moving opposite the direction of the current. The result is the same in either case.

Consider first the change in voltage as a charge q moves from position A through the battery. Chemical energy must be expended to pull the charge through the battery away from the negative terminal and toward the positive terminal. For a positive charge, the trip across the battery is like being pulled up a hill whose height in units of volts is $\mathcal{E}$, the emf from the negative to positive terminal of the battery.

During the rest of the trip, the charge moves through resistors. The potential difference across a resistor of resistance R through which a current I flows is given by Ohm's law: $V = IR$. If the charge is moving in the direction of the current, the potential decreases as the charge moves through the resistor. The motion of the charge is analogous to the fall of a skydiver in the earth's gravitational field. The skydiver's elevation and gravitational potential energy decrease as the skydiver is pulled downward in the gravitational field. The electrical potential and potential energy of a positive electric charge also decreases as it is pulled through a resistive material by an electric field.

In the case of the circuit shown in Fig. 25.2b, there is no change in potential along the wires since they have almost zero resistance, and according to Ohm's law, $V = IR = I(0) = 0$. However, the tungsten filament of the light bulb does have resistance, and the change in voltage across this resistance is $V_{CD} = -IR$, the negative sign being used to indicate a lower voltage at D than at C.

The charge then moves along the second wire from position D back to position A where it started. The net voltage change for the whole trip is zero. The charge is at the same position and electrical elevation as it was when it started the trip. Thus, the sum of the changes in voltage for each step along the trip must add to zero:

	Voltage changes		
Through battery	Along wire	Through resistor	Along wire
$\mathcal{E}$ $+$	0 $-$	IR $+$	0 $= 0.$

The algebraic sum of the voltage changes around any closed loop in an electric circuit is zero.

The rule applies to circuits in which a steady current flows. If abrupt changes in current occur, such as when a switch is closed, the rule may not be applicable.

When using Kirchhoff's loop rule, we must be careful to use the correct signs for the voltage change across each element. These voltage-change conventions are summarized in Fig. 25.3. Note that the potential is higher by an amount $\mathcal{E}$ at the positive terminal of a battery compared to the potential at the

FIG. 25.3. Potential difference across batteries and resistors.

negative terminal. * In the case of a resistor, the potential is IR higher on the side that a current I enters a resistor than on the side it leaves. In other words, the drop in potential across the resistor in the direction of the current is $-IR$.

(a)

EXAMPLE 25.1 (a) Write a loop-rule equation for the circuit shown in Fig. 25.4a. (b) Solve the equation for the current that is flowing.

SOLUTION (a) To start the problem, we first draw an arrow indicating the direction of expected current flow (Fig. 25.4b). For this circuit the current flows counterclockwise, since the 10-V battery "pushes harder" than the 4-V battery, which opposes the current as drawn. We will discover later that if the direction is incorrectly chosen, the equations will indicate our error.

Starting at A (shown in Fig. 25.4b) and moving counterclockwise, we find the following voltage changes:

$$+10\text{ V} - I(2\,\Omega) - 4\text{ V} - I(6\,\Omega) - I(4\,\Omega) = 0.$$

Notice that the voltage change across the 4-V battery was negative; we moved from the positive to the negative terminal. (b) The above equation is now solved for the current:

$$I = \frac{10\text{ V} - 4\text{ V}}{2\,\Omega + 6\,\Omega + 4\,\Omega} = \frac{6\text{ V}}{12\,\Omega} = \underline{0.5\text{ A}}. \qquad \blacksquare$$

(b)

FIG. 25.4. (a) An electric circuit and (b) the same circuit with the current identified (including its direction).

25.2 Kirchhoff's Junction Rule

In the last section we looked at circuits with one loop. However, many electric circuits contain several loops. The current moving through a wire in one loop may come to a junction where it divides to travel in two or more wires, or the currents from several wires may join at a junction to move through a single wire. The flow of electric charge into and out of these *junctions* is much like the flow of water in a pipe. If the pipe divides into several pipes at a junction, the water flowing in the original pipe must equal the water leaving the junction in the other pipes.

For electric charge, the current flowing along one or more wires into a junction must equal the current flowing out of the junction along other wires. Consider point A of the circuit shown in Fig. 25.5. The currents in each wire are represented by the symbols I, I_1, and I_2, and their directions are indicated by the arrows. For point A, the current I entering that junction must equal the sum of the currents I_1 and I_2 leaving that junction:

$$I = I_1 + I_2$$

or

$$I - I_1 - I_2 = 0.$$

FIG. 25.5. The current divides into two branches at point A.

*We are ignoring the internal resistance of the battery.

The above equation is one example of the application of **Kirchhoff's junction rule:**

The algebraic sum of the electric currents flowing into and out of any junction is zero. That is,

$$\Sigma I = I_1 + I_2 + \cdots = 0. \tag{25.1}$$

A current is positive if entering the junction and negative if leaving the junction.

25.3 Use of Kirchhoff's Rules

You can apply Kirchhoff's rules most easily when analyzing electric circuits by using this procedure:

1. Assign a direction and symbol to the current in each independent branch of a circuit. If you mistakenly assign the wrong direction to a current, do not worry. When your equations are solved, you will find that the current has a negative sign. This indicates that the current actually moves in the opposite direction to the one you assigned.

2. Apply Kirchhoff's junction rule to enough junctions so that each current appears in at least one equation.

3. Apply Kirchhoff's loop rule to enough closed loops so that each current appears in at least one loop equation. Remember to use the sign conventions listed in Figure 25.3.

4. Solve the equations for the desired unknowns.

EXAMPLE 25.2 Solve for the currents in each branch of the circuit shown in Fig. 25.6a.

SOLUTION First, the circuit shown in Fig. 25.6a has three branches, whose currents are represented in Fig. 25.6b by the symbols I, I_1, and I_2. We expect that positive charge flows away from the positive terminal of the battery in a counterclockwise direction. The arrows for I, I_1, and I_2 are drawn in this direction.

Next, the junction rule applied to point A shown in Fig. 25.6b is

$$I - I_1 - I_2 = 0. \tag{25.2}$$

The junction rule could also be applied to point B. However, the equation we get is identical to Eq. (25.2), only with positive and negative signs reversed.

FIG. 25.6. (a) An electric circuit and (b) the same circuit with the currents in each branch identified.

(a)

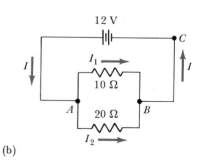

(b)

Next, we use the loop rule for the two different loops shown in Fig. 25.7. In each case, we start at point C and move around the loop in a counterclockwise direction.

$$+12 - 10I_1 = 0,$$
$$+12 - 20I_2 = 0.$$

The last step is to solve the equations for the unknowns. The above equations can be solved for I_1 and I_2:

$$I_1 = \tfrac{12}{10} = \tfrac{6}{5}\,\text{A},$$
$$I_2 = \tfrac{12}{20} = \tfrac{3}{5}\,\text{A}.$$

Substituting these results in Eq. (25.2), we find that

$$I = I_1 + I_2 = \tfrac{6}{5}\,\text{A} + \tfrac{3}{5}\,\text{A} = \tfrac{9}{5}\,\text{A}. \qquad \blacksquare$$

EXAMPLE 25.3 Calculate the currents flowing in each branch of the circuit shown in Fig. 25.8a. Also, calculate the potential difference across the 18-Ω resistor.

(a)

(b)

FIG. 25.7. Two loops of the circuit shown in Fig. 25.6.

(a)

(b)

FIG. 25.8

SOLUTION The circuit is redrawn in Fig. 25.8b with the current symbols and directions identified. We have intentionally drawn the direction of current I_3 incorrectly to show that the proper application of Kirchhoff's rules will indicate our error. The junction rule applied to point A results in the equation

$$I_1 - I_2 + I_3 = 0. \qquad (25.3)$$

Notice that I_3 has a plus sign in front of it because the arrow shows current I_3 entering point A. The loop rule is applied for the two different loops shown in Fig. 25.9 as follows:

$$+3\,\text{V} - (18\,\Omega)I_2 = 0$$

or

$$I_2 = \frac{3\,\text{V}}{18\,\Omega} = \frac{1}{6}\,\text{A}.$$

For the loop in Fig. 25.9b we find that

$$+9\,\text{V} - (12\,\Omega)I_1 - 3\,\text{V} = 0$$

or

$$I_1 = \frac{9\,\text{V} - 3\,\text{V}}{12\,\Omega} = \frac{1}{2}\,\text{A}.$$

(a)

(b)

FIG. 25.9. Two loops of the circuit shown in Fig. 25.8.

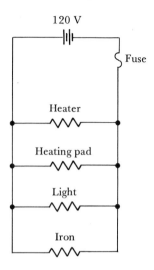

120 V

Fuse

Heater

Heating pad

Light

Iron

FIG. 25.10. The circuit for one outlet of a home wiring system.

Then, using Eq. (25.3), we find that

$$I_3 = I_2 - I_1 = \tfrac{1}{6}\,\mathrm{A} - \tfrac{1}{2}\,\mathrm{A} = \underline{-\tfrac{1}{3}\,\mathrm{A}}.$$

The negative sign indicates that the direction of I_3 was chosen incorrectly. The 9-V battery actually forces charge through the 3-V battery from left to right. The voltage across the 18-Ω resistor is

$$V = -I_2 R = -(\tfrac{1}{6}\,\mathrm{A})(18\ \Omega) = \underline{-3\ \mathrm{V}}.$$

The voltage is 3 V lower at point B than at point A. ∎

EXAMPLE 25.4 A 1380-watt electric heater, a 180-watt heating pad, a 120-watt light bulb, and a 1200-watt iron are connected in parallel across the terminals of a single wall outlet (Fig. 25.10). If the current through the wire from the fuse box to the wall outlet exceeds 20 A, a fuse is blown. Will the fuse be blown when all these appliances are in operation at the same time? The voltage across each appliance is 120 V.

SOLUTION Equation (24.11) can be used to calculate the current through each appliance when connected to the 120-V wall outlet:

$$\text{Heater:}\qquad I = \frac{P}{V} = \frac{1380\ \text{watts}}{120\ \text{V}} = 11.5\ \text{A},$$

$$\text{Heating pad:}\qquad I = \frac{180\ \text{watts}}{120\ \text{V}} = 1.5\ \text{A},$$

$$\text{Light:}\qquad I = \frac{120\ \text{watts}}{120\ \text{V}} = 1.0\ \text{A},$$

$$\text{Iron:}\qquad I = \frac{1200\ \text{watts}}{120\ \text{V}} = 10\ \text{A}.$$

In order to provide the current for each appliance, the wires from the fuse box to the wall outlet must carry the total current needed for all of the appliances, or $11.5 + 1.5 + 1.0 + 10 = 24\ \text{A}$. If the current exceeds 20 A, the fuse, a thin piece of copper wire, becomes hot and melts. Thus, when all the appliances are operating, the fuse melts or blows and the current stops. If a fuse were not included in the circuit for the wire, then when the current exceeded 20 A the wires that carry the current through the walls of the house would become very warm and could possibly start a fire. ∎

25.4 Equivalent Resistance

When analyzing an electric circuit, it is often convenient to replace several of its resistors with a single equivalent resistor that produces the same effect as the resistors it replaces. These replacements can be made for several resistors in series and for several resistors in parallel.

Resistors in Series

A circuit containing three resistors in series is shown in Fig. 25.11a. Resistors are in series if the same current flows through each with no branching between resistors. If a source of voltage $\mathcal{E}$ is placed across the resistors, a current I flows through them. We wish to replace the three resistors in series by a single equivalent resistor R_{eq} that will cause the same current to flow when attached to an identical voltage source $\mathcal{E}$ (Fig. 25.11b).

The loop rule applied to the circuit shown in Fig. 25.11a results in the equation

$$\mathcal{E} - IR_1 - IR_2 - IR_3 = 0$$

or

$$I = \frac{\mathcal{E}}{R_1 + R_2 + R_3}.$$

When applied to the circuit shown in Fig. 25.11b, the loop rule results in the equation

$$\mathcal{E} - IR_{eq} = 0$$

or

$$I = \frac{\mathcal{E}}{R_{eq}},$$

where R_{eq} is the resistor that is equivalent to R_1, R_2, and R_3. Since the current is the same in both circuits,

$$I = \frac{\mathcal{E}}{R_1 + R_2 + R_3} = \frac{\mathcal{E}}{R_{eq}}.$$

This will be true if

$$R_{eq} = R_1 + R_2 + R_3.$$

In general, any group of **series resistors** can be replaced by a single equivalent resistor whose resistance is

$$R_{eq} = R_1 + R_2 + R_3 + \cdots. \tag{25.4}$$

Resistors in Parallel

Three parallel resistors are shown in Fig. 25.12a. Resistors are in parallel if one side of each resistor is connected to a common junction and the other side of each resistor is connected to another common junction. If a source of voltage $\mathcal{E}$ is connected across these resistors, the current I leaving it splits when it reaches the first junction (point A in Fig. 25.12a). The voltage across each resistor in Fig. 25.12a is the same and has a value $\mathcal{E}$. Thus, the current through the resistors is given by

$$I_1 = \frac{\mathcal{E}}{R_1}, \qquad I_2 = \frac{\mathcal{E}}{R_2}, \qquad I_3 = \frac{\mathcal{E}}{R_3}.$$

When the junction rule is applied to point A, the total current I is seen to be

$$I = I_1 + I_2 + I_3.$$

Substituting for I_1, I_2, and I_3, we find that

$$I = \frac{\mathcal{E}}{R_1} + \frac{\mathcal{E}}{R_2} + \frac{\mathcal{E}}{R_3} = \mathcal{E}\left(\frac{1}{R_1} + \frac{1}{R_2} + \frac{1}{R_3}\right).$$

(a)

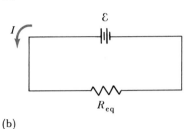

(b)

FIG. 25.11. (a) A circuit with three series resistors and (b) an equivalent circuit.

(a)

(b)

FIG. 25.12. (a) A circuit with three parallel resistors and (b) an equivalent circuit.

If the three parallel resistors are replaced by a single equivalent resistor (Fig. 25.12b), the current passing through it is

$$I = \frac{\mathcal{E}}{R_{eq}} = \mathcal{E}\left(\frac{1}{R_{eq}}\right).$$

In order for the current I to be the same for each circuit, the equivalent resistance must be given by the equation:

$$\frac{1}{R_{eq}} = \frac{1}{R_1} + \frac{1}{R_2} + \frac{1}{R_3}.$$

In general, any group of **parallel resistors,** $R_1, R_2, R_3, \ldots$, can be replaced by an equivalent resistor whose resistance is calculated using the equation:

$$\frac{1}{R_{eq}} = \frac{1}{R_1} + \frac{1}{R_2} + \frac{1}{R_3} + \cdots. \tag{25.5}$$

These two rules for replacing series and parallel resistors by an equivalent resistor can now be applied to replacing, step by step, a group of many resistors by one or more equivalent resistors that produce the same effect as all the resistors. Each step involves replacing either several series resistors or several parallel resistors by an equivalent resistance. We cannot, however, simultaneously replace combinations of series and parallel resistors by a single equivalent resistor. Consider the next examples.

EXAMPLE 25.5 (a) Calculate the equivalent resistance of the group of resistors shown in Fig. 25.13a. (b) Calculate the total current I flowing through the power supply.

SOLUTION (a) It is simpler to calculate the current I by using equivalent resistors rather than by using Kirchhoff's rules. The procedure is done in steps. The first step involves replacing the 3-Ω and 6-Ω parallel resistors by an equivalent resistor:

$$\frac{1}{R_{eq}} = \frac{1}{3\,\Omega} + \frac{1}{6\,\Omega} = \frac{2}{6\,\Omega} + \frac{1}{6\,\Omega} = \frac{3}{6\,\Omega}$$

or

$$R_{eq} = \frac{6\,\Omega}{3} = 2\,\Omega.$$

The new circuit with the parallel resistors replaced by a single 2-Ω resistor is shown in Fig. 25.13b.

We are now left with three resistors in series. The equivalent resistance of these series resistors is:

$$R_{eq} = 5\,\Omega + 2\,\Omega + 8\,\Omega = \underline{15\,\Omega}.$$

(b) Our new circuit, shown in Fig. 25.13c, has only one 15-Ω resistor attached to the 12-V power source. The current I through this resistor and power source is given by

$$\mathcal{E} - IR = 0.$$

(a)

(b)

(c)

FIG. 25.13. (a) A circuit with four resistors is reduced in steps to one with a single equivalent resistor (c).

Thus,

$$I = \frac{\mathcal{E}}{R} = \frac{12\ \mathrm{V}}{15\ \Omega} = \underline{0.80\ \mathrm{A}}.$$ ■

EXAMPLE 25.6 A more complicated step-by-step reduction of a group of resistors is shown in Fig. 25.14. You should follow each step to see how a single resistor replaces two or more resistors that were part of the previous circuit.

■

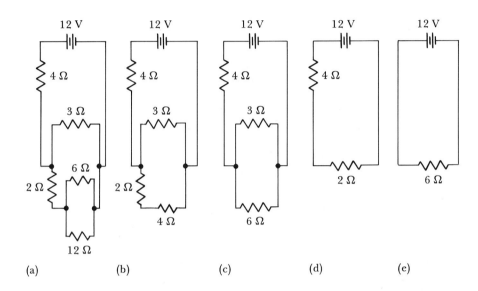

FIG. 25.14. A series of steps is used to calculate the equivalent resistance of a group of resistors shown in (a).

25.5 Measuring Current and Voltage

Kirchhoff's laws and the ideas of equivalent resistance are used to design electrical circuits. Once built, the circuits can be tested using meters or other electrical instruments that measure the voltage at various parts of the circuit and the current flowing through different elements in the circuit.

Care must be taken when making any type of measurement that the measuring device does not alter the quantity being measured. A ruler used to measure a person's height does not change the height. A meter used to measure current or voltage should not change the current or voltage that is being measured. The effect of a meter on a circuit depends on what is called its **internal resistance**—the net electrical resistance of all the parts inside the meter. We will find that meters used to measure current must have a very low internal resistance and meters used to measure voltage must have a large internal resistance.

Measuring Current

To measure the current flowing in one part of a circuit, one must open the circuit (Fig. 25.15) and insert into it a current-measuring meter called an **ammeter.** The current being measured must now pass through the ammeter, which indicates the value of the current.

FIG. 25.15. (a) To measure the current in a circuit, it should first be opened, as in (b). (c) The current is measured by inserting an ammeter into the circuit.

In order that the meter does not alter the current being measured, the meter's internal resistance must be much less than the resistance of other resistors in the circuit through which the current flows. For example, in the circuit shown in Fig. 25.15a, the current is $I = \mathcal{E}/R$, where R is the resistance of the circuit before the meter is added. After the meter is placed in the circuit, the total resistance of the circuit increases to $R + R_{int}$, where R_{int} is the internal resistance of the meter. With the meter in the circuit, the current is reduced to $I' = \mathcal{E}/(R + R_{int})$. If $R_{int} \ll R$, then we can ignore R_{int} in the expression for I', and the measured current I' is for practical purposes equal to I, the current when the meter is not present. The internal resistance of ammeters is typically 1 ohm or less.

The fact that the internal resistance of an ammeter is small can result in damage to the meter if it is used incorrectly. For instance, if the wires (called **leads**) that are used to connect the meter with the circuit being tested are accidentally placed across a resistor or across a battery, the voltage across the resistor or battery may be large enough to produce a very large current through the meter. The current is likely to damage the ammeter. An ammeter should always be inserted into a circuit in a way that allows the normal current of the circuit to pass through it without producing a voltage across the leads.

FIG. 25.16. Voltage is measured by placing the leads from a voltmeter across the section of the circuit for which you wish to measure the voltage change.

Measuring Voltage

The potential difference or voltage between two different points in a circuit, such as from one side of a resistor or battery to the other side, is measured by connecting the leads from a **voltmeter** to these points (Fig. 25.16). Unlike an ammeter, a voltmeter should have a much higher internal resistance than the resistance of the section of the circuit across which the voltage is being measured. Suppose, for example, that you wish to measure the voltage across a resistor of resistance R that has a current I flowing through it. When the voltmeter is not connected (Fig. 25.16a), the voltage across the resistor is, according to Ohm's law, $V = IR$. If a voltmeter having internal resistance R_{int} less than R is connected in parallel to R (Fig. 25.26b), then most of the current flows through the voltmeter rather than through the resistor R. If the current through the resistor is reduced, the voltage drop (IR) across the resistor is also reduced and an incorrect voltage reading will be made. The internal resistance of the meter should be roughly 100 times or more greater than the resistance of the elements in the circuit across which the voltage is being measured.

The internal resistance of a voltmeter is usually stated on the voltmeter. For example, if the maximum or full-scale reading of the voltmeter is 30 V and its internal resistance is 20,000 Ω per volt full-scale, then its internal resistance is $30 \times 20,000 \ \Omega = 6 \times 10^5 \ \Omega$.

Summary

1. **Kirchhoff's loop rule:** The algebraic sum of the changes in voltage around any closed loop in an electric circuit is zero. The positive terminal of a battery is at $+\mathcal{E}$ higher potential than the negative terminal (we ignore the internal resistance of the battery). The voltage drop across a resistor in the direction of a current is $-IR$.

2. **Kirchhoff's junction rule:** The algebraic sum of the electric currents flowing into or out of any junction in an electric current is zero:

$$\Sigma I = I_1 + I_2 + \cdots = 0. \qquad (25.1)$$

A current is positive if entering the junction and negative if leaving the junction.

3. **Equivalent resistors:** A group of **series resistors,** R_1, R_2, R_3, . . . , can be replaced by an equivalent resistor R_{eq} of resistance

$$R_{eq} = R_1 + R_2 + R_3 + \cdots. \qquad (25.4)$$

A group of **parallel resistors,** R_1, R_2, R_3, . . . , can be replaced by an equivalent resistor whose resistance R_{eq} is determined using the equation

$$\frac{1}{R_{eq}} = \frac{1}{R_1} + \frac{1}{R_2} + \frac{1}{R_3} + \cdots. \qquad (25.5)$$

4. **Meters:** An **ammeter** is a device used to measure the current flowing through a part of an electric circuit. Ammeters have low internal resistance and are inserted in the circuit when measuring current. A **voltmeter** is a device used to measure the voltage or potential difference between two different points of a circuit. A voltmeter has high internal resistance and is connected in parallel with the part of the circuit across which the voltage is being measured.

Questions

1. You wish to be able to switch a light in the center of a hallway on and off from either end of the hall. Draw an electronic circuit that will allow you to do this.

2. It is possible to make Christmas tree lights that work when connected in parallel. If lights having a different resistance are used, they can also be connected in series. (a) Describe one advantage of lights connected in parallel. (b) Should the parallel lights or the series lights have a greater resistance so that each light will use the same electrical power in each situation? Explain.

3. Three 9-Ω resistors are to be connected to a 12-V battery. In order to produce the most thermal energy in the resistors, should they be connected in series or in parallel? Justify your answer.

4. Two light bulbs, one 30 watt and the other 120 watt, are built to be connected across a potential difference of 120 V. Which bulb uses the most electrical power when the bulbs are connected in series to a 240-V potential difference? Explain.

5. Describe one of the differences between an ammeter and a voltmeter.

Problems

25.1, 25.2, and 25.3
Kirchhoff's Rules

1. Calculate the current in each branch of the circuit shown in Fig. 25.17 where $R_1 = 0\,\Omega$, $R_2 = 18\,\Omega$, $R_3 = 9\,\Omega$, and $\mathcal{E} = 6.0$ V.

FIG. 25.17

2. The currents through resistors R_1, R_2, and R_3 shown in Fig. 25.17 are I_1, I_2, and I_3, respectively. (a) Use Kirchhoff's junction rule to write an equation relating the currents to each other. (b) Use Kirchhoff's loop rule for three different loops to write three equations that relate these quantities and the emf of the battery.

3. Calculate the current in each branch of the circuit shown in Fig. 25.17 where $R_1 = 50\,\Omega$, $R_2 = R_3 = 20\,\Omega$, and $\mathcal{E} = 120$ V.

4. The currents through resistor R_1 in Fig. 25.17 is 2.0 A. Calculate the currents through resistors R_2 and R_3, and the emf of the battery. $R_1 = 4\,\Omega$, $R_2 = 10\,\Omega$, and $R_3 = 40\,\Omega$.

5. Calculate the current through the circuit shown in Fig. 25.18 and the voltage from A to B, where $\mathcal{E}_1 = 12$ V, $\mathcal{E}_2 = 3$ V, $R_1 = R_2 = 1\,\Omega$, and $R_3 = 16\,\Omega$.

FIG. 25.18

6. Calculate the value of $\mathcal{E}_1$ so that a current of 1 A flows clockwise in the circuit shown in Fig. 25.18, where $\mathcal{E}_2 = 12$ V, $R_1 = 2\,\Omega$, $R_2 = 1\,\Omega$, and $R_3 = 12\,\Omega$.

7. Write the loop rule for two different loops in the circuit shown in Fig. 25.19 and the junction rule for point A. Solve the equations to find the current in each loop when $\mathcal{E}_1 = 3$ V, $\mathcal{E}_2 = 6$ V, $R_1 = 10\ \Omega$, $R_2 = 20\ \Omega$, and $R_3 = 30\ \Omega$.

FIG. 25.19

8. Determine the value of R_2, shown in Fig. 25.19, so that the current through R_3 equals twice that through R_2. The values of the other circuit elements are: $\mathcal{E}_1 = 12$ V, $\mathcal{E}_2 = 15$ V, $R_1 = 15\ \Omega$, and $R_3 = 30\ \Omega$.

9. A simplified electrical circuit for a home is shown in Fig. 25.20. Calculate the currents through the fuse, light bulb, electric crock, and toaster.

FIG. 25.20

10. The voltage across wall sockets in Europe is twice that in the United States (220 V rather than 110 V). If you plugged your electric clock radio into an outlet in a European hotel, it might be burned out. To prevent this you place a light bulb in series with one of the wires from the wall socket to the radio (see Fig. 25.21). Explain carefully what the resistance of the bulb should be relative to that of the radio so that the voltage across the radio will be 110 V.

FIG. 25.21

11. (a) Write Kirchhoff's rules for two loops and one junction for the circuit shown in Fig. 25.22. (b) Solve the equations for the current in each branch of the circuit when $\mathcal{E}_1 = 10$ V, $\mathcal{E}_2 = 2$ V,

FIG. 25.22

$R_1 = 50\ \Omega$, $R_2 = 200\ \Omega$, and $R_3 = 20\ \Omega$. (c) Calculate the voltage across resistor R_3 from point B to point A.

12. The circuit shown in Fig. 25.23 is an example of a voltage divider. The 20-V potential difference is divided by R_1 and R_2, and only a fraction of the voltage appears across the load resistor, R_L. Calculate the value of R_2 so that the voltage across R_L is 4 V when $R_1 = 80\ \Omega$ and $R_L = 600\ \Omega$.

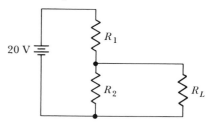

FIG. 25.23.   A voltage divider.

13. A Wheatstone bridge circuit used to measure resistance is shown in Fig. 25.24. Resistances R_1 and R_2 equal $20\ \Omega$, and R_3 is adjusted so that no current flows through the *galvanometer* at the center. The current I_g is zero when R_3 is $67.5\ \Omega$. Calculate the value of the unknown resistance R_x. Be sure to show clearly your calculation technique.

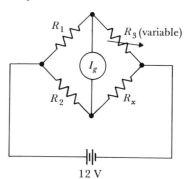

FIG. 25.24.   A Wheatstone bridge.

14. A battery has a small **internal resistance** r to charge flowing through it. The battery can be represented by a circuit such as shown in Fig. 25.25a where points A and B represent the battery terminals. If a current flows through the battery, the voltage across the terminals is

$$V_{AB} = \mathcal{E} - Ir.$$

For most purposes the small Ir drop can be ignored. A potentiometer is a device that can measure the value of the battery's emf $\mathcal{E}$ when no current flows. The slide wire of the potentiometer circuit shown in Fig. 25.25b is moved along the resistor until the current I_g through the battery stops. Derive an equation showing how the unknown emf $\mathcal{E}$ of the battery is related to $\mathcal{E}_p$, R_1, and R_2 when I_g is zero.

15. One 120-V electrical line in a home is connected to a 60-watt light bulb, a 1050-watt toaster, a 600-watt heater, and the 2000-watt burner of a stove. The devices, connected in parallel in a circuit similar to that shown in Fig. 25.10, cause the circuit's fuse to blow when all are on at the same time. However, if all but the light are on, the fuse does not blow. Roughly what current causes the fuse to blow?

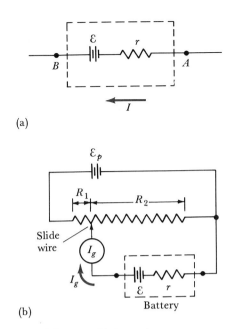

(a)

(b)

FIG. 25.25. (a) A battery with internal resistance. (b) A potentiometer is used to measure the emf of the battery.

16. Three resistors are connected as shown in Fig. 25.26. Each resistor can dissipate no more than 0.5 watt of electrical power. Calculate the maximum current that can flow through the circuit.

FIG. 25.26

17. A battery has an emf of 12 V and an internal resistance of $3 \, \Omega$ (see the discussion with Problem 14). (a) Calculate the power delivered to a resistor R connected to the battery terminals for values of R equal to 1, 2, 3, 4, 5, and $6 \, \Omega$. (b) Plot on a graph the calculated values of P versus the different values of R. Connect the points by a smooth curve. You should confirm that the maximum power is delivered when R has the same resistance as the internal resistance of the power source ($3 \, \Omega$ in this example).

25.4 Equivalent Resistance

18. Calculate the equivalent resistance of the resistors shown in Fig. 25.26.

19. Calculate the equivalent resistance between points 1 and 2 for the circuit shown in Fig. 25.27.

20. Calculate the equivalent resistance between points 1 and 2 for the circuit shown in Fig. 25.28. The 1200-Ω resistor on the bottom is divided in half by its contact with the 300-Ω resistor at the arrow.

FIG. 25.27

FIG. 25.28

21. Repeat Problem 20 for the situation in which the arrow attached to the 300-Ω resistor contacts the 1200-Ω resistor one fourth of the distance from its left edge (the 1200-Ω resistance is divided into 300-Ω and 900-Ω resistances).

22. Calculate the resistance between points A and B for the circuit shown in Fig. 25.29 if $R_1 = 5 \, \Omega$, $R_2 = 4 \, \Omega$, $R_3 = R_4 = 12 \, \Omega$, $R_5 = 3 \, \Omega$, and $R_6 = 6 \, \Omega$.

FIG. 25.29

23. Calculate the resistance between points A and B for the circuit shown in Fig. 25.29 if $R_1 = 10 \, \Omega$, $R_2 = 20 \, \Omega$, $R_3 = 80 \, \Omega$, $R_4 = 100 \, \Omega$, $R_5 = 400 \, \Omega$, and $R_6 = 120 \, \Omega$.

24. A battery produces a 2.0-A current when connected to an unknown resistance R. When a 10-Ω resistor is connected in series with R, the current drops to 1.2 A. Calculate the emf $\mathcal{E}$ of the battery and the resistance R.

25. A fuse for one line in your home's electrical system melts if more than 30 A of current flows. Determine if the fuse will melt if the following appliances are all connected in parallel to that line: a 13-Ω toaster, an 18-Ω dishwasher, a 24-Ω refrigerator, and a 15-Ω heater. The potential difference across the devices is 120 V.

26. Your high-fidelity amplifier has one output for a speaker whose resistance is 8 Ω. How can you arrange two 8-Ω speakers, one 4-Ω speaker, and one 12-Ω speaker so that all are powered by

the amplifier and their equivalent resistance when connected together in this way is 8 Ω.

27. Nine Christmas tree lights are connected in parallel across a 120-V potential difference. The cord to the wall socket carries a current of 0.36 A. (a) Calculate the resistance of one of the bulbs. (b) What would the current be if the bulbs were connected in series?

28. Calculate the current through the circuit shown in Fig. 25.30.

FIG. 25.30

29. (a) Calculate the current I through the circuit shown in Fig. 25.31 when $R_1 = 200$ Ω and $R_2 = 200$ Ω. (b) Suppose that R_2 is broken and its resistance becomes infinite. By what factor must the voltage increase in order for I to be the same as in part (a). (c) The circuit is analogous in many ways to the circulatory system of the body. Develop the analogy. For example, what parts of the circulatory system are analogous to the current, voltage, and resistance? What happens in the circulatory system when the resistance of one of its branches increases?

FIG. 25.31

30. Two light bulbs use 30 watts and 60 watts, respectively, when connected in parallel to a 120-V power source. How much power does each bulb use when connected in series across the 120-V power source assuming that their resistances remain the same?

31. A bus has twenty 15-watt light bulbs connected in parallel to a 12-V battery. The battery is rated at 120 A · hr—that is, it will provide 1 A of current for 120 hours, 2 A of current for 60 hours, and so on. How long will the battery last with all of the lights operating while the bus is parked?

32. Three equal resistors, when connected in series, together consume electrical energy at a rate of 15 watts. Calculate the power consumed by the resistors when connected in parallel to the same potential difference.

25.5. Measuring Current and Voltage

33. A voltmeter with an internal resistance of 10,000 Ω is used to measure the voltage across R_2 shown in Fig. 25.32. Calculate the voltage that will be measured when $R_1 = R_2 = 1000$ Ω.

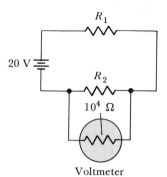

FIG. 25.32   Voltmeter

34. Repeat the calculations in Problem 33 when $R_1 = R_2 = 20,000$ Ω. Explain why the calculated voltage when the meter is connected across R_2 does not agree with the calculated voltage across R_2 when no meter is present.

35. A voltmeter measures a voltage V across an ammeter of internal resistance R_A and an unknown resistance R that is in series with the ammeter. A current I flows through the ammeter and through R. Show that

$$R = \frac{V}{I} - R_A.$$

36. Show that for the meter arrangement shown in Fig. 25.33,

$$R = \left(\frac{V}{IR_V - V}\right) R_V,$$

where R_V is the voltmeter resistance.

FIG. 25.33

37. A galvanometer is a device whose needle deflection is proportional to the current passing through its coil (Fig. 25.34). Suppose that the deflection is maximum with a 1-mA current through the coil. Calculate the value of a parallel resistance, called a **shunt resistance** R_S, which causes a 1-mA current through a 10-Ω coil when a 10-A current flows through the galvanometer.

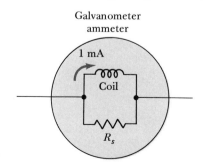

FIG. 25.34

38. The galvanometer described in the last problem can also be used to measure voltage. In this case a large resistance R_V is placed in series with the galvanometer coil (Fig. 25.35). Calculate the value of R_V in order for a 1-mA current to flow when the voltage across the galvanometer voltmeter is 1000 V.

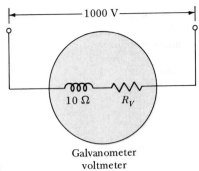

Galvanometer
voltmeter

FIG. 25.35

26 Capacitors

In the last two chapters we studied electric circuits and some of the elements commonly found in them, such as direct-current power supplies and resistors. Another device commonly found in electronic circuits is a capacitor, a device that stores electric charge on its conducting surfaces. Capacitors are used in radio and television sets, in automobile ignition systems, in electronic flash attachments for cameras, in the circuits of heart pacemakers, in the starting circuits for electric motors, as electronic filters, and as parts of radiation detection devices.

In this chapter, we will consider the structure of capacitors, how they store electric charge and electric energy, and how they are used in electric circuits with resistors, batteries, and switches.

26.1 Capacitors

A **capacitor** consists of two electrical conductors separated by a nonconducting material. For example, a parallel-plate capacitor has two parallel metal plates separated by air, rubber, paper, or some other material that does not conduct electric charge (Fig. 26.1a). When the conducting plates are connected to the terminals of a voltage source, as in Fig. 26.1b, negative charge is forced onto one plate, leaving the other plate with a net positive charge.

While some capacitors are made of parallel metal plates, most (especially those used in electronic equipment) consist of strips of metal foil separated by a thin piece of paper or some other nonconducting material. The strips are then rolled in a cylinder so that the unit occupies little volume (Fig. 26.2).

Not all capacitors are manufactured. Nerve cells and other cells in the body have properties of a capacitor (Fig. 26.3), the conductors of which are the fluids on the inside and the outside of the cell. The cell membrane serves as a moderately nonconductive medium that separates the two conducting fluids. In the membrane chemical processes, which are only partly understood, cause ions to be "pumped" across the membrane. The membrane's inner surface accumulates a small excess of negative charge, and the outer surface accumulates a small excess of positive charge. Approximately 10 percent of our metabolic energy is used to pump electric charge across the membranes of body cells. In nerve cells the energy is released when a nerve impulse travels along the cell. For other

FIG. 26.1. (a) A parallel-plate capacitor. (b) When the capacitor is attached to a voltage source, opposite-sign charges move onto the two plates.

FIG. 26.2. A capacitor used in electronic equipment.

510

body cells the energy stored by these separated charges becomes involved in a complex way with the chemical reactions that provide energy for our body.

All capacitors, when drawn in an electric circuit, are represented by the following symbol: ─┤├─ . Occasionally one of the lines is drawn slightly curved (─┤)─).

26.2 Capacitance

In earlier chapters we learned that the current flowing through a resistor depends on the voltage across the resistor and on its resistance. In the case of capacitors we are usually interested in the positive and negative electric charge that accumulates on its plates when a voltage is placed across them. The positive charge that accumulates on one plate and the negative charge on the other plate depends on the voltage across the plates and on another quantity called the capacitance of the capacitor.

Capacitance is an indicator of the amount of charge a particular capacitor will store per volt of potential difference across its plates; the larger the capacitance, the more the charge that is stored. The capacitance of a capacitor is analogous to the resistance of a resistor in that capacitance depends on the materials of which the capacitor is made and on its dimensions. Specifically, the capacitance depends on the size of its conducting plates or surfaces, on the separation of these plates, and on the type of material separating them.

To understand how these factors affect capacitance, look at the parallel-plate capacitors shown in Fig. 26.4. First, capacitance is greater for capacitors with plates of large area A; the more space available on which to store charge,

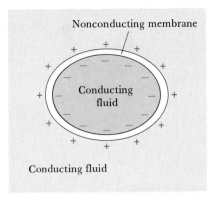

FIG. 26.3. A body cell has properties of a capacitor. The inside and outside fluids serve as conducting plates separated by the nonconducting membrane.

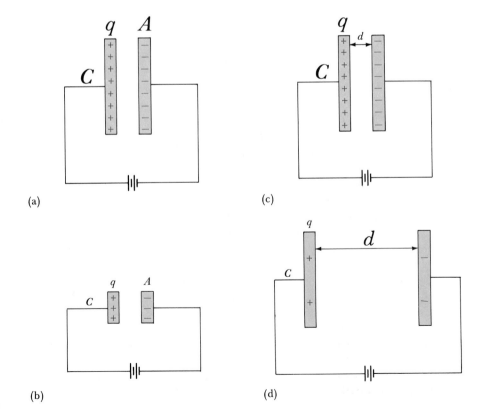

(a)

(b)

(c)

(d)

FIG. 26.4. (a) A capacitor with plates of large area can store more charge than (b) one with plates of small area. More charge is stored if (c) the plates are close to each other rather than (d) far apart.

the greater the capacitance (Fig. 26.4a). Similarly, the smaller the plate area, the less the ability to store charge and the less the capacitance (Fig. 26.4b).

Capacitance also depends upon the separation d of the plates (Fig. 26.4c and d). When the plate separation is small, as in (c), more charge is drawn onto the plates because of the greater mutual attraction between the opposite-sign charges. When the plates are separated by a large distance, as in (d), little charge is attracted to them. Negative charge on the right plate feels little attraction for positive charge on the left one. The ability of capacitor plates to attract and store charge is inversely related to the separation d of the plates; as d increases, the capacitance decreases, and vice versa.

The third quantity on which the capacity to store electric charge depends is the dielectric constant K of the insulating material between the plates. Recall from Section 22.4 that materials with high dielectric constant are often made of polar molecules (molecules with an excess positive charge on one side and an equal excess negative charge on the other side). When capacitor plates become charged, polar molecules between them become oriented so that their negative side points toward the positively charged plate and the positive side points toward the negatively charged plate. The polar molecules near the plate tend to cancel the effect of some of the charge on the plate and make it easier for more charge to move onto the plates. Thus, more charge moves onto the capacitor plates for capacitors whose plates are separated by a material of high dielectric constant. The capacitance increases in proportion to the dielectric constant of the material. We will have more to say about this in Section 26.3.

The capacitance of a parallel-plate capacitor is calculated using the equation

$$C = \frac{K}{4\pi k}\left(\frac{A}{d}\right), \tag{26.1}$$

where k is the constant in Coulomb's law $(k = 9.0 \times 10^9\,\text{N} \cdot \text{m}^2/\text{C}^2)$. For other types of capacitors (cylindrical or those shaped like a body cell), different equations must be used. In most of our work, capacitance will be a given quantity. Keep in mind, though, that capacitance depends on intrinsic geometrical and material properties of the capacitor.

Charge on a Capacitor

If a capacitor of capacitance C is attached to a voltage source with voltage V across its terminals, the magnitude of charge q on each plate (positive on one plate and negative on the other) is

$$q = CV. \tag{26.2}$$

Equation (26.2) is often used to define capacitance:

$$C = \frac{q}{V}.$$

However, the capacitance C of a capacitor is an intrinsic property of its dimensions and the material separating the conducting plates. Equation (26.2) simply relates the three quantities C, q, and V and can be used to calculate the value of any one of the three quantities if the other two are known.

Unit of Capacitance

We see by rearranging Eq. (26.2) that $C = q/V$. Thus, the unit of capacitance must equal the units of the quantity q/V—that is, coulomb/volt. The unit of capacitance called the **farad** (named after Michael Faraday) is defined as one coulomb/volt:

$$1 \text{ farad} = 1 \frac{\text{coulomb}}{\text{volt}} \quad \text{or} \quad 1 \text{ F} = 1 \frac{\text{C}}{\text{V}}.$$

A one-farad capacitor attached to a one-volt power supply would have a charge of $+1$ C on one plate and a charge of -1 C on the other plate. Most capacitors store much less than 1 C of charge when attached to a one-volt power supply. Capacitors found in electronic equipment usually have capacitances between 5000 μF and 1 pF, where $1 \mu\text{F} = 10^{-6}$ F and $1 \text{ pF} = 10^{-12}$ F.

EXAMPLE 26.1 (a) Estimate the capacitance of your physics book (its capacity to store electric charge) if it had metal covers. The dielectric constant of paper pages is approximately 6. (b) What voltage must be placed across its covers in order to get 10^{-6} C of charge on each cover (positive on one cover and negative on the other)?

SOLUTION (a) The covers of the book have an area of about 0.20 m $\times$ 0.25 m = 0.050 m². The covers are separated by about 3.6 cm $= 3.6 \times 10^{-2}$ m. The material between the covers is paper, which has a dielectric constant of about 6. The capacitance of the book with metal covers is calculated using Eq. (26.1):

$$C = \frac{(6)(0.050 \text{ m}^2)}{(4\pi 9.0 \times 10^9 \text{ N} \cdot \text{m}^2/\text{C}^2)(3.6 \times 10^{-2} \text{ m})} = \underline{74 \times 10^{-12} \text{ F.}} \, *$$

(b) The answer to (a) means that our book capacitor will have 74×10^{-12} C of charge on each cover (positive on one cover and negative on the other) for each volt of potential difference across the plates of the capacitor. Using Eq. (26.2), we find the voltage needed to get 10^{-6} C of charge on each cover:

$$V = \frac{q}{C} = \frac{10^{-6} \text{ C}}{74 \times 10^{-12} \text{ F}} = 14 \times 10^3 \frac{C}{F} = \underline{14 \times 10^3 \, V.}$$

Even with this large voltage, we still have only 10^{-6} coulombs of charge on the metal book covers. ∎

EXAMPLE 26.2 Calculate (a) the total capacitance C and (b) the charge q on all the membranes of a body having 10^{13} cells if each cell has a surface area of 1.8×10^{-9} m²; and if the membranes are 8.0×10^{-9} m (8.0 nm) thick and have a 0.090-V potential difference across them and a dielectric constant of 8.0.

SOLUTION The cell diameters are much greater than the thickness of their membranes. Thus, each cell can be considered as a parallel-plate capacitor. The plates are the inside and outside cell fluids. The membrane is a relatively non-

*Note that the unit $\text{C}^2/\text{N} \cdot \text{m} = \text{C}^2/\text{J} = \text{C/V} = \text{F}$.

conducting material separating the fluids. (a) The total membrane area is the number of cells multiplied by the area of each cell:

$$A = (10^{13} \text{ cells})(1.8 \times 10^{-9} \text{ m}^2/\text{cell}) = 1.8 \times 10^4 \text{ m}^2.$$

This is the area of about three football fields. Using Eq. (26.1), we find that the total capacitance of these cells is

$$C = \frac{K}{4\pi k}\left(\frac{A}{d}\right) = \frac{8(1.8 \times 10^4 \text{ m}^2)}{(4\pi 9.0 \times 10^9 \text{ N} \cdot \text{m}^2/\text{C}^2)(8.0 \times 10^{-9} \text{ m})} = \underline{160 \text{ F}}.$$

(b) The total charge separated by the membranes of the cells, when a 0.090-V potential difference exists across the membranes, is calculated using Eq. (26.2):

$$q = CV = (160 \text{ F})(0.090 \text{ V}) = \underline{14 \text{ C}}.$$

Although these calculations are approximate, it is clear that the separation of electric charge across cell membranes must be an important part of our metabolic processes. The 14-C charge stored on our membrane capacitors is about the same as the charge on a cloud before a lightning flash. Discharging our cell capacitors in one surge would be an event of no small significance. ∎

(a)

(b)

FIG. 26.5. (a) An electric field exists between the plates of a charged capacitor. (b) Dipole molecules between the plates are oriented by the field. The charge inside the circle in (b) is the same as that inside the circle in (a), but the capacitor plate in (b) holds more charge because the polar molecules are present.

26.3 Dielectrics

In Chapter 22, we found that materials with large dielectric constant can greatly reduce the force between charges immersed in them. These materials are often made of polar molecules. If placed between the plates of a capacitor, a material with large dielectric constant can greatly increase the charge on the plates of a capacitor—assuming the voltage across the plates remains constant. This can be understood easily by referring to Fig. 26.5.

Two parallel plates separated by air are attached to a voltage source. The charges on the plates create an electric field between the plates (Fig. 26.5a). If a material made of polar molecules replaces the air, as in Fig. 26.5b, the molecules become oriented by the electric field. The negative ends of the polar molecules point toward the positive plate and effectively neutralize some of its positive charge. Similarly, the positive ends of the molecules point toward the negative plate and neutralize some of its negative charge. Because part of the electric charge on each plate is neutralized by the polar molecules, more charge is pushed by the battery onto the plate. The total charge on the plates increases by a factor K, where K is the dielectric constant of the material between the plates.

The capacitance of any capacitor increases by a factor K when a material of dielectric constant K is placed between the conductors. If its capacitance is C_0 when the conductors are separated by air, its capacitance is $C = KC_0$ when they are separated by the dielectric material. When a voltage V is placed across the plates of the capacitor, the magnitude of the charge on the plates increases by a factor K:

$$q = CV = KC_0V.$$

EXAMPLE 26.3 A 5.0-μF capacitor with air separating the plates is charged by a 100-V power supply. The capacitor is disconnected from the supply, and a sheet of Plexiglas with dielectric constant 3.40 is placed between the plates so that it fills the space. Calculate the voltage across the plates.

SOLUTION When the capacitor is removed from the power supply, charge cannot leave the plates. Thus, the charge on the plates remains constant as the Plexiglas is inserted between them.

Because Plexiglas has a greater dielectric constant than air, the capacitor's capacitance increases as the Plexiglas is inserted between its plates. But, since q is constant, we see from Eq. (26.2) that V must decrease as the capacitance increases. Thus,

$$q = C_0 V_0 = CV = (KC_0)V,$$

where C_0 is the initial capacitance with air between the plates, $V_0 = 100$ volts is the initial voltage, C is the capacitance when the Plexiglas of dielectric constant $K = 3.40$ is inserted in the air gap, and V is the voltage after the Plexiglas is inserted. We find that

$$V = \frac{C_0 V_0}{C} = \frac{C_0 V_0}{(KC_0)} = \frac{V_0}{K} = \frac{100 \text{ volts}}{3.40} = \underline{29 \text{ volts}}.$$

A device such as just described could be used to detect motion. A burglar alarm, for example, could be constructed so that a Plexiglas sheet was inserted between the plates of a capacitor when a door or window was opened. The resulting decrease in voltage would activate an alarm. ∎

26.4 Dielectric Breakdown

The magnitude of the electric field between the plates of a parallel-plate capacitor is given by Eq. (23.8):

$$E = \frac{V}{d},$$

where V is the voltage across the plates and d is the distance between them.

Normally, the material between the plates does not conduct electric charge, but if the electric field in the material is very large, the atoms and molecules in this material may become ionized. The electrons are then pulled toward the positive capacitor plate while the positive ions left behind attract electrons from the negatively charged plate. As a result, a very intense current flows for a brief time as the charge on the capacitor plates flows or discharges through the material between them.

This phenomenon, called **dielectric breakdown,** occurs when a nonconducting material is subjected to a very large electric field, becomes ionized, and conducts electricity. In such situations the onset of conduction occurs suddenly and often produces a spark. The **dielectric strength** of a material is defined as the value of the electric field for which breakdown of the material occurs. Nonconducting materials, such as plastic, rubber, paraffin, and transformer oils, have dielectric strengths of 8×10^6 to 20×10^6 V/m.

If the dielectric strength of a material is 8×10^6 V/m, then the voltage across a 1-m thick slab of the material would have to be 8×10^6 V or greater in order for breakdown to occur. If the slab were 0.1 m thick, then a voltage of 8×10^5 V or greater would cause breakdown. (The electric field in the material is still 8×10^6 V/m or greater since, according to Eq. (23.8), $E = V/d = 8 \times 10^5 \text{ V}/0.1 \text{ m} = 8 \times 10^6 \text{ V/m}$.)

EXAMPLE 26.4 The dielectric strength of air is 3×10^6 V/m. As you walk across a synthetic rug, your body accumulates electric charge, causing a potential difference of 6000 V between it and a door knob. Estimate the separation of your finger from the doorknob when dielectric breakdown of air occurs causing a spark of electric charge to jump between your body and the knob.

SOLUTION In this problem the two conducting surfaces are the doorknob and your body. Air is the nonconducting material separating them. Breakdown occurs when the electric field between your finger and the door exceeds 3×10^6 V/m. Rearranging Eq. (23.8) to determine the separation d of your finger and the knob when the field exceeds this amount, we find that

$$d = \frac{V}{E} = \frac{6000 \text{ V}}{3 \times 10^6 \text{ V/m}} = 2 \times 10^{-3} \text{ m} = \underline{2 \text{ mm}}. \qquad \blacksquare$$

Explosions in grain elevators have been ignited by sparks caused by dielectric breakdown. Similar explosions of anesthetics could occur in operating rooms. To prevent such explosions, physicians and nurses wear shoes with soles that conduct electricity. This prevents the accumulation of charge on their bodies and hence prevents sparks.

EXAMPLE 26.5 Calculate the electric field across the membrane of a body cell if its thickness is 7.0 nm and the potential difference across it is 90 mV.

SOLUTION The magnitude of the electric field in the membrane is determined using Eq. (23.8):

$$E = \frac{V}{d} = \frac{0.090 \text{ V}}{7.0 \times 10^{-9} \text{ m}} = \underline{13 \times 10^6 \text{ V/m}}.$$

If our cell membranes were made of a material with dielectric strength less than 13×10^6 V/m, dielectric breakdown would occur and the 14 C of charge separated by these membranes (calculated in Example 26.2) would be discharged. Fortunately, the dielectric strength of our cell membranes is about double that of the electric field in the membranes. $\qquad \blacksquare$

Lightning

Lightning is a complex phenomenon that depends on dielectric breakdown. Perhaps the most common type of lightning (there are many types) is cloud-to-ground lightning. As shown in Fig. 26.6a, a cloud becomes positively charged at the top (the P-region) and negatively charged at the bottom (the N-region). A small portion of the cloud at the bottom also has a positive charge (the p-region). The reasons for this charge distribution are complex and only partially understood.

If the electric field caused by this charge distribution is large enough, dielectric breakdown occurs, and electrons leap from the N-region toward the p-

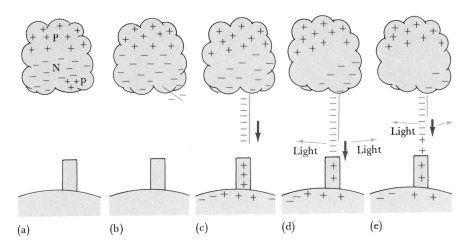

FIG. 26.6. (a) Charge distribution in a cloud just before cloud-to-ground lightning. (b) A discharge between the N- and p-regions of the cloud starts the lightning. (c) The step leader moves toward the ground. (d) Ionization of air occurs near the ground, causing light and an intense downward flow of electrons. (e) Electric charge from farther up flows down.

(a)   (b)   (c)   (d)   (e)

region. The freed electrons continue to move down and away from the rest of the negative charge in the cloud (Fig. 26.6b). The electrons make successive 50-m jumps toward the earth at intervals of one-millionth of a second. This downward trip of the electrons is called a **stepped leader** (Fig. 26.6c).

As the electrons approach the earth, positive charge on the earth's surface is attracted toward the leader, while negative charge is repelled (Fig. 26.6c). The electric field in the air just above the surface becomes so intense that it ionizes atoms and molecules in the air. The air experiences dielectric breakdown (Fig. 26.6d), and an intense flash of light is produced. Breakdown often occurs at elevated parts of the earth or at the tops of buildings or trees, since the electric field is strongest at places that are nearest the approaching leader (recall that $E = V/d$; thus E is greatest for small d).

After breakdown occurs, negative charge in the stepped leader farther above the earth can now rush down through the region of ionized air (Fig. 26.6e). This intense flow of electrons originating farther above the earth causes a flash of light that, when photographed, appears to move upward because electron flow starts from successively higher and higher positions. The flash, which appears to move up, and the large downward electron flow, make up what is called the **return stroke.** Eventually, a large portion of electrons originally on the cloud discharge from the cloud to the earth. About five coulombs of electrons move down during the stepped leader, and another 20 to 30 coulombs of electrons move down during the return stroke.

In the United States, about 100 to 200 deaths are caused each year by lightning. In eighteenth-century Europe, church bells were often rung to protect against lightning. Some persons thought, erroneously, that the sound of the bells disturbed the path of the lightning and protected those persons near the bells. Unfortunately, the metal bells were usually at high elevations and thus were likely spots for lightning to strike. In one thirty-three-year period during the eighteenth century, 386 bell towers were struck, and 103 bell ringers were killed at their ropes.

What should you do during a lightning storm? You should not make a lightning rod of yourself or stand beneath one (such as a tree). If you are in an open space, find a ravine, valley, or depression in the ground. Crouch so that you do not project above the surrounding landscape. Lightning usually strikes objects farthest above the earth's surface. Do not lie on the ground. A large

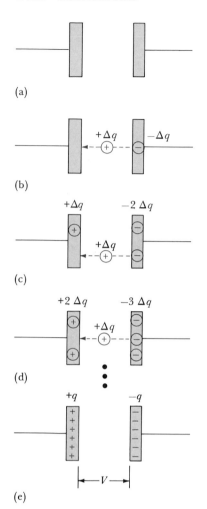

(a)

(b)

(c)

(d)

(e)

FIG. 26.7. The charging of a capacitor starts with the transfer of a small charge Δq from the right to the left plate. This transfer is fairly easy. However, as more and more charge is transferred, additional transfer is more difficult.

voltage may develop between your head and feet, causing an undesirable flow of electric charge through your body. When crouched, keep your feet close together.

26.5 Energy in a Charged Capacitor

As we learned in Chapter 23, whenever positive and negative electric charges move apart, electrical potential energy increases. In Fig. 26.7a two parallel plates with no net charge are shown. We will charge the capacitor by moving small increments of charge Δq from one plate to the other; one plate gains a charge $+\Delta q$, leaving the other plate with a charge deficiency $-\Delta q$.

Little energy is used to move the first Δq, since there is no charge on the left plate to oppose its motion (Fig. 26.7b). The next small charge Δq is harder to move (Fig. 26.7c), since there is now a $+\Delta q$ charge on the left plate that repels its motion while a $-2\Delta q$ charge on the right plate pulls it back. The more charge that is transferred, the harder it becomes to move another increment, Δq. Eventually, we will have moved a total charge q from one plate to the other.

Some external energy source, such as a battery, usually provides the increased electrical potential energy needed to charge a capacitor. The energy needed can be calculated using Eq. (23.5):

$$\Delta PE_q = qV_{average}$$

where $V_{average}$ is the average potential difference between the capacitor plates while being charged. Since the initial voltage across the plates is zero and the final voltage is V, the average potential difference between the plates is

$$V_{average} = \frac{0 + V}{2} = \tfrac{1}{2}V.$$

Substituting this expression for $V_{average}$ in Eq. (23.5), we find that

$$\Delta PE_q = qV_{average} = q(\tfrac{1}{2}V).$$

The above equation for the *change in electrical potential energy* that occurs when a capacitor receives a charge q from a potential difference V can be rewritten in three different forms using Eq. (26.2), $q = CV$:

$$\Delta PE_q = \tfrac{1}{2}qV = \tfrac{1}{2}CV^2 = \tfrac{1}{2}q^2/C. \tag{26.3}$$

EXAMPLE 26.6 Estimate the increase in electrical potential energy needed to charge all the membrane capacitors of a person's body. In Example 26.2 we estimated that the total capacitance of the membranes of 10^{13} cells was 160 F, the total charge on these cells was 14 C, and the voltage across the membranes was 0.090 V.

SOLUTION The increased electrical potential energy when the membrane capacitors of the body are charged can be calculated using any of the three expressions in Eq. (26.3):

$$\Delta PE_q = \tfrac{1}{2}qV = \tfrac{1}{2}(14 \text{ C})(0.090 \text{ V}) = \underline{0.6 \text{ J}},$$
$$\Delta PE_q = \tfrac{1}{2}CV^2 = \tfrac{1}{2}(160 \text{ F})(0.090 \text{ V})^2 = \underline{0.6 \text{ J}},$$
$$\Delta PE_q = \tfrac{1}{2}q^2/C = \tfrac{1}{2}(14 \text{ C})^2/(160 \text{ F}) = \underline{0.6 \text{ J}}. \qquad ■$$

26.6 Charging and Discharging a Capacitor Through a Resistor

A charged capacitor stores electrical potential energy; positive charge on one plate is separated from negative charge on the other. This energy is released if the negative charge on one plate returns to the positively charged plate by flowing through a circuit attached to the plates. The electrical potential energy lost by the capacitor is transferred to elements in the circuit through which the charge flows. For example, the light from a camera's flashlamp is produced by discharging a capacitor through a xenon lamp. The energy for pulsed lasers is supplied by discharging large capacitors. Even the electrical energy that a heart pacemaker uses to initiate a heartbeat can be supplied by a repetitive charging and discharging of a capacitor.

For many of these applications the time needed to charge or discharge the capacitor is important. To cause a bright flash of light in your camera's flash-lamp, its capacitor must discharge almost completely in less than one millisec-ond (10^{-3} s). The discharge time depends on both the capacitance of the capaci-tor and the resistance of the circuit through which the charge flows.

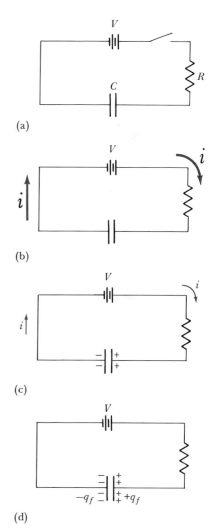

(a)

(b)

(c)

(d)

FIG. 26.8. (a) An RC series circuit. (b) When the switch is closed, a large current starts to flow and charge begins to accumu-late on the capacitor plates. (c) As more charge accumulates, the current is reduced because of its repulsion from the charge al-ready on the plates. (d) Finally, the plates become completely charged and no more cur-rent flows.

Charging a Capacitor

Figure 26.8a shows the charging of the capacitor through a circuit consisting of a battery attached to a resistor R in series with a capacitor C (called an **RC series circuit**). An open switch ($-\frown-$) prevents the battery from charging the capacitor.

When the switch is closed (Fig. 26.8b), the battery forces current around the circuit in a clockwise direction. However, the charge cannot flow between the capacitor plates. Instead, opposite-sign charges start to accumulate on the plates. This charge makes it difficult for more charge to flow onto the plates, and the current decreases (Fig. 26.8c). When the capacitor has become fully charged with a final charge q_f, the current stops (Fig. 26.8d).

We can better understand the time variation of the current i through the resistor and the charge q on the capacitor by analyzing the voltage across the three parts of the circuit shown in Fig. 26.8. (We will use lower-case letters to denote quantities that change with time.) The voltage across the battery is con-stant and equals V. The voltage across the resistor varies with time and is $v_R = iR$ at the instant of time that a current i flows through the resistor. The voltage across the capacitor also varies with time and depends on the instanta-neous charge on its plates according to Eq. (26.2): $v_C = q/C$. The sum of the voltages across the resistor and capacitor equals the constant voltage across the battery:

$$V = v_R + v_C = iR + \frac{q}{C}.$$

Rearranging, we find that

$$i = \frac{V}{R} - \frac{q}{RC}.$$

At the instant the switch is closed, $q = 0$ and the initial current is $i_0 = V/R$. As charge accumulates on the capacitor, q/RC increases and the current de-

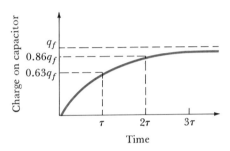

FIG. 26.9. The variation of (a) current and (b) charge with time as a capacitor is charged through a resistor.

creases. Eventually, when the capacitor is fully charged, $i = 0$ and

$$0 = \frac{V}{R} - \frac{q_f}{RC}$$

or

$$q_f = VC.$$

The final charge q_f on the capacitor is just the charge we expect using Eq. (26.2) when a voltage V is across the capacitor.

The variation with time of the current through the resistor and charge on the capacitor are shown in Fig. 26.9. The current starts abruptly with an initial value i_0 and decreases with time. The charge on the capacitor increases from zero toward a final value q_f. Equations for the variation of i and q with time can be derived using techniques involving calculus. We present the results without derivation:

$$i = i_0 e^{-t/RC}, \tag{26.4}$$

$$q = q_f(1 - e^{-t/RC}). \tag{26.5}$$

We see that the current decreases exponentially with time and that the charge also varies in a way that depends on an exponential function of time.

One very important feature of Eqs. (26.4) and (26.5) is the appearance of the quantity RC, called the **time constant** τ of the circuit:

$$\tau = RC. \tag{26.6}$$

At a time $t = RC$, the current has decreased to e^{-1} of its initial value. Since $e = 2.72$, we find that at $t = RC$, $i = i_0 e^{-1} = i_0 0.37$ and that the charge on the capacitor has increased to $q = q_f(1 - e^{-1}) = 0.63 q_f$. At twice this time ($t = 2RC$), $i = i_0 e^{-2} = i_0 0.14$ and $q = q_f(1 - e^{-2}) = q_f 0.86$. At $t = 3RC$, $i = i_0 e^{-3} = i_0 0.05$ and $q = q_f(1 - e^{-3}) = q_f 0.95$. We see that the time needed to charge a capacitor through a resistor depends on the values of R and C. At a time $t = 2RC$, the capacitor is 86 percent charged; at a time $t = 3RC$, the capacitor is 95 percent charged, and so forth.

Discharging a Capacitor

The discharge of a fully charged capacitor through a resistor is represented by the series of steps shown in Fig. 26.10. Initially, when the switch is open (Fig. 26.10a), the capacitor has a charge q_0. When the switch is closed, charge leaves the capacitor and flows through the resistor. The charge q on the plate decreases exponentially with time according to the equation

$$q = q_0 e^{-t/RC}, \tag{26.7}$$

where q_0 is the initial charge on the plates (Fig. 26.11).

Once again, the time constant $\tau = RC$ appears in the equation for the discharging of a capacitor through a resistor. At a time $t = RC$, $q = q_0 e^{-1} = q_0 0.37$; at $t = 2RC$, $q = q_0 e^{-2} = q_0 0.14$; at $t = 3RC$, $q = q_0 e^{-3} = q_0 0.05$; and so forth. Thus, the capacitor is 95 percent discharged at a time of $3RC$.

Although it is not immediately obvious, the quantity RC which has the units of resistance times capacitance, is also a unit of time, as the following

C

(a)

i

(b)

(c)

i

(d)

FIG. 26.10. (a) A charged capacitor. (b) When the switch is closed, current flows through the resistor, and (c) the capacitor eventually becomes discharged (d).

calculations confirm:

$$R = \frac{V}{I} \quad \text{or} \quad \text{ohm} = \frac{\text{volt}}{\text{ampere}}$$

and

$$I = \frac{\Delta q}{\Delta t} \quad \text{or} \quad \text{ampere} = \frac{\text{coulomb}}{\text{second}}.$$

Thus,

$$\text{ohm} = \frac{\text{volt}}{\text{ampere}} = \text{volt}\left(\frac{\text{second}}{\text{coulomb}}\right).$$

Also note that

$$C = \frac{q}{V} \quad \text{or} \quad \text{farad} = \frac{\text{coulomb}}{\text{volt}}.$$

The unit of the time constant τ is the same as the unit of RC. Thus,

$$\text{ohm} \cdot \text{farad} = \left[\text{volt}\left(\frac{\text{second}}{\text{coulomb}}\right)\right]\left(\frac{\text{coulomb}}{\text{volt}}\right) = \text{second}.$$

EXAMPLE 26.7 If you take a picture using a flash attachment, there is a delay of 5 to 10 seconds before you can take the next. The delay is caused by the time needed to charge a capacitor through a resistor (see Fig. 26.12).* Once charged, the capacitor is ready to release its charge through a xenon lamp, producing light. Estimate the series resistance R of the charging circuit if the capacitor has a capacitance of 450 μF.

SOLUTION Since you must wait 5 to 10 seconds between pictures, let us estimate that the time τ needed to get the capacitor 63 percent charged is 5 s.

*The actual circuit in a flash attachment is somewhat more complicated than this.

Charge on capacitor

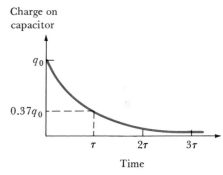

q_0

$0.37 q_0$

τ 2τ 3τ

Time

FIG. 26.11. The variation of charge with time as a capacitor is discharged through a resistor.

Switch

V C Xenon flashtube

R

(a)

V C Light

R

Electron flow

(b)

FIG. 26.12. (a) A circuit for a camera flashtube. When not in use, a capacitor is charged by a battery. (b) When the switch is closed, the capacitor charge flows through a xenon lamp, thus producing light.

Charging circuit

(a)

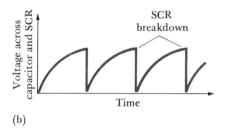

(b)

FIG. 26.13. (a) A circuit such as this could be used for a heart pacemaker. (b) The variation of voltage across the capacitor as a function of time.

Then, according to Eq. (26.6),

$$R = \frac{\tau}{C} = \frac{5 \text{ s}}{450 \times 10^{-6} \text{ F}} = 1.1 \times 10^4 \ \Omega = \underline{11 \text{ k}\Omega}.$$ ∎

The charging and discharging of capacitors is also used in the electrical circuits of some heart pacemakers. A simplified circuit for a pacemaker is shown in Fig. 26.13a. A battery charges the capacitor C through a resistor R. The values of R and C are chosen so that the capacitor becomes almost fully charged in about 1 s.

Another device, called a silicon-controlled rectifier (SCR), is placed in parallel with the capacitor. Little current flows through the rectifier until the voltage across it reaches some predetermined value where breakdown occurs. At this breakdown voltage the rectifier becomes a conductor with very low resistance, and the charge on the capacitor is discharged almost instantaneously through the rectifier.

In summary, then, the voltage across the capacitor builds up slowly in about one second as it is charged by the battery through the resistor R. When the rectifier breaks down, the capacitor discharges quickly, and the process restarts. The variation of voltage with time is shown in Fig. 26.13b.

When a pacemaker is connected to the heart, the voltage produced stimulates the heart to beat at a fixed frequency. A new heartbeat is initiated by each increase in voltage.

EXAMPLE 26.8 A 0.40-μF capacitor is used in the circuit for a heart pacemaker (Fig. 26.13a). If the time constant for the RC charging circuit is to be 0.70 s, what should the resistance of the resistor be?

SOLUTION According to Eq. (26.6),

$$R = \frac{\tau}{C} = \frac{0.70 \text{ s}}{0.40 \times 10^{-6} \text{ F}} = 1.8 \times 10^6 \ \Omega = \underline{1.8 \text{ M}\Omega}.$$ ∎

26.7 Capacitors in Series and in Parallel

We have been considering electronic circuits that contain only one capacitor. Often, it is convenient or necessary to use several capacitors in an electronic circuit. For example, the power for a pulse laser is provided by discharging a large capacitor through a flashlamp, thereby exciting the laser. The flash can be made greater by simultaneously discharging several capacitors through the lamp. In this section we consider the total capacitance when several capacitors are connected in parallel or in series.

Capacitors in Parallel

Several capacitors connected in parallel are shown in Fig. 26.14. The potential difference V across each capacitor is the same, so the charge on any one capacitor depends on its capacitance: $q_i = C_i V$, where i identifies a particular capacitor. The sum of the charges on the left side of all of the capacitors is

$$q = q_1 + q_2 + q_3 + \cdots$$
$$= C_1 V + C_2 V + C_3 V + \cdots$$

or

$$q = (C_1 + C_2 + C_3 + \cdots)V.$$

A single, equivalent capacitor, built to replace the parallel capacitors, should hold the same amount of charge when a voltage V is placed across its plates (Fig. 26.14b). Thus,

$$q = C_{eq} V.$$

By comparing the last two equations, we see that the *equivalent capacitance of one capacitor that replaces several parallel capacitors* is

$$C_{eq} = C_1 + C_2 + C_3 + \cdots. \qquad (26.8)$$

Capacitors in Series

Now, consider capacitors attached in series, as shown in Fig. 26.15a. The battery will push a charge $-q$ to the right plate of capacitor C_3. An equal charge will be repelled from its left plate onto the right plate of capacitor C_2. The net charge in the region between capacitors C_2 and C_3 is zero, as it was before the capacitors were charged (see the region inside the dashed lines in Fig. 26.15a). A similar phenomenon occurs between capacitors C_1 and C_2. Thus, the left plate of each capacitor has a charge $+q$, and the right plate of each capacitor has a charge $-q$. The potential difference V across all the capacitors is just the sum of the potential differences across each capacitor:

$$V = V_1 + V_2 + V_3 + \cdots.$$

For each capacitor, $V_i = q/C_i$ [Eq. (26.2)]; thus,

$$V = \frac{q}{C_1} + \frac{q}{C_2} + \frac{q}{C_3} + \cdots$$
$$= q\left(\frac{1}{C_1} + \frac{1}{C_2} + \frac{1}{C_3} + \cdots\right).$$

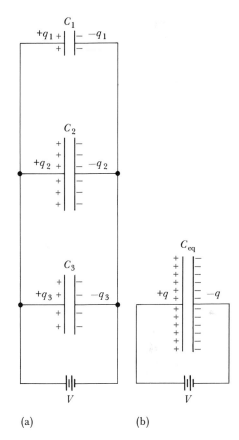

FIG. 26.14. (a) Three capacitors in parallel and (b) an equivalent capacitor.

(a)

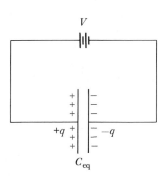

(b)

FIG. 26.15. (a) Three capacitors in series and (b) an equivalent capacitor.

If a single equivalent capacitor replaces the series capacitors, its capacitance should produce the same charge on each plate when the same voltage is applied across its plates (Fig. 26.15b). Thus,

$$V = q\left(\frac{1}{C_{eq}}\right).$$

For V and q to be the same in the last two equations, the *equivalent capacitance of a group of series capacitors* must satisfy the following equation:

$$\frac{1}{C_{eq}} = \frac{1}{C_1} + \frac{1}{C_2} + \frac{1}{C_3} + \cdots. \tag{26.9}$$

Notice that the expressions for equivalent capacitance are just the opposite of those for equivalent resistances—series resistors are added linearly, whereas series capacitors are added by first taking their inverses.

EXAMPLE 26.9 Calculate the total energy stored in three 10–μF capacitors when attached in parallel and when attached in series. A 100-V power supply is used to charge the capacitors.

SOLUTION First, calculate the equivalent capacitance for each situation using Eqs. (26.8) and (26.9):

Parallel: $C_{eq} = C_1 + C_2 + C_3 = 3(10 \ \mu\text{F}) = 30 \ \mu\text{F},$

Series: $\dfrac{1}{C_{eq}} = \dfrac{1}{C_1} + \dfrac{1}{C_2} + \dfrac{1}{C_3} = \dfrac{1}{10 \ \mu\text{F}} + \dfrac{1}{10 \ \mu\text{F}} + \dfrac{1}{10 \ \mu\text{F}} = \dfrac{3}{10 \ \mu\text{F}}$

or

$$C_{eq} = \frac{10 \ \mu\text{F}}{3} = 3.3 \ \mu\text{F}.$$

The energy stored in the capacitors is calculated using Eq. (26.3):

Parallel: $\Delta PE_q = \frac{1}{2}C_{eq}V^2 = \frac{1}{2}(30 \times 10^{-6} \ \text{F})(100 \ \text{V})^2 = \underline{0.15 \ \text{J}},$

Series: $\Delta PE_q = \frac{1}{2}C_{eq}V^2 = \frac{1}{2}(3.3 \times 10^{-6} \ \text{F})(100 \ \text{V})^2 = \underline{0.033 \ \text{J}}.$

As you can see, much more energy can be stored in parallel capacitors. Consequently, if we wished to release a large amount of electrical potential energy by simultaneously discharging several capacitors through the flashlamp of a laser, the capacitors should be combined in parallel, not in series. ∎

Summary

1.   A **capacitor** is a device that stores electric charge. It consists of two electrical conductors separated by a nonconducting material. A capacitor is given the symbol —||— in diagrams of electronic circuits.

2.   The **capacitance** C of a capacitor depends on the area A of its conducting surfaces, the separation d of these surfaces, and the dielectric constant K of the nonconducting material between the conductors. For a parallel-plate capacitor

$$C = \frac{KA}{(4\pi k)d}, \tag{26.1}$$

where k is a constant equal to $9.0 \times 10^9 \ \text{N} \cdot \text{m}^2/\text{C}^2$.

3.   **Charging a capacitor:** When a capacitor is attached to a power source of potential difference V, a charge $+q$ accumulates

on one plate and $-q$ on the other. The magnitude of the charge on each plate is

$$q = CV. \tag{26.2}$$

4. The **energy** needed to separate the charges onto the plates of a capacitor is given by any of the equations:

$$\Delta PE_q = \tfrac{1}{2}qV = \tfrac{1}{2}CV^2 = \tfrac{1}{2}q^2/C. \tag{26.3}$$

5. **RC series circuit:** If a capacitor C is charged through a series resistor R, the charge on the capacitor increases with time according to the equation:

$$q = q_f(1 - e^{-t/RC}), \tag{26.5}$$

where q_f is the final charge on the capacitor and RC is called the **time constant** τ of the circuit. If a capacitor is discharged through a resistor, the variation of the charge with time is given by the equation

$$q = q_0 e^{-t/RC}, \tag{26.7}$$

where q_0 is the initial charge on the capacitor.

6. The **equivalent capacitance** C_{eq} of a group of capacitors connected in **parallel** is

$$C_{eq} = C_1 + C_2 + C_3 + \cdots. \tag{26.8}$$

The equivalent capacitance of a group of capacitors connected in **series** is determined by solving Eq. (26.9) for C_{eq}:

$$\frac{1}{C_{eq}} = \frac{1}{C_1} + \frac{1}{C_2} + \frac{1}{C_3} + \cdots. \tag{26.9}$$

Questions

1. A dielectric material vibrates in and out of the space between two parallel plates. The plates are connected to a small resistor and a battery, as shown in Fig. 26.16. What happens to the charge on the plates as the dielectric material is pushed between them? As the material is pulled out, what happens to the charge? Explain why a voltage is measured across the resistor as the dielectric material moves in and out between the plates. Devices such as this are used to measure vibrations.

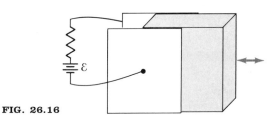

FIG. 26.16

2. Would you expect the dielectric constant of a material, such as water, to increase, decrease, or remain the same as the temperature of the material increases? Explain. (*Hint:* When water mole-

cules are placed in an electric field, more molecules (a small fraction) point in the direction of the field than in any other direction. Consider how temperature would affect that fraction.)

3. How does the state of a polar dielectric material change as the electric field in which it is placed increases?

4. A 10-μF capacitor and a 0.01-μF capacitor are both charged by a 100-V potential difference. From which capacitor will you get the bigger shock if it is discharged through your body? Explain how you arrive at your answer.

5. Two capacitor plates are charged by a battery. The plates are then disconnected from the battery and pulled apart so that their separation is doubled. Explain how each of the following quantities changes: the capacitance of the capacitor, the charge on the capacitor plates, the voltage across the plates, and the energy stored by the capacitor. Justify each of your answers.

6. A capacitor is connected in series to a resistor and a battery. (a) Does the time constant for charging the capacitor increase, decrease, or remain the same if a second resistor is connected in parallel with the first resistor? Explain. (b) Explain what happens if a second resistor is added to the circuit in series.

Problems

26.2 and 26.3
Capacitance and Dielectrics

1. Two strips of aluminum foil, separated by a strip of paper 0.070 mm thick and with dielectric constant 6.0, are wrapped together in a cylinder. Calculate the area of the aluminum sheets needed to make (a) a 5.0-μF capacitor and (b) a 500-pF capacitor.

2. Two round metal plates 0.40 m in diameter are separated by a sheet of mica 0.30 mm thick, whose dielectric constant is 6.3. Calculate the capacitance of the plates.

3. A parallel-plate "trimmer" capacitor has a screw adjustment that can be used to vary the separation of its plates. The plates have an area of 8.0 cm^2 and are separated by a 0.100-mm air gap. (a) Calculate the capacitor's capacitance. (b) To what separation should the plates be moved to change the capacitance to 100 pF?

4. Two square plates, 1.0 m long on each side, are separated by 1.0 cm of air. (a) Calculate the capacitance of the plates. (b) If the plates are cut into fourths to make four capacitors, what must the plate separation be so that each of the four new capacitors will have the same capacitance as the original? Explain. (c) Transformer oil with dielectric constant 2.1 is placed between the plates of one of the small capacitors. To what separation must its plates now be adjusted so that its capacitance will be the same as in part (b)?

5. Charge can move about with relative ease on your body. Thus, your body can be considered as one plate of a capacitor. *Estimate* the capacitance between your body and that of your physics professor during physics class. Justify any numbers used in your calculations. We will assume that charge cannot leave the bodies through their contact with chairs or the floor.

6. A cylindrical capacitor consists of two cylindrical conducting materials, one inside the other. The capacitance of a cylindrical capacitor is calculated using the equation:

$$C = \frac{KL}{2k \ln(R_2/R_1)}, \qquad (26.10)$$

where $k = 9.0 \times 10^9 \text{ N} \cdot \text{m}^2\text{C}^{-2}$, K is the dielectric constant of the material between the two cylinders, L is the length of the capacitor, R_1 is the radius of the smaller cylinder, and R_2 is the radius of the larger cylinder. Calculate the capacitance of a coaxial cable with an air dielectric, a 1.0-mm inner radius, a 2.0-mm outer radius, and a 1.5-m length.

7. A 100-V potential difference is placed across two round metal plates, each 20 cm in diameter, separated by a 2.0-mm air gap. Calculate the magnitude of the charge on the plates.

8. Aluminum foil is glued on the top and bottom surfaces of a desk in your physics lecture room. Roughly, how much voltage is needed to separate $+10^{-7} \text{ C}$ of charge onto one sheet of foil and -10^{-7} C of charge onto the other? Justify any numbers used in your estimation.

9. Two square metal plates of area 9.0 cm^2 are separated by a 0.40-cm air gap. The plates are then connected to a 200-V power supply. (a) Calculate the capacitance of the plates. (b) Calculate the magnitude of the charge on the plates if wax with dielectric constant 2.25 is placed between the plates while the 200-V power supply remains connected.

10. A parallel-plate capacitor with plates of area 100 cm^2 has a capacitance of 800 pF. Calculate the number of excess electrons on a 1.0-cm^2 area of the negatively charged plate when the capacitor is charged by a 12-V battery.

11. A 4.2-μF capacitor has a charge of $+7.6 \times 10^{-4} \text{ C}$ on one plate and a charge of $-7.6 \times 10^{-4} \text{ C}$ on the other. Charge slowly "leaks" through the dielectric material separating the plates so that after one hour the magnitude of charge on the plates is $5.4 \times 10^{-4} \text{ C}$. Calculate the change in voltage across the plates during that hour.

12. Two plates of 200-cm^2 area and separated by air are charged by a battery with equal but opposite charges of magnitude $3.6 \times 10^{-7} \text{ C}$. When the battery is disconnected and the space between the plates is filled with a dielectric material, the electric field between the plates has a magnitude of $4.2 \times 10^5 \text{ V/m}$. Calculate the dielectric constant of the dielectric material. (*Hint:* Write down all equations that seem to be involved in this problem and see if they can be combined to calculate the desired unknown.)

13. A 2.0-μF capacitor and a 4.0-μF capacitor are both charged by a 100-V power supply. The capacitors are then disconnected from the supply. Each plate of each capacitor is connected by a wire to a plate of the opposite charge on the other capacitor. (a) Calculate the total charge on two plates connected together. (b) Calculate the magnitude of the charge on each capacitor. (*Hint:* The voltage is the same across each capacitor.) (c) What is the voltage across the capacitors?

14. Two square parallel metal plates that are 20 cm long on each side are separated by a 3.0-mm air gap. The plates are charged by a 100-V power supply and then disconnected from it. A sheet of rubber of dielectric constant 3.0 is placed between the plates. (a) Does the charge on the plates change? Explain. If your answer is yes, then consider where the charge originates. (b) Does the

voltage change? If it does, then calculate the change. (c) Does the electric field between the plates change? If it does, calculate the change.

15. A 2.0-μF parallel-plate capacitor with an air gap separating the plates is connected to a 12-V battery. The air gap is then filled with polyethylene, which has a dielectric constant of 2.25. Calculate the magnitude of the charge added to or removed from a plate when the polyethylene is placed between them. The capacitor remains connected to the battery.

26.4 Dielectric Breakdown

16. Your body becomes charged as you shuffle across a carpet. A spark leaps 0.15 cm from your finger to another person's nose. Estimate the potential difference between your body and the other person's body just before the spark jumped.

17. You have probably seen demonstrations in which a large metal sphere becomes charged by a Van de Graaff generator. Sparks leap from the sphere to the fingers of a person's hand held 5 or 10 cm away. Calculate the potential difference between the sphere and the person's hand just before a spark leaps a 10-cm distance.

18. Electrons in the stepped-leader portion of a lightning flash jump a distance of 50 m. Estimate the potential difference between the two parts of the cloud across which the jump occurs.

19. Calculate the voltage at which dielectric breakdown occurs between the plates described in Problem 2. The dielectric strength of mica is $2.0 \times 10^8 \text{ V/m}$.

26.5 Energy in a Charged Capacitor

20. A 120-μF capacitor, when charged, stores 2.4 J of energy. Calculate the voltage across the capacitor and the magnitude of charge on its plates.

21. The flashlamp for a laser must discharge 8.0 J of energy in order to provide an intense flash to excite the laser. A capacitor, charged by a 10,000-V power supply, releases this much energy when discharged. (a) Calculate the capacitance of the capacitor. (b) How much charge does the capacitor store?

22. A parallel-plate capacitor with air between its plates has a capacitance of 0.060 μF. When paper is inserted between the plates, the capacitance increases to 0.300 μF. (a) Calculate the dielectric constant of the paper. (b) The capacitor with a paper gap receives a charge of $1.5 \times 10^{-8} \text{ C}$. Calculate the stored energy. (c) The voltage source is removed, and the paper is pulled out of the gap. The charge on the plates remains the same. Calculate the stored energy. (d) Account for any gain or loss of energy.

23. A 0.100-μF parallel-plate capacitor with a variable air gap is charged by a 100-V potential difference. The voltage source is removed from the capacitor, leaving it with a fixed charge. (a) Calculate the energy stored in the capacitor. (b) The capacitor plates are forced apart, reducing its capacitance to 0.080 μF, while leaving it with the same charge. Calculate the voltage across the plates. (c) Calculate the stored energy. (d) Using the work-energy equation, account for the change in electrical potential energy.

24. Suppose a large capacitor charged at the beginning of your physics lecture could provide all of the metabolic energy needed by the students during the class. *Estimate* the capacitance of a capacitor that, when charged by 100,000 V, would provide the needed energy. Justify any numbers used in your calculations.

25. A 1.6-μF capacitor will break down (spark) if the voltage across it exceeds 200 V. (a) Calculate the maximum electrical energy the capacitor can store. (b) If the capacitor plates are separated by polystyrene insulation whose dielectric strength is 2.0×10^7 V/m, calculate the separation of the plates.

26. A defibrillation unit used to electrically stimulate the heart muscles following a heart attack consists of a 20-μF capacitor charged by a 6000-V potential difference. (a) Calculate the energy released when the defibrillator is discharged through the body. (b) How much charge is released?

27. A cloud has -25 C excess charge. The potential difference from the cloud to the earth is 6.0×10^6 V. (a) Calculate the capacitance of this capacitor (the cloud is considered as one plate and the earth as the other). (b) How much electrical potential energy is released during a lightning flash that discharges the capacitor? (c) Is the energy enough to boil all of the water in a person's body? (*Hint:* The body is about two-thirds water.)

26.6 Charging and Discharging a Capacitor Through a Resistor

28. A 10-μF capacitor is connected in series to a 2.0-MΩ resistor and a 100-V power source. (a) Calculate the time constant of this circuit. (b) What is the final charge on the capacitor? How many seconds are required to get the capacitor (c) 63 percent charged and (d) 95 percent charged?

29. A 1.5-μF capacitor in a television set is charged by a 22,000-V potential difference. In order to prevent harm to the television repair technician, a large "bleeder" resistor is connected across the capacitor's ends. When the television set is turned off, the capacitor discharges through the bleeder resistor rather than through the technician, who might accidentally touch its ends. (a) Calculate the time needed to discharge 95 percent of the charge on the capacitor through a 1.0-MΩ resistor. (b) What is the total charge that must be discharged?

30. The circuit for a flashing light is shown in Fig. 26.17. A battery charges a capacitor through a resistor. When the voltage across the capacitor exceeds 0.86 times the voltage that is present when the capacitor is fully charged, the neon lamp breaks down and becomes a conductor of electricity. The capacitor discharges through the neon lamp, causing a flash of light. The recharging of the capacitor starts again. Calculate three different pairs of values for the resistance and capacitance in order to make a circuit that will flash five times a second.

FIG. 26.17

31. You are to construct a circuit similar to that described in the previous problem. Your circuit should produce one flash of light each second in order to keep time with the ticks of a clock. The neon lamp breaks down (offers zero resistance) when the voltage across it exceeds 86 V. The capacitor discharge must provide 0.2 J of energy. Calculate suitable values for V, C, and R.

32. A defibrillation unit used to resuscitate a heart-attack victim was discussed in Problem 26. Calculate the time needed for 86 percent of the charge on the defibrillator's capacitor plates to discharge through the body if the resistance across the body is 10,000 Ω.

33. A heart pacemaker has a capacitor that is charged by a 6-V potential difference before it discharges. When discharged, the capacitor releases 8.0×10^{-3} J of energy. Calculate the resistance of the resistor in the charging circuit if the pacemaker time constant is 0.8 s.

26.7 Capacitors in Series and in Parallel

34. (a) Three capacitors of capacitances 2.0 μF, 3.0 μF, and 6.0 μF are connected in parallel across a 120-V potential difference. Calculate the equivalent capacitance of a single capacitance that can replace the three capacitors, and calculate the magnitude of the charge on its plates. (b) Repeat the problem for the same three capacitors arranged in series.

35. (a) A 10-μF capacitor and a 6-μF capacitor are connected in parallel. Find the energy stored by the capacitors when charged by a 1000-V potential difference. (b) Repeat the problem for a circuit in which the capacitors are connected in series.

36. A 0.080-μF capacitor is given a charge of 9.6 μC. The capacitor is removed from the voltage source and connected in parallel to an uncharged capacitor. The potential difference across the capacitors is now 80 V. Calculate the capacitance of the second capacitor. (*Hint:* A single equivalent capacitor has a charge of 9.6 μC when charged by an 80-V potential difference.)

37. Using techniques similar to those used when calculating equivalent resistance, calculate the equivalent capacitance of the capacitors shown in Fig. 26.18.

FIG. 26.18

38. Calculate the energy stored by the capacitors in the circuit shown in Fig. 26.19.

FIG. 26.19

Magnetic Forces and Fields

Our study of magnetism in this chapter focuses on two basic ideas: (1) an electric charge experiences a magnetic force when moving through a magnetic field, and (2) a moving electric charge produces a magnetic field. With these two ideas we can understand many of the fascinating phenomena that depend on magnetic forces and fields, such as the operation of a mass spectrometer and of electric motors, the use of magnetic fields to contain extremely hot gases, the way in which the earth's magnetic field shields the earth from high-energy particles flying toward us from the sun, the cause of the northern lights, and the way in which a permanent magnet works. Hopefully, you will find that magnetism is a surprising and interesting subject.

27.1 Magnetic Fields

We have all seen a compass needle rotate in the earth's magnetic field. Perhaps you have seen iron filings oriented along the magnetic field lines that emanate from the north pole of a bar magnet and terminate at the south pole (Fig. 27.1).

At this point we will not examine the cause of these magnetic-field lines—their existence is implied by the simple experiments just mentioned. However, before we begin our study of magnetism, we must define magnetic fields in quantitative terms. The symbol **B** represents the magnitude and direction of a *magnetic field* at a certain position. The direction of the field is the direction in which a compass needle points when in the field. The magnitude of the magnetic field might be determined by measuring the torque exerted on a compass needle by the field when the needle is held perpendicular to the field direction. The greater the torque, the greater is the magnitude of the field. However, other techniques are usually used to measure the magnitude of the field, as we learn later.

The magnitude of the field is given in a unit called the **tesla** (T). A magnetic field with a magnitude of one tesla is quite large—by comparison, the earth's magnetic field is about 5×10^{-5} T. Representative values of the magnetic fields produced by different magnets are illustrated in Fig. 27.2. Magnetic fields can also be measured in terms of weber/m² and gauss. The weber/m² is an older name for a tesla. The three units are related to each other as shown below:

$$1 \text{ gauss} = 10^{-4} \text{ T} = 10^{-4} \text{ weber/m}^2,$$
$$1 \text{ T} = 1 \text{ weber/m}^2 = 10^4 \text{ gauss}.$$

FIG. 27.1. Iron filings line up along the magnetic field lines of a bar magnet. (Fundamental Photographs, New York)

27.2 Magnetic Force on a Moving Charge

The force caused by a magnetic field is different in a number of ways from the forces caused by gravitational and electric fields. (1) While an electric field exerts a force on an electric charge and a gravitational field exerts a force on a mass (a gravitational "charge"), there seems to be no magnetic charge on which a magnetic field exerts a force. Instead, a magnetic field exerts a force on an electric charge. (2) Although electric and gravitational forces are exerted on both stationary and moving objects, a magnetic force is exerted only on an electric charge when it moves. No magnetic force is exerted on a charge at rest. (3) Gravitational and electric forces do not depend on the direction in which objects move, whereas a magnetic force is exerted only if a component of the motion of the charge is perpendicular to the field. If the charge moves parallel to the field, no magnetic force is exerted. (4) Finally, while gravitational and electric fields always exert force in the direction of the field, the magnetic force does not point in the direction of the field, nor does it point in the direction of the charge's velocity. The magnetic force is perpendicular to both the magnetic field and to the charge's velocity.

The unusual nature of the magnetic force might be contrasted to driving a car along a highway across which a wind blows (Fig. 27.3a). If the "wind field" acted like a magnetic field, the car would be lifted rather than blown off the highway (Fig. 27.3b). The lifting force would be perpendicular to both the car's velocity and the direction of the wind. If the car were at rest in the field, no force would be exerted on it. If the car drove into the wind or with the wind, it would feel no force from the wind, since no force is exerted on an object moving parallel to the field.

The nature of the magnetic force produces many interesting phenomena and also provides special opportunities for useful applications. Before discussing these phenomena and applications, let us carefully define the magnetic force.

Magnetic force on a moving charge: If an electric charge q moves with a velocity **v** through a magnetic field **B**, a magnetic force $\mathbf{F}_m$ is exerted on the charge. The magnitude of the force is

$$F_m = qvB \sin \theta, \qquad (27.1)$$

where θ is the angle (180° or less) between the directions of **v** and **B.** The force is exerted perpendicularly to both **v** and **B**, as discussed shortly.

3–5 × 10⁻⁵ T

0.01–0.1 T

1–100 T

FIG. 27.2. Representative values of magnetic fields.

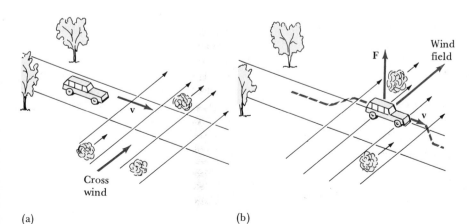

Cross wind

Wind field

F

v

(a)   (b)

FIG. 27.3. If a car entered a region where air blew across a highway and if the blowing air (a wind field) acted like a magnetic field, an upward force would be exerted by the air on the car.

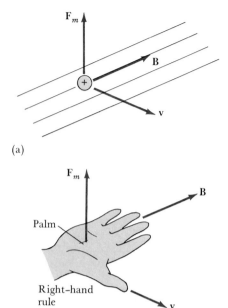

(a)

(b)

FIG. 27.4. (a) The magnetic force on a moving charge is perpendicular to **B** and **v**. (b) If your fingers point in the direction of **B** and your thumb toward **v**, your palm faces the direction of **F**$_m$.

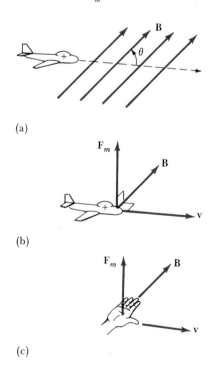

(a)

(b)

(c)

FIG. 27.5. (a) An airplane passes across the earth's magnetic field. (b) An upward magnetic force is exerted on the airplane, as shown by using the right-hand rule (c).

The vectors, **v** and **B**, lie in a plane (see Fig. 27.4a). The magnetic force **F**$_m$ on a moving charge is perpendicular to that plane. From what we have said, the force could be up, as in Fig. 27.4a, or down, since both directions are perpendicular to the plane of **B** and **v**. The right-hand rule illustrated in Fig. 27.4b is used to determine which is the correct choice for the direction of the force.

Right-hand rule: Point the fingers of your open right hand in the direction of the magnetic field (Fig. 27.4b). Orient your hand so that the thumb points along **v**, as though you were "thumbing" a ride (hitchhiking) in the direction of motion. The direction of the magnetic force on a positive charge is the direction in which your open palm would push. The force on a negative charge is in the opposite direction.

When Eq. (27.1) is used, the unit of **B** is the tesla, the unit of **v** is the meter per second, the unit of q is the coulomb, and the unit of force is the newton. We see that

$$1\,\text{N} = 1\,\text{C(m/s)}\,\text{T} \quad \text{or} \quad 1\,\text{T} = \frac{\text{N} \cdot \text{s}}{\text{C} \cdot \text{m}} \tag{27.2}$$

EXAMPLE 27.1 An airplane, shown in Fig. 27.5a, accumulates an excess positive charge of $2.0 \times 10^{-5}\,\text{C}$. Calculate the magnetic force on the airplane when passing at $240\,\text{m/s}$ through and perpendicular to the earth's magnetic field ($B = 5.0 \times 10^{-5}\,\text{T}$).

SOLUTION By using Eq. (27.1), we find that the magnitude of the magnetic force is

$$F_m = qvB \sin \theta = (2.0 \times 10^{-5}\,\text{C})(240\,\text{m/s})(5.0 \times 10^{-5}\,\text{T}) \sin 90°$$
$$= \underline{2.4 \times 10^{-7}\,\text{N}}.$$

This is about one-thirtieth the weight of one of the hairs on your head—hardly a factor in aviation. The direction of the force is up, as shown in Fig. 27.5c.

∎

Magnetohydrodynamics

Magnetohydrodynamics (MHD) is a new technology for efficient production of electricity by power-generating plants. The process depends on the magnetic force exerted on moving charges. Gaseous fuel, such as pulverized coal, enters a combustion chamber and burns there at high temperature (1000–2000°C) and pressure. If alkali metals, such as potassium, are injected into the burning gas, free electrons and ions are formed. The electrons and ions pass from the combustion chamber through a nozzle and into a magnetic field directed perpendicular to their motion (Fig. 27.6a). The magnetic force caused by the field deflects positive ions in one direction and electrons (negative charges) in the other direction.

The opposite-sign charges are collected on large plates that face each other and serve much the same role as the terminals of a battery. As the plates become charged, they produce an electric field that opposes the accumulation of more charge (Fig. 27.6b). When the opposing electric force **F**$_q$ on a charged particle

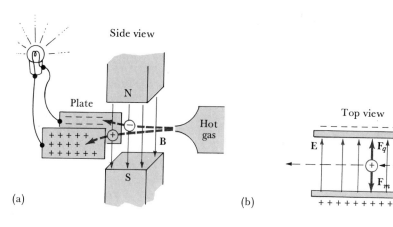

Side view

(a)   (b)

FIG. 27.6. (a) A magnetohydrodynamic generator. (b) Opposite-sign charge accumulates on the plates. Further charge is prevented from accumulating because of the electric field caused by charges already on the plates.

Top view

between the plates just balances the magnetic force F_m on the particle, the net force on the particle is zero,

$$F_m - F_q = 0, \qquad (27.3)$$

and the particle moves straight ahead rather than toward a plate. This condition allows us to calculate the voltage across the plates when they stop accumulating more charge.

The electric force on the moving charges is given by Eq. (22.4):

$$F_q = qE,$$

where E is the electric field between the plates. The electric field depends on the voltage V across the plates and on the plate separation d according to Eq. (23.8):

$$E = \frac{V}{d}.$$

Therefore,

$$F_q = q\frac{V}{d}.$$

From Eq. (27.1), we know that the magnetic force on the charge has a magnitude

$$F_m = qvB.$$

Substituting these expressions for F_q and F_m into Eq. (27.3), we find that

$$qvB - q\frac{V}{d} = 0 \qquad (27.4)$$

or

$$V = vBd.$$

The voltage of the magnetohydrodynamic generator depends on the speed v of the charges, the strength of the magnetic field B, and the separation of the plates d.

Acting alone, an MHD generator is only about 20 to 30 percent efficient at producing electric power whereas a conventional steam power plant is about 30 to 40 percent efficient. However, if an MHD generator is used in combination with a conventional steam plant, overall efficiency is expected to be 50 to 60 percent. The combination plant, besides producing power more efficiently, also produces much less environmental pollution.

This process of producing a potential difference by separating positive and negative charges as they pass through a magnetic field is used in other technologies besides magnetohydrodynamic-power generation. The speed of blood flow is also measured in a similar manner. A magnetic field is created perpendicular to the vessel carrying the blood. As positive and negative ions pass through the field, they are deflected by a magnetic force in opposite directions. Opposite-sign charges accumulate on opposite walls of the vessel. Tiny electrodes measure the voltage between these walls. With the use of Eq. (27.4), the speed of the ions can be calculated if the voltage, magnetic field, and charge of the ions are known.

27.3 Circular Motion in a Magnetic Field

You might recall from Chapter 6 that an object moves in a circular path if a centripetal force continually pulls the object toward the center of a circle. A ball swung at the end of a string is pulled inward by the tension force of the string. The moon, as it circles the earth, is pulled inward by its gravitational attraction toward the earth. These centripetal forces are exerted perpendicular to the motion of the object and do not change the object's speed. The forces change only the object's direction of motion—instead of flying straight ahead, the object moves in a circle.

Because it is directed perpendicular to the velocity of an electric charge, a magnetic force can act as a centripetal force and cause circular motion. To understand how this happens, lay your physics book on a flat surface. Suppose that a uniform magnetic field points downward into the book from the ceiling. A charge entering the field in a horizontal plane will move in a circle. A top view of the charge's path is pictured in Fig. 27.7.

Suppose, for example, that a positive charge moves along the right edge of the open book toward the top of the page (position a in Fig. 27.7). Using the right-hand rule, you can determine the direction of the force on the charge: your fingers should point down into the book in the direction of **B;** your thumb should point toward the top of the book in the direction of **v;** and the force on a positive charge is in the direction that your palm faces—toward the center of the book. Since the force is perpendicular to the charge's velocity, its speed does not change—only its direction. As the charge moves upward, it is deflected slightly to the left by the magnetic force and follows the curved path indicated by the dashed line in Fig. 27.7.

Eventually the charge reaches the top, middle of the open book (position b in Fig. 27.7). The charge is now moving to the left. Using the right-hand rule again, you will find that the magnetic force again points toward the center of the page. If you do this for every point on the circle shown in Fig. 27.7, you find that the magnetic force always points toward the center of the circle. In Chapter 6 we learned that Newton's second law, when applied to circular motion, was

$$\sum_{\substack{\text{radial} \\ \text{direction}}} F = ma_c,$$

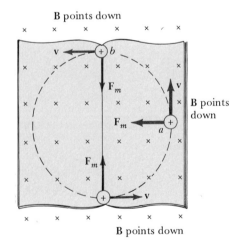

FIG. 27.7. A magnetic field (the crosses) points down into the page. The magnetic force on a charge moving in a circle is shown for several points on the circle.

where the components of all forces acting on the particle in the radial direction are added. The centripetal acceleration a_c caused by these forces has a value v^2/r [Eq. (6.7)], where r is the radius of the circle. In the present situation, the magnetic force provides the centripetal force. Since the angle θ between **v** and **B** is 90°, the magnitude of the force is

$$F_m = qvB \sin 90° = qvB.$$

Substituting in Newton's second law for the force and centripetal acceleration, we find that

$$qvB = m\frac{v^2}{r}. \qquad (27.5)$$

Equation (27.5) can be solved for one unknown quantity if the other quantities are known.

EXAMPLE 27.2 An electron from the sun enters the earth's magnetic field high above its surface and undergoes circular motion (Fig. 27.8). Calculate the radius of the circle. At this high elevation the earth's magnetic field is about 3×10^{-7} T. The electron's speed v is 2×10^8 m/s, the magnitude of its charge q is 1.6×10^{-19} C, and its mass m is 9.11×10^{-31} kg.

SOLUTION Equation (27.5) is solved for the radius of the circle:

$$r = \frac{mv}{qB} = \frac{(9.11 \times 10^{-31}\,\text{kg})(2 \times 10^8\,\text{m/s})}{(1.6 \times 10^{-19}\,\text{C})(3 \times 10^{-7}\,\text{T})} = 4 \times 10^3\,\text{m} = \underline{4\,\text{km}}. \qquad \blacksquare$$

FIG. 27.8. An electron in the solar wind moves in a circle after entering the earth's magnetic field.

Magnetic Bottles

This last example leads us to a very interesting phenomenon involving the circular motion of electric charges in magnetic fields. A charged particle can enter a magnetic field with its velocity **v** not completely perpendicular to the field (Fig. 27.9). The perpendicular component of the velocity $\mathbf{v}_\perp$ causes the charge to move in a circle, as we have just seen. However, the component of the velocity that is parallel to the field $\mathbf{v}_\parallel$ is unaffected by the field, since the angle between $\mathbf{v}_\parallel$ and **B** is 0°, and sin 0° is zero—hence no force. The particle undergoes a helical motion. It rotates in a circle at speed $v_\perp$ while moving along the field at speed $v_\parallel$ (Fig. 27.9).

If the magnitude of the magnetic field is not constant but increases, the particle is in for a real surprise. A field of increasing magnitude is represented by

FIG. 27.9. If a charged particle enters a magnetic field with its velocity not completely perpendicular to the field, the charge undergoes a helical motion.

FIG. 27.10. (a) When a charge undergoes helical motion in a converging magnetic field, the magnetic force has a backward component that can reverse the particle's direction (b).

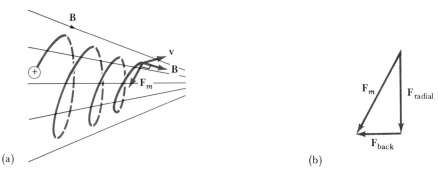

(a)   (b)

more closely spaced magnetic-field lines (Fig. 27.10a). As the lines come together, the plane in which **v** and **B** are located tilts. For example, at the top of the loop shown in Fig. 27.10a, the **B**-field points slightly down, whereas **v** is horizontal. Thus, the plane of **v** and **B** tilts down at the front. Since the magnetic force F_m is perpendicular to this plane, it points down and slightly back, as shown in Fig. 27.10a. F_m consists of two components shown in Fig. 27.10b: a radial component F_{radial} pointing toward the center of the circle, and a backward component F_{back} that opposes its forward motion. If strong enough, F_{back} can actually reverse the particle's direction. The place where the particle's direction is reversed by a converging magnetic field is called a **magnetic mirror** (the particle is reflected as if it hit a mirror).

Magnetic fields can be constructed to confine hot, highly ionized gases, called **plasmas.** The ions are reflected by the "pinched" magnetic field at the ends of the magnetic container—called a **magnetic bottle** (Fig. 27.11). Unfortunately, a magnetic bottle is somewhat leaky. Ions moving along its center axis are not deflected at its ends. Efforts are being made to construct magnetic bottles to confine the hot gas in which nuclear fusion reactions occur (discussed in Chapter 35).

One of the most interesting magnetic bottles occurs high above the earth's atmosphere. The earth is continually "showered" by ions that have escaped from the sun. These showers are especially intense following solar flares. When the ions reach the earth's magnetic field (Fig. 27.12), they are deflected and move in helical paths in two regions at distances of about 3000 km and 16,000 km above the earth's equator (called the Van Allen belts). As the earth's magnetic-field lines converge near the earth's poles, they form a magnetic mirror that reflects the ions back in the opposite direction. The ions are trapped in the magnetic bottle of the earth's own magnetic field.

Like man-made magnetic bottles, the earth's magnetic bottle leaks, and some of the ions reach the earth's atmosphere near the north and south poles. When these ions collide with atoms in our atmosphere, the atoms are ionized. As the ionized atoms recombine with their missing electrons, light is emitted. The aurora or northern (and southern) lights are produced. A picture of these dramatic lights is shown in Fig. 27.13.

FIG. 27.11. A "magnetic bottle."

FIG. 27.12. Electrons and protons from the sun are trapped in the earth's magnetic field—a giant magnetic bottle.

Mass Spectrometers

The circular motion of charged particles in magnetic fields helps us understand natural phenomena, such as the production of the earth's northern lights. This circular motion has also been put to practical use by the development of instruments, such as mass spectrometers, that are used to determine the mass of atom-sized particles.

In a **mass spectrometer,** an atom or molecule is first ionized by removal of one of its electrons (Fig. 27.14). The ion is then accelerated across a voltage and passes through a velocity selector (see Problem 7 in this chapter) that allows only those ions moving at a predetermined speed to pass. The ions then enter a magnetic field, where they move in one-half of a circular path. The radius r of the circle can be measured by observing the place where the ions strike a detector after moving halfway around the circle. The mass m of the ion can be calculated using Eq. (27.5):

$$m = \frac{qrB}{v},$$

where q is the charge of the ion, r is the radius of the semicircle, v is the ion's speed, and B is the magnitude of the magnetic field.

FIG. 27.13. The northern lights. (Wide World Photos)

27.4 Torque on a Current Loop

Another important application involving magnetic force occurs when electric charge moves through a loop of wire in a magnetic field. The magnetic force acting on these moving charges can exert a torque on the loop that causes it to rotate. An electric motor turns because of such a torque.

In order to calculate this torque, consider the magnitude of the magnetic force acting on all charges moving along one side of the loop of wire shown in Fig. 27.15. (When calculating the force, it is easier to assume that positive charges move in the direction of the current rather than that negative charges move opposite the current; the result is the same in each case.) Suppose that there are n moving charges in this wire. Each charge q travels perpendicular to the magnetic field. Hence the net magnetic force on the charges and on the wire is

$$F_m = nqvB. \tag{27.6}$$

If a charge moves the length b of the wire in time t, its speed is

$$v = \frac{b}{t}.$$

In time t, all n charges in the wire will pass a cross section at its end. According to Eq. (24.1), the current is

$$I = \frac{nq}{t}.$$

Using the above equations to substitute for v in Eq. (27.6), we find that

$$F_m = nqvB = nq\left(\frac{b}{t}\right)B = \left(\frac{nq}{t}\right)bB = IbB.$$

The force is the same magnitude on each side wire shown in Fig. 27.15, but the force on one side wire is directed up, whereas that on the other wire is directed down. Although the net force on the two wires is zero, together they produce a torque on the loop of the wire.

FIG. 27.14. A mass spectrometer.

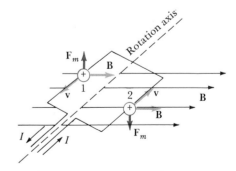

FIG. 27.15. The magnetic force on positive charges moving along the side of a loop of wire causes a torque that will rotate the loop clockwise.

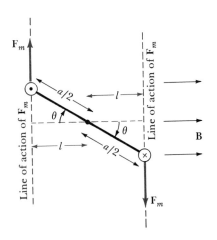

FIG. 27.16. The magnetic force on each side wire of a loop has a moment arm l. The torque due to each force has a magnitude of $\tau = lF_m$.

To calculate the torque, consider the end view of the loop shown in Fig. 27.16. If the plane of the loop is oriented at an angle θ with respect to the field, then the moment arm l of the force about an axis along the center of the loop is

$$l = \left(\frac{a}{2}\right) \cos \theta,$$

where a is the width of the loop. The torque due to both wires is calculated using Eq. (2.4):

$$\tau = 2(l)(F_m) = 2\left(\frac{a}{2} \cos \theta\right)(IbB).$$

The inside area A of the loop is the product of its width and length; that is, $A = ab$. Hence, the **torque** is

$$\tau = IAB \cos \theta. \tag{27.7}$$

(Magnetic forces exerted on moving charges at the ends of the loop stretch or compress the loop but cause no torque.)

The torque is greatest when the plane of the loop is oriented parallel to the direction of the magnetic field (Fig. 27.17a). The angle θ is then $0°$, and since $\cos 0° = 1$, the torque is a maximum. When the loop is perpendicular to the field, as in Fig. 27.17b, two equal and opposite forces are produced, each having a zero moment arm. Thus, no torque is produced. (Note that when θ is $90°$, $\cos 90° = 0$, and the torque is zero.)

Although Eq. (27.7) was derived for a rectangular loop, it applies equally well to a loop of any shape, as long as the loop is flat.

FIG. 27.17. (a) The torque on the loop is maximum when the plane of the loop is parallel to the **B** field. The forces have the greatest moment arm. (b) The torque is zero when the loop is perpendicular to the field, since the moment arms of the forces are zero.

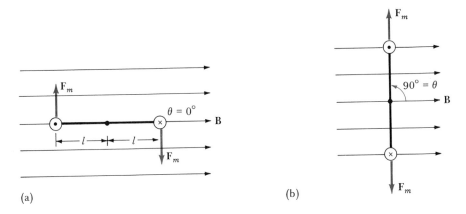

27.5 A Direct-Current Electric Motor

The torque just described is used to turn the shaft of a direct-current electric motor. The motor has a large number of loops of wire, called the **armature,** wound around an iron core. An armature with one loop is shown in Fig. 27.18. In a motor the armature is placed between the poles of a magnet. When electric current flows through the loops of wire, a magnetic force is exerted on these moving charges. The force causes a torque that turns the armature and its core. The core is attached to a shaft that causes some other object, such as a fan, to rotate.

The motor shown in Fig. 27.18 has one serious problem: when the loop has rotated a little less than one-quarter turn, the torque on the loop becomes counterclockwise rather than clockwise. Look at the side view of a loop shown in Fig. 27.19a. The circle with a cross ($\otimes$) indicates a current moving into the paper along a side wire of the loop. A circle with a dot ($\odot$) indicates a current moving out of the paper on the other side wire of the loop. When in the horizontal position with the currents flowing as shown, a clockwise torque is exerted on the loop. As the loop passes the vertical orientation, the torque becomes counterclockwise. However, if the current is reversed, as in Fig. 27.19b, the torque remains clockwise. Consequently, to make the armature continue to rotate clockwise, the current must change direction each time the loop passes the vertical orientation.

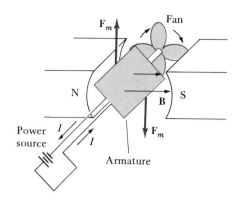

FIG. 27.18. A dc electric motor.

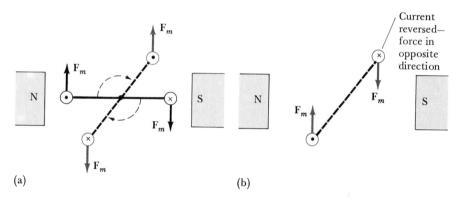

(a) (b)

FIG. 27.19. (a) When the loop is horizontal (the solid lines), a clockwise torque is exerted on the loop. As the loop passes the vertical orientation, the torque becomes counterclockwise. (b) If the current direction in the loop is reversed as it passes the vertical, the torque continues to be clockwise.

This current reversal is made possible using a commutator (Fig. 27.20). The **commutator** consists of two semicircular rings that are attached to the armature and rotate with it. Sliding contacts or brushes from the power source contact the commutator rings. The current direction is reversed in the middle of each rotation, as the contact of the brushes pass from one commutator ring to the next.

27.6 The Cause of a Magnetic Field

All the effects we have discussed—magnetic forces, circular motion of charges in a magnetic field, torques on current loops—depend on the presence of a magnetic field. At the start of the chapter we assumed that these fields exist because we saw their effects. But what causes a magnetic field? We know that electric fields are caused by electric charges and that gravitational fields are caused by masses. Is there an analogous object that causes magnetic fields? Some scientists say that there is such an object, called a magnetic monopole, that is analogous to an electric charge. Whereas most magnets have a north and a south pole, a magnetic monopole is a single pole—either north or south. A north pole might be likened to a positive charge and a south pole to a negative charge. Either one supposedly can produce a magnetic field. While the search for a magnetic monopole has, to date, been unsuccessful, some experiments seem to hint at its existence.

There are, nonetheless, ways to produce a magnetic field that do not require the existence of a magnetic monopole. For example, putting an electric charge in motion produces a magnetic field. You have probably performed experi-

FIG. 27.20. A simple dc motor used to turn a fan.

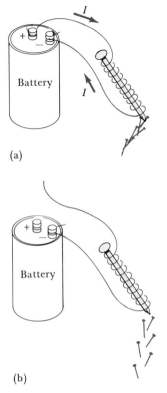

(a)

(b)

FIG. 27.21. (a) When a current passes through a wire wrapped around a nail, a magnetic field is produced that allows the nail to become a magnet and attract metal pins. (b) If the current stops, the pins fall, since a magnetic field is no longer being produced.

FIG. 27.22. (a) The arrangement used by Oersted to show that currents did not produce magnetic fields. (b) If the compasses are placed under the wire, they both swing to the right. (c) The magnetic-field lines produced by a wire are concentric circles.

ments, such as shown in Fig. 27.21, in elementary school that confirm this idea. When a wire is wrapped around a nail and attached to a battery, current flows through the wire. The moving charges produce a magnetic field, and as a result, the nail acts like a magnet. The nail attracts pins and other objects made of iron when the current is flowing but does not when the current stops.

The fact that moving charges create magnetic fields was discovered in 1820 by Hans Christian Oersted (1777–1851), a Danish physicist. Before that time other scientists had surmised that electricity—that is, moving charges—and magnetism were related in some unknown way. In fact, prizes were offered for a discovery of the connection. It is said that Oersted, who often gave public lectures, was demonstrating a variety of electric and magnetic phenomena. He had demonstrated many times that there was no observable interaction between electric currents and nearby magnets.

One experiment that Oersted devised to demonstrate this noninteraction involved initiating a current in a long, straight wire by closing a switch that connected it to a battery. Two compass needles were placed near the wire, as shown in Fig. 27.22a. Oersted expected that if a magnetic field was created by the wire, it would be directed parallel to the current. This field should deflect the compass needles so that both would point along the direction of the current. When the current was initiated, though, neither compass needle moved.

After one lecture, the compasses, which were usually on a platform at the same level as the wire, were moved below the wire. When Oersted closed the switch, the needles, to his amazement, swung around and both pointed perpendicular to the wire (Fig. 27.22b). Repeated experiments showed that the current was producing a magnetic field and that the field could be represented by concentric circles around the wire (Fig. 27.22c).

At the time it seemed unreasonable that the field should point at right angles to the current. Oersted's contemporaries "reacted violently to his demonstrations in subsequent lectures. The audience 'disliked' the force at right angles, just as later audiences 'disliked' relativity; for example, that mass changed with velocity, as Einstein asserted in lectures at the University of Chicago . . . ."*

Oersted's work eventually gained acceptance. Scientists of his time developed techniques to calculate the magnetic field caused by moving charges. One of these techniques will be discussed briefly next.

*This is an excerpt from an interesting book about magnetism by Francis Bitter, *Magnets: The Education of a Physicist* (Garden City, N.Y.: Doubleday Anchor Books, 1959).

(a)

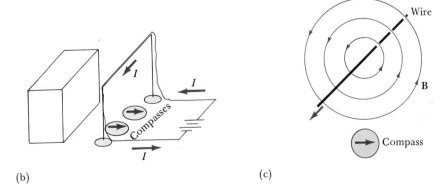

(b)

(c)

27.7 Calculating the Magnetic Field Caused by Moving Charges

Two French scientists, Jean Baptiste Biot (1774–1862) and Felix Savart (1791–1841), formulated an equation to calculate the magnetic field caused by moving charges. The equation is difficult to use without the assistance of integral calculus, so we will simply outline the Biot-Savart technique and emphasize some results of its use. Perhaps the most important idea to remember is that moving charges produce magnetic fields. The technique is illustrated with reference to the current moving in a long, straight wire, shown in Fig. 27.23a.

Biot and Savart found experimentally that the magnetic field **B** caused by a charge q moving at a speed v has a magnitude

$$B = \left(\frac{\mu}{4\pi}\right)\frac{qv\sin\theta}{r^2}. \tag{27.8}$$

The equation is called the **Biot-Savart law.** The quantity r is the distance from the moving charge to the point where the field is being calculated. The quantity μ is called the **permeability** of the medium in which the magnetic field exists. In free space the permeability μ_0 has a value $\mu_0 = 4\pi \times 10^{-7}\,\text{T}\cdot\text{m/A} = 1.257 \times 10^{-6}\,\text{T}\cdot\text{m/A}$; thus

$$\frac{\mu_0}{4\pi} = 10^{-7}\,\text{T}\cdot\text{m/A}.$$

The permeability of other materials is a complex subject. Even for a particular type of material, such as iron, the permeability depends on its past history: has it been magnetized or demagnetized, and to what degree? Although more will be said about permeability later, we will note that the permeability of iron is about 1000 times that of air. Thus, magnetic fields produced by currents are much stronger in iron than in air.

The angle θ that appears in Eq. (27.8) is the angle between the direction of the charge's velocity **v** and a vector **r** drawn from the moving charge to the field point, as shown in Fig. 27.23a. The direction of **B** is perpendicular to the plane containing **v** and **r.** The orientation of **B** relative to that plane is determined using a different right-hand rule than that used to find the force on a moving charge.

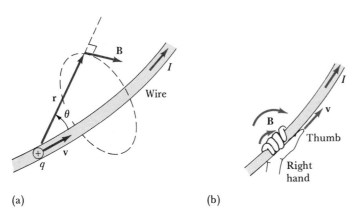

(a) (b)

FIG. 27.23. (a) The magnetic field produced by a moving charge is perpendicular to the planes of **v** and **r.** (b) The magnetic field circles the wire in the direction of your fingers when your right thumb points in the direction of the current.

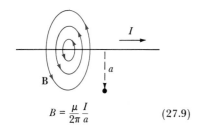

$$B = \frac{\mu}{2\pi} \frac{I}{a} \qquad (27.9)$$

(a)

$$B = \frac{\mu I}{2a} \qquad \text{At center of loop} \qquad (27.10)$$

$$B = \frac{N\mu I}{2a} \qquad \text{At center of loop with } N \text{ turns} \qquad (27.11)$$

(b)

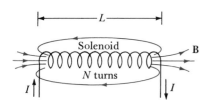

$$B = \mu \frac{N}{L} I \qquad \text{Inside solenoid} \qquad (27.12)$$

(c)

FIG. 27.24. The magnetic fields created by currents in (a) a straight wire, (b) a single loop of wire, and (c) a solenoid.

Right-hand rule to find direction of B field: Point the thumb of your right hand in the direction of the current—that is, in the direction of a moving, positive charge. The field circles around the current in the same direction as your fingers curl around the current (Fig. 27.23b).

Since most magnetic fields are produced by a very large number of charges moving in a wire, the net magnetic field at a point is the vector sum of the fields caused by each moving charge. Integral calculus must be used to add these fields. Usually the magnitude of the resultant field is expressed in terms of the electric current I causing the field. The direction of the field at different locations is shown by drawing magnetic-field lines. The lines are closely spaced at regions where the field is greatest. Several examples are shown in Fig. 27.24.

EXAMPLE 27.3 A straight wire carries a current of 5 A. At what distance from the wire is the magnetic field equal in magnitude to the earth's magnetic field (5×10^{-5} T)?

SOLUTION The magnetic field at distance a from a straight wire carrying a current I is given by Eq. (27.9), which is listed in Fig. 27.24a:

$$B = \frac{\mu_0 I}{2\pi a}.$$

Rearranging to solve for a, we find that

$$a = \frac{\mu_0 I}{2\pi B} = \frac{(4\pi \times 10^{-7}\,\text{T} \cdot \text{m/A})5\,\text{A}}{2\pi(5 \times 10^{-5}\,\text{T})} = 2 \times 10^{-2}\,\text{m} = \underline{2\,\text{cm}}.$$

Thus, we would not want to put the compass of a ship close to electrical wires. The compass might respond to the magnetic field caused by currents in the wires rather than to the earth's magnetic field. ∎

EXAMPLE 27.4 We wish to observe the effect of a magnetic field on the muscles in a frog's leg. The measurements can be made by placing the frog's leg in a solenoid (Fig. 27.25) that has a diameter of 1.0 cm, is 5.0 cm long, and is made of 200 turns of wire. Calculate the strength of the field inside the solenoid if a current of 4.0 A flows in it. The permeability of the frog's leg (and of all living matter) is about equal to that of air.

SOLUTION The magnetic field at the center of a solenoid is given by Eq. (27.12), shown in Fig. 27.24c:

$$B = \mu_0 \frac{N}{L} I = (4\pi \times 10^{-7}\,\text{T} \cdot \text{m/A}) \times \frac{200}{0.050\,\text{m}}(4.0\,\text{A}) = \underline{2.0 \times 10^{-2}\,\text{T}}.$$

It is unlikely that this field will produce any easily observable effects on the frog's leg. ∎

EXAMPLE 27.5 An electron in the hydrogen atom can be thought of as circling the proton nucleus in a circular orbit of radius 0.53×10^{-10} m. The electron completes one trip around the orbit each 1.5×10^{-16} s. Calculate the magnitude of the magnetic field at the center of the circle where the proton is located.

SOLUTION The electron's motion produces the same effect as if a circular loop of wire carried an electric current around the proton (Fig. 27.24b). The field at the center of the loop is given by Eq. (27.10): $B = \mu_0 I/2r$. To determine B, we must first calculate the current I due to the electron's motion. Since one electron passes around the loop or orbit each 1.5×10^{-16} s, we can calculate the "current" that is flowing by using Eq. (24.1):

$$I = \frac{q}{t} = \frac{(1.6 \times 10^{-19}\,\text{C})}{(1.5 \times 10^{-16}\,\text{s})} = 1.1 \times 10^{-3}\,\text{A}.$$

This single electron, because of its small orbit and its high speed, produces a current of 1.1 mA.

The magnetic field at the center of the loop is now determined using Eq. (27.10):

$$B = \frac{\mu_0 I}{2r} = \frac{(4\pi \times 10^{-7}\,\text{T}\cdot\text{m/A})(1.1 \times 10^{-3}\,\text{A})}{2(0.53 \times 10^{-10}\,\text{m})} = \underline{13\,\text{T}}.$$

This is a very strong magnetic field. ∎

FIG. 27.25. The current through a solenoid creates a magnetic field in a frog's leg.

27.8 Magnetic Materials

Now that we have examined (1) the force of a magnetic field on a moving charge, and (2) the magnetic field produced by moving charges, we can use this information to see how a common bar magnet works. A bar magnet, like all magnetic materials, derives its properties from the magnetic fields created by atoms and the forces (magnetic and others) acting on atoms.

An atom consists of a very small nucleus surrounded by clouds of electrons, which can be thought of as circling the nucleus in elliptical orbits. The motion of each electron is equivalent to a loop of wire carrying one electron. The magnetic field produced by this electron current loop is identical to that produced by a tiny bar magnet (see Fig. 27.26). The electron also produces a magnetic field because it has an intrinsic *magnetic dipole moment* with a north and south pole and is like a very tiny bar magnet. Thus, an electron in an atom creates a magnetic field because of its orbital motion and because of its intrinsic magnetic moment.

In atoms with more than one electron, the fields created by different electrons often cancel each other. The electrons pair off so that the intrinsic magnetic moment of one electron points opposite that of another electron, and their magnetic fields cancel. The field caused by the orbital motion of electrons also cancel since some electron currents move clockwise while others move counterclockwise. If the cancellation is not complete, the atom has a net magnetic dipole moment and produces a magnetic field like that of a small bar magnet. Atoms with nonzero magnetic dipole moments such as iron, aluminum, sodium, oxygen, and nickel are called *paramagnetic* atoms.

Normally the paramagnetic atoms in a material are oriented randomly. The magnetic field produced by one atom is canceled by the field of another

(a)

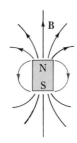

(b)

FIG. 27.26. The field caused by an electron current loop in an atom (a) looks just like that produced by a tiny bar magnet (b).

atom that is oriented in the opposite direction. The net field produced by all atoms is zero. However, if this material that is made of paramagnetic atoms is placed in an externally generated field, the magnetic dipole moments of the atoms have a tendency to align with the external field (like a compass needle aligning with the earth's magnetic field). The magnetic fields caused by the atoms no longer cancel each other completely but add together to produce their own field that adds to the external field. In most materials the paramagnetic atoms contribute little additional field because the thermal agitation by neighboring atoms causes the atoms to remain randomly oriented. For example, the field produced by a unit volume of aluminum atoms is only 2×10^{-5} times the field used to orient the atoms.

In certain materials such as iron, nickel, cobalt, and gadolinium, the alignment of their atoms is not random. If iron, for example, is placed in an externally generated magnetic field, the atomic magnetic moments of the iron atoms may become aligned so that the net field (the external field plus that caused by the magnetic moments of the iron atoms) is 100 to 1000 times greater than the external field causing the alignment. In these materials, called *ferromagnetic* materials, the magnetic moments of neighboring atoms are locked together and oriented in the same direction by a complicated force that is described by theories of atomic physics.

Even when not in an external magnetic field, the alignment of neighboring atoms in these materials is uniform in small regions called **domains.** In a particular domain the magnetic dipoles of all atoms point in the same direction. Each domain may include 10^{15} to 10^{16} atoms and occupies a space less than a millimeter on a side. In a neighboring domain the magnetic dipoles of the atoms may all point in a different direction. In a piece of iron that is unmagnetized, like a nail, the domains are oriented randomly, and the magnetic field produced by one domain is cancelled by that produced by another.

If an unmagnetized piece of iron is placed in an externally produced magnetic field caused by a loop of wire with current flowing through it, the domains oriented in the direction of the field increase in size or grow, and those oriented in other directions decrease in size or shrink. The material is then said to be magnetized, and it produces its own strong magnetic field. In a permanent magnet, the domains remain oriented in the same direction, and the material continues to produce a strong magnetic field even when removed from the external field.

We can now understand why the magnetic permeability μ of a material affects the magnetic field produced by a current. Picture in your mind a solenoid with air inside its coils and another with iron inside. The magnetic field produced by the solenoid with air is caused entirely by the current in the loops of wire, but for a solenoid with iron inside, the field produced by the current loops causes the domains oriented in the direction of the field to grow. Their field adds to that of the coil. If most of the iron magnetic moments become aligned in the direction of the field of the coil, the field of the iron and coil together may be a thousand times greater than if the iron were not present.

This increase in magnetic field caused by ferromagnetic materials has many practical applications. The coating on magnetic tape provides a record of information by the degree to which it becomes permanently magnetized. Airport metal detectors, transformers, electric motors, loudspeakers, and electric generators all depend on the magnetization of ferromagnetic materials.

Summary

1. **Magnetic force on a moving charge:** If an electric charge q moves with velocity $\mathbf{v}$ through a magnetic field $\mathbf{B}$, a magnetic force $\mathbf{F}_m$ is exerted on the charge. The magnitude of the force is

$$F_m = qvB \sin \theta, \tag{27.1}$$

where θ is the angle between the directions of $\mathbf{v}$ and $\mathbf{B}$. The direction of the magnetic force is perpendicular to both $\mathbf{v}$ and $\mathbf{B}$. The right-hand rule is used to determine the orientation of $\mathbf{F}_m$. Point the fingers of your open right hand in the direction of $\mathbf{B}$ and orient your hand so that the thumb points along $\mathbf{v}$. Your open palm then faces in the direction of the magnetic force on a positive charge. The force on a negative charge is in the opposite direction.

2. **Circular motion in a magnetic field:** When an object of mass m and charge q moves at speed v perpendicular to a uniform magnetic field $\mathbf{B}$, the magnetic force ($F_m = qvB$) acting on the charge causes a centripetal acceleration ($a_c = v^2/r$) that keeps the charge moving in a circle. Newton's second law relating the magnetic force and centripetal acceleration is as follows:

$$qvB = m\frac{v^2}{r}. \tag{27.5}$$

3. **Torque on a current loop:** If a current I flows through a loop of wire of area A while the loop is in a magnetic field B, the magnetic force on the charges moving in the wire causes a torque τ on the loop of magnitude

$$\tau = IAB \cos \theta, \tag{27.7}$$

where θ is the angle between the plane of the loop and the direction of the magnetic field.

4. **Magnetic field produced by moving charges:** Moving electric charges produce magnetic fields. The Biot-Savart law, along with integral calculus, can be used to calculate the field caused by these moving charges. The fields produced by electric currents in wires of different shape are shown in Fig. 27.24.

5. **Magnetic materials:** A material that can be magnetized is made of atoms whose electrons produce magnetic fields because of the orbits and intrinsic magnetic moments of their electrons. The atoms act like tiny bar magnets. In ferromagnetic materials, the atomic bar magnets in a region of the material fall in line; a **domain** is produced. If domains become aligned, the material is magnetized and produces its own large magnetic field.

Questions

1. A beam of electrons is not deflected as it moves through a region of space in which a magnetic field exists. Give two explanations why the electrons might move along a straight path.

2. A "magnetic lens" is used to focus electrons in an electron microscope. We can gain a little understanding of how the lens works by considering the path of electrons moving out of the electron beam at a small angle. Describe the motion of these electrons if a magnetic field points in the direction of desired motion. Compare the motion to that which would occur if the magnetic field were not present.

3. A U-shaped wire through which a current flows hangs with the bottom of the U between the poles of an electromagnet (see Fig. 27.27). When the field in the magnet is increased, does the U swing toward the right or the left? Explain.

4. Why does the torque on the current loop of the motor shown in Fig. 27.19 change, even though the force on the wires is constant?

5. A magnetic field is said to exist in a region of space. Describe three experiments you could do to confirm this statement.

6. An electron enters a solenoid at a small angle relative to the magnetic field inside. Describe the electron's motion.

FIG. 27.27

7. Two parallel wires carry electric current in the same direction. Does the moving charge in one wire cause a magnetic force to be exerted on the moving charge in the other wire? If so, in what direction is the force relative to the wires? Explain. Repeat for currents moving in opposite directions.

8. Why is the magnetic field so much greater at the end of a solenoid that has an iron core than at the end of a solenoid of similar size with an air core?

Problems

27.2 Magnetic Force on a Moving Charge

1. Each of the lettered dots shown in Fig. 27.28 has $+10^{-8}$ C of charge that moves at speed 2×10^7 m/s. Calculate the magnitude of the force on each charge and indicate carefully, in a drawing, the direction of the force. A uniform 0.5-T magnetic field points in the positive z direction.

FIG. 27.28

2. A duck accumulates a positive charge of 3.0×10^{-8} C while flying north at speed 18 m/s. The earth's magnetic field at the duck's location has a magnitude of 5.3×10^{-5} T and points in a direction 62° below a horizontal line pointing north. Calculate the magnitude and direction of the magnetic force on the duck.

3. A 1000-kg automobile drives west along the equator near Lake George in Uganda. At this location the earth's magnetic field is 3.5×10^{-5} T and points north parallel to the earth's surface. If the automobile carries a $-(2.0 \times 10^{-6})$-C charge, how fast must it move so that the magnetic force balances 1 percent of its weight?

4. An electron flies toward the screen of a television set at a speed of 1.0×10^7 m/s. The electron moves perpendicular to the (5.0×10^{-5})-T magnetic field of the earth. Calculate the ratio of the magnitude of the gravitational force acting on the electron and the magnitude of the magnetic force.

5. A hydroxide ion (OH^-) in a glass of water has an average speed of about 600 m/s. (a) Calculate the electrical force between the hydroxide ion (charge $-e$) and a positive ion (charge $+e$) that is 1.0×10^{-8} m away (about the separation of 30 atoms). (b) Calculate the maximum magnetic force that can act on the ion when in the earth's (5.0×10^{-5})-T magnetic field. (c) On the basis of these two calculations, does it seem likely that the earth's magnetic field has much effect on the biochemistry of the body?

6. An automobile with a charge of $+2.0$ C is in a (4.0×10^{-5})-T magnetic field that is parallel to the ground and points north. Calculate the magnetic force (magnitude and direction) on the automobile when (a) at rest, (b) moving south at 120 m/s, (c) moving west at 120 m/s, (d) moving 30° east of north at 120 m/s, and (e) moving 37° west of south at 120 m/s.

7. An electron moves between two parallel plates, as shown in Fig. 27.29. A 480-V/m electric field points from the upper plate toward the lower. A 0.12-T magnetic field points into the paper (represented by the crosses ($\times$) between the plates). At what

FIG. 27.29

speed must the electron move so that it is not deflected while passing between the plates? (If narrow slits are placed at the entrance and exit of the plates, the device allows charges of only one speed to pass through both slits. Such a device is called a **velocity selector.**)

8. A velocity selector (discussed in Problem 7) allows charges moving at 3.0×10^6 m/s to pass through it. If the selector has a magnetic field of 0.15 T and its plates are separated by 2.5 cm, how much voltage should be placed across the plates?

9. A magnetohydrodynamic generator develops a potential difference of 1.0×10^4 V when ions pass through a 10-T magnetic field and accumulate on plates separated by 10 cm. (a) What is the average speed of the ions? (b) What must their speed be if the plates are separated by 30 cm?

10. A blood-flow meter produces a 0.020-T magnetic field perpendicular to the flow direction. (a) Calculate the magnetic force on an ion of charge e while moving with the blood at a speed of 0.30 m/s. (b) Positive and negative ions are forced in opposite directions, producing an electric field and voltage across the blood vessel. The ion separation stops when the electric force caused by this voltage balances the magnetic force. What voltage will be developed across a (3.0×10^{-3})-m diameter blood vessel?

11. A wire, shown in Fig. 27.30, moves perpendicular to a magnetic field. (a) In which direction are the electrons in the wire forced to move? (b) After the electrons are forced in one direction, leaving positive charges behind, an electric field and voltage V are developed from one end of the wire to the other. Electrons no longer move, since the electric and magnetic forces balance. Show that this happens when the voltage from one end of the wire to the other is

$$V = vLB,$$

where v is the speed of the wire, L is its length, and B is the magnitude of the magnetic field.

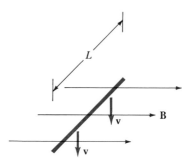

FIG. 27.30

12. Calculate the voltage developed across the wings of a Boeing 747 jet airliner when traveling north at a speed of 240 m/s. The wing span is 60 m. The earth's (5.0×10^{-5})-T magnetic field points in a direction 60° below a horizontal line directed north. In

solving this problem, follow the outline developed in the previous problem. Note that the magnetic field is not perpendicular to the plane's velocity.

27.3 Circular Motion in a Magnetic Field

13. An electron moves in a circular path perpendicular to the earth's (3.5×10^{-5})-T magnetic field. The radius of the circle is 0.70 m. Calculate the electron's speed.

14. An electron and a proton, moving side by side at the same speed, enter a 0.020-T magnetic field. The electron moves in a circular path of radius 7.0 mm. Calculate the radius of the adjacent circle in which the proton moves.

15. (a) Calculate the speed of electrons accelerated through a 28-kV potential difference in a color television set. (b) Calculate the radius of curvature of one of these electrons if moving perpendicular to the earth's (3.5×10^{-5})-T magnetic field at the equator. (c) If the television set faces east, are the electrons deflected up or down? Explain your answer.

16. A lithium ion with charge e and speed 8.0×10^5 m/s enters the 0.35-T magnetic field of a mass spectrometer. The ion moves in a circular path of radius 0.17 m. Calculate the mass of the ion.

17. A proton in the solar wind enters the earth's magnetosphere where the magnetic field is about 3×10^{-7} T. The proton moves in a helical path of radius 3.38 km. (a) Calculate the magnitude of the component of the proton's velocity that is perpendicular to the field. (b) If the proton's velocity makes an angle of 88.7° with the field, what is the magnitude of the parallel component of velocity?

18. Carbon ions of charge e leave the velocity selector of a mass spectrometer at a speed of 1.6×10^5 m/s. They then move in a semicircular path in a 0.80-T magnetic field. The carbon consists of two isotopes, one with six protons and six neutrons in its nucleus and the other with six protons and seven neutrons in its nucleus; the first is called carbon 12 and the second, carbon 13. Carbon 12 has a mass of 1.99×10^{-26} kg and carbon 13 a mass 13/12 greater. Calculate the distance apart of the ions after they move in a semicircular path in the mass spectrometer.

19. E. O. Lawrence built one of the first cyclotrons (particle accelerators) in 1932. Protons were accelerated while moving in a circular path perpendicular to a magnetic field. The protons attained a speed of 1.5×10^7 m/s while moving in a circle of radius 0.14 m. (a) Calculate the magnitude of the magnetic field. (b) How many rotations in a circle could one proton make each second?

27.4 Torque on a Current Loop

20. A wire 0.50 m long is supported at its ends in a north-south direction. A 5.0-A current flows north in the wire. (a) In which direction should a magnetic field point in order that an upward magnetic force be exerted on the wire? Explain. (b) If a 2.0-T field points in that direction, calculate the magnitude of the magnetic force exerted on the wire.

21. An east-west power line at the equator carries 100 A of current toward the east. At this location the earth's magnetic field has a magnitude of 3.5×10^{-5} T and points north. (a) Calculate the magnitude and direction of the magnetic force on a 250-m length of the wire whose mass is 160 kg. (b) Calculate the fraction of the wire's weight that this force equals.

22. (a) Calculate the magnetic force (magnitude and direction) that acts on segments a, b, and c of wire shown in Fig. 27.31. Segments a and c are 2.0 cm long, and segment b is 10.0 cm long. A 3.0-A current flows through the wire, which is in a 0.15-T magnetic field. (b) Calculate the torque caused by forces $\mathbf{F}_a$ and $\mathbf{F}_c$ together, and the torque caused by force $\mathbf{F}_b$.

FIG. 27.31

23. (a) What is the maximum torque on a 500-turn circular coil of radius 6.0 cm when carrying a 0.30-A current in a 0.10-T magnetic field? (b) How is the coil oriented relative to the field when this maximum torque is exerted on it? (c) What should its orientation be when the torque is reduced by 50 percent?

24. A 500-turn coil of wire is hinged to the side of a table, as shown in Fig. 27.32. Each side of the movable part of the coil has a length of 0.50 m. (a) In which direction should a magnetic field point in order to help lift the free end of the loop off the table? (b) Calculate the torque caused by a 0.70-T field pointing in the direction described in (a) when a 0.80-A current flows through the wire and the coil is parallel to the table.

FIG. 27.32

25. A 20-turn coil, 5.0 cm by 10.0 cm, hangs with the plane of the coil parallel to a 0.12-T magnetic field. (a) Calculate the torque on the coil when a 0.5-A current flows through it. (b) The support for the coil exerts an opposing torque, which increases as the coil is deflected. The opposing torque is calculated using the equation $\tau = -0.016\,\theta$ (in units of N · m), where the angle θ is in units of radians (Fig. 27.33). At approximately what angle does

FIG. 27.33

the current torque balance the torque caused by the supporting wire? An ammeter can be calibrated to measure currents in this manner.

27.5 A Direct-Current Electric Motor

26. An electric motor has a square armature with 500 turns. Each side of the coil is 12 cm long and carries a current of 4.0 A. (a) Calculate the magnetic force on one side of the coil when it is perpendicular to a 1.2-T magnetic field. (b) Calculate the torque on the coil when the armature's surface is in the plane of the field.

27. An electric motor has a circular, 100-turn armature of radius 7.0 cm and carries a current of 50 A when the switch is first turned on. The surface of the coil is parallel to a 0.80-T magnetic field. An 8.0-N · m friction torque initially opposes the rotation of the armature. Calculate the net torque on the armature when the switch is turned on.

27.7 Calculating the Magnetic Field Caused by Moving Charges

28. An east-west electric power line carries a 500-A current toward the east. The line is 10 m above the earth's surface. Calculate the magnitude and direction of the magnetic field at the earth's surface directly under the wire.

29. Two long, parallel wires are separated by 2.0 m. Each wire carries a 30-A current, but the currents flow in opposite directions. (a) Calculate the magnitude of the net magnetic field midway between the wires. (b) Calculate the net magnetic field at a point 1.0 m to the side of one wire and 3.0 m from the other wire.

30. During World War II, explosive mines were dropped by the Nazis in the harbors of England. The mines, which lay at the bottom of the harbors, were activated by the changing magnetic field that occurred when a large metal ship passed above them. Small English boats called minesweepers would tow long, current-carrying coils of wire around the harbors. The field created by the coils activated the mines, causing them to explode under the coils rather than under ships. (a) Calculate the current that must flow in one long, straight wire in order to create a 0.0050-T magnetic field at a depth of 20 m under the water. The magnetic permeability of water is about the same as that of air. (b) How might the field be created using a smaller current?

31. A solenoid of radius 1.0 m with 750 turns and a length of 5.0 m is wound around a pigeon cage. What current must flow so that the solenoid field just cancels the earth's (4.2×10^{-5})-T magnetic field, which is occasionally used by the pigeons to determine the direction they move.

32. Two concentric solenoids carry currents in opposite directions. The inner solenoid has 50 turns per centimeter, the outer solenoid 20 turns per centimeter. The inner coil carries a current of 0.80 A. (a) Calculate the current needed in the outer coil so that its field cancels the field produced by the inner coil. (b) How large a field is produced by the inner coil?

33. An excited electron in a hydrogen atom moves with a speed of 1.09×10^6 m/s in a circular orbit whose radius is 2.12×10^{-10} m. (a) Calculate the time required for one trip around the circle. (b) Calculate the current caused by the electron's motion. (c) Calculate the magnetic field at the center of the circular orbit.

General Problems

G1. A singly charged ion of unknown mass is accelerated from rest through a 6000-V potential difference. The ion enters perpendicular to a 1.2-T magnetic field and moves in a circular path of radius 10.7 cm. Calculate the ion's mass. (*Hint:* You will need to solve two equations simultaneously.)

G2. A mass spectrometer has a velocity selector (see Problem 7) that allows ions traveling at only one speed to pass undeflected through slits at the ends. While moving through the velocity selector, the ions pass through a 60,000-V/m electric field and a 0.0500-T magnetic field. The three vectors, **v**, **E**, and **B**, are mutually perpendicular. (a) Calculate the speed of undeflected ions. (b) After leaving the velocity selector, the ions continue to move in the 0.0500-T magnetic field. Calculate the radius of curvature of a singly charged lithium ion whose mass is 1.16×10^{-26} kg.

G3. Two electrons move past each other along lines separated by 2.0×10^{-8} m. Each electron moves at a speed of 3.0×10^7 m/s. (a) Calculate the magnetic field at one electron caused by the other when the electrons are next to each other. (b) Calculate the magnitude of the magnetic force caused by this field. (c) Calculate the magnitude of the electric force of one electron on the other when separated by 2.0×10^{-8} m.

G4. An alien from a planet in the galaxy M31 (Andromeda) has a ray gun that shoots protons at a speed of 2.0×10^5 m/s. Design a magnetic shield that will deflect the protons away from your body. Using rough *estimates*, show that your magnetic shield will, in fact, protect you. Indicate the orientation of magnetic field in front of you and the direction in which the ions are deflected.

Induced Voltage

28

Two fundamental ideas dominated our study of magnetism in Chapter 27: (1) a force is exerted on a moving charge as it passes through a magnetic field, and (2) a moving electric charge produces a magnetic field. In this chapter another important magnetic phenomenon is introduced: *electromagnetic induction* is the generation of electrical voltage by the variation of the magnetic field in a loop of wire. In addition to describing how electromagnetic induction occurs, we will examine some of the ways it is used—for example, in airport metal detectors and in electric generators and transformers.

28.1 Electromagnetic Induction

A simple example of electromagnetic induction is illustrated in Fig. 28.1. In (a) a loop of wire is attached to a sensitive voltmeter, which measures the voltage across the ends of the loop. Since there are no batteries or other sources of electromotive force (emf) in the circuit, the voltage across the ends of the loop is normally zero.

(a)

(c)

(e)

(b)

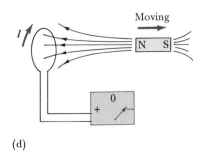

(d)

FIG. 28.1. The voltage across a loop changes as the magnet is moved toward the loop (b) and away from the loop (d). The voltage is zero when the number of magnetic-field lines through the loop is constant, as in (a), (c), and (e).

Strong B field

(a)

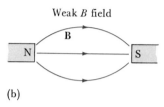

Weak B field

(b)

FIG. 28.2. (a) A large-magnitude magnetic field is represented by closely spaced lines, whereas (b) a weak field has sparsely spaced lines.

Zero flux

(a)

Intermediate flux

(b)

Maximum flux

(c)

FIG. 28.3. The magnetic flux through a loop of wire depends on its orientation in a magnetic field.

If a bar magnet is moved toward the loop of wire, as in Fig. 28.1b, a voltage is induced across the ends of the wire and a small current flows around the loop. If the magnet's motion is stopped (Fig. 28.1c), the voltage returns to zero and the current stops. If the magnet is withdrawn (Fig. 28.1d), a voltage is again produced and a current flows. However, this time the polarity of the voltage is reversed, and the current flows in the opposite direction. The voltage and current return to zero as the magnet is moved far away (out of sight in Fig. 28.1e).

The most important observation relative to this experiment is the fact that a voltage is produced only when the magnet is moving. The voltage $\mathcal{E}$ produced by moving the magnet relative to the loop of wire is an example of a process called **electromagnetic induction**—the induction of a voltage by a *varying* magnetic field inside a loop of wire. Electromagnetic induction was discovered in 1831 after a seven-year search by the English scientist Michael Faraday. The theoretical description of electromagnetic induction is based on what is called Faraday's law.

28.2 Faraday's Law

Before we can examine Faraday's law, we must first define a quantity called magnetic flux. Magnetic flux depends on the magnetic field in a region of space, such as the region inside a loop of wire. Usually the field is depicted by lines, as in Fig. 28.2. If the field is strong, the lines are more closely spaced (Fig. 28.2a); if it is weak, the lines are farther apart (Fig. 28.2b).

If a coil or loop of wire is placed in a magnetic field, some of the field lines pass through the loop. The **magnetic flux** through the loop equals the number of magnetic-field lines that pass through the coil and depends on the magnitude B of the field, on the area A of the loop, and on the orientation of the loop in the field. For example, when the plane of the loop is parallel to the field (Fig. 28.3a), no lines pass through the loop, and the magnetic flux through the loop is zero. If the loop is rotated so that its plane becomes oriented more nearly perpendicular to the field, as in Fig. 28.3b and c, more lines pass through the loop and the magnetic flux increases.

The **magnetic flux** Φ through a loop is calculated using the equation

$$\Phi = AB \cos \phi, \qquad (28.1)$$

where A is the area inside the loop, B is the magnitude of the magnetic field inside the loop, and ϕ is the angle between the direction of the field and the direction of a line drawn *perpendicular* to the area inside the loop (Fig. 28.3b). When the perpendicular line and the field are parallel (Fig. 28.3c), ϕ is zero and $\cos 0° = 1$; then the flux through the loop is maximum. When ϕ is 90° as in Fig. 28.3a, $\cos 90° = 0$, and the flux is zero. Magnetic flux is measured in units of tesla-meter^2 (T $\cdot$ m^2).

The magnetic flux through a loop is analogous to the mass of air that blows through a loop held in a wind. The stronger the wind, the greater is the mass of air blowing through the loop. Also, more air passes through a loop of large area than through one of small area. Finally, the mass of air passing through the loop depends on its orientation. If the loop's surface is parallel to the direction of motion of the wind, the mass of air blows above and below the loop, but not through it—the air flux would be zero. As the loop is rotated perpendicular to the wind's direction, the mass of air blowing through the loop increases.

Magnetic flux plays an important role in electromagnetic induction. Notice that as the bar magnet moves closer to the loop shown in Fig. 28.1b, the flux through the loop increases. Similarly, when the magnet is withdrawn, as in Fig. 28.1d, the flux decreases. In both cases, voltage and current are induced in the coil. However, if the magnet is not moving, the flux does not change, and the voltage and current are zero.

Since Faraday's original experiments, a great many other experiments have shown that the voltage induced in a loop of wire is equal to the time rate of change of the magnetic flux through the loop. This is called Faraday's law.

Faraday's law: The induced emf (induced voltage) $\mathcal{E}$ through a loop of wire or a coil of wire with N loops (or turns) equals the change of magnetic flux $\Delta\Phi$ through the loops divided by the time Δt needed for that change in flux:

$$\mathcal{E} = -N\frac{\Delta\Phi}{\Delta t} = -N\frac{\Phi - \Phi_0}{t - t_0}. \tag{28.2}$$

The unit of $\mathcal{E}$ is volts if $\Delta\Phi$ is in units of tesla-meter^2 and Δt is in seconds.

In Eq. (28.2), the expression $-N\,\Delta\Phi/\Delta t$ is used to calculate the instantaneous induced voltage during a very short time Δt, and the expression $-N(\Phi - \Phi_0)/(t - t_0)$ is used to calculate the average induced emf over a longer time $(t - t_0)$. The negative sign in the equation will be discussed shortly. Regardless of the sign, it is evident from this equation that voltage is induced only while flux is changing and that the voltage is greatest when there is large flux change in a short period of time.

EXAMPLE 28.1 A circular loop of wire with one turn ($N = 1$) and a radius of 2.0 cm is moved from outside the poles of a horseshoe magnet to between the poles in a time of 0.10 s (Fig. 28.4). The field between the poles is 0.12 T. Calculate the average induced voltage as the loop moves into the magnetic field.

SOLUTION The area of the loop is $A = \pi r^2 = \pi(0.020\text{ m})^2 = 1.3 \times 10^{-3}\text{ m}^2$, and a line perpendicular to the plane in which the loop lies is parallel to the field; thus, $\phi = 0$. When outside the field at $t_0 = 0$, the field B_0 through the loop is zero. After moving into the field at $t = 0.10$ s, the field B through the loop increases to 0.12 T. Thus, by combining Eqs. (28.1) and (28.2), we find that the average induced voltage across the loop as it moves into the field is

$$
\begin{aligned}
\mathcal{E} &= -N\frac{\Phi - \Phi_0}{t - t_0} = -\frac{AB\cos\phi - AB_0\cos\phi}{t - t_0} \\
&= -\frac{A\cos\phi\,(B - B_0)}{t - t_0} = -\frac{(1.3 \times 10^{-3}\text{ m}^2)(\cos 0°)(0.12\text{ T} - 0)}{(0.10\text{ s} - 0)} \\
&= -1.6 \times 10^{-3}\text{ V} = \underline{-1.6\text{ mV}}.
\end{aligned}
$$

FIG. 28.4

EXAMPLE 28.2 A circular coil of wire with ten turns ($N = 10$) and a 4.0-cm radius is located between the poles of a large electromagnet that produces a steady magnetic field of 0.50 T. The surface of the loop, which is originally parallel to the field, is rotated in 0.70 s so that its surface is perpendicular to the field. Calculate the average induced voltage across the ends of the loop as the coil rotates.

SOLUTION The area of one loop of the coil is $\pi r^2 = \pi(0.04 \text{ m})^2 = 5.0 \times 10^{-3} \text{ m}^2$. The magnetic field B in the region of the coil has a constant value 0.50 T. However, the orientation of the loop changes. The surface of the loop is originally parallel to the magnetic field; thus, a line drawn perpendicular to the loop's surface makes an angle of 90° with the magnetic field ($\phi_0 = 90°$) at $t_0 = 0$. At $t = 0.70$ s, the loop has rotated so that a line perpendicular to its surface is parallel to the field ($\phi = 0°$). By combining Eqs. (28.1) and (28.2), we find that the average induced voltage as the coil rotates is

$$\mathcal{E} = -N\frac{\Phi - \Phi_0}{t - t_0} = -10\frac{AB\cos\phi - AB\cos\phi_0}{t - t_0}$$

$$= -\frac{10AB(\cos\phi - \cos\phi_0)}{t - t_0}$$

$$= -\frac{10(5.0 \times 10^{-3} \text{ m}^2)(0.50 \text{ T})(\cos 0° - \cos 90°)}{(0.70 \text{ s} - 0)}$$

$$= \underline{-3.6 \times 10^{-2} \text{ V}}. \qquad\blacksquare$$

28.3 The Polarity of Induced Voltage

We have seen that a changing magnetic flux inside a loop of wire produces an induced voltage. We now turn to the polarity of this induced voltage; that is, which end of the loop of wire is at a higher voltage? The polarity of the induced voltage has an important effect on the operation of electric motors and on devices called induction coils, which are used in electronic equipment such as television sets and in the electric circuit for the spark plugs of your car.

To determine the polarity of the induced voltage, we must first consider the magnetic field produced by this induced current. As we learned in Section 27.7, a current flowing in a loop of wire produces a magnetic field such as the one shown in Fig. 28.5. The direction of the field is determined using the right-hand rule: if the thumb of the right hand points in the direction of the current, then the fingers wrap around the wire in the direction of the field.

When the magnetic flux through a loop changes, a current is induced in the loop. The induced current produces an "induced" magnetic field inside the loop that opposes the change in magnetic flux that caused current. For example, if the flux through a loop points toward the left and is increasing, then the induced current will produce a flux through the coil that points to the right. However, if the magnetic flux is toward the left and decreasing, the induced current will produce a flux through the coil toward the left in an attempt to prevent the original decrease in flux.

The relationship between the changing flux through a coil and the polarity of the induced voltage is summarized by **Lenz's law:**

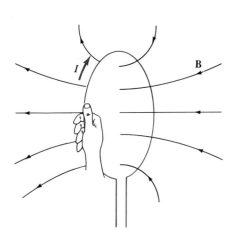

FIG. 28.5. The magnetic field caused by a current in a loop.

The **polarity** of an induced voltage is such that it produces a current whose own magnetic field opposes the change in flux that caused the induced voltage.

The minus sign in Faraday's law simply reminds us that the induced voltage $\mathcal{E}$ *opposes* the changing flux that caused the voltage. The negative sign has no other meaning.

To determine the polarity of the induced voltage, we simply determine in which direction the induced field should point so that it opposes the change in flux. Using the right-hand rule, we curl the fingers of the right hand in the direction of the induced field. The thumb then points in the direction of the induced current. The induced current flows around the coil toward the higher-voltage terminal of the coil and away from the lower-voltage terminal, just as current passes through a battery (another emf source) from its low voltage terminal to its high voltage terminal.

EXAMPLE 28.3 In Fig. 28.6a, a bar magnet moves toward a loop of wire. Which lead of the wire is at a higher voltage, *a* or *b*?

SOLUTION The flux of magnetic-field lines passing through the loop from right to left increase as the magnet moves closer to the loop. Hence, the induced current should produce a magnetic field that opposes this increase, and to do so, the induced field should pass through the loop from left to right, as shown in Fig. 28.6b. The right-hand rule shows that the current must flow counterclockwise in order to produce this field from left to right inside the loop (Fig. 28.6c). Since current flows through a source of emf toward the higher-voltage terminal, <u>*a* must be at higher voltage than *b*.</u> ∎

EXAMPLE 28.4 A rectangular loop that is 10 cm long on each side is moved into a uniform magnetic field of 1.5 T at a speed of 3.0 m/s (Fig. 28.7). (a) Calculate the induced emf across the ends of the loop as the loop enters the field. (b) Is point *a* or point *b* at higher voltage?

SOLUTION (a) In this problem the magnetic field is nonzero in one region inside the coil and zero elsewhere. As more of the loop enters the field, the nonzero region increases. The induced emf is caused by the increasing area ΔA of the region occupied by the field.

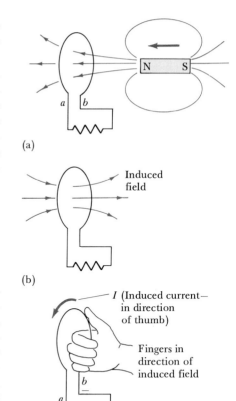

(a)

(b)

(c)

FIG. 28.6. (a) As a magnet moves toward the loop, the flux through the loop increases and (b) induces a field that opposes the increasing flux. (c) The induced field is caused by an induced current that moves from *b* to *a*. Thus, *a* is at a higher voltage than *b*, since current moves through a source of emf toward higher voltage.

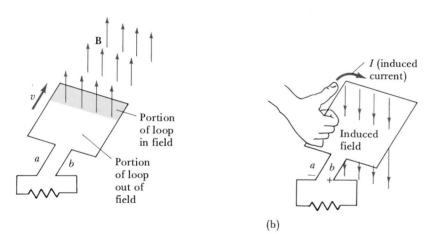

(a)   (b)

FIG. 28.7. (a) A loop enters a field. (b) The induced current flows from *a* to *b*, causing a downward induced field.

In general, for a rectangular loop whose front edge has a length L, and that moves at a speed $v = \Delta x / \Delta t$, the area increases by

$$\Delta A = L \, \Delta x$$

in a time Δt. Thus, the changing flux per time Δt is

$$\mathcal{E} = -\frac{\Delta \Phi}{\Delta t} = -\frac{(\Delta A)B \cos \theta}{\Delta t} = -\frac{(L \, \Delta x)B \cos \theta}{\Delta t}.$$

Substituting $\Delta x / \Delta t = v$, we find that

$$\mathcal{E} = -LvB \cos \theta.$$

We are given that $L = 0.10$ m, $v = 3.0$ m/s, $B = 1.5$ T, and $\theta = 0°$. Thus,

$$\mathcal{E} = -(0.10 \text{ m})(3.0 \text{ m/s})(1.5 \text{ T})1 = \underline{-0.45 \text{ V}}.$$

(b) Because the induced voltage is caused by an increasing magnetic flux that points up, the induced field should point down. A downward field is caused by a current flowing clockwise from a to b (Fig. 28.7b). Hence, <u>point b must be at a higher voltage than point a.</u> ■

28.4 The Electric Generator

An electric generator is an important device whose operation we can understand by using Faraday's law. A homemade version of an electric generator is shown in Fig. 28.8a. Water in a flask is heated by a bunsen burner. Steam leaves a tube at the top of the flask and strikes the blades of a turbine, causing the turbine to rotate. A loop of wire attached to the turbine also rotates in a steady magnetic field. The changing magnetic flux through the loop causes an induced voltage. In the simple power plant shown in Fig. 28.8a, the voltage drives a current through a light bulb. The power plant of an electric utility company represented schematically in Fig. 28.8b works in much the same manner, only on a much larger scale.

Our concern in this section is with the rotating coil and the induced emf caused by this coil. A top view of a coil with one turn is shown in Fig. 28.9. The magnetic flux through the coil depends on the orientation of the coil: if the coil is parallel to the magnetic field (Fig. 28.9a), the field lines pass along the sides of the coil but not through it and the flux is zero. If the coil's surface is perpendicular to the field direction (Fig. 28.9b), then the flux through the coil is $\Phi = BA$. If a line perpendicular to the coil's surface is oriented at an angle ϕ relative to the field (Fig. 28.9c), then the flux through the coil is

$$\Phi = BA \cos \phi. \tag{28.3}$$

Suppose that the coil is oriented so that $\phi = 0$ at $t = 0$ and that a short time t later the coil makes an angle ϕ with the field. The angular velocity ω of the coil is then

$$\omega = \frac{\phi}{t}. \tag{6.1}$$

If the coil rotates through 2π radians in one second, it will turn at a frequency of one rotation per second. If rotating through 4π radians per second, its frequency

(a)

(b)

FIG. 28.8. (a) A homemade version of an electric power plant, and (b) a representation of an actual electric power plant.

is two rotations per second. In general, the frequency f of rotation and the angular velocity ω are related by the equation

$$\omega = 2\pi f.$$

Combining the last two equations, we find that

$$\phi = \omega t = 2\pi f t.$$

This expression for ϕ is substituted into Eq. (28.3) to obtain an equation for the changing flux through a rotating coil as a function of time:

$$\Phi = BA \cos 2\pi f t. \qquad (28.4)$$

The above equation for flux is an example of a cosinusoidal function and is plotted against time on the graph in Fig. 28.10a.

Each time the coil makes one-half rotation, the magnitude of the flux through the coil is the same as it was one-half rotation earlier. However, the orientation of the coil has reversed relative to the field. The alternating positive and negative values of flux shown in Fig. 28.10a reflect the changing orientation of the coil relative to the field.

The voltage induced in the coil by the changing flux is given by Eq. (28.2): $\mathcal{E} = -N\Delta\Phi/\Delta t$. As the flux changes, so does the voltage induced across its ends. If you are familiar with differential calculus, then you can take the negative time derivative of Φ as given in Eq. (28.4) to determine an equation for $\mathcal{E}$. For those unfamiliar with calculus, we present the result without proof:

$$\mathcal{E} = -\frac{\Delta\Phi}{\Delta t} = (NBA2\pi f)\sin 2\pi f t = \mathcal{E}_0 \sin 2\pi f t. \qquad (28.5)$$

The induced voltage varies sinusoidally with time. The alternating voltage plotted in Fig. 28.10b changes polarity at the same frequency as the generator's coil rotates. If one terminal is at higher voltage than the other at one instant of time, then one-half turn later the polarity is reversed—the high-voltage terminal has become the low-voltage terminal. The alternating positive and negative voltages shown in Fig. 28.10b reflect the changing polarity of the voltage across the terminals of the generator.

The **peak voltage** $\mathcal{E}_0$ produced by a generator equals the quantity $(NBA2\pi f)$ that appears in front of the sine function in Eq. (28.5). For example, the electric voltage provided by a power plant for your home has a frequency of 60 Hz and a peak voltage of 170 V. The voltage alternates between $+170$ V and -170 V 60 times each second. The following symbol is used for an alternating current (ac) electric power source that produces an alternating voltage: $-\!\bigcirc\!-$.

(a)

(b)

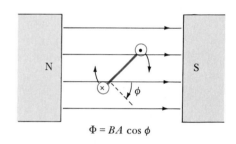

(c)

FIG. 28.9. The flux through a coil depends upon its orientation in the field.

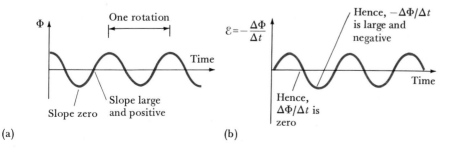

(a)

(b)

FIG. 28.10. The time variation of (a) the flux through and (b) the induced voltage across the leads of a coil rotating in a magnetic field.

FIG. 28.11. A person produces electricity by rotating a coil in a magnetic field.

EXAMPLE 28.5 A small generator has a coil with 100 loops each of area 80 cm^2. The generator is turned by a gear system attached to the pedals of a stationary bicycle (Fig. 28.11). A student pedals the bicycle so that the generator rotates at a frequency of 20 Hz in a magnetic field of 0.30 T. Find an expression for the voltage produced by the generator.

SOLUTION We are given the following information: $N = 100$, $A = 8.0 \times 10^{-3}$ m^2, $B = 0.30$ T, and $f = 20$ Hz $= 20$ s^{-1} (20 rotations/second). Thus, the generated voltage is

$$\mathcal{E} = (NBA2\pi f) \sin 2\pi ft = [(100)(0.30 \text{ T})(80 \times 10^{-4} \text{ m}^2)(2\pi 20 \text{ s}^{-1})] \sin (40\ \pi t)$$
$$= \underline{30 \sin (40\pi t)} \quad \text{(in units of volts)}.$$

The voltage produced from one terminal of the generator to the other will vary 20 times a second from $+30$ volts to -30 volts, and then back to $+30$ volts. ∎

28.5 Mutual Inductance

Now we encounter a different example of electromagnetic induction. When a current passes through the loops of a coil, a magnetic field is produced. If a second coil is held near the first, some of the magnetic-field lines caused by current in the first coil may pass through the second coil. If the current in the first coil changes, the magnetic field and flux through the second coil will also change, and a voltage will be induced in the second coil because of the changing flux. This process is called **mutual inductance**—a voltage is induced in one coil by a changing current in the other.

The mutual inductance of two coils is defined with reference to Fig. 28.12. A current i_1 passing through coil 1, which has N_1 turns, produces a magnetic field.* Some of the magnetic-field lines produced by coil 1 pass through coil 2 with N_2 turns. By convention, we designate the magnetic flux through each loop of coil 2 by the symbol Φ_{21}. (The subscript 21 refers to the flux through the loops of coil 2 that are caused by the current in coil 1.) The total flux through all N_2 turns of coil 2 is $N_2\Phi_{21}$.

The mutual inductance M of the coils is defined as

$$M = \frac{N_2\Phi_{21}}{i_1}$$

or

$$N_2\Phi_{21} = Mi_1. \tag{28.6}$$

The quantity M depends on the size, orientation, and separation of the coils. If they are close to each other, M is greater, since a current in coil 1 will produce a greater flux through the loops of coil 2. As we will see later, the value of M also depends on the magnetic permeability of the material in the region between and through the two coils.

If the current in coil 1 varies with time, Eq. (28.6) becomes

$$N_2 \frac{\Delta\Phi_{21}}{\Delta t} = M \frac{\Delta i_1}{\Delta t}.$$

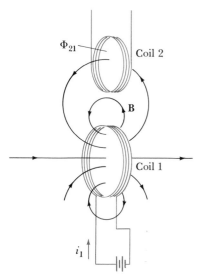

FIG. 28.12. A current in coil 1 produces a magnetic flux Φ_{21} through each loop of coil 2.

*Varying currents are usually represented by a lower-case i and steady currents by a capital I.

The left side of this equation is just the negative of the voltage induced in coil 2 by the changing current in coil 1 ($\mathcal{E} = -N_2 \Delta\Phi/\Delta t$). Consequently,

$$\mathcal{E}_2 = -M \frac{\Delta i_1}{\Delta t}. \qquad (28.7)$$

A similar equation applies to coil 1. If the current in coil 2 changes at a rate $\Delta i_2/\Delta t$, then a voltage is induced in coil 1:

$$\mathcal{E}_1 = -M \frac{\Delta i_2}{\Delta t}. \qquad (28.8)$$

The same M is used in both Eqs. (28.7) and (28.8). The unit of mutual inductance is called a **henry** (H). Rearranging and substituting for the units in Eq. (28.7), we see that

$$1 \text{ H} = \frac{1 \text{ V}}{1 \text{ A/s}}.$$

A 1-H mutual inductance between two coils means that a 1-V potential difference is induced in one coil by a 1-A/s change in current in the other coil.

The unit *henry* was named in honor of the American physicist Joseph Henry (1797–1878), who independently discovered the law of electromagnetic induction, which was named in honor of Michael Faraday. Henry published his findings just a little later than Faraday and had to settle for a unit named in his honor rather than a law.

EXAMPLE 28.6 The mutual inductance between two coils is 0.10 H. Calculate the induced emf in coil 2 caused by a current changing at a rate of 50 A/s in coil 2.

SOLUTION Using Eq. (28.7), we see that

$$\mathcal{E}_2 = M \frac{\Delta i_1}{\Delta t} = -(0.10 \text{ H})(50 \text{ A/s}) = \underline{-5.0 \text{ V}}. \qquad \blacksquare$$

Mutual Inductance and Metal Detectors

A metal detector like those through which you walk at airports can detect metals with high magnetic permeability. The detector could be built to detect a change in the mutual inductance of two large coils, such as shown in Fig. 28.13. An alternating current i_1 in coil 1 produces a magnetic field whose magnitude at each instance of time is given approximately by Eq. (27.11):

$$B = \frac{N_1 \mu i_1}{2r},$$

where N_1 is the number of loops in the first coil, μ is the permeability of the material inside the coil, and r is its radius (the coil is assumed to be circular in shape).

Since the second coil is so near the first, the field through each of its N_2 loops is about the same as that through the loops of the current-carrying coil.

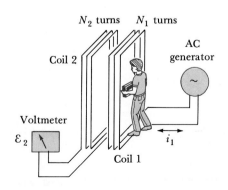

FIG. 28.13. An airport metal detector.

Thus, the flux in the second coil caused by the current in the first is approximately:

$$N_2\Phi_{21} = N_2 BA = N_2 \left(\frac{N_1\mu i_1}{2r}\right) A,$$

where A is the area inside the second coil. By comparing the above equation to Eq. (28.6), $N_2\Phi_{21} = Mi_1$, we see that the mutual inductance between the coils is

$$M = \frac{N_2 N_1 \mu A}{2r}.$$

The magnitude of the voltage induced in coil 2 by the oscillating current in coil 1 is

$$\mathcal{E}_2 = -M\frac{\Delta i_1}{\Delta t}.$$

One way that the airport metal detector could work is by producing an alternating current in the first coil. If the mutual inductance between the first and second coils changes as a person walks through the detector, the induced voltage $\mathcal{E}_2$ changes, causing an alarm to go off. Such a change in mutual inductance will occur if a person carries a metal object of high magnetic permeability through the loops of wire.

When a person carrying no metal objects on his body walks through the alarm, the mutual inductance is

$$M = \frac{N_2 N_1 \mu_0 A}{2r},$$

since the magnetic permeability of the body is the same as that of air: $\mu_{\text{body}} \approx \mu_0$. Should a person carry a piece of iron, however, the permeability may change. For example, if the object has a permeability 300 times that of air and occupies 1/300 of the area inside the coil, the average permeability of the material in the loop doubles, the mutual inductance between the coil doubles, and the induced voltage in coil 2 doubles, causing the alarm to turn on. In the next chapter we will learn how the metal detector can be made even more sensitive.

(a)

(b)

FIG. 28.14. (a) No current flows through the coil when the switch is open. (b) When the switch is closed, the current produces a magnetic field and flux through the coil.

28.6 Self-Inductance

In the preceding section the induced emf in one coil was caused by a changing current in another coil. A voltage can also be induced in a coil because of a change in its own current. Such a voltage is called a self-induced voltage, or **self-induced emf.**

The circuit shown in Fig. 28.14a illustrates self-inductance. When the switch is open, no current flows and no magnetic field passes inside the coil. When the switch is closed, a current flows and produces a magnetic field in the coil (Fig. 28.14b). The magnetic field and flux through the coil increase from zero (with the switch open) to a finite value (with the switch closed). The changing magnetic field and flux through the coil cause a voltage to be induced in the coil, and this induced voltage initiates an induced current that opposes the change in the original current.

The self-induced emf of a coil is defined in terms of the self-inductance L of the coil. Suppose that a current i flowing through the coil shown in Fig. 28.14b produces a magnetic field whose flux through each loop is represented by the symbol Φ. The total flux through the N loops of the coil is $N\Phi$. The **self-inductance** L of the coil is defined by the following equation:

$$L = \frac{N\Phi}{i}. \qquad (28.9)$$

The greater a coil's self-inductance, the greater is the voltage induced in the coil if the current passing through it changes.

The self-inductance of a coil depends on the size and shape of the coil, on the number of turns in it, and also on the magnetic permeability of material inside it. For example, the self-inductance of a coil with an iron core is several hundred times that of a similar coil with an air core. The reason is that the magnetic field caused by the current through the coil aligns the tiny "bar magnets" of the iron atoms, adding their fields to the field produced by the current. As a consequence, there is a much greater flux through a coil that surrounds iron than through one that surrounds air.

Equation (28.9) can be used to determine the induced emf in a coil when the current flowing through it changes. Since $N\Phi = Li$, the changing flux is related to a changing current by the equation

$$N\frac{\Delta\Phi}{\Delta t} = L\frac{\Delta i}{\Delta t}.$$

The left side of this equation is just the negative of the voltage induced in the coil ($\mathcal{E} = -N\,\Delta\Phi/\Delta t$). Consequently,

$$\mathcal{E} = -L\frac{\Delta i}{\Delta t}. \qquad (28.10)$$

The **induced emf** (voltage) can be calculated if the rate of current change and self-inductance of a coil are known. The induced emf is sometimes called a **back emf**, since it opposes the changing current that caused the induced emf.

EXAMPLE 28.7 (a) Calculate the self-inductance of a solenoid with an air core of length 8.0 cm with 520 turns and an inside area of $7.2 \times 10^{-3}\,\text{m}^2$ (Fig. 28.14). (b) What is the average induced voltage in the solenoid if the current passing through it changes from zero to 1.5 A in 0.20 s?

SOLUTION (a) The inductance of the coil is $L = N\Phi/i$, where the flux through each coil is $\Phi = BA$ [Eq. (28.1)], and (for a solenoid) $B = N\mu i/l$ [Eq. (27.12)], where l is the length of the solenoid. Thus,

$$L = \frac{N\Phi}{i} = \frac{N(BA)}{i} = \frac{N(N\mu i/l)A}{i} = \frac{N^2\mu A}{l}.$$

For the coil in this problem, $N = 520$, $\mu = \mu_0 = 4\pi \times 10^{-7}$ (the solenoid has an air core), $A = 7.2 \times 10^{-3}\,\text{m}^2$, and $l = 8.0 \times 10^{-2}\,\text{m}$. Thus,

$$L = \frac{N^2\mu A}{l} = \frac{(520)^2(4\pi \times 10^{-7}\,\text{T}\cdot\text{m/A})(7.2 \times 10^{-3}\,\text{m}^2)}{8.0 \times 10^{-2}\,\text{m}}$$

$$= 3.1 \times 10^{-2}\,\text{H} = \underline{31\,\text{mH}}.$$

(b) The induced emf in the solenoid is determined using Eq. (28.10):

$$\mathcal{E} = -L\frac{\Delta i}{\Delta t} = -(31 \times 10^{-3}\,\text{H})\left(\frac{1.5\,\text{A}}{0.20\,\text{s}}\right) = \underline{-0.23\,\text{V}}.$$

The negative sign implies that the current caused by the induced voltage flows in the opposite direction of the increasing current that induced the voltage.

∎

28.7 Transformers

A **transformer** is a device that uses the mutual inductance between two coils to convert an alternating voltage across one coil into a larger or smaller alternating voltage across the other coil. This increase or decrease in voltage is much like the increase or decrease in force that is caused by the use of a lever; a small force applied to the long end of a lever can produce a large force at the short end, and vice versa. Similarly, a small voltage at the input of a transformer can be converted to a large voltage at its output, and vice versa.

A transformer (Fig. 28.15) consists of two coils of wire wound around a core, or yoke, that is made of a material that can be magnetized. The coil that is attached to the alternating voltage source $\mathcal{E}_1$ is called the **primary coil.** The alternating current flowing through the primary coil produces a fluctuating magnetic field B in the transformer core. This field passes around the core and through the secondary coil. Since the field alternates direction, an induced voltage $\mathcal{E}_2$ is produced in the **secondary coil.** If the secondary coil is attached to a light bulb, current is forced by the induced voltage through the light's filament, causing it to glow.

Transformers are useful mainly because of the difference between the voltage going into the primary coil and that coming out of the secondary coil. The secondary voltage can be larger than the primary voltage if the secondary has more turns than the primary, smaller if it has fewer turns than the primary. The cause of this behavior is outlined below.

The alternating current in the primary coil produces a self-induced voltage of magnitude

$$\mathcal{E}_1 = -N_1\frac{\Delta\Phi}{\Delta t},$$

where Φ is the magnetic flux through each loop of the coil, and N_1 is the number of primary loops.

In a good transformer nearly all of the magnetic flux through the loops of the primary coil also passes through the loops of the secondary coil; hence the induced emf in this coil is

$$\mathcal{E}_2 = -N_2\frac{\Delta\Phi}{\Delta t},$$

where N_2 is the number of loops in the secondary coil, and Φ is the same flux that passed through the primary coil.

If we divide the first equation by the second, we find that

$$\frac{\mathcal{E}_1}{\mathcal{E}_2} = \frac{N_1}{N_2}. \tag{28.11}$$

FIG. 28.15. A transformer.

The secondary coil voltage is

$$\mathcal{E}_2 = \frac{N_2}{N_1}\,\mathcal{E}_1.$$

We see that $\mathcal{E}_2$ depends on the ratio of the number of turns in each coil and also on the voltage in the primary coil.

EXAMPLE 28.8 The 120-V alternating voltage $\mathcal{E}_1$ from a wall socket feeds into the primary coil of a toy electric train transformer, which has 800 turns (N_1). Calculate the number of turns in the secondary coil so that the output voltage $\mathcal{E}_2$ from the transformer is 12 V—enough to run a toy electric train safely.

SOLUTION Equation (28.11) can be rearranged to solve for the number of turns in the secondary coil:

$$N_2 = \frac{\mathcal{E}_2}{\mathcal{E}_1}\,N_1 = \frac{12\text{ V}}{120\text{ V}}\,800 = \underline{80}. \qquad \blacksquare$$

Transformers have a number of uses in everyday life. As we saw in the last example, they can reduce voltages to safe levels for toys. Another important use is related to the economics of transmitting electricity from power plants to homes. Suppose that you live in a community with 100,000 people, which is located fifty miles from the electric power plant serving the city. In 1980 the average electric-power use rate per person in the United States and Canada for a 24-hour period was about 1000 watts. A power plant would have to supply 100,000 × 1000 watts or 1×10^8 watts of power to your city continuously (somewhat more during hours of peak use).

Recall from Chapter 24 that electric power used is the product of the voltage difference at its usage point and the current that is flowing:

$$P = VI. \qquad (24.11)$$

If the power is delivered at 120 V, then the current to this city must average

$$I = \frac{P}{V} = \frac{1 \times 10^8\text{ watt}}{120\text{ V}} = 8 \times 10^5\text{ A}.$$

The wires in a home should carry no more than about 20 A of current. To supply electricity to your city, 8×10^5 A/20 A = 40,000 of these wires must run from the power plant to the city. The cost of the wires alone would prohibit this transmission system. In addition, since such long wires have considerable electrical resistance, the power lost due to resistive heating (I^2R) would be great. If a way can be found to deliver the electric power at low current, considerable energy is saved and less wire is needed to deliver the power.

Power companies have solved this problem by using a "step-up" transformer. The electrical generator in a power plant produces electricity at about 12,000 V. The voltage is placed across the primary coil of a large transformer. The secondary coil has about ten times as many turns as the primary coil, so that the voltage is raised to about 120,000 V. Since the voltage is 1000 times greater than the 120 V used in your home, the same power ($P = VI$) can be transmitted by $\frac{1}{1000}$ the current. Only $\frac{1}{1000}$ as much wire is needed to transmit the power, and the heating is greatly reduced, since the current is reduced. After reaching the city, the voltage is stepped down to lower voltage (120 or 240 V) in local transformers for use in homes.

Summary

1. **Electromagnetic induction** is the generation of an electrical voltage in a loop of wire by the variation of the magnetic field through the area inside the loop. The **magnetic flux** Φ through the loop is

$$\Phi = AB \cos \phi, \tag{28.1}$$

where A is the loop's area, B is the magnetic field inside the loop, and ϕ is the angle between $\mathbf{B}$ and a line drawn perpendicular to the loop.

2. **Faraday's law:** The voltage $\mathcal{E}$ induced in a coil with N turns depends upon the time rate of change of magnetic flux through the coil:

$$\mathcal{E} = -N\frac{\Delta\Phi}{\Delta t} = -N\frac{\Phi - \Phi_0}{t - t_0}. \tag{28.2}$$

The polarity of the induced emf is such that it produces a current whose own field opposes the change in flux that caused the induced voltage.

3. **Electric generator:** An electric generator produces an alternating voltage by rotating a coil with N loops in a steady magnetic field B. The induced voltage is

$$\mathcal{E} = (NBA2\pi f) \sin 2\pi ft = \mathcal{E}_0 \sin 2\pi ft, \tag{28.5}$$

where A is the area of each loop, f is the frequency of rotation, and $\mathcal{E}_0$ is the peak voltage produced by the generator.

4. **Mutual inductance:** A mutual inductance exists between two coils if an emf is induced in one coil by a changing current in the other. The **mutual inductance** M between the coils is

$$M = \frac{N_2\Phi_{21}}{i_1}, \tag{28.6}$$

where N_2 is the number of loops in coil 2, and Φ_{21} is the magnetic flux through coil 2 caused by a current i_1 in coil 1. The emf induced in coil 2 by a changing current in coil 1 is

$$\mathcal{E}_2 = -M\frac{\Delta i_1}{\Delta t}. \tag{28.7}$$

5. **Self-inductance:** If the current in a coil changes, the magnetic field and flux produced by the current also change. An emf is induced in the coil by its own changing flux. This emf is called a **self-induced electromotive force.** The **self-inductance** L of a coil indicates the coil's ability to induce a voltage when its current changes. The self-inductance is defined by the equation

$$L = \frac{N\Phi}{i}, \tag{28.9}$$

where Φ is the flux through each loop of a coil with N turns when carrying a current i. The **self-induced emf** due to a changing current is

$$\mathcal{E} = -L\frac{\Delta i}{\Delta t}. \tag{28.10}$$

6. **Transformer:** A transformer is a device that uses the mutual inductance between a primary coil with N_1 turns and a secondary coil with N_2 turns to convert a primary voltage $\mathcal{E}_1$ to a different secondary voltage $\mathcal{E}_2$. The four quantities are related by the equation

$$\frac{\mathcal{E}_1}{N_1} = \frac{\mathcal{E}_2}{N_2}. \tag{28.11}$$

Questions

1. A coil of wire can detect the vibrations or movements of a nearby magnet. How is this accomplished? Describe one possible application of this detection system.

2. Describe three different ways in which the flux through a loop of wire can change, thus inducing a voltage in the loop.

3. A circular metal plate swings past the north pole of a permanent magnet, as shown in Fig. 28.16. Consider the metal as consisting of a series of rings of increasing radius. (a) Indicate the direction of the current in one ring as the metal swings down from the left. (b) Indicate the direction of the current in the same ring as it swings back up toward the left. Using Lenz's law, justify your answers. Similar currents, called eddy currents, are induced in metal sheets exposed to changing magnetic fields.

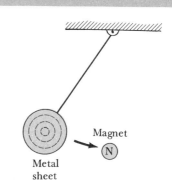

Metal sheet

Magnet

FIG. 28.16

4. A 12-V automobile battery provides the thousands of volts needed to produce a spark across the gap of a spark plug. How might this be done?

5. (a) Show the orientation of a generator coil relative to a constant magnetic field when the *flux* through the coil is maximum. (b) If the coil rotates at constant angular velocity, show the orientation of the coil when the *emf* generated by the coil is maximum, and (c) when the *emf* is minimum.

6. A coil connected to a voltmeter can be used to detect alternating currents in other circuits. Such a device can detect currents as small as 5 μA. Carefully explain how one of these devices might work.

7. The induction method of geophysical prospecting depends upon the alternating magnetic field produced by an alternating current in a source coil. Refer to Fig. 28.17 and carefully explain how the induced emf in a receiver coil might be affected by a high-permeability mass (iron, for example) under the earth's surface.

8. You wear a coil of wire on your T-shirt. The coil is attached to a switch and battery. When the switch is closed, current flows clockwise as seen from your front. A friend standing opposite you also has a coil on his or her T-shirt. Does the voltage induced in that coil when the switch to your coil is closed cause a clockwise or counterclockwise current, as seen by you? Explain carefully.

9. A respiration detector consists of a coil placed on a person's chest and another placed on the person's back. A current flows

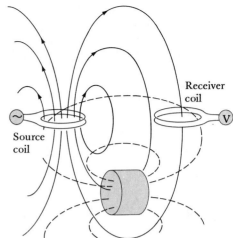

FIG. 28.17 High-permeability mass

through one coil, producing a magnetic flux through the other coil. Why does the flux change as the person breathes, thus inducing a voltage in the other coil?

10. Suppose Lenz's law says that the voltage induced in a coil supports the changing flux through the coil rather than opposing it. If this is true, the current in a coil, once it starts, will continue to increase until the coil melts. Carefully explain in words why this will happen.

Problems

28.1–28.3 Electromagnetic Induction

1. The magnetic field in a region has a magnitude of 0.40 T and points in the positive z direction shown in Fig. 28.18. Calculate the magnetic flux through (a) surface *abcd*, (b) surface *bcef*, and (c) surface *adef*. (*Hint:* The calculation in part (b) can be simplified by careful reasoning, but be sure to justify any simplifications you make.)

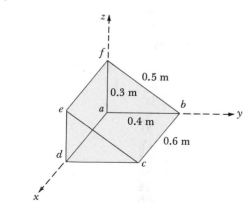

FIG. 28.18

2. The earth's magnetic field in the southern region of the United States has a magnitude of 5.1×10^{-5} T and points north with a downward inclination of about 60° below the horizontal. Estimate the magnetic flux through one side window of a car

when (a) the car is facing east, and when (b) the car is facing north.

3. The earth's magnetic field in British Columbia has a magnitude of 5.8×10^{-5} T and points roughly north with a downward inclination of 72°. Calculate the magnetic flux through a 1.0 m $\times$ 1.8 m window that faces (a) south, (b) west, and (c) southwest.

4. A large circular loop of wire is held so that its surface is parallel to the earth's (5.5×10^{-5})-T magnetic field. The loop is rotated for 0.50 s through a 90° angle so that its surface is perpendicular to the field. Calculate the radius of the loop needed to develop an average induced emf of 12 V. Is this a practical way to produce voltage? Explain your answer.

5. A circular coil with 25 turns, each of radius 1.2 cm, is in a 0.35-T magnetic field. The coil is rotated in 0.18 s from an orientation with its surface parallel to the field to a perpendicular orientation. Calculate the average induced emf in the coil while it is being rotated.

6. The earth's magnetic field at one location in the magnetosphere high above the earth's atmosphere is 3.0×10^{-7} T. A coil whose inside area is 5000 m² rotates in the magnetosphere so that the flux varies from the maximum possible value to the minimum possible value in 1.2 s. How many turns must the coil have in order to develop an average induced emf of 120 V for use by a nearby space platform?

7. A circular coil of 0.30-m diameter with 25 loops rests between the poles of a large electromagnet. The magnet initially produces

a 0.50-T field, which passes through the coil perpendicular to its surface. The magnetic field is reduced to zero in 1.5 s. Calculate the magnitude of the induced voltage in the coil.

8. A 0.20 m × 0.40 m coil of wire with 12 turns is pulled into a 0.25-T magnetic field at a speed of 1.8 m/s. The 0.20-m side enters first. Calculate the induced emf as the coil enters the field, which is perpendicular to the coil's surface.

9. A person wearing wire-framed glasses rotates his head quickly in the earth's magnetic field. *Estimate* the induced emf in the metal ring encircling one glass lens while the head is being turned. Justify any numbers used in your estimate.

10. A race car at the Indianapolis Speedway, driving north, rounds a 90° curve in 8.0 s, so that the car is moving west. Calculate the average induced emf in the metal frame for one of the car's windows caused by the change in the earth's magnetic flux through the window whose dimensions are 0.60 m by 0.40 m. The earth's field has a magnitude of 5.8×10^{-5} T, points north, and tilts downward from the horizontal at a 72° angle.

11. The surface of a circular wire loop of radius 9.0 cm is located perpendicular to a uniform 0.35-T magnetic field. The loop is pulled from two sides into a long, thin shape with the two sides next to each other. The change of shape occurs in 0.11 s. Calculate the average induced emf while the loop is being pulled.

12. A pair of rails, separated by 1.5 m, are connected at one end to form a long, thin U-shape. An axle with wheels on each side rolls along the rails at a speed of 25 m/s. The area of the loop formed by the rails and axle increases as the wheels roll. Calculate the induced voltage across the axle if the earth's magnetic field is 5.4×10^{-5} T and tilts downward 68° below the horizontal.

13. A bar magnet is dropped through a loop of wire (Fig. 28.19). (a) As the magnet approaches the loop from above, is position *a* or position *b* at a higher voltage? (b) As the magnet falls away from the loop from below, is *a* or *b* at a higher voltage? Justify each answer with a careful explanation based on Lenz's law.

FIG. 28.19

14. The cardboard tube in Fig. 28.20 has two insulated wires wrapped around it. The switch in circuit A is closed, causing current to flow through loop A. During the short time that the current develops in A, a voltage is induced in loop B. (a) Does the current flow through resistor *R* from left to right or from right to left? Justify your answer, using Lenz's law. (b) In which direction does the current flow through the resistor when the switch is reopened? Expain.

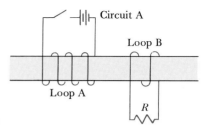

FIG. 28.20

15. The coil in Fig. 28.21 has 18 turns and an inside area of 0.060 m². The coil is pulled from a 0.42-T magnetic field in 0.080 s. (a) Calculate the average induced field in the coil. (b) Is position *a* or position *b* at a higher voltage while the coil is being withdrawn? Explain your answer.

FIG. 28.21

16. The surface of the coil of wire shown in Fig. 28.21 and discussed in problem 15 is initially oriented parallel to the field with lead *a* on the left. (a) Calculate the average induced emf if the coil is rotated 90° in the clockwise direction in 0.050 s. (b) Is lead *a* or lead *b* at a higher voltage? Explain. (c) Calculate the average induced emf if the coil is rotated another 90° in the clockwise direction in 0.050 s. (d) Is lead *a* or lead *b* at a higher voltage during the latter 90° rotation? Explain.

28.4 The Electric Generator

17. A generator has a 200-turn coil with each turn having an area of 400 cm². The coil rotates in a 0.40-T magnetic field. (a) At what frequency must it rotate to produce a peak voltage of 120 V? (b) Write an expression for the voltage as a function of time. (c) What is the voltage across the generator terminals at a time of 0.021 s (assuming the voltage is zero at time zero)?

18. A generator has a 450-turn coil that is 10 cm long and 12 cm wide. The coil rotates at 8 revolutions per second in a 0.10-T magnetic field. Calculate the generator's peak voltage.

19. A 50-Hz generator produces a peak voltage of 310 V. The variation of voltage with time is represented by the expression $310 \sin(2\pi 50t)$. (a) Calculate the voltage across the generator terminals at the following times: 0 s, $\frac{1}{200}$ s, $\frac{2}{200}$ s, $\frac{3}{200}$ s, $\frac{4}{200}$ s, $\frac{5}{200}$ s, $\frac{6}{200}$ s, $\frac{7}{200}$ s, and $\frac{8}{200}$ s. (b) Plot these values of voltage versus time on a graph and connect the points in a smooth curve. Does the curve resemble a sinusoidal curve?

20. You want to make a generator for your bicycle light that will provide an alternating voltage whose peak value is 4.2 V. The generator coil has 55 turns and rotates in a 0.080-T magnetic field. If the coil rotates at 12 revolutions per second, what must the area of the coil be to develop this voltage?

21. A football quarterback, in order to obtain good grades while strengthening his throwing arm, is asked to vibrate a coil in and out of a 0.18-T magnetic field. The coil has 350 turns, which are 20 cm long and 18 cm wide. At what frequency must the coil move in and out of the magnetic field to provide a peak voltage of

30 V (assuming the motion is sinusoidal)? The voltage will be used to operate a light bulb for his studying.

28.5 Mutual Inductance

22.   The current in one coil decreases from 4.2 A to zero in 0.80 s. A 0.035-V emf is induced in a neighboring coil. (a) Calculate the mutual inductance between the two coils. (b) Calculate the emf induced in the second coil if the current in the first coil now increases from zero to 1.2 A in 0.050 s. (c) Why is the magnitude of the voltage calculated in part (b) greater than that given in part (a), even though the change in current is less?

23.   The mutual inductance between two coils of wire is 0.025 H. (a) Calculate the voltage induced in one coil by a 1.2-A current decrease in another coil in a time of 0.036 s. (b) By how much must the current decrease in 0.108 s in order to produce the same average induced emf? Explain.

24.   A heart pacemaker has a small coil in which a voltage is induced by an alternating current in a larger coil placed on a person's chest. The mutual inductance between the two coils is 0.040 H. (a) Calculate the current change per unit time in the chest coil to produce a 0.24-V induced emf in the pacemaker coil. (b) Calculate the induced emf if a steady current of 1.2 A flows in the chest coil.

25.   The spark coil in an automobile ignition system converts an abrupt change in current in one coil into a large induced voltage (25,000 V) in a second coil whose leads are across the spark plug. This large voltage produces dielectric breakdown and a spark in the spark plug. If the mutual inductance between the coils is 1.2 H and the current changes from 16 A to zero, during what time must this current change in order to produce the 25-kV induced emf in the second coil?

26.   An airport metal detector consists of two coils whose mutual inductance, when unoccupied, is 8.2 H. An alternating current through one coil induces a 120-mV peak alternating voltage in the second coil. A person wearing an iron medallion walks through the detector. The medallion occupies $\frac{1}{400}$ of the space in the second coil and has a magnetic permeability 600 times that of air. Estimate the peak voltage induced as the person passes through the coil. Justify in words your calculation procedure.

27.   A current i flows through a long, straight wire. (a) Write an expression for the magnetic field at a distance a from the wire. (b) If a small coil of radius r with N turns is held perpendicular to the field at distance a from the wire, what is the flux through the coil? (c) Now, derive an expression for the mutual inductance between the wire and coil. (d) If the current in the wire decreases from 15 A to zero in $\frac{1}{240}$ s and a coil of 1.2 cm radius with 10 turns is held 15 cm from the wire, what is the average emf induced in the coil?

28.6 Self-Inductance

28.   The current in a circuit changes from 12 A to zero in 0.0040 s, causing a self-induced emf of 105 V. (a) Calculate the self-inductance of the circuit. (b) Calculate the induced emf if the current increases from zero to 4.0 A in 0.0020 s.

29.   The current passing through a 0.40-H coil increases from 2.0 A to 10.0 A in 0.12 s. (a) Calculate the voltage induced in the coil. (b) If the current then decreases to 8.0 A in the next 0.020 s, what is the induced voltage? Is its polarity the same as in part (a)? Explain.

30.   The armature coil of a motor has an inductance of 0.055 H. Calculate the self-induced voltage when the current increases from zero to 8.0 A in $\frac{1}{240}$ s. Does the induced voltage aid or oppose the increasing current?

31.   The current through the armature of a motor changes from 3.0 A to -3.0 A in 0.0030 s. A 104-V induced emf is produced by the changing current. Calculate the inductance of the armature.

32.   The average magnetic field inside a flat coil of area A with N turns around which a current i flows is approximately $B = N\mu_0 i/2\sqrt{A/\pi}$. (a) Determine an expression for the flux through the coil when a current i flows. (b) Determine an expression for the coil's inductance. (c) If the current through a coil of area 45 cm^2 with 20 turns decreases from 4.0 A to 2.5 A in 0.030 s, calculate the induced emf in the coil.

33.   (a) Calculate the self-inductance of a 0.10-m long solenoid with 200 turns, each of 8.0×10^{-5} m^2 area, in which a solid iron core of magnetic permeability 1.0×10^{-3} resides. (b) If a steady current flows through the coil, what is the self-induced emf in the coil? Explain. (*Note:* If the iron core vibrates in and out of the coil, a voltage is induced even though the current is constant. The voltage is caused by the changing flux through the coil as the iron is pushed in and out. The iron core surrounded by the solenoid acts as a vibration detector.)

28.7 Transformers

34.   Calculate the ratio of the secondary to primary turns of a transformer that can step the voltage up from 120 V to 12,000 V in order to operate a neon sign.

35.   A home's electric bell operates on 10 V. Should a step-up or step-down transformer be used to convert the home's 120 V to 10 V? Calculate the ratio of the secondary to primary turns needed for the bell's transformer.

36.   A transformer has 100 turns in its primary coil, which is connected across 120 V. Calculate the number of turns in the secondary coil needed to operate (a) an 18-V toy racetrack, (b) a 200-V welder.

37.   The power for a city of 50,000 people is to be carried by one large transmission line. The line can carry a maximum current of 180 A. (a) Calculate the voltage at which 1000 watts of power per person must be transmitted. (b) Calculate the ratio of secondary to primary turns in a transformer used to step down the voltage to 120 V.

General Problems

G1.   (a) Two coils are held near each other. A steady current flows through one coil. By using the definition of mutual inductance, show that the voltage induced in the second coil by a change of mutual inductance between the coils is

$$\mathcal{E} = -i\frac{\Delta M}{\Delta t}.$$

(b) Draw a rough design for a device used to detect metal during food processing that is based on the change in mutual inductance between two coils as given by the above equation. Explain in words why it operates as it does—especially why M changes.

G2. A power line 12 m above the ground carries a current of 200 A. (a) Calculate the magnetic field at the ground level below the wire. (b) A hermit builds a 0.50 m $\times$ 3.0 m coil with 100 turns and places it below the wire. The coil lies with a 3.0-m side on the ground. Indicate in a drawing the orientation of the coil relative to the power line so that a maximum flux passes through the coil. (c) If the current decreases from 200 A to zero in $\frac{1}{240}$ second, calculate the average emf induced in the coil.

G3. (a) A steady current flows from a battery through a solenoid. Show, using the definition of inductance and Faraday's law, that if the inductance of the solenoid changes, a back emf is induced whose magnitude is

$$\mathcal{E} = i \frac{\Delta L}{\Delta t}.$$

(b) The back emf opposes the current and causes it to decrease or increase, depending on whether L increases or decreases. Describe carefully how this solenoid might be used to build a burglar alarm.

G4. A transmission line is to carry 1.5×10^8 watts of power at 750 kV over a distance of 800 km. Calculate the cross-sectional area of aluminum wires in the transmission line that will cause a 1 percent I^2R power loss.

Alternating-Current Circuits

29

We learned in the last chapter that a coil rotating in a magnetic field produces an alternating voltage. At first, one terminal of the alternating-current generator is at a positive voltage relative to the other terminal; when the generator's coil rotates one-half turn, the voltage is reversed. The terminals of the generator are said to change polarity each time the voltage across the terminals is reversed. If a wire is connected from one terminal to the other, electrons in the wire oscillate back and forth, always moving toward the positive terminal. Since the terminals change polarity as the coil rotates, the electrons change direction each time the polarity changes. The back-and-forth motion of the electrons is called an **alternating current** (ac).

About 99 percent of electric power is generated and transmitted by ac circuits. Most appliances in your home operate on alternating current, as do radio, television, high-fidelity, telephone, and all other communications systems.

In our brief discussion of ac circuits, we will consider only a few basic ideas concerning the role of resistors, induction coils, and capacitors in these circuits. A relationship between the current through and voltage across each device will be examined. The chapter ends with an analysis of a simple ac circuit containing each of these elements in series.

29.1 AC Generator

An alternating-current (ac) generator is a voltage source whose polarity changes in a repetitive pattern. The generator, symbolized by a circle with a wavy line inside (–⊙–), first has positive voltage on one side and negative on the other, as depicted in Fig. 29.1a. A short time later the positive voltage side becomes negative, and the negative side positive.

Electronic generators have been built that produce a variety of time-varying voltage patterns, called voltage waveforms. Several examples are shown in Fig. 29.1b–d. Notice that in each case the polarity of the voltage across the generator reverses in a repetitive manner. The change in voltage can be gradual, as in the sinusoidal waveform (Fig. 29.1b), or abrupt, as in the square-wave and sawtooth waveforms in (c) and (d). In this chapter we will examine circuits powered only by sinusoidal generators.

A **cycle** or period (shown in Fig. 29.1b–d) is the time needed for the voltage waveform to make one complete vibration pattern. The frequency f of the ac

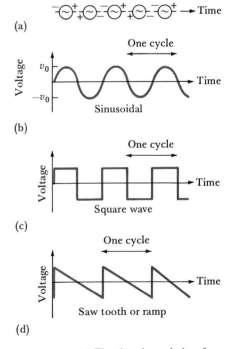

FIG. 29.1. (a) The changing polarity of an alternating voltage. (b)–(d) Different ac voltage waveforms.

generator is the number of cycles per unit time. The voltage generated by electric power plants in the United States has a frequency of 60 Hz; the voltage completes 60 cycles in one second, and during each cycle the polarity changes from plus to minus and back to plus again.

The voltage across a sinusoidal generator varies with time, according to the equation

$$v = v_0 \sin (2\pi f t), \qquad (29.1)$$

where v_0 is called the **peak voltage**—that is, the magnitude of the maximum voltage across the generator during each cycle (see Fig. 29.1b). Note that lowercase letters such as v and i are used to represent voltages and currents that change with time. A capital V or I represents constant voltages or currents that do not change with time.

29.2 Alternating Current Through Resistors

The current produced in a resistor when connected to an ac generator (Fig. 29.2) is calculated using Ohm's law: $i = v/R$. Since the voltage varies with time according to Eq. (29.1), we find that the current through the resistor also varies with time:

$$i = \frac{v}{R} = \frac{v_0}{R} \sin (2\pi f t)$$

or

$$i = i_0 \sin (2\pi f t), \qquad (29.2)$$

where $i_0 = v_0/R$ is called the **peak current.**

The ac voltage across the resistor as well as the current through the resistor are plotted against time in Fig. 29.2b and c, respectively. The current fluctuates in exactly the same pattern as the voltage—first becoming positive, then decreasing and becoming negative, and then completing a cycle by becoming positive again.

What does a negative current mean? If we arbitrarily say that a current moving through the resistor toward the right is positive, then a negative current is one that moves toward the left. The fluctuating current indicates that the electric charge in the resistor first moves in one direction and then in the opposite direction.

EXAMPLE 29.1 The voltage across the wall socket in your home has a peak voltage v_0 of 170 V and a frequency f of 60 Hz. Write an expression for the current through a toaster whose resistance R when hot is 12 Ω.

SOLUTION Substituting $i_0 = v_0/R$ into Eq. (29.2), we find that current varies with time as follows:

$$i = i_0 \sin (2\pi f t) = \frac{v_0}{R} \sin (2\pi f t) = \frac{170 \text{ V}}{12 \text{ Ω}} \sin (2\pi 60 t)$$

$$= \underline{14 \sin (2\pi 60 t)} \qquad \text{(in units of amperes).} \qquad \blacksquare$$

(a)

(b)

(c)

FIG. 29.2. (a) An ac generator produces an alternating current in a resistor. The variations (b) of voltage with time and (c) of current with time.

The alternating current through a resistor, such as the heating coil in a toaster, produces thermal energy in much the same way as a direct current produces thermal energy, as we see next.

29.3 RMS Voltage and Current

You will recall from past chapters that the word *power* describes a change in energy per unit time. In electricity, power is defined as the electrical potential energy added to an electric circuit per unit time. If a direct current I flows through a resistor of resistance R, the electrical energy added to the resistor per unit time is

$$P = \frac{\Delta PE_q}{\Delta t} = I^2 R. \qquad (24.11)$$

As electrons move through a resistor, they bump into atoms and ions in the resistor, which in turn gain thermal energy, causing the resistor to become hot. The hot resistor transfers thermal energy to its surroundings by radiation, conduction, and/or convection. Eventually, an equilibrium is reached where the resistor gains electrical energy just as fast as the energy is lost by this heat transfer to the surrounding environment.

The same process occurs in a resistor through which an alternating current flows. However, the rate of energy transfer differs from one instant to the next because the current passing through the resistor continually changes. At the instant the current has a value $i = i_0 \sin(2\pi f t)$, the power transferred to the resistor is

$$p = \frac{\Delta PE_q}{\Delta t} = i^2 R = [i_0^2 \sin^2(2\pi f t)]R. \qquad (29.3)$$

This expression for power is plotted versus time in Fig. 29.3b. Notice that all values of p are positive because the square of a negative number (the negative value of current in this case) is always positive. Thus, the power (thermal-energy generation rate) varies between zero and a maximum of $i_0^2 R$. Usually we are interested only in the average rate of thermal-energy production. For example, a slice of bread in a toaster does not respond to the 120 power peaks each second (2 per cycle from the 60-Hz ac voltage); instead, the rate of toasting depends on the average power to the heating elements. The average of the $\sin^2(2\pi f t)$ is $\frac{1}{2}$. *

(a)

*The average of the square of the sine of a function can be calculated with the aid of the trigonometric identity,

$$\sin^2 \theta + \cos^2 \theta = 1.$$

The $\sin^2 \theta$ and $\cos^2 \theta$ functions have exactly the same shape when plotted against θ. The only difference is that one is shifted along the θ axis relative to the other. However, over one cycle of θ, both functions have two full humps, such as shown in Fig. 29.3b, and both have the same average. Thus,

$$(\sin^2 \theta)_{av} = (\cos^2 \theta)_{av}$$

and

$$(\sin^2 \theta + \cos^2 \theta)_{av} = (2\sin^2 \theta)_{av} = 1$$

or

$$(\sin^2 \theta)_{av} = \tfrac{1}{2}.$$

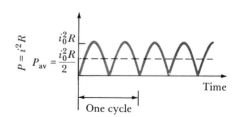

(b)

FIG. 29.3. (a) An oscillating current through a resistor causes electric power absorption that varies with time as shown in (b).

Referring back again to Eq. (29.3), we see that the average power P to a resistor through which an alternating current flows is

$$P = [i_0^2 \sin^2 (2\pi ft)]_{\text{av}} R = [\sin^2 (2\pi ft)]_{\text{av}} i_0^2 R = \tfrac{1}{2} i_0^2 R.$$

Usually we are most interested in the average power used by a device in an electric circuit. Because of this interest in average power, it is convenient to define what is called a root-mean-square (rms) current that relates in a direct way to average power. The **root-mean-square current** I_{rms} is the magnitude of a direct current that produces thermal energy in a resistor at the same rate as the energy is produced by an alternating current of peak value i_0. By equating the expressions for power production by a direct current I_{rms} and by an alternating current of peak value i_0 $(P = I_{\text{rms}}^2 R = \tfrac{1}{2} i_0^2 R)$, we find that $I_{\text{rms}}^2 = \tfrac{1}{2} i_0^2$ or

$$I_{\text{rms}} = \frac{i_0}{\sqrt{2}} = 0.707 i_0. \tag{29.4a}$$

A 0.707-A direct current passing through a resistor produces thermal energy at the same rate as a sinusoidal alternating current of peak value 1.0 A. The rms current is sometimes called the **effective current**; it produces the same effect as the alternating current.

A similar expression can be used to define root-mean-square (rms) or effective voltage. The **root-mean-square voltage** V_{rms} is the magnitude of a dc voltage that produces thermal energy in a resistor at the same rate as an ac voltage of peak value v_0. V_{rms} and v_0 are related by an equation similar to that relating I_{rms} and i_0:

$$V_{\text{rms}} = \frac{v_0}{\sqrt{2}} = 0.707 v_0. \tag{29.4b}$$

Many ac voltmeters and ammeters are calibrated to read root-mean-square voltages and currents rather than peak values. In fact, most of our future discussion of ac circuits will involve root-mean-square values rather than peak values because they relate directly to power consumption. The same equation $(P = I^2 R = I_{\text{rms}}^2 R)$ can be used to calculate either ac or dc power used by a resistor. Also, Ohm's law applies equally well for instantaneous voltage and current, peak voltage and current, and rms voltage and current:

$$\frac{v}{i} = \frac{v_0}{i_0} = \frac{V_{\text{rms}}}{I_{\text{rms}}} = R. \tag{29.5}$$

In the rest of this chapter, rms values of current and voltage will be used, unless stated otherwise, and the subscript rms will be dropped. This notation for ac circuits is summarized in Table 29.1. Note that $I = 0.707 i_0$, $v = 0.707 v_0$, and $P = I^2 R = \tfrac{1}{2} i_0^2 R$.

EXAMPLE 29.2 The peak value v_0 of the alternating voltage provided by a power company to your home is 170 V. Calculate the rms value V of this voltage.

SOLUTION We solve Eq. (29.4b) for V as follows:

$$V = 0.707 v_0 = 0.707(170 \text{ V}) = \underline{120 \text{ V}}. \qquad \blacksquare$$

TABLE 29.1 Notation for AC Circuits

	Current	Voltage	Power
Direct current	I	V	P
Alternating current:			
Instantaneous value	i	v	p
Peak value	i_0	v_0	p_0
Effective (rms) value	I	V	P

As far as producing thermal energy in a toaster or light bulb, an alternating voltage of peak value 170 V is equivalent to a constant dc voltage of 120 V. When the leads of an ac voltmeter are placed across the terminals of a wall socket, they measure 120 V. The voltage from the wall socket is almost always referred to as 120 V, since it effectively provides power like a 120-V dc power supply.

EXAMPLE 29.3 A 60-watt light bulb is connected to a 120-V ac power source (the rms voltage V provided to a home by a power company). Calculate the resistance R of the light bulb.

SOLUTION We are given that $P = 60$ watts and $V = 120$ V. The substitution of Eq. (29.5) into Eq. (24.11) results in the expression:

$$P = I^2R = (V/R)^2R = V^2/R$$

or

$$R = V^2/P = (120 \text{ V})^2/(60 \text{ watts}) = \underline{240\ \Omega}. \qquad ■$$

29.4 Apparent Resistance of a Coil

The current through a resistor caused by an alternating voltage is calculated using Ohm's law ($I = V/R$). A similar expression relates the current through and voltage across an induction coil such as those described in Section 28.6. However, the expression for the resistance of the coil is quite different from what we would expect!

To get a better understanding of the resistance of a coil to alternating current, let us connect a coil of wire to several different voltage sources, as shown in Fig. 29.4. In each case the current through the coil will be measured, and the ratio of the measured voltage and measured current will indicate the coil's resistance to the current. In the future we will use the word *impedance* rather than resistance to describe the opposition to current flow through an ac circuit. Thus, for the coil we define its ac impedance as

$$\text{impedance} = \frac{V}{I}.$$

First, when the coil is connected to a 12-V dc voltage source (Fig. 29.4a), an ammeter measures a current of 12 A. Using Ohm's law, we find that the coil's dc

(a)

(b)

(c)

FIG. 29.4. The current through a coil decreases as the frequency of the voltage source increases.

resistance (the word resistance is used when direct current flows) is $1\,\Omega$, since

$$R = \frac{V}{I} = \frac{12\,\text{V}}{12\,\text{A}} = 1\Omega \qquad \text{(dc resistance)}.$$

Next, the coil is connected to a 12-V, 60-Hz alternating voltage source (Fig. 29.4b). We find, surprisingly, that the alternating current is only 0.12 A, implying an impedance of $100\,\Omega$, not $1\,\Omega$:

$$\text{impedance} = \frac{V}{I} = \frac{12\,\text{V}}{0.12\,\text{A}} = 100\,\Omega \qquad \text{(at 60 Hz)}.$$

If the frequency is tripled to 180 Hz (Fig. 29.4c) while the voltage is kept constant, the current is reduced by one-third to 0.04 A, and the impedance is now $300\,\Omega$:

$$\text{impedance} = \frac{V}{I} = \frac{12\,\text{V}}{0.04\,\text{A}} = 300\,\Omega \qquad \text{(at 180 Hz)}.$$

Finally, we try one last experiment (not depicted in Fig. 29.4). The coil is stretched so that it is a long, straight wire. Then, when the voltage of the different frequency ac generators is placed across the stretched coil, they all produce a current of 12 A. The stretched coil's impedance at all frequencies is $1\,\Omega$, the same as its dc resistance:

$$\text{impedance} = \frac{12\,\text{V}}{12\,\text{A}} = 1\,\Omega \qquad \text{(stretched coil)}.$$

Evidently, the coil's impedance or opposition to alternating current is a consequence of its shape, since the impedance decreases dramatically when the coil is stretched straight, even though the current travels through the same length of wire.

This puzzling behavior depends on two different types of opposition to current—one related to Ohm's law and the other to Faraday's law. First, consider the Ohm's law resistance. As electrons move through the coil, they bump into atoms and molecules in the wire. This collision-type resistance R is the same for both dc and ac and equals $1\,\Omega$ for the example just discussed.

A second type of opposition to current flow becomes dominant as the frequency of the voltage source is increased. The frequency-dependent impedance is given the ominous name *inductive reactance* and is represented by the symbol X_L. It has nothing to do with electrons colliding with atoms and ions but is caused instead by a voltage induced in the coil by its own changing current. This frequency-dependent impedance is a result of the coil's self-inductance, a subject discussed in Section 28.6. We found that the voltage $\mathcal{E}$ induced in a coil of inductance L when the current in the coil was changing with time at a rate $\Delta i/\Delta t$ was

$$\mathcal{E} = -L\frac{\Delta i}{\Delta t}. \tag{28.10}$$

If the induced voltage $\mathcal{E}$ is large, the opposition to current trying to pass through the coil is also large. We see that the induced voltage and opposition to current depends on two quantities: the inductance L of the coil and the rate of change of current $\Delta i/\Delta t$ through the coil.

Inductance L is a geometrical property of the coil and depends on the number of turns, the inside area of the coil, and the type of material inside the coil. Because a straight wire has no turns, its inductance is essentially zero. Hence, no opposing voltage is induced by an alternating current through a straight wire. On the other hand, a coil with large inductance produces a large induced voltage and offers considerable opposition to a high-frequency alternating current.

The induced voltage that opposes the changing current also depends on $\Delta i / \Delta t$. If a direct current flows through a coil, no opposing voltage is induced, since $\Delta i / \Delta t$ is zero—the current is steady and does not change with time. However, a high-frequency alternating current fluctuates back and forth rapidly ($\Delta i / \Delta t$ is large), and the induced voltage opposing the current is also large. The impedance to alternating current caused by the induced voltage in the coil increases in direct proportion to the frequency of the current.

The frequency-dependent impedance, called **inductive reactance** X_L, of a coil is calculated using the equation

$$X_L = 2\pi f L, \tag{29.6}$$

where f is the frequency of the current, and L is the self-inductance of the coil. When an alternating voltage source is connected across a coil of inductance L, the rms values of the current and voltage are related by the equation

$$I = \frac{V}{X_L} = \frac{V}{2\pi f L}. \tag{29.7}$$

This equation is analogous to Ohm's law with the reactance of the coil replacing the resistance of the resistor. The peak values of current and voltage are related by a similar equation ($i_0 = v_0 / X_L$).

EXAMPLE 29.4 Calculate the current in a coil of inductance 0.05 H when driven by a 120-V, 60-Hz voltage source.

SOLUTION The inductive reactance of the coil is calculated using Eq. (29.6):

$$X_L = 2\pi f L = 2\pi(60 \text{ s}^{-1})(0.05 \text{ H}) = 18.8 \ \Omega.$$

The ac current through the coil is

$$I = \frac{V}{X_L} = \frac{120 \text{ V}}{18.8 \ \Omega} = \underline{6.4 \text{ A}}. \qquad \blacksquare$$

EXAMPLE 29.5 A coil, called a "choke," is used to reduce the amplitude of high-frequency signals while allowing low frequencies to pass. (a) Calculate the inductance of a coil that causes a 1200-Ω reactance for a 20-kHz signal. (b) What is the coil's reactance to a 60-Hz signal?

SOLUTION (a) The following information is given: $X_L = 1200 \ \Omega$ when $f = 20 \times 10^3 \text{ Hz} = 20 \times 10^3 \text{ s}^{-1}$. We are to find the value of the coil's inductance L that produces this reactance. Rearranging Eq. (29.6), we find that

$$L = \frac{X_L}{2\pi f} = \frac{1.2 \times 10^3}{2\pi(20 \times 10^3 \text{ s}^{-1})} = 9.5 \text{ mH}.$$

Metal

FIG. 29.5. A metal detector detects a change in alternating current if metal is carried through the coil. The reactance of the coil increases, and the current is reduced.

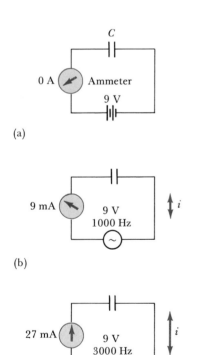

(a)

(b)

(c)

FIG. 29.6. The apparent resistance of a capacitor decreases as the frequency of the voltage source increases.

(b) The same coil offers much less than 1200-Ω resistance to a 60-Hz signal, since

$$X_L \text{ (at 60 Hz)} = 2\pi f L = 2\pi(60 \text{ s}^{-1})(9.5 \times 10^{-3} \text{ H}) = 3.6 \text{ Ω}.$$

The coil effectively "chokes" high-frequency signals, allowing only lower-frequency signals to pass. ∎

In the last chapter we discussed an airport metal detector that depended on the mutual inductance between two coils. The detector can also work using the inductive reactance of a single coil. This type of detector consists of a single coil across which an alternating voltage is connected (Fig. 29.5). An ac ammeter measures the current in the coil. If the current changes, an alarm rings.

When a person walks through the coil, the coil's inductance changes little, since the magnetic susceptibility of a person's body is the same as that of air or a vacuum. If the person carries a piece of iron, though, the inductance of the coil changes dramatically. Even a piece of iron that occupies only a tiny fraction of the area inside the coil can double the coil's inductance L. If L doubles, so does the coil's reactance, since $X_L = 2\pi f L$. If X_L doubles, the alternating current through the coil is reduced by half, since $I = V/X_L$. The alarm goes off.

29.5 Apparent Resistance of a Capacitor

In the last section we learned that the reactance—that is, opposition of a coil to the flow of alternating current—depends on the frequency of the current and the inductance of the coil. A capacitor also has a reactance that opposes an alternating current. The capacitive reactance depends on the frequency of the current and the capacitance of the capacitor, as illustrated by the following simple experiments.

A capacitor is first connected to a 9-V dc power source, such as a battery (Fig. 29.6a). At the instant the capacitor is connected, a current flows. Once the capacitor is charged, the current stops. At this point the capacitor is like an open switch that offers infinite resistance to direct current.

Next, connect the capacitor to a 9-V 1000-Hz alternating voltage source (Fig. 29.6b). A small 0.009-A alternating current is measured. The capacitor has an ac impedance (opposition to current flow) of 1000 ohms, since

$$\text{impedance} = \frac{V}{I} = \frac{9 \text{ V}}{0.009 \text{ A}} = 1000 \text{ Ω}.$$

If the frequency of the power source is tripled, the current also triples to 0.027 A. Evidently, the impedance has been reduced by one-third:

$$\text{impedance} = \frac{V}{I} = \frac{9 \text{ V}}{0.027 \text{ A}} = 333 \text{ Ω}.$$

As the frequency increases, the impedance of the capacitor decreases. This frequency-dependent impedance of a capacitor is called **capacitive reactance.** Like inductive reactance, it has nothing to do with electrons colliding with atoms and ions in a wire but is caused by the voltage across the charged capacitor. As the capacitor accumulates a charge q on one plate and $-q$ on the other plate, a voltage given by

$$V = \frac{q}{C} \qquad (26.2)$$

opposes the further accumulation of charge on the plates. If the plates become fully charged, current stops.

Capacitive reactance depends on how fully the capacitor becomes charged during each one-half cycle of an alternating current. At low frequency, more time is available to charge the capacitor during each flow of current in one direction and before flow in the other direction starts, and the capacitor offers a large average resistance at low frequency. At high frequency, however, there is little time for charge to accumulate on the plates before the capacitor discharges as current flows in the opposite direction. As a result, the capacitor offers little resistance to high-frequency current.*

The opposing voltage ($V = q/C$) also depends on the capacitance C of the capacitor. The greater is C, the less the capacitor becomes fully charged during any time period. Hence, a capacitor with large capacitance produces much less resistance to alternating-current flow than does one with small capacitance.

The frequency-dependent impedance (**capacitive reactance X_C**) caused by a capacitor is given by the equation

$$X_C = \frac{1}{2\pi f C}, \qquad (29.8)$$

where f is the frequency of the current, and C is the capacitance of the capacitor. When an alternating voltage is connected across a capacitor of capacitance C, the rms values of the current and voltage are related by the equation

$$I = \frac{V}{X_C} = \frac{V}{1/(2\pi f C)} = (2\pi f C)V. \qquad (29.9)$$

This equation is analogous to Ohm's law with the capacitive reactance replacing resistance.

EXAMPLE 29.6 Calculate the capacitive reactance of a 0.03-μF capacitor (a) at 60 Hz, and (b) at 90 MHz.

SOLUTION (a) The value of X_C at each frequency is determined using Eq. (29.8):

$$X_C = \frac{1}{2\pi f C} = \frac{1}{2\pi (60 \text{ Hz})(0.03 \times 10^{-6} \text{ F})} = \underline{8.8 \times 10^4 \, \Omega} \qquad \text{(at 60 Hz)}.$$

(b)

$$X_C = \frac{1}{2\pi (90 \times 10^6 \text{ Hz})(0.03 \times 10^{-6} \text{ F})} = \underline{0.059 \, \Omega} \qquad \text{(at 90 MHz)}.$$

---

*The idea of current flowing "through" a capacitor may bother you. The charge does not actually flow from one plate to the other. The accumulation of charge on the plates allows current to flow in other parts of a circuit attached to a capacitor. An alternating current flows in a circuit with a capacitor, even if charge does not cross from one plate to the other.

A 0.03-μF capacitor offers negligible resistance to an FM radio signal transmitted at 90 MHz. The same capacitor will offer considerable opposition to a 60-Hz noise signal transmitted from fluorescent lights or motors operated by the 60-Hz power used in a home. Consequently, the capacitor reduces the current caused by low-frequency electronic "noise" signals, much of which is at a frequency of 60 Hz, but allows high-frequency radio and television signals to pass. ■

29.6 Phase Angles of Voltages and Currents in AC Circuits

In the last two sections we have ignored a subtle feature that affects the power used by an ac circuit. This ignored feature is the phase of the current and voltage. Often, current and voltage do not peak at the same times and are said to be *out of phase* with each other. The difference in phase is expressed in terms of a *phase angle*.

Before examining phase angles, let us compare the two different sinusoidal functions shown in Fig. 29.7. The first function (Fig. 29.7a) is a sinusoidal current expressed by the function $i = i_0 \sin(2\pi ft)$. Plotted below the sinusoidal current is the sin θ versus θ. The two functions, $i_0 \sin(2\pi ft)$ versus t and sin θ versus θ, have exactly the same shape. Notice that one cycle of the sin θ function corresponds to a change in θ of 2π radians or 360°. One-half cycle is 180°, and one-quarter cycle is 90°.

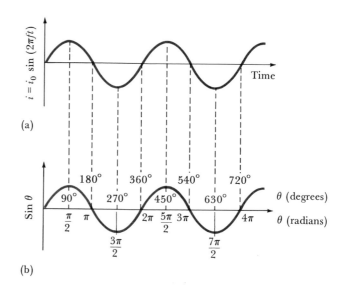

FIG. 29.7. (a) A sinusoidal current varies with time in a way that appears identical to the variation of sin θ versus θ shown in (b). One cycle corresponds to 360°.

Now let us return to voltage and current. In Fig. 29.8a we have plotted a sinusoidal current. Plotted below the current in Fig. 29.8b is a sinusoidal voltage that is in phase with the current—the current and voltage reach their peaks simultaneously.

Other sinusoidal voltage functions are plotted in Fig. 29.8c and d. These voltage functions are different from that shown in (b) in that they lead or lag the current by what is called a phase angle ϕ. Compare, for example, (a) and (c) in Fig. 29.8. Notice that at time zero the current shown in (a) is just starting to rise from zero. However, the voltage shown in (c) is zero and rising at an earlier

FIG. 29.8. (a) A sinusoidal current. (b)–(d) A variety of voltages that differ in phase from the current.

time; by time zero it has already reached its positive peak. Since the voltage has already completed one-fourth of the cycle (90°) by time zero, the voltage is said to "lead" the current by 90°. The voltage shown in Fig. 29.8d "lags" the current by 90°, since it does not start rising from zero until a time when the current has already completed one-fourth cycle.

In the rest of this section we will talk about the phase angle between the current and the voltage in an ac circuit with one resistor, with one inductor, or with one capacitor. In the next section we consider a circuit with all three elements in the same circuit.

Circuit with One Resistor

For an ac circuit containing a single resistor (Fig. 29.9a), the instantaneous current i and the voltage v are related by Ohm's law:

$$v = Ri = Ri_0 \sin(2\pi ft) = v_0 \sin(2\pi ft),$$

where $v_0 = Ri_0$. *The current and voltage are in phase,* as shown in Fig. 29.9b and c.

Circuit with One Inductor

The phase between the current through an inductor and the voltage across it must be determined carefully. Let us start by applying Kirchhoff's loop rule (Section 25.1) to the inductor circuit shown in Fig. 29.10. As we move clockwise around the circuit, the voltage change across each element is added. We first pass the ac generator, across which there is an unspecified voltage change v. Next we pass through the inductor. The induced emf across the inductor is

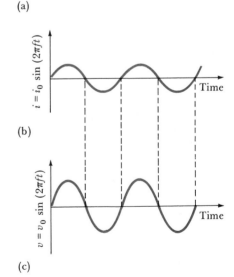

FIG. 29.9. (a) An ac circuit with a resistor. (b) The sinusoidal current passing through the resistor is in phase with (c) the voltage that causes the current.

(a)

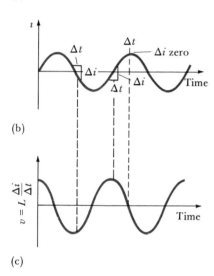

(b)

(c)

FIG. 29.10. (a) An ac circuit with an inductor. (b) If a sinusoidal current flows, the voltage shown in (c) must lead the current by 90° in order that $v = L(\Delta i/\Delta t)$.

$-L(\Delta i/\Delta t)$ (Eq. 28.10). Since these are the only two elements in the circuit, the two voltages must add to zero:

$$v + \left(-L\frac{\Delta i}{\Delta t}\right) = 0$$

or

$$v = L\frac{\Delta i}{\Delta t}.$$

Notice that the voltage across the coil is proportional to $\Delta i/\Delta t$ and not to i. A sinusoidal current is plotted in Fig. 29.10b. In (c) the change of current per unit time ($\Delta i/\Delta t$) is plotted. The $\Delta i/\Delta t$ curve leads the current by 90°. Since the generator voltage is proportional to $\Delta i/\Delta t$, it also leads the current by 90°. *The voltage across an inductor leads the current by 90°.*

Circuit with One Capacitor

Finally, examine the voltage across the capacitor shown in Fig. 29.11a. The voltage across the ac generator has an unspecified value v. If at a given instant the voltage v across the generator is positive, then the voltage across the capacitor must be negative, and it has a value given by Eq. (26.2): $-q/C$. Applying Kirchhoff's loop rule to the circuit shown in Fig. 29.11, we find that

$$v - \frac{q}{C} = 0. \tag{26.2}$$

The current flowing through the circuit is related to the charge q on the capacitor plates by the equation

$$i = \frac{\Delta q}{\Delta t}. \tag{24.2}$$

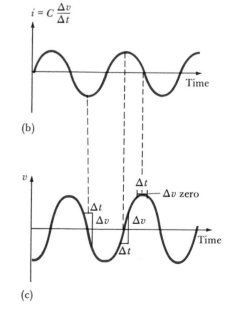

(b)

FIG. 29.11. (a) An ac circuit with a capacitor. (b) If a sinusoidal current flows, the voltage in (c) must lag the current by 90° in order that $i = C(\Delta v/\Delta t)$.

(a)

(c)

Solving Eq. (26.2) for q and substituting into Eq. (24.2), we find that

$$i = \frac{\Delta q}{\Delta t} = \frac{\Delta(Cv)}{\Delta t} = C\frac{\Delta v}{\Delta t},$$

where $\Delta v/\Delta t$ is the change in voltage in a short time Δt. If the current has a sinusoidal shape (Fig. 29.11b), then the voltage curve whose change with time $\Delta v/\Delta t$ produces that sinusoidal current curve is as shown in Fig. 29.11c. *The voltage across a capacitor lags the current by 90°.*

29.7 RLC Series Circuit

In this section we examine a circuit, called an **RLC series circuit**, that contains a resistor, induction coil, and capacitor in series with each other. Our analysis will lead to an understanding of the idea of the electrical impedance of an ac circuit and will develop the background we need in order to understand electrical resonance and electrical power transfer in ac circuits.

An RLC series circuit is shown in Fig. 29.12a. An ammeter in the circuit measures the current flowing through the circuit. The current in a series circuit is the same at each point in the circuit and is represented by the sinusoidal curve shown in Fig. 29.12b.

We learned in the preceding section that the voltage across each element in the circuit is not necessarily in phase with the current. In fact, if we represent the instantaneous voltages across the resistor, inductor, and capacitor by the symbols v_R, v_L, and v_C, respectively, and the instantaneous current by the symbol i, then

1. v_R is in phase with i.
2. v_L leads the current by 90°.
3. v_C lags the current by 90°.

The net voltage v across all three circuit elements at each instant of time is the sum of the instantaneous voltages across each element at that time:

$$v = v_R + v_L + v_C.$$

The net voltage v produced by adding v_R, v_L, and v_C at each instant is shown in Fig. 29.12f. Notice that the voltages v_L and v_C tend to cancel each other and that both are out of phase with v_R. Thus, the amplitude of the net voltage is a complicated function of the amplitudes of v_R, v_L, and v_C, as is the phase of the net voltage. For the example shown in Fig. 29.12, the net voltage shown in (f) leads the current in (b) by a phase angle ϕ of about 30° to 40°.

Instead of adding the instantaneous voltages across each element to find the amplitude and phase of the resultant voltage, as in Fig. 29.12, another technique involving vectors is often used. We can represent the voltages across each circuit element by a vector (Fig. 29.13). The length of the vector represents the magnitude of the voltage (the peak voltage), and the direction of the vector represents the phase of the voltage relative to the current. The positive x axis is arbitrarily chosen as the direction of a voltage in phase with the current. Thus, $\mathbf{v}_R$ points in the positive x direction (see Fig. 29.13a). The voltage across the inductor coil leads the current by 90° and is represented by a vector $\mathbf{v}_L$ pointing in the

(a)

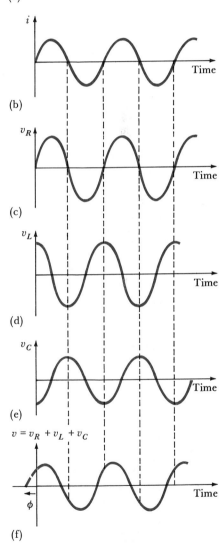

(b)

(c)

(d)

(e)

$v = v_R + v_L + v_C$

(f)

FIG. 29.12. (a) An RLC series circuit. (b) The current flowing in the circuit. (c)–(e) The instantaneous voltages are shown across R, L, and C, respectively. (f) The resultant voltage across the whole circuit.

(a)

(b)

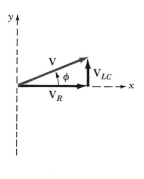

(c)

FIG. 29.13. (a) The rms voltages across a resistor, inductor, and capacitor are represented by vectors. (b) $\mathbf{V}_L$ and $\mathbf{V}_C$ are added to find the voltage across the inductor and capacitor, together. (c) $\mathbf{V}_{LC}$ and $\mathbf{V}_R$ are added to find the resultant voltage across the circuit.

positive y direction. The voltage across the capacitor lags the current by 90° and is represented by a vector $\mathbf{v}_C$ pointing in the negative y direction.

The net or resultant voltage $\mathbf{v}$ is the vector sum of $\mathbf{v}_R$, $\mathbf{v}_L$, and $\mathbf{v}_C$. To find this sum, let us first add $\mathbf{v}_L$ and $\mathbf{v}_C$ to find $\mathbf{v}_{LC}$, the voltage across the inductor and capacitor:

$$\mathbf{v}_{LC} = \mathbf{v}_L + \mathbf{v}_C.$$

The voltages $\mathbf{v}_L$ and $\mathbf{v}_C$ point in opposite directions; when added, the two vectors tend to cancel each other. The magnitude of the combined voltage of $\mathbf{v}_L$ and $\mathbf{v}_C$ is the difference of their amplitudes:

$$v_{LC} = v_L - v_C.$$

The vector $\mathbf{v}_{LC}$ points up or down, depending on whether v_L is greater or less than v_C (see Fig. 29.13b).

Next, $\mathbf{v}_{LC}$ is added to $\mathbf{v}_R$ (Fig. 29.13c). Since the vectors are at right angles, the resultant voltage has a magnitude that is the square root of the sum of the squares of the voltages:

$$v = \sqrt{v_R^2 + v_{LC}^2} = \sqrt{v_R^2 + (v_L - v_C)^2}.$$

The phase angle ϕ that $\mathbf{v}$ makes with the positive x axis is calculated using the tangent of the angle ϕ, shown in Fig. 29.13c:

$$\tan \phi = \frac{v_{LC}}{v_R} = \frac{v_L - v_C}{v_R}.$$

The last two equations apply equally well for the rms voltages across the various circuit elements since $V_{rms} = V = 0.707v$. Substituting rms voltages in these expressions and using the fact that $V_R = IR$, $V_L = IX_L$, and $V_C = IX_C$, we find that

$$V = \sqrt{V_R^2 + (V_L - V_C)^2} = I\sqrt{R^2 + (X_L - X_C)^2} \qquad \textbf{(29.10)}$$

and

$$\tan \phi = \frac{V_L - V_C}{V_R} = \frac{IX_L - IX_C}{IR} = \frac{X_L - X_C}{R}. \qquad \textbf{(29.11)}$$

The quantity under the square root sign in Eq. (29.10) is called the **imped-ance** of the RLC series circuit and is represented by the symbol Z:

$$Z = \sqrt{R^2 + (X_L - X_C)^2}. \qquad \textbf{(29.12)}$$

The impedance of an ac circuit plays a role analogous to the resistance of a dc circuit—both represent the opposition to the flow of current. The big difference between impedance and resistance is that the X_L and X_C parts of impedance are caused by the opposing induced voltage across the inductor coil and the opposing voltage across the charged capacitor. No heating occurs because of these opposing voltages. Thus, the X_L and X_C parts of Z result in no net energy loss due to heating. However, the R part of impedance is caused by collisions of electrons with atoms and ions in a resistor, and it results in thermal energy creation and in heat loss in the electronic circuit.

The equations just developed apply for rms currents and voltages and are summarized below.

The rms current through and voltage across an RLC series circuit are related by the equation

$$I = \frac{V}{Z}, \tag{29.13}$$

where

$$Z = \text{IMPEDANCE} = \sqrt{R^2 + (X_L - X_C)^2}. \tag{29.12}$$

The voltage leads or lags the current by a phase angle ϕ calculated by solving the equation

$$\tan \phi = \frac{X_L - X_C}{R}. \tag{29.11}$$

The phase angle ϕ is positive if the voltage leads the current and negative if it trails the current. The rms voltages across the individual circuit elements are

$$V_R = IR, \tag{29.5}$$
$$V_L = IX_L, \tag{29.7}$$
$$V_C = IX_C. \tag{29.9}$$

EXAMPLE 29.7 A 0.020-H inductor with 3.0-Ω resistance is connected in series with a 1.2-μF capacitor. A 15-V, 1000-Hz power source is connected across the RLC circuit. (a) Calculate the current in the circuit. (b) Calculate the phase angle between the voltage and the current. (c) Calculate the voltage measured across each element. (d) The voltages calculated in (c) add to more than 15 V. Explain.

SOLUTION (a) The current is given by $I = V/Z$. The circuit's resistance R is 3.0 Ω. To find Z, we must also know the values of X_L and X_C, the circuit's inductive and capacitive reactances. Using Eqs. (29.6) and (29.8), we find that

$$X_L = 2\pi f L = 2\pi (1000 \text{ Hz})(0.020 \text{ H}) = 126 \, \Omega$$

and

$$X_C = \frac{1}{2\pi f C} = \frac{1}{2\pi (1000 \text{ Hz})(1.2 \times 10^{-6} \text{ F})} = 133 \, \Omega.$$

The impedance of the RLC series circuit is

$$Z = \sqrt{R^2 + (X_L - X_C)^2} = \sqrt{3^2 + (126 - 133)^2} = \sqrt{9 + 49} = 7.6 \, \Omega.$$

The magnitude of the current through the circuit is:

$$I = \frac{V}{Z} = \frac{15 \text{ V}}{7.6 \, \Omega} = \underline{2.0 \text{ A}}.$$

(b) The phase angle between the voltage and the current is calculated using Eq. (29.11):

$$\tan \phi = \frac{X_L - X_C}{R} = \frac{126 - 133}{3} = \frac{-7}{3} = -2.33$$

or

$$\phi = \underline{-66°}.$$

The voltage lags the current by 66°. (c) The rms voltages across each element in the circuit are:

$$V_R = IR = (2.0 \text{ A})(3.0\,\Omega) = \underline{6\,\text{V}},$$
$$V_L = IX_L = (2.0 \text{ A})(126\,\Omega) = \underline{252\,\text{V}},$$
$$V_C = IX_C = (2.0 \text{ A})(133\,\Omega) = \underline{266\,\text{V}}.$$

(d) The large voltages across the inductor and capacitor are 180° out of phase with each other and tend to cancel. The net voltage across L and C is their difference or $(252 - 266) = -14$ V. This voltage is 90° out of phase with the voltage across the resistor. Together, they produce a net voltage

$$V = \sqrt{14^2 + 6^2} = 15 \text{ V}. \qquad\blacksquare$$

29.8 Power in AC Circuits

In Section 24.5 we learned that the electric power used by a dc circuit is given by Eq. (24.5), $P = VI$, where V is the voltage across the circuit, and I is the current through the circuit. The electric power used in a dc circuit results in the production of thermal energy as current passes through a resistor.

In an ac circuit containing resistors, inductors, and capacitors, the power is used in three ways: (1) thermal energy is produced in the resistor; (2) a magnetic field builds up and collapses in an inductor; and (3) a capacitor becomes charged and discharged. The last two alternatives result in momentary energy storage or energy release rather than energy loss. For example, as a capacitor is charged, its electrical potential energy increases. During discharge, that energy is returned to the circuit or to the source doing the charging. A similar rise and fall of energy occurs in an inductor (a process not discussed in this text). Averaged over one cycle of alternating current flow, the inductor and capacitor consume or absorb no net energy.

Since the only energy absorbed by an ac circuit is that due to heating in its resistors, the average power absorbed in an ac circuit is

$$P = V_R I, \qquad (24.11)$$

where V_R is the rms voltage across the resistor, and I is the rms current through the resistor.

For an RLC series circuit, $V_R = IR$ and $I = V/Z$. Substituting for V_R and I in the above equation, we find that

$$P = (IR)\left(\frac{V}{Z}\right) = VI\frac{R}{Z},$$

where V is the rms voltage across the whole ac circuit, I is the rms current through the circuit, Z is the circuit's impedance, and R is the resistance of the circuit. We see from Fig. 29.14 that $R/Z = \cos\phi$, where ϕ is the phase angle between the voltage and current. Thus,

$$P = VI\cos\phi, \qquad (29.14)$$

where $\cos\phi = R/Z$ is called the **power factor** of the circuit.

The power factor has an important effect on the efficiency of electric power transmission and use. Suppose, for instance, that you have an electric motor that must use electric energy at a rate of 1200 watts when connected to the 120-V rms

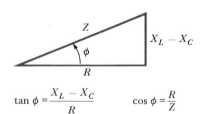

$$\tan\phi = \frac{X_L - X_C}{R} \qquad \cos\phi = \frac{R}{Z}$$

FIG. 29.14. An impedance triangle relating resistance, inductances, impedance, and phase angle.

voltage from the wall socket in your home. The wires wound around the arma-ture of the motor cause it to have a self-inductance L and resistance R but little capacitance C. Thus, the capacitive reactance X_C of the motor is zero, but the inductive reactance X_L and the resistance R are not zero. The power factor can be determined by substituting for Z using Eq. (29.12) and setting $X_C = 0$:

$$\cos \phi = \frac{R}{Z} = \frac{R}{\sqrt{R^2 + (X_L - X_C)^2}} = \frac{R}{\sqrt{R^2 + X_L^2}}.$$

Since X_L is not zero, the denominator of the above equation is greater than the numerator, and $\cos \phi$ is less than 1 (for an electric motor, the power factor is typically 0.8). We can now use Eq. (29.14) to determine the current that must flow through the motor (assuming $\cos \phi = 0.8$) to provide it with the desired power:

$$I = \frac{P}{V \cos \phi} = \frac{1200 \text{ watts}}{(120 \text{ volts})(0.8)} = 12.5 \text{ amperes}.$$

If the power factor were 1.0, the same power would be provided by a current of only 10 amperes:

$$I = \frac{1200 \text{ watts}}{(120 \text{ volts})(1.0)} = 10 \text{ amperes}.$$

The electric power company must charge more for electric power, since they need more transmission lines and have greater loss of energy in these lines when delivering extra current to our homes because of the less-than-optimum power factor of our motors and other appliances.

Although fluorescent light bulbs are much more efficient at producing light than are incandescent light bulbs, they too have inductive circuits that reduce their power factor. Factories or universities using fluorescent light bulbs are often charged by the electric power company at a slightly higher rate because of the extra current that must be delivered to provide the needed power for their lights and other appliances that have reduced power factors.

The power factor of a motor or fluorescent light bulb can be increased to 1.0 by connecting a capacitor in series with the motor or bulb. The same power is provided by less current, as we see in the next example.

EXAMPLE 29.8 An electric motor has an inductance of 0.040 H and a resistance of 20 Ω. (a) Calculate the current through the motor when connected to a 120-V, 60-Hz power source. (b) Calculate the power used by the motor. (c) Calculate the series capacitance that, when added to the motor, increases its power factor to 1.0. (d) What current must now flow through the motor to provide the same power as calculated in part (b)?

SOLUTION (a) The current depends on the voltage and impedance of the circuit. The impedance depends on the resistance and the inductive and capaci-tive reactances:

$$R = 20 \text{ } \Omega,$$
$$X_L = 2\pi f L = 2\pi(60 \text{ Hz})(0.040 \text{ H}) = 15 \text{ } \Omega,$$
$$X_C = 0 \quad \text{since the circuit has no capacitor.}$$

Thus, the impedance is

$$Z = \sqrt{R^2 + (X_L - X_C)^2} = \sqrt{20^2 + (15 - 0)^2} = 25\,\Omega.$$

The current through the circuit is:

$$I = \frac{V}{Z} = \frac{120\,\text{V}}{25\,\Omega} = \underline{4.8\,\text{A}}.$$

(b) The power factor for the circuit is

$$\cos\phi = \frac{R}{Z} = \frac{20\,\Omega}{25\,\Omega} = 0.8.$$

The power used by the circuit is, then,

$$P = IV\cos\phi = (4.8\,\text{A})(120\,\text{V})(0.8) = \underline{460\,\text{watts}}.$$

(c) The power factor $\cos\phi = R/Z$ can be increased to 1.0 if a capacitor is added to the circuit whose capacitive reactance X_C equals the inductive reactance X_L of the motor. If $X_C = X_L$, then

$$Z = \sqrt{R^2 + (X_L - X_C)^2} = \sqrt{R^2 + 0} = R,$$

and $\cos\phi = R/Z = R/R = 1.0$. Thus, for this example, we must add a capacitor whose reactance is $X_C = X_L = 15\,\Omega$. The value of the needed capacitance is determined by solving Eq. (29.8) for C:

$$C = \frac{1}{2\pi f X_C} = \frac{1}{2\pi(60\,\text{Hz})(15\,\Omega)} = 177 \times 10^{-6}\,\text{F} = \underline{177\,\mu\text{F}}.$$

(d) With this capacitance, the motor's impedance is

$$Z = \sqrt{R^2 + (X_L - X_C)^2} = \sqrt{20^2 + (15 - 15)^2} = 20\,\Omega.$$

The power factor of the motor is

$$\cos\phi = \frac{R}{Z} = \frac{20}{20} = 1.0.$$

If 460 watts of power are to be used, the current through the motor and capacitor is given by Eq. (29.14):

$$I = \frac{P}{V\cos\phi} = \frac{460\,\text{watts}}{(120\,\text{V})(1.0)} = \underline{3.8\,\text{A}}.$$

Instead of delivering 4.8 A, the power company can now deliver 3.8 A and still provide the same power for the motor. ■

29.9 Resonance in RLC Series Circuits

The current measured in the RLC series circuit shown in Fig. 29.15 is

$$I = \frac{V}{Z},$$

where the circuit's impedance is

$$Z = \sqrt{R^2 + (X_L - X_C)^2}.$$

FIG. 29.15. An RLC series circuit.

The circuit's impedance depends on the frequency of the power source, since both X_L and X_C depend on frequency. At one particular frequency, called the **resonant frequency** f_0, the impedance of the circuit is a minimum and the current is a maximum. This current maximum occurs at a frequency when

$$X_L - X_C = 0, \tag{29.15}$$

in which case the impedance is a minimum, since

$$Z = \sqrt{R^2 + 0^2} = R.$$

The capacitor and inductor together offer no net opposition to current flow; their effects cancel each other.

To determine an expression for the resonant frequency, we substitute for X_L and X_C into Eq. (29.15) and rearrange:

$$2\pi f_0 L - \frac{1}{2\pi f_0 C} = 0$$

or

$$f_0 = \frac{1}{2\pi \sqrt{LC}}. \tag{29.16}$$

At this frequency the current in the RLC circuit is

$$I = \frac{V}{R} \qquad \text{(at resonance)}.$$

At frequencies higher than the resonant frequency, $X_L > X_C$, since X_L increases with increasing frequency, whereas X_C decreases with increasing frequency. The current

$$I = \frac{V}{Z} = \frac{V}{\sqrt{R^2 + (X_L - X_C)^2}}$$

is less than the current at the resonant frequency, since the $(X_L - X_C)^2$ term contribtures to the impedance (see Fig. 29.16). The current is also less at frequencies less than f_0. For these frequencies, $X_C > X_L$, and the $(X_L - X_C)^2$ part again adds to the impedance. The impedance is least at resonance, and the current is greatest.

At the resonant frequency, a large ac current oscillates back and forth, first charging the capacitor, then discharging through the inductor. The increasing current in the inductor induces a back emf that sends the charge back toward the capacitor. The charge oscillates back and forth in much the way a person oscillates on a swing.

The resonant curve shown in Fig. 29.16 is broader for circuits with larger resistance (like a swing with lots of friction in its bearings). To construct a circuit with a very narrow resonant curve, the circuit should have large capacitive and inductive reactances and as little resistance as possible. (This circuit oscillates easily, like a swing with frictionless bearings.)

An RLC circuit, such as just discussed, is very responsive at its resonant frequency to small input voltages. The tuning circuit on your radio depends on this responsiveness. Usually your radio has a variable capacitor whose capacitance can be changed (Fig. 29.17). You change the capacitor surface area by

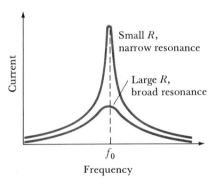

FIG. 29.16. The current is maximum at the resonant frequency of an RLC circuit. The breadth of the resonance depends upon how large the resistance is.

FIG. 29.17. A variable capacitor.

rotating one set of plates into the region between another stationary set of plates. As the area between plates changes, so does the capacitor's capacitance. To tune the radio for a particular station, we change C so that the resonant frequency of the tuning circuit is at the transmission frequency of that station.

EXAMPLE 29.9 The tuning circuit of a radio consists of a variable capacitor and a 20-mH coil of 8 Ω resistance. (a) Calculate the capacitance needed to tune the circuit to a station broadcasting at 920 kHz. (b) If the radio wave arriving at the antenna produces a (2.0×10^{-5})-V ac signal, how large is the current in the antenna circuit? (c) What is the ac voltage across the coil?

SOLUTION (a) Equation (29.16) is solved for the unknown capacitance:

$$C = \frac{1}{(2\pi f_0)^2 L} = \frac{1}{4\pi^2(9.2 \times 10^5 \text{ Hz})^2(20 \times 10^{-3} \text{ H})} = 1.5 \times 10^{-12} \text{ F} = \underline{1.5 \text{ pF}}.$$

With a 1.5-pF capacitor, the circuit will have a resonant frequency of 920 kHz. (b) At resonance, $X_L - X_C = 0$. Hence, $Z = R$, and the current is

$$I = \frac{V}{R} = \frac{2.0 \times 10^{-5} \text{ V}}{8 \text{ Ω}} = \underline{0.25 \times 10^{-5} \text{ A}}.$$

(c) The voltage across the inductor is

$$V_L = IX_L = I(2\pi f L) = (0.25 \times 10^{-5} \text{ A})2\pi(9.2 \times 10^5 \text{ Hz})(20 \times 10^{-3} \text{ H})$$
$$= \underline{0.29 \text{ V}}.$$

This fairly large voltage across the inductor provides a relatively large input signal to a radio amplifier. ∎

You might wonder how such a small input signal, such as the (2×10^{-5})-V signal discussed in the last example, can produce such a large voltage (0.29 V) across the inductor and across the capacitor. The key to answering this question is the word *resonance*. Let us return to the analogy of a swing. If we continually give the swing a small push at its swinging frequency, it eventually develops large-amplitude oscillations. The same phenomenon occurs at electrical resonance in an RLC series circuit. The current oscillates back and forth between the capacitor and the inductor. The frequency of this to-and-fro motion is such that each device gets an optimum current or charge. The external voltage signal gives the vibrating charge a little push each time it returns to the capacitor or inductor for the next swing. Large-amplitude ac current vibrations are caused by many of these small voltage pushes by the external signal that is being detected.

Summary

1. **Alternating current:** An alternating voltage source (such as an ac generator) first pushes charge in one direction and then pulls it back in the opposite direction, thus producing an alternating current (ac). The alternating currents, voltages, and power are characterized in terms of their instantaneous values (i, v, and p) that change with time, their peak values (i_0, v_0, and p_0), and their rms or effective values (I, V, and P). The rms values of current and voltage are related to the peak values by the equations $I = 0.707i_0$ and $V = 0.707v_0$.

2. **Alternating current through resistors, induction coils, and capacitors:** The rms alternating current I through a resistor, in-

duction coil, or capacitor when an rms voltage V is placed across the device is given by the following equations:

RESISTOR: $$I = \frac{V_R}{R},$$ (29.5)

where R is the resistance of the resistor. The instantaneous voltage is in phase with the current.

COIL: $$I = \frac{V_L}{X_L},$$ (29.7)

where X_L, the **inductive reactance** of the coil, is

$$X_L = 2\pi f L.$$ (29.6)

The instantaneous voltage leads the current by $90°$.

CAPACITOR: $$I = \frac{V_C}{X_C},$$ (29.9)

where X_C, the **capacitive reactance** of the capacitor, is

$$X_C = \frac{1}{2\pi f C}.$$ (29.8)

The instantaneous voltage lags the current by $90°$.

3. **RLC series circuit:** The rms alternating current I through a series circuit containing a resistor of resistance R, a coil of inductive reactance X_L, and a capacitor of capacitive reactance X_C

across which an rms voltage V exists is given by the equation

$$I = \frac{V}{Z},$$ (29.13)

where Z, the **impedance** of the circuit, is

$$Z = \sqrt{R^2 + (X_L - X_C)^2}.$$ (29.12)

The voltage leads or lags the current by an angle ϕ that is determined by solving the equation

$$\tan \phi = \frac{X_L - X_C}{R}.$$ (29.11)

The voltage leads the current if ϕ is positive and lags the current if ϕ is negative.

4. **Power:** The average power absorbed by an ac circuit is

$$P = VI \cos \phi,$$ (29.14)

where the **power factor,** $\cos \phi$, is equal to R/Z.

5. **RLC series resonance:** An RLC series circuit has a **resonant frequency** f_0 at which its impedance is a minimum and the alternating current in the circuit is a maximum. Resonance occurs when $X_L - X_C = 0$ and at a frequency

$$f_0 = \frac{1}{2\pi} \sqrt{\frac{1}{LC}}.$$ (29.16)

Questions

1. A light has a filament made of a very thin wire that is wrapped in loops. To dim the light, an iron core is inserted into the coil. Why does this dim the light?

2. Explain why a coil of wire placed in series in a circuit allows low-frequency current to pass but not high-frequency current.

3. Explain why a capacitor placed in series in a circuit allows high-frequency current to pass but not low-frequency current.

4. A parallel-plate capacitor is connected in series to a light. An ac voltage source is connected across the two elements. Will the light be bright or dimmer when a material with dielectric constant greater than 1.0 is placed between the capacitor plates? Explain your answer.

5. Under what conditions is the power factor for an RLC series circuit 1.0? Draw three simple circuits that have a power factor of zero.

6. A light, a resistor, an inductor, and a capacitor are all connected in series to an ac power source. If an extra inductor is added to the circuit, the light dims, whereas if an extra capacitor is added, the light brightens. Explain why this occurs.

7. An ac generator is connected across an RLC series circuit. The frequency of the generator is less than the resonant frequency of the circuit. Does the voltage lead or lag the current? Explain your answer.

8. An RLC series circuit has a variable capacitor. If the circuit's resonant frequency is higher than that of a radio station, should the capacitance of the capacitor be increased or decreased in order to tune the circuit to the radio's frequency? Explain.

Problems

All values of voltage and current given in these problems are rms values unless stated otherwise.

29.1 AC Generator

1. A citizen's band radio transmitter operates at a frequency of 27.1 MHz. The peak voltage across the transmitter is 10 V. Write

an expression for the variation of voltage with time. What is the time for one cycle?

2. While exploring for ore, a geophysicist uses an electrical generator that produces a sinusoidal voltage with a peak value of 150 V at a frequency of 20 Hz. (a) Write an expression for the

variation of voltage with time. (b) Determine two different times at which the voltage is +150 V, two different times at which it is zero, and two different times at which it is −150 V.

3. The voltage across the terminals of an ac generator varies with time according to the equation

$$v = 30 \sin(2\pi 1000t) \text{ volts.}$$

(a) What are the peak voltage and frequency of the generator? (b) Calculate the value of the voltage at each of the following times: 0 s, 0.25×10^{-3} s, 0.50×10^{-3} s, 0.75×10^{-3} s, and 1.0×10^{-3} s.

29.2 Alternating Current Through Resistors

4. A 60-Hz ac generator develops a peak voltage of 170 V. The generator is connected to a light bulb whose resistance when operating is 240 Ω. Calculate an expression for the variation of current through the bulb with time.

5. The generator discussed in Problem 2 is connected to two electrodes placed in the ground 1000 m from each other. The ac current flowing through the ground between the electrodes has a peak value of 250 mA. (a) Calculate the resistance of the ground. (b) Write an expression for the current as a function of time. (c) Calculate the instantaneous current at a time of 0.04 s.

6. A 50-Hz generator with a peak voltage of 310 V (such as used in Europe) is connected to a toaster whose resistance when operating is 12 Ω. (a) Calculate an expression for the current through the toaster as a function of time. (b) Calculate the instantaneous current at the following times: 0 s, 0.005 s, 0.010 s, 0.015 s, and 0.020 s.

29.3 RMS Voltage and Current

7. If you placed an ac voltmeter across the outlets of a wall socket in Europe, you would measure voltage of about 220 V. (a) Calculate the peak voltage. (b) Calculate the rms current and peak current through a toaster whose resistance when operating is 15 Ω. (b) Calculate the power used by the toaster.

8. An alternating current passes through a 5000-Ω resistor and causes thermal energy to be generated at a rate of 0.50 watt. Calculate (a) the rms current through the resistor, (b) the peak current, (c) the rms voltage across the resistor, and (d) the peak voltage.

9. You hold in your hands terminals connected to the 120-V output from a wall socket. (a) Calculate the rms current through your body if the resistance between electrodes is 10,000 Ω. (b) Calculate the peak current. (c) Will any health hazards be caused by this current? (See Table 24.2.)

10. (a) A number 12 copper wire used for home wiring can safely carry a current of 25 A. Calculate the peak value of the current. (b) Why is the wire rated in terms of rms current rather than peak current?

11. (a) Calculate the resistance of a light bulb that absorbs electrical energy at a rate of 60 watts when connected to a 120-V power source. (b) Calculate the rms and peak currents through the bulb. (c) and (d) Repeat parts (a) and (b) for a 120-watt light bulb.

12. You wish to produce an electric heater by running current through a large resistor. The resistor is connected to a 120-V, 60-Hz power source. (a) Calculate the desired resistance in order to produce 1000 watts of heat. (b) Calculate the rms and peak current through the resistor.

29.4 Apparent Resistance of a Coil

13. A 0.35-H inductor is connected across a 12-V ac power source whose frequency can be varied. Calculate the current through the inductor when the frequency is (a) 10 Hz, (b) 100 Hz, (c) 1000 Hz, and (d) 100,000 Hz.

14. Inductors are connected, one at a time, to the terminals of a 60-Hz, 12-V power source. Calculate the current through inductors of inductances (a) 0.010 H, (b) 0.10 H, (c) 1.0 H, and (d) 10 H.

15. A coil has inductive reactance of 5000 Ω when placed in an ac circuit with a 27.1-MHz citizen's band radio signal. (a) Calculate the inductance of the coil. (b) What is its reactance to a 60-Hz signal?

16. The primary of a transformer for a toy car racetrack has an inductance of 3.5 H. Calculate the current that flows when the primary is connected to a 120-V, 60-Hz power source.

17. A 230-V, 60-Hz voltage is connected across the primary of a transformer for an electric welder. The ac voltage across the secondary is 30 V, and a current of 250 A flows. Assume that the current is limited by the reactance of the secondary coil. (a) Calculate the inductance of the secondary. (b) Calculate the primary to secondary turns ratio.

29.5 Apparent Resistance of a Capacitor

18. A 1000-pF capacitor is connected across a 12-V ac power source whose frequency can be varied. Calculate the capacitor current when the frequency is (a) 10 Hz, (b) 10 kHz, and (c) 10 MHz.

19. Capacitors are connected, one at a time, to the terminals of a 60-Hz, 12-V power source. Calculate the current through capacitors of capacitance (a) 10 pF, (b) 1000 pF, (c) 0.1 μF and (d) 10 μF.

20. Static type noise can occur on automobile radios because of ignition sparking in the motor. The noise can be reduced by connecting a 0.50 μF capacitor from the primary coil of the distributor to ground. Some of the high-frequency electrical noise is effectively shorted to ground. Calculate the reactance of the capacitor to 20-kHz noise.

21. A 1-μF and a 2-μF capacitor are connected in series across a 0.50-V, 10-kHz voltage. (a) Calculate the current in the circuit. (b) Repeat the problem for capacitors connected in parallel.

22. In Example 26.1 we learned that your physics book, if it had metal covers, would have a capacitance of 74 pF. (a) Calculate the current through the book when it is connected to the voltage from a wall socket (120 V at 60 Hz). (b) Does current actually flow through the pages of the book? Explain.

29.7 RLC Series Circuit

23. A 250-Ω resistor, 1.2-H inductor, and 1.8-μF capacitor are connected in series across a 60-Hz, 120-V power supply. Calculate (a) the circuit's impedance, (b) the current, (c) the phase angle of the voltage relative to the current, and (d)–(f) the measured voltages across the resistor, inductor, and capacitor.

24. A 420-Ω resistor, 0.12-H inductor, and 0.48-μF capacitor are connected in series to a 20-V, 60-Hz power source. (a) Calculate the phase angle and magnitude of the measured current through the circuit. (b) Repeat the problem for a 20-V, 1000-Hz power source.

25. An ac circuit with a resistor and an inductor has an impedance of 100 Ω at 50 Hz and an impedance of 600 Ω at 500 Hz. Calculate the values of R and L. What is the phase angle between the voltage across and the current through the 50-Hz circuit?

26. An ac circuit with a 1200-Ω resistor and a capacitor of unknown capacitance has an impedance of 1600 Ω at 100 Hz. (a) Calculate the capacitor's capacitance and (b) the circuit's impedance at 500 Hz.

27. A solenoid has an inductance L due to its coiled wire and a resistance R due to the resistivity of the wire. When connected to a 12-V battery, the solenoid draws 4.0 A of current. When connected to a 12-V, 100-Hz power source, the current is 1.8 A. Calculate the values of R and L.

28. A feedthrough capacitor allows alternating current to pass through a circuit but blocks direct current. A 3000-pF capacitor in series with a 5000-Ω resistor is connected to a 10-mV, 9.2-MHz voltage at the input of an amplifier. (a) Calculate the circuit's impedance to the 9.2-MHz input signal and its impedance to direct current. (b) What fraction of the 9.2-MHz impedance is caused by the capacitor?

29.8   Power in AC Circuits

29. A 6.0-μF capacitor, a 0.12-H inductor, and an 85-Ω resistor are connected in series across a 10-V, 300-Hz voltage source. Calculate (a) the current in the circuit, (b) the phase angle between the voltage and current, and (c) the power delivered to the circuit.

30. A 240-Ω resistor is connected across a 120-V, 60-Hz voltage source. (a) Calculate the current through the resistor and the electric power used by the resistor. (b) A 0.053-H inductor and 13-μF capacitor are connected in series with the resistor. Calculate the power used by this series circuit when connected to the same power source.

31. A large motor has an armature whose inductance is 0.12 H. The motor has no capacitance. The series resistance R of the armature's wire is 25 Ω. (a) Calculate the power used by the motor when connected to a 120-V, 60-Hz power source. (b) Calculate the capacitance of a series capacitor that maximizes the power used by the motor. (c) Calculate the power used when this capacitance

is added. (*Warning:* Note that both I and ϕ change when the capacitor is added to the circuit.)

32. Your electric lawn mower does not have enough power to cut long grass. In order to draw more current through the motor and produce more power when it is connected to a 120-V, 60-Hz voltage source, a 540-μF capacitor is connected in series to the mower. The motor's armature has a resistance R of 12 Ω and an inductance of 0.016 H. Calculate the ratio of the power after the capacitor is connected to that before it is connected.

33. In the "new" West, sheep-shearing contests are conducted in which electric shears are used. An inventive physics student connects a variable capacitor in series to her electric shears in order to increase its power. When connected to a 120-V, 60 Hz voltage source, the capacitor causes the current through the shears to increase from 3.0 A to 4.2 A. (a) Calculate the shears' impedance with and without the capacitor. (b) If the shears' resistance R is 1.5 Ω, what is the percentage increase in power when the capacitor is added?

29.9   Resonance in RLC Series Circuits

34. Television Channel 2 broadcasts at a frequency of about 65 MHz. Calculate the inductance of an inductor in series with a 0.024-pF capacitor such that the circuit resonates at this frequency.

35. A citizen's band radio operator has a receiver circuit with a 0.23-mH inductor and a 0.15-pF capacitor. (a) Calculate the resonant frequency of the receiver circuit. (b) If the circuit also has a resistance R of 9.0 Ω, calculate the current through the receiver when a 0.10-mV input voltage is placed across the circuit at its resonant frequency. (c) Calculate the voltage across the inductor.

36. A radio station broadcasts at a frequency of 9.2 MHz. (a) Calculate the capacitance of a capacitor placed in series with a 0.71-mH inductor such that the circuit resonates at this frequency. (b) What is the wavelength of the radio waves produced by the station? The waves travel at 3.0×10^8 m/s.

37. A 2.0-Ω resistor, 1.9-μF capacitor, and 0.12-H inductor are connected in series. (a) Calculate the resonant frequency of the circuit. (b) The inductor has a metal core that can move in and out of the inductor's coil. Calculate the ratio of the circuit's original impedance at its resonant frequency and its impedance when the inductance is increased by 10 percent because the core has moved. The device is a sensitive detector of movement and vibration.

General Problems

G1. The voltage across your shoulders, caused by the changing dipole charge on your heart, varies between about 2 mV and −2 mV once each heartbeat. The resistance in the tissue between the shoulders is about 300 Ω. Calculate the peak current. *Estimate* the distance that an ion in the tissue moves in one direction before moving in the opposite direction. The number of ions per unit volume in the tissue is approximately 10^{25} to 10^{26} molecules/m³. Justify other numbers used in your estimate. (*Hint:* You might consider the typical drift velocity of charges, discussed in Chapter 24.)

G2. The high-pass filter circuit shown in Fig. 29.18 can be used to "pass" signals of high frequency while reducing signals of low

FIG. 29.18

frequency. The circuit could for example be used to reduce 60-Hz "noise" that enters the amplifier of a radio from its antenna. (a) Calculate a general expression for the alternating current I in terms of V_1, f, C, and R. (b) The output voltage V_2 across the resistor connects to an amplifier circuit in the radio. Show that

$$\frac{V_2}{V_1} = \frac{2\pi f RC}{\sqrt{1 + (2\pi f RC)^2}}.$$

(c) Show that for $2\pi f RC \gg 1$, $V_2/V_1 = 1$ (high frequencies are not attenuated). (d) Show that for $2\pi f RC \ll 1$, $V_2/V_1 \ll 1$ (at low frequencies, the output voltage is greatly attenuated).

G3. A low-pass filter is shown in Fig. 29.19 (see discussion for the previous problem). (a) Calculate the current caused by an input voltage V_1. (b) Show that the ratio of the output and input voltages is

$$\frac{V_2}{V_1} = \frac{1}{\sqrt{1 + (2\pi f RC)^2}}.$$

(c) Calculate the limit of this ratio for very high frequencies and for very low frequencies ($f = 0$).

FIG. 29.19

G4. Your neighbor has a citizen's band radio that broadcasts at 27.1 MHz. His communications are being picked up on your television antenna. You can effectively draw his signal to ground by connecting a series capacitor and inductor, which resonate at 27.1 MHz, to ground (see Fig. 29.20). (a) Calculate the impedance of the circuit at 27.1 MHz. (b) Calculate the impedance at 60 MHz, the frequency of the nearest television channel.

(c) Explain why the circuit keeps his conversations out of your television but still allows you to have good television reception.

FIG. 29.20

G5. A pressure-measuring device consists of a 0.15-μF capacitor, a 2.0-Ω resistor, and a 0.17-H inductor connected in series across a 6-V ac voltage source whose frequency matches the resonant frequency of the RLC series circuit. (a) Calculate the current through the circuit. (b) A pressure increase causes the inductor core to be pushed farther into the inductor. Calculate the current at the same frequency as used in part (a) if the inductance increases by 1 percent.

G6. An airport metal detector consists of a large coil of wire through which you walk. The inductance of the coil is 83 mH and its resistance due to the resistivity of the wire is 0.50 Ω. (a) Calculate the capacitance of a capacitor that when placed in series with the coil results in a resonant frequency of 200 Hz. (b) Calculate the current that flows when the circuit is connected to a 1.0-V power source oscillating at the resonant frequency of the circuit. (c) A person carries a small piece of metal through the coil so that its inductance increases by 3 percent. Calculate the current in the coil. (d) Calculate the ratio of the current after the inductance change and before the change.

INTERLUDE Impedance Matching

Impedance is a useful concept whenever power transfers from an object that provides the power (the source) to another object that receives the power (the load). The source and load might be the amplifier and loudspeakers of a high-fidelity system, the electrodes and recorder of an electrocardiogram apparatus, or the vibrating air caused by a sound wave and our inner ear where nerves are located to detect that sound.

The source and load can each be characterized in terms of their **impedance,** a measure of the resistance they offer to the flow of power. In general, the maximum power transfer from a source to a load occurs when they have the same impedance; their impedances are then said to be **matched.**

As a simple example of power transfer and the effect of impedance on this transfer, imagine a battery that produces an emf $\mathcal{E}$. All sources of power, including batteries, have an internal resistance that resists the flow of power through the source. In Fig. 1 the internal resistance of the battery (its impedance) is represented by R_S, the source resistance. When connected to a load resistor of resistance R_L, a current flows through the circuit.

The electric power used by the load is given by Eq. (24.11): $P = I^2 R_L$. For the circuit shown in Fig. 1, the current is

$$I = \frac{\mathcal{E}}{R_S + R_L}.$$

Thus, the power used by the load resistor is

$$P = I^2 R_L = \frac{\mathcal{E}^2 R_L}{(R_S + R_L)^2}$$

and is plotted for different values of R_L in Fig. 2. Instead of specifying particular values of R_L, we have shown the value of P for different values of R_L relative to the value of R_S. For example, the value 2 on the abscissa corresponds to $R_L = 2R_S$, the value 3 corresponds to $R_L = 3R_S$, and so forth. We see that the maximum power transfer occurs at 1 when $R_L = R_S$—that is, when the load has the same resistance as the source; the impedances are then matched. For values of R_L not equal to R_S, less power is transferred to the load; the impedances are then **mismatched** (not equal).

FIG. 1. A battery with internal resistance R_S is connected to a load resistor of resistance R_L.

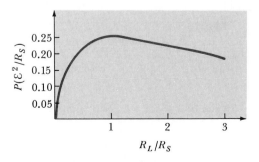

FIG. 2. The variation of power to the load resistor is optimum when $R_L/R_S = 1$.

We considered another example of impedance mismatch in Chapter 17 (see Figs. 17.8 and 17.9) in terms of pulses traveling on ropes or strings. If two ropes of different mass per unit length are connected and a pulse is initiated in one rope, the pulse is partially reflected when it reaches the boundary to the second rope. However, if the ropes have the same mass per unit length, their impedances are then matched, and 100 percent of the incident pulse is transmitted to the second rope.

In Section 29.7 we considered another example of impedance and found that a series RLC circuit has an impedance given by Eq. (29.12):

$$Z = \sqrt{R^2 + (X_L - X_C)^2},$$

where R is the resistance of the circuit, and X_L and X_C are its inductive and capacitive reactances, respectively. All electrical circuits have impedance given by Eq. (29.12), where the values of R, X_L, and X_C depend on the arrangement and values of the resistances, inductances, and capacitances in the circuit, and on the frequency at which the impedance is calculated. In order to have maximum power transfer from an electrical source to a load, their resistances must be equal and the reactance parts of their impedance $(X_L - X_C)$ must have opposite signs:

$$R_{\text{source}} = R_{\text{load}},$$
$$(X_L - X_C)_{\text{source}} = -(X_L - X_C)_{\text{load}}.$$

Similar expressions for impedance can be developed for mechanical systems that include masses, springs, and friction. In these systems, mass is similar to an electrical inductor, a spring is analogous to an electrical capacitor, and the viscous friction force is similar to an electrical resistor. The analog between these mechanical and electrical quantities is illustrated in Table 1.

The human ear is a mechanical system whose impedance can be characterized in terms of the mechanical analog of Eq. (29.12). The ear has a number of different masses, elastic elements, and sources of viscous friction (see Fig. 3).

TABLE 1 Analogous Quantities and Equations for Electrical and Mechanical Impedance

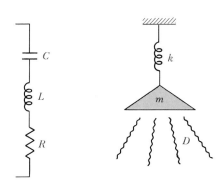

	Electrical		Mechanical	
	Voltage v		Force F	
	Current i		Velocity v	
Resistance	$v = Ri$	Drag force	$F = Dv$	
Inductance	$v = L(\Delta i/\Delta t)$	Inertia	$F = m(\Delta v/\Delta t)$	
Capacitance	$v = \dfrac{1}{C}q$, where $I = \Delta q/\Delta t$	Elastic force	$F = kx$, where $v = \Delta x/\Delta t$	
	$V = ZI,$		$\bar{F} = Z\bar{v},$	
	where $Z = \sqrt{R^2 + (X_L - X_C)^2}$		where $Z = \sqrt{D^2 + (X_m - X_k)^2}$	

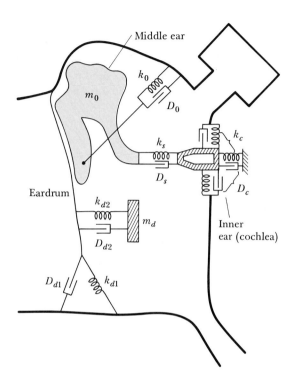

Middle ear

m_0

k_0

D_0

k_s

D_s

k_c

D_c

Eardrum

k_{d2}

m_d

D_{d2}

D_{d1}

k_{d1}

Inner
ear (cochlea)

FIG. 3. The human ear in terms of masses, springs, and friction devices called dashpots (—⊏—). (Adapted from Marquet et al., *Acta Oto-Rhino-Laryngologica Belgica* 27:137, 1973. Reproduced with permission.)

Various hearing disorders of the middle ear can be diagnosed by measuring the ear's acoustic impedance—that is, the impedance of the ear to the absorption of the energy of sound. For example, if the middle ear has a massive adhesion, its inductive-type reactance is greatly increased.

Transformers

Often we wish to transfer power from a source to a load that has a different impedance. If the two are connected directly, an impedance mismatch occurs, and less than optimum power is transferred from the source to the load. This mismatch can be partially or completely corrected using a device called a **transformer**—a two-sided device that has different impedances on each side and is built so that most power entering one side is transferred to the other side. The transformer, when connected between the source and load, matches the source impedance on one side and the load impedance on the other. Power is transferred efficiently from the source to the transformer and from the transformer to the load. Since the power flows efficiently through the transformer, the flow of power from source to load is greatly improved. The outer and middle ear serve as a transformer that improves the efficiency of transfer of sound from air to the fluid in the inner ear. If sound passed directly from air to the inner ear, the

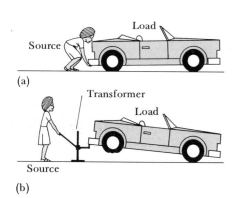

(a)

(b)

FIG. 4. A lever is a transformer that matches the impedance of a person to that of the car.

impedance mismatch at their interface would cause most of the sound to be reflected. But the outer ear and eardrum have a relatively low impedance that matches the impedance of air, whereas the bones of the middle ear have a higher impedance that matches the impedance of the fluid in the inner ear. This transformer produces approximately a thirty-fold improvement in the transfer of sound to the inner ear.

A lever is another form of transformer that provides efficient transfer of energy from a source to a load. A person trying to lift a car so that a flat tire can be fixed is depicted in Fig. 4. Although expending much energy, the person in (a) is unable to lift the car. Using a lever as in (b), the person lifts the car easily. The lever is a transformer that provides an efficient transfer of energy from the person to the car.

As we have seen, the concept of impedance and impedance matching can be applied to a variety of studies. Whenever power, or information, or both are to be transferred efficiently between different objects or between different pieces of equipment, efforts must be made to match the impedances of the objects to each other, either by choosing the equipment correctly or by placing impedance-matching transformers between the source and the load.

Questions and Problems

1. A 10-V source of emf has an internal resistance of $50\,\Omega$. (a) Calculate the power dissipated to a load resistor with the following values of resistances: $10\,\Omega$, $30\,\Omega$, $50\,\Omega$, $100\,\Omega$, and $500\,\Omega$. (b) Plot a graph of power used by the load resistor for these different values of resistance and draw a smooth curve through the points.

2. The variation of the reactive part of the impedance $(X_L - X_C)$ of a normal ear is shown in Fig. 5. (a) Based on the analogy between electrical and mechanical impedance and your knowledge of the frequency dependence of X_L and X_C, explain why the ear's elastic constant dominates its impedance at low frequency. (b) Why does the reactive impedance decrease as the sound frequency increases?

3. For each of the following transformers, indicate the source, load, and type of energy or power that is transmitted through each transformer: (a) a flyswatter, (b) a megaphone, and (c) stairsteps.

4. Describe three everyday objects that act as transformers (do not use the transformers mentioned elsewhere in this interlude).

5. An ac source of electric power operates at a frequency of 300 Hz and has a resistance R of $100\,\Omega$, an inductance L of 0.40 H, and a capacitance C of $10\,\mu F$. The source is connected to a load whose inductance is 0.10 H. Calculate the values of R and C for the load in order that its impedance matches that of the source.

FIG. 5. The variation of a normal ear's resistance and reactance with frequency.

Modern Physics

In 1875 a bright, 17-year-old student named Max Planck (1858–1947) entered the University of Munich. He was told he should not pursue a career in physics because "all the important discoveries in physics have been made." The laws of mechanics, thermodynamics, electricity, and magnetism accounted for the observed behavior of the universe. Despite this warning, Planck chose physics and eventually became a physics professor. In the last week of 1899 at a meeting of the German Physical Society, Planck presented a paper having to do with the nature of light that ushered in the most exciting thirty-year period in the development of our understanding of the physical universe. It was, according to Albert Einstein, "a marvelous time to be alive."

The experiments and theories of that short period brought us into realms seldom experienced in daily living. Einstein's theory of relativity considered the world of objects moving near the speed of light. Another new principle, called quantum mechanics, offered an extraordinary view of the behavior of electrons and atoms. Revolutionary as these two theories were, they are now well established among the principles of modern physics.

At present, an explanation is being sought for the behavior of atomic nuclei and the particles of which they are made. New words and ideas are being introduced—gluons, strangeness, charm, leptons, quarks, and neutrinos. This search will certainly open a new chapter of progress in physics.

Special Relativity

30

Albert Einstein (1879–1955) became internationally famous in the fall of 1919. A group of British scientists, after studying a solar eclipse, announced that their study confirmed Einstein's prediction that light passing the sun is bent by the sun's gravitational field. Newspapers around the world carried headlines proclaiming Einstein's new theory of relativity and decrying its complexity and incomprehensibility. The *New York Times* on November 25, 1919, reported that "so complicated has this revolutionary theory proved that even some of the most learned have been confounded."

Early in the theory's history the ideas of relativity seemed to be hidden in a cloud of mathematics. Years of discussion and debate have clarified these ideas, and relativity can now be presented concisely with little mathematical sophistication.

Perhaps the greatest problem encountered in studying relativity concerns our willingness to accept some of its predictions—predictions that conflict with our intuition. For example, the hands of a clock will appear to turn more slowly if the clock is moving past an observer than if the clock is stationary in front of the observer. An arrow will appear to be shorter when moving at high speed than it does when at rest. An object's mass will increase as its speed increases. These paradoxical predictions have been proven correct by a variety of experiments. In these experiments, however, the objects under investigation are moving very fast—near the speed of light (3.0×10^8 m/s). At the normal speeds we encounter in our daily lives these phenomena—variations in time, length, and mass—are so small that we do not observe them. But they do occur.

There is one important relativistic phenomenon whose effects are apparent in our lives each day: mass is a form of energy and can be converted to other forms of energy, just as other forms of energy can be converted to mass. The sun's radiative energy is provided by the conversion of mass to energy—about four million tons each second. Nuclear power plants provide electricity by converting mass to energy.

Our discussion in this chapter will center on two main topics: (1) a determination of the way in which an object's appearance varies as a result of its motion relative to an observer and (2) a reformulation of some of our ideas concerning energy.

30.1 Events
and Inertial Reference Frames

In relativity we are concerned with differences in the observations made by two observers when examining the same set of events. An **event** is anything that happens—a person's departure and return from a trip, or two flashes of light, one at the nose of a rocket and the other at its tail. We will find that if one observer is moving relative to the second observer, the time between the events and their separation in space seem different for each observer. One observer measures a different time for a trip in space than the other observer and measures a different distance for the length of the spaceship.

Before discussing these differences in time intervals and in distances as measured by two observers, we must define carefully the reference frames in which the events are measured. A **reference frame** consists of a clock and a set of coordinate axes. Each observer has a clock that sits at rest relative to the observer and is used to measure the time at which events occur. Each observer also uses a set of coordinate axes that are at rest relative to the observer. These are used to record the location of events and to indicate the size or length of objects. Since one observer moves relative to the other observer, their coordinate systems move relative to each other.

Relativity is divided into two broad categories called special relativity and general relativity. In general relativity no restrictions are placed on the reference frames, whereas in special relativity we can only use reference frames that are not accelerating. In special relativity, for example, we could not use a point on the dashboard of an accelerating car as the origin of a set of coordinates to describe the location of events occurring either inside or outside the car.

A reference frame that is not accelerating is called an **inertial reference frame.** If a body is placed at rest in an inertial reference frame and if no forces act on it, then it remains at rest. For example, if a piece of dry ice is placed at rest on a horizontal, smooth surface in a car that moves at constant velocity, the dry ice remains at rest, but if the car accelerates, the dry ice slides backward. When accelerating, the car is no longer an inertial reference frame; only when moving at constant velocity is the car such an inertial reference frame.

Earth observer

(a)

Spaceship observer

(b)

FIG. 30.1. (a) Light leaves the earth in pursuit of a fleeing spaceship. The light's speed relative to an earth observer is c. (b) To an observer on the spaceship moving at speed v, the light's speed is also c and not $(c - v)$.

30.2 Postulates of Special Relativity

Einstein formulated the special theory of relativity using only two postulates:

1. The basic laws of nature are the same in all inertial reference frames; that is, basic laws such as Newton's second law ($\Sigma \mathbf{F} = m\mathbf{a}$) should have the same mathematical form in all inertial reference frames.

2. The speed of light in a vacuum is always measured to be 3.0×10^8 m/s and does not depend on whether or not the light source or the observer that measures the light speed are moving.

While the first postulate seems reasonable, the second seems less so. For example, suppose that an intense pulse of light is shot from a laser resting on the earth at a fleeing spaceship. A stationary observer on the earth measures the light's speed c as 3.0×10^8 m/s (Fig. 30.1a).

Now, let us imagine the speed of the light as it passes the spaceship. Suppose that the spaceship moves away from the earth at a speed v that is just a little less than the speed of light. We might expect that the light would have a difficult time reaching the fleeing ship since, according to the earth observer, the light's speed is only $(c - v)$ greater than that of the spaceship. However, an observer on the spaceship measures the light's speed to be $c = 3.0 \times 10^8$ m/s as the light passes the ship (Fig. 30.1b). If a second spaceship approaches the earth at a speed v that is nearly the speed of light, an observer in it also measures the light's speed relative to the spaceship as $c = 3.0 \times 10^8$ m/s, not as $c + v$.

The fact that the speed of light is the same in all reference frames contradicts our intuition. At one time scientists thought that an invisible substance called ether* filled space and vibrated as light passed through it. (Ether served the same function for a light wave as air does for a sound wave.) Scientists thought that light traveled at speed c through the ether and if an observer moved relative to the ether, light's speed would appear slower or faster than c, depending on whether the observer moved with or against the light.

A famous experiment devised by American physicists A. A. Michelson and E. W. Morley in 1887 showed that the measured speed of light is the same in all reference frames and that it does not depend on the motion of the observer. The experiment resulted in the demise of the ether theory and is the experimental basis for the second postulate of relativity.

Some remarkable consequences result from these two postulates. One of these consequences is called time dilation.

30.3 Time Dilation

We usually think of the time between two events as being independent of the motion of an observer. The time between two flashes of light on a passing fire truck appears the same to the truck driver as to a person standing on the sidewalk nearby. If the truck or a spaceship could move at almost the speed of light, however, the flashes of light from the truck or spaceship would appear less frequent to a person standing on the earth. We say that there is a **time dilation**—a lengthening of the time between events that occur on an object moving relative to the observer. A moving clock runs more slowly than the same clock when sitting at rest in front of the observer.

We now derive an equation that compares the time between two events as measured by two different observers. For making this comparison, we analyze the special situation in which one observer is in what is called the **proper reference frame**—the frame where the events occur at the same place relative to the observer. An observer on a spaceship in which a light flashes is in the proper reference frame, since the flashes occur at the same place relative to the observer. The time between events in the proper reference frame is called the **proper time.** Relative to a second observer, such as a person standing on the earth, the events (flashes of light on the spaceship) occur at different places. Thus, the earth is not the proper reference frame for this example, and the time between flashes is delayed for this observer.

To derive an equation relating the time between events as measured in the proper reference frame and in another reference frame, consider the special "clock" on the spaceship shown in Fig. 30.2. An observer on the ship sees a short

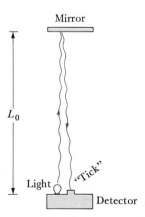

FIG. 30.2. One "tick" of the clock is the time needed for light to travel up to a mirror and back down—a total distance of $2L_0$.

*The ether that supposedly filled space should not be confused with the substance used as an anesthetic.

pulse of light leave the lower surface and travel a distance L_0 to a mirror, which reflects it back to the bottom where it is detected and produces a "tick" sound. We will call the time interval for a round trip one "tick." Since the light travels a distance $2L_0$ at speed c, the time Δt_0 for one tick as observed by a person on the spaceship is

$$\Delta t_0 = \frac{2L_0}{c}.$$

Suppose a person on the earth could observe the tick as the spaceship passed. Since the clock moves relative to the earth observer, the bottom of the clock is in a different location at the end of a tick than it is at the beginning (see Fig. 30.3). The light seems to follow a triangular path of length $2L$. Because $2L$ is longer than $2L_0$, the time for one tick seems longer to the earth observer (remember that light travels at the same speed for both observers).

The distance L can be determined using the Pythagorean theorem. Note that L is the hypotenuse of a right triangle of height L_0 and base $\frac{1}{2}v\,\Delta t$ (see Fig. 30.3). Thus,

$$L = \sqrt{L_0^2 + (\tfrac{1}{2}v\,\Delta t)^2}.$$

The time for one tick according to the earth observer is

$$\Delta t = \frac{2L}{c} = \frac{2\sqrt{L_0^2 + (\tfrac{1}{2}v\,\Delta t)^2}}{c}.$$

Substituting $L_0 = \frac{1}{2}c\,\Delta t_0$ into the above equation and rearranging to solve for Δt, we find that

$$\Delta t = \frac{\Delta t_0}{\left(1 - \dfrac{v^2}{c^2}\right)^{1/2}}. \tag{30.1}$$

This is the time-dilation equation we have been seeking. The procedure for using this equation is summarized as follows:

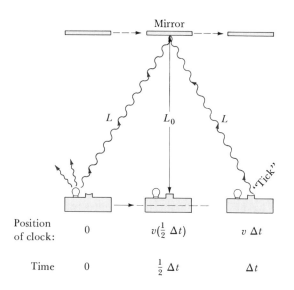

FIG. 30.3. An earth observer sees a triangular light path when the clock shown in Fig. 30.2 moves. A tick lasts longer according to the earth observer because the light travels farther.

Time dilation: Carefully define your events. Then determine in what reference frame the events occur at the same place. This is the **proper reference frame.** The time between the events as measured by an observer in the proper reference frame is the **proper time** Δt_0. The time Δt between the same two events as measured by an observer moving at speed v relative to the proper reference frame is

$$\Delta t = \frac{\Delta t_0}{\left(1 - \dfrac{v^2}{c^2}\right)^{1/2}}. \qquad (30.1)$$

The words "moving at speed v relative to the proper reference frame" are especially important. Recall our example of the spaceship on which a light flashes. The time between flashes as measured by a passenger on the ship is the proper time, since the flashes occur at the same place relative to the passenger. For an observer on the earth, the flashes occur at different places as the spaceship moves past the earth. The earth is not the proper reference frame. We say that the earth moves at speed v relative to the proper reference frame on the spaceship.

You may feel uncomfortable saying that the earth moves relative to the spaceship, but, in relativity we do not attach any special significance to the reference frame on the earth. When using Eq. (30.1), we must first determine which inertial reference frame is the proper one. Then, all other reference frames are regarded as moving relative to that frame. The particular reference frame that is the proper frame depends on the events being considered and may differ from one problem to the next.

EXAMPLE 30.1 As a spaceship moves past the earth at a speed of 2.00×10^8 m/s, its light flashes at 10-s intervals (according to the clock of an observer on the spaceship). How much time elapses between flashes as measured by an earth observer?

SOLUTION The events we are considering are two flashes that are separated in time by 10 s according to an observer on the spaceship. These events occur at the same place on the spaceship (the proper reference frame) and hence

$$\Delta t_0 = 10 \text{ s}.$$

Relative to the spaceship, the earth moves at a speed of 2.00×10^8 m/s. The time Δt between the two events, as measured by the earth observer, is

$$\Delta t = \frac{\Delta t_0}{\sqrt{1 - \dfrac{v^2}{c^2}}} = \frac{10 \text{ s}}{\sqrt{1 - \left(\dfrac{2.00 \times 10^8 \text{ m/s}}{3.00 \times 10^8 \text{ m/s}}\right)^2}} = \underline{13.4 \text{ s}}. \qquad \blacksquare$$

EXAMPLE 30.2 (Muon decay) Direct confirmation of time dilation is provided by an experiment involving particles called muons. A muon has about 200 times as much mass as an electron and is produced during nuclear reactions in high-energy particle accelerators. Once produced, muons have a half-life of 2.2×10^{-6} s. This means that if 1000 muons are produced, then 2.2×10^{-6} s later only 500 remain (the other 500 have disintegrated to produce electrons and energy). Suppose that a beam of 1000 muons is produced by a source and emitted at a speed of 0.95c. How far from the source will the muons have traveled before their number is reduced to 500?

SOLUTION The muons have a half-life of 2.2×10^{-6} s in their own proper reference frame (one in which they are produced and decay at the same location). The proper reference frame for the muons when they are flying through the laboratory is a reference frame that moves through the laboratory with them at a speed of 0.95c. The muons at rest *relative to this reference frame* are produced and disintegrate at the same place in this frame as it moves through the laboratory. In this proper reference frame, their half-life is

$$\Delta t_0 = 2.2 \times 10^{-6} \text{ s}.$$

Relative to the proper reference frame, the laboratory seems to move past the muons at a speed of 0.95c. In the laboratory frame, which is considered to move relative to the rest frame of the muons, half the muons live a time Δt given by Eq. (30.1):

$$\Delta t = \frac{\Delta t_0}{\left(1 - \dfrac{v^2}{c^2}\right)^{1/2}} = \frac{2.2 \times 10^{-6} \text{ s}}{\left(1 - \dfrac{0.95^2 c^2}{c^2}\right)^{1/2}} = 7.0 \times 10^{-6} \text{ s}.$$

By the laboratory clock, 50 percent of the muons will remain after 7.0×10^{-6} s. During this time they move a distance (as measured in the laboratory frame) of

$$d = v \, \Delta t = (0.95c)(7.0 \times 10^{-6} \text{ s}) = \underline{2000 \text{ m}}.$$

The distance they travel is over three times longer than the distance they would travel if their half-life was only 2.2×10^{-6} s in the laboratory frame. ■

Many experiments have shown that muons and other short-lived particles travel much farther than would be possible during their proper half-lives. Their extended lives, as measured by a laboratory observer, agrees well with theories of relativity.

The physiological aging of humans is also affected by motion. The rate of a traveler's heartbeat slows relative to an observer on earth. If the traveler is moving near the speed of light, his or her heart may appear (to the observer on earth) to beat only once in several seconds while to an observer moving with the traveler, it appears to beat about once each second. Some unusual family relationships could result from this slowed physiology, as we see in the next example.

EXAMPLE 30.3 Suppose that a 25-year-old mother leaves on a rocket trip. She travels at a speed of 0.99c relative to the earth and is gone for thirty years as measured on an earth clock. How old is the mother, in terms of physiological age, when she returns? How old is her daughter, who was 2 when the mother left?

SOLUTION A person's physiological age is meaningful only in his or her own proper reference frame. The earth is the proper reference frame for the daughter.

During thirty earth years, the daughter will age thirty years. Since she was 2 years old when the mother left, the daughter will be 32 years old at the end of the mother's trip.

The mother's physiological age change must be determined by a different clock in her own proper reference frame, the spaceship. In that frame she has aged by an unknown amount Δt_0—the time on the spaceship's clock from the start to the finish of the trip. Relative to the spaceship, the earth's clock moves at a speed $v = 0.99c$ and measures a time $\Delta t =$ thirty years for the trip. The mother's age change is calculated using Eq. (30.1):

$$\Delta t_0 = \Delta t \left(1 - \frac{v^2}{c^2}\right)^{1/2} = (30 \text{ years}) \left[1 - \left(\frac{0.99c}{c}\right)^2\right]^{1/2} = 4.2 \text{ years.}$$

Since the mother was 25 years old when she left, on returning she is 25 + 4 or 29 years old. The mother is now three years younger than her daughter! ∎

The trip discussed in Example 30.3 is an inappropriate subject for special relativity since a traveler must accelerate and decelerate while making a space trip, and measurements in special relativity are restricted to inertial reference frames—frames that do not accelerate. Over the years careful discussion and argument have led most scientists to believe that such a phenomenon as described in Example 30.3 is possible. A person in an accelerating reference frame would age more slowly than a person in a nonaccelerating frame, and the difference in aging would be the same as that predicted by applying the equations of special relativity.

A test of this idea was performed by J. C. Hafele and Richard Keating in 1971. They transported four very precise cesium atomic clocks around the world on commercial airlines. The clocks were in the air for approximately forty-five hours. The time change on these clocks was compared to reference clocks that remained at rest on the earth. The scientists observed a shift in time, a time dilation, that agreed well with the theory of special relativity. Because the airplanes flew at speeds much less than the speed of light, the time dilation was very small—several hundred nanoseconds during forty-five hours of flight.

30.4 Length Contraction

Time dilation is linked closely to another phenomenon called **length contraction:** the lengths of objects moving relative to us seem to shorten. The faster an object moves, the shorter is its length. Length contraction occurs only along the directions of motion. An object's size perpendicular to the direction of motion does not appear to change.

We can derive an expression for the dependence of an object's length on its speed by considering an arrow that moves past a stationary clock (Fig. 30.4). As the arrowhead passes the clock, it trips a switch that starts the clock (Fig. 30.4a). As the arrow's feathers pass the clock, the switch is again tripped, and the clock stops (Fig. 30.4b). The two events occurred in the same place in the reference frame of the clock (the proper reference frame for this time measurement), so the clock's reading is the proper time Δt_0. If a small fly rides on the arrow, the fly

FIG. 30.4. (a) As the arrowhead passes a clock, it starts the clock. (b) The back of the arrow stops the clock Δt_0 later.

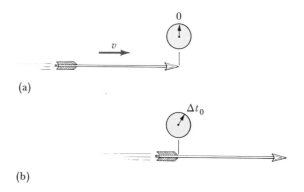

(a)

(b)

would say that the time interval Δt between the events is, according to Eq. (30.1),

$$\overbrace{\text{Arrow}\atop\text{time}}\ \Delta t = \frac{\overbrace{\Delta t_0}}{\left(1 - \dfrac{v^2}{c^2}\right)^{1/2}},\ \ \text{Stationary}\atop\text{clock time}$$

where v is the speed of the fly on the arrow relative to the clock.

Next, how is the arrow's length affected by its motion relative to an observer? The **proper length** L_0 of the arrow is its length in a coordinate system in which the arrow is stationary. Since the arrow is at rest relative to the fly, the fly says the arrow's proper length is L_0. To an observer at rest in the clock's reference frame, the length L of the moving arrow is different. We can relate L to L_0 by noting that the arrow's speed appears the same to an observer on the clock as the clock's speed does to the fly on the arrow.

The arrow, according to the clock observer, moves a distance L (not a proper length) in a time Δt_0. For the fly, the arrow moves its proper length L_0 past the clock in a time Δt. Thus,

$$\underset{\substack{\text{Clock}\\\text{observer}}}{}\quad \underset{\substack{\text{Fly on}\\\text{arrow}}}{}$$

$$v = \frac{L}{\Delta t_0} = \frac{L_0}{\Delta t}. \tag{30.2}$$

Rearranging Eq. (30.2) and substituting Eq. (30.1) for $\Delta t_0/\Delta t$ leads to the following equation, which relates the length L_0 of an object observed in its proper reference frame to its length L in a frame moving at speed v relative to the proper frame:

$$\frac{L}{L_0} = \frac{\Delta t_0}{\Delta t} = \left(1 - \frac{v^2}{c^2}\right)^{1/2}.$$

Length contraction: An object has a length L_0 when measured in a reference frame at rest relative to the object. This frame is the object's **proper reference frame,** and the length L_0 is called the **proper length.** In a reference frame moving at speed v relative to the proper reference frame, the object's length L is measured to be

$$L = L_0\left(1 - \frac{v^2}{c^2}\right)^{1/2}. \tag{30.3}$$

EXAMPLE 30.4 An arrow flies past a person. When at rest, the arrow is 1.00 m long. Calculate the arrow's length L when moving (a) at a speed of $0.90c$, and (b) at a speed of 300 m/s.

SOLUTION (a) To a person at rest relative to the arrow, the arrow's length L_0 is 1.00 m. If the arrow moves at speed v past an observer, then we can also say that the observer moves at a speed v relative to the arrow. The arrow is measured by this observer to have a length L (when moving at $v = 0.90c$) given by Eq. (30.3):

$$L = L_0\left(1 - \frac{v^2}{c^2}\right)^{1/2} = (1.00 \text{ m})\left[1 - \left(\frac{0.90c}{c}\right)^2\right]^{1/2} = \underline{0.44 \text{ m}}.$$

(b) When moving at 300 m/s, the arrow's length is measured to be

$$L = (1.00 \text{ m})\left[1 - \left(\frac{3 \times 10^2 \text{ m/s}}{3 \times 10^8 \text{ m/s}}\right)^2\right]^{1/2}$$

$$= (1.00 \text{ m})[1 - 10^{-12}]^{1/2} \simeq \underline{1.00 \text{ m} - \tfrac{1}{2}(10^{-12} \text{ m})}.$$

(The binomial expansion has been used for the last approximation.) The arrow appears to be shortened by less than $\frac{1}{100}$ the size of an atom when moving at 300 m/s. Relativistic effects are unobservable at the ordinary speeds of daily life.

◼

30.5 Relativistic Mass

In the past two sections we learned that the time between two events and the length of an object depends on the motion of the observer. In this section, we turn to a third relativistic phenomenon: an object's mass increases as its speed relative to an observer increases. Like the changes in time and length, a change in mass is not noticeable at the low speeds of daily living. However, large mass changes occur for objects moving near the speed of light. For example, the mass of an electron increases by a factor of about 40,000 as it is accelerated to nearly the speed of light at the Stanford Linear Accelerator.

Einstein, by performing a thought experiment, was able to accurately predict the dependence of an object's mass on its speed. The thought experiment involved the collision of two identical objects as viewed by two observers. In order for momentum to be conserved in the reference frame of each observer, the object's mass when moving had to be related to its mass when at rest by the following equation (which we will not derive):

$$m = \frac{m_0}{\left(1 - \frac{v^2}{c^2}\right)^{1/2}}, \tag{30.4}$$

where m_0 = object's mass when at rest (called its proper mass or **rest mass**),
  m = object's mass when moving at a speed v.

Einstein's relativistic mass equation has been confirmed many times using particle accelerators. The variation of an object's mass with speed is shown in Fig. 30.5. At low speeds, an object's mass is nearly constant and equals its rest

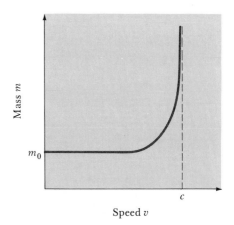

FIG. 30.5. A particle's mass increases dramatically as it approaches the speed of light.

mass m_0. As the object's speed approaches the speed of light c, the denominator of Eq. (30.4) becomes very small, and its mass increases dramatically.

We will see in the next section that the object's mass increase is a result of energy gained as forces pushing and pulling the object cause it to accelerate to higher speed. However, as its speed approaches the speed of light c, its mass approaches an infinite value. To increase the object's speed to the speed of light would require an infinite force and infinite energy. Since this is not possible, *the speed of an object with nonzero rest mass can never equal or exceed the speed of light.*

EXAMPLE 30.5 An electron is accelerated to a speed of $0.995c$. How much greater is its mass when moving at this speed than its mass when at rest?

SOLUTION Rearranging Eq. (30.4), we find that the ratio of its mass when moving at $v = 0.995c$ and its mass m_0 when at rest is

$$\frac{m}{m_0} = \frac{1}{\left(1 - \frac{v^2}{c^2}\right)^{1/2}} = \frac{1}{\left[1 - \left(\frac{0.995c}{c}\right)^2\right]^{1/2}} = \frac{1}{(1 - 0.990)^{1/2}} = \underline{10}. \quad \blacksquare$$

EXAMPLE 30.6 You leave on a spaceship bound for a planet in the Crab Nebula. If your acceleration from rest is $10g$ (that is, 10 times the acceleration of gravity), how long will you have traveled before your mass increases by 1 percent (as noted by an observer on the earth)?

SOLUTION Let us first use Eq. (30.4) to calculate your speed after your mass has increased by 1 percent:

$$\frac{m}{m_0} = 1.01 = \frac{1}{\left(1 - \frac{v^2}{c^2}\right)^{1/2}}$$

or

$$\left(1 - \frac{v^2}{c^2}\right) = \left(\frac{1}{1.01}\right)^2 = 0.980.$$

Hence

$$v = (1 - 0.980)^{1/2}c = 0.14c = 4.2 \times 10^7 \text{ m/s}.$$

If your acceleration is constant and equal to $10g$, the time needed to reach this speed from rest is given by Eq. (3.7) from kinematics:

$$v - v_0 = at.$$

or

$$t = \frac{v - v_0}{a} = \frac{v - v_0}{10g} = \frac{4.2 \times 10^7 \text{ m/s} - 0}{10(9.8 \text{ m/s}^2)} = 4.3 \times 10^5 \text{ s} = \underline{5.0 \text{ days}}. \quad \blacksquare$$

Relativity and Newtonian Mechanics

The principles of Newtonian mechanics were built on the idea that one can measure time, length, and mass accurately and uniquely when describing the changing motion of an object. Must we reject Newtonian mechanics now that we have learned that measurements of time, length, and mass vary from one observer to another? The answer is "yes," if we are talking about objects that move near the speed of light. But at speeds of normal living, the relativistic changes of time, length, and mass are infinitesimal. A person's mass increases by less than a factor of 10^{-13} when the person's speed increases from zero to 100 m/s. Consequently, we can ignore this mass change when using Newtonian mechanics to analyze the person's motion.

As a rough rule of thumb, we can usually ignore relativistic changes in mass, time, and length when the speeds of objects or moving reference frames are less than about one-tenth the speed of light. At this speed ($v = 0.1c$) the factor $(1 - v^2/c^2)^{1/2}$ that appears in the equations for time dilation, length contraction, and mass change equals 0.995. Thus, at $v = 0.1c$ there is only a 0.5 percent change in a time interval, length, or mass. At lower speeds the changes are even smaller, and the equations of Newtonian mechanics provide an adequate description of nature.

30.6 Relativistic Energy

A particle accelerator causes the speed of a particle such as an electron or proton to increase by moving the particle across a large voltage (or across a large number of smaller voltages). In the case of an electron its decrease in electrical potential energy when crossing a voltage V is $\Delta PE_q = -eV$. This decrease is balanced by an increase in kinetic energy. In classical physics, the change in kinetic energy of a particle that starts at rest is

$$\Delta KE = \tfrac{1}{2}m_0 v^2 - \tfrac{1}{2}m_0 0^2 = \tfrac{1}{2}m_0 v^2,$$

where m_0 is the particle's rest mass and is constant. Balancing the particle's kinetic and electrical potential energy changes results in the equation

$$0 = \tfrac{1}{2}m_0 v^2 - eV.$$

Solving for the particle's speed, we find that

$$v = \sqrt{\frac{2eV}{m_0}}.$$

If an electron accelerates across about 10^4 V or less, the above equation provides a fairly accurate expression for its speed. For greater voltages, the equation predicts incorrect results. For example, if the electron crosses 10^6 V, its speed is calculated to be 6×10^8 m/s, or twice the speed of light! Electrons never move faster than the speed of light, even when accelerated through voltages much greater than 10^6 V. Evidently, our classical expression for kinetic energy does not apply to particles that move very fast.

Suppose an electron is already traveling near the speed of light when it passes across a large voltage. The decreased electrical potential energy ($-eV$) must be balanced by increased kinetic energy, yet the electron cannot move much faster, since it is already moving near the speed of light. If it experiences a small increase in speed, its mass increases dramatically (see the upward-curved portion of the line near the point $v = c$ in Fig. 30.5). Since the decreased electrical potential energy seems to be balanced by increased mass rather than by increased speed, mass must be a form of energy.

Albert Einstein, using a moderately complicated thought experiment, was able to show that a particle's total energy is proportional to its relativistic mass:

The **total energy** E of a particle moving at speed v is

$$E = mc^2, \tag{30.5}$$

where

$$m = \frac{m_0}{\left(1 - \dfrac{v^2}{c^2}\right)^{1/2}}$$

and m_0 is the particle's rest mass. If the particle sits at rest, its energy is

$$E_0 = m_0 c^2, \tag{30.6}$$

where E_0 is called the particle's **rest-mass energy.**

The total energy in Eq. (30.5) includes both the particle's energy due to its rest mass and the particle's energy due to its motion (kinetic energy). When the particle is at rest, its energy is $E_0 = m_0 c^2$. If the particle accelerates to speed v, its energy increases to $E = mc^2$. The difference in the initial energy E_0 and the final energy E is the kinetic energy gained by the particle.

The relativistic **kinetic energy** of a mass moving at speed v is the difference of its total energy E and its rest-mass energy E_0:

$$KE = E - E_0 = mc^2 - m_0 c^2, \tag{30.7}$$

where m is the particle's mass when moving at speed v and m_0 is its mass when at rest.

As depicted in Fig. 30.6, the increasing kinetic energy of an electron as it moves across a large voltage first appears as an increase in speed; then, as the electron's speed approaches c, continued energy gain results in a noticeable increase in mass.

The relativistic equation for a particle's kinetic energy looks nothing like the equation used in classical physics ($KE = \frac{1}{2}m_0 v^2$). We find, though, that if the relativistic expression for mass m [Eq. (30.4)] is substituted into Eq. (30.7) for speeds much less than the speed of light ($v \ll c$), the relativistic equation for kinetic energy reduces to the classical equation. For speeds somewhat less than the speed of light, the classical equation for kinetic energy can be used; for speeds near the speed of light, Eq. (30.7) should be used to calculate a particle's kinetic energy. If the particle's speed is unknown, Eq. (30.7) can be used, since it provides the correct answer at all speeds.

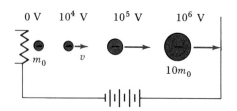

FIG. 30.6. As an electron passes across greater voltages, its mass and energy increase.

EXAMPLE 30.7 Protons are accelerated from rest to a speed of $0.993c$ in a "small" 150-m diameter booster accelerator that is part of the Fermi National Accelerator Laboratory in Batavia, Illinois (Fig. 30.7). (a) Calculate the rest-mass energy E_0 of a proton whose rest mass is 1.67×10^{-27} kg. (b) Calculate the proton's final energy E as it leaves the booster accelerator. (c) Calculate the increase in kinetic energy of the proton in the booster accelerator.

SOLUTION (a) The proton's initial energy when at rest is determined using Eq. (30.6):

$$E_0 = m_0 c^2 = (1.67 \times 10^{-27}\,\text{kg})(3 \times 10^8\,\text{m/s})^2 = \underline{1.50 \times 10^{-10}\,\text{J}}.$$

(b) The proton's energy as it leaves the booster accelerator at a speed of $0.993c$ is determined using Eq. (30.5):

$$E = mc^2 = \frac{m_0 c^2}{\left(1 - \dfrac{v^2}{c^2}\right)^{1/2}} = \frac{1.50 \times 10^{-10}\,\text{J}}{\left(1 - \dfrac{0.993^2 c^2}{c^2}\right)^{1/2}} = \underline{12.70 \times 10^{-10}\,\text{J}}.$$

The proton's energy has increased by a factor of 8.5. (c) The proton's kinetic energy as it leaves the booster accelerator is the difference between its total energy and its rest-mass energy:

$$KE = E - E_0 = 12.70 \times 10^{-10}\,\text{J} - 1.50 \times 10^{-10}\,\text{J} = \underline{11.20 \times 10^{-10}\,\text{J}}. \quad \blacksquare$$

EXAMPLE 30.8 An electron is accelerated from rest across the (2.5×10^4)-V potential difference from the cathode to the screen of a television set. Calculate the percentage increase in the electron's mass. An electron's rest mass m_0 is 9.1×10^{-31} kg.

SOLUTION As the electron moves from the cathode to the screen, its electrical potential energy decreases and its kinetic energy increases by an equal amount. The mass increase can be calculated using the work-energy equation [Eq. (8.10)]:

$$0 = \Delta KE + \Delta PE_q,$$

where

$$\Delta KE = mc^2 - m_0 c^2 = (m - m_0)c^2$$

and

$$\Delta PE_q = qV = (-e)V = (-1.6 \times 10^{-19}\,\text{C})(2.5 \times 10^4\,\text{V}) = -4.0 \times 10^{-15}\,\text{J}.$$

We find, then, that

$$0 = (m - m_0)c^2 + (-4.0 \times 10^{-15}\,\text{J})$$

or

$$m - m_0 = \frac{4.0 \times 10^{-15}\,\text{J}}{c^2} = 4.4 \times 10^{-32}\,\text{kg}.$$

The electron's percentage increase in mass is

$$\left(\frac{m - m_0}{m_0}\right) 100 = \frac{4.4 \times 10^{-32}\,\text{kg}}{9.1 \times 10^{-31}\,\text{kg}} \times 100 = \underline{4.9\ \text{percent.}} \qquad \blacksquare$$

30.7 Equivalence of Mass and Energy

Einstein derived his equation relating mass and energy ($E = mc^2$) using a thought experiment that we will not duplicate here. Nevertheless, a variety of experiments confirm the correctness of the equation. These experiments have convinced physicists that mass, whether stationary or moving, is equivalent to energy.

Electron-Positron Annihilation

A positron is a small particle whose rest mass equals the rest mass of an electron, but it has a charge of $+e$ rather than the $-e$ charge of the electron. Positrons are occasionally produced when high-energy particles from an accelerator collide with the nucleus of an atom. Suppose that a positron created by one of these collisions is near an electron. Because they have opposite electric charges, the positron and electron are pulled together by their electrical attraction. As they collide, a burst of electromagnetic radiation is produced and the electron and positron cease to exist. Their masses have been converted entirely into the energy of electromagnetic radiation—a form of energy (such as light) that has no rest mass.

Nuclear Fusion

A free proton on the sun collides with, and is captured by, the nucleus of a nitrogen atom. Electromagnetic radiation sprays outward as the new nucleus is formed. Following this reaction, the mass of the new nucleus is found to be less than the masses of the original proton and the nitrogen nucleus. The lost mass has been converted to the energy of electromagnetic radiation (light, ultraviolet radiation, x-rays, and so forth) and into kinetic energy. On the sun this capture of protons by nitrogen nuclei is just one of many nuclear processes that are responsible for producing the radiative energy of the sun. Each second the sun's mass decreases by about 4×10^9 kg because of these processes, and an equivalent amount of radiative energy is produced. This mass-energy conversion has occurred for over a billion years. Yet because of the sun's huge initial mass (about 2×10^{30} kg), over 99.99 percent of its mass still remains!

Nucleus

γ ray

(a)

Pair Production

In the last two examples we saw that mass could be converted to electromagnetic energy. The reverse can also occur; electromagnetic energy can be converted to mass. The most familiar form of electromagnetic radiation is light; other forms include radio waves, microwaves, infrared radiation, ultraviolet radiation, x-rays, and gamma rays. All these forms of radiation travel at the speed of light in a vacuum and have no rest mass (they do not exist at rest). The types of radiation differ in their frequency and energy, gamma rays being the most energetic form of electromagnetic radiation. A gamma ray has enough energy to create the mass energy of an electron and a positron. If a gamma ray passes near another atom, it occasionally disappears and an electron and a positron appear in its place (Fig. 30.8). The nearby atom is unchanged. The energy of the gamma ray has been converted to the mass energy of the two particles, a process called **pair production.**

These are just a few of the many experiments that indicate that energy, in any form, is equivalent to mass and that mass is equivalent to energy.

Nucleus

Positron

Electron

(b)

(c)

EXAMPLE 30.9 About 2×10^{10} J of energy are needed to heat and cool your home each year. If the energy could be obtained by converting mass to other forms of energy, how much mass would be needed?

SOLUTION According to Eq. (30.5),

$$m = \frac{E}{c^2} = \frac{2 \times 10^{10}\,\text{J}}{(3 \times 10^8\,\text{m/s})^2} = 2.2 \times 10^{-7}\,\text{kg} = \underline{0.22\,\text{mg}}. \qquad \blacksquare$$

FIG. 30.8. (a) A gamma ray passes a nucleus. (b) It disappears and in its place are an electron and a positron. The gamma ray's energy has been converted to mass. (c) The tracks of an electron and positron in a cloud chamber after their production from a gamma ray (the gamma ray produces no track). The particles move apart because of the force of a magnetic field. (Courtesy of Brookhaven National Laboratory)

Summary

Our discussion of relativity emphasized two main ideas: (1) a determination of the way an object's appearance varies depending on the motion of an observer relative to the object and (2) a reformulation of some of our ideas concerning energy.

1. **Time dilation:** The time between two events as observed in a reference frame where the events occur at the same location is called the **proper time** Δt_0, and the reference frame is called the **proper reference frame.** When the same events are observed in a reference frame moving at speed v relative to the proper reference frame, the time Δt between the events is

$$\Delta t = \frac{\Delta t_0}{\left(1 - \dfrac{v^2}{c^2}\right)^{1/2}}, \qquad (30.1)$$

where c, the speed of light, equals 3.0×10^8 m/s.

2. **Length contraction:** The **proper length** L_0 of an object is its length as measured in a reference frame in which the object is at rest. Its length L as seen in a reference frame in which the object moves at speed v is

$$L = L_0 \left(1 - \frac{v^2}{c^2}\right)^{1/2}. \qquad (30.3)$$

3. **Mass increase:** An object's mass m increases as its speed relative to an observer increases:

$$m = \frac{m_0}{\left(1 - \dfrac{v^2}{c^2}\right)^{1/2}}, \qquad (30.4)$$

where $m_0 =$ the object's mass when measured at rest (its **rest mass**) and $m =$ the object's mass when moving at speed v.

4. **Relativistic energy:** An object's **total energy** E when moving at a speed v is

$$E = mc^2 = \frac{m_0 c^2}{\left(1 - \dfrac{v^2}{c^2}\right)^{1/2}}. \qquad (30.5)$$

The above includes both the object's **rest-mass energy** E_0,

$$E_0 = m_0 c^2, \qquad (30.6)$$

and its **kinetic energy** KE,

$$KE = E - E_0 = mc^2 - m_0 c^2. \qquad (30.7)$$

The equivalence of mass and energy, as predicted by Einstein's famous equation $E = mc^2$, has been confirmed by a variety of experiments in which mass is converted to other forms of energy and other forms of energy are converted to mass.

Questions

1. Explain the difference between a proper reference frame and other inertial reference frames in which the time between two events is measured.

2. Explain how it is theoretically possible for a daughter or son to be physiologically older than her or his natural parents.

3. You move toward a star at a speed of $0.9c$. At what speed does light from the star pass you? What if you move away from the star?

4. You pass the earth in a spaceship that moves at $0.9c$ relative to the earth. Do you notice a change in your heartbeat rate? Does an observer on the earth think your heart is beating at the normal rate? Explain both answers.

5. It takes light approximately 10^{10} years to reach the earth from the edge of the observable universe. Would it be possible for a person to travel this distance during a lifetime? Explain.

6. A person holds a meter stick in a spaceship that travels at $0.95c$ past the earth. The person rotates the meter stick so that it is first parallel and then perpendicular to the ship's velocity. Describe its changing appearance to (a) an observer at rest on the earth and (b) to the person in the spaceship.

7. Name several ways in which your life would be different if the speed of light were 20 m/s rather than its actual value.

8. If the speed of light were infinity, how would time dilation, length contraction, and mass increase be affected. Justify your answer carefully.

9. *Estimate* the amount of energy of all types you use during a lifetime. How much mass is equivalent to this energy?

Problems

You may need to know the following masses in order to work some of these problems: $m_{electron} = 9.11 \times 10^{-31}$ kg and $m_{proton} = 1.673 \times 10^{-27}$ kg.

30.3 Time Dilation

1. A particle called Σ^+ lives for 0.80×10^{-10} s in its proper reference frame before breaking apart into two other particles. How long does the particle seem to live according to a laboratory observer when the particle moves past at a speed of 2.4×10^8 m/s?

2. A Σ^+ such as the one discussed in Problem 1 appears to a laboratory observer to live for 1.0×10^{-10} s. How fast is it moving relative to the observer?

3. A light flashes on a billboard once each 2.0 seconds according to a person standing beside the billboard. What is the time between flashes for a person in a spaceship moving at a speed of 2.95×10^8 m/s?

4. A spaceship moves away from the earth at a speed of $0.990c$. The pilot looks back and measures the time for the earth to rotate once on its axis. How much time does the pilot measure?

5. A free neutron lives about 1000 s before breaking apart to become an electron and a proton. If a neutron leaves the sun at a speed of $0.999c$, (a) how long does it live according to an earth observer, and (b) will the neutron reach Pluto (5.9×10^{12} m from the sun) before breaking apart? Explain your answers.

6. A K^+-meson lives only 1.2×10^{-8} s in its proper reference frame. If traveling at a speed of $0.95c$, how far will it move through a laboratory before it disintegrates?

7. A Σ^+ particle that lives only 0.80×10^{-10} s in its proper reference frame moves 0.17 m through a bubble chamber before decaying to other particles. What is the speed at which the Σ^+ particle travels?

8. Suppose the speed of light were 15 m/s. You run a 100-m dash in 10 s according to the timer's clock. How long did the race last according to your clock?

9. Suppose you have roughly 50 more years to live. At what speed must your spaceship move in order to travel from the earth to the edge of the observable universe (a distance of 10^{10} light-years) before you die?

10. An explorer travels at speed 2.90×10^8 m/s from the earth to a planet of Alpha Centauri, a distance of 4.3 light-years. (a) How long does the trip last according to an earth observer? (b) How much older physiologically is the explorer at the end of the trip?

30.4 Length Contraction

11. A spaceship moving at a speed of $0.95c$ relative to the earth appears to be 30.0 m long according to an earth observer. What is the spaceship's length according to the ship's pilot?

12. At what speed must a meter stick move past an observer in order that it will appear to be 0.50 m long?

13. A billboard is 10 m high and 15 m long according to a person standing in front of it. At what speed must the person drive past in a fast car in order that the billboard will appear square?

14. An astronomer on earth measures the distance to the star Arcturus as 3.8×10^{17} m (40 light-years). What is the earth-Arcturus separation according to a traveler moving at a speed of $0.995c$ from the earth to Arcturus?

15. A spaceship moving at a speed of 3.0×10^7 m/s appears to be 50 m long. Does the spaceship appear longer or shorter than 50 m when at rest, and if so, how much longer or shorter?

30.5 Relativistic Mass

16. At what speed must an object move in order for its mass to be twice its rest mass?

17. Calculate the ratio of an electron's mass and rest mass when moving at the following speeds: (a) 300 m/s, (b) 3.0×10^6 m/s, (c) 3.0×10^7 m/s, (d) 1.0×10^8 m/s, (e) 2.0×10^8 m/s, and (f) 2.9×10^8 m/s.

18. In order to escape the gravitational pull of the sun, protons (which are part of the solar wind) must have a speed of at least 6.2×10^5 m/s. By what factor is the proton's mass increased when moving at this speed?

19. At what speed does an object move so that its mass is 1 percent greater than its rest mass? 10 percent greater?

20. A 50-kg space traveler starts at rest and accelerates at $5g$ for 30 days. Calculate the person's mass after 30 days.

30.6 Relativistic Energy

21. A proton's energy after passing through the accelerator at Fermilab is 500 times its rest energy. Calculate the proton's speed.

22. A 1000-kg rocket starts at rest and accelerates to a speed of $0.90c$. Calculate the kinetic energy needed for this change in speed.

23. Calculate the mass, total energy, rest-mass energy, and kinetic energy of a person with 60-kg rest mass moving at speed $0.95c$.

24. At what speed must you travel in order that your total energy will be ten times your rest-mass energy?

25. An electron is accelerated from rest across 50,000 V in a machine used to produce x-rays. Calculate the percentage increase in the electron's mass.

26. An electron is accelerated from rest across a potential difference of 9.0×10^9 V. Calculate the electron's speed (a) using the nonrelativistic kinetic-energy equation and (b) using the relativistic kinetic-energy equation.

27. A proton is given a kinetic energy of 200 GeV (1 GeV = 1.6×10^{-10} J) at the Fermilab. Calculate the ratio of the proton's mass and rest mass.

28. A particle of rest mass m_0 initially moves at speed $0.4c$. (a) If the particle's speed is doubled, calculate the ratio of its final kinetic energy and its initial kinetic energy. (b) If the particle's kinetic energy increases by a factor of 100, by what factor does its speed increase?

29. The relativistic momentum p of an object of mass m moving at speed v is $p = mv$. Use this and the relativistic energy equations to verify that $E = (p^2c^2 + m_0^2c^4)^{1/2}$.

30.7 Equivalence of Mass and Energy

30. Calculate the amount of mass whose rest energy equals the total yearly energy consumption of the world (4×10^{20} J).

31. A city of about 100,000 people consumes about 2×10^{15} J of electrical energy each year. How much mass converted to electrical energy at 33 percent efficiency will provide this 2×10^{15} J?

32. An electric utility company charges a customer 1.5 cents for 10^6 J of electrical energy. At this rate, calculate the cost of a 1-gram mass if converted entirely to energy.

33. A nuclear power plant produces 10^9 watt of electric power and 2×10^9 watt of waste heat. (a) At what rate must mass be converted to energy in the reactor? (b) What is the total mass converted to energy each year?

34. Separating a carbon monoxide molecule CO into a carbon and an oxygen atom requires 1.76×10^{-18} J of energy. (a) Calculate the mass equivalent of this energy. (b) Calculate the ratio of this mass and the original mass of a CO molecule (4.65×10^{-26} kg).

35. A hydrogen-oxygen fuel cell combines 2 kg of hydrogen with 16 kg of oxygen to form 18 kg of water, thus releasing 2.5×10^5 J of energy. What fraction of the mass has been converted to energy?

36. (a) Calculate the energy radiated by the sun each second by conversion of 4×10^9 kg of mass to energy. (b) Determine the fraction of this energy intercepted by the earth. The earth is 1.50×10^{11} m from the sun and has a radius of 6.38×10^6 m.

General Problems

G1. Suppose the speed of light is 100 m/s and that you are driving a racecar at a speed of 90 m/s. (a) How much time is required for you to travel 900 m along a track's straightaway, according to a timer on the track? (b) According to your own clock? (c) How long does the straightaway appear to you? (d) Notice that the speed at which the track moves past is your answer to (c) divided by your answer to (b). Does this speed agree with the speed as measured by the stationary timer?

G2. An observer on a track measures the passage of an extraordinary rocketship that travels 100 m in 5.0×10^{-7} s. (a) Calculate the ship's speed as determined by the ground observer. (b) How long a distance does the 100 m seem to a man in the rocketship? (c) What time has elapsed on his watch during this passage? (d) How fast does the earth seem to pass the rocketship?

G3. A rod when at rest is 1.0 m long and has a cross section of 1.0×10^{-4} m^2 and a mass of 1.0 kg. (a) Calculate the rod's density. (b) The rod is now made to move at a speed of $0.995c$ in a direction parallel to its long dimension. Calculate the rod's mass, length, cross section, and density as determined by a stationary observer.

G4. A pilot and her spaceship have a mass of 400 kg. She expects to live 50 years and wishes to travel to a galaxy that requires 100 years to reach if traveling at the speed of light. (a) Determine the average speed she must travel to reach the galaxy during her next 50 years. (b) In order to attain this speed, a certain mass m of matter is consumed and converted to the spaceship's kinetic energy. How much mass is needed? (We will ignore the energy needed to accelerate the fuel that has not yet been consumed.)

G5. (a) A container holding 4 kg of water is heated from 0°C to 60°C. Calculate its change in mass as a result of this increased energy. (b) If the water, initially at 0°C, is converted to ice at 0°C, calculate its change in mass.

Wave-Particle Duality

31

We learned in the last chapter about the modifications in classical physics needed to describe objects moving near the speed of light. Now we learn about modifications needed to describe very small objects, such as electrons. This new description of the way small objects behave incorporates portions of wave theory and portions of the classical theory used to describe particles with mass. This dual wave-particle description applies not only to objects with mass (electrons, protons, and atoms) but also to a massless form of energy called electromagnetic radiation. Our study begins with the wavelike nature of electromagnetic radiation.

31.1 Electromagnetic Waves

An **electromagnetic wave** is an electrical and magnetic disturbance that moves through space at the speed of light ($c = 3.0 \times 10^8$ m/s). You probably recognize the names of most of the different types of electromagnetic waves—radio waves, microwaves, infrared waves, light, ultraviolet radiation, x-rays, and gamma rays. The waves differ from each other in their frequency and wavelength and in the way they are produced and interact with matter. The waves are similar in that they all move at the speed of light and consist of moving electric and magnetic fields.

The production of an electromagnetic wave by an oscillating electric charge in an antenna is illustrated in Fig. 31.1. The antenna consists of two vertical metal rods connected to an ac generator that causes charge to oscillate up and down from one rod to the other. In Fig. 31.1a, an excess negative charge resides at the top of the antenna, leaving a positive charge (a deficiency of electrons) at the bottom. In Fig. 31.1b, about one-eighth of a cycle of oscillation later ($\tau/8$), some electrons have moved down from the top rod to the bottom rod causing the excess charge on each rod to be reduced. As electrons continue to move down, the rods are neutralized (Fig. 31.1c), and eventually the lower rod becomes negatively charged (Fig. 31.1d). One-half cycle of oscillation is completed when the charge distribution shown in Fig. 31.1a is reversed as in Fig. 31.1e. The second half of oscillation involves an upward movement of electrons and a return to the original charge distribution (Fig. 31.1i).

The charge on the antenna produces an electric field directly in front of the antenna. The field fluctuates up and down in phase with the charge distribution

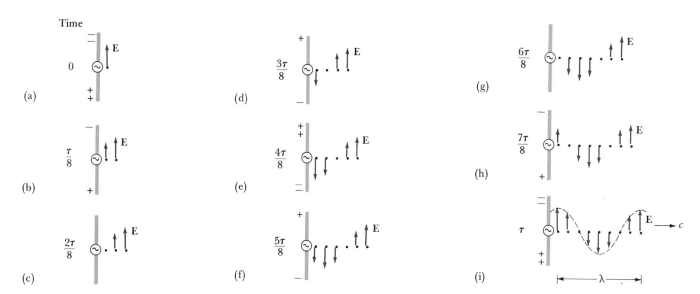

FIG. 31.1. As electrons oscillate up and down on the antenna, an electric field is produced in front of it. As the field moves away at a speed of 3.0×10^8 m/s, the antenna continues to produce new fields.

on the antenna. In Fig. 31.1a the field points up. When the charge on the antenna is reduced by one-half, as in Fig. 31.1b, the field directly in front of the antenna is reduced by one-half. When the charge distribution is reversed, as in Fig. 31.1e, the field directly in front of the antenna is reversed.

While the field near the antenna oscillates up and down in phase with the oscillating charge distribution on the antenna, the electric field created at earlier times moves away from the antenna at the speed of light. The symbol **E** traces the position of the original field (produced at time zero) at successive time intervals. A pattern of alternating up-and-down electric field lines leaves the antenna as its charge oscillates up and down. One wavelength of the electric field pattern (Fig. 31.1i) is caused by one cycle of oscillation of charge on the antenna. As the antenna charge continues to oscillate, the wave grows longer and the whole pattern continues to move away from the antenna at the speed of light.

An electric charge some distance from the antenna and in the wave's path is forced by the electric field of the passing wave to vibrate up and down. For example, electrons in the antenna of a car radio are pushed up and down by passing electromagnetic waves generated by a radio station antenna that is miles from the car.

The moving charges in the antenna shown in Fig. 31.1 also produce a magnetic field that encircles the antenna rods as shown in Fig. 31.2. (See Section 27.7 for a discussion of how magnetic fields are produced by moving charges.) If the current moves up the antenna, a magnetic field such as shown in Fig. 31.2a is produced. If the current moves down, the field is reversed (Fig. 31.2b). As the electrons oscillate up and down in the antenna, an alternating magnetic field pattern is produced (Fig. 31.2c). This field also moves away from the antenna at the speed of light. The combined magnetic and electric fields produced by the oscillating charge on the antenna is called an **electromagnetic (EM) wave** and is shown in Fig. 31.2d. The magnetic and electric fields are perpendicular to each other and are both perpendicular to the direction in which the wave moves. While the pattern shown in Fig. 31.2d represents the varying fields in one direction, similar patterns can be drawn along other directions since the EM wave spreads in many directions.

(a)

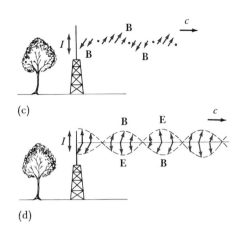

(c)

(b)

(d)

FIG. 31.2. (a) and (b) The magnetic fields produced in front of an antenna by its electric current. (c) A magnetic wave moving at a speed c leaves the antenna. (d) The combined electric and magnetic field moving at a speed c is called an electromagnetic wave.

31.2 Electromagnetic Spectrum

If we could watch an EM wave as it passed a point in space, the directions of the electric and magnetic fields at that point would alternate back and forth. The frequency f of the wave is the number of complete vibrations per second of the field at a point along the path of the passing wave. The frequency also equals the vibration frequency of the wave's source.* For example, if the electrons in a radio station antenna move up and down at a frequency of 90 MHz, then the electromagnetic waves produced by the antenna have a frequency of 90 MHz.

As with other waves, the wavelength of an electromagnetic wave depends on its frequency and speed. In a vacuum, the wavelength of all electromagnetic waves is given by Eq. (17.8):

$$\lambda = \frac{c}{f},$$

where c, the wave's speed, equals 3.0×10^8 m/s. The different types of waves move at about the same speed in air but at different speeds in other media. (See Table 19.1 for some representative values of the speed of light in different media.)

Electromagnetic waves are often classified by frequency in a scheme called the **electromagnetic spectrum** (Fig. 31.3). A particular frequency region in the EM spectrum includes waves produced in a similar manner. Infrared waves are produced by molecular vibrations, whereas radio waves are produced by charges vibrating back and forth in antennas. The boundaries between regions are not sharp.

The low radio-frequency region from 0 to 10^9 Hz consists of waves used mostly for communications—the lower frequencies being using for AM radio and telegraphy and the higher frequencies for FM radio and television communications. Microwaves, the next part of the electromagnetic spectrum, extend from about 10^9 Hz to more than 10^{11} Hz. Microwaves are used for radar tracking, radio navigation, communication, heating in microwave ovens, drying, and other industrial purposes.

---

*We assume that the wave's source and the device detecting the wave are at rest relative to each other. If not, the frequency of the wave is shifted because of the Doppler effect.

FIG. 31.3. The electromagnetic spectrum.

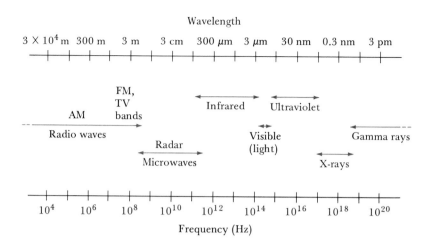

The infrared portion of the spectrum from about 5×10^{11} Hz to 4×10^{14} Hz comes next. Infrared waves, when absorbed by your skin, increase the amplitude of molecular vibrations in the skin. Your skin is warmed. For this reason infrared waves are sometimes called heat or thermal radiation.

The portion of the electromagnetic spectrum that includes visible light waves starts with red light at about 4×10^{14} Hz and ends with blue light at about 8×10^{14} Hz. This tiny portion of the whole EM spectrum is important to humans because of the sensitivity of our eyes to light. Light waves are caused by the motion of electrons in atoms as they jump between different orbits.

Higher-frequency ultraviolet radiation (8×10^{14} to 10^{17} Hz) and x-rays (10^{17} to 10^{19} Hz), the next parts of the EM spectrum, are also produced by the motion of electrons in atoms. Ultraviolet radiation darkens the skin. X-rays are used for examining the interiors of objects that are opaque to light, such as the human body. The spectrum is completed by high-frequency gamma rays (10^{19} Hz and higher) produced by the motion of charged particles in nuclei.

Light waves and many other forms of electromagnetic waves are often described in terms of wavelength instead of frequency. The metric unit used for the wavelength of light waves is the nanometer (1 nm = 10^{-9} m). Red light has a wavelength of about 700 nm, blue light a wavelength of about 400 nm, and x-rays have wavelengths of about 0.1 nm, roughly the radius of an atom.

31.3 Photons

Isaac Newton and his contemporaries in the eighteenth century believed that light consisted of many separate bundles of energy. In 1802 Thomas Young's famous double-slit interference experiment started an era in which the wave theory of light dominated. Approximately a hundred years after Young's experiment, in December 1899, the German physicist Max Planck read a paper to a meeting of the German Physical Society in Berlin. This paper, along with the research of other scientists that followed, led to a resurrection of the particle theory of light. Light, and all forms of EM radiation, were shown to consist of tiny bundles of energy called quanta. These quanta eventually became known as *photons*. During the first twenty years of the twentieth century, much experimental evidence was collected in support of the photon theory of EM radiation.

A **photon** is a massless bundle of electromagnetic energy. The energy E of a single photon depends on its frequency f and is given by the equation

$$E = hf, \tag{31.1}$$

where h, a fundamental constant called Planck's constant, has a value $h = 6.63 \times 10^{-34}$ J·s.

FIG. 31.4. A flashlight sends out many photons of light. The frequencies f and energies ($E = hf$) of the photons differ.

In this definition, the word *massless* is used to distinguish photons from objects that have measurable rest mass, such as electrons and atoms. A photon cannot be captured and weighed on a scale. When a photon is caught, such as happens when light is absorbed by your skin, it ceases to exist. Its energy is converted to some other form of energy—for example, into thermal energy of atoms and molecules in your skin. Photons exist only while moving and do not have rest mass. Their speed is 3.0×10^8 m/s in air or in a vacuum and is usually slower when passing through other materials.

According to the photon theory, light from a flashlight might be pictured as shown in Fig. 31.4—over 10^{20} individual photons leave the light each second. The brightness of the light depends on the number of photons passing along the light beam each second.

The energy of photons in units of joules is very small. A photon of red light has an energy of about 3×10^{-19} J. A more convenient energy unit has been developed to describe photons and other phenomena related to atoms: **one electron volt** (eV) is the magnitude of the change in electrical potential energy that occurs when an electron crosses a potential difference of 1 V [$\Delta PE_q = qV = (1.6 \times 10^{-19}$ C$)(1$ V$) = 1.6 \times 10^{-19}$ J]. We see that

$$1 \text{ eV} = 1.6 \times 10^{-19} \text{ J.} \tag{31.2}$$

The energy of a photon of red light is about 2 eV. The energies in units of eV associated with other atomic phenomenon are listed in Table 31.1.

TABLE 31.1 Energy of Several Atomic Occurrences and Photons in Units of Electron-Volts (eV)

Ionizing a hydrogen atom	13.6 eV
Breaking a hydride bond	1 eV
Causing a DNA mutation	3–10 eV
Energy of light photon	2–3 eV
Energy of UV photon	3–300 eV
Energy of x-ray photon	$\sim$40,000 eV
Thermal energy of one atom at room temperature	$\frac{1}{40}$ eV

EXAMPLE 31.1 Calculate the energy of a photon of red light with wavelength 660 nm in units of (a) joules and (b) electron volts.

SOLUTION (a) Using Eqs. (31.1) and (17.8), we see that

$$hf = h\left(\frac{c}{\lambda}\right) = (6.63 \times 10^{-34} \text{ J·s})\left(\frac{3.0 \times 10^8 \text{ m/s}}{660 \times 10^{-9} \text{ m}}\right) = \underline{3.0 \times 10^{-19} \text{ J.}}$$

(b) The solution to (a) in joules is converted to electron volts using Eq. (31.2):

$$3.0 \times 10^{-19} \text{ J} \left(\frac{1 \text{ eV}}{1.6 \times 10^{-19} \text{ J}}\right) = \underline{1.9 \text{ eV.}} \quad \blacksquare$$

EXAMPLE 31.2 An x-ray photon produced during a medical chest x-ray has an energy of about 40,000 eV. (a) Calculate the photon's wavelength. (b) During a chest x-ray, a person's body typically absorbs x-ray photons whose total energy is about 1.5×10^{-3} J. Calculate the number of x-ray photons absorbed.

SOLUTION (a) The wavelength is determined by combining and rearranging Eqs. (31.1) and (17.8):

$$E = hf = h\left(\frac{c}{\lambda}\right),$$

$$\lambda = \frac{hc}{E} = \frac{(6.63 \times 10^{-34}\,\mathrm{J \cdot s})(3.0 \times 10^8\,\mathrm{m/s})}{(40{,}000\,\mathrm{eV} \times 1.6 \times 10^{-19}\,\mathrm{J/eV})} = 3.1 \times 10^{-11}\,\mathrm{m} = \underline{0.031\,\mathrm{nm}}.$$

Note that one must convert the energy from electron volts to joules before calculating the wavelength in meters. (b) The energy of each photon is converted to joules with the assistance of Eq. (31.2):

$$\text{energy per photon} = 40{,}000\,\mathrm{eV}\left(\frac{1.6 \times 10^{-19}\,\mathrm{J}}{1\,\mathrm{eV}}\right) = 6.4 \times 10^{-15}\,\mathrm{J}.$$

Since the total energy of all x-ray photons absorbed is $1.5 \times 10^{-3}\,\mathrm{J}$, the number of photons absorbed is approximately

$$\frac{1.5 \times 10^{-3}\,\mathrm{J}}{6.4 \times 10^{-15}\,\mathrm{J\ per\ photon}} = \underline{2.3 \times 10^{11}\,\text{photons}}. \qquad \blacksquare$$

31.4 Photon Experiments

The photon theory of EM radiation is supported by a variety of experiments. The first success of the photon theory involved an explanation of the frequency distribution of electromagnetic radiation that comes from hot objects. The hot filament of a light bulb emits most of its energy in the infrared and visible portions of the EM spectrum, as predicted by the photon theory; using the wave theory, however, most of the filament's energy was predicted incorrectly to be emitted in the ultraviolet and x-ray regions. Other successes of the photon theory included the photoelectric effect and the Compton effect.

Photoelectric Effect

The photoelectric effect involves the ejection of electrons from the surface of a metal when it is irradiated with EM radiation. It was first observed experimentally in 1887 by Heinrich Hertz and was explained in terms of the photon theory by Albert Einstein in 1905. In 1921, Einstein was awarded the Nobel Prize for this and other work.

The apparatus for the photoelectric effect is shown in Fig. 31.5. Light of one frequency and wavelength, called monochromatic light, strikes a metal plate. If an electron in the metal absorbs enough energy from the light, the electron can escape the metal's surface. The minimum energy needed for an electron to escape, called the metal's **work function** ϕ, is similar to the energy needed by a water molecule to escape the surface of liquid water (to evaporate). The value of the work function differs for various types of metals.

If an electron escapes the metal, it passes across a vacuum and is captured by a collector (another piece of metal). The electron moving from the plate to the collector constitutes what is called a **photoelectric current.** This current, after reaching the collector, continues through a wire back to the original metal plate and is measured by an ammeter along the way.

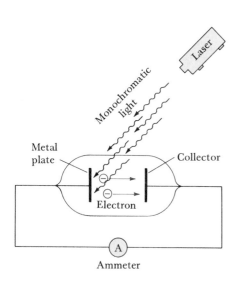

FIG. 31.5. An apparatus used to observe the photoelectric effect.

The wave theory of EM radiation could not explain the photoelectric current. In this theory, the energy of a light wave is spread out over the wavefronts of the light. Calculations have shown that a single electron intercepts only a small fraction of the wave's energy, and considerable time would be needed for an electron to absorb enough energy from the light to escape the metal's surface.

However, when a photoelectric-effect experiment is performed, the current is observed at the instant the light is turned on; there is no delay. This instantaneous current can be explained if one assumes that light consists of photons. An electron gains all the energy (hf) of one photon. Part of the energy is used to escape the metal, and any energy remaining after the electron escapes the surface is available as kinetic energy. In terms of energy, the absorption of a photon and the subsequent escape of an electron from the metal's surface is represented by the equation

$$\underbrace{-hf}_{} + \underbrace{(\underbrace{\phi}_{\text{to escape}} + \underbrace{\tfrac{1}{2}mv^2}_{\text{leftover}})}_{} = 0. \qquad (31.3)$$

$$\overbrace{}^{\substack{\text{Energy lost by}\\\text{deceased photon}}} \qquad \overbrace{\phantom{(\phi + \tfrac{1}{2}mv^2)}}^{\substack{\text{Energy gained}\\\text{by electron}}}$$

Energy lost by deceased photon; Energy gained by electron; to escape metal; leftover kinetic energy

If the light has a frequency such that hf is less than ϕ, the electron will not gain enough energy from an absorbed photon to escape the surface. Thus, no photoelectric current is observed if low-frequency EM radiation illuminates the metal plate, even if the radiation is intense.

The work function of a metal can be determined by increasing the frequency of the EM radiation to a value where the photoelectric current starts. The minimum frequency f_0 of the EM radiation needed to start the photoelectric current is such that

$$-hf_0 + \phi = 0. \qquad (31.4)$$

At this frequency the electron leaves the surface of the metal plate but has no excess energy. For most metals the photoelectric current is initiated by EM radiation in the visible portion of the EM spectrum. But, if the work function of a metal is large, ultraviolet frequency photons may be needed to initiate the current.

EXAMPLE 31.3 Light strikes the surface of a piece of aluminum whose work function ϕ is 4.20 eV. (a) Calculate the minimum-frequency photon that will supply enough energy for an electron to escape the surface of the aluminum. (b) If a 200-nm ultraviolet-wavelength photon is absorbed by an electron in the metal's surface, how much kinetic energy does the electron have when it leaves the metal?

SOLUTION (a) For the electron to gain just enough energy to escape the aluminum surface, the photon's energy hf_0 must equal the energy ϕ needed to remove the electron from the metal, or $hf_0 = \phi$. Thus,

$$f_0 = \frac{\phi}{h} = \frac{(4.20 \text{ eV})(1.6 \times 10^{-19} \text{ J/eV})}{6.63 \times 10^{-34} \text{ J} \cdot \text{s}} = 1.01 \times 10^{15} \text{ s}^{-1} = \underline{1.01 \times 10^{15} \text{ Hz}}.$$

(b) The energy of a 200-nm photon is determined by combining Eqs. (31.1) and (17.8):

$$E = hf = h\frac{c}{\lambda} = (6.63 \times 10^{-34}\,\text{J}\cdot\text{s})\left(\frac{3.0 \times 10^8\,\text{m/s}}{200 \times 10^{-9}\,\text{m}}\right)$$

$$= 9.95 \times 10^{-19}\,\text{J} = 6.22\,\text{eV}.$$

Since an electron needs only 4.20 eV of energy to escape the aluminum, the energy that remains after absorbing a photon is in the form of kinetic energy:

$$\tfrac{1}{2}mv^2 = hf - \phi = 6.22\,\text{eV} - 4.20\,\text{eV} = \underline{2.02\,\text{eV}}. \qquad \blacksquare$$

Several devices used to detect light and ultraviolet radiation rely on the photoelectric effect. Photocells in burglar alarms, automatic door openers, and smoke detectors measure the flow of electric charge from a metal surface that is exposed to the light. If the light striking the surface is interrupted, the current leaving the surface stops.

A **photomultiplier tube** is one of the most sensitive light detectors and is found in many scientific instruments that measure the properties of atoms and molecules as indicated by the wavelengths of light they absorb or emit. The tube consists of a series of metal surfaces called stages or dynodes (Fig. 31.6). A potential difference of about 50 to 100 V is connected across adjacent stages. When a single photon strikes the first stage, a single electron is released, which is accelerated by the voltage toward the second stage. By the time the electron hits the second stage, the electron has gained enough energy to knock several electrons out of its surface. These electrons accelerate across a voltage toward the third stage where each electron frees several more electrons. This may continue through ten or more stages. At the last stage a huge number of electrons are released. For example, if each electron releases three electrons as it hits the next stage, then the number released at the tenth stage due to the absorption of a single photon at the first stage is $3^9 = 19{,}683$. Thus, one photon can cause a fairly large pulse of electric charge that is easily detected.

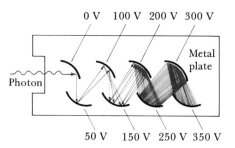

0 V   100 V   200 V   300 V

Metal plate

Photon

50 V   150 V   250 V   350 V

FIG. 31.6. A photomultiplier tube.

The Compton Effect

A. H. Compton (1892–1962), an American physicist from Washington University, provided additional support for the photon theory of EM radiation in 1923. In his experiment Compton irradiated a graphite block with monochromatic x-rays and observed that some of the scattered x-rays had a frequency lower than that of the original x-rays. Just as we would not expect a sound wave to change frequency when scattered from a wall, Compton did not expect x-rays to change frequency when scattered from graphite—that is, if the x-rays were considered to be waves. However, the frequency change could be explained nicely using the photon theory of EM radiation.

To see how the photon theory was used, suppose that an x-ray photon of frequency f_0 collides with a stationary electron in the graphite (Fig. 31.7). During the collision the photon imparts kinetic energy to the electron, causing its speed to increase. The increased kinetic energy of the electron must be balanced by a decrease in the photon's energy from an original value hf_0 to a final value

Before collision

Incident photon

hf_0

Electron

After collision

v

hf Scattered photon

FIG. 31.7. The frequency of a photon changes after it is scattered from an electron.

hf, where f is the frequency of the scattered photon. Thus,

$$\underbrace{(hf_0 - hf)}_{\substack{\text{Energy lost} \\ \text{by photon}}} = \underbrace{\Delta KE_{\text{electron}}}_{\substack{\text{Energy gained} \\ \text{by electron}}}.$$

We see that the photon's frequency must decrease as a result of the collision (f is less than f_0, as observed by Compton).

Since the electron's speed has increased, its momentum has also increased. Recall from Chapter 5 that momentum is a conserved quantity; thus, a momentum increase by the electron must be balanced by a momentum decrease of the photon. But the photon must have momentum in order to lose it.

Compton deduced an expression for a photon's momentum by equating the expression for its energy ($E = hf$) to the relativistic expression for the energy of a particle of relativistic mass m ($E = mc^2$). Thus, $mc^2 = hf$, or $mc = hf/c$. The quantity mc represents the momentum of a particle of mass m moving at speed c. The photon moves at speed c and, because of the equivalence of mass and energy, is considered to have mass $m = E/c^2 = hf/c^2$. Thus, Compton deduced that the momentum of the photon must be

$$\text{momentum of a photon} = p = mc = \left(\frac{hf}{c^2}\right)c = \frac{hf}{c} = \frac{h}{\lambda}, \qquad (31.5)$$

where $\lambda = c/f$ [Eq. (17.8)] is the photon's wavelength. Compton combined equations for the conservation of momentum and the conservation of energy during the photon-electron collision, and he was able to predict exactly the change in wavelength of x-ray photons that were scattered at different angles by the electrons in the graphite. The experiment and its interpretation provided another important support for the photon theory of electromagnetic radiation.

31.5 The Wave-Particle Nature of EM Radiation

While many experiments imply that EM radiation is made of photons and exhibits a particlelike behavior, other experiments, such as those discussed in Chapter 21, can be explained only by the use of a wave theory for EM radiation. As a result, we have two theories for the nature of EM radiation, a wave theory and a photon theory, and each theory is supported by its own experimental evidence. What, then, is EM radiation—a wave or a photon?

Our present conception of EM radiation draws from both theories. The dual nature of EM radiation is illustrated by Young's double-slit interference experiment discussed in Section 21.2 and depicted in Fig. 31.8a. Coherent light of very low intensity irradiates a pair of slits. After passing through the slits, the light moves toward a photographic plate.

If the plate is exposed to the light for a very short time and if the light intensity is low, the photographic plate becomes exposed as in Fig. 31.8b. Each little spot represents the place where a photon has hit the plate and caused it to become exposed. If the light is allowed to expose the plate for a longer time, a pattern of fringes begins to emerge as in Fig. 31.8c and d. In Chapter 21, we explained these fringes as being caused by the interference of light waves coming

(a)

(b)

(c)

(d)

FIG. 31.8. (a) An apparatus for observing the double-slit interference pattern of light. (b), (c), and (d) As more photons hit the photographic plate, a pattern of interference fringes appears.

from the two slits. If the light is intense, the fringe pattern is produced rapidly and the fact that individual photons are causing the pattern is not apparent. However, by reducing the light intensity, we see that photons and not waves are hitting the plate (as in Fig. 31.8b). The photons hit the plate at places where we would expect to see constructive interference of waves.

Based on this and many other experiments, scientists now believe that EM radiation has both a wave and a particle (that is, photon) behavior—a **wave-particle duality.** Whenever radiation is detected by an experimental apparatus or interacts with matter, it behaves like a photon. For example, when light strikes the plate of a photomultiplier tube or a photographic plate, it behaves like photons. The wavelike behavior of EM radiation determines the probability of detecting photons at different places. When light passes through two slits and then strikes a photographic plate, we must use a wavelike description of the light to decide which are the probable places on the plate for the photons to strike. For the double-slit experiment, photons hit those places on the plate where waves from the slits undergo constructive interference. The dual wave-particle behavior is sometimes summarized by saying that EM radiation behaves like photons when interacting with matter and behaves like a wave when propagating through space from a source to a place where it is detected (the wave determines probable places for detecting the photon).

This dual wave-particle behavior of EM radiation differs from the behavior of objects we are used to seeing with our eyes. A baseball, when moving from a bat to the glove of an outfielder, follows a well-defined path and exhibits no wavelike behavior. On the other hand, an ocean wave acts like a wave as it moves toward and collides with a beach; the wave does not strike the beach as discrete bundles of energy at particular locations. The fact that EM radiation exhibits both wave and particle behavior is one of the interesting surprises that physicists have discovered during this century. Perhaps even more surprising has been the discovery that objects with mass also exhibit a dual wave-particle behavior.

31.6 The Wave-Particle Nature of Matter

In 1923 a 31-year-old French graduate student named Louis de Broglie (1892–) made the unusual proposal that if EM radiation had a dual wave-particle behavior, then objects with mass such as electrons and protons might also exhibit this dual behavior. The idea that an electron might exhibit wave behavior seemed strange. Waves are characterized by their wavelength, frequency, speed, amplitude, and tendency to spread as they move. Waves also undergo interference. None of these properties resemble our usual conception of a particle such as an electron, which we think of as a small, well-defined ball of mass.

Nevertheless, de Broglie thought that particles with mass might also exhibit a wavelike behavior. He even deduced an expression for the wavelength λ that should be associated with the motion of an object of mass m when moving at speed v. First he equated the equations for the momentum of a photon ($p = hf/c = h/\lambda$) and the momentum of a particle of mass m when moving at speed v ($p = mv$) to obtain the equation $mv = h/\lambda$. By rearranging, he obtained an expression for the wavelength (called the **de Broglie wavelength**) associated with the motion of the moving mass:

$$\lambda = \frac{h}{mv}. \tag{31.6}$$

Three years after de Broglie's proposal, two experiments were performed that showed clearly the wave behavior of electrons. Since that time, many other experiments have been performed that show that all tiny masses—electrons, protons, neutrons, atoms, and ions—behave somewhat like waves and that the wavelength associated with their motion is given by Eq. (31.6). Instead of reviewing these experiments, we will again use a modified version of Young's double-slit experiment to illustrate the wave nature of mass particles.

The experiment is illustrated in Fig. 31.9a. A beam of electrons moves at speed v toward a pair of slits. After passing through the slits, the electrons fly toward a fluorescent screen. Each time an electron hits the screen, a flash of light occurs.

If the electrons behave only like particles, we expect to see electrons hitting the screen in two bands (Fig. 31.9b) that lie along a line from the electron source through the slits and to the screen. The pattern that is actually observed is nothing like this. After only a few electrons have hit the screen, the distribution appears quite random, as shown in Fig. 31.9c. As more and more electrons reach the screen, a pattern consisting of alternating bright and dark fringes emerges (Fig. 31.9d and e). This pattern is exactly like the pattern produced when light passes through a pair of slits.

We see that electrons, when they hit the screen, appear as discrete particles; but the places where the electrons hit the screen are places where waves from the two slits undergo constructive interference. The angular deflection to these bright bands where many electrons hit the screen agrees perfectly with calculations using de Broglie's equation for the wavelength of an electron ($\lambda = h/mv$) and the double-slit interference equation developed in Section 21.2 [Eq. (21.1)].

The experiment just described is difficult to perform because the wavelengths of electrons are very short, even at low speed, and the slits must be very close to each other in order to produce a fringe pattern that can be seen. However, the bands are easily observed if electron beams pass through crystals. The space between adjacent rows of atoms serve as closely spaced slits. The actual experiments that confirmed the wavelike nature of electrons used such solid materials as the "slits" to cause interference patterns of the electron waves. These experiments were first performed in 1927 by C. J. Davisson and L. H. Germer of Bell Telephone Laboratories and independently by G. P. Thomson, an English physicist whose father had discovered the electron in 1897.

By 1930 the wave nature of matter had become the heart of a completely new theoretical description of the behavior of atomic-size masses. The new theory, called wave mechanics or quantum mechanics, has been fine-tuned for the last fifty years to describe a wide variety of properties of atoms, molecules, and the atomic properties of liquids and solids. In this theory the wave that is calculated for a particle represents the probability of detecting the particle at different locations. For the electrons passing through the double slit, the wave determines the probability of observing an electron at different places on the fluorescent screen. The electrons hit the screen only at those places where a wave of wavelength $\lambda = h/mv$ undergoes constructive interference after passing through the slits.

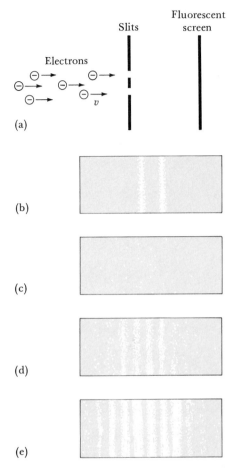

FIG. 31.9. (a) An apparatus for observing the double-slit interference pattern of electrons. (b), (c), and (d) As more electrons hit the fluorescent screen, a pattern of interference fringes appears.

We see that objects with mass exhibit a wave-particle duality just as EM radiation does. The objects behave like particles when interacting with a measuring device or with matter, and they behave like waves when moving from a source to the place where they are detected. The wave determines their probability of being observed at different locations.

EXAMPLE 31.4 Calculate the wavelength of the wave associated with an electron of mass 9.11×10^{-31} kg that has been accelerated from rest across a 3000-V potential difference. Ignore any relativistic effects.

SOLUTION In order to calculate λ, we first need to know the electron's speed. The speed is determined from the work-energy equation:

$$0 = \Delta KE + \Delta PE_q = \tfrac{1}{2}mv^2 + (-e)V.$$

Thus,

$$v = \sqrt{\frac{2eV}{m}} = \sqrt{\frac{2(1.6 \times 10^{-19}\,\text{C})(3000\,\text{V})}{(9.11 \times 10^{-31}\,\text{kg})}} = 3.2 \times 10^7\,\text{m/s}.$$

The electron's wavelength is given by Eq. (31.6):

$$\lambda = \frac{h}{mv} = \frac{6.63 \times 10^{-34}\,\text{J}\cdot\text{s}}{(9.11 \times 10^{-31}\,\text{kg})(3.2 \times 10^7\,\text{m/s})} = 2.2 \times 10^{-11}\,\text{m} = \underline{0.022\,\text{nm}}.$$

This wavelength is about 0.5×10^{-4} times shorter than the wavelength of light. The ability of electron microscopes to view small objects that cannot be seen with light depends on this short wavelength, as we learn in Section 31.8. ■

You might be wondering why the wave nature of matter is unobserved for large objects such as people and cars. Consider the next example.

EXAMPLE 31.5 Calculate the de Broglie wavelength of a 1-kg ball rolling at a speed of 1 m/s.

SOLUTION

$$\lambda = \frac{h}{mv} = \frac{6.63 \times 10^{-34}\,\text{J}\cdot\text{s}}{(1\,\text{kg})(1\,\text{m/s})} = \underline{6.6 \times 10^{-34}\,\text{m}}.$$

The ball's wavelength is about 10^{-20} times smaller than the size of the nucleus of an atom. The crests and troughs of the wave are so close to each other that they cannot possibly be used to indicate the probability of finding the bowling ball at one place compared to another. The wave nature of an object is observed only for very small masses about the size of atoms and the particles of which they are made. ■

31.7 The Uncertainty Principle

We have seen that both EM radiation and objects with mass exhibit a dual wave-particle behavior. This behavior has led to interesting discussions about the limits of our ability to accurately measure the properties of small objects.

The discussion revolves around the idea that a measuring device must affect the quantity being measured.

Suppose, for example, that you wish to measure the temperature of a cup of tea. If a thermometer is placed in the tea, the tea cools a little as the thermometer warms. The final temperature of the thermometer and tea is less than the tea's initial temperature. The measuring device has altered the quantity being measured.

We can usually correct for errors in measurement if we know the characteristics of our measuring device (such as the temperature of the thermometer before entering the tea, and its mass and specific heat capacity). There is, however, an intrinsic limit in our ability to make accurate measurements. This limit depends on the dual wave-particle nature of matter.

To see how this limit occurs, consider an experiment in which the position of a small object is determined by scattering electromagnetic radiation from the object. After passing the object, the radiation produces a diffraction pattern on a screen behind the object and indicates its position. A detailed analysis such as performed in the interlude on page 434 indicates that the object's uncertainty in position Δx equals roughly the wavelength λ of the radiation used to determine its location:

$$\text{uncertainty in position} = \Delta x \sim \lambda.$$

The shorter the wavelength, the better we can determine the object's position.

However, short-wavelength electromagnetic radiation has a disadvantage. Remember that this radiation has a particlelike behavior as well as a wavelike behavior. The electromagnetic "particles" called photons have momentum $p = h/\lambda$ and can transfer momentum to an object that they hit. If a photon has a direct hit, the object moves backward; if the photon has a glancing blow, the object is bumped to the side. The object's sideways momentum after the collision can vary from about zero to $\pm h/\lambda$, where λ is the wavelength of the scattered photon. The uncertainty in the object's momentum after the photon is scattered is

$$\text{uncertainty in momentum} = \Delta p \sim \frac{h}{\lambda}.$$

To disturb the object's momentum only slightly, long-wavelength radiation should be used.

We are now in a predicament. To locate the object's position most accurately, short-wavelength radiation should be used; but this radiation when scattered from the object creates considerable uncertainty in its momentum. If we use long-wavelength radiation, the object's momentum is only slightly disturbed, but its position is more uncertain. We can express the predicament quantitatively by taking the product of the object's uncertainties in position Δx and momentum Δp:

$$\Delta x \, \Delta p \sim (\lambda) \frac{h}{\lambda} = h.$$

The product of these uncertainties equals Planck's constant h and is independent of the type of radiation used for the measurement. This means that no matter what we do, we cannot simultaneously determine an object's exact position and the exact value of its momentum.

A careful derivation of this uncertainty principle, as done by the German physicist Werner Heisenberg (1901–1976) in 1927, shows that the constant h should be replaced by $h/2\pi$. The Heisenberg uncertainty principle is stated, then, as follows:

Uncertainty principle: It is impossible to know simultaneously with certainty the position and momentum of a particle. At best, the product of the uncertainties of these two quantities must exceed $h/2\pi$, where $h = 6.63 \times 10^{-34}$ J · s. Thus,

$$\Delta x\,\Delta p \geq \frac{h}{2\pi}. \qquad (31.7)$$

A similar relationship applies to our ability to know the value of a particle's energy and the length of time during which the energy is measured. Suppose that a photon of wavelength λ is used to locate a small object. As we said earlier, the object's position is uncertain by an amount $\Delta x \sim \lambda$. The photon travels at speed c; hence, it takes a time

$$\Delta t = \frac{\Delta x}{c} \sim \frac{\lambda}{c}$$

for the photon to pass through a distance the length of the object's position uncertainty. During that time, the photon can transfer all or none of its energy to the object, depending on the way in which it scatters from the object. The object's energy is uncertain by an amount

$$\Delta E \sim hf = h\,\frac{c}{\lambda}.$$

The product of these two uncertainties is

$$\Delta E\,\Delta t \sim h.$$

Heisenberg's more careful derivation of this uncertainty equation showed that the constant h should be replaced by $h/2\pi$.

A measurement on an object requiring a time Δt produces an uncertainty in its energy ΔE that can be no less than is consistent with the equation

$$\Delta E\,\Delta t \geq \frac{h}{2\pi}. \qquad (31.8)$$

EXAMPLE 31.6 An electron in an atom is bumped into an orbit farther from the nucleus by a collision with another atom. The electron moves around this "excited" orbit for 10^{-8} s and then falls back to the orbit in which it moved before the collision. Calculate the minimum uncertainty in the energy of the excited orbit.

SOLUTION Since the electron resided in this orbit only 10^{-8} s, this is the maximum time available to measure the energy of an electron when in the orbit. Using the best equipment we could ever devise, the uncertainty in the measurement of the electron's energy would still be equal to or greater than that given by Eq. (31.8):

$$\Delta E \geq \frac{h}{2\pi \, \Delta t} = \frac{6.63 \times 10^{-34} \, \mathrm{J \cdot s}}{2\pi (10^{-8} \, \mathrm{s})} = \underline{1.1 \times 10^{-26} \, \mathrm{J}}.$$ ∎

EXAMPLE 31.7 A free electron in a copper wire moves randomly with a speed of 1000 m/s. Suppose that we can devise an experiment to measure the electron's velocity with a 10 percent uncertainty. Calculate the uncertainty in the electron's position. The mass of an electron is 9.11×10^{-31} kg.

SOLUTION The electron's speed is uncertain by 10 percent of 1000 m/s, or $\Delta v = 100$ m/s. The electron's position in the wire is determined using Eq. (31.7) where we substitute $m \, \Delta v$ for the momentum uncertainty:

$$\Delta x \geq \frac{h}{2\pi m \, \Delta v} = \frac{6.63 \times 10^{-34} \, \mathrm{J \cdot s}}{2\pi (9.11 \times 10^{-31} \, \mathrm{kg})(100 \, \mathrm{m/s})} = \underline{1.2 \times 10^{-6} \, \mathrm{m}}.$$

This position uncertainty is greater than the length of 1000 atoms lined in a row. It is impossible to pinpoint the location of an electron in a wire. In modern physics the free electron is described by a wave that extends over long distances in the wire and whose amplitude at each location represents the probability of finding the electron at that location.   ∎

31.8 Electron Microscopes

One of the interesting and useful technological applications of the wavelike behavior of matter is the **electron microscope,** a device that views small objects, such as human body cells, by scattering electrons off them. Electrons have considerable advantage over light waves for viewing small objects because of the short wavelength associated with their motion. If the wavelength is shorter, then details of smaller structures can be seen. When we use the best microscope that transmits light, two objects must be farther apart than about 100 nm in order that their images be distinct. When we use an electron microscope, however, two objects separated by 0.1 nm form distinct images. The superior resolution of the electron microscope is due to the fact that the scattered electrons have a wavelength ($\lambda = h/mv$) of about 0.005 nm rather than the 500-nm wavelength of light. The resolution of the electron microscope is limited not by the electrons' wavelength but by other factors in the construction of the microscope and in the preparation of the sample being viewed.

A schematic representation of a transmission electron microscope is shown in Fig. 31.10. Electrons are emitted from a hot filament and focused in a parallel beam by a condensing lens that consists of a coil of wire through which a current flows. The current creates a magnetic field that bends and focuses the electron beam, much as glass lenses bend and focus light. After leaving the condensing lens, the electron beam passes through a thin sample (50 nm or less) that is being viewed. The electron beam scattered off the sample is then focused and magnified by two more magnetic lenses—the objective lens and the projection lens. A final image of the sample is formed on a fluorescent screen or on a photographic plate.

FIG. 31.10. A transmission electron microscope.

FIG. 31.11. A scanning electron microscope.

The focal length of a magnetic lens depends on the speed of the electrons. In order to have a uniform focal length for all electrons, the electrons must all move at about the same speed, which is why the sample must be so thin. If it were thicker, some electrons would be slowed more than others while passing through the sample, and a blurry image would be formed. Thus, sample preparation is a problem when using transmission electron microscopes. In addition, the depth of field is very small in this type of microscope, and only a flat, two-dimensional object can be viewed.

A scanning electron microscope presents neither of these problems. In this microscope a focused electron beam scans across the specimen along a series of adjacent lines (Fig. 31.11). As the electron beam strikes each point on the specimen, electrons are knocked out of the specimen. These electrons are then collected by a detector. The number of electrons released from each point on the specimen determines the brightness of a spot on the screen of a cathode ray tube (CRT), such as found in a television set. As the microscope electron beam scans across the sample, the spot moves across the CRT. Each point on the specimen corresponds to a point on the screen of the CRT.

If the microscope electron beam strikes the side of an upward projection of the specimen, more electrons are released from the sample and a brighter spot is produced on the CRT screen. In this way the orientation of different parts of the specimen can be determined. If two pictures are taken at different angles, a stereoscopic view of the specimen is seen (Fig. 31.12).

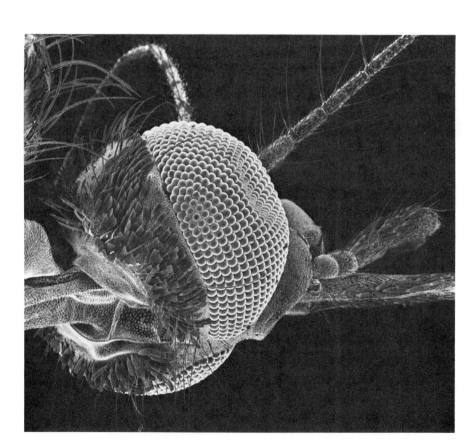

FIG. 31.12. The underside of the head of a female mosquito (*Aedes*) as seen with a scanning electron microscope. Shown are the neck (left), eyes, and proboscis (right), magnified 200 times. (Copyright © David Scharf/Peter Arnold, Inc. All rights reserved.)

Although the resolution of a scanning electron microscope is only about 10 nm, as opposed to 0.1 nm for the transmission electron microscope, the decreased resolution is balanced by increased depth of field and ease of sample preparation.

Summary

1. **Wave-particle nature of electromagnetic radiation:** Electromagnetic radiation behaves like a wave when propagating through space and like a particle called a photon when it interacts with matter. An **electromagnetic wave** is an alternating electric and magnetic field that moves through space at a speed of 3.0×10^8 m/s. The various forms of electromagnetic waves, such as radio waves, microwaves, infrared waves, light, ultraviolet radiation, x-rays, and gamma rays, differ from each other in frequency and wavelength. The frequency f, wavelength λ, and speed c of the waves are related by the equation $f\lambda = c$.

A **photon** is a bundle of electromagnetic energy with zero rest mass. A photon's energy depends on its frequency as given by the equation

$$E = hf, \tag{31.1}$$

where h, Planck's constant, equals 6.63×10^{-34} J · s. A photon's momentum is given by the equation

$$p = \frac{h}{\lambda}. \tag{31.5}$$

2. **Wave-particle nature of mass:** Particles with rest mass also exhibit a dual wave-particle behavior. The energy of a particle with mass m is

$$E = mc^2 \tag{30.5}$$

and its momentum is

$$p = mv. \tag{5.1}$$

The particle exhibits a wavelike behavior when propagating through space. The **de Broglie wavelength** associated with the particle's motion when moving at speed v is

$$\lambda = \frac{h}{p} = \frac{h}{mv}. \tag{31.6}$$

The wave is used to determine the probability of observing the particle at different locations.

3. **Uncertainty principle:** It is impossible to know simultaneously with certainty the position and momentum of a particle. At best, the product of the uncertainties of these two quantities must exceed $h/2\pi$, where $h = 6.63 \times 10^{-34}$ J · s. Thus,

$$\Delta x \, \Delta p \geq h/2\pi. \tag{31.7}$$

Similarly, a measurement of an object's energy during a time Δt must be uncertain by an amount ΔE consistent with the following equation:

$$\Delta E \, \Delta t \geq h/2\pi. \tag{31.8}$$

Questions

1. Why are x-ray photons more hazardous than microwave photons?

2. Why do you think that ultraviolet radiation causes a suntan, whereas infrared radiation does not?

3. Why is visible radiation able to remove electrons from sodium but not from zinc?

4. When film is being developed, a special red light is turned on in the developing room, whereas the regular light is not. Explain.

5. Why is the photon theory supported by the fact that light below a certain frequency, when striking a metal plate, does not cause electrons to be freed from the metal?

6. Light from a relatively powerful laser has lifted and supported the weight of glass spheres that are 20×10^{-6} m in diameter. How is this possible?

7. Is the radiation pressure of light greater on a mirror or on a black surface of the same size? Explain.

8. Suppose that Planck's constant were 100 J · s rather than 6.63×10^{-34} J · s. Describe several ways in which the world would appear different. Be specific.

9. Why is the wavelike behavior of matter not observed in daily life?

10. A proton and an electron have the same speeds. Calculate the ratio of their de Broglie wavelengths.

11. Students are seldom motionless during a physics lecture. (a) *Estimate* the uncertainty in the speed due to wiggling. The number may be small but is not zero. (b) Based on the answer to (a), estimate the person's uncertainty in momentum. (c) Now, suppose that the world were changed, with Planck's constant much larger. Calculate the value of Planck's constant if a person's uncertainty in position while sitting in class was 10 m, given the momentum uncertainty estimated above. With this large position uncertainty, the person might be found almost anywhere in a typical lecture room.

12. Your boyfriend or girlfriend walks through a doorway. *Estimate* the least uncertainty with which you think his or her momentum could be measured. Suppose that the world had been created with Planck's constant equal to 1.0×10^{-2} J · s. Would your friend's shape be discernible? Explain.

Problems

You may need to know the following masses in order to work some of these problems: $m_{electron} = 9.11 \times 10^{-31}$ kg, $m_{proton} = 1.673 \times 10^{-27}$ kg, and $m_{neutron} = 1.675 \times 10^{-27}$ kg.

31.1 and 31.2 Electromagnetic Waves and Spectrum

1. (a) X-rays used in a medical examination have a wavelength of about 2.5×10^{-11} m. Calculate the frequency of this radiation. (b) A microwave oven produces microwaves of frequency 2.45 GHz. Calculate the wavelength of the microwaves.

2. You change your AM radio from a station broadcasting at frequency 54 kHz to one broadcasting at 140 kHz. Calculate the ratio of the wavelength of the latter station to that of the former.

31.3 Photons

3. (a) Calculate the energy in electron volts of a 300-nm ultraviolet photon partly responsible for the production of a suntan. (b) A vibrating carbon monoxide molecule (CO) produces infrared photons of energy 0.37 eV. Calculate the frequency of CO vibration, which is the same as the frequency of the infrared radiation the molecule emits.

4. Suppose the bond in a biologically important molecule is broken by photons of energy 5.0 eV. Calculate the frequency and wavelength of these photons and the region of the electromagnetic spectrum in which they are located.

5. A 1-eV photon's wavelength is 1240 nm. Using a ratio technique, calculate the wavelength of a 5-eV photon.

6. A laser pulse produces 2.0×10^{18} photons of wavelength 694 nm. Calculate the energy in one pulse.

7. On a bright day with the sun directly overhead, the intensity of light striking the earth is about 1400 watt/m^2. Calculate the number of photons striking a 1.0-m^2 surface each second, assuming the average photon wavelength is 700 nm.

8. Roughly 5 percent of the power of a 100-watt light bulb is emitted as light, the rest being emitted as heat and longer-wavelength radiation. Calculate the number of photons of light coming from a bulb each second. Assume that the wavelength of an average photon appears in the middle of the visible part of the spectrum.

9. For an object to be seen with the unaided eye, the light intensity coming to the eye must be about 5×10^{-12} J/$m^2 \cdot$s or greater. Calculate the minimum number of photons that must enter the eye's pupil each second in order for the object to be seen. Assume that the pupil's radius is 0.20 cm and the wavelength of the light is 550 nm.

10. Is the ear better at detecting sound power than the eye is at detecting light power, or vice versa? Explain. (*Note:* Refer to Problem 9 concerning the eye and to Section 18.5 concerning the ear's threshold of audibility.)

11. A small (1.0×10^{-5})-g piece of dust falls in the earth's gravitational field. Calculate the distance it must fall so that its change in gravitational potential energy equals the energy of a 0.10-nm x-ray photon.

31.4 Photon Experiments

12. The work function of cesium is 1.9 eV. (a) Calculate the lowest-frequency photon that can eject an electron from cesium. (b) Calculate the maximum possible kinetic energy in electron volts of a photoelectron that absorbs a 400-nm photon.

13. The work function of calcium is 2.7 eV. Determine if an electron is ejected from calcium by absorbing a photon of wavelength (a) 400 nm, (b) 500 nm, and (c) 600 nm.

14. Visible light shines on the metal surface of a photocell having a work function of 1.30 eV. The maximum kinetic energy of the electrons leaving the surface is 0.92 eV. Calculate the light's wavelength.

15. Light of wavelength 430 nm strikes a metal surface, releasing electrons with kinetic energy equal to 0.58 eV or less. Calculate the metal's work function.

16. Suppose that light of intensity 1.0×10^{-2} watt/m^2 is made of waves rather than photons and that the waves strike a sodium surface with a work function of 2.2 eV. (a) Determine the power in watts incident on the area of a single sodium atom at the metal's surface (the radius of a sodium atom is approximately 1.7×10^{-10} m). (b) How long will it take for an electron in the sodium to accumulate enough energy to escape the surface, assuming it collects all of the light incident on the atom?

17. A film becomes exposed when light striking it initiates a complex chemical reaction. A particular type of film does not become exposed if struck by light of wavelength longer than 670 nm. Calculate the minimum energy in electron volts needed to initiate the chemical reaction.

18. During photosynthesis, eight photons of 670-nm wavelength can cause the following reaction: $CO_2 + H_2O \longrightarrow \frac{1}{6}C_6H_{12}O_6 + O_2$. During respiration, when the human body metabolizes sugar, the reverse reaction releases 4.9 eV of energy per CO_2 molecule. Calculate the ratio of the energy released (respiration) and the energy absorbed (photosynthesis)—a measure of the photosynthetic efficiency.

19. A photomultiplier tube has twelve stages, each of which emits six electrons when struck by an electron released from a previous stage. Calculate the number of electrons emitted after the last stage. Notice that one electron is emitted from the first stage after absorption of a photon, six from the second, and so on.

20. In a Compton-effect scattering experiment, an incident photon's frequency is 2.0×10^{20} Hz; the scattered photon's frequency is 1.4×10^{20} Hz. Calculate the kinetic energy increase of the electron, in units of electron volts, when the photon is scattered.

21. An electron hit by an x-ray photon of energy 5.0×10^4 eV gains 3.0×10^3 eV of energy. Calculate the wavelength of the scattered photon leaving the site of the collision.

22. A laser produces a short pulse of light whose energy equals 0.20 J. The wavelength of the light is 694 nm. (a) How many photons are produced? (b) Calculate the total momentum of the emitted light pulse.

23. A microwave transmitter produces short pulses of radiation, each with 10 J of energy. The radiation consists of photons with

10.0-cm wavelength. (a) How many photons are emitted per pulse? (b) Calculate the total momentum of each pulse.

31.6 The Wave-Particle Nature of Matter

24. Calculate the de Broglie wavelengths of an electron and a proton when moving at speed 1.0×10^3 m/s.

25. The United States Golf Association requires that a golf ball have a mass of 45.6 g and reach a speed of 76 m/s when tested on one of their machines. Calculate the de Broglie wavelength of this ball. How does the wavelength compare with the size of a proton (approximately 10^{-15} m in diameter)?

26. An electron and a proton have equal kinetic energy (nonrelativistic). Calculate the ratio of the electron's and the proton's de Broglie wavelengths.

27. An electron starting at rest crosses a 22-kV potential difference in a television set. Calculate its de Broglie wavelength. (Ignore relativistic effects.)

28. In order to study the wave nature of electrons, some physics students propose to accelerate a group of electrons through a voltage so that the de Broglie wavelength of the electrons will equal that of light (about 600 nm). Interference of the electron waves can then be observed easily by passing the electrons through double slits used to study light interference. (a) Calculate the speed of electrons having this wavelength. (b) Calculate the voltage through which the electrons should be accelerated. (c) Does the experiment seem feasible? Explain.

29. (a) Estimate your de Broglie wavelength as you walk between classes. (b) Suppose that the world were restructured so that Planck's constant h could be much larger than its present value. How large would it have to be in order for your de Broglie wavelength, when you move between classes, to be 1 m? If this happened, students entering the door of a classroom would be diffracted into a pattern, such as shown in Fig. 31.13.

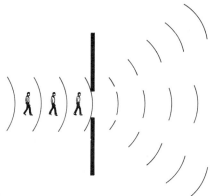

FIG. 31.13

30. Calculate the de Broglie wavelength of a thermal neutron—that is, one whose kinetic energy equals $\frac{3}{2}kT$, where k, Boltzmann's constant, is 1.38×10^{-23} J/K, and T is the temperature in degrees Kelvin of the neutron's environment (room temperature, or about 20°C, for this problem).

31. Suppose that Planck's constant were 2.0×10^4 J · s. A person running on a track would be guided by his or her de Broglie wave. The most probable places for the person on the track would depend on the amplitude of his or her de Broglie wave at these places. In order for the wave to "fit" around the track's circumference, an integral number of wavelengths must equal the length of the circumference. (a) Calculate the speed that a 50-kg woman must run on a 400-m circular track so that one of her waves will fit around the track. (b) Calculate the speed for two of her waves. A similar phenomenon occurs to electrons in atoms.

32. Show that the de Broglie wavelength of an electron accelerated to nonrelativistic speed through a potential difference V equals $1.23 \times 10^{-9}/\sqrt{V}$ in units of meters.

31.7 The Uncertainty Principle

33. A proton travels at speed (4.66 ± 0.08) m/s. Calculate the least uncertainty with which its position can be known.

34. Suppose the speeds of an electron and of a 45.9-g golf ball are each measured with an uncertainty of $\Delta v = 1.0 \times 10^{-3}$ m/s. Calculate the minimum uncertainties in their positions.

35. The uncertainty in the momentum of an electron and of a 5.0-g mass are both $\Delta p = 10^{-3}p$, where p is the object's momentum. Calculate the minimum uncertainty in the position of the electron if moving at speed 1.0×10^7 m/s and of the 5.0-g mass if moving at 0.60 m/s.

36. A 60-kg runner's instantaneous speed can be determined with an uncertainty of 0.01 m/s. Calculate the minimum uncertainty of the runner's position as limited by the uncertainty principle. Does the uncertainty principle limit the accuracy of times recorded for track races?

37. A light microscope can determine a particle's position with an uncertainty of about 250 nm. (a) Calculate the uncertainty in momentum of a (2.0×10^{-13})-kg red blood cell viewed by the microscope. (b) Calculate the cell's uncertainty in speed.

38. At one time, scientists thought that an atom's nucleus was made of protons and electrons. However, the electron cannot exist in the nucleus, as is apparent using the uncertainty principle. (a) Calculate the uncertainty in an electron's momentum if restricted to a region the size of a nucleus (10^{-14} m). (b) Assuming that the electron's momentum must be at least as large as its uncertainty, calculate its energy, using the relativistic equation $E = \sqrt{p^2c^2 + m_0^2c^4}$ (convert your answer to units of electron volts). An electron with such great energy could not reside in the nucleus.

31.8 Electron Microscopes

39. Calculate the wavelength of electrons accelerated across a 40,000-V potential difference. Compare their wavelength to the wavelength of light.

40. The smallest distance that can be resolved by any microscope under optimum conditions, called its resolving power, is approximately equal to the wavelength used in the microscope. A typical electron microscope uses electrons accelerated through a 50,000-V potential difference. Calculate the microscope's theoretical resolving power.

41. In order for an atom to be "seen" with an electron microscope, the wavelength of the electrons should be no bigger than the size of the atom—roughly 0.1 nm. Through what potential difference must the electrons be accelerated to have this wavelength?

General Problems

G1. Calculate the number of 650-nm photons whose energy equals the rest energy of an electron.

G2. A beam of light of wavelength 560 nm is reflected perpendicularly from a mirror. If 10^{20} photons hit the mirror each second, calculate the force of the light exerted on the mirror. (*Hint:* Refer to the impulse-momentum equation from Chapter 5. You may assume that the magnitude of the photons' momenta are unchanged by the collision but their directions are reversed.)

G3. Electrons starting at rest accelerate across a 30-V potential difference. The electrons then pass through a pair of slits and strike a screen. (a) Calculate the slit separation that causes the $n = 1$ interference maximum to have an angular deflection of 10°. (b) Using the same slits, calculate the angular deflection if the electrons accelerate across a 60-V potential difference.

G4. An electron that resides by itself in an open region of space is struck by a photon of light. Using nonrelativistic formulas, show that the electron cannot absorb the photon's energy and simultaneously absorb its momentum. In order to conserve both energy and momentum, the photon must be absorbed by the electron near another mass, which carries away some of the momentum but little of the energy.

G5. We wish to calculate the net force on the earth due to absorption of light from the sun. (a) Calculate the net area of the earth exposed to sunlight (the earth's radius is 6.38×10^6 m). (b) The solar radiation intensity is 1400 J/m$^2 \cdot$s. Calculate the momentum of photons hitting the earth's surface per second. (c) Use the impulse-momentum equation to calculate the average force of this radiation on the earth ($F =$ change in momentum per unit time).

The Bohr Model of the Atom

32

At the beginning of this century, scientists knew that matter was made of basic building blocks called atoms. The blocks were extremely small (for example, 6×10^{23} helium atoms had a mass of only 4 grams, and the same number of carbon atoms had a mass of 12 grams). The masses of the various types of atoms differed, as did other of their properties. For example, carbon atoms combined with a variety of atoms to form molecules (CO, CO_2, CH_4, . . .) or solids (graphite and diamond), while helium atoms were inert—they would not combine easily with other atoms. Sodium atoms would combine with chlorine atoms to make a crystalline solid but would not combine with carbon atoms. There was also variation in the wavelengths of light and of other forms of EM radiation emitted and absorbed by different types of atoms. While all atoms of a particular type had identical properties, atoms of different types differed greatly from each other.

Near the beginning of the twentieth century, scientists began to search for a structure of the atom that would account for this variation from one atom to the next. The first atomic theory that could be used to calculate the wavelengths of light emitted and absorbed by an atom was presented by Niels Bohr in 1913 and became known as the Bohr model of the atom. Bohr's model, which applied only to atoms or ions with one electron, was shown after its development to draw on both the particle and wave behavior of electrons. In 1925 a comprehensive atomic theory, called quantum mechanics, was devised. In this theory the dual wave-particle behavior of matter was brought together in one consistent theory.

In this chapter we will examine the method by which Bohr developed his model and the properties of atoms that can be determined using the model. In the next chapter we will examine the quantum mechanical theory of the atom.

32.1 The Nuclear Model of the Atom

Near the end of the nineteenth century, it was well known that matter was made of atoms. Little was known, however, of the structure of the atom. One of the first atomic models, called the "plum-pudding" model (Fig. 32.1), was presented by J. J. Thomson (1856–1940) soon after he discovered the electron in 1897. He proposed that negatively charged electrons (the "plums") were imbedded in a sphere of uniform positive charge (the "pudding").

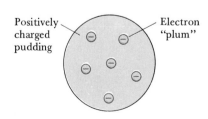

FIG. 32.1. The "plum-pudding" model of the atom.

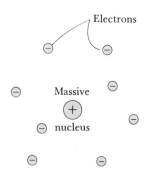

FIG. 32.2. Rutherford's nuclear atom consists of a massive, positively charged nucleus surrounded by electrons with little mass. (Not to scale.)

In 1911 Ernest Rutherford (1871–1937) and his colleagues in England performed an interesting experiment that led to the development of a different atomic model. Rutherford shot a beam of **alpha (α) particles** (the nuclei of helium atoms—helium atoms with their electrons removed) at a thin foil of gold atoms. Most α particles passed straight through the gold foil. This result was consistent with the plum-pudding model, since there were no concentrations of positive charge and mass in the pudding to reflect the α particles. The mass of the negatively charged electrons was about $\frac{1}{8000}$ the mass of the α particles and was too small to deflect an α particle backward.

Rutherford and his colleagues also found that while most alpha particles passed through the foil undeflected, a few were scattered back toward their source as if they had hit a mass bigger than their own. This result was inconsistent with the plum-pudding model because there was nothing in the "pudding" to cause this back-scattering. This observation led Rutherford to propose the **nuclear model** of the atom. In this model the atom consisted of a tiny, massive core called the nucleus, which contained all of the atom's positive charge and most of its mass. Electrons occupied the space outside the core (Fig. 32.2). The electrons outside the nucleus had too little mass to deflect an α particle backward, but the massive nucleus with its large positive charge could cause back-scattering during a direct hit by the α particle.

To account for the small fraction of back-scattered α particles, the nucleus could be only about 10^{-14} m in diameter. Since the whole atom occupied a space that was over 10^{-10} m in diameter, the atom seemed to consist mostly of empty space dotted with electrons and a very small nucleus at the center.

To develop a feeling for the small size of the nucleus compared to the region of space occupied by electrons, suppose that your thumbnail were the atom's nucleus. The electrons would then occupy a cross-sectional area a little larger than the size of a football field. Most α particles would be undeflected while traveling through this open space, but when an α particle hit the "thumbnail" nucleus, a dramatic collision would occur.

Although Rutherford's nuclear model of the atom agrees well with our present conception of the atom, it lacks the detail needed to explain many other properties of atomic behavior, such as the wavelengths of light emitted and absorbed by atoms and the reasons why some atoms combine to form molecules and others do not.

32.2 Atomic Spectra

More than thirty years before Rutherford introduced his nuclear model of the atom, scientists had begun to record light spectra emitted by different types of atoms. A **spectrum** consists of the various wavelengths of EM radiation that are emitted (an emission spectrum) or absorbed (an absorption spectrum) by a particular type of atom.

One technique for recording an emission spectrum is illustrated in Fig. 32.3. Atoms in the form of a gas are placed in a tube with glass walls. A large voltage across the metal ends of the tube causes electrons to flow through the gas. As the electrons collide with the atoms in the tube, the atoms gain energy; they become "excited." The excited atoms quickly lose their extra energy by emitting light and other forms of EM radiation.

FIG. 32.3. A spectrometer for measuring the wavelength of light coming from a lamp.

The light leaving the tube is collimated as it passes through consecutively placed single slits. After the slits, light passes through a diffraction grating or a prism that bends the light at angles that depend on the wavelength of the light. If a film is placed beyond the grating, it becomes exposed only in narrow bands or lines that correspond to the wavelengths of light emitted from the atoms. A spectrum recorded in this manner is called a **band spectrum** or a **line spectrum.**

The line spectra of several types of atoms are shown in Fig. 32.4 and on the color insert facing page 432. The spectrum of each type of atom or molecule is like a fingerprint in that it differs from those of other atoms or molecules. Astronomers can determine the types and relative amounts of atoms that are on distant stars by analyzing their spectra. Geologists and soil scientists use atomic absorption spectroscopy to analyze the types and concentrations of different

FIG. 32.4. The emission spectra of several gases (wavelengths are in units of nanometers). (Courtesy of Bausch & Lomb)

types of atoms in samples of soil. Using spectral techniques, a blood analyzer can identify the concentrations of sixty different atoms, ions, and molecules in the blood.

During the latter part of the nineteenth century, scientists were looking for one theory that would account for the variety of spectral lines emitted from different atoms. Much of the effort centered on hydrogen, the simplest atom. In 1885 Johann Balmer, a mathematics teacher in a girls' high school in Switzerland, developed (in his spare time) a mathematical equation for calculating the wavelength of a group of these hydrogen lines. The equation became known as **Balmer's equation:**

$$\frac{1}{\lambda} = R\left(\frac{1}{2^2} - \frac{1}{n^2}\right) \qquad \text{where } n = 3, 4, 5, \ldots.$$

The constant R, later called the **Rydberg constant,** has a value of $1.097 \times 10^7 \text{ m}^{-1}$. The wavelength of each line is calculated by substituting an appropriate integer n into the equation.

EXAMPLE 32.1 Calculate the wavelength of a hydrogen emission line that corresponds to the integer $n = 4$.

SOLUTION

$$\frac{1}{\lambda} = (1.097 \times 10^7 \text{ m}^{-1})\left(\frac{1}{2^2} - \frac{1}{4^2}\right) = 2.057 \times 10^6 \text{ m}^{-1}$$

or

$$\lambda = \frac{1}{2.057 \times 10^6 \text{ m}^{-1}} = 486.2 \times 10^{-9} \text{ m} = \underline{486.2 \text{ nm}}.$$

Light with this wavelength appears blue to the human eye.   ■

In the twentieth century it was found that hydrogen emits electromagnetic radiation in two wavelength series besides those given by Balmer's equation. The wavelengths of these three series of lines emitted by hydrogen atoms are calculated using the following equations:

Lyman series: $\dfrac{1}{\lambda} = R\left(\dfrac{1}{1^2} - \dfrac{1}{n^2}\right)$ where $n = 2, 3, \ldots,$ **(32.1)**

Balmer series: $\dfrac{1}{\lambda} = R\left(\dfrac{1}{2^2} - \dfrac{1}{n^2}\right)$ where $n = 3, 4, \ldots,$ **(32.2)**

Paschen series: $\dfrac{1}{\lambda} = R\left(\dfrac{1}{3^2} - \dfrac{1}{n^2}\right)$ where $n = 4, 5, \ldots,$ **(32.3)**

where $R = 1.097 \times 10^7 \text{ m}^{-1}$ for each series. The **Lyman series** consists mostly of radiation emitted by hydrogen in the ultraviolet region; the **Paschen series** consists of radiation in the infrared region.

The next big step in the history of atomic theory was made by the Danish physicist Niels Bohr. Relying mostly on basic principles, he devised a theory that accounted for all the spectral lines emitted by hydrogen atoms.

32.3 The Bohr Model

Niels Bohr (1885–1962) was a man of many talents. As a college student he was selected for the All-Danish Soccer Team. He went on to attain a doctorate in physics from the University of Copenhagen and in 1912 spent a few months with Ernest Rutherford at Manchester University in England. Rutherford was so pleased with his new colleague that he told a friend: "This young Dane is the most intelligent chap that I have ever met." In turn, Bohr had a high opinion of Rutherford, who had just developed the nuclear model of the atom. Soon after Bohr returned to Copenhagen in 1913, he extended Rutherford's nuclear model so that all the spectral emissions of a hydrogen atom could be calculated with great accuracy. In 1922 at age 37 Bohr received the Nobel Prize in Physics for his model of the atom.

The Bohr model is restricted to atoms and ions with one electron. Bohr postulated that the electron can move in circular orbits about the atom's nucleus, as pictured in Fig. 32.6, but that only certain radii orbits are allowed. The energy of the atom depends on the radius of the orbit its electron occupies. While in a particular orbit, Bohr postulated that the atom's energy was stable. The electron could jump from one stable orbit to another, in which case the atom's change in energy was balanced by the emission or absorption of a photon. If the electron moved to an orbit of lower energy, it emitted a photon. In order that energy be conserved, the energy lost by the atom equaled the *energy* (*hf*) *of the newly created photon:*

$$E_i - E_f = hf, \tag{32.4}$$

where E_i is the atom's initial energy and E_f is its final energy. If an atom absorbed a photon, its electron moved to an orbit of higher energy.

Using this model of the atom and basic principles of physics, Bohr was able to derive an expression for the radii of the allowed orbits and the energy of the atom when its electron was in these stable orbits. This energy equation, when substituted in Eq. (32.4), could be used to calculate the frequency of photons that hydrogen atoms were expected to emit. When compared to the experimentally observed frequencies of emitted photons, Bohr's predictions agreed perfectly.

FIG. 32.5. Niels Bohr and his wife, Copenhagen, 1931. (AIP Niels Bohr Library, Uhlenbeck Collection)

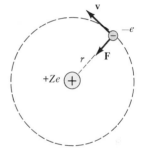

FIG. 32.6. Bohr's model of a one-electron ion.

Energy of One-Electron Atoms

Much of Bohr's theory relies on principles of classical physics. For example, Bohr knew that when an object moves with speed v in a circular orbit of radius r, it experiences a centripetal acceleration $a_c = v^2/r$ [Eq. (6.7)]. For an electron moving in a circle in a one-electron atom, Bohr assumed that the acceleration was caused by the electrical force of attraction between the electron of charge $-e$ and the nucleus of charge $+Ze$. (The quantity Z, called the **atomic number,** represents the number of protons in the nucleus. For hydrogen, $Z = 1$; for lithium, $Z = 2$, and so forth.) The electrical force keeps the electron in its circular orbit just as the gravitational force of the earth on the moon keeps the moon in its orbit. The magnitude of the electrical force between the electron and nucleus is given by Coulomb's law [Eq. (22.1)]: $F_q = k(e)(Ze)/r^2$. Substituting the Coulomb force and the centripetal acceleration into **Newton's second law** when written for circular motion ($\Sigma F_{\text{radial}} = ma_c$), we find that

$$\frac{kZe^2}{r^2} = m\frac{v^2}{r}, \tag{32.5}$$

where m is the electron's mass. This equation will be used shortly.

The **atom's energy** E when its electron is in an orbit of radius r consists of the kinetic energy of the electron plus the electrical potential energy due to the electric charges of the electron and nucleus:

$$E = KE + PE_q = \frac{1}{2}mv^2 + \frac{k(Ze)(-e)}{r}$$

$$= \frac{1}{2}mv^2 - \frac{kZe^2}{r}. \tag{32.6}$$

The electrical potential energy is defined to be zero when the electron is infinitely far from the nucleus ($r = \infty$). When the oppositely charged electron and nucleus are closer together, the electrical potential energy is less than zero—hence the negative sign for the PE_q part of Eq. (32.6).

This equation for the atom's energy depends on two unknown quantities: the electron's speed and the radius of its orbit. By combining Eqs. (32.5) and (32.6), we get an expression for the atom's energy that depends only on the radius of the electron orbit. Canceling an r from the denominator of each side of Eq. (32.5), we find that $kZe^2/r = mv^2$. If this expression for mv^2 is substituted in Eq. (32.6), we find that the atom's energy is

$$E = -\frac{kZe^2}{2r}. \tag{32.7}$$

Notice that as the radius of the electron's orbit decreases, the atom's energy lessens (becomes more negative). This is because the atom's electrical potential energy is less when the electron is closer to the nucleus.

We see that according to Eq. (32.7), the atom's energy depends only on the radius of the orbit of its electron. Bohr's next task was to determine the radii of the orbits in which the electron was allowed to move. The energy of these allowed orbits should be such that when an electron jumps from one orbit to another, the change in energy of the atom is balanced by the energy of a photon that is known to be emitted or absorbed by the atom.

After several failures, Bohr found a suitable *condition for determining the allowed orbits*. An electron, according to Bohr, can move in only those orbits in which its angular momentum mvr [Eq. (7.18)] equals a positive integer n times the constant $h/2\pi$, where h is Planck's constant and equals 6.63×10^{-34} J · s:

$$mv_n r_n = n\frac{h}{2\pi} \qquad \text{for } n = 1, 2, 3, 4, \ldots. \tag{32.8}$$

Each possible value of n corresponds to a different allowed orbit. For example, if $n = 3$, the radius r_3 of the third orbit and the electron's speed v_3 when in that orbit must satisfy the condition $mv_3 r_3 = 3(h/2\pi)$. Bohr had no justification for imposing this condition other than the fact that when combined with his other equations, it allowed him to calculate the energy of an atom when its electron was in one of these orbits and the correct frequencies of photons emitted from the atom when its electron jumped from one allowed orbit to another.

To calculate the radii of allowed orbits, Eq. (32.8) is solved for v_n and substituted into Eq. (32.5). We find that

$$r_n = \left(\frac{h^2}{4\pi^2 k e^2 m}\right)\frac{n^2}{Z}.$$

Notice that all the quantities inside the parentheses of the above equation are constants ($h = 6.63 \times 10^{-34}\,\text{J}\cdot\text{s}$, $k = 9.0 \times 10^9\,\text{N}\cdot\text{m}^2/\text{C}^2$, $e = 1.6 \times 10^{-19}\,\text{C}$, and $m = 9.11 \times 10^{-31}\,\text{kg}$). When evaluated, the term in parentheses equals $0.53 \times 10^{-10}\,\text{m}$. Thus, the radii of allowed orbits in a one-electron atom or ion are given by the equation

$$r_n = \frac{(0.53 \times 10^{-10})n^2}{Z}\ \text{m}. \tag{32.9}$$

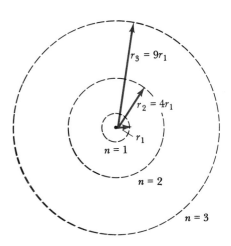

FIG. 32.7. The allowed orbits for an electron in a hydrogen atom ($r_1 = 0.53 \times 10^{-10}\,\text{m}$).

EXAMPLE 32.2 Calculate the radii of the smallest three orbits in the hydrogen atom ($Z = 1$). The orbits are pictured schematically in Fig. 32.7.

SOLUTION

$n = 1$: $\quad r_1 = (0.53 \times 10^{-10}\,\text{m})1 = \underline{0.53 \times 10^{-10}\,\text{m}}$,

$n = 2$: $\quad r_2 = (0.53 \times 10^{-10}\,\text{m})4 = \underline{2.12 \times 10^{-10}\,\text{m}}$,

$n = 3$: $\quad r_3 = (0.53 \times 10^{-10}\,\text{m})9 = \underline{4.77 \times 10^{-10}\,\text{m}}$. ∎

If the expression for the radii of the allowed orbits is substituted for r in the equation for the atom's energy [Eq. (32.7)], we find that the energy of the atom when its electron is in an orbit that corresponds to a particular value of n is

$$E_n = -\left(\frac{2\pi^2 k^2 e^4 m}{h^2}\right)\frac{Z^2}{n^2}. \tag{32.10}$$

Once again, all the quantities inside the parentheses are constants. Substituting for these constants, we find that the energy of a one-electron atom or ion when its electron is in the nth orbit is

$$E_n = -\frac{(2.2 \times 10^{-18})Z^2}{n^2}\ \text{J}.$$

The energy unit electron volt (eV) is much more convenient for calculations involving atoms. Converting the above equation from units of joules to units of electron volts ($1\,\text{eV} = 1.6 \times 10^{-19}\,\text{J}$) results in the following equation for the energy of a one-electron atom or ion:

$$E_n = -\frac{13.6Z^2}{n^2}\ \text{eV}. \tag{32.11}$$

EXAMPLE 32.3 Calculate the energy of the hydrogen atom ($Z = 1$) when its electron is in orbits $n = 1$, 2, and 3.

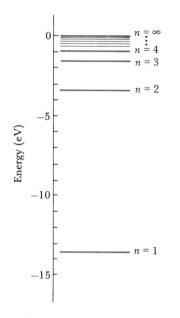

FIG. 32.8. An energy-level diagram for hydrogen.

SOLUTION Using Eq. (32.11) for the energy of an atom having a nucleus with one proton ($Z = 1$), we find that the energies of the atoms with electrons in these different orbits are as follows:

$$n = 1: \quad E_1 = \underline{-13.6 \text{ eV}},$$
$$n = 2: \quad E_2 = -13.6/4 = \underline{-3.4 \text{ eV}},$$
$$n = 3: \quad E_3 = -13.6/9 = \underline{-1.5 \text{ eV}}. \quad ■$$

Energy-Level Diagrams

The energy of the atom when its electron is in these different orbits is often represented by means of an **energy-level diagram** (Fig. 32.8). A vertical axis is drawn whose scale is in units of energy. A horizontal line, called an **energy level,** is drawn for each possible value of energy for the atom. Since a hydrogen atom can exist in states with energy -13.6 eV, -3.4 eV, -1.5 eV, and so on, horizontal lines appear at these energies in Fig. 32.8. Usually the value of n corresponding to a particular energy level is placed at its side.

The integer n can assume any positive integral value from one to infinity, and an infinite number of lines could be drawn in the diagram. Usually, however, lines appear for only the first few energy levels. (The rest are crowded into a very small energy region near zero energy and are not shown.)

Ground States and Excited States

The different allowed orbits of a one-electron atom, often called its states, are labeled by the integer n, which is called a **quantum number.** The state with lowest energy (the $n = 1$ state for hydrogen) is called the **ground state.** Higher energy states ($n = 2, 3, 4, \ldots$) are called **excited states.**

In a group of hydrogen atoms, most atoms have their electron in the ground state (the $n = 1$ orbit). As the atoms move above and collide with each other, an electron of one atom is occasionally excited from the ground state to an allowed state of higher energy. The energy gained by the excited atom is balanced by losses of translational kinetic energy of the atoms involved in the collision.

An excited atom remains excited for only about 10^{-8} s; its electron then falls back to the ground state in one step or in a series of steps, and during each step a photon is emitted from the atom.

While one atom is becoming deexcited by emitting a photon, another atom is excited by a collision. At any one time a small fraction of atoms are in excited states. On the sun's surface, for example, approximately six out of every billion hydrogen atoms are in the $n = 2$ state. A smaller fraction are in higher excited states. As the temperature of a group of atoms increases, their collisions become more violent, and a larger fraction of them are bumped by collisions into excited states. Since there are more excited atoms, more photons are being emitted as the atoms become deexcited. Thus, a hot object emits more light than a cool object.

Wavelength of Emitted and Absorbed Photons

We can now use Bohr's equations to calculate the wavelengths (or frequencies) of photons that are emitted and absorbed by one-electron atoms. As an atom emits a photon, an electron in the atom falls from an initial excited state of energy E_{n_i} to a final state of energy E_{n_f}. The energy lost by the atom equals the energy of the newly created photon:

$$hf = E_{n_i} - E_{n_f}.$$

Substituting Eq. (17.8) ($f = c/\lambda$) for the photon's frequency and rearranging, we find that

$$\frac{1}{\lambda} = -\frac{1}{hc}(E_{n_f} - E_{n_i}).$$

If we substitute for E_{n_f} and E_{n_i} using Bohr's equation for the energy of the atom [Eq. (32.10)], we find that

$$\frac{1}{\lambda} = \left(\frac{2\pi^2 k^2 e^4 m}{h^3 c}\right) Z^2 \left(\frac{1}{n_f^2} - \frac{1}{n_i^2}\right).$$

All the quantities inside the first term in parentheses of the above equation are constants, and when their values are substituted, the term equals the Rydberg constant R, which was determined empirically by Balmer. Substituting for R, we now have an equation for calculating the *wavelength of a photon emitted by a one-electron atom or ion* whose electron falls from state n_i to state n_f:

$$\frac{1}{\lambda} = RZ^2 \left(\frac{1}{n_f^2} - \frac{1}{n_i^2}\right), \tag{32.12}$$

where R equals $1.097 \times 10^7 \text{ m}^{-1}$ and for hydrogen $Z = 1$. Notice that this equation is in exactly the same form as Balmer's equation [Eq. (32.2)], which was an experimentally based equation for the wavelength of emitted photons.

The wavelengths of photons that can be absorbed by the atom can be calculated by equating the photon's energy ($hf = hc/\lambda$) to the change in energy of the atom ($E_{n_f} - E_{n_i}$) as its electron jumps from an initial orbit n_i to a final orbit n_f.

Bohr had for the first time applied basic physical principles to a model of the atom and had successfully calculated the wavelengths of light emitted and absorbed by an atom. Even with this success, the model had some faults. First, the Bohr model could not be used to calculate the spectra of atoms or ions with more than one electron. Second, Bohr could not justify in terms of physical reasoning his assumption that an electron's angular momentum equaled an integral multiple of $h/2\pi$. Finally, his assumption that the energy of an atom was stable when its electron moved in one of the allowed orbits conflicted with classical theories of electricity and magnetism. When viewed from the side, the electron's orbital motion resembled that of an electron oscillating in an antenna. An oscillating charge emits electromagnetic radiation (see Fig. 31.1) and loses energy as it moves back and forth. But if the electron loses energy, its orbit is not stable; its separation from the nucleus would decrease and it would spiral into the nucleus.

It is now known that we cannot picture atoms in the way that Bohr did. Electrons in atoms exhibit a wavelike behavior, and their paths as they move around the nucleus are uncertain. We can determine the probability of finding an electron at different locations relative to the nucleus, but we cannot determine the electron's exact orbit as it moves.

Even though Bohr's model is now regarded as inappropriate, it still serves a useful purpose in helping us visualize many of the interesting properties of atoms.

EXAMPLE 32.4 Two hydrogen atoms $(Z = 1)$ are excited into the $n = 4$ state by collisions with other atoms. One atom emits one photon as it falls back to the ground state $(n = 1)$. The other atom emits a photon as it falls to the $n = 2$ state, then emits another photon as it falls from the $n = 2$ to the $n = 1$ state. Calculate the wavelengths of each of these three photons.

SOLUTION The electron of the first atom falls from the state with $n_i = 4$ to the state with $n_f = 1$. The wavelength of the photon is determined by solving Eq. (32.12):

$$\frac{1}{\lambda} = (1.097 \times 10^7 \, \text{m}^{-1})(1^2)\left(\frac{1}{1^2} - \frac{1}{4^2}\right) = 1.03 \times 10^7 \, \text{m}^{-1}$$

or

$$\lambda = 97.2 \times 10^{-9} \, \text{m} = \underline{97.2 \, \text{nm}}.$$

The second atom returns to the ground state in two transitions. First, its electron falls from the state with $n_i = 4$ to the state with $n_f = 2$:

$$\frac{1}{\lambda} = (1.097 \times 10^7 \, \text{m}^{-1})(1^2)\left(\frac{1}{2^2} - \frac{1}{4^2}\right) = 2.06 \times 10^6 \, \text{m}^{-1}$$

or

$$\lambda = 486 \times 10^{-9} \, \text{m} = \underline{486 \, \text{nm}}.$$

This is a visible-wavelength photon (it appears bluish-green to the eyes) in the Balmer series. In the second transition the electron falls from the $n_i = 2$ state to the ground state $(n_f = 1)$:

$$\frac{1}{\lambda} = (1.097 \times 10^7 \, \text{m}^{-1})(1^2)\left(\frac{1}{1^2} - \frac{1}{2^2}\right) = 8.23 \times 10^6 \, \text{m}^{-1}$$

or

$$\lambda = 122 \times 10^{-9} \, \text{m} = \underline{122 \, \text{nm}}.$$

This is an ultraviolet wavelength photon in the Lyman series. ∎

The Lyman series corresponds to photons emitted when the electron in a hydrogen atom $(Z = 1)$ falls from any initial excited state $(n_i = 2, 3, 4, \ldots)$ directly to the ground state $(n_f = 1)$; the Balmer series consists of photons emitted when excited electrons fall to the $n_f = 2$ state from excited states of higher energy; the Paschen series consists of photons emitted when electrons fall from excited states of higher energy $(n_i = 4, 5, 6, \ldots)$ to the $n_f = 3$ state. These different transitions are indicated by arrows in the energy-level diagram shown

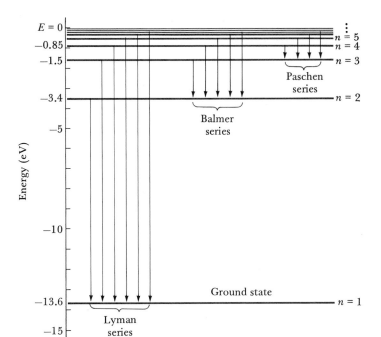

FIG. 32.9. The vertical lines represent electron transitions in the hydrogen atom that produce the different spectral series.

in Fig. 32.9. The Lyman and Paschen series were discovered after development of Bohr's theory.

EXAMPLE 32.5 Calculate (a) the radii and (b) the energy of the three smallest allowed orbits for the electron in the Li^{2+} ion. (Li^{2+} consists of the nucleus of a lithium atom ($Z = 3$) orbited by a single electron.) (c) What are the energy and wavelength of a photon that, when absorbed, causes an electron in Li^{2+} to be excited from the $n = 1$ to $n = 3$ state?

SOLUTION (a) We determine the radii of the $n = 1$, 2, and 3 orbits using Eq. (32.9) with $Z = 3$:

$$n = 1: \quad r_1 = \frac{0.53 \times 10^{-10} \text{ m}}{3} 1^2 = \underline{0.18 \times 10^{-10} \text{ m}},$$

$$n = 2: \quad r_2 = \frac{0.53 \times 10^{-10} \text{ m}}{3} 2^2 = \underline{0.71 \times 10^{-10} \text{ m}},$$

$$n = 3: \quad r_3 = \frac{0.53 \times 10^{-10} \text{ m}}{3} 3^2 = \underline{1.59 \times 10^{-10} \text{ m}}.$$

(b) Substituting $Z = 3$ into the energy equation [Eq. (32.11)], we find that the energies of the $n = 1$, 2, and 3 orbits are

$$n = 1: \quad E_1 = \frac{-13.6(3^2)}{1^2} = \underline{-122.4 \text{ eV}},$$

$$n = 2: \quad E_2 = \frac{-13.6(3^2)}{2^2} = \underline{-30.6 \text{ eV}},$$

$$n = 3: \quad E_3 = \frac{-13.6(3^2)}{3^2} = \underline{-13.6 \text{ eV}}.$$

(c) The photon's energy must equal exactly the energy $(E_3 - E_1)$ needed to excite the electron. Thus, the photon's energy hf is

$$hf = E_3 - E_1 = (-13.6 \text{ eV}) - (122.4 \text{ eV}) = \underline{108.8 \text{ eV}}.$$

Substituting $f = c/\lambda$ in the above equation, rearranging, and converting the energy to units of joules, we find that

$$\lambda = \frac{hc}{E_3 - E_1} = \frac{(6.63 \times 10^{-34} \text{ J} \cdot \text{s})(3.0 \times 10^8 \text{ m/s})}{(108.8 \text{ eV})(1.6 \times 10^{-19} \text{ J/eV})}$$

$$= 11.4 \times 10^{-9} \text{ m} = \underline{11.4 \text{ nm}}. \qquad ∎$$

32.4 De Broglie's Quantum Condition

In developing his atomic model, Bohr had proposed that electrons could move only in those orbits in which their angular momentum equaled an integral multiple of $h/2\pi$; that is, $mvr = n(h/2\pi)$. His only justification for this seemingly arbitrary proposal was the fact that he was then able to derive an equation that correctly predicted the wavelengths of light emitted and absorbed by hydrogen atoms. This condition on the angular momentum of electrons in orbit about the nucleus was made less arbitrary ten years later in 1923 by de Broglie's proposal that an electron of mass m moving at speed v behaves like a wave of wavelength

$$\lambda = \frac{h}{mv}. \qquad (31.5)$$

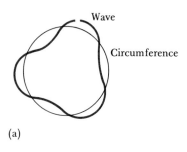

To see how these two ideas are related, picture in your mind the circular orbit of an electron in a hydrogen atom. We now regard the electron as a wave propagating around the circle. The situation is very similar to the formation of standing waves on a stretched string, such as discussed in Section 18.1. The wave on the string, after reflection from the end of the string, interferes with itself. If the total distance traveled by the wave up the string and back equals an integral number of wavelengths, the interference is constructive and large-amplitude standing waves are formed on the string. For all other wavelengths the interference is destructive, and waves of these other wavelengths cannot resonate on the string.

A similar situation occurs for electron waves moving in orbits about the nucleus of a one-electron atom. If the total distance around the orbit (the circumference $2\pi r$ of a circle) equals an integral number of electron wavelengths, the wave always returns to a particular position on the string with the same amplitude (Fig. 32.10a). The amplitude of vibration at each point on the orbit increases to form a standing-wave vibration—a resonance. If the circumference of the orbit is not an integral number of wavelengths, then destructive interference occurs as the wave makes a trip around the loop (Fig. 32.10b); the wave cannot resonate on these other orbits.

(b)

FIG. 32.10. (a) The de Broglie wave of the electron returns to its starting position in phase with the wave that started earlier, and constructive interference occurs. (b) The wave is out of phase with itself, and destructive interference occurs.

Constructive interference of electron waves on circular orbits of radius r takes place when the circumference $(2\pi r)$ equals an integral number of wavelengths:

$$2\pi r = n\lambda \qquad \text{for } n = 1, 2, 3, \ldots. \qquad (32.13)$$

When we substitute de Broglie's expression for the wavelength of an electron of

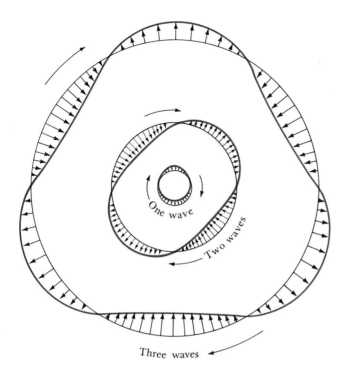

FIG. 32.11. Electron standing-wave patterns in the first three orbits of a hydrogen atom.

One wave

Two waves

Three waves

mass m moving at speed v ($\lambda = h/mv$), we find that

$$2\pi r = n\,\frac{h}{mv}$$

or

$$mvr = n\,\frac{h}{2\pi} \qquad \text{for } n = 1,\ 2,\ 3,\ \ldots.$$

This is exactly the condition used by Bohr [Eq. (32.8)] to select those orbits on which an electron was allowed to move. The situation is pictured schematically in Fig. 32.11. Notice that the inner orbit with the smallest circumference has room for only one wavelength. The second orbit, longer than the first, holds two waves, and so forth.

De Broglie's justification for Bohr's condition for selecting orbits foreshadowed the role that waves would soon play in the description of atomic phenomena. Two years after de Broglie's wave proposal, a whole new formulation of atomic physics called wave mechanics or quantum mechanics was introduced. The wavelike description of electrons in quantum mechanics was somewhat different from Bohr's orbits, as we learn in the next chapter.

Summary

1. **Nuclear model of atom:** By scattering alpha particles from gold atoms in a gold foil, Ernest Rutherford determined that an atom has a nucleus that is small in size but large in mass (about 10^{-14} m in diameter) and is made of protons and neutrons. (Rutherford did not know about the neutrons.) Electrons move around the nucleus in a variety of different orbits whose radii are roughly 10^{-10} m.

2. **Atomic spectra:** Atoms emit and absorb electromagnetic radiation of different wavelengths (called the atom's **spectrum**). The spectrum of each type of atom differs from that of other atoms.

3. **Bohr model of one-electron atom:** Niels Bohr in 1913 developed the first model of the atom that could be used to calculate the spectra of radiation emitted and absorbed by an atom. Ac-

cording to his model, which applies only to one-electron atoms and ions, the electron can move around the nucleus only in certain allowed orbits. The electrical force between the electron and nucleus provides the centripetal force to keep the electron in orbit $[kZe^2/r^2 = mv^2/r]$. When the electron is in an allowed orbit, the atom's energy is stable and consists of the electron's kinetic energy and the electrical potential energy due to the charges of the electron and nucleus $[E = \frac{1}{2}mv^2 + k(Ze)(-e)/r^2]$. Bohr postulated that the angular momentum of an electron when in an allowed orbit must equal an integral multiple of $h/2\pi$; that is, $mvr = n(h/2\pi)$, where $n = 1, 2, 3, \ldots$. Finally, when an electron fell from an initial orbit of higher energy E_{n_i} to a final orbit of lower energy E_{n_f}, a photon was emitted whose energy equaled the difference in energy of the atom:

$$hf = E_{n_i} - E_{n_f}.$$

An electron could jump to an orbit of higher energy by absorbing a photon whose energy equaled the energy gained by the atom.

By combining these equations, Bohr was able to derive expressions for the radii of allowed orbits, the energy of the atom when its electron was in these orbits, and the wavelength of radiation emitted as an electron fell from one orbit to another:

$$r_n = \frac{(0.53 \times 10^{-10})n^2}{Z} \, \text{m} \qquad \text{for } n = 1, 2, 3, \ldots, \quad \textbf{(32.9)}$$

$$E_n = -\frac{13.6Z^2}{n^2} \, \text{eV} \qquad \text{for } n = 1, 2, 3, \ldots, \quad \textbf{(32.11)}$$

$$\frac{1}{\lambda} = RZ^2 \left(\frac{1}{n_f^2} - \frac{1}{n_i^2} \right), \qquad \textbf{(32.12)}$$

where Z is the number of protons in the nucleus and R, called the Rydberg constant, equals $1.097 \times 10^7 \, \text{m}^{-1}$.

4. **De Broglie's electron waves:** Bohr's condition that the angular momentum of an electron in an allowed orbit must equal $n(h/2\pi)$ was given physical meaning by de Broglie's proposal that electrons exhibit a wavelike behavior. Bohr's allowed orbits were found to be those in which an electron wave formed a resonant standing-wave vibration pattern. Resonance occurred if the circumference of the orbit equaled an integral number of de Broglie wavelengths:

$$2\pi r = n\lambda = n(h/mv) \qquad \text{for } n = 1, 2, 3, \ldots. \quad \textbf{(32.13)}$$

Questions

1. How can you determine if the sun contains iron?

2. If we know the value of n for the orbit of an electron in a hydrogen atom, we can determine the values of three other quantities related to either the electron's motion or the atom as a whole. List these three quantities.

3. Explain carefully why hydrogen gas at room temperature emits little or no light whereas hydrogen gas on the sun emits considerable light.

4. How can an atom with only one electron have so many spectral lines?

5. A group of hydrogen atoms are in a container at room temperature. Infrared radiation, light, and ultraviolet rays are passed through the tube, but only the ultraviolet rays are absorbed at the wavelengths of the Lyman series. Explain why this is true.

6. Why cannot an electron move in an orbit whose circumference is 1.5 times the electron's de Broglie wavelength?

7. If Planck's constant were larger, would atoms be larger or smaller? Explain your answer.

Problems

32.1 The Nuclear Model of the Atom

1. The electron in a hydrogen atom spends most of its time $5.3 \times 10^{-11} \, \text{m}$ from the proton nucleus, whose radius is about $1.4 \times 10^{-15} \, \text{m}$. If each dimension of an atom were increased by the same factor and the proton's radius were increased to the size of a tennis ball, how far from the proton would the electron be?

2. A high-energy α particle scattered from the nucleus of an atom helps determine the size and shape of the nucleus. For best results, the de Broglie wavelength of the α particle should be the same size as the nucleus (approximately $10^{-14} \, \text{m}$) or smaller. If the mass of the α particle is $6.7 \times 10^{-27} \, \text{kg}$, at what speed must it travel to produce a wavelength of $10^{-14} \, \text{m}$?

3. A 10 m × 10 m board has twenty holes in it, each with a 3.0-cm radius. If one million BBs are shot randomly at the board, how many will pass through the holes? This experiment is analogous to shooting α particles at a gold foil. However, in that experiment, α particles pass through in most places and are deflected when they hit the gold nuclei.

4. A single layer of gold atoms lies on a table. The radius of each gold atom is about $1.5 \times 10^{-10} \, \text{m}$, and the radius of each gold nucleus is about $6 \times 10^{-14} \, \text{m}$. A particle much smaller than the nucleus is shot at the layer of gold atoms. Roughly, what is its chance of hitting a nucleus and being scattered? (The projectile is not affected by the electrons around the atom.)

5. (a) Calculate the mass of a gold foil that is 0.010 cm thick and whose area is 1 cm × 1 cm. The density of gold is 19,300 kg/m³.

(b) Calculate the number of gold atoms in the foil if the mass of each atom is 3.27×10^{-25} kg. (c) The radius of a gold nucleus is 6.0×10^{-15} m. Calculate the area of a circle with this radius. (d) Calculate the chance that an α particle passing through the gold foil will hit a gold nucleus. Ignore the α particle's size and assume that all gold nuclei are exposed to it; that is, no gold nuclei are hidden behind other nuclei.

32.2 Atomic Spectra

6. Calculate the wavelengths of the first three lines in the Balmer series. In what part of the electromagnetic spectrum do these lines appear?

7. Calculate the wavelengths of the first three lines in the Lyman series. In what part of the electromagnetic spectrum do these lines appear?

8. Calculate the longest-wavelength and shortest-wavelength lines in the Paschen series. In what part of the electromagnetic spectrum do these lines appear?

9. Calculate the wavelength, frequency, and photon energy of the line with $n = 5$ in the Balmer series.

10. An imaginary atom is observed to emit electromagnetic radiation at the following wavelengths: 250 nm, 2250 nm, 6250 nm, 12,250 nm, ... Invent an empirical equation for calculating these wavelengths; that is, determine $\lambda = f(n)$ where f is the unknown function of an integer n, which can have values 1, 2, 3, 4, ...

32.3 The Bohr Model

11. (a) Calculate the radii and energy of the $n = 1$, 2, and 3 orbits in the He$^+$ ion ($Z = 2$). (b) Construct an energy-level diagram for this ion.

12. (a) Calculate the radii and energy of the $n = 1$, 2, and 3 orbits of a sodium ion in which ten of its eleven electrons have been removed. (b) Construct an energy-level diagram for the ion.

13. A uranium atom ($Z = 92$) has two electrons in an $n = 1$ orbit. *Estimate* the radius of this orbit.

14. Calculate the least energy needed to remove an electron from the $n = 1$ orbit in He$^+$ to infinity.

15. *Estimate* the energy needed to remove an electron from (a) the $n = 1$ orbit to the $n = \infty$ orbit of iron ($Z = 26$) and (b) from the $n = 1$ to the $n = \infty$ orbit of hydrogen ($Z = 1$).

16. Calculate the electrical potential energy and kinetic energy in electron volts of an electron in the $n = 1$ orbit of hydrogen.

17. A 1000-kg satellite circles the earth once every 24 hours in an orbit of radius 4.23×10^7 m. Assuming the orbit satisfies Bohr's quantum condition, calculate the value of n for this orbit.

18. Some of the energy states of a hypothetical one-electron atom are, in units of eV, $E_1 = -18.72$, $E_2 = -6.36$, $E_3 = -3.70$, $E_4 = -1.74$, $E_5 = -0.96$, . . . , $E_\infty = 0.00$. (a) Draw an energy-level diagram for this atom. (b) An atom in the $n = 1$ state is involved in a collision in which it gains 15.02 eV of energy. In what state is its electron following the collision? (c) After the collision, the electron falls to the next lower state. What is the energy of the photon created by this transition?

19. Calculate the speed and frequency of the revolution of an electron around the first Bohr orbit in hydrogen. According to

classical physics, the atom should emit electromagnetic radiation at this frequency. In what portion of the electromagnetic spectrum is this frequency?

20. Show that the frequency of revolution of an electron around the nucleus of a hydrogen atom is $f = (4\pi^2 k^2 e^4 m/h^3)(1/n^3)$.

21. An electron in a hydrogen atom falls from the $n = 4$ to the $n = 3$ state. Calculate the wavelength of the emitted photon.

22. An electron falls from the $n = 15$ to the $n = 5$ state of hydrogen. Calculate the wavelength, frequency, and energy of the emitted photon.

23. An electron in a He$^+$ ion falls from the $n = 3$ to the $n = 1$ state. Calculate the wavelength, frequency, and energy of the emitted photon.

24. Calculate the energy and wavelength of a photon whose absorption raises the electron in a hydrogen atom from the $n = 3$ to the $n = 8$ state.

25. Calculate the energy, frequency, and wavelength of a photon whose absorption raises the electron in a He$^+$ atom from the $n = 1$ to the $n = 6$ state.

26. The average thermal energy due to the random transitional motion of a hydrogen atom at room temperature is $\frac{3}{2}kT$ where k is the Boltzmann constant. Would a typical collision between two hydrogen atoms be likely to transfer enough energy to one of the atoms to raise its electron from the $n = 1$ to the $n = 2$ energy state? Explain your answer. (*Note:* Actually, free hydrogen on the earth is in the molecular form H$_2$. However, the above reasoning still explains why a gas of hydrogen molecules at room temperature is seldom excited so that it can emit light.)

27. An electron in the excited state of an atom usually remains in that state only about 10^{-8} s before falling to a lower state. How many times will an electron in the $n = 3$ state of hydrogen rotate around the orbit before falling to the $n = 2$ or $n = 1$ state?

28. A gas of hydrogen atoms in a tube is excited by collisions with free electrons. If the maximum excitation energy gained by an atom is 12.5 eV, calculate the wavelengths of light emitted from the tube as atoms return to the ground state.

29. A hypothetical one-electron atom emits radiation in a series of lines whose wavelengths are given by the equation $1/\lambda = 300[1 - (1/n^2)]$ in units of nanometers for $n = 2$, 3, 4, . . . , ∞. Calculate (a) the most energetic photon emitted and (b) the least energetic photon emitted in the series.

32.4 De Broglie's Quantum Condition

30. (a) Use Eq. (32.5) to calculate the speed of the electron in a hydrogen atom when in the $n = 1$ orbit. The radius of the orbit is 0.529×10^{-10} m. (b) Calculate the electron's de Broglie wavelength. (c) Confirm that the circumference of the orbit equals one de Broglie wavelength.

31. Repeat Problem 30 for the $n = 3$ orbit, whose radius is 4.76×10^{-10} m. Three de Broglie wavelengths should fit around the $n = 3$ orbit.

32. A 70-kg person runs around a circular track whose radius is 64 m. Calculate the lowest two speeds at which the person can run so that the track's circumference will equal an integral number of the person's de Broglie wavelengths. For this problem assume that Planck's constant equals 1.4×10^5 J · s.

General Problems

G1. Calculate the ratio of the electric force between a proton and an electron in the ground state of the hydrogen atom and the gravitational force between the two particles. Based on your answer, are we justified in ignoring the gravitational force in the Bohr theory?

G2. Are we justified in using nonrelativistic energy equations in the Bohr theory? (Is the electron's speed less than $0.1c$?)

G3. A mass M moving at speed v_0 has a direct elastic collision with a second mass m which is at rest. Using the conservation-of-energy and momentum principles, we showed in Chapter 8 that the final velocity of mass M is $v = (M - m)v_0/(M + m)$. Using this result, determine the final velocity of an alpha particle following a head-on collision with (a) an electron, and (b) a gold nucleus, each initially at rest. The alpha particle's velocity before the collision is $0.010c$; $m_\alpha = 6.6 \times 10^{-27}\,\text{kg}$; $m_{el} = 9.11 \times 10^{-31}\,\text{kg}$; and $m_{gold} = 3.3 \times 10^{-25}\,\text{kg}$. (c) Based on your answers, could an alpha particle be deflected backward by hitting an electron in a gold atom?

Quantum Mechanics and Atoms

33

In 1925, twelve years after Bohr developed his model for one-electron atoms, Erwin Schrödinger (1887–1961) and Werner Heisenberg (1901–1976) independently presented different but equivalent forms of a new theory for describing atoms. The theory, called **quantum mechanics,** combines the wave and particle nature of matter in one equation. The solutions to this equation, called wave functions, can be used to determine the energy of electrons in atoms and the wavelength and intensity of spectral emissions from the atoms. The theory of quantum mechanics is not restricted to atoms but has also been used to describe the properties of molecules (small and large) and of solid materials such as those used to make transistors and superconductors. In short, quantum mechanics has proved so successful at describing a broad range of phenomena that most physicists regard it as the fundamental theory underlying the behavior of nature at the atomic and molecular level. In this chapter we will describe the results of the quantum theory as applied to atoms.

33.1 Quantum Mechanics

In the approach to quantum mechanics developed by Schrödinger, calculations are performed by solving an equation named in his honor. The solutions to Schrödinger's equation are called **wave functions** and are given the symbol ψ (psi). For atoms, the wave functions are analogous to the orbits in the Bohr theory, in that they can be used to determine possible locations for electrons in an atom. In the Bohr theory, however, the electron moves in a well-defined orbit, whereas the path of an electron described by a wave function is not well defined. Instead, the square of the wave function ψ^2 indicates the *probability* of finding the electron at different locations. This probability (ψ^2) for electrons in atoms is often pictured as in Fig. 33.1, looking something like a cloud; the more dense the cloud is, the greater the probability of finding the electron at that position.

Just as there are many allowed electron orbits in the Bohr theory, many wave functions are solutions of the Schrödinger equation. Each wave function represents a possible state for a particle—a place for it to reside. The energy of a particle depends on the state it occupies. For a one-electron atom the energies of the different states that are calculated using the Schrödinger equation are the same as the energies of electrons when in different Bohr orbits. Whereas the Bohr

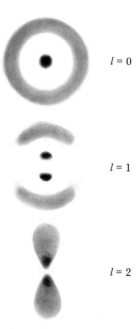

$l = 0$

$l = 1$

$l = 2$

FIG. 33.1. Three different $n = 3$ orbital distributions. The density of the pattern equals the probability of finding an electron at a point.

theory applies only to one-electron atoms, the theory of quantum mechanics has been applied with success to almost every type of atom, to many molecules, and even to electrons in solids and in biological molecules. Unfortunately, the solution of quantum-mechanics problems is often complex and usually requires the use of large computer facilities. In the rest of the chapter we will review qualitatively some of the main results of quantum-mechanical calculations with atoms.

33.2 Quantum Numbers for Atomic Wave Functions

In the Bohr model of the one-electron atom, the different allowed orbits for the electron were labeled using a quantum number n. The number could have any integral value from one to infinity. The radius of an orbit and the energy of an electron in that orbit depended on the value of n for the orbit.

A similar situation occurs in quantum mechanics. The wave functions that are solutions of the Schrödinger equation differ from each other in a variety of ways, such as the energy of an electron when in a state described by the wave function, and the average distance, shape, and orientation of the electron's probability cloud when in that state. The wave functions are labeled using four **quantum numbers,** and the values of these quantum numbers for a particular wave function give a rough indication of the energy and type of orbit of the electron when in that state compared to when in other states. (For convenience, we will occasionally use the word *orbit* to indicate an electron's probability distribution.) The quantum numbers can each have different integral or half-integral values; they are described below along with an indication of the way in which they are related to the electron and its orbit.

Principal Quantum Number n

The **principal quantum number** n is analogous to the quantum number n in Bohr's theory. It can have any positive integral value

$$n = 1, 2, 3, 4, \ldots \tag{33.1}$$

and is an indicator of the energy and average separation of the orbit from the nucleus. The larger the value of n, the greater is the energy of an electron when in that state and the greater the average separation of the electron from the nucleus.

Orbital Angular Momentum Quantum Number l

The orbital angular-momentum quantum number l indicates orbits of different shape and also the orbital angular momentum of an electron when in a state. For a particular value of n there are several l-type states of different shape, which are numbered using the following values of l:

$$l = 0, 1, 2, \ldots, n - 1. \tag{33.2}$$

For example, three differently shaped electron distributions are possible for $n = 3$ states (these differently shaped orbits have l-values equal to 0, 1, and 2).

Five differently shaped orbits are available to electrons in $n = 5$ states (the l-values of these states are 0, 1, 2, 3, or 4). An $l = 0$ state is spherically symmetric (Fig. 33.1a); the electron's probability distribution is the same along any direction away from the nucleus. When in states with higher values of l, the electron distribution is not symmetric. Examples of $l = 1$ and 2 states are shown in Fig. 33.1b and c. The angular momentum L of an electron when in a state with quantum number l is $L = \sqrt{l(l + 1)}\, h/2\pi$.

Magnetic Quantum Number m_l

Angular momentum is a vector quantity, and the **magnetic quantum number** m_l indicates the direction of the angular momentum of an electron when in a particular state and also the orientation of the electron's orbit when in that state. In general, for a state with orbital angular momentum quantum number l, $2l + 1$ differently oriented states exist corresponding to the following values of m_l:

$$m_l = 0, \pm 1, \pm 2, \ldots, \pm l. \tag{33.3}$$

FIG. 33.2. An $l = 1$ type orbit can assume three different orientations in a magnetic field.

For a state with $l = 0$, the angular momentum L of an electron is zero, and the m_l quantum is also zero. For an $l = 1$ state, however, the angular momentum is not zero, and it can have three different orientations corresponding to the three allowed values of m_l ($-1, 0, +1$). The orbits shown in Fig. 33.2 represent schematically the three possible orientations of an $l = 1$ electron orbit. For an $l = 2$ state, five orbits of different orientation exist corresponding to the five allowed values of m_l ($-2, -1, 0, +1, +2$).

The word *magnetic* is chosen for the m_l quantum number because of its importance relative to the magnetic properties of the atom. When an electron moves in an orbit around a nucleus, it creates a magnetic field much like that produced by an electron current in a circular loop of wire. The atom is said to have a magnetic moment because of this orbital motion of the electron and the magnetic field it produces. The m_l quantum number indicates the orientation of the magnetic moment due to the orbital motion of the electron; it is important in a variety of experiments involving the magnetic properties of atoms.

Spin Magnetic Quantum Number m_s

The fourth and last quantum number is related not to the motion of the electron about the nucleus (as are n, l, and m_l) but to an intrinsic property of the electron called its spin angular momentum. An electron seems to spin about an axis as it moves in orbit about the nucleus, much the way the earth spins about its axis as it orbits the sun. Any spinning or rotating mass has angular momentum. For an electron, the magnitude of its spin angular momentum is $S = \sqrt{s(s + 1)}\,(h/2\pi)$, where $s = \frac{1}{2}$.

If the electron is pictured as a solid sphere (Fig. 33.3), its electric charge must spin as the mass spins, and each portion of charge Δq moves in a loop and produces an electric current. The current, in turn, produces a magnetic field.* We see that an atom's magnetic field and magnetic moment are caused by two

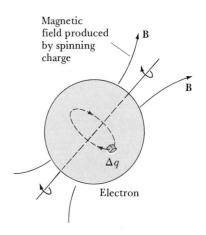

FIG. 33.3. A model of a spinning electron.

*This model provides an intuitive idea of how an electron's intrinsic magnetic moment is produced. Unfortunately, detailed calculations are not consistent with this model.

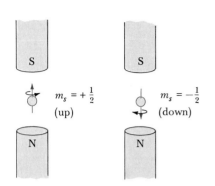

$m_s = +\frac{1}{2}$ (up) $m_s = -\frac{1}{2}$ (down)

FIG. 33.4. An electron's spin axis can assume only two orientations in a magnetic field.

TABLE 33.1 Quantum Numbers for Electron States in Atoms

Name	Symbol	Allowed values
Principal	n	1, 2, 3, 4, ...
Orbital	l	For a given n, l can have the values 0, 1, 2, ..., $n-1$.
Magnetic	m_l	For given n and l, m_l can have the values 0, ± 1, ± 2, ..., $\pm l$.
Spin magnetic	m_s	For each set n, l, and m_l, the spin magnetic quantum number can have two values: $m_s = +\frac{1}{2}$ or $-\frac{1}{2}$.

different motions: (1) the orbital motion of its electrons about the nucleus, and (2) the spinning motion of the electrons about their own axes. Many careful experiments have shown that the electron's spin magnetic moment can assume only two orientations relative to an externally applied magnetic field—one in the direction of the field and the other opposite the direction of the field (Fig. 33.4). The spin magnetic quantum number m_s is used to indicate the orientation of an electron's spin angular momentum and its magnetic moment. The quantum number can have two values,

$$m_s = +\tfrac{1}{2} \quad \text{and} \quad -\tfrac{1}{2}, \tag{33.4}$$

corresponding to the two possible orientations of the electron's spin angular momentum and magnetic moment. The two orientations are sometimes called "up" and "down."

The allowed values for each of the four quantum numbers for electron states in atoms are summarized in Table 33.1.

33.3 Multielectron Atoms

Together, the four quantum numbers introduced in the last section (n, l, m_l, and m_s) are used to label the wave functions for states occupied by electrons in atoms. The symbol $|nlm_lm_s\rangle$ will be used to represent one of these wave functions. For example, $|300 - \frac{1}{2}\rangle$ represents a wave function with $n = 3$, $l = 0$, $m_l = 0$, and $m_s = -\frac{1}{2}$. The probability of finding an electron at different locations around the nucleus when in a state ("orbit") described by this wave function is shown in Fig. 33.1a.

Early in the study of the light emitted and absorbed by atoms, a special vernacular was developed for the states of different orbital angular momentum quantum number l. If an electron fell from an excited state with $l = 0$ to a state of lower energy, the spectral line appeared sharp or well defined. Thus, an $l = 0$ state became known as an s-state. Transitions from $l = 1$ states were often the principal bands of light observed, and $l = 1$ states became known as p-states. The letter designations for these and other l-states are summarized in Table 33.2. This letter designation is used extensively in the rest of this chapter. An $n = 1$ and $l = 0$ state is called a $1s$ state; an $n = 4$ and $l = 2$ state is called a $4d$ state; and so forth.

The energy of an electron when in a particular state of the atom depends on the values of n and l but not on the values of m_l and m_s. The smaller the values of n and l, the less the electron's energy. (The energy depends more on n than on l.)

The electrons in atoms usually occupy the lowest energy states that are

TABLE 33.2 Letter Designations of Different l-States

l-value	0	1	2	3	4	5	6
Letter	s	p	d	f	g	h	j

available. The electron in a hydrogen atom almost always resides in a $1s$ state, the state in which $n = 1$ and $l = 0$. In lithium, a three-electron atom, two of its three electrons are in the $1s$ state. Its third electron normally resides in a $2s$ state. The orbits of the $1s$ states seem to be "filled" by the first two electrons, and the third electron, finding no room, must reside in an orbit of greater energy and one that is farther from the nucleus (the $n = 2$, $l = 0$ orbit).

Having observed that electrons can exclude other electrons from the orbits they occupy, we might ask ourselves how many electrons a particular orbit can hold. Why, for example, does it take two electrons to fill the orbits with $n = 1$, $l = 0$ ($1s$ states) instead of one or three electrons? In 1925 the Austrian physicist Wolfgang Pauli (1900–1958) developed an empirical principle to answer this question:

Pauli exclusion principle: Each electron state $|nlm_lm_s\rangle$ of an atom can hold no more than one electron.

For example, for the $n = 1$, $l = 0$ state, m_l can only be zero and m_s can be $+\frac{1}{2}$ or $-\frac{1}{2}$. Thus, there are two different states with $n = 1$ and $l = 0$; the $|nlm_lm_s\rangle = |100\frac{1}{2}\rangle$ and $|100 - \frac{1}{2}\rangle$ states. According to Pauli's exclusion principle, each state can hold one electron; so two electrons can occupy the $n = 1$, $l = 0$ states. These two states are said to form a **subshell**—a group of states with the same values of n and l. If each state in a subshell is occupied by an electron, the subshell is said to be **filled.**

The number of states in a subshell depends on the number of different allowed combinations of m_l and m_s for the given values of n and l. The rules restricting the values of n, l, m_l, and m_s are summarized in Table 33.1.

EXAMPLE 33.1 Determine the number of states and the quantum number designation of each state for a $3d$ subshell.

SOLUTION For a $3d$ subshell, $n = 3$ and $l = 2$ (see Table 33.2). For an $l = 2$ state, the m_l quantum number can have the values $m_l = -2, -1, 0, 1, 2$. For each of these five values of m_l, the m_s quantum number can have two values: $m_s = -\frac{1}{2}$ or $+\frac{1}{2}$. Thus, there are ten unique $|nlm_lm_s\rangle$ states for the $3d$ subshell:

$$|32 - 2\tfrac{1}{2}\rangle, \quad |32 - 1\tfrac{1}{2}\rangle, \quad |320\tfrac{1}{2}\rangle, \quad |321\tfrac{1}{2}\rangle, \quad |322\tfrac{1}{2}\rangle,$$
$$|32 - 2 - \tfrac{1}{2}\rangle, \quad |32 - 1 - \tfrac{1}{2}\rangle, \quad |320 - \tfrac{1}{2}\rangle, \quad |321 - \tfrac{1}{2}\rangle, \quad |322 - \tfrac{1}{2}\rangle.$$

If one electron resides in each of these states, such as occurs in a copper atom, the atom is said to have a filled $3d$ subshell. ■

Electron Configurations

The electrons in an atom normally occupy the states with lowest energy. The atom as a whole is then said to be in its **ground state.** An electron configuration for the ground state of an atom indicates the subshells occupied by its electrons and the number of electrons in each subshell. For example, a sodium atom has eleven electrons, distributed as follows: $1s^22s^22p^63s$. The superscript represents the number of electrons in each subshell: $1s^2$ indicates two electrons occupy the $1s$ subshell, $2p^6$ indicates that six electrons occupy the $2p$ subshell, and so forth.

TABLE 33.3 Atomic Subshells Listed in Order from Lowest to Highest Energy

Subshell	Maximum number of electrons in subshell	Total number of electrons in atom to fill subshell
1s	2	2
2s	2	4
2p	6	10
3s	2	12
3p	6	18
4s	2	20
3d	10	30
4p	6	36
5s	2	38
4d	10	48
5p	6	54
6s	2	56
4f	14	70
5d	10	80
6p	6	86
7s	2	88
5f	14	102
6d	10	112

The maximum number of electrons that can be in a subshell and their approximate order of filling are listed in Table 33.3. We see that for sodium, ten electrons are needed to fill the 1s, 2s, and 2p subshells. The eleventh electron in sodium occupies a state in the next available subshell—the 3s subshell. When sodium is ionized to form Na^+, it loses the electron from the 3s shell.

33.4 The Periodic Table

Many of the properties of atoms (their ability to form positive or negative ions and to combine with other atoms to form molecules) depend on the number of electrons in the subshell of the atom that is only partially filled and on the type of subshell it is. These "outer" electrons are usually farthest from the nucleus and interact most with other atoms in their vicinity. Also, these outer electrons usually require the least energy to be removed from the atom. When sodium is ionized to form Na^+, it loses the electron that was in the 3s subshell before it was ionized.

Groups of atoms with the same number of outer electrons in similar subshells usually exhibit similar properties. This leads naturally to the classification of the atoms as presented in the periodic table of elements (Table 33.4). (The word *element* indicates a substance made of one type of atom.) In the table the different types of atoms are grouped together in columns so that atoms in the same column have similar electron configurations for their outer electrons.

Group I. Atoms in this group all have one outer electron in an *s* state (a state with $l = 0$). Examples are hydrogen (1s), lithium (2s), sodium (3s),

TABLE 33.4 Periodic Table of the Elements*

I	II											III	IV	V	VI	VII	0
1 **H** 1.0080																	2 **He** 4.0026
3 **Li** 6.941	4 **Be** 9.0122											5 **B** 10.81	6 **C** 12.011	7 **N** 14.0067	8 **O** 15.9994	9 **F** 18.9984	10 **Ne** 20.179
11 **Na** 22.9898	12 **Mg** 24.305				Transition elements							13 **Al** 26.9815	14 **Si** 28.086	15 **P** 30.9738	16 **S** 32.06	17 **Cl** 35.453	18 **Ar** 39.948
19 **K** 39.102	20 **Ca** 40.08	21 **Sc** 44.956	22 **Ti** 47.90	23 **V** 50.941	24 **Cr** 51.996	25 **Mn** 54.9380	26 **Fe** 55.847	27 **Co** 58.9332	28 **Ni** 58.71	29 **Cu** 63.54	30 **Zn** 65.37	31 **Ga** 69.72	32 **Ge** 72.59	33 **As** 74.9216	34 **Se** 78.96	35 **Br** 79.909	36 **Kr** 83.80
37 **Rb** 85.467	38 **Sr** 87.62	39 **Y** 88.906	40 **Zr** 91.22	41 **Nb** 92.906	42 **Mo** 95.94	43 **Tc** (99)	44 **Ru** 101.07	45 **Rh** 102.906	46 **Pd** 106.4	47 **Ag** 107.870	48 **Cd** 112.40	49 **In** 114.82	50 **Sn** 118.69	51 **Sb** 121.75	52 **Te** 127.60	53 **I** 126.9045	54 **Xe** 131.30
55 **Cs** 132.906	56 **Ba** 137.34	57 **La** 138.906	72 **Hf** 178.49	73 **Ta** 180.948	74 **W** 183.85	75 **Re** 186.2	76 **Os** 190.2	77 **Ir** 192.2	78 **Pt** 195.09	79 **Au** 196.967	80 **Hg** 200.59	81 **Tl** 204.37	82 **Pb** 207.2	83 **Bi** 208.981	84 **Po** (210)	85 **At** (210)	86 **Rn** (222)
87 **Fr** (223)	88 **Ra** (226)	89 **Ac** (227)	104 **Rf** (261)	105 **Ha** (262)													

Lanthanide series

58 **Ce** 140.12	59 **Pr** 140.908	60 **Nd** 144.24	61 **Pm** (147)	62 **Sm** 150.4	63 **Eu** 151.96	64 **Gd** 157.25	65 **Tb** 158.925	66 **Dy** 162.50	67 **Ho** 164.930	68 **Er** 167.26	69 **Tm** 168.934	70 **Yb** 173.04	71 **Lu** 174.97

Actinide series

90 **Th** (232)	91 **Pa** (231)	92 **U** (238)	93 **Np** (237)	94 **Pu** (242)	95 **Am** (243)	96 **Cm** (248)	97 **Bk** (249)	98 **Cf** (249)	99 **Es** (254)	100 **Fm** (257)	101 **Md** (258)	102 **No** (259)	103 **Lr** (260)

*Atomic weights of stable elements are those adopted in 1969 by the International Union of Pure and Applied Chemistry. For those elements having no stable isotope, the mass number of the "most stable" isotope is given in parentheses.

and potassium ($4s$). This outer electron is easily removed, leaving behind a positive ion: H^+, Li^+, Na^+, and K^+.

Group II. Atoms in this group have two electrons in an outer s state. Examples are beryllium ($2s^2$), magnesium ($3s^2$), and calcium ($4s^2$). These atoms become ions with a double positive charge when their outer electrons are removed (Be^{2+}, Mg^{2+}, Ca^{2+}).

Group III. Atoms in this group have two outer electrons in s states and a third outer electron in a p state. Examples are boron ($2s^2 2p$), aluminum ($3s^2 3p$), and gallium ($4s^2 4p$).

Groups IV–VII. Each atom to the right of the preceding atom in the periodic table has one more p electron than the neighboring atom to the left. Thus, Groups IV, V, VI, and VII consist of atoms with outer electron configurations $s^2 p^2$, $s^2 p^3$, $s^2 p^4$, and $s^2 p^5$, respectively. Notice that the Group VII atoms are one p electron short of having a filled p subshell. Group VII atoms often capture an extra electron to fill their p subshell, since filled subshells are usually very stable. Negative ions, such as F^-, Cl^-, Br^-, and I^-, are formed in this way.

Group 0. All atoms in this group have outer electrons in a filled p subshell, with the exception of helium, which has a $1s^2$ configuration. Examples are neon ($2p^6$), argon ($3p^6$), krypton ($4p^6$), and xenon ($5p^6$). Since filled subshells are very stable, Group 0 atoms are less likely to be ionized and are relatively inert.

Transition atoms. This category includes atoms that do not fit neatly into the other groups of the periodic table. The atoms in the first row have eighteen electrons in filled $1s$, $2s$, $2p$, $3s$, and $3p$ subshells. Additional electrons occupy the $4s$ and $3d$ subshells; for example, titanium with 22 electrons has its outer electrons in a $4s^2 3d^2$ configuration, while copper with 29 electrons has its outer electrons in a $4s^1 3d^{10}$ configuration. When ionized, transition atoms tend to lose their $4s$ electrons rather than their $3d$ electrons. Thus, these transition atoms tend to form positive ions with partially filled $3d$ subshells.

Second-row transition atoms starting with yttrium involve atoms whose $5s$ and $4d$ states are being filled. The lanthanides include atoms whose $6s$, $5d$, and $4f$ subshells are partially or totally filled. The actinides include atoms whose $7s$, $6d$, and $5f$ subshells are partially or totally filled.

33.5 Absorption and Emission of Radiation

Many of the interesting properties of atoms are related to an electron's jump between two different orbits—called an electron **transition.** The absorption of a photon by an atom results in an electron transition from a state of lower energy to one of higher energy, an event that is called **excitation.** To be absorbed the photon must have an energy equal to the difference in energy of the atom when in an excited state and when in the ground state. The absorption of photons in the visible and ultraviolet wavelength regions usually involves the excitation of an outer electron from the subshell in which it normally resides to a vacant subshell of higher energy (Fig. 33.5). Excitation can also occur as a result of collisions of atoms with each other.

Atoms and molecules can often be identified by their absorption **spectra**—the wavelengths of light they absorb. Absorption spectroscopy is used for purposes such as monitoring the type and concentration of atoms and molecules in samples of blood, air, soil, and water, and for measuring the changing concentration of molecules involved in chemical and biochemical reactions.

Photon **emission** occurs when atoms fall from an excited state to a state of lower energy. Deexcitation usually involves the transition of a single electron (Fig. 33.6) and occurs within 10^{-8} s after the atom becomes excited. The energy of the photon equals the atom's change in energy.

Fluorescence occurs when an atom that has been excited by absorbing a photon returns to the ground state in two or more steps, where each step involves the emission of a photon (Fig. 33.7). The energy of the emitted photons together equals the energy of the absorbed photon. As a result, each emitted photon has less energy than the absorbed photon that caused excitation.

Fluorescent light bulbs produce light by fluorescence. A gas inside the bulb

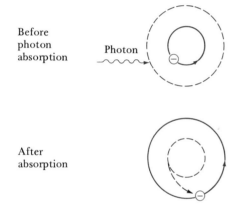

FIG. 33.5. Photon absorption resulting in the excitation of an atom.

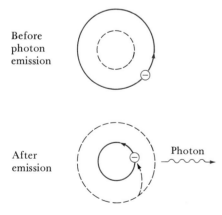

FIG. 33.6. Spontaneous photon emission by deexcitation of an atom.

is excited by collisions with electrons that have been accelerated by an applied voltage. The excited atoms return to their ground state by emitting ultraviolet-wavelength photons. The ultraviolet-wavelength photons strike a coating of fluorescent material on the inside surface of the bulb, causing these atoms to become excited. These fluorescent atoms return to the ground state in more than one step, and at least one step involves the emission of a photon of visible light.

Phosphorescence is similar to fluorescence except that the intermediate state (E_2 in Fig. 33.7) to which an excited atom falls is a "metastable" state ("almost stable"). The atom, when in the **metastable state,** remains there somewhat longer than the 10^{-8} s that an atom usually remains in an excited state. The lifetime of an excited state is the time needed for half the atoms to fall from the excited state. For a metastable state the lifetime could be as short as 10^{-5} seconds or as long as an hour or so, depending on the particular type of phosphorescent atom. A watch that glows in the dark is sometimes made of a phosphorescent material. Some of the light absorbed during the day causes electrons to be excited to the metastable state. Light is emitted by this material long after the exciting light is turned off; the phosphorescent material glows in the dark.

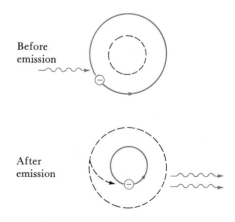

FIG. 33.7. An energy-level diagram for a material made of phosphorescent atoms. The state with energy E_2 is a metastable state. Electrons are excited from E_1 to E_3 and then fall to E_2. Because E_2 is metastable, the fall of the electrons back to E_1 is delayed.

33.6 Stimulated Emission and Lasers

The emission of radiation that was discussed in the last section is called **spontaneous emission:** an atom spontaneously falls from an excited state to a state of lower energy and emits a photon in the process. No external inducements are needed to encourage the transition.

An excited atom can fall to a state of lower energy by another process called stimulated emission. **Stimulated emission** occurs when a photon passes an excited atom and induces or stimulates the deexcitation of the atom. A new photon is created as the atom falls to a state of lower energy. This photon and the photon that stimulated the deexcitation move off together (Fig. 33.8). Stimulated emission occurs only if the energy lost by the atom as it falls from its excited state equals the energy of the photon that induced the transition. This photon has exactly the same energy and wavelength as the newly created photon. The two photons are also in phase, and their waves add constructively to form a wave twice the energy as that of the original photon.

A **laser** is a device that relies on stimulated emission to produce an intense, narrow beam of light. Lasers can be used to align roads through mountain tunnels, to perform surgery on the eye's retina, to drill holes through metal, to carry thousands of telephone messages along a thin glass fiber, to measure the distance to the moon, and to help keep the blade of a grader at the same elevation while leveling a field for irrigation.

A laser consists of a specially chosen group of atoms (or molecules) that can be excited to a metastable state, an apparatus for exciting these atoms, and a container to hold the atoms. Partially or totally reflecting mirrors are usually placed at the two ends of the container. The atoms to be excited may be in a gaseous, liquid, or solid form. Excitation occurs in a variety of ways. For example, in a helium-neon gas laser, such as used in many physics laboratories, electrons are accelerated through a gas of helium and neon atoms. Collisions between the electrons and atoms cause the atoms to become excited. In other lasers, a flash of light may cause excitation. The excitation of an atom to a state

FIG. 33.8. Stimulated emission. A photon stimulates an atom to fall from an excited state, creating a second photon and a deexcited atom.

Excitation

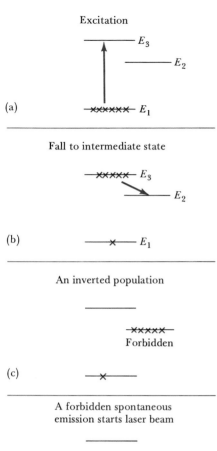

(a)

Fall to intermediate state

(b)

An inverted population

(c)

A forbidden spontaneous
emission starts laser beam

(d)

A stimulated transition
adds another photon to beam

(e)

FIG. 33.9. Steps leading to the production of a laser beam.

of higher energy is represented in Fig. 33.9a by the heavy arrow. Each cross (✗) in this figure represents an atom (or a group of atoms) in the energy state where the cross occurs. Originally almost all atoms are in the ground state with energy E_1. After excitation many of these atoms have been excited to state E_3 (Fig. 33.9b).

The excited atoms fall quickly to an intermediate, metastable state of energy E_2; the arrow in (b) indicates this transition. The atoms have now experienced a population inversion, a situation in which more atoms are in the excited state of energy E_2 than in the ground state of energy E_1 (c). The transition from the metastable state back to the ground state is called a **forbidden transition,** since it occurs spontaneously at a much slower rate than most atomic deexcitation transitions. For this reason the atoms remain much longer in state E_2 than they normally remain in other excited states—the state is "almost stable" or metastable.

The deexcitation of this inverted population begins slowly (compared to the normal deexcitation time). An atom falls spontaneously back to the ground state and emits a photon (d). This photon, as it passes another atom in the metastable state, can cause it to fall to the ground state by stimulated emission. Two photons of equal energy and in phase with each other leave the second atom (e). These two photons, in turn, stimulate deexcitation of two more excited atoms. We now have four photons moving together, each of which can stimulate more transitions. Before long, a huge cascade of atoms is returning to the ground state by stimulated emission, and each atom contributes one photon whose energy and phase equals that of the other photons. The process, represented schematically in Fig. 33.10, is called light amplification by the stimulated emission of radiation; hence the acronym **laser.**

The deexcitation of the inverted population is not complete after one pass of the beam through the excited atoms. Mirrors placed at the end of the laser reflect the beam back and forth through the excited atoms many times. If the population remains inverted, the intensity of the beam grows during each round trip. However, the mirror at one end of the laser reflects only part of the beam and allows the other part to leave the laser. In a **continuous laser,** an equilibrium is reached where energy lost by the beam leaving the laser equals energy gained by the partially reflected beam as it makes another round trip through the excited atoms. A continuous narrow beam of light leaves the laser.

A beam of light from a laser is unique for several reasons. (1) The light is **monochromatic;** that is, all the photons have the same wavelength. (2) The photons in the beam are in phase with each other (the light is said to be **coherent**) rather than a group of independent photons whose vibrations are independent of each other. The coherence of a laser beam allows it to be used to transmit information (such as telephone conversations) in much the way that high-frequency radio waves transmit information for television broadcasting. (3) Ordinary sources of light such as tungsten bulbs emit light in all directions,

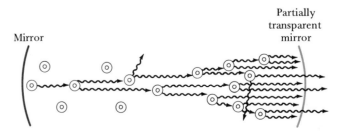

FIG. 33.10. The increasing intensity of a laser beam as it passes through a group of excited atoms in a laser.

and the light intensity drops as the light spreads after leaving the source. On the other hand, a laser beam is very narrow, being limited only by diffraction at the mirror where it leaves the laser. If the laser is well designed, the angular spread of the beam is approximately λ/D (see Section 21.5 and the interlude on page 434), where λ is the wavelength of the light and D is the diameter of the mirror. Even though the power of the laser may be low, the small spread of the beam causes the light to be relatively intense (Fig. 33.11).

FIG. 33.11. A narrow laser beam illuminates London's Big Ben from almost three miles away. The laser light on Big Ben is much brighter than the light illuminating the tower from more powerful lamps on the nearby bridge. (United Press International Photo)

Laser Use in Eye Surgery

The fact that a laser beam is very narrow has led to applications requiring a significant amount of energy in very localized places, such as when drilling small holes in metal sheets. Another of these applications is in eye surgery. As you may recall from Section 20.4, the retina of the eye contains about 130 million cones and rods, each of which serves as a light detector. The cones and rods connect to nerves that transmit visual information to the brain. If a disease or a severe blow to the head causes part of the retina to detach from the underlying tissue (Fig. 33.12a), the path of nerve impulses to the brain is interrupted, and the affected area becomes blind.

Retina

Detached retina

(a)

Laser weld

(b)

FIG. 33.12. (a) The eye showing a small retinal detachment. (b) A magnified cross section of a laser weld to hold the retina in place. (Courtesy of International Research and Development Co. Ltd., Newcastle-upon-Tyne, England)

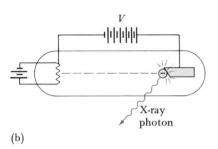

FIG. 33.13. (a) An electron starts its trip across an x-ray tube. (b) When the electron crashes into the anode, x-ray photons are created.

Modern treatment of a detached retina involves welding it to the underlying tissue by using the heat produced by a short laser flash. The cross section of one of these welds is shown in Fig. 33.12b. Lasers are very useful in eye surgery for several reasons. (1) The surgery is noninvasive; other parts of the eye are not invaded by foreign objects such as scalpels, clamps, sponges, or tweezers. (2) The weld spots can be extremely small, an important factor when the detachment occurs in the central part of the retina that is used for seeing fine detail. (3) Welds can be made in a fraction of a second, so the eye does not have to be clamped open. (4) The laser weld causes no pain.

33.7 X-Ray Emission

Usually, the outer electrons in atoms are the only electrons involved in transitions from one state to another. The outer electrons are bumped into excited states when atoms collide with each other, and they jump between different states when photons of light and ultraviolet radiation are absorbed and emitted by atoms. Inner electrons can also change states, but much more energy is needed to excite an inner electron to a vacant state of higher energy. Consequently, these inner electrons usually remain in their orbits close to the nucleus. One opportunity for leaving an inner orbit occurs if an inner electron of an atom is struck by an energetic electron in an x-ray tube.

An x-ray tube, such as shown in Fig. 33.13, has a filament that becomes hot when an electric current flows through it. The extra thermal energy of the atoms and ions in the filament causes electrons to be bumped from the filament. These electrons are accelerated by a large voltage toward a metal anode commonly made of tungsten. In the x-ray tube used to produce x-rays for medical and dental examinations, the voltage from the filament to the anode is typically about 40,000 V. The speed of an electron released from the filament increases to over 10^8 m/s, almost the speed of light, just before it crashes into the anode.

The electron, as it strikes the anode, can be stopped by a single dramatic collision or by a series of collisions. During each collision the electron loses kinetic energy, which is balanced by the creation of a photon. The photons produced during these collisions are called **Bremsstrahlung radiation** (from the German word meaning "braking radiation") and are a consequence of the electron's deceleration. Most of the photons are in the x-ray portion of the electromagnetic spectrum.

If the electron stops in one direct hit, it emits one photon whose energy equals the energy gained by the electron as it accelerated across the tube. Thus, the maximum energy of a photon created by a collision is

$$hf_{max} = eV. \qquad (33.5)$$

$$\begin{array}{cc} \text{photon} & \text{energy gained as} \\ \text{energy} & \text{electron crosses} \\ & \text{x-ray tube} \end{array}$$

If the electron stops in a series of collisions, a group of lesser-energy photons is created.

A typical x-ray spectrum produced by these electron collisions is shown in Fig. 33.14. Notice that no photons are produced with frequencies greater than $f_{max} = eV/h$, as predicted by Eq. (33.5). The smooth part of the curve in Fig.

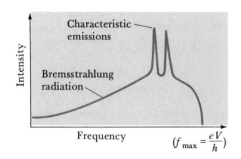

FIG. 33.14. The intensity of radiation emitted by an x-ray tube as a function of frequency.

33.14 is caused by less energetic collisions, which produce Bremsstrahlung radiation.

The pointed lines on the curve in Fig. 33.14, called **characteristic lines,** result from collisions of the electron flying across the tube with inner electrons of atoms in the anode. If a collision causes the $1s$ electron to be knocked from a tungsten atom, then outer electrons of the atom fall into the vacancy left in the inner subshell. The characteristic lines in Fig. 33.14 are the result of photons emitted by $2p \longrightarrow 1s$ and $3p \longrightarrow 1s$ transitions of tungsten atoms in the anode.

We can estimate the energy needed to knock an inner electron from a many electron atom by using Bohr's equation for the energy of an electron when in different orbits [Eq. (32.11)]:

$$E_n = -\frac{13.6Z^2}{n^2} \text{ eV.}$$

Notice that the energy is proportional to Z^2, the number of protons in the nucleus. For tungsten, $Z = 74$, and the $n = 1$ electrons in tungsten should have energy of approximately

$$E_1 \simeq -\frac{13.6(74^2)}{1^2} = -70{,}000 \text{ eV.}$$

To remove an $n = 1$ electron from tungsten would require about $70{,}000$ eV of energy.

Bohr's equation was developed only for one-electron atoms, and it overestimates the energy needed to remove an $n = 1$ electron from an atom with many electrons. Using Eq. (32.11), we estimate the energy of an $n = 1$ electron in iron ($Z = 26$) to be -9200 eV, whereas the experimentally determined value is -7130 eV. Nevertheless, the equation does provide a rough indication of the great energy needed to dislodge an inner electron from an atom. The energy depends very much on the nuclear charge [the Z^2 dependence in Eq. (32.11)].

Summary

1. **Quantum mechanics** is a theory used to determine the energy, distribution in space, and other properties of particles such as electrons when in different environments. **Schrödinger's equation** is solved for **wave functions,** which, when squared, indicate the probability of finding an electron or some other particle at different locations. The wave functions can also be used to determine the electron's energy when in a state described by a particular wave function and its chance of jumping to another state of the atom.

2. **Quantum numbers:** Usually, many different wave functions are solutions to Schrödinger's equation, and each wave function defines a place for an electron to reside—an electron **state.** For atoms the states are labeled using four quantum numbers, which have integral or half-integral values and indicate various properties about the electron and its orbit.

The **principle quantum number** n has positive integral values

$$n = 1, 2, 3, 4, \ldots \tag{33.1}$$

and indicates the energy and average distance of the orbit from the nucleus. The larger the value of n, the greater the electron's average separation from the nucleus and the greater its energy.

The **orbital angular momentum quantum number** l indicates the shape of the electron distribution and the angular momentum of an electron. For a particular value of n, states exist for the following values of l:

$$l = 0, 1, 2, 3, \ldots, n - 1 \tag{33.2}$$
$$s \quad p \quad d \quad f$$

The **orbital magnetic quantum number** m_l indicates the orientation of the orbit and the orientation of its magnetic moment due to its orbital motion. For a particular value of n and l, states exist for the following values of m_l:

$$m_l = 0, \pm 1, \pm 2, \ldots, \pm l \tag{33.3}$$

The **spin magnetic quantum number** m_s indicates the orientation of the axis of an electron's spinning motion. The two allowed values of m_s are

$$m_s = +\tfrac{1}{2} \text{ (up)} \quad \text{or} \quad -\tfrac{1}{2} \text{ (down)} \qquad (33.4)$$

3. **Multielectron atoms:** The electrons in an atom usually occupy the states with lowest energy, but subject to the **Pauli exclusion principle:** an electron state of an atom, represented by the symbol $|nlm_lm_s\rangle$, can hold no more than one electron. The lowest-energy electron states are those with (1) the smallest value of n, and (2) the smallest value of l. An electron configuration indicates the electron distribution among different orbits (for example, $1s^2 2s^2 2p^6 3s^2 3p^6 4s$, and so forth), where the superscript indicates the number of electrons in a subshell with a particular value of n and l. The letters $s, p, \ldots$, indicate the value of l for a subshell (see Table 33.2).

4. **Periodic table:** Atoms whose outer electrons are in similar electron configurations are grouped together in the periodic table. These atoms exhibit similar properties.

5. **Electron transitions:** An atom can **absorb** a photon if the photon's energy equals the difference in energy of the state in which an electron resides and a higher-energy vacant state. An excited atom can emit a photon by falling **spontaneously** to a lower-energy state. An excited atom can also emit a photon by a process called **stimulated emission:** another photon passes the excited atom and induces or stimulates its deexcitation, thus causing the production of a second photon whose energy and phase equal that of the first photon. The coherent narrow beam of light from a laser is caused by stimulated emission.

6. **X-rays** are produced by the deceleration of fast-moving electrons as they hit the metal anode of an x-ray tube. X-ray photons are also created as outer electrons in atoms fall into a vacant inner orbit.

Questions

1. Discuss the similarities and differences in the way a hydrogen atom is pictured in Bohr's model and in quantum mechanics.

2. Explain why each of the following is an unlikely ground-state electron configuration for an atom: (a) $1s^2 2s^2 2p^6 3s^2 4s^2$; (b) $1s^2 2s^2 2p^6 3s^2 4p^3$; (c) $1s^2 2s^2 2p^4 3s^2 4s$; (d) $1s^2 2s^2 2p^6 3s^2 3p^6 4s^2 4d^2$.

3. Why do fluorine and chlorine exhibit similar properties?

4. Why do atoms in the first column of the periodic table exhibit similar properties?

5. Explain how astronomers might determine the different types of atoms that are on the sun and the amount of each type.

6. Can a phosphorescent material absorb visible radiation and emit, after some delay, ultraviolet radiation? Explain.

7. Indicate the difference between spontaneous emission and stimulated emission.

8. What property of a laser beam makes it useful for (a) cutting metal, and (b) carrying information?

Problems

33.2 Quantum Numbers for Atomic Wave Functions

1. (a) An electron is in an $n = 5$ orbit. List the possible values of its l quantum number. (b) If the electron happens to be in an $l = 2$ state, list the possible values of its m_l quantum number. (c) List the possible values of its m_s quantum number.

2. (a) An electron in an atom is in a state with $m_l = 3$ and $m_s = +\tfrac{1}{2}$. What are the smallest possible values of l and n for that state? (b) Repeat for an $m_l = 2$ and $m_s = -\tfrac{1}{2}$ state.

3. (a) An electron resides in an $n = 7$ orbit. List the possible values of its l quantum number. (b) Of these different l states, the electron occupies the $l = 4$ state. List the possible values of the m_l quantum number.

4. Using arrows, represent schematically the possible orientations of an $l = 2$ state in a magnetic field. Indicate the value of m_l for each orientational state.

33.3 Multielectron Atoms

5. List the $|nlm_lm_s\rangle$ states available to an electron in a $4p$ subshell.

6. List the $|nlm_lm_s\rangle$ states available to an electron in a $4f$ subshell.

7. Identify the values of n and l for each of the following subshell designations: $3s$, $2p$, $4d$, $5f$, and $6s$.

8. Determine the electron configuration for oxygen ($Z = 8$).

9. Determine the electron configuration for chlorine ($Z = 17$).

10. Determine the electron configuration for iron ($Z = 26$).

11. Manganese ($Z = 25$) has two $4s$ electrons. How many $3d$ electrons does it have? Explain your answer.

33.4 The Periodic Table

12. Determine the electron configurations of four elements in Group I of the periodic table. Explain why these elements are likely to have similar properties.

13. Determine the electron configurations of three elements in Group VI of the periodic table. Explain why these elements are likely to have similar properties.

33.5 Absorption and Emission of Radiation

14. A sodium atom in its ground state absorbs a 589-nm photon when its $3s$ electron jumps to the $4p$ state. Calculate the energy difference of the $3s$ and $4p$ states.

15. A fluorescent molecule, oxyluciferin, absorbs a photon of wavelength 370 nm and emits two photons, one of wavelength 480 nm. Determine the wavelength of the second emitted photon using the fact that the energy of the absorbed photon equals the sum of the energies of the emitted photons.

33.6 Stimulated Emission and Lasers

16. (a) A laser pulse emits 2.0 J of energy during a time of 1.0×10^{-9} s. Calculate the power emitted during that short time. (b) Calculate the light intensity (power per area) in the laser beam if its cross-sectional area is 8.0×10^{-6} m^2.

17. A pulsed argon laser of 456.5-nm wavelength emits 3.0×10^{-3} J of energy to produce a tiny weld in the retina of a person's eye. How many photons are in the laser pulse?

18. A pulsed laser used for welding produces 100 watts of power during a time of 10 ms. Calculate the energy delivered to the weld.

19. A continuous helium-neon laser emits a beam that diverges at an angle of 4.2×10^{-5} radians. Calculate the width of a cross section of the laser beam after it has traveled (a) 1.0 km, and (b) to the moon.

33.7 X-Ray Emission

20. Electrons are accelerated across a 40,000-V potential difference. Calculate the frequency and wavelength of the maximum-energy x-ray photons created when the electrons crash into the anode of an x-ray tube.

21. An x-ray tube emits photons of frequency 1.33×10^{19} Hz or less. Calculate the electrical potential difference across the tube.

22. Estimate the energy of a $1s$ electron in gold ($Z = 79$).

23. Estimate the minimum voltage across an x-ray tube that results in the removal of $1s$ electrons from molybdenum atoms ($Z = 42$) in the anode of the tube.

General Problems

G1. One of the easiest problems involving quantum mechanics involves the determination of the wave functions for an electron inside an infinite square-well potential (Fig. 33.15). The electron is prohibited from leaving the well but moves freely inside it. Its allowed wave functions appear exactly like the waves on a string fixed at both ends. For a well of width a, the allowed wavelengths

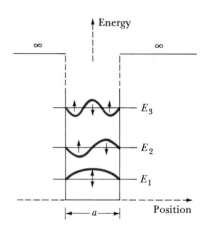

FIG. 33.15. The wave functions for an electron inside an infinitely deep square-well potential look like standing waves on a string fixed at both ends.

are $\lambda = 2a/n$ where $n = 1, 2, 3, \ldots$ [see Eq. (18.3)]. Using the expression for the de Broglie wavelength and kinetic energy of an electron (its potential energy in the well is zero), show that the electron's energy when in a state with quantum number n is $E_n = (h^2/8a^2m)n^2$.

G2. The mass of one mole of iron (6.0×10^{23} atoms) is 0.056 kg and iron's density is 7.9×10^3 kg/m^3. (a) Show that the volume of space occupied by one iron atom is 1.2×10^{-29} m^3. (b) Calculate the volume of space occupied by a lithium atom, the metal of least density 0.53×10^3 kg/m^3. The mass of one mole of lithium atoms is 0.0069 kg. (c) Repeat the calculations in (b) for an osmium atom, the metal of greatest density 22.5×10^3 kg/m^3. The mass of one mole of osmium atoms is about 0.192 kg.

G3. A laser used to weld the retina of an eye emits 20 mW of power for 100 ms. The light is focused on a spot 0.1 mm in diameter. Assume that the laser's energy is deposited in a small sheet of water of 0.10 mm diameter and 0.30 mm thickness. (a) Calculate the energy deposited. (b) Calculate the mass of this water. (c) Calculate the increase in temperature of the water (assume that it does not boil and that its heat capacity is 4180 J/kg · C°).

G4. An x-ray tube operates at 40,000 V. The beam of electrons flowing from the filament to the anode produces a current of 10 mA. (a) Calculate the rate of heat production in the anode by the collisions of these electrons with the anode. (b) If the specific heat capacity of the 0.10-kg anode is 460 J/kg · C°, calculate its temperature increase in 1.0 second if no cooling water is used.

34

The Nucleus
and Radioactivity

An atom's nucleus occupies only about 10^{-12} percent of the atom's volume. If an atom's size increased so that its cross section equalled that of a football field, the nucleus would be smaller than your thumbnail. Although it is so small, the nucleus contains 99.97 percent of the atom's mass because the constituents of the nucleus, protons and neutrons, are packed closely together and each particle has a mass roughly 2000 times greater than that of an electron, the third particle of which atoms are made. The protons and neutrons in a nucleus are held together by a strong, though poorly understood, nuclear force. Disturbances of these tightly packed particles in the nucleus may cause reactions in which the nucleus breaks apart, ejects one or a group of its constituent particles, or absorbs a new particle.

In this chapter we will see how to determine the amount of energy released or absorbed during nuclear reactions, with emphasis on nuclei undergoing radioactive decay. We will also learn of several applications of radioactive decay, such as determining the age of ancient objects unearthed by archeologists and the effects that particles emitted during radioactive decay have on biological materials.

34.1 The Nucleus

The protons in a nucleus of an atom each have a charge $+e$ of 1.6×10^{-19} C and a mass m of 1.673×10^{-27} kg. The neutrons found in a nucleus have no net electric charge and a mass of 1.675×10^{-27} kg—slightly greater than the mass of a proton. The protons and neutrons in a nucleus are referred to as **nucleons.**

A nucleus is labeled by indicating its number of protons and neutrons:

Z = atomic number = number of protons in nucleus,
N = neutron number = number of neutrons in nucleus,
A = mass number = $Z + N$
  = the total number of nucleons (protons and neutrons) in nucleus.

The values of A, Z, and N for several nuclei appear in Table 34.1. The symbol $^A_Z X$, shown in the first column of this table, combines the information about nuclear constituency with the chemical symbol for an atom. For example, the lithium atom, with $Z = 3$, $A = 7$, and chemical symbol Li, is written as $^7_3$Li. Although the number of neutrons is not given in this symbol, it can be determined easily by subtracting Z from A.

664

TABLE 34.1 Values of A, Z, and N for the Nuclei of Several Atoms

Nucleus	Mass number $(Z + N)$ A	Atomic number (Number of protons) Z	Neutron number N
^1_1H	1	1	0
$^2_1\text{H} = \text{D}$	2	1	1
^4_2He	4	2	2
^6_3Li	6	3	3
^7_3Li	7	3	4
$^{12}_6\text{C}$	12	6	6
$^{13}_6\text{C}$	13	6	7
$^{14}_6\text{C}$	14	6	8
$^{16}_8\text{O}$	16	8	8
$^{235}_{92}\text{U}$	235	92	143
$^{238}_{92}\text{U}$	238	92	146

Providing both the chemical symbol Li and the value of Z is unnecessary, since lithium is the name of an atom with three electrons whose nucleus has three protons, and thus $Z = 3$ for lithium. For this reason we often omit the value of Z when writing the symbol for an atom or nucleus (^7Li instead of ^7_3Li).

You will notice in Table 34.1 that several nuclei (such as carbon) have the same number of protons but different numbers of neutrons: $^{12}_6\text{C}$, $^{13}_6\text{C}$, and $^{14}_6\text{C}$. Nuclei with the same number of protons but different numbers of neutrons are called **isotopes**—$^{12}_6\text{C}$ is a carbon isotope. Each isotope has the same number of orbital electrons; consequently, their chemical properties are very similar. Because the isotopes have a different number of neutrons in their nuclei, however, the properties of the nuclei may be quite different. For example, the uranium isotope, $^{235}_{92}\text{U}$, can take part in the nuclear reactions that provide energy for power plants, while the $^{238}_{92}\text{U}$ isotope cannot.

Isotopes also vary in **natural abundance**—the percentage of an element found in a particular isotopic form. The abundances of the three carbon isotopes mentioned earlier are as follows: $^{12}_6\text{C}$ (98.89 percent), $^{13}_6\text{C}$ (1.11 percent), and $^{14}_6\text{C}$ (10^{-10} percent).

The size of a nucleus can be determined by scattering high-energy electrons, protons, or alpha particles from the nucleus. The nucleus is found to be roughly spherical in shape (Fig. 34.1), and it has a radius in meters given by the equation

$$R = 1.2 \times 10^{-15} A^{1/3} \text{ m},$$

where A is the number of nucleons in the nucleus.

FIG. 34.1. The nucleus consists of protons ($\oplus$) and neutrons ($\bullet$) held close to each other by a strong nuclear force.

34.2 Nuclear Force

Holding a group of protons together within a volume the size of a nucleus is no small task. The protons, having like electrical charges and being so close together, repel each other with an electrical force about 100 million times greater than the force that holds electrons in orbit about the nucleus. Without some strong attractive force to balance this electrical repulsion, nuclei would quickly disintegrate as protons were repelled from the nucleus by their positively charged neighbors.

The attractive force that keeps the protons and neutrons together in a nucleus is called the **strong nuclear force.** A precise mathematical description of this force has not been formulated, but enough is known about the force to realize that it is quite different from the two other fundamental forces we have already studied, the gravitational force and the electrical force. The strong nuclear force is a short-range force: it provides a very strong attraction between two nucleons that are separated by about 5×10^{-15} m or less; but if they are separated farther, the force is essentially zero. By comparison, the gravitational and electrical forces decrease as masses and electric charges are separated farther, but the forces do not decrease to zero until the separation is infinite. The strong nuclear force is the same between any two nucleons (two protons, two neutrons, or a proton and a neutron) and does not depend on their charge. Finally, the magnitude of the strong nuclear force of attraction between two nucleons separated by 5×10^{-15} m or less is much greater than the electrical force and very much greater than the gravitational force between the nucleons. It is the strength of the nuclear force that keeps protons and neutrons bound together tightly in the nucleus.

Because of the short range of the strong nuclear force, a particular proton or neutron in the nucleus is attracted by the strong nuclear force only to its nearest neighbors. Yet, a proton feels an electrical repulsion from all other protons in a nucleus. For nuclei with many protons, the electrical repulsion of a proton from all other protons may exceed the strong nuclear attraction of the proton for its neighbors. The nucleus may be unstable and eject one or a group of its particles; the nucleus is said to undergo **radioactive decay.** Surprisingly, the nucleus emits not a proton but an alpha particle (a small nucleus that consists of two protons and two neutrons) or an electron. The reason for this behavior is discussed in Section 34.5.

34.3 Binding Energy per Nucleon

Discussing the stability of nuclei in terms of the strong nuclear force is a cumbersome problem because of our incomplete understanding of the strong nuclear force and because of the unknown distribution of protons and neutrons in nuclei. We can address the stability of nuclei and many other nuclear phenomena more easily using the ideas of mass defect, binding energy, and binding energy per nucleon.

The mass of every stable nucleus (and even of unstable nuclei) is less than the mass of the nucleons of which it is made. The nucleus is said to have a *mass defect;* it cannot break apart into free protons and neutrons, since the nucleus lacks the mass needed for these particles. We can think of the stability of the nucleus as resulting from this mass deficiency. The greater the mass defect, the more stable the nucleus.

The mass defect usually constitutes only a small fraction of the total mass of the nucleus and is calculated conveniently using a unit called the atomic mass unit.

One **atomic mass unit** (u) equals one-twelfth the mass of a $^{12}_{6}$C atom, including the mass of its six electrons. In terms of kilograms,

$$1 \, u = 1.660566 \times 10^{-27} \, kg.$$

TABLE 34.2 The Rest Masses of Several Particles and Atoms

Particle or atom	Mass		
	kg	u	MeV/c²
Electron	9.11×10^{-31}	0.000549	0.511
Proton	1.673×10^{-27}	1.007276	938.3
Neutron	1.675×10^{-27}	1.008665	939.6
^1_1H	1.673×10^{-27}	1.007825	938.8
^4_2He	6.647×10^{-27}	4.002604	3,728.4
^7_3Li	11.651×10^{-27}	7.016005	6,535.4
$^{23}_{11}\text{Na}$	38.176×10^{-27}	22.989770	21,414.9

A list of the masses of several particles and atoms appears in Table 34.2. Appendix C contains a more comprehensive list.

In order to understand how the mass of a nucleus is used to determine the energy binding it together, we can perform a simple thought experiment using a lithium atom, ^7_3Li, whose mass, according to Table 34.2, is 7.016005 u. We pull the atom apart to obtain the separate particles of which it is made: three electrons that had been orbiting the nucleus plus three protons and four neutrons from the nucleus (Fig. 34.2). If we now combine each proton with an electron to form a hydrogen atom, we are left with three hydrogen atoms (^1_1H) and four neutrons ($^1_0 n$).* The mass of the separated parts is

$$3m_{^1_1\text{H}} = 3(1.007825 \text{ u}) = 3.023475 \text{ u}$$
$$4m_{^1_0 n} = 4(1.008665 \text{ u}) = 4.034660 \text{ u}$$
$$\text{total mass of constituents} = 7.058135 \text{ u}.$$

The total mass of the constituents (7.058135 u) is 0.042130 u greater than the mass of the lithium atom itself (7.016005 u). Almost all of this missing mass is associated with the nucleus and the particles of which it is made. The lithium nucleus lacks 0.042130 u of mass that is needed to spontaneously break apart to form its separate protons and neutrons. Lacking the mass to break apart, the particles must stay together.

The "missing mass" or **mass defect** of a nucleus is defined as follows:

$$\text{mass defect} = \Delta m = [Z m_{^1_1\text{H}} + (A - Z)m_{^1_0 n}] - m_{^A_Z \text{X}}, \qquad \textbf{(34.1)}$$

*The symbol for a neutron is $^1_0 n$, since its charge is zero and its nucleon number is 1.

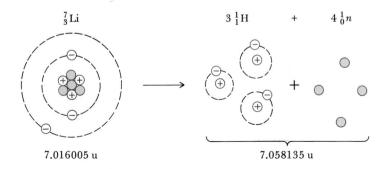

FIG. 34.2. A ^7_3Li atom has less mass than the particles of which it is made.

where Z is the number of protons in the nucleus, $m_{_Z^A X}$ is the mass of the atom whose mass defect is being calculated, $m_{_1^1 H}$ is the mass of a hydrogen atom, and $m_{_0^1 n}$ is the mass of a neutron.

The nuclear **binding energy** (BE) of a nucleus equals the energy that is equivalent to the mass defect Δm. This energy is determined using Einstein's famous equation that relates mass and energy:

$$BE = \Delta m \, c^2. \tag{34.2}$$

The binding energy represents the total energy needed to separate the atom and its nucleus into hydrogen atoms and neutrons. All but an infinitesimal amount of this energy is needed to pull the nucleus apart.

The binding energy of a nucleus is usually calculated in units of million electron volts (MeV), where

$$1 \, \text{MeV} = 10^6 \, \text{eV} = 1.602 \times 10^{-13} \, \text{J}.$$

The calculation of binding energy from mass defect [Eq. (34.2)] is made easier using a conversion that relates a 1-u mass to its equivalent energy:

$$mc^2 = E$$

or

$$(1 \, \text{u})c^2 = (1.660566 \times 10^{-27} \, \text{kg})(3.0 \times 10^8 \, \text{m/s})^2$$
$$= 1.492 \times 10^{-10} \, \text{J}$$
$$= 931.5 \, \text{MeV}.$$

Rearranging the last equation, we see that

$$1 \, \text{u} = 931.5 \, \text{MeV}/c^2. \tag{34.3}$$

EXAMPLE 34.1 Calculate the binding energy of a $_{11}^{23}$Na nucleus.

SOLUTION First we calculate the mass defect using Eq. (34.1):

$$\Delta m = [11m_{_1^1 H} + 12m_{_0^1 n}] - m_{_{11}^{23} Na}$$
$$= 11(1.007825 \, \text{u}) + 12(1.008665 \, \text{u}) - 22.989770 \, \text{u} = 0.200285 \, \text{u}.$$

Substituting this value for Δm into Eq. (34.2), we find that

$$BE = \Delta m \, c^2 = \left(0.200285 \, \text{u} \times \frac{931.5 \, \text{MeV}/c^2}{1 \, \text{u}}\right) c^2 = \underline{186.6 \, \text{MeV}}.$$

Notice that the conversion $931.5 \, \text{MeV}/c^2 = 1 \, \text{u}$ has been used to convert the mass defect in atomic mass units to its equivalent energy in MeV; the c^2 canceled, as did the atomic mass unit u. ∎

The stability of a nucleus depends more on the average binding energy of each nucleon that on the total binding energy of the whole nucleus. For sodium, which has 23 nucleons and a total binding energy of 186 MeV, the average binding energy per nucleon is 186 MeV/23 nucleons, or 8.1 MeV/nucleon. The binding energy per nucleon for most stable nuclei ranges from about 6 to 8 MeV/nucleon. To calculate the binding energy per nucleon, divide the total

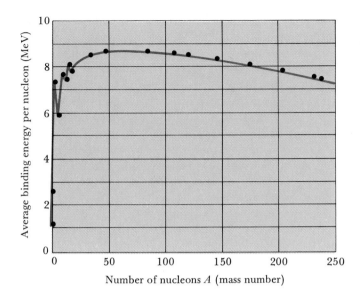

FIG. 34.3. The average binding energy per nucleon is plotted versus the number of nucleons for stable nuclei.

binding energy by the atomic mass number A:

$$\text{binding energy per nucleon} \equiv \frac{BE}{A}. \qquad \textbf{(34.4)}$$

The binding energy per nucleon for stable nuclei is plotted as a function of A in Fig. 34.3. Notice that nuclei with $A \simeq 60$ have the greatest binding energy per nucleon and are also the most stable nuclei, while very small and very large nuclei have less binding energy per nucleon and are less stable. The binding-energy-per-nucleon curve shown in Fig. 34.3 is used to analyze several nuclear processes, such as fission (the splitting of large nuclei) and fusion (the combining of small nuclei).

34.4 Nuclear Reactions

Nuclear reactions, like chemical reactions, involve the conversion or transmutation of nuclear reactants to different nuclei that are products of the reaction. In these reactions two nuclei may interact to form one or more new nuclei, or a single nucleus may divide into two or more nuclei or emit a small particle, thus leaving behind a different nucleus. The nuclear reactions are often written using the atomic symbols rather than special nuclear symbols. Consider the reaction of a proton with a lithium nucleus to form two helium nuclei (Fig. 34.4):

$$^1_1\text{H} + {}^7_3\text{Li} \longrightarrow {}^4_2\text{He} + {}^4_2\text{He} + 17.3\,\text{MeV}. \qquad \textbf{(34.5)}$$

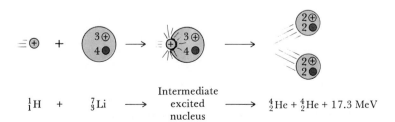

FIG. 34.4. A nuclear reaction in which a proton ($^1_1$H) combines with a lithium nucleus to form two helium nuclei.

Notice that the proton, the nucleus of a hydrogen atom, is represented by the symbol for the hydrogen atom ($_1^1H$). The lithium nucleus is represented by $_3^7Li$, the symbol for a lithium atom, and so forth. This convention causes no problem, since an equal number of orbital electrons are included on each side of Eq. (34.5)—one for the hydrogen plus three for the lithium equals two each for the two helium atoms. The advantage of using this convention for writing nuclear reactions is that atomic masses found in atomic mass tables can be used to calculate the energy changes during the reaction. Even though atomic masses are used, the energy changes are associated almost entirely with differences in the nuclear masses of the reactants and the products of the reaction.

These nuclear reactions must satisfy several rules, such as those used to balance chemical reactions:

Rule 1. The sum of the atomic numbers (Z) must be the same on both sides of the equation.

Notice, for example, that the atomic numbers for the lithium-proton reaction described by Eq. (34.5) add to four on each side of the equation ($1 + 3 = 2 + 2$). This rule is needed to conserve the electric charge of the nuclei; we must start and finish with the same amount of electric charge.

Rule 2. The sum of the atomic mass numbers (A) must be the same on both sides of the equation.

For the reaction described by Eq. (34.5), the mass numbers add to eight on each side of the reaction ($1 + 7 = 4 + 4$). This rule indicates that the number of nucleons (protons plus neutrons) must be conserved. Reactions in which neutrons are converted to protons (or vice versa) can occur, but the total number of nucleons of either type must not change during the reaction.*

Rule 3. The energy released or consumed during a reaction can be determined by subtracting the mass of the products from the mass of the reactants and converting this mass difference to energy:

$$\Delta E = \left(\sum m_{\text{reactants}} - \sum m_{\text{products}} \right) c^2. \qquad \textbf{(34.6)}$$

This energy difference is sometimes called the **Q of the reaction.** If Q is positive, energy is released. If Q is negative, energy must be absorbed to produce the reaction products.

EXAMPLE 34.2   Calculate the Q of the following reaction:

$$_1^1H + {}_3^7Li \longrightarrow {}_2^4He + {}_2^4He.$$

SOLUTION

$$\sum m_{\text{reactants}} = m_{_1^1H} + m_{_3^7Li} = 1.007825 \text{ u} + 7.016005 \text{ u} = 8.023830 \text{ u}$$
$$\sum m_{\text{products}} = 2m_{_2^4He} = 2(4.002604 \text{ u}) = 8.005208 \text{ u}.$$

---

*In the next chapter we will find that this rule is stated differently, using a quantity called baryon number. For now Rule 2 will be satisfactory as stated, even though it is not strictly correct.

Since the products have less mass than the reactants, energy is released during the reaction. Using Eq. (34.6) and the conversion $1 \text{ u} = 931.5 \text{ MeV}/c^2$, we find that the energy released is

$$\Delta E = [(8.023830 \text{ u} - 8.005208) \text{ u}]c^2$$

$$= \left[0.018622 \text{ u} \times \frac{931.5 \text{ MeV}/c^2}{1 \text{ u}}\right]c^2 = \underline{17.3 \text{ MeV}}. \qquad \blacksquare$$

EXAMPLE 34.3 Use Rules 1 and 2 and the periodic table (Table 33.4) to determine the missing product or reactant for each of the nuclear reactions shown below:

(a) $\qquad\qquad {}^{6}_{3}\text{Li} + {}^{1}_{0}n \longrightarrow ? + {}^{3}_{1}\text{H} + \text{energy}.$

(b) $\qquad\qquad\quad {}^{7}_{4}\text{Be} + {}^{0}_{-1}e \longrightarrow ? + \text{energy}.$

(c) $\qquad {}^{1}_{0}n + {}^{235}_{92}\text{U} \longrightarrow {}^{147}_{56}\text{Ba} + ? + 3\,{}^{1}_{0}n + \text{energy}.$

(d) $\qquad\qquad\quad {}^{4}_{2}\text{He} + ? \longrightarrow {}^{17}_{8}\text{O} + {}^{1}_{1}\text{H} - \text{energy}.$

(e) $\qquad\qquad\qquad {}^{14}_{6}\text{C} \longrightarrow ? + {}^{0}_{-1}e + \text{energy}.$

SOLUTION (a) For the values of Z and A to be equal on each side of this reaction, the missing nucleus must have $Z = 2$ and $A = 4$. We see from the periodic table that helium has $Z = 2$. Thus, the missing product is ${}^{4}_{2}\text{He}$. (b) An electron is represented by the symbol ${}^{0}_{-1}e$. The value of Z equals -1, since its charge is $-e$, and A equals 0, since it is not a nucleon. The missing nucleus must have $Z = 4 - 1 = 3$ and $A = 7 + 0 = 7$ and is lithium (${}^{7}_{3}\text{Li}$). (c) You should confirm that the missing nucleus is ${}^{86}_{36}\text{Kr}$. (Do not forget to include *three* neutrons in your calculations.) (d) and (e) The missing nucleus in both reactions is ${}^{14}_{7}\text{N}$.

$\qquad\qquad\qquad\qquad\qquad\qquad\qquad\qquad\qquad\qquad\qquad\qquad\qquad\qquad\blacksquare$

34.5 Radioactive Decay

The nuclei of some unstable atoms undergo a process called **radioactive decay** in which a small particle is emitted from the nucleus, resulting in the formation of a new nucleus. Radioactive decay was discovered quite by accident in 1896 by a French physicist, Henri Becquerel (1852–1908). Becquerel had been investigating a phenomenon called phosphorescence—the excitation of an atom followed by a delayed emission of light from the atom.

Becquerel's equipment consisted of a photographic plate wrapped carefully in two sheets of heavy black paper so that no light could enter. Becquerel placed some uranium salt crystals on the plate, then set the plate in the sun. When he developed the plate, Becquerel found that it was exposed at the places over which the crystals had been lying. Becquerel thought that the sunlight's interaction with the crystals was somehow involved in producing the radiation that passed through the paper and exposed the plates.

The next day, when Becquerel tried to repeat his experiment, cloudy weather intervened, so he stored his salt sample on top of a box of unexposed photographic plates in a drawer. Several days later, while checking to be sure his plates were still good, he found that they had been exposed in a shape resembling the pile of uranium crystals. Becquerel now realized that the sunlight had nothing to do with this exposure; his plates were being exposed by some unusual emission from the uranium crystals. He had accidentally discovered radioactivity, the discovery for which he received the Nobel Prize in Physics in 1903.

We now know that these unusual radioactive emissions are caused by the disintegration or decay of unstable nuclei. The decay usually involves the ejection of a small particle from the "parent" nucleus. A new "daughter" nucleus is left behind. In some cases the newly formed daughter nucleus is unstable and also undergoes radioactive decay.

A nucleus may be unstable for several reasons. If there are too many protons in relation to the number of neutrons, the repulsive electric force between protons exceeds the nuclear force holding the protons together. It would seem that a nucleus could correct this excess-proton instability problem by ejecting one or more protons from the nucleus; yet, surprisingly, we seldom see a radioactive nucleus emit a proton. Instead, a nucleus that spontaneously disintegrates emits two other types of particles: (1) helium nuclei [called **alpha** (α) **particles**], and (2) electrons or positrons [called **beta** (β) **particles**]. A positron has the same mass as an electron and is similar to an electron in all respects, except that its charge is $+e$ rather than $-e$. The positron is called β^+ and the electron β^-. These different decay processes are examined next.

Alpha Decay

Of the small nuclei, helium ($^4_2$He) is one of the most stable. In a large nucleus, small numbers of nucleons tend to group together as helium nuclei (like friends separated in small groups in a large crowd). If the nucleus is unstable, one of these groups may be ejected as a unit—that is, as a helium nucleus. This emission reduces the number of protons in the original nucleus and reduces the electrical repulsive force acting on the protons that remain. Since a helium nucleus also contains two neutrons, the total number of nucleons A in the original nucleus decreases by 4.

Figure 34.5 illustrates the α decay of radium-226 to radon-222 and an α particle. The daughter nucleus and alpha particle together have less mass than the parent nucleus, and energy is released in the form of kinetic energy.

EXAMPLE 34.4 (a) Calculate the kinetic energy of the product nuclei when $^{212}_{84}$Po undergoes alpha decay. (b) Could ^{212}Po decay by emitting a proton from its nucleus? The masses of the nuclei are $m_{212_{Po}} = 211.9889$, $m_{208_{Pb}} = 207.9766$, $m_{4_{He}} = 4.0026$, $m_{211_{Bi}} = 210.9873$, and $m_{1_H} = 1.0078$.

FIG. 34.5. The alpha decay of a radium-226 to form a radon-222 and an alpha particle ($^4_2$He).

SOLUTION (a) The alpha decay of polonium-212 is represented by the equation:

$$^{212}_{84}\text{Po} \longrightarrow \, ^{208}_{82}\text{Pb} + \, ^{4}_{2}\text{He} + \text{energy}.$$

The Q of this reaction is given by Eq. (34.6):

$$Q = (m_{^{212}\text{Po}} - m_{^{208}\text{Pb}} - m_{^{4}\text{He}})c^2$$
$$= (211.9889 \text{ u} - 207.9766 \text{ u} - 4.0026 \text{ u})c^2$$
$$= \left(0.0097 \text{ u} \times \frac{931.5 \text{ MeV}/c^2}{1 \text{ u}}\right)c^2 = \underline{9.0 \text{ MeV}}.$$

Of this 9.0 MeV of released energy, 8.5 MeV is converted to the kinetic energy of the α particle (^{4}He). Most of the remaining 0.5 MeV is converted to the kinetic energy of the recoiling lead-208 nucleus. In addition, a small fraction of the energy may be released as a gamma ray, a high-energy photon. (b) The decay of polonium-212 by proton emission would be represented by the following reaction:

$$^{212}_{84}\text{Po} \xrightarrow{\,?\,} \, ^{211}_{83}\text{Bi} + \, ^{1}_{1}\text{H}.$$

The total mass of the products together equals 210.9873 u + 1.0078 or 211.9951 u. Since this product mass exceeds the 211.9889-u mass of the original polonium atom, polonium-212 does not have enough mass to spontaneously decay by proton emission from the nucleus. Most decay schemes involving proton emission are prohibited for the same reason. ∎

Beta Decay

Beta decay, another nuclear decay process, occurs by electron or positron emission. For example, a carbon-14 nucleus decays by electron emission (Fig. 34.6a), whereas sodium-21 decays by positron emission (Fig. 34.6b).

Beta decay is unusual in several respects. First, electrons and positrons are not constituents of nuclei, yet are emitted from many nuclei during radioactive decay. The electron is created in the nucleus when a neutron changes into a

FIG. 34.6. (a) The beta-minus decay of a carbon-14 nucleus to form nitrogen-14 and an electron ($_{-1}^{0}e$). (b) The beta-plus decay of a sodium-21 nucleus to form neon-21 and a positron ($_{+1}^{0}e$).

proton and electron:

$$^1_0 n \longrightarrow {}^1_1 H + {}^{\ 0}_{-1} e.$$

The electron is emitted and the proton is left behind in the nucleus. Positron emission occurs when a proton converts to a neutron and a positron ($^0_1 e$):

$$^1_1 H \longrightarrow {}^1_0 n + {}^0_1 e.$$

The positron is emitted from the nucleus, and the neutron remains behind.

A second intriguing aspect of beta decay concerns the conservation of energy during the reaction. Careful measurements and calculations indicated that energy was not conserved in beta decay reactions. In 1931, Wolfgang Pauli proposed that an as yet undiscovered particle carried away the missing energy. The particle, called a **neutrino,** was postulated to have zero mass and zero charge and to travel at the speed of light. Although this sounds like the description of a photon, there is a subtle difference between a photon and a neutrino. Photons can be absorbed by atoms as their electrons jump between different energy states. These allowed transitions occur only between states that differ in angular momentum quantum number by $\Delta l = \pm 1$. The photon's angular momentum equals that lost by the atom. On the other hand, a neutrino has only half as much angular momentum as a photon and cannot cause an electron transition in an atom. Neutrinos were discovered twenty-five years after Pauli proposed their existence.

Beta decay reactions always result in the production of a neutrino ν or antineutrino $\bar{\nu}$ (we will not discuss their differences):

$$^{14}_6 C \longrightarrow {}^{14}_7 N + {}^{\ 0}_{-1} e + \bar{\nu},$$
$$^{22}_{11} Na \longrightarrow {}^{22}_{10} Ne + {}^0_1 e + \nu.$$

Gamma Decay

A number of experiments indicate that protons and neutrons occupy discrete energy states in a nucleus, similar to the energy states occupied by electrons in atoms. If a nucleus becomes excited as a result of a collision with another particle, one of the protons or neutrons in the nucleus may be excited to a higher energy state. When the proton or neutron falls back to its normal state, a **gamma-ray photon** (γ) is emitted. These photons are similar to light or x-ray photons, but they have much greater energy.

Gamma rays are sometimes emitted from a daughter nucleus after the emission of an alpha or beta particle from the parent nucleus. The daughter

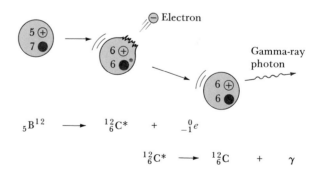

FIG. 34.7. A beta decay leads to the formation of an excited carbon-12 nucleus ($^{12}C^*$). It settles into its unexcited state by emitting a gamma-ray photon.

$$_5 B^{12} \longrightarrow {}^{12}_6 C^* + {}^{\ 0}_{-1} e$$

$$^{12}_6 C^* \longrightarrow {}^{12}_6 C + \gamma$$

may be created with one of its protons or neutrons in an excited state. As the proton or neutron falls into the lowest possible energy state, a gamma ray is emitted. This can happen when boron-12 undergoes beta decay to form carbon-12. An excited $^{12}_{6}C*$ (the asterisk indicates a nucleus excited to a higher energy state) returns to its lowest energy state by gamma decay (Fig. 34.7).

34.6 Half-Life and Decay Rate

In a sample of radioactive nuclei, each time a nucleus emits an alpha or a beta particle, a different nucleus is formed. The number of the original type of nuclei decreases by one. As time progresses, more and more radioactive decays occur in the sample, and the number of radioactive nuclei continues to decrease.

The **half-life** T of a particular type of radioactive nucleus is the time required for the number of these nuclei in a given sample to be reduced by one half.

The half-lives of several nuclei are listed in Table 34.3.

The exponentially decreasing function shown in Fig. 34.8 illustrates the decreasing number of radioactive nuclei in a material. Notice that the material starts with N_0 radioactive nuclei at time $t = 0$. After one half-life at time T, the number has decreased to $N_0/2$. During the next half-life, the number decreases again by one half; thus, at time $2T$, $N_0/4$ radioactive nuclei remain. After n half-lives, the number of radioactive nuclei is reduced to $N_0/2^n$.

EXAMPLE 34.5 Carbon-11 with a half-life of 21 minutes was used in pioneering experiments to trace the absorption of carbon dioxide gas into barley plants. Investigators found that radioactive carbon ^{11}C incorporated in carbon dioxide gas ($^{11}CO_2$) eventually become part of carbohydrate molecules produced by the barley in the presence of the gas. Suppose that transporting the radioactive $^{11}CO_2$ gas from the place where it is made to the barley plants required 2 hours and 6 minutes. Calculate the fraction of the original $^{11}CO_2$ molecules that would still be radioactive after this delay.

TABLE 34.3 Half-Lives and Decay Constants of Some Common Nuclei

Isotope	Half-life	Decay constant (s^{-1})
$^{87}_{37}Rb$	4.88×10^{10} yr	4.50×10^{-19}
$^{238}_{92}U$	4.5×10^{9} yr	4.9×10^{-18}
$^{40}_{19}K$	1.28×10^{9} yr	1.72×10^{-17}
$^{239}_{94}Pu$	2.44×10^{4} yr	9.00×10^{-13}
$^{14}_{6}C$	5730 yr	3.84×10^{-12}
$^{226}_{88}Ra$	1602 yr	1.37×10^{-11}
$^{137}_{55}Cs$	30.0 yr	7.32×10^{-10}
$^{90}_{38}Sr$	28.1 yr	7.82×10^{-10}
$^{3}_{1}H$	12.4 yr	1.77×10^{-9}
$^{60}_{27}Co$	5.26 yr	4.18×10^{-9}
$^{131}_{53}I$	8.05 day	9.96×10^{-7}
$^{11}_{6}C$	21 min	5.5×10^{-4}

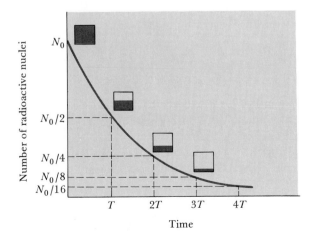

Number of radioactive nuclei

N_0
$N_0/2$
$N_0/4$
$N_0/8$
$N_0/16$

T $2T$ $3T$ $4T$

Time

FIG. 34.8. For every time period T the number of radioactive nuclei decreases by one half.

SOLUTION The delay lasts for 126 minutes, which equals six 21-minute half-lives. Thus, the number of radioactive nuclei has been reduced by the fraction

$$\frac{1}{2^6} = \frac{1}{64} = \underline{0.0156}.$$

When this technique is used to determine the fraction of radioactive nuclei remaining after a certain time, the delay can be any length of time—not just an integral number of half-lives. For example, if the transport of the $^{11}CO_2$ gas had taken 120 minutes, or $120/21 = 5.7$ half-lives, the fraction remaining would be

$$\frac{1}{2^{5.7}} = 2^{-5.7} = 0.0192. \qquad \blacksquare$$

Most applications using radioactive nuclei involve measuring the decay rate or activity of a sample of these nuclei.

The **decay rate,** or **activity** (A), of a sample of radioactive material is the number of disintegrations ΔN divided by the time Δt during which those disintegrations occur:

$$\text{decay rate} \equiv \frac{\Delta N}{\Delta t}. \qquad (34.7)$$

The decay rate, or activity, of a sample is usually stated in units of **curies,** where a sample with an activity of one curie (Ci) experiences 3.70×10^{10} decays per second.

This unusual number (3.70×10^{10}) was chosen because it represents roughly the activity of 1 g of pure radium. A recently defined SI unit of activity, called the **becquerel** (Bq), equals one disintegration per second: 1 Bq = 1 decay/s. We see that $1 \text{ Ci} = 3.70 \times 10^{10}$ Bq. The curie and its submultiples—the mCi (10^{-3} Ci) and μCi (10^{-6}Ci)—are still the most commonly used units of activity. Radioactive tracers used in medicine usually have activities of μCi. The luminous dial on a watch also has an activity of approximately 1 μCi.

Measurements have indicated that the decay rate depends on the number N of radioactive nuclei in a sample and on a quantity called the **decay constant** λ that is different for each type of radioactive nucleus (see Table 34.3). The relationship between the decay rate and these quantities is as follows:

$$\text{decay rate} = \frac{\Delta N}{\Delta t} = -\lambda N. \qquad (34.8)$$

The negative sign in Eq. (34.8) reflects the fact that the number of radioactive nuclei decreases by amount ΔN during time Δt.

We can see from Eq. (34.8) that if λ is large, $\Delta N/\Delta t$ is large, and that a large number of decays occur in a short time. In turn, this implies that the half-life of the decaying nuclei is short. In fact, the decay constant and half-life are related by the equation,

$$\lambda = \frac{0.693}{T}. \qquad (34.9)$$

A nucleus with short half-life has a large decay constant, and vice versa.

The proportionality between N and $\Delta N/\Delta t$, as expressed by Eq. (34.8), provides a convenient means for measuring the number of radioactive nuclei in a sample.

EXAMPLE 34.6 A person whose thyroid gland is suspected of being hyperactive is given a small dose of radioactive iodine ($^{131}_{53}$I) dissolved in a liquid. Twenty-four hours later the thyroid gland has enough iodine nuclei to cause an activity of 1.2×10^3 disintegrations per second. Calculate the number of radioactive iodine atoms absorbed in the person's thyroid at that time. The half-life of ^{131}I is 8.05 days (see Table 34.3).

SOLUTION We are given that the decay rate $\Delta N/\Delta t = -1.2 \times 10^3$ disintegrations/s $= -1.2 \times 10^3\,\text{s}^{-1}$ and that $T = 8.05$ days. Using Eq. (34.9), we find that the decay constant for ^{131}I is

$$\lambda = \frac{0.693}{T} = \frac{0.693}{(8.05 \text{ days})(24 \text{ hr/day})(3600 \text{ s/hr})} = 9.96 \times 10^{-7}\,\text{s}^{-1}.$$

According to Eq. (34.8), the number of radioactive iodine atoms in the thyroid causing the given decay rate is

$$N = -\frac{\Delta N/\Delta t}{\lambda} = -\frac{(-1.2 \times 10^3\,\text{s}^{-1})}{(9.96 \times 10^{-7}\,\text{s}^{-1})} = \underline{1.2 \times 10^9}.$$

In the diagnosis of thyroid problems by this method, the number of radioactive iodine atoms absorbed from a given dose by people with normal thyroids is compared to the number of iodine atoms absorbed by the person being tested. A hyperactive thyroid normally absorbs twice as much iodine as a normal thyroid. ∎

34.7 Radioactive Dating

The activity of a radioactive sample decreases with time in direct proportion to the number of radioactive nuclei in the sample [Eq. (34.8)]. Using calculus and this equation, it can be shown that the number of nuclei varies with time, according to the equation:

$$N = N_0 e^{-\lambda t}, \tag{34.10}$$

where N_0 is the number of radioactive nuclei at time $t = 0$, N is the number at time t, λ is the decay constant of the nucleus, and $e = 2.718\ldots$ is the base of the natural logarithm system.

Equation (34.10) is often used by archeologists and geologists to determine the length of time that a radioactive sample has been decaying. Dividing both sides of the equation by N_0 and taking the natural logarithm of each side leads to the equation

$$\ln \frac{N}{N_0} = -\lambda t.$$

Substituting $\lambda = 0.693/T$ (Eq. 34.9) and rearranging, we get an expression for the age of a radioactive sample:

$$t = -\frac{\ln(N/N_0)}{0.693} T. \tag{34.11}$$

In this equation T is the half-life of the decaying nucleus, and t is the sample's age when N radioactive nuclei are left. The sample started at time zero with N_0 radioactive nuclei.

EXAMPLE 34.7 Tritium, an isotope of hydrogen ($^{3}_{1}H$), is produced in the atmosphere by the cosmic-ray-induced nuclear reaction:

$$^{14}_{7}N + ^{1}_{0}n \longrightarrow ^{12}_{6}C + ^{3}_{1}H.$$

Tritium undergoes β^- decay with a half-life of 12.4 years. Fresh rainwater contains a small amount of tritium. If isolated, the tritium in the water decays, and 12.4 years later contains only one half as much tritium. A bottle of wine recovered from a sunken ship contains only 4.0×10^{-4} times the normal amount of tritium. How old is the wine?

SOLUTION Water in the wine at the time the wine was made had the same amount of tritium as an equal volume of fresh rainwater. However, as the wine aged, the tritium underwent beta decay and its concentration decreased. By the time the wine was discovered, the number of tritium nuclei had been reduced by a factor $N/N_0 = 4.0 \times 10^{-4}$. Substituting this and the 12.4-year half-life of tritium into Eq. (34.11), we find that the wine's age is

$$t = -\frac{\ln(4.0 \times 10^{-4})}{0.693}(12.4 \text{ yr}) = -\frac{(-7.8)}{0.693}(12.4 \text{ yr}) = \underline{140 \text{ years}}. \quad \blacksquare$$

An important type of radioactive dating, called **carbon dating,** is used to determine the age of objects that are less than 40,000 years old. Most of the carbon in our environment consists of the isotope carbon-12, but about 1 in 10^{12} carbon atoms is radioactive carbon-14 with a half-life of 5700 years. The carbon-14 is continuously produced in our atmosphere by collisions of neutrons from the solar wind with nitrogen nuclei in the atmosphere, resulting in the reaction $^{1}_{0}n + ^{14}_{7}N \longrightarrow ^{14}_{6}C + ^{1}_{1}H$. At their present concentration, the carbon-14 isotopes disintegrate by radioactive decay just as fast as they are created by the above reaction. An equilibrium is established.

Any plant or animal that metabolizes carbon incorporates about one carbon-14 isotope into its structure for each 10^{12} carbon-12 isotopes that are present. When an animal dies, this is the concentration of radioactive carbon-14 in its bones. Since carbon is no longer absorbed and metabolized after death, the carbon-14 in the bones starts to disintegrate by radioactive decay. After 5700 years the concentration decreases by one-half. Two half-lives, or 11,400 years after death, the concentration dwindles to one-fourth the concentration at death. A measurement of the carbon-14 concentration indicates the age of the bone (see Fig. 34.9).

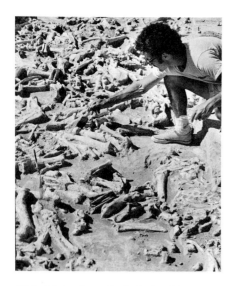

FIG. 34.9. An archeologist brushing debris from a layer of butchered bison bones of a Paleo-Indian site near Crawford, Nebraska. The age of the site can be determined by carbon-14 dating of the bones. (© 1976 Tom McHugh/Photo Researchers, Inc.)

EXAMPLE 34.8 A bone found by an archeologist contains a small amount of radioactive carbon-14. The radioactive emissions from the bone produce a measured decay rate of 3.3 decays per second. The same mass of fresh cow bone produces 30.8 decays per second. Estimate the age of the sample.

SOLUTION The 30.8 decays per second reflects the activity or decay rate of a sample at death. The bone of unknown age must have had that activity at the animal's death. Using this information and the fact that the decay rate and number of radioactive nuclei are proportional to each other ($\Delta N/\Delta t = -\lambda N$), we can determine the ratio of radioactive carbon now in the bone to that in the bone when the animal died:

$$\frac{(\Delta N/\Delta t)_{\text{now}}}{(\Delta N/\Delta t)_{\text{at death}}} = \frac{3.3 \text{ s}^{-1}}{30.8 \text{ s}^{-1}} = \frac{-\lambda N_{\text{now}}}{-\lambda N_{\text{at death}}}$$

or

$$\frac{N_{\text{now}}}{N_{\text{at death}}} = \frac{N}{N_0} = \frac{3.3}{30.8} = 0.107.$$

Substituting this value into Eq. (34.11), we find the age of the bone:

$$t = -\frac{\ln (N/N_0)}{0.693} T = -\frac{\ln (0.107)}{0.693} (5700 \text{ yr})$$

$$= -\frac{-2.23}{0.693} (5700 \text{ yr}) = \underline{18,400 \text{ yr}}. \qquad \blacksquare$$

34.8 Radioactive Decay Series

Many of the daughter nuclei formed by radioactive decay are themselves radioactive. In fact, the decay of the original parent nucleus may start a chain of successive decays called a **decay series.** An example of a decay series is illustrated in Fig. 34.10. First, the alpha decay of $^{238}_{92}$U leads to the formation of $^{234}_{90}$Th. Thorium-234 then undergoes beta decay to form $^{234}_{91}$Pa. The series continues until $^{218}_{84}$Po is formed. Polonium-218 can decay by either alpha or beta emission. The series branches at this point and at several other points as well. Eventually, the stable lead isotope ($^{206}_{52}$Pb) is formed and the series ends.

Radioactive series, such as illustrated in Fig. 34.10, replenish our environment with nuclei that would normally have disappeared long ago. For example, radium ($^{226}_{88}$Ra) has a half-life of 1600 years. During the 5×10^9 years of our solar system's existence, the original supply of radium-226 would have been depleted by radioactive decay. However, the supply is continually replenished as uranium-238, with a (4.5×10^9)-year half-life, decays and leads to the production of ^{226}Ra via a series of reactions such as shown in Fig. 34.10.

34.9 Ionizing Radiation

Ionizing radiation consists of photons and moving particles whose energy exceeds the 1 to 10 eV of energy needed to knock an electron out of an atom or molecule, thus forming an ion. There are a variety of forms of ionizing radiation, such as photons in the ultraviolet, x-ray, and gamma-ray regions of the electromagnetic spectrum; alpha and beta particles emitted during the radioactive decay of nuclei; and cosmic rays (high-energy particles) that reach the earth from outer space.

FIG. 34.10. A radioactive decay series beginning with uranium-238 and ending with lead-206.

Life has evolved on the earth in a steady background of this radiation, most of it coming from emissions of radioactive nuclei in the earth's crust and from cosmic rays passing down through the earth's atmosphere. The technological use of ionizing radiation *supposedly* began over eighty years ago (see Fig. 34.11). Today, our exposure from man-made sources of ionizing radiation accounts for about 40 percent of our total exposure (see Table 34.6 in the next section). Before discussing the sources of this radiation and the biological damage it causes, we must first become familiar with the special terminology used to describe ionizing radiation.

Quantities and Units Used to Describe Ionizing Radiation

Five different quantities and their units of measure have been defined to describe the activity of a source of ionizing radiation, the intensity of radiation incident on a target, and the effect of the radiation on the matter in which the radiation is absorbed. The first quantity, the **decay rate** of a radioactive source, was defined in Section 34.6 and indicates the number of alpha particles, beta

FIG. 34.11. An advertisement from the December 1898 issue of *The Cosmopolitan* magazine claimed that doses of x-rays could be sent anywhere in the United States to cure a headache. X-rays had been discovered just three years earlier by W. C. Roentgen.

particles, or gamma rays emitted by a radioactive source per second. The decay rate is measured in units of curies.

The amount of radiation incident on a medium is known as the **exposure** and is measured by the number of ions produced as the radiation passes through dry air. The unit of exposure called the **roentgen** (R) constitutes the amount of radiation that will produce a number of ions whose electric charge is 2.58×10^{-4} coulombs.

The **dose** of radiation is the amount of energy absorbed by a medium per unit mass and is measured in a unit called the **radiation absorbed dose, or rad.** One rad is the absorption of 10^{-2} joules of energy per kilogram of mass:

$$1 \text{ rad} = 10^{-2} \text{ J/kg}. \tag{34.12}$$

A fourth quantity, the **relative biological effectiveness, RBE** (also called the quality factor), indicates the relative damage caused by different types of ionizing radiation. A mass receiving a dose of 1 rad of x-rays will experience a different amount of damage than the same mass when receiving a dose of 1 rad of alpha particles. Alpha particles move more slowly through mass and interact with a large number of atoms. On the other hand, x-rays and gamma rays deposit a large fraction of their energy with a small number of atoms. The net damage—the number of bonds broken and the number of ions formed—is about ten times greater in the case of the alpha particles than for x-rays. The RBE of alpha particles is ten times that of x-rays. For the RBE scale, x-rays whose photons have an energy of 2×10^5 eV are used as a standard. The RBE of some other form of ionizing radiation is the damage caused by a certain amount of absorbed energy of that type of radiation divided by the damage caused by an equal amount of absorbed x-ray energy. The RBE of several different types of radiation are listed in Table 34.4.

This brings us to our last quantity, the **effective dose,** an indicator of the net biological effect of a dose of radiation. The effective dose of radiation equals the product of the dose and the relative biological effectiveness of the radiation and is reported in units of **roentgen equivalent man (rem):**

$$\text{effective dose (in rem)} = \text{dose (in rad)} \times \text{RBE}. \tag{34.13}$$

We see, using Eq. (34.13) and Table 34.4, that a dose of 1 rad of alpha particles (RBE = 10) results in an effective dose of 10 rem, whereas a dose of 1 rad of x-rays (RBE = 1) results in an effective dose of 1 rem.

TABLE 34.4 Relative Biological Effectiveness (RBE) of Different Types of Radiation

Type	RBE
X-rays and γ rays	1.0
Electrons	1.0–1.7
Slow neutrons	4–5
Protons	10
Alpha particles	10
Heavy ions	20

TABLE 34.5 Summary of Radiation Units

Quantity	Unit	Value
Source activity	Ci	$3.7 \times 10^{10}\,\text{s}^{-1}$
Exposure	R	$2.58 \times 10^{-4}\,\text{C/kg}$
Dose	rad	$10^{-2}\,\text{J/kg}$
Effective dose	rem	rad $\times$ RBE

These various quantities used to describe ionizing radiation and its effects on a medium are summarized in Table 34.5.

EXAMPLE 34.9 During a chest x-ray, you receive an effective dose of 20 mrem in about 5 kg of body tissue. Calculate the number of x-ray photons absorbed if each x-ray photon has an energy of 50,000 eV.

SOLUTION The RBE of x-rays is 1. Thus, the effective dose of x-rays in rem equals the dose in rad. For a dose of 20 mrad, the energy absorbed per kilogram of exposed tissue is determined using the conversion given by Eq. (34.12):

$$20\,\text{mrad} = (20 \times 10^{-3}\,\text{rad})\left(\frac{10^{-2}\,\text{J/kg}}{1\,\text{rad}}\right) = 20 \times 10^{-5}\,\frac{\text{J}}{\text{kg}}.$$

Since 5 kg of tissue receive this dose, the total absorbed energy is

$$(5\,\text{kg})\left(20 \times 10^{-5}\,\frac{\text{J}}{\text{kg}}\right) = 1.0 \times 10^{-3}\,\text{J}.$$

The energy of each x-ray photon in units of joules is determined using the conversion $1\,\text{eV} = 1.6 \times 10^{-19}\,\text{J}$:

$$\text{energy of each photon} = 50,000\,\frac{\text{eV}}{\text{photon}}\left(\frac{1.6 \times 10^{-19}\,\text{J}}{1\,\text{eV}}\right)$$

$$= 8.0 \times 10^{-15}\,\frac{\text{J}}{\text{photon}}.$$

Dividing the total absorbed energy by the energy of each photon gives us the number of absorbed photons:

$$\frac{1.0 \times 10^{-3}\,\text{J}}{8.0 \times 10^{-15}\,\text{J/photon}} = 1.3 \times 10^{11}\,\text{photons.} \qquad \blacksquare$$

34.10 Sources and Effects of Ionizing Radiation

The U.S. Environmental Protection Agency estimated that the average effective dose of ionizing radiation received by a person in the United States in the year 1980 was about 222 millirem per year (mrem/yr). This represents the average effective dose for each kilogram of body mass. The dose varies somewhat for different body parts. Also, the effective dose varies greatly from one person to another, depending on where they live, the number of medical and dental x-rays taken during a year, the person's occupation, and so on. The sources of this radiation, listed in Table 34.6, can be divided into two broad categories: natural and man-made.

Natural Sources

The largest natural source consists of radioactive elements in the earth's crust, such as uranium-238, potassium-40, radon-226 and 228, thorium-232, and the radioactive daughters of these nuclei. These radioactive elements are even incorporated into building materials, such as pumice stone, granite, light concrete

TABLE 34.6 Estimated Annual Exposure to Ionizing Radiation in 1980 Averaged over the U.S. Population[a]

Natural sources	Dose (mrem/yr)
Radioactive elements:	
External from surroundings	60
Internal (absorbed in diet and breathing)	25
Cosmic rays	45
	130

Man-made sources	
Medical and dental:	
X-rays	72
Radiopharmaceuticals	14
Nuclear fallout	4
Miscellaneous (TV, fire alarms, clocks, etc.)	2
Occupational	0.8
Nuclear power production	0.04
	92
Total dose/yr	222

[a] Adapted from "Estimates of Ionizing Radiation Doses in the United States 1960–2000," U.S. Environmental Protection Agency, August 1972.

derived from alum shale, and phosphogypsum. The earth's radioactive crust contributes an effective dose of 60 mrem/year, or 27 percent of the total average dose absorbed by a United States citizen. The dose varies greatly for different locations in the United States (from 30 mrem/yr in some places to 90 mrem/yr in others). In several inhabited regions of the world the dose due to radioactive elements in rocks, soil, or building materials is considerably higher. In Kerala in India, for example, over 100,000 people receive a background dose due to radioactive nuclei of about 1300 mrem/yr. One-sixth of the population of France receive a dose from radioactive granite that ranges from 180 to 350 mrem/yr.

Another large natural source results from the inhalation of radioactive nuclei in the air and the ingestion of foods that contain radioactive isotopes. Potassium-40, which is present in small concentration in fertilizers containing potassium, finds its way from fields into our food products. These nuclei, which are absorbed into the body and are called internal sources, contribute an average effective dose of 25 mrem/yr, or 11 percent of the total dose.

Cosmic rays comprise the third natural source. A **cosmic ray** is a charged atomic particle moving near the speed of light. Examples of cosmic rays are electrons, positrons, protons, alpha particles, and other heavier nuclei. Protons account for almost 90 percent of all cosmic radiation; alpha particles represent about 9 percent, and the other particles make up the remaining 1 percent. These particles appear to have been produced during huge explosive reactions of supernovas in our galaxy and seem to permeate all space. The mass of cosmic rays in space equals roughly the mass of the stars.

Our exposure from cosmic rays would be well over 100 mrem/yr. However, two-thirds of the rays are diverted away from the earth by the earth's magnetic field. The net dose from cosmic rays is about 45 mrem/yr, or 20 percent of the total dose. The cosmic-ray exposure is greater in locations at high elevation—three times greater at an elevation of 2000 m than at sea level.

Man-Made Sources

Our man-made exposure to ionizing radiation results primarily from the use of x-rays and radioactive nuclei as diagnostic tools in medicine and dentistry. Together, these sources account for a dose of 86 mrem/yr—38 percent of our total exposure and over 90 percent of the man-made exposure. The dose due to medical x-rays is slowly increasing from year to year. The medical use of radioactive nuclei (radionuclides) has been increasing much more rapidly—about 25 percent per year for the last decade.

The second largest source of man-made radiation is the fallout of radioactive nuclei produced in atmospheric testing of nuclear bombs (4 mrem/yr). Miscellaneous sources of radiation (2 mrem/yr) include radiation from television sets, radioactive clock dials and increased cosmic-ray exposure due to air travel. The occupational dose listed in Table 34.6 (0.8 mrem/yr) is associated primarily with extra exposure received by radiologists who give medical and dental x-rays.

In 1980 the world's nuclear power industry accounted for about 0.04 mrem/yr. This includes radiation caused by uranium mining, fuel processing, power plant operation, and waste disposal facilities.

Effects of Ionizing Radiation on Humans

The effects of ionizing radiation on living organisms are divided into two categories: genetic damage and somatic damage. **Genetic damage** occurs when the DNA molecules in the genes of a person's reproductive organs are altered, causing a mutation. These genetic changes are passed on to future generations. **Somatic damage** involves cellular changes caused by ionizing radiation in all other parts of the body except the reproductive organs. Of major concern is the induction of various forms of cancer.

It is difficult to assess the risk of cancer or of other forms of somatic or genetic damage as a result of exposure to ionizing radiation. Populations in regions where the background radiation is high, such as in Kerala, India, show no apparent effect of their exposure. Yet it is nearly impossible to make an accurate comparison between the 100,000 people living in Kerala and other population groups because of differences in their diet, social habits, and ethnic origin. At present, the attitude of health agencies such as the U.S. Bureau of Radiological Health and the World Health Organization is that small doses of ionizing radiation pose a small risk of genetic and somatic damage. For example, the risk of inducing leukemia that is associated with a single chest x-ray is of the order of 2 in 10 million; for an x-ray examination of the skull it is 1.6 in 1 million. The chance of getting thyroid cancer is approximately 12 in 10 million for a chest x-ray.* Approximately three people per one million population die each year as a result of cancer caused by medical and dental x-rays.†

These risks are partly justified because of the medical benefits that result from x-ray diagnosis. However, the dosage could be reduced significantly with

---

*Eric J. Hall, *Radiobiology for the Radiologist,* 2nd ed. (New York: Harper and Row, 1978), p. 425.

†R. L. Gotchy, "NRC Estimates of Health Risks Associated with Low-Level Radiation Exposure," 10th Annual National Conference on Radiation Control, HEW Publication (FDA)79-8054 (June 1979).

no adverse effect on the diagnostic benefits of the x-rays.* For exposure were consistently limited to the size of the film, one-third of the administered radiation received would be avoided. During skull examinations using x-rays, a 75 to 80 percent reduction in the number of films results in equally good diagnosis. The same type of analysis shows that a reduction of intravenous urograms reduces the annual gonadal x-ray dose to 38 percent with no loss of information. Routine chest x-rays as an entrance procedure to hospitals reveal no information for people under 20 years of age and should not be given. Another study showed that physicians with their own x-ray machines use twice as many x-ray examinations as physicians who do not own a machine. Many of these x-rays are probably unneeded. As we can see, much of this x-ray exposure could be reduced without losing any of the diagnostic value of these examinations. A medical or dental x-ray should confer to the exposed person an immediate benefit that more than balances the risk. If it does not, then the risk should not be taken.

34.11 Radiation Detection Devices

A variety of devices have been built for detecting ionizing radiation. These devices can be used for making radioactive dating measurements, for calibrating x-ray machines, for detecting particles emitted by radioactive nuclei used in medical diagnosis, for detecting the background radiation at different locations on the earth, and for many other purposes. Most of the devices depend on the fact that ionizing radiation produces ions as it passes through matter.

Geiger Counter

Probably the most widely used radiation detection device, a geiger counter consists of a gas-filled cylindrical metal tube with a positively charged wire at high voltage ($\sim 10^3$ V) down the center (Fig. 34.12). If a high-energy particle (or photon) enters the window at one end of the tube, the particle's collisions with atoms in the gas inside cause several ions to be formed. The electrons that are freed accelerate toward the positively charged wire and cause other atoms in their path to be ionized. These secondary electrons that are freed strike other atoms, and soon a cascade of electrons rushes toward the wire. When they reach it, they produce a pulse of electric current that passes down the wire into an electric circuit attached across the wire and metal tube. If a loudspeaker is placed in the circuit, a "click" sound is heard each time a cascade is initiated by the entrance of an ionizing particle or photon into the tube.

Each pulse of current is called a count. The counting rate is the number of counts per second. Geiger counters are very sensitive to all forms of ionizing radiation but cannot be used to measure high counting rates, nor do they indicate the type of radiation being detected or the energy of the particles or photons in the beam. A geiger counter is relatively inexpensive and is rugged enough for use as a portable detector.

FIG. 34.12. A geiger tube.

*H. L. Abrams, "Factors Underlying the Overutilization of Radiological Examinations," National Conference on Referral Criteria for X-ray Examinations, HEW Publication 79-8083 (April 1979).

Scintillation Detector

This detector is made of a material that emits a flash of light when struck by an energetic particle or photon. A photomultiplier tube (see Section 31.4) near the scintillation material detects the flash of light, thus counting the particle or photon that caused the flash. Scintillation materials usually consist of a special type of plastic, crystals such as sodium iodide, or liquids. Radioactive samples whose age is to be determined are often placed directly in a liquid scintillation material. In this way all radioactive emissions from the sample must pass through the scintillation material and thus produce a flash of light.

Semiconductor Diode Detector

A semiconductor diode is a device that prohibits the flow of electric current in one direction through the diode while allowing it to flow with little resistance in the opposite direction. If an energetic particle passes into the diode, ionization occurs, and a pulse of current can flow in the direction in which current normally does not flow. The diode detects ionizing radiation by a pulse of current.

Photographic Emulsion

The three detectors just described are used to indicate the number of high-energy particles or photons entering the device per unit time (the counting rate). The next three ionizing radiation detection devices produce a trail along the path followed by individual charged particles. If a high-energy charged particle enters a photographic emulsion, the particle ionizes atoms along its path. The chemical changes caused by this ionization become apparent when the photographic emulsion is developed. We see the particle's path as it passed through the emulsion.

Cloud Chamber

A cloud chamber contains a gas vapor cooled to a temperature just below its condensation temperature. Any disturbance of the gas initiates condensation in the region of the disturbance. A high-energy charged particle passing through the cloud causes atoms along its path to be ionized, and condensation is initiated on these ions. The condensation track, which can be photographed, indicates the path of the particle. If a magnetic field exists in the cloud, negative and positive charges are deflected in opposite directions by the magnetic field.

Bubble Chamber

A bubble chamber consists of a liquid that is heated to a temperature near its boiling point. Charged particles passing through the liquid produce a trail of bubbles that can be photographed to show the particle's path (Fig. 34.13).

FIG. 34.13. A bubble-chamber photo (top) shows electron-positron pairs produced by the annihilation of a gamma-ray photon. (Courtesy of the Lawrence Radiation Laboratory, University of California)

Summary

1. **Nucleus.** A nucleus consists of protons and neutrons held together by the strong nuclear force. An atom with Z protons (Z is called the atomic number) and N neutrons is represented by the symbol $_Z^A X$, where the mass number A equals $Z + N$, and X represents the chemical symbol for the element. Nuclei with the same number of protons but different numbers of neutrons are called **isotopes.**

2. **Binding energy per nucleon.** The **mass defect** of a nucleus equals the difference between the mass of its separated parts and the mass of the nucleus itself:

$$\text{mass defect} = \Delta m = [Z m_{^1\text{H}} + (A - Z) m_{^0_0\text{n}}] - m_{^A_Z\text{X}}. \quad \textbf{(34.1)}$$

If a nucleus has less mass than the particles of which it is made, it cannot spontaneously separate into these parts. It is bound together because of a small amount of missing mass (a mass defect). The energy of this mass defect is called the **nuclear binding energy:**

$$\text{nuclear binding energy} = BE = \Delta m\, c^2. \quad \textbf{(34.2)}$$

When the binding energy is divided by A, the number of nucleons in the nucleus, we have the

$$\text{binding energy per nucleon} = \frac{BE}{A}. \quad \textbf{(34.4)}$$

When calculating mass defect, we use the atomic mass unit u, where $1\text{ u} = 1.660566 \times 10^{-27}\text{ kg} = 931.5\text{ MeV}/c^2$ and $1\text{ MeV} = 10^6\text{ eV} = 1.6 \times 10^{-13}\text{ J}$.

3. **A nuclear reaction** results in the transmutation of one or more nuclei into other product nuclei. The reactions must satisfy the rules listed in Section 34.4. The energy released or absorbed during the reaction is called the Q of the reaction:

$$Q = \left(\sum m_{\text{reactants}} - \sum m_{\text{products}} \right) c^2. \quad \textbf{(34.6)}$$

4. **Radioactive decay.** Unstable nuclei can spontaneously decay by emitting an **alpha particle** ($_2^4\text{He}$), a **beta particle** (an energetic electron or positron), or a **gamma ray** (a high-energy photon). The **half-life** T of a particular type of radioactive nucleus is the time required for one half of the nuclei in a sample to experience radioactive decay. If the sample starts at time zero with N_0 radioactive nuclei, then at a time nT the number remaining is $N_0/2^n$. The **decay rate,** or **activity** (A), of a radioactive material is the number of decays ΔN per time Δt and equals

$$\text{decay rate} \equiv \frac{\Delta N}{\Delta t} = -\lambda N, \quad \textbf{(34.8)}$$

where N is the number of radioactive nuclei in the sample and λ, the **decay constant,** is related to the half-life by the equation

$$\lambda = \frac{0.693}{T}. \quad \textbf{(34.9)}$$

5. **Radioactive dating.** The number of radioactive nuclei in a sample decreases exponentially ($N = N_0 e^{-\lambda t}$), thus allowing us to determine the sample's age by comparing the number of radioactive nuclei now in the sample to the number in the sample at some previous time.

6. **Ionizing radiation.** Any particle or photon that has enough energy to ionize an atom or molecule is called ionizing radiation. Several quantities are used to describe this radiation and its effect on living matter. The **curie** (Ci) is a unit for the decay rate of a radioactive sample: $1\text{ Ci} = 3.70 \times 10^{10}$ decays per second. The amount of radiation incident on a medium is the **exposure** and is measured in units of **roentgen** (R): an exposure of 1 R produces 2.58×10^{-4} coulombs of charged ions when passing through dry air. The **dose** of radiation is the energy absorbed by a medium per unit mass and is measured in **rads:** $1\text{ rad} = 10^{-2}\text{ J/kg}$. The **effective dose** is the product of the dose and the **relative biological effectiveness** (RBE) of the radiation and is measured in **rem:** $1\text{ rem} = 1\text{ rad} \times \text{RBE}$.

Questions

1. *Estimate* the radius of a sphere that would hold all of your mass if its density equaled the density of a nucleus.

2. Determine the chemical elements represented by the symbol X in each of the following: (a) $_{10}^{20}\text{X}$; (b) $_{24}^{52}\text{X}$; (c) $_{27}^{59}\text{X}$; and (d) $_{41}^{93}\text{X}$.

3. Explain why very heavy nuclei are unstable.

4. The half-life of radium-226 is 1622 years. Since the earth is about 4 billion years old, essentially all of the radium-226 should have decayed. How can you explain the fact that radium-226 still remains in our environment in moderate abundance?

5. Design an experiment to determine whether O_2 emitted from plants comes from H_2O or from CO_2, the basic input molecules that lead to plant growth.

6. The age of the Dead Sea Scrolls was found by carbon dating. Are they made of stone or wood? Explain.

7. Design an experiment in which radioactive nuclei are used to test the ability of different detergents to remove dirt from clothes.

Problems

34.1 The Nucleus

1. Determine the number of protons, neutrons, and nucleons in each of the following nuclei: $^9_4$Be, $^{16}_8$O, $^{27}_{13}$Al, $^{56}_{26}$Fe, ^{64}Zn, and ^{107}Ag.

2. (a) Calculate the radius of a $^{56}_{26}$Fe nucleus. (b) Do the same for a $^{112}_{48}$Cd nucleus. (c) Even though the cadmium nucleus has twice as many nucleons as iron, its radius is not twice as large. Explain.

3. (a) Calculate the radius of a copper nucleus ($A = 63$). (b) Using the answer to (a) and the fact that a copper atom has a radius of about 0.1 nm, *estimate* the atom's size if the nucleus grew until it was the size of a tennis ball.

4. (a) *Estimate* the density of a nucleus. (b) Calculate what the earth's radius would be if it had this same density.

34.2 Nuclear Force

5. *Estimate* the magnitude of the repulsive electrical force between two protons in a helium nucleus.

6. *Estimate* the magnitude of the repulsive electrical force on a proton near the edge of a gold nucleus ($^{197}_{79}$Au) due to all of the other protons in the nucleus. State any assumptions used in your calculation. This is an order-of-magnitude estimate, so do not become bogged down in details.

34.3 Binding Energy per Nucleon

7. Calculate the rest-mass energies of an electron, a proton, and a neutron in units of MeV.

8. Calculate the total binding energy and the binding energy per nucleon for helium ($^4_2$He).

9. Calculate the total binding energy and the binding energy per nucleon for arsenic ($^{75}_{33}$As).

10. Calculate the binding energies per nucleon for $^{238}_{92}$U and $^{120}_{50}$Sn. Based on these numbers, which nucleus is more stable? Explain.

11. A 5.5-g marble is dropped from rest in the earth's gravitational field. How far must it fall before its decrease in gravitational potential energy is 938 MeV, the same as the rest-mass energy of a proton?

34.4 Nuclear Reactions

12. Insert the missing symbol:

(a) $\qquad ^4_2$He $+ ^{12}_6$C $\longrightarrow ^{15}_7$N $+ ?$

(b) $\qquad ^2_1$H $+ ^3_1$H $\longrightarrow ^4_2$He $+ ?$

(c) $\qquad ^1_0 n + ^{235}_{92}$U $\longrightarrow ^{140}_{54}$Xe $+ ? + 2\,^1_0 n$

(d) $\qquad ^3_1$H $\longrightarrow ? + ^0_{-1}e$

13. Which of the following reactions violate one or more of the rules for nuclear reactions? Explain why.

(a) $\qquad ^1_0 n + ^{238}_{94}$Pu $\longrightarrow ^{140}_{54}$Xe $+ ^{96}_{40}$Zr $+ 3\,^1_0 n$

(b) $\qquad ^{14}_6$C $\longrightarrow ^{14}_7$N $+ ^0_{+1}e$

(c) $\qquad ^4_2$He $+ ^{27}_{13}$Al $\longrightarrow ^{32}_{15}$P $+ ^1_0 n$

(d) $\qquad ^2_1$H $+ ^1_1$H $\longrightarrow ^4_2$He $+ ^1_1$H

14. Explain why the following reaction will not occur spontaneously: $^4_2$He $\longrightarrow ^3_1$H $+ ^1_1$H.

15. The following reaction occurs on the sun: $^3_2$He $+ ^4_2$He $\longrightarrow ^7_4$Be. How much energy is released?

16. Oxygen is produced on stars by the following reaction: $^{12}_6$C $+ ^4_2$He $\longrightarrow ^{16}_8$O. How much energy is absorbed or released in units of MeV by the reaction?

17. One part of the carbon-nitrogen cycle that provides energy for the sun is the reaction $^{12}_6$C $+ ^1_1$H $\longrightarrow ^{13}_7$N $+ 1.943$ MeV. Using the known masses of ^{12}C and ^{1}H and the results of this reaction, determine the mass of ^{13}N.

18. Determine the missing nucleus in the following reaction and calculate the energy released: $^{232}_{92}$U $\longrightarrow ? + ^4_2$He $+$ energy.

19. Determine the missing nucleus in the following reaction and calculate its mass: $? \longrightarrow ^{211}_{83}$Bi $+ ^4_2$He $+ 8.20$ MeV.

34.5 Radioactive Decay

20. The following nuclei produced in a nuclear reactor each undergo radioactive decay. Determine the daughter nucleus formed by each decay reaction. (a) $^{239}_{94}$Pu alpha decay; (b) $^{144}_{58}$Ce beta-minus decay; (c) $^{129}_{53}$I beta-minus decay; and (d) $^{65}_{30}$Zn beta-plus decay.

21. Potassium-40 ($^{40}_{19}$K) undergoes both beta-plus and beta-minus decay. Determine the daughter nucleus in each case.

22. Radon-222 ($^{222}_{86}$Rn) is released into the air during uranium mining and undergoes alpha decay to form $^{218}_{84}$Po of mass 218.0089 u. Calculate the energy released by the decay reaction. Most of this energy is in the form of alpha-particle kinetic energy.

23. Carbon-11 ($^{11}_6$C) undergoes beta-plus decay. Determine the product of the decay and the energy released.

24. (a) Calculate the total binding energy of radium-226 ($^{226}_{88}$Ra). (b) Calculate and add together the binding energies of a radon-222 ($^{222}_{86}$Rn) and an alpha particle. (c) Calculate the difference of these numbers, which equals the energy released during alpha decay of ^{226}Ra.

25. The body contains about 7 mg of radioactive $^{40}_{19}$K, absorbed with foods we eat. Each second, about 2.0×10^3 of these potassium nuclei undergo beta decay (either β^- or β^+). Assuming that all decays are beta-plus, determine the energy in MeV added to the body each second by ^{40}K decay. What is the rate of energy addition in watts?

34.6 Half-Life and Decay Rate

26. Strontium-90, a waste product of nuclear reactors, has a half-life of 25 years. Calculate the fraction of ^{90}Sr remaining in a reactor fuel rod (a) 100 years after it is removed from the reactor, (b) 200 years after, (c) 1000 years after.

27. How many years are required for the amount of krypton-85 ($^{85}_{36}$Kr) in a spent nuclear-reactor fuel rod to be reduced by a factor of $\frac{1}{8}$? $\frac{1}{32}$? $\frac{1}{128}$? The half-life of ^{85}Kr is 10.8 years.

28. In 1898 Madame Curie collected 200 mg of radium-226. Roughly, how much of this radium will be left in the year 8306?

29. What fraction of the $^{235}_{92}$U present at the creation of the universe 10 billion years ago still survives? The half-life of ^{235}U is 2.25×10^{16} s.

30. A radioactive sample initially undergoes 4.8×10^4 decays per second. Twenty-four hours later its activity is 1.2×10^4 decays per second. Determine the half-life of the radioactive species in the sample.

31. An unlabeled container of radioactive material has an activity of 90 decays per minute. Four days later the activity is 72 decays per minute. Calculate the half-life of the material. When will its activity reach nine decays per minute?

32. A student nicknamed "Hothead" accidentally swallows $0.10 \mu g$ of iodine-131 while pipetting the radioactive material. (a) Calculate the number of ^{131}I atoms swallowed (the mass of ^{131}I is approximately 131 u). (b) Calculate the activity of this material. The half-life of ^{131}I is 8.08 days.

33. A 100-mCi dose of radioactive gold-198 is administered to a person for treatment of a soft-tissue tumor. *Estimate* the value of this gold at present market prices. The gold-198 has a half-life of 2.7 days and a mass of 198 u.

34. Calculate the ratio of the number of atoms of iodine-129 (half-life of 1.7×10^7 years) and cobalt-58 (half-life of 71 days), whose activities are equal. Explain in words why so many more atoms of the one material are needed in order for their activities to be equal.

35. In order to estimate the number of ants in a nest, 100 ants are removed and fed sugar made from radioactive carbon of long half-life. The ants are returned to the nest. Several days later, it is found that of 200 ants tested, only 5 are radioactive. Roughly, how many ants are in the nest? Explain your calculation technique.

36. A small amount of radioactive material placed in $1 cm^3$ of water has an activity of 7500 decays per minute. An identical amount of radioactive material is injected into a person's blood system. A short time later, after the material has mixed throughout the blood, a 1-cm^3 sample of blood has an activity of 16 decays per minute. Calculate the person's blood volume.

34.7 Radioactive Dating

37. A mallet found at an archeological excavation site has $\frac{1}{16}$ the normal carbon-14 decay rate. Use both the $(\frac{1}{2})^n$ technique and Eq. (34.11) to calculate the mallet's age.

38. Scientists have estimated that the uranium-238 in the earth's crust has decreased by a factor of 0.50 during the earth's life. Roughly, how old is the earth? The half-life of uranium-238 is 4.51×10^9 years.

39. A sample of water from a deep well contains only 30 percent as much tritium as fresh rainwater. How old is the water in the well?

40. Charcoal from an ancient Indian campsite has a carbon-14 decay rate of 11.4 decays per minute per gram of carbon in the sample. The rate for fresh charcoal is 15.3 decays per minute per gram. How old is the Indian camp?

41. A 10-g sample of tree uprooted and buried by the Wisconsin glacier has a carbon-14 decay rate of 15.3 decays per minute. One gram of fresh tree has a decay rate of 65.6 decays per minute. How many years ago was the tree uprooted by the glacier?

42. The ^{235}U in a rock decays with a half-life of 7.07×10^8 years. A geologist finds that for each ^{235}U now remaining in the rock, 2.6 ^{235}U have already decayed to form daughter nuclei. Calculate the age of the rock.

34.8 Radioactive Decay Series

43. A different radioactive series than that shown in Fig. 34.9 begins with $^{232}_{90}Th$ and undergoes the following series of decays: $\alpha \beta^- \beta^- \alpha \alpha \alpha \alpha \beta^- \alpha \beta^-$. Determine each nucleus in the series.

44. After a series of alpha- and beta-minus decays, plutonium-239 ($^{239}_{94}Pu$) becomes lead-207 ($^{207}_{82}Pb$). Determine the number of alpha and beta particles emitted in the complete decay process. Explain your method for determining these numbers.

34.9 and 34.10 Ionizing Radiation, Sources and Effects of Ionizing Radiation

45. An x-ray examination as a screening procedure for breast cancer results in a dosage of 2.0 rads to the breast tissue. Calculate the number of 50-keV x-ray photons absorbed per kilogram of tissue.

46. (a) How much energy is absorbed by a 70-kg person receiving a 250-mrad whole-body dose of radiation? (b) Is it better to absorb 250 mrad of x-rays or 250 mrad of beta rays? Explain. (c) What is the effective dose in each case?

47. (a) X-rays passing through dry air produce 6.68×10^{-10} coulomb of electron charge per cubic centimeter of air. What is the x-ray exposure per cubic centimeter of air? (b) The average energy needed to produce one electron ion pair is 33.7 eV. How much energy in joules is deposited per cubic centimeter of dry air by the x-ray exposure in part (a)?

48. Calculate the effective dose caused by a 70-mrad dose of the following types of radiation: (a) x-rays; (b) beta rays; (c) protons; (d) alpha particles; and (e) heavy ions.

49. A sample of human serum albumin containing radioactive ^{131}I has an activity of 8.5 μCi. (a) How many iodine atoms are in the sample? (b) If each atom has a mass of 131 u, what is the mass of ^{131}I in the sample?

50. The yearly whole-body effective dose caused by radioactive ^{40}K absorbed in our tissues is 17 mrem. (a) Assuming that ^{40}K undergoes beta decay with RBE of 1.4, calculate the dose in units of rads. (b) How much beta-ray energy is absorbed by an 80-kg person? (*Note:* ^{40}K also emits gamma rays, many of which leave the body before being absorbed. Since fatty tissue has low potassium concentration and muscle has higher concentration, gamma-ray emissions indicate indirectly the amount of fat in a person.)

51. A pilot who flies 18 hours per week receives an additional whole-body effective dose of 0.70 mrem per hour of flight due to increased cosmic radiation. Calculate the pilot's yearly effective dose due to flying.

52. A body cell of 1.0×10^{-5} m radius and density $1000 kg/m^3$ receives a radiation dose of 1 rad. (a) Calculate the mass of the cell (assume it is shaped like a sphere). (b) Calculate the energy absorbed by the cell. (c) If the average energy needed to produce one positively charged ion is 100 eV, calculate the number of positive ions produced in the cell. (d) Repeat the procedure for the (3.0×10^{-6})-m radius nucleus of the cell, where its genetic information is stored.

53. While working with ^{14}C, you accidentally swallow through a pipette 10 μCi. (a) If each ^{14}C decay produces a 0.155-MeV β^- particle, which is absorbed by your tissue, calculate the energy in joules absorbed per year (ignore the decrease in ^{14}C activity).

(b) If your body mass is 50 kg, calculate your radiation dose from the ^{14}C for one year, assuming the radiation is distributed uniformly throughout the body.

General Problems

G1. *Estimate* the total number of (a) nucleons and (b) electrons in your body. (c) Indicate roughly the fraction of the body's volume occupied by the nucleons.

G2. A radioactive ^{60}Co nucleus emits a gamma ray of wavelength 0.93×10^{-12} m. If the cobalt was initially at rest, use the conservation-of-momentum equation to calculate its speed following the gamma-ray emission.

G3. One gram of pure, radioactive radium produces 130 J of heat per hour and has an activity of 1 Ci. Calculate the average energy in electron-volts released by each radioactive decay.

G4. Calculate the number of ^{40}K nuclei in the body of an 80-kg person using the information provided in Problem 50 and the fact that ^{40}K has a radioactive half-life of 1.28×10^9 years, and assuming that each ^{40}K beta decay results in 1.4 MeV of energy that is deposited in the person's tissue.

We live in a world where growth has become a way of life: our population grows by a small percentage each year as does our use of energy, the number of cars we drive, the area of the earth covered by asphalt and concrete, the cost of a loaf of bread, the national debt, and the money in our savings account (assuming there are no withdrawals). If the growth is steady, these quantities can be described by the exponential function. In this interlude we will examine the properties of this function and some of the surprising consequences we might expect from exponential growth.*

Growth Rate $\Delta N / \Delta t$

Exponential growth (or decay) occurs when the rate of change of a quantity is proportional to the quantity itself. Suppose that you have some money in a savings account. The money ΔN that is added to your account during a time Δt depends on three quantities: (1) the amount N of money already in the account, (2) the interest rate being paid, and (3) the time Δt that the money is collecting interest. These quantities are related by the equation

$$\Delta N = kN \Delta t$$

or

$$\frac{\Delta N}{\Delta t} = kN, \tag{1}$$

where k is a proportionality constant that depends on the interest rate being paid, and $\Delta N/\Delta t$ is called the **growth rate.**

The meaning of the proportionality constant k becomes apparent when we rearrange Eq. (1):

$$k = \frac{\Delta N/N}{\Delta t}.$$

We see that k equals the fractional change in N—that is $\Delta N/N$— per time Δt. If the money in a savings account increases by 6 percent per year, then the fractional change in the money during the year is $\Delta N/N = 0.06$ and the proportionality constant

$$k = \frac{\Delta N/N}{\Delta t} = \frac{0.06}{1 \text{ yr}} = 0.06 \text{ yr}^{-1}.$$

Notice that k has dimensions of time^{-1}.

---

*Many of the ideas and examples presented here are from a series of articles, referenced at the end of the interlude, written by Professor Albert Bartlett of the Department of Physics at the University of Colorado, Boulder. We would like to thank Professor Bartlett for his permission to use this material.

When discussing quantities that grow, it is common to indicate their **percentage growth rate** P. For the money in a savings account, $P = 6$ percent/yr. We relate the percentage growth rate to k by multiplying by 100:

$$P = \text{percentage growth rate} = 100k. \qquad (2)$$

The Exponential Function

It is quite easy, by means of calculus, to derive from Eq. (1) another equation that can be used to calculate the value of N at some future time t if its value N_0 is known at time zero. The right side of this equation, which we will not derive, is called the **exponential function:**

$$N = N_0 e^{kt}, \qquad (3)$$

where $e = 2.7182\ldots$ is the base of the natural logarithm system.

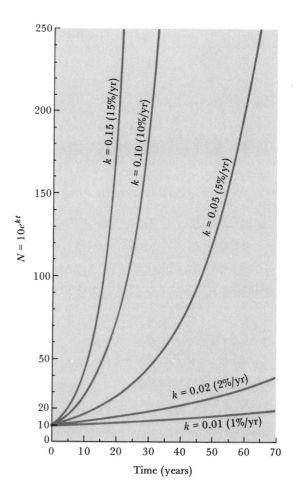

FIG. 1. Graphs of five different exponential functions are plotted on linear graph paper. $N_0 = 10$ for each graph, and the value of k is different for each graph.

If k is a positive number, exponential growth occurs. If k is a negative number (or if a negative sign appears in front of a positive-valued k), exponential decay occurs. We studied examples of exponential decay in Sections 34.6 and 34.7 (radioactive decay) and Section 26.6 (the decay of the electric charge on the plates of a capacitor as it discharges through a resistor). In the rest of this interlude we will restrict our attention to examples of exponential growth ($k > 0$).

The general characteristics of exponential growth are illustrated by means of the linear graphs shown in Fig. 1, where a steadily growing quantity N is plotted as a function of time for five different values of k. In each curve shown in Fig. 1 the value of N at time zero equals 10 ($N_0 = 10$). The curve that increases least rapidly is for $k = 0.01$ yr^{-1} (or a percentage growth rate of 1 percent/yr). The curve that increases most rapidly is for $k = 0.15$ yr^{-1} (a percentage growth rate of 15 percent/yr).

These same graphs are plotted on semilogarithmic graph paper in Fig. 2.

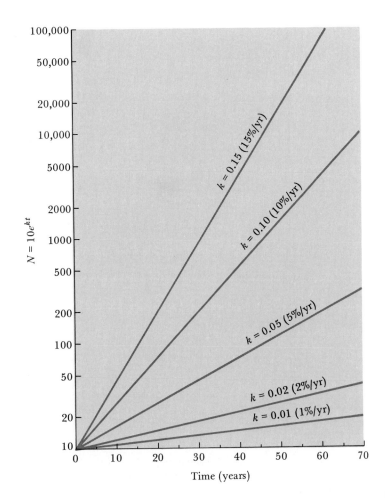

FIG. 2. The same exponential functions that were plotted in Fig. 1 on linear graph paper are plotted again on semilogarithmic graph paper.

693

Note that the exponentially increasing functions appear as straight lines on semilog graph paper (see Appendix B). These straight-line graphs can be extended easily to determine the value of an exponentially increasing quantity at some time in the future. (Remember that "exponentially increasing" is synonymous with "steadily growing.") This has been done in Fig. 3, where we show the increasing costs of several consumer products that are inflating at a rate of 10 percent/yr. Notice that with 10 percent/yr inflation, the cost of a movie will

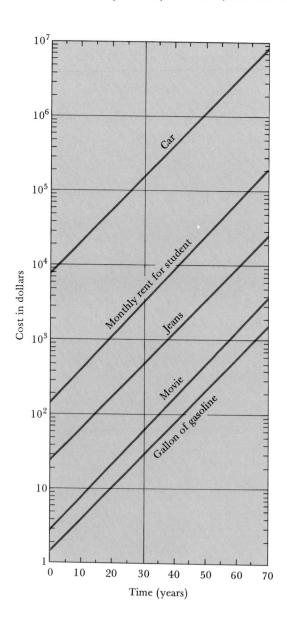

FIG. 3. The changing cost of several different consumer products that experience 10 percent per year inflation are plotted on semilogarithmic graph paper.

increase from \$3 at present to \$60 at a time 30 years in the future. An economy car's price will increase during the next 30 years from about \$8000 to \$160,000.

We do not need to rely on graphs alone to estimate the way in which constant growth rate affects the value of a quantity, as we learn next.

Doubling Time

Equation (3) can be used to derive a simple expression for calculating the time needed for a quantity's value to double. Rearranging Eq. (3) and taking the natural logarithm of each side, we find that

$$t = \frac{\ln (N/N_0)}{k}.$$

The growing quantity has a value N_0 at time zero, and increases to a value $N = 2N_0$ in a time T, called the **doubling time,** given by the expression

$$T = \frac{\ln (2N_0/N_0)}{k} = \frac{\ln 2}{k} = \frac{0.693}{k}.$$

If we now substitute $k = P/100$ [Eq. (2)] into the above equation, we find that

$$\text{doubling time} = T = \frac{69.3}{P} \simeq \frac{70}{P}, \tag{4}$$

where P is the percentage growth rate of the quantity. For example, if the world's population growth rate is 1.9 percent/yr, then the population doubling time is

$$T = \frac{70}{1.9 \text{ yr}^{-1}} = 37 \text{ yr.}$$

If the world's yearly energy usage is increasing at a rate of 4 percent/yr, then the energy used per year will double in

$$T = \frac{70}{4 \text{ yr}^{-1}} = 18 \text{ yr.}$$

The units need not be years. For example, if a colony of bacteria grows at a rate of 3.5 percent/min, then the doubling time is

$$T = \frac{70}{3.5 \text{ min}^{-1}} = 20 \text{ min.}$$

Growth in One Human Lifetime

We often hear people talking about the good old days when a cup of coffee cost a nickel, a gallon of gasoline 30 cents, and a nice home \$10,000. When we use the exponential function to help us look into the future, though, we may find that the good old days of low prices are now. Equation (3) can be used to derive a simple expression for calculating the factor by which a quantity will increase in the next seventy years (about one human lifetime) if it experiences constant growth. Substituting $t = 70$ years and $k = P/100$ into Eq. (3) and rearranging, we find that

$$\frac{N}{N_0} = e^{kt} = e^{(P/100)(70)} = e^{0.7P}.$$

But $e^{0.7P} = (e^{0.7})^P = 2^P$. Thus, *in approximately one lifetime of 70 years we can expect an exponentially increasing quantity to increase by a factor.*

$$\frac{N}{N_0} = 2^P \text{ (change in 70 years)}, \tag{5}$$

where P is the percentage growth rate per year. If the price of a home increases by 10 percent/yr, then in 70 years its price will increase by $2^{10} = 1024$; a home costing \$100,000 today would cost \$102,400,000 70 years from now. If the population of a city grows at a rate of 2 percent/yr, then 70 years from now, the city will have to increase its water, sewage, and waste disposal facilities by a factor of $2^2 = 4$. On the other hand, if the city grows by 5 percent/yr, these facilities must be increased by a factor of $2^5 = 32$. Few cities in the United States should experience this type of growth, since the percentage growth rate for the country is now about 0.6 percent/yr. In Mexico, the percentage growth rate in 1977 was 3.5 percent/yr. If this growth continues, then in 70 years the population of Mexico will have increased by a factor of $2^{3.5} = 11.3$.

Use During Doubling Period Exceeds All Previous Use

Another interesting property of a quantity that increases exponentially is that the change in the quantity during any doubling period is greater than the sum of the changes in the quantity in all the time preceding that doubling period. Suppose that you have \$100 in a savings account and that you are paid 7 percent/yr interest. The doubling time for your money is ten years [Eq. (4)]. The amount of money in the account increases with time as shown in Table 1. The money added to the account during the last doubling period (\$1600) exceeds the sum of the money added during all earlier doubling periods (\$100 + \$200 + \$400 + \$800 = \$1500).

For quantities whose yearly rate of use increases exponentially, we can make a similar statement: *the amount of the quantity used during any doubling period exceeds the amount of that quantity used during all the time preceding that doubling period.* For example, the rate of electrical energy use in the United States has increased at about 7 percent/yr (a doubling period of ten years). If this continues, we will use more electrical energy during the next ten years than was used in all the previous history of our country!

TABLE 1 Increase in Money in a Savings Account That Collects 7 Percent Annual Interest

Time	Money in account	Money added in last doubling period
0	\$ 100	—
10 yr	\$ 200	\$ 100
20 yr	\$ 400	\$ 200
30 yr	\$ 800	\$ 400
40 yr	\$1600	\$ 800
50 yr	\$3200	\$1600

696

It is clear that exponential growth cannot continue forever. In the real world we do not have resources that are infinitely large. Unfortunately, we do not realize the scarcity of a resource until it is nearly all gone, a condition that arrives abruptly. The idea is illustrated by considering bacteria that grow on a tray. The growth is started at 11:00 A.M. and the number of bacteria double each minute for one hour, at which time the tray has become completely covered. At 11:58 A.M. the tray is only one-fourth full of bacteria, and considerable room seems still to be available. One minute later, at 11:59 A.M., the bacteria have doubled and the tray is one-half full; there still seems to be enough room. One minute later, at noon, the bacteria double one last time and the tray is completely filled. It seems foolish to wait until the tray is full before halting growth.

Supplementary Reading

Highly recommended as supplementary reading are a number of articles by Albert A. Bartlett published in the *Physics Teacher,* which are cited below by volume and page.

1. **14:**393 (1976).   4. **15:**98 (1977).   7. **16:**158 (1978).
2. **14:**485 (1976).   5. **15:**225 (1977).   8. **17:**23 (1979).
3. **15:**37 (1977).   6. **16:**92 (1978).

Problems

1.   A community decides that its water facilities can be increased by no more than a factor of 2.0 in 24 years. (a) Calculate the maximum percentage growth rate in population of the community. (b) At this rate, by what factor will the water facilities need to be increased in seventy years?

2.   Suppose that the number of nuclear fission reactions in a nuclear power plant increased by 0.035 percent every 0.10 seconds. Calculate (a) the percentage growth rate and (b) the doubling time.

3.   The energy in the beam of a laser increases by a factor of 4.0×10^8 in 2.0×10^{-6} s as the beam induces stimulated emission while passing through the laser cavity. Use Eq. (3) to calculate the value of k for this growth process, assuming the growth is exponential.

4.   (a) The world usage of crude oil is increasing at a rate of about 7 percent per year. What is the doubling time? (b) If we are now using 3×10^9 barrels/yr, how many barrels can we expect to use per year twenty years from now (assuming the usage rate continues to grow at 7 percent)? (c) *Estimate* the number of barrels that have been used by the world in all of its history preceding this year.

5.   If the number of cars registered in the United States is increasing at a rate of 3.9 percent, how many years are required for the number to (a) double, (b) quadruple, (c) increase by a factor of ten (approximately), and (d) increase by a factor of 100 (approximately)?

6.   The cost of mailing a letter has varied with time as shown in the table below. (a) Is the cost increasing exponentially (approximately)? (b) What will be the cost approximately in the year 2000 if the cost continues to increase at the same rate? (c) What is the percentage increase per year? (d) What is the doubling time? (*Note:* You should use semilog graph paper for this problem.)

Year	Cost (cents)	Year	Cost (cents)
1968	6	1975	13
1971	8	1978	15
1974	10	1981	18

7.   The cost of a gall bladder operation has varied as follows: 1965 ($839), 1970 ($1397), 1975 ($2208). (a) Does the rate seem to be increasing exponentially? (b) What do you expect will be the cost in the year 1990? (c) What is the doubling time? (*Note:* You should use semilog graph paper for this problem.)

697

35

Fission, Fusion, and Elementary Particles

In Chapter 34 we learned about the binding energy that holds nuclei together and the radioactive emissions of nuclei that are unstable. Now we will encounter two types of nuclear reactions called fission and fusion. *Fission* occurs when a heavy nucleus splits into two smaller nuclei; *fusion* takes place when two small nuclei combine to form a larger nucleus. Fission reactions release energy for use in nuclear power plants, while fusion reactions release energy that warms the sun and other stars.

Later in the chapter we turn from nuclear physics to a study of the particles found in nuclei and of a large number of other particles that have been discovered in the last fifty years. We will discuss a relatively new theory of particle physics that suggests that protons, neutrons, and many other particles are made of smaller objects called quarks.

35.1 Nuclear Fission

Nuclear **fission** occurs when a heavy nucleus, such as $^{235}_{92}U$, splits to form two daughter nuclei of roughly equal mass. This process is called fission because it resembles the fission or division of a biological cell. Fission does not usually occur spontaneously; instead, it must be induced by collision and absorption of a low-energy neutron into the heavy nucleus (Fig. 35.1). After a short delay, the excited nucleus splits into two pieces. In addition to forming two medium-sized daughter nuclei, fission frees several neutrons. These neutrons may move on to induce fission reactions in other nuclei.

A typical fission reaction involving a uranium nucleus proceeds as follows:

$$^{1}_{0}n + ^{235}_{92}U \longrightarrow ^{236}_{92}U^* \longrightarrow ^{141}_{56}Ba + ^{92}_{36}Kr + 3\,^{1}_{0}n + \text{energy}. \qquad (35.1)$$

The excited $^{236}U^*$ nucleus that is formed after neutron absorption by ^{235}U can split in many different ways; in all, about 90 different daughter nuclei have been formed by uranium fission. For all of these possible uranium fission reactions we find that, on the average, 2.5 neutrons are produced per fission.

In addition to releasing daughter nuclei and neutrons, fission reactions also release energy; in fact, a single fission reaction releases about 10^7 times more energy than does an ordinary chemical reaction, such as the combustion of a molecule of coal. This energy release during fission can be sustained, since neu-

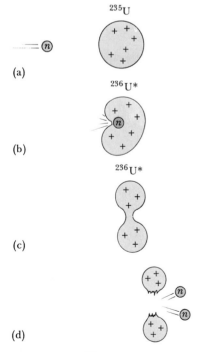

FIG. 35.1. A ^{235}U fission reaction induced by neutron capture.

698

FIG. 35.2. The average binding energy per nucleon is 0.9 MeV greater for the products of uranium-235 fission ($A \simeq 112$) than it is for uranium-235. The fission of ^{235}U releases about 0.9 MeV of energy for each of the 235 uranium nucleons.

trons released by one reaction can induce further reactions. Energy from controlled fission is now used in nuclear power plants to produce electricity.

The amount of energy released during a fission reaction can be analyzed using the graph of binding energy per nucleon shown in Fig. 35.2 (see Section 34.3 for a review of binding energy). Notice that the binding energy per nucleon for ^{235}U is about 7.6 MeV. Thus, in order to separate ^{235}U into its individual protons and neutrons, each nucleon needs 7.6 MeV of energy in the form of mass. This mass defect—that is, missing mass per nucleon—for ^{235}U is

$$\Delta m = \frac{\Delta E}{c^2} = \frac{7.6 \,\text{MeV} \left(\dfrac{1\,\text{u}}{931.5\,\text{MeV}/c^2} \right)}{c^2} = 8.16 \times 10^{-3}\,\text{u}.$$

Each of the two products of fission has about half as many nucleons as uranium; the average of their mass numbers is about 112. According to Fig. 35.2, a nucleus with $A = 112$ has a binding energy per nucleon of about 8.5 MeV, or a mass defect per nucleon of 9.13×10^{-3} u. During a fission reaction, the average mass defect per nucleon increases from about 8.16×10^{-3} u to 9.13×10^{-3} u, or each nucleon loses 0.97×10^{-3} u of mass. The mass lost in fission converts to energy—0.9 MeV per nucleon. Since uranium fission involves 235 nucleons, the total energy released by one fission is

$$235 \,\text{nucleons} \left(\frac{0.9\,\text{MeV}}{\text{nucleon}} \right) \simeq 200\,\text{MeV}.$$

In comparison, the combustion of one molecule of octane (C_8H_{18}) in the engine of your automobile releases 57×10^{-6} MeV of energy.

35.2 Nuclear Power Plants

Electric power plants fueled by the combustion of coal, oil, and nuclear fission all operate in much the same way. In a coal or oil plant, shown schematically in Fig. 35.3a, heat supplied by burning fuel converts water to steam. The steam, under high pressure, passes the blades of a turbine and causes it to rotate. A

Fossil–fuel power plant

(a)

Nuclear boiling–water reactor

(b)

Nuclear pressurized–water reactor

(c)

FIG. 35.3. (a) A fossil-fuel electric power plant, (b) a nuclear boiling-water reactor, and (c) a nuclear pressurized-water reactor.

generator coil attached to the turbine turns in a magnetic field, and electricity is produced. After passing the turbine, the steam is condensed at low pressure by cooling water and returns to the boiler for another trip around the power cycle. A nuclear boiling-water reactor BWR (Fig. 35.3b) works in the same way as a fossil-fuel plant except that the heat necessary to boil water is supplied by nuclear fission rather than by combustion of fossil fuel.

In a nuclear pressurized-water reactor (PWR), such as the one shown in Fig. 35.3c, fission reactions that occur in the reactor core supply heat to water that is maintained at high pressure so that it does not boil. The hot water is then pumped through a heat exchanger, which transfers heat from the pressurized water to a separate supply of water, thus converting it to steam. The steam then drives a turbine. In this system, the steam used to turn the turbine is isolated from the nuclear fuel, and there is less chance of carrying radioactive nuclei from the core of the reactor to other parts of the power plant.

In the rest of this section we will focus on the reactions that occur in the reactor core and on the mechanisms used to control the heating of water in the core.

Nuclear Fuel

Uranium, the fuel for nuclear reactors, comes in two isotopic forms: $^{238}_{92}$U (99.3 percent natural abundance) and $^{235}_{92}$U (0.7 percent natural abundance). While ^{238}U does not undergo fission, ^{235}U does. For most nuclear power plants the ^{235}U concentration must be increased from 0.7 percent to about 3 percent; the uranium is said to be "enriched" in its ^{235}U concentration. At this concentration, at least one of the neutrons produced by a fission reaction has a good chance of causing fission in another ^{235}U nucleus. The fission reactions are sustained, one following the other. At lower ^{235}U concentrations, the rate of reactions may slow down or even stop.

Once enriched, the uranium is placed in ceramic pellets, which retain their solid form even at the high temperature inside the core. Each pellet is about the size of the tip of your little finger. Pellets are stacked end-to-end in tubes 4 m long (called fuel rods) made of zirconium alloy. About 100 fuel rods are precisely arranged into bundles, called fuel assemblies; a few hundred assemblies provide the fuel for a reactor. A moderate-sized reactor may require 40,000 fuel rods weighing 100 tons or more. All of these fuel rods together are called the reactor **core.**

Water flowing past the fuel rods in the core is heated by contact with the rods, whose average temperature is about 300°C. The water serves three purposes: (1) it is the working fluid that, when heated and converted into steam, drives the turbine; (2) it prevents the reactor core from overheating and melting; and (3) it serves as a moderator for neutrons.

Moderators

A neutron released by a fission reaction normally moves near the speed of light. When moving at this speed, its chance of being captured by another ^{235}U nucleus is small. If the neutron can be slowed, its chance of capture is much better.

The function of a moderating material is to slow neutrons. To best accomplish this, the moderating material must have atoms that are approximately the same mass as a neutron. One direct hit between a moving neutron and a proton will completely stop the neutron. (This happens to any moving object that hits head-on another stationary object of equal mass—such as a cue ball hitting another billiard ball.)

Water serves as the moderator for nuclear power plants in the United States. The two nuclei of the hydrogen atoms in water (H_2O) quickly slow neutrons by collisions. Unfortunately, some collisions result in the combining of a proton and neutron to form deuterium ($_1^2H$):

$$_1^1H + _0^1n \longrightarrow _1^2H.$$

This combination is unfortunate because an absorbed neutron will not produce future fission reactions. The absorption of neutrons by water is one reason why uranium-235 must be enriched to a concentration of 3 percent. At lower concentrations the fission reactions cannot be sustained because too many neutrons are absorbed as they move between ^{235}U nuclei.

Reactors that use water as a moderator are called light-water reactors (LWR). The word *light* indicates that the hydrogen nuclei in the water molecules are protons instead of a heavier hydrogen isotope, $_1^2H$, known as deuterium.

Deuterium nuclei also serve as excellent moderators for neutrons and have the extra advantage of absorbing few neutrons. If deuterated water (that is, water in which the protons are replaced by deuterium (D) to form D_2O) is used as the working fluid, the reactor is called a heavy-water reactor. Although expensive, heavy water has an advantage over light water because the fission reactions can be sustained using uranium that has not been enriched. Nuclear reactors in Canada use heavy water.

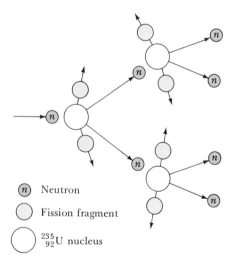

n Neutron

◯ Fission fragment

◯ $_{92}^{235}U$ nucleus

FIG. 35.4. A chain reaction. Neutrons released by one fission reaction are absorbed by other uranium nuclei and cause them to undergo fission.

Chain Reaction

For a reactor core to continue producing a steady supply of energy, each nuclear fission reaction must cause a subsequent fission reaction. Before considering this situation, let us see what happens if each fission reaction releases two neutrons (the average number is actually 2.5) that are absorbed by other ^{235}U nuclei and thus cause two more fission reactions (Fig. 35.4). The new fissions each release two more neutrons for a total of four neutrons. These four neutrons cause four more fusions, which release eight neutrons, and so on. If each neutron produces a fission, the number of nuclei undergoing fission per unit time increases exponentially; the reaction goes out of control and an explosion may occur. Such uncontrolled explosive reactions occur with uranium enriched to contain about 90 percent ^{235}U, the concentration of ^{235}U in atomic bombs. A fission explosion cannot occur when we use uranium enriched with only 3 percent ^{235}U. At this concentration the majority of the neutrons released by one fission reaction escape the reactor core, are absorbed by a proton, or undergo some other nonfission fate. But even at the low ^{235}U concentration, the reaction rate can increase to the point where the reactor core becomes so hot it melts.

Controlling Fission

For each ^{235}U fission, an average of 2.5 neutrons are released. In order to supply heat at a steady rate, exactly one of these neutrons should cause another fission reaction and the other 1.5 neutrons should not cause fission. With ^{235}U enriched to 3 percent or more, slightly more than one neutron released per fission causes future fission reactions. If uncontrolled, the fission rate would increase slowly and would eventually produce enough heat to melt the core. To prevent this increase, rods made of cadmium are inserted into the core. These rods absorb neutrons and slow the reaction. When the control rods are completely inserted, the reaction stops. If the rods are withdrawn, fewer neutrons are absorbed, and the rate of fission increases.

Spent Fuel Rods

When the fuel rods have been in a reactor for about three years, the ^{235}U concentration decreases so that efficient, sustained reactions are no longer possible. The fuel rods must then be removed. They continue to produce heat long after fission reactions have stopped because many of the products of fission are radioactive. The krypton-92 and barium-141 formed by the reaction shown in Eq. (35.1) both undergo beta decay with half-lives of 3.0 s and 18 min, respectively. The radioactive decay of these fission products in the fuel rods releases energy and causes the rods to remain hot for days.

A large number of fissionable uranium nuclei still remain in a fuel rod after it is removed from a reactor. The rods could be sent to a fuel-recyling plant to recover and incorporate this material into new fuel rods. At present, fuel-recycling plants are not allowed in the United States. All of the spent fuel is stored in pools of water at the reactor sites.

Nuclear Power Safety

Nuclear power production has been widely criticized as a threat to human health. Nuclear power production does involve a health hazard, as do most forms of human activity—from the hazard of slipping in the bathtub to that of driving to the store in an automobile. In order to put the nuclear power generation hazard in perspective, we will compare it to the hazard of producing electricity by a coal-fired power plant. These two power-generation techniques are chosen for comparison because they are both expected to play an increasing role in future electric-power production as our supplies of oil and gas become depleted.

The economic and health aspects of coal and nuclear power production have been analyzed in a variety of comprehensive studies. * A study published in 1977 indicated that on the average, 1.7 occupational deaths occur each year in the mining, transportation, and operation of a 1000-megawatt coal-fired power

---

*A Report of Advisory Committee on the Biological Effects of Ionizing Radiations, "Considerations of Health Benefit-Cost Analysis for Activities Involving Ionizing Radiation Exposure and Alternatives," U.S. Environmental Improvement Agency, EPA 520/4-77-03 (1977); Report of the Nuclear Energy Policy Study Group, *Nuclear Power Issues and Choices* (Cambridge, Mass.: Ballinger, 1977).

plant.* In addition, roughly 0.4 to 25 deaths per year result from the emission of sulfur-related pollutants from coal combustion. This number can be reduced by using low-sulfur coal and lime scrubbers to absorb sulfur emissions. In summary, for a new coal-fired 1000-megawatt plant meeting present safety standards, analysis indicates premature deaths for workers and the general public in the range of 2 to 25 per year.† In 1975 the numbers were larger (from 20 to 60) because fewer emission control devices were used on coal plants.

Major health concerns related to the use of nuclear fuel include the possibility of a major accident and the effects of low-level radiation emitted into the environment. Estimates used to calculate these hazards are even more uncertain than those applied to power generation using coal. Low-level radiation exposure resulting from all aspects of the operation of a 1000-megawatt nuclear reactor for one year is expected to cause 0.6 to 1.0 deaths per year. The WASH-1400 Report, a reactor safety study published in 1975,‡ concluded that a serious nuclear accident had a very small probability but would result in about 3300 prompt fatalities and approximately 45,000 cancer fatalities in the 40 years following. To compare this risk to that related to coal power production, the accident risk has been averaged over the number of years estimated to pass between these serious accidents, none of which has ever occurred. The WASH-1400 Report projected 0.02 accident-related fatalities per year as a result of the operation of a 1000-megawatt reactor. This report has been criticized as underestimating the risk; at the extreme upper limit, if all uncertainties are viewed pessimistically, the risk might be as high as ten deaths per year per 1000-megawatt reactor. In summary, the hazards from low-level radiation and

*EPA 520/4-77-03 (1977), cited above.

†EPA 520/4-77-03 (1977), cited above.

‡The Reactor Safety Study, Nuclear Regulatory Commission (1975). This report is often referred to as the Rasmussen Report or the WASH-1400 Report.

TABLE 35.1 Comparisons of Mortality Risks[a]

Cause of death	Number of deaths
One 1000-megawatt coal-fired power plant[b]	2–25
One 1000-megawatt nuclear reactor[b]	0.6–10
Lightning	200
Medical x-rays	900
Electrocution	2,000
Air travel	3,000
Drowning	9,000
Fires and hot substances	10,000
Falls	30,000
Motor vehicles	90,000
All causes	3,000,000

[a] Adapted from "NRC Estimates of Health Risks Associated with Low-Level Radiation Exposure," 10th Annual Conference on Radiation Control, Bureau of Radiological Health, HEW Publication (FDA) 79-8054 (1979), and from the WASH-1400 Report. The numbers, except for the first two, are for a population of 300 million.

[b] The generation of electricity for a population of 300 million would require about 600 1000-megawatt power plants. If 100 percent of our electric power in the year 2000 were produced by nuclear power plants, the expected number of deaths per year would be from 0.6 to 10 times 600, or 360 to 6000. For coal power plants producing 100 percent of our electricity, the number of deaths per year would be from 1200 to 15,000.

possible accidents at a 1000-megawatt reactor are estimated to produce approximately 0.6 to 10 fatalities per year, slightly fewer than the estimates for a coal-fired plant.

To help place this mortality risk in perspective, the risks of other types of accidental deaths are shown in Table 35.1. The numbers represent the expected fatalities each year for a population of 300 million, roughly the expected United States population by the year 2000.

35.3 Breeder Reactors

Another type of nuclear reactor, called a breeder reactor, converts nonfissionable ^{238}U with 99.3 percent natural abundance to plutonium-239, a fissionable material. Some of the neutrons emitted by the fission of ^{235}U are absorbed by ^{238}U, and after the series of reactions shown below, plutonium-239 ($^{239}_{94}$Pu) is formed:

$$_{0}^{1}n + {}^{238}_{92}\text{U} \longrightarrow {}^{239}_{92}\text{U}^* \longrightarrow {}^{239}_{93}\text{Np} + {}_{-1}^{0}e,$$
$$_{93}^{239}\text{Np} \longrightarrow {}^{239}_{94}\text{Pu} + {}_{-1}^{0}e.$$

The ^{239}Pu isotope is a fissionable nucleus that releases about three neutrons per fission and can serve as a nuclear fuel for sustained fission reactions.

In a breeder reactor, one neutron released by ^{239}Pu fission produces another fission. A second neutron is absorbed by a ^{238}U nucleus and produces a new ^{239}Pu nucleus. In this way, the concentration of the ^{239}Pu fuel remains constant. If the third neutron is also absorbed by ^{238}U, two ^{239}Pu nuclei are produced for each ^{239}Pu fission. Since more fuel is produced than is consumed, the reactor is said to "breed" nuclear fuel. Breeding fissionable plutonium-239 from nonfissionable uranium-238 increases the supply of fissionable fuel in the earth's crust by a factor of more than 100: without breeding, only 0.7 percent of natural uranium, the ^{235}U, is fissionable; with breeding, all of the uranium is fissionable.

35.4 Fusion

A proposed technique for producing electrical power that should have less impact on the environment than do present methods is based on a process called nuclear fusion. **Fusion** occurs when two small nuclei combine or fuse together to form a larger nucleus.

The simplest fusion reaction is the combination of a proton ($^{1}_{1}$H) and a neutron ($^{1}_{0}n$) to form a deuterium ($^{2}_{1}$H):

$$_{1}^{1}\text{H} + {}_{0}^{1}n \longrightarrow {}_{1}^{2}\text{H} + \gamma. \qquad (35.2)$$

The mass of the deuterium is smaller than the masses of the proton and neutron. Energy is released by the reaction; part of the energy results in the production of a gamma ray (γ) and the rest is converted to kinetic energy of the deuterium.

EXAMPLE 35.1 Calculate the mass-energy released by the fusion of a neutron and proton.

SOLUTION The energy released equals c^2 times the difference in mass of the reactants and the product of the reaction [Eq. (34.6)]:

$$Q = \text{energy released} = (m_{{}^1_1\text{H}} + m_{{}^1_0 n} - m_{{}^2_1\text{H}})\,c^2$$

$$= (1.007825\ \text{u} + 1.008665\ \text{u} - 2.014102\ \text{u})\,c^2$$

$$= \left(0.002388\ \text{u} \times \frac{931.5\ \text{MeV}/c^2}{1\ \text{u}}\right) c^2 = \underline{2.22\ \text{MeV}}. \qquad \blacksquare$$

Many fusion reactions release relatively large amounts of energy. In fact, fusion provides the energy for the sun and other stars. Since the sun's mass is approximately three-fourths hydrogen, most of the fusion occurring on the sun involves reactions combining protons—the nuclei of hydrogen atoms—with each other and with other particles and nuclei. Helium is also important in these reactions, since it accounts for approximately one-fourth of the sun's mass.

An example of a series of fusion reactions that provides some of the sun's energy is the **proton-proton cycle** summarized below:

$$\begin{array}{lll} {}^1_1\text{H} + {}^1_1\text{H} \longrightarrow {}^2_1\text{H} + {}^0_1 e + \nu & (0.42\ \text{MeV}), & \textbf{(35.3a)} \\[4pt] {}^2_1\text{H} + {}^1_1\text{H} \longrightarrow {}^3_2\text{He} + \gamma & (5.49\ \text{MeV}), & \textbf{(35.3b)} \\[4pt] {}^3_2\text{He} + {}^3_2\text{He} \longrightarrow {}^4_2\text{He} + {}^1_1\text{H} + {}^1_1\text{H} & (12.86\ \text{MeV}). & \textbf{(35.3c)} \end{array}$$

Net reaction:

$$6\,{}^1_1\text{H} \longrightarrow {}^4_2\text{He} + 2\,{}^1_1\text{H} + 2\,{}^0_1 e + 2\nu \qquad (24.68\ \text{MeV}). \qquad \textbf{(35.3)}$$

We see that a product of one reaction is involved in the subsequent reaction. Because two heliums (${}^3_2\text{He}$) are needed for the last reaction [Eq. (35.3c)], the first two reactions must each occur twice before the last reaction can proceed. Besides helium-4, two neutrinos (ν) and two positrons (${}^0_1 e$) are produced by the reaction. The positrons annihilate quickly with electrons to produce an additional 2 MeV of energy.

The net energy released by the proton-proton cycle (including the electron-positron annihilation) is 26.68 MeV. One gram of hydrogen undergoing the reactions summarized by Eq. (35.3) releases enough energy to heat an average home for over 40 years! Although the abundance of hydrogen in our environment makes these reactions seem like attractive sources of energy, one major obstacle stands in the way of their use. In order for two protons to react as represented by Eq. (35.3a), they must be brought to a separation of about 10^{-14} m or less so that the strong nuclear force between them exceeds their electrical repulsion. This close encounter occurs only if the protons have a large amount of kinetic energy that is converted to electrical potential energy as they approach each other (see Fig. 35.5). Let us calculate the initial kinetic energy needed to bring the protons within 10^{-14} m of each other. From this calculation we will be able to estimate the temperature at which we can expect protons to fuse.

Suppose that the protons are moving straight toward each other with an initial speed v_0 and that they are initially separated by a very large distance. As they draw closer, their speed decreases while their electrical potential energy increases. Finally, when separated by about 10^{-14} m or less, the protons become "locked" together by the strong nuclear force. For the protons to get this close, their initial kinetic energy must balance their final electrical potential energy

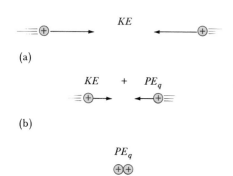

(a)

(b)

(c)

FIG. 35.5. As two protons approach each other, their initial kinetic energy converts to electrical potential energy.

when separated by a distance $R \simeq 10^{-14}$ m:

$$\text{initial } KE = \text{final } PE_q$$

$$2\left(\frac{1}{2}mv_0^2\right) = \frac{ke^2}{R}.$$

Solving for the initial kinetic energy of one of the protons leads to the expression

$$\frac{1}{2}mv_0^2 = \frac{ke^2}{2R} = \frac{(9 \times 10^9 \,\text{N} \cdot \text{m}^2/\text{C}^2)(1.6 \times 10^{-19}\,\text{C})^2}{2(10^{-14}\,\text{m})} = 1.15 \times 10^{-14}\,\text{J}.$$

In Chapter 12 we learned that the temperature of a gas reflects the average random kinetic energy of the atoms in the gas; the temperature T is related to the average kinetic energy by Eq. (12.12):

$$\tfrac{3}{2}kT = \tfrac{1}{2}mv^2,$$

where k is Boltzmann's constant (1.38×10^{-23} J/K). For a gas of protons to have sufficient kinetic energy for fusion to occur during an average collision, the temperature must be

$$T = \frac{\tfrac{1}{2}mv^2}{\tfrac{3}{2}k} = \frac{1.15 \times 10^{-14}\,\text{J}}{\tfrac{3}{2}(1.38 \times 10^{-23}\,\text{J/K})} = 5.6 \times 10^8\,\text{K}.$$

The sun's temperature is about 0.15×10^8 K at its core. However, even at this "low" temperature, enough protons have the kinetic energy needed to cause fusion. Obviously, fusion cannot be expected to occur in a home furnace in which the flame temperature is only about 600 K. In fact, even the sun's outer shell, the photosphere, is too cool (5800 K) for fusion to occur.

It is expected that the commercial production of energy by controlled fusion reactions will become a reality at some time in the future. At present, more energy is expended in initiating fusion than is produced by it. One fusion technique that shows promise consists of shooting tiny pellets of fusionable material into the center of a large number of ion beams (Fig. 35.6). When focused on the pellets, the ions produce a great amount of thermal energy, resulting in pellet temperatures of 10^8 K; fusion is induced. Other fusion techniques, such as shooting the pellets into a large number of focused laser beams, are also being tested.

The production of energy by fusion has several advantages over energy production by fission. Fusion reactions produce few radioactive isotopes, and those formed have short lifetimes. The fusion-energy release can be stopped by simply turning off the ion beams or by stopping the flow of pellets into the beam. Also, the earth contains an unlimited supply of fusionable material. Fusion, if it can be made to work, offers a somewhat cleaner source of energy than either coal or nuclear fission.

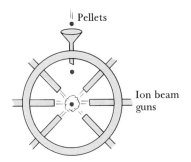

FIG. 35.6. Ion beams focused on fuel pellets containing deuterium cause intense heat, resulting in fusion.

35.5 Elementary-Particle Physics

In this and the next two sections we conclude our study of physics by considering one of the most interesting and challenging problems physicists have studied in many years. The significance of this endeavor parallels that of the development of Newtonian mechanics and the atomic theory of quantum mechanics. These great mental achievements of the past have led to many new technologies, such as communications satellites, electronic devices of all sorts, lasers, and various ways of using and providing energy. The technological impact of new theories such as quantum mechanics could not have been predicted or even imagined at the time of their development. When J. J. Thomson discovered the electron in 1897, he could not have envisioned the many modern devices that rely on information and energy transfer by electrons. Scientists working today in the field of elementary-particle physics find themselves in a similar situation. They are trying to understand the basic nature of particles such as electrons, protons, and neutrons, but they cannot envision what new applications will come of this understanding.

To most of us, electrons, protons, and neutrons seem like unchangeable balls of mass—some with electric charge and others with none. They were supposedly part of the universe when it was born and will supposedly be with us, unchanged, to the end. But some scientists now think that these basic building blocks of nature may not be so basic after all. Physicists currently suggest that each proton and neutron consists of three smaller, more basic particles called **quarks.** The quarks—not the proton and the neutron—are the elementary particles, the basic building blocks.

The electron, along with several other particles, still retains its cherished role of being "elementary"—an indivisible, single piece of matter. But even the electron's structure is a puzzle. How can this tiny mass and charge possibly stick together? Its diameter is less than 10^{-15} m. If we think of its mass and electric charge as occupying a sphere of that diameter and draw an imaginary slice through its middle, as depicted in Fig. 35.7a, the electrical repulsive force of one

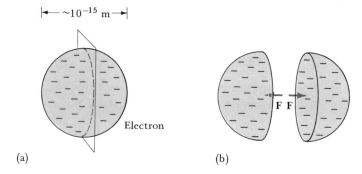

$$\vdash\!\!-\!\!- \sim\!10^{-15}\text{ m}-\!\!\dashv$$

Electron

F F

(a)   (b)

FIG. 35.7. (a) An imaginary slice is drawn through an electron. (b) The repulsive electrical force of one half on the other is over 50 lb.

half of the sphere on the other half is over 50 lb (Fig. 35.7b). The strong nuclear force has no effect on the electron, and evidently some other force is needed to hold an electron together. An interesting proposal suggests that an electron is a very small, stable black hole and that the great density of mass inside the electron holds its pieces together, even though they are being pushed apart by their electrical repulsion.

The intense interest in elementary particles in recent years has resulted partly from the discovery of over 200 new particles that in many ways seem as elementary as protons and neutrons. All these particles have masses greater than the electron, and many are more massive than protons and neutrons.

A list of several stable particles appears in Table 35.2. A particle is considered stable if it reaches the ripe old age of about 10^{-12} s or older before decaying into other particles. This time is long compared to the life of most of the other 200 particles—about 10^{-23} s. The decay of an elementary particle often leads to the formation of another particle, which in turn decays. The decay process eventually stops when a truly stable particle, such as a proton or electron, is

TABLE 35.2 "Stable" Elementary Particles

Category	Particle name	Symbol	Anti-particle	Rest-mass energy (MeV)	B	S	L_e	L_μ	L_τ	Life-time (s)	Decay mode
Photon	Photon	γ	Self	0	0	0	0	0	0	Stable	
Leptons	Electron	e^-	e^+	0.51	0	0	1	0	0	Stable	
	Electron neutrino	ν_e	$\bar{\nu}_e$	0	0	0	1	0	0	Stable	
	Muon	μ^-	μ^+	105.7	0	0	0	1	0	2.2×10^{-6}	$\mu^- \longrightarrow e^- + \bar{\nu}_e + \nu_\mu$
	Muon neutrino	ν_μ	$\bar{\nu}_\mu$	0	0	0	0	1	0	Stable	
	Tau	τ^-	τ^+	1782	0	0	0	0	1	$\sim 10^{-13}$	$\tau^- \longrightarrow e^- + \bar{\nu}_e + \nu_\tau$
	Tau neutrino	ν_τ	$\bar{\nu}_\tau$	0	0	0	0	0	1	Stable	
Hadrons Mesons	Pion	π^+	π^-	139.6	0	0	0	0	0	2.6×10^{-8}	$\pi^+ \longrightarrow \mu^+ + \nu_\mu$
		π^0	Self	135.0	0	0	0	0	0	0.8×10^{-16}	$\pi^0 \longrightarrow \gamma + \gamma$
	Kaon	K^+	K^-	493.8	0	+1	0	0	0	1.2×10^{-8}	$K^+ \longrightarrow \mu^+ + \nu_\mu$; $\pi^+ + \pi^0$
		K^0_S	$\overline{K}^0_S$	497.8	0	+1	0	0	0	0.9×10^{-10}	$K^0_S \longrightarrow \pi^+ + \pi^-$; $\pi^0 + \pi^0$
		K^0_L	$\overline{K}^0_L$	497.8	0	+1	0	0	0	5.7×10^{-8}	$K^0_L \longrightarrow \pi^\pm + e^\mp + \left(\frac{\nu_e}{\bar{\nu}_e}\right)$; $\pi^\pm + \mu^\mp + \left(\frac{\nu_\mu}{\bar{\nu}_\mu}\right)$; $\pi^0 + \pi^0 + \pi^0$
	Eta	η^0	Self	548.8	0	0	0	0	0	$<10^{-16}$	$\eta^0 \longrightarrow \gamma + \gamma$
Baryons	Proton	p	$\bar{p}$	938.3	+1	0	0	0	0	Stable	
	Neutron	n	$\bar{n}$	939.6	+1	0	0	0	0	10^3	$n \longrightarrow p + e^- + \bar{\nu}_e$
	Lambda	Λ^0	$\bar{\Lambda}^0$	1116	+1	−1	0	0	0	2.5×10^{-10}	$\Lambda^0 \longrightarrow p + \pi^-$; $n + \pi^0$
	Sigma	Σ^+	$\bar{\Sigma}^-$	1189	+1	−1	0	0	0	0.8×10^{-10}	$\Sigma^+ \longrightarrow p + \pi^0, n + \pi^+$
		Σ^0	$\bar{\Sigma}^0$	1192	+1	−1	0	0	0	$<10^{-14}$	$\Sigma^0 \longrightarrow \Lambda^0 + \gamma$
		Σ^-	$\bar{\Sigma}^+$	1197	+1	−1	0	0	0	1.5×10^{-10}	$\Sigma^- \longrightarrow n + \pi^-$
	Xi	Ξ^0	$\bar{\Xi}^0$	1315	+1	−2	0	0	0	3×10^{-10}	$\Xi^0 \longrightarrow \Lambda^0 + \pi^0$
		Ξ^-	$\bar{\Xi}^+$	1321	+1	−2	0	0	0	1.7×10^{-10}	$\Xi^- \longrightarrow \Lambda^0 + \pi^-$;
	Omega	Ω^-	$\bar{\Omega}^+$	1672	+1	−3	0	0	0	1.3×10^{-10}	$\Omega^- \longrightarrow \Xi^0 + \pi^-$, $\Lambda^0 + K^-$

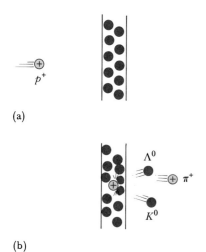

(a)

(b)

FIG. 35.8. (a) A proton flying at high speed crashes (b) into a target nucleus, causing several new particles to spray out of the target material.

formed. The decay modes of the unstable "stable" particles appear in Table 35.2, along with their lifetimes and a variety of other information about them.

The particles are produced by accelerating protons or electrons to a very high energy. When the moving particle hits some target nucleus, new particles spray out from the collision (Fig. 35.8). The particles produced by the collision hit other target nuclei or decay, thus leading to the formation of new particles.

The different particles produced by the collision can be detected by the trails they leave in bubble chambers, cloud chambers, or nuclear emulsion plates. Bubble-chamber tracks similar to those shown in Fig. 35.9 represent a series of events started by a collision of a moving kaon (K^-) with a stationary proton. The reaction that follows is represented by the equation

$$K^- + p^+ \longrightarrow \Xi^- + K^0 + \pi^+.$$

Charged particles produce tracks in the bubble chamber, whereas particles with no charge, such as the K^0, produce no tracks. From the point of the collision, we see only the short upward track of the Ξ^- and the track toward the right of the π^+.

The Ξ^- lives only about 10^{-10} s before decaying into a Λ^0 and π^-, the latter producing a track directed up toward the right. Both the Λ^0 and the K^0, produced earlier, move undetected through the bubble chamber until they decay, each causing the creation of two new charged particles:

$$K^0 \longrightarrow \pi^+ + \pi^-,$$
$$\Lambda^0 \longrightarrow p^+ + \pi^-.$$

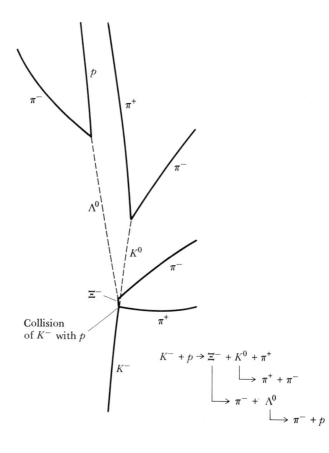

FIG. 35.9. A K^- hits a proton, resulting in the production of a Ξ^-, K^0, and π^+. The tracks are similar to those produced in a bubble chamber.

The charged products of each decay produce tracks that together look like the letter V. V-tracks appear frequently in these experiments.

Through elaborate analysis of these and many other tracks, it becomes possible to identify the mass and electric charge of the different particles involved in the collisions. The masses are not reported in atomic mass units; instead, we list the rest-mass energy of the particles in units of MeV. For example, a proton of mass 1.0072766 u has a rest-mass energy of 938.3 MeV.

Each particle listed in Table 35.2 has an **antiparticle** that has the same mass but an opposite electric charge. A proton's antiparticle has the same mass as a proton but has charge $-e$; an electron's antiparticle, the positron, has charge $+e$ and the same mass as the electron. Some scientists speculate that, given the great amount of symmetry found in the world, there is probably a part of the universe made of antiparticles. A hydrogen atom in this antiworld would consist of a positron circling a negatively charged antiproton. Antiparticles do not last long in our part of the universe, since they quickly join with their oppositely charged particles, causing an annihilation that releases all the mass of each particle in the form of electromagnetic radiation. A more recent theory says that there is no antiworld and that most of the antiparticles were annihilated during the explosive beginning of our universe, the big bang. The excess of particles that were not annihilated by antiparticles constitute the remaining mass of the universe.

Having discovered over 200 of these elementary particles leaves us with an impression of confusion. By examining a large number of particle collisions, scientists have devised several classification schemes that bring a little order to this chaos. These schemes have led to the development of several new conservation laws for particles.

35.6 Classification Schemes and Conservation Laws

Much of our present understanding of elementary particles is summarized in classification schemes that group together particles with similar behaviors and properties. One of the most useful schemes groups particles according to the forces they exert on each other. For example, protons and neutrons are subject to the strong nuclear force, whereas electrons are not. In Table 35.2, the **hadron** group consists of particles that exert strong nuclear forces on each other; the proton and neutron are the best known of this group. The **leptons,** including the electron, do not interact strongly with other particles.

The hadrons are subdivided into two groups, the mesons and the baryons. **Baryons** are distinguished by the fact that their number must be conserved in any nuclear process. The conservation, however, occurs in a very special way. Each baryon listed in Table 35.2 has a baryon number, $B = +1$. The antiparticle of a baryon has $B = -1$. Other particles have baryon number $B = 0$. The **conservation-of-baryon-number law** is stated as follows:

In any nuclear reaction or decay process, the sum of the baryon numbers of the reactants must equal the sum for the products of the reaction.

The law was developed merely by noting that no exceptions have been observed in any particle-reaction experiment performed so far.

EXAMPLE 35.2 Determine whether each reaction listed below obeys the baryon number conservation law:

(a) $$p^+ + n \longrightarrow p^+ + n + p^- + p^+.$$

(b) $$p^+ + n \longrightarrow p^+ + p^+ + p^-.$$

(c) $$K + p^+ \longrightarrow \Xi^- + K^0 + \pi^+.$$

SOLUTION

(a) B: $1 + 1 = 1 + 1 - 1 + 1.$

(b) B: $1 + 1 \neq 1 + 1 - 1.$

(c) B: $0 + 1 = 1 + 0 + 0.$

Reactions (a) and (c) satisfy the law, whereas (b) violates the baryon-number-conservation law. Reaction (b) has never been observed and is said to be forbidden because it violates the baryon-number conservation law. ∎

On the basis of observation of many particle reactions, other laws have been proposed. A group of particles discovered in the 1950s exhibited what was regarded as strange behavior; these became known as **strange** particles, examples being the K, Λ, and Σ. One example of their "strange" behavior is that they are always produced in pairs. For example, if a π^--meson hits a proton, a K^0 and Λ^0 are produced:

$$\pi^- + p \longrightarrow K^0 + \Lambda^0.$$

According to conservation laws known at that time, a neutron should be able to replace the Λ^0 in the above equation, but such a neutron reaction never occurred. Because of this and other strange behavior, another new conservation law was hypothesized, the **conservation-of-strangeness law:**

During any nuclear reaction or decay process involving particles that exert strong nuclear forces on each other, the strangeness of the reactants must equal that of the reaction products.

The strangeness number S of the elementary particles listed in Table 35.2 appears in the column labeled "S". The strangeness of an antiparticle has the opposite sign to that of the particle.

EXAMPLE 35.3 Determine whether the following reactions satisfy the conservation-of-strangeness law:

(a) $$\pi^- + p \longrightarrow K^0 + \Lambda^0.$$

(b) $$\pi^- + p \longrightarrow K^0 + n.$$

SOLUTION

(a) S: $0 + 0 = 1 - 1$ allowed.

(b) S: $0 + 0 \neq 1 + 0$ forbidden.

Reaction (b) is forbidden by the conservation-of-strangeness law. ∎

In recent years, particles have also been classified in other ways: electron lepton number (L_e), muon lepton number (L_μ), and tau lepton number (L_τ). Most recently, the particles have been postulated as having charm, truth, and beauty; such words reflect the whimsy of physicists more than the actual particle properties.

35.7 Quarks

The classification and conservation laws just described are analogous to the development of the periodic table of elements by Dmitri Mendeleev in 1869. Not until almost sixty years later, after the formulation of quantum mechanics, could scientists explain why certain groups of atoms had similar behavior. Physicists now are searching for a basic theory of elementary particles that will explain why they are classified as baryons, strange, and so forth.

The theory that has gone farthest in explaining these properties of elementary particles is based on a **quark model.** This theory, proposed independently in 1963 by Murray Gell-Mann and George Zweig of the California Institute of Technology, suggests that hadrons are composed of several smaller particles called quarks. In the original 1963 theory, three quarks (named "up," "down," and "strange") and their antiquarks were needed to construct all the hadrons then known. Since that time, three other quarks named "charm," "truth," and "beauty" have been added to the list.* These quarks have properties such as mass, charge, baryon number, and strangeness (see Table 35.3), and each quark

TABLE 35.3 Properties of Quarks and Antiquarks

Name	Quarks					Antiquarks				
	Symbol	Charge	Baryon number	Strangeness	Charm	Symbol	Charge	Baryon number	Strangeness	Charm
Up	u	$+\frac{2}{3}e$	$\frac{1}{3}$	0	0	$\bar{u}$	$-\frac{2}{3}e$	$-\frac{1}{3}$	0	0
Down	d	$-\frac{1}{3}e$	$\frac{1}{3}$	0	0	$\bar{d}$	$\frac{1}{3}e$	$-\frac{1}{3}$	0	0
Strange	s	$-\frac{1}{3}e$	$\frac{1}{3}$	-1	0	$\bar{s}$	$\frac{1}{3}e$	$-\frac{1}{3}$	$+1$	0
Charm	c	$+\frac{2}{3}e$	$\frac{1}{3}$	0	1	$\bar{c}$	$-\frac{2}{3}e$	$-\frac{1}{3}$	0	-1
Truth (top)	t	$+\frac{2}{3}e$	$\frac{1}{3}$	0	0	$\bar{t}$	$-\frac{2}{3}e$	$-\frac{1}{3}$	0	0
Beauty (bottom)	b	$-\frac{1}{3}e$	$\frac{1}{3}$	0	0	$\bar{b}$	$\frac{1}{3}e$	$-\frac{1}{3}$	0	0

has an antiquark whose numbers have opposite signs. Baryons are made of three quarks; mesons of two. A particle's net charge, baryon number, strangeness, and so on equal the sum of the numbers for the quarks that compose the particle. The quark compositions of several particles are listed in Table 35.4.

EXAMPLE 35.4 Show that the charge, baryon number, and strangeness of the proton equal the sum of these numbers for its three quarks (*uud*).

*Truth and beauty are often called "top" and "bottom." We will use their more poetic names.

TABLE 35.4 Quark Composition of Several Stable Hadrons

Mesons	Constituent quarks
π^+	$u\bar{d}$
π^-	$\bar{u}d$
K^+	$u\bar{s}$
K^-	$\bar{u}s$
J/ψ	$c\bar{c}$
γ	$b\bar{b}$

Baryons	Constituent quarks
Proton	uud
Neutron	udd
Λ^0	uds
Σ^+	uus
Σ^0	uds
Σ^-	dds
Ξ^0	uss
Ξ^-	dss

SOLUTION Using Table 35.3, we find that:

$$\begin{array}{ccc} u & u & d \end{array}$$
$$q = \tfrac{2}{3}e + \tfrac{2}{3}e - \tfrac{1}{3}e = e,$$
$$B = \tfrac{1}{3} + \tfrac{1}{3} + \tfrac{1}{3} = 1,$$
$$S = 0 + 0 + 0 = 0.$$

We see from Table 35.2 that these numbers agree with the values of q, B, and S listed for a proton. ∎

In recent years researchers have extended the quark model in an effort to devise a theory for the way quarks interact with each other. The theory in favor at present is called **quantum chromodynamics.** This theory is analogous in many ways to the one used to describe the electrical interactions between charged particles. In electrical theory particles are said to have electric charge whose value can be $+e$, 0, or $-e$. The concept of electric charge was invented not because electric charge could be seen stuck to a particle but because it was needed to explain the electrical force that one particle exerts on another.

A similar situation exists in particle physics. In order to explain the fact that quarks must exert very strong forces on each other when held together in a particle, the quarks are said to have a property called *color*. A quark can be either red, blue, or yellow. In quantum chromodynamics, the word *color* does not mean that the quarks actually look red, blue, or yellow. Instead, the color represents a physical property of a quark that causes it to exert a strong force on another quark (just as the electric charge and mass of particles cause them to exert electrical and gravitational forces on each other). This strong force is related in some as yet undetermined way to the strong nuclear force that hadrons exert on each other.

Another analogy between electrical theories and quantum chromodynamics is the ability of charged particles and of quarks to emit and absorb other forms of energy. Particles with electric charge emit and absorb photons; quarks emit and absorb particles called *gluons*. The gluons are supposedly small objects (never observed) that are exchanged or shared by quarks and bind them together to form protons, neutrons, and the other hadrons.

Knowledge in the field of elementary particle physics is changing rapidly. Old theories are regularly refined and revised to fit the results of new experiments. New ideas emerge. In 1968 Abraham Pais, a particle theorist, likened the state of particle physics to a symphony hall some time before the start of the concert.

> On the podium one will see some but not yet all of the musicians. They are tuning up. Short brilliant passages are heard on some instruments; improvisations elsewhere; some wrong notes, too. There is a sense of anticipation for the moment when the symphony starts. *

Some physicists think that the symphony has now started—that it consists of an interesting combination of colored quarks. Others think that the orchestra is still tuning up and that we have yet to see all the musicians. We hope the final symphony will include truth and beauty—if not as quarks, then in other ways.

*Abraham Pais, "Particles," *Physics Today*, May 1968, p. 28.

Summary

1. **Fission** occurs when a heavy nucleus (such as $^{235}_{92}U$) splits, forming two smaller daughter nuclei. Fission is normally induced by absorption of a neutron. Along with the daughter nuclei, several neutrons are emitted. These neutrons can induce more fissions, thus creating a series of fission reactions. Roughly 200 MeV of energy is released by the fission of one ^{235}U nucleus. This energy can be used to produce electricity in nuclear power plants.

2. **Fusion** is the combination of two small nuclei to form a larger nucleus. Many fusion reactions release a large amount of energy. In order for fusion to occur, the nuclei must move to a distance of about 10^{-14} m or less from each other. This requires large initial kinetic energy that is converted to electrical potential energy as the nuclei with like charges approach each other. Because of the large kinetic energy needed to cause a close approach of the nuclei, fusion usually occurs only at very high temperature, such as in the interior of the sun. Fusion reactors are being developed for controlled release of energy for use in generating electricity.

3. **Elementary particles** are the basic building blocks of nature, electrons, protons, and neutrons being the most familiar examples.

At the present time more than 200 elementary particles have been discovered. Most of these particles have very short lifetimes—10^{-6} s or less. These particles are now classified according to certain properties that seem to be conserved during their reactions: electric charge, lepton number, baryon number, strangeness, charm, and so on.

4. **Quarks.** One of the popular new theories for explaining the properties of elementary particles is based on the idea that many of the elementary particles are not elementary, but instead are constructed from other particles called quarks. Particles such as protons and neutrons are made of three quarks, and mesons are made of two quarks. The properties of a particle depend on the properties of the particular quarks of which it is made. The six different quarks (each with an antiquark) are named "up," "down," "strange," "charm," "truth" ("top"), and "beauty" ("bottom"). Each of these six quarks can have a different color, which is a fundamental property (like mass and electric charge) that affects the force quarks exert on each other.

Questions

1. Why does a reactor core remain hot long after fission has stopped?

2. Explain carefully how control rods manage to increase or decrease the rate of fission in a nuclear reactor.

3. Suppose that each married couple had three or more children. How would this affect the world's population? Explain the analogy between this situation and the control of nuclear fission by absorbing more or fewer neutrons with the control rods in a reactor core.

4. Discuss the advantages and disadvantages of producing electricity by (a) burning coal, (b) nuclear fission, and (c) nuclear fusion.

5. Why is a high temperature usually needed in order for fusion to occur?

6. A neutron and a proton will fuse at low temperature to form deuterium, causing the release of energy. Why is this not a suitable way to produce energy?

Problems

You may need to use the masses given in Appendix C and the unit conversion 1 u = 1.6606×10^{-27} kg = 931.5 MeV/c^2 for some of these problems.

35.1 Nuclear Fission

1. Calculate the energy released in the following fission reaction: $^{1}_{0}n + {}^{235}_{92}U \longrightarrow {}^{141}_{56}Ba + {}^{92}_{36}Kr \longrightarrow 3\,{}^{1}_{0}n$. The masses of the nuclei are: $m_n = 1.0087$ u, $m_U = 235.0439$ u, $m_{Ba} = 140.9141$ u, and $m_{Kr} = 91.8981$ u.

2. Calculate the energy released in the following fission reaction: $^{1}_{0}n + {}^{235}_{92}U \longrightarrow {}^{88}_{38}Sr + {}^{136}_{54}Xe + 12\,{}^{1}_{0}n$. The masses of the nuclear fragments are: $m_{Sr} = 87.9056$ u and $m_{Xe} = 135.9072$ u.

3. Calculate the energy released in the following fission reaction: $^{1}_{0}n + {}^{235}_{92}U \longrightarrow {}^{92}_{37}Rb + {}^{140}_{55}Cs + 4\,{}^{1}_{0}n$. The masses of the nuclear fragments are: $m_{Rb} = 91.9191$ u and $m_{Cs} = 139.9171$ u.

4. Calculate the mass of uranium-235 needed to supply all the yearly electrical energy for the United States (approximately

2×10^{19} J). Assume that the energy released per fission is 200 MeV and that one-third of the energy produces electricity, the other two-thirds producing waste heat.

5. *Estimate* the number of ^{235}U nuclei that must undergo fission in order to provide the energy to lift an ant 1 cm.

6. Convert the 200 MeV per nucleus energy that is released by ^{235}U fission to units of joules per kilogram. By comparison, the energy release of coal is 3.3×10^{7} J/kg.

7. (a) Calculate the energy release in MeV of gasoline per molecule of *n*-heptane burned. The molecular mass of *n*-heptane is 100 u, and it releases energy at a rate of 4.8×10^{7} J/kg. (b) Calculate the ratio of energy released by one uranium-nucleus fission (approximately 200 MeV) and one *n*-heptane molecule combustion.

8. Calculate the mass decrease due to uranium fission of the first atomic bomb, whose energy release was equivalent to twenty tons of TNT. (The energy release of TNT is 1.4×10^{10} J/ton.)

35.2 Nuclear Power Plants

9. Calculate the number of ^{235}U fission reactions per second in a nuclear power plant that produces 1000 megawatts of electricity and 2000 megawatts of waste heat. Each fission releases 200 MeV of energy.

10. *Estimate* the mass of the ^{235}U nuclei that undergo fission while operating a 500-megawatt electric power plant for one year. Most plants are approximately 33 percent efficient. Be sure to show your estimation procedure.

11. The world's uranium supply is approximately 10^9 kg (10^6 tons), 0.7 percent of which is ^{235}U. (a) How much energy is available from the fission of this ^{235}U? (b) The world's energy consumption rate for production of electricity is about 10^{20} joules per year. At this rate, how many years would the uranium last if it were used to provide all our electrical energy?

12. A 1000-megawatt nuclear reactor requires about 1.3 g of ^{235}U per day per megawatt of electricity generated. Eighty-five percent of this mass is converted to fission products. (a) Calculate the mass of fission products produced in one year by a 1000-megawatt plant. (b) Calculate the volume of the fission products if their average density is 10,000 kg/m^3.

13. When an object of mass m_1 collides elastically and head-on with a stationary object of mass m_2, the final and initial kinetic energies of m_1 are related by the equation $KE_f = [(m_1 - m_2)/(m_1 + m_2)]^2 KE_i$. Use this to show that hydrogen is a better neutron moderator than deuterium. (We ignore the disadvantage that hydrogen absorbs neutrons more than deuterium.)

14. Think of a chain reaction in which each fission, on the average, produces 1.1 future fissions. If at time zero ten fissions occur, how many occur during the next generation (the second generation)? During the third? During the tenth? During the nth?

15. Suppose that each fission produces an average of 1.0001 neutrons that initiate a subsequent fission and that the time between fissions is 10^{-3} s. Calculate the factor by which the rate of fission reactions increases in a 60-s time interval.

35.3 Breeder Reactors

16. (a) A ^{239}Pu fission releases 180 MeV of energy per nucleus. Convert this energy yield to joules per kilogram. (b) A breeder reactor converts 1 kg of ^{238}U to 1 kg of ^{239}Pu. If the world supply of ^{238}U is 10^9 kg, calculate the energy available from its conversion to ^{239}Pu and subsequent ^{239}Pu fission. (c) If 10^{20} J of this energy is used each year to produce the world's electricity, how many years will the uranium last?

35.4 Fusion

17. Calculate the energy released by the following fusion reaction: $^2_1H + ^3_1H \longrightarrow ^4_2He + ^1_0n$.

18. A series of reactions on the sun leads to the combination of three helium nuclei (^4_2He) to form one carbon nucleus ($^{12}_6C$). (a) Calculate the net energy released by the reactions. (b) What fraction of the total mass of the three helium nuclei is converted to energy?

19. Deuterium nuclei have an approximately equal chance of reacting in either of the following ways: (a) $^2_1H + ^2_1H \longrightarrow ^1_0n + ^3_2He$ and (b) $^2_1H + ^2_1H \longrightarrow ^1_1H + ^3_1H$. Calculate the energy released in each reaction.

20. A series of reactions that might provide power for a nuclear fusion reactor are summarized by the equation: $6\,^2_1H \longrightarrow 2\,^1_1H + 2\,^1_0n + 2\,^4_2He$. (a) Calculate the net energy released by the reaction. (b) Convert this answer to units of joules per kilogram of deuterium (^2_1H).

21. The synthesis of oxygen on the sun involves the combination of ^{12}C and ^4He to form ^{16}O. Does the reaction consume or release energy, and how much?

22. Approximately 4×10^9 kg of mass is converted to energy each second on the sun. If all the energy is a result of the proton-proton series of reactions, including the positron-electron annihilation, calculate the number of proton-proton series reactions occurring each second.

23. (a) Determine the number of water molecules in 1 kg of water (about one quart). One mole of water has a mass of 18 g. (b) For every 67,000 hydrogen nuclei (^1_1H) in water, there is one deuterium (^2_1H) nucleus. Calculate the number of ^2_1H nuclei in the 1 kg of water. (c) The fusion of two deuterium nuclei releases about 4 MeV of energy. Calculate the energy, in units of joules, that could be released by fusion of the deuterons in 1 kg of water. By comparison, burning 1 kg of coal releases 2×10^7 J of energy.

24. World energy consumption in 1980 was about 2.5×10^{20} J. (a) Calculate the number of deuterium nuclei that would be needed to produce this energy. (b) Calculate the volume of water of density 1000 kg/m^3 needed to supply this energy. Use the numbers given in the preceding problem.

35.5 Elementary-Particle Physics

25. A Ξ^+ particle is produced by a high-energy accelerator. Using the decay schemes shown in Table 35.2, develop a sequence of decays that lead to a group of stable particles with lifetimes longer than 10^{-5} s.

26. Do the same as in the previous problem, only for the K^+ particle.

27. A K^+ enters a bubble chamber and decays to a μ^+ and ν_μ. Draw a set of lines on a piece of paper representing the tracks you expect to be formed by the K^+ and its decay products and their decay products.

28. Shown in Fig. 35.10 is a set of bubble-chamber tracks caused by the following particles and their reactions: $\Omega^- \longrightarrow \Lambda^0 + K^-$, followed by $\Lambda^0 \longrightarrow p + \pi^-$ and $K^- \longrightarrow \mu^- + \bar{\nu}_\mu$. Redraw the tracks on a separate paper and identify which track is caused by which particle. Also, add dashed lines for particles that cause no tracks.

FIG. 35.10. A hypothetical set of bubble-chamber tracks.

29. A proton and antiproton with negligible kinetic energy annihilate one another to produce two photons. Calculate the energy in electron volts of each photon and its frequency and wavelength.

30. Calculate the wavelength of the lowest-energy photon capable of producing an electron-positron pair. Ignore the question of conserving momentum (a poor assumption).

35.6 Classification Schemes and Conservation Laws

31. Show that the laws of conservation of charge and baryon number are not violated by the π^+, K^+, Λ^0, and Ξ^0 decay reactions shown in Table 35.2.

32. Show that the following reactions are forbidden by the conservation of charge and/or baryon number laws:

(a) $$\Lambda^0 \longrightarrow p + e^- + \nu_e + n.$$

(b) $$\pi^0 + p \longrightarrow \Sigma^0 + \pi^0.$$

(c) $$\Omega^- \longrightarrow \Sigma^- + p + \pi^- + e^+ + e^-.$$

(d) $$\pi^- + p \longrightarrow \bar{K}^0 + \eta^0 + \pi^0.$$

33. Use the conservation of charge, baryon number, and strangeness laws to show that the following reactions are not allowed:

(a) $$\pi^- + p \longrightarrow \Xi^0 + K^0 + \pi^0.$$

(b) $$\Lambda^0 + n \longrightarrow p^+ + p^- + 3\pi^+ + 3\pi^-.$$

(c) $$\pi^- + p \longrightarrow K^+ + \Omega^-.$$

(d) $$p + p \longrightarrow p + \pi^0 + \Lambda^0 + K^-.$$

34. A moving proton strikes another proton that is at rest, causing the reaction $p + p \longrightarrow p + n + \pi^+$. What is the least kinetic energy of the incident proton in order for this reaction to occur? Ignore momentum conservation.

35.7 Quarks

35. Using Tables 35.2 and 35.3, confirm that the charge, baryon number, and strangeness of the following particles equals the sum of the numbers for the quarks of which the particle is made: (a) n (udd), (b) Σ^+ (uus); (c) Ξ^- (dss).

36. Repeat Problem 35 for the following particles: (a) Σ^- (dds), (b) π^- $(\bar{u}d)$, and (c) $K^0(d\bar{s})$.

37. A new meson is postulated to have the following quantum numbers: $q = S = +1$ and $B = 0$. What combination of quarks produces these numbers?

General Problems

G1. An excited $^{236}_{92}\text{U}^*$ nucleus decays into $^{144}_{56}\text{Ba}$ and $^{92}_{36}\text{Kr}$ fragments. Suppose that these fragments are spherical and barely touch just after formation. (a) Calculate the radii of each fragment and their separations. (b) Calculate their electrical potential energy compared to that when infinitely far apart.

G2. ^{235}U and ^{238}U can be separated by allowing molecules of UF_6 to diffuse through filters. Since the $^{235}\text{UF}_6$ is less massive, it moves faster and becomes more concentrated. Calculate the ratio of the speeds of $^{235}\text{UF}_6$ and $^{238}\text{UF}_6$ at room temperature.

G3. A 1000-megawatt nuclear power plant uses about 400 kg of ^{235}U per year. Approximately 0.3 percent of the ^{235}U fission reactions produce a ^{85}Kr nucleus. (a) Calculate the number of ^{85}Kr nuclei produced per year. (b) Calculate the activity in curies of this krypton, whose decay constant is 2.0×10^{-9} s^{-1} (half-life of 10.8 yr). If these ^{85}Kr nuclei were released to the atmosphere, the ^{85}Kr radiation exposure from all nuclear plants would reach about 1 mrem/yr per person by the year 2000.

G4. (a) Calculate the average kinetic energy of a proton at the center of a star whose temperature is 2×10^7 K. (b) If two of these protons have a head-on collision, calculate their closest separation.

G5. A Σ^+ baryon with a lifetime of 0.8×10^{-10} s in its proper inertial frame of reference and with a rest mass energy of 1189 MeV has an equal amount of kinetic energy. (a) Calculate its speed. (b) Calculate its lifetime as determined by a laboratory observer. (c) Calculate the length of its path in a bubble chamber before decay occurs.

APPENDIX A Mathematics Review

A1. Significant Figures

The uncertainty in the value of a measured quantity can be indicated conveniently by using significant figures. Suppose you use a ruler to measure the length of the cover of a book and find that it is 26.0 cm long. This number has three **significant figures,** the number of digits whose values have been ascertained reliably by your measurement. By convention, the last digit to which a value is assigned may be uncertain by one or two numbers on either side of the assigned value. By stating that the book is 26.0 cm long, we mean that the book's length is 26.0 ± 0.1 cm, or that its length lies between 25.9 and 26.1 cm. We cannot say that the book's length is 26.00 cm unless we can actually measure it with a certainty of ± 0.01 cm. Such certainty is usually not possible using rulers, since the smallest division on most rules is 0.1 cm.

When adding or subtracting, the result should not be carried beyond the first column having a doubtful figure. For example, $8.16 + 74 = 82$, not 82.16. *When multiplying or dividing, the result should have the number of digits as the least significant number used in the calculation.* Suppose, for example, that you wish to calculate the area of a poster that is 5.3 cm high and 11.2 cm wide. Your answer should have only two significant figures since 5.3 has only two significant figures: $A = 5.3 \times 11.2 = 59.36 = 59$ cm^2. To see why we should not retain the .36 in our answer, we must realize that the calculated value of A can range from $5.2 \times 11.1 = 57.72$ cm^2 to $5.4 \times 11.3 = 61.02$ cm. Our answer of 59 cm^2 reflects the fact that the calculated number is uncertain by about ± 1 cm. If we stated our answer as 59.36 cm^2, we would be implying an unjustifiable uncertainty of ± 0.01 cm^2.

A2. Powers of Ten

Powers of ten are used to write the large and small numbers in science. For instance, the speed of light, 300,000,000 m/s, is written in powers of ten notation as 3×10^8 m/s, where $10^8 = 100,000,000$. Other examples of numbers written in the powers of ten notation are given below:

$$100 = 10 \times 10 = 10^2,$$
$$1000 = 10 \times 10 \times 10 = 10^3,$$
$$5,700,000 = 5.7 \times 10^6,$$
$$49,200 = 4.92 \times 10^4.$$

Numbers less than one can also be written in the powers of ten notation. For instance,

$$0.1 = \frac{1}{10} = 10^{-1},$$

$$0.001 = \frac{1}{1000} = \frac{1}{10^3} = 10^{-3},$$

$$0.0000073 = 7.3 \times 10^{-6},$$
$$0.000000012 = 1.2 \times 10^{-8}.$$

We can see from these examples that

$$\frac{1}{10^n} = 10^{-n}. \tag{A.1}$$

In order to multiply and divide numbers that are written in the powers of ten notation, we use the following rules:

$$10^n \times 10^m = 10^{n+m}, \tag{A.2}$$

$$\frac{10^n}{10^p} = 10^n \times 10^{-p} = 10^{n-p}. \tag{A.3}$$

where n, m, and p can be any number. The following examples use these rules for the case where n, m, and p are integers:

$$10^2 \times 10^4 = 10^6,$$

$$\frac{10^4}{10^2} = 10^4 \times 10^{-2} = 10^2,$$

$$\frac{10^{-2}}{10^{-3}} = 10^{-2} \times 10^3 = 10,$$

$$(5 \times 10^2)(4 \times 10^4) = (5 \times 4)(10^{2+4}) = 20 \times 10^6 = 2 \times 10^7,$$

$$(6.6 \times 10^{-34})(5.0 \times 10^{15}) = (6.6 \times 5.0)(10^{-34+15}) = 33 \times 10^{-19} = 3.3 \times 10^{-18},$$

$$\frac{(6.6 \times 10^{-34})(3 \times 10^8)}{(5 \times 10^{-7})} = \left(\frac{6.6 \times 3}{5}\right)(10^{-34+8+7}) = 4 \times 10^{-19}.$$

The powers of ten notation is preferred in science for a variety of reasons: (1) numbers are more easily read and compared in powers of ten notation than if written with a large number of zeros before or after a decimal point; (2) the numbers can be transcribed from one place to another more easily; (3) calculations are easier and less prone to error; and (4) the notation provides a satisfactory method for indicating the number of significant figures in a number.

A3. Exponents and Logarithms

The expression a^n means that a should be multiplied by itself n times. For example, $2^3 = 2 \times 2 \times 2$ and $10^4 = 10 \times 10 \times 10 \times 10$. The superscript n is called an **exponent.**

The **logarithm** of a number N is the exponent n to which a number a must be raised in order to equal that number ($a^n = N$). If

$$a^n = N, \quad \text{then} \quad n = \log_a N, \tag{A.4}$$

where a is called the **base number.** Two examples of the use of Eq. (A.4) are given below:

$$10^4 = 10,000 \quad \text{or} \quad 4 = \log_{10} 10,000,$$
$$2^5 = 32 \quad \text{or} \quad 5 = \log_2 32.$$

The decimal system of numbers has the base 10, a common base number for calculations involving logarithms. In fact, when the base number 10 is used, we omit the subscript 10 and write $n = \log N$ which implies that $10^n = N$. This system with base number 10 is called the **common logarithm system.**

Another commonly used system of logarithms has the base number $e = 2.718\ldots$ and is called the **natural logarithm system.** In this system logarithms are written as $n = \ln N$ which implies that $e^n = N$. A logarithm of a number N in one system can be converted to a logarithm in the other system by using one of the following:

$$\ln N = 2.3026 \log N, \tag{A.5}$$

$$\log N = 0.43429 \ln N. \tag{A.6}$$

We often take logarithms of numbers that are multiplied or divided by each other. The following rules are useful for this purpose:

$$\log AB = \log A + \log B, \tag{A.7}$$

$$\log \frac{A}{B} = \log A - \log B, \tag{A.8}$$

$$\log A^n = n \log A. \tag{A.9}$$

Let us prove the first rule. Suppose that

$$10^x = A \quad \text{and} \quad 10^y = B.$$

Then,

$$x = \log A \qquad y = \log B,$$

and

$$x + y = \log A + \log B.$$

Also,

$$AB = 10^x 10^y = 10^{x+y}.$$

Using the definition of logarithm [Eq. (A.4)], we find that

$$x + y = \log AB.$$

We now have two equations that each have $x + y$ on the left side. Equating the right sides of these equations, we find that

$$\log AB = \log A + \log B,$$

which is the desired result. The other rules can be proved easily in a similar manner.

APPENDIX B Graphing

A **graph** is a diagram that illustrates by means of points and lines a relationship between different quantities whose values are plotted on the graph. Graphical analysis, the drawing and interpreting of graphs, helps us visualize the way a change in one variable affects the values of another variable. Graphical analysis also helps us determine the type of mathematical equation that relates the quantities being graphed.

As an example of graph construction, consider the changing position of a ball that rolls on a smooth, frictionless surface at constant speed. As the ball crosses a starting line, we start a timing clock. We have arranged our experiment so that the ball is at position zero when the time is zero. The observed values of position x and time t for later times are recorded in Table B.1. Notice that the time measurement is so precise that the experimental error in this quantity can be ignored. However, we estimate that the position of the ball can be determined no better than ± 0.1 m. This is an estimate of the uncertainty in the measurement of the ball's position.

Let us now construct a graph of the data recorded in Table B.1.

1. Choose the graph paper: There are many types of graph paper such as linear, semilog, and log-log. We will find that lines passing through our data points have different shapes when plotted on these various types of graph paper. *If the data can be fit by a straight line on one type of graph paper, this allows us to determine the type of equation that relates the variables plotted on that graph.* For example, equations of the form $y = ax + b$ produce straight lines on linear graph paper. Exponential equations of the form $y = y_0 e^{kx}$ produce straight lines on semilog graph paper, and power laws of the form $y = ax^n$ produce straight lines on log-log graph paper. If our data produce a straight line on one of these types of graph paper, we are able to write an equation that relates the variables x and y. For now, we will use only linear graph paper to develop our graphing techniques.

2. Choose the axes: You must decide what variable to put on each axis. You may be told in the instructions for an experiment to plot position versus time. In this case the variable before the "versus" is plotted on the vertical axis (the **ordinate**) and the variable after the versus is plotted on the horizontal axis (the **abscissa**). If you were asked to plot x versus t, the x would go on the vertical axis and the t on the horizontal axis. The variable on the vertical axis is usually called the **dependent variable,** as its values often depend on the value of the variable plotted on the horizontal axis, the **independent variable.** Occasionally you may wish to plot some function of a variable rather than the variable itself (for example, plot x^2 versus t or $1/x$ versus t).

Having chosen the variables for each axis, the axes must be labeled with the name of the variable followed by its unit of measure in parentheses.

3. Choose the scales: The **scale** of a graph is the number of ruled lines on the graph that correspond to a particular change in the value of the variable being plotted on that axis. If plotting time, for instance, we may choose one large division to be 0.25 second. Ten large divisions would be 2.5 seconds.

The scale is chosen by considering the range of values of the variable to be

TABLE B.1 Position of a Ball as a Function of Time

Time in s (± 0.0 s)	Position in m (± 0.1 m)
0.0	0.0
0.5	1.0
1.0	2.3
1.5	3.3
2.0	4.2
2.5	5.3
3.0	6.7

plotted and the total number of divisions on the graph paper. Try to use most of the graph paper. If the values of the time variable range from zero to 3.0 seconds and there are 18 large divisions (18 cm) on the graph paper, we might choose the scale so that five large divisions equal one second. Fifteen large divisions equal three seconds, and the data will be spread over most of the paper (see Fig. B.1). Having chosen the scale, numbers should be placed along this axis to indicate the value of the variable at different places along the axis. A scale must also be chosen for the vertical axis.

4. Plot the data: Having determined the scale and labeled the axes, we next plot the data points on the graph. Notice that the point for the distance measurement at time 2.0 seconds (Table B.1) is a point in Fig. B.1 above the time scale at 2.0 seconds and is horizontally across from the place on the distance scale where $x = 4.2$ m.

It is important to include error bars on your graph. An **error bar** is a line passing through the data point. The length of the error bar indicates the uncertainty in the value of the variable plotted in that direction. The position variable plotted in Fig. B.1, for example, has an uncertainty of ± 0.1 m. As a result, an error bar for a point on the graph is parallel to the position coordinate axis, is 0.2 m long, and extends 0.1 m above and below the data point.

5. Draw a line through your data points: There are a variety of ways to draw lines through the data points. Suppose that your data points can be fit by a straight line. Often, a computer can be used to choose the best straight line, assuming of course that a straight line seems to fit the data. Even with the best straight line, it is unlikely that all of the data points will lie on the line. In fact, none of the data points need lie on the line. A computer chooses the straight line so that the sum of the squares of the deviations of the actual data points from the line is a minimum. This procedure produces a line that passes through the center of the data points and hopefully most data points lie near the line. Usually an equal number of points will be above the line as below.

If you cannot use a computer for choosing the line, then an "eyeball" procedure must be tried. You will have to draw your line so that it is not too far from any point and so that the points above the line balance the deviations of

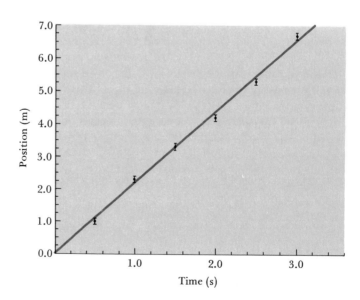

FIG. B.1. A straight line graph on linear graph paper.

points below the line. In the example shown in Fig. B.1, one point lies on the line, two above the line, and three below.

Although your line should pass within the error bars of most points, this is not necessary. The lengths of the error bars are estimates of the uncertainty of the measurements. The line passing through the data points should be within the uncertainty limits of most data points. It is not unusual, though, for the error bars of some data points not to lie on the line. (For a so-called normal distribution of data points, we expect about one out of three error bars to be off the line.)

6. Analyze the graph: If the data points can be fit by a straight line, we can use this line to write an equation that relates the independent variable to the dependent variable. We will review this procedure for each type of graph paper in the following sections.

What if our data points vary in a way that cannot be fit by a straight line? Then we must draw a curved line that moves smoothly through the data points. This curved line will help you develop a feel for the way the data are changing. Does the curved line represent an accelerating or decelerating object? Does it look like an exponentially decreasing or increasing function? The curved line may also help us guess the kind of mathematical equation that relates the independent variable to the dependent variable. This will help us determine whether another type of graph paper should be used to plot the data.

7. Linear graph paper and linear equations: **Linear graph paper** has equal spacing between the lines along each axis. A **linear equation** is of the form

$$y = ax + b, \tag{B.1}$$

where x is the independent variable, y is the dependent variable, and a and b are constants. A linear equation produces a straight line when plotted on linear graph paper. Conversely, if a set of data, when plotted on linear graph paper, can be fit by a straight line, then we know that the dependent variable is related to the independent variable by a linear equation.

The constant a in Eq. (B.1) is called the **slope** of the plotted line. To find the slope, we choose two points on the line. (We do not choose actual data points unless they lie on the line.) It is best to select points that are far from each other; that is, near the ends of the line. The uncertainty in the slope is less when calculated using distant points. The slope is given by the equation

$$\text{slope} = a = \frac{y_2 - y_1}{x_2 - x_1}, \tag{B.2}$$

where (x_1, y_1) and (x_2, y_2) are the values of the two variables at the chosen points on the line. The units of the slope are the units of the y variable divided by the units of the x variable.

The constant b is called the **y intercept** and equals the value of y at the point where the plotted line crosses the y axis (the value of y when $x = 0$).

EXAMPLE B.1 Determine the equation that fits the data plotted in Fig. B.2.

SOLUTION The data produce a straight line on linear graph paper. The dependent variable v must be related to the independent variable t by an equation of the form $v = at + b$. Using Eq. (B.2), we see that

$$a = \frac{v_2 - v_1}{t_2 - t_1} = \frac{(2.22 - 0.80)\text{m/s}}{(0.95 - 0.05)\text{s}} = \frac{1.42\,\text{m/s}}{0.90\,\text{s}} = 1.6\,\text{m/s}^2.$$

FIG. B.2. A linear graph.

The value of v where the straight line crosses the vertical axis is $0.7\,\text{m/s}$; thus, $b = 0.7\,\text{m/s}$. The equation that fits this data is

$$v = (1.6\,\text{m/s}^2)t + 0.7\,\text{m/s}.$$

FIG. B.3. On a log scale, the variable doubles in value for equal length steps up the scale.

8. Semilog graph paper and exponential functions: **Semilog graph** paper has a linear scale (equal spacing between lines) on the horizontal axis and a logarithmic scale on the vertical axis (Fig. B.3). To make a logarithmic scale, we place a group of equally spaced horizontal marks on the vertical axis. These marks do not represent the numbers 1, 2, 3, 4, 5, $\cdots$ as in linear graph paper; instead, they represent the numbers 1, 2, 4, 8, 16, $\cdots$. Each equally spaced step up the vertical axis represents a doubling in the value of the variable plotted on that axis. Exponential functions of the form

$$y = y_0 e^{kx} \tag{B.3}$$

have the property that the magnitude of the change in the y variable, Δy, doubles as the x variable is changed in equal steps, Δx. If we plot the values of y versus x for exponential functions on semilog graph paper, a straight line connects the data points. Conversely, if the data points for the variables y versus x produce a straight line on semilog graph paper, the variables are related by an exponential equation [Eq. (B.3)].

Semilog graph paper comes in different types that vary in the range of values that can be plotted on the logarithmic scale. Each range of 10 is called a **cycle.** A two-cycle log scale can accommodate data that range in value over two powers of ten (for example, from 1–100 or from 10^3–10^5). If the data that you wish to plot on the logarithmic scale range from 2 to 6400, you will need a four-cycle logarithmic scale. The first cycle is for points from 1–10, the second cycle for 10–100, the third cycle for 100–1000, and the fourth for 1000–10,000.

If the data when plotted on semilog graph paper can be fit by a straight line, you know that the variables are related by an equation like

$$y = y_0 e^{kx}.$$

The constant y_0, the **y intercept,** is the value of y when $x = 0$ (it is also the place where the straight line crosses the y axis). The constant k is determined by choosing two points on the line, (y_1, x_1) and (y_2, x_2). The value of k is determined by taking the natural logarithm of y_2/y_1 and substituting in the equation

$$k = \frac{\ln(y_2/y_1)}{x_2 - x_1}. \qquad \textbf{(B.4)}$$

EXAMPLE B.2 Determine an equation that fits the data shown in Fig. B.4. The graph shows the change in cost of gasoline (y) versus time (t) assuming a 10 percent per year rate of inflation.

SOLUTION Since the data produce a straight line on semilog graph paper, we know that the dependent variable y is related to the independent variable t by an equation of the form $y = y_0 e^{kt}$. We see that when $t = 0$, $y = y_0 e^{k0} = y_0 = \$2.00$. The value of k is determined using Eq. (B.4) and the points (\$32.00, 28 years) and (\$2.00, 0 years):

$$k = \frac{\ln(32.00/2.00)}{28 \text{ yr} - 0 \text{ yr}} = \frac{2.77}{28 \text{ yr} - 0} = 0.10 \text{ yr}^{-1}.$$

The equation relating y and t is

$$y = \$2.00 e^{(0.10 \text{ yr}^{-1})t}. \qquad \blacksquare$$

9. Log-log graph paper and power laws: Log-log graph paper has a logarithmic scale on each axis. Power law equations of the form $y = ax^n$ produce

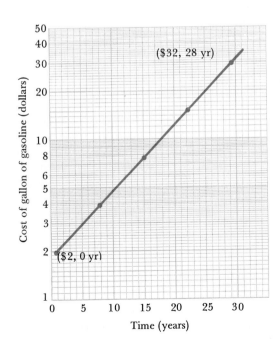

FIG. B.4. A semilog graph.

straight lines on log-log graph paper. The constant n (an exponent) can be determined by choosing two points on the line [(x_1, y_1) and (x_2, y_2)] and substituting the values of x and y for these points into the equation

$$n = \frac{\ln(y_2/y_1)}{\ln(x_2/x_1)}. \qquad \textbf{(B.5)}$$

Having found n, the constant a can be determined by substituting the values of n, x_2, and y_2 into the equation

$$a = \frac{y_2}{x_2^n}. \qquad \textbf{(B.6)}$$

EXAMPLE B.3   Determine an equation that fits the data shown in Fig. B.5.

SOLUTION   Since the data produce a straight line on log-log graph paper, the variables y and t must be related by an equation of the form $y = at^n$. Choosing the points (2 m, 0.5 s) and (200 m, 5 s) that are on the line, we find using Eq. (B.5) that

$$n = \frac{\ln(200 \text{ m}/2 \text{ m})}{\ln(5 \text{ s}/0.5 \text{ s})} = \frac{\ln 100}{\ln 10} = \frac{4.6}{2.3} = 2.$$

FIG. B.5.   A log-log graph.

Substituting $n = 2$, $y_2 = 200$ m, and $t_2 = 5$ s into Eq. (B.6), we find that

$$a = \frac{200 \text{ m}}{(5 \text{ s})^2} = 8 \text{ m/s}^2.$$

The equation relating y and t is

$$y = (8 \text{ m/s}^2)t^2.$$ ∎

(1) Atomic number Z	(2) Element	(3) Symbol	(4) Mass number, A	(5) Atomic mass†	(6) Percent abundance, or decay mode if radioactive	(7) Half-life (if radioactive)
0	(Neutron)	n	1	1.008665	β^-	10.6 min
1	Hydrogen	H	1	1.007825	99.985	
	Deuterium	D	2	2.014102	0.015	
	Tritium	T	3	3.016049	β^-	12.33 yr
2	Helium	He	3	3.016029	0.00014	
			4	4.002604	99.99986	
3	Lithium	Li	6	6.015123	7.5	
			7	7.016005	92.5	
4	Beryllium	Be	7	7.016930	EC, γ	53.3 days
			9	9.012183	100	
5	Boron	B	10	10.012938	19.8	
			11	11.009305	80.2	
6	Carbon	C	11	11.011433	β^+, EC	20.4 min
			12	12.000000	98.89	
			13	13.003355	1.11	
			14	14.003242	β^-	5730 yr
7	Nitrogen	N	13	13.005739	β^+	9.96 min
			14	14.003074	99.63	
			15	15.000109	0.37	
8	Oxygen	O	15	15.003065	β^+, EC	122 s
			16	15.994915	99.76	
			18	17.999159	0.204	
9	Fluorine	F	19	18.998403	100	
10	Neon	Ne	20	19.992439	90.51	
			22	21.991384	9.22	
11	Sodium	Na	22	21.994435	β^+, EC, γ	2.60 yr
			23	22.989770	100	
			24	23.990964	β^-, γ	15.0 h
12	Magnesium	Mg	24	23.985045	78.99	
13	Aluminum	Al	27	26.981541	100	
14	Silicon	Si	28	27.976928	92.23	
			31	30.975364	β^-, γ	2.62 h
15	Phosphorus	P	31	30.973763	100	
			32	31.973908	β^-	14.28 days
16	Sulfur	S	32	31.972072	95.0	
			35	34.969033	β^-	87.4 days
17	Chlorine	Cl	35	34.968853	75.77	
			37	36.965903	24.23	
18	Argon	Ar	40	39.962383	99.60	
19	Potassium	K	39	38.963708	93.26	
			40	39.964000	β^-, EC, γ, β^+	1.28×10^9 yr
20	Calcium	Ca	40	39.962591	96.94	
21	Scandium	Sc	45	44.955914	100	

*Data are taken from C. M. Lederer and V. S. Shirley, eds., *Table of Isotopes,* 7th ed. (New York: John Wiley & Sons, Inc., 1978).
†The masses given in column (5) are those for the neutral atom, including the Z electrons.

(1) Atomic number Z	(2) Element	(3) Symbol	(4) Mass number, A	(5) Atomic mass†	(6) Percent abundance, or decay mode if radioactive	(7) Half-life (if radioactive)
22	Titanium	Ti	48	47.947947	73.7	
23	Vanadium	V	51	50.943963	99.75	
24	Chromium	Cr	52	51.940510	83.79	
25	Manganese	Mn	55	54.938046	100	
26	Iron	Fe	54	53.939612	5.8	
			56	55.934939	91.8	
27	Cobalt	Co	59	58.933198	100	
			60	59.933820	β^-, γ	5.27 yr
28	Nickel	Ni	58	57.935347	68.3	
			60	59.930789	26.1	
29	Copper	Cu	63	62.929599	69.2	
			65	64.927792	30.8	
30	Zinc	Zn	64	63.929145	48.6	
			66	65.926035	27.9	
31	Gallium	Ga	69	68.925581	60.1	
32	Germanium	Ge	72	71.922080	27.4	
			74	73.921179	36.5	
33	Arsenic	As	75	74.921596	100	
34	Selenium	Se	80	79.916521	49.8	
35	Bromine	Br	79	78.918336	50.69	
36	Krypton	Kr	84	83.911506	57.0	
37	Rubidium	Rb	85	84.911800	72.17	
38	Strontium	Sr	86	85.909273	9.8	
			88	87.905625	82.6	
			90	89.907746	β^-	28.8 yr
39	Yttrium	Y	89	88.905856	100	
40	Zirconium	Zr	90	89.904708	51.5	
41	Niobium	Nb	93	92.906378	100	
42	Molybdenum	Mo	98	97.905405	24.1	
43	Technetium	Tc	98	97.907210	β^-, γ	4.2×10^6 yr
44	Ruthenium	Ru	102	101.904348	31.6	
45	Rhodium	Rh	103	102.90550	100	
46	Palladium	Pd	106	105.90348	27.3	
47	Silver	Ag	107	106.905095	51.83	
			109	108.904754	48.17	
48	Cadmium	Cd	114	113.903361	28.7	
49	Indium	In	115	114.90388	95.7; β^-	5.1×10^{14} yr
50	Tin	Sn	120	119.902199	32.4	
51	Antimony	Sb	121	120.903824	57.3	
52	Tellurium	Te	130	129.90623	34.5; β^-	2×10^{21} yr
53	Iodine	I	127	126.904477	100	
			131	130.906118	β^-, γ	8.04 days
54	Xenon	Xe	129	128.90478	26.4	
			131	130.90508	21.2	
			132	131.90415	26.9	
			136	135.90722	8.9	
55	Cesium	Cs	133	132.90543	100	
56	Barium	Ba	137	136.90582	11.2	
			138	137.90524	71.7	
57	Lanthanum	La	139	138.90636	99.911	
58	Cerium	Ce	140	139.90544	88.5	
59	Praseodymium	Pr	141	140.90766	100	
60	Neodymium	Nd	142	141.90773	27.2	
61	Promethium	Pm	145	144.91275	EC, α, γ	17.7 yr

(1) Atomic number Z	(2) Element	(3) Symbol	(4) Mass number, A	(5) Atomic mass†	(6) Percent abundance, or decay mode if radioactive	(7) Half-life (if radioactive)
62	Samarium	Sm	152	151.91974	26.6	
63	Europium	Eu	153	152.92124	52.1	
64	Gadolinium	Gd	158	157.92411	24.8	
65	Terbium	Tb	159	158.92535	100	
66	Dysprosium	Dy	164	163.92918	28.1	
67	Holmium	Ho	165	164.93033	100	
68	Erbium	Er	166	165.93031	33.4	
69	Thulium	Tm	169	168.93423	100	
70	Ytterbium	Yb	174	173.93887	31.6	
71	Lutecium	Lu	175	174.94079	97.39	
72	Hafnium	Hf	180	179.94656	35.2	
73	Tantalum	Ta	181	180.94801	99.988	
74	Tungsten	W	184	183.95095	30.7	
75	Rhenium	Re	187	186.95577	62.60, β^-	4×10^{10} yr
76	Osmium	Os	191	190.96094	β^-, γ	15.4 days
			192	191.96149	41.0	
77	Iridium	Ir	191	190.96060	37.3	
			193	192.96294	62.7	
78	Platinum	Pt	195	194.96479	33.8	
79	Gold	Au	197	196.96656	100	
80	Mercury	Hg	202	201.97063	29.8	
81	Thallium	Tl	205	204.97441	70.5	
82	Lead	Pb	206	205.97446	24.1	
			207	206.97589	22.1	
			208	207.97664	52.3	
			210	209.98418	α, β^-, γ	22.3 yr
			211	210.98874	β^-, γ	36.1 min
			212	211.99188	β^-, γ	10.64 h
			214	213.99980	β^-, γ	26.8 min
83	Bismuth	Bi	209	208.98039	100	
			211	210.98726	α, β^-, γ	2.15 min
84	Polonium	Po	210	209.98286	α, γ	138.38 days
			214	213.99519	α, γ	164 μs
85	Astatine	At	218	218.00870	α, β^-	≈ 2 s
86	Radon	Rn	222	222.017574	α, γ	3.8235 days
87	Francium	Fr	223	223.019734	α, β^-, γ	21.8 min
88	Radium	Ra	226	226.025406	α, γ	1.60×10^3 yr
89	Actinium	Ac	227	227.027751	α, β^-, γ	21.773 yr
90	Thorium	Th	228	228.02873	α, γ	1.9131 yr
			232	232.038054	100, α, γ	1.41×10^{10} yr
91	Protactinium	Pa	231	231.035881	α, γ	3.28×10^4 yr
92	Uranium	U	232	232.03714	α, γ	72 yr
			233	233.039629	α, γ	1.592×10^5 yr
			235	235.043925	0.72; α, γ	7.038×10^8 yr
			236	236.045563	α, γ	2.342×10^7 yr
			238	238.050786	99.275; α, γ	4.468×10^9 yr
			239	239.054291	β^-, γ	23.5 min
93	Neptunium	Np	239	239.052932	β^-, γ	2.35 days
94	Plutonium	Pu	239	239.052158	α, γ	2.41×10^4 yr
95	Americium	Am	243	243.061374	α, γ	7.37×10^3 yr
96	Curium	Cm	245	245.065487	α, γ	8.5×10^3 yr
97	Berkelium	Bk	247	247.07003	α, γ	1.4×10^3 yr
98	Californium	Cf	249	249.074849	α, γ	351 yr
99	Einsteinium	Es	254	254.08802	α, γ, β^-	276 days

(1) Atomic number Z	(2) Element	(3) Symbol	(4) Mass number, A	(5) Atomic mass†	(6) Percent abundance, or decay mode if radioactive	(7) Half-life (if radioactive)
100	Fermium	Fm	253	253.08518	EC, α, γ	3.0 days
101	Mendelevium	Md	255	255.0911	EC, α	27 min
102	Nobelium	No	255	255.0943	EC, α	3.1 min
103	Lawrencium	Lr	257	257.0998	α	$\approx$35 s
104			261	261.1087	α	1.1 min
105			262	262.1138	α	0.7 min
106			263	263.1184	α	0.9 s
107			261	261	α	1–2 ms

Answers to Odd-Numbered Problems

Chapter 1

1. 70 km 38° S of E
3. 17 km 83° S of W
5. 398 km 59° N of W
7. 199 N 59° above $+x$ axis
9. 156 N 23° below $+x$ axis
11. 246 N 6.8° above $+x$ axis
13. 18 N 27° below $+x$ axis
15. (a) 18 km 35° S of W, (b) 18 km 35° S of W, (c) 26 km 85° N of E, (d) 26 km 85° N of E
17. $A_x = 10$ km, $A_y = 0$, $C_x = -6.0$ km, $C_y = 8.0$ km
19. $F_{1x} = 173$ N, $F_{1y} = 100$ N, $F_{3x} = -150$ N, $F_{3y} = 0$
21. (4.1 km, -11.3 km) and (-6.1 km, -5.1 km)
23. 70 km 38° S of E
25. 17 km 83° S of W
27. 398 km 59° N of W
29. 199 km 59° above $+x$ axis
31. 156 N 23° below $+x$ axis
33. 0.85 km 21° S of E
35. (a) 8.06 km 30° above $-x$ axis, (b) 36 N 56° below $+x$ axis, (c) 10 N along $+y$ axis, (d) 20 N along $-x$ axis
37. 113 N 77° above $-x$ axis

Chapter 2

1. 56 lb 30° below $-x$ axis
3. 43 N 9.8° above $+x$ axis
5. $T_1 = 10$ lb, $T_2 = 16.6$ lb, $T_3 = 13.3$ lb
7. $T_1 = w = 76.6$ N, $T_3 = 64.3$ N
9. 50 N
11. $\theta_1 = 84.3°$, $\theta_2 = 2.9°$
13. 143 lb
15. Yes
17. $\tau_1 = 0$, $\tau_2 = -153$ N · m, $\tau_3 = +200$ N · m, $\tau_4 = -200$ N · m
19. $\tau_{ball} = -14$ N · m, $\tau_{biceps} = +17$ N · m, $\tau_{triceps} = -3$ N · m
23. 500 N
25. (a) 1000 N · m, (b) 250 N, (c) 750 N

27. 233 N
29. $T = 220$ lb, $F = 200$ lb
31. $T = 1500$ N, $F = 1550$ N 15° above horizontal
33. $T = 200$ N, $F = 173$ N along horizontal toward right
35. 15 N
37. $F_1 = -270$ lb, $F_2 = +450$ lb
39. 173 N
41. $w = 600$ N 5.8 m from left cable

Chapter 3

1. 1.3 km, 0.82 mile
3. (a) 9.5×10^{15} m, (b) 4.3 yr
5. 0.69 m
7. 83 km/hr
9. 0.24 m
11. (a) 29 m/s, (b) 11 m/s, (c) 62 mph
13. (a) 1.6×10^4 cm^2, (b) 17.2 ft^2
15. 0.033 m/s, 0.074 mph
17. (a) 0.710 μm/s, (b) 0.760 μm/s, (c) 3.6×10^4 s = 10 hr
19. (a) 4.5 ft/s, 8.5 ft/s, 5.0 ft/s, -4.0 ft/s, -8.5 ft/s, -5.5 ft/s, (b) 9 s
21. (a) 30.3 s, (b) 2.18 m/s^2
23. 5.6×10^{11} m/s^2
25. (a) 1.6 m/s^2, (b) 1.6 m/s^2, 1.6 m/s^2, 1.7 m/s^2, 1.6 m/s^2, 1.6 m/s^2
31. (a) 400 s, (b) 4800 m
33. 1.71 m/s^2, 21.6 s
35. (a) 185 ft/s^2, (b) -616 ft/s^2, (c) 2310 ft, (d) 693 ft, (e) 19.3 g
37. 0.95 yr, 4.5×10^{13} m
39. No
41. 123 m
43. 14.3 m/s
45. 12 ft/s
47. 5.3 m/s
49. 76.7 m/s
51. 10 barrels
53. 0.4 m
55. (a) 0.1 s, 4.04 s, (b) 8.8 ft, 355 ft, (c) yes

G1. (a) 2.85, (b) 136 lb
G3. 3.84 m/s^2

Chapter 4

1. (a) 90 N, (b) 2.3 m/s^2
3. 1000 kg
5. (a) 2.3 m/s^2, (b) 0.20 N
7. (a) 2500 m/s^2, (b) 100 N
9. (a) 125 m/s^2, (b) 2.3×10^5 N
11. 167 lb, 742 N
13. 0.094 slug, 1.4 kg, 13 N
21. 1.7 m/s^2
23. 1400 N
25. 1.9×10^4 N
27. (a) 3.1×10^5 m/s^2, (b) 970 N, (c) yes
29. (a) 125 lb, (b) 75 lb, (c) 0
31. 1350 N, 1690 N
33. 0.86 m/s^2
35. 65 N, 3.3 m/s^2
37. 0.67 m/s^2, 105 N
39. 0.26
41. (a) 193 lb, (b) 36 ft/s^2, (c) 11 ft/s^2
43. (a) 24 N, (b) 36 N, (c) -1.2 m/s^2
45. (a) 343 N, (b) 1.96 m/s^2
47. (a) 1.0×10^4 N, (b) -6.9 m/s^2, (c) 36 m
49. 74 lb and 54 lb
G1. (a) 240 lb, (b) 1700 lb
G3. (a) 46 lb, (b) 53 lb

Chapter 5

1. (a) 6.7 m/s, (b) 2.7×10^3 m/s
5. (a) 10.9 kg · m/s, (b) 2.8 kg · m/s
7. (a) 7.5×10^5 N, (b) 2.5×10^5 N
9. 0.034 s
11. (a) 640 N, (b) 190 N
13. (a) 267 N, (b) 0.040 m/s
15. (a) 1500 N, (b) yes, (c) 0.167 s
17. (a) 5.7×10^{-3} s, (b) 6.7×10^4 N
19. (a) 0.060 s, (b) 0.45 m

21. 75 ft

23. (a) 130 kg $\cdot$ m/s, (b) 130 N

25. 0.5 m/s

27. 54 m/s away from ship

29. 3.7×10^5 m/s

31. 0.080 mph

33. 1.8 m/s

35. 7.1 m/s

37. 1.31 kg

G1. (a) 4300 N, (b) 1600 N

G3. (a) 0.90 m/s, (b) 0.14 m

Chapter 6

1. 0.87 rad, 2.79 rad, 270°, 22.9°, 13.2 rad, 1080°

3. (a) 1.7 m, (b) 22 m

5. 1.9 m

7. 157 m, 0.90 rad

9. 9.36×10^5 rev

11. 1.2×10^4 rad/s, 1.15×10^5 rpm

13. (a) 8π rad/s, (b) 8.29 m/s

15. (a) 6.60×10^{15} rev, (b) 2.20×10^6 m/s

17. 0.28 rad/s, 141 m/s

19. 2.2 m/s^2, 2.3 g

23. 3.0 rad/s

25. 5.3×10^3 N

27. 57 km/hr

29. (b) 5830 N

31. (a) 4.3×10^{20} N, (b) 2.0×10^{20} N

33. 2.67×10^7 m

35. 2.33×10^6 s = 27 days

37. 1.34

39. (a) 77.6 kg, (b) 820 N

41. (c) 322 N

43. 0.313 rad/s

G1. $\omega \leq 1.50$ rad/s

G3. (a) 2960 N, (b) 0.20

G5. (b) 47.8° above horizontal tilted toward axis of rotation

Chapter 7

1. (a) -0.058 rad/s^2, (b) 0.99 s

3. 200 m/s^2, 133 rad/s^2

5. (a) 4 m/s^2, (b) 0.31 m

7. (a) 60 rad/s, (b) 18 m/s

9. 3.8 rad/s^2

11. (a) 0.25 m/s^2, (b) 0.83 rad/s^2

13. (a) 0.17 m/s^2, (b) 0.014 rad/s^2

15. (a) 9.4 N $\cdot$ m, (b) 12 rad/s^2

17. (a) 400 rad/s^2, (b) 24 N $\cdot$ m

19. 8 kg $\cdot$ m^2

21. 4 kg $\cdot$ m^2

23. 0.71 m from pivot point

25. -6.9×10^{-4} N $\cdot$ m

27. (a) -4.3×10^{-22} rad/s^2, (b) -4.2×10^{16} N $\cdot$ m, (c) 6.6×10^9 N

29. 5.9×10^4 N $\cdot$ m

31. (a) 4.8×10^7 kg $\cdot$ m^2/s, (b) 2.18×10^6 m/s

33. 30 s

35. 13.3 kg $\cdot$ m^2

37. 0.5 rad/s

Chapter 8

1. (a) 2180 J, (b) 2400 J

3. (a) 4000 J, (b) -4000 J

5. 1.16×10^5 J

9. (a) 250 J, (b) 0.05 m, (c) 250 J

11. (a) 90 J, (b) 180 J

13. 14.6 m/s

17. (a) 2.5×10^3 J

19. 0.036 J

21. (a) 3.9×10^7 J, (b) 230

23. 3.8×10^6 J

25. -1.8×10^3 J

27. (a) -1.8×10^5 J, (b) 42 m/s

29. (a) 22 J, (b) -12 J, (c) -10 J

31. 0.55 m

33. (a) -8.8×10^5 J, (b) 1.2 m

35. 4.0 m/s

37. 2.3×10^4 lb

39. Yes, 3.7 m/s

41. 4.7×10^4 N

43. (a) 8.9 m/s, 0.10 m

45. 3100 N/m

47. (a) 1.5 m (b) 7.9×10^3 N/m

49. 2940 J

51. 1050 m

53. 11.6 m/s

55. 0.40

57. (a) 0.635 m/s, (b) 0.103 m

59. (a) 6.7 m/s, (b) 0.33

61. $v_1 = +1.2$ m/s, $v_2 = +3.2$ m/s

63. 1.9 m/s

65. (a) 3.9×10^4 J, (b) 0.91 watt

67. 5.9×10^3 watts

69. 930 kcal/day assuming no exercise

71. (a) 1.14×10^7 J, (b) 240 g, (c) 210 g

G1. (a) 30.7 m/s, (b) 0.11 C°

G3. 870 N

Chapter 9

1. (a) 41.7°C, (b) 103°F

3. (a) 1.9 F°, (b) -88°C, 185 K

5. 29.8°F and 86°F

7. -321°F, 77 K

9. 4.0 C°

11. 0.34 kg

13. (a) 2.8 J (b) 168 J, (c) 57 C°

15. (a) 8.7×10^7 J, (b) 1.1×10^4 s or 3.1 hr

17. (a) 585 kg, (b) 0.081 kg/s, (c) 3100 J/s

19. 173 C°

21. 38.9°C

23. 6.1 C°

25. 50 kg

27. 0.69 C°

29. 2.1×10^5 J

31. 0.36 kg

33. 48°C

35. (a) 2.26×10^5 J, (b) 0.854×10^5 J

37. 5.6×10^3 s or 1.6 hr

39. Gaseous state at 90.2°C

41. -0.97 ft

43. 0.067 ft

45. 0.71 m

47. (a) 30.0054 m, (b) 252.55 m

49. 5750 cm^3

51. 0.21 liters

G1. Steam at about 105°C

G3. (a) 4.18×10^5 J, (b) 0.185 kg

G5. (a) 35 mm Hg, (b) -2.7×10^3 J, (c) -6.5 C°

Chapter 10

1. (a) 2.7×10^6 J, (b) 8.1 kg

3. (a) 473 watts, (b) 4.1×10^7 J

5. 173 watts

7. -14 watts

9. -370 watts

11. -2.4×10^{-2} watts

13. 3.6×10^{-5} m^2

15. (a) 297 K, (b) 277 K

17. −63.0 watts

19. 28.6 m²

21. −16 watts

23. (a) 2.5×10^{-2} g/s,
(b) 2.2×10^3 g/day

25. (a) 8.3×10^{-3} g/s, (b) 60 g

27. (a) +900 watts, (b) −140 watts,
(c) −880 watts (d) 0.36 g/s

29. (a) −850 watts,
(b) 1.8×10^{-4} kg/s, (c) 1.9 kg

31. (a) −140 watts, −143 watts,
(b) 283 watts

33. 5.8 kW

35. 0.55 kW

G1. −10.8°C

G3. 105 s

Interlude, p. 205

1. 4.3 s

3. 7810 N

5. $327

7. 8.6 in

Chapter 11

1. (b) 1, 5, 10, 10, 5, 1
(c) 0, 1.6 k, · · ·

3. (a) 972, (b) 25

5. (a) 2.52, (b) 1.17×10^{10}, (c) 52.7

7. (b) There are three, (c) there are six.

9. (a) −609 J/K, (b) 3020 J/K

11. (a) 0.078×10^3 J/K,
(b) 1.22×10^3 J/K,
(c) 1.29×10^3 J/K,
(d) 6.05×10^3 J/K,
(e) 0.10×10^3 J/K, (f) 8.74×10^3 J/K

13. 3.3 J/K

15. (a) 2.6°C, (b) +8 J/K, yes

17. 1.9×10^4 J/K

19. 0.55 J/K

21. (a) 5.7×10^4 J, (b) 0.57

23. 620°C, 7.8×10^7 J

25. (a) 0.047, (b) 2.1×10^8 J/s

27. 0.37 or 37 percent

29. (a) 8.9, (b) 8940 J, (c) 2500 J

31. (a) 6.66×10^5 J,
(b) 0.49×10^5 J, (c) 7.15×10^5 J

Chapter 12

1. 3.2×10^{-9} m, 11

3. (a) 5×10^{-8} m, (b) 17

5. (a) 1290 lb, (b) 220 lb

7. 1.75 in

9. (a) 4.9 atm, (b) 73 lb/in²,
(c) 3750 mm Hg, (d) 5.1×10^3 cm H_2O

11. -0.10×10^5 N/m²,
0.91×10^5 N/m²

13. 2.13×10^5 N

15. 0.72 cm³

17. (a) 7.24×10^{-3} m³,
(b) 2.90×10^{-3} m³, (c) 0.088 m

19. 18.9

21. 102°C

23. 0.19 m³

25. (a) 8.5×10^{22}, (b) 2.13×10^{23},
(c) 1870 g

27. (a) 2.5×10^{25} molecules, (b) 41.7 moles

29. (a) 2.05 moles,
(b) 1.03×10^5 N/m²

31. 1.5×10^3 m/s

33. 510 m/s, 0.88 m/s, 5.3×10^{-10} m/s

35. (a) 1.61×10^5 K,
(b) 1.00×10^4 K

37. (b) 343 m/s

41. 2000

43. (a) −168 J, (b) +168 J

45. 2000 J, expand, increase

47. (a) 0, (b) 5000 J

49. 9.4×10^3 J

51. (a) −9.0 J, (b) 0, (c) +6.0 J

53. −9 J, 19 J; 0, −9 J; 6 J, −7 J

55. 9.98×10^5 N/m² toward salt water

57. (a) 6.4 moles/m³, (b) 8980 N/m² toward soil

59. (a) 6.1×10^2 mole/m³

Chapter 13

1. 5400 N, 1200 lb

5. 630 kg

7. (a) 69 lb/ft³, (b) 2.2 slugs/ft³

9. Decreases by 7 percent

11. 3.9×10^5 N/m², 57 lb/in², 3.9 atm

13. 2.0×10^3 N/m²

15. 0.34 m

17. 1140 kg/m³

19. 0.022 m²

21. (a) 6.3×10^5 N/m²,
(b) 1.0×10^4 N

23. 0.96

25. 0.095 N

27. 0.0156 m³

29. 9.5 N

31. 172 N

33. (a) 229 N, (b) 6.6 m/s²

35. 2.4×10^{-3} N

37. 1.05×10^{-4} kg

39. 29 stitches

41. 9.5×10^{-5} m, 6.1 cm

43. (a) 1.5×10^{-4} m,
(b) 3.0×10^{-5} m, (c) 7.4×10^{-6} m,
(d) 1.5×10^{-6} m, (e) 3.0×10^{-7} m

45. 3.0 cm

G1. 0.77 N

Chapter 14

1. (a) 32 cm/s, (b) 350 cm/s

3. 0.060 m

5. 2.8 m/s

7. (a) 3.1 m/s, (b) 1.8×10^4 N/m²

9. 1.51×10^5 N/m²

11. 21 m/s

13. (a) 1.1×10^5 N,
(b) 2.2×10^3 N/m², (c) 212 m/s

15. (a) 406 N/m², (b) 1070 N pushing out

17. 4.1×10^{-3} m

19. (a) 2.0×10^3 N/m², (b) 5.9 cm

21. 1.15×10^3 N/m²

23. (a) 5.1×10^{-4} m/s,
(b) 1600 N/m²

25. (a) 0.66, (b) 1.52

27. Laminar

29. Turbulent

31. 630 N, 1050 N

35. (a) 5.3×10^{-12} m/s,
(b) 5.2×10^6 hr

37. 0.25 m/s

39. 33 m²

G3. 52 percent

Chapter 15

1. (a) 180 N, (b) 0.16 m

3. 21 cm

5. (a) 1.5×10^5 N/m²,
(b) 7.8×10^{-7} m², 5.0×10^{-4} m

7. (a) 0.0030, (b) 0.12 m

9. 6.3×10^{-4} m

11. 1.4×10^{-2} m

13. (a) 0.040, (b) $4 \times 10^7 \, \text{N/m}^2$, (c) 1200 N

15. 7.8×10^{-3}

19. $1.7 \times 10^{-4} \, \text{m}$

21. $0.40 \, \text{m}^2$

23. $3 \times 10^9 \, \text{m}^2$

25. $-3.2 \times 10^{-6} \, \text{m}^3$, $-6.6 \times 10^{-6} \, \text{m}^3$, $-2.5 \times 10^{-4} \, \text{m}^3$, $-5.6 \times 10^{-4} \, \text{m}^3$

27. 190 m

29. $-3.3 \times 10^{-5} \, \text{m}^3$

31. 0.8 J

G1. 8.0 m

Interlude p. 306

1. 80 kg

3. (a) 116 nails, (b) 76.5 kg, (c) mother

7. (a) $1240 \, \text{lb/in}^2$, (b) 29 ft

9. $40 \, \text{N/cm}^2$

Chapter 16

1. (a) $2.27 \times 10^{-3} \, \text{s}$, (b) $4.74 \times 10^{13} \, \text{Hz}$

3. (a) 20 Hz, (b) $5.0 \times 10^{-5} \, \text{s}$

5. $20 \, \text{N/cm} = 2000 \, \text{N/m}$

7. (a) 1/4, 3/4, (b) $0.71 \, A$

9. (a) 250 N/m, (b) 0.45 J

11. $1 \times 10^{-10} \, \text{m}$

13. $1.0 \times 10^4 \, \text{N/m}$, 12.6 Hz

15. 1.2 N/m

17. (a) 6860 N/m, (b) 8.6 J, 1.6 Hz

19. 1, 1, 1.4

21. 0.070 Hz, 1.0 m

23. 0.25 m

25. 2.2 m/s

G1. (a) 235 N/m, (b) 0.29 J

G3. $f = (1/2\pi)\sqrt{2\,k/m}$

G5. (a) $2.0 \times 10^8 \, \text{N/m}$, (b) $2 \times 10^{-2} \, \text{m}$

Chapter 17

1. 17 m, 0.017 m

3. $7.3 \times 10^{-3} \, \text{m}$

5. (a) $2.1 \times 10^9 \, \text{N/m}^2$, (b) 2980 m/s

7. 25

9. (b) 820 m

11. $8.1 \times 10^{10} \, \text{N/m}^2$, $4.9 \times 10^{10} \, \text{N/m}^2$

13. 7.1 m, out of phase

15. 0.5 to 0.6 cm left

29. (a) 266 Hz, (b) 236 Hz, (c) 265 Hz, (d) 235 Hz

31. (a) 4.3 m/s, (b) 4.2 m/s

33. (a) $4.28 \times 10^5 \, \text{Hz}$, (b) $4.65 \times 10^5 \, \text{Hz}$

35. (a) 3 plates/min, (b) 1 plate/min

G1. (a) 0.35 m/s, (b) $7.0 \times 10^{-4} \, \text{m}$

G5. 43.6 m/s

Interlude, p. 351

1. $Cg^{1/2}L^{-1/2}$

3. $CL^{-1/2}m^{-1/2}T^{1/2}$

Chapter 18

1. (a) 541 m/s

3. 0.14 m

5. 219 Hz

7. 2.25

9. 2.2 Hz

11. (a) 425 Hz, 850 Hz, 1275 Hz, (b) 212 Hz, 638 Hz, 1063 Hz

13. (a) 0.061 Hz, (b) ~330

15. (a) 3270 Hz

19. 350 m/s

23. 710 m

25. (a) 3.8×10^{-6} watts (b) 6.9 cm

27. 40 dB, 100 dB

29. (a) 113 dB, (b) 107 dB

31. 72 dB

33. 21 s

35. 32

37. 70.1

G1. 1.009

G3. 1.3×10^5 watts

Chapter 19

1. 14 m, $4.7 \times 10^{-8} \, \text{s}$

3. 9.7 cm

5. (a) 1.43, (b) 1.30

7. (a) $2.33 \times 10^{-6} \, \text{m}$, (b) $3.11 \times 10^{-6} \, \text{m}$

11. (a) 39°, (b) 1.44

13. 21°

15. 4980 m/s

17. 13.7°, 16.3°

19. 58.1° to right of vertical line

21. 13°

23. No

27. 90°

29. 1.41

31. (a) 29.6°, (b) 30.4°, (c) 56.0°, (d) 49.8°

Chapter 20

1. (a) 15 cm, (b) 35 cm, (c) -10 cm

3. (a) -8 cm, (b) -6 cm, (c) -3 cm

5. (a) -7.5 cm, (b) -5.8 cm, (c) -3.3 cm

7. (a) 18.5 cm, (b) 20.4 cm

9. The film must be moved 1.0 cm farther from the lens.

11. (a) $h = -1.25$ cm, $s' = 12.5$ cm, real, inverted, (b) $h = 0.83$ cm, $s' = -8.3$ cm, virtual, upright

13. -2.5 m

15. 0.16 m

17. (a) 2.10 cm, (b) 2.09 cm, (c) 1.94 cm

19. (a) fp $= \infty$, np $= 30$ cm; (b) fp$= 300$ cm, np $= 25$ cm

21. 28.6 cm

23. 5.8

25. (a) 5.7 cm, (b) 18

27. (a) 67 cm to left of diverging lens, (b) upright, (c) virtual

29. (a) 15.6 cm to right of diverging lens, (b) -1.5 cm

31. -16.7 cm

33. 70 cm to right of last lens

35. (a) 82 cm to left of diverging lens, (b) 0.20, (c) 5

37. (a) 105 cm to left of eyepiece, (b) -0.15 cm, (c) 271

39. (a) 2.44 cm to left of lens, (b) 0.41 cm, (c) -439

41. (a) 9.16 cm, (b) -102

43. (a) 8.1 cm, (b) -20.5

45. (a) -40 cm, virtual, upright, (b) -27 cm, virtual, upright, (c) -19 cm, virtual upright

47. 16.5 m in front of mirror, 0.15 m in diameter

49. $f = 0.51$ m, $s = 0.56$ m

51. (a) -0.86 m (behind mirror), (b) 0.49 m

53. (a) $f_1 = -18$ m, $f_2 = +0.95$ m, (b) $m_1 = 0.78$, $m_2 = -0.23$, (c) The $f_1 = -18$ m mirror

Chapter 21

5. (a) $0°$, $0.090°$, $0.180°$; (b) 1.6 cm

7. 640 nm

9. (a) 330 m, (b) 170 m

11. 656 nm: $20.3°$, $44.1°$; 410 nm: $12.6°$, $25.8°$

13. 530 nm

15. 1.7×10^5 lines/m

17. 1.5

19. (a) $0.84°$, (b) $3.1°$

21. 130 nm, 389 nm

23. 118 nm

25. (a) 600 nm, (b) 300 nm

27. (a) $1.8°$, (b) $0.18°$, (c) $0.018°$

29. 3.3 cm

31. (a) 240 cm, (b) 48 cm, (c) 4.8 cm

Interlude, p. 438

1. 1.8×10^{-6} m

3. (a) 1.3×10^{-5} m, (b) 9.0×10^{-6} m

5. $d \geq 5.2$ mm

Chapter 22

1. (a) -9.6×10^{-14} C, (b) 9.9×10^7 electrons, -1.6×10^{-11} C

3. 3.6×10^{-10} N

5. 3.6×10^{32} electrons, 320 kg

7. (a) 2.3×10^{-8} N, (b) 1.9×10^{-44} N, (c) 1.2×10^{36}

9. 3.7×10^{21} N

11. 160 N toward positive charges

13. 2.2×10^{-9} N toward Na$^+$ ion

15. $-0.444 \, q$ between charges and 0.333 m from q

17. 1.56×10^9 N $85°$ below $-x$ axis

19. 43 N pointing away from charge at opposite corner

21. 4.5×10^{-11} N

23. 2.2×10^{-12} N toward right

25. (a) 2.2×10^{-11} N toward left, (b) 2.2×10^{-11} N toward right

27. 1.0×10^{-2} C

29. 2.3×10^{-19} N toward right

31. (a) 5.6×10^7 N/C away from charge, (b) 1.4×10^7 N/C away from charge, (c) 5.6×10^7 N/C toward charge, (d) 1.4×10^7 N/C toward charge

33. 1.1×10^9 N/C toward left at A, 1.4×10^9 N/C toward right at B

35. 57 N/C

37. 15 N/C toward shield

G1. (a) 8.2×10^{-8} N, (b) 9.0×10^{22} m/s^2, 2.2×10^6 m/s

G3. 3.0×10^{-27} N $\cdot$ m

Chapter 23

1. (a) 8.0×10^{-20} J, (b) 1.0×10^{-21} J

3. (a) -810 J, (b) $+945$ J

5. (a) 4.32×10^{-6} N toward sphere, (b) 3.57×10^{-6} N toward sphere, (c) 3.95×10^{-10} J, (d) 3.93×10^{-10} J

7. (a) -6.1×10^{-16} J, (b) $+6.1 \times 10^{-16}$ J

9. -3.4×10^{10} J

11. -3.4×10^{-15} J

13. 1.2×10^{-14} m

15. 5.5×10^{-3} J

17. 7.3 m/s

19. 20.4 m/s

21. -1.2×10^9 J

23. 8.8×10^7 m/s

25. (a) -1.9 J, (b) 1.9 J

27. (a) 60 J, (b) 2.2×10^5 J

29. (a) -5.4×10^5 V, (b) -2.0×10^5 V

31. (a) 0 V, (b) 7.5×10^4 V

33. (a) 2.4×10^4 V/m, (b) 4×10^{-5} m

35. 6000 V

G1. (a) 3.1×10^6 m/s, (b) 9.3×10^6 kg $\cdot$ m/s, (c) 2.3×10^3 m/s

G3. 1.6×10^4 s

Chapter 24

1. (a) 6.3×10^{20} electrons/s, (b) 6.3×10^{21} electrons

3. 3.4×10^{22} electrons

5. (a) 3.1×10^{19} electrons/s, (b) 9.4×10^{-10} m^3/s

7. 3.4×10^7 s or 1.06 years

9. 1.2 mA and 24 mA

13. 1.0×10^{-2} m

15. $32.4 \, \Omega$

17. 21 percent

19. 10 mV, no

21. $25.60 \, \Omega$

23. $0.0175 \, \Omega$

25. $3100°$C

27. 2.4 watts

29. 2.0×10^4 hr

31. (a) 65 kW $\cdot$ h, 2.3×10^8 J; (b) $\$4.50$

33. 192 watts

35. (a) 9×10^{11} kg, (b) 2×10^{19} J

G1. 6.0 m

G3. 9.4 min

G5. $4.3°$C

Chapter 25

1. $I_1 = 1.00$ A, $I_2 = 0.33$ A, $I_3 = 0.67$ A

3. $I_1 = 2.0$ A, $I_2 = I_3 = 1.0$ A

5. 0.5 A, 11 V

7. $I_1 = 0.03$ A cw, $I_2 = 0.14$ A cw, $I_3 = 0.11$ A ccw

9. $I = 7.50$ A, $I_{bulb} = 0.47$ A, $I_{crock} = 1.41$ A, $I_{toaster} = 5.63$ A

11. (a) $I_1 = 0.149$ A cw, $I_2 = 0.023$ A cw, $I_3 = 0.127$ A down, (c) 2.54 V

13. $67.5 \, \Omega$

15. Greater than 30.4 A

17. (a) 9.0 watts, 11.5 watts, 12.0 watts, 11.8 watts, 11.3 watts, 10.7 watts

19. $12.5 \, \Omega$

21. $454 \, \Omega$

23. $50 \, \Omega$

25. 28.9 A, no

27. (a) $3000 \, \Omega$, (b) 4.4 mA

29. (a) 1 A, (b) 2

31. 4.8 hr

33. 9.52 V

37. $1.0 \times 10^{-3} \, \Omega$

Chapter 26

1. (a) 6.6 m^2, (b) 6.6×10^{-4} m^2

3. (a) 71 pF, (b) 0.071 mm

7. 1.4×10^{-8} C

9. (a) 2.0 pF, (b) 9.0×10^{-10} C

11. 52 V

13. (a) 2×10^{-4} C, (b) 0.67×10^{-4} C and 1.33×10^{-4} C, (c) 33 V

15. 3×10^{-5} C

17. 3×10^5 V

19. 6×10^4 V

21. (a) $0.16 \, \mu$F, (b) 1.6×10^{-3} C

23. (a) 5.0×10^{-4} J, (b) 125 V, (c) 6.3×10^{-4} J

25. (a) 0.032 J, (b) 0.01 mm

27. (a) 4.2×10^{-6} F, (b) 7.5×10^7 J, (c) no

29. (a) 4.5 s, (b) 0.033 C
31. 100 V, 5.4×10^{-5} F, 9200 Ω
33. 1.8 kΩ
35. (a) 8 J, (b) 1.9 J
37. 4 pF

Chapter 27

1. 0.10 N, 0, 0.10 N, 0.071 N
3. 1.4×10^{12} m/s (This speed could not be attained according to the theory of relativity.)
5. (a) 2.3×10^{-12} N, (b) 4.8×10^{-21} N, (c) No
7. 4000 m/s
9. (a) 10^4 m/s, (b) 3.3×10^3 m/s
13. 4.3×10^6 m/s
15. (a) 9.9×10^7 m/s, (b) 16 m, (c) down
17. (a) 9.7×10^4 m/s, (b) 2.2×10^3 m/s
19. (a) 1.1 T, (b) 1.7×10^7 rotations/second
21. 0.88 N up, (b) 5.6×10^{-4}
23. (a) 0.17 N $\cdot$ m, (b) Plane of coil is parallel to the field. (c) Plane of coil is 60° from field direction.
25. (a) 0.0060 N, (b) $\theta \simeq 20°$
27. 54 N $\cdot$ m
29. (a) 1.2×10^{-5} T, (b) 0.4×10^5 T
31. 0.22 A
33. (a) 1.2×10^{-15} s, (b) 1.3×10^{-4} A, (c) 0.39 T
G1. 2.20×10^{-25} kg
G3. (a) 1.2×10^{-3} T, (b) 5.8×10^{-15} N, (c) 5.8×10^{-13} N

Chapter 28

1. (a) 0.096 T $\cdot$ m^2, (b) 0.096 T $\cdot$ m^2, (c) 0
3. (a) 3.2×10^{-5} T $\cdot$ m^2, (b) 0, (c) 2.3×10^{-5} T $\cdot$ m^2
5. 0.022 V
7. 0.59 V
9. 1 to 3×10^{-7} V
11. 0.081 V
13. (a) Position b
15. (a) 5.7 V, (b) position a
17. (a) 6.0 Hz, (c) 85 V
19. 0 V, 310 V, 0 V, -310 V, 0 V, 310 V, 0 V, -310 V, 0 V
21. 2.1 Hz

23. (a) 0.83 V, (b) 3.6 A
25. 7.7×10^{-4} s
27. (c) $N\mu r^2/2a$, (d) 2.2×10^{-5} V
29. (a) 27 V, (b) 40 V
31. 0.052 H
33. (a) 0.032 H, (b) 0
35. 0.083
37. (a) 2.8×10^5 V, (b) 4.3×10^{-4}

Chapter 29

5. (a) 600 Ω, (c) -0.24 A
7. (a) 311 V, (b) 14.7 A and 20.7 A, (c) 3230 watts
9. (a) 0.012 A, (b) 0.017 A, (c) pain
11. (a) 240 Ω, (b) 0.50 A and 0.71 A, (c) 120 Ω, (d) 1.0 A and 1.4 A
13. (a) 550 mA, (b) 55 mA, (c) 5.5 mA, (d) 0.055 mA
15. (a) 2.9×10^{-5} H, (b) 0.011 Ω
17. (a) 3.2×10^{-4} H, (b) 7.7
19. (a) 4.5×10^{-8} A, (b) 4.5×10^{-6} A, (c) 4.5×10^{-4} A, (d) 4.5×10^{-2} A
21. (a) 2.1×10^{-2} A, (b) 9.4×10^{-2} A
23. (a) 1050 Ω, (b) 0.114 A, (c) $-76°$, (d) 29 V, (e) 52 V, (f) 168 V
25. 80 Ω, 0.19 H, 36°
27. 3.0 Ω, 9.5×10^{-3} H
29. (a) 0.062 A, (b) 58°, (c) 0.32 watts
31. (a) 27 watts, (b) 59 μF, (c) 580 watts
33. (a) 29 Ω, 40 Ω; (b) 96 percent
35. (a) 27 MHz, (b) 1.1×10^{-5} A, (c) 0.43 V
37. (a) 333 Hz, (b) 0.079
G1. 0.7×10^{-5} A
G3. (a) $V/\sqrt{R^2 + (1/2\pi f C)^2}$, (c) for high f, $V_2/V_1 \to 0$; for low f, $V_2/V_1 \to 1$
G5. (a) 3.0 A, (b) 0.55 A

Chapter 30

1. 1.3×10^{-10} s
3. 11 s
5. (a) 2.2×10^4 s, (b) yes
7. 2.97×10^8 m/s
9. $(1 - 2.5 \times 10^{-17})^{1/2}$ c
11. 96 m

13. 2.2×10^8 m/s
15. Longer, 0.25 m
17. (a) 1.00000, (b) 1.00005, (c) 1.005, (d) 1.06, (e) 1.34, (f) 3.9
19. 4.2×10^7 m/s, 1.25×10^8 m/s
21. 2.999994×10^8 m/s
23. (a) 190 kg, (b) 1.7×10^{19} J, (c) 5.4×10^{18} J and 1.2×10^{19} J
25. 9.8 percent
27. 214
31. 0.067 kg
33. 3.3×10^{-8} kg/s
35. 1.5×10^{-13}
G1. (a) 10 s, (b) 4.4 s, (c) 390 m, (d) yes
G3. (a) 1.0×10^4 kg/m^3, (b) 10 kg, 0.10 m, 1.0×10^{-4} m^2, 1.0×10^6 kg/m^3
G5. (a) 1.1×10^{-11} kg, (b) -1.5×10^{-11} kg

Chapter 31

1. (a) 1.2×10^{19} Hz, (b) 0.122 m
3. (a) 4.1 eV, (b) 8.9×10^{13} Hz
5. 248 nm
7. 4.9×10^{21}
9. 170
11. 2.0×10^{-8} m
13. (a) yes, (b) no, (c) no
15. 2.31 eV
17. 1.86 eV
19. 3.6×10^8
21. 0.026 nm
23. (a) 5.0×10^{24}, (b) 3.3×10^{-8} kg $\cdot$ m/s
25. 1.9×10^{-34} m or about 1.9×10^{-15} smaller than the diameter of a proton
27. 8.3×10^{-12} m
29. (a) $\sim 10^{-35}$ m, (b) 10 to 100 J $\cdot$ s
31. 1.0 m/s
33. $\pm 7.9 \times 10^{-7}$ m
35. 1.2×10^{-5} m, 3.5×10^{-29} m
37. (a) 4.2×10^{-28} kg $\cdot$ m/s, (b) 2.1×10^{-15} m/s
39. 6.1×10^{-12} m, 10^{-5} times the wavelength of light
41. 150 V
G1. 2.7×10^5
G3. (a) 1.3×10^{-9} m, (b) 7.1°
G5. (a) 1.28×10^{14} m^2, (b) 6.0×10^8 kg $\cdot$ m/s, (c) 6.0×10^8 N

Chapter 32

1. 1200 m

3. 565

5. (a) 1.9×10^{-4} kg, (b) 5.9×10^{20}, (c) 1.1×10^{-28} m^2, (d) 6.7×10^{-4}

7. 122 nm, 103 nm, 97 nm

9. 434 nm, 6.91×10^{14} Hz, 4.58×10^{-19} J or 2.86 eV

11. 0.27×10^{-10} m, 1.06×10^{-10} m, 2.39×10^{-10} m; -54.4 eV, -13.6 eV, -6.0 eV

13. 5.8×10^{-13} m

15. (a) 9190 eV, (b) 13.6 eV

17. 1.23×10^{48}

19. 2.19×10^6 m/s, 6.6×10^{15} Hz

21. $1.88\ \mu$m

23. 25.7 nm, 1.17×10^{16} Hz, 48.4 eV

25. 23.5 nm, 1.28×10^{16} Hz, 52.9 eV

27. 2.4×10^6

29. (a) 5.97×10^{-14} J, (b) 4.47×10^{-14} J

31. (a) 7.29×10^5 m/s, (b) 9.98×10^{-10} m

G1. 2.3×10^{39}, yes

G3. (a) $0.010\ c$, (b) $-0.0096\ c$

Chapter 33

15. 1620 nm

17. 6.9×10^{15}

19. (a) 4.2 cm, (b) 1.6×10^4 m

21. 5.51×10^4 V

23. 2.4×10^4 V

G3. (a) 2.0×10^{-3} J, (b) 2.4×10^{-9} kg, (c) 200 C°

Chapter 34

3. (a) 4.8×10^{-15} m, (b) ~ 700 m

5. 100 N

7. 0.511 MeV, 938 MeV, 939 MeV

9. 653 MeV, 8.70 MeV/nucleon

11. 2.8×10^{-9} m

15. 1.59 MeV

17. 13.005739 u

19. $^{215}_{85}$At, 214.99867 u

23. $^{11}_{5}$B, 1.47 MeV

25. 2.0×10^3 MeV, 3.2×10^{-10} watts

27. 32.4 yr, 54.0 yr,

29. 6.0×10^{-5}

31. 12.4 days, 41.3 days

33. 4.1×10^{-10} kg or 1.4×10^{-8} oz at $600 per oz costs 9×10^{-4} cents.

35. 4000

37. 22,920 yr

39. 21.5 yr

41. 12,040 yr

45. 2.5×10^{12}

47. (a) 2.59×10^{-6} R/cm^3, (b) 2.25×10^{-8} J

49. (a) 3.2×10^{11}, (b) 6.9×10^{-14} kg

51. 660 mrem

53. (a) 0.29 J, (b) 0.58 rad

G3. 6.1 MeV

Interlude, p. 697

1. (b) 7.5

3. 9.9×10^6 s^{-1}

5. (a) 18 yr, (c) 60 yr

7. (a) yes, (b) $\sim$$10,000

Chapter 35

1. 199.6 MeV

3. 169 MeV

7. (a) 5.0×10^{-5} MeV/molecule, (b) 4.0×10^6

9. 9.4×10^{19} reactions/second

11. (a) 6×10^{20} J, (b) 6 yr

15. 403

17. 17.6 MeV

19. (a) 3.3 MeV, (b) 4.0 MeV

21. 7.16 MeV released

23. (a) 3.3×10^{25}, (b) 1.0×10^{21}, 3.2×10^8 J

29. 938 MeV, 2.26×10^{23} Hz, 1.32×10^{-15} m

G1. (a) $R_{Ba} = 6.3 \times 10^{-15}$ m, $R_{Kr} = 5.4 \times 10^{-15}$ m, $r = 11.7 \times 10^{-15}$ m (b) 4.0×10^{-11} J = 250 MeV

G3. (a) 3.1×10^{24}, (b) 1.7×10^5 Ci

G5. (a) 0.866 c = 2.60×10^8 m/s, (b) 1.6×10^{-10} s, (c) 4.2 cm

Index

Boldface type is used to indicate the pages on which a term, principle, or law is defined.

Aberrations, **406–408**
Abscissa, 721
Absolute potential, **471–472**
Absolute pressure, **229**
Absolute temperature scale, 163
Absolute zero, 162–163
Absorption of wave energy, 332n
Absorption spectra, 634, 645, 656
Acceleration, 40–54
    average, **40–41**
    graphical analysis of, 43–45
    gravitational, **50–52**
    instantaneous, **41–42**
    linear motion with constant acceleration, 45–50
    units of, 41
Accommodation, 395
Actinides, 656
Action-reaction forces, 61–62, 87
Activity of radioactive material, **676,** 687
Alpha decay, **672–673,** 687
Alpha particles, **634,** 645, 672–673, 687
Alternating current, 481, 553–554, **565–585**
Alternating current generator, 565–566, 584
Ammeter, **503,** 505
Ampere (unit), 480, 492
Amplitude of vibration, **310,** 322
Amplitude of wave, **329,** 343
Angular acceleration, **115–116,** 117–128
Angular magnification, **397–400,** 411
Angular momentum, 125–127, 650–653, 661–662
    in atoms, 650–652, 661–662
    quantization of, 650–653, 661–662
    spin, 651–653, 661–662
Angular position coordinate, **98–99,** 111
Angular size, **398,** 411
Angular velocity, **101–102,** 111
Antenna, electric field production, 456–457, 613–615
Antinodes, 354

Antiparticles, **711**
Antiproton, 711
Archimedes, 260
Archimedes' principle, **259–262,** 266
Arc-length coordinate, **98–99**
Aristotle, 59
Armature, **536–537**
Astigmatism, 407–408
Atmosphere (unit), 229
Atomic absorption spectroscopy, 635–636
Atomic mass unit, **666–668,** 687
Atomic number, 637, **664–**665, 687
Atomic spectra, **634–636,** 645
Atomic structure, 633–662
    Bohr model, 637–646, 649
    of hydrogen, 637–646, 649–662
    multielectron atoms, 649–662
    nuclear model of, 634
    plum-pudding model, 633
Atomic wave functions, 649–652
Atoms, 444, 637–646, 649–662
    energy levels, **640**
    excited states of, **640–**644
    ground state of, **640**
    multielectron, 649–662
    one-electron, 637–646
    states of, 649–654
    subshells in, **653,** 662
Average acceleration, **40–41**
Average speed, **36,** 54
Average velocity, **37–38**
Avogadro's number, **233,** 248
Axis of lens, **388**

Back emf, **557**
Balmer, Johann, 636
Balmer's equation, **636**
Balmer series, 636, 642–643
Band spectrum, **635–636**
Bar (unit), **229**
Baryon number, 670n, 711
Baryons, **711**
Basal metabolic rate, **168**
Base number, **719–**720
Basic quantities, **4**
Battery, 478
Beat frequency, **338–339**

Becquerel, Henri, 671
Becquerel (unit), **676**
Bernoulli, Daniel, 273
Bernoulli's equation, 273–280, 285
Beta decay, **673–674,** 687
Beta particles, 673–674, 687
Big bang theory, **220**
Big crunch, **220**
Bilateral impedance rheography, **488–489**
Binding energy, nuclear, 666–669, 687
Binding energy per nucleon, 668–**669,** 687
Biot, Jean Baptiste, 539
Biot-Savart law, **539,** 543
Blood flow, 272–273, 278–279, 281, 282–283
    measured with ultrasound, 342
Blood pressure, 255, 281
Body temperature control, 195–196
Bohr, Niels, 633, 636, 637
Bohr model of the atom, 637–644, 649–650
    emission and absorption of photons, 640–644, 646
    quantization condition, 638, 646
Boiling, 172–173
Boiling temperature, **172**
Boltzmann's constant, 232, 248
Boyle, Robert, 231–232
Boyle's law, **231–232,** 248
Breeder reactors, **705**
Bremsstrahlung radiation, **660–661**
Bubble chamber, 686
Bulk modulus, **296–297,** 299
Buoyant force, **259–262,** 266

Calorimetry, 166–168
Cameras, 393–394
Capacitance, 511–514, 573
Capacitive reactance, 572–574, 585
Capacitors, **510–525**
    charge on, 512
    in circuit with resistance, 519–522
    energy stored by, 518
    impedance of, 572–574
    in parallel, 522–523
    in series, 523–524
Capillary action, 265–266

Carbon dating, **678**–679
Cells, light diffraction by, 436
Celsius, Anders, 162
Celsius temperature scale, 162–163
Center of gravity, **28**
Centigrade temperature scale, 162
Centrifuge, 105
Centripetal acceleration, 102–105, 111
Centripetal force, 105–107
Chain reaction, 702
Change of state, 169–174
   boiling and condensation, 172–174
   melting and freezing, 169–172
   sublimation, 173n
Charles' Law, 232
Choke coil, 571–572
Chromatic aberration, **407**
Circle of least confusion, **407**
Circular motion, 98–111, 532–535, 543
   in magnetic field, 532–535, 543
Closed pipe, resonant frequencies, 357–
   358, 365
Cloud chamber, **686**
Coal power production, safety of, 703–
   705
Coefficient of kinetic friction, 72–74
Coefficient of performance
   heat pump, **217**–218
   refrigerator, **219**
Coefficient of static friction, 72
Coefficient of viscosity, **280–281**, 286
Coherent light, 658
Coherent wave, **372**
Coil
   impedance of, 569–572
   mutual inductance of, 554–556, 560
   self-inductance of, 556–558, 560, 569–
   572
Collisions
   elastic, **152**
   inelastic, **151**
Color (quarks), 714, 715
Common logarithms, 720
Commutator, **537**
Complex waves, 335–337
Components
   of a vector, 9–13
   vector addition by, 8–13
Compound microscope, 404–406
Compressibility, **296**–297
Compressive strength, **294**
Compressive stress, **292**
Compton, A. H., 620
Compton effect, 620–621
Condensation, 173–174, 191

Condensation temperature, 173
Conduction, heat, 182–184, 199
Conductive heat-transfer rate, **183**
Conductivity, **485**
Conductor, electrical, **485**
Conservation laws
   angular momentum, **125–127**
   baryon number, **711**
   energy, **142–143**, 193
   mass number, 670
   momentum, 87–90
   strangeness, 712
Conserved quantities, 85–86, 90, 142–143
Constant acceleration and linear motion,
   45–54
Constructive interference, **335**, 424
Contact force, **16**
Continuity equation, **273**, 285
Continuous laser, 658
Control rods, 703
Convection coefficient, **185**
Convection of heat, **184**–187
   forced, 185
   natural, 185
Convective heat-transfer rate, **185**
Converging lens, **387**
Converging mirror, **408**
Convex mirror, **409**
Core of nuclear reactor, 701
Corpuscular theory, 371–372
Cosmic rays, **683**
Coulomb, Charles Augustin de, 445
Coulomb's law, **445**–446
Coulomb (unit), **445**
Crests, **327**
Critical angle, **379**
Critically damped oscillator, **319**
Curie (unit), **676**, 687
Current. *See* Electric current
Current loop, 540
Cycle, 310, 565
Cycle on log scale, **725**

Damped oscillation, 318–319, 322
Dating, radioactive, 677–679, 687
Davisson, C. J., 623
Davisson-Germer experiment, 623
De Broglie, Louis, 622, 644
De Broglie wavelength, **622–624**, 629,
   644–646
Decay, nuclear. *See* Radioactive decay
Decay constant, **676–677**, 687
Decay rate, **676**, 680–681, 687
Decay series, **679**

Deceleration, **40**
Decibel (unit), 362
Deformations, types of, 290
Degree (unit), 99
Density, **253–255**, 266
Dependent variable, 721
Derived quantity, **4**
Destructive interference, **335**, 424
Determinism, 434
Dielectric breakdown, 515–518
Dielectric constant, **449**–450
Dielectrics, 514–515
Dielectric strength, 515–516
Diffraction, **417**, 426–428, 431, 435–438,
   659
   through a hole, 435–438
   through a slit, 426–428
Diffraction grating, 421–423, 430
Diffraction limit of circular aperture,
   435–438
Diffuse reflection, **373**
Diffusion, **244**
Dimensional analysis, **347–350**
Diopter, **397**
Dipole moment, 451–452
Direct current, **478–482**
Direct-current electric motor, 536–537
Discharging a capacitor, 520–522
Disordered states, **209**
Dispersion, 381–383
Displacement, **5**
Distance, **5**
Diverging lens, **388**
Diverging mirror, **409**
Domains, magnetic, **542**, 543
Doppler effect, 339–343
Doppler radar, 342
Dose of radiation, **681**, 687
Dot product, 134n
Double-slit experiment, **418**
Doubling time, **695–697**
Drag force, viscous, **284–285**, 286, 484–
   485
Drift velocity, 481–482
Dynamic equilibrium, **16**
Dynamics, 59–76

Ear, human, 364–365
Earthquakes, 295–296
Effective current, **568**, 584
Effective dose, **681**, 687
Effective voltage, **568**, 584
Efficiency, thermodynamic, **215–217**, 221
Einstein, Albert, 595, 616

Elastic collisions, **152**
Elastic deformation, 289–290, 298
Elastic membranes, 297–298
Elastic potential energy, 135, **140–142,** 154
Electric charge, **443**–445
Electric circuits
  ac, 565–585
  dc, 495–509
  RC, 519–522
  resonant, 582–584
  RLC, 577–585
Electric current
  ac, **565**–585
  dc, **478**–492
  effective, 568
  hazards, 487–488
  peak, 566
Electric dipole, 451–452
Electric dipole moment, **451–452,** 457
Electric field, **452**–455, 455–456, 456–457, 473–474
Electric field lines, 455–456
Electric force, 445–449, 450
Electric generator, 552–554, **565**–566
Electric motor
  (ac), 581–582
  (dc), 536–537
Electric potential difference, **469–473,** 474
Electric potential energy, **461–466,** 467–474
Electric power, **490**–491
  in ac circuits, 567–569, 580–582, 585
  in dc circuits, 490–491, 492
  production by fission, 699–705
  production by fusion, 707
Electric resistance, **485–486**
Electrocardiogram, 452
Electrodes, 478–479
Electrolyte, 478–479
Electromagnetic induction, **547**–552, 560
Electromagnetic radiation, 187, 457, **613–616**
Electromagnetic spectrum, 615–616
Electromagnetic wave, 457, **613–614,** 615–616
Electromotive force, 479, 492
  back emf, 557
  battery, 479
  of generator, 552–554, 560
  induced, 548–552, 560
  self-induced, **556**–558, 560
Electron
  de Broglie wavelength of, 644–646

Electron [*cont.*]
  as an elementary particle, 708
  orbital angular momentum in atom, 650–651, 661–662
  orbits in atom, 637–640, 649–653, 661–662
  spin angular momentum, 651–653, 661–662
Electron configuration, **653**–654, 662
Electron microscope, 627–629
Electron-position annihilation, 608
Electron state, 640, 649–652, 662
Electron transition, **656**–658, 660–662
Electron volt (unit), **617**
Elementary particles, 707–715
  classification of, 708–713
  and conservation laws, 711–713
  production of, 710
  quarks, **708,** 713–715
  stability of, 708–713
Elements, 654
Emission
  photon, 656
  spontaneous, 656, **657,** 662
  stimulated, **657,** 662
Emission spectrum, 634–636, 645
Emissivity, **188**
Energy, 132–154
  in a capacitor, 518
  conservation of, 142–143
  elastic potential, 135, **140–142**
  electrical potential, **461–466,** 467–474
  gravitational potential, **138–140**
  kinetic, **136–138**
  kinetic, rotational, **137–138**
  vibrational, 311–312
Energy-level diagram, **640**
Energy states in atoms
  in Bohr model, 637–644
  excited, **640**–644
  ground, **640**
  in multielectron atoms, 649–662
  in one-electron atoms, 637–646
Energy transfer between oscillators, 319–321, 322
English system, 62–63
Entropy, **206**–221
Environment, **86,** 90, 142
Equations of state, 227, 230–232, 248
Equilibrium, 16–34
  first condition, 16–18
  second condition, 24–27
Equilibrium position, **289,** 308
Equilibrium state, **210**
Equipotential surface, **472**

Equivalence of mass and energy, 608–609
Error bar, **722**
Estimates, procedures for making, 55, 94–97
Ether, 597
Evaporation, **172,** 189–193
Evaporation rate, **190**
Evaporative heat transfer rate, 191
Event, **596**
Eye's optical system, 395, 411
Eye surgery with lasers, 659–660
Excitation, **656**
Excited states, **640–644**
Expansion, thermal. *See* Thermal expansion
Exponential function, 691–697, 725–726
Exponents, 719–720
Exposure to radiation, **681,** 687
Extended body, 28, **121–122**
External forces, **87–88,** 90

Fahrenheit, Gabriel, 162
Fahrenheit temperature scale, 162–163
Faraday, Michael, 455, 548
Faraday's law, 57, 548–552
Farad (unit), 513–514
Far point of eye, **395,** 411
Farsightedness, 396–397
Ferromagnetic materials, 542, 543
Fiber optics, 380
First law of thermodynamics, 193–195
Fission, nuclear. *See* Nuclear fission
Flow rate, **272**–273, 285
Fluid, 252–266
Fluid flow, 271–285
Fluorescence, **656**
Fluorescent light bulbs, 581, 656
Focal length
  of a lens, **388, 391,** 411
  of a mirror, **410**
Focal point, **388**
Forbidden transition, **658**
Force, 6
  buoyant, **259–262**
  contact, **16**
  drag, **284–285**
  friction, **71–75**
  gravitational, **107–109**
  resultant, **12**
  of spring, 290
  strong nuclear, **665–666**
  tension, **19**
  units of, 4, **62**
  weight, 16, **64–65**

Force constant of spring, **310,** 322
Formants, **360**
Free-electron density, **482**
Freezing, 168–172
Freezing temperature, 170
Frequency
    of standing waves, 354, 357
    of traveling wave, **331,** 343
    of vibration, **310**–311, 312–314, 315–
        317, 322
Frequency spectrum, **337**
Friction force, 71–75, 76
Friction in fluids, 280–283, 286
Fuel rods, 701–703
Fulcrum, **22**
Fundamental frequency of vibration, **353**
Fusion
    heat of, **169**–172
    nuclear. *See* Nuclear fusion

Galileo Galilei, 35
Galvanometer, 506, 508, 509
Gamma decay, **674**–675, 687
Gamma rays, 616, **674**–675, 687
Gases, 226–248
    work done by, 239–242, 243–244
Gauge pressure, 229–**230**
Gauss, Karl Friedrich, 454
Gauss's law, 454
Gauss (unit), 528
Geiger counter, 685
General relativity, 596
Generator, electric, 552–554, 560, **565**–
    566, 584
Genetic damage, **684**
Glaucoma, 259
Gluons, 714
Graphical addition of vectors, 6–8
Graphical analysis, 721–727
Gratings, **421**–423, 430
Gravitation, universal law of, **107**–**109,**
    110
Gravitational acceleration, **50**–**52,** 54
Gravitational force, **107**–**109,** 110
Gravitational potential energy, 135,
    138–**140,** 154, 275
Gravitational torques, 27–30
Ground point, electric, 471
Ground state, **640,** **653,** 656–658
Groups of atoms, 654–656
Growth rate, **691**–692

Hadrons, **711**
Half-life, **675**–677, 687

Half-life [*cont.*]
    and decay constant, 676–677
    and radioactive dating, 678–679
    table of for nuclei, 675, 728–731
Harmonics, 353–358, 359–360
Heart, dipole moment of, 452
Heart attacks, 278–279
Heat, 164–174, 616
    and boiling and evaporation, 172–174
    and calorimetry, 166–168
    and melting and freezing, 168–172
    specific heat capacity, 164–166
Heat capacity, specific, **164**–166
Heat death of universe, **220**
Heat engines, 214–217
Heat of condensation, **173**
Heat of fusion, **169**–172
Heat of vaporization, **172**–174
Heat pump, 217–218
Heat transfer, **164**–174, 182–193, 199
    by conduction, **182**–184
    by convection, **184**–187
    by evaporation, **189**–193
    by radiation, **187**–189
Heat-transfer rate, **182**
Heavy water, 702
Heisenberg, Werner, 626, 649
Heisenberg uncertainty principle, 624–
    626, 627
Henry, Joseph, 555
Henry (unit), **555**
Hertz, Heinrich, 616
Hertz (unit), **310**
Holography, 428–430, 431
Home temperature control, 196–198
Hooke, Robert, 404
Hooke's law, **290**
Hooke's law-type force, **309**–**310**
Horsepower (unit), 153
Human ear, 364–365
Human eye, 394–397
Huygens, Christian, 372, 416
Huygens' principle, **416**–**417,** 430
Hydraulic brakes, 258
Hydraulic lift, 258
Hydraulic press, 258
Hydrogen atom and Bohr theory, 636–
    646

Ideal gas law, **232**–**235,** 248
Image distance
    lenses, 389, 391
    mirrors, 410
Impedance, **333,** 569–574, 577–584,
    589–590

Impedance [*cont.*]
    of capacitor, 572–574
    of coil, 569–572
    matched, 589–590
    mismatched, 589–590
    of RLC series circuit, 577–580
Impedance matching, 589–592
Impulse, 80, **81**–**85,** 90
Impulse-momentum equation, **81**–**85,** 90
Impulsive wave, **326**
Inches of mercury (unit), **229**
Incoherent wave, **373**
Independent variable, 721
Index of refraction, **374**–375
Induced emf, 547–560
    of electric generator, 552–554
    polarity of, 550–552
    in transformers, 558–560
Induced voltage, 547–560
Inductance
    mutual, **554**–556, 560
    self, **556**–**558,** 570–572
Induction coil. *See* Coil
Inductive reactance, 569–572
Inductor. *See* Coil
Inelastic collision, **151**
Inertia, 61
Inertial reference frame, **596**
Infiltration, **198**
Infrared waves, 187, 616
Instantaneous acceleration, **41**–42
Instantaneous speed, **38**–**40,** 54
Instantaneous velocity, 38–40
Insulator, electrical, **485**
Insulators, thermal, **183**
Intensity, 360–364, 365
Intensity level, 362–364, 365
Intercept, 724, 725
Interference, 335, 416–431
    constructive, 335
    destructive, 335
    diffraction, 426–428, 435–437
    double-slit, 417–421
    gratings, 421–423
    thin-film, 423–426
Internal energy, 160
Internal resistance, 503–504
Inversion, population, 658
Ion current, **480**
Ionizing radiation, 679–686
    detection of, 685–686
    effects on humans, 684–685
    medical uses, 684–685
    sources of, 682–684
    units for, 680–682
Ions, 444

Irreversible processes, 207, 221
Isolated system, **86,** 90
Isotopes, **665,** 677–679, 687
    radioactive, 683–684
    table of, 728–731

Jeans, James, 233
Joule, James, 135
Joule (unit), 135
Junction rule, **497**

Kelvin temperature scale, **163**
Kelvin (unit), **163**
Kilocalorie (unit), **153**
Kilogram (unit), **62**
Kilowatt-hour (unit), **491**
Kinematics, 35–54
    for constant angular acceleration,
      117–118
    for linear motion with constant accel-
      eration, 45–52
    for simple harmonic motion, 317–318
Kinetic energy, 136–138, 154
    relativistic, **606–608**
    rotational, **137–138,** 154
    translational, **136–137,** 154
Kinetic friction, **73–75**
Kinetic theory of gases, 235–239
Kirchhoff's junction rule, **497–498,** 505
Kirchhoff's loop rule, **495–497,** 498–500,
    505

Laminar fluid flow, 271, 285
Lanthanides, 656
Lasers, **657–660,** 662
Latent heats
    of fusion, 169–172
    of vaporization, **172–175**
Law of reflection, **374**
Length contraction, **601–603**
Length deformation, 292–294, 299
Lenses, 387–408
Lenz's law, **550–552,** 560
Leptons, **711**
Light, 371–372, 616, 658
    coherent, **658**
    monochromatic, **658**
Lightning, 516–519
Light pipe, **380**
Linear equation, **723**
Linear expansion, **175–176**
Linear graph paper, **723**
Linear magnification, **393,** 411

Linear motion with constant accelera-
    tion, 45–54
Line of action, **22**
Lines of force, **455–456**
Line spectrum, **635–636**
Logarithms, 719–720
Log-log graph, 726–727
Longitudinal wave, **327–329,** 343
Loop rule, **495–497,** 498–500, 505
Loudness of sound, 360–364, 365
Lyman series, **636,** 642–643

Macrostate, **208–210,** 221
Magnetic bottles, 533–534
Magnetic dipole moment, **541**
Magnetic domains, **542,** 543
Magnetic fields, 528–543
    path of charged particle in, 532–535
    produced by moving charge, 537–541
    units of, **528**
Magnetic flux, **548–560**
Magnetic force, 529–535, 543
Magnetic lens, 627–628
Magnetic materials, 541–542, 543
Magnetic mirror, 534
Magnetic moment
    orbital, **651,** 661
    spin, **651–652,** 661
Magnetic permeability, **539**
Magnetic quantum number, **651,** 661
Magnetohydrodynamics, 530–532
Magnifying glass, 397–400
Magnifying power, **400**
Magnitude of vector, **5,** 13
Manometer, **229–230**
Mass, **61,** 64, 76
Mass and weight, 64–65, 76
Mass defect, **667–668**
Mass-energy equivalence, 608–609
Mass increase, 603–605
Mass number, **664–665,** 687
Mass spectrometer, 534–535
Matched impedances, 589
Maxwell, James Clerk, 372
Measurable quantities, 3–5, 13
Measuring current, 503–504
Measuring voltage, 504
Melting, 168–172
Melting temperature, **169**
Metabolic rate, **153–154**
Metal detector, 555–556
Metastable state, **657**
Meters. *See* Ammeter; Voltmeter
Meter (unit), 4
Metric system, 62

Michelson, A. A., 597
Michelson–Morley experiment, 597
Microstate, **208–210,** 221
Microwave oven, 452
Microwaves, 615
Millimeters of mercury (unit), **229**
Mirages, 378
Mirror equation, 409–411
Mirrors, 408–409, 411
Mismatched impedances, 589–590
Moderators, 701–702
Modulus of elasticity, 290, **291–292,** 299
Molar concentration, **246,** 248
Mole, **233–235**
Moment arm, **23**
Moment of inertia, **120–122**
Momentum, 80–91, 621
    of a mass, **80**
    of a photon, **621**
Momentum conservation, 87–90
Monochromatic light, 618, 658
Morley, E. W., 597
Motion, 35–54, 98–111
    circular, 98–111
    graphical analysis of, 43–45
    linear with constant acceleration,
      45–52
    projectile, 52–54
Motion and Newton's laws, 59–76
Multielectron atoms, 649–656, 662
Muon decay, **599–600**
Mutual inductance, **554–556,** 560
    and metal detectors, 555–556

Nanometer (unit), **616**
Natural abundance, **665**
Natural logarithms, 720
Near point of eye, **395,** 411
Nearsightedness, 395–396
Negative ion, **444**
Neutrino, **674**
Neutron number, **664–665,** 687
Newton, Isaac, 4, 59, 107, 371, 616
Newtonian mechanics, 1, 59–76
Newton's first law, **59–60,** 76
Newton's laws of motion, 59–76
Newton's second law, **60–62,** 66–71, 76,
    637
Newton's third law, **65–66,** 76
Newton's universal law of gravitation,
    107–109, 110
Newton (unit), 4, 6, **62**
Node, **354**
Nuclear binding energy, 666–669, 687
Nuclear decay. *See* Radioactive decay

Nuclear energy
  released in fission, 698–699
  released in fusion, 705–706
Nuclear fission, 698–705, 715
  and chain reactions, 702
  control of, 703
  energy released during, 698–699
Nuclear force, 665–666
Nuclear fusion, 608–609, **705**–707, 715
  energy released during, 705–706
  reactions, 706–707
Nuclear model of the atom, **634,** 645
Nuclear power, 699–707
  safety of, 703–705
Nuclear power plants, 699–705
Nuclear radiation, 666, 671–677, 679–687
  activity of source, **676,** 687
  decay rate of, **676,** 687
  effects on humans, 684–685
  half-life of, **675–677,** 687
  sources of, 682–684
Nuclear reactions, 669–675, 687
  Q of, **670,** 673
Nuclear reactors, 699–705
  boiling-water, 700–701
  breeder, 705
  core of, 701
  fuel for, 701, 703
  heavy-water, 702
  light-water, 702
  moderators for, 701–702
  pressurized-water, 700–701
  safety of, 703–705
Nucleon number, 664–665, 667n
Nucleons, **664**–671
Nucleus, **444,** 664–680

Object distance
  of a lens, 389–390, **391–394,** 411
  of a mirror, 410, 412
Object height
  of lenses, 410
  of mirrors, 393
Oboe reed, 277
Oersted, Hans Christian, 538
Ohm, Georg Simon, 482–486
Ohm's law, **482–492,** 566, 568, 569–570
  applications of, 486–489
Ohm (unit), **485,** 492
Optics
  geometrical, 387–415
  physical, 416–433
Optics of the eye, 395–397

Orbital angular momentum, 650–651, 661
Orbital angular momentum quantum number, **650–651,** 661
Orbital magnetic quantum number, **651,** 661
Orbits, electron, 637–640, 649–653, 661–662
Ordered states, 209–211
Order of interference pattern, **418–421**
Order of magnitude estimate, 94–97
Ordinate, 721
Oscillators, energy transfer between, 319–321
Osmosis, 245–247, 248
Osmotic pressure, 246–247, 248
Overdamped oscillator, 319
Overtones, 353–358, 359–360

Pacemaker for hearts, 552
Pair production, 609
Parachutist, 148
Parallel capacitors, **522–523,** 525
Parallel resistors, **501–503,** 505
Paramagnetic materials, 541
Particles
  elementary, 707–715
  table of stable, 709
Pascal's principle, 255–259, 266
Paschen series, **636,** 642–643
Pauli, Wolfgang, 142, 653, 674
Pauli exclusion principle, **653,** 662
Peak current, 566, 584
Peak voltage, **533,** 560, 566, 584
Pendulum, 315–317
  effective force constant of, 316
  physical, 316n
  simple, 315–317
Percentage growth rate, **692**
Performance coefficients, 217–219
Period
  of ac voltage, 565
  of simple harmonic motion, 313
  of vibration, **310–311**
Periodic table, **654–656,** 662
Periodic waves, 326, 343
Permeability, **539**
Permittivity, 450n
Phase
  in ac circuits, 574–577, 585
  of waves, 333
Phosphorescence, 657
Photoelectric current, 618
Photoelectric effect, 618–620

Photographic emulsion, 686
Photomultiplier tube, 620, 686
Photons, 372, **616–623,** 629
  absorption, 656, 662
  emission, 656
  experiments supporting concept of, 618–621
  momentum, 621
Physical pendulum, 316n
Physical quantities, 3–4
Pitch of a sound, 359, 365
Planck, Max, 593, 616
Planck's constant, 617
Plane mirror, 408
Plasma, 534
Plate tectonics, 295–296
Plum-pudding model of the atom, 633
Poiseuille, Louis Marie, 282
Poiseuille's law, **282,** 286
Poise (unit), 281
Polarizability, 460
Polar molecules, 451
Population inversion, 658
Position coordinate, angular **98–101,** 111
Positive ion, 444
Positron, 711
Positron-electron annihilation, 608
Postulates of special relativity, 596
Potential, electrical, **471–472**
Potential difference, electrical, **469–472,** 474
  and electric field, 473–474
Potential energy
  elastic, 135, **140–142**
  electrical, **461–464,** 465–466
  gravitational, **138–140,** 275
  minimization of, 466
Potentiometer, 506
Pound (unit), 4, **62**
Power, 152–153, 154
  electrical, 490–491, 492
  generated by fission, 699–705
  generated by fusion, 707
  and work, 152–153
Power factor, **580–582,** 585
Power laws, 726–727
Power of a lens, 397
Power plants, nuclear, 699–705
Powers of ten, 718–719
Pressure, **228–230,** 247–248, 252–253, 255–259, 266
  absolute, 229
  atmospheric, 229
  blood, 255, 281
  gauge, 229–230

Pressure [*cont.*]
 hydraulic, 258–259
 osmotic, 246–247, 248
 units for, 229
Pressure variation with depth, 255–257
Primary coil, **558**–560
Principal quantum number, **650,** 661
Prisms, **381,** 382–383
Probability
 in quantum mechanics, 649–650
 and second law of thermodynamics,
  208–210
 wave-particle duality, 622, 624
Problem-solving techniques
 dynamics, 66–67
 optics with two lenses, 400–401
 rotational motion, 122–123
 statics, 19
 work and energy, 144, 466–467
Projectile motion, 52–54
Proper length, 601–603
Proper reference frame, 597–601, 602–
  603, 609
Proper time, **597–601,** 609
Proportions, 203–205
Proton-proton cycle, 706
P-V diagrams, 242–244

Q of nuclear reaction, **670,** 673
Quality factor of ionizing radiation, 681
Quality of sound, 359–360, 365
Quanta, 616
Quantization condition, 638, 646
Quantum chromodynamics, 714
Quantum condition, de Broglie's, 644–
  646
Quantum mechanics, 623, 633, **649–662**
Quantum mechanics, and determinism,
  434
Quantum number, **640,** 649–653
 magnetic, **650–651,** 661–662
 orbital angular momentum, **651,** 661
 principal, **650,** 661
 spin magnetic, **651–653,** 661–662
Quantum theory of atoms, 649–662
Quark model, 613–715

Radar, doppler systems, 342
Radians, angular measure, **99,** 111
Radiation
 absorption and emission of, 656–658,
  660–662
 detection devices, 620, 685–686

Radiation [*cont.*]
 effects on humans, 684–685
 and heat transfer, 187–189
 ionizing. *See* Ionizing radiation
Radiation absorbed dose, rad (unit), 681
Radiative heat transfer rate, 188
Radioactive dating, 677–679, 687
Radioactive decay, 666, 671–679, 687
 activity of, **676,** 687
 by alpha particle emission, **672–673,**
  687
 by beta particle emission, **673–674,**
  687
 decay rate of, **676, 680–681,** 687
 by gamma ray emission, **674–675,** 687
 half-life of, **675–677,** 687
Radioactive decay constant, **676,** 687
Radioactive decay series, 679
Radioactive fallout, 684
Radioactive isotopes, 683–684
Radioactivity, 671–679
Radio waves, 615
Rad (unit), **681,** 687
Rainbows, 383
Rarefaction, 328
Ray diagrams, **388–391,** 409, 411
Rayleigh, Lord, 350
Rays, **373,** 384
RBE (Relative biological effectiveness),
  of ionizing radiation, **681,** 687
RC series circuit, 519–522, 525
Reactance
 capacitative, 572–574
 inductive, 569–572
Reactions, nuclear, 669–675, 687
Reactor. *See* Nuclear reactors
Real image, **389–390,** 411
Recurrence paradox, 219–220
Reference frame, 596
Reflection, 332–334, 343, 371, **373–374,**
  378–381, 384
 diffuse, 373, 384
 law of, **373–374,** 384
 phase change during, 333
 of sound waves, 333–334
 specular, 373, 384
 total internal, 378–381, 384
 of waves on a string, 332–334, 343
Refraction, **374–380,** 381–383, 384
Refrigerator, 218–219
Relative biological effectiveness (RBE),
  681, 687
Relative density, 255
Relativistic energy, **605–608,** 609
Relativistic kinetic energy, 606

Relativistic mass, 603–604
Relativity, special theory of, 595–612
 and Newtonian mechanics, 604–605
 postulates of, 596
Rem (unit), **681,** 687
Resistance, electrical, 485–492
Resistance and temperature, 489–490
Resistance to fluid flow, 288
Resistive drag force, 481–482
Resistivity, **485–489,** 492
Resistors
 in ac circuits, 566–569, 584–585
 in dc circuits, 482–489
 in parallel, 501–503
 in series, 500–501
Resonance
 of electron waves in atoms, 644–646
 in RLC series circuit, 582–584, 585
 of sound in air columns, 356–358
 of waves on a string, 352–356
Resonant frequencies
 of closed air columns, 358
 of open air columns, 357
 in RLC series circuit, 583–584
 of waves on string, 352–356
Response function, 320–321
Rest mass, **603–604,** 609
Rest-mass energy, 606
Restoring force, **309–311,** 322
Resultant, 7
Reversible process, 207, 212, 221
Revolutions (angular measure), 99
Reynolds number, **283, 286**
Right-hand rule
 for direction of B-field, 540–541
 for force on moving charge, 530
 for polarity of induced emf, 551
RLC series circuit, 577–585
 impedance of, 577–580, 585
 phase angles of, 578–580, 585
Roentgen equivalent mass, rem (unit),
  681, 687
Roentgen (unit), **681,** 687
Root-mean-square values (rms)
 of current, 567–569, 584
 of speed, 239
 of voltage, 567–569, 584
Rotational kinetic energy, 137–138
Rotational motion, **36,** 115–128
Rutherford, Ernest, 634
Rydberg constant, 636

Savart, Felix, 539
Scalars, 3, **5,** 13

Scale of a graph, 721
Scaling, 302–306
Scaling factor, 302
Scanning electron microscope, 628–629
Schrödinger, Erwin, 649
Scintillation detector, 686
Secondary coil, 558–560
Second condition of equilibrium, **24–27,** 30
Second law of thermodynamics, 213–221
Self-induced emf, **556–558,** 560
Self-inductance, **556–558,** 560
Semiconductor diode detector, 686
Semilog graph, 693–695, 724–726
Series capacitors, **523–524,** 525
Series resistors, **500–501,** 505
Shear deformation, **294–296,** 299
Shear force, 294
Shear modulus, **294–296,** 299
Shear strain, 294
Shear stress, 294
Shunt resistance, 508
Sign conventions for lenses, 391
Significant figures, 718
Silicon-controlled rectifier, 522
SI metric system of units, 4, 62
Simple harmonic motion, 310, 311–318
    applied to pendulum, 315–317
    energy of, 311–312
    frequency of, 312–314
    kinematics of, 317–318
    period of, 313
Simple pendulum, 315–317
Single-slit diffraction, **426–428,** 431
Sinusoidal motion, 318
Sinusoidal wave, 334–335
Slope of graph, 45, 723
Slug (unit), 63
Snell, Willebrord, 376
Snell's law, **376–378,** 384
Snoring, 277
Solenoid, 540
Solids, types of deformations, 290
Somatic damage, 684
Sound, 352–365
    intensity of, 360–364, 365
    intensity level of, 362–364, 365
    loudness of, 360–364, 365
    pitch of, 359, 365
    quality of, 359–360, 365
Special theory of relativity, 595–612
Specific gravity, 255
Specific heat capacity, **164**–166
Spectrograph, 422–423

Spectroscopy, atomic absorption, 635–636
Spectrum
    absorption, 634, 645, 656
    band, 635–636
    electromagnetic, 615–616
    emission, 634–636, 645
    line, 635–636
    x-ray, 660–661
Specular reflection, **373,** 384
Speed
    average, 36
    instantaneous, 38–40
    tangential, 101–102, 111
Spherical aberration, 407
Spin angular momentum, **651–653,** 661–662
Spin magnetic quantum number, **651–653,** 661–662
Spontaneous emission, 656, **657,** 662
Spring, force constant of, 290
Standards of measurement, 3–4
Standing waves, 352–358, 365
    in a closed pipe, 357–358, 365
    frequencies of, 357–358, 365
    in an open pipe, 356–357, 365
    on a string, 352–356, 365
    wavelengths of, 354–355
State
    disordered, 209–211
    equilibrium, 210
    ordered, 209–211
State variables, 208–210, 227, 230–232
Static electric force, 446–449
Static electricity, 443–457
Static equilibrium, 16–34
Static friction, **71–73,** 76
Statics, 16–34
Statistics and entropy, 208–211, 221
Stefan-Boltzmann constant, 188
Stefan's law, 188
Stimulated emission, **657,** 662
Stoke's law, 284
Strain, 290–292, 299
Strangeness, 712
Strange particles, 712
Streamline flow, 271
Stress, 290–292, 298
Strong nuclear force, **666,** 687
Sublimation, heat of, 173n
Subshells in atoms, **653,** 662
Superposition, principle of, 334–335, 343
Surface tension, **262–265,** 266
    importance of, 264–265

Surface tension [*cont.*]
    of soap bubbles, 263
Synchronous earth satellite, 109
System, 86, 142
    isolated, 86
Systems of units, 4, 62–64

Tacoma Narrows bridge, 319
Tangential acceleration, 116–117
Tangential speed, 101–102, 111
Telescopes, 402–404
Temperature, **161**–163
Temperature and volume change of a gas, 241–242
Temperature coefficient of resistance, 489
Temperature control, 193–199
Temperature interval, 162
Temperature point, 162
Temperature scales
    absolute, 163
    Celsius, 162–163
    Fahrenheit, 162–163
    Kelvin, 163
Tensile stress, 292
Tension force, 19–21
Terminals of a battery, 479
Terminal speed, 285
Tesla (unit), 528
Thermal conductivity, 183
Thermal energy, 135, **149–151,** 154, **160–161,** 163–164, 177, 238, 248
    average, 238
Thermal expansion, 174–177, 178
    of air, 177
    coefficients of, 175, 176
    linear expansion, 174–176
    volume expansion, 176–177
    of water, 177
Thermal radiation, 616
Thermistors, 490
Thermodynamic probability, 208–210
Thermodynamics, 206–221
    and entropy, 211–213, 221
    first law of, 193–199
    of gases, 242–244
    second law of, 213–221
Thermogram, 189
Thermometric property, 161
Thin-film interference, **423–426,** 431
Thin-lens equation, **391–394,** 411
    sign conventions for, **391**
Thomson, G. P., 623

Thomson, J. J., 633
Threshold of audibility, 360
Time constant, 520, 525
Time dilation, **597-601,** 609
  test of, 601
Toricelli's theorem, 277-278
Torque, **22-24,** 30, 118-120
  on a current loop, 535-536, 543
  gravitational, 27-30
Total internal reflection, **378-381,** 384
Total relativistic energy, 606
Transformer, **558-560,** 591-592
Transmission electron microscope, 627-629
Transition atoms, 656
Transitions in atoms, **656**-658, 660-662
Translational motion, 35
Transverse waves, **327-329,** 343
Traveling waves, 326-343
Troughs, 327
Tuning circuit, 583-584
Turbulent fluid flow, **271-272,** 283, 285

Ultimate compressive strength, 294
Ultimate tensile strength, 294
Ultrasound, 334
  measurement of blood flow, 342
Ultraviolet radiation, 616
Uncertainty in measurement, 434-438
Uncertainty principle, **624-627,** 629
  philosophical comments on, 434-438
Units of measurement, 3-5, 62-64
Universal gas constant, 234

Vaporization, heat of, 172-174
Vectors, 3, **5,** 6-13
  addition of, by components, 8-9, 12-13
  addition by graphical techniques, 6-8
  components of, **9-11,** 12-13
  graphical representation of, 5-6

Vectors [*cont.*]
  magnitude of, 5
  resultant, 7
  subtraction by graphical technique, 6-8
Vector symbols, 5
Velocity
  angular, **101-102,** 111
  average, **37-38**
  drift, 481
  instantaneous, **38**
Velocity selector, 544
Vibration, **309-**322
Vibrational motion, 36
Virtual image, **390-391,** 411
Viscosity, coefficient of, **280,** 286
Viscous force, 280
Visible light, 616
Voltage
  effective, **568-569,** 584
  induced, 547-560
  peak, 566
  self-induced, **556-558,** 560
Voltage divider, 506
Voltmeter, 469, **504,** 505
Volt (unit), 469
Volume
  change, 290
  deformation, 296-297, 299
  expansion, 176-177
  strain, 296
  stress, 296

Watt (unit), 153
Waveforms, 328-329, 337
Wavefronts, **372-373,** 384
Wave functions, **649-**662
Wave interference, 334-339, 416-431
Wavelength, **331-332,** 343
  de Broglie, 644-646
Wavelets, 417
Wave mechanics, 623

Wave-particle duality, 613-629
  and electromagnetic radiation, 621-623
  and matter, 622-624
Waves, 326-343
  complex, 335-338, 343
  electromagnetic (EM), 613-614
  periodic, 326, 343
  properties of, 326-327
  speed of, 329-331, 343
  standing, 352-356, 365
Wave speed, 329-331
Wave theory of light, 371-372
Weakest-link rule, 96
Weakly damped oscillator, 318
Weber per meter^2 (unit), 528
Weight, **16, 64-65,** 76
Weight density, 255
Weightlessness, 109-111
Wheatstone bridge, 506
Wind-chill temperature, 186-187
Wing of airplane, lift on, 279-280
Work, **133-135,** 154, 275
  done by a gas, 239-242, 248
  done on a gas, 240-242, 248
  units of, 135
Work-energy calculations, 466-468, 474
Work-energy equation, **143-148,** 154, 274
Work-energy principle, 164
Work function, 618

X-rays, 660-662
  Bremsstrahlung, 660-661
  characteristic lines, 661
  medical uses of, 684-685
  risks of, 684-685
X-ray spectrum, 660-661
X-ray tube, 660

Young, Thomas, 372, 417-418, 616
Young's modulus, **293,** 299

POWER OF TEN PREFIXES

Prefix	Abbrev.	Value
Exa	E	10^{18}
Peta	P	10^{15}
Tera	T	10^{12}
Giga	G	10^9
Mega	M	10^6
Kilo	k	10^3
Hecto	h	10^2
Deka	da	10^1
Deci	d	10^{-1}
Centi	c	10^{-2}
Milli	m	10^{-3}
Micro	μ	10^{-6}
Nano	n	10^{-9}
Pico	p	10^{-12}
Femto	f	10^{-15}
Atto	a	10^{-18}

SOME USEFUL FACTS

Area of circle (radius r)	πr^2
Area of sphere (radius r)	$4\pi r^2$
Volume of sphere	$4\pi r^3/3$
Density of dry air (STP)	1.293 kg/m³
Density of water (4° C)	1000 kg/m³
Speed of sound in air (0° C)	331.4 m/sec

Trig definitions:

$\sin \theta = $ (opposite side)/(hypotenuse)

$\cos \theta = $ (adjacent side)/(hypotenuse)

$\tan \theta = $ (opposite side)/(adjacent side)

Quadratic equation:

$$0 = ax^2 + bx + c,$$

where $x = \dfrac{-b \pm \sqrt{b^2 - 4ac}}{2a}$